ANNUAL REVIEW OF PLANT PHYSIOLOGY AND PLANT MOLECULAR BIOLOGY

ANNUAL REVIEW OF PLANT PHYSIOLOGY AND PLANT MOLECULAR BIOLOGY

VOLUME 49, 1998

RUSSELL L. JONES, *Editor*
University of California, Berkeley

CHRISTOPHER R. SOMERVILLE, *Associate Editor*
Carnegie Institution of Washington, Stanford, California

VIRGINIA WALBOT, *Associate Editor*
Stanford University

http://www.AnnualReviews.org   science@annurev.org   650-493-4400

ANNUAL REVIEWS   4139 EL CAMINO WAY   P.O. BOX 10139   PALO ALTO, CALIFORNIA 94303-0139

ANNUAL REVIEWS
Palo Alto, California, USA

International Standard Serial Number: 1040-2519
International Standard Book Number: 0-8243-0649-X
Library of Congress Catalog Card Number: 50-13143

⊗ The paper used in this publication meets the minimum requirements of American
National Standards for Information Sciences—Permanence of Paper for Printed Library
Materials, ANZI Z39.48-1992.

TYPESET BY TECHBOOKS, FAIRFAX, VA
PRINTED AND BOUND IN THE UNITED STATES OF AMERICA

Annual Review of Plant Physiology and Plant Molecular Biology
Volume 49 (1998)

CONTENTS

(*continued*)   v

SOME RELATED ARTICLES IN OTHER *ANNUAL REVIEWS*

From the *Annual Review of Biochemistry*, Volume 67

Base Flipping, Richard J. Roberts and Xiaodong Cheng

The DNA Replication Fork in Eukaryotic Cells, Shou Waga and Bruce Stillman

Structure and Function in GroEL-Mediated Protein Folding, Paul B. Sigler, Zhaohui Xu, Hays S. Rye, Steven G. Burston, Wayne A. Fenton, and Arthur L. Horwich

The Ubiquitin System, Avram Hershko and Aaron Ciechanover

The AMP-Activated/SNF1 Protein Kinase Subfamily: Metabolic Sensors of the Eukaryotic Cell? D. Grahame Hardie, David Carling, and Marian Carlson

Enzymatic Transition States and Transition State Analog Design, Vern L. Schramm

Ribonucleotide Reductases, Peter Reichard and Albert Jordan

Phosphoinositide Kinases, David A. Fruman, Rachel E. Meyers, and Lewis C. Cantley

Sphingolipid Functions in Saccharomyces Cerevisiae: *Comparison to Mammals*, Robert C. Dickson

Role of Small G Proteins in Yeast Cell Polarization and Wall Biosynthesis, Enrico Cabib, Jana Drgonová, and Tomás Drgon

Transporters of Nucleotide Sugars, ATP, and Nucleotide Sulfate in the Endoplasmic Reticulum and Golgi Apparatus, Carlos B. Hirschberg, Phillips W. Robbins, and Claudia Abeijon

G Protein–Coupled Receptor Kinases, Julie Pitcher, Neil Freedman, and Robert Lefkowitz

Nucleocytoplasmic Transport: The Soluble Phase, Ludwig Englmeier and Iain W. Mattaj

From the *Annual Review of Biophysics & Biomolecular Structure*, Volume 26

Flexibility of RNA, Paul J. Hagerman

Modular Peptide Recognition Domains in Eukaryotic Signaling, John Kuriyan and David Cowburn

Protein Folds in the All-β and All-α Classes, Cyrus Chothia, Tim Hubbard, Steven Brenner, Hugh Barns, and Alexey Murzin

Calcium in Close Quarters: Microdomain Feedback in Excitation-Contraction Coupling and Other Cell Biological Phenomena, Eduardo Rios and Michael D. Stern

Eukaryotic Transcription Factor–DNA Complexes, G. Patikoglou and S. K. Burley

Lessons from Zinc-Binding Peptides, Jeremy Berg and Hilary Arnold Godwin

From the *Annual Review of Cell & Developmental Biology*, Volume 13

From the *Annual Review of Genetics*, Volume 31

From the *Annual Review of Phytopathology*, Volume 36

For the convenience of readers, a detachable order form/envelope is bound into the back of this volume.

Len Sussex

Annu. Rev. Plant Physiol. Plant Mol. Biol. 1998. 49:xiii–xxii

THEMES IN PLANT DEVELOPMENT

Ian Sussex

Department of Plant and Microbial Biology, University of California, Berkeley, California 94720; e-mail: sussex@nature.berkeley.edu

KEY WORDS: embryogeny, flower development, leaf development, meristem organization

CONTENTS

INTRODUCTION

Recently I was reading an alumni magazine in which the editor had asked a number of retired or soon-to-be retired professors how they got into their disciplines. I was surprised that for over half of them it had been a seemingly trivial or incidental event such as "I was planning to be a physics major, but I picked up my roommate's introductory biology text and became fascinated by the subject" that had precipitated the initial decision.

In reflecting on my own life, I am also surprised at the number of times when seemingly chance events, such as being in the right place at the right time, had major consequences for my career as a biologist. I think that the realization that there is no one true path to be followed has also influenced how science has been done in our lab by myself and by the people who came to study with me. Instead of defining a biological problem and following it to its end, with everyone contributing a part, my approach has been to find a question that is interesting to me, and to investigate it as long as it remains interesting, then to move on to a new question.

1040-2519/98/0601-xiii$08.00

So my research has involved organisms as diverse as fungi, algae, lycopsids, ferns, monocots, and dicots, and has examined questions as varied as gametophyte-sporophyte relations; the initiation of leaves, roots, and flowers; embryonic polarity; somatic embryogenesis; hormonal regulation of storage proteins and dormancy; auxin metabolism; and some others.

Most of the people who worked in our lab chose their own projects, and the only restriction was that it should be something on which I could give them encouragement and some advice. This was what I learned from Claude Wardlaw at Manchester during my PhD studies. Unfortunately, it seems to be a way of doing science that is dying out as the pressure of large lab groups, competitive grants, and professors who prefer to be called PIs increases.

PLACES

I was born and grew up in a semirural suburb of Auckland, New Zealand, where we had access to pastures, salt marshes, mud flats, beaches, and native bush. It was easy in this environment to develop an interest in plants.

In the 1940s, there was no career advising in New Zealand schools, and I probably would not have attended university had not a teacher suggested this to my parents. At Auckland University College, my undergraduate degree was a BSc in botany. This required all science and no humanities or social science courses but gave me an excellent education in what are now called the classical fields. While I was there, VJ Chapman came from England to be head of the department, and his enthusiasm inspired a generation of students, me included, to study marine and shoreline biology. After completing an MSc degree in 1949 on the ecology of a shoreline plant, I was fortunate enough to obtain a three-year scholarship to Manchester University in England.

The 1950s were exciting times to be in Manchester. It had a large botany department. Professor Eric Ashby was doing research on control of leaf shape and aging in *Lemna*. HE Street, with whom I worked for three months after completing my PhD, was using excised tomato root cultures to study metabolism, and Wardlaw's work on morphogenesis was attracting international attention. In addition, Alan Turing, after picking up a pine cone on a Sunday afternoon walk and being intrigued by the spirals of scales that he observed, began developing his diffusion-reaction theory of morphogenesis. During this time he consulted frequently with Wardlaw, who helped to translate his mathematical models into terms that would be understandable to biologists.

At Manchester I discovered the joys of having access to a good library, and I read a lot. I was especially influenced by Joseph Needham's *Biochemistry and Morphogenesis*, Paul Weiss's *Principals of Development* and D'Arcy

Thompson's *On Growth and Form*. Although these books were principally or entirely devoted to animal questions, they had a major influence on the way I thought about developmental questions in plants.

Taylor Steeves came to Wardlaw's lab for a year on a Sheldon Traveling Scholarship to continue his PhD studies, and he and I began a personal and scientific friendship that has lasted through two editions of *Patterns in Plant Development* and still continues today. After Taylor returned to Harvard, he convinced Ralph Wetmore, his advisor, to invite me to postdoc in his lab, which I did for eight months. Taylor and I then spent four months together working in France with Georges Morel at CNRS, which was located on the palace grounds at Versailles.

Finally, in 1954, it was time to face reality, and I got a position as junior lecturer in the botany department of Victoria University College in Wellington. Although I had intended to remain in New Zealand, I stayed only one year, because the department was unable to support the level of research activity that interested me and there were no external funding sources available. Happily, the situation is very much better in New Zealand universities now.

I applied for and got (without an interview) an assistant professorship at the University of Pittsburgh where I stayed for five years. When Yale offered Taylor Steeves and myself positions in the botany department, it seemed like an ideal opportunity to resume our research collaboration. However, Taylor had only the year previously moved to the University of Saskatchewan and felt that he could not move again. So I went to New Haven in 1960, pleased to be in a coastal environment again. Within two years of my arrival, the botany and zoology departments merged, and I had the opportunity to teach plant development in courses collaboratively with animal developmental biologists. My years at Yale were very happy and productive ones, largely because of the succession of excellent graduate students, postdocs, and sabbatical visitors who worked with me.

In 1989, another of those unanticipated events occurred when I was serving on an NSF/USDA/DOE panel to evaluate biological science centers' grant applications. Interactions with panel members led to my being invited to apply for a position at Berkeley in the newly organized plant biology department. Berkeley presented a great opportunity: a new building, a new lab, and a much larger number of plant colleagues than I had been used to. The lab quickly filled with graduate students and postdocs, and we have had an extended period where people worked together without too many comings and goings. Now that I have decided to retire at age 70, after 43 years of teaching and writing grant proposals, the lab is emptying out as people leave for jobs. I am getting back to doing research with my own hands.

RESEARCH

Above, I commented that my research has involved working on a problem until it was no longer interesting to me and then beginning something new. Often "no longer interesting" meant that there were no methods currently available to advance the problem further. However, in many cases, new approaches did become available later, and I frequently found myself returning to an old problem but using new approaches and new methods. So, although I have used many different organisms in my research, and have investigated many different questions, there are themes that run through these studies of plant development connecting work that was done many years apart. I discuss three of these where the theme is particularly strong.

Meristem Organization

Anyone who has dissected the bud of a vascular plant under a stereomicroscope must surely have been thrilled by the translucent, glistening beauty of the apical meristem and the surrounding leaf primordia. There are subtleties that are lost in the starkness of an SEM image. As well as being thrilled by the appearance of the meristem, one must surely also be awed, thinking, "How does it work and how can I find out?" My interest in meristem organization began in Claude Wardlaw's lab. He was using microsurgical procedures to investigate morphogenesis in fern apical meristems. He agreed that I should work on similar problems, but in angiosperms.

In those days in England there were no course requirements for the PhD, hence the need to read independently, and graduate students did not prepare a thesis prospectus or have a dissertation committee. Wardlaw simply said to me, "Go away and do something." For the next nine months, I wrestled with what to do and what to do it on. One day Frank Cusick, another Wardlaw graduate student, brought me some sprouted potato tubers that he had found in his vegetable bin. These turned out to be ideal experimental material. I could punch out "eyes" on plugs of tuber tissue that were easy to orient on the stage of a stereomicroscope.

The buds contained only about 10 developing scale leaves with no obscuring trichomes and could be dissected easily and quickly. New leaf primordia were initiated at about 24-hour intervals, so development was very rapid. Within a few months, I had done a series of experiments that involved surgically bisecting the shoot apical meristem, isolating the meristem or parts of it from lateral tissue by four incisions, and puncturing it in terminal and subterminal positions. These experiments pointed to the importance of the slowly dividing, distally located cells in maintaining the integrity of the meristem. However, this was about as far as I could go with the surgical approach, and when I left the lab I moved on to other questions.

My next venture into meristem organization came in the 1970s when I became interested in how meristem fate is established. The angle meristem of *Selaginella* was ideal for study of this question. In intact plants, these meristems form rhizophores, root-like structures that initiate roots when they contact the soil surface. If the main shoot apex of the plant is removed, the angle meristem produces a leafy shoot. Thus the meristem is formed before its fate as rhizophore or shoot apex is determined. By the time we began these studies, organ sterile culture systems were quite well developed, so we excised angle meristems devoid of surrounding tissue and found that when we cultured them on a basal medium they developed as shoots, but when cultured on an IAA-containing medium they developed as rhizophores. Furthermore, by culturing excised angle meristems first on basal medium then transferring to the IAA medium, we found that meristem fate was determined several days after the meristem began to grow.

The idea of meristem fate reappeared in our research in the 1980s and 1990s when we produced fate maps of shoot apical meristems. In my reading of the animal literature I had been impressed by the fate maps that embryologists had produced, and how these had formed a foundation for experimental analysis of embryonic development. In 1978, Coe & Neuffer published a fate map of the shoot apical meristem of maize. This showed that specific parts of the meristem gave rise to specific parts of the shoot. This idea of meristem cells being partitioned out in a modular fashion was at odds with anatomical descriptions of shoot development, in which cell lineages from the apex appear to "flow" seamlessly into the developing shoot. In order to investigate whether the model was restricted to maize or was more general, we made fate maps of the shoot apical meristem of *Helianthus* and *Arabidopsis*. These are both dicots, but, like maize, flower terminally. These fate maps were strikingly similar to the Coe & Neuffer fate map of maize, suggesting that in the shoot meristem, cells are partitioned out as components of morphological units consisting of node, leaf, internode, and axillary bud. The beauty of this concept is that it tells us that the morphological units that we identify visually in a plant have a developmental reality. Now we need to find out how that reality is achieved.

One of the most fascinating questions concerning meristems is how cells in the different layers integrate their activities to maintain meristem organization, and function together in organ and tissue formation. In most dicotyledons, the shoot apical meristem consists of three cell layers (L1, L2, and L3) that generate discrete lineages in the plant. L1 forms the epidermis, L2 several underlying cell layers, and L3 the core of the plant. Since there is no exchange of cells between these layers within the meristem, how do the cells of different layers interpret their positions and integrate their functions? Animal developmental biologists were able to analyze similar questions by generating chimeric organisms. In these, embryonic cells of two genetically identifiable organisms were disaggregated and allowed to reassociate as a single embryo in which the cellular contribution of each partner to the new organism could be assessed. Comparable experiments have not been possible in plants because disaggregation and reassociation of tissue cells has not been achieved consistently. However, chimeric plants can be produced from graft regions of two genetically different plants if one of the meristem cell layers in a regenerated apex originates from one of the graft partners and the other layers from the other. By exploiting this approach, we were able to show that many features of meristem function such as meristem size and number of floral organs per whorl are controlled by the L3 layer, the L1 and L2 cell layers behaving as though they respond to signals from the L3 layer. Now the nature of these signals and how they are transferred between layers remain to be discovered.

Most recently, we have taken a new direction in the study of plant meristems, and this is to examine the molecular and cellular events involved in the initiation of a meristem. This work was begun before the discovery of PCR, so it was necessary to use an experimental system that would provide large amounts of developmentally synchronous material for extraction of mRNA. Lateral root initiation turned out to be excellent for this work. Lateral roots can be induced synchronously along the whole length of seedling radish roots that have been exposed to IAA. From these, subtracted cDNA libraries were made that were enriched in genes expressed at specific times in meristem initiation and development, and we identified many such genes. Continuation of this work in *Arabidopsis* focused on the cellular origin of the meristem, and in vitro culture experiments revealed that there was formation of an initial primordium within which a subset of cells became organized to function as the root apical meristem.

Leaf Development

While I was carrying out my graduate studies on the apical meristem of potato, I noticed that if one of the surgical incisions was located between the shoot apical meristem and the presumptive site of the next leaf to be initiated, that leaf frequently would develop as a radially symmetrical and not a dorsiventral organ. This result was strikingly similar to results that Wardlaw had obtained on the fern apex, but in his experiments the radial organ developed as a new shoot apex. Evolutionary morphologists had earlier suggested that the leaves of vascular plants are modifications of branch systems, and these experimental results seemed to indicate that the shift from radial to dorsiventral symmetry resulted from an influence of the shoot apical meristem on the developing lateral organ. However, at that time, molecular or genetic approaches to examine this question were not available.

My next venture into leaf development was begun in collaboration with Taylor Steeves, who had been working on leaf development in the fern *Osmunda cinnamomea*. We found that leaf primordia excised from the bud and placed on quite simple culture media would continue to develop and mature as small replicas of normal leaves even to the extent of initiating sporangia. Fern leaves were ideal for study of the effects of nutrition on leaf shape because, in contrast to dicot leaves that typically have limited or no apical growth, fern leaves develop from an apical meristem that functions for a long time, forming a succession of leaflets along the axis. Later, we showed that by modifying the sucrose level of the medium we could reproduce all the leaf forms from juvenile to adult in excised leaves, thus providing support for Goebel's idea that the simple juvenile leaf form of ferns results from carbohydrate starvation.

Above, I referred to the absence of apical growth in typical dicot leaves. This was not generally realized at the time, and the model that had been developed from anatomical studies invoked apical and subapical initial cells that generated the leaf axis, and marginal and submarginal initial cells that generated the leaf blade. Even at the time, it could have been seen that this model was incorrect because leaves of chimeric plants in which L1 is genetically green, L2 is albino, and L3 is green have white margins and green centers, indicating that all three layers of the meristem contribute to formation of the leaf. But at that time, chimeric plants were thought to be anomalies, and it was not until we carried out a clonal analysis of leaf development in tobacco that experimental proof of the absence of extensive apical and marginal activity in this leaf was provided. Clonal analysis also showed that each leaf is initiated from many founder cells and not from the very few initial cells that earlier anatomical studies had suggested.

A final proof that all three cell layers from the meristem contribute to leaf shape came from studies of tomato/nightshade chimeras in which leaves were large, compound, and tomato-like if L3 was from tomato, and were small, simple, and nightshade-like if L3 was from nightshade, regardless of the genotype of L1 and L2.

Embryogeny

After my early studies on meristems had been frustrated by their small size and the absence of biochemical methods to carry the work further, we decided in 1970 that embryo development would be an interesting way to investigate developmental programs. However, because the earliest stages of a plant embryo are as small as or smaller than its meristems, we decided to focus on later events when the embryo is larger, and to choose plants in which the embryo becomes very large. For this reason, we selected *Phaseolus vulgaris*, the kidney bean, and *P. coccineus*, the scarlet runner bean. These were ideal for our studies. *P. vulgaris*, in particular, is a spontaneously self-pollinating, day-neutral, determinate plant in which the stage of embryo development can be accurately determined from morphological features of the pod and seed. An interesting question in plant embryogeny is the developmental arrest of the embryo and formation of the seed. These phenomena, which were crucial in the evolution of gymnosperms and angiosperms by providing dispersal and survival mechanisms, are in marked contrast to the situation in ferns where embryogeny and postembryonic development occur without interruption. So, how did this new event, developmental arrest, intervene between two previously continuous stages? Our investigations led to the conclusion that the synthesis of abscisic acid, possibly in response to the changing osmotic environment of the embryo, brought about growth cessation and arrest. This conclusion was supported by work on the viviparous *vp1* mutant of maize in which arrest of the embryo does not occur and the embryo germinates precociously on the cob because it fails to respond to the presence of ABA in the seed.

However, the role of ABA was even more profound than this, because in excised embryos we showed that application of exogenous ABA resulted in the activation of synthesis of seed-specific storage proteins and the suppression of expression of genes required for chloroplast function. Thus, ABA appeared to play a pivotal role in the shift from embryo growth and development to arrest and seed maturation.

A way to study developmental events in very early stages of the embryo became available when it became possible to generate a large number of embryo lethal mutants in *Arabidopsis*. Continuation of this line of work has resulted in molecular and biochemical analysis of embryogeny in many labs.

Flower Development

At about the same time that we began work on meristem fate determination in *Selaginella*, we also became interested in how meristem fate is changed in the transition from vegetative to reproductive function so that the lateral organs are now the various kinds of floral organs and not leaves, and the meristem is determinate, not indeterminate, in its growth. Our first approach to this question was to combine surgical and in vitro culture methods. We bisected and excised tobacco floral meristems at different stages of development and grew them in culture where they completed organogenesis. The new organs were floral and were formed in the appropriate whorls so we concluded that floral meristem fate was fixed irreversibly in tobacco and that the factors that determined organ identity were located within the meristem.

Much later, we returned to examine the question of floral meristem determinacy. This was after the ABC model of floral organ identity had been established and when progress was being made in defining the functions of the floral homeotic genes by means of mutant analysis. We were able to show that the AP1 and AP2 genes are required for determinate development of the floral meristem and that they also suppress the formation of axillary buds in first whorl floral organs.

CONCLUSIONS

What are some of the changes I have seen during my academic career? First is the huge increase in the number of scientists who are working in the fields that can broadly be defined as developmental plant biology. In my early years, it was rare to find another person working on the same organism or the same question that I was. Now it is usual to find several labs working on the same genes in the same plant. This has had many consequences. The level of competition has increased, but so has the rate of progress. Two of the great advantages of this population explosion are that it has provided opportunities for more women to become scientists, and it has virtually overwhelmed the "old boy network."

Next is the increased frequency of scientific meetings. I used to go to two meetings a year, the Growth Society, the forerunner of the Society for Developmental Biology, and the AIBS meetings. These were held on a university campus and were cheap to attend. They were small, and it was possible to know essentially all the people attending because they returned year after year. Now there are Gordon and Keystone conferences and FASEB meetings as well as many others that compete for our time and attention. These are usually held in expensive resorts. Most of these meetings are large or their subject matter changes year-to-year so it gets harder to know even a good fraction of the attendees.

Small science versus big science. By small science I mean a project on which one or two people in a lab are working. By big science I do not mean the genome project, valuable as it undoubtedly is, but "the lab project" where the PI has obtained one or more federal grants, and essentially everyone in the lab works on a part of it. It seems to me that the latter way of doing science is a disservice to graduate students because their focus is necessarily narrowed. Because of this, granting agencies have repeatedly urged interdisciplinary collaborations between labs that may be physically distant. But, with small science, which I like to think our lab practiced, at any one time the lab might contain a cell biologist, a biochemist, a molecular biologist, and a geneticist each working on their own projects, but also able to contribute to other projects. In this way, our lab seemed to have many of the aspects of an interdisciplinary approach.

ACKNOWLEDGMENTS

I thank all the graduate students and postdocs who have worked in our lab. They helped me define questions more clearly, and they brought new ideas into the lab. I have not given the names of students identified with the specific studies I have described. It would have been impossible to name everybody and their individual contributions. Next, I thank NSF, USDA, NIH, and other agencies for the financial support that allowed us to carry out our research. Without their aid most of the results discussed here would have remained ideas. Finally, my thanks to Nancy Kerk, my wife, for keeping my feet to the fire when I languished during the writing of this article.

Annu. Rev. Plant Physiol. Plant Mol. Biol. 1998. 49:1–24

GENETIC ANALYSIS OF OVULE DEVELOPMENT

C. S. Gasser, J. Broadhvest, and B. A. Hauser

Section of Molecular and Cellular Biology, Division of Biological Sciences, University of California, One Shields Avenue, Davis, California 95616; e-mail: csgasser@ucdavis.edu

KEY WORDS: embryo sac, morphogenesis, plant evolution, seed, Arabidopsis

ABSTRACT

Ovules are the direct precursors of seeds and thus play central roles in sexual plant reproduction and human nutrition. Extensive classical studies have elucidated the evolutionary trends and developmental processes responsible for the current wide variety of ovule morphologies. Recently, ovules have been perceived as an attractive system for the study of genetic regulation of plant development. More than a dozen regulatory genes have now been identified through isolation of ovule mutants. Characterization of these mutants shows that some aspects of ovule development follow independent pathways, while other processes are interdependent. Some of these mutants have ovules resembling those of putative ancestors of angiosperms and may help in understanding plant evolution. Clones of several of the regulatory genes have been used to determine expression patterns and putative biochemical functions of the gene products. Newly constructed models of genetic regulation of ovule development provide a framework for interpretation of future discoveries.

CONTENTS

1

INTRODUCTION

Function and Basic Structure of Ovules

As the precursors to seeds, ovules play a central role in sexual reproduction of angiosperms and gymnosperms, the currently dominant groups of land plants. Seeds represent a stable, readily dispersed propagule, which commonly includes stored materials for rapid and efficient establishment of seedlings. These properties result in part from the fact that seeds consist of a cooperative association between three plant generations—the parental sporophyte, the gametophyte, and the progenal sporophytes (embryo and endosperm). Evolution of this cooperation between generations, and hence evolution of the ovule, was the key event in the origin of seed plants.

The clear agronomic and evolutionary importance of ovules led to extensive study of ovule structure (reviewed in 5). Ovules of angiosperms contain only three or four morphologically distinct structures. The nucellus is the terminal region of the ovule and is the site of embryo sac formation (Figure 1*a*). Surrounding the nucellus are one or two integuments, lateral structures that usually tightly encase the nucellus (Figure 1*a,b*). The integuments are not fused at the apex of the nucellus but have an opening, the micropyle, through which a pollen tube can gain access to the embryo sac. The basal part of the ovule is the funiculus, a supporting stalk that attaches the ovule to the placental region within the carpel.

Recently, because of its relative morphological simplicity, the ovule has been perceived as an attractive structure for the study of regulation of morphogenesis. Despite this simplicity, ovule development embodies nearly all of the processes that characterize plant development in general: primordium initiation, directional division and cell expansion, asymmetric growth, and cellular differentiation. Thus, an understanding of ovule development has the potential to illuminate many aspects of plant development.

Ovule Evolution

While ontogeny does not actually recapitulate phylogeny, information on evolution of a structure can often contribute to understanding morphogenesis. Seeds (and hence ovules) are first observed in the fossil record in the Upper Devonian or Lower Carboniferous, approximately 330 to 350 million years ago. Fossils

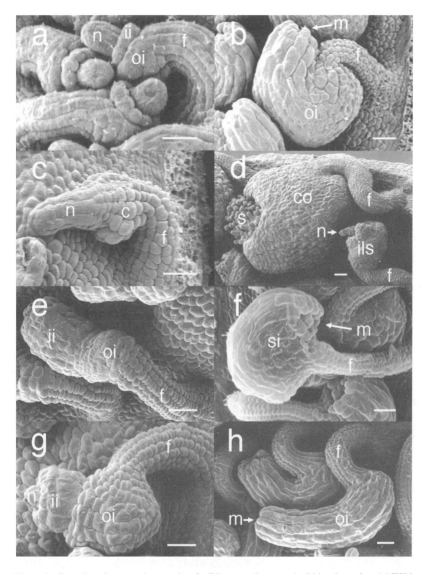

Figure 1   Scanning electron micrographs of wild-type and mutant Arabidopsis ovules. (*a*) Wild-type ovules at the time of integument initiation; (*b*) wild-type ovules at anthesis; (*c*) *ant-5* ovules at anthesis; (*d*) *bel1-4* ovules showing the most common phenotype (*lower right*) and a carpelloid ovule (co); (*e*) *ino* at anthesis; (*f*) *ats* at anthesis; (*g*) *sin1-1* at anthesis; (*h*) *sup-5* at anthesis. Bars = 20 μm in all panels. c, Collar of tissue; f, funiculus; ii, inner integument; ils, integument-like structure; m, micropyle; n, nucellus; oi, outer integument; s, stigma; si, single integument. Photos are courtesy of K Robinson-Beer (*a*, *b*, *d*, and *e*) and JM Villanueva (*f*) or are reproduced from References 14 (*g*) or (40) (*h*), with permission.

from this time period show a series of putative evolutionary intermediate forms that suggest the origin of the first integument. These fossils show fusion of appendages (telomes or sporangiophores) surrounding a megasporangium (or megasporangiophore) to form a sheathlike integument (1, 17). Ovules of such plants were erect, had clearly defined micropyles, and closely resembled ovules of many extant gymnosperms. The current interpretation is that the first integuments originated directly from fused appendages and not from modification of leaves; in fact, leaves are also hypothesized to have derived from fusion of telomes (for example, see 16).

The details of evolution of angiosperms from their gymnospermous predecessors remain largely obscure because of major gaps in the fossil record. Despite this, several firm conclusions can be drawn regarding the evolution of the angiosperm ovule. Extant and fossil gymnosperms have unitegmic (single integument) ovules (16, 49). Bitegmy is the primitive morphology within the angiosperms, because it is the primary condition in all putatively basal groups (5; CS Gasser, unpublished information). Unitegmy is clearly a derived state that has arisen several times within the angiosperms and is largely confined to crown groups within several of the larger clades (5). Thus, the presence of a second integument is a key character separating ovules of angiosperms and gymnosperms.

The second integument has often been discussed as deriving from a cupule, a structure surrounding one or more ovules; the cupule is found in a number of fossil gymnosperms (16, 48, 49). While firm evidence supporting a relationship between the cupule and the outer integument is lacking, it is clear that the origin of the second integument occurred sometime in the Upper Jurassic or Lower Cretaceous, close to 200 million years after the origin of the first (and likely inner) integument.

Ovule Development

The development of a seed is a continuous process. For this discussion we consider ovule development to be those processes occurring prior to fertilization, with further development constituting seed development. Angiosperm ovule development was comprehensively reviewed by Bouman (5) and has also been the subject of more recent reviews (2, 38). A brief summary is presented here.

The sites of ovule initiation are referred to as placentas and are at various locations within the carpels, depending on the species. Ovules originate subdermally through periclinal divisions within the L2 or L3 layers of the placenta. Ongoing division of these cells, in combination with anticlinal divisions in the epidermis, results in formation of an ovule primordium. One or two integuments are initiated from the chalaza, commonly in a ring of cells around the circumference of the primordium (Figure 1a). The inner integument is usually initiated first and is almost invariably of dermal (L1) origin (5). The outer integument can be of dermal or subdermal (L2, or L2 and L3) origin (5). The integuments grow around and are tightly appressed to the nucellus and to each other. In the majority of angiosperms, asymmetric growth of the funiculus, the chalaza, the nucellus, or the integuments, or a combination of these structures, results in curvature of the ovules such that the micropyle is adjacent to the funiculus (Figure 1*b*).

Concomitant with the above processes, a subdermal cell within the nucellus differentiates to form an enlarged archesporial cell. A megaspore mother cell will differentiate directly from this cell or from a mitotic product of this cell (5). Four megaspores result from meiosis of the megaspore mother cell and, depending on the species, one, two, or all four will go on to form the megagametophyte or embryo sac. Most commonly, the embryo sac derives from a single megaspore. Subsequent cell divisions often produce eight nuclei separated into seven cells, but there are many species with a different cellular constitution (for reviews see 38, 53).

GENETIC ANALYSIS AND MOLECULAR CLONING

Recently, a number of groups have initiated classical and molecular genetic analysis of ovule development. This research has primarily focused on *Arabidopsis thaliana* and *Petunia hybrida*. Arabidopsis has the advantages of relatively well-developed genetic tools and extensive information on flower development. A large body of existing molecular work and a very simple transformation system make petunia a useful system for many types of molecular genetic studies. In this section, we use progression of ovule development as a framework to describe results of such genetic studies that provide a useful extension of prior morphological and anatomical studies.

Ovule Initiation and Identity

One would expect mutants in ovule initiation to lack ovules or to have a dramatic alteration in ovule placement or number. To date, such mutants have not been described. In contrast, tobacco mutant plants and transgenic petunia plants have been described that exhibit apparent alterations in ovule identity.

TOBACCO OVULE MUTANTS   Two tobacco mutants (MGR3 and MGR9), which may be defective in ovule identity, were regenerated from cultured cells selected for resistance to a polyamine biosynthesis inhibitor (13, 28). The ovaries of both mutants produced apparently normal ovules and also style- or carpel-like structures in place of some ovules (13). It is possible that these mutants are defective in a function that promotes ovule identity. Both mutants had elevated levels of polyamines, but the relationship between ovule defects and the polyamine phenotype remains unclear, as further work on these lines has not been published.

ECTOPIC EXPRESSION OF *AGAMOUS*   To learn more about the function of *AGAMOUS* (*AG*) genes, Mandel et al (29) produced transgenic tobacco plants ectopically expressing *Brassica napus AG* (*BAG*) under control of the cauliflower mosaic virus (CaMV) 35S promoter. A significant fraction of the resulting plants exhibited floral abnormalities including conversion of sepals to carpel-like structures that develop ectopic ovules, and conversion of petals to stamens (29). These features are similar to those observed in *APETALA2* (*AP2*) mutants of Arabidopsis (7, 23). Within the gynoecium, a subset of ovules of the transgenic plants are converted to style-like structures (29). This phenotype closely resembled that observed by Evans et al (12) in their tobacco mutants. This indicates that ectopic *AG* expression can cause ovule primordia to deviate from their normal developmental fate. As discussed below, however, ectopic expression of *AG* genes in other species can produce different results.

FLORAL BINDING PROTEINS 7 AND 11   Floral binding proteins (FBP) 7 and 11 are encoded by petunia cDNAs that were isolated on the basis of homology to the MADS box domain common to several genes that regulate flower development (3). These proteins share 90% amino acid identity with each other (3). Interestingly, both appear to be single copy genes in allotetraploid *Petunia hybrida* even though both are present in each of the ancestors of this species (3). RNA blot analyses and in situ hybridizations showed that the two genes are expressed in similar patterns with expression confined to the gynoecium (3). Initial expression is in the cells at the center of the flower that will give rise to the placenta. Expression persists in the developing placenta but is later confined to the ovule primordia and then to the funiculus and emerging integuments (3). In mature ovules, expression is strong in the endothelium (3).

Transgenic petunia plants were generated in which expression of both FBP7 and FBP11 was reduced because of the presence of the transgene (3). In homozygous progeny of one such co-suppression line, both FBP7 and FBP11 mRNAs were undetectable. Ovaries of these plants contained a few sterile ovules with wild-type morphology, but the majority of ovules were partially or completely converted to style-like structures (3; Figure 2*b*). These structures emerged directly from the placenta, not from any visible ovular structure (3). The style-like organs closely resemble those observed in the tobacco mutants of Evans et al (13) and in tobacco plants ectopically expressing BAG (29) A similar, but much weaker phenotype was observed in plants hemizygous for the transgene. The levels of mRNA of putative petunia orthologs of *AG* were slightly higher than wild-type levels in ovaries of homozygous plants (3).

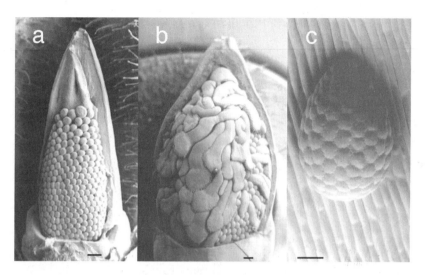

Figure 2 Scanning election micrographs of petunia ovules. Dissected ovaries of (*a*) wild-type petunia and (*b*) transgenic plant with reduced expression of FBP7 and FBP11. (*c*) Ectopic ovule on corolla of a plant ectopically expressing FBP11. Bars = 500 μm (*a* and *b*) and 50 μm (*c*). Reproduced from References 3 (*a* and *b*) and 10 (*c*), with permission.

Effects of ectopic expression of FBP11 were examined by generating transgenic petunia plants with this coding region under control of a CaMV 35S promoter (10). One primary transformant with an extreme phenotype had alterations in sepal development that included the absence of trichomes and the formation of placenta-like tissues bearing ovules at fusion points of the sepals (10). Rarely, ovules were also observed on the abaxial side of the tubular corolla, which showed no other apparent alterations (10; Figure 2*c*).

The carpelloid ovule phenotype of plants with suppressed expression of FBP7 and FBP11 implicates at least one of these genes as an important determinant of identity of ovules or of the placenta (2, 3). The reduced effect of the transgene in hemizygous versus homozygous plants may reflect the need for a threshold of activity as a switch for selecting the developmental fate of the meristems. A role for FBP11 in ovule and placenta identity is further supported by the formation of ectopic placental regions bearing ovules on sepals, and of occasional ovules on petals, of plants ectopically expressing this gene (10). While together these results present a strong case for FBP11 and FBP7 involvement in ovule and placental identity, other interpretations of some results are possible. The coincidence of an absence of trichomes on the ovule-bearing sepals of the FBP11-overexpressing plants suggests at least partial conversion of these organs to carpels, rather than merely production of ectopic ovules. Conversion of sepals to carpel-like structures is also seen in *AP2* mutants of Arabidopsis and in tobacco (20, 29) or Arabidopsis (30) plants ectopically expressing *AG* or other related MADS-box genes (MF Yanofsky, unpublished information). Because FPB11 is in the *AG* group of MADS-box genes (3, 32, 34), the effects of ectopic expression of FBP11 on sepals may occur because FBP11 weakly mimicks *AG*.

AP2   The Arabidopsis *ap2* mutation was originally identified because it reduces floral organ number and causes homeotic changes in the first two whorls of floral organs (7, 23). More recently, two *ap2* alleles, *ap2-6* and *ap2-7*, were shown also to affect ovule development. Normal ovules are produced by *ap2-6* plants, but filamentous structures and structures with features of both carpels and sepals also arise from the end of a funiculus or directly from the placenta (31). These aberrations suggest that *AP2* may have a role in the promotion of ovule or placental identity (31). However, these mutants produce many normal ovules, and other strong alleles of *ap2* produce only morphologically normal ovules (19). *AP2*, is therefore, not essential for ovule development; the precise role of this gene in ovule development remains unclear.

Integument Initiation, Identity, and Development

The first morphological change to occur in the initially featureless ovule primordium is the emergence of the integuments. Extension of the integuments results from anticlinal cell divisions in combination with directional cell elongation. A number of Arabidopsis mutants have been identified that affect initiation, identity, or development of integuments.

ANT   As shown in Figure 1*c*, strong *aintegumenta* (*ant*) mutants fail to develop integuments (4, 11, 21). Even in putative null alleles, elimination of the integuments is not always complete, and some expansion of the chalazal region may occur late in ovule development (4, 11, 21). This expansion is at least partly under control of other genetic loci, as the degree of expansion varies in different genetic backgrounds (11). Strong *ant* mutants also fail to form embryo sacs (4, 11, 21).

Several weaker alleles of *ant* have been described (4, 21). The *ant-8* allele forms a single asymmetric structure in place of the two integuments (4). This structure can grow to resemble an outer integument that partially encloses the nucellus. The *ant-3* allele has even more extensive integument growth and forms two integuments, but the separation between the two integument primordia is less distinct than in wild type (21). Initiation of the inner integument is asymmetrical, and the integuments do not grow to their normal size. In rare cases a functional embryo sac develops in *ant-3* ovules (21).

Klucher et al (21) showed that the early expression of the *BEL1* gene (see below) was not altered in *ant* mutants. Because *BEL1* expression specifically marks the chalazal region of the ovule, this observation shows that a proximal-distal zonation of the ovule primordium occurs in *ant* mutants. Baker et al (4) noted that cells on the surfaces of the nucellus, chalaza, and funiculus of Ant⁻ ovules take on different shapes and textures as the primordia age. This indicates continuation of some aspects of ovule development. Together, these observations show that strong *ant* mutations result in failure in integument primordia formation or enlargement rather than failure in formation of a chalazal region or an arrest of ovule development in general.

ANT is thus necessary for normal initiation of integument primordia, for formation of two separate primordia, and for subsequent expansion of primordia, with larger amounts of activity necessary for each of these progressive steps. Whether the absence of embryo sac development is a direct effect of the *ant* mutations or an indirect effect of the lack of integuments is unknown. However, the absence of *ANT* expression in the nucellus during embryo sac formation (11) argues against a direct role for ANT protein in megagametophyte development.

In addition to altered ovule development, *ant* mutants exhibit a consistent decrease in the size and numbers of other floral organs (4, 11, 21). As for the effects on ovule development, these effects indicate that *ant* mutations act to inhibit initiation or expansion of lateral structures.

Two groups have independently isolated clones of the *ANT* gene using insertional mutants (11, 21). The ANT protein is homologous to the AP2 protein of Arabidopsis. In addition to the presence of a two-fold repeat of a putative DNA binding domain and a conserved spacer sequence (like AP2), ANT also includes a serine-rich region, a glutamine-rich region, and a putative nuclear localization signal (11, 21). Together, these properties make it highly likely that ANT is a transcriptional regulator.

The pattern of expression of *ANT* as determined by in situ hybridization, is largely consistent with the visible effects of *ant* mutations. *ANT* mRNA was found in the primordia of all floral organs early in their development and was persistent in the margins of petals (11). *ANT* mRNA was present throughout ovule primordia early in their development but was restricted to the chalazal region by the time of integument initiation (Figure 3*a*). *ANT* mRNA was also found in meristematic regions of the shoot despite a lack of phenotypic effects of *ant* mutations on these structures (11). The *ANT* gene does not appear to be autoregulatory as the pattern of accumulation of *ANT* mRNA was similar in ovules of wild-type and *ant-9* (apparently null) mutant plants (11).

Analysis of *ant-9 ap2-2* double mutant plants revealed strong synergism between these two mutations. Strong *ap2* mutations led to decreased numbers of organs in the first three whorls of flowers and caused homeotic transformations

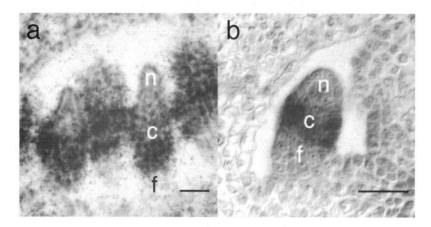

Figure 3 Localizaton of *ANT* and *BEL1* mRNA. Sections of ovule primordia hybridized with (*a*) antisense *ANT* cDNA and (*b*) antisense *BEL1* cDNA. In both cases, *dark regions* indicate hybridization signal. Bars = 10 μm. Symbols are as in Figure 1. Reproduced from References 11 (*a*) and 39 (*b*), with permission.

of organs that do develop in the first two whorls (8). All first, second, and third whorl organs are usually absent from flowers of double mutant plants, and such flowers consist of only a gynoecium and occasional subtending rudimentary filaments or bract-like organs (11). Thus, the effects of *ant* and *ap2* mutations on organ number are additive. This observation, and the homology between these two genes, suggest that each gene may partially compensate for the absence of the other in promoting floral organ initiation and growth. The ovules of the double mutant plants do not differ significantly from those of *ant* single mutants, indicating that *AP2* cannot compensate for loss of *ANT* function in ovule development.

BEL1   Phenotypic effects of *bell* (*bel1*) mutations are confined to the ovules where integument identity appears to be largely lost. Significant growth does occur at the chalazal regions of Bel1⁻ ovules in the form of a single relatively amorphous collar of tissue [the "integument-like structure" (ILS); Figure 1*d*] (31, 40). While the asymmetric shape of the ILS primordium resembles that of a normal outer integument, subsequent growth is irregular, and the ILS does not resemble either integument. Embryo sac development rarely proceeds beyond the formation of megaspores in *bell* mutants, and further steps are aberrant when they occur (17, 31, 40). Growth of the ILS is not always evenly distributed, and a variable number of protuberances often extend from the edges of this structure (17, 31, 40). While the cells of the protuberances are usually parynchymatous, a subapical cell can take on the appearance of a megasporo-cyte (17). Herr (17) has hypothesized that these protuberances may represent ectopic nucelli.

While the formation of a collar-like ILS is the terminal condition for the majority of Bell⁻ ovules, the ILS of a significant subset of ovules can expand dramatically forming carpelloid structures (31, 36; Figure 1d). These structures can include ovary and stylar regions, stigmatic regions, and secondary ovules (which reiterate the Bell⁻ phenotype; 31, 36).

Reiser et al (39) cloned the *BEL1* gene using a T-DNA (transferred DNA of *Agrobacterium tumefaciens*) tagged line. The deduced BEL1 protein includes a homeodomain DNA-binding motif and is therefore likely to be a DNA-binding transcriptional regulator. The sequence of BEL1 differs significantly from that of most previously described homeodomain proteins, thus BEL1 represents a member of a new class of such proteins (39). Within flowers, *BEL1* mRNA is found exclusively in ovules. *BEL1* mRNA is initially present throughout an ovule primordium but becomes restricted to the chalazal region before emer-gence of the integument primordia. Thus, the pattern of *BEL1* expression (like that of *ANT*) demonstrates that the chalazal domain is established before emer-gence of the integuments (39) and is a molecular marker for this region.

The ILS of a Bell⁻ ovule has been interpreted as replacing only the outer integument, with the inner integument being absent (31, 36, 40). Thus, like *ANT*, *BEL1* may be necessary for initiation of the inner integument. *BEL1* would then have a different role in the outer integument—directing it to its normal developmental fate (39). However, it is also possible that the ILS derives from cells that would normally give rise to both integuments and thus would represent a fusion of these two structures (15). In this model, *BEL1* has a single role—determining the identity of the region giving rise to both integuments.

Ray et al (36) focused on the homeotic conversion of integuments to carpels late in development of Bell⁻ ovules. They observed that the expression of *AG*, a gene closely associated with carpel development (7), appeared to be higher in late stage Bell⁻ ovules than in wild-type ovules. In addition, they found that Arabidopsis plants containing a transgene that should lead to overexpres-sion of *AG* had ovules similar to those of *bell* mutants. On the basis of these observations, they hypothesized that the carpelloid nature of *bell* ovules re-sulted from ectopic *AG* expression and that *BEL1* was a negative regulator of *AG*. Subsequently, Reiser et al (39) showed that levels and distribution of *AG* mRNA were unaltered in Bell⁻ ovules early in development when the mutant phenotype was first manifest, indicating that if such negative regulation exists, it must be indirect. On the basis of his observation of putative ectopic nucelli on some Bell⁻ ovules, Herr (17) hypothesized that the *bell* mutation may be atavistic, converting ovules to structures resembling unfused sporangiophores homologous to precursors of the first integuments.

Both these models, as well as other discussions on this gene (31, 39, 40), include the concept that *BEL1* plays an important role in determining integument identity. In fact, it is possible to explain all these phenotypes if *BEL1* simply directs the meristematic cells in the chalazal region to form integument primordia. In this model, the chalazal cells maintain their meristematic properties and continue to divide and expand under control of *ANT* and possibly other genes. BEL1 activity causes this growth to be directed toward integument formation. In the absence of BEL1 activity, this region has three possible fates. The most common is simple maintenance of the undifferentiated state, producing the parynchymatous ILS of most Bel1$^-$ ovules. Less frequently, this region reverts to the identities of the meristematic regions from which it has derived—either the placenta, where it then gives rise to ectopic ovule primordia, or the central floral meristem, where it gives rise to carpels. In the absence of the strong directive influence of *BEL1*, there may be a delicate balance between these three fates, in which stochasitc deviation from the undifferentiated state leads to a self-reinforcing commitment to the carpel primordium or placental developmental pathway.

INO   Effects of the Arabidopsis *inner no outer* (*ino*) mutation appear to be confined to the outer integument where both organ initiation and subsequent development are affected (4, 14, 45). Following normal initiation of an inner integument, the outer integument of an Ino$^-$ ovule appears to initiate on the opposite side from that of wild-type ovules (4). The rotation appears to be specific to the outer integument primordium because other bilaterally symmetrical aspects of ovule development are unaltered. The funiculus bends in the normal direction toward the base of the pistil, and the nucellus bends normally toward the apex (stigma) of the pistil (4). Thus, the effect of the *ino* mutation on initiation of the outer integument appears to be a 180° displacement of the region producing this structure around the axis of the ovule primordium.

The aberrantly oriented outer integument primordium of *ino* mutants undergoes minimal further development following initiation (Figure 1e). *INO* may have two roles, orientation of the outer integument primordium and promotion of growth of this structure. Alternatively, *INO* may only affect orientation of the outer integument primordium, and the absence of further development may be a secondary effect of misorientation.

ALTERED TESTA SHAPE   In ovules of the Arabidopsis *altered testa shape* (*ats*) mutant ovules, the inner and outer integuments are replaced by a single integumentary structure (25). In wild-type ovules, the inner and outer integuments consist of three- and two-cell layers, respectively. The innermost layer of the inner integument differentiates to form an endothelium (40). In mature seeds of Arabidopsis, the external layer of the outer integument produces columellae, characteristic central elevations in the desiccated cells (25). The integument of an *ats* ovule consists of only three cell layers that include both an inner endothelium and an outer layer that will form columellae (25). Thus, Ats⁻ ovules have a single integument with properties of both inner and outer integuments.

One interpretation of this phenotype is that ats ovules fail to form the furrow separating the two integuments (25). *ats* integuments remain fused together, but cell layers within the compound structure maintain their identities and differentiate appropriately.

SIN1   The Arabidopsis *short integuments 1* (*sin1*) mutation affects growth of the integuments and general growth of the plant. *sin1* was originally identified in an *erecta* (*er*) background where it resulted in reduced apical dominance, short internodes, late flowering time, reduced pollen production, and infertile ovules with short integuments as a result of reduced cell elongation (Figure 1g; 24, 40). Sin1⁻ plants are infertile because meiosis does not occur (40). Lang et al (24) found that in an *ER* background *sin1* internodes were of normal length (*SIN1 er* plants have internodes of intermediate length), and normal pollen production was restored. The majority of ovules of *sin1 ER* plants have short outer integuments, inner integuments much longer than in wild type, and arrested megasporogenesis. Following pollination, some ovules develop a normal outer integument, but the inner integument enlarges into a hollow structure that can be even larger than a normal seed. No morphological changes were observed after pollination of the *sin1 er* ovules (24). Several publications (24, 35, 37, 40) provide more details on these and other aspects of the Sin1⁻ phenotype.

The above data indicate that ER can mask the effects of *sin1* on internode elongation, and that SIN1 can partially compensate for the absence of ER (24). Thus, while both ER and SIN1 can contribute to internode elongation, ER is more critical for this process. Effects of *er* mutations on ovules are only visible in a *sin1* background, indicating that SIN1 can completely compensate for absence of ER in developing ovules. One of several possible explanations for these effects is that SIN1 and ER are similar proteins and that their different effects reflect levels of expression of one or the other gene in specific structures.

SUP   Arabidopsis *superman* (*sup*; also referred to as *floral mutant 10, flo10*) mutants were originally identified by their effects on gross floral morphology. Sup⁻ flowers have supernumerary stamens and a corresponding reduction in the amount of carpel tissue, sometimes leading to a complete absence of a gynoecium (6, 47). Gaiser et al (14) noted that Sup⁻ ovules are aberrant. Formation of the asymmetric outer integument primordium is normal in Sup⁻ plants. However, subsequent growth is approximately equal on all sides of the primordium, resulting in a long tubular outer integument (Figure 1*h*). The radially symmetrical inner integument is visible in ovules of *sup ino* double mutants and does not appear to be affected by *sup* mutations (14). The specific role of the *SUP* gene in ovule development thus appears to be to suppress growth of the outer integument on the adaxial side of the ovule (14).

Effects of *sup* mutations on stamen number and carpel development have been shown to largely or completely result from expansion of expression of a floral homeotic gene, *APETALA3* (*AP3*), outside the third whorl of primordia and into the region normally giving rise to the gynoecium (7, 44, 47). Ovules of *sup ap3* double mutants are indistinguishable from those of *sup* single mutants (14). This indicates that *AP3* does not play a role in the effect of *sup* mutations on ovule development and that there are significant differences between the mechanisms by which *SUP* mediates floral and ovule development.

Cloning and sequencing of the *SUP* gene showed it to encode a protein with properties of a DNA-binding transcription factor (44). *SUP* mRNA was detected in the innermost region of the third whorl of floral organs, and in the funiculus adjacent to the outer integument, but was not detected in the fourth whorl primordia, or in the integument primordia—the structures most affected by the mutation (44). One simple mechanism that could explain both apparently non-cell-autonomous effects of *SUP* is that the function of the SUP gene product is to create a boundary that prevents expansion of the zone of expression of some factor beyond the region of cells where SUP is present. In the floral apex, this factor could be AP3, while an as yet unknown growth-promoting factor would be regulated by SUP in the outer integument of ovules (44).

TOUSLED   The *tousled* (*tsl*) mutation was originally identified by its floral aberrations (43). This mutation leads to a decrease in the number of organs in the three outer whorls, a slight increase in the number of carpels in the fourth whorl, and altered morphology of all floral organs (42). The Tsl⁻ gynoecium exhibits reduced development in apical tissues, leads to failure of postgenital fusion of the style and septum (42). Tsl⁻ ovules have a protruding inner integument as a result of abnormal elongation of this structure, and variable but usually reduced development of the outer integument (42). The opposite effects of TSL activity on growth of the gynoecium and the inner integument indicate that the serine/threonine kinase activity of this protein (41, 43) regulates different aspects of cell proliferation in different plant structures.

LEUNIG   The *leunig* (*lug*) mutation was identified by its pleiotropic effects on leaves and floral organs (22, 26). Lug⁻ plants have narrow serrated leaves, slightly carpelloid sepals, and stamenoid petals. Genetic experiments indicate that a primary role of LUG may be to negatively regulate expression of *AG* and that aberrant *AG* expression may be responsible for many of the effects of the *lug* mutation (26). Lug⁻ and Tsl⁻ plants have similar gynoecia and ovules, with ovules of both having a protruding inner integument (42, 45). The observation that ovules of *tsl lug* double mutants have similar phenotypes to either single mutant suggests that both genes could act on a single process in ovule development (42). The effects of *lug* mutations on ovules are distinct from the phenotypes observed in plants overexpressing *AG*; it is unlikely that *AG* is responsible for the Lug⁻ ovule phenotype.

OTHER MUTANTS   In their recent analysis of a large set of Arabidopsis ovule mutants, Schneitz et al (45) provide initial descriptions of six new mutations affecting integument development. In *huellenlos* (*hll*) mutants, integuments appear to initiate, but their development is limited. The inner integument primordia undergo only a few cell divisions, and the region from which the outer integument would form usually undergoes only minimal cell expansion and no cell division. This lack of development can be followed by precocious degeneration of the nucellar region. The *unicorn* (*unc*) mutation acts relatively early in ovule development and leads to formation of a protrusion at the base of the outer integument. Other mutations act later in development and lead to dissected outer integuments (*strubbelig, sub*), highly irregular integuments (*blasig, bag*), integuments with enlarged cells (*mollig, mol*), or a protruding inner integument (*laelli, lal*). Further characterization of these mutations will allow more complete understanding of their roles in ovule development. Many additional genes likely remain to be identified.

Embryo Sac Development

As noted above, several ovule mutants affect both the sporophytic parts of ovules and the embryo sac. Strong alleles of *bell*, *sin1*, and *ant* lead to early arrest of embryo sac development (4, 11, 21, 40). The absence of detectable expression of *ANT* or *BEL1* in cells giving rise to the embryo sac led to the hypothesis that failure in embryo sac formation may be an indirect result of the absence of integuments (11, 21, 39). The recently described *hll* mutation, which results in highly reduced integuments, also fails to form an embryo sac (45). *ats* and *ino* mutants produce ovules that have one integument around the nucellus, and both mutants produce at least some functional embryo sacs (4, 25). *ats ino* double mutants have a naked nucellus and fail to form embryo sacs (4). In every case in which integuments do not enclose the nucellus, an embryo sac fails to form, and at least one integument may be essential for normal embryo sac formation. It is clear that a sheathing integument is not sufficient for this process, however, as numerous mutants have been described that have aberrant embryo sacs despite the presence of normal integuments (18, 45, 51).

MODELS FOR GENE ACTION

A number of conclusions about the regulation of this process can be drawn from observations of wild-type ovule development (31, 40, 46). The regular arrangement of ovules on the placenta demonstrates the existence of a patterning process to define the locations for ovule initiation. After initiation of an ovule primordium from the placenta, proximal-distal patterning of the primordium defines three zones that will give rise to the funiculus, integuments, and nucellus (46). In addition, the bilateral symmetry of most ovules indicates that lateral patterning must also occur at this time (4). Ovule mutants will aid characterization of genetic interactions to formulate more detailed models of regulation of ovule development.

Ovule Genes and the ABC Model

Two groups have attempted to interrelate regulation of ovule development with the now well-established "ABC" model of floral organ identity (9, 27, 33, 52). As noted above, the simple model of Ray et al (36), in which *BEL1* acts directly as an inhibitor of *AG* action in ovules, was not supported by more recent molecular studies (39). Transgenic petunia plants under- or overexpressing FBP11 show an apparent loss of ovule identity and formation of ectopic ovules, respectively (3, 10). Based on these observations, and the fact that the petunia placenta arises from the floral meristem, it was proposed that FBP11 is a member of a "D" class of genes regulating placenta and ovule identity (2, 10). Because no specific interactions between D genes and ABC genes are proposed, additional work will be required to see whether this model adds to our understanding of ovule and placental development.

Ovule Gene Interactions

In contrast to the ABC classes of genes regulating floral organ identity, where numerous different interactions were found (9, 27, 33, 52), interactions among ovule genes have primarily been shown to be either strict epistasis or simple additivity. With respect to effects on ovules, *ant* was found to be epistatic to *bel1* (4, 21), *ap2* (11), *ino* (4), *sin1* (4), and *sup* (4). *bel1* was shown to be epistatic to both *ino* (4) and *sup* (14), and *ino* was shown to be epistatic to *sup* (4). These relationships imply that these genes act in a common developmental (but not necessarily biochemical or signal transduction) pathway, and help to define their order of action (see below). The epistasis of *ant* over all these mutations is easily explained by the fact that they affect the integuments, which are absent in *ant* mutants.

While *ant* was epistatic to *sin1*, *sin1* showed apparent simple additivity with *bel1* (40), and *ino* (4). The additivity with *bel1* and *ino* may indicate that the effects of *sin1* are relatively independent of the actions of these other two genes. Because *ino* simply eliminates the outer integument, it is not surprising that the effects of *sin1* on the inner integument are still readily apparent in the double mutant. That *sin1* has an effect on the ILS of *bel1* mutant ovules is somewhat more surprising. Most other evidence indicates that the ILS forms as a result of loss of integument identity. However, the fact that *sin1* still has an effect on this structure is an indication that the loss of integument identity may not be complete. Additive effects with several different ovule mutations may be an indication that *SIN1* has a general role in cellular function and that its effects on integument development may be due to its pattern of expression. That *sin1* mutations can also have effects on elongation of other parts of plants is also consistent with this hypothesis.

The *ino* and *ats* mutations also show additive effects (4). In double mutant plants, the aborted outer integument (resulting from the *ino* mutation) is fused to the inner integument (the result of the *ats* mutation) supporting the hypothesis that *ats* causes integument fusion (4). It further appears that the fusion to the abortive outer integument prevents full development of the inner integument, and the nucellus remains uncovered. As noted above, this absence of a sheathing integument is associated with failure in embryo sac development (4).

Comprehensive Models of Ovule Development

In their recent review, Angenent & Colombo (2) integrated information from studies on Arabidopsis and petunia to describe ovule development as a linear series of steps, and associated a total of seven genes with regulation of specific steps. According to their model, FBP7 and FPB11 participate in ovule initiation, *ANT* participates in integument initiation, *BEL1* participates in integument identity, *INO* and *SIN1* participate in integument elongation, and *SUP* participates in asymmetric proliferation.

On the basis of new and previous analysis of ovule mutants, Baker et al (4) propose a similar order of gene action in a more complex, branched model for genetic regulation of ovule development. In this model, following the patterning of the ovule primordium into the funiculus, chalaza, and nucellus, each of these regions develops in at least partial independence of the others. The Arabidopsis ovule genes described to date are proposed to act primarily within the chalazal region, where they govern the development of the integuments. Origination of the integument primordia as two separate structures is viewed as a specific genetically regulated event under the control of *ATS*, *ANT*, and *BEL1*. The pathway branches further with separate developmental pathways for the inner and outer integuments, and for cell division and cell expansion in these structures. Effects of the ovule mutations on embryo sac development are proposed to be indirect, resulting from the absence of sheathing integuments (as also proposed by others for *BEL1* 39; and *ANT* 11, 21).

Schneitz et al (45) independently formulated a branched model of regulation of ovule development that shares many features with that proposed by Baker et al (4). The model includes branches to indicate the relative independence of development of different parts of ovules but parses the overall process into types of developmental processes (pattern formation, initiation of morphogenesis, and morphogenesis) rather than into ovule-specific developmental events. This model also incorporates additional genes newly identified by the authors (45).

Using information from all the above models, we propose a consensus model representing the current state of knowledge of the order of action and roles of genes known to be involved in ovule development (Figure 4). The model describes Arabidopsis ovule development because most currently identified genes are from this species. As in other models, we hypothesize that there may be ortholog(s) of the petunia FBP7/FBP11 genes in Arabidopsis that play similar roles in this species. The model also includes establishment of a lateral pattern, necessary for the bilaterally symmetrical aspects of ovule development, which was not addressed in previous models.

OVULE GENES, FLOWER GENES, AND PLEIOTROPIC EFFECTS

As noted in the detailed descriptions of the genes affecting ovule development, mutations in a number of these genes result in pleiotropic effects, with many of these effects confined to flowers. In some cases, enough information is available to provide a mechanistic explanation for the pleiotropy. For example, the data on sequence similarity and overlapping expression patterns for *ANT* and *AP2* provide a reasonable explanation for the partial functional redundancy of these genes in floral organ formation (11, 21). This redundancy raises questions about the "original" role of *ANT*, the origin of *AP2*, and the role of their common ancestor gene. Because ovules precede flowers in the ancestors of angiosperms, one possibility is that the earliest role of *ANT* was in promotion of integument formation. *AP2* could be a diverged duplicate of *ANT* that evolved to its current role during the evolution of flowers. Because Arabidopsis contains a large family of other related genes, *AP2* could also derive from another member of this family that had a role in other aspects of development in the ancestors to angiosperms. *lug* mutations also appear to interact synergystically with *ap2* mutations, but *lug* has entirely different effects on ovule development than does *ant*. The effects of *lug* mutations on ovules are, however, not as

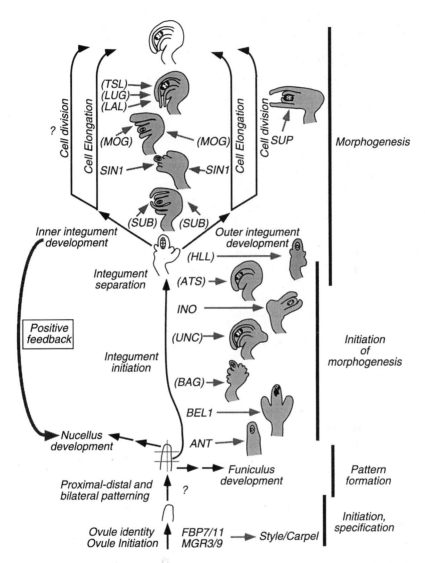

Figure 4 Integrated model of genetic regulation of ovule development. Features of published models for regulation of ovule development (2, 4, 39, 45, 46) were assembled to summarize current knowledge of genetic regulation of ovule development. Progress of ovule development is indicated from *bottom* to *top* by *black arrows*, and intermediate structures of wild-type ovule development are shown *unshaded*. Gene designations are adjacent to the process they are believed to affect. Parentheses indicate greater uncertainty in placement. ?, Predicted genes that have not yet been identified. *Shaded* structures represent phenotypes at anthesis of ovules of mutants in each of the indicated genes. Bars at the *right* of the figure relate the pathway to developmental process descriptions of Schneitz et al (45).

profound as those of *ant* mutations, and thus *LUG* may play a less essential and more recently derived role in ovule development. The dual roles of *SUP* in flower development and in outer integument asymmetry have been hypothesized to be two manifestations of regulation of cellular proliferation by this gene (44). Because *SUP* acts through *AP3* in flower development and not in ovule development, it is also possible that two different mechanisms are involved in these processes. It is unknown whether evolution of integument asymmetry, or even a second integument, preceded flower evolution; thus, it is not possible to tell which of the two roles of *SUP* is primary and which is derived.

Both *tsl* and *sin1* mutations have multiple pleiotropic effects on flower and plant development. TSL is a serine/threonine protein kinase (41) and thus may have impacts on a variety of different process that can be regulated by protein phosphorylation. Plants contain many protein kinases that may differentially mask the effects of *tsl* mutations in different parts of plants. TSL may, therefore, play roles in plant development in addition to those indicated by the observed phenotypes. The variety of effects of *sin1* on ovule, flower, and plant development indicate that, like TSL, the product of this gene may affect a variety of processes through a single mechanism. The only clue we have to the possible nature of SIN1 is its apparent partial overlap in function with ER, making it possible that these proteins are homologous or that they interact in some way. ER is proposed to be a receptor kinase that includes extracellular domains (50). If SIN1 is indeed homologous to ER, then its pattern of expression, in combination with differential presence of specific phosphorylation targets, could explain the wide range of processes affected by *sin1* mutations.

OVULE GENES AND OVULE EVOLUTION

The evolution of ovules was briefly described at the beginning of this review. Because ovule development is ultimately under genetic control (although this control can manifest through other downstream mechanisms—wall tension, cytoskeletal arrangements, diffusible gradients, plasmodesmatal trafficking, etc), evolutionary changes in ovule morphology must coincide with changes in genes regulating this process. Several of the currently identified ovule regulatory genes may have been important participants in initial evolution of ovules and subsequent radiation of angiosperm ovule morphology.

Ovules of strong *ant* mutants completely lack integuments. In this respect, they resemble the sporangiophores of the precursors to seed plants. The essential role of *ANT* in formation of integuments implies that evolution of this function for this gene was concomitant with evolution of integuments. Investigation of conservation of function of *ANT* orthologs in other species will help to verify or refute this hypothesis.

Herr (17) interprets one of the phenotypes of Bel1⁻ ovules as a branching axis with multiple terminal nucelli. On this basis, he has speculated that *bell* mutants are atavistic mimics of an even earlier evolutionary precursor to ovules—a fertile axis comprising multiple sporangiophores. However, this is only one of three fates of the ILS of *bell* mutants, the others being an amorphous collar of tissue or a carpelloid structure. We therefore favor the hypothesis, outlined in our description of this mutant, that the loss identity resulting from *bell* mutations leaves the cells in a meristematic state without a firm direction to a specific fate. In either case, the role of *BEL1* in determination of integument identity argues for coincidence of evolution of this gene and evolution of integuments.

Ovules of *ino* mutants, which have only a single (inner) integument, resemble those of extant and fossil gymnospermous plants, including putative progenitors of angiosperms. This implicates the *INO* gene as a critical component in both development and evolution of the outer integument. The origin of the outer integument remains largely obscure, and molecular analysis of this gene—and of potential orthologous genes in conifers and Gnetales—will provide a new avenue to address the previously untractable question of this origin. Unitegmy is clearly a derived state in the majority of angiosperms displaying this trait (see the section on Ovule Evolution, above). The unitegmic ovules of some angiosperms appear to result from loss of the outer integument (5) and therefore represent phenocopies of *ino* mutants. It will be of great interest to examine *INO* orthologs in these species. The most common alteration leading to unitegmy is, however, congenital fusion of the two integuments into a single structure (5). This is the apparent effect of the *ats* mutation. Molecular studies on *ATS* orthologs will enable testing of the obvious hypothesis that evolution of this type of unitegmy results from alteration in the nature or expression patterns of such genes.

The asymmetric shape of ovules can be seen to result from several different genetically separable steps including curvature of the funiculus, the initial orientation and asymmetric shape of the outer integument primordium, bending of the base of the nucellus, curvature of the nucellus and embryo sac, and asymmetric expansion of the outer integument (4, 45). *SUP* is clearly a critical determinant in the last of these processes; this gene is likely to be involved in the evolutionary changes separating amphitropous and orthotropous ovules. Studies on the nature and expression of *SUP* in a variety of angiosperms may allow the determination of the molecular mechanism for some of the evolutionary changes in ovule morphology.

PERSPECTIVE

The past five years have seen rapid progress on genetic regulation of ovule development. Before this period, no systematic attempts were made to identify genes involved in this process. Now more than a dozen loci have been identified, and ongoing efforts should uncover more. While some of these genes apparently function only in ovules, many also regulate aspects of development in other floral and even vegetative structures. In retrospect, this is not surprising given that mutations were selected for both alterations of identity of ovules and their substructures, but also for effects on ovule morphology. Thus, genes regulating general aspects of cell division and directional expansion, which would be expected to be important in morphology of all parts of plants, could be identified. Ovules provide a simple system in which it may be possible to determine the specific roles of such genes. The isolation of ovule regulatory genes has only just begun but has already produced interesting results. Studies using the cloned genes will eventually provide an understanding of regulation of ovule development at the molecular level. Of equal importance, cloned genes allow ongoing studies to cross taxonomic lines by facilitating isolation of orthologous genes from different species. Examining the presence or absence of such genes, or the nature and regulation of genes once found, may provide a new avenue for understanding the evolutionary origin of angiosperm ovules and the evolution of the wide variety of ovule morphologies within the angiosperms.

ACKNOWLEDGMENTS

We wish to thank GC Angenent, RL Fischer, DR Smyth, K Robinson-Beers, and JM Villanueva for providing photographs; MF Yanofsky for communication of data before publication; and all members of the Gasser laboratory for intellectual contributions. Supported by grants 96-35305-3707 from the NRI Competitive Grants Program/USDA and IBN 95-07157 from the National Science Foundation.

Visit the *Annual Reviews home page* at
http://www.AnnualReviews.org.

Literature Cited

1. Andrews HNJ. 1963. Early seed plants. *Science* 142:925–31
2. Angenent GC, Colombo L. 1996. Molecular control of ovule development. *Trends Plant Sci.* 1:228–32
3. Angenent GC, Franken J, Busscher M, Van Dijken A, Van Went JL, et al. 1995. A novel class of MADS box genes is involved in ovule development in petunia. *Plant Cell* 7:1569–82
4. Baker SC, Robinson-Beers K, Villanueva JM, Gaiser JC, Gasser CS. 1997. Interactions among genes regulating ovule development in *Arabidopsis thaliana*. *Genetics* 145:1109–24
5. Bouman F. 1984. The ovule. See Ref. 19a, pp. 123–57
6. Bowman JL, Sakai H, Jack T, Weigel D, Mayer U, et al. 1992. *SUPERMAN*, a regulator of floral homeotic genes in Arabidopsis. *Development* 114:599–615
7. Bowman JL, Smyth DR, Meyerowitz EM. 1989. Genes directing flower development in Arabidopsis. *Plant Cell* 1:37–52
8. Bowman JL, Smyth DR, Meyerowitz EM. 1991. Genetic interactions among floral

homeotic genes of *Arabidopsis*. *Development* 112:1–20

9. Coen ES, Meyerowitz EM. 1991. The war of the whorls: genetic interactions controlling flower development. *Nature* 353:31–37

10. Colombo L, Franken J, Koetje E, Van Went J, Dons HJM, et al. 1995. The petunia MADS box gene FBP11 determines ovule identity. *Plant Cell* 7:1859–68

11. Elliott RC, Betzner AS, Huttner E, Oakes MP, Tucker WQJ, et al. 1996. *AINTEGUMENTA*, an *APETALA2*-like gene of Arabidopsis with pleiotropic roles in ovule development and floral organ growth. *Plant Cell* 8:155–68

12. Evans PT, Holaway BL, Malmberg RL. 1988. Biochemical differentiation in the tobacco flower probed with monoclonal antibodies. *Planta* 175:259–69

13. Evans PT, Malmberg RL. 1989. Alternative pathways of tobacco placental development: time of commitment and analysis of a mutant. *Dev. Biol.* 136:273–83

14. Gaiser JC, Robinson-Beers K, Gasser CS. 1995. The Arabidopsis *SUPERMAN* gene mediates asymmetric growth of the outer integument of ovules. *Plant Cell* 7:333–45

15. Gasser CS. 1996. Homeodomains ring a *BELL* in plant development. *Trends Plant Sci.* 1:134–36

16. Gifford EM, Foster AS. 1989. *Morphology and Evolution of Vascular Plants*, pp. 537–45. New York: Freeman. 3rd ed.

17. Herr JM. 1995. The origin of the ovule. *Am. J. Bot.* 82:547–64

18. Hülskamp M, Schneitz K, Pruitt RE. 1995. Genetic evidence for a long-range activity that directs pollen tube guidance in Arabidopsis. *Plant Cell* 7:57–64

19. Jofuku KD, den Boer BGW, Van Montagu M, Okamuro JK. 1994. Control of Arabidopsis flower and seed development by the homeotic gene *APETALA2*. *Plant Cell* 6:1211–25

19a. Johri BM, ed. 1984. *Embryology of the Angiosperms*. New York: Springer-Verlag

20. Kempin SA, Mandel MA, Yanofsky MF. 1993. Conversion of perianth into reproductive organs by ectopic expression of the tobacco floral homeotic gene NAG1. *Plant Physiol.* 103:1041–46

21. Klucher KM, Chow H, Reiser L, Fischer RL. 1996. The *AINTEGUMENTA* gene of Arabidopsis required for ovule and female gametophyte development is related to the floral homeotic gene *APETALA2*. *Plant Cell* 8:137–53

22. Komaki MK, Okada K, Nishino E, Shimura Y. 1988. Isolation and characterization of novel mutants of *Arabidopsis thaliana* defective in flower development. *Development* 104:195–204

23. Kunst L, Klenz JE, Martinez-Zapater J, Haughn GW. 1989. *AP2* gene determines the identity of perianth organs in flowers of *Arabidopsis thaliana*. *Plant Cell* 1:1195–208

24. Lang JD, Ray S, Ray A. 1994. *sin1*, a mutation affecting female fertility in Arabidopsis, interacts with *mod1*, its recessive modifier. *Genetics* 137:1101–10

25. Leon-Kloosterziel KM, Keijzer CJ, Koornneef M. 1994. A seed shape mutant of Arabidopsis that is affected in integument development. *Plant Cell* 6:385–92

26. Liu ZC, Meyerowitz EM. 1995. *LEUNIG* regulates *AGAMOUS* expression in Arabidopsis flowers. *Development* 121:975–91

27. Ma H. 1994. The unfolding drama of flower development—recent results from genetic and molecular analyses. *Genes Dev.* 8:745–56

28. Malmberg RL, McIndoo J. 1983. Abnormal floral development of a tobacco mutant with elevated polyamine levels. *Nature* 305:623–25

29. Mandel MA, Bowman JL, Kempin SA, Ma H, Meyerowitz EM, et al. 1992. Manipulation of flower structure in transgenic tobacco. *Cell* 71:133–43

30. Mizukami Y, Ma H. 1992. Ectopic expression of the floral homeotic gene *AGAMOUS* in transgenic Arabidopsis plants alters floral organ identity. *Cell* 71:119–31

31. Modrusan Z, Reiser L, Feldmann KA, Fischer RL, Haughn GW. 1994. Homeotic transformation of ovules into carpel-like structures in Arabidopsis. *Plant Cell* 6:333–49

32. Münster T, Pahnke J, Di Rosa A, Kim JT, Martin W, et al. 1997. Floral homeotic genes were recruited from homologous MADS-box genes preexisting in the common ancestor of ferns and seed plants. *Proc. Natl. Acad. Sci. USA* 94:2415–20

33. Okamuro JK, den Boer BGW, Jofuku KD. 1993. Regulation of Arabidopsis flower development. *Plant Cell* 5:1183–93

34. Purugganan MD. 1997. The MADS-box floral homeotic gene lineages predate the origin of seed plants: phylogenetic and molecular clock estimates. *Mol. Evol.* 45:392–96

35. Ray A, Lang JD, Golden T, Ray S. 1996. *SHORT INTEGUMENT (SIN1)*, a gene required for ovule development in

Arabidopsis, also controls flowering time. *Development* 122:2631–38

36. Ray A, Robinson-Beers K, Ray S, Baker SC, Lang JD, et al. 1994. The *Arabidopsis* floral homeotic gene BELL (*BEL1*) controls ovule development through negative regulation of AGAMOUS gene (*AG*). *Proc. Natl. Acad. Sci. USA* 91:5761–65

37. Ray S, Golden T, Ray A. 1996. Maternal effects of the short integument mutation on embryo development in Arabidopsis. *Dev. Biol.* 180:365–69

38. Reiser L, Fischer RL. 1993. The ovule and embryo sac. *Plant Cell* 5:1291–301

39. Reiser L, Modrusan Z, Margossian L, Samach A, Ohad N, et al. 1995. The *BELL1* gene encodes a homeodomain protein involved in pattern formation in the Arabidopsis ovule primordium. *Cell* 83:735–42

40. Robinson-Beers K, Pruitt RE, Gasser CS. 1992. Ovule development in wild-type Arabidopsis and two female-sterile mutants. *Plant Cell* 4:1237–49

41. Roe JL, Durfee T, Zupan JR, Repetti PP, McLean BG, et al. 1997. TOUSLED is a nuclear serine-threonine protein kinase that requires a coiled-coil region for oligomerization and catalytic activity. *J. Biol. Chem.* 272:5838–45

42. Roe JL, Nemhauser JL, Zambryski PC. 1997. *TOUSLED* participates in apical tissue formation during gynoecium development in Arabidopsis. *Plant Cell* 9:335–53

43. Roe JL, Rivin CJ, Sessions RA, Feldmann KA, Zambryski PC. 1993. The *TOUSLED* gene in *A. thaliana* encodes a protein kinase homolog that is required for leaf and flower development. *Cell* 75:939–50

44. Sakai H, Medrano LJ, Meyerowitz EM. 1995. Role of *SUPERMAN* in maintaining Arabidopsis floral whorl boundaries. *Nature* 378:199–203

45. Schneitz K, Hülskamp M, Kopczak S, Pruitt R. 1997. Dissection of sexual organ ontogenesis: a genetic analysis of ovule development in *Arabidopsis thaliana*. *Development* 124:1367–76

46. Schneitz K, Hülskamp M, Pruitt RE. 1995. Wild-type ovule development in *Arabidopsis thaliana*—a light microscope study of cleared whole-mount tissue. *Plant J.* 7:731–49

47. Schultz EA, Pickett FB, Haughn GW. 1991. The *FLO10* gene product regulates the expression domain of homeotic genes *AP3* and *PI* in Arabidopsis flowers. *Plant Cell* 3:1221–37

48. Stebbins GL. 1974. *Flowering Plants: Evolution Above the Species Level,* pp. 199–245. Cambridge, MA: Harvard Univ. Press

49. Stewart WN. 1983. *Paleobotany and the Evolution of Plants.* New York: Cambridge Univ. Press

50. Torii KU, Mitsukawa N, Oosumi T, Matsuura Y, Yokayama R, et al. 1996. The arabidopsis *ERECTA* gene encodes a putative receptor protein kinase with extracellular leucine-rich repeats. *Plant Cell* 8:735–46

51. Vollbrecht E, Hake S. 1995. Deficiency analysis of female gametogenesis in maize. *Dev. Genet.* 16:44–63

52. Weigel D, Meyerowitz EM. 1994. The ABCs of floral homeotic genes. *Cell* 78:203–9

53. Willemse MTM, van Went JL. 1984. The female gametophyte. See Ref. 19a, pp. 159–96

Annu. Rev. Plant Physiol. Plant Mol. Biol. 1998. 49:25–51

POSTTRANSLATIONAL ASSEMBLY OF PHOTOSYNTHETIC METALLOPROTEINS

Sabeeha Merchant and Beth Welty Dreyfuss

Department of Chemistry and Biochemistry, University of California, Los Angeles, Los Angeles, California 90095-1569; e-mail: merchant@chem.ucla.edu; bdreyfus@ucla.edu

KEY WORDS: FeS (iron-sulfur), chloroplast, cofactor, mutant, cytochrome

ABSTRACT

The assembly of chloroplast metalloproteins requires biochemical catalysis. Assembly factors involved in the biosynthesis of metalloproteins might be required to synthesize, chaperone, or transport the cofactor; modify or chaperone the apoprotein; or catalyze cofactor-protein association. Genetic and biochemical approaches have been applied to the study of the assembly of chloroplast iron-sulfur centers, cytochromes, plastocyanin, and the manganese center of photosystem II. These have led to the discovery of NifS-homologues and cysteine desulfhydrase for iron-sulfur center assembly, six loci (*CCS1–CCS5, ccsA*) for *c*-type cytochrome assembly, four loci for cytochrome b_6 assembly (*CCB1–CCB4*), the CtpA protease, which is involved in pre-D1 processing, and the *PCY2* locus, which is involved in holoplastocyanin accumulation. New assembly factors are likely to be discovered via the study of assembly-defective mutants of *Arabidopsis*, cyanobacteria, *Chlamydomonas*, maize, and via the functional analysis of candidate cofactor metabolizing components identified in the genome databases.

CONTENTS

INTRODUCTION

Plastids contain many diverse metalloproteins that function in metabolic pathways (e.g. fatty acid desaturase, acetyl-CoA carboxylase, choline monooxygenase), electron transfer reactions (e.g. cytochromes, iron-sulfur proteins), metabolic regulation (e.g. ferredoxin, metalloproteases), or metal storage (e.g. ferritin) (Table 1). Metalloproteins as a class are generally well studied and were among the first proteins to be purified and characterized, perhaps because the spectroscopic properties of the metal center could be exploited for purification, structural analysis, or mechanistic studies of function. Today, a topical area of research on metalloproteins concerns the mechanics of assembly of these proteins in vivo. The present emphasis is on deducing the biochemical mechanism of metallocluster assembly and cofactor-protein ligation, and on understanding the cell biology of cofactor and apoprotein metabolism in the context of compartmentalized assembly.

Metalloproteins and Metal Binding Sites

Metalloproteins are found in a variety of flavors. In this review, they are divided into three categories: those in which the metal cofactor(s) is coordinated only by residues from the protein (e.g. carbonic anhydrase, plastocyanin; Table 1A), those in which the metal ion is associated with other inorganic ions in a stable "cluster" that, in turn, is bound to specific functional groups in the protein through the metal (e.g. ferredoxin, F_X of PSI; Table 1B), and those containing a stable (tetrapyrrole)-chelated metal center (e.g. cytochromes, chlorophyll proteins; Table 1C). Regardless of the category, each metal cofactor binding site is defined by functional groups, generally amino acid side chains, which serve as ligands to the metal. In some cases, the functional group can be generated by posttranslational modification of an amino acid—as occurs during the biosynthesis of urease where the nickel binding ligand is generated by CO_2-dependent modification of a lysine residue (54, 86), or during the activation of clotting factors where a Glu residue is carboxylated to generate a bidentate ligand for phospholipid bound Ca^{2+} (108). Metals (Lewis acids) bind to these functional groups (Lewis bases) according to preferences determined by the

Table 1 Examples of metalloproteins in the chloroplast

Metal cofactor	Holoprotein/holocomplex	Location	Assembly factors[a]
A. Coordinated directly with apoprotein			
Cu	1 Plastocyanin	Lumen	Pcy2
	2 Polyphenol oxidase	Lumen	
Fe	3 PS II reaction center (D1, D2)	Thylakoid membrane	
	4 Plastid superoxide dismutase	nd/stroma	
	5 Hydrogenase	Stroma	
	6 Lipoxygenase[b]	Envelope membrane	
	7 Ferritin	Stroma	
Zn	8 Carbonic anhydrase	Stroma	
	9 FtsH protease[b]	Thylakoid membrane	
	10 MGDG synthase[b]	Envelope membrane	
	11 Acetyl-CoA carboxylase[b]	nd	
	12 Chloroplast precursor processing enzyme (CPE)[b,c]	Stroma	
	13 EP1 endopeptidase	Stroma	
B. Inorganic cluster			
Fe_2-S_2	14 Rieske protein of cyt $b_6\,f$ complex	Thylakoid membrane	
	15 Ferredoxin	Stroma	C-DES
	16 Ferrochelatase	Thylakoid membrane associated	
	17 Choline monooxygenase	Stroma	
	18 Envelope membrane electron carriers[d]	Envelope membrane	
Fe_4-S_4	19 F_X of PS I reaction center (PsaA, PsaB)	Thylakoid membrane	
	20 F_A/F_B of PS I reaction center (PsaC)	Thylakoid membrane	NifS homologue(s)
	21 Chl L (FrxC) subunit of light-independent protochlorophyllide reductase		
	22 Ferredoxin-thioredoxin reductase	Stroma	
	23 FrxB/Ndh I subunit of NADH-dehydrogenase	Thylakoid membrane	
	24 Envelope membrane electron carriers[d]	Envelope membrane	
Fe_4-S_4, siroheme	25 Ferrodoxin-nitrite reductase	Stroma	
Fe_3-S_4, FAD, FMN	26 Ferredoxin-glutamate synthase	Stroma	
Mn_4 cluster (4Mn, 1Ca, 1Cl)	27 PS II oxygen-evolving complex (D1, D2)	Lumen	MntCAB, CtpA autocatalytic assembly
C. Organic cofactors			
b-type cytochrome	28 Cyt b_{559} of PSII reaction center (α, β subunits)	Thylakoid membrane	
	29 Cyt b_6 (b_p and b_n) of cyt b_6/f complex	Thylakoid membrane	Ccb1–4
c-type cytochrome	30 Cyt f of cyt b_6/f complex	Thylakoid membrane	Ccs1–5, CcsA
	31 Cyt c_6	Lumen	Ccs1–5, CcsA
heme	32 Catalase	Lumen	

nd = not determined.

[a]In general, function assigned based on genetic analysis or homology to other known factors.

[b]Identity of metal cofactor not directly determined.

[c]*Chlamydomonas* processing protease activity is not inhibited by metal chelators.

[d]Suggested to be fatty acid desaturase components.

Table 2 Ligand preferences for metals commonly found in metalloproteins[a]

Metals	Ligands	
Hard		
Mn^{2+}	H_2O	OH^-
Fe^{3+}	ROH	RO^-
Mg^{2+}	PO_4^{3-}	$ROPO_3^{2-}$
Ca^{2+}	$R\text{-}CH_2COO^-$	CO_3^{2-}
	$RNH\text{-}COO^-$	RNH_2
Intermediate		
Fe^{2+}	Imidazolyl nitrogen	
Ni^{2+}		
Zn^{2+}		
Co^{2+}		
Cu^{2+}		
Soft		
Cu^+	R_2S	RS^-
	RSH	

[a]The indicated ligand preferences are according to the hard-soft classification (adapted from Reference 68), which is intended to serve only as a guideline for thinking about metal binding sites in proteins. The nature of the ligand affects dramatically the chemical properties of the metal. Nature uses different ligands and geometries to create unique metal catalysts in vivo.

"hard-soft" theory of acids and bases, which states that hard acids bind to hard bases and soft acids to soft bases (Table 2; 68). Thus, Ca^{2+}, Mg^{2+}, Mn^{2+}, and Fe^{3+} are generally found bound to oxygen ligands (carboxylates, phenolates, carbonates, and phosphates), while Zn^{2+}, Ni^{2+}, and Fe^{2+} have an affinity for imidazolyl nitrogen, and Cu^+ has a strong preference for sulfur ligands (thiols, thiolates, thioethers). In the redox proteins where the metal binding site must accommodate different valence states of a metal (Fe^{2+}/Fe^{3+} or Cu^+/Cu^{2+}), the binding site is optimized for the electron transfer function of the protein. In the case of metal-porphyrin complexes, the axial ligands have important effects on the stability of the metal-porphyrin interaction. Proteins with functionally important Mg^{2+}-binding sites are not usually referred to as metalloproteins because of the loose association of Mg^{2+} with its binding site. In addition, other divalent ions (e.g. Mn^{2+}) can often substitute quite well for Mg^{2+}.

In some metalloproteins, such as the cytochromes, the metal cofactor is an essential determinant of the structure (27, 105), while in others, such as plastocyanin and carbonic anhydrase, the metal binding site is preformed in the apoprotein (11, 33); regardless, the holoprotein is generally more thermodynamically stable than the apoprotein. The difference in stability or structure between apo- and holoforms of some metalloproteins has been exploited in studies of metalloprotein assembly (see below). With the exception of the FeS-dependent regulatory proteins where reversible cluster formation in vivo is key to function (10, 45, 94), cofactor-protein assembly in vivo is considered unidirectional. Even if the cofactor and polypeptide are associated through noncovalent interactions, the resulting complex can be so stable that the association is practically irreversible under physiological conditions. In other cases, the cofactor is solvent inaccessible within a tightly folded structure so that there is a kinetic barrier to cofactor dissociation. For some metalloproteins, incorporation of the metal is coupled with changes in the redox state of the metal or apoprotein, which serves to "trap" the metal within the complex, as exemplified by copper binding to laccase, iron incorporation into ferritin, or assembly of the manganese center of PSII (reviewed in 20, 42, 116).

Although knowledge of the structure of the metal center and the nature of its association with the protein allows certain inferences to be drawn regarding the chemistry essential for assembly, the complexity of the in vivo mechanisms is just beginning to be appreciated. This review focuses on the assembly of metalloproteins in chloroplasts, the most abundant and well-characterized of which are those that function in photosynthesis. The contributions from the cyanobacterial model systems are included, since they are immediately relevant to chloroplast processes, but contributions from other model systems are not discussed in detail except for the purpose of illustration and comparison. Informative discussions of specific pathways of metalloprotein assembly may be found in the following articles (40, 42, 58, 75, 118). Although the chlorophyll proteins are the most abundant cofactor-containing proteins in the chloroplast, a discussion of chlorophyll protein assembly is excluded from this work. The interested reader is referred to recent reviews of this special topic (77, 87).

Cell Biology of Metalloprotein Assembly in the Chloroplast

A nucleus-encoded chloroplast protein is synthesized outside the chloroplast, generally (although not always) as a higher molecular weight precursor, which is imported into the organelle after translation (reviewed in 99). The imported protein is processed to its mature form within the plastid, sorted to its final destination (envelope or thylakoid membranes, stroma, intermembrane space, or lumen) via specific sorting mechanisms, and (if it is a subunit of a multicomponent complex) assembled with other proteins or cofactors into a functional complex. Some of these postimport steps are required also for the biosynthesis of organelle-encoded proteins.

The determination of the temporal sequence of each of the above posttranslational events is key to defining the substrate for metalloprotein assembly in vivo and the suborganellar site of assembly. For ferredoxin, plastocyanin, cytochromes c_6 and f, and the PSII Mn cluster, association of the apoprotein with the cofactor is a near-terminal event in the biosynthetic pathway, occurring in the same compartment in which the mature protein functions with the processed polypeptide serving as the substrate for assembly (21, 43, 51, 66, 67, 111). This tends to be the case in other biological systems as well—exemplified by heme insertion into c-type cytochromes in the bacterial periplasm or mitochondrial intermembrane space (37, 83, 118), periplasmic insertion of copper into nitrous oxide reductase (136), and postsecretion insertion of copper into melanin (15). Late assembly of the metalloprotein in the same compartment in which it functions can be rationalized on the basis that protein translocation across membranes requires partial unfolding of the polypeptide. Not only might unfolding be incompatible with maintenance of the (often noncovalent) cofactor-protein association, but the increased stability of the folded metalloprotein relative to the apoprotein could conceivably inhibit translocation. Nevertheless, it should be noted that while some posttranslational modifications involving organic cofactors (e.g. flavinylation, pantothenylation) also tend to occur in the organelle after import in vivo (e.g. 26, 92), these modifications do not seem to affect import in vitro (e.g. 102, 115, 130).

Since the same type of metallocenter (e.g. FeS center) is found in different compartments within the plastid (Table 1B), it is possible that distinct "assembly factors" are required in each compartment—particularly if cofactor insertion occurs after intraorganellar sorting. The determination of the substrate specificity and suborganellar location of candidate assembly factors is, therefore, relevant to an understanding of its function in vivo (see discussion of NifS homologues).

Many thylakoid membrane metalloproteins are found in multisubunit complexes. In fact, in some cases the metal binding site is created by ligands provided by two different subunits—as in the F_X binding site in PSI or the heme binding site of cytochrome b_{559} (Table 1B, line 19; 1C, line 28). In these cases, assembly of PsaA-PsaB or the α and β subunits of cyt b_{559} must be intimately coordinated with assembly of the F_X center into PSI or the heme(s) into PSII, respectively, which makes it difficult to resolve apoprotein-cofactor assembly from the process of subunit association. This difficulty can prevent specific functional assignment to genetically defined assembly factors in the absence of biochemical assays for cofactor insertion (see below).

Biochemistry of Metalloprotein Assembly

For metalloproteins like plastocyanin (Table 1A) where the metal cofactor is simply coordinated by residues provided by the polypeptide, the holoprotein can often be formed in vitro without catalysis (43, 65, 67, 113). However, the reaction is slow in vitro and does not display the metal selectivity noted in vivo. Indeed, when the related blue copper protein azurin was expressed in *Escherichia coli*, a significant fraction of the expressed protein contained bound zinc at the copper site (78). Selective insertion of a metal ion, present normally in trace amounts, in the presence of much higher concentrations of other ions is a key feature of metalloprotein biosynthesis in vivo. This may be accomplished by concentration of the cofactor in the compartment in which assembly occurs or by delivery of the metal ion to the assembly factor via a "metal chaperone" (14, 62, 70, 136). Thus, an appreciation of trace metal metabolism and its regulation is integral to the study of metalloprotein biosynthesis.

For metalloproteins containing inorganic clusters (Table 1*B*), the substrates required for cluster synthesis (e.g. sulfur and ferrous ion for the formation of FeS centers) must be generated in vivo by mobilization/reduction of stored iron and enzymatic desulfurylation of cysteine (19, 28, 134). Assembly of the cluster can require the function of a scaffold protein as for the FeMo cofactor of nitrogenase, which is preassembled on a scaffold consisting of the *nifE* and *nifN* gene products, and transferred after assembly to the active site of the enzyme (reviewed in 19). When the cofactor is covalently associated with the polypeptide as in the *c*-type cytochromes (Table 1*C*), enzymes are required for formation of the linkages (114). The importance of biological catalysis of this type of reaction is underscored by the observation that uncatalyzed chromophore attachment to cysteinyl residues of the phycobilins occurs without discrimination for the chemical identity of the bilin, whereas the in vivo reactions show strict specificity for the chromophore (25). Maintenance or protection of the cofactor-binding cysteinyl residues in the apoproteins requires thiol reductants in vitro (64, 65), and can be mediated by thioredoxin-like proteins or protein dithiol-disulfide oxidoreductases in vivo. Indeed, a family of such proteins is required for *c*-type cytochrome synthesis in the bacterial periplasm (8, 18, 96, 125). Alternatively, this function could be provided by chaperones. Chaperone activity may also be necessary for presentation of the apoprotein substrate so that only the cofactor-binding ligands are presented for interaction with the cofactor. Other biochemical activities necessary for metalloprotein biosynthesis include modification of the polypeptide prior to or concomitant with metal insertion, such as introduction of a metal-binding carbamate at the Rubisco active site (41).

For metalloproteins like the cytochromes (Table 1*C*), the cofactor is usually synthesized in one compartment in the cell and must be transported to other compartments for assembly with various apoproteins. This might require the function of specific transporters and also, perhaps, carrier proteins or cofactor chaperones (analogous to the metal chaperones mentioned above) to deliver the cofactor to the apocytochrome. The "carrier" protein that binds the molybdocofactor and delivers it to aponitrate reductase would be an example of such a cofactor chaperone (2). Finally, the cofactor may need to be processed (e.g. by reduction) before it is a suitable substrate (80).

Some of the assembly factors discussed above might be required specifically for the assembly of a single metalloprotein in a particular compartment, and distinct isozymes may function in different compartments, whereas others may be more general factors required for the synthesis of all members of a particular class of metalloprotein. In most cases, the assembly factor is expected to be several orders of magnitude less abundant in a mature cell than its cognate metalloprotein product (see below).

EXPERIMENTAL APPROACHES

In Vitro Studies

The types of questions that can be addressed by the application of biochemical methods include definition of the apoprotein substrates for assembly—e.g. proteolytically processed vs precursor proteins (43, 51, 67), definition of the substrates for cofactor synthesis—e.g. the source of "acid labile S" in the FeS centers (111), and definition of the compartment in which assembly occurs (51, 67, 111). Naturally, the best-studied systems are those where the holo-protein can be distinguished readily from the apoprotein; for example, on the basis of a spectroscopic signal, enzyme activity or a significant conformational or other structural change. The use of nondenaturing gel electrophoresis to separate apo- from holo- forms of various proteins is a common tool for the study of metalloprotein biosynthesis (64, 66, 67, 111), whereas spectroscopy and measurement of catalytic activity have been exploited primarily for the study of iron-sulfur center assembly (28, 48, 71, 131, 132; M Antonkine, F Yang, S Parkin, MP Scott, JM Bollinger Jr & JH Golbeck, manuscript in preparation), and separation by denaturing gel electrophoresis or HPLC has been used only for proteins that carry covalently attached cofactors (49, 79, 83). The experimental strategy used to deduce the pathway of metalloprotein maturation involves monitoring the fate of radiolabeled precursors either in vivo or in organello. Once the pathway is elaborated and the substrates for cofactor-protein association are deduced, the door is open to the establishment of in vitro assays for cofactor insertion, which is a prerequisite for the purification of assembly factors.

Isolation of Mutants

Classical genetic approaches have been extraordinarily useful in defining *trans*-acting assembly factors required for *c*-type cytochrome formation in bacteria, mitochondria, and chloroplasts (reviewed by 37, 50, 58, 118), molybdocofactor biosynthesis in prokaryotes and eukaryotes (32, 90, 106, 119), cytochrome oxidase assembly in *Saccharomyces cerevisiae* (40), and nitrogenase assembly (19). These factors were defined originally on the basis of mutations that affected the accumulation of functional forms of specific enzymes but did not affect the expression of structural genes nor synthesis and processing of the apoprotein. In some cases, the apoproteins accumulated in the mutant strains and the function of the *trans*-acting loci were deduced readily (19, 75, 83, 136). In other cases, the phenotype could be suppressed by feeding the mutant with an excess of the cofactor (6, 35, 70, 124), which suggested a defect in cofactor transport or delivery. Biochemical activities for some of the cloned candidate assembly factors have been proposed on the basis of conserved sequence motifs; included among these assembly factors are candidate transporter components (6, 9, 91, 136), thiol metabolizing proteins (8, 69), enzymes involved in cofactor biosynthesis (123), and metal-metabolizing proteins (30, 62). The present emphasis is on demonstrating that the gene products exhibit these activities in vitro and on discerning the relevance of the activity to the assembly process in vivo.

For many of the metalloproteins in the photosynthetic complexes, a fundamental problem is the definition of the phenotype of a cofactor insertion mutant, which makes it difficult to screen specifically for an assembly defect of interest. For instance, loss of heme binding to cyt b_{559} results in the destabilization of the entire PSII complex in *Synechocystis* sp. 6803 (85). Likewise, *ccs* strains of *Chlamydomonas* (see below) that are defective in heme attachment to cyt f lack the entire cytochrome b_6/f complex (49). Thus, a cyt b_{559} assembly mutant exhibits the same general PSII-minus phenotype as a mutant with a defect in some other aspect of PSII assembly, for example, subunit-subunit associations in the reaction center. And a c-heme attachment mutant with a defect in cyt f assembly exhibits the same general cytochrome b_6/f-minus phenotype as a b-heme insertion mutant (34). Similarly, loss of the F_A and F_B FeS centers of PsaC results in a pleiotropic deficiency in several subunits of the PSI complex (84, 131), while loss of the F_X center prevents assembly of PsaA and PsaB, and hence the entire PSI complex (128; reviewed in 16). The same phenomenon prevented specific assignment of functions to many loci defined by respiratory mutants of *S. cerevisiae* whose phenotypes indicated defective assembly of particular multisubunit complexes, but tentative functional roles for the affected genes are now being proposed by recognition of conserved sequence motifs as the wild-type alleles are cloned (reviewed in 40, 89, 122).

Given the complexity of each of the photosynthetic complexes, the task of refining the assembly-defective phenotypes to resolve distinct categories of defects at the biochemical level appears daunting. Nevertheless, this has been possible occasionally. For instance, c-heme attachment mutants in *Chlamydomonas* were defined on the basis of a pleiotropic c-type cytochrome deficiency (49). Recently, b-heme insertion mutants were recognized on the basis of a unique phenotype exhibited by strains carrying site-directed changes at the b-heme binding ligands of apocyt b_6, and this was exploited to define a pathway for the assembly of hemes into chloroplast cyt b_6 and to identify four nuclear loci required for this process (see below).

There are a number of *Arabidopsis*, *Chlamydomonas*, maize and cyanobacterial mutants that are defective in the accumulation of specific plastid complexes. Some of these are categorized as assembly defective (5, 7, 74, 121, 126) because expression of the polypeptide components of the affected complex(es) is normal. While some of the mutants are known to define general assembly factors (e.g. 107), a subset of them might be defective in plastid metalloprotein assembly or metallocofactor metabolism. The recognition of defects in this category is straightforward when apoproteins accumulate (reviewed in 19, 63, 75, 83). Unfortunately, this is often not the case in the chloroplast, particularly in *Chlamydomonas* chloroplasts where unassembled or misassembled proteins are degraded rapidly (55, 73, 98; reviewed in 1, 93). In these instances, it might be possible to classify such mutants on the basis of pleiotropic deficiencies in proteins with common cofactors (e.g. 32). For instance, iron-sulfur center assembly mutants might display deficiencies in more than one iron-sulfur protein, or a b-heme insertion mutant could be deficient in both PSII and cytochrome b_6/f function. Ultimately, more detailed characterization of the phenotype of assembly mutants, particularly in the context of biochemical studies of metalloprotein assembly, should result in the recognition of cofactor processing defects.

Molecular Genetics

The functional relationship between pathways in the chloroplast and analogous ones in other organisms can be exploited in order to identify chloroplast homologues of well-characterized metal metabolizing factors in other systems. One method relies on complementation of appropriate *S. cerevisiae* or *E. coli* mutants. This approach is used widely and has led to the identification of plant genes corresponding to molybocofactor synthesis components and also to various metal transporters (24, 47, 56, 104). In principle, this method could be extended also to the isolation of intracellular and/or intraorganellar metal transporters and specific chloroplast metalloprotein assembly factors. Candidate homologues can be identified also on the basis of sequence relationships. This method, whose usefulness is increasing daily as the genome databases grow, is effective—and accordingly, quite popular. A few examples of candidate chloroplast cofactor metabolizing components identified via the databases include CutA, a putative chloroplast-targeted copper transporter in the *Arabidopsis* dbEST; NifS, a sulfur-mobilizing enzyme required for iron-sulfur center assembly identified in the *Arabidopsis* dbEST and the *Synechocystis* sp. 6803 genome (M Antonkine, F Yang, S Parkin, MP Scott, JM Bollinger Jr & JH Golbeck, manuscript in preparation); and several cytochrome assembly components (129b). In some cases (e.g. NifS homologues), the sequence relationship can be quite high, while for other candidate homologues (e.g. CcsA), the prediction of function was made on the basis of only a conserved sequence motif (9). Regardless, the function of the candidate homologues must be demonstrated— either by reverse genetic approaches, where loss of function results in a specific metabolic defect, or by demonstration of a specific biochemical activity for the expressed gene product.

ASSEMBLY OF SPECIFIC CHLOROPLAST PROTEINS

Iron-Sulfur Centers in PSI and Ferredoxin

One of the first examples of reconstitution of a metalloprotein was the in vitro chemical synthesis of holoferredoxin from ferrous ion, inorganic sulfide, and apoferredoxin in the presence of thiol reducing reagents (71). Variations of this procedure have been used to reconstitute FeS centers into other plastid proteins like the Fe_4S_4 centers, F_A and F_B, of PsaC (72) and the Fe_2S_2 Rieske center (48). The ability to reconstitute metalloproteins in vitro is of great practical value because it facilitates functional analysis of the metal center by site-directed mutagenesis of metal-binding ligands. This approach, applied to PsaC, allowed Golbeck and coworkers to assign specific cysteine residues to the F_A vs the F_B center of PsaC and to undertake functional analysis of reconstituted Fe_3S_4 clusters or mixed ligand Fe_4S_4 clusters (reviewed by 16).

While the facility with which FeS centers can be reassembled in vitro had led some to wonder whether the process might occur without catalysis in vivo, it is clear today that formation of holo-FeS proteins in vivo does require specific enzyme-catalyzed steps (discussed in 28, 29). The in vitro reaction is performed under nonphysiological conditions and requires concentrations of Fe^{2+} and sulfide at much higher levels than are known to occur in cells. In addition, the rate of uncatalyzed metallocenter formation is not sufficient to keep up with biosynthetic demands. Plastids contain a variety of FeS-proteins (Table 1B, lines 14–26), some of which are rather abundant—like ferredoxin. Formation of holoferredoxin occurs in the chloroplast stroma after import and processing of the preprotein (67, 88, 109, 111), and this is probably the case for the other FeS proteins as well (although their biosyntheses have not been studied in this context). Many of the other abundant FeS centers, like F_X, F_A, and F_B, are plastid-encoded, and their assembly must occur in the organelle. Therefore, the substrates (Fe^{2+} and sulfur) for FeS cluster formation have to be generated in the plastid at an appreciable rate to support the biosynthesis of the various FeS proteins.

Studies of holoferredoxin synthesis in organello identified cysteine as the source of "acid-labile" sulfur in the ferredoxin FeS cluster (111) and established also that assembly of the cluster in extracts from lysed chloroplasts required ATP hydrolysis and NADPH (110, 112). NADPH (but not NADH) was required for generation of inorganic sulfur from cysteine, whereas ATP hydrolysis was required subsequently at the step of cluster formation or cluster incorporation into apoferredoxin. The enzymes required for these processes were noted to be soluble and localized in the stroma (110) and are active in etioplasts as well as chloroplasts (109). The observed ATP and NADPH requirement could not be attributed to iron mobilization processes (110).

In the meantime, studies of various *nif* mutants of *Azotobacter vinelandii* suggested that the *nifS* gene product was required in *trans* for formation of the Fe_4S_4 cluster of the iron protein of nitrogenase (reviewed in 19). NifS is a pyridoxal phosphate containing enzyme that catalyzes the desulfurization of L-cysteine to form L-alanine and S^0 via a NifS-bound persulfide (133, 134), and the formation of the holo Fe protein from L-cysteine, Fe^{2+} and apo Fe protein in most presence of a reductant (132). In the absence of apoprotein, S^0 can be converted to S^{2-} by dithiothreitol. Recombinant *A. vinelandii* NifS can also catalyze the reconstitution of either Fe_4S_4 or Fe_2S_2 clusters in heterologous FeS proteins, including the transcription factors FNR and SoxR of *E. coli* (39, 44).

NifS homologues are found in the genomes of many bacteria including nondiazotrophs, and it was hypothesized that these homologues might be involved in the biosynthesis of iron-sulfur proteins in these organisms (e.g. 29). Indeed, a NifS homologue was purified from *E. coli* extracts on the basis of its ability to reconstitute an FeS center into dihydroxyacid dehydratase (28). Mechanistic studies of the *A. vinelandii* and *E. coli* NifS enzymes indicate that the cysteinyl sulfur is transferred to an active-site sulfhydryl on NifS yielding a cysteine-persulfide. A carrier protein is proposed to transfer S^0 from NifS to the apoprotein where it is "stored" for FeS cluster formation (Scheme 1).

transfer of S^0 to carrier   transfer of S^0 to apoprotein   incorporation of Fe

NifS—S-SH $\longrightarrow$ Carrier—S-SH $\longrightarrow$ Apoprotein—S-SH $\longrightarrow$ [FeS] protein

Scheme 1

The *Synechocystis* sp. 6803 genome (57) contains three NifS homologues (sll0704, slr007, slr0387). Two of these have been expressed in *E. coli*, purified and demonstrated to catalyze the reconstitution of Fe_4S_4 centers into apo-PsaC in vitro and also into a synthetic peptide corresponding to the F_X site of PSI (M Antonkine, F Yang, S Parkin, MP Scott, JM Bollinger Jr & JH Golbeck, manuscript in preparation). Thus, the biochemical activity of the NifS homologues is consistent with a function in sulfur mobilization; however, genetic evidence for a specific function in the assembly of a particular subset of iron-sulfur proteins is being sought presently. A number of enzymes can generate sulfide or elemental sulfur from organic or inorganic compounds, including *O*-acetylserine sulfhydrylase, β-cystathionase, and rhodanese, and these function also as catalysts of FeS center formation in vitro, but their relevance to FeS cluster synthesis in vivo is doubtful (13, 29). Definitive demonstration of a physiological role for candidate FeS center assembly proteins in FeS center formation in vivo is, therefore, critical.

Recently, a distinct *cysteine-desulfyhdrase* activity (C-DES) was purified from *Synechocystis* PCC 6714 on the basis of its ability to catalyze the reconstitution of holoferredoxin from apoferredoxin, cysteine, and Fe^{2+} in the presence of a thiol reductant (glutathione) and under anaerobic conditions (64). Cysteinyl sulfur is quantitatively assembled into the Fe_2S_2 cluster of holoferredoxin by C-DES. Like NifS, C-DES is a pyridoxal phosphate containing enzyme, and like NifS, the presence of the apoprotein is not required for removal of sulfur from cysteine. But unlike the NifS-catalyzed reaction, the product of the C-DES catalyzed reaction is pyruvate and ammonia rather than alanine. In addition, C-DES can catalyze sulfur elimination from cystine as well as cysteine. Based on the abundance of the protein (1:150 stoichiometry relative to ferredoxin), its ability to catalyze quantitative transfer of cysteinyl sulfur to ferredoxin (via a cysteine persulfide), and its turnover number in vitro, Leibrecht & Kessler (64) suggest that C-DES might be the physiological equivalent of NifS for ferredoxin formation in cyanobacteria and, by extension, chloroplasts. Nevertheless, this needs to be demonstrated genetically.

Flint (28) noted that the purified *E. coli* NifS homologue is not as active in reconstituting the FeS center of dihydroxyacid dehydratase as is the crude extract from which it was purified. On this basis he suggests, as do Zheng & Dean (132), that the function of NifS lies only in release of cysteinyl sulfur and proposes that another enzyme is required for assembly of the iron-sulfur cluster from Fe^{2+} and S^0. This view is supported by the in vitro studies of Takahashi et al (110), who distinguished at least three steps in holoferredoxin biosynthesis in chloroplast extracts: (*a*) NADPH-dependent mobilization of sulfur from cysteine, (*b*) mobilization of iron from ferritin, and (*c*) ATP-dependent cluster formation, and the genetic studies of nitrogenase assembly that indicate that the *nifU* and *nifM* gene products are required in addition to NifS for full activation of the Fe protein (19). NifU and NifM could participate in Fe mobilization (see below) or cluster assembly on the protein. The fact that NifS- and C-DES-catalyzed release of sulfur from cysteine and incorporation into an FeS cluster are not coupled, and the lack of specificity of NifS for various apoprotein substrates, are consistent with the idea that the role of NifS is restricted to provision of activated S^0. The wide distribution of NifS-like proteins suggests that they may serve a general cellular function for mobilization of sulfur for metallocluster formation.

Iron is stored in the plastid in the form of ferritin, where it is mineralized with phosphate (127). The abundance of ferritin in the plastid is high in etioplasts and decreases dramatically during plastid development when synthesis of most of the FeS-containing metalloproteins is occurring, suggesting that ferritin is the source of iron for these proteins. The biochemistry of iron mobilization in the plastid is completely unknown; presumably, reduction of Fe^{3+} is required. NifU might be a candidate catalyst for this process. Like NifS, NifU is required for the formation of the holoform of the Fe protein (19). Purified NifU is a homodimer and contains one Fe_2S_2 center per subunit (31). Besides the four cysteines required to bind the FeS center, another four cysteines are conserved in the primary sequence and are speculated to function in mobilizing iron for FeS cluster formation. The NifU-like sequence (ssl12667) present in *Synechocystis* sp. 6803 (which does not fix nitrogen) might be a good candidate for catalyzing iron mobilization during the formation of FeS centers.

Cytochromes c

Owing to the covalent nature of the interaction between the cofactor and the apoprotein, the need for catalysis of holocyt c formation has long been appreciated (114). Accordingly, the maturation of cytochromes c has been studied for some time, and the factors required for holocytochrome c formation in mitochondria and bacteria are well defined. In fungal and mammalian mitochondria, enzymes called cyt c (or c_1)/heme lyases are required in the intermembrane space for formation of the thioether linkages between the apoprotein and heme. Two such enzymes have been identified genetically, one specific for holocyt c formation and the other specific for holocyt c_1 formation (22, 23, 135). These appear to be the only factors required for mitochondrial c-type cytochrome biogenesis in this system. Nevertheless, since the in vitro reaction catalyzed by cyt c and c_1/heme lyases requires reductants for maintenance of heme in the reduced form (80, 81), it is possible that additional "general assembly factors" are involved (see below, discussion of bacterial oxidoreductases).

In bacteria, genetic analysis led to the definition of multiple loci whose products are required for the periplasmic assembly of all c-type cytochromes (58, 118). This pathway is quite distinct from the one discovered in fungal mitochondria and appears to operate in the proteobacteria and also plant mitochondria (12, 38, 100, 101). In this system, an ABC-type transporter (HelABCD in *Rhodobacter* sp. or CcmABCD in *E. coli*) is required for transport of heme from its site of synthesis to the site of cytochrome assembly, a cyt c/heme lyase (Ccl1 in *Rhodobacter* sp. or CcmF in *E. coli*) is required for formation of the thioether linkages, a cyt c–specific thioredoxin (HelX or CcmG) plus a series of thioredoxin-like or protein disulfide isomerase-like molecules that function also in the maturation of many periplasmic proteins (DsbAD, DipZ in *E. coli*) are required for maintenance and presentation of the substrate thiols to the putative cyt c/heme lyase by sequential oxidation and reduction of substrate thiols (18, 76, 95, 96). The above cyt-specific assembly components are thought to be associated in a membrane "cytochrome assembly complex." Other components of this complex include CycJ and CycH (corresponding to CcmE and the C-terminal portion of CcmH in *E. coli*) originally identified in *Bradyrhizobium japonicum* and Ccl2/CcmH (58, 61, 91, 117). The same pathway is thought to operate in plant mitochondria because homologues of some of the components have been identified in plant mitochondrial genomes (e.g. HelBC, Ccl1) or in the dbEST (CycJ). However, candidate homologues are not found in the yeast genome, which suggests that the biochemistry of the two pathways is different.

Chloroplasts contain up to two c-type cytochromes (Table 1C, lines 30 & 31), both with the signature CxxCH heme-binding motif near the N terminus. Cyt c_6 is found only in algae, while cyt f is required in all chloroplasts. Both, proteins are translocated across the thylakoid membrane, and their N-terminal presequences are processed on the lumen-side where assembly with heme occurs (50–52). The order of heme attachment vs processing does not appear to be obligatory (4, 59), and the sequence may be determined by the rate of the two reactions in vivo. For cyt c_6, heme attachment succeeds processing in vivo (51), and this is likely for cyt f as well, since processing occurs very rapidly—perhaps even before translation is completed. Certainly, processing is a prerequisite for complete assembly of cyt f because the α-NH2 group of the mature protein serves as one of the two axial ligands to the heme iron (17).

Holocyt c assembly mutants were identified readily in *Chlamydomonas* on the basis of a pleiotropic c-type cytochrome deficiency (49), and these were shown by biochemical methods to be blocked at the step of heme attachment (49, 52, 129b). Since cyt c_6 is nucleus encoded and cyt f is plastid encoded, the recognition of the pleiotropic phenotype facilitated a screen for additional assembly mutants that were named *ccs* strains for *c*-type cytochrome synthesis (B Dreyfuss & S Merchant, unpublished results; 129b). A minimum of six loci (nuclear *CCS1* through *CCS5* and plastid *ccsA*) are required for heme attachment in the chloroplast, and these function independently of genes required for mitochondrial *c*-cytochrome synthesis. The *ccsA* gene (formerly *ycf5*) encodes a hydrophobic protein that probably spans the membrane multiple times. The conserved C-terminal region contains a sequence motif—WGxxWxWDxxE— called WWD motif, which is present also in Ccl1/CcmF and HelC/CcmC (58, 129a). However, the protein does not appear to be a true homologue of any of the bacterial cytochrome assembly components (see above). Homologues of CcsA are present instead in cyanobacteria, *Helicobacter pylori*, and gram-positive bacteria (129b). The *CCS1* locus also encodes a membrane-associated protein (53a) which is proposed to interact with CcsA in a putative "cytochrome assembly complex" (129b). As for CcsA, homologues of Ccs1 (formerly Ycf44) are found in cyanobacteria, gram-positive bacteria and other algae (129b, 53a) but not in most proteobacteria or *Saccharomyces cerevisiae*. In *Porphyra purpurea*, immediately proximal to *ccs1* is found another open reading frame whose product is a candidate homologue of CcdA—a protein required for holocytochrome *c* synthesis in *Bacillus subtilis* and present also in other gram-positive bacteria (97). On the basis of the distribution of CcsA, Ccs1, and CcdA sequences in the databases, it appears that the chloroplast system for *c*-type cytochrome synthesis is distinct from the well-characterized ones operating in mitochondria or in most proteobacteria (129b).

The function of CcsA, Ccs1, and CcdA remains a key question. Mutagenesis of conserved residues in CcsA indicates that (*a*) the protein contains essential histidine residues in the putative *trans*-membrane segments, and (*b*) some of the tryptophans and the aspartic acid of the WWD motif are essential (Z Xie, N Yu, B Dreyfuss & S Merchant, unpublished results). However, besides the WWD motif [which is hypothesized to be a heme binding motif (9)], CcsA does not display any similarity to proteins of known function. Thus, the specific biochemical function of CcsA is not known. The essential W and D residues of the WWD motif are predicted to lie in extramembrane segments on the lumen side of the thylakoid membrane, which is not inconsistent with a function for CcsA directly in the lyase reaction or in heme delivery to the putative lyase. The functions of Ccs1 and CcdA are also not predictable from sequence analysis and will have to be deduced by biochemical analysis of the gene products. The *ccsA* and *Ccs1* genes are only weakly expressed at the RNA level (129b, 53a); therefore, the proteins are not expected to be abundant. This is not surprising for catalysts of assembly. As noted above, C-DES is 10^2- to 10^3-fold less abundant than ferredoxin whose assembly it catalyzes (64). By considering the functions required for cytochrome biogenesis in the bacterial periplasm (topologically analogous to the thylakoid lumen), the products of the *CCS* loci are hypothesized to encode heme transport/delivery components, the heme attachment enzyme, and oxidoreductases, which keep the substrates reduced (discussed in 50).

The b Cytochromes

Cytochrome b_{559} consists of two subunits, α and β, each of which provides histidine ligands for axial coordination of heme within the membrane (reviewed in 129). Mutation of either one of the histidines results in disruption of heme binding and destabilization of the entire PSII complex (85), which is not surprising considering that cytochrome b_{559} is an integral component in this complex and has been proposed to serve as a nucleation site for PSII assembly (reviewed in 129). The pleiotropic assembly-defective phenotype does not facilitate the discrimination of heme insertion defects from defects in other early steps of PSII assembly, while the intersubunit nature of the heme-protein interaction and the hydrophobic properties of the subunits are not conducive to the application of biochemical methods. The study of cytochrome b_{559} assembly has, therefore, not received much more attention.

The study of cytochrome b_6 (Table 1*C*, line 29) assembly faced some of the same barriers. The protein, which spans the membrane four times, binds two hemes (b_p and b_n on the lumen and stromal side, respectively) by interhelix bis-histidyl ligation (17). Mutation of any of the four histidines prevents heme binding. Apocytochrome b_6 does not accumulate, nor do the other subunits of the complex (59a). This phenotype is similar to that of $\Delta petA$, $\Delta petB$, or $\Delta petD$ mutants, which carry disruptions in the structural genes encoding cytochrome *f*, cytochrome b_6, or Subunit IV, respectively, and other b_6/*f* assembly mutants, like the *ccs* strains discussed above (53, 60). However, analysis of cytochrome b_6 synthesis in *Chlamydomonas* by pulse-radiolabeling techniques allowed Wollman and coworkers to distinguish between holocyt b_6 (which migrated as a broad diffuse band), apocytochrome b_6 (which migrated as a single sharp band), and a b_p-dependent biosynthetic intermediate (which was resolved as a doublet of sharp bands) after electrophoretic separation of the products under denaturing conditions. By monitoring the synthesis of these distinct forms of cytochrome b_6 in strains carrying mutations at the b_p- or b_n-heme-binding sites or in gabaculine-treated wild-type cells, the following pathway was deduced (Scheme 2), where the two hemes are inserted sequentially into the protein. This work further demonstrated that the association of heme(s) with apocytochrome b_6 is independent of its association with other subunits, which allowed putative heme assembly mutants to be distinguished from those affected in some other aspect of cytochrome b_6/f assembly.

$$\text{apocyt } b_6 + b_p \text{ heme} \longrightarrow \text{intermediate} + b_n \text{ heme} \longrightarrow \text{holocyt } b_6$$

Ccb1-Ccb4

Scheme 2

This phenotype was also used to distinguish b_n-heme association defects from a collection of b_6/f-deficient strains of *Chlamydomonas*, and genetic analysis indicated that they correspond to four nuclear loci, *CCB1–CCB4*, which are distinct from the *CCS* loci discussed above. Curiously, b_p-heme association mutants were not found. One possibility is that b_p-heme association is uncatalyzed; another is that such mutants might exhibit a pleiotropic cytochrome deficiency and the initial screen used in that work would have bypassed such strains. Since mutants with defects in b-cytochrome assembly have not been recognized in other systems, it is likely that functional analysis of these genes and their products will lead to the discovery of some novel aspects of cofactor-protein assembly. Candidate functions for these loci include cofactor chaperoning/delivery/transport, apoprotein processing (e.g. a membrane protein prolyl *cis-trans* isomerase), or catalysis of the unusually tight cofactor-protein association.

Manganese Center of PSII

The Mn cluster is a complex structure containing four manganese in various oxidation states, calcium, and chloride (Table 1*B*, line 27). The D1 protein and to a lesser extent the D2 protein of PSII provide most of the ligands for the cluster. Although the chemical structure of the cluster is not yet known, studies of cluster formation have been ongoing for almost three decades, owing to the interesting autocatalytic process of assembly called photoactivation (reviewed in 20). In brief, the steps of assembly, which occur on the lumen side after assembly of the reaction center polypeptides, involve photooxidation of a ligated Mn^{2+} to form a Mn^{3+} intermediate, ligation of a second Mn^{2+} to form a binuclear intermediate, photooxidation of the second Mn^{2+} to yield a metastable Mn^{3+}-Mn^{3+} complex, followed by coordination of two more Mn^{2+} ions that must also be oxidized to yield a stable Mn_4 cluster. Ca^{2+} and chloride ions are required for functional assembly of the Mn_4 complex, but the chemical sequence and roles are not understood precisely, particularly owing to the absence of a final structural model.

Biochemical characterization of D1 synthesis in *Scenedesmus obliquus* indicated that the protein was synthesized in precursor form, inserted into the thylakoid membrane, and processed with a half-time of under 2 min to yield the mature 34-kDa form of the protein (21). When processing is blocked, as in the LF1 mutant of *Scenedesmus* or by directed mutagenesis of the processing site (82), D1 integrates into the thylakoid membrane, but a stable Mn_4 cluster cannot be formed. The protease has been purified (or partially purified) from various sources including cyanobacteria, algae, and plants, and was cloned recently from *Scenedesmus* (120). The gene (*ctpA*) was identified also in *Synechocystis* sp. 6803 on the basis of its ability to complement mutant strain (SK18), which displayed a phenotype similar to that of *Scenedesmus* LF1 (103). Reverse genetics confirmed that CtpA is required for processing of D1 (3), and hence for assembly of the Mn_4 cluster.

It is likely that further exploitation of genetic approaches will lead to the identification of additional factors required for specific steps in metallocenter assembly in PSII. Indeed, a manganese transporter, required (under certain conditions) for Mn uptake by cyanobacteria, was identified on the basis of a Mn_4 cluster deficiency in the BP13 strain that carries a mutation in one subunit (MntA) of this ABC-type transporter encoded by *mntCAB* (6). Although the Mn transporter identified in that work was found to be a plasma membrane rather than a thylakoid membrane transporter, the same approach could be applied to identify manganese-metabolizing components required for cluster assembly in the lumen. However, it is possible that manganese is transported via a broad specificity metal transporter (as is the case for Ni transport in *Bradyrhizobium japonicum*), which could complicate the genetic screen.

Plastocyanin

Studies of plastocyanin synthesis in vivo and in organello established that the import and processing of plastocyanin occur independently of copper (73). Copper-protein assembly is a near-terminal step in the biosynthetic pathway and occurs in the thylakoid lumen after proteolytic processing by the thylakoid peptidase (43, 67). The fact that metal incorporation into plastocyanin is much more selective in vivo than in vitro (46) suggests that some aspect of metal metabolism is catalyzed in vivo. Genetic analysis of plastocyanin biosynthesis in *Chlamydomonas* revealed only one *trans*-acting locus, called *PCY2*, whose function is required in the thylakoid lumen and is specific for holoplastocyanin formation (66).

Strain *pcy2-1* exhibits a weak nonphotosynthetic phenotype in copper-supplemented, but not copper-deficient, medium (when plastocyanin function is not essential for photosynthesis), which is attributed to accumulation of apoplastocyanin in the thylakoid lumen at the expense of holoplastocyanin. Curiously, apoplastocyanin accumulates in copper-supplemented *pcy2-1* cells, whereas apoplastocyanin created by copper deficiency is normally degraded rapidly

(66, 73). Since the locus has not been cloned, the function of Pcy2 can only be speculated upon. By analogy to functions required for copper insertion into tyrosinase, nitrous oxide reductase and cytochrome oxidase (14, 36, 136), one might predict that Pcy2 is a copper or apoprotein chaperone in the thylakoid lumen, or perhaps a component of a *trans*-thylakoid membrane copper transporter. Recently, a gene with significant similarity (52%) to CutA (involved in copper tolerance in *E. coli*) was identified in the *Arabidopsis* genome. The 152-codon open reading frame includes an N-terminal putative transit peptide, which suggests that the protein might be localized to the chloroplast (K Keegstra, personal communication). However, the gene has not been characterized with respect to expression or localization of its product; its role in chloroplast copper metabolism therefore remains hypothetical.

CONCLUSIONS AND FUTURE PROSPECTS

With the complete genome sequence of so many organisms already available in the databases, and with several more expected in the next few years, many components required for the assembly of chloroplast metalloproteins might be identified on the basis of sequence relationships with well-characterized proteins from other systems. NifS and CcdA, discussed above, are good examples of assembly factors that were recognized by this approach. Demonstration of function for a candidate gene of interest will require parallel biochemical approaches, where the activity of the expressed protein is assessed in a reconstituted in vitro system, and reverse genetic approaches, where the phenotype of a "knock out" can be studied in the context of the pathway of interest. The latter approach has received considerably more attention relative to the former, but ultimately, functional analysis of specific gene products depends also on the dissection of mechanism, structure, and activity in vitro.

For the discovery of novel (unanticipated) biochemical functions, the traditional methods of mutant identification and phenotypic characterization have enormous potential. This is exemplified by the *ccb* mutants of *Chlamydomonas* that are affected in a process that has not been studied in any organism. The biochemical functions of the *CCB* loci are not likely to be identified immediately by genome analysis. It is also possible that the mechanisms operating for certain metalloprotein assembly processes in the chloroplast might be distinct from those studied in other systems—as noted already for the novel *CCS* loci required for chloroplast *c*-type cytochrome maturation.

Acknowledgments

Our own work in this area has been supported by grants (to SM) from the NIGMS at the NIH and the National Research Initiative of the USDA. BD is grateful for fellowship support from the NIH. We thank John Golbeck for communicating results before publication, and Joan Valentine, Elizabeth Theil, Richard Debus, and Parag Chitnis for helpful discussions.

> Visit the *Annual Reviews home page* at
> http://www.AnnualReviews.org.

Literature Cited

1. Adam Z. 1996. Protein stability and degradation in chloroplasts. *Plant Mol. Biol.* 32:773–83
2. Aguilar M, Kalakoutskii K, Cardenas J, Fernandez E. 1992. Direct transfer of molybdopterin cofactor to aponitrate reductase from a carrier protein in *Chlamydomonas reinhardtii*. *FEBS Lett.* 307:162–63
3. Anbudurai PR, Mor TS, Ohad I, Shestakov SV, Pakrasi HB. 1994. The *ctpA* gene encodes the C-terminal processing protease for the D1 protein of the photosystem II reaction center complex. *Proc. Natl. Acad. Sci. USA* 91:8082–86
4. Anderson CM, Gray JC. 1991. Effect of gabaculine on the synthesis of heme and cytochrome *f* in etiolated wheat seedlings. *Plant Physiol.* 96:584–87
5. Barkan A, Voelker R, Mendel-Hartvig J, Johnson D, Walker M. 1995. Genetic analysis of chloroplast biogenesis in higher plants. *Physiol. Plant.* 93:163–70
6. Bartsevich VV, Pakrasi HB. 1995. Molecular identification of an ABC transporter complex for manganese: analysis of a cyanobacterial mutant strain impaired in the photosynthetic oxygen evolution process. *EMBO J.* 14:1845–53
7. Bartsevich VV, Pakrasi HB. 1997. Molecular identification of a novel protein that regulates biogenesis of photosystem I, a membrane protein complex. *J. Biol. Chem.* 272:6382–87
8. Beckman DL, Kranz RG. 1993. Cytochromes *c* biogenesis in a photosynthetic bacterium requires a periplasmic thioredoxin-like protein. *Proc. Natl. Acad. Sci. USA* 90:2179–83
9. Beckman DL, Trawick DR, Kranz RG. 1992. Bacterial cytochromes *c* biogenesis. *Gen. Dev.* 6:268–83
10. Beinert H, Kiley P. 1996. Redox control of gene expression involving iron-sulfur proteins. Change of oxidation-state or assembly/disassembly of Fe-S clusters? *FEBS Lett.* 382:218–19

11. Bertini I, Luchinat C. 1994. The reaction pathways of zinc enzymes and related biological catalysts. In *Bioinorganic Chemistry*, ed. I Bertini, HB Gray, SJ Lippard, JS Valentine, pp. 37–106. Mill Valley, CA: Univ. Sci. Books
12. Bonnard G, Grienenberger J-M. 1995. A gene proposed to encode a transmembrane domain of an ABC transporter is expressed in wheat mitochondria. *Mol. Gen. Genet.* 246:91–99
13. Cerletti P. 1986. Seeking a better job for an under-employed enzyme. *Trends Biochem. Sci.* 11:369–72
14. Chen LY, Chen MY, Leu WM, Tsai TY, Lee YH. 1993. Mutational study of *Streptomyces* tyrosinase trans-activator MelC1. MelC1 is likely a chaperone for apotyrosinase. *J. Biol. Chem.* 268:18710–16
15. Chen LY, Leu WM, Wang KT, Lee YH. 1992. Copper transfer and activation of the *Streptomyces* apotyrosinase are mediated through a complex formation between apotyrosinase and its *trans*-activator MelC1. *J. Biol. Chem.* 267:20100–7
16. Chitnis PR. 1996. Photosystem I. *Plant Physiol.* 111:661–69
17. Cramer WA, Martinez SE, Furbacher PN, Huang D, Smith JL. 1994. The cytochrome $b_6 f$ complex. *Curr. Opin. Struct. Biol.* 4:536–44
18. Crooke H, Cole J. 1995. The biogenesis of *c*-type cytochromes in *Escherichia coli* requires a membrane-bound protein, DipZ, with a protein disulphide isomerase-like domain. *Mol. Microbiol.* 15:1139–50
19. Dean DR, Bolin JT, Zheng L. 1993. Nitrogenase metalloclusters: structures, organization, and synthesis. *J. Bacteriol.* 175:6737–44
20. Debus RJ. 1992. The manganese and calcium ions of photosynthetic oxygen evolution. *Biochim. Biophys. Acta* 1102: 269–352
21. Diner BA, Ries DF, Cohen BN, Metz

JG. 1988. COOH-terminal processing of polypeptide D1 of the photosystem II reaction center of *Scenedesmus obliquus* is necessary for the assembly of the oxygen-evolving complex. *J. Biol. Chem.* 263:8972–80

22. Drygas ME, Lambowitz AM, Nargang FE. 1989. Cloning and analysis of the *Neurospora crassa* gene for cytochrome *c* heme lyase. *J. Biol. Chem.* 264:17897–906

23. Dumont ME, Ernst JF, Hampsey DM, Sherman F. 1987. Identification and sequence of the gene encoding cytochrome *c* heme lyase in the yeast *Saccharomyces cerevisiae*. *EMBO J.* 6:235–41

24. Eide D, Broderius M, Fett J, Guerinot ML. 1996. A novel iron-regulated metal transporter from plants identified by functional expression in yeast. *Proc. Natl. Acad. Sci. USA* 93:5624–28

25. Fairchild CD, Glazer AN. 1994. Nonenzymatic bilin addition to the γ subunit of an apophycoerythrin. *J. Biol. Chem.* 269:28988–96

26. Fernandez MD, Lamppa GK. 1990. Acyl carrier protein (ACP) import into chloroplasts does not require the phosphopantetheine: evidence for a chloroplast holo-ACP synthase. *Plant Cell* 2:195–206

27. Fisher WR, Taniuchi H, Anfinsen CB. 1973. On the role of heme in the formation of the structure of cytochrome *c*. *J. Biol. Chem.* 248:3188–95

28. Flint DH. 1996. *Escherichia coli* contains a protein that is homologous in function and N-terminal sequence to the protein encoded by the *nifS* gene of *Azotobacter vinelandii* and that can participate in the synthesis of the Fe-S cluster of dihydroxy-acid dehydratase. *J. Biol. Chem.* 271:16068–74

29. Flint DH, Tuminello JF, Miller TJ. 1996. Studies on the synthesis of the Fe-S cluster of dihydroxy-acid dehyratase in *Escherichia coli* crude extract. *J. Biol. Chem.* 271:16053–67

30. Fu CL, Javedan S, Moshiri F, Maier RJ. 1994. Bacterial genes involved in incorporation of nickel into a hydrogenase enzyme. *Proc. Natl. Acad. Sci. USA* 91:5099–103

31. Fu WG, Jack RF, Morgan TV, Dean DR, Johnson MK. 1994. *nifU* gene product from *Azotobacter vinelandii* is a homodimer that contains two identical [2Fe–2S] clusters. *Biochemistry* 33:13455–63

32. Gabard J, Marion-Poll A, Cherel I, Meyer C, Muller A, Caboche M. 1987. Isolation and characterization of *Nico-tiana plumbaginifolia* nitrate reductase-deficient mutants: genetic and biochemical analysis of the *NIA* complementation group. *Mol. Gen. Genet.* 209:596–606

33. Garrett TPJ, Clingeleffer DJ, Guss JM, Rogers SJ, Freeman HC. 1984. The crystal structure of poplar apoplastocyanin at 1.8-Å resolution. The geometry of the copper-binding site is created by the polypeptide. *J. Biol. Chem.* 259:2822–25

34. Girard-Bascou J, Choquet Y, Gumpel N, Culler D, Purton S, et al. 1995. Nuclear control of the expression of the chloroplast *pet* genes in *Chlamydomonas reinhardtii*. In *Photosynthesis: From Light to Biosphere*, ed. P Mathis, 3:683–86. Dordrecht: Kluwer

35. Glerum DM, Shtanko A, Tzagoloff A. 1996. Characterization of *COX17*, a yeast gene involved in copper metabolism and assembly of cytochrome oxidase. *J. Biol. Chem.* 271:14504–9

36. Glerum DM, Shtanko A, Tzagoloff A. 1996. *SCO1* and *SCO2* act as high copy suppressors of a mitochondrial copper recruitment defect in *Saccharomyces cerevisiae*. *J. Biol. Chem.* 271:20531–35

37. Gonzales DH, Neupert W. 1990. Biogenesis of mitochondrial *c*-type cytochromes. *J. Bioenerg. Biomembr.* 22:753–68

38. Gonzalez DH, Bonnard G, Grienenberger J-M. 1993. A gene involved in the biogenesis of cytochromes is co-transcribed with a ribosomal protein gene in wheat mitochondria. *Curr. Genet.* 24:248–55

39. Green J, Bennett B, Jordan P, Ralph ET, Thomson AJ, Guest JR. 1996. Reconstitution of the [4Fe–4S] cluster in FNR and demonstration of the aerobic-anaerobic transcription switch *in vitro*. *Biochem. J.* 316:887–92

40. Grivell LA. 1995. Nucleo-mitochondrial interactions in mitochondrial gene expression. *Crit. Rev. Biochem. Mol. Biol.* 30:121–64

41. Hartman FC, Harpel MR. 1994. Structure, function, regulation, and assembly of D-ribulose-1, 5-bisphosphate carboxylase/oxygenase. *Annu. Rev. Biochem.* 63:197–234

42. Hausinger RP. 1990. Mechanisms of metal ion incorporation into metalloproteins. *Biofactors* 2:179–84

43. Hibino T, deBoer AD, Weisbeek PJ, Takabe T. 1991. Reconstitution of mature plastocyanin from precursor apoplastocyanin expressed in *Escherichia*

coli. Biochim. Biophys. Acta 1058:107–12

44. Hidalgo E, Demple B. 1996. Activation of SoxR-dependent transcription *in vitro* by noncatalytic or NifS-mediated assembly of [2Fe–2S] clusters into apo-SoxR. *J. Biol. Chem.* 271:7269–72

45. Hidalgo E, Ding H, Demple B. 1997. Redox signal transduction via iron-sulfur clusters in the SoxR transcription activator. *Trends Biochem. Sci.* 22:207–10

46. Hill KL, Li HH, Singer J, Merchant S. 1991. Isolation and structural characterization of the *Chlamydomonas reinhardtii* gene for cytochrome c_6. Analysis of the kinetics and metal specificity of its copper-responsive expression. *J. Biol. Chem.* 266:15060–67

47. Hoff T, Schnorr KM, Meyer C, Caboche M. 1995. Isolation of two *Arabidopsis* cDNAs involved in early steps of molybdenum cofactor biosynthesis by functional complementation of *Escherichia coli* mutants. *J. Biol. Chem.* 270:6100–7

48. Holton B, Wu XN, Tsapin AI, Kramer DM, Malkin R, Kallas T. 1996. Reconstitution of the 2Fe–2S center and $g = 1.89$ electron paramagnetic resonance signal into overproduced *Nostoc* sp. PCC 7906 Rieske protein. *Biochemistry* 35:15485–93

49. Howe G, Merchant S. 1992. The biosynthesis of membrane and soluble plastidic *c*-type cytochromes of *Chlamydomonas reinhardtii* is dependent on multiple common gene products. *EMBO J.* 11:2789–801

50. Howe G, Merchant S. 1994. The biosynthesis of bacterial and plastidic *c*-type cytochromes. *Photosynth. Res.* 40:147–65

51. Howe G, Merchant S. 1994. Role of heme in the biosynthesis of cytochrome c_6. *J. Biol. Chem.* 269:5824–32

52. Howe G, Mets L, Merchant S. 1995. Biosynthesis of cytochrome *f* in *Chlamydomonas reinhardtii*: analysis of the pathway in gabaculine-treated cells and in the heme attachment mutant B6. *Mol. Gen. Genet.* 246:156–65

53. Howe G, Quinn J, Hill K, Merchant S. 1992. Control of the biosynthesis of cytochrome c_6 in *Chlamydomonas reinhardtii. Plant Physiol. Biochem.* 30:299–307

53a. Inoue K, Dreyfuss BW, Kindle KL, Stern DB, Merchant S, Sodeinde OA. 1997. *Ccs1*, a nuclear gene required for the post-translational assembly of chloroplast *c*-type cytochromes. *J. Biol. Chem.* 272:31747–54

54. Jabri E, Carr MB, Hausinger RP, Karplus PA. 1995. The crystal structure of urease from *Klebsiella aerogenes. Science* 268:998–1004

55. Jensen KH, Herrin DL, Plumley FG, Schmidt GW. 1986. Biogenesis of photosystem II complexes: transcriptional, translational, and posttranslational regulation. *J. Cell. Biol.* 103:1315–25

56. Kampfenkel K, Kushnir S, Babiychuk E, Inze D, Van Montagu M. 1995. Molecular characterization of a putative *Arabidopsis thaliana* copper transporter and its yeast homologue. *J. Biol. Chem.* 270:28479–86

57. Kaneko T, Sato S, Kotani H, Tanaka A, Asamizu E, et al. 1996. Sequence analysis of the genome of the unicellular cyanobacterium *Synechocystis* sp. strain PCC6803. II. Sequence determination of the entire genome and assignment of potential protein-coding regions. *DNA Res.* 3:109–36

58. Kranz RG, Beckman DL. 1995. Cytochrome biogenesis. In *Anoxygenic Photosynthetic Bacteria*, ed. RE Blankenship, MT Madigan, CE Bauer, pp. 709–23. Dordrecht: Kluwer

59. Kuras R, Buschlen S, Wollman F-A. 1995. Maturation of preapocytochrome *f in vivo*. A site-directed mutagenesis study of *Chlamydomonas reinhardtii. J. Biol. Chem.* 270:27797–803

59a. Kuras R, de Vitry C, Choquet Y, Girard-Bascou J, Culler D, et al. 1997. Molecular genetic identification of a pathway for heme binding to cytochrome b_6. *J. Biol. Chem.* 272:32427–35

60. Kuras R, Wollman F-A. 1994. The assembly of cytochrome b_6/f complexes: an approach using genetic transformation of the green alga *Chlamydomonas reinhardtii. EMBO J.* 13:1019–27

61. Lang SE, Jenney FE, Daldal F. 1996. *Rhodobacter capsulatus* CycH: a bipartite gene product with pleiotropic effects of the biogenesis of structurally different *c*-type cytochromes. *J. Bacteriol.* 178:5279–90

62. Lee MH, Pankratz HS, Wang S, Scott RA, Finnegan MG, et al. 1993. Purification and characterization of *Klebsiella aerogenes* UreE protein: a nickel-binding protein that functions in urease metallocenter assembly. *Protein Sci.* 2:1042–52

63. Lee Y-H, Chen B-F, Wu S-Y, Leu W-M, Lin J-J, et al. 1988. A *trans*-acting gene is required for the phenotypic expression of a tyrosinase gene in *Streptomyces. Gene* 65:71–81

64. Leibrecht I, Kessler D. 1997. A novel L-cysteine/cystine C-S-Lyase directing [2Fe-2S] cluster formation of Synechocystis ferredoxin. J. Biol. Chem. 272:10442–47

65. Li HH, Merchant S. 1995. Degradation of plastocyanin in copper-deficient Chlamydomonas reinhardtii. J. Biol. Chem. 270:23504–10

66. Li HH, Quinn J, Culler D, Girard-Bascou J, Merchant S. 1996. Molecular genetic analysis of plastocyanin biosynthesis in Chlamydomonas reinhardtii. J. Biol. Chem. 271:31283–89

67. Li H-M, Theg SM, Bauerle CM, Keegstra K. 1990. Metal-ion-center assembly of ferredoxin and plastocyanin in isolated chloroplasts. Proc. Natl. Acad. Sci. USA 87:6748–52

68. Lippard SJ, Berg JM. 1994. Principles in Bioinorganic Chemistry. Mill Valley, CA: Univ. Sci. Books

69. Loferer H, Hennecke H. 1994. Protein disulphide oxidoreductases in bacteria. Trends Biochem. Sci. 19:169–71

70. Maier T, Lottspeich F, Bock A. 1995. GTP hydrolysis by HypB is essential for nickel insertion into hydrogenases of Escherichia coli. Eur. J. Biochem. 230:133–38

71. Malkin R, Rabinowitz JC. 1966. The reconstitution of clostridial ferredoxin. Biochem. Biophys. Res. Commun. 23:822–27

72. Mehari T, Qiao F, Scott MP, Nellis DF, Zhao J, et al. 1995. Modified ligands to F_A and F_B in photosystem I. I. Structural constraints for the formation of iron-sulfur clusters in free and rebound PsaC. J. Biol. Chem. 270:28108–17

73. Merchant S, Bogorad L. 1986. Rapid degradation of apoplastocyanin in Cu(II)-deficient cells of Chlamydomonas reinhardtii. J. Biol. Chem. 261:15850–53

74. Meurer J, Meierhoff K, Westhoff P. 1996. Isolation of high-chlorophyll-fluorescence mutants of Arabidopsis thaliana and their characterisation by spectroscopy, immunoblotting and northern hybridisation. Planta 198:385–96

75. Mobley HL, Island MD, Hausinger RP. 1995. Molecular biology of microbial ureases. Microbiol. Rev. 59:451–80

76. Monika EM, Goldman BS, Beckman DL, Kranz RG. 1997. A thioreduction pathway tethered to the membrane for periplasmic cytochrome c biogenesis; in vitro and in vivo studies. J. Mol. Biol. 271:679–92

77. Morishige DT, Dreyfuss BW. 1997. The Light-harvesting complexes of higher plants. In Photosynthesis—A Comprehensive Treatise, ed. AS Raghavendra, pp.18–28. Cambridge: Cambridge Univ. Press

78. Nar H, Huber R, Messerschmidt A, Filippou AC, Barth M, et al. 1992. Characterization and crystal structure of zinc azurin, a by-product of heterologous expression in Escherichia coli of Pseudomonas aeruginosa copper azurin. Eur. J. Biochem. 205:1123–29

79. Nicholson DW, Kohler H, Neupert W. 1987. Import of cytochrome c into mitochondria. Cytochrome c heme lyase. Eur. J. Biochem. 164:147–57

80. Nicholson DW, Neupert W. 1989. Import of cytochrome c into mitochondria: reduction of heme, mediated by NADH and flavin nucleotides, is obligatory for its covalent linkage to apocytochrome c. Proc. Natl. Acad. Sci. USA 86:4340–44

81. Nicholson DW, Stuart RA, Neupert W. 1989. Biogenesis of cytochrome c_1. Role of cytochrome c_1 heme lyase and of the two proteolytic processing steps during import into mitochondria. J. Biol. Chem. 264:10156–68

82. Nixon PJ, Trost JT, Diner BA. 1992. Role of the carboxy terminus of polypeptide D1 in the assembly of a functional water-oxidizing manganese cluster in photosystem II of the cyanobacterium Synechocystis sp. PCC 6803: assembly requires a free carboxyl group at C-terminal position 344. Biochemistry 31:10859–71

83. Page MD, Ferguson SJ. 1990. Apo forms of cytochrome c_{550} and cytochrome cd_1 are translocated to the periplasm of Paracoccus denitrificans in the absence of haem incorporation caused by either mutation or inhibition of haem synthesis. Mol. Microbiol. 4:1181–92

84. Pakrasi HB. 1995. Genetic analysis of the form and function of photosystem I and photosystem II. Annu. Rev. Genet. 29:755–76

85. Pakrasi HB, De Ciechi P, Whitmarsh J. 1991. Site directed mutagenesis of the heme axial ligands of cytochrome b559 affects the stability of the photosystem II complex. EMBO J. 10:1619–27

86. Park IS, Hausinger RP. 1995. Requirement of carbon dioxide for in vitro assembly of the urease nickel metallocenter. Science 267:1156–58

87. Paulsen H. 1995. Chlorophyll a/b-binding proteins. Photochem. Photobiol. 62:367–82

88. Pilon M, de Kruijff B, Weisbeek PJ. 1992. New insights into the import mechanism of the ferredoxin precursor into chloroplasts. *J. Biol. Chem.* 267:2548–56

89. Poyton RO, McEwen JE. 1996. Crosstalk between nuclear and mitochondrial genomes. *Annu. Rev. Biochem.* 65:563–607

90. Rajagopalan KV, Johnson JL. 1992. The pterin molybdenum cofactors. *J. Biol. Chem.* 267:10199–202

91. Ramseier TM, Winteler HV, Hennecke H. 1991. Discovery and sequence analysis of bacterial genes involved in the biogenesis of *c*-type cytochromes. *J. Biol. Chem.* 266:7793–803

92. Robinson KM, Lemire BD. 1996. Covalent attachment of FAD to the yeast succinate dehydrogenase flavoprotein requires import into mitochondria, presequence removal, and folding. *J. Biol. Chem.* 271:4055–60

93. Rochaix JD. 1996. Post-transcriptional regulation of chloroplast gene expression in *Chlamydomonas reinhardtii*. *Plant Mol. Biol.* 32:327–41

94. Rouault TA, Klausner RD. 1996. Iron-sulfur clusters as biosensors of oxidants and iron. *Trends Biochem. Sci.* 21:174–77

95. Sambongi Y, Ferguson SJ. 1994. Specific thiol compounds complement deficiency in *c*-type cytochrome biogenesis in *Escherichia coli* carrying a mutation in a membrane-bound disulphide isomerase-like protein. *FEBS Lett.* 353:235–38

96. Sambongi Y, Ferguson SJ. 1996. Mutants of *Escherichia coli* lacking disulphide oxidoreductases DsbA and DsbB cannot synthesise an exogenous monohaem *c*-type cytochrome except in the presence of disulphide compounds. *FEBS Lett.* 398:265–68

97. Schiott T, von Wachenfeldt C, Hederstedt L. 1997. Identification and characterization of the *ccdA* gene, required for cytochrome *c* synthesis in *Bacillus subtilis*. *J. Bacteriol.* 179:1962–73

98. Schmidt GW, Mishkind ML. 1983. Rapid degradation of unassembled ribulose 1, 5-bisphosphate carboxylase small subunits in chloroplasts. *Proc. Natl. Acad. Sci. USA* 80:2632–36

99. Schnell D. 1998. Protein targeting to the thylakoid membrane. *Annu. Rev. Plant Physiol. Plant Mol. Biol.* 49:97–126

100. Schuster W. 1994. The highly edited *orf206* in *Oenothera* mitochondria may encode a component of a heme transporter involved in cytochrome *c* biogenesis. *Plant Mol. Biol.* 25:33–42

101. Schuster W, Combettes B, Flieger K, Brennicke A. 1993. A plant mitochondrial gene encodes a protein involved in cytochrome *c* biogenesis. *Mol. Gen. Genet.* 239:49–57

102. Serra EC, Krapp AR, Ottado J, Feldman MF, Ceccarelli EA, Carrillo N. 1995. The precursor of pea ferredoxin-NADP$^+$ reductase synthesized in *Escherichia coli* contains bound FAD and is transported into chloroplasts. *J. Biol. Chem.* 270:19930–35

103. Shestakov SV, Anbudurai PR, Stanbekova GE, Gadzhiev A, Lind LK, Pakrasi HB. 1994. Molecular cloning and characterization of the *ctpA* gene encoding a carboxyl-terminal processing protease. Analysis of a spontaneous photosystem II-deficient mutant strain of the cyanobacterium *Synechocystis* sp. PCC 6803. *J. Biol. Chem.* 269:19354–59

104. Stallmeyer B, Nerlich A, Schiemann J, Brinkmann H, Mendel RR. 1995. Molybdenum co-factor biosynthesis: the *Arabidopsis thaliana* cDNA *cnx1* encodes a multifunctional two-domain protein homologous to a mammalian neuroprotein, the insect protein Cinnamon and three *Escherichia coli* proteins. *Plant J.* 8:751–62

105. Stellwagen E, Rysavy R, Babul G. 1972. The conformation of horse heart apocytochrome *c*. *J. Biol. Chem.* 247:8074–77

106. Stewart V. 1988. Nitrate respiration in relation to facultative metabolism in enterobacteria. *Microbiol. Rev.* 52:190–232

107. Sundberg E, Slagter JG, Fridborg I, Cleary SP, Robinson C, Coupland G. 1997. *ALBINO3*, an Arabidopsis nuclear gene essential for chloroplast differentiation, encodes a chloroplast protein that shows homology to proteins present in bacterial membranes and yeast mitochondria. *Plant Cell* 9:717–30

108. Suttie JW. 1988. Vitamin K-dependent carboxylation of glutamyl residues in proteins. *Biofactors* 1:55–60

109. Suzuki S, Izumihara K, Hase T. 1991. Plastid import and iron-sulfur cluster assembly of photosynthetic and nonphotosynthetic ferredoxin isoproteins in maize. *Plant Physiol.* 97:375–80

110. Takahashi Y, Mitsui A, Fujita Y, Matsubara H. 1991. Roles of ATP and NADPH in formation of the Fe-S cluster of spinach ferredoxin. *Plant Physiol.* 95:104–10

111. Takahashi Y, Mitsui A, Hase T, Matsubara H. 1986. Formation of the iron-sulfur cluster of ferredoxin in isolated chloroplasts. *Proc. Natl. Acad. Sci. USA* 83:2434–37

112. Takahashi Y, Mitsui A, Matsubara H. 1991. Formation of the Fe-S cluster of ferredoxin in lysed spinach chloroplast. *Plant Physiol.* 95:97–103

113. Tamilarasan R, McMillin DR. 1986. Absorption spectra of d^{10} metal ion derivatives of plastocyanin. *Inorg. Chem.* 25:2037–40

114. Taniuchi H, Basile G, Taniuchi M, Veloso D. 1983. Evidence for formation of two thioether bonds to link heme to apocytochrome *c* by partially purified cytochrome *c* synthetase. *J. Biol. Chem.* 258:10963–66

115. Taroni F, Rosenberg LE. 1991. The precursor of the biotin-binding subunit of mammalian propionyl-CoA carboxylase can be translocated into mitochondria as apo- or holoprotein. *J. Biol. Chem.* 266:13267–71

116. Theil EC. 1987. Ferritin: structure, gene regulation, and cellular function in animals, plants and microorganisms. *Annu. Rev. Biochem.* 56:289–315

117. Thony-Meyer L, Fischer F, Kunzler P, Ritz D, Hennecke H. 1995. *Escherichia coli* genes required for cytochrome *c* maturation. *J. Bacteriol.* 17:4321–26

118. Thony-Meyer L. 1997. Biogenesis of respiratory cytochromes in bacteria. *Microbiol. Mol. Biol. Rev.* 61:337–76

119. Tomsett AB, Garrett RH. 1980. The isolation and characterization of mutants defective in nitrate assimilation in *Neurospora crassa*. *Genetics* 95:649–60

120. Trost JT, Chisholm DA, Jordan DB, Diner BA. 1997. The D1 C-terminal processing protease of photosystem II from *Scenedesmus obliquus*. Protein purification and gene characterization in wild type and processing mutants. *J. Biol. Chem.* 272:20348–56

121. Turner M, Gumpel N, Ralley L, Lumbreras V, Purton S. 1996. Biogenesis of the electron-transfer complexes in *Chlamydomonas reinhardtii*. *Biochem. Soc. Trans.* 24:733–38

122. Tzagoloff A, Dieckman CL. 1990. *PET* genes of *Saccharomyces cerevisiae*. *Microbiol. Rev.* 54:211–25

123. Tzagoloff A, Nobrega M, Gorman N, Sinclair P. 1993. On the functions of the yeast COX10 and COX11 gene products. *Biochem. Mol. Biol. Int.* 31:593–98

124. Ugalde RA, Imperial J, Shah VK, Brill WJ. 1985. Biosynthesis of the iron-molybdenum cofactor and the molybdenum cofactor in *Klebsiella pneumoniae*: effect of sulfur source. *J. Bacteriol.* 164:1081–87

125. Vargas C, Wu G, Davies AE, Downie JA. 1994. Identification of a gene encoding a thioredoxin-like product necessary for cytochrome *c* biogenesis and symbiotic nitrogen fixation in *Rhizobium leguminosarum*. *J. Bacteriol.* 176:4117–23

126. Voelker R, Barkan A. 1995. Nuclear genes required for post-translational steps in the biogenesis of the chloroplast cytochrome $b_6 f$ complex in maize. *Mol. Gen. Genet.* 249:507–14

127. Waldo GS, Wright E, Whang ZH, Briat JF, Theil EC, Sayers DE. 1995. Formation of the ferritin iron mineral occurs in plastids. *Plant Physiol.* 109:797–802

128. Webber AN, Gibbs PB, Ward JB, Bingham SE. 1993. Site-directed mutagenesis of the photosystem I reaction center in chloroplasts. The proline-cysteine motif. *J. Biol. Chem.* 268:12990–95

129. Whitmarsh J, Pakrasi HB. 1996. Form and function of cytochrome b–559. In *Oxygenic Photosynthesis: The Light Reactions*, ed. DR Ort, CF Yocum, 4:249–64. Dordrecht: Kluwer

129a. Xie Z, Merchant S. 1996. The plastid-encoded *ccsA* gene is required for heme attachment to chloroplast *c*-type cytochromes. *J. Biol. Chem.* 271:4632–39

129b. Xie ZY, Culler D, Dreyfuss BW, Girard-Bascou J, Kuras R, Wollman FA, Merchant S. 1998. Genetic analysis of chloroplast *c*-type cytochrome assembly: One chloroplast locus and at least four nuclear loci are required for heme attachment. *Genetics.* In press

130. Yang LM, Fernandez MD, Lamppa GK. 1994. Acyl carrier protein (ACP) import into chloroplasts. Covalent modification by a stromal holoACP synthase is stimulated by exogenously added CoA and inhibited by adenosine $3'$, $5'$-bisphosphate. *Eur. J. Biochem.* 224:743–50

131. Yu JP, Vassiliev IR, Jung Y-S, Golbeck JH, McIntosh L. 1997. Strains of *Synchocystis sp.* PCC 6803 with altered PsaC. I. Mutations incorporated in the cysteine ligands of the two [4Fe-4S] clusters F_A and F_B of photosystem I. *J. Biol. Chem.* 272:8032–39

132. Zheng L, Dean DR. 1994. Catalytic formation of a nitrogenase iron-sulfur cluster. *J. Biol. Chem.* 269:18723–26

133. Zheng L, White RH, Cash VL, Dean DR. 1994. Mechanism for the desulfurization of L-cysteine catalyzed by the *nifS* gene product. *Biochemistry* 33:4714–20

134. Zheng L, White RH, Cash VL, Jack RF, Dean DR. 1993. Cysteine desulfurase activity indicates a role for NIFS in metallocluster biosynthesis. *Proc. Natl. Acad. Sci. USA* 90:2754–58

135. Zollner A, Rodel G, Haid A. 1992. Molecular cloning and characterization of the *Saccharomyces cerevisiae CYT2* gene encoding cytochrome-c_1-heme lyase. *Eur. J. Biochem.* 207:1093–100

136. Zumft WG, Viebrock-Sambale A, Braun C. 1990. Nitrous oxide reductase from denitrifying *Pseudomonas stutzeri*. Genes for copper-processing and properties of the deduced products, including a new member of the family of ATP/GTP-binding proteins. *Eur. J. Biochem.* 192:591–99

Annu. Rev. Plant Physiol. Plant Mol. Biol. 1998. 49:53–75

BIOSYNTHESIS AND FUNCTION OF THE SULFOLIPID SULFOQUINOVOSYL DIACYLGLYCEROL

Christoph Benning

Department of Biochemistry, Michigan State University, East Lansing, Michigan 48824

KEY WORDS: glycerolipid, photosynthesis, phosphate limitation, *sqd* genes, thylakoid
               membrane, UDP-sulfoquinovose

ABSTRACT

The sulfolipid sulfoquinovosyl diacylglycerol is an abundant sulfur-containing nonphosphorous glycerolipid that is specifically associated with photosynthetic membranes of higher plants, mosses, ferns, algae, and most photosynthetic bacteria. The characteristic structural feature of sulfoquinovosyl diacylglycerol is the unique head group constituent sulfoquinovose, a derivative of glucose in which the 6-hydroxyl is replaced by a sulfonate group. While there is growing evidence for the final assembly of the sulfolipid by the transfer of the sulfoquinovosyl moiety from UDP-sulfoquinovose to the *sn*-3 position of diacylglycerol, very little is known about the biosynthesis of the precursor UDP-sulfoquinovose. Recently, a number of mutants deficient in sulfolipid biosynthesis and the corresponding *sqd* genes have become available from different organisms. These provide novel tools to analyze sulfolipid biosynthesis by a combination of molecular and biochemical approaches. Furthermore, the analysis of sulfolipid-deficient mutants has provided novel insights into the function of sulfoquinovosyl diacylglycerol in photosynthetic membranes.

CONTENTS

53

INTRODUCTION

The sulfolipid sulfoquinovosyl diacylglycerol (SQDG) of higher plants and other photosynthetic organisms is characterized by its unique sulfonic acid head group, a 6-deoxy-6-sulfo-glucose, referred to as sulfoquinovose (Figure 1). Due to its strong amphipathic character, this anionic lipid is commonly thought to have excellent detergent properties (11) that may explain its apparent antiviral and antitumor activities (24, 71). It has been estimated that SQDG is one of the most abundant sulfur-organic compounds in the biotic world following glutathione, cysteine, and methionine, and thus plays an important role in the global sulfur cycle (27, 28).

Since the discovery of SQDG (12) and the elucidation of its structure by Benson and coworkers (11) over 30 years ago, a plenitude of largely theoretical discussions has been provided on the biosynthesis of SQDG (for the most recent reviews, refer to 28, 38, 43, 53). In the past, experimental progress has been hampered by the lack of a suitable in vitro system for SQDG biosynthesis

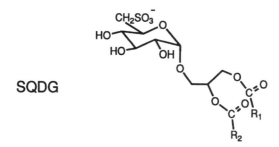

SQDG

Figure 1   Structure of the sulfolipid sulfoquinovosyl diacylglycerol (SQDG). A more precise designation of this compound according to the International Union of Pure and Applied Chemistry (IUPAC) is: 1,2-di-O-acyl-3-O-(6-deoxy-6-sulfo-α-D-glucopyranosyl)-sn-glycerol. R_1 and R_2 indicate acyl chains of different length and degree of unsaturation.

and a large number of sulfurous water-soluble compounds in extracts of higher plants. Only recently has it become feasible to rigorously test the different hypotheses for the pathway of SQDG biosynthesis by a combination of biochemical and genetical approaches. A brief historic overview is provided, but the focus of this review is on the "sugar-nucleotide pathway" hypothesis for the head group biosynthesis of SQDG, beginning with the synthesis of UDP-sulfoquinovose (UDP-SQ) from UDP-glucose (UDP-Glc) followed by a transfer of sulfoquinovose to diacylglycerol (DAG). This pathway, discussed in detail by Pugh et al (57), seems to be best supported by the current experimental evidence. Less attention is paid to the origin of the DAG moiety, an aspect of SQDG biosynthesis that has already been extensively covered (38, 43, 53). Because genes essential for the biosynthesis of SQDG in photosynthetic bacteria and higher plants have recently been identified, it has become feasible to begin a comparison of the pathways for SQDG biosynthesis in different photosynthetic organisms. Already the limited data available suggest that some aspects of SQDG biosynthesis are conserved among organisms containing SQDG, while other aspects diverged during evolution.

The assumption that all photosynthetic organisms contain the sulfolipid SQDG, in conjunction with the unusual structure of the head group, led to the speculation that SQDG may play a specific role in photosynthesis (e.g. 5). However, the correlation between photosynthesis and the presence of SQDG has its exceptions, at least in bacteria. Furthermore, SQDG-deficient mutants of different organisms are now available, and their analysis has resulted in a revised hypothesis on the function of SQDG.

OCCURRENCE AND ABUNDANCE OF SULFOLIPID

Photosynthetic Organisms

It is generally accepted that the sulfolipid SQDG is present in all higher plants, mosses, ferns, and algae. This assumption is based on an extensive survey of different photosynthetic organisms following the discovery of the sulfolipid SQDG as summarized by Haines (26). More recently, a large number of marine algae have been shown to contain SQDG (1, 17, 18). However, it should be emphasized that the identification of SQDG was rarely based on rigorous structural analysis but, rather, on cochromatography with standards. It is known that other sulfur-containing lipids are present in some algae (53), and a new survey relying more on structural analysis may be desirable. As summarized by Heinz (28), some marine algae contain more than 40 mol% of SQDG in lipid extracts compared with roughly 4 mol% found in lipid extracts of higher plants. Fractionation of subcellular membranes from higher plants reveals that SQDG is exclusively associated with plastid membranes (19).

While there seems to be a tight correlation between the presence of SQDG and the capability to perform photosynthesis in plants, many photosynthetic bacteria lack SQDG or may contain it only in small amounts. Cyanobacteria conduct oxygenic photosynthesis like higher plants and typically show a membrane lipid composition similar to that of plant thylakoid membranes. Thus, a large number of cyanobacterial species isolated from different habitats contain approximately 10 mol% SQDG (e.g. 50, 74, 79). However, there has been a recent report that the cyanobacterium *Gloeobacter violaceus* sp. PCC7421 lacks SQDG (68). Among the anoxygenic photosynthetic bacteria are at least two genera, *Ectothiorhodospira* and *Heliobacterium*, that are completely devoid of glycolipids and, thus, SQDG (3, 34). Within the purple nonsulfur bacteria, some species like *Rhodobacter sphaeroides* contain SQDG, as confirmed by extensive structural analysis (21, 58), while only traces of SQDG can be detected in the closely related species *Rhodobacter capsulatus* (33), and no SQDG seems to be present in *Rhodopseudomonas viridis* (48). Furthermore, there is some uncertainty whether SQDG or a structurally related sulfolipid is present in photosynthetic bacteria of the family Chromatiaceae and the green sulfur bacteria (35). Therefore, previous statements that "SQDG is present in all photosynthetic organisms" are clearly misleading. Even the assumption that "SQDG is present in all oxygenic photosynthetic organisms" may no longer be valid following the discovery of a SQDG-deficient cyanobacterium.

Nonphotosynthetic Organisms

Another widespread but erroneous assumption has been that "the occurrence of SQDG is restricted to photosynthetic organisms." The recent identification and complete structural elucidation of SQDG isolated from the nonphotosynthetic bacterium *Rhizobium meliloti* (15) clearly demonstrate beyond any doubt that SQDG is present in some nonphotosynthetic organisms. However, it should be pointed out that rhizobia are closely related to photosynthetic purple nonsulfur bacteria (78) and that there are even photosynthetic species of rhizobia (75). Thus it may be more surprising that SQDG is apparently present in a gram-positive extreme thermoacidophilic nonphotosynthetic bacterium of the genus *Alicyclobacillus* formerly called *Bacillus acidocaldarius* (46). The identification was accomplished by cochromatography with authentic standards, IR spectroscopy, and isotope labeling, leaving little doubt that SQDG is, indeed, present in this bacterium. Moreover, there is a report on the presence and function of SQDG in eggs and sperm cells of the sea urchin *Pseudocentrotus depressus* (36).

The Effect of Phosphate Availability

In general, the membrane lipid composition of photosynthetic tissues in higher plants is highly constant between different species and is, to a large extent,

Table 1 Polar lipid composition of the purple bacterium *Rhodobacter sphaeroides* and the cyanobacterium *Synechococcus* sp. PCC7942[a]

Bacterium	Pi[b] (mM)	Phospholipids[c]			Glycolipids[d]			Others[e]	
		PC	PE	PG	SQDG	MHD	DHD	OL	BL
R. sphaeroides	1.0	27.7	39.9	22.8	2.2	—	Tr[f]	5.5	Tr
R. sphaeroides	0.1	2.9	6.8	12.2	16.6	Tr	31.1	11.2	31.1
Synechococcus	0.18	—	—	16.6	10.3	60.6	12.5	—	—
Synechococcus	0.018	—	—	7.2	22.3	42.5	28.0	—	—

[a]Data from Benning et al (7) and Güler et al (23).
[b]Concentration of inorganic phosphate (Pi) in the medium.
[c]PC, Phosphatidylcholine; PE, phosphatidylethanolamine; PG, phosphatidylglycerol. In mol percent.
[d]SQDG, Sulfoquinovosyl diacylglycerol; MHD, monohexosyl diacylglycerol (monogalactosyl diacylglycerol for *Synechococcus*); DHD, dihexosyl diacylglycerol (glucosylgalactosyl diacylglycerol for *R. sphaeroides* or digalactosyl diacylglycerol for *Synechococcus*). In mol percent.
[e]OL, Ornithine lipid; BL, betaine lipid. In mol percent.
[f]Tr, traces detected, but not quantified.

indifferent to varying growth conditions at the lipid class level. However, with regard to bacteria it has long been known that the availability of phosphate can drastically affect the membrane lipid composition. Thus, phosphate-limiting growth conditions result in a replacement of acidic phospholipids by acidic glycolipids in *Pseudomonas diminuta* (51). Likewise, it has been reported that in the purple bacterium *R. sphaeroides* and the cyanobacterium *Synechococcus* sp. PCC7942 drastic changes in the complex lipid composition occur during phosphate-limiting growth (7, 8, 23). Table 1 shows that the relative amounts of phospholipids are reduced under phosphate limitation and that the relative amounts of nonphosphorous lipids such as SQDG drastically increase. Moreover, in *R. sphaeroides* novel glycolipids, as well as a betaine lipid previously thought to be present only in algae and lower plants, are newly synthesized under phosphate limitation (8). It should also be pointed out that there is an inverse relationship between the abundance of the two anionic lipids phosphatidylglycerol and SQDG such that the total amount of anionic lipids remains constant under optimal and phosphate-limiting conditions.

It has been recently discovered that phosphate availability also affects the membrane lipid composition in higher plants. Seedlings of *Arabidopsis thaliana* grown on agar-solidified medium with decreasing concentrations of inorganic phosphate showed similar changes in the lipid composition, as observed for the cyanobacterium *Synechococcus* sp. PCC7942 (C Benning, unpublished data). The relative amount of SQDG increases up to fivefold, and the relative amounts of phosphatidyl glycerol decrease accordingly with decreasing phosphate concentrations in the medium. The increase in the relative amount of SQDG in response to phosphate limitation appears to be due to an active regulatory process rather than a simple dilution of phospholipids, because in parallel the expression of a gene essential for SQDG biosynthesis is induced in *A. thaliana* (B Essigmann & C Benning, unpublished data).

GENES ESSENTIAL FOR SULFOLIPID BIOSYNTHESIS

Approximately 10 years ago, we began our efforts to search for genes essential for the biosynthesis of SQDG in different photosynthetic organisms. Initially, the purple bacterium *R. sphaeroides* was chosen as a model organism for three reasons: (*a*) It contains SQDG like higher plants (21), (*b*) it can grow photoheterotrophically allowing the isolation of SQDG-deficient mutants even if SQDG deficiency may have had adverse effects on photosynthesis, and (*c*) common bacterial genetic techniques are available for *R. sphaeroides*. Screening a chemically mutagenized population by thin-layer chromatography of lipid extracts, a large number of lipid mutants of *R. sphaeroides* were isolated, including those deficient in phosphatidylcholine (2) and SQDG (9, 10). The first three genes essential for SQDG biosynthesis were isolated by genetic complementation of three mutants (9, 10). The organization of the sulfolipid genes of *R. sphaeroides*, designated *sqdA*, *sqdB*, and *sqdC* in two transcription units, is depicted in Figure 2. Subsequently, a fourth sulfolipid gene (*sqdD*) was identified by insertional inactivation of an open reading frame flanked by *sqdB* and *sqdC* (61). Predicted amino acid sequences of two of the sulfolipid genes

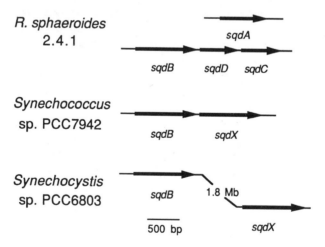

Figure 2   Organization of sulfolipid genes in purple bacteria and cyanobacteria. The *sqd* genes of *R. sphaeroides* and *Synechococcus* were confirmed by mutational analysis, while those of *Synechocystis* were identified based on only sequence similarity. bp, base pairs; Mb, megabase pairs.

showed considerable similarity to sequences of enzymes with known function: *sqdB* to UDP-glucose-4-epimerases and nucleotide-hexose-4,6,-dehydratases, and *sqdD* to glycogenins, a class of UDP-glucose-dependent glycosyltransferases (10). Thus in two cases, the sequence analysis of the sulfolipid genes provided a promising basis for further analysis of the gene function and suggested that the *sqdB* gene product may be a sugar nucleotide modifying enzyme, possibly involved in the biosynthesis of UDP-SQ, and that the *sqdD* gene product may encode the glycosyltransferase catalyzing the last reaction of SQDG biosynthesis in *R. sphaeroides*.

To study SQDG function in an organism with oxygenic photosynthesis, we isolated and inactivated a homolog of the *R. sphaeroides sqdB* gene from the cyanobacterium *Synechococcus* sp. PCC7942 (23). Functional homology between the *sqdB* genes of the two bacteria is inferred based on an amino acid sequence identity of 67% and SQDG deficiency following inactivation in *Synechococcus*. Furthermore, a presumed *sqdB* gene from *Synechocystis* sp. PCC6803 has been identified by complete genome analysis of this cyanobacterium (39), but other *sqd* genes found in *R. sphaeroides* are less conserved or not present in *Synechocystis*. Thus far, the limited species comparison suggests that the *sqdB* genes are highly conserved among different photosynthetic bacteria, while other *sqd* genes are not. However, we identified a novel sulfolipid gene in *Synechococcus* sp. PCC7942, designated *sqdX*, by inactivation of an open reading frame following directly the 3′ end of *sqdB* in this bacterium (S Güler & C Benning, unpublished data). This open reading frame encodes a putative protein with sequence similarity to sucrose synthase type glycosyl transferases, and its inactivation causes SQDG deficiency. Thus, it seems likely that in *Synechococcus* the last reaction of SQDG biosynthesis is catalyzed by a different type of glycosyl transferase than in *R. sphaeroides*. Moreover, in *Synechocystis* sp. PCC6803, the putative *sqdB* gene is not followed by an open reading frame similar to *sqdX*. However, a putative coding sequence with 73% identity to *sqdX* is located approximately 1.8 Mb away from the putative *sqdB* gene in the genome of *Synechocystis* sp. PCC6803, suggesting a different organization of sulfolipid genes in related cyanobacteria. A comparison between the organization of sulfolipid genes in *R. sphaeroides, Synechococcus* sp. PCC7942, and *Synechocystis* sp. PCC6803 is shown in Figure 2.

Comparing bacterial *sqd* gene DNA sequences against DNA databases, we were able to identify a partial cDNA from rice (D46477; Reference 63) with high sequence similarity to bacterial *sqdB* genes. We used the rice cDNA to isolate a corresponding cDNA from *Arabidopsis thaliana* encoding a putative protein with a 42% amino acid sequence identity to the bacterial *sqdB* gene products (S Güler & C Benning, unpublished data). Expression of this cDNA in antisense orientation under the control of a strong promoter caused SQDG deficiency, confirming that the gene indeed encodes a protein involved in SQDG biosynthesis (RA Narang & C Benning, unpublished data). It appears that the *sqdB* gene class has been highly conserved among photosynthetic bacteria and even higher plants, suggesting that the respective proteins catalyze a reaction unique to SQDG biosynthesis.

THE PATHWAY OF SULFOLIPID BIOSYNTHESIS

A Brief Historic Overview

Despite nearly forty years of efforts, our current knowledge about the reactions leading toward the formation of the sulfoquinovose head group still falls short of the claim that the pathway of SQDG biosynthesis has been solved. The complexity of this classic problem of plant biochemistry becomes apparent if one considers that three major pathways contribute to the biosynthesis of SQDG: carbohydrate metabolism, sulfur assimilation, and fatty acid biosynthesis. Of these three metabolic processes, carbohydrate metabolism has been most intensely studied, and many ideas regarding the possible pathway for SQDG biosynthesis have been derived by analogy to common reactions of carbohydrate biosynthesis or degradation.

An early suggestion by Benson (11) was the "sulfoglycolytic pathway" for SQDG biosynthesis, which was thought to involve sulfonated analogs of intermediates of glycolysis. Experimental support seemed to be provided by the discovery of different sulfonated compounds in extracts of the unicellular algae *Chlorella*, including sulfoacetic acid, sulfolactic acid, cysteic acid, and a nucleoside diphosphate sulfoquinovose (69). Based on labeling experiments with the unicellular alga *Euglena*, Davis et al (16) outlined a version of the sulfoglycolytic pathway starting with the synthesis of 3-L-sulfolactic acid from phosphoenolpyruvate and 3'-phosphoadenosyl-5'-phosphosulfate (PAPS), which is converted to 3-sulfo-D-lactaldehyde via a proposed epimerase and a dehydrogenase. Sulfolactaldehyde and dihydroxyacetone phosphate could condense via an aldolase reaction, forming 6-sulfo-6-deoxy-D-fructose-1-phosphate, which was suggested to be a direct precursor of 6-sulfoquinovose. Activated sulfoquinovose in the form of UDP-SQ was thought to be the substrate for a glycosyl transferase catalyzing the last reaction of SQDG biosynthesis, as first suggested by Benson (11). It was shown that *Euglena* incorporates cysteic acid into SQDG via an assumed side reaction of the sulfoglycolytic pathway (16), but Haines proposed that cysteic acid feeds directly into the pathway via a transaminase reaction to give rise to 3-sulfo-L-lactic acid (26). However, labeling experiments with spinach seedlings provided no evidence for a direct precursor role of cysteic acid in higher plants (52). Furthermore, it has been shown that *Euglena* is somewhat unusual, because sulfite derived from sulfur assimilation in mitochondria serves as precursor for SQDG biosynthesis in chloroplasts of this alga (62).

The observation that isolated spinach chloroplasts are able to incorporate sulfate into SQDG initially provided a promising new model system for the analysis of sulfolipid biosynthesis (25, 37). Incorporation rates of up to 700 pmol/mg chlorophyll were observed in the light (40), and incorporation of sulfate into SQDG was detected in the dark following the supply of chloroplasts with ATP (41). Recently, isolated chloroplasts from pea and lettuce were also shown to incorporate sulfate into SQDG (56). The general conclusion drawn from these experiments is that all the enzymes required for the biosynthesis of SQDG must be associated with chloroplasts in higher plants. Unfortunately, chloroplasts have limitations as a system for investigating the biosynthesis of the SQDG head group, because they are quite selective in their uptake of small molecules, a property not suitable for precursor and inhibitor studies. In the past, the biggest drawback to conventional biochemical analysis of SQDG biosynthesis has been the lack of enzyme activity following the rupture of spinach chloroplasts, with the exception of the enzyme catalyzing the last reaction of SQDG biosynthesis (see below). The lack of a suitable in vitro system has been one of the reasons for the slow progress in the elucidation of SQDG biosynthesis in higher plants. However, it should be noted that incorporation of PAPS, sulfite, and sulfate into SQDG by extracts of *Chlamydomonas reinhardtii* has been observed (31), an experiment that has not been repeated to date. Furthermore, Pugh et al (57) recently observed SQDG biosynthesis using broken pea chloroplasts, a promising system that may be used in the future to study SQDG biosynthesis in vitro.

The Sugar-Nucleotide Pathway Hypothesis of Sulfolipid Biosynthesis

An alternative to the sulfoglycolytic pathway outlined above is a two-step reaction sequence starting with UDP-glucose as the direct precursor for UDP-SQ, which one can call the sugar-nucleotide pathway for SQDG biosynthesis. Our current working hypothesis and the proposed involvement of *sqd* gene products in SQDG biosynthesis in *R. sphaeroides* and *Synechococcus* sp. PCC7942 are outlined in Figure 3. In this model, the sulfur is introduced at the level of hexose and not triose as predicted by the sulfoglycolytic pathway. The similarity of the *R. sphaeroides sqdB* gene product to nucleotide-modifying enzymes such as UDP-glucose-4-epimerases and nucleotide-hexose-4,6,-dehydratases (10) provided a first piece of experimental evidence in support of the sugar-nucleotide pathway hypothesis for SQDG biosynthesis. Furthermore, SQDG-deficient mutants of *R. sphaeroides* inactivated in *sqdB* do not accumulate a sulfur-labeled water-soluble compound (C Benning, unpublished data), suggesting that the

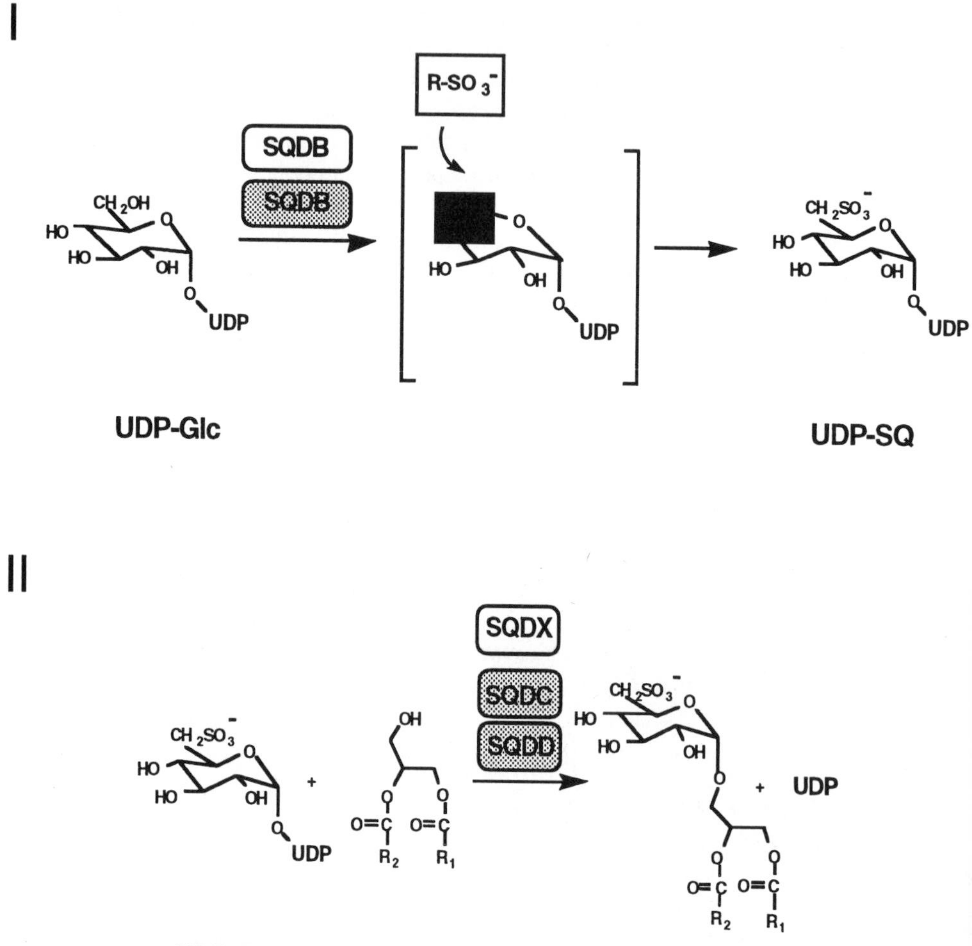

Figure 3 Sugar-nucleotide pathway hypothesis for SQDG biosynthesis. Step I, biosynthesis of UDP-SQ; Step II, UDP-SQ:DAG sulfoquinovosyltransferase reaction. Gene products thought to be involved in cyanobacteria (*open symbols*) and in *R. sphaeroides* (*stippled symbols*) are indicated above the respective arrows. A predicted intermediate involved in step I, of which the exact structure is unknown (*black box*), is placed in brackets, and a predicted sulfur donor of unknown structure ($R\text{-}SO_3^-$) is indicated. Abbreviations are as defined in the text.

sqdB gene product is acting before or during the incorporation of sulfur into the head-group moiety. Recently, we observed that the *sqdB* gene product of *R. sphaeroides* accepts UDP-glucose and converts it to a new compound in vitro (B Essigmann & C Benning, unpublished data). Activated reaction intermediates such as UDP-4-ketoglucose-5-ene produced during the biosynthesis of 6-deoxyhexoses by sugar nucleotide-hexose-4,6-dehydratases (20) could serve as a sulfur acceptor. That 6-sulfoquinovose found in SQDG may be derived from such an intermediate was first suggested by Barber (4). Recently Pugh et al (57) discussed the possible mechanism involved in the conversion of UDP-Glc to UDP-SQ in detail and observed the incorporation of UDP-[^{14}C]-Glc into SQDG by broken pea chloroplasts. Furthermore, a stimulation of sulfate incorporation into SQDG by α-methyl glucosenide addition to broken and intact pea chloroplasts was reported by the same authors. Although it is unknown how α-methyl glucosenide is metabolized to give rise to a suitable precursor for SQDG biosynthesis, the two observations seem to support the sugar-nucleotide pathway for SQDG biosynthesis.

All pathway proposals outlined above are based on UDP-SQ as the head group donor for SQDG biosynthesis. Although the involvement of UDP-SQ was first suggested by Benson over 30 years ago (11) following the observation of a sulfur-containing sugar nucleotide in extracts of *Chlorella* (69), the existence of UDP-SQ remained uncertain until recently. Indirect evidence for the involvement of UDP-SQ in SQDG biosynthesis was obtained by Heinz et al (29), demonstrating that chemically synthesized UDP-SQ could stimulate SQDG biosynthesis by membranes of spinach chloroplasts. Taking advantage of *R. sphaeroides* mutants deficient in SQDG biosynthesis, we were able to isolate a sulfate-labeled water-soluble compound accumulating in a mutant inactivated in *sqdD*, which we could identify as UDP-SQ (61). Furthermore, labeled UDP-SQ isolated from the bacterial mutant was directly converted to SQDG by isolated spinach chloroplasts. Recently, UDP-SQ has been identified also in extracts of different unicellular algae and a moss (C Tietje & E Heinz, personal communication). In conclusion, there is indirect and direct evidence that UDP-SQ is the crucial intermediate in SQDG biosynthesis in bacteria as well as plants, and there is circumstantial evidence based on *sqdB* gene sequences from different organisms suggesting that carbohydrate and sulfur metabolism converge at the level of UDP-hexose to give rise to UDP-SQ.

UDP-Sulfoquinovose:Diacylglycerol Sulfoquinovosyltransferase

According to step II depicted in Figure 3, the final assembly of SQDG is catalyzed by an enzyme designated SQDG synthase or, more precisely, UDP-sulfoquinovose:diacylglycerol sulfoquinvosyltransferase. The most likely candidate gene to encode this activity in *R. sphaeroides* is *sqdD*, because the predicted protein is similar to glycosyl transferases like glycogenin (10), and a mutant specifically inactivated in *sqdD* accumulates UDP-SQ (61), as would be expected following a metabolic block in the last reaction of SQDG biosynthesis. The enzymatic activity of the recombinant protein has not yet been demonstrated in vitro (M Rossak & C Benning, unpublished data). A possible reason may be that *sqdD* encodes only a subunit of the SQDG synthase in *R. sphaeroides* and that a fully functional enzyme requires additional proteins. Supporting evidence for this assumption may be based on our observation that a mutant of *R. sphaeroides* inactivated in *sqdC* accumulates a novel compound, sulfoquinovosyl-1-*O*-dihydroxyacetone (SQ-DHA) (60). Because labeled SQ-DHA was not incorporated into SQDG by extracts of *R. sphaeroides* or by spinach chloroplasts, we concluded that this compound is not an intermediate of SQDG biosynthesis but a side product synthesized in the absence of a functional *sqdC* gene product. It should also be pointed out that *sqdC* mutants of *R. sphaeroides*, including the one carrying an insertion in *sqdC*, still contain small amounts of SQDG (10, 60), suggesting either that a second protein can substitute for the *sqdC* gene product or that the *sqdC* protein has not an essential but an auxiliary function in SQDG biosynthesis. The interpretation of our observations for *sqdC* null mutants of *R. sphaeroides* is summarized in Figure 4. Based on our results we proposed that the *sqdC* gene product interacts with the *sqdD* gene product encoding the sulfoquinvosyltransferase, thereby providing substrate specificity for diacylglycerol (DAG), or promoting membrane association of the SQDG synthase. In the absence of a *sqdC* gene product, the sulfoquinvosyltransferase can still act to a small extent on DAG, but most of the sulfoquinovosyl moiety is transferred onto a soluble acceptor, dihydroxyacetone (DHA), or a similar compound, giving rise to SQ-DHA (Figure 4). Further experiments, like the demonstration that the *sqdD* and *sqdC* gene products directly interact, will be required to test this hypothesis.

In cyanobacteria, the most likely candidate gene encoding the SQDG synthase seems to be *sqdX* because of the sequence similarity of the predicted gene product to glycosyltransferases (see above). Unlike the *sqdD* gene product of *R. sphaeroides*, the cyanobacterial *sqdX* gene product contains a membrane-spanning domain and thus may be membrane associated (S Güler & C Benning, unpublished data). Further experiments will be required to test whether *sqdX* indeed encodes the SQDG synthase in cyanobacteria.

Although the gene for the UDP-galactose:dicacylglycerol galactosyltransferase involved in the biosynthesis of galactolipids has recently been isolated (70), a gene for the equivalent SQDG synthase has not yet been identified in higher plants. However, by employing chemically synthesized UDP-SQ it has been possible to characterize the SQDG synthase of spinach chloroplasts in

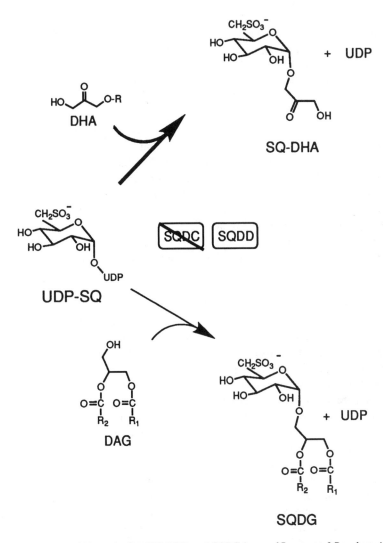

Figure 4 Proposed biosynthesis of SQ-DHA and SQDG in a *sqdC* mutant of *R. sphaeroides*. The corresponding reaction occurring in the wild type is shown in step II in Figure 3. The flux through two competing reactions in the *sqdC* mutant is indicated by the thickness of the arrows. Abbreviations are as defined in the text.

some detail (29, 67). Intact chloroplasts were biochemically preloaded with labeled DAG to provide an acceptor for the sulfoquinovose moiety. Biosynthesis of SQDG following the addition of unlabeled UDP-SQ was only observed for osmotically ruptured chloroplasts. Furthermore, the activity was associated with the envelope membranes and has recently been shown to exclusively copurify with the inner envelope membrane (C Tietje & E Heinz, personal communication). Of the different nucleotide sulfoquinovoses tested, UDP-SQ was the most active, while the GDP derivative showed some, but lower, activity (29). The enzyme exhibited maximal activity at pH 7.5 with a K_m for UDP-SQ of 10 μM in the presence of 5 mM magnesium ions (67). Testing the substrate specificity of the enzyme with regard to different molecular species of DAG, it was found to be much less discriminatory than the equivalent enzyme involved in galactolipid biosynthesis. In summary, while molecular information on the presumed SQDG synthase is available for photosynthetic bacteria, the biochemical characterization of this enzyme is more advanced in higher plants. Already based on the limited molecular information on putative SQDG synthases in bacteria, one has to expect differences in the biochemical properties of the SQDG synthase in different bacteria and possibly also between bacterial and higher plant SQDG synthases.

Biosynthesis of UDP-Sulfoquinovose

The essential problem in understanding SQDG biosynthesis is to determine how the head group sugar becomes sulfonated. Because there is no parallel to the synthesis of sulfoquinovose known in the biotic world, it appears that the capability to form a sulfonated hexose has evolved only once. Of all the *sqd* genes isolated thus far, the bacterial *sqdB* genes as well as the *sqdB*-like gene from *A. thaliana* are the most conserved. Thus it seems plausible that an *sqdB*-like gene is present in all SQDG-containing organisms. Based on our current knowledge of the *sqdB* gene product (outlined above), it seems most likely that the respective enzyme acts on UDP-Glc, converting it to an activated UDP-hexose intermediate that can accept sulfur from an appropriate donor at the carbon 6 position of the hexose (Figure 3, step I). Free sulfite with its oxidation state +4 like that of the sulfonate group in SQDG, a protein-bound thiosulfate similar or identical to that proposed to be involved in the "adenosyl-5'-phosphosulfate (APS)-bound" pathway of sulfate reduction (30, 66), or the sulfur of the mixed acid anhydrates APS or PAPS, may serve as the sulfur donor (28). In view of differences in sulfate reduction pathways, it is questionable whether the same sulfur donor is used in bacteria, algae, and higher plants. However, a mutant approach targeting different genes involved in sulfur assimilation in a photosynthetic bacterium, similar to experiments done with *Salmonella* (32), may provide an answer to this question—at least for bacteria. Because an increasing number of genes encoding proteins involved in sulfur metabolism are becoming available from higher plants (30, 47), a reverse genetics approach seems feasible to address this question in higher plants as well.

The assumption that UDP-Glc may be providing the carbohydrate backbone for the head group of SQDG is posing an interesting problem with regard to higher plants. From sulfate-labeling experiments with intact plastids it must be assumed that all enzymes required for SQDG biosynthesis are associated with plastids. Furthermore, the *sqdB*-like gene from *A. thaliana* encodes a protein imported into plastids (D Linke & C Benning, unpublished data). However, ^{31}P-NMR analysis of isolated spinach chloroplasts revealed only small amounts of UDP-Glc, if anything at all (14). Thus it is necessary to postulate that either a carrier for UDP-Glc exists in the chloroplast envelope or that the biosynthesis of UDP-SQ is associated with the inner envelope membrane permitting access to UDP-Glc synthesized in the cytosol. A close association of UDP-SQ biosynthesis and SQDG synthase, known to be located in the inner envelope (see above), would allow substrate channeling, a possible reason why it may have not been possible to isolate UDP-SQ from higher plants thus far and only osmotically ruptured spinach plastids are accessible to UDP-SQ (29, 67). In this context it is also interesting to note that the sulfur of labeled APS added to isolated spinach chloroplasts was incorporated into SQDG (42), indicating that UDP-SQ biosynthesis has access to cytosolic precursors at least in spinach. Cytosolic forms of the ATP-sulfurylase involved in APS biosynthesis are known for spinach (49, 59) and potatoes (44) but may be absent from *A. thaliana* (30).

CLUES ABOUT SULFOLIPID FUNCTION DERIVED FROM MUTANT ANALYSIS

Anoxygenic Photosynthetic Bacteria

The availability of *sqd* genes from *R. sphaeroides* allowed the construction of SQDG-deficient null mutants isogenic to the wild type. In general, these are ideally suited to test the function of SQDG in a photosynthetic organism because they permit the establishment of a firm correlation between the primary phenotype, SQDG deficiency, and secondary phenotypes, e.g. photosynthetic performance and growth. Unlike chemical or radiation mutagenesis, specific disruption of individual *sqd* genes by gene replacement techniques avoids secondary mutations that may bias the interpretation of the results. Accordingly, a SQDG-deficient null mutant of *R. sphaeroides* inactivated in *sqdB* showed no impairment in growth or photosynthetic electron transport when grown under optimal conditions (7). This result strongly suggests that SQDG has no essential role in anoxygenic photosynthesis. Nevertheless, SQDG appears to be of conditional importance to *R. sphaeroides*, e.g. under limited phosphate availability, because the growth of the SQDG-deficient null mutant inceased earlier compared with the wild type during incubation in medium with reduced phosphate (7).

Cyanobacteria, Algae, and Higher Plants

Because there are many anoxygenic photosynthetic bacteria that lack SQDG or contain only very small amounts (see above), it was not too surprising that SQDG is not essential for photosynthetic growth of *R. sphaeroides*. Furthermore, *R. sphaeroides*, like other anoxygenic photosynthetic bacteria, is characterized by a less complex photosynthetic apparatus compared with oxygenic cyanobacteria, algae, or plants that may exhibit less stringent lipid requirements. The latter contain two photosystems and a water-splitting, oxygen-evolving complex associated with photosystem II, while anoxygenic photosynthetic bacteria harbor only a single photosystem and do not release oxygen (6, 13, 22). Thus, the properties of the *R. sphaeroides* mutants did not exclude the possibility that SQDG may play an essential role in oxygenic photosynthesis. To clarify this issue, we isolated and inactivated the *sqdB* gene from the cyanobacterium *Synechococcus* sp. PCC7942, thereby creating a SQDG-deficient organism with oxygenic photosynthesis (23). Detailed analysis of this null mutant revealed only subtle changes in the photosynthesis-related biochemistry of molecular oxygen and a slight increase in the variable room temperature chlorophyll fluorescence yield. However, the light response curves for oxygen evolution were nearly identical for the mutant and the wild type, and photosynthetic growth of the mutant was not impaired under optimal growth conditions. Thus, the lack of SQDG did not affect the overall electron transport rate nor the optical cross section of oxygen evolution, although the components of the electron transport chain are intrinsic or peripheral membrane proteins directly in contact with thylakoid membrane lipids, including SQDG. Taken together, these results suggested that SQDG plays no essential role in oxygenic photosynthesis but did not exclude the possibility that SQDG may be of conditional importance. In fact, we observed that under phosphate-limiting growth conditions, the cyanobacterial mutant ceased to grow earlier than the wild type (23), as was observed for the equivalent mutant of *R. sphaeroides* (see above). Likewise, the relative amount of PG remained much higher in the mutant compared with the wild type under phosphate-limiting conditions.

Because of the close structural relationship of photosynthetic reaction centers and other components of the photosynthetic electron-transport chain, it seems legitimate to extrapolate from cyanobacteria to plants (6) and to assume that the thylakoid lipid SQDG also plays no essential role in reactions of the photosynthetic electron transport of plants. However, there are two fundamental differences with regard to the photosynthetic apparatus and the thylakoid membranes of cyanobacteria and eukaryotic photosynthetic organisms. First, the different light-harvesting antenna systems, with phycobilisomes in cyanobacteria (72) and chlorophyll *a/b*-binding protein complexes in higher plants (45), and second, the presence of plastids in eukaryotic photosynthetic organisms. While the cyanobacterial phycobilisomes are extramembranous antenna structures with presumably few direct contacts to thylakoid lipids, this is not the case for the light-harvesting antenna of green algae and plants. For example, SQDG was reported to be tightly associated with light harvesting complex II of *C. reinhardtii* (64, 73). However, PG but not SQDG was found in preparations of light-harvesting complex II trimers of pea chloroplasts, and contrary to PG, SQDG was not required for in vitro crystallization of the complex (54). Based on these observations, it is still a matter of discussion whether SQDG has a specific function in the light-harvesting antenna of algae or plants. From the organization of the photosynthetic apparatus in plastids of eukaryotic photosynthetic organisms, one might infer additional roles for SQDG during plastid development or the assembly of the photosynthetic apparatus. Indeed, a specific interaction between chloroplast envelope lipids, in particular SQDG, and the transit peptide of a chloroplast precursor protein has been observed (76), suggesting that charged lipids like SQDG are required for proper protein import into chloroplasts and thus the assembly of the photosynthetic apparatus. One valuable approach to test this idea would be the analysis of SQDG-deficient mutants of algae such as *C. reinhardtii* or higher plants such as *A. thaliana*. Sato et al (65) were the first to obtain a SQDG-deficient eukaryotic mutant following the mutagenesis of the unicellular alga *C. reinhardtii*. The mutant was identified based on a high chlorophyll fluorescence phenotype, which, by itself, is an indication of disturbances in the photosynthetic apparatus. Closer examination of this mutant revealed an altered light-response curve for oxygen evolution and a 42% reduction in the activity of photosystem II, but photosystem I activity was not affected (64). In agreement with this observation, a preferential association of SQDG with preparations of photosystem II and the light-harvesting complexes was observed. The relative amounts of the different pigment-binding complexes and their subunit composition remained unchanged. The growth of this SQDG mutant was slightly reduced by 17%. While the observed secondary phenotype of the *C. reinhardtii* mutant is more drastic than that of the SQDG-deficient cyanobacterial null mutant (23), it is difficult to assume an essential role for SQDG in photosynthesis of *C. reinhardtii* because the effect of possible secondary mutations in the mutant cannot be excluded. Transgenic plants of *A. thaliana* with reduced amounts of SQDG expressing a *sqdB*-like cDNA in antisense orientation have been isolated, and their preliminary analysis showed no obvious growth impairment (RA Narang & C Benning, unpublished data).

The Sulfolipid-Phospholipid Substitution Hypothesis

The absence of SQDG from a large number of anoxygenic photosynthetic bacteria and the preliminary analysis of SQDG-deficient mutants of two different photosynthetic bacteria, a unicellular green alga, and a higher plant clearly suggest that SQDG is not essential for photosynthetic growth in bacteria and eukaryotes. However, a more subtle function of SQDG in the photosynthetic apparatus of oxygenic photosynthetic organisms based on the phenotype of the mutants of *Synechococcus* and *Chlamydomonas* cannot be ruled out at this time. Taken together, these observations raise the question about the nature of the selective pressure presumably required to establish and maintain the pathway of SQDG biosynthesis in many photosynthetic, and even a few nonphotosynthetic, organisms. One appealing idea would be that the capability to synthesize SQDG is part of the repertoire of photosynthetic organisms to adapt to specific environmental conditions, e.g. a limitation in the availability of essential nutrients. In support of this idea we observed dramatic changes in lipid composition of photosynthetic bacteria (7, 8, 23) and the plant *A. thaliana* (C Benning, unpublished data) in response to phosphate limitation in the medium. Particularly intriguing was an inverse relationship between the relative amounts of the two anionic thylakoid lipids PG and SQDG. Their biophysical properties are thought to be similar, because both adopt a cylindrical shape in membranes due to their large head groups and low unsaturation of fatty acid constituents, and both are thought to stabilize the lamellar lipid phase (77). Furthermore, of all phospholipids, PG was the only one not to be reduced in an SQDG-deficient mutant of *R. sphaeroides* (7) and was also maintained in the corresponding cyanobacterial mutant (23) when grown under phosphate limitation. Based on these observations, one may conclude that SQDG can substitute for PG under phosphate limitation, a condition frequently encountered in natural habitats. Thereby a certain amount of anionic, bilayer-forming lipids is maintained that is required for the proper function of the thylakoid membrane. A suitable designation for this concept of SQDG function would be the sulfolipid-phospholipid substitution hypothesis. Taking into consideration that approximately one third of organic phosphorus in a plant cell is bound in phospholipids (55), the replacement of phospholipids by nonphosphorous lipids seems to be a reasonable strategy for photosynthetic organisms to conserve phosphate. Corroborating evidence has been provided by the observation that the expression of a gene involved in the biosynthesis of SQDG in *A. thaliana* is increased under phosphate limitation (B Essigmann & C Benning, unpublished data).

PERSPECTIVES

Although the isolation of the first SQDG-deficient mutants and genes essential for SQDG biosynthesis represents a significant step forward, our understanding of the biosynthesis and function of SQDG is far from complete. We cannot be certain that the pathway of SQDG biosynthesis has been saturated with mutations, and we have yet to demonstrate the function of a purified *sqd* gene product in vitro. The most elusive aspect of SQDG biosynthesis is still the origin of UDP-SQ, the head group precursor in bacteria and plants. Based on our current knowledge on *sqd* genes in different organisms, it seems reasonable to assume that an understanding of *sqdB* gene function will lead to an understanding of UDP-SQ biosynthesis. Another emerging aspect is the low conservation between the glycosyltransferases catalyzing the last reaction of sulfolipid biosynthesis in different organisms, which is in stark contrast to the high conservation of *sqdB*-like genes in bacteria and higher plants. In conjunction with the spotty occurrence of SQDG in photosynthetic bacteria and some nonphotosynthetic bacteria, this aspect makes the pathway of SQDG biosynthesis an interesting subject of study in the context of evolution. Furthermore, the availability of SQDG genes provides the means to study the regulation of thylakoid membrane lipid composition at the level of lipid classes.

The fact that SQDG-deficient mutants are not at all, or only mildly, impaired in photosynthetic growth suggests that SQDG has no specific function in photosynthesis. However, it is difficult to accept that there may not be a specific reason for the establishment and maintenance of an elaborate pathway for an unusual compound like SQDG that is preferentially, but not exclusively, associated with photosynthetic organisms. A partial answer to this puzzling question has been provided by SQDG-deficient mutants, which are at a disadvantage under phosphate-limiting conditions. In addition, the strong increase in relative amounts of SQDG in bacteria and plants and the increase in the expression of sulfolipid genes under phosphate limitation support the sulfolipid-phospholipid substitution hypothesis for SQDG function discussed above. Based on this hypothesis, one would predict that an increase in the relative amounts of SQDG in crop plants may reduce their dependency on phosphate fertilizer, a cost-saving factor that may also benefit the environment. In principle, all means are available to test this idea. Algae are known with relative amounts of close to 40% SQDG in their membranes (17, 18), and there is no reason to assume that these high amounts could not be achieved in a crop plant by genetic engineering. Crop plants with increased amounts of SQDG could also serve as an inexpensive source of this lipid. In view of reports on antitumor and antiviral properties (24, 71) and the excellent detergent properties of SQDG, biomedical applications may be developed that require an inexpensive and reliable source of SQDG.

ACKNOWLEDGMENT

I would like to thank the members of my laboratory for critically reading the manuscript and Professor E Heinz for sharing unpublished data. The work on sulfolipid conducted in my laboratory was financially supported in part by the Deutsche Forschungsgemeinschaft (Grants Be 1591/1-1 and Be 1591/1-2) and the Bundesministerium für Bildung und Forschung (Grant 0311024).

> Visit the *Annual Reviews home page* at
> http://www.AnnualReviews.org.

Literature Cited

1. Araki S, Eichenberger W, Sakurai T, Sato N. 1991. Distribution of Diacylgly-cerylhydroxymethyltrimethyl-β-alanine (DGTA) and phosphatidylcholine in brown algae. *Plant Cell Physiol.* 32:623–28
2. Arondel V, Benning C, Somerville CR. 1993. Isolation and functional expression in *Escherichia coli* of a gene encoding phosphatidylethanolamine methyl transferase (EC 2.1.1.17) from *Rhodobacter sphaeroides*. *J. Biol. Chem.* 268:16002–8
3. Asselineau J, Trüper HG. 1982. Lipid composition of six species of the phototrophic bacterial genus *Ectothiorhodospira*. *Biochim. Biophys. Acta* 712:111–16
4. Barber GA. 1963. The formation of uridine diphosphate L-rhamnose by enzymes of the tobacco leaf. *Arch. Biochem. Biophys.* 103:276–82
5. Barber J, Gounaris K. 1986. What role does sulpholipid play within the thylakoid membrane? *Photosynth. Res.* 9:239–49
6. Barry BA, Boerner RJ. 1994. The use of cyanobacteria in the study of the structure and function of photosystem II. See Ref. 14a, pp. 217–57
7. Benning C, Beatty JT, Prince RC, Somerville CR. 1993. The sulfolipid sulfoquinovosyl diacylglycerol is not required for photosynthetic electron transport in *Rhodobacter sphaeroides* but enhances growth under phosphate limitation. *Proc. Natl. Acad. Sci. USA* 90:1561–65
8. Benning C, Huang ZH, Gage DA. 1995. Accumulation of a novel glycolipid and a betaine lipid in cells of *Rhodobacter sphaeroides* grown under phosphate limitation. *Arch. Biochem. Biophys.* 317:103–11
9. Benning C, Somerville CR. 1992. Isolation and genetic complementation of a sulfolipid-deficient mutant of *Rhodobacter sphaeroides*. *J. Bacteriol.* 174:2352–60
10. Benning C, Somerville CR. 1992. Identification of an operon involved in sulfolipid biosynthesis in *Rhodobacter sphaeroides*. *J. Bacteriol.* 174:6479–87
11. Benson AA. 1963. The plant sulfolipid. *Adv. Lipid Res.* 1:387–94
12. Benson AA, Daniel H, Wiser R. 1959. A sulfolipid in plants. *Proc. Natl. Acad. Sci. USA* 45:1582–87
13. Blankenship RE. 1994. Protein structure, electron transfer and evolution of procaryotic photosynthetic reaction centers. *Antoine van Leeuwenhoek* 65:311–29
14. Bligny R, Gardestrom P, Roby C, Douce R. 1990. ^{31}P NMR studies of spinach leaves and their chloroplasts. *J. Biol. Chem.* 265:1319–26
14a. Bryant DA, ed. 1994. *The Molecular Biology of Cyanobacteria*. Dordrecht: Kluwer
15. Cedergreen RA, Hollingsworth RI. 1994. Occurrence of sulfoquinovosyl diacylglycerol in some members of the family Rhizobiaceae. *J. Lipid Res.* 35:1452–61

16. Davies WH, Mercer EI, Goodwin TW. 1966. Some observations on the biosynthesis of the plant sulfolipid by *Euglena gracilis*. *Biochem. J.* 98:369–73
17. Dembitsky VM, Pechenkina-Schubina EE, Rozentsvet OA. 1991. Glycolipids and fatty acids of some sea weeds and marine grasses from the black sea. *Phytochemistry* 30:2279–83
18. Dembitsky VM, Rozentsvet OA, Pechenkina EE. 1990. Glycolipids, phospholipids and fatty acids of brown algae species. *Phytochemistry* 29:3417–31
19. Douce R, Joyard J. 1990. Biochemistry and function of the plastid envelope. *Annu. Rev. Cell Biol.* 6:173–216
20. Gabriel O. 1987. Biosynthesis of sugar residues for glycogen, peptidoglycan, lipopolysaccharide, and related systems. In *Escherichia coli and Salmonella typhimurium: Cellular and Molecular Biology*, ed. FC Neidhardt, JL Ingraham, KB Low, B Magasanik, M Schaechter, HE Umbarger, 1:504–11. Washington, DC: Am. Soc. Microbiol.
21. Gage DA, Huang ZH, Benning C. 1992. Comparison of sulfoquinovosyl diacylglycerol from spinach and the purple bacterium *Rhodobacter sphaeroides* by fast atom bombardment tandem mass spectrometry. *Lipids* 27:632–36
22. Golbeck JH. 1993. Shared thematic elements in photochemical reaction centers. *Proc. Natl. Acad. Sci. USA* 90:1642–46
23. Güler S, Seeliger S, Härtel H, Renger G, Benning C. 1996. A null mutant of *Synechococcus* sp. PCC7942 deficient in the sulfolipid sulfoquinovosyl diacylglycerol. *J. Biol. Chem.* 271:7501–7
24. Gustafson KR, Cardellina JH II, Fuller RW, Weislow OS, Kiser RF, et al. 1989. AIDS-antiviral sulfolipids from cyanobacteria (blue-green algae). *J. Natl. Cancer Inst.* 81:1254–58
25. Haas R, Siebertz HP, Wrage K, Heinz E. 1980. Localization of sulfolipid labeling within cells and chloroplasts. *Planta* 148:238–44
26. Haines TH. 1973. Sulfolipids and halosulfolipids. In *Lipids and Biomembranes of Eucaryotic Microorganisms*, ed. JA Erwin, pp. 197–232. New York: Academic
27. Harwood JL, Nicholls RG. 1979. The plant sulfolipid—a major component of the sulphur cycle. *Biochem. Soc. Trans.* 7:440–47
28. Heinz E. 1993. Recent investigations on the biosynthesis of the plant sulfolipid. In *Sulfur Nutrition and Assimilation in Higher Plants*, ed. LJ De Kok, pp. 163–78. Den Haag, Netherlands: SPB Acad.
29. Heinz E, Schmidt H, Hoch M, Jung K-H, Binder H, Schmidt RR. 1989. Synthesis of different nucleoside 5′-diphosphosulfoquinovoses and their use for studies on sulfolipid biosynthesis in chloroplasts. *Eur. J. Biochem.* 184:445–53
30. Hell R. 1997. Molecular physiology of plant sulfur metabolism. *Planta* 202:138–48
31. Hoppe W, Schwenn JD. 1981. In vitro biosynthesis of the plant sulpholipid: on the origin of the sulphonate head group. *Z. Naturforsch. Teil C* 36:820–26
32. Hulanicka MD, Hallquist SG, Kredich NM, Mojica-A T. 1979. Regulation of *O*-acetylserine sulfhydrylase B by L-cysteine in *Salmonella typhimurium*. *J. Bacteriol.* 140:141–46
33. Imhoff JF. 1984. Sulfolipids in phototrophic purple nonsulfur bacteria. In *Structure, Function and Metabolism of Plant Lipids*, ed. PA Siegenthaler, W Eichenberger, pp. 175–78. Amsterdam: Elsevier
34. Imhoff JF. 1988. Lipids, fatty acids and quinones in taxonomy and phylogeny of anoxygenic phototrophic bacteria. In *Green Photosynthetic Bacteria*, ed. JM Olson, JG Ormerod, J Amesz, E Stackebrandt, HG Trüper, pp. 223–32. New York: Plenum
35. Imhoff JF, Bias-Imhoff U. 1995. Lipids, quinones and fatty acids of anoxygenic phototrophic bacteria. In *Anoxygenic Photosynthetic Bacteria*, ed. RE Blankenship, MT Madigan, CE Bauer, pp. 179–205. Dordrecht: Kluwer
36. Isono Y, Mohri H, Nagai Y. 1967. Effect of egg sulpholipid on respiration of sea urchin spermatozoa. *Nature* 214:1336–38
37. Joyard J, Blee E, Douce R. 1986. Sulfolipid synthesis from $^{35}SO_4^{2-}$ and [1-^{14}C] acetate in isolated spinach chloroplast. *Biochim. Biophys. Acta* 879:78–87
38. Joyard J, Block MA, Malherbe A, Maréchal E, Douce R. 1993. Origin and synthesis of galactolipid and sulfolipid head groups. In *Lipid Metabolism in Plants*, ed. TS Moore Jr, pp. 231–58. Boca Raton, FL: CRC
39. Kaneko T, Sato S, Kotani H, Tanaka A, Asamizu E, et al. 1996. Sequence analysis of the genome of the unicellular cyanobacterium *Synechocystis* sp. strain PCC6803. II. Sequence determination of the entire genome and assignment of potential protein-coding regions. *DNA Res.* 3:109–36
40. Kleppinger-Sparace KF, Mudd JB,

Bishop DG. 1985. Biosynthesis of sulfoquinovosyl diacylglycerol in higher plants: the incorporation of $^{35}SO_4$ by intact chloroplasts. *Arch. Biochem. Biophys.* 240:859–65

41. Kleppinger-Sparace KF, Mudd JB. 1987. Biosynthesis of sulfoquinovosyldiacylglycerol in higher plants. The incorporation of $^{35}SO_4$ by intact chloroplast in darkness. *Plant Physiol.* 84:682–87

42. Kleppinger-Sparace KF, Mudd JB. 1990. Biosynthesis of sulfoquinovosyldiacylglycerol in higher plants. Use of adenosine-5'-phosphosulfate and adenosine-3'-phosphate-5'-phosphosulfate as precursors. *Plant Physiol.* 93:256–63

43. Kleppinger-Sparace KF, Mudd JB, Sparace SA. 1990. Biosynthesis of plant sulfolipids. In *Sulfur Nutrition and Sulfur Assimilation in Higher Plants*, ed. H Rennenberg, C Brunold, LJ De Kok, I Stulen, pp. 77–88. Den Haag, Netherlands: SPB Acad.

44. Klonus D, Höfgen R, Willmitzer L, Riesmeier JW. 1994. Isolation and characterization of two cDNA clones encoding ATP-sulfurylases from potato by complementation of a yeast mutant. *Plant J.* 6:105–12

45. Kühlbrandt W. 1994. Structure and function of the plant light-harvesting complex, LHC-II. *Curr. Opin. Struct. Biol.* 4:519–28

46. Langworthy TA, Mayberry WR, Smith PF. 1976. A sulfonolipid and novel glucosamidyl glycolipids from the extreme thermoacidophile *Bacillus acidocaldarius*. *Biochim. Biophys. Acta* 431:550–69

47. Leustek T. 1996. Molecular genetics of sulfate assimilation in plants. *Physiol. Plant.* 97:411–19

48. Linscheid M, Diehl BWK, Övermöhle M, Riedl I, Heinz E. 1997. Membrane lipids of *Rhodopseudomonas viridis*. *Biochim. Biophys. Acta* 1347:151–63

49. Lunn JE, Droux M, Martin J, Douce R. 1990. Localization of ATP-sulfurylase and O-acetylserine(thiol)lyase in spinach leaves. *Plant Physiol.* 94:1345–52

50. Merrit MV, Rosenstein SP, Loh C, Chou RH, Allen MM. 1991. A comparison of the major lipid classes and fatty acid composition of marine unicellular cyanobacteria with freshwater species. *Arch. Microbiol.* 155:107–13

51. Minnikin DE, Abdolrahimzadeh H, Baddiley J. 1974. Replacement of acidic phospholipids by acidic glycolipids in *Pseudomonas diminuta*. *Nature* 249:268–69

52. Mudd JB, Dezacks R, Smith J. 1980. Studies on the biosynthesis of sulfoquinovosyldiacylglycerol in higher plants. In *Biogenesis and Function of Plant Lipids*, ed. P Mazliak, P Beneviste, C Costes, R Douce, pp. 57–66. Amsterdam: Elsevier

53. Mudd JB, Kleppinger-Sparace KF. 1987. Sulfolipids. In *The Biochemistry of Plants. A Comprehensive Treaty. Lipids: Structure and Function*, ed. PK Stumpf, 9:275–89. New York: Academic

54. Nußberger S, Dörr K, Wang DN, Kühlbrandt W. 1993. Lipid-protein interactions in crystals of plant light-harvesting complex. *J. Mol. Biol.* 234:347–56

55. Poirier Y, Thoma S, Somerville C, Schiefelbein J. 1991. A mutant of *Arabidopsis* deficient in xylem loading of phosphate. *Plant Physiol.* 97:1087–93

56. Pugh CE, Hawkes T, Harwood JL. 1995. Biosynthesis of sulphoquinovosyl diacylglycerol by chloroplast fractions from pea and lettuce. *Phytochemistry* 39:1071–75

57. Pugh CE, Roy AB, Hawkes T, Harwood JL. 1995. A new pathway for the synthesis of the plant sulpholipid, sulphoquinovosyldiacylglycerol. *Biochem. J.* 309:513–19

58. Radunz A. 1969. Über das Sulfoquinovosyl-Diacylglycerin aus höheren Pflanzen, Algen und Purpurbakterien. *Hoppe-Seyler's Z. Physiol. Chem.* 350:411–17

59. Renosto F, Patel HC, Martin RL, Thomassian C, Zimmermann G, Segel IH. 1993. ATP-sulfurylase from higher plants: kinetic and structural characterization of the chloroplast and cytosol enzymes from spinach leaf. *Arch. Biochem. Biophys.* 307:272–85

60. Rossak M, Schäfer A, Xu NX, Gage DA, Benning C. 1997. Accumulation of sulfoquinovosyl-1-O-dihydroxyacetone in a sulfolipid-deficient mutant of *Rhodobacter sphaeroides* inactivated in *sqdC*. *Arch. Biochem. Biophys.* 340:219–30

61. Rossak M, Tietje C, Heinz E, Benning C. 1995. Accumulation of UDP-sulfoquinovose in a sulfolipid-deficient mutant of *Rhodobacter sphaeroides*. *J. Biol. Chem.* 270:25792–97

62. Saidha T, Schiff JA. 1989. The role of mitochondria in sulfolipid biosynthesis by *Euglena* chloroplasts. *Biochim. Biophys. Acta* 1001:268–73

63. Sasaki T, Song JY, Koga-Ban Y, Matsui E, Fang F, et al. 1994. Toward cataloging all rice genes: large-scale sequencing of randomly chosen rice cDNAs from a callus cDNA library. *Plant J.* 6:615–24

64. Sato N, Sonoike K, Tsuzuki M,

Kawaguchi A. 1995. Impaired photosystem II in a mutant of *Chlamydomonas reinhardtii* defective in sulfoquinovosyl diacylglycerol. *Eur. J. Biochem.* 234:16–23

65. Sato N, Tsuzuki M, Matsuda Y, Ehara T, Osafune T, Kawaguchi A. 1995. Isolation and characterization of mutants affected in lipid metabolism of *Chlamydomonas reinhardtii*. *Eur. J. Biochem.* 230:987–93

66. Schmidt A, Jäger K. 1992. Open questions about sulfur metabolism in plants. *Annu. Rev. Plant Physiol. Mol. Biol.* 43:325–49

67. Seifert U, Heinz E. 1992. Enzymatic characteristics of UDP-sulfoquinovose:-diacylglycerol sulfoquinovosyltransferase from chloroplast envelopes. *Bot. Acta* 105:197–205

68. Selstam E, Campbell D. 1996. Membrane lipid composition of the unusual cyanobacterium *Gloeobacter violaceus* sp. PCC7421, which lacks sulfoquinovosyl diacylglycerol. *Arch. Microbiol.* 166:132–35

69. Shibuya I, Yagi T, Benson AA. 1963. Sulfonic acids in algae. In *Microalgae and Photosynthetic Bacteria*, ed. Jpn. Soc. Plant Physiol., pp. 627–36. Tokyo: Univ. Tokyo Press

70. Shimojima M, Ohta H, Iwamatsu A, Masuda T, Shioi Y, Takamiya K. 1997. Cloning of the gene for monogalactosyldiacylglycerol synthase and its evolutionary origin. *Proc. Natl. Acad. Sci. USA* 94:333–37

71. Shirahashi H, Murakami N, Watanabe M, Nagatsu A, Sakakibara J, et al. 1993. Isolation and identification of anti-tumor-promoting principles from the fresh water cyanobacterium *Phormidium tenue*. *Chem. Pharmacol. Bull.* 41:1664–66

72. Sidler WA. 1994. Phycobilisomes and phycobilliprotein structure. See Ref. 14a, pp. 139–216

73. Sigrist M, Zwillenberg C, Giroud CH, Eichenberger W, Boschetti A. 1988. Sulfolipid associated with the light-harvesting complex associated with photosystem II apoproteins of *Chlamydomonas reinhardtii*. *Plant Sci.* 58:15–23

74. Skallal AK, Nimer NA, Radwan SS. 1990. Lipid and fatty acid composition of fresh water cyanobacteria. *J. Gen. Microbiol.* 136:2043–48

75. So RB, Ladha JK, Young JP. 1994. Photosynthetic symbionts of *Aeschymone* spp. from a cluster with bradyrhizobia on the basis of fatty acid and rRNA analyses. *Int. J. Syst. Bacteriol.* 44:392–403

76. van't Hof R, van Klompenburg W, Pilon M, Kozubek A, de Korte-Kool G, et al. 1993. The transit sequence mediates the specific interaction of the precursor of ferredoxin with chloroplast envelope lipids. *J. Biol. Chem.* 268:4037–42

77. Webb MS, Green BR. 1991. Biochemical and biophysical properties of thylakoid acyl lipids. *Biochim. Biophys. Acta* 1060:133–58

78. Woese CR. 1987. Bacterial evolution. *Microbiol. Rev.* 51:221–71

79. Yagi H, Miyari S, Nakahara T, Ichimura T, Matsui I, Honda K. 1992. Lipid and fatty acid composition of the cytoplasmic and thylakoid membranes from the thermophilic bacterium *Synechococcus elongatus*. *J. Jpn. Oil Chem. Soc.* 41:1025–28

Annu. Rev. Plant Physiol. Plant Mol. Biol. 1998. 49:77–95

SPLICE SITE SELECTION IN PLANT PRE-mRNA SPLICING

J. W. S. Brown and C. G. Simpson

Department of Cell and Molecular Genetics, Scottish Crop Research Institute,
Invergowrie, Dundee DD2 5DA, Scotland, United Kingdom;
e-mail: jbrown@scri.sari.ac.uk; csimps@scri.sari.ac.uk

KEY WORDS: pre-mRNA splicing, intron, exon

ABSTRACT

The purpose of this review is to highlight the unique and common features of splice site selection in plants compared with the better understood yeast and vertebrate systems. A key question in plant splicing is the role of AU sequences and how and at what stage they are involved in spliceosome assembly. Clearly, intronic U- or AU-rich and exonic GC- and AG-rich elements can influence splice site selection and splicing efficiency and are likely to bind proteins. It is becoming clear that splicing of a particular intron depends on a fine balance in the "strength" of the multiple intron signals involved in splice site selection. Individual introns contain varying strengths of signals and what is critical to splicing of one intron may be of less importance to the splicing of another. Thus, small changes to signals may severely disrupt splicing or have little or no effect depending on the overall sequence context of a specific intron/exon organization.

CONTENTS

77

1040-2519/98/0601-0077$08.00

INTRODUCTION

Splicing is an integral step in the overall process of gene expression and is one level at which gene expression may be regulated. The excision of intron sequences from precursor nuclear messenger RNA (pre-mRNA) transcripts and ligation of exons is a dynamic yet orderly process that involves the assembly of a large ribonucleoprotein complex, called the spliceosome on pre-mRNAs. The spliceosome is formed by the sequential assembly of small nuclear ribonucleoprotein particles (snRNPs) and other non-snRNP spliceosomal proteins. It catalyzes the precise recognition and cleavage of the intron and subsequent ligation of the exons. Failure to remove introns or any inaccuracy in intron excision would alter the open reading frame of the mRNA leading to altered protein size, sequence, and probably function. Inaccurate splicing may affect mRNA stability or transport. The accuracy of splicing is mediated by conserved sequences in the pre-mRNA and by the spliceosomal snRNPs and other factors that recognize these sequences (reviewed by 5, 46, 47, 71).

The availability of in vitro splicing extracts from animal and yeast cells along with yeast RNA processing mutants has facilitated the dissection of the splicing process in these organisms. The lack of an in vitro splicing extract for plants has meant that investigations into plant pre-mRNA splicing have progressed at a much slower pace. Nevertheless, splicing in plants is particularly interesting. First, there are clear differences in plant splicing compared with fungi and animals. Second, splicing differences exist between monocotyledonous (monocot) and dicotyledonous (dicot) plants (29, 41). More recently, examples of splicing differences between different species of dicots have been found, suggesting yet another level of subtlety in plant intron signal recognition (39, 65). Third, differentiation in animals often requires alternative or differential splicing to produce cell-type specific mRNAs from a single transcript; this selection of alternative splice sites produces proteins with different activities (89). A number of examples of alternative splicing have been described in plants, but currently little information is available on the functional significance of the alternative products. As more genes are analyzed, we anticipate more examples. If alternative splicing is an important factor in the control of plant development and differentiation, an understanding of intron sequence and structure, the factors involved in splice site recognition, and spliceosome assembly will be a prerequisite to studying the regulation of differential splicing.

Over the past 10 years, various aspects of plant intron structure, spliceosomal components, and splicing have been investigated. Much emphasis was placed initially on the differences between plant and animal introns and their splicing behavior in reciprocal assays. In particular, the main distinguishing feature of plant introns is AU and U richness in relation to flanking exons, a feature which has received a great deal of attention. The first decade of analysis of plant splicing has been reviewed thoroughly by Simpson & Filipowicz (85).

Because of their comprehensive review, we do not wish to reiterate large areas of research. Although the review was published relatively recently, a number of new and interesting observations have been made, allowing current hypotheses of the mechanism of splice site selection to be reassessed. In particular, we discuss the changing perceptions of the role of various splicing signals in splice site selection, the importance of exon sequences (particularly with regard to exon scanning in plant intron splicing), and how these signals may interact with the important AU-rich sequences in the intron. Other advances have been made in the areas of alternative splicing and splicing regulation, communication between introns, and in the discovery of a novel second class of pre-mRNA introns with (normally) noncanonical 5' and 3' splice sites.

THE SPLICING PROCESS

Spliceosome assembly and the spliceosome cycle have been determined mainly using vertebrate and yeast systems. Introns are excised in a two-step cleavage-ligation reaction, where the first step involves cleavage at the 5' splice site with formation of an intron lariat at an adenosine nucleotide (the branchpoint), usually 18–40 nucleotides (nt) upstream of the 3' splice site. In the second step, following cleavage at the 3' splice site, the exons are ligated, and the intron is released as a lariat, which is then debranched and degraded (46, 47, 71). The accuracy of splicing depends on the ability of the splicing machinery to recognize various intron signals. This recognition is mediated by spliceosomal U-type small nuclear ribonucleoprotein particles (snRNPs) and numerous non-snRNP protein factors such as hnRNP proteins (reviewed in 20, 42), SR proteins (a family of proteins with arginine/serine-rich domains) (reviewed in 22, 61), and DEAD- or DEAH-box–containing proteins (RNA-dependent ATPases or ATP-dependent RNA helicases) (reviewed in 5, 23). HnRNP and SR proteins both contain RNA-binding domains and auxiliary domains, often glycine-rich in hnRNP proteins, and serine-arginine-rich in SR proteins, which are involved in protein-protein interactions (22, 42). The DEAD- and DEAH-box proteins are thought to mediate the conformational changes that occur in spliceosome assembly and disassembly (14).

The spliceosome cycle has been reviewed in detail (5, 46, 47, 71). Briefly, the pre-mRNA transcript becomes associated with hnRNP proteins, which may define intron/exon regions by virtue of sequence-specific binding and aid the association and stability of other splicing factors. U1snRNP is recruited to the 5' splice site where the 5' end of U1snRNA interacts with the 5' splice site through base pairing. The polypyrimidine tract (lying between the branchpoint and 3' splice site) is bound by U2AF^{65} (vertebrates)/Mud2p (yeast) and interacts with the U1-5' splice site complex, through cross-intron bridging interactions, bringing the 5' and 3' splice sites into close proximity. In yeast, the branchpoint binding protein (BBP) interacts specifically with the branchpoint and with the proteins, Mud2p at the polypyrimidine tract, and Prp40, a component of the U1snRNP (1, 7). This commitment complex is the earliest functional intermediate in spliceosome assembly and targets the pre-mRNA to the splicing pathway. U2AF^{65} and ATP are required for the association of U2snRNP, which also involves base pairing between the branchpoint sequence and U2snRNA, resulting in bulging of the branchpoint nucleotide, usually an adenosine (78). Formation of the presplicing complex requires a number of splicing factors, including SR proteins, that may function in promoting or stabilizing interactions between components (5, 46, 47). The addition of the U4/U6-U5 tri-snRNP forms the splicing complex in which a number of conformational changes occur to bring the interactive splice sites together. These changes require ATP and splicing factors such as RNA helicases/ATPases and involve a series of interactions between the pre-mRNA, snRNAs, and protein components (5, 46, 47, 60, 71). Ultimately, the reactive sites of the pre-mRNA are arranged into the correct conformation [catalytic center(s)] in which the splicing reactions occur. Following catalysis, the spliced exons are released, the postsplicing complex containing U2, U4, U5, and U6snRNPs is disassembled and the snRNPs recycled, and the intron lariat debranched and degraded.

PLANT INTRON SEQUENCES

Plant intron 5' and 3' splice site consensus sequences are very similar to those of vertebrate introns, although individual introns exhibit great variation around the highly conserved :GU and AG: dinucleotides (11, 13, 15, 21, 33, 35, 55, 58, 85, 86; Figure 1). In yeast, the internal branchpoint sequence is absolutely conserved—UACUAAC, where the underlined adenosine is the branchpoint nucleotide. In animals, the branchpoint sequence is variable, with the consensus YURAY (45), usually positioned 18–40 nt upstream of the 3' splice site. Only recently have branchpoints been mapped in plant introns and shown to be an important splicing signal (49, 83). Again, the plant intron branchpoint consensus is very similar to that of vertebrates (49, 83, 99). In vertebrate introns, the fourth splicing signal required for efficient splicing is a polypyrimidine tract, found directly upstream of the 3' splice site. In some yeast introns, a poly U sequence in this region is necessary for splicing. In plant introns, this region is often

U-rich and potentially may be involved in branchpoint definition and 3' splice site selection as in vertebrate introns (4).

The major difference between plant introns and those of vertebrates and yeast is that plant, and particularly dicot, introns are AU rich. In fact, in an analysis of base composition of 271 dicot and 146 monocot plant introns in the regions extending 50 nucleotides upstream and downstream of the 5' and 3' splice sites, the A content of introns matches that of the flanking exons. Introns are ~15% more U-rich than the flanking exons, which are ~15% more GC-rich (85). This property and the associated difference in nucleotide composition between introns and exons are important factors in pre-mRNA splicing in plants (28, 29, 105). The emphasis of many splicing studies in both dicot and monocot systems has been aimed at defining the role of AU richness in general, and at identifying the functional elements of this characteristic, such as U- and AU-rich elements and AU/GC intron/exon border sequences in particular, and is discussed in detail below (4, 15, 16, 26, 28, 29, 53–55, 57, 58, 66–68, 82, 85; Figure 1).

Recently, the AT-AC introns were identified in vertebrates as a minor class of introns with nonconventional splice site sequences (31, 107). AT-AC introns

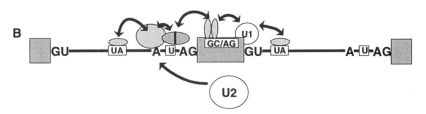

Figure 1 Model of early spliceosomal complex formation in plants. (*A*) Important intron signals: GU, 5' splice site; AG, 3' splice site; A, branchpoint; UA, UA-rich elements; U, U-rich elements between the branchpoint and 3' splice site and GC/AG-rich exonic elements. (*B*) Exon definition: By analogy to assembly of the vertebrate commitment complex, U1snRNP is recruited to the 5' splice site. In plants, this may be directed or the interaction may be stabilized by factors that bind to UA-rich elements in the introns. At the 3' splice site/branchpoint, factors presumably similar to factors such as U2AF^{65} assemble in the branchpoint region. This may also be mediated by factors binding to UA- or U-rich elements which aid definition of the branchpoint and 3' splice site. Once factors are assembled and stabilized across an exon, interactions between the 5' splice site and branchpoint region across the intron will form the commitment complex before U2snRNP addition and functional spliceosome assembly.

have an invariant :AUAUCCUY sequence at the 5' splice site, YAC: at the 3' splice site, and a presumptive branchpoint (UCCUURAY) 16–19 nt upstream of the 3' splice site (31, 107). The mechanism of splicing of these introns is a two-step transesterification reaction, as for conventional introns, but recognition of the splicing signals requires a unique set of snRNAs (32, 95, 96). U11 and U12 basepair with the 5' splice site and branchpoint of AT-AC introns respectively in an analogous manner to U1 and U2 of the major class of spliceosomes (32, 43, 95). U5 is common to both spliceosomes (95), and U4atac/U6atac snRNAs are found as a di-snRNP and are essential to AT-AC spliceosome function (96).

In plants, two examples of AT-AC introns have been described: intron 14 of the RecA-like protein gene and intron 7 of the G5 gene from *Arabidopsis* are AT-AC introns, although the latter contains an AA: dinucleotide at the 3' splice site (106). Other examples of such introns have been proposed based on computer analysis. A second class of AT-AC introns with canonical splice sites (GU-AG) but with other AT-AC intron features (e.g. the proximity of the branchpoint to the 3' splice site) have also been proposed (P Rouzé, personal communication). Whether these introns have a function in regulating levels of their host mRNA remains an important question.

AU RICHNESS

Splicing in plants has progressed, for the most part, using transient expression analysis. Plasmid constructs consisting of a constitutive promoter, the test intron with complete or partial flanking exon sequences, and/or intron-containing reporter genes are introduced into plant protoplasts. The accuracy and efficiency of splicing is determined by Northern, RNase A/T_1 protection mapping; RT-PCR; or reporter gene expression. While an analysis of splicing in vivo rather than in vitro may well be an advantage for assessing true splice site choice and efficiency, there are limitations that need to be considered. First, strong constitutive promoters designed to give high levels of the transcript of interest can mean abnormal levels of pre-mRNA transcript, which may overload the splicing process. Second, most analyses to date of plant splicing use single intron constructs, taken out of their authentic gene context, and often tested in a heterologous species. There is evidence that introns are influenced by the presence of their neighboring intron and exon sequences (12, 15, 28, 66) and that splicing variation exists among plant species (29, 39, 41, 65). These limitations have the potential to give information that may not be a precise representation of splicing in a normal genetic background. Despite these weaknesses, studies so far have established the key principles for splicing in plants. By analyzing the increasing number of splicing mutants, especially in *Arabidopsis*, many of these limitations are removed. This, along with the ability to prepare transgenic plants, will permit us to test these principles and establish a more detailed understanding of plant splicing processes and mechanisms.

The differences between plant and animal splicing systems were highlighted by variable splicing of animal or plant/animal hybrid introns in plant cells. For example, the only animal introns that are efficiently spliced in plants to date are the human β-globin intron 1 assayed in maize (29), the SV40 small-t intron (37), and the mouse hsc70 intron 5 (JWS Brown, unpublished data) assayed in tobacco. These animal introns are all AU rich.

Goodall & Filipowicz (28) first demonstrated the requirement for AU richness for efficient plant intron splicing using synthetic introns assayed in tobacco cells. They proposed roles for the AU sequences either as binding sites for putative AU-binding proteins or sites to reduce secondary structure formation. Of great influence in the formulation of models for plant intron splicing were the findings that intron splice sites bordering AU-rich sequences were accurately and efficiently used despite the lack of a conserved branchpoint or polypyrimidine tract (28). The suggestion that a conserved branchpoint was not required and that presumably any adenosine (or other nucleotide) in the AU-rich intron could be used led to the idea of plant introns being defined by virtue of binding of specific proteins to AU-rich elements within the intron (28). This model was developed further with a series of experiments using pea and maize introns in *Nicotiana benthamiana* cells that suggested that 5' and 3' splice sites were selected because of their proximity to the transitions between the AU-rich intron and AU-poor exon. In these experiments, the positions of AU/GC borders were moved relative to splice sites by insertion and deletion of AU-rich and GC-rich sequences and mutation of authentic splice sites (53, 54, 67). The data suggested activation of splice sites at AU/GC transitions and "masking" of splice sites embedded in AU-rich regions. Similar insertional experiments with monocot introns tested in maize cell transient assays also supported the transition model, but it was recognized that in some cases an "internal" signal appeared to be needed (58). Splicing of nonintronic AU-rich regions (82) also supported the hypothesis that the only requirements for efficient plant intron splicing were AU-richness with adjacent splice sites (28, 53, 54, 58, 67). The importance of plant exon sequences was demonstrated by analyzing splicing of introns where GC- and AG-rich elements were inserted or were present in flanking exons, altering splicing efficiencies (15, 68; Figure 1). The key elements of AU-rich intron sequences were first defined as AU islands of 4–7 nt and later as U-rich elements with the ability to bind nuclear proteins (26, 53–55, 67). It is assumed that U-binding proteins associate with intron sequences and presumably a different set of proteins with different specificity bind to exon sequences, thereby aiding the delimitation of the exon/intron (21, 28, 29, 53–55, 67, 68, 85).

Despite the weight of experimental evidence, some observations of plant intron sequence and splicing behavior are not entirely consistent with intron definition on the basis of AU sequences and AU/GC transitions. First, while consensus plots of AU/GC content highlight intron/exon borders (85, 105), as with all consensus sequences, there is a great deal of variation on the individual gene sequence level. Some plant introns do not exhibit a significant or, in some cases, any AU/GC differential, and some exons contain AU islands in close proximity to splice sites. For example, a comparison of neighboring exons and introns in 209 *Arabidopsis* genes found 10% of the genes to contain at least one exon/intron transition where the exon has between 0% and 8% higher AU content than the intron (CG Simpson & JWS Brown, unpublished data) that, if bound by proteins, might be expected to interfere with splicing. In addition, splice sites found within AU-rich regions or GC-rich regions can be selected. An authentic splice site was surrounded by GC-rich sequences and utilized 85% of the time (77), and a splice site embedded in AU-rich sequence was efficiently selected when a competing distal 5' splice site was mutated (68). On the other hand, when dicot/monocot (AU-rich/AU-poor) hybrid introns were introduced into dicot cells, the authentic monocot intron splice sites were efficiently selected despite being some distance from the AU/GC transition formed by the intron fusions, and potential splice site–like sequences more proximal to the transitions were not selected (83). Thus, local AU/GC transitions around splice sites may not be major determining signals but reflect the method by which plant equivalents of hnRNP proteins exert their sequence specificity.

SIMILARITIES WITH OTHER SYSTEMS

The similarities among eukaryotes are highlighted by the strong conservation of splicing signals, snRNAs in important regions of primary sequence or secondary structure (91), the ever-increasing number of plant homologues to yeast and animal spliceosomal proteins (27, 79, 84, 87; JWS Brown & CG Simpson, unpublished data), SR proteins (48, 51, 52) (in particular, homologs of U2AF^{65}; C Domon & W Filipowicz, personal communication), a putative hnRNP C1/C2 (J Turner, personal communication), and the evidence for similar mechanisms of splice site selection operating in plants (see below). At the whole cell or nucleus level, there is also extensive organizational conservation with nuclear spliceosomal components being distributed throughout the fibrous interchromatin nuclear network and in coiled bodies directly paralleling that seen in animals (8, 18, 25, 85). Altogether, these similarities suggest that fundamental differences in splicing between plants and animals lie at the level of intron sequence recognition early in spliceosome formation. It seems likely that the distinguishing features of plant pre-mRNAs and proteins specific to them function in establishing the commitment or presplicing complexes after which spliceosome assembly and splicing will be very similar to other systems.

3' Splice Site Selection

Mutation of the conserved :GU and AG: dinucleotides abolishes correct splicing and optimizing splice site sequences can increase splicing efficiency (16, 26, 29, 83). The most important aspects of splice site selection in terms of this review are the similarities of the splicing behavior of many plant introns to that of vertebrate introns. For example, in mammalian intron splicing, the scanning model for 3' splice site selection proposes that the spliceosome scans from the branchpoint and usually selects the first AG: dinucleotide downstream of the branchpoint (90). If more than one AG: lies in close proximity, local scanning in this region may select the best 3' splice site on the basis of sequence context (88). In addition, when branchpoint, 3' splice site and associated exon fragments are found in duplicate, the downstream-most splice site was generally selected (80). There is strong evidence that such selection mechanisms operate in plant splicing (12, 83 and references therein). From experiments with a 3' splice site sequestered in a stem-loop structure, early recognition of the 3' splice site in plant spliceosome assembly was demonstrated and was suggested to be involved in commitment complex formation or branchpoint selection (50). These examples parallel yeast and vertebrate splicing where the 3' splice site is recognized early in spliceosomal assembly and aids identification of the internal branchpoint sequence. This is further supported by the need for a conserved branchpoint sequence for efficient splicing (83). Early recognition of the 3' splice site may involve the positioning of AU-binding proteins at the exon/intron border and subsequent recruitment of splicing factors analogous to vertebrate factors (e.g. U2AF) that recruit the U2snRNP to the branchpoint allowing formation of a commitment complex. In terms of exon definition (see below), interactions between an upstream 5' splice site and factors defining the preferred branchpoint sequence may be a key process in correct 3' splice site selection. This model, which parallels the vertebrate and yeast models, would allow accurate selection of 3' splice sites for introns that lack a AU/GC transition and are located within AU-rich or GC-rich regions (Figure 1).

Exon Definition

Exon sequences have been known to be important in vertebrate intron splicing for many years. In mammals exons tend to be small (<300 nt) and introns very large. A model for exon definition or scanning in splice site selection has been proposed whereby interactions between factors at a 3' splice site and a downstream 5' splice site define the exon (reviewed in 6, 9). In some cases of regulated splicing, interactions can occur between 5' or 3' splice site factors and proteins bound to sequences in an adjacent intron (intron enhancers) or exon (exon enhancers) to define the splice site (100). For example, stable U1snRNP binding to a 5' splice site can promote U2AF binding to an upstream 3' splice site (36), and binding of SR proteins or U1snRNP to purine-rich exon enhancer sequences enhances splicing of the upstream intron (98, 101). Mutation in splice site sequences can disrupt these interactions, often leading to exon skipping (removal of both introns and the intervening exon from the pre-mRNA). In contrast, lower eukaryotes with usually short introns may not require exon definition, and factors may interact directly across intron sequences (intron definition; 6).

In plants, although exon sequences have long been recognized as important in splicing by virtue of the contrast in base composition to adjacent introns, only recently have more active roles been discovered. Analysis of mutants in various *Arabidopsis* genes, with mutations in and around splice sites, has uncovered a number of examples of exon skipping in plant splicing (12). To date, six such mutations have been described: *ag-4, cop1-1, cop1-2, cop1-8, spy-1,* and *spy-2* (38, 70, 108; CG Simpson & JWS Brown, manuscript in preparation; Figure 2).

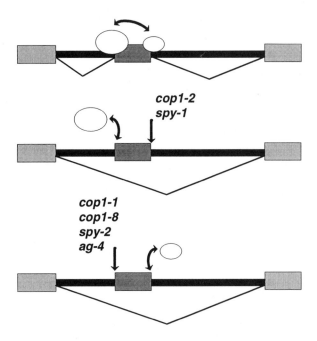

Figure 2 Exon definition in plant splicing. In multi-intron transcripts, factors assembled across an exon aid selection of the splice sites. Mutations to either the 5' or 3' splice sites flanking the central exon will disrupt assembly of the splicing components at these sites leading to a loss of interaction/stabilization across the exon and exon skipping.

The two mutant alleles of *SPINDLY, spy-1* and *spy-2*, both give rise to skipping of exon 8 (38). The *spy-2* allele contains a G → A mutation at the 3' splice site -AG of intron 7, and the *spy-1* mutation is to the last nucleotide in exon 8. The *cop1-1* and *cop1-2* mutants cause exon skipping of exon 6. *Cop1-1* has a mutation in the 3' splice site of the upstream intron, *cop1-2* is mutated in the 5' splice site of the downstream intron, and *cop1-8* carries a mutation in the 3' splice site of intron 10, which leads to removal of introns 10 and 11 and exon 11 (CG Simpson & JWS Brown, unpublished data). Finally, in the *ag-4* mutant, the majority of transcripts undergo exon skipping of exon 6 (108). In the exon skipping mutants, it would be expected that splicing factors would associate with the adjacent unmutated splice site, but the lack of, or unstable binding of, proteins to the mutated splice site would impair cross exon interactions and block correct splicing.

The examples of exon skipping in the *Arabidopsis* intron mutants provide strong evidence for exon definition (12; Figure 1). If the intron is the unit of definition in plant splicing, mutations to splice sites would not be expected to affect splicing of neighboring introns (6, 9). Therefore, splicing of at least some plant introns may involve exon definition and require direct interactions across the intron to form the spliceosome. One implication of exon definition in plant splicing is that splicing of individual introns in a multi-intron pre-mRNA may be affected by other introns in the transcript. Such cooperation or communication could occur by protein interactions where factors on one intron promote factor assembly at other introns. Evidence for such cooperation among vertebrate introns comes mainly from human genetic mutations where mutation of one intron can affect processing of adjacent or even distal introns (2, 76) and by in vivo splicing analyses where the presence of one intron is required for maximal splicing efficiency of other introns (72, 73).

SPECIES DIFFERENCES IN SPLICING

Dicots need ~60% AU for efficient splicing, while monocots are more flexible and can splice introns with as little as 30% AU. This difference between the AU-content requirement for splicing in monocots and dicots may reflect poorer affinity of the dicot AU-binding proteins for introns with few or only short AU-rich sequences. The apparently more flexible monocot splicing machinery would suggest that monocot hnRNP protein analogs have a broader sequence specificity that is reflected in the range of AU content seen in monocot introns (29). Besides the monocot/dicot difference, differences in splicing behavior of pre-mRNA transcripts among different species are beginning to emerge. For example, the relative usage of combinations of different splice sites in the Ds transposable element of maize differs between tobacco and *Arabidopsis* (75).

Similarly, cryptic splicing of the fourth intron of the *Ac* transposase gene in *Arabidopsis* has been postulated as the reason for poor levels of *Ac* excision in transgenic *Arabidopsis* (39, 65), compared with other dicotyledonous plants, such as tomato and tobacco, and in its host plant, maize (3, 40). Finally, expression of the *Aequoria victoria* GFP gene was severely hampered by a cryptic splicing event that removed an 84-nt intron with the GFP coding region (34). The cryptic intron (68% AU) contained 5' and 3' splice site and branchpoint sequences with a good match to plant intron consensus sequences. In *Arabidopsis*, this intron was efficiently spliced in almost 100% of transcripts, giving little evidence of fluorescence in transgenic plants. However, in tobacco, splicing efficiency of the cryptic intron was 40%, resulting in expression of the protein. Thus, considerable variation exists in the ability of the splicing apparatus of different dicot species to recognize and excise the same intron sequence.

ALTERNATIVE SPLICING

In animals, mechanisms exist to regulate splice site choice via alternative splicing, which can result in either retention or inclusion of introns in the open reading frame; selection of alternative 5' and 3' splice sites, which leads to increases or decreases in the size of particular exons; selective inclusion of mutually exclusive exons; and exon skipping (69, 89, 98). Alternative splicing results in the production of different mRNAs that encode proteins with functional differences often in a tissue-specific or developmental stage-specific manner and is therefore an important regulatory mechanism in gene expression.

In plants, a number of examples of alternative splicing have emerged, but little functional or developmental information is available. A common form of alternative splicing in plants is intron retention (27, 74, 85, 93, 97), which probably reflects poor recognition of the intron rather than active processes inhibiting the splicing reaction. The presence of a poorly spliced intron in a pre-mRNA can potentially regulate transcript levels because of competition with other RNA-processing events such as polyadenylation. Polyadenylation sites present within such an intron may compete with the splicing apparatus with the result that many transcripts are prematurely polyadenylated (39, 59, 85). In vertebrates, the presence of in-frame premature stop codons within intron sequences can lead to effects on mRNA stability, transport from the nucleus, translation, and subsequently levels of gene expression (17, 62). Alternative splicing of the *wxG* allele (G retrotransposon insertion in *waxy*) in maize shows a down-regulation of expression in endosperm but not in pollen. This tissue-specific effect may be related to the presence of premature termination codons within exon 13 (63). Alternatively, levels of *Arabidopsis* U1 70k mRNA transcripts that retain intron

7 showed variable levels of expression in different tissues, but in this case there is no premature termination codons or polyadenylation within the intron (27), suggesting that other mechanisms linked to splicing may operate to alter levels of gene expression.

Selection of alternative 5' and 3' splice sites occurs in a number of plant introns. In some cases, both splice sites are selected in a similar ratio, such as in rubisco activase from a number of species (103). In others, the relative usage displays different cell- or tissue-specific splicing patterns. For example, alternative 5' splice site selection in the third intron of chorismate synthase (LeCS2) from tomato and alternative 3' splice site selection in the first intron of the H-protein subunit of glycine decarboxylase from *Flaveria trinervia* show an alteration toward more distal and weaker splice sites in floral and root tissue, while stronger proximal sites are predominantly selected in leaf, stem, and cotyledon (30, 44). In animal systems, particular splicing factors can alter the splice site choice of competing splice sites (109), such that the tissue or organ-specific splicing pattern differences may reflect different splicing factor levels. Similarly, in animals, splicing patterns can be altered in response to stress; however, in plants, splicing patterns tend to be unaffected by stress situations (64, 85). Recently, however, an example of exon skipping has been described in which a mini-exon is skipped at low levels in response to cold storage conditions (10). This effect may reflect lower splicing accuracy or changes in splicing factors under these physiological conditions, but because the mini-exon encodes part of the most conserved region of the protein, the possibility remains that the exon skip produces a functionally modified peptide.

Although the functional significance of alternative splicing systems in plants is mostly unknown, the effect of alternative splicing events can be extensive on gene expression. For example, intron retention can lead to alterations in cellular location for functional peptides when they are found separating a signal sequence from the functional peptide. Use of a translation start codon in the intron can produce cytosolic forms of otherwise targeted proteins (19, 81, 97). The different forms of alternative splicing can also lead to the production of truncated peptides due to premature termination either by activation of intronic polyadenylation signals or by bringing translation stops in frame (24, 59, 92, 93). The role of these truncated peptides, with the exception of secreted and membrane-bound *SLG* (94), is not known, but in some cases significant levels of the truncated peptide accumulate (59). It is feasible that such peptides are involved in regulation, possibly acting as *trans*-dominant inhibitors. Alternatively, competition between processing signals such as splicing and polyadenylation could regulate the levels at which the functional peptide is found. Overexpression of the flowering control gene, *FCA*, resulted in only a limited acceleration of flowering time, reflected by only a small increase in fully spliced mRNA transcript. However, the levels of the short polyadenylated transcript, which terminates in intron 3, showed a 150-fold increase over the wild type. Thus, the default pathway for polyadenylation in intron 3 may keep the levels of FCA at the correct level to keep flowering time constant (59). Finally, it has long been known that the insertion of transposable elements into gene sequences can result in gross changes in the processing of the pre-mRNA transcript (55, 56, 102, 104). Often the transposable element inserts into intron sequences and is then removed by splicing (using cryptic splice sites within the transposable element) but often inaccurately such that pre-mRNA encodes a protein that differs from the wild type and often has reduced functionality. Although the number of alternative splicing systems in plants is increasing, the importance of this area of gene regulation is still in its infancy, and a better understanding of the functions of different peptides and the underlying splicing mechanisms is readily awaited.

CONCLUDING REMARKS

As more information on animal, yeast, and plant splicing and their splicing components accrues, it is becoming clear that greater similarity exists than was previously thought. It is, for example, inconceivable that a protein such as PRP8, known to be essential in animal and yeast systems for many stages of spliceosome assembly and which is so highly conserved in plants, should not carry out the same functions in plant splicing. The recent demonstrations of the importance of branchpoint sequences and exons in plant splicing again draw closer parallels with other eukaryotic systems in terms of mechanisms of splice site selection. This degree of similarity would suggest that the distinguishing feature of UA-rich sequences in plant introns appears to be involved early in intron recognition and prespliceosomal complex formation. Its particular role and the factors that recognize this intron signal still remain to be resolved fully. The increasing number of alternatively spliced gene systems, the characterization of splicing mutants, the examples of exon skipping and recent experiments on test introns all underline the subtle complexity of the splicing process and how splicing of any particular intron depends on the balance of a number of signals and factors. An appreciation of this balance will be necessary in understanding how gene expression can be regulated at the level of pre-mRNA splicing. The study of splicing of specific gene systems in their own context may provide new insights into our knowledge of plant splicing.

Literature Cited

1. Abovich N, Rosbash M. 1997. Cross-intron bridging interactions in the yeast commitment complex are conserved in mammals. *Cell* 89:403–12

2. Antoniou M. 1995. Clinical defects in pre-mRNA processing. See Ref. 47a, 12:187–200

3. Baker B, Schell J, Lorz H, Federoff N. 1986. Transposition of the maize controlling element "*Activator*" in tobacco. *Proc. Natl. Acad. Sci. USA* 83:4814–44

4. Baynton CE, Potthoff SJ, McCullough AJ, Schuler MA. 1996. U-rich tracts enhance 3'-splice-site recognition in plant nuclei. *Plant J.* 10:703–11

5. Beggs JD. 1995. Yeast splicing factors and genetic strategies for their analysis. See Ref. 47a, 5:79–95

6. Berget SM. 1995. Exon recognition in vertebrate splicing. *J. Biol. Chem.* 270:2411–14

7. Berglund JA, Chua K, Abovich N, Reed R, Rosbash M. 1997. The splicing factor BBP interacts specifically with the pre-mRNA branchpoint sequence UACUAAC. *Cell* 89:781–87

8. Beven AF, Simpson GG, Brown JWS, Shaw PJ. 1995. The organisation of spliceosomal components in the nuclei of higher plants. *J. Cell Sci.* 108:509–18

9. Black DL. 1995. Finding splice sites within a wilderness of RNA. *RNA* 1:763–71

10. Bournay A-S, Hedley PE, Maddison A, Waugh R, Machray GC. 1996. Exon skipping induced by cold stress in a potato invertase gene transcript. *Nucleic Acids Res.* 24:2347–51

11. Brown JWS. 1986. A catalogue of splice junction and putative branchpoint sequences from plant introns. *Nucleic Acids Res.* 14:9549–59

12. Brown JWS. 1996. *Arabidopsis* intron mutations and pre-mRNA splicing. *Plant J.* 10:771–80

13. Brown JWS, Smith P, Simpson CG. 1996. *Arabidopsis* consensus intron sequences. *Plant Mol. Biol.* 32:531–35

14. Burgess SM, Guthrie C. 1993. Beat the clock–paradigms for NTPases in the maintenance of biological fidelity. *Trends Biochem. Sci.* 18:381–84

15. Carle-Urioste JC, Brendel V, Walbot V. 1997. A combinatorial role for exon, intron and splice site sequences in splicing in maize. *Plant J.* 11:1253–63

16. Carle-Urioste JC, Ko CH, Benito M-I, Walbot V. 1994. In vivo analysis of intron processing using splicing-dependent reporter gene assays. *Plant Mol. Biol.* 26:1785–95

17. Carter MS, Li S, Wilkinson MF. 1996. A splicing-dependent regulatory mechanism that detects translation signals. *EMBO J.* 15:5965–75

18. Chamberland H, Lafontaine JG. 1993. Localisation of snRNP antigens in nucleolus associated bodies: study of plant interphasic nuclei by confocal and electron microscopy. *Chromosoma* 102:220–26

19. Domon C, Everard J-L, Pillay DTN, Steinmetz A. 1991. A 2.6 kb intron separates the signal peptide codong sequence of an anther-specific protein from the rest of the gene in sunflower. *Mol. Gen. Genet.* 229:238–44

20. Dreyfuss G, Matunis MJ, Piñol-Roma SN, Burd CG. 1993. HnRNP proteins and the biogenesis of mRNA. *Annu. Rev. Biochem.* 1962:289–321

21. Filipowicz W, Gniadkowski M, Klahre U, Liu H-X. 1995. Pre-mRNA splicing in plants. See Ref. 47a, 4:65–77

22. Fu X-D. 1995. The superfamily of arginine/serine-rich splicing factors. *RNA* 1:663–80

23. Fuller-Pace V. 1994. RNA helicases: modulators of RNA structure. *Trends Cell Biol.* 4:271–74

24. Giranton J-L, Ariza MJ, Dumas C, Cock JM, Gaude T. 1995. The *S* locus receptor kinase gene encodes a soluble glycoprotein corresponding to the SRK extracellular domain in *Brassica oleracea*. *Plant J.* 8:827–34

25. Glyn MCP, Leitch AR. 1995. The distribution of a spliceosomal protein in cereal Triticeae: interphase nuclei from cells with different metabolic activities and through the cell cycle. *Plant J.* 8:531–40

26. Gniadkowski M, Hemmings-Mieszcak M, Klahre U, Liu H-X, Filipowicz W. 1996. Characterisation of intronic uridine-rich sequence elements acting as possible targets for nuclear proteins during pre-mRNA splicing in *Nicotiana plumbaginifolia*. *Nucleic Acids Res.* 24:619–27

27. Golovkin M, Reddy ASN. 1996. Structure and expression of a plant U1 snRNP 70K gene: alternative splicing of U1 snRNP 70K pre-mRNAs produces two different transcripts. *Plant Cell* 8:1421–35

28. Goodall GJ, Filipowicz W. 1989. The AU-rich sequences present in the introns of

plant nuclear pre-mRNAs are required for splicing. *Cell* 58:473–83

29. Goodall GJ, Filipowicz W. 1991. Different effects on intron nucleotide composition and secondary structure on pre-mRNA splicing in monocot and dicot plants. *EMBO J.* 10:2635–44

30. Gorlach J, Raesecke H-R, Abel G, Wehrli R, Amrhein N, et al. 1995. Organ-specific differences in the ratio of alternatively spliced chorismate synthase (*LeCS2*) transcripts in tomato. *Plant J.* 8:451–56

31. Hall SL, Padgett RA. 1994. Conserved sequences in a class of rare eukaryotic nuclear introns with nonconsensus splice sites. *J. Mol. Biol.* 239:357–65

32. Hall SL, Padgett RA. 1996. Requirements of U12 snRNA for in vivo splicing of a minor class of eukaryotic nuclear pre-mRNA introns. *Science* 271:1716–18

33. Hanley BA, Schuler MA. 1988. Plant intron sequences: evidence for distinct groups of introns. *Nucleic Acids Res.* 16:7159–76

34. Haseloff J, Siemering KR, Prasher DC, Hodge S. 1997. Removal of a cryptic intron and subcellular localization of green fluorescent protein are required to mark transgenic *Arabidopsis* plants brightly. *Proc. Natl. Acad. Sci. USA* 94:2122–27

35. Hebsgaard SM, Kornin PG, Tolstrup N, Engelbrecht J, Rouze P, et al. 1996. Splice-site prediction in *Arabidopsis thaliana* pre-messenger-RNA by combining local and global sequence information. *Nucleic Acids Res.* 24:3439–52

36. Hoffman BE, Grabowski PJ. 1992. U1snRNP targets an essential splicing factor, U2AF, to the 3' splice site by a network of interactions spanning the exon. *Genes Dev.* 6:2554–68

37. Hunt AG, Mogen BD, Chu NM, Chua N-H. 1991. The SV40 small t intron is accurately and efficiently spliced in tobacco cells. *Plant Mol. Biol.* 16:375–79

38. Jacobsen SE, Binkowski K, Olszewski NE. 1996. *SPINDLY*—a tetratricopeptide repeat protein involved in gibberellin signal transduction in *Arabidopsis*. *Proc. Natl. Acad. Sci. USA* 93:9292–96

39. Jarvis P, Belzile F, Dean C. 1997. Inefficient and incorrect processing of the *Ac* transposase transcript in *iae1* and wild-type *Arabidopsis thaliana*. *Plant J.* 11:921–31

40. Jones JDG, Carland FM, Maliga P, Dooner MK. 1989. Visual detection of the maize element *Activator* (*Ac*) in tobacco seedlings. *Science* 244:204–7

41. Keith B, Chua NH. 1986. Monocot and dicot pre-mRNAs are processed with different efficiencies in transgenic tobacco. *EMBO J.* 5:2419–25

42. Kiledjian M, Burd CG, Gorlach M, Portman DS, Dreyfuss G. 1994. Structure and function of hnRNP proteins. In *RNA-Protein Interactions*, ed. K Nagai, IW Mattaj, 6:127–49. Oxford: IRL Press

43. Kolossova I, Padgett RA. 1997. U11sn-RNA interacts in vivo with the 5' splice site of U12–dependent (AU-AC) pre-mRNA introns. *RNA* 3:227–33

44. Kopriva S, Cossu R, Bauwe H. 1995. Alternative splicing results in two different transcripts for H-protein of the glycine cleavage system in the C_4 species *Flaveria trinervia*. *Plant J.* 8:435–41

45. Krainer AR, Maniatis T. 1988. RNA splicing. In *Transcription and Splicing*, ed. BD Hames, DM Glover, 4:131–206. Oxford: IRL Press

46. Krämer A. 1995. The biochemistry of pre-mRNA splicing. See Ref. 47a, 3:35–64

47. Krämer A. 1996. The structure and function of proteins involved in mammalian pre-mRNA splicing. *Annu. Rev. Biochem.* 65:367–409

47a. Lamond AI, ed. 1995. *Pre-mRNA Processing.* Georgetown, TX: Landes

48. Lazar G, Schaal T, Maniatis T, Goodman HM. 1995. Identification of a plant serine-arginine-rich protein similar to mammalian splicing factor SF2/ASF. *Proc. Natl. Acad. Sci. USA* 92:7672–76

49. Liu H-X, Filipowicz W. 1996. Mapping of branchpoint nucleotides in mutant pre-mRNAs expressed in plant cells. *Plant J.* 9:381–89

50. Liu H-X, Goodall GJ, Kole R, Filipowicz W. 1995. Effects of secondary structure on pre-mRNA splicing: hairpins sequestering the 5' but not the 3' splice site inhibit intron processing in *Nicotiana plumbaginifolia. EMBO J.* 14:377–88

51. Lopato S, Mayeda A, Krainer AR, Barta A. 1996. Pre-mRNA splicing in plants: characterization of Ser/Arg splicing factors. *Proc. Natl. Acad. Sci. USA* 93:3074–79

52. Lopato S, Waigmann E, Barta A. 1996. Characterization of a novel Arginine/Serine-rich splicing factor in *Arabidopsis. Plant Cell* 8:2255–64

53. Lou H, McCullough AJ, Schuler MA. 1993. 3' splice selection in dicot plant nuclei is position dependent. *Mol. Cell. Biol.* 13:4485–93

54. Lou H, McCullough AJ, Schuler MA. 1993. Expression of maize *Adh1* intron mutants in tobacco. *Plant J.* 3:393–403

55. Luehrsen KR, Taha S, Walbot V. 1994. Nuclear premessenger-RNA processing in higher plants. *Prog. Nucleic Acid Res. Mol. Biol.* 47:149–93

56. Luehrsen KR, Walbot V. 1990. Insertion of *Mu1* elements in the first intron of the *Adh1–S* gene of maize results in novel RNA processing events. *Plant Cell* 2:1225–38

57. Luehrsen KR, Walbot V. 1991. Intron enhancement of gene expression and the splicing efficiency of introns in maize cells. *Mol. Gen. Genet.* 225:81–93

58. Luehrsen KR, Walbot V. 1994. Addition of A- and U-rich sequence increases the splicing efficiency of a deleted form of a maize intron. *Plant Mol. Biol.* 24:449–63

59. Macknight R, Bancroft I, Page T, Lister C, Schmidt R, et al. 1997. *FCA*, a gene controlling flowering time in *Arabidopsis*, encodes a protein containing RNA-binding domains. *Cell* 89:737–45

60. Madhani HD, Guthrie C. 1994. Dynamic RNA-RNA interactions in the spliceosome. *Annu. Rev. Genet.* 28:1–26

61. Manley JL, Tacke R. 1996. SR proteins and splicing control. *Genes Dev.* 10:1569–79

62. Maquat LE. 1995. When cells stop making sense: effects of nonsense codons on RNA metabolism in vertebrate cells. *RNA* 1:453–65

63. Marillonnet S, Wessler SR. 1997. Retrotransposon insertion into the maize *waxy* gene results in tissue-specific RNA processing. *Plant Cell* 9:1–12

64. Marrs KA, Walbot V. 1997. Expression and RNA splicing of the maize glutathione *S*-transferase *Bronze 2* gene is regulated by Cadmium and other stresses. *Plant Physiol.* 113:93–102

65. Martin DJ, Firek S, Moreau E, Draper J. 1997. Alternative processing of the maize *Ac* transcript in *Arabidopsis*. *Plant J.* 11:933–43

66. McCullough AJ, Baynton CE, Schuler MA. 1996. Interactions across exons can influence splice site recognition in plant nuclei. *Plant Cell* 8:2295–307

67. McCullough AJ, Lou H, Schuler MA. 1993. Factors affecting authentic 5′ splice site selection in plant nuclei. *Mol. Cell. Biol.* 13:1323–31

68. McCullough AJ, Schuler MA. 1997. Intronic and exonic sequences modulate 5′ splice site selction in plant nuclei. *Nucleic Acids Res.* 25:1071–77

69. McKeown M. 1992. Alternative mRNA splicing. *Annu. Rev. Cell Biol.* 8:133–55

70. McNellis TW, von Arnim AG, Araki T, Komeda Y, Misera S, et al. 1994. Genetic and molecular analysis of an allelic series of *cop1* mutants suggests functional roles for multiple protein domains. *Plant Cell* 6:487–500

71. Moore MJ, Query CC, Sharp PA. 1993. Splicing of precursors to mRNA by the spliceosome. In *The RNA World*, ed. RF Gesteland, JF Atkins, 13:303–57. Cold Spring Harbor, NY: Cold Spring Harbor Lab. Press

72. Neel H, Weil D, Giansante C, Dautry F. 1993. In vivo cooperation between introns during premessenger RNA processing. *Genes Dev.* 7:2194–205

73. Nesic D, Maquat LE. 1994. Upstream introns influence the efficiency of final intron removal and RNA 3′-end formation. *Gene. Dev.* 8:363–75

74. Nishihama R, Banno H, Kawahare E, Irie K, Machida Y. 1997. Possible involvement of differential splicing in regulation of the activity of *Arabidopsis* ANP1 that is related to mitogen-activated protein kinase kinase kinases (MAPKKKs). *Plant J.* 12:39–48

75. Nussaume L, Harrison K, Klimyuk V, Martienssen R, Sundarsan V, et al. 1995. Analysis of splice donor and acceptor site function in a transposable gene trap derived from the maize element Activator. *Mol. Gen. Genet.* 249:91–101

76. Ohno K, Suzuki K. 1988. Multiple abnormal β-hexosaminidase of chain mRNAs in a compound-heterozygous Ashkenazi Jewish patient with Tay-Sachs disease. *J. Biol. Chem.* 263:18563–67

77. Okagaki RJ, Sullivan TD, Schiefelbein JW, Nelson OE. 1992. Alternative 3′ splice acceptor sites modulate enzymic activity in derivative alleles of the maize *bronze1–mutable 13* allele. *Plant Cell* 4:1453–62

78. Query CC, Moore MJ, Sharp PA. 1994. Branch nucleophile selection in pre-mRNA splicing: evidence for the bulged duplex model. *Genes Dev.* 8:587–97

79. Reddy ASN, Czernik AJ, An G, Poovaiah BW. 1992. Cloning of the cDNA for U1 small nuclear ribonucleoprotein particle 70K protein from *Arabidopsis thaliana*. *Biochim. Biophys. Acta* 1171:88–92

80. Reed R, Maniatis T. 1986. A role for exon sequences and splice site proximity in splice site selection. *Cell* 46:681–90

81. Rosche E, Westhoff P. 1995. Genomic structure and expression of the pyruvate, orthophosphate dikinase gene of the dicotyledonous C_4 plant *Flaveria trinervia*

(Asteraceae). *Plant Mol. Biol.* 29:663–78

82. Simpson CG, Brown JWS. 1993. Efficient splicing of an AU-rich antisense intron sequence. *Plant Mol. Biol.* 21:253–65

83. Simpson CG, Clark G, Davidson D, Smith P, Brown JWS. 1996. Mutation of putative branchpoint consensus sequences in plant introns reduces splicing efficiency. *Plant J.* 9:369–80

84. Simpson GG, Clark G, Rothnie H, Boelens W, van Venrooij W, et al. 1995. Molecular characterisation of the spliceosomal proteins U1A and U2B″ from higher plants. *EMBO J.* 14:4540–50

85. Simpson GG, Filipowicz W. 1996. Splicing of precursors to messenger RNA in higher plants: mechanism, regulation and sub-nuclear organisation of the spliceosomal machinery. *Plant Mol. Biol.* 32:1–41

86. Simpson CG, Leader DJ, Brown JWS. 1993. Characteristics of plant pre-mRNA introns. In *Plant Molecular Biology Labfax*, ed. RRD Croy, 6:183–251. Oxford: BIOS Sci.

87. Simpson GG, Vaux P, Clark G, Waugh R, Beggs JD, et al. 1991. Evolutionary conservation of the spliceosomal protein U2B″. *Nucleic Acids Res.* 19:5213–17

88. Smith CWJ, Chu TT, Nadal-Ginard B. 1993. Scanning and competition between AGs are involved in 3′ splice site selection in mammalian introns. *Mol. Cell. Biol.* 13:4939–52

89. Smith CWJ, Patton JG, Nadal-Ginard B. 1989. Alternative splicing in the control of gene expression. *Annu. Rev. Genet.* 23:437–49

90. Smith CWJ, Porro EB, Patton JG, Nadal-Ginard B. 1989. Scanning from an independently specified branchpoint defines the 3′ splice site of mammalian introns. *Nature* 342:243–47

91. Solymosy F, Pollák T. 1993. Uridylate-rich small nuclear RNAs UsnRNAs, their genes and pseudogenes, and UsnRNPs in plants—structure and function—a comparative approach. *Crit. Rev. Plant Sci.* 12:275–369

92. Tamaoki M, Tsugawa H, Minami E, Kayano T, Yamamoto N, et al. 1995. Alternative RNA products from rice homeobox gene. *Plant J.* 7:927–38

93. Taniguchi M, Sugiyama T. 1997. The expression of 2-Oxoglutarate/Malate translocator in the bundle-sheath mitochondria of *Panicum miliaceum*, a NAD-malic

enzyme-type C_4 plant, is regulated by light and development. *Plant Physiol.* 114:285–93

94. Tantikanjana T, Nasrallah ME, Stein JC, Chen C-H, Nasrallah JB. 1993. An alternative transcript of the *S* locus glycoprotein gene in class II pollen-recessive self-incompatibility haplotype of *Brassica oleracea* encodes a membrane-anchored protein. *Plant Cell* 5:657–66

95. Tarn W-Y, Steitz JA. 1996. A novel spliceosome containing U11, U12, and U5snRNPs excises a minor class (AT-AC intron in vitro). *Cell* 84:801–11

96. Tarn W-Y, Steitz JA. 1996. Highly diverged U4 and U6 small nuclear RNAs required for splicing rare AT-AC introns. *Science* 273:1824–32

97. Thorbjornsen T, Villand P, Kleczkowski LA, Olsen OA. 1996. A single-gene encodes two different transcripts for the ADP-glucose pyrophosphorylase small-sub-unit from barley (*Hordeum vulgare*). *Biochem. J.* 313:149–54

98. Tian M, Maniatis T. 1993. A splicing enhancer complex controls alternative splicing of *doublesex* pre-mRNA. *Cell* 74:105–14

99. Tolstrup N, Rouzé P, Brunak S. 1997. A branch point concensus from *Arabidopsis* found by noncircular analysis allows for better prediction of acceptor sites. *Nucleic Acids Res.* 25:3159–63

100. Valcárcel J, Sing R, Green MR. 1995. Mechanisms of regulated pre-mRNA splicing. See Ref. 47a, 6:97–112

101. Watakabe A, Tanaka K, Shimura Y. 1993. The role of exon sequences in splice site selection. *Genes Dev.* 7:407–18

102. Weil CF, Wessler SR. 1990. The effects of plant transposable element insertion on transcription initiation and RNA processing. *Annu. Rev. Plant Physiol. Plant Mol. Biol.* 41:521–52

103. Werneke JM, Chatfield JM, Ogren WL. 1989. Alternative mRNA splicing generates the two ribulosebisphosphate carboxylase/oxygenase activase polypeptides in spinach and *Arabidopsis*. *Plant Cell* 1:815–25

104. Wessler SR. 1989. The splicing of maize transposable elements from pre-mRNA—a mini review. *Gene* 82:127–33

105. Wiebauer K, Herrero JJ, Filipowicz W. 1988. Nuclear pre-mRNA processing in plants: distinct modes of 3′ splice site selection in plants and animals. *Mol. Cell. Biol.* 8:2042–51

106. Wu H-J, Gaubier-Comella P, Delseny M, Grellet F, Van Montagu M, et al. 1996. Noncanonical introns are at least 10^9

years old. *Nat. Genet.* 14:383–84
107. Wu Q, Krainer AR. 1997. Splicing of a divergent subclass of AT-AC introns requires the major spliceosomal snRNAs. *RNA* 3:586–601
108. Yanofsky MF, Ma H, Bowman JL, Drews GN, Feldmann KA, Meyerowitz EM. 1990. The protein encoded by the *Arabidopsis* homeotic gene *agamous* resembles transcription factors. *Nature* 346:35–39
109. Zahler AM, Roth MB. 1995. Distinct functions of SR proteins in recruitment of U1 small nuclear ribonucleoprotein to alternative 5′ splice sites. *Proc. Natl. Acad. Sci. USA* 92:2642–46

Annu. Rev. Plant Physiol. Plant Mol. Biol. 1998. 49:97–126

PROTEIN TARGETING
TO THE THYLAKOID MEMBRANE

Danny J. Schnell

Department of Biological Sciences, Rutgers, The State University of New Jersey, Newark, New Jersey 07102; e-mail: schnell@andromeda.rutgers.edu

KEY WORDS: chloroplast, protein translocation, organelle biogenesis, signal peptide, transit sequence

ABSTRACT

The assembly of the photosynthetic apparatus at the thylakoid begins with the targeting of proteins from their site of synthesis in the cytoplasm or stroma to the thylakoid membrane. Plastid-encoded proteins are targeted directly to the thylakoid during or after synthesis on plastid ribosomes. Nuclear-encoded proteins undergo a two-step targeting process requiring posttranslational import into the organelle from the cytoplasm and subsequent targeting to the thylakoid membrane. Recent investigations have revealed a single general import machinery at the envelope that mediates the direct transport of preproteins from the cytoplasm to the stroma. In contrast, at least four distinct pathways exist for the targeting of proteins to the thylakoid membrane. At least two of these systems are homologous to translocation systems that operate in bacteria and at the endoplasmic reticulum, indicating that elements of the targeting mechanisms have been conserved from the original prokaryotic endosymbiont.

CONTENTS

1040-2519/98/0601-0097$08.00

INTRODUCTION

The thylakoid membrane has evolved an exquisitely organized set of intrinsic and extrinsic protein complexes that function in concert to couple the photo-oxidation of water with electron transport and chemiosmosis, thereby providing NADPH and ATP for the carbon fixation reactions of photosynthesis. The biogenesis and maintenance of the photosystems, electron carriers, and ATP-synthase relies on the contributions of protein subunits encoded by both nuclear and plastid genomes. As a consequence of this dual genetic origin, the assembly of the thylakoid components begins with the targeting of stromally and cyto-plasmically synthesized proteins to the thylakoid and their integration into or translocation across the membrane. The general features of protein targeting to the thylakoid membrane adhere to similar principles as those that govern protein targeting to other cellular membrane systems. For example, thylakoid proteins are synthesized with intrinsic targeting signals that are necessary and sufficient to target a protein to the thylakoid membrane. Plastid-encoded proteins possess a single thylakoid targeting signal, whereas nuclear-encoded proteins require dual signals that first direct import of the protein across the double membrane of the chloroplast envelope and subsequently target the protein to the thylakoids. Furthermore, experimental evidence supports the existence of highly specific soluble and membrane-bound recognition systems that decode the targeting signals and initiate translocation through protein conducting channels.

Although the import of nuclear-encoded proteins at the chloroplast enve-lope appears to use a common general import machinery, recent investigations into the molecular mechanism of protein targeting to the thylakoid membrane have revealed a multitude of targeting systems that function with specific sub-classes of thylakoid proteins. Remarkably, at least two of these pathways are homologous to known translocation systems that operate in bacteria and the endoplasmic reticulum (ER). Two other pathways represent previously uniden-tified translocation systems, one of which may also have a bacterial counterpart. The discovery of these multiple pathways has contributed significantly to our understanding of the initial stages in the assembly of the photosynthetic appa-ratus and has provided new insights into the evolutionary relationships between membrane targeting systems at different organelles.

The majority of information on the thylakoid targeting processes has been obtained through the elegant manipulation of in vitro assays in which radio-labeled preproteins are incubated with intact isolated chloroplasts or thylakoid membranes. Following the in vitro targeting reaction, the fate of the preprotein is determined by analyzing its association with the thylakoid membrane or another chloroplast subfraction and quantifying the extent of appropriate proteolytic processing. In addition to these biochemical systems, in vivo models for thylakoid targeting in maize and the green alga *Chlamydomonas* are now available, and the combination of in vitro and in vivo approaches has contributed to the identification of a number of translocation components. The intent of this review is to summarize the recent developments that have led to the discovery of multiple pathways for protein targeting to the thylakoids in plants. It will focus on recent advances in identifying the components of the targeting pathways at the thylakoid membranes. Several current reviews of protein import across the chloroplast envelope have been published (26, 32, 38, 40, 72, 78, 127). Therefore, a complete review of envelope translocation is not presented, but the salient features of the import mechanism are summarized to provide the background for understanding this critical first step in the localization of nuclear-encoded thylakoid proteins. Additional information and perspectives are available in several recent reviews on chloroplast protein import and targeting to the thylakoid (26, 30, 32, 63, 72, 78, 109, 127).

TARGETING SIGNALS OF THYLAKOID PROTEINS

The Dual Targeting Signals of Nuclear-Encoded Thylakoid Proteins

Nuclear-encoded thylakoid proteins are synthesized on cytoplasmic ribosomes as precursors containing a cleavable N-terminal extension designated the transit sequence. Targeting to the thylakoid occurs posttranslationally by a two-step process requiring dual targeting signals that direct import across the double membrane of the chloroplast envelope and subsequent transport to the thylakoid membrane (Figure 1). The analysis of the targeting signals of a variety of nuclear-encoded thylakoid preproteins suggests that the envelope and thylakoid targeting signals occur in two configurations. In the first configuration, the transit sequence is bipartite (122), containing the information for import across the chloroplast envelope as well as information for targeting to the thylakoid. The two targeting signals reside in separate domains of the transit sequence with the stromal targeting domain (STD) located at the N-terminal region and the thylakoid lumenal targeting domain (LTD) located at the C-terminal region of the transit sequence (26). Representatives of this class that have been studied most extensively are the lumenal proteins, preplastocyanin (prePC), and the subunits of the oxygen evolving complex, preOE16, preOE23, and preOE33. The bipartite nature of the transit sequences of thylakoid lumenal proteins was first proposed by Smeekens et al (122) when they observed that replacement of the preplastocyanin transit sequence with that of the stromal protein, preferredoxin, redirected plastocyanin to the stroma in in vitro import studies. These results indicated that the preplastocyanin transit sequence contained information in

	STROMAL TARGETING DOMAIN	LUMEN TARGETING DOMAIN
LHCP	MAASSSSSMALSSPTLAGKQLKLNPSSQELGAARFT	
CP24	MAAATSATAIVNGFTSPFLSGGKKSSQSLLFVNSKVGAGVSTTSRKLVVVA	
CFoII		MANMLVASSSKTLPTTTTTTITPKPKFPLLKTPLLKLSPPQLPPLKHLNLSVLKSAAITATPLTLSFLLPYPSL
PII-X		MASTSAMSLVTPLNQTRSSPFLKPLPLKPSKALVATGGRAQRLQVRALKMDKALITGISAAALTASMVI
PII-W		MATTTASSSASLVARASLVHNSRVGVSSSPPILGLPSMTKRSKVTCSIENKPSTTETTTTTNKSMGASLLAAAAAATISNPAM

	A	N	H	C

	STROMAL TARGETING DOMAIN	LUMEN TARGETING DOMAIN
OE33	MAASLQASTTFLQPTKVASRNTLQLRSTQNVCKAFGVESASSGGRLSLSLQSDLKELANKCVDATKLAGLALATSALIASGANA	
PC	MATVASSAAVAVPSFTGLKASGSIKPTTAKIIPTTTAVPRLSVKASLKNVGAAVVATAAGLLAGNAMA	
OE23	MASTQCFLHHQYAITTPTRTLSQRQVVTTKPNHIVCKAQKQDDVVDAVV**SRR**LAISVLIGAAAVGS**K**VSPADA	
OE16	MAQAMASMAGLRGASQAVLEGSLQISGSMRLSGPTTSRVAVPKMGLNIRAQQVSAEAET**SRR**AMLGFVAAGLASGSFV**K**AVLA	
PSI-N	MAGVNTSVVGLKPAAAVPQSASPAAAKRVQVAPAKD**RR**SALIGLAAVFAATAASAGSA**R**A	

Figure 1 Structures of the transit peptides of representative nuclear-encoded thylakoid proteins. A detailed description of the structural elements of the transit sequences is provided in the text. The primary structures of the transit peptides of the precursors for pea light-harvesting a/b binding protein (LHCP) (16), spinach apoprotein of the CP24 complex (CP24) (124), spinach CF$_0$II subunit of the ATP synthase (CF$_0$II) (48), Arabidopsis photosystem II (PSII) subunit X (PII-X) (60), spinach PSII subunit W (PII-W) (77), spinach 33-kDa subunit of the oxygen-evolving complex (OE33) (125), spinach plastocyanin (PC) (111), pea 23-kDa subunit of the oxygen-evolving complex (OE23) (142), spinach 16-kDa subunit of the oxygen-evolving complex (OE16) (56), and barley PSI subunit N (PSI-N) (65) are shown. The positions of the A region (A), N region (N), H region (H), and C region (C) of the lumenal targeting domains are indicated. The hydrophobic core of the H region is *underlined*. The twin arginine motif of the N-regions and the basic residue within the C regions of OE23, OE16, and PSI-N are in *boldface* type.

addition to an import signal that directed transport to the thylakoids. Several integral thylakoid membrane proteins, such as the CF_0II subunit of the ATP synthase (88) and the X and W proteins of photosystem II (PSII) (60, 77), also appear to contain bipartite transit sequences based on structural similarities to the transit sequences of lumenal proteins.

In the second configuration, the transit sequences of integral membrane proteins, such as the precursor to the light-harvesting chlorophyll *a/b* binding protein (preLHCP) and the precursor to the 20-kDa subunit of the CP24 complex (15), contain only information for envelope transport (Figure 1). The signals for targeting of these proteins to the thylakoid appear to reside within the primary structures of the mature polypeptides (75).

ENVELOPE TRANSLOCATION AND THYLAKOID TARGETING ARE INDEPENDENT PROCESSES   The discovery that nuclear-encoded thylakoid proteins possess separate targeting signals for chloroplast import and thylakoid targeting led to the hypothesis that these two targeting processes proceed independent of one another (42). This hypothesis has been supported by several subsequent experimental observations. First, the demonstration of protein transport into isolated thylakoids confirmed that import at the envelope and thylakoid transport were not obligatorily coupled processes (23, 61). Second, time course analysis of the import and thylakoid targeting of preLHCP (105), prePC (122), preOE23 (7), preOE33 (55, 61), and preOE16 (55) in in vitro assays using intact chloroplasts revealed soluble stromal intermediates. These intermediate forms accumulated if envelope translocation was normal, but thylakoid transport was inhibited (25, 28, 47, 66, 71, 90). The intermediates represented a productive step in the targeting process because they were subsequently targeted to thylakoids if the inhibitor was removed (27, 28, 31, 47, 105). The existence of a stromal intermediate has been confirmed for the targeting pathway for cytochrome c_6 using pulse/chase labeling studies in *Chlamydomonas* cultures (52). These results provide in vivo evidence for the two-step pathway of thylakoid targeting. A final set of evidence for the independence of import and thylakoid targeting was provided by the observation that mature LHCP (28, 134) and intermediate size forms of preOE23 (31, 54, 90) and preOE33 (54) that retain their LTDs, but lack STDs, are transported into isolated thylakoid membranes. Therefore, the stromal targeting signal is not required for thylakoid targeting, although its removal is not necessarily a prerequisite for thylakoid transport (8, 12, 23, 61).

THE TRANSIT SEQUENCE DIRECTS IMPORT ACROSS THE CHLOROPLAST ENVELOPE Evidence for the role of the transit sequence in stromal targeting was provided by early in vitro import experiments that demonstrated that N-terminal deletion mutants of thylakoid integral and lumenal preproteins were not imported into isolated intact chloroplasts (21, 42, 45, 69, 134). Structural analysis of the transit sequences of different preproteins revealed no similarities in primary structure; however, the overall characteristics of the STDs share similarity to the transit sequences of nuclear-encoded stromal proteins (139). These characteristics include a size range from approximately 30 to 70 amino acids, an overall basic charge, a high proportion of hydroxylated amino acids, and a deficiency in acidic amino acids. Subsequent studies indicated that the transit sequences of different thylakoid and stromal proteins are functionally identical. For example, interchange of the presequences of prePC, and preOE16, preOE23, and preOE33 by the construction of recombinant chimeric preproteins, had little effect on the import of these proteins into the stroma (22). Likewise, the transit sequences of thylakoid proteins direct the efficient import of the mature domains of stromal preproteins (45, 122) or foreign passenger proteins (87) into isolated chloroplasts. The functional similarity of the transit sequence was confirmed by the demonstration that thylakoid and stromal proteins compete for import into isolated chloroplasts (28, 97). These results, coupled with the observation that the energy requirements for the import of stromal and thylakoid proteins are similar (32), have led to the hypothesis that all proteins destined for internal compartments of the chloroplast use a similar general import pathway at the envelope.

LUMENAL TARGETING DOMAINS RESEMBLE BACTERIAL SIGNAL PEPTIDES   The C-terminal LTDs of lumenal thylakoid proteins exhibit significantly more hydrophobic characteristics than STDs or the transit sequences of stromal preproteins (6a, 139). This observation provided the first suggestion that these domains may function as separate targeting signals. The function of the LTD in thylakoid targeting was confirmed by the demonstration that deletion of this domain from preOE33 (69) and prePC (42) resulted in accumulation of these preproteins in the stroma in in vitro chloroplast import assays. Conversely, transfer of the LTD to stromal preproteins or foreign passenger proteins results in localization of these proteins to the thylakoid membrane (22, 42, 69, 70, 87). However, it should be noted that chimeric proteins containing fusions between LTDs and passenger domains derived from different polypeptides vary widely in the efficiency of thylakoid targeting (22, 42). In addition, truncations of the C-termini of the mature domains of preOE23 (110) and preOE33 (70) can influence the efficiency of thylakoid targeting. These results suggest that although the LTDs are necessary and sufficient for thylakoid targeting, they may have evolved in the context of their corresponding mature sequences to optimize thylakoid transport of the specific passenger polypeptide.

LTDs have been divided into four regions based on the clustering of particular classes of amino acids within their primary structures (46) (Figure 1). Three of these domains display a remarkable similarity in characteristics to the domain structures proposed for the signal peptides of proteins targeted to the

ER or the bacterial plasma membrane (44). LTDs and signal peptides both contain a short, positively charged N-terminal region (N-domain), a hydrophobic core region of 12–18 amino acids (H-region), and a polar C-terminal region (C-domain) (137, 138). These structural similarities led to the first proposals that the LTDs of lumenal preproteins were equivalent to signal peptides and that the pathway of thylakoid targeting for this class of proteins was conserved from the original bacterial endosymbiont (139). The fourth domain of LTDs is not found in signal peptides. This A-domain consists of an extended acidic domain of approximately 12–15 amino acids that separates the LTD from the STD of the transit sequence.

Additional support for the conservative origin of LTDs is provided by the observation that the thylakoid lumenal processing peptidase (TPP) (43) that cleaves the LTD following membrane transport is similar to the bacterial leader (signal) peptidase (LPP). *Escherichia coli* LPP and the TPP correctly cleave the LTDs of lumenal proteins as well as signal peptides of both bacterial and eukaryotic proteins (2, 44), indicating a similar site specificity for both enzymes. Small neutral residues are invariably found at positions −3 and −1 in the C-domain of the LTD relative to the N-terminal amino acid of the mature protein, and these residues are necessary for processing (121). A relatively conserved motif of Ala-X-Ala^X can be assigned to the cleavage site (138).

THYLAKOID TARGETING SIGNALS FOR INTEGRATION OF LHCP ARE INTRINSIC TO THE MATURE PROTEIN   Unlike lumenal proteins, the transit sequence of pre-LHCP is unable to direct a foreign passenger protein to the thylakoid membrane. The mature sequences of the Rubisco small subunit (RBCS) (45) or neomycin phosphotransferase II (131) were targeted to the stroma in in vitro import experiments when fused to the transit sequence of preLHCP. In contrast, chimeric proteins consisting of mature LHCP fused to the transit sequence of preRBCS are efficiently imported into isolated chloroplasts and properly assembled into light-harvesting complexes (75). Therefore, the preLHCP transit sequence only contains an STD, and the targeting information for the thylakoid membrane resides within the mature sequence of LHCP. The analysis of deletion mutants of preLHCP suggests that the signals for thylakoid integration reside within one or more of the three transmembrane domains of the protein (45, 53, 75). Deletion of any portion of these domains dramatically decreases the productive association of LHCP with the thylakoid, indicating that integration of this protein may require the cooperation of a complex set of signals.

The Targeting Signals of Plastid-Encoded Thylakoid Proteins

Approximately 50% of thylakoid membrane proteins are encoded by the plastid genome in plants and green algae. However, there is less information available on the signals and pathways for thylakoid targeting of these proteins than their nuclear-encoded counterparts. The best studied of these proteins is the cytochrome f component of the cytochrome $b_6 f$ complex that couples electron transport between PSII and PSI. Cytochrome f is synthesized as a preprotein containing an N-terminal extension that resembles the LTD of nuclear-encoded lumenal proteins and the signal peptides of bacterial and eukaryotic secreted proteins. In vitro import experiments confirm that the N-terminal presequence is necessary for targeting to the thylakoids and indicate that the presequence functions as an LTD (95, 150). Integration of cytochrome f into the thylakoid membrane must rely on a secondary signal that is likely to reside within the mature sequence of the protein. The function of this signal is presumably similar to the stop-transfer sequences of membrane proteins of the secretory pathway.

Two additional components of the cytochrome $b_6 f$ complex, cytochrome b_6 and subunit IV, are plastid-encoded. Both proteins are synthesized as mature polypeptides without cleavable signal sequences, indicating that their thylakoid targeting signals reside within the mature polypeptide. Zak et al (150) fused both proteins to the transit sequences of nuclear-encoded stromal proteins to target the proteins into chloroplasts in in vitro import assays and studied their subsequent integration into the thylakoid membrane. They demonstrated that the intrinsic signals of cytochrome b_6 and subunit IV function like uncleaved versions of the cytochrome f signal peptide rather than the intrinsic thylakoid targeting domains of preLHCP. The exact location of targeting signals for these proteins remains to be defined.

A number of other thylakoid integral membrane proteins are plastid encoded and appear to be synthesized on thylakoid membrane-bound ribosomes. Targeting of this class of proteins is difficult to study because in vitro assays for the targeting of nascent chain-ribosome complexes to the thylakoid have not been developed. The D1 protein of PSII is representative of this class of membrane proteins. D1 is synthesized without an N-terminal presequence. Targeting of the nascent chain-ribosome complex to the membrane presumably is mediated by an intrinsic noncleavable signal (62), but the nature of this signal has not been investigated.

THE MECHANISM OF PROTEIN IMPORT
AT THE CHLOROPLAST ENVELOPE

The Stages of Import

The import of nuclear-encoded proteins across the chloroplast envelope has been investigated largely using stromal proteins as substrates. However, import competition studies (14, 103) and transit sequence swapping experiments (45, 122) indicate that stromal and thylakoid preproteins are transported from the cytoplasm into the organelle by a common recognition and translocation machinery at the chloroplast envelope. This section is a brief summary of current knowledge of this process with a focus on the components of the import machinery. Although different names have been assigned to the same components of the import machinery by various groups, these investigators recently have agreed on a uniform nomenclature for these proteins (117). This nomenclature has been adopted for the descriptions in this article in accordance with the recently published guidelines.

The import of nuclear-encoded proteins into the chloroplast stroma is mediated by the coordinate interactions of import complexes (translocons) in the outer and inner chloroplast envelope membranes (38, 78). The Toc complex (translocon at the outer envelope membrane of chloroplasts) mediates the recognition of cytoplasmic preproteins at the chloroplast surface and their subsequent translocation across the outer membrane (117). Translocation across the inner membrane is mediated by the Tic complex (translocon at the inner envelope membrane of chloroplasts) (117).

The import reaction can be divided into three steps based on analysis of the energetics of preprotein binding and translocation at the envelope. First, the cytoplasmic preprotein specifically associates with the envelope membrane via the interaction of the transit sequence with proteinaceous receptors at the surface of the outer membrane (29, 37, 80, 104). This interaction appears to be reversible and energy independent. Second, the preprotein inserts across the protein-conducting machinery of the outer membrane. This step requires the hydrolysis of low concentrations of both ATP and GTP (less than 100 μM) in the cytoplasm or in the intermembrane space of the envelope (59, 80, 99, 100). Insertion across the outer envelope membrane triggers the association of the outer and inner import machineries and brings the transit sequence near to import components of the inner membrane (80, 116). The functional association of the outer and inner membrane import machineries at this step in import occurs at contact sites where the two membranes are held in close proximity (116). Outer membrane insertion is likely to represent the committed step in the import reaction because the interaction of preproteins with the envelope at this stage is irreversible (37). Preproteins bound to the envelope at this stage in import have been designated early import intermediates (104, 116).

Finally, the preprotein inserts across the inner membrane, and translocation into the stroma proceeds simultaneously across both envelope membranes at contact sites (116). This step requires the hydrolysis of ATP in the stromal compartment (101, 126). Proteins in transit across both envelope membranes have been isolated and are referred to as late import intermediates. Unlike mitochondrial protein import or bacterial protein export, chloroplast protein import does not require a membrane potential (101, 126). Upon translocation across the envelope, the transit sequences of stromal and thylakoid preproteins undergo proteolytic processing by a specific soluble metalloendopeptidase, designated the general stromal processing peptidase (SPP) (1, 108, 132). Proteolytic processing of lumenal thylakoid proteins by SPP yields soluble stromal intermediates that have lost the STD of their transit sequence but retain their LTDs (55), whereas SPP processing of preLHCP yields a mature-size stromal intermediate (20).

Components of the Import Machinery

A TRIMERIC COMPLEX FORMS THE CORE OF THE OUTER MEMBRANE TRANSLOCON   Candidates for components of the preprotein recognition and membrane translocation machinery of the envelope have been identified, and several lines of evidence provide clues to their functions in the import process (Figure 2).

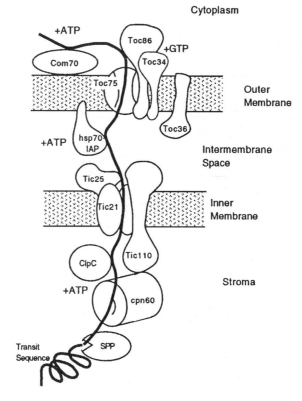

Figure 2   Schematic representation of the protein import machinery of the chloroplast envelope. Protein import is mediated by the coordinated action of translocon complexes in the outer membrane (*Toc complex*) and inner membrane (*Tic complex*). The Toc and Tic complexes associate at envelope contact sites to facilitate direct transport of preproteins from the cytoplasm to the stromal compartment (see text for discussion). The names of the known import components are indicated.

Remarkably, these studies have revealed that the structures of the import components and the mechanism of envelope translocation is unrelated to protein import into mitochondria. The three integral membrane proteins, Toc34 (IAP34, OEP34), Toc86 (IAP86, OEP86), and Toc75 (IAP75, OEP75) (117), form a trimeric complex at the core of the outer membrane translocon. Toc86 and Toc34 are related in primary structure and contain cytoplasmic GTP-binding domains (50, 59, 119). The topology of these two components and the observation that GTP hydrolysis is required at the early stages of import (59, 99) suggest that Toc34 and Toc86 may function in regulating the recognition of the cytoplasmic preprotein at the outer membrane. Support for the roles of Toc34 and Toc86 as a preprotein receptor system is twofold. First, covalent cross-linking studies indicate that Toc86 and Toc34 are in intimate contact with the precursors during the initial binding to the outer envelope membrane (80, 104; Kouranov & Schnell, unpublished data). Second, anti-Toc86 IgGs block the binding of preproteins to the envelope (50). It has been proposed that Toc34 and Toc86 act together in a manner analogous to the functions of the signal recognition particle (SRP) and its receptor in targeting nascent polypeptides to the translocation machinery of the ER (143). In this hypothesis, Toc34 and Toc86 would regulate the presentation of the cytoplasmic preprotein to the protein-conducting machinery of the translocon through cycles of GTP binding and hydrolysis (118).

Toc75 has been shown to cross-link directly to an early intermediate in import when the preprotein is inserted across the outer membrane (80, 104). In addition, antibodies to Toc75 block the import of precursor proteins into intact chloroplasts (128). Toc75 is deeply embedded in the outer membrane (118, 128). These observations indicate a direct role for Toc75 in protein import and suggest that the protein may function as a constituent of the protein-conducting channel in the outer membrane translocon.

CHAPERONES PARTICIPATE IN OUTER MEMBRANE TRANSLOCATION   One notable similarity between the translocation machinery of the outer envelope membrane and other membrane translocation systems that have been studied is the presence of molecular chaperones of the hsp70 family. Two hsp70 homologues, Com70 (67, 73, 145) and hsp70 IAP (84, 118, 141), are localized at the outer and inner face of the outer envelope membrane, respectively. The presence of two hsp70 homologues at the outer membrane has been proposed to account for the requirement of ATP at the early stages of import (118); however, this has not been demonstrated directly.

Com70 is bound to the cytoplasmic surface of the outer membrane and is closely related in primary structure to the major cytoplasmic hsp70s (67). It can be covalently cross-linked to a chimeric envelope-bound preprotein consisting of preOE33 fused to dihydrofolate reductase (145), and its role in import has been confirmed by the observation that antibodies to Com70 inhibit protein import into the chloroplast (73). Although the exact function of Com70 remains to be determined, its location and activities suggest that it may act to maintain the import competence of the preproteins at the cytoplasmic face of the outer membrane by binding to preproteins at the early stages of import. This role would be similar to that proposed for the cytoplasmic hsp70s in translocation into mitochondria and the ER (115). Unlike Com70 and other cytosolic chaperones, the hsp70 IAP is tightly anchored to the inner face of the outer membrane (84, 118). It has been proposed that the hsp70 IAP binds to the precursor protein as it emerges across the outer membrane, thereby maintaining the import competence of the polypeptide until the transit sequence engages the inner membrane import machinery (118).

Evidence also has been presented demonstrating that preLHCP requires a cytoplasmic hsp70 for targeting and import to the chloroplast envelope (140). Cytoplasmic hsp70s are essential for protein translocation at the ER and mitochondria because they maintain the unfolded import competent state of preproteins (18, 33). A similar mechanism is likely for their role in preLHCP targeting.

SEVERAL COMPONENTS OF THE INNER MEMBRANE TRANSLOCON HAVE BEEN IDENTIFIED   The structure of the inner membrane translocon has not been studied in detail. However, several components of the import machinery have been identified. Two components of the chloroplast envelope, Tic(21) and Tic22, have been covalently cross-linked to preproteins at an intermediate stage in import (80). Tic22 is peripherally associated with the outer face of the inner membrane, suggesting that it may serve as a receptor for preproteins at the inner membrane (Kouranov & Schnell, unpublished data). Tic(21) is an integral inner membrane protein, making it a candidate for a component of the translocation channel in the inner membrane (80). An additional integral inner membrane component, Tic110 (IAP100, IEP110, Cim97), has been detected in complexes containing proteins at late stages in import (79, 118, 145). Anti-Tic110 coprecipitates two stromal chaperones, the hsp60 homologue, cpn60 (58), and the chloroplast ClpC homologue (93). These observations suggest that Tic110 may function as a docking site for molecular chaperones that participate in driving transport across the inner membrane or in folding of newly imported proteins in the stroma.

An additional envelope component, Toc36 (Bce44B, Com44), is present in a covalently cross-linked complex containing an envelope-bound chimeric preprotein (145). Toc36 appears to be one member of a family of immunologically related envelope proteins that is localized in both the outer and inner envelope membranes (68). The functions of the Toc36 family of proteins in import remains to be investigated.

MULTIPLE MECHANISMS FOR THYLAKOID TARGETING

The Energy Requirements for Thylakoid Targeting

Early in vitro targeting experiments using intact chloroplasts established that thylakoid targeting was an energy-dependent process. Newly imported thylakoid proteins were shown to accumulate in the stroma in the presence of protonophores that dissipate the pH gradient (ΔpH) across the thylakoid membrane (25, 90). Unfortunately, in organello assays could not distinguish whether the inhibition of transport by protonophores was due to a direct effect on the proton gradient or an indirect effect due to the uncoupling of electron transport and ATP synthesis. The advent of targeting assays with isolated thylakoid membranes allowed the energy requirements for targeting to be determined for a number of thylakoid proteins. These experiments clearly showed that the effects of protonophores were direct for certain preproteins and indirect in other cases.

The possibility of a single thylakoid targeting mechanism similar to that established for envelope translocation was quickly challenged because different proteins were shown to have distinct energy requirements for targeting to this membrane. Surprisingly, even proteins that possessed structurally similar LTDs were shown to fall into three targeting classes based on the energy requirements for translocation. The existence of multiple targeting mechanisms was further substantiated by targeting competition studies that separated various proteins into distinct transport classes. It is now clear that at least four distinct pathways for targeting to the thylakoid exist and that these differences reflect fundamentally distinct targeting mechanisms (Figure 3).

THE ATP-DEPENDENT AND ΔPH-DEPENDENT PATHWAYS   The first pathway for thylakoid targeting is represented by the nuclear-encoded lumenal proteins, prePC, and preOE33, the nuclear-encoded precursor to the F subunit of PSI (prePSI-F), an integral membrane protein, and the plastid-encoded membrane protein, precytochrome *f*. Translocation of these proteins into the thylakoid lumen is absolutely dependent upon ATP and is stimulated by the transthylakoidal ΔpH (23, 54, 57, 61, 82, 91, 147). This pathway is referred to as the ATP-dependent pathway. The second pathway is characterized by a singular requirement for the ΔpH. This pathway does not require nucleoside triphosphates and is represented by the nuclear-encoded lumenal proteins, preOE23, and preOE16, and the precursor to subunit N of PSI (prePSI-N) (13, 24, 64, 82, 90, 94).

The apparent similarity in LTDs for the proteins of the ATP- and ΔpH-dependent pathways raised the immediate question of the nature of the determinant for pathway specificity. Was pathway specificity conferred by the LTD or the passenger polypeptide? To address this question, the energy requirements of

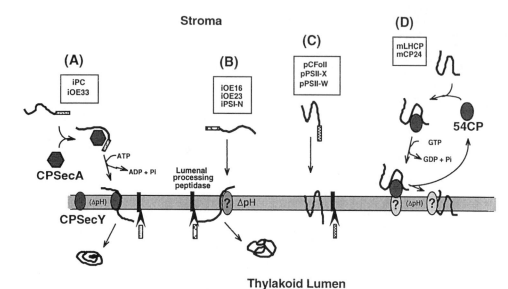

Figure 3   Schematic representation of the four known pathways of protein targeting to the thylakoid. Representatives of proteins that are transported on each pathway are indicated in the boxed regions (i, intermediate; m, mature; p, precursor protein). (*A*) The Sec pathway is a homologue of the bacterial preprotein translocase. Transport requires ATP hydrolysis and is stimulated by a pH gradient across the thylakoid membrane. (*B*) The ΔpH-dependent pathway requires proteinaceous components at the thylakoid membrane and relies exclusively on the transthylakoidal pH gradient. (*C*) The spontaneous pathway does not appear to require proteinaceous receptors or an energy source for transport. (*D*) The SRP pathway involves a soluble targeting factor (54CP) that is similar to the mammalian and bacterial signal recognition particles. Targeting is driven by a GTP hydrolysis cycle at the 54CP.

thylakoid import were determined for chimeric constructs in which the LTDs of preproteins from the ΔpH- and ATP-dependent pathways were switched. In one scenario, the transit sequences of the ΔpH-dependent preproteins, preOE23 or preOE16, were fused to the mature sequence of plastocyanin (47, 107). The plastocyanin passenger was transported into the thylakoid in both cases, and transport was dependent only on the ΔpH. In the converse approach, the transit sequences of preOE33 (47, 83) or prePC (83) were shown to switch the transport of mature OE16 to the ATP-dependent pathway, albeit with low efficiency. The results of these transit sequence swapping experiments clearly demonstrated that pathway specificity is determined by the transit sequence. It is interesting to note that the transport of the preOE33-OE16 and prePC-OE16 fusions was more sensitive to the protonophore, nigericin, than that of authentic preOE33 or prePC, suggesting that the passenger protein can influence the contribution of the ΔpH on the ATP-dependent pathway (47, 83).

How could similar targeting signals direct targeting along two different pathways? A reexamination of the LTDs of preproteins from both pathways revealed the presence of a unique twin arginine motif within the N-domain of the LTD of preproteins targeted via the ΔpH-dependent pathway that was not found in other thylakoid transit sequences (Figure 1) (11, 17). Substitution of the Arg-Arg sequence with Gln-Gln, Arg-Lys, or Lys-Arg blocked or drastically inhibited thylakoid targeting. It is apparent from these results that the specific dipeptide and not simply a basic cluster are essential for the targeting signal.

Two groups (10, 46) attempted to further define the pathway specific motifs of the LTDs of preproteins for the ATP- and ΔpH-dependent pathways by studying the transport of preproteins containing mutant and chimeric LTDs. These studies demonstrated that both N- and H-domains are essential for targeting to both pathways. Transport on the ΔpH pathway required an N-domain with a twin arginine motif, but was tolerant of changes within the H-domain. In contrast, transport on the ATP-dependent pathway required the H-domain of a preprotein from the ATP-dependent pathway, but the sequence requirements of the N-domain were flexible with the only requirement being an overall positive charge. The investigators also noted that the ATP-dependent pathway was less tolerant of passenger proteins from the ΔpH pathway although the ΔpH pathway efficiently transported plastocyanin or OE33 as a passenger protein. This phenomenon previously had been observed with other chimeric constructs (47, 83), and Henry et al (46) speculated that the ΔpH pathway may have evolved to transport proteins that were incompatible with the other transport system. Bogsch et al (10) have termed these basic amino acids that flank the H-domain a "Sec-avoidance" motif and suggested that they have evolved to avoid nonproductive interactions of passenger proteins, such as OE23, with the ATP-dependent system. This hypothesis is reinforced by the observation that removal of the twin arginine motif and a second lysine in the C-domain of preOE23 generates an LTD that is completely compatible with the ATP-dependent system (10).

Additional support for the role of the passenger protein in the evolution of the ΔpH- and ATP-dependent pathways is provided by the observation that the two pathways differ in the substrate conformations required for translocation. In general, the translocation of preproteins across membranes is thought to require an unfolded conformation that allows linear transport of the polypeptide through the protein-conducting machinery in the membrane bilayer (115). This hypothesis is supported for translocation at the chloroplast envelope and on the ATP-dependent pathway for thylakoid transport by the observations that tightly folded preproteins block membrane transport (1a, 34a, 40a). In contrast, evidence has been presented that the ΔpH-dependent pathway is tolerant of partially or fully folded translocation substrates. The stromal intermediate form of OE23 assumes a tightly folded conformation, and a specific, folded conformation is required for efficient membrane translocation at the thylakoid membrane (31). Recently, Clark & Theg (21a) showed that a chimeric preprotein consisting of preOE16 fused to bovine trypsin inhibitor (BPTI) was efficiently transported into thylakoids even when the tertiary structure of the BPTI moiety was stabilized by internal disulfide bridges. These data suggest that the protein-conducting machinery of the ΔpH-dependent pathway can accommodate folded proteins. The molecular basis of translocation mechanism remains to be defined, but the protein-conducting machinery of the ΔpH-dependent pathway appears to maintain a tight seal because transport does not disrupt the electrochemical potential at the thylakoid membrane (SA Teter & S Theg, personal communication).

SEVERAL INTEGRAL MEMBRANE PROTEINS INTEGRATE INDEPENDENT OF ENERGY INPUT   The CF_0II subunit of the thylakoid ATP synthase and the PSII-X and -W proteins of PSII are nuclear-encoded and synthesized with bipartite transit sequences similar to the transit sequences of lumenal proteins. Interestingly, transport of these proteins to the thylakoids does not require ATP and is insensitive to protonophores, suggesting that their integration represents a distinct import pathway (60, 77, 88). The lack of an energy requirement for their integration is suggestive of the spontaneous mechanisms proposed for the targeting of several proteins to the chloroplast and mitochondrial outer membranes (115). Additional studies will be necessary to determine whether CF_0II, PSII-X, and PSII-W use a comparable mechanism for thylakoid integration. The PSI-H, -K, and -L subunits of PSI also appear to undergo energy-independent integration into the thylakoid membrane (82). These components are nuclear-encoded but do not possess cleavable LTDs. The targeting of these proteins has not been studied in detail, and therefore it is not clear whether they may share a targeting pathway with CF_0II, PSII-X, and PSII-W.

LHCP TARGETING REQUIRES GTP   The original studies on the targeting of LHCP indicated that ATP was required for integration into isolated thylakoid membranes and that the ΔpH had a stimulatory effect (23–25, 134). Upon reexamination, however, the nucleoside triphosphate requirement was demonstrated to be specific for GTP. Hoffman & Franklin (51) showed that low concentrations of exogenous GTP were more effective than ATP in promoting LHCP targeting after the components of in vitro targeting assays were pretreated to ensure the removal of endogenous free nucleotides. Furthermore, slowly hydrolyzable analogs of GTP blocked the ability of ATP to promote transport, whereas slowly hydrolyzable analogs of ATP had a minor effect on GTP-stimulated targeting. The role of GTP in LHCP targeting clearly distinguishes this pathway from the ATP-dependent pathway for proteins with cleavable LTDs and brings to four the potential mechanisms for targeting to the thylakoid membrane.

The Components of the Thylakoid Targeting Pathways

COMPETITION STUDIES CONFIRM TARGETING SUBCLASSES   The existence of subclasses of thylakoid proteins with distinct energy requirements for transport presented a strong argument for the existence of distinct transport pathways. Definitive evidence for multiple mechanisms was provided by the development of an in organello competition assay for thylakoid transport (28). The relatively high rate of preprotein import across the envelope compared to the rate of thylakoid targeting results in the accumulation of saturating levels of stromal intermediates for nuclear-encoded thylakoid proteins in in vitro assays. Under these conditions, it was demonstrated that the stromal intermediate of OE23 selectively inhibited OE16 targeting, and the OE33 stromal intermediate competed for the targeting of intermediate prePC. LHCP targeting was not competed by either intermediate OE23 or OE33. In a separate study, intermediate OE23 was shown not to compete for CF_oII integration (88). Thus, the results of competition assays correlated directly with the distinct energy requirements for each class of thylakoid protein and supported the existence of multiple targeting pathways.

THE ATP-DEPENDENT PATHWAY IS A HOMOLOG OF THE BACTERIAL SEC PATHWAY The structural similarity of thylakoid LTDs to bacterial signal peptides and the marked similarity in energy requirements between the ATP-dependent thylakoid targeting pathway and bacterial protein export provided compelling indications of an evolutionary relationship between thylakoid targeting and bacterial protein export. Additional evidence for a bacterial origin of this pathway was provided by two observations. First, homologues of two components of the E. coli preprotein translocase, SecA and SecY, are encoded by the plastid genomes of certain algae (35, 106, 113, 114, 130), suggesting that a "Sec" pathway exists in plastids. Second, preOE33 or forms of OE33 and plastocyanin containing only the LTD were efficiently exported to the periplasmic space and properly processed to their mature sizes when expressed in E. coli (41, 86, 120). Likewise, the N-terminal region of precytochrome f was capable of targeting β-galactosidase to the E. coli cytoplasmic membrane (112). These results provided unequivocal evidence that the LTD was functionally equivalent to the E. coli signal sequence.

The functional correlations among LTDs and signal peptides provided the impetus for experiments to directly test whether the ATP-dependent targeting pathway represented a homologue of the E. coli Sec system. In E. coli, the SecA ATPase is a major cytoplasmic factor that binds preproteins, targets the complex to the cytoplasmic membrane, and facilitates membrane translocation (144). A hallmark of Sec-dependent export is the ability of azide to inhibit the reaction by blocking the SecA ATPase (98). Azide also inhibits the ATP-dependent pathway of thylakoid targeting (47, 66). Early investigations established that transport of preproteins into isolated thylakoids via the ATP-dependent pathway requires the presence of stromal extract in addition to ATP (54, 61, 91), and given the azide sensitivity of transport, a SecA homologue was a logical candidate for the stromal factor. Yuan et al (149) purified a stromal factor that could completely substitute for stromal extract in the in vitro transport of preplastocyanin and preOE33. Antibodies to algal SecA cross-reacted with the stromal factor, providing direct evidence that it corresponded to a chloroplast SecA homologue. Concomitantly, Nakai and colleagues (92, 96) cloned a SecA homologue from pea and demonstrated that anti-SecA inhibited the transport of OE33, but not OE23, into thylakoids. Like the *E. coli* protein, the chloroplast SecA homologue, cpSecA, is a homodimer of approximately 110 kDa subunits (149). It is nuclear-encoded, contains a typical envelope transit sequence, and exhibits 43–49% identity to bacterial SecA proteins and 60–65% identity to cyanobacterial SecA (9, 96). These results confirmed that the ATP-dependent pathway for thylakoid targeting represents a chloroplast Sec system.

A cDNA for a homologue of the *E. coli* integral membrane protein, SecY, has been identified in Arabidopsis (74). SecY is a component of the Sec YEG complex that forms the protein-conducting channel of the *E. coli* preprotein translocase (144). In *E. coli*, the preprotein-SecA complex binds to this membrane complex during preprotein targeting to initiate translocation. Like cpSecA, cpSecY is nuclear-encoded. Laidler et al (74) demonstrated that Arabidopsis SecY was imported into isolated chloroplasts and targeted to the thylakoid membrane, providing compelling evidence that it functions at this membrane. A role for cpSecY in thylakoid targeting has not been demonstrated directly, but its discovery contributes to the growing evidence for the function of a Sec translocase in the transport of nuclear-encoded lumenal proteins, such as prePC and preOE33, and the plastid-encoded membrane protein, cytochrome *f*.

LHCP IS TARGETED BY A PATHWAY HOMOLOGOUS TO THE SRP PATHWAY FOR TARGETING TO THE ER   The targeting of LHCP to isolated thylakoids also requires a stromal extract (19, 39). The first step toward identifying the stromal factor(s) for the LHCP pathway was the observation that the stromal intermediate of LHCP was present in a large soluble complex of approximately 120 kDa (102). Two clues pointed to the identity of a stromal component of the transit complex: LHCP targeting required GTP (51), and a stromal homologue of the GTP-binding subunit of the SRP had been identified in pea chloroplasts (36). SRP is the cytoplasmic ribonucleoprotein particle that targets the nascent chain-ribosome complex of secretory proteins to the ER during cotranslational protein translocation (143). The chloroplast SRP homologue was designated 54CP.

The presence of the chloroplast 54CP in the LHCP transit complex was confirmed by immunoprecipitation and covalent cross-linking experiments (76). The 54CP was shown to be an essential component of the transport pathway because immunodepletion of the polypeptide from stromal extracts inhibited formation of the transit complex and blocked LHCP integration into isolated thylakoids. The 54CP binds tightly to the third transmembrane domain of LHCP (49), consistent with the observation that the targeting signals for LHCP integration are located within the transmembrane domains. The mature 54CP is a 53-kDa polypeptide that is encoded by a nuclear gene. It exhibits 27% identity with the 54-kDa subunit of mammalian SRP and 44% identity with the bacterial SRP homologue, Ffh.

High et al (49) have investigated the specificity of 54CP binding to a variety of thylakoid preproteins by covalent cross-linking. As expected, 54CP binds strongly to LHCP but does not interact detectably with OE23 or OE33. These results are consistent with the hypothesis of distinct targeting pathways. However, cross-linking also was observed between 54CP and two additional thylakoid polypeptides, precytochrome f and the Rieske FeS protein, a nuclear-encoded preprotein with an uncleaved thylakoid targeting domain (34, 81). The targeting pathway of the Rieske FeS protein has not been clearly defined, but cytochrome f previously had been assigned to the Sec pathway. High et al (49) suggested that cytochrome f may be able to use both SRP and Sec pathways or that certain components of each pathway may act sequentially during targeting. However, 54CP has not been shown to participate in cytochrome f targeting, and therefore the significance of the cross-linking results must await the demonstration that the cytochrome f-54CP interaction is relevant to the cytochrome f targeting reaction.

The exact function of 54CP is not known, but recent data suggest parallels between its activities and those of mammalian SRP (49). For example, the nascent chain-ribosome complex of preprolactin can be covalently cross-linked to 54CP when added to a stromal extract. In addition, the increased affinity of 54CP binding for targeting signals is correlated with increased hydrophobicity of the targeting signal, a property shared with mammalian SRP. On the basis of these observations, it is likely that the binding determinants for 54CP and mammalian SRP are similar.

Mammalian SRP acts as a multisubunit soluble targeting factor that binds to the signal sequences of nascent polypeptide chains and delivers the translation complex to the protein-conducting channel at the membrane of the rough ER (143). The targeting cycle of SRP is regulated by GTP binding and hydrolysis at its 54-kDa subunit (3, 89). It has been proposed that 54CP acts as a specific targeting factor in LHCP integration by a similar mechanism (26, 36). However, comparisons of 54CP with mammalian SRP or bacterial Ffh raise important questions. For example, does 54CP act alone or as a component of a macromolecular complex? Both SRP and Ffh contain an essential 7S RNA component, and SRP contains five additional protein subunits (143). An RNA component of 54CP has not been discovered, but Cline & Henry (26) have reported that native 54CP exists as a large complex and not a monomer. Furthermore, SRP and Ffh act cotranslationally to target nascent chain-ribosome complexes to the membrane. In the case of LHCP, the activity of 54CP is clearly posttranslational. Perhaps the lack of a need to interact with the ribosome has streamlined the evolution of 54CP into a targeting factor that lacks the structural features that were required for the ribosome binding and elongation arrest functions of its mammalian and *E. coli* counterparts. It will be of interest to determine whether 54CP plays a role in the targeting of proteins that are cotranslationally integrated into the thylakoid, such as the D1 protein of PSII (62, 133).

A role for 54CP in thylakoid targeting is clear. However, it also is clear that the LHCP transit complex requires the presence of additional stromal factors to target to the thylakoid membrane. What are these additional factors? CpSecA does not appear to play a role because the purified cpSecA protein is unable to substitute for the requirement of stromal extract in in vitro targeting reactions, and LHCP targeting is azide insensitive (149). It has been proposed that the additional requirement may be the stromal hsp70 chaperone (146). However, stromal hsp70s do not appear to be components of the transit complex, and immunodepletion of hsp70 from stromal extracts does not affect the ability of the extract to support targeting (148). The nature of the additional factors for LHCP targeting remains to be defined, and the availability of reconstituted transit complex and thylakoid targeting systems provides the biochemical assay for their isolation.

THE ΔpH PATHWAY IS MEDIATED BY PROTEINACEOUS RECEPTORS   The singular requirement for the proton gradient makes the ΔpH-dependent pathway unique among known membrane translocation systems, but little information is available on the components of this pathway. Treatment of isolated thylakoids with exogenous protease before in vitro transport assays inhibits the transport of preOE23 (31, 108a). In addition, the transport of OE23 and OE16 are competitive and saturable (28). These results suggest that targeting and/or translocation are mediated by proteinaceous components at the thylakoid surface. Recently, a maize mutant, *hcf106,* has been identified that appears to selectively affect the targeting of proteins that use the ΔpH-dependent pathway (135). The isolation of the hcf106 gene is beginning to shed light on one candidate for a component of this pathway (see the following section).

In Vivo Models Support the Assignment of Multiple Targeting Pathways

The recent identification of mutations that affect thylakoid biogenesis in higher plants (6) and green algae (123) supports the assignment of distinct thylakoid targeting mechanisms and is beginning to aid in the identification of transport components. A class of thylakoid assembly (*tha*) mutants are deficient in the assembly of a number of photosynthetic complexes, but the phenotypes do not appear to be due to defects in gene expression (5). The phenotype of the *tha1* mutant is consistent with a defect in the Sec pathway because it selectively affects the OE33, plastocyanin, and cytochrome *f* targeting pathway (135). The steady state levels of plastocyanin, OE33, and PSI-F are reduced in this mutant, and these proteins, as well as cytochrome *f*, were shown to accumulate in the stroma as their intermediate-size forms. The *tha1* gene has recently been isolated and shown to be the maize CPSecA (136), providing in vivo evidence for the existence of the conserved Sec pathway for thylakoid targeting.

A second maize mutant, *hcf106*, also exhibits a defect in the accumulation of photosystem complexes. In contrast to *tha1*, the *hcf106* defect results in the reduced accumulation of OE16 and OE23 and the appearance of higher molecular weight stromal intermediates of these two proteins (135). This mutant also exhibits an unusual thylakoid membrane morphology consistent with a role for *hcf106* in protein transport and membrane biogenesis (85). Analysis of the *hcf106* gene indicates that it encodes a thylakoid membrane protein that exhibits similarity to genes in a variety of bacterial species (120a). The bacterial genes appear to play a role in the secretion of a subset of proteins that use a pathway exclusively dependent on the ΔpH at the plasma membrane. Thus, the ΔpH-dependent pathway also may represent a derivative of a targeting pathway that existed in the original prokaryotic endosymbiont.

The analysis of a set of mutations in the LTD of cytochrome *f* has led to a genetic selection system for the study of thylakoid transport in *Chlamydomonas* (4, 123). One of these LTD mutants, *A15E*, also affects the accumulation of LHCP and the D1 protein but does not affect plastocyanin, OE16, OE23, or OE33 accumulation. These results suggest that the pathways of cytochrome *f*, D1, and LHCP integration may converge at a common element. Although cytochrome *f* appears to utilize the Sec pathway for targeting, Smith & Kohorn (123) have suggested that it may share a component for membrane translocation with other integral membrane proteins. Extragenic suppressors of the cytochrome *f* mutants have been identified, and the genes for several of these thylakoid insertion proteins (tip proteins) have been isolated (KK Bernd & BD Kohorn, personal communication). Their analysis should provide important in vivo evidence to test this hypothesis (4, 123).

PERSPECTIVES

Although the relationships of the various pathways for thylakoid targeting are not yet clear, the identification of components of each pathway provides the necessary tools to unravel the complexity of these elaborate protein targeting systems. To date, studies of the targeting components have focused on soluble targeting factors. With the identification of a SecY homologue and hcf106, the components of the membrane translocation machinery are beginning to reveal their secrets. Other Sec proteins and proteins such as a 54CP receptor are predicted to exist. The in vivo molecular genetic approaches hold particular promise for identifying components of the targeting machinery. Proteins such as hcf106 are candidates for previously unidentified factors that may contribute to the elucidation of the apparently unique ΔpH-dependent pathway. In addition, at least four additional *tha* mutants have been identified that disrupt either the Sec or ΔpH pathways. One of these mutants has been assigned to a gene, tha4, that encodes an integral thylakoid membrane protein similar to hcf106 and its bacterial homologues (A Barkan, personal communication). These data provide additional evidence for the role of this family of proteins in thylakoid targeting, and indicate that a similar, previously unknown pathway exists in prokaryotes.

How independent are the multiple targeting pathways? If the bacterial and ER systems are good precedents, the membrane translocation systems are likely to be the sites of convergence for multiple targeting pathways. For example, the SecA and Ffh targeting pathways operate in parallel in bacteria, but both feed into a common membrane-bound SecYEG translocase. Likewise, SRP-dependent and SRP-independent pathways for targeting to the ER exist, but both pathways utilize the Sec61p translocation complex at the ER membrane. In fact, some evidence suggests that the multiple pathways for thylakoid targeting may overlap at common components or that single proteins may be able to utilize more than one pathway. One of the cytochrome *f* LTD mutants in *Chlamydomonas* shows a dominant negative effect on the integration of LHCP and the D1 protein (123). Although in vitro data indicate separate targeting pathways for these proteins, the analysis of this mutant suggests that they may share a common element at one stage in transport.

Perhaps the most striking revelation from investigations over the past 10 years is the complexity of mechanisms for protein targeting to the thylakoids. As with most biological processes, what could have been modeled as a single stepwise pathway actually requires the operation of a number of distinct parallel mechanisms that reflect the diversity of thylakoid proteins themselves. Why do these multiple pathways exist? The accumulated evidence is consistent with the proposal that the characteristics of the passenger protein (e.g. hydrophobicity or folding properties) may require distinct transport mechanisms (26).

For example, ΔpH-dependent proteins are inefficiently transported by the Sec pathway (47, 83), suggesting an incompatibility between the Sec machinery and these polypeptides. With regard to the role of the 54CP pathway, the Ffh protein has recently been shown to operate in targeting a specific subset of polytopic membrane proteins to the *E. coli* inner membrane (129). By analogy, it is conceivable that 54CP may have the ability to serve both as a targeting signal and as a specialized chaperone for very hydrophobic polypeptides such as LHCP. The definition of these specificities, in concert with the elucidation of the transport components of these systems, should provide knowledge of the essential elements of membrane translocation in these systems and their bacterial and mammalian counterparts.

Visit the *Annual Reviews home page* at
http://www.AnnualReviews.org.

Literature Cited

1. Abad MS, Clark SE, Lamppa GK. 1989. Properties of a chloroplast enzyme that cleaves the chlorophyll a/b binding protein precursor. *Plant Physiol.* 90:117–24

1a. America T, Hageman J, Guéra A, Rook F, Archer K, et al. 1994. Methotrexate does not block import of a DHFR fusion protein into chloroplasts. *Plant Mol. Biol.* 24:283–94

2. Anderson CM, Gray J. 1991. Cleavage of the precursor of pea chloroplast cytochrome f by leader peptidase from *Escherichia coli. FEBS Lett.* 280:383–86

3. Bacher G, Lütcke H, Jungnickel B, Rapaport TA, Dobberstein B. 1996. Regulation by the ribosome of the GTPase of the signal-recognition particle during protein targeting. *Nature* 381:248–51

4. Baillet B, Kohorn BD. 1996. Hydrophobic core but not amino-terminal charged residues are required for translocation of an integral thylakoid membrane protein in vivo. *J. Biol. Chem.* 271:18375–78

5. Barkan A, Miles D, Taylor W. 1986. Chloroplast gene expression in nuclear photosynthetic mutants of maize. *EMBO J.* 5:1421–27

6. Barkan A, Voelker R, Mendel-Hartvig J, Johnson D, Walker M. 1995. Genetic analysis of chloroplast biogenesis in higher plants. *Physiol. Plant.* 93:163–70

6a. Bassham DC, Bartling D, Mould RM, Dunbar B, Weisbeek P, et al. 1991. Transport of proteins into chloroplasts: delineation of envelope "transit" and thylakoid "transfer" signals within the presequences of three imported thylakoid lumen proteins. *J. Biol. Chem.* 266:23606–10

7. Bauerle C, Dorl J, Keegstra K. 1991. Kinetic analysis of the transport of thylakoid lumenal proteins in experiments using intact chloroplasts. *J. Biol. Chem.* 266:5884–90

8. Bauerle C, Keegstra K. 1991. Full-length plastocyanin precursor is translocated across isolated thylakoid membranes. *J. Biol. Chem.* 266:5876–83

9. Berghöfer J, Karnauchov I, Herrmann RG, Klösgen RB. 1995. Isolation and characterization of a cDNA encoding the SecA protein from spinach chloroplasts. *J. Biol. Chem.* 270:18341–46

10. Bogsch E, Brink S, Robinson C. 1997. Pathway specificity for a delta pH-dependent precursor thylakoid lumen protein is governed by a 'Sec-avoidance' motif in the transfer peptide and a 'Sec-incompatible' mature protein. *EMBO J.* 16:3851–59

11. Brink S, Bogsch EG, Mant A, Robinson C. 1997. Unusual characteristics of amino-terminal and hydrophobic domains in nuclear-encoded thylakoid signal peptides. *Eur. J. Biochem.* 245:340–48

12. Brock IW, Hazell L, Michl D, Nielsen WS, Møller BL, et al. 1993. Precursors of one integral and five lumenal thylakoid proteins are imported by isolated

pea and barley thylakoids: optimisation of in vitro assays. *Plant Mol. Biol.* 23:717–25

13. Brock IW, Mills JD, Robinson D, Robinson C. 1995. The delta pH-driven, ATP-independent protein translocation mechanism in the chloroplast thylakoid membrane: kinetics and energetics. *J. Biol. Chem.* 270:1657–62

14. Buvinger WE, Michel H, Bennett J. 1989. A truncated analog of a prelight-harvesting chlorophyll a/b protein II transit peptide inhibits protein import into chloroplasts. *J. Biol. Chem.* 264:1195–202

15. Cai D, Herrmann RG, Klösgen RB. 1993. The 20 kDa apoprotein of the CP24 complex of photosystem II: an alternative model to study import and intra-organellar routing of nuclear-encoded thylakoid proteins. *Plant J.* 3:383–92

16. Cashmore AR. 1984. Structure and expression of a pea nuclear gene encoding a chlorophyll a/b-binding polypeptide. *Proc. Natl. Acad. Sci. USA* 81:2960–64

17. Chaddock AM, Mant A, Karnauchov I, Brink S, Herrmann RG, et al. 1995. A new type of signal peptide: central role of a twin-arginine motif in transfer signals for the delta-pH-dependent thylakoidal protein translocase. *EMBO J.* 14:2715–22

18. Chirico WJ, Waters MG, Blobel G. 1988. 70K heat shock related proteins stimulate protein translocation into microsomes. *Nature* 322:805–10

19. Chitnis PR, Nechushtai R, Thornber JP. 1987. Insertion of the precursor of the light-harvesting chlorophyll a/b-protein into the thylakoids requires the presence of a developmentally regulated stromal factor. *Plant Mol. Biol.* 10:3–11

20. Clark SE, Abad MS, Lamppa GK. 1989. Mutations at the transit peptide-mature protein junction separate two cleavage events during chloroplast import of the chlorophyll a/b-binding protein. *J. Biol. Chem.* 264:17544–50

21. Clark SE, Oblong JE, Lamppa GK. 1990. Loss of efficient import and thylakoid insertion due to N- and C-terminal deletions in the light-harvesting chlorophyll a/b binding protein. *Plant Cell* 2:173–84

21a. Clark SA, Theg SM. 1997. A folded protein can be transported across the chloroplast envelope and thylakoid membranes. *Mol. Biol. Cell* 8:923–34

22. Clausmeyer S, Klösgen RB, Herrmann RG. 1993. Protein import into chloroplasts: the hydrophilic lumenal proteins exhibit unexpected import and sorting specificities in spite of structurally conserved transit peptides. *J. Biol. Chem.* 268:13869–76

23. Cline K. 1986. Import of proteins into chloroplasts: membrane integration of a thylakoid precursor protein reconstituted in chloroplast lysates. *J. Biol. Chem.* 261:14804–10

24. Cline K, Ettinger WF, Theg SM. 1992. Protein-specific energy requirements for protein transport across or into thylakoid membranes: two lumenal proteins are transported in the absence of ATP. *J. Biol. Chem.* 267:2688–96

25. Cline K, Fulsom DR, Viitanen PV. 1989. An imported thylakoid protein accumulates in the stroma when insertion into thylakoids is inhibited. *J. Biol. Chem.* 264:14225–32

26. Cline K, Henry R. 1996. Import and routing of nucleus-encoded chloroplast proteins. *Annu. Rev. Cell Dev. Biol.* 12:1–26

27. Cline K, Henry R, Li CJ, Yuan JG. 1992. Pathways and intermediates for the biogenesis of nuclear-encoded thylakoid proteins. In *Research in Photosynthesis*, ed. N Murata, pp. 149–56. Dordrecht: Kluwer

28. Cline K, Henry R, Li CJ, Yuan JG. 1993. Multiple pathways for protein transport into or across the thylakoid membrane. *EMBO J.* 12:4105–14

29. Cline K, Werner-Washburne M, Lubben TH, Keegstra K. 1985. Precursors to two nuclear-encoded chloroplast proteins bind to the outer envelope membrane before being imported into chloroplasts. *J. Biol. Chem.* 260:3691–96

30. Cohen Y, Yalovsky S, Nechushtai R. 1995. Integration and assembly of photosynthetic protein complexes in chloroplast thylakoid membranes. *Biochim. Biophys. Acta* 1241:1–30

31. Creighton AM, Hulford A, Mant A, Robinson D, Robinson C. 1995. A monomeric, tightly folded stromal intermediate on the delta pH-dependent thylakoid protein transport pathway. *J. Biol. Chem.* 270:1663–69

32. De Boer AD, Weisbeek PJ. 1991. Chloroplast protein topogenesis: import, sorting and assembly. *Biochim. Biophys. Acta* 1071:221–53

33. Deshaies R, Koch B, Werner-Washburne M, Craig E, Schekman R. 1988. A subfamily of stress proteins facilitates translocation of secretory and mitochondrial precursor polypeptides. *Nature* 332:800–5

34. De Vitry C. 1994. Characterization of the gene of the chloroplast Rieske iron-sulfur protein in *Chlamydomonas reinhardtii*. *J. Biol. Chem.* 269:7603–9

34a. Endo T, Kawakami M, Gogo A, America T, Weisbeek P, Nakai M. 1994. Chloroplast protein import: chloroplast envelopes and thylakoids have different abilities to unfold proteins. *Eur. J. Biochem.* 225:403–9

35. Flachmann R, Michalowski CB, Löffelhardt W, Bohnert HJ. 1993. SecY, an integral subunit of the bacterial preprotein translocase, is encoded by a plastid genome. *J. Biol. Chem.* 268:7514–19

36. Franklin AE, Hoffman NE. 1993. Characterization of a chloroplast homologue of the 54-kDa subunit of the signal recognition particle. *J. Biol. Chem.* 268:22175–80

37. Friedman AL, Keegstra K. 1989. Chloroplast protein import: quantitative analysis of precursor binding. *Plant Physiol.* 89:993–99

38. Fuks B, Schnell DJ. 1997. Mechanism of protein-transport across the chloroplast envelope. *Plant Physiol.* 114:405–10

39. Fulson DR, Cline K. 1988. A soluble protein factor is required in vitro for membrane insertion of the thylakoid precursor protein, pLHCP. *Plant Physiol.* 88:1146–53

40. Gray J, Row PE. 1995. Protein translocation across chloroplast envelope membranes. *Trends Cell Biol.* 5:243–47

40a. Guéra A, America T, van Waas M, Weisbeek PJ. 1993. A strong protein unfolding activity is associated with the binding of precursor chloroplast proteins to chloroplast envelopes. *Plant Mol. Biol.* 23:309–34

41. Haehnel W, Jansen T, Gause K, Klösgen RB, Stahl B, et al. 1994. Electron transfer from plastocyanin to photosystem I. *EMBO J.* 13:1028–38

42. Hageman J, Baecke C, Ebskamp M, Pilon R, Smeekens S, Weisbeek P. 1990. Protein import into and sorting inside the chloroplast are independent processes. *Plant Cell* 2:479–94

43. Hageman J, Robinson C, Smeekens S, Weisbeek P. 1986. A thylakoid-located processing protease is required for complete maturation of the lumen protein plastocyanin. *Nature* 324:567–69

44. Halpin C, Elderfield PD, James HE, Zimmermann R, Dunbar B, Robinson C. 1989. The reaction specificities of the thylakoid processing peptidase and *Escherichia coli* leader peptidase are iden-tical. *EMBO J.* 8:3917–21

45. Hand JM, Szabo LJ, Vasconcelos AC, Cashmore AR. 1989. The transit peptide of a chloroplast thylakoid membrane protein is functionally equivalent to a stromal-targeting sequence. *EMBO J.* 8:3195–206

46. Henry R, Carrigan M, McCaffery M, Ma XY, Cline K. 1997. Targeting determinants and proposed evolutionary basis for the Sec and delta-pH protein transport systems in chloroplast thylakoid membranes. *J. Cell Biol.* 136:823–32

47. Henry R, Kapazoglou A, McCaffery M, Cline K. 1994. Differences between lumen targeting domains of chloroplast transit peptides determine pathway specificity for thylakoid transport. *J. Biol. Chem.* 269:10189–92

48. Herrmann RG, Steppuhn J, Herrmann GS, Nelson N. 1993. The nuclear-encoded polypeptide CF_0II from spinach is a real ninth subunit of chloroplast ATP synthase. *FEBS Lett.* 326:192–98

49. High S, Henry R, Mould RM, Valent Q, Meacock S, et al. 1997. Chloroplast SRP54 interacts with a specific subset of thylakoid precursor proteins. *J. Biol. Chem.* 272:11622–28

50. Hirsch S, Muckel E, Heemeyer F, von Heijne G, Soll J. 1994. A receptor component of the chloroplast protein translocation machinery. *Science* 266:1989–92

51. Hoffman NE, Franklin AE. 1994. Evidence for a stromal GTP requirement for the integration of a chlorophyll a/b-binding polypeptide into thylakoid membranes. *Plant Physiol.* 105:295–304

52. Howe G, Merchant S. 1993. Maturation of thylakoid lumen proteins proceeds post-translationally through an intermediate in vivo. *Proc. Natl. Acad. Sci. USA* 90:1862–66

53. Huang LQ, Adam Z, Hoffman NE. 1992. Deletion mutants of chlorophyll a/b binding proteins are efficiently imported into chloroplasts but do not integrate into thylakoid membranes. *Plant Physiol.* 99:247–55

54. Hulford A, Hazell L, Mould RM, Robinson C. 1994. Two distinct mechanisms for the translocation of proteins across the thylakoid membrane, one requiring the presence of a stromal protein factor and nucleotide triphosphates. *J. Biol. Chem.* 269:3251–56

55. James HE, Bartling D, Musgrove JE, Kirwin PM, Herrmann RG, Robinson C. 1989. Transport of proteins into chloroplasts: import and maturation of precursors to the 33-, 23-, and 16-kDa

proteins of the photosynthetic oxygen-evolving complex. *J. Biol. Chem.* 264:19573–76

56. Jansen T, Rother C, Steppuhn J, Reinke H, Bayreuther K, et al. 1987. Nucleotide sequence of cDNA clones encoding complete "23 kDa" and "16 kDa" precursor proteins associated with the photosynthetic oxygen-evolving complex from spinach. *FEBS Lett.* 216:234–40

57. Karnauchov I, Cai D, Schmidt I, Herrmann RG, Klösgen RB. 1994. The thylakoid translocation of subunit 3 of photosystem I, the psaF gene product, depends on a bipartite transit peptide and proceeds along an azide-sensitive pathway. *J. Biol. Chem.* 269:32871–78

58. Kessler F, Blobel G. 1996. Interaction of the protein import and folding machineries in the chloroplast. *Proc. Natl. Acad. Sci. USA* 93:7684–89

59. Kessler F, Blobel G, Patel HA, Schnell DJ. 1994. Identification of two GTP-binding proteins in the chloroplast protein import machinery. *Science* 266:1035–39

60. Kim SJ, Robinson D, Robinson C. 1996. An *Arabidopsis thaliana* cDNA encoding PS II-X, a 4.1 kDa component of photosystem II: a bipartite presequence mediates SecA/delta pH-independent targeting into thylakoids. *FEBS Lett.* 390:175–78

61. Kirwin PM, Meadows JW, Shackleton JB, Musgrove JE, Elderfield PD, et al. 1989. ATP-dependent import of a lumenal protein by isolated thylakoid vesicles. *EMBO J.* 8:2251–55

62. Klein RR, Mason HS, Mullet JE. 1988. Light-regulated translation of chloroplast proteins. I. Transcripts of PsaA-PsaB, psbA, and RbcL are associated with polysomes in dark-grown and illuminated barley seedlings. *J. Cell Biol.* 106:289–301

63. Klösgen RB. 1997. Protein transport into and across the thylakoid membrane. *J. Photochem. Photobiol.* 38:1–9

64. Klösgen RB, Brock IW, Herrmann RG, Robinson C. 1992. Proton gradient-driven import of the 16 kDa oxygen-evolving complex protein as the full precursor protein by isolated thylakoids. *Plant Mol. Biol.* 18:1031–34

65. Knoetzel J, Simpson DJ. 1993. The primary structure of a cDNA for PsaN encoding an extrinsic lumenal polypeptide of barley photosystem I. *Plant Mol. Biol.* 22:337–45

66. Knott TG, Robinson C. 1994. The SecA inhibitor, azide, reversibly blocks the translocation of a subset of proteins across the chloroplast thylakoid membrane. *J. Biol. Chem.* 269:7843–46

67. Ko K, Bornemisza O, Kourtz L, Ko ZW, Plaxton WC, Cashmore AR. 1992. Isolation and characterization of a cDNA clone encoding a cognate 70-kDa heat shock protein of the chloroplast envelope. *J. Biol. Chem.* 267:2986–93

68. Ko K, Budd D, Wu CB, Seibert F, Kourtz L, Ko ZW. 1995. Isolation and characterization of a cDNA clone encoding a member of the Com44/Cim44 envelope components of the chloroplast protein import apparatus. *J. Biol. Chem.* 270:28601–8

69. Ko K, Cashmore AR. 1989. Targeting of proteins to the thylakoid lumen by the bipartite transit peptide of the 33-kd oxygen-evolving protein. *EMBO J.* 8:3187–94

70. Ko K, Ko ZW. 1992. Carboxyl-terminal sequences can influence the in vitro import and intraorganellar targeting of chloroplast protein precursors. *J. Biol. Chem.* 267:13910–16

71. Konishi T, Watanabe A. 1993. Transport of proteins into the thylakoid lumen: stromal processing and energy requirements for the import of the precursor to the 23-kDa protein of PSII. *Plant Cell Physiol.* 34:315–19

72. Kouranov A, Schnell DJ. 1996. Protein translocation at the envelope and thylakoid membranes of chloroplasts. *J. Biol. Chem.* 271:31009–12

73. Kourtz L, Ko K. 1997. The early stage of chloroplast protein import involves Com70. *J. Biol. Chem.* 272:2808–13

74. Laidler V, Chaddock AM, Knott RF, Walker D, Robinson C. 1995. A SecY homolog in *Arabadopsis thaliana*. *J. Biol. Chem.* 270:17664–67

75. Lamppa GK. 1988. The chlorophyll a/b binding protein inserts into the thylakoids independent of its cognate transit peptide. *J. Biol. Chem.* 263:14996–99

76. Li XX, Henry R, Yuan JG, Cline K, Hoffman NE. 1995. A chloroplast homologue of the signal recognition particle subunit SRP54 is involved in the posttranslational integration of a protein into thylakoid membranes. *Proc. Natl. Acad. Sci. USA* 92:3789–93

77. Lorkovíc ZJ, Schröder WP, Pakrasi HB, Irrgang KD, Herrmann RG, Oelmüller R. 1995. Molecular characterization of PsbW, a nuclear-encoded component of the photosystem II reaction center complex in spinach. *Proc. Natl. Acad. Sci. USA* 92:8930–34

78. Lübeck J, Heins L, Soll J. 1997. Protein import into chloroplasts. *Physiol. Plant.* 100:53–64

79. Lübeck J, Soll J, Akita M, Nielsen E, Keegstra K. 1996. Topology of IEP110, a component of the chloroplastic protein import machinery present in the inner envelope membrane. *EMBO J.* 15:4230–38

80. Ma YK, Kouranov A, LaSala SE, Schnell DJ. 1996. Two components of the chloroplast protein import apparatus, IAP86 and IAP75, interact with the transit sequence during the recognition and translocation of precursor proteins at the outer envelope. *J. Cell Biol.* 134:315–27

81. Madueño F, Bradshaw SA, Gray JC. 1994. The thylakoid-targeting domain of the chloroplast Rieske iron-sulfur protein is located in the N-terminal hydrophobic region of the mature protein. *J. Biol. Chem.* 269:17458–63

82. Mant A, Nielsen VS, Knott TG, Møller BL, Robinson C. 1994. Multiple mechanisms for the targeting of photosystem I subunits F, H, K, L, and N into and across the thylakoid membrane. *J. Biol. Chem.* 269:27303–9

83. Mant A, Schmidt I, Herrmann RG, Robinson C, Klösgen RB. 1995. Sec-dependent thylakoid protein translocation: delta pH requirement is dictated by passenger protein and ATP concentration. *J. Biol. Chem.* 270:23275–81

84. Marshall JE, DeRocher AE, Keegstra K, Vierling E. 1990. Identification of heat shock protein hsp70 homologues in chloroplasts. *Proc. Natl. Acad. Sci. USA* 87:374–78

85. Martienssen RA, Barkan A, Scriven A, Taylor WC. 1987. Identification of a nuclear gene involved in thylakoid structure. In *Plant Membranes: Structure, Function, Biogenesis*, ed. H Sze, pp. 181–92. New York: Liss

86. Meadows JW, Robinson C. 1991. The full precursor of the 33-kDa oxygen-evolving complex protein of wheat is exported by *Escherichia coli* and processed to the mature size. *Plant Mol. Biol.* 17:1241–43

87. Meadows JW, Shackleton JB, Hulford A, Robinson C. 1989. Targeting of a foreign protein into the thylakoid lumen of pea chloroplasts. *FEBS Lett.* 253:244–46

88. Michl D, Robinson C, Shackleton JB, Herrmann RG, Klösgen RB. 1994. Targeting of proteins to the thylakoids by bipartite presequences: CF_0II is imported by a novel, third pathway. *EMBO J.* 13:1310–17

89. Miller JD, Wilhelm H, Gierasch L, Gilmore R, Walter P. 1993. GTP binding and hydrolysis by the signal recognition particle during initiation of protein translocation. *Nature* 366:351–54

90. Mould RM, Robinson C. 1991. A proton gradient is required for the transport of two lumenal oxygen-evolving proteins across the thylakoid membrane. *J. Biol. Chem.* 266:12189–93

91. Mould RM, Shackleton JB, Robinson C. 1991. Transport of proteins into chloroplasts: requirements for the efficient import of two lumenal oxygen-evolving complex proteins into isolated thylakoids. *J. Biol. Chem.* 266:17286–89

92. Nakai M, Goto A, Nohara T, Sugita D, Endo T. 1994. Identification of the SecA protein homolog in pea chloroplasts and its possible involvement in thylakoidal protein transport. *J. Biol. Chem.* 269:1338–41

93. Nielsen E, Akita M, Davila-Aponte J, Keegstra K. 1997. Stable association of chloroplastic precursors with protein-translocation complexes that contain proteins from both envelope membranes and a stromal Hsp100 molecular chaperone. *EMBO J.* 16:935–46

94. Nielsen VS, Mant A, Knoetzel J, Møller BL, Robinson C. 1994. Import of barley photosystem I subunit N into the thylakoid lumen is mediated by a bipartite presequence lacking an intermediate processing site. *J. Biol. Chem.* 269:3762–66

95. Nohara T, Asai T, Nakai M, Sugiura M, Endo T. 1996. Cytochrome f encoded by the chloroplast genome is imported into thylakoids via the SecA-dependent pathway. *Biochem. Biophys. Res. Commun.* 224:474–78

96. Nohara T, Nakai M, Goto A, Endo T. 1995. Isolation and characterization of the cDNA for pea chloroplast SecA: evolutionary conservation of the bacterial-type SecA-dependent protein transport within chloroplasts. *FEBS Lett.* 364:305–8

97. Oblong JE, Lamppa GK. 1992. Precursor for the light-harvesting chlorophyll a/b-binding protein synthesized in *Escherichia coli* blocks import of the small subunit of ribulose-1,5-bisphosphate carboxylase/oxygenase. *J. Biol. Chem.* 267:14328–34

98. Oliver DB, Cabelli RJ, Dolan KM, Jarosuk GP. 1990. Azide-resistant mutants of

Escherichia coli alter the SecA protein, an azide-sensitive component of the protein export machinery. *Proc. Natl. Acad. Sci. USA* 87:8227–31

99. Olsen LJ, Keegstra K. 1992. The binding of precursor proteins to chloroplasts requires nucleoside triphosphates in the intermembrane space. *J. Biol. Chem.* 267:433–39

100. Olsen LJ, Theg SM, Selman BR, Keegstra K. 1989. ATP is required for the binding of precursor proteins to chloroplasts. *J. Biol. Chem.* 264:6724–29

101. Pain D, Blobel G. 1987. Protein import in chloroplasts requires a chloroplast ATPase. *Proc. Natl. Acad. Sci. USA* 84:3288–92

102. Payan LA, Cline K. 1991. A stromal protein factor maintains the solubility and insertion competence of an imported thylakoid membrane protein. *J. Cell Biol.* 112:603–13

103. Perry SE, Buvinger WE, Bennett J, Keegstra K. 1991. Synthetic analogues of a transit peptide inhibit binding or translocation of chloroplastic precursor proteins. *J. Biol. Chem.* 266:11882–89

104. Perry SE, Keegstra K. 1994. Envelope membrane proteins that interact with chloroplastic precursor proteins. *Plant Cell* 6:93–105

105. Reed JE, Cline K, Stephens LC, Bacot KO, Viitanen PV. 1990. Early events in the import/assembly pathway of an integral thylakoid protein. *Eur. J. Biochem.* 194:33–42

106. Reith M, Mulholland J. 1993. A high-resolution gene map of the chloroplast genome of the red alga *Porphyra purpurea*. *Plant Cell* 5:465

107. Robinson C, Cai D, Hulford A, Brock IW, Michl D, et al. 1994. The presequence of a chimeric construct dictates which of two mechanisms are utilized for translocation across the thylakoid membrane: evidence for the existence of two distinct translocation systems. *EMBO J.* 13:279–85

108. Robinson C, Ellis RJ. 1984. Transport of proteins into chloroplasts: partial purification of a chloroplast protease involved in the processing of imported precursor polypeptides. *Eur. J. Biochem.* 142:337–42

109. Robinson C, Klösgen RB. 1994. Targeting of proteins into and across the thylakoid membrane—a multitude of mechanisms. *Plant Mol. Biol.* 26:15–24

109a. Robinson D, Karnauchov I, Herrmann RG, Klösgen RB, Robinson C. 1996. Protease-sensitive thylakoidal import

machinery for the Sec-, delta pH- and signal recognition particle-dependent protein targeting pathways, but not for CF_0II integration. *Plant J.* 10:149–55

110. Roffey RA, Theg SM. 1996. Analysis of the import of carboxyl-terminal truncations of the 23-kilodalton subunit of the oxygen-evolving complex suggests that its structure is an important determinant for thylakoid transport. *Plant Physiol.* 111:1329–38

111. Rother C, Jansen T, Tyagi A, Tittgen J, Herrmann RG. 1986. Plastocyanin is encoded by an uninterrupted nuclear gene in spinach. *Curr. Genet.* 11:171–76

112. Rothstein SJ, Gatenby AA, Willey DL, Gray JC. 1985. Binding of pea cytochrome f to the inner membrane of *Escherichia coli* requires the bacterial secA gene product. *Proc. Natl. Acad. Sci. USA* 82:7955–59

113. Scaramuzzi CD, Hiller RG, Stokes HW. 1992. Identification of a chloroplast-encoded secA gene homologue in a chromophytic alga: possible role in chloroplast protein translocation. *Curr. Genet.* 22:421–27

114. Scaramuzzi CD, Stokes HW, Hiller RG. 1992. Characterization of a chloroplast-encoded secY homologue and atpH from a chromophytic alga: evidence for a novel chloroplast genome. *FEBS Lett.* 304:119–23

115. Schatz G, Dobberstein B. 1996. Common principles of protein translocation across membranes. *Science* 271:1519–26

116. Schnell DJ, Blobel G. 1993. Identification of intermediates in the pathway of protein import into chloroplasts and their localization to envelope contact sites. *J. Cell Biol.* 120:103–15

117. Schnell DJ, Blobel G, Keegstra K, Kessler F, Ko K, Soll J. 1997. A nomenclature for the protein import components of the chloroplast envelope. *Trends Cell Biol.* 7:303–4

118. Schnell DJ, Kessler F, Blobel G. 1994. Isolation of components of the chloroplast protein import machinery. *Science* 266:1007–12

119. Seedorf M, Waegemann K, Soll J. 1995. A constituent of the chloroplast import complex represents a new type of GTP-binding protein. *Plant J.* 7:401–11

120. Seidler A, Michel H. 1990. Expression in *Escherichia coli* of the psbO gene encoding the 33 kD protein of the oxygen-evolving complex from spinach. *EMBO J.* 9:1743–48

120a. Settles M, Yonetani A, Baron A, Bush

D, Cline K, Martienssen R. 1997. The maize gene Hcf106 encodes an ancient conserved protein required for Sec-independent protein translocation. *Science*. In press

121. Shackleton JB, Robinson C. 1991. Transport of proteins into chloroplasts. The thylakoidal processing peptidase is a signal-type peptidase with stringent substrate requirements at the −3 and −1 positions. *J. Biol. Chem.* 266:12152–56

122. Smeekens S, Bauerle C, Hageman J, Keegstra K, Weisbeek P. 1986. The role of the transit peptide in the routing of precursors toward different chloroplast compartments. *Cell* 46:365–75

123. Smith TA, Kohorn BD. 1994. Mutations in a signal sequence for the thylakoid membrane identify multiple protein transport pathways and nuclear suppressors. *J. Cell Biol.* 126:365–74

124. Spangfort M, Larsson UK, Ljungberg U, Ryberg M, Andersson B, et al. 1990. The 20-kDa apo-polypeptide of the chlorophyll a/b protein complex CP24. Characterization and complete primary amino acid sequence. In *Current Research in Photosynthesis*, ed. M Bakscheffsky, 2:253–56. The Hague: Kluwer

125. Tagi A, Hermans J, Steppuhn J, Jansson C, Vater F, et al. 1987. Nucleotide sequence of cDNA clones encoding the complete "33 kDa" precursor protein associated with the photosynthetic oxygen-evolving complex from spinach. *Mol. Gen. Genet.* 207:288–93

126. Theg SM, Bauerle C, Olsen LJ, Selman BR, Keegstra K. 1989. Internal ATP is the only energy requirement for the translocation of precursor proteins across chloroplastic membranes. *J. Biol. Chem.* 264:6730–36

127. Theg SM, Scott SV. 1993. Protein import into chloroplasts. *Trends Cell Biol.* 3:186–90

128. Tranel PJ, Froehlich J, Goyal A, Keegstra K. 1995. A component of the chloroplastic protein import apparatus is targeted to the outer envelope membrane via a novel pathway. *EMBO J.* 14:2436–46

129. Ulbrandt ND, Newitt JA, Bernstein HD. 1997. The *E. coli* signal recognition particle is required for the insertion of a subset of inner membrane proteins. *Cell* 88:187–96

130. Valentin K. 1993. SecA is plastid-encoded in a red alga: implications for the evolution of plastid genomes and the thylakoid protein import apparatus. *Mol. Gen. Genet.* 236:245–50

131. Van Den Broeck G, Van Houtven A, Van Montague M, Herrera-Estrella L. 1988. The transit peptide of a chlorophyll a/b-binding protein is not sufficient to insert neomycin phosphotransferase II in the thylakoid membrane. *Plant Sci.* 58:171–76

132. VanderVere PS, Bennett TM, Oblong JE, Lamppa GK. 1995. A chloroplast processing enzyme involved in precursor maturation shares a zinc-binding motif with a recently recognized family of metalloendopeptidases. *Proc. Natl. Acad. Sci. USA* 92:7177–81

133. Van Wijk K, Knott TG, Robinson C. 1995. Evidence for SecA- and delta pH-independent insertion of D1 into thylakoids. *FEBS Lett.* 368:263–66

134. Viitanen PV, Doran ER, Dunsmuir P. 1988. What is the role of the transit peptide in thylakoid integration of the light-harvesting chlorophyll a/b protein. *J. Biol. Chem.* 263:15000–7

135. Voelker R, Barkan A. 1995. Two nuclear mutations disrupt distinct pathways for targeting proteins to the chloroplast thylakoid. *EMBO J.* 14:3905–14

136. Voelker R, Mendel-Hartvig J, Barkan A. 1997. Transposon-disruption of a maize nuclear gene, tha1, encoding a chloroplast SecA homologue: in vivo role of cp-SecA in thylakoid protein targeting. *Genetics* 145:467–78

137. Von Heijne G. 1985. Signal sequences: the limits of variation. *J. Mol. Biol.* 184:99–105

138. Von Heijne G. 1986. A new method for predicting signal sequence cleavage sites. *Nucleic Acids Res.* 14:4683–90

139. Von Heijne G, Steppuhn J, Herrmann RG. 1989. Domain structure of mitochondrial and chloroplast targeting peptides. *Eur. J. Biochem.* 180:535–45

140. Waegemann K, Paulsen H, Soll J. 1990. Translocation of proteins into isolated chloroplasts requires cytosolic factors to obtain import competence. *FEBS Lett.* 261:89–92

141. Waegemann K, Soll J. 1991. Characterization of the protein import apparatus in isolated outer envelopes of chloroplasts. *Plant J.* 1:149–58

142. Wales R, Newman BJ, Rose SA, Pappin D, Gray JC. 1989. Characterization of cDNA clones encoding the extrinsic 23 kDa polypeptide of the oxygen-evolving complex of photosystem II in pea. *Plant Mol. Biol.* 13:573–82

143. Walter P, Johnson AE. 1994. Signal sequence recognition and protein targeting to the endoplasmic reticulum

membrane. *Annu. Rev. Cell Biol.* 10:87–119

144. Wickner W, Leonard MR. 1996. *Escherichia coli* preprotein translocase. *J. Biol. Chem.* 271:29514–16

145. Wu CB, Seibert FS, Ko K. 1994. Identification of chloroplast envelope proteins in close proximity to a partially translocated chimeric precursor protein. *J. Biol. Chem.* 269:32264–71

146. Yalovsky S, Paulsen H, Michaeli D, Chitnis PR, Nechushtai R. 1992. Involvement of a chloroplast HSP70 heat shock protein in the integration of a protein (light-harvesting complex protein precursor) into the thylakoid membrane. *Proc. Natl. Acad. Sci. USA* 89:5616–19

147. Yuan JG, Cline K. 1994. Plastocyanin and the 33-kDa subunit of the oxygen-evolving complex are transported into thylakoids with similar requirements as predicted from pathway specificity. *J. Biol. Chem.* 269:18463–67

148. Yuan JG, Henry R, Cline K. 1993. Stromal factor plays an essential role in protein integration into thylakoids that cannot be replaced by unfolding or by heat shock protein Hsp70. *Proc. Natl. Acad. Sci. USA* 90:8552–56

149. Yuan JG, Henry R, McCaffery M, Cline K. 1994. SecA homolog in protein transport within chloroplasts: evidence for endosymbiont-derived sorting. *Science* 266:796–98

150. Zak E, Sokolenko A, Unterholzner G, Altschmied L, Herrmann RG. 1997. On the mode of integration of plastid-encoded components of the cytochrome bf complex into thylakoid membranes. *Planta* 201:334–41

Annu. Rev. Plant Physiol. Plant Mol. Biol. 1998. 49:127–50

PLANT TRANSCRIPTION FACTOR STUDIES

C. Schwechheimer, M. Zourelidou, and M. W. Bevan

Molecular Genetics Department, John Innes Centre, Norwich, Norfolk, NR4 7UH, United Kingdom; e-mail: bevan@bbsrc.ac.uk

KEY WORDS: domain, interaction, localization, modification, technique

ABSTRACT

Major advances have been made in understanding the role of transcription factors in gene expression in yeast, Drosophila, and man. Transcription factor modification, synergistic events, protein-protein interactions, and chromatin structure have been successfully integrated into transcription factor studies in these organisms. While many putative transcription factors have been isolated from plants, most of them are only poorly characterized. This review summarizes examples where molecular biological techniques have been successfully employed to study plant transcription factors. The functional analysis of transcription factors is described as well as techniques for studying the interactions of transcription factors with other proteins and with DNA.

CONTENTS

127

1040-2519/98/0601-0127$08.00

INTRODUCTION

Regulated gene expression is one of the most complex activities in cells because it involves the integration of signal transduction pathways, the movement of proteins between cellular compartments, alterations in chromosome structure, RNA synthesis, and RNA processing (10). To understand plant growth and development at the molecular level, a detailed knowledge of the mechanisms of transcription is required. To achieve this, a comprehensive set of techniques and approaches must be assembled and used in a complementary manner. Here, using selected examples, systems are described that are presently available for studying transcription factors in vivo and in vitro. Techniques are outlined that allow the identification of transcription factor domains and of proteins that interact with transcription factors. In the closing section, techniques are summarized for studying the interactions between transcription factors and DNA.

TRANSCRIPTION STUDIES IN VIVO AND IN VITRO

Transient assays can provide prompt information about transcription factor function or their DNA-binding specificity. Test systems for plant transcription factors range from transient transformation of plant cells (Table 1) to in vitro transcription systems (107, 135) as well as transformation of yeast (49, 70, 73, 87, 88, 91, 92, 106, 122) and human cells (56, 97).

Transient Transformation of Plant Cells

Plant tissue can be transiently transformed by particle bombardment, electroporation, or polyethylene-glycol (PEG)/$CaCl_2$–mediated procedures. Table 1 summarizes transient assays from different plants that have been used for the study of plant transcription factors. For in vivo experiments, a DNA construct, directing the expression of the transcription factor of interest (the effector), is cotransfected with a suitable promoter/reporter construct (67). Depending on the nature of the transcription factor, reporter gene expression should be activated or repressed in its presence. Reporter gene activity can be measured within hours or days of transformation. To ensure reproducibility and to account for variations between individual transformations, a second reporter gene is usually cotransfected that is not affected by the transcription factor under study (67). In several cases, transcription factor action could be stimulated

Table 1 Summary of examples for transient assay systems that have been used for transcription factor studies

Method	Plant species	Plant material	Transcription factor	Reference
Particle bombardment	*Hordeum vulgare*	Aleurone	GAMyb, Viviparous1	41, 48, 103
	Nicotiana tabacum	Leaves	activating sequences	29
	Petunia hybrida	Floral organs	Lc, C1	83
	Phaseolus vulgaris	Cotyledons, leaves	PvAlf, ROM1, ROM2	8, 17, 18
	Zea mays	Cotyledons, endosperm	C1, R, Opaque2	9, 118
		Embryo, aleurone, embryogenic cellus	B, C1, Viviparous1	35, 36, 48, 54, 69, 86
			activating sequences	29
		A636, L6 maize suspension cells	B, C1, P, Opaque2	93, 132
Protoplast transformation	*Arabidopsis thaliana*	Leaf protoplasts	ATMYB2	119
	Nicotiana tabacum	Mesophyll protoplasts	HSF, PosF21, VSF-1, SPA, Myb.Ph3	3, 4, 106, 111, 112
	Zea mays	BY2	GAL4-fusions	74
		Black Mexican Sweet suspension cells	SPA	4
Electroporation	*Glycine max*	Cell culture	GBF1	97
	Daucus carota	WOO1C	GBF, GAL4-fusions	117
		Cell suspension culture	Viviparous1	71
	Nicotiana tabacum	Suspension culture (line XD)	GT-2	24
	Oriza sativa	Suspension culture (line Oc)	OSH42, OSH44, OSH45	108
			Osvp1, RITA-1	45, 52
	Zea mays	Black Mexican Sweet suspension cells	Opaque2, C1, R	9, 116
		Endosperm-derived suspension culture cells	Opaque2, Viviparous1	46, 54, 71, 116, 125

by the addition of phytohormones, elicitors, or other inducers of gene activity (25, 41, 46, 48, 54, 95, 102, 117, 125, 128).

The choice of assay can be crucial for accurate assessment of transcription factor action because the requirements for cofactors or posttranscriptional modifications should be considered. This was shown in the case of C1 and P, two maize Myb-like transcription factors. In vitro, both proteins can bind the same promoter element of the dehydroflavonol reductase (*A1*) promoter (58, 93, 114). But while P is sufficient to activate reporter gene expression from the *A1* promoter in a maize suspension culture system, the closely related C1 protein requires the presence of the Myc-like protein B (or R) for transcriptional activation. C1 cannot activate transcription in tissues unless either B or R is expressed (58, 93).

Activator studies require the target promoter to be silent or of low activity in the absence of ectopic effector. It is difficult to test stress- or wounding-induced promoters in transient assay systems, because these promoters can be activated during tissue preparation (90, 91). The tissue from where the effector originates is generally not suitable for transient assays because it contains endogenous transcription factor. Consequently, the ideal test tissue would be derived from a mutant where the transcription factor is not expressed. This strategy has been employed in studies of the maize transcription factors C1, P, R, B, and Viviparous1 (VP1). The complex developmental and interactive regulation of these proteins was elucidated using transient transformation experiments in mutant maize tissue and also in different suspension cultures of known genetic composition (46, 48, 71, 125). It was shown that the VP1 protein is a transcriptional activator of the anthocyanin regulator C1 (46) and the seed maturation–associated EM protein (71). Application of abscisic acid to the VP1 transient assays led to a synergistic increase in the activation of EM (71) but had only an additive effect on the expression of C1 (46). During seed maturation, VP1 represses the expression of the seed germination–specific α-amylase gene, and overexpression of VP1 during seed germination can even reverse the activating effects of gibberellic acid on α-amylase gene expression (48).

Yeast as an Alternative Assay

Studies in yeast can provide important information about plant transcription factors and can be useful in defining their DNA-binding specificity. These studies are possible because certain classes of transcriptional regulators activate transcription through activation domains that can mediate transcriptional activation in plants and in yeast, e.g. acidic activation domains (73, 91, 92, 106). Other classes of activation domains, e.g. the glutamine-rich class, are not active in yeast (82) and cannot be productively modeled in this heterologous host.

For the study of Myb-like transcription factors, DNA-binding specificity and activation potential have been characterized from a range of promoter elements for genes of the phenylpropanoid biosynthetic pathway (73, 91, 92, 106). Activation studies in yeast were used to confirm results obtained from a random binding site selection (RBSS) with Myb.Ph3 from petunia (106). Myb.Ph3, but not mutant forms deficient in DNA binding, could activate reporter gene expression from reporter constructs bearing consensus binding sites derived from RBSS, thus confirming its DNA-binding specificity. The same binding sites could be identified in a number of chalcone synthase promoters from various plant species, and it was shown that Myb.Ph3 can transcriptionally activate expression from a petunia chalcone synthase (*chsJ*) promoter in tobacco protoplasts (106).

The activity of the cauliflower mosaic virus 35S (CaMV 35S) promoter was the subject of several studies in yeast. This promoter is repressed in *Saccharomyces cerevisiae* but can be activated under nitrogen-limiting conditions and by cAMP. The regulatory elements of the CaMV 35S promoter involved in this activation were mapped to the *as*-1 element, which contains binding sites for the bZIP transcription factor TGA1 (87). Subsequently, it was shown that co-expression of TGA1 and a reporter gene regulated by either the CaMV 35S promoter or consensus *as*-1 elements can confer high levels of transcriptional activation in yeast cells (88, 122).

An interference between plant and endogenous yeast activators was observed in studies with the maize bZIP protein OPAQUE-2 (O2). O2 recognizes promoter elements that confer high levels of endosperm-specific storage protein expression in maize. Its cognate promoter elements share high sequence homology to the DNA-binding sites of the yeast bZIP transcription factor GCN4. Although it had been demonstrated in yeast that O2 can recognize and activate transcription from the yeast GCN4 DNA-binding site (70) and from the related plant promoter elements (49, 99), the results with the plant promoter elements also indicated that the activator function depended on an intact yeast GCN4 protein. O2 could activate transcription in a *gcn4⁻* yeast mutant background only to a minor extent (49, 70); it was postulated that heterodimerization between maize O2 and yeast GCN4 is required for the formation of an active transcription factor (49).

In Vitro Transcription Assays

In vitro transcription is a powerful tool for studying general and activated transcription and in defining the requirements for initiation, elongation, and termination. In vitro transcription offers a number of advantages over in vivo studies in that transcription from the desired template can be studied using heterologously expressed or biochemically purified transcription factors. The lack of reliable plant in vitro transcription systems in the past has meant that to date only one heterologously expressed plant transcription factor has been studied in in vitro assays. There, it could be shown in a wheat germ–based system that the tobacco bZIP protein TGA1a activates transcription by increasing the number of active pre-initiation complexes (129).

Several attempts have been made to establish reliable and reproducible in vitro transcription systems for plants using a variety of nuclear and cell extracts (107, 135). The most promising transcription system makes use of nuclear extracts from tobacco BY-2 cells (30, 31, 51, 131), a rapidly dividing cell culture. The BY-2 system supports transcription of RNA polymerase I– (31), RNA polymerase II– (30), and RNA polymerase III–dependent genes (30, 131). Accurate initiation of transcription was reported in all cases. Primer extension is generally used for the detection of transcript, but in the case of the tRNA[Ser] analysis it was possible to detect the transcript directly by the addition of radioactively labeled nucleotides to the transcription reaction (131). In the most intriguing report, transcription from the light-inducible tomato *RbcS* promoter, which is not active in BY-2 nuclear extracts, is restored by the addition of leaf nuclear extract from light-grown tomatoes (30). This demonstrated that plant in vitro systems can be used to study plant-specific regulatory mechanisms by complementation and that the addition of transcription factors and cofactors can mediate transcriptional activation in a nonresponsive extract. A second in vitro transcription system is based on whole cell extracts derived from rice or tobacco. Using this system, it has been demonstrated that accurate transcription can be initiated from a rice phenylalanine ammonia-lyase gene and from a tobacco sesquiterpene cyclase gene promoter (136, 137). The main test for both in vitro transcription systems will be whether they can be reproduced independently in a number of laboratories.

TRANSCRIPTION FACTOR ANALYSIS IN PLANTA

In the absence of genetic analysis, identification of transcription factor target genes is one of the most demanding tasks in transcription factor studies. The high conservation between transcription factors has led to the identification of many orphan transcription factors of unknown function. Where mutants are available, downstream genes can be securely identified using genetic analysis and differential display or related techniques.

Two methods have principally been used to study the role of transcription factors in plants: overexpression and antisense technology. Overexpression, where a gene is expressed from a high-level constitutive or tissue-specific promoter in transgenic plants, can produce either plants that accumulate high levels of transcription factor or knock-out plants through inactivation of the transgene and/or the endogenous gene by cosuppression (5, 94, 115). Antisense technology, where an RNA is expressed that is complementary to a target mRNA, is used to suppress expression of the endogenous transcription factor gene (11, 81, 121). Both approaches can cause lethal or strong pleiotropic effects in transgenic plants that cannot easily be differentiated from the desired phenotypes (94, 115). Overexpression and inactivation of the transgene do not necessarily occur in all plant cells and the extent of overexpression or cosuppression is difficult to assess. Transgene inactivation by antisense requires homology over as little as 50 bp, which makes it difficult to generate antisense plants for highly conserved transcription factors such as the class of MADS-box or the Myb-like transcription factors (11). High-level expression of a transcription factor in a plant cell might favor the binding of the transcription factor to low affinity binding sites and the activation of gene expression from noncognate promoters. In addition, it is difficult to assess whether genes, which are upregulated in an overexpressing line, are directly upregulated by the transcription factor under study or indirectly as a consequence of the expression of other genes.

An alternative technique, which allows the identification of mutants in specific genes, employs the polymerase chain reaction to screen large populations of plants containing T-DNA or transposon insertions (61). This approach is currently being used for the identification of mutant *Myb* loci in Arabidopsis (62), a family of transcription factors that comprises around 100 members in this plant species. The outcome of this research should provide valuable information regarding the role of the many different Myb proteins in Arabidopsis and their functional and genetic redundancy.

Inducible Gene Expression

Inducible gene expression is used to avoid problems associated with overexpression. It allows the temporal, spatial, and quantitative control of gene expression in a mutant for the transcription factor or in a heterologous host plant or tissue (33). To prevent interference from endogenous plant genes, inducible systems are based on nonplant components. In uninduced conditions, the transgene is not expressed or not active, should not interfere with normal plant development, and therefore should not cause pleiotropic effects. Gene activation or expression should occur rapidly after induction and be stable for a defined period of time. The inducing signal should be readily perceived, taken up, distributed within the plant, and be active in the whole plant or a tissue of interest in a dosage-dependent manner.

Several inducible gene expression systems have been developed for plants that fulfill one or several of these criteria (33). To study plant transcription factors, posttranslational induction makes use of animal steroid–inducible receptors (6, 7, 65, 95, 105). The steroid-binding domain of the glucocorticoid receptor is fused to a plant transcription factor. The transcription factor accumulates in the inactive unliganded state (79, 80). It is thought that the unliganded hormone-binding domain represses nuclear localization, DNA binding, and perhaps other activities of the transcription factor. After induction by application of a steroid, repression is relieved, and active protein can rapidly enter the nucleus and exert its transcription factor function. Glucocorticoid receptor fusion proteins were first tested in transiently transformed tobacco protoplasts, and induction was achieved using the glucocorticoid derivatives dexamethasone and progesterone (95). A glucocorticoid-responsive GAL4-VP16 fusion protein has been used to induce the activation of a luciferase reporter gene in transgenic Arabidopsis and tobacco plants, either by growing the plants on nutrient agar containing dexamethasone or by spraying the plants with the inducing compound (6). Induction of the target gene was dosage-dependent and could be observed within 1 h after application of the chemical. Four different glucocorticoid derivatives were tested for their induction levels and sustainability of induction. In this and other studies, glucocorticoids had no visible impact on wild-type plants.

The maize regulatory protein R was studied using a glucocorticoid-based system by inducing its overexpression in the Arabidopsis mutant transparent testa glabra (*ttg*) (65); *ttg* mutant Arabidopsis plants lack trichomes, anthocyanins, and seed coat pigment and produce excess root hairs. Production of trichomes and anthocyanins was restored by overexpression of the maize transcription factor R in a constitutive and inducible manner. Trichomes started to form 24 h after immersion of plants in inducing solution. The number of trichomes formed on a leaf and anthocyanin accumulation in plants depended on the concentration of the glucocorticoid.

In another example, the Arabidopsis gene *Constans* (*Co*) was expressed as a glucocorticoid fusion in a *co* mutant background (105). CO is a protein with homology to the GATA class of transcription factors, and mutations in CO delay flowering under long day conditions but have almost no effect under short days. The onset of flowering is CO dosage–dependent. After induction by dexamethasone, Arabidopsis plants carrying a CO-glucocorticoid transgene flowered earlier than the untransformed *co* mutant plant. Promotion of flowering could be observed at any stage in plant development, from early in germination until the time when the mutant plant would form flowers. Transcripts of the floral meristem-identity gene *Leafy* and of *Terminal Flower* could be detected by in situ hybridization 24 h after CO induction, indicating that these genes are downstream of CO.

Using inducible expression of transcription factors in combination with differential display should make a large contribution to the identification of target genes activated by a transcription factor. To date, at least two reports exist where this approach has been successful (16, 89).

A Viral Vector for the Ectopic Expression of Transcription Factors

Studying the interaction between transcription factors and their promoter binding sites is difficult. Ideally, it requires the generation of independent transgenic lines carrying either the reporter or the effector gene. These lines are crossed to study the effects of the transcription factor. This problem was circumvented in one study by using a potato virus X (PVX)-based vector (14, 90). Leaves from transgenic tobacco plants carrying different versions of the bean phenylalanine ammonia-lyase 2 (PAL2) promoter regulating a GUS reporter gene were inoculated with a PVX-construct expressing Myb305 (90). Myb305 from snapdragon is a putative homologue of the tobacco transcriptional activator that regulates the PAL2 promoter in tobacco petals (91). Ectopic expression of Myb305 using PVX produced high concentrations of transcription factor in the infected tissue and resulted in the expression of the GUS reporter gene in plants carrying the wild-type but not a mutant PAL2 promoter element (90). The relative instability of the vectors and the restricted host range of the virus may, however, limit its use.

TRANSCRIPTION FACTOR DOMAINS

Traditionally, transcription factors have been described as modular proteins containing a variety of domains for DNA binding, activation, binding of signaling molecules, and interaction with other proteins. Nuclear localization motifs regulate the import of transcription factors into the nucleus. Modularity permits the combinations of different domains to form transcription factors with discrete functions from a relatively small number of components.

Gene Fusions for the Identification of Protein Domains

DNA-binding domains are usually highly conserved and can often be identified from the primary amino acid sequence. In contrast, activation domains are not so conserved and can only be classified by their overall amino acid composition as being rich in acidic, glutamine, or proline residues (113).

Fusion of a putative activation domain to a known DNA-binding domain can define an activation domain even for a transcription factor with no defined target promoters. The DNA-binding domain of the yeast transcriptional activator GAL4 is most frequently chosen to "host" the putative activation domain. GAL4-transcription factor fusion constructs can be tested in transient assays for the activation of GAL4 responsive reporter genes. GAL4 fusions have been used to identify the activation domains of the Arabidopsis G-box binding factor GBF1 (97); the maize activators O2 (118), C1 (35), and VP1 (71); the bean protein PvAlf (8); and the group of alternatively spliced rice transcription factors OSH42, OSH44, and OSH45 (108). Although there are two reports where the GAL4 DNA-binding domain alone could confer reporter gene activation (74, 98), generally this has not posed a problem. Addition of GAL4 binding sites to a minimal promoter increased the background activity of the reporter construct in the GBF1 study, but this was negligible when compared to the expression levels obtained with the activator fusion proteins (97).

Auxin-responsive elements (AuxREs) of the soybean *GH3* promoter are required but not sufficient for auxin-inducible gene expression (117). Promoter sequences flanking the AuxREs were constitutively active when the AuxRE element was deleted or mutated. Promoters containing the AuxRE element and its flanking sequences were silent, but gene expression could be activated through the addition of auxins. Using a composite promoter containing GAL4 binding sites and an auxin-responsive element (AuxRE), the heterologous transcriptional activator GAL4-cRel can only activate transcription from this promoter when it is derepressed by the addition of auxin (117).

Not all DNA-binding proteins are transcriptional activators. To show that a DNA-binding protein can recognize a putative promoter target sequence, fusions of the DNA-binding protein to a strong activation domain can be tested. The activation domains of the herpes simplex virus protein VP16 and the yeast activator GAL4 both act as strong activation domains in plants. VP16 fusions have been used to show DNA recognition of the parsley bZIP protein CPRF1 (common plant regulatory factor 1) (32) and the maize regulator VP1 (71). CPRF1 could not activate a GUS reporter gene on its own but yielded high levels of activation once fused to VP16 (32). When the VP16 activation domain was used to replace the endogenous activation domain of VP1, the resulting VP1/VP16 fusion protein was significantly less active than the original VP1 protein but was still a strong transcriptional activator (71). In the case of C1, the endogenous activation domain was replaced by the activation domain of GAL4. Activation by the C1-GAL4(AD) fusion protein from the *Bronze1* target promoter was still dependent on the presence of the Myc-like cofactor B-Peru, suggesting that B-Peru interacts with C1 at its DNA-binding domain (35).

Nuclear Localization of Transcription Factors

The import of transcription factors from the cytoplasm into the nucleus is a necessary and important step in posttranslational control (64). For small proteins (<40–60 kDa), this import could take place by diffusion, but most proteins require nuclear localization signals (NLSs) for selective import into the nucleus (26, 84). These NLSs can be recognized by transporter proteins that shuttle between the cytoplasm and the nucleus or by import receptors located at nuclear pores (26, 64, 84). NLSs have been described as rich in arginine and lysine residues and have been classified into three groups: those resembling the SV40 large T type antigen NLS (PKKKRKV), those resembling the NLS of the yeast mating type factor Matα2 (KIPIK), and NLSs with a bipartite structure, which are usually a combination of two basic protein regions separated by approximately 10 amino acids (e.g. nucleoplasmin SPP-KAVKRPAATKKAGQAKKKKLDKEDES) (26, 84).

The nuclear localization signals in a range of plant transcription factors have been identified mainly in transient transformation assays using GUS-fusion constructs (1, 72, 104, 120, 124). After histochemical staining of transformed tissue, GUS enzymatic activity is identified in the cytoplasm or the nucleus. C- and N-terminal OPAQUE-2 (O2)/GUS reporter gene fusions showed that O2 possesses two NLSs, designated NLS A (SV40-like motif) and NLS B (bipartite motif) as observed in stable tobacco transformants and in transiently transformed onion epidermis cells (124). Both NLSs conferred nuclear localization to the O2/GUS protein independently, with NLS B being more efficient than NLS A. The nuclear localization function of the bipartite NLS B is located at the basic domain of the bZIP domain of O2, which is involved in DNA binding. That the nuclear localization function in O2 is independent from its ability to bind DNA was demonstrated with an O2 mutant deficient in DNA binding (123). More recently, the presence of NLS-binding proteins (NBPs) for the O2 NLS B peptide at the nuclear pores of tobacco cells has been suggested (47).

The MADS-box transcription factors APETALA3 (AP3) and PISTILLATA (PI) belong to the B class of organ identity genes and are involved in the formation of petals and pistils in Arabidopsis (130). Several lines of evidence suggested that AP3 and PI form a heterodimer (37, 85, 130). Using C- and N-terminal GUS fusions, it was found that nuclear import of AP3 and PI was mutually interdependent (72). Expression of AP3/GUS or PI/GUS fusion proteins alone resulted in cytoplasmic localization of the fusion proteins. However, when one fusion protein was expressed together with the other transcription factor in its native form, nuclear localization of the GUS fusion protein was observed. Primary results from transiently transformed onion epidermis cells were confirmed by stable Arabidopsis transformation.

The use of green fluorescent protein (GFP) as a reporter gene for nuclear localization could allow the observation of nuclear import and export as a dynamic process. Attempts to use GFP for nuclear localization studies have, however, only been successful when a full-length transcription factor was fused to GFP (110). GFP fused to short NLS sequences appears to diffuse freely in and out of the nucleus, perhaps because of its small size (39). To circumvent this, GFP and GUS were jointly fused to an NLS for nuclear localization studies (39).

Apart from the use of reporter genes, a number of immunochemical techniques have been employed to study nuclear import of transcription factors.

The cellular distribution of the tomato heat shock proteins HSFA1 and HSFA2 was studied in tobacco protoplasts using immunofluorescence (68). While under stable temperature conditions, the heat shock factors appeared to be present in the cytoplasm and the nucleus, a clear shift in distribution from the cytoplasm to the nucleus could be observed after temperature stress. Light-dependent transport of the G-box binding factor (GBF) to the nucleus was examined using an in vitro GBF antibody cotranslocation assay (44). The rationale of these experiments was that only under conditions that cause the nuclear import of GBF would the cotransported GBF antibody also be present in the nucleus. Therefore, dark-grown evacuolated parsley protoplasts were incubated with antisera against GBF and subsequently irradiated by light. Nuclear extracts were prepared and the presence of antibody was evaluated by western blot analysis. Antibody was detected in the nuclear fraction only after light treatment, indicating that the transport of GBF from the cytoplasm to the nucleus was light-dependent. Conditions reported to inhibit nuclear import, e.g. addition of lectins and decrease of the incubation temperatures, could significantly decrease GBF import into the nucleus.

PROTEIN-PROTEIN INTERACTIONS

Regulation of transcription factors through heterodimerization (63) or modifications by protein kinases have been shown to be important in transcription factor activation or inactivation and in determining their DNA-binding specificity (50, 55).

Protein-Protein Interaction Studies In Vivo

Traditionally, biochemical techniques such as immunoprecipitation (37, 85, 132) or expression library screening with radiochemically labeled proteins (132) were used to identify protein-protein interactions. The yeast two-hybrid system can now be used for finding proteins that interact with a protein of interest (15, 19, 28). The protein of interest is translationally fused to the DNA-binding domain of the yeast transcription factor GAL4 or the bacterial repressor LexA in a yeast expression vector. This "bait" construct is cotransformed into yeast with "prey" constructs bearing translational fusions that combine either a known protein or a cDNA expression library with a strong activation domain. If the expressed bait and prey proteins can interact, they reconstitute a functional transcription factor recognized as reporter gene activation from a promoter bearing GAL4 or LexA binding sites (15, 19, 28).

The two-hybrid assay faces some potential difficulties in that many transcriptional activators, e.g. those belonging to the class of acidic activators, are also transcriptional activators in yeast (34, 73, 91, 106). This will lead to auto-activating bait proteins where reporter gene expression is activated in the absence of a prey construct. Nevertheless, many plant transcription factors do not activate transcription in yeast, and in several cases it has been possible to use the two-hybrid system for the identification of proteins that interact with transcription factors (23, 27, 34, 78, 93, 100).

The interaction between the maize Myc-like factor B and the Myb-like factor C1 was studied with the two-hybrid assay (34, 93). Fusions of both proteins to the GAL4 DNA-binding (DBD) and activation (AD) domains were prepared, but the GAL4(DBD)/C1 fusion strongly activated reporter gene expression in yeast (34). In contrast, GAL4(DBD) fusions of B and a fragment of B did not function as activators. When the GAL4(DBD)/B fusions were cotransformed with expression constructs encoding C1 fused to the GAL4(AD) or even C1 alone, high levels of reporter gene activity were obtained with the bait construct containing the B fragment but not with the full-length B protein. This suggested a physical interaction between B and C1 and also that C1 confers transcription activation potential to the B/C1 heterodimer (34). The fact that the B/C1 interaction could only be observed with a fragment of B and not with the full-length protein highlights that proteins can be misfolded or that interacting domains are not always displayed properly in the context of protein fusions. In a further study in yeast, it was shown that a deletion mutant of C1, C1-I, which is deficient in DNA binding, can still interact with the B protein, indicating that interaction with B and DNA binding are independent functions of C1(93). These findings were also of interest because C1 has similar DNA-binding specificity to the Myb-like regulator P, which regulates phlobaphene synthesis (40, 93). Both proteins can bind to the *A1* promoter in vitro. In vivo P alone can activate *A1* but C1 requires co-expression of B or R to activate transcription in plant cells. C1 alone can act as a strong transcriptional activator in yeast when fused to the GAL4 DNA-binding domain (40, 93). Therefore, it has been suggested that B relieves an inhibitory masking of the C1 activation domain by the C1 protein itself (93).

Several new transcription factors have been identified in a two-hybrid screen using a flower-specific cDNA expression library with the floral homeotic MADS-box transcription factors PLENA, DEFICIENS, and GLOBOSA from *Antirrhinum majus* as bait (23). Using northern blotting analysis and in situ hybridization, it was confirmed that the cDNAs encoding interacting domains were expressed in the same tissues as the bait constructs. The screen with DE-FICIENS detected only GLOBOSA cDNAs, and reciprocally the screen with GLOBOSA detected only DEFICIENS cDNAs, suggesting that both transcription factors interact exclusively with each other. This screen also confirmed biochemical data obtained from the Arabidopsis homologues APETALA3 and PISTILLATA, which suggested that these proteins also form a heterodimer (37, 85). SQUAMOSA and three novel MADS-box factors were identified with the PLENA construct. The interaction with SQUAMOSA was the most interesting because SQUAMOSA is also expressed in tissues where PLENA cannot be detected. This indicated that SQUAMOSA might also exert a function that is independent from the interaction with PLENA, perhaps by interaction with other MADS-box factors. The novel MADS-box factors, DEFH42 and DEFH200, have putative orthologues in petunia and tomato, FBP2 and TM5, and studies in these plants underline their role in PLENA action (5, 81). A third MADS-box factor, DEFH49, showed high homology to the Arabidopsis protein AGAMOUS-LIKE 2 but had a different expression pattern from the Arabidopsis protein.

Posttranslational Modifications

Many stimuli that affect gene expression also activate protein kinases. Some transcription factors are directly regulated by phosphorylation (50, 55) at three levels: import of transcription factors or associated proteins into the nucleus, enhancement or repression of the DNA-binding affinity, and positive or negative regulation of the activation potential. Regulation at several distinct levels could be achieved by phosphorylation at different sites by different protein kinases.

Casein kinase II (CKII) is hypothesized to be involved in central cellular functions such as cell division and growth, gene expression, and DNA replication (50). Several casein kinases have been identified from plants (38, 59, 60, 133) that potentially participate in the regulation of signal transduction pathways by counteracting the activity of protein phosphatases. It has been shown that the DNA-binding activity of the Arabidopsis GBF1, which interacts with a conserved element in the promoter of several light-inducible genes, is regulated by phosphorylation (59, 60). Purified nuclear CKII from broccoli (60) and a recombinant and reconstituted CKII from Arabidopsis (59) have been used to phosphorylate GBF1 and enhance its DNA-binding activity in an ATP- or GTP-dependent manner. DNA-binding activity was lost upon treatment with calf alkaline phosphatase and could then be rescued by rephosphorylation with CKII (60). In a different study on GBFs, cytosolic and nuclear protein extracts from parsley were tested for their ability to bind to the G-box (44). Upon dephosphorylation with alkaline phosphatase, DNA-binding activity in both fractions was almost completely lost. By adding ATP to the protein extracts, G-box binding activity could be restored in the cytosolic fraction but not in the nuclear fraction, suggesting the presence of a protein kinase in the cytosol that can regulate GBF DNA-binding activity. A different study revealed that the maize transcription factor O2 is phosphorylated in vivo and can be phosphorylated in vitro by CKII (20). Only nonphosphorylated and hypophosphorylated forms could bind target DNA sequences. The different phosphorylated forms of O2 were separated by isoelectric focusing. When these gels were blotted and probed with an O2 target DNA sequence, no DNA binding was observed for the phosphorylated forms unless the protein blot had been treated with potato acidic phosphatase. It was also shown that the profile of phosphorylated to hypophosphorylated forms changed diurnally, suggesting that storage protein synthesis might be slowed down at night and that this event is regulated by phosphorylation of O2 (20).

DNA-PROTEIN INTERACTIONS

Most promoters contain an array of *cis*-elements that can be recognized by transcription factors. Promoter studies are usually initiated by fusing promoter regions to a reporter gene and assaying reporter gene activity in transiently or stably transformed plants. Defined promoter regions can then be further characterized by electrophoretic mobility shift assays (EMSAs) where short promoter fragments are used in binding reactions with a nuclear protein extract or a heterologously expressed DNA-binding protein. More detailed analysis can be carried out using footprinting techniques or random binding site selection (RBSS), which allow the identification of bases important for transcription factor binding.

Electrophoretic Mobility Shift Assay

EMSA, the most widely applied technique for studying DNA-protein interactions, is based on the ability of a DNA-binding protein to alter the mobility of DNA in a nondenaturing acrylamide gel (12). EMSA is a reliable and simple technique for demonstrating the specific binding of a protein to a particular DNA sequence, but it provides only limited information about the bases that are directly involved in protein binding. Using antibodies, it is possible to confirm the identity of a protein present in a DNA-protein complex because the addition of the antibody to the binding reaction can supershift the complex or inhibit complex formation (17, 73, 91).

EMSAs can be used for the identification of DNA-binding domains. Several lines of evidence suggested the presence of multiple DNA-binding domains in the rice transcription factor GT-2, which can recognize three different GT-boxes in the rice phytochrome A gene promoter (24). Using a range of truncated GT-2 polypeptides, the boundaries of two minimal DNA-binding domains could be defined by EMSA. These two DNA-binding domains bound preferentially to the GT-2 and the GT-3 box respectively, and their binding specificity was further enhanced by flanking peptide sequences (75).

EMSAs have also been used to study the ability of DNA-binding proteins to form homo- and heterodimers, e.g. in the case of the Arabidopsis bZIPs GBF1, GBF2, and GBF3 (96). While addition of one GBF to the binding reaction resulted in a single band representing the homodimer, addition of a second GBF led to the formation of two or three bands representing homo- and heterodimeric complexes. For each GBF protein, a different potential for homo- and heterodimer formation was observed.

The effect of covalent modifications on the DNA-binding ability of a protein can also be studied by EMSA. Phosphorylation by CKII of the bZIP protein GBF1 was shown to stimulate its DNA-binding activity (59), while it has been reported for other transcription factors that their DNA-binding ability can be enhanced by dephosphorylation (22, 73).

DNA Footprinting Techniques

Footprinting techniques have been used extensively to determine transcription factor binding sites on promoter fragments. Footprinting methods are based on either chemical modification and subsequent cleavage of the promoter DNA or on DNase I treatment (2, 127). When the resulting DNA fragments are separated on denaturing acrylamide gels, residues protected by the protein are identified as missing or underrepresented bands.

In vitro footprinting with DNase I exploits the ability of DNase I to nick randomly either strand of DNA. Treatment with DNase I can take place before (interference) or after (protection) protein binding, but the latter technique is far more frequently used. DNase I footprinting was used to delineate the binding sites for the bean repressor proteins ROM1 and ROM2 (17, 18). ROM1 and ROM2 antagonize the activating function of the bean VP1-homologue PvAlf in the promoters of the storage protein genes phytohemagglutinin and β-phaseolin. Using recombinant proteins, it was shown that the bZIPs ROM1 and ROM2 recognize almost identical elements in the promoters of both target genes, but expression studies indicate that they do so at different times during seed maturation. While ROM1 expression precedes the onset of phytohemagglutinin and β-phaseolin expression (18), ROM2 expression coincides with a decrease in the transcription of the storage proteins at later stages of seed maturation (17).

In a number of cases, it has been possible to link chromatin structure and DNA binding using footprinting techniques. In the case of the wheat bZIP Em-binding protein (EmBP-1), Em promoter DNA was assembled in vitro with nucleosome cores from wheat and HeLa cells (76). It was shown that EmBP-1 binding is reduced in packaged DNA compared with naked DNA and that the position of the binding site within the nucleosome contributes to the extent of DNA binding. Although this study showed that histone octamers inhibit DNA binding by EmBP-1, linker histone 1 (which is not a component of the histone octamer complex) can enhance DNA binding of EmBP-1 in EMSAs (101).

Dimethyl sulfate (DMS) (91, 96, 97, 126), diethyl pyrocarbonate (DEPC) (40, 126), or 1,10-phenanthroline-copper (53, 57, 126) are used in chemical footprinting. Treatment with the chemical agents modifies specific nucleotides, and subsequent piperidine treatment breaks the DNA strand at the site of modification. The different modifying reagents react with different nucleotides, and therefore the choice of reagent depends on the base composition of the DNA fragment under study. A combination of three different footprinting techniques has been used to delineate the DNA-binding sites of the Arabidopsis Myb-like transcription factor CCA1 of the light-harvesting chlorophyll *a/b* protein gene (126). Only limited information about the cognate binding sites of CCA1 was obtained from DMS and DEPC footprints because of their AT-rich nature, as DMS and DEPC modify only guanine and adenine residues. 1,10-phenanthroline footprinting, however, which can modify every base, allowed the delineation of the CCA1-binding site (126).

Protein-DNA interactions over extensive promoter fragments can be studied in vivo in the context of intact, transcriptionally active chromatin using footprinting techniques. In vivo footprinting allows the detection of temporal and spatial interactions between DNA-binding proteins and their cognate elements. In vivo footprinting was used to confirm binding of the Arabidopsis bZIP transcription factor GBF3 to the G-box of the alcohol dehydrogenase promoter (66). Comparison of the footprints detected in vivo and in vitro suggests that GBF3 binds to the protected site. The extended sequence homology between different members of the bZIP family over their DNA-binding domain results, however, in similar DNA-binding signatures in vitro and makes it difficult to discriminate between interactions of different GBFs with a particular promoter element.

While the standard in vivo footprinting techniques are generally carried out using cell cultures, in vivo footprinting coupled with ligation-mediated polymerase chain reaction (LM-PCR) allows the analysis of transcription factors in the tissues from where they originate, when a suitable cell culture is not available (42, 43). In this technique, linkers are ligated to cleaved genomic DNA fragments that can then be used to amplify the resulting DNA fragments by PCR (42). Using LM-PCR, it was shown that the endosperm box of a wheat glutenin promoter is protected during the early stages of endosperm development in wheat grains (43). This observation was consistent with results that show that this element confers endosperm-specific expression in transgenic tobacco (21). The footprint was detected only in endosperm tissue. It was shown that the endosperm box consists of two distinct sites that are occupied sequentially by different proteins just before glutenin expression reaches its maximum levels (43).

Random Binding Site Selection

The range and degree of the DNA-binding specificity of a transcription factor can be determined by random binding site selection (RBSS). RBSS is based on the amplification of specific protein binding sites from a pool of randomized DNA sequences (77). This technique is particularly useful for defining differential DNA binding and the role of sequences flanking a core consensus site for different members of the same transcription factor family. It is difficult to interpret the results from RBSS when nothing is known about the cognate binding site of the transcription factor.

RBSS has been used successfully to determine the binding specificity of Arabidopsis GBF1 (97), which has been cloned by means of its binding to the tomato *RbcS*-3A G-box-like element (**ACACGTGG**) (96). GBF1 was also shown to bind to the "perfect" G-box (**CCACGTGG**) in the promoters of Arabidopsis *RbcS*-1A and alcohol dehydrogenase, and of parsley chalcone synthase, and to several other G-box–like elements, which raised the question of its DNA-binding specificity. RBSS, together with EMSAs, revealed that only sequences that contained the core motif ACGTG were bound by GBF1 with high affinity. In addition, specific bases flanking the G-box core motif were found to be required for high-affinity binding. Substitutions of bases in the core motif resulted in low-affinity binding (97).

RBSS has been employed to define binding specificity in the case of the maize Myb-like proteins P and C1, which regulate the expression of several genes in the flavonoid biosynthetic pathway (58, 93, 114). Both C1 and P recognize the same binding site ($CC^T/_ACC$) in the *A1* promoter (93). RBSS with C1 and P revealed that C1 has a more diverse binding spectrum than P, suggesting that the broader DNA-binding specificity of C1 results in an increased number of target promoter sites (93).

PERSPECTIVES

In this review, techniques were described that can be employed for transcription factors studies in plants. However, it is apparent that certain aspects of regulated gene expression cannot as yet be studied by the techniques currently available. Major advances are needed in the development of reliable in vitro systems and protein purification procedures from plants. This should allow the investigation of processes that are downstream of transcription factor–promoter recognition, such as the mechanisms of transcriptional activation and repression, and transcription factor–mediated changes in chromatin structure. Several recent reports describe also the integration of signal transduction cascades and gene expression (13, 109, 134). This research together with other approaches will soon provide us with an accurate understanding of the function of plant transcription factors.

ACKNOWLEDGMENTS

We are particularly grateful to Roderick Card and Gordon Simpson for helpful discussions on the manuscript. We also thank Manuela Costa, Giovanni Murtas, Manolo Pineiro, and Michael Hammond-Kosack for their critical proofreading. Because of length restrictions and the availability of other reviews, we apologize in advance for being unable to cite all the publications on this topic. Our work is supported by the Biotechnology and Biological Sciences Research Council (BBSRC); the Ministry of Agriculture, Fisheries, and Food (MAFF); and the State Scholarships Foundation of Greece.

Visit the *Annual Reviews* home page at
http://www.AnnualReviews.org.

Literature Cited

1. Abel S, Theologis A. 1995. A polymorphic bipartite motif signals nuclear targeting of early auxin-inducible proteins related to PS-IAA4 from pea (*Pisum sativum*). *Plant J.* 8:87–96
2. Adhya S, Becker MM, Burkjoff AM, Corman N, Fogerty S, et al. 1991. *A Laboratory Guide to In Vitro Studies of Protein-DNA Interactions*. Basel: Birkhäuser
3. Aeschbacher RA, Schrott M, Potrykus I, Saul MW. 1991. Isolation and molecular characterization of *PosF21*, an *Arabidopsis thaliana* gene which shows characteristics of a b-ZIP class transcription factor. *Plant J.* 1:303–16
4. Albani D, Hammond-Kosack MCU, Smith C, Conlan S, Colot V, et al. 1997. The wheat transcriptional activator SPA: a seed-specific bZIP protein that recognizes the GCN4-like motif in the bifactorial endosperm box of prolamin genes. *Plant Cell* 9:171–84
5. Angenent GC, Franken J, Busscher M, Weiss D, van Tunen AJ. 1994. Cosuppression of the petunia homeotic gene *fbp2* affects the identity of the generative meristem. *Plant J.* 5:33–44
6. Aoyama T, Chua N-H. 1997. A glucocorticoid-mediated transcriptional induction system in transgenic plants. *Plant J.* 11:605–12
7. Aoyama T, Dong C-H, Wu Y, Carabelli M, Sessa G, et al. 1995. Ectopic expression of the *Arabidopsis* transcriptional activator Athb-1 alters leaf fate in tobacco. *Plant Cell* 7:1773–85
8. Bobb AJ, Eiben HG, Bustos MM. 1995. PvAlf, an embryo-specific acidic transcriptional activator enhances gene expression from phaseolin and phytohemagglutinin promoters. *Plant J.* 8:331–43
9. Bodeau JP, Walbot V. 1992. Regulated transcription of the maize *Bronze-2* promoter in electroporated protoplasts requires the *C1* and *R* gene products. *Mol. Gen. Genet.* 233:379–87
10. Calkhoven CF, Geert AB. 1996. Multiple steps in the regulation of transcription-factor level and activity. *Biochem. J.* 317:329–42
11. Cannon M, Platz J, O'Leary M, Sookdeo C, Cannon F. 1990. Organ-specific modulation of gene expression in transgenic plants using antisense RNA. *Plant Mol. Biol.* 15:39–47
12. Carey J. 1991. Gel retardation. *Methods Enzymol.* 208:103–17
13. Chandra S, Martin GB, Low PS. 1996. The Pto kinase mediates a signalling pathway leading to the oxidative burst in tomato. *Proc. Natl. Acad. Sci. USA* 93:13393–97
14. Chapman S, Kavanagh T, Baulcombe D. 1992. Potato virus X as a vector for gene expression in plants. *Plant J.* 2:549–57
15. Chen C-T, Bartel PL, Sternglanz R, Fields S. 1991. The two-hybrid system: a method to identify and clone genes from proteins that interact with a protein of interest. *Proc. Natl. Acad. Sci. USA* 88:9578–82
16. Chen X, Meyerowitz E. 1997. Identification of a gene negatively regulated by the floral homeotic gene *AGAMOUS*. *Int. Conf. Arabidopsis Res., 8th*, Madison, Wis. (Abstr.)

17. Chern M-S, Bobb AJ, Bustos MM. 1996. The regulator of *MAT2* (ROM2) protein binds to early maturation promoters and represses PvALF-activated transcription. *Plant Cell* 8:305–21
18. Chern M-S, Eiben HG, Bustos MM. 1996. The developmentally regulated bZIP factor ROM1 modulates transcription from lectin and storage protein genes in bean embryos. *Plant J.* 10:135–48
19. Chevray PM, Nathans D. 1992. Protein interaction cloning in yeast: identification of mammalian proteins that react with the leucine zipper of Jun. *Proc. Natl. Acad. Sci. USA* 89:5789–93
20. Ciceri P, Gianazza E, Lazzari B, Lippoli G, Genga A, et al. 1997. Phosphorylation of Opaque2 changes diurnally and impacts its DNA binding activity. *Plant Cell* 9:97–108
21. Colot V, Robert LS, Kavanagh TA, Bevan MW, Thompson RD. 1987. Localization of sequences in wheat endosperm protein genes which confer tissue-specific expression in tobacco. *EMBO J.* 6:3559–64
22. Datta N, Cashmore AR. 1989. Binding of a pea nuclear protein to promoters of certain photoregulated genes is modulated by phosphorylation. *Plant Cell* 1:1069–77
23. Davies B, Egea-Cortines M, de Andrade Silva E, Saedler H, Sommer H. 1996. Multiple interactions amongst floral homeotic MADS box proteins. *EMBO J.* 15:4330–43
24. Dehesh K, Hung H, Tepperman JM, Quail PH. 1992. GT-2: a transcription factor with twin autonomous DNA-binding domains of closely related but different target sequence sepcificity. *EMBO J.* 11:4131–44
25. Dietrich A, Mayer JE, Hahlbrock K. 1990. Fungal elicitor triggers rapid, transient, and specific protein phosphorylation in parsley cell suspension cultures. *J. Biol. Chem.* 265:6360–68
26. Dingwall C, Laskey RA. 1991. Nuclear targeting sequences—a consensus? *Trends Biochem. Sci.* 16:478–81
27. Dubois C, Chern M-S, Bustos M. 1997. Temporal regulation of gene expression during embryogenesis: the role of ABI3 complexes in transcriptional control of early and late maturation genes of *Arabidopsis. Int. Conf. Arabidopsis Res., 8th,* Madison, Wis. (Abstr.)
28. Durfee T, Becherer K, Chen P-L, Yeh S-H, Yang Y, et al. 1993. The retinoblastoma protein associates with the protein phosphatase type 1 catalytic subunit. *Genes Dev.* 7:555–69
29. Estruch JJ, Crossland L, Goff SA. 1994. Plant activating sequences: positively charged peptides are functional as transcriptional activation domains. *Nucleic Acids Res.* 22:3983–89
30. Fan H, Sugiura M. 1995. A plant basal *in vitro* system supporting accurate transcription of both RNA polymerase II- and III-dependent genes: supplement of green leaf component(s) drives accurate transcription of a light-responsive *rbcS* gene. *EMBO J.* 14:1024–31
31. Fan H, Yakura K, Miyanishi M, Sugita M, Sugiura M. 1995. *In vitro* transcription of plant RNA polymerase I-dependent rRNA genes is species-specific. *Plant J.* 8:295–98
32. Feldbrügge M, Sprenger M, Dinkelbach M, Yazaki K, Harter K, et al. 1994. Functional analysis of a light-responsive plant bZIP transcriptional regulator. *Plant Cell* 6:1607–21
33. Gatz C. 1997. Chemical control of gene expression. *Annu. Rev. Plant Physiol. Plant Mol. Biol.* 48:89–108
34. Goff SA, Cone KC, Chandler VL. 1992. Functional analysis of the transcriptional activator encoded by the maize B gene: evidence for a direct functional interaction between two classes of regulatory proteins. *Genes Dev.* 6:864–75
35. Goff SA, Cone KC, Fromm ME. 1991. Identification of functional domains in the maize transcriptional activator C1: comparison of wild-type and dominant inhibitor proteins. *Genes Dev.* 5:298–309
36. Goff SA, Klein TM, Roth BA, Fromm ME, Cone KC, et al. 1990. Transactivation of anthocyanin biosynthetic genes following transfer of *B* regulatory genes into maize tissues. *EMBO J.* 9:2517–22
37. Goto K, Meyerowitz EM. 1994. Function and regulation of the *Arabidopsis* floral homeotic gene *PISTILLATA. Genes Dev.* 8:1548–60
38. Grasser KD, Maier U-G, Feix G. 1989. A nuclear casein type II kinase from maize endosperm phosphorylating HMG proteins. *Biochem. Biophys. Res. Commun.* 162:456–63
39. Grebenok RJ, Pierson E, Lambert GM, Gong F-C, Afonso CL, et al. 1997. Green-fluorescent protein fusions for efficient characterization of nuclear targeting. *Plant J.* 11:573–86
40. Grotewold E, Drummond BJ, Bowen B, Peterson T. 1994. The *myb*-homologous *P* gene controls phlobaphene pigmentation in maize floral organs by directly activating a flavonoid biosynthetic gene subset. *Cell* 76:543–53

41. Gubler F, Kalla R, Roberts JK, Jacobsen JV. 1995. Gibberellin-regulated expression of a *myb* gene in barley aleurone cells: evidence for Myb transactivation of a high-pI α-amylase gene promoter. *Plant Cell* 7:1879–91

42. Hammond-Kosack MCU, Bevan MW. 1993. A pratical guide to ligation-mediated PCR footprinting and *in-vivo* DNA analysis using plant tissues. *Plant Mol. Biol. Rep.* 11:249–72

43. Hammond-Kosack MCU, Holdsworth MJ, Bevan MW. 1993. *In vivo* footprinting of a low molecular weight glutenin gene (LMWG-1D1) in wheat endosperm. *EMBO J.* 12:545–54

44. Harter K, Kircher S, Frohnmeyer H, Krenz M, Nagy F, et al. 1994. Light-regulated modification and nuclear translocation of cytosolic G-box factors in parsley. *Plant Cell* 6:545–59

45. Hattori T, Terada T, Hamasuna S. 1995. Regulation of the *Osem* gene by abscisic acid and the transcriptional activator VP1: analysis of *cis*-acting promoter elements required for regulation by abscisic acid and VP1. *Plant J.* 7:913–25

46. Hattori T, Vasil V, Rosenkrans L, Hannah LC, McCarty DR, et al. 1992. The *Viviparous-1* gene and abscisic acid activate the *C1* regulatory gene for anthocyanin biosynthesis during seed maturation in maize. *Genes Dev.* 6:609–18

47. Hicks GR, Raikhel NV. 1995. Nuclear localization signal binding proteins in higher plant nuclei. *Proc. Natl. Acad. Sci. USA* 92:734–38

48. Hoecker U, Vasil IK, McCarty DR. 1995. Integrated control of seed maturation and germination programs by activator and repressor functions of Viviparous-1 of maize. *Genes Dev.* 9:2459–69

49. Holdsworth MJ, Munoz-Blanco J, Hammond-Kosack M, Colot V, Schuch W, et al. 1995. The maize transcription factor Opaque-2 activates a wheat glutenin promoter in plant and yeast cells. *Plant Mol. Biol.* 29:711–20

50. Hunter T, Karin M. 1992. The regulation of transcription by phosphorylation. *Cell* 70:375–87

51. Iwataki N, Hoya A, Yamazaki K-I. 1997. Restoration of TATA-dependent transcription in a heat-inactivated extract of tobacco nuclei by recombinant TATA-binding protein (TBP) from tobacco. *Plant Mol. Biol.* 34:69–79

52. Izawa T, Foster R, Nakajima M, Shimamoto K, Chua N-H. 1994. The rice bZIP transcriptional activator RITA-1 is highly expressed during seed development. *Plant Cell* 6:1277–87

53. Jordano J, Almoguera C, Thomas TL. 1989. A sunflower helianthinin gene upstream sequence ensemble contains an enhancer and sites of nuclear protein interaction. *Plant Cell* 1:855–66

54. Kao C-Y, Cocciolone SM, Vasil IK, McCarty DR. 1996. Localization and interaction of the *cis*-acting elements for abscisic acid, VIVIPAROUS1, and light activation of the *C1* gene of maize. *Plant Cell* 8:1171–79

55. Karin M, Hunter T. 1995. Transcriptional control by protein phosphorylation: signal transmission from the cell surface to the nucleus. *Curr. Biol.* 5:747–57

56. Katagiri F, Yamazaki K-I, Horikoshi M, Roeder RG, Chua N-H. 1990. A plant DNA-binding protein increases the number of active preinitiation complexes in a human in vitro transcription system. *Genes Dev.* 4:1899–909

57. Kehoe DM, Degenhardt J, Winicov I, Tobin EM. 1994. Two 10-bp regions are critical for phytochrome regulation of a *Lemna gibba Lhcb* gene promoter. *Plant Cell* 6:1123–34

58. Klein TM, Roth BA, Fromm ME. 1989. Regulation of anthocyanin biosynthetic genes introduced into intact maize tissues by microprojectiles. *Proc. Natl. Acad. Sci. USA* 86:6681–85

59. Klimczak LJ, Collinge MA, Farini D, Giuliano G, Walker JC, et al. 1995. Reconstitution of *Arabidopsis* casein kinase II from recombinant subunits and phosphorylation of transcription factor GBF1. *Plant Cell* 7:105–15

60. Klimczak LJ, Schindler U, Cashmore AR. 1992. DNA binding activity of the *Arabidopsis* G-box binding factor GBF1 is stimulated by phosphorylation by casein kinase II from broccoli. *Plant Cell* 4:87–98

61. Koes R, Souer E, van Houwelingen A, Mur L, Spelt C, et al. 1995. Targeted gene inactivation in petunia by PCR-based selection of transposon insertion mutants. *Proc. Natl. Acad. Sci. USA* 92:8149–53

62. Kranz H, Denekamp M, Greco R, Jin H-L, Leyva A, et al. 1997. Analysis of MYB transcription factors in *Arabidopsis thaliana*. *Int. Conf. Arabidopsis Res., 8th*, Madison, Wis. (Abstr.)

63. Lamb P, McKnight SL. 1991. Diversity and specificity in transcriptional regulation: the benefits of heterotypic dimerization. *Trends Biochem. Sci.* 16:417–22

64. Laskey RA, Dingwall C. 1993. Nuclear shuttling: the default pathway for nuclear proteins? *Cell* 74:585–86

65. Lloyd AM, Schena M, Walbot V, Davis RW. 1994. Epidermal cell fate determination in *Arabidopsis*: patterns defined by a steroid-inducible regulator. *Science* 266:436–39

66. Lu GH, Paul A-L, McCarty DR, Ferl RJ. 1996. Transcription factor veracity: is GBF3 responsible for ABA-regulated expression of *Arabidopsis Adh*? *Plant Cell* 8:847–57

67. Luehrsen KR, De Wet JR, Walbot V. 1992. Transient expression analysis in plants using firefly luciferase reporter gene. *Methods Enzymol.* 216:397–414

68. Lyck R, Harmening U, Höhfeld I, Treuter E, Scharf K-D, Nover L. 1997. Intracellular distribution and identification of the nuclear localization signals of two plant heat-stress transcription factors. *Planta* 202:117–25

69. Marrs KA, Alfenito MR, Lloyd AM, Walbot V. 1995. A glutathione S-transferase involved in vacuolar transfer encoded by the maize gene *Bronze-2. Nature* 375:397–400

70. Mauri I, Maddaloni M, Lohmer S, Motto M, Salamini F, et al. 1993. Functional expression of the transcriptional activator Opaque-2 of *Zea may* in transformed yeast. *Mol. Gen. Genet.* 241:319–26

71. McCarty DR, Hattori T, Carson CB, Vasil V, Lazar M, et al. 1991. The *Viviparous-1* developmental gene of maize encodes a novel transcriptional activator. *Cell* 66:895–905

72. McGonigle B, Bouhidel K, Irish VF. 1996. Nuclear localization of the *Arabidopsis* APETALA3 and PISTILLATA homeotic gene products depends on their simultaneous expression. *Genes Dev.* 10:1812–21

73. Moyano E, Martínez-Garcia JF, Martin C. 1996. Apparent redundancy in *myb* gene function provides gearing for the control of flavonoid biosynthesis in *Antirrhinum* flowers. *Plant Cell* 8:1519–32

74. Nakayama T, Okanami M, Meshi T, Iwabuchi M. 1997. Dissection of the wheat transcription factor HBP-1a(17) reveals a modular structure of the activation domain. *Mol. Gen. Genet.* 253:553–61

75. Ni M, Dehesh K, Tepperman JM, Quail PH. 1996. GT-2: in vivo transcriptional activation activity and definition of novel twin DNA binding domains with reciprocal target sequence selectivity. *Plant Cell* 8:1041–59

76. Niu XP, Adams CC, Workman JL, Guiltinan MJ. 1996. Binding of the wheat basic leucine zipper protein EmBP-1 to nucleosomal binding sites is modulated by nucleosome positioning. *Plant Cell* 8:1569–87

77. Ouellette MM, Wright WE. 1995. Use of reiterative selection for defining protein-nucleic acid interactions. *Curr. Opin. Biotechnol.* 6:65–72

78. Pelaz S, Gustafson-Brown C, Kohalmi S, Crosby B, Yanofsky MF. 1997. Isolation of genes whose products interact with *APETALA1* and *CAULIFLOWER* using the yeast 2-hybrid system. *Int. Conf. Arabidopsis Res., 8th*, Madison, Wis. (Abstr.)

79. Picard D. 1993. Steroid-binding domains for regulating the functions of heterologous proteins in *cis. Trends Cell Biol.* 3:278–80

80. Picard D, Salser SJ, Yamamoto KR. 1988. A movable and regulable inactivation function within the steroid binding domain of the glucocorticoid receptor. *Cell* 54:1073–80

81. Pnueli L, Hareven D, Broday L, Hurwitz C, Lifschitz E. 1994. The *TM5* MADS box gene mediates organ differentiation in the three inner whorls of tomato flowers. *Plant Cell* 6:175–86

82. Ponticelli AS, Pardee TS, Struhl K. 1995. The glutamine-rich activation domains of human Sp1 do not stimulate transcription in *Saccharomyces cerevisiae. Mol. Cell. Biol.* 15:983–88

83. Quattrocchio F, Wing JF, Leppen HTC, Mol JNM, Koes RE. 1993. Regulatory genes controlling anthocyanin pigmentation are functionally conserved among plant species and have distinct sets of target genes. *Plant Cell* 5:1497–512

84. Raikhel N. 1992. Nuclear targeting in plants. *Plant Physiol.* 100:1627–32

85. Riechmann JL, Krizek BA, Meyerowitz EM. 1996. Dimerization specificity of *Arabidopsis* MADS domain homeotic proteins APETALA1, APETALA3, PISTILLATA, and AGAMOUS. *Proc. Natl. Acad. Sci. USA* 93:4793–98

86. Roth BA, Goff SA, Klein TM, Fromm ME. 1991. *C1-* and *R*-dependent expression of the maize *Bz1* gene requires sequences with homology to mammalian *myb* and *myc* binding sites. *Plant Cell* 3:317–25

87. Rüth J, Hirt H, Schweyen RJ. 1992. The cauliflower mosaic virus 35S promoter is regulated by cAMP in *Saccharomyces cerevisiae. Mol. Gen. Genet.* 235:365–72

88. Rüth J, Schweyen RJ, Hirt H. 1994. The plant transcription factor TGA1 stimulates expression of the CaMV 35S promoter in *Saccharomyces cerevisiae. Plant Mol. Biol.* 25:323–28

89. Sablowski R, Meyerowitz E. 1997. Beyond floral homeotic genes: finding their target genes. *Int. Conf. Arabidopsis Res., 8th*, Madison, Wis. (Abstr.)

90. Sablowski RWM, Baulcombe DC, Bevan M. 1995. Expression of a flower-specific Myb protein in leaf cells using a viral vector causes ectopic activation of a target promoter. *Proc. Natl. Acad. Sci. USA* 92:6901–5

91. Sablowski RWM, Moyano E, Culianez-Macia FA, Schuch W, Martin C, et al. 1994. A flower-specific Myb protein activates transcription of phenylpropanoid biosynthetic genes. *EMBO J.* 13:128–37

92. Sainz MB, Goff SA, Chandler VL. 1997. Extensive mutagenesis of a transcriptional activation domain identifies single hydrophobic and acidic amino acids important for activation in vivo. *Mol. Cell. Biol.* 17:115–22

93. Sainz MB, Grotewold E, Chandler VL. 1997. Evidence for direct activation of an anthocyanin promoter by the maize C1 protein and comparison of DNA binding by related Myb domain proteins. *Plant Cell* 9:611–25

94. Samach A, Kohalmi SE, Motte P, Datla R, Haughn GW. 1997. Divergence of function and regulation of class B floral organ identity genes. *Plant Cell* 9:559–70

95. Schena M, Lloyd AM, Davis RW. 1991. A steroid-inducible gene expression system for plant cells. *Proc. Natl. Acad. Sci. USA* 88:10421–25

96. Schindler U, Menkens AE, Beckmann H, Ecker JR, Cashmore AR. 1992. Heterodimerization between light-regulated and ubiquitously expressed *Arabidopsis* GBF bZIP proteins. *EMBO J.* 11:1261–73

97. Schindler U, Terzaghi W, Beckmann H, Kadesch T, Cashmore AR. 1992. DNA-binding site preferences and transcriptional activation properties of the *Arabidopsis* transcription factor GBF1. *EMBO J.* 11:1275–89

98. Schläppi M, Raina R, Fedoroff N. 1996. A highly sensitive plant hybrid protein assay system based on the *Spm* promoter and TnpA protein for detection and analysis of transcription activation domains. *Plant Mol. Biol.* 32:717–25

99. Schmidt RJ, Ketudat M, Aukerman MJ, Hoschek G. 1992. Opaque-2 is a transcriptional activator that recognizes a specific target site in 22-kD zein genes. *Plant Cell* 4:689–700

100. Schultz TF, Quatrano RS. 1997. Characterization and expression of a rice RAD23 gene. *Plant Mol. Biol.* 34:557–62

101. Schultz TF, Spiker S, Quatrano RS. 1996. Histone H1 enhances the DNA binding activity of the transcription factor EmBP-1. *J. Mol. Biol.* 271:25742–45

102. Schulze-Lefert P, Dangl JL, Becker-André M, Hahlbrock K, Schulz W. 1989. Inducible *in vivo* DNA footprints define sequences necessary for UV light activation of the parsley chalcone synthase gene. *EMBO J.* 8:651–56

103. Shen QX, Zhang PH, Ho T-HD. 1996. Modular nature of abscisic acid (ABA) response complexes: composite promoter units that are necessary and sufficient for ABA induction of gene expression in barley. *Plant Cell* 8:1107–19

104. Shieh MW, Wessler SR, Raikhel NV. 1993. Nuclear targeting of the maize R protein requires two nuclear localization sequences. *Plant Physiol.* 101:353–61

105. Simon R, Igeno MI, Coupland G. 1996. Activation of floral meristem identity genes in *Arabidopsis*. *Nature* 384:59–62

106. Solano R, Nieto C, Avila J, Canas L, Diaz I, Pazares J. 1995. Dual DNA-binding specificity of a petal epidermis-specific MYB transcription factor (MYB.Ph3) from *Petunia hybrida*. *EMBO J.* 14:1773–84

107. Sugiura M. 1997. Plant in vitro transcription systems. *Annu. Rev. Plant Physiol. Plant Mol. Biol.* 48:383–98

108. Tamaoki M, Tsugawa H, Minami E-I, Kayano T, Yamamoto N, et al. 1995. Alternative RNA products from a rice homeobox gene. *Plant J.* 7:927–38

109. Tang X, Frederick RD, Zhou J, Halterman DA, Jia Y, et al. 1996. Initiation of plant disease resistance by physical interaction of AvrPto and Pto kinase. *Science* 274:2060–65

110. Terzaghi WB, Bertekap RL Jr, Cashmore AR. 1997. Intracellular localization of GBF proteins and blue light-induced import of GBF2 fusion proteins into the nucleus of cultured *Arabidopsis* and soybean cells. *Plant J.* 11:967–82

111. Torres-Schumann S, Ringli C, Heierli D, Amrhein N, Keller B. 1996. *In vitro* binding of the tomato bZIP transcriptional activator VSF-1 to a regulatory element that controls xylem-specific gene expression. *Plant J.* 9:283–96

112. Treuter E, Nover L, Ohme K, Scharf K-D. 1993. Promoter specificity and deletion analysis of three heat stress transcription factors of tomato. *Mol. Gen. Genet.* 240:113–25

113. Triezenberg SJ. 1995. Structure and function of transcriptional activation domains. *Curr. Opin. Genet. Dev.* 5:190–96

114. Tuerck JA, Fromm ME. 1994. Elements of the maize *A1* promoter required for transactivation by the anthocyanin *B/C1* or phlobaphene *P* regulatory genes. *Plant Cell* 6:1655–63

115. Ueberlacker B, Klinge B, Werr W. 1996. Ectopic expression of the maize homeobox genes *ZmHox1a* or *ZmHox1b* causes pleiotropic alterations in the vegetative and floral development of transgenic tobacco. *Plant Cell* 8:349–62

116. Ueda T, Waverczak W, Ward K, Sher N, Ketudat M, et al. 1992. Mutations of the 22- and 27-kD zein promoters affect transactivation by the Opaque-2 protein. *Plant Cell* 4:701–9

117. Ulmasov T, Liu Z-B, Hagen G, Guilfoyle TJ. 1995. Composite structure of auxin responsive elements. *Plant Cell* 7:1611–23

118. Unger E, Parsons RL, Schmidt RJ, Bowen B, Roth BA. 1993. Dominant negative mutants of Opaque2 suppress transactivation of a 22-kD zein promoter by Opaque2 in maize endosperm cells. *Plant Cell* 5:831–41

119. Urao T, Noji M-A, Yamaguchi-Shinozaki K, Shinozaki K. 1996. A transcriptional activation domain of ATMYB2, a drought-inducible *Arabidopsis* Myb-related protein. *Plant J.* 10:1145–48

120. van der Krol AR, Chua N-H. 1991. The basic domain of plant B-ZIP proteins facilitates import of a reporter protein into plant nuclei. *Plant Cell* 3:667–75

121. van der Krol AR, Mur LA, de Lange P, Gerats AGM, Mol JNM, et al. 1990. Antisense chalcone synthase genes in petunia: visualization of variable transgene expression. *Mol. Gen. Genet.* 220:204–12

122. van der Zaal BJ, Droog FNJ, Pieterse FJ, Hooykaas PJJ. 1996. Auxin-sensitive elements from promoters of tobacco GST genes and a consensus as-1-like element differ only in relative strength. *Plant Physiol.* 110:79–88

123. Varagona MJ, Raikhel NV. 1994. The basic domain in the bZIP regulatory protein Opaque2 serves two independent functions: DNA binding and nuclear localization. *Plant J.* 5:207–14

124. Varagona MJ, Schmidt RJ, Raikhel NV. 1992. Nuclear localization signal(s) required for nuclear targeting of the maize regulatory protein Opaque-2. *Plant Cell* 4:1213–27

125. Vasil V, Marcotte WR Jr, Rosenkrans L, Cocciolone SM, Vasil IK, et al. 1995. Overlap of Viviparous1 (VP1) and abscisic acid response elements in the *Em* promoter: G-box elements are sufficient

but not necessary for VP1 transactivation. *Plant Cell* 7:1511–18

126. Wang Z-Y, Kenigsbuch D, Sun L, Harel E, Ong MS, Tobin EM. 1997. A Myb-related transcription factor is involved in the phytochrome regulation of an *Arabidopsis Lhcb* gene. *Plant Cell* 9:491–507

127. Wissmann A, Hillen W. 1991. DNA contacts probed by modification protection and interference studies. *Methods Enzymol.* 208:365–79

128. Xiang CB, Miao Z-H, Lam E. 1996. Coordinated activation of *as-1*-type elements and a tobacco glutahione S-transferase gene by auxins, salicylic acid, methyl-jasmonate and hydrogen peroxide. *Plant Mol. Biol.* 32:415–26

129. Yamazaki K-I, Katagiri F, Imaseki H, Chua N-H. 1990. TGA1a, a tobacco DNA-binding protein, increases the rate of initiation in a plant *in vitro* transcription system. *Proc. Natl. Acad. Sci. USA* 87:7035–39

130. Yanofsky MF. 1995. Floral meristems to floral organs: genes controlling early events in *Arabidopsis* flower development. *Annu. Rev. Plant Physiol. Plant Mol. Biol.* 46:167–88

131. Yukawa Y, Sugita M, Sugiura M. 1997. Efficient *in vitro* transcription of plant nuclear tRNA^{Ser} genes in a nuclear extract from tobacco cultured cells. *Plant J.* 12:965–70

132. Zhang B, Chen W, Foley RC, Büttner M, Singh KB. 1995. Interactions between distinct types of DNA binding proteins enhance binding to *ocs* element promoter sequences. *Plant Cell* 7:2241–52

133. Zhang SQ, Jin C-D, Roux SJ. 1993. Casein kinase II-type protein kinase from pea cytoplasm and its inactivation by alkaline phosphatase in vitro. *Plant Physiol.* 103:955–62

134. Zhou JM, Tang XY, Martin GB. 1997. The Pto kinase conferring resistance to tomato bacterial speck disease interacts with proteins that bind a *cis*-element of pathogenesis-related genes. *EMBO J.* 16:3207–18

135. Zhu Q. 1996. RNA polymerase II-dependent plant *in vitro* transcription systems. *Plant J.* 10:185–88

136. Zhu Q, Chappell J, Hedrick SA, Lamb C. 1995. Accurate in vitro transcription from circularized plasmid templates by plant whole cell extracts. *Plant J.* 7:1021–30

137. Zhu Q, Dabi T, Lamb C. 1995. TATA box and initiator function in the accurate transcription of a plant minimal promoter in vitro. *Plant Cell* 7:1681–89

Annu. Rev. Plant Physiol. Plant Mol. Biol. 1998. 49:151–71

LESSONS FROM SEQUENCING OF THE GENOME OF A UNICELLULAR CYANOBACTERIUM, *SYNECHOCYSTIS* SP. PCC6803

H. Kotani and S. Tabata

Kazusa DNA Research Institute, 1532-3 Yana, Kisarazu, Chiba 292, Japan;
e-mail: tabata@kazusa.or.jp

KEY WORDS: cyanobacteria, genome sequencing, photosynthesis, plastid genomes, CyanoBase

---

ABSTRACT

The nucleotide sequence of the entire genome of the unicellular cyanobacterium, *Synechocystis* sp. PCC6803, has been determined. The length of the circular genome was 3,573,480 bp, and a total of 3168 protein-coding genes were assigned to the genome by a computer-assisted analysis. The functions of approximately 45% of the genes were deduced based on sequence similarity to known genes. Here are distinctive features of genetic information carried by the cyanobacteria, which have a phylogenetic relationship to both bacteria and plants.

CONTENTS

151

1040-2519/98/0601-0151$08.00

INTRODUCTION

Cyanobacteria, also called blue-green algae, are one of the 11 major eubacterial phyla on the basis of 16S rRNA, 23S rRNA, and protein sequences (73). They are capable of oxygenic photosynthesis. Cyanobacteria are clearly distinct from other photosynthetic bacteria such as purple and green bacteria, because cyanobacteria utilize H_2O as an electron donor but the others do not. Cyanobacteria, algae, and multicellular plants share striking similarity in the machinery and mechanism of photosynthesis. Cyanobacteria have long been used as model organisms for the study of oxygenic photosynthesis of higher plants, in which a more complex genetic system regulates this process.

From an evolutionary viewpoint, cyanobacteria are ancient, proposed to have evolved some 2.5 billion years ago. It is generally accepted that progenitors of present-day unicellular cyanobacteria were also the progenitors of plant plastids through endosymbiotic events. Today, because of a variety of physiological, morphological, and developmental characteristics, over 1500 species of cyanobacteria constitute an extremely diverse group of prokaryotes. Cyanobacteria colonize a wide range of habitats.

Although cyanobacteria are one of the largest groups of gram-negative bacteria, only a few strains for which genetic engineering technology is applicable, have been used for physiological and genetic studies. Among such strains, *Synechocystis* sp. PCC6803 is a unicellular cyanobacterium that is naturally transformable (10, 19, 56). It is capable of both photoautotrophic growth and, in the absence of light, heterotrophic growth. This allows analysis of the mechanism of oxygenic photosynthesis through the characterization of mutants deficient in both photosystems I and II.

Recent progress in DNA sequencing technology has allowed the generation of large quantities of nucleotide sequence data in a short period of time. This facilitated the sequencing of complete genomes: *Haemophilus influenzae* Rd (16), *Mycoplasma genitalium* (15), *Methanococcus jannaschii* (6), *Mycoplasma pneumoniae* (23), *Helicobacter pylori* (63), and *Saccharomyces cerevisiae* (72), and sequencing of other bacterial and eukaryotic genomes is under way. In 1996, nucleotide sequencing of the entire genome of *Synechocystis* sp. PCC6803 was completed (30), and this was the first report of the complete genomic information of a photoautotrophic organism. In this review, distinctive features of the genome and the genes of *Synechocystis*, revealed mostly by intensive analysis of the nucleotide sequence data of the entire genome, will be described in relation to the characteristics of the genomes of other cyanobacteria and of plastids.

GENOME STRUCTURES OF CYANOBACTERIA

Physical and Gene Maps of the Cyanobacterial Genomes

Although the structure and function of individual genes of diverse cyanobacteria have been intensively studied, comparative information on overall genome and gene organization is still inadequate. With the development of pulsed-field gel electrophoresis (PFGE), physical maps of the *Anabaena* sp. PCC7120 genome (6.4 Mb, G+C; 42.5%), the *Synechococcus* sp. PCC7002 genome (2.7 Mb, G+C; 49.1%), and the *Synechococcus* sp. PCC6301 (2.7 Mb, G+C; 55.1%) genome have been constructed (1, 11, 22, 29). Genes have been localized to specific large restriction fragments of DNA, thus merging the gene and physical maps into an accurate representation of the genome. In addition, the map is useful in constructing a set of overlapping clones representing the entire genome for sequence analysis. A map of the 6.4-Mb genome of *Anabaena* sp. PCC7120 has been constructed from *Avr*II, *Pst*I, and *Sal*I restriction fragments, and 30 genes and gene clusters have been localized to particular restriction fragments by Southern hybridization. In a map of the 2.7-Mb genome of *Synechococcus* sp. PCC7002, restriction fragments from *Asc*I, *Not*I, and *Sfi*I were analyzed by cross-hybridization, and 23 genes and gene clusters were located. In addition, a PCC6301 map has been constructed using the restriction endonucleases *Swa*I, *Pme*I, and I-*Ceu*I, and 52 genes and gene clusters have been mapped.

For construction of the physical map of the *Synechocystis* sp. PCC6803 genome, several approaches were combined (34, 35). A primary map was created by the conventional restriction analysis of total genomic DNA using three restriction endonucleases, *Asc*I, *Mlu*I, and *Spl*I. The map was confirmed by Southern hybridization of the genomic DNA digested with three enzymes using the gel-purified restriction fragments as probes. Then, 1500 clones from a cosmid genomic library spotted on nitrocellulose membranes were hybridized with the restriction fragments to roughly localize the clones on the map. The genome size of *Synechocystis* was estimated to be approximately 3.6 Mb judging by the combined lengths of the restriction fragments (34). *Synechocystis* has a medium-sized genome compared with other cyanobacteria. More detailed physical and gene maps were then constructed by the ordering of cosmid clones for the entire genome and by localizing on the map genes and gene clusters, which had been reported in this strain, by hybridization (35). Most of the multicopy cosmid clones carrying *Synechocystis* DNA were stably maintained in *Escherichia coli* cells, and about 90% of the entire genome was successfully covered with 11 contigs, consisting of 1100 cosmid clones. Then, the neighboring contigs were connected by clones from the lambda genomic library and by long polymerase chain reaction (PCR). Finally, the entire genome was completely covered with 107 cosmid clones, 15 lambda clones, and 17 long PCR products. Using a set of ordered clones, approximately 80 genes and gene clusters were localized on the map, to less than a 40-kb resolution.

Comparison of Genome Structure Among Cyanobacteria

The physical and gene maps of the *Synechocystis* genome are presented in Figure 1. The *Asc*I site, connecting *Asc*I-F and *Asc*I-E fragments, was designated

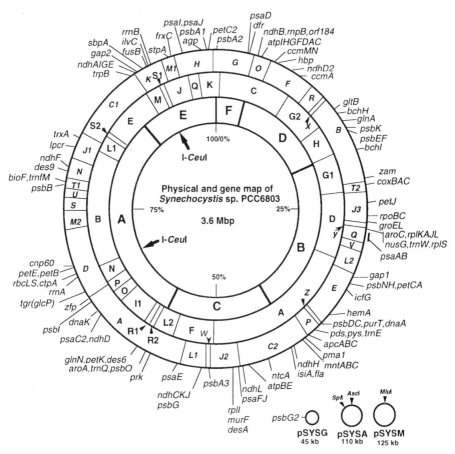

Figure 1   Physical and gene map of the genome of *Synechocystis* sp. PCC6803. Restriction sites for *Asc*I, *Mlu*I, and *Spl*I are indicated by *bars* in *inner, middle,* and *outer* circles, respectively. I-*Ceu*I sites that recognize the rRNA gene are also shown by *arrowheads*. Locations of 82 genes and gene clusters are indicated outside the circular map. *Small circles* at the bottom represent three plasmid DNAs.

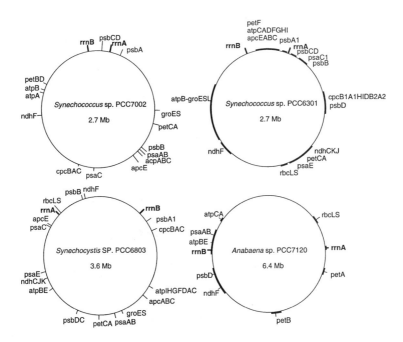

Figure 2 Comparison of gene organization among four cyanobacterial genomes. Locations of ribosomal RNA gene clusters and positions or regions where major protein-coding genes locate are indicated on the circular physical maps of four cyanobacterial strains.

as the 0/100 map position of the circular genome (34). Because the I-*Ceu*I cleavage sites specifically contained in the regions for the *rrn* operon (37) were identified at two locations at the positions 68.5 and 93 on the *Synechocystis* map, these sites were designated as *rrn*A and *rrn*B, respectively.

To compare the gene organization with that of other cyanobacteria, the physical genome maps of four cyanobacterial strains are illustrated in Figure 2. The locations of some common genes are identified. While the presence of two *rrn* operons has also been reported for three other cyanobacteria (1, 11, 29), relative locations vary. No apparent correlation is observed when the relative positions of various protein-coding genes, along with those of two *rrn* operons, are compared among four cyanobacteria, although some of the gene clusters—*psaAB, atpBE,* and *apcABC*—seem to be conserved among some strains. Common features of gene organization are not obvious, even between two *Synechococcus* strains PCC6301 and PCC7002. All bacteria able to photosynthesize oxygenically are classified as cyanobacteria; it is not clear if all species are related by descent. Genome structures may vary as a result of extensive rearrangement.

Plasmids in Cyanobacteria

Since the discovery of small extrachromosomal covalently closed circular DNA in a *Synechococcus* species, much effort has been put into characterizing the plasmids in cyanobacteria (13). In *Synechocystis* sp. PCC6803, plasmids ranging from 2.1 Kb to about 100 Kb have been reported (8). During the course of physical map construction, three large extrachromosomal DNA elements of approximate sizes 45, 110, and 125 kb, and designated as pSYSG, pSYSA, and pSYSM, respectively, were identified. The probes for *psbG2*, the gene for the subunit of NAD(P)H dehydrogenase, hybridized to the cosmid clones derived from the 45-kb plasmid and not to any restriction fragments or clones from the genomic DNA. This indicated that the gene is located on a plasmid, in accordance with the observation of Steinmuller & Bogorad (59).

GENERAL FEATURES OF THE *SYNECHOCYSTIS* SP. PCC6803 GENOME

Sequencing of the Entire Genome of Synechocystis sp. PCC6803

Sequencing was initiated from multiple points of the genome using as the templates the cosmid clones, the lambda clones, and the long PCR products of known map positions. The modified shotgun method was adapted to facilitate assembly of the raw sequence data (30, 31). Briefly, two shotgun libraries cloned into an M13 vector were prepared, one with about 700-bp inserts (element shotgun clones) and the other with 2.5-kb inserts (bridge shotgun clones). The element clones were sequenced from only one end, but the bridge clones were sequenced from both ends to assist computer assembly. For the cosmid clones with 40-kb inserts, typically approximately 650 element clones and 200 bridge clones were analyzed by the Applied Biosystems 373XL and 377XL type DNA sequencers. This yielded sequence data of about 10 times the original insert length. Then, the raw sequence data were assembled and finished using the autoassembler software of Applied Biosystems. To exclude any ambiguity, the entire genome of both strands was sequenced, and additional sequencing was performed where ever necessary.

To confirm the authenticity of the assembled sequences, every 20-kb region of the genome was amplified directly from the genomic DNA by long PCR, and the restriction patterns of the amplified fragments were compared with those expected from the finished sequences. This allowed for the detection of any possible rearrangements, insertions, or deletions during cloning and clone maintenance in *E. coli* or computer errors in the finishing step as a result of repetitive sequences, within the limits of resolution of agarose gel electrophoresis. In fact, for the entire genome, no inconsistency was found between the restriction patterns obtained by the two methods.

The total length of the circular genome of *Synechocystis* sp. PCC6803 was 3,573,470 bp (30). The percentage of nucleotides was A: 26.1%, C: 23.8%, G: 23.9%, and T: 26.2%, which results in a G+C content of 47.7%. The overall accuracy of the final sequence was estimated to be better than 99.999%, based on a comparison with previously reported sequences and on conformity with overlapped sequences that had been independently deduced. The nucleotide position was numbered from the junction of *AscI*-E and -F fragments according to the physical map.

Repetitive Sequences in the Genome of Synechocystis

One of the notable features of the nucleotide sequence of the *Synechocystis* genome is the high content of two types of repetitive sequences, insertion sequence (IS)-like sequences and HIP1 sequences.

IS-LIKE SEQUENCES   Bacterial IS are DNA elements, typically containing terminal inverted repeats. ISs are transposable elements with concomitant mutagenesis of interrupted genes and rearrangement of the genome by recombination between the elements (44). Transposition of an IS element typically causes a short duplication of the target sequence. In the *Synechocystis* genome, a total of 99 open reading frames (ORFs), derived from perhaps 77 IS-like elements. These IS elements showed significant similarity to putative bacterial transposases and were dispersed throughout the genome (31, 30). These ORFs could be classified into nine groups on the basis of family-specific 17–36-bp inverted repeats on element termini (T Kaneko, personal communication), and each of them appeared to be located within a stretch that constitutes an IS-like element. Interestingly, only 26 IS appeared to encode a functional transposases. The remaining ORFs have been disrupted by mutations such as frame-shifts and deletions and by insertions of other IS-like elements (44, 64, 65). The G+C contents of IS-like elements (32–45%) are clearly different from that of the chromosome, suggesting that the elements were transferred horizontally by another source. The presence of many intact and disrupted IS-like elements strongly suggests that rearrangement of the genome occurred frequently during the evolution of this microorganism. In *Anabaena* sp. PCC7120, frequent transposition of IS892 during prolonged culture was observed by genomic hybridization (7).

SHORT REPETITIVE SEQUENCES   Various types of short repetitive sequences have been reported in cyanobacteria. Three classes of short tandemly repeated repetitive elements consisting of 7-bp sequences, first identified in *Calothrix* sp. PCC7601, were found in many filamentous and heterocystous cyanobacteria

(41). An octameric palindromic sequence (5′-GCGATCGC-3′) designated as HIP1 (highly iterated palindrome) occurs abundantly in many cyanobacterial genomes (53). HIP1 was identified in a deletion mutant of the *smtB* gene, encoding a transcriptional repressor of the metallothionein gene, in *Synechococcus* sp. PCC6301 (20). Structural analysis of the mutation point revealed that the palindromic sequence, 5′-GCGATCGC-3′, was located at the deletion termini. In the *Synechocystis* genome, the HIP1D sequence, a decamer (5′-GGCGATCGCC-3′) which is 2 bp longer than HIP1, can be defined as a consensus sequence. A total of 2818 copies of HIP1D are found distributed fairly evenly throughout the genome, on the average every 1268 bp (A Tanaka, personal communication). About 86% of HIP1D sequences are found in protein-coding regions, and the remaining 14% are found in intergenic regions. HIP1D distribution reflects the ratio of the coding (87%) and noncoding (13%) regions in the genome, indicating that HIP1D sequences are randomly distributed. Furthermore, about 70% of the HIP1D sequences located in the protein-coding regions are translated as Ala-Ile-Ala. The origin and function of HIP1D in the *Synechocystis* genome are is not yet known. A nine-base pair sequence (USS), which is repeated 1465 times in the genome of *H. influenzae* (1 copy/1249 bp), is involved in the uptake of DNA into the cells (57). Although *Synechocystis* can take up DNA, there is no evidence that HIP1D participates in this process. Because HIP1 is located at the borders of gene deletions, this may indicate that HIP1(D) is a site for DNA recombination. It is tempting to consider two alternative possibilities for the function of HIP1(D). The first possibility is that HIP1(D), by providing sites for genetic recombination, enhances genome plasticity by increasing the pool of genetic diversity. The second possibility is that the mechanism that represses homologous recombination among HIP1(D) sequences may also work for the maintenance of polyploidy (up to 30 copies per cell) (39) by repressing recombination among the multicopy genomes.

ASSIGNMENT OF RNA AND PROTEIN-CODING GENES IN THE *SYNECHOCYSTIS* GENOME

Potential Structural RNA Genes in the Synechocystis Genome

Assignment of the potential RNA coding genes was performed by a computer-assisted analysis, which included a similarity search and prediction by a computer program. As a consequence, two copies of an rRNA gene cluster, 42 tRNA genes, and a gene for an RNA subunit of RNase P were located in the *Synechocystis* genome.

Two identical rRNA gene clusters were identified in the genome. Each consisted of a 5028-bp sequence containing in order 16S RNA, Ile-tRNA, 23S

RNA, and 5S RNA genes. The clusters are approximately 870 kb apart in reverse orientation (see the physical map in Figures 1 and 2). Other cyanobacteria, and most of algal and higher plant plastids, contain two copies of an rRNA gene cluster in their genomes (Figure 2), whereas other eubacteria, whose complete genome structures have been determined, contain varying numbers of copies (1–7 copies) of rRNA gene clusters, suggesting a phylogenetic relationship among cyanobacteria and plant plastids.

Based on the sequence similarity to known tRNAs and their genes and by computer prediction with the tRNAscan-SE program, 42 tRNA genes representing 41 tRNA species were identified in the *Synechocystis* genome (Y Nakamura, personal communication), sufficient for recognition of all the codons used in this species (Y Nakamura & S Tabata, submitted for publication). Only the gene for trnI-GAU was duplicated; two identical copies were in the rRNA gene cluster. There are fewer tRNA genes in *Synechocystis* compared with *E. coli* (86 tRNA genes), where multiplication of genes is commonly observed (T Ikemura, personal communication). In *Synechocystis*, the tRNA genes are scattered throughout the genome. Most of the genes seem to be transcribed separately as individual units, though the genes for trnY-GUA and trnT-GGU are tandemly aligned on the same strand of DNA with an 8-bp interval. This is quite different from *E. coli*, where 70% of the tRNA genes form clusters consisting of two to nine genes. It is also notable that many of the tRNA genes in *Synechocystis* show a high degree of sequence similarity to those in plant plastids when compared with those in other bacteria (Y Nakamura, personal communication).

Potential Protein Genes in the Synechocystis Genome

Potential protein-coding regions were assigned by combination of similarity search and computer prediction. The ORFs longer than 50 sense codons in the six translation frames were translated into amino acid sequences, and they were subjected to a similarity search with the MPSRCH program (58) against the public protein database. The ORFs with similarity to known genes were preferentially analyzed, and ORFs that showed no apparent similarity were assigned to unoccupied spaces. If two ORFs overlapped, the longer one was chosen, unless the function of the shorter one could be anticipated. Then, the short DNA stretches between the assigned ORFs were searched for similarity to known short genes. As a consequence, 3001 potential protein-coding regions were assigned in the *Synechocystis* genome.

In parallel with the similarity search, the nucleotide sequence of the entire genome was subjected to computer analysis for prediction of the coding region. GeneMark calculates the frequency of the regional appearance of oligonucleotides along the DNA region and calculates the difference of frequency between the coding and the noncoding regions (3–5, 25). The GeneMark analysis predicted 167 additional short coding regions in the *Synechocystis* genome, while 282 ORFs assigned by similarity search were not included in the GeneMark prediction. By combining the results of the similarity search with the GeneMark analysis, a total of 3168 potential protein genes were assigned in the *Synechocystis* genome (30). The average length of the putative gene products was 326 amino acid residues, and the potential protein-coding regions occupied 87.0% of the genome. The gene density was 1 gene per 1.1 kb, a typical value for bacterial genomes.

Of the 3168 potential protein genes assigned, 145 (4.6%) were identical to the reported *Synechocystis* genes, 935 (29.4%) were homologous to the genes whose function had been deduced, 324 (10.2%) showed lower similarity to known genes, 340 genes (10.8%) were similar to hypothetical genes of other organisms, and the remaining 1424 (45.0%) did not match any sequences in the public DNA databases. The results indicate that the functions of approximately 56% of the genes remain undetermined.

Classification of the Potential Protein Genes in Synechocystis

The 1402 genes with deduced functions were classified into 15 categories according to their biological function. The categories were originally adopted by Riley for the classification of *E. coli* genes (52) and were modified for this cyanobacterium (Table 1). The sequenced genes of other bacterial genomes were also classified to highlight distinctive features of gene composition in the *Synechocystis* genome. The notable features of the *Synechocystis* genome deduced from the comparison to *Haemophilus influenzae* is as follows. 1. The difference between the numbers of common genes in each category occurs because there are multiple hits of *Synechocystis* genes against a single *Haemophilus* gene, which suggests that gene multiplication has occurred in the *Synechocystis* genome. 2. More genes are in common in the categories involved in basic cellular activities: DNA replication, restriction, modification, recombination and repair, transcription, and translation. This indicates that these genes are conserved between the two species; however, there are also a number of species-specific genes in these categories. 3. The categories where fewer genes are in common contain more genes unique to the *Synechocystis* genome. These categories include photosynthesis and respiration, regulatory functions, and other.

CHARACTERISTIC FEATURES OF *SYNECHOCYSTIS* GENES

Seven bacterial genomes, including those of a cyanobacterium and an archaebacterium, and 10 plastid genomes have been sequenced. Using nucleotide

Table 1   Classification of the *Synechocystis* and *Haemophilus* genes according to biological function

Categories	*Synechocystis* PCC6803	Number of *Haemophilus* genes homologous to *Synechocystis* genes[c]	*H. influenzae*[d]
Amino acid biosynthesis	84	53 (49)	68
Biosynthesis of cofactors, prosthetic groups, and carriers	108 (2)[a]	34 (34)	54
Cell envelope	64	25 (25)	84
Cellular processes	68	27 (20)	53
Central intermediary metabolism	31	16 (13)	30
Energy metabolism	86 (7)[b]	36 (33)	105
Fatty acid, phospholipid, and sterol metabolism	35	20 (18)	25
Photosynthesis and respiration	138 (7)[b]	21 (21)	
Purines, pyrimidines, nucleosides, and nucleotides	39 (1)[a]	26 (26)	53
Regulatory functions	135 (1)[a]	39 (18)	64
DNA replication, restriction, modification, recombination, and repair	49	34 (34)	87
Transcription	24	20 (16)	27
Translation	144	112 (99)	141
Transport and binding proteins	158	69 (45)	123
Other categories	248	26 (24)	93

[a,b]Values in parentheses are numbers of genes classified into two categories.
[c]Values in parentheses are numbers of genes after removal of redundancy.
[d]Values are taken from Reference 15.

sequence and gene annotation information, various studies on gene and genome structure, and gene organization are under way with the help of computers. One of the most effective ways to deduce distinctive features of genes in a given genome is to compare the gene complements between various genomes. The following is an outline of several characteristics of the genes in the *Synechocystis* genome revealed by such an analysis.

Photosynthetic Genes

A total of 128 genes were deduced in the genome on the basis of their sequence similarity to previously reported photosynthetic genes of cyanobacteria and plants (Table 2). The photosynthetic processes and the numbers of genes anticipated to be involved were ATP synthase (9 genes), CO_2 fixation (24 genes), cytochrome *b6f* complex (8 genes), cytochrome oxidase (7 genes), NADH dehydrogenase (17 genes), photosystem I (12 genes), photosystem II (26 genes), phycobillisome (17 genes), and soluble electron carriers (8 genes). Most of

Table 2 Photosynthetic genes in *Synechocystis*

ATP synthase

sll1326	ATP synthase a subunit	atpA
slr1329	ATP synthase b subunit	atpB
sll1327	ATP synthase g subunit	atpC
sll1325	ATP synthase d subunit	atpD
slr1330	ATP synthase e subunit	atpE
sll1324	ATP synthase subunit b	atpF
sll1323	ATP synthase subunit b'	atpG
ssl2615	ATP synthase subunit c	atpH
sll1322	ATP synthase subunit a	atpI

CO₂ fixation

sll1028	carbon dioxide concentrating mechanism protein CcmK	ccmK
sll1029	carbon dioxide concentrating mechanism protein CcmK	ccmK
sll1030	carbon dioxide concentrating mechanism protein CcmL	ccmL
sll1031	carbon dioxide concentrating mechanism protein CcmM	ccmM
slr0436	carbon dioxide concentrating mechanism protein CcmK	ccmK
sll0934	carboxysome formation protein	ccmA
slr1838	carbon dioxide concentrating mechanism protein CcmK	ccmK
slr1839	carbon dioxide concentrating mechanism protein Ccmk	ccmK
slr0051	carbonic anhydrase	icfA
slr1347	carbonic anhydrase	icfA
slr0009	ribulose bisphosphate carboxylase large subunit	rbcL
slr0012	ribulose bisphosphate carboxylase small subunit	rbcS
slr0394	phosphoglycerate kinase	pgk
sll1342	glyceraldehyde-3-phosphate dehydrogenase (NADP+)	gap2
slr0783	triosephosphate isomerase	tpi
sll0018	fructose-1,6-bisphosphate aldolase	cbbA
slr0943	fructose-bisphosphate aldolase	fda
slr0952	fructose 1,6-bisphosphatase	fbp
sll1070	transketolase	tktA
slr0194	ribose 5-phosphate isomerase	rpiA
sll0807	pentose-5-phosphate-3-epimerase	cfxE
sll1525	phosphoribulokinase	prk
sll0030	rubisco operon transcriptional regulator	rbcR
sll1594	rubisco operon transcriptional regulator	rbcR

Photosystem I

slr1834	P700 apoprotein subunit Ia	psaA
slr1835	P700 apoprotein subunit Ib	psaB
ssl0563	photosystem I subunit VII	psaC
slr0737	photosystem I subunit II	psaD
ssr2831	photosystem I subunit IV	psaE
sll0819	photosystem I subunit III	psaF
smr0004	photosystem I subunit VIII	psaI
sml0008	photosystem I subunit IX	psaJ
sll0629	photosystem I subunit X	psaK
ssr0390	photosystem I subunit X	psaK
slr1655	photosystem I subunit XI	psaL
smr0005	photosystem I PsaM subunit	psaM

Photosystem II

slr1181	photosystem II D1 protein	psbA1
slr1311	photosystem II D1 protein	psbA2
sll1867	photosystem II D1 protein	psbA3
slr0906	photosystem II CP47 protein	psbB
sll0851	photosystem II CP43 protein	psbC
sll0849	photosystem II D2 protein	psbD
slr0927	photosystem II D2 protein	psbD2
ssr3451	cytochrome b_{559} a subunit	psbE
smr0006	cytochrome b_{559} b subunit	psbF
slr1280	NADH-ubiquinone oxidoreductase subunit PsbG	psbG
ssl2598	photosystem II PsbH protein	psbH
sml0001	photosystem II PsbI protein	psbI
smr0008	photosystem II PsbJ protein	psbJ
sml0005	photosystem II PsbK protein	psbK
smr0007	photosystem II PsbL protein	psbL
sml0003	photosystem II PsbM protein	psbM
smr0009	photosystem II PsbN protein	psbN
sll0427	photosystem II manganese-stabilizing polypeptide	psbO
smr0001	photosystem II PsbT protein	psbT
sll1194	photosystem II 12 kD extrinsic protein	psbU
sll0258	cytochrome c_{550}	psbV

sml0002	photosystem II PsbX protein	psbX
sll1398	photosystem II 13-kD protein	
slr1645	photosystem II 11-kD protein	
slr1739	photosystem II 13-kD protein (PsbW) homologue	
sll0247	iron-stress chlorophyll-binding protein	isiA

Phycobilisome

slr2067	allophycocyanin a chain	apcA
slr1986	allophycocyanin b chain	apcB
ssr3383	phycobilisome LC linker polypeptide	apcC
sll0928	allophycocyanin-B	apcD
slr0335	phycobilisome LCM core-membrane linker polypeptide	apcE
slr1459	phycobilisome core component	apcF
sll1578	phycocyanin a subunit	cpcA
sll1577	phycocyanin b subunit	cpcB
sll1579	phycocyanin associated linker protein	cpcC
sll1580	phycocyanin associated linker protein	cpcC
ssl3093	phycocyanin associated linker protein	cpcD
slr1878	phycocyanin alpha phycocyanobilin lyase CpcE	cpcE
sll1051	phycocyanin alpha phycocyanobilin lyase CpcF	cpcF
sll1471	phycobilisome rod-core linker polypeptide CpcG	cpcG
sll2051	phycobilisome rod-core linker polypeptide CpcG	cpcG
ssl0452	phycobilisome degradation protein NblA	nblA
ssl0453	phycobilisome degradation protein NblA	nblA

Soluble electron carriers

sll0199	plastocyanin	petE
sll1382	ferredoxin	petF
slr0150	ferredoxin	petF
slr1828	ferredoxin	petF
ssl0020	ferredoxin	petF
slr1643	ferredoxin-NADP oxidoreductase	petH
sll1796	cytochrome c_{553}	petJ
sll0248	flavodoxin	isiB

Cytochrome $b_6 f$ complex

sll1317	apocytochrome f	petA
slr0342	cytochrome b_6	petB
sll1182	cytochrome $b_6 f$ complex iron-sulfur subunit	petC
sll1316	plastoquinol—plastocyanin reductase	petC
slr1185	cytochrome $b_6 f$-complex iron-sulfur protein	petC
slr0343	cytochrome $b_6 f$ complex subunit 4	petD
smr0010	PetG subunit of the cytochrome $b_6 f$ complex	petG
smr0003	cytochrome $b_6 f$ complex subunit PetM	petM

Cytochrome oxidase

sll1899	cytochrome c oxidase folding protein	ctaB
sll0813	cytochrome c oxidase subunit II	ctaC
slr1136	cytochrome c oxidase subunit II	ctaC
slr1137	cytochrome c oxidase subunit I	ctaD
slr2082	cytochrome c oxidase subunit I	ctaD
slr1138	cytochrome c oxidase subunit III	ctaE
slr2083	cytochrome c oxidase subunit III	ctaE

NADH dehydrogenase

sll0519	NADH dehydrogenase subunit 1	ndhA
sll0223	NADH dehydrogenase subunit 2	ndhB
slr1279	NADH dehydrogenase subunit 3	ndhC
sll0027	NADH dehydrogenase subunit 4	ndhD
sll1733	NADH dehydrogenase subunit 4	ndhD
slr0331	NADH dehydrogenase subunit 4	ndhD
slr2007	NADH dehydrogenase subunit 4	ndhD2
slr1291	NADH dehydrogenase subunit 4	ndhE
sll0522	NADH dehydrogenase subunit 4L	ndhF
sll0026	NADH dehydrogenase subunit 5	ndhF
sll1732	NADH dehydrogenase subunit 5	ndhF
slr0844	NADH dehydrogenase subunit 5	ndhG
slr2009	NADH dehydrogenase subunit 5	ndhH
sll0521	NADH dehydrogenase subunit 6	ndhI
slr0261	NADH dehydrogenase subunit 7	ndhJ
sll0520	NADH dehydrogenase subunit NdhI	
slr1281	NADH dehydrogenase subunit I	

the genes were single copy, although gene multiplication was observed for the genes *psbA, psbD, psbC* (*isiA*), *cpcC, cpcG, psaK, ndhD, ndhF, petC,* and *petF.* The genes of small subunits of photosystems I and II components, namely, *psaG* (47), *psaH* (48), *psaN* (33), *psbP* (26), *psbQ* (40), *psbR* (17), *psbS* (69), *psbT* (32), and *psbW* (62), which are commonly present in the nuclear genomes of higher plants, were not found in the *Synechocystis* genome.

Relationship to Plant Plastids

It is generally accepted that progenitors of unicellular cyanobacteria also evolved to plastids after a long period of endosymbiosis. If this is the case, evolutionary traces of such a relationship may remain in the genomes of extant cyanobacteria and plastids. Presently, the complete nucleotide sequences of *Nicotiana tabacum* (55), *Zea mays* (38), *Oryza sativa* (24), *Pinus thunbergii* (67), *Marchantia polymorpha* (45), *Epifagus virginiana* (74), *Chlorella ellipsoidea* (66), *Euglena gracilis* (21), *Cyanophora paradoxa* (60), *Odentella sinensis* (36), and *Porphyra purpurea* (51) are available. The complete lists of potential gene complements of each plastid have been compiled. The genomes of plastids from green plants are 120 to 156 kb circles that contain 100 to 200 genes and gene candidates (62). In contrast, the genome of *E. virginiana,* a nongreen plant that has lost the ability to photosynthesize, is only 70 kb long with just 38 genes (74). Comparison of 3168 genes of the *Synechocystis* genome with those of 10 plastid genomes has shown that a total of 224 genes in the *Synechocystis* genome have homologues in the plastid genomes. The *Synechocystis* genes common to each plastid were grouped according to biological function, and the number of genes in each category is listed in Table 3. It seems that genes with a variety of functions are conserved in cyanobacterial and plastid genomes; fewer genes are conserved in the plastids of higher plants, despite the fact that fewer genes are contained in the genomes of higher plant plastids compared with lower plants and algae. The most significant similarity in genomes was observed between *Synechocystis* and the plastid of a red alga, *P. purpurea* (51), where nearly 95% of the plastid genes showed significant homology to *Synechocystis* genes.

Genes of Two-Component Signal Transduction Systems

Bacteria acclimate to a changing environment (49) using two-component signal transduction systems containing a sensory kinase and a response regulator (61). Sensory kinases monitor environmental conditions and phosphorylate regulators; response regulators modulate transcription of the genes that are involved in cell response. Eighty genes for two component signal transducers, including 26 genes for sensory kinases, 38 genes for response regulators, and 16 genes for hybrid sensory kinases containing both transmitter and receiver domains

Table 3 Comparison of gene complements among *Synechocystis* and 10 plastids

Functional categories	*Nicotiana tabacum*	*Zea mays*	*Oryza sativa*	*Pinus thunbergii*	*Marchantia polymorpha*	*Euglena gracilis*	*Epifagus virginiana*	*Cyanophora paradoxa*	*Odontella sinensis*	*Porphyra purpurea*
Photosynthesis and respiration	41	41	41	31	42	26	—	47	42	54
Amino acid biosynthesis	—	—	—	—	—	—	—	2	—	7
Biosynthesis of cofactors, prosthetic groups, and carriers	—	—	—	3	3	1	—	8	2	9
Cell envelope	—	—	—	—	—	—	—	1	—	6
Cellular processes	—	—	—	—	—	—	—	4	6	4
Energy metabolism	—	—	—	—	—	—	—	—	—	5
Fatty acid, phospholipid, and sterol metabolism	1	—	—	1	1	—	1	1	1	—
Purines, pyrimidines, nucleosides, and nucleotides	—	—	—	—	—	—	—	3	1	6
DNA replication, restriction, modification, recombination, and repair	—	—	—	—	—	—	—	—	1	1
Transcription	4	4	4	4	4	3	—	4	4	5
Translation	23	23	23	22	23	22	17	39	46	54
Transport and binding proteins	1	1	1	1	2	—	—	4	2	2
Other categories	—	—	—	—	1	—	—	2	1	4
Hypothetical	5	5	5	6	7	2	—	18	16	37
Total	75	74	74	68	83	54	18	133	122	194

were identified in *Synechocystis* (43). The genes are scattered throughout the genome, and many of the genes exist as a single unit. This feature of gene organization makes pairing sensory kinases with specific response regulators difficult. In contrast, most *E. coli* two-component system genes that interact are in operons (42).

It is natural to speculate that genes for two-component systems in the progenitors of cyanobacteria have been transmitted to plastids. Actually, genes coding for sensors and regulators were found in the plastid genomes of the red alga *P. purpurea* (51) and the glaucocystophyte *C. paradoxa* (60). Some plastid genes for two-component systems show sequence similarity to those of the *Synechocystis* genome. In addition, the gene products of *ycf26, ycf27,* and *ycf29* are similar to those of sll0698 (drug sensory protein A) (2), slr0947, which codes for a regulator belonging to OmpR subfamily, and slr1783 and slr1909, regulators of NarL subfamily, respectively (43).

The genes for a putative two-component system were reported in the nuclear genome of the higher plant, *Arabidopsis thaliana*, and their homologues were also found in tomato (9, 28, 71). This observation may suggest a common lineage of the two-component system genes in cyanobacteria and plants. Portions of several putative gene products of sensory kinase genes in *Synechocystis* showed a high degree of sequence similarity to a chromophore-binding domain commonly found in phytochromes of green plants. The translated amino acid sequence of slr0473 is similar to that of phytochrome C in higher plants (12). The gene product of slr0473 produced in *E. coli* can bind phycocyanobilin efficiently in vitro. The gene product also showed the spectra characteristic to a plant phytochrome after irradiation with red and far-red light (27). These results indicate that the gene product of slr0473 fulfills the criteria of a functional phytochrome. A putative product of another sensory kinase gene, sll1124, also contained the chromophore-binding domain (70). Introduction of mutations in sll1124 had no effect on photoautotrophic growth under white or far-red light; a drastic effect occurred in blue light, suggesting that the gene may function as one of the light receptors involved in photoautotrophic growth.

PERSPECTIVES

Postsequencing

For organisms with entirely sequenced genomes, the most urgent goal is to understand gene functions. In *Synechocystis*, even though the structures of 3168 genes have been deduced from nucleotide sequence, the functional role of approximately 55% of these genes has yet to be been ascertained. To achieve this goal, several strategies were examined, and some of them proved to be useful.

Utilizing the natural transformation ability of *Synechocystis* (19), targeted gene disruption, in which a DNA fragment containing a gene for a selective marker is inserted into a given gene by homologous recombination, has been carried out for a variety of genes, including many that are involved in photosynthesis. The availability of sequence information for the entire genome should accelerate a large-scale gene disruption project. Random gene disruption by insertion of *E. coli* transposons may also be useful because technology for the insertion of *E. coli* transposons into a cloned DNA segment using in vivo or in vitro transposition systems, followed by integration into the genome by transformation, has been developed (68). Another promising strategy would be to express *Synechocystis* genes in *E. coli* so that proteins can be characterized in vitro. One example discussed above is of slr0473. The gene products of slr0088 and slr0611 were shown to be β-carotene ketolase, which produces echinenone in the carotenoid biosynthesis pathway (14), and solanesyl diphosphate synthase, which is involved in the biosynthesis of ubiquinone 9 (46), respectively.

Because the primary structures of genes and their putative products are known, it may be possible to monitor the expression of many or even all the genes in the genome using sequence information as tags. Transcriptional levels of each of the gene constituents can be measured by microarray technology (18) using a panel with DNA segments corresponding to the individual genes. Translational levels of genes can be assessed by the two-dimensional gel patterns of soluble proteins in cell extracts. By analyzing the partial amino acid sequence of the N terminus of a given protein spot, and by comparing the back-translated sequence with the genome sequence, the corresponding gene can be easily identified. In *Synechocystis*, the N termini of 96 major proteins were sequenced, and each of them was correlated to the corresponding gene in the genome (54). These new technologies are useful for observing changes of transcript or protein levels on a genome-wide basis at the same time under different conditions; for example, under various culture conditions, in the presence and absence of drugs, and before and after disruption of genes.

Database for the Genome Information of Synechocystis sp. PCC6803

Genome information contained in the 3,573,470-bp sequence of *Synechocystis* is too much to process at one time. The nucleotide sequence and information about the assigned 3168 protein-coding genes in the *Synechocystis* genome have been posted in the web database, CyanoBase, at http://www.kazusa.or.jp/cyano, in such a way that pages containing various kinds of information are closely linked. Presented in CyanoBase are the graphical images of a circular physical map and a linear gene map, gene information such as the translated amino acid sequences and the nucleotide sequences of the coding and adjacent regions, the results of a similarity search, and a gene category list in which genes are grouped into 15 functional categories. CyanoBase also provides a keyword search and similarity search engines, and an FTP site that makes available the nucleotide sequence file and the results of similarity search for all the genes. To extract as much information as possible from the nucleotide sequence of the entire genome of *Synechocystis* and to fully utilize the sequence data for the functional analysis of the genes in the future, careful examination of the data from various viewpoints is necessary. CyanoBase, therefore, is one of the most powerful tools for such examination and is useful for studies that may lead to a better understanding of the genetic systems in cyanobacteria.

> Visit the *Annual Reviews home page* at
> http://www.AnnualReviews.org.

Literature Cited

1. Bancroft I, Wolk CP, Oren EV. 1989. Physical and genetic maps of the genome of the heterocyst-forming cyanobacterium *Anabaena* sp. strain PCC 7120. *J. Bacteriol.* 171:5940–48
2. Bartsevich VV, Shestakov SV. 1995. The dspA gene product of the cyanobacterium *Synechocystis* sp. strain PCC 6803 influences sensitivity to chemically different growth inhibitors and has amino acid similarity to histidine protein kinases. *Microbiology* 141:2915–20
3. Borodovsky M, Koonin EV, Rudd KE. 1994. New genes in old sequences: a strategy for finding genes in a bacterial genome. *Trends Biochem. Sci.* 19:309–13
4. Borodovsky M, McIninch JD. 1993. GENMARK: parallel gene recognition for both DNA strands. *Comput. Chem.* 17:123–33
5. Borodovsky M, Rudd KE, Koonin EV. 1994. Intrinsic and extrinsic approaches for detecting genes in a bacterial genome. *Nucleic Acids Res.* 22:4756–67
6. Bult CJ, White O, Olsen GJ, Zhou LX, Fleischmann RD, et al. 1996. Complete genome sequence of the Methanogenic archaeon, *Methanococcus jannaschii*. *Science* 273:1058–73
7. Cai Y. 1991. Characterization of insertion sequence IS892 and related elements from the cyanobacterium *Anabaena* sp. strain PCC 7120. *J. Bacteriol.* 173:5771–77
8. Castets AM, Houmard J, de Marsac NT. 1986. Is cell motility a plasmid-encoded function in the cyanobacterium *Synechocystis* 6803? *FEMS Microbiol. Lett.* 37:277–81
9. Chang C, Kwok SF, Bleecker AB, Meyerowitz EM. 1993. *Arabidopsis* ethylene-response gene ETR1: similarity of product to two-component regulators. *Science* 262:539–44
10. Chauvat F, Rouet P, Bottin H, Boussac A. 1989. Mutagenesis by random cloning of an *Escherichia coli* kanamycin resistance gene into the genome of the cyanobacterium *Synechocystis* PCC 6803: selection of mutants defective in photosynthesis. *Mol. Gen. Genet.* 216:51–59
11. Chen XG, Widger WR. 1993. Physical genome map of the unicellular cyanobacterium *Synechococcus* sp. strain PCC7002. *J. Bacteriol.* 175:5106–16
12. Cowl JS, Hartley N, Xie DX, Whitelam GC, Murphy GP, et al. 1994. The PHYC gene of *Arabidopsis*. Absence of the third intron found in *PHYA* and *PHYB*. *Plant Physiol.* 106:813–14
13. de Marsac NT, Houmard J. 1987. Advance in cyanobacterial molecular genetics. In *The Cyanobacteria*, ed. P Fay, CV Baalen, pp. 251–302. Amsterdam: Elsevier
14. Fernandez-Gonzalez B, Sandmann G, Vioque A. 1997. New type of asymmetrically acting beta-carotene ketolase is required for the synthesis of echinenone in the cyanobacterium *Synechocystis* sp. PCC 6803. *J. Biol. Chem.* 272:9728–33
15. Fraser CM, Gocayne JD, White O, Adams MD, Clayton RA, et al. 1995. The minimal

gene complement of *Mycoplasma genitalium*. *Science* 269:397–403
16. Freischmann RD, Adams MD, White O, Clayton RA, Kiekness EF, et al. 1995. Whole-genome random sequencing and assembly of *Haemophilus influenzae* Rd. *Science* 269:496–512
17. Gil-Gomez G, Marrero PF, Haro D, Ayte J, Hegardt FG. 1991. Characterization of the gene encoding the 10-kDa polypeptide of photosystem II from *Arabidopsis thaliana*. *Plant Mol. Biol.* 17:517–22
18. Goffeau A. 1997. Molecular fish on chips. *Nature* 385:202–3
19. Grigorieva G, Shestakov S. 1982. Transformation in the cyanobacterium *Synechocystis* 6803. *FEMS Microbiol. Lett.* 13:367–70
20. Gupta A, Morby AP, Turner JS, Whitton BA, Robinson NJ. 1993. Deletion within the metallothionein locus of cadmium-tolerant *Synechococcus* PCC 6301 involving a highly iterated palindrome (HIP1). *Mol. Microbiol.* 7:189–95
21. Hallick RB, Hong L, Drager RG, Favreau MR, Montfort A, et al. 1993. Complete sequence of the *Euglena gracilis* chloroplast DNA. *Nucleic Acids Res.* 21:3537–44
22. Herdman M, Janvier M, Waterbury JB, Rippka R, Stanier RY. 1979. Deoxyribonucleic acid base composition of cyanobacteria. *J. Gen. Microbiol.* 111:63–71
23. Himmelreich R, Hilbert H, Plagens H, Pirkl E, Li BC, et al. 1996. Complete sequence analysis of the genome of the bacterium *Mycoplasma pneumoniae*. *Nucleic Acids Res.* 24:4420–49
24. Hiratsuka J, Shimada H, Whittier R, Ishibashi T, Sakamoto M, et al. 1989. The complete sequence of the rice (*Oryza sativa*) chloroplast genome: intermolecular recombination between distinct tRNA genes account for a major plastid inversion during the evolution of cereals. *Mol. Gen. Genet.* 217:185–94
25. Hirosawa M, Kaneko T, Tabata S, McIninch JD, Hayes WS, et al. 1995. Computer survey for likely genes in the one megabase contiguous genomic sequence data of *Synechocystis* sp. strain PCC6803. *DNA Res.* 2:239–46
26. Hua SB, Dube SK, Barnett NM, Kung SD. 1991. Nucleotide sequence of a cDNA clone encoding 23-kDa polypeptide of the oxygen-evolving complex of photosystem II in tobacco, *Nicotiana tabacum* L. *Plant Mol. Biol.* 16:749–50
27. Hughes J, Lamparter T, Mittmann F, Hartmann E, Gärtner W, et al. 1997. A prokaryotic phytochrome. *Nature* 386:663

28. Kakimoto T. 1996. CKI1, a histidine kinase homolog implicated in cytokinin signal transduction. *Science* 274:982–85
29. Kaneko T, Matsubayashi T, Sugita M, Sugiura M. 1996. Physical and gene maps of the unicellular cyanobacterium *Synechococcus* sp. strain PCC6301 genome. *Plant Mol. Biol.* 31:193–201
30. Kaneko T, Sato S, Kotani H, Tanaka A, Asamizu E, et al. 1996. Sequence analysis of the genome of the unicellular cyanobacterium *Synechocystis* sp. strain PCC6803. II. Sequence determination of the entire genome and assignment of potential protein-coding regions. *DNA Res.* 3:109–36
31. Kaneko T, Tanaka A, Sato S, Kotani H, Sazuka T, et al. 1995. Sequence analysis of the genome of the unicellular cyanobacterium *Synechocystis* sp. strain PCC6803. I. Sequence features in the 1 Mb region from map positions 64% to 92% of the genome. *DNA Res.* 2:153–66
32. Kapazoglou A, Sagliocco F, Dure L III. 1995. PSII-T, a new nuclear encoded lumenal protein from photosystem II. Targeting and processing in isolated chloroplasts. *J. Biol. Chem.* 270:12197–202
33. Knoetzel J, Simpson DJ. 1993. The primary structure of a cDNA for *PsaN*, encoding an extrinsic lumenal polypeptide of barley photosystem I. *Plant Mol. Biol.* 22:337–45
34. Kotani H, Kaneko T, Matsubayashi T, Sato S, Sugiura M, et al. 1994. A physical map of the genome of a unicellular cyanobacterium *Synechocystis* sp. strain PCC6803. *DNA Res.* 1:303–7
35. Kotani H, Tanaka A, Kaneko T, Sato S, Sugiura M, et al. 1995. Assignment of 82 known genes and gene clusters in the genome of the unicellular cyanobacterium *Synechocystis* sp. strain PCC6803. *DNA Res.* 2:133–42
36. Kowallik KV, Stoebe B, Schaffran I, Kroth-Pancic P, Freier U. 1995. The chloroplast genome of a chlorophyll a+c-containing alga, *Odontella sinensis*. *Plant Mol. Biol. Report.* 13:336–42
37. Liu SL, Hessel A, Sanderson KE. 1993. The *XbaI-BlnI-CeuI* genomic cleavage map of *Salmonella enteritidis* shows an inversion relative to *Salmonella typhimurium* LT2. *Mol. Microbiol.* 10:655–64
38. Maier RM, Neckermann K, Igloi GL, Kossel H. 1995. Complete sequence of the maize chloroplast genome: gene content, hotspots of divergence and fine tuning of genetic information by transcript editing. *J. Mol. Biol.* 251:614–28

39. Mann N, Carr NG. 1974. Control of macromolecular composition and cell division in the blue-green algae *Anacystis nidulans*. *J. Gen. Microbiol.* 83:399–405

40. Mayfield SP, Schirmer-Rahire G, Frank H, Zuber H, Rochaix JD. 1989. Analysis of the genes of the oee1 and oee3 proteins of the photosystem II complex of *Chlamydomonas reinhardtii*. *Plant Mol. Biol.* 12:683–93

41. Mazel D, Houmard J, Castets AM, de Marsac NT. 1990. Highly repetitive DNA sequences in cyanobacterial genomes. *J. Bacteriol.* 172:2755–61

42. Mizuno T. 1997. Compilation of all genes encoding two-component phosphotransfer signal transducers in the genome of *Escherichia coli*. *DNA Res.* 4:161–68

43. Mizuno T, Kaneko T, Tabata S. 1996. Compilation of all genes encoding bacterial two-component signal transducers in the genome of the cyanobacterium, *Synechocystis* sp. strain PCC6803. *DNA Res.* 3:407–14

44. Ohtsubo E, Sekine Y. 1996. Transposable elements. *Curr. Top. Microbiol. Immunol.* 204:1–26

45. Ohyama K, Fukuzawa H, Kohchi T, Shirai H, Sano T, et al. 1986. Chloroplast gene organization deduced from complete sequence of liverwort *Marchantia polymorpha*. *Nature* 322:572–74

46. Okada K, Minehira M, Zhu XF, Suzuki K, Nakagawa T, et al. 1997. The *ispB* gene encoding octaprenyl diphosphate synthase is essential for growth of *Escherichia coli*. *J. Bacteriol.* 179:3058–60

47. Okkels JS, Nielsen VS, Scheller HV, Moller BL. 1992. A cDNA clone from barley encoding the precursor from the photosystem I polypeptide PSI-G: sequence similarity to PSI-K. *Plant Mol. Biol.* 18:989–94

48. Okkels JS, Scheller HV, Jepsen LB, Moller BL. 1989. A cDNA clone encoding the precursor for a 10.2 kDa photosystem I polypeptide of barley. *FEBS Lett.* 250:575–79

49. Parkinson JS. 1993. Signal transduction schemes of bacteria. *Cell* 73:857–71

50. Deleted in proof

51. Reith M, Munholland J. 1995. Complete nucleotide sequence of the *Porphyra purpurea* chloroplast genome. *Plant Mol. Biol. Report.* 13:333–35

52. Riley M. 1993. Functions of the gene products of *Escherichia coli*. *Microbiol Rev.* 57:862–952

53. Robinson NJ, Robinson PJ, Gupta A, Bleasby AJ, Whitton BA, et al. 1995. Singular over-representation of an octameric

palindrome, HIP1, in DNA from many cyanobacteria. *Nucleic Acids Res.* 23:729–35

54. Sazuka T, Ohara O. 1996. Sequence features surrounding the translation initiation sites assigned on the genome sequence of *Synechocystis* sp. strain PCC6803 by amino-terminal protein sequencing. *DNA Res.* 3:225–32

55. Shinozaki K, Ohme M, Tanaka M, Wakasugi T, Hayashida N, et al. 1986. The complete nucleotide sequence of the tobacco chloroplast genome: its gene organization and expression. *EMBO J.* 5:2043–49

56. Smart LB, McIntosh L. 1993. Genetic inactivation of the *psaB* gene in *Synechocystis* sp. PCC 6803 disrupts assembly of photosystem I. *Plant Mol. Biol.* 1:177–80

57. Smith HO, Tomb J, Drougherty BA, Fleischmann RD, Venter JC. 1995. Frequency and distribution of DNA uptake signal sequences in the *Haemophilus influenzae* Rd genome. *Science* 269:538–40

58. Smith TF, Waterman M. 1981. Identification of common molecular subsequences. *J. Mol. Biol.* 147:195–97

59. Steinmuller K, Bogorad L. 1990. Identification of a psbG-homologous gene in *Synechocystis* sp. PCC6803. In *Current Research in Photosynthesis*, ed. M Baltsscheffsky, 3:12557–60. Dordrecht: Kluwer

60. Stirewalt VL, Michalowski CB, Loffelhardt W, Bohnert HJ, Bryant DA. 1995. Nucleotide sequence of the cyanelle genome from *Cyanophora paradoxa*. *Plant Mol. Biol. Report.* 13:327–32

61. Stock JB, Surette MG, Levit M, Stock AM. 1995. Two-component signal transduction systems: structure-function relationships and mechanisms of catalysis. In *Two-Component Signal Transduction*, ed. JA Hoch, TJ Silhavy, pp. 25–51. Washington, DC: Am. Soc. Microbiol

62. Sugiura M. 1992. The chloroplast genome. *Plant Mol. Biol.* 19:149–68

63. Tomb JF, White O, Kerlavage AR, Clayton RA, Sutton GG, et al. 1997. The complete genome sequence of the gastric pathogen *Helicobacter pylori*. *Nature* 388:539–47

64. Umeda M, Ohtsubo E. 1989. Mapping of insertion elements IS1, IS2 and IS3 on the *Escherichia coli* K-12 chromosome. Role of the insertion elements in formation of Hfrs and F′ factors and in rearrangement of bacterial chromosomes. *J. Mol. Biol.* 208:601–14

65. Umeda M, Ohtsubo E. 1990. Mapping of insertion element IS5 in the *Escherichia coli* K-12 chromosome. Chromosomal rearrangements mediated by IS5. *J. Mol. Biol.* 213:229–37

66. Wakasugi T, Nagai T, Kapoor M, Sugita M, Ito M, et al. 1997. Complete nucleotide sequence of the chloroplast genome from the green alga *Chlorella vulgaris*: the existence of genes possibly involved in chloroplast division. *Proc. Natl. Acad. Sci. USA* 94:5967–72

67. Wakasugi T, Tsudsuki J, Ito S, Nakashima K, Tsudsuki T, et al. 1994. Loss of all *ndh* genes as determined by sequencing the entire chloroplast genome of the black pine *Pinus thunbergii*. *Proc. Natl. Acad. Sci. USA* 91:9794–98

68. Wang G, Blakesley RW, Berg DE, Berg CM. 1993. pDUAL: a transposon-based cosmid cloning vector for generating nested deletions and DNA sequencing templates in vivo. *Proc. Natl. Acad. Sci. USA* 90:7874–78

69. Wedel N, Klein R, Ljungberg U, Andersson B, Herrmann RG. 1992. The single-copy gene *psbS* codes for a phylogenetically intriguing 22 kDa polypeptide of photosystem II. *FEBS Lett.* 314:61–66

70. Wilde A, Churin Y, Schubert H, Börner T. 1997. Disruption of a *Synechocystis* sp. PCC 6803 gene with partial similarity to phytochrome genes alters growth under changing light qualities. *FEBS Lett.* 406:89–92

71. Wilkinson JQ, Lanahan MB, Yen HC, Giovannoni JJ, Klee HJ. 1995. An ethylene-inducible component of signal transduction encoded by never-ripe. *Science* 270:1807–9

72. Williams N. 1996. Yeast genome sequence ferments new research. *Science* 272:481

73. Wilmotte A. 1994. Molecular evolution and taxonomy of the cyanobacteria. In *The Molecular Biology of Cyanobacteria*, ed. DA Bryant, pp. 1–25. Dordrecht: Kluwer

74. Wolfe KH, Morden CW, Palmer JD. 1992. Function and evolution of a minimal plastid genome from a nonphotosynthetic parasitic plant. *Proc. Natl. Acad. Sci. USA* 89:10648–52

Annu. Rev. Plant Physiol. Plant Mol. Biol. 1998. 49:173–98

ELABORATION OF BODY PLAN AND PHASE CHANGE DURING DEVELOPMENT OF *ACETABULARIA*: How Is the Complex Architecture of a Giant Unicell Built?

Dina F. Mandoli

Department of Botany, Box 355325, University of Washington, Seattle, Washington 98195-5325; e-mail: mandoli@u.washington.edu

KEY WORDS: morphogenesis, grafting, amputation, mutants, translation

ABSTRACT

While uninucleate and unicellular, *Acetabularia acetabulum* establishes and maintains functionally and morphologically distinct body regions and executes phase changes like those in vascular plants. Centimeters tall at maturity, this species has allowed unusual experimental approaches. Amputations revealed fates of nucleate and enucleate portions from both wild type and mutants. Historically, graft chimeras between nucleate and enucleate portions suggested that morphological instructions were supplied by the nucleus but resided in the cytoplasm and could be expressed interspecifically. Recently, graft chimeras enabled rescue of mutants arrested in vegetative phase. Since the 1930s, when *Acetabularia* provided the first evidence for the existence of mRNAs, a dogma has arisen that it uses long-lived mRNAs to effect morphogenesis. While the evidence favors translational control, the postulated mRNAs have not been identified, and the mechanism of morphogenesis remains unknown. Amenable to biochemistry, physiology, and both classical and molecular genetics, *Acetabularia* may contribute yet new insights into plant development and morphogenesis.

CONTENTS

173

1040-2519/98/0601-0173$08.00

WHY USE *ACETABULARIA?*

Species of marine green algae in the order Dasycladales present opportunities to study extreme cases of organismal body plan (patterning) and development without cellularization. Despite dependence on a single nucleus for most of their life cycle, these evolutionarily successful species reach heights of 1–10 cm at reproductive maturity and undergo elaborate morphogenesis concomitant with development (7). The most studied member of this order (cf suggested elevation to class Dasycladophyceae, 109) is the type species, *Acetabularia acetabulum* (= *mediterranea*). When vegetative (Figure 1), *A. acetabulum* superficially resembles the "horsetail" *Equisetum.* When reproductive, *A. acetabulum* resembles a miniature goblet, hence the nickname, The Mermaid's Wineglass. Many aspects of growth of and development in this unicell are reminiscent of multicellular vascular plants. For example, during vegetative growth, whorls of hairs and interwhorls alternate and are stacked one on the other, just like the phyllodes of vascular plants. In addition, in preparation for reproductive phase, *A. acetabulum* goes through juvenile and adult vegetative phases (87). In any unicell, consideration of cell-cell interactions within the organism is moot, so being unicellular eliminates a layer of developmental complexity. Compared with other unicellular models for development such as oocytes and early embryos of *Fucus*, flies, and frogs, *Acetabularia* has added appeal because all of vegetative development and the switch to reproductive phase is orchestrated by a single cytoplasmic compartment in conjunction with a single nucleus.

Technical advantages made this a developmental model of choice from circa 1930–1970, as one publication history indicates (69). Simple amputations led to an accurate description of the origin and role of mRNAs in eukaryotes (35, 36) thirty years before the elucidation of the chemical nature of mRNAs in prokaryotes (45). The discovery that this alga tolerated grafting (reviewed in 40, 41) led to experiments that mixed and matched disparate nucleate and enucleate body parts within and between species and genera (13, 37, 112). Nuclear transplantations were performed to infer the contributions of the nucleus and cytoplasm to development, morphogenesis, and circadian rhythmicity (e.g. 8, 37, 55, 96, 98),

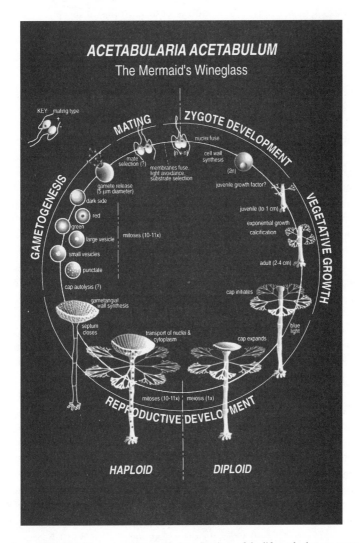

Figure 1 A view of the life cycle of *A. acetabulum*. Portions of the life cycle that are speculative or supported only by observation are flagged with *question marks*. More extensive discussion of the life cycle can be found elsewhere (69). The arrangement of the flagella during mating has not been clearly documented, but it is 11 and 5 o'clock in free-swimming gametes (see references in 109). A color rendition of this image resides at http://weber.washington.edu/~mandoli.

and to transform this plant (85). Some biochemical assays demanded as few as 1–4 plants (3, 4, 98), and the isolation of crude cytoplasm for those assays was easy: The stalk was snipped off, and the cytoplasm was stripped out just as one squeezed toothpaste from a tube. Previous researchers were cognizant that the basal portion left behind retained the ability to make progeny and potentially could serve as a renewable source of cytoplasm because it readily regrew after each amputation of the previous apical portion (see references in 5, 12, 50, 93). However, this feature of the organism was not fully exploited. Microinjection of nuclei with foreign DNA (19, 20) led to stable transformation in 70% of the plants (85). Finally, there was a hint of the potential for genetic analysis (59, 60): *A. acetabulum* has a reasonable genome size [20 chromosomes (24), 0.92 pg per haploid genome (105), small intervening sequences (106)], each plant produces millions of progeny per generation (see discussion and references in 69), and it can be selfed and outcrossed (33, 34, 59, 60). Despite its classical contributions, technical versatility, and obvious appeal, use of this taxon has decreased since the 1970s (69).

Three major roadblocks contributed to reduction in use of the system (69): the life cycle was too long, it was labor intensive to obtain and keep clean monocultures for routine use, and large-scale genetic analyses were not feasible. Significant headway has been made on all three fronts. The life cycle, which runs 1–2 years in the wild and 6 months in most laboratories, was reduced to 94 days for heterogeneous wild type (69) by improving the physiological conditions in which the species was grown. The first significant reductions in the duration of the life cycle were obtained when cultures were rendered axenic (42, 74), when plants were grown in an artificial seawater tailor-made for the type species (44), and when the effect of population density on the rate of development was understood (22, 69, 113). Later, a method that yielded single-cell suspensions of zygotes (73) made it possible to adjust the population densities of cultures (69). This method was important because zygotes tend to attach to each other, creating sheets or balls of plants, a physical arrangement that seems detrimental to growth. In contrast, populations of plants that are not attached to each other are fairly synchronous in development and are healthier and faster growing than siblings that are clumped (see further discussion in 69). An initial foray into genetics was made feasible by developing a mating matrix that allowed virtually all progeny to be recovered (73) and by beginning to study phenotypes defective in development and morphology (68, 70). In sum, improved understanding of the physiology of the plant has enabled these three roadblocks to be removed or substantively reduced.

Now, large-scale genetic analyses are practical in this species (69). Inbreeding and selection have decreased the genetic load carried by wild-type laboratory strains, promising uniform genetic background for mutant analyses, and have resulted in homogeneous wild-type lineages of plants with uniform wild-type morphology that self well (BE Hunt & DF Mandoli, unpublished data). We have defined conditions for long-term storage of haploid germplasm [15 months with 89% recovery (43)] and have begun to adapt microparticle bombardment to *A. acetabulum* (C Geil & DF Mandoli, unpublished data). Although further improvements in culture conditions, strains, and genetic manipulations can and will be made, the minor problems with culture that remain do not preclude full use of the system (69), and genetic analyses of interesting developmental mutations are under way (e.g. 72).

APICES ESTABLISH THE BODY PATTERN DURING PHASE CHANGE

Images and description of the life cycle of this organism abound in the literature (23, 54, 61, 69, 93, 100), and reviews covering diverse topics are numerous (2, 5, 12, 15, 16, 32, 39–41, 50, 54, 68, 69, 93, 97, 108). Details of the life cycle of *A. acetabulum* as we now understand it (Figure 1) are offered because, in my opinion, a deep understanding of development throughout the life cycle can only increase the utility of any model system.

The body plan of *A. acetabulum* is elaborated by three types of growth regions that are regulated in number and location on the body during development and that have functions, behaviors, and morphologies that distinguish them from each other (Figure 2). I suggest that the term *apex* is applicable to each of these localized regions of growth. The *stalk apex* (1 per plant) grows upward and generates the stalk and the whorls of hairs during vegetative growth and generates whorls of the cap during reproductive growth (Figure 2). The *hair apices* that are arranged in rings (10–19 whorls of hairs per stalk with >4 more arising from the cap) initially grow upward then settle perpendicular to the stalk axis and, except the first few made during juvenile growth, bifurcate as they grow. The *rhizoid apices* (~8–10 per plant) grow downward and generate the digits of the rhizoid. The stalk and hair apices make straight cell walls, and rhizoid apices make curvilinear ones. To the best of my knowledge, the mechanism(s) of growth at these apices are not defined for any species in the Dasycladales, and this is a fundamental gap in our knowledge. Identifying the borders of and potential subdomains within the apices of *A. acetabulum* with molecular and genetic markers may also be useful, especially in light of the insights gained by understanding functional domains and organogenesis (see 21) within the shoot apical meristem of vascular plants (see 51). The biology of the apices of *A. acetabulum* suggests that these growing regions are biochemically distinct and spatially restricted, but how such regional specificity is established and maintained given the lack of cellularization is entirely unknown.

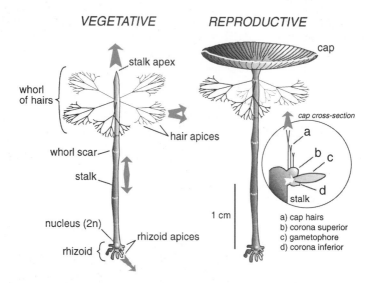

Figure 2   Anatomy of vegetative and reproductive plants are shown with each region of growth on *A. acetabulum* highlighted with an *arrow* indicating the major direction(s) of growth (modified from 86). In the cap cross section, the *white arrow* indicates the position of continued growth of the stalk apex, and the *gray arrow* indicates growth of the hair apices of the cap. For the sake of clarity, only one whorl of hairs is depicted on the vegetative plant that would normally have 10–19 whorls of hairs (see Figure 4). During reproductive phase, the whorls of hairs fall off as shown.

As it generates the majority of the body plan of the organism, the stalk apex of *A. acetabulum* progresses through discrete and predictable changes in shape during vegetative and reproductive development (Figure 3). The shoot apical meristem of a multicellular vascular plant partitions itself into spatially distinct stem and leaves, whereas the stalk apex of *A. acetabulum* alternately produces regions of stalk and whorls of sterile hairs during vegetative growth. The patterns of interwhorl and whorl production in *A. acetabulum* bear an uncanny resemblance to the patterns of internodes and nodes of multicellular plants. Like internodes, the interwhorls of *A. acetabulum* continue to grow in length (87) but do not alter the body plan per se. Like leaf arrangements and structures in vascular plants, the hair apices generate branched, needle-like hairs that are arranged in a ring. Unlike the leaves of vascular plants, hairs of *A. acetabulum* lack dormant growing regions adjacent to the stalk that will grow if the primary hair apex is lost, i.e. they lack regions that function like axillary buds. Akin to flower production in vascular plants, during reproductive growth the stalk apex makes a morphologically distinct, trilobed-whorl which is called

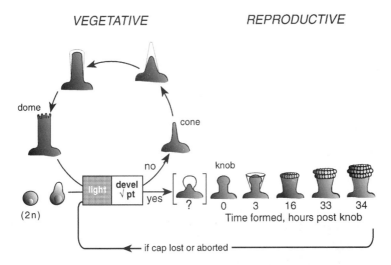

Figure 3 Shape changes in the stalk apex during vegetative and reproductive phases (adapted from 62). Vegetative shapes are based on data of Schmid (95). Reproductive onset depends on completion of vegetative phases, i.e. the plant must pass some "developmental checkpoint" and is marked by "*knob*." Knob (t = 0) is the only light-regulated shape change in the switch from vegetative to reproductive growth (62). The time below each *apex shape* indicates the hour at which 50% of a population of plants had a stalk apex of that shape.

the "cap" (Figures 1 and 2). The shape changes characteristic of the hair and rhizoidal apices have not been documented. However, especially given the paucity of data on the apices of *A. acetabulum*, it is premature to speculate whether any of the resemblances to vascular plants noted above are any more than superficial.

On a larger temporal and spatial scale, *A. acetabulum* progresses through juvenile and adult phases in development (69, 80, 87) as do vascular plants (63). These phases are morphologically distinct (Figure 4), temporally sequential and predictable, spatially stacked as they are in *Zea*, and physiologically distinct (Table 1). The basic features of juvenile and adult phases are similar in genetically heterogeneous wild type (87) and in two inbred wild-type strains (Table 2; 80). In general, the borders between the phases or "phase transitions" have been poorly characterized, and the molecular basis of phase change is unknown in *A. acetabulum* and has only begun to be elucidated in vascular plants (18).

Juvenile phase (Figure 4) comprises the first 1 cm of growth, represents 25–28% of the life cycle, and includes about 5–6 whorl-interwhorl units (87).

Table 1   Summary of phase characteristics of *Acetabularia acetabulum*[a]

	Juvenile phase	Adult phase	Reproductive phase
Physiology[b]			
Population density required	100 per mL	0.4 per mL	0.4 per mL
Growth rates[c]			
Rhizoids, mm per day[d]	0.04	0.02	?
Stalk, mm per day	0.17	0.61	0.19
Whorls of hairs, # per day	0.16	0.35	4–5 sets of cap hairs
Morphogenetic potential[c] of			
Apical portion			
Whorls of hairs, mean	0.2	2	<1 set of cap hairs
Cap	0	1	Not applicable
Basal portion			
Whorls of hairs, mean	13	10	10
Cap	1	1	1

[a]Values pertain to axenic growth and development in iterations of an artificial seawater (44).
[b]See References 69, 113.
[c]Hunt & Mandoli, unpublished data; 80.
[d]See Reference 86.

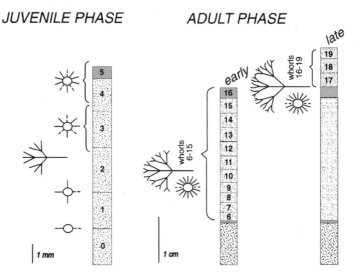

Figure 4   Summary of the spatial and morphological features that distinguish the vegetative phases from each other (derived from original data in 87). Note the separate scales relevant to juvenile and adult phases. Interwhorl lengths are drawn to scale and cross-referenced between the cartoons by common *shading textures* with likely regions of phase transitions indicated with a distinct texture. Positions of the whorls of hairs on each stalk are demarcated by *light horizontal lines*—these *lines* do not represent crosswalls. To the left of each stalk the number of hairs in a whorl and the branching patterns of those hairs is indicated for groups of whorls. *Dashed lines* indicate the standard errors for the hair arrangements in a group of whorls.

Juvenile stalks are threadlike and make whorls of just a few hairs that are clear (87). Juvenile rhizoids grow rapidly (Table 1; 86). In addition, juveniles may be uniquely able to attach readily to substrata (93), may retract their cytoplasm when overcrowded (93), and, unlike adults, may tend not to calcify (75). Physiologically, juvenile phase initiates well only in crowded conditions; zygotes differentiate best at a population density one million times the optimum for reproductive onset (22, 113), a population density at which adults die (113). One interpretation of the need for crowding of zygotes is that zygotes or juveniles make a growth factor that must accumulate in the medium to trigger differentiation to the siphonous growth habit, but this remains conjectural. In addition, juveniles may be more cold tolerant than adults. In the wild, observations suggest that "young algae overwinter" by retracting the cytoplasm into the rhizoid (93), and laboratory-grown juveniles, but not adults, can be stored in the cold often without adverse effects (7, 69). Apical portions amputated from juveniles survive but cannot initiate a cap without the nucleus (Table 2; 80), supporting application of the term juvenile for this portion of development.

Adult phase (Figure 4) comprises the remaining 2–3 cm of growth, represents 20–21% of the life cycle, and includes the next 10–14 whorl-interwhorls, i.e. the sixth to nineteenth ones made (87). Adult stalks grow rapidly (Table 1), average ~0.3 mm in width (71, 72), and make hairs that contain chloroplasts (87). Adult phase was split into early and late portions based on the morphology of the whorls of hairs: Hairs are more ramified and live twice as long in the late portion of adult phase (87). In general, the apical region of an early adult looks bald, and that of a late adult looks tufted because there are more whorls of hairs close to the stalk apex late in adult phase. Adults grow well only at low population densities, conditions in which zygotes fail to differentiate and juveniles grow poorly (22, 113). The stalk apex becomes competent to make a

Table 2 How phases in *Acetabularia acetabulum* might relate to morphogenetic potential

Primary bioassay	Interpretation of bioassay	Juvenile	Adult Early	Adult Late	Reproductive
When the intact plant makes a cap	Cytosolic inhibitor prevents premature reproductive onset	0%	0%	0%	100%
Apical portion (enucleate) makes a cap	Stable, cap-specific mRNAs are localized in the apical portion	0%[a]	15%[b]	60%[b]	Not applicable

[a]See Reference 80.
[b]See Reference 94.

cap without the nucleus during adult phase (Table 2). Adult phase terminates with initiation of the cap by the stalk apex.

Possible Functional Significance of Phases in A. acetabulum

Juvenile growth may serve to embed the sole nucleus rapidly in a protected locale, and, to a lesser degree, juvenile responses to body loss and adverse physiological conditions suggest that this phase might aid species survival. After the nascent zygote finds and adheres to a solid substratum, in the wild, the digits of the juvenile rhizoid grow into crevices in shells and pebbles (92, 93), and in the laboratory they spread out and flatten so as to grip the bottom of the Petri dishes. Because it houses the nucleus, rapid growth of the rhizoid (86) may quickly sequester the sole nucleus in a protected location away from grazers such as the ascoglossan, *Elysia timida* (75). While the extent to which these attributes are confined to juvenile phase and are missing in adult phase is not yet clear, assessing their functional significance using genetic analyses is now possible and looks promising.

Adult phase may function to increase the volume and surface area of the plant in preparation for the prolific reproduction of this unicell. Adult growth accounted for the major increase in plant height with a concomitant increase in the number of whorls of hairs (80, 87), implying that both the calculated surface area (31) and the volume of the plant increased more dramatically during adult phase. It is plausible that an increase in surface area may be important for nutrient uptake and for photosynthesis: Adult whorls of hairs contain chloroplasts (see Figure 2 in 31), each has a calculated surface area equal to that of the entire stalk (31), and each incorporates radioactive DNA and protein precursors (e.g. 11). The number of chloroplasts increases during development (67), with the main increase probably occurring during what we now call adult phase, an inference based on plant size. The response of the adult basal portion to loss of the apical portion is to repeat adult growth (Figure 5), which may reflect an important function: Given the fecundity of the organism and the presumed need for each gamete to have mitochondria and a chloroplast to be viable, successful reproduction may demand a full complement of the cytoplasm made in adult phase.

Responses to Removing and Adding Body Regions

In a unicell, recovery from loss of body parts cannot be achieved by replacing cells or recruiting new cells to fulfill the role of the lost body regions as a multicellular species would. The primary response to wounding by *A. acetabulum* is to limit loss of body contents (cytoplasm, periplasmic space, vacuole, etc) using the cytoskeleton and wound plug formation to heal the breach rapidly and efficiently (29, 77, 79). The secondary response of the body regions of

A. acetabulum to body loss depends on whether the nucleus is present, when in development that body loss was sustained, and how much of the body was lost (Figure 5). Hence, although *A. acetabulum* lacks structural redundancy, i.e. it lacks cell populations that can de-differentiate to heal a wound, it has functional redundancy because the basal portion can repeat development.

Responses to body loss have been interpreted to reflect both the morphogenetic capacity of each of the amputated body regions and the influence that that body region has on the morphogenetic potential of the remainder of the organism. As both portions of the organism survive, it is trivial to build simple and elegant internal controls into such amputation experiments. Interactions between the chloroplast and nuclear genomes have been inferred by comparing the numbers of organelles (e.g. 67, 99) or enzymatic activity (e.g. 25) in intact and amputated body portions (see other references in 12). Although nuclear implantations (e.g. 8, 37, 55, 96, 98) allow the roles of the nucleus and the cytoplasm to be further explored, what Hämmerling (40) called "nucleocytoplasmic"

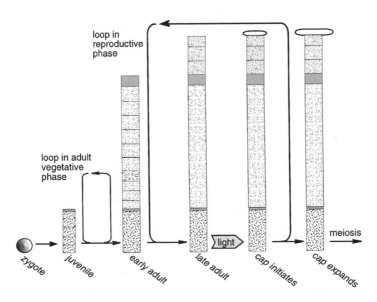

Figure 5 Amputation of the stalk apex prior to meiosis induces the rhizoid to repeat portions of development, i.e. to loop back through development (derived from 80, 94). The portion of development repeated depends on the volume of the apical portion removed and on when in development the amputation occurred. The vegetative loop shown was induced by removing all the apical portion, whereas the reproductive loop shown was triggered by removing just the cap and the whorl of hairs immediately below it. Textures of the phases and interwhorl lengths are as in Figure 4.

interactions, the responses of body portions that result from amputation are not synonymous with the responses of the cytoplasm to the loss of the nucleus and vice versa because there is always cytoplasm left with the rhizoid.

The responses to amputation of body portions during different phases in development suggest that the strategy of the intact plant is to render the stalk apex morphologically competent to reproduce before the nucleus commits to meiosis (33, 59, 61, 94) and that the nucleus then regulates the timing of reproductive onset. Here, *basal portion* means the entire rhizoid that contains the nucleus plus a variable amount of stalk and whorls depending on the experimental design, and *apical portion* means the stalk and whorls but does not include the nucleus or rhizoid. Note that these terms do not imply a time in development, e.g. an apical portion from a plant in reproductive phase would include a cap. Apical portions removed from juveniles grew, but few made a whorl of hairs (Figure 5), and none made a cap [0 out of 100 plants in Table 2 (80)]. In contrast, many apical portions removed from adults morphologically skipped the rest of vegetative development, making just 1–2 whorls of hairs before making a cap (36, 80). Taken together, these data suggest that the juvenile stalk apex has a limited ability to undergo morphogenesis or development without the nucleus and that the adult stalk apex is poised to reproduce but in some way is inhibited by the nucleus. In contrast, basal portions removed from juveniles finished vegetative growth without repeating vegetative development and without a delay in cap initiation [i.e. juvenile phase has no loop in Figure 5 (80)], as if loss of this part of the juvenile body was inconsequential. In contrast, basal portions removed from plants in adult or reproductive phase recapitulated the growth characteristic of adult phase both morphologically and temporally (*small loop* in Figure 5). These behaviors are reminiscent of the responses of portions excised from vascular plants (see references in 76, 76a–c). The response of the adult apical portion to loss of the basal portion implies that if nuclear function were blocked in an intact adult plant, then the organism would form a cap prematurely (Table 2); this is exactly what happened when transcriptional inhibitors were applied (114).

The responses of *A. acetabulum* to adding or replacing body parts, achieved by amputation followed by grafting, show that it responds to changes in body make-up and hint at how the enucleate and nucleate portions of the organism interact during development (Table 3). Interspecific grafts that combined the apical portion of one species with the basal portion from another could make viable progeny (reviewed in 12), whereas intergeneric grafts frequently died and failed to make progeny as if the apex and rhizoid were incompatible (13). In preliminary graft chimeras using the apical portion of one developmentally arrested (*da*) phenotype to the basal portion of another, interactions ranged from kill of one partner by the other to full acceptance and use of the donated

Table 3 Opportunities with graft chimeras of *Acetabularia* species

Nature of grafts	Purposes	References
Basal portion with other basal portion(s)[a]	Rescue putative mutants	71
	Infer interactions of nuclei in trans	86
	Infer nuclear dosage effects	14, 86
Apical portion with a basal portion	Define contribution of apical and basal portions to morphogenesis both intra- and interspecifically	10, 37, 41, 112
	Compensate putative mutants	71
	Characterize mutant biology (dependence on the polarity of the graft partner, etc)	70
	Compensation analysis (i.e. interactions of rhizoid of one mutant with apical portion of another or of wild type)	DF Mandoli & BE Hunt, unpublished data
	Intergenomic interactions (i.e. between organelle populations)	

[a]Number of rhizoids per graft chimera can be varied from 2 to 9 (14).

apical portion, suggesting that the graft partners both sensed and responded to changes in body composition (BE Hunt & DF Mandoli, unpublished data). These graft chimeras frequently form functional ectopic stalk apices, a response to replacement of body parts that has not been well studied and the significance of which is not known.

When a wild-type apical portion from a reproductive plant was grafted to the basal portion of a vegetative one from the same species, the graft chimeras generated haploid nuclei sooner than the nucleus in the basal portion would have been expected to had it been left intact (39) and much sooner than if the apical portion had simply been removed. Electron microscopy indicated that once associated with an older enucleate graft partner, a young nucleus would take on the appearance of an older one; it was smaller in volume and had reduced periplasmic space (6, 8), as if it were from an older plant about to enter meiosis. These results suggest that the reproductive apical portion hastened the vegetative nucleus to assume a reproductive morphology and induced karyokinesis, but this result has not been repeated or corroborated by other means, e.g. with molecular markers. Taken together, data from graft chimeras suggest that both the apical and basal portions mutually influence each other to orchestrate age progression and reproductive onset and have the potential to provide insight about how the identities of body regions are established and maintained.

The ability to compare genes *in trans* and *in cis* exists at several levels in *A. acetabulum*. For example, graft chimeras that combine from 2–9 rhizoids, producing bi- or multinucleate heterokaryons, enable rescue of mutants or nuclear dosage experiments (Table 3). Given that genetic analysis is just beginning, it is hard to predict what comparison of complementation analysis and the cell biological cognate, *compensation analysis* (interactions of a cytoplasmic genotype with a different nuclear genotype; Table 3), will reveal. The utility and power of such genetic and cell biological comparisons will become clear only once the organelle and nuclear genomes of *A. acetabulum* are better characterized.

A MECHANISM OF MORPHOGENESIS?

Words change meaning as a field or idea evolves, so when the language used to describe events is questioned, it signals increased understanding. Conversely, when word choices are not challenged, they can inadvertently hamper understanding as much as missing or poor controls, preconceived notions, or hidden (therefore untested) assumptions (see discussion of "homologous" versus "analogous" in 49). In the case of *Acetabularia*, use of the term "mRNA" in relation to morphogenesis of the cap is a word choice that needs revisiting. In the 1930s, when this unicell first made its mark, the complexity of the RNA world—that is, the diversity of types, functions, and catalytic abilities of RNA was unknown. Accordingly, the criteria for applying the term mRNA as it is used today have dramatically changed. Curtailing use of the phrase "translational control" in view of the current criteria that term now implies and the limited data available on *Acetabularia* species seems similarly apropos. Hence, in the section that follows, every attempt has been made to stick closely to what was actually done in the experiments that originally engendered the somewhat entrenched belief that morphogenesis of the hair and cap whorls in this unicell is under translational control and that cap-specific mRNAs exist.

The Case for Translational Control of Morphogenesis During Development

The concept that morphogenesis is directed by unique information concentrated in the stalk apex arose from four kinds of data: fates of body regions post amputation, distributions of biochemical activities, interspecific grafting, and inhibitor studies.

The first important result was that the nucleus was the source of morphological information: An implanted nucleus alone could confer on a middle portion the ability to make a rhizoid or a stalk apex (this and other experiments on this point are reviewed in 40). Three laboratories used grafts to show that the nucleus contained morphological information that was species-specific (9, 13, 37, 111). A vegetative apical portion and a vegetative basal portion of two species that make caps of different shapes were joined. Usually, cap morphology was detailed (e.g. 111), and then the cap was amputated each time one formed. In succession, the cap shape followed that of the donor of the apical portion, then was intermediate in shape, and finally was that of the donor of the basal portion (37). Note that while these data suggest that some aspects of cap shape are conserved enough to function across species lines, they do not indicate the chemical nature of the information. While the role of the nucleus as a source of species-specific morphological information may seem obvious now, in the 1930s it was revolutionary.

The basic finding that the enucleate apical portion can undergo morphogenesis is robust. Several independent studies corroborated that apical portions removed from older plants (based on their size and age, these were probably adults) could make one or two whorls of hairs and then a cap after amputation (9, 35, 36, 80), but apical portions removed from juveniles could make one whorl of hairs at best (80). The enucleate middle portion had a lower probability of making a cap at the apical pole than an apical portion did, and middle portions occasionally made a rhizoid-like structure at their basal poles (37). Primarily based on these amputations and the morphology and development of the intact plant, Hämmerling (reviewed in 37, 38, 66) proposed that there were two gradients of morphogenetic information in this species such that instructions for differentiating the rhizoid and stalk apices were concentrated at the basal and apical poles respectively.

There is independent biochemical evidence for apical-basal gradients of protein synthesis (30, 66, 89, 90), of ribosomal and mRNA (30, 40, 46, 47, 56–58, 65, 84, 97), of thiol groups (110), and even of specific enzymatic activities (e.g. 116) in this alga. These gradients differ widely in degree or strength of the gradient, in the crudeness of the assay used, and in direction, i.e. some constituents of the plant are concentrated at one pole and some at the other. It is possible that comparing gradients of total proteins, ribosomes, and redox potential in diverse wild-type and mutant backgrounds might provide insight into the strongly directional growth of the organism. However, the relevance of any gradient to pole differentiation and morphogenesis can only begin to be addressed using specific molecular markers and can be tested only by manipulation of the most important molecules in defined genetic backgrounds.

Two additional kinds of experiments suggest that at least one morphogenetic event, cap initiation, entails translation of relatively few proteins. For example, stalk apices treated with ultraviolet radiation are half as likely to make a cap (17, 91)—results that are consistent with RNA playing an important role in cap morphogenesis. However, such fairly nonspecific treatments do not rule out alternate explanations, e.g. preventing the growth needed to fashion the shape can account equally well for these results and similar types of experiments that prevent morphogenesis (reviewed in 12). When analyzed on 2-D gels, in vitro translation of mRNAs from plants poised to make caps revealed just 29 proteins made de novo at cap initiation (101–104). Taken together, these data argue that cap initiation does not entail general up-regulation of translation. However, to the best of my knowledge, these experiments on cap morphogenesis have not been independently corroborated, and the question of the specificity of translation for morphogenesis of the whorls of hairs has not been addressed even to the limited extent it has been for cap morphogenesis.

The suggestion that the timing of cap initiation is regulated by an inhibitor that is made by the nucleus has support from three kinds of studies: amputation, grafting, and inhibitor studies in wild type and now in one mutant, *nightstick*. Since an apical portion taken from a heterogeneous wild-type plant formed a cap faster than an intact plant did (Table 2, Figure 5), Beth (10) surmised that "the nucleus" actively prevents premature cap initiation in the intact plants that, based on their size (about 2 cm tall at amputation), were probably adults. Because this result has not been extended with nuclear implantations, it is safest to state that the rhizoid actively prevents cap initiation by the (adult) apex. Consistent with this hypothesis, the transcriptional inhibitor, actinomycin D, turned on cap initiation in intact (probably adult) wild-type plants (115), implying that cessation of transcription turned on cap morphogenesis once the stalk apex had gained competence to make a cap. Intact adults make an average of 2.2 whorls of hairs in actinomycin D (80), suggesting that the information for making hairs may also be preloaded into the stalk apex. Although the lack of temporal delays in cap onset in interspecific grafts led to the suggestion that this inhibitor may be species-specific in nature (1), it is not clear that this effect, a whole plant bioassay, can be solely attributed to a molecule that remains putative. Finally, the amputation behavior of the *da* mutant *nightstick* (*nst*; Figure 6) supports the existence of a cytosolic inhibitor of cap initiation. In brief, *nst* is a recessive trait with a terminal morphology that arrests late in adult phase, i.e. it is capless when intact or wounded. After amputation of the apical portion, the *nst* basal portion made a cap. Furthermore, the probability that a *nst* basal portion would make a cap after amputation of its own cytoplasm was directly proportional to the volume of the apical portion that had been removed (72). In sum, the hypothesis that the nucleus produces a cytosolic inhibitor of cap initiation that plays a role in the timing of reproductive onset (9) deserves further consideration.

Taking a Genetic Approach: Are the Controls Spatial, Temporal, or Both?

Clearly, the solid foundation of knowledge about morphogenesis in *Acetabularia* species provides a springboard for further advances in understanding how

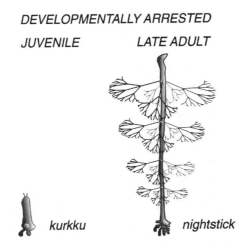

DEVELOPMENTALLY ARRESTED

JUVENILE   LATE ADULT

kurkku   nightstick

Figure 6   Phenotypes of two developmentally arrested mutants of *A. acetabulum* (68), both of which are recessive in outcrosses to wild type (71, 72). *kurkku* makes 0–1 whorl of hairs per plant consisting of 1–2 hairs each (70, 71). *nightstick* progresses normally through development, making an average of 10 ± 0.5 whorls of hairs until it arrests late in adult phase (72). Classes of *da* defects can be distinguished by their amputation behavior (68).

body plan is established and maintained in a unicellular context. However, a wealth of questions remain: How distinct are vegetative and reproductive whorls?—e.g. are hair and cap morphogenesis effected by distinct or overlapping sets of mRNAs, proteins, etc? How are juvenile and adult phases related—i.e. how abrupt is the switch and how is it timed? How is localized differentiation achieved without cellular partitioning?

The naiveté of the regulation of *Acetabularia* morphogenesis (e.g. Table 2) can be most succinctly illustrated by an example from a unicellular system about which more is known, the fly egg. As background, the gene *oskar* is central to posterior pole determination. When *oskar* activity is missing, *Drosophila* females make oocytes that lack a germ line or abdomen (64). If *oskar* is expressed in the wrong place, oocytes form posterior pole features ectopically (28). Three aspects of *oskar*'s role in germ plasm determination are of particular interest to thinking about *Acetabularia*. *oskar* RNA is found throughout the early oocyte, it becomes physically localized in several steps during oocyte development (27, 53). *oskar* mRNA that is not at the posterior pole is translationally repressed (52). oskar protein is needed to keep *oskar* RNA localized (27, 53, 75a, 93a; see also *staufen* e.g. 107). In sum, *oskar* activity becomes tightly restricted to the posterior pole of the oocyte by two mechanisms: by localization of *oskar* RNA and by spatially confined translation of *oskar*. It seems reasonable to anticipate that pole determination and differentiation may be as complex in *Acetabularia* as it is in *Drosophila.*

One powerful aspect of a genetic approach is that prior knowledge of the target is unnecessary. This statement comes with the proviso that the phenotypes sought must be sufficiently broad to encompass all likely targets. Since it is premature to predict whether spatial or temporal cues will prove more important or experimentally tractable in identifying factors relevant to morphogenesis in *A. acetabulum,* we study two broad groups of phenotypes. Developmentally arrested (*da*) phenotypes (Figure 6) fail to make a cap but are normal in morphology until the point of arrest. In contrast, morphologically altered (*ma*) phenotypes are aberrant in body plan (Figure 7). Study of these phenotypes with genetic and cell biological manipulations is under way (71, 72). One way to explain development in this unicell would be to alter the apices (*ma* genes) as the organism ages (*da* genes) (Figure 8). That is, it is theoretically possible to regulate all of the patterning of the elaborate body plan of the Dasycladales just by locally controlling differentiation of the critical subcellular regions—the apices—and to regulate phase change by globally controlling development of the organism as a whole. Clearly, the first part of this working concept

MORPHOLOGICALLY ALTERED

ESTABLISHMENT   PROPORTIONS   MAINTENANCE

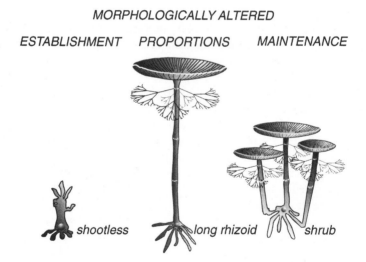

Figure 7 Morphologically altered phenotypes of *A. acetabulum* fall into three classes: those that fail to establish a pole, those that make too much of one body region, and those that fail to maintain the identity of a region after it is made (drawn based on photographs in 68). These phenotypes have been assigned a brief, descriptive name, but it is not yet known whether these represent genes.

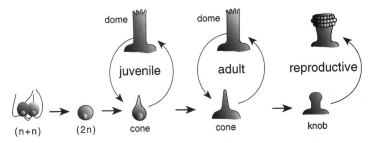

Figure 8   One idea of how morphological changes at the stalk apex and phase change might interface in *A. acetabulum.* All the whorl-interwhorl units made in juvenile or adult phases are represented by a simple alternation between a cone-shaped stalk apex or a dome-shaped stalk apex that has just initiated a whorl of hairs. It has not been proved which of these apical shapes is the default or uninduced shape, and it is not known how *ma* and *da* defects might be related.

has resonance with the emerging picture that cellular position is important to morphogenesis in multicellular vascular plants (21, 51, 81). My bias is that genes that effect the temporal sequence of development and genes that alter the spatial definition of body regions are needed if we are to begin to understand morphogenesis and phase change as well as the interface between these processes.

POTENTIAL FOR FUTURE INSIGHTS INTO DEVELOPMENT AND MORPHOGENESIS

It is striking that phase change is shared by this ancient unicellular protist, *A. acetabulum,* and by modern multicellular plants. Perhaps the study of phase change in a unicellular context will lend insight about why it evolved and into how it is accomplished. Of paramount interest would be aspects of phase change that have been evolutionarily conserved independent of whether the body plan of the organism is partitioned into one or many cells. Why is competence to reproduce, a feature of adult phase (63), acquired so early in development in *A. acetabulum* and in multicellular plants (e.g. 18)? Is the strategy to prevent premature reproductive onset rather than to delay competence an ancient one? This said, it would be folly to ignore the alternate view that the very concept of phase change is not useful in thinking about development but an artificial construct without real biological relevance (76) until we have more than an inkling of the genetic circuitry (e.g. 82) that embodies vegetative growth.

How shape and function of localized regions are established and then maintained within cells, a question of fundamental importance in development, is one to which *A. acetabulum* is well suited. Given the central role of localized determinants to patterning of the body plan in animals, in the long term it will be interesting to compare the mechanisms for localizing determinants in species as diverse as fly oocytes (88) and *Acetabularia*. The extent to which features of the diplophase body plan are parentally derived or imprinted, e.g. maternal mRNAs in *Drosophila* or structural elements of body plan like surgically altered cilia in *Paramecium* (48), cannot be ruled out and will be important to distinguish from those which are made de novo.

The physical height of the organism may have implications for the crosstalk between the nucleus and the major site of morphogenesis, the shoot apex. Given this long-distance relationship, it is hard to envision that the cytoskeleton (78) and molecular motors would not be intimately involved in moving and targeting information for aspects of pole establishment, differentiation, or maintenance of body regions. *Acetabularia* have sizable internal currents (see references in 83), but whether such currents play any role in morphogenesis in this plant is unknown. Perhaps the physical distance between the poles of *A. acetabulum* provides a clue about why the data support translational rather than transcriptional regulation of reproductive onset. For example, if speed of execution is important to changing the shape of the apex or in coordinating apical and nuclear crosstalk at specific times in development, then the physical distance between the nucleus and the shoot apex may preclude transcriptional regulation. In sum, being a giant unicell with a single nucleus may in itself have had ramifications for the way this algal species functions and, perhaps, for how it has or has not evolved.

THE HALLMARKS OF A CLASSIC DEVELOPMENTAL SYSTEM

In these days of limited funding, technical access to cross phyla comparisons, and well-established model systems, what are the most compelling reasons to use *A. acetabulum*? For development and morphogenesis, the size of its unicellular body that still retains the architectural and developmental complexity of "higher" plants physically enables the network of internal and external cues to be partly unraveled. For molecular evolution, the Dasycladales offers a 570-M year fossil record [186 total with 11 extant genera (7)] that will help address whether the unicellular nature of *A. acetabulum* and other species is primary or derived from a multicellular ancestor, and that will be useful to analyses of the conservation of homeotic genes. For cell biology, the ability to manipulate the body of the organism (to amputate, to mix-and-match the cytoplasm of one genotype with the nucleus of another) and the chemical composition of the environment or the organism (to deliver drugs and inhibitors exogenously in an axenic, defined growth medium or via microinjection) presents tantalizing opportunities. For routine classical genetics, the species offers self- and out-crosses, packaged mating types, amplification of nuclei, millions of progeny per plant per generation and the possibility of making stable haploids (26). For molecular genetics, it offers abundant DNA per generation, haploid and diploid transformation of high yield and stable expression. However, to me, the special appeal of *A. acetabulum* lies not just in the ability to address important questions in developmental and structural biology in the context of a physically large and architecturally complex unicell, but in being able to do so with access to a diverse and robust toolkit because this means that if one avenue of attack does not work, another probably will.

ACKNOWLEDGMENTS

Thanks to the talented undergraduates who have populated my lab over the years, to my graduate and postdoctoral students, especially Kyle Serikawa, who challenge my thinking at every step, to my wonderful technician, Brenda E Hunt, and to my colleagues here and around the globe whose faith powered (and probably funded) me through the early stages of this work and whose hard-nosed critiques cajoled me into doing better. Our work discussed here was supported in part by National Science Foundation grant IBN-9305473 and IBN-9630618, by the Graduate Research Fund of the University of Washington, and by Project R/B-8, of the Washington Sea Grant Program, with funding from grant NA89AA-D-SG022 from the National Oceanic and Atmospheric Administration, US Department of Commerce.

Visit the *Annual Reviews home page* at
http://www.AnnualReviews.org.

Literature Cited

1. Bannwarth H, de Groot EJ. 1991. Temporal control of differentiation by the cell nucleus: evidence for a species-specific nuclear suppression of enzymes at the early generative phase in *Acetabularia*. *J. Cell Sci.* 100:863–68
2. Bannwarth H, Ikehara N, Schweiger H-G. 1977. Nucleo-cytoplasmic interactions in the regulation of thymidine phosphorylation in *Acetabularia*. *Proc. R. Soc. London Ser. B* 198:177–90
3. Bannwarth H, Ikehara N, Schweiger H-G. 1982. Deoxyctidine monophosphatase deaminase in *Acetabularia*: properties and regulation in the early generative phase. *Eur. J. Cell Biol.* 27:200–5
4. Bannwarth H, Schweiger H-G. 1983. The influence of the nucleus on the regulation of the dCMP deaminase in *Acetabularia*. *Cell Biol. Int. Rep.* 7:859–68
5. Berger S, de Groot EJ, Neuhaus G, Schweiger M. 1987. *Acetabularia*: a giant single cell organism with valuable advantages for cell biology. *Eur. J. Cell Biol.* 44:349–70
6. Berger S, Herth W, Franke WW, Falk H, Spring H, Schweiger HG. 1975. Morphology of the nucleo-cytoplasmic interactions during the development of *Acetabularia* cells. II. The generative phase. *Protoplasma* 84:223–56
7. Berger S, Kaever MJ. 1992. *Dasyclad-*

ales: An Illustrated Monograph of a Fascinating Algal Order. London: Elsevier. 247 pp.

8. Berger S, Schweiger HG. 1975. Cytoplasmic changes in the ultrastructure of the *Acetabularia* nucleus and perinuclear cytoplasm. *J. Cell Sci.* 17:517–29

9. Beth K. 1953. Über den Einfluss des kernes auf die Formbildung von *Acetabularia* in verschiedenen Entwicklungstadien. *Z. Naturforsch. Teil* B8:771–75

10. Beth K, Hämmerling J. 1943. Ein- und zweikernige Transplantate zwischen *Acetabularia mediterranea* und *Acicularia schenckii. Z. induckt. Abstamm. Vererb. Lehre* 81:271–312

11. Bonotto S. 1969. Quelques observations sur les verticilles d'*Acetabularia mediterranea. Bull. Soc. R. Bot. Belg.* 102:165–79

12. Bonotto S. 1988. Recent progress in research on *Acetabularia* and related Dasycladales. In *Progress in Phycological Research*, ed. FE Round, DJ Chapman, 6:59–235. Bristol: BioPress

13. Bonotto S. 1989. Graft incompatibility between *Acetabularia* and *Batophora* (Dascladales, Rhodophyta). *Giorn. Bot. Ital.* 123:55–62

14. Bonotto S, Kirchmann R, Manil P. 1971. Cell engineering in *Acetabularia*: a graft method for obtaining large cells with two or more reproductive caps. *Giorn. Bot. Ital.* 105:1–9

15. Borghi H, Puiseux-Dao S, Durand M, Dazy AC. 1983. Morphogenesis, bioelectric polarity and intracellular streaming in a giant cell, *Acetabularia mediterranea*: studies on their recovery after prolonged dark period. *Plant Sci. Lett.* 31:75–86

16. Brachet J. 1981. A comparison of nucleocytoplasmic interactions in *Acetabularia* and in eggs. In *Fortschritte der Zoologie*, ed. HW Sauer, 26:15–32. New York: Gustav Fisher Verlag

16a. Brachet J, Bonotto S. 1970. *Biology of Acetabularia.* New York: Academic. 300 pp.

17. Brachet J, Olszewska M. 1960. Influence of localized ultra-violet irradiation on the incorporation of adenine-8-^{14}C and D, L-methionine-^{35}S in *Acetabularia mediterranea. Nature* 187:954–55

18. Bradley D, Ratcliffe O, Vincent C, Carpenter R, Coen E. 1997. Inflorescence commitment and architecture in *Arabidopsis. Science* 275:80–83

19. Cairns E, Doerfler W, Schweiger H-G. 1978. Expression of a DNA animal virus genome in a plant cell. *FEBS Lett.* 96:295–97

20. Cairns E, Sarkar S, Schweiger H-G. 1978. Translation of tobacco mosaic virus RNA in *Acetabularia* cell cytoplasm. *Cell Biol. Int. Rep.* 2:573–78

21. Clark SE. 1997. Organ formation at the vegetative shoot meristem. *Plant Cell* 9:1067–76

22. Cooper J, Mandoli DF. 1998. *Physiological optimization of early differentiation events in the life cycle of* Acetabularia acetabulum *(Dasycladales, Chlorophyta).* http://weber.washington. edu/~mandoli

23. Crawley JCW. 1970. The fine structure of the gametes and zygotes of *Acetabularia.* See Ref. 16a, pp. 73–83

24. De DN, Berger S. 1990. Karyology of *Acetabularia mediterranea. Protoplasma* 155:19–28

25. de Groot EJ, Schweiger H-G. 1985. Regulation of a ribonucleosidase reductase during the early generative phase in *Acetabularia. J. Cell Sci.* 73:1–5

26. Dübel S, Neuhaus G, Berger S, Schweiger H-G. 1985. Characterization of haploid *Acetabularia* cells. *Eur. J. Cell Biol.* 38:328–34

27. Ephrussi A, Dickinson LK, Lehmann R. 1991. *oskar* oganizes the germ plasm of the posterior determinant *nanos. Cell* 66:37–50

28. Ephrussi A, Lehmann R. 1992. Induction of germ line cell formation by *oskar. Nature* 358:387–92

29. Fester R, Hopkins C, Mandoli DF. 1993. Wounds, incurred during routine cell culture, prolong the life cycle of *Acetabularia acetabulum* and require K$^+$ to heal. *Protoplasma* 177:123–31

30. Garcia E, Dazy A-C. 1986. Spatial distribution of poly(A)$^+$RNA and protein synthesis in the stalk of *Acetabularia acetabulum. Biol. Cell* 58:23–30

31. Gibor A. 1973. Observations on the sterile whorls of *Acetabularia. Protoplasma* 78:195–202

32. Gradmann D. 1984. Electrogenic pump in the marine alga *Acetabularia.* In *Chloride Transport Coupling in Biological Membrane and Epithelia*, ed. GA Teregerencser, pp. 13–61. Amsterdam: Elsevier

33. Green BR. 1973. Evidence for the occurrence of meiosis before cyst formation in *Acetabularia mediterranea* (Chlorophyceae, Siphonales). *Phycologia* 12:233–35

34. Green BR. 1976. Approaches to the genetics of *Acetabularia.* In *The Genetics of Algae*, ed. RA Lewin, pp. 236–56. Berkeley: Univ. Calif. Press

35. Hämmerling J. 1931. Entwicklung und Formbildungsvermögen von *Acetabularia mediterranea*. I. Die normale Entwicklung. *Biol. Z.* 51:633–47

36. Hämmerling J. 1932. Entwicklung und Formbildungsvermögen von *Acetabularia mediterranea*. *Biol. Zentralbl.* 52:42–61

37. Hämmerling J. 1934. Über formbildende Substanzen bei *Acetabularia mediterranea*, ihre räumliche und zeitliche Verteilung und ihre Herkunft. *Wilhelm Roux Arch. Entwicklungsmech. Org.* 131:1–81

38. Hämmerling J. 1936. Studien zum Polaritätsproblem I-III. *Zool. Jahrb. Abt. Allg. Zool. Physiol.* 56:440–86

39. Hämmerling J. 1953. Nucleo-cytoplasmic relationships in the development of *Acetabularia*. *Int. Rev. Cytol.* 2:475–98

40. Hämmerling J. 1963. Nucleo-cytoplasmic interactions in *Acetabularia* and other cells. *Annu. Rev. Plant Physiol.* 14:65–92

41. Hämmerling J. 1963. The role of the nucleus in differentiation especially in *Acetabularia*. *Symp. Soc. Exp. Biol.* 17:127–37

42. Hunt BE, Mandoli DF. 1992. Axenic cultures of *Acetabularia* (Chlorophyta): a decontamination protocol with potential application to other algae. *J. Phycol.* 28:407–14

43. Hunt BE, Mandoli DF. 1998. *Conditions for maturation and long term storage* Acetabularia acetabulum *(Chlorophyta)*. http://weber.washington. edu/~mandoli

44. Hunt BE, Mandoli DF. 1996. A new artificial seawater that facilitates growth of large numbers of cells of *Acetabularia acetabulum* (Chlorophyta) and reduces the labor inherent in cell culture. *J. Phycol.* 32:483–95

44a. Irish EE, Jegla D. 1997. Regulation of extent of vegetative development of the maize shoot meristem. *Plant J.* 11:63–71

44b. Irish EE, Nelson TM. 1988. Development of maize plants from cultured shoot apices. *Planta* 175:9–12

44c. Irish EE, Nelson TM. 1991. Vegetative to floral conversion occurs in multiple steps in maize tassel formation. *Development* 112:891–98

45. Jacob F, Monod J. 1961. Genetic regulatory mechanisms in the synthesis of proteins. *J. Mol. Biol.* 3:318–56

46. Janowski M. 1963. Incorporation de phosphore radioactif dans les acides ribonucléiques de fragments nucléés et anucléés d'*Acetabularia mediterranea*. *Arch. Int. Physiol. Biochim.* 71:819–20

47. Janowski M. 1966. Detection of ribosomes and polysomes in *Acetabularia mediterranea*. *Life Sci.* 5:2113–16

48. Jerka-Dziadosz M, Beisson J. 1990. Genetic approaches to ciliate pattern formation: from self-assembly to morphogenesis. *Trends Genet.* 6:41–45

49. Kaplan DR, Hagemann W. 1991. The relationship of cell and organism in vascular plants. *BioScience* 41:693–703

50. Keck K. 1964. Culturing and experimental manipulations of *Acetabularia*. In *Methods in Cell Physiology*, ed. DM Prescott, 1:189–213. New York: Academic

51. Kerstetter RA, Hake S. 1997. Shoot meristem formation in vegetative development. *Plant Cell* 9:1001–10

52. Kim-Ha J, Kerr K, Macdonald PM. 1995. Translational regulation of *oskar* mRNA by bruno, an ovarian RNA-binding protein, is essential. *Cell* 81:403–12

53. Kim-Ha J, Smith JL, Macdonald PM. 1991. *oskar* mRNA is localized to the posterior pole of the *Drosophila* oocyte. *Cell* 66:23–35

54. Kloppstech K. 1982. *Acetabularia*. In *The Molecular Biology of Plant Development*, ed. H Smith, D Grierson, pp. 136–58. London: Blackwell Sci.

55. Kloppstech K, Schweiger H-G. 1973. Nuclear genome codes for chloroplast ribosomal proteins in *Acetabularia*. I. Isolation and characterization of chloroplast ribosomal particles. *Exp. Cell Res.* 80:63–68

56. Kloppstech K, Schweiger H-G. 1975. Polyadenylated RNA from *Acetabularia*. *Differentiation* 4:115–23

57. Kloppstech K, Schweiger H-G. 1977. The rate of synthesis of poly(A) RNA in *Acetabularia mediterranea*. In *Progress in Acetabularia Research*, ed. CLF Woodcock, pp. 19–32. New York: Academic

58. Kloppstech K, Schweiger H-G. 1982. Stability of poly(A)$^+$ RNA in nucleate and anucleate cells of *Acetabularia*. *Plant Cell Rep.* 1:165–67

59. Koop H-U. 1977. Genetic aspects of *Acetabularia mediterranea*. In *Progress in Acetabularia Research*, ed. CLF Woodcock, pp. 7–18. London: Academic

60. Koop H-U. 1977. Selection of a morphologically changed strain of *Acetabularia mediterranea*. *Protoplasma* 91:224–25

61. Koop H-U. 1979. The life cycle of *Acetabularia (Dasycladales, Chlorophyceae)*: a compilation of evidence for meiosis in the primary nucleus. *Protoplasma* 100:353–66

62. Kratz R, Young PY, Mandoli DF. 1997. Reproductive onset in *Acetabularia*: light dependence of shape change in the context of a model for developmental progression. *J. Phycol.* 34:In press

63. Lawson EJR, Poethig RS. 1995. Shoot development in plants: time for a change. *Trends Genet.* 11:263–68

64. Lehmann R, Nüsslein-Volhard C. 1986. Abdominal segmentation, pole cell formation, and embryonic polarity require the localized activity of *oskar*, a maternal gene in *Drosophila*. *Cell* 47:141–52

65. Li-Weber M, Schweiger H-G. 1985. Evidence for and mechanism of translational control during cell differentiation in *Acetabularia*. *Eur. J. Cell Biol.* 38:73–78

66. Lüttke A. 1985. Morphogenesis of *Acetabularia*: a new concept. *Acetabularia* 1984, pp. 83–90. Mol, Belgium: Belg. Nucl. Cent., CEN-SNK

67. Lüttke A, Bonotto S. 1981. Chloroplast DNA of *Acetabularia mediterranea*: cell cycle related changes in distribution. *Planta* 153:536–42

68. Mandoli DF. 1996. Establishing and maintaining the body plan of *Acetabularia acetabulum* in the absence of cellularization. In *Seminars in Cell and Developmental Biology*, ed. I Sussex, pp. 891–901. London: Academic

69. Mandoli DF. 1997. What ever happened to *Acetabularia*? Bringing a once-classic model system into the age of molecular genetics. *Int. Rev. Cytol.* In press

70. Mandoli DF, Hunt BE. 1996. *kurkku*, a phenotype of *Acetabularia acetabulum* that is arrested in vegetative growth, can be rescued with wildtype cytoplasm. *Plant Cell* 8:323–32

71. Mandoli DF, Hunt BE. 1998. *Development and compensation (graft) analysis of* kurkku, *a recessive trait that results in juvenile arrest in* A. acetabulum. http://weber.washington.edu/~mandoli

72. Mandoli DF, Hunt BE. 1998. Nightstick, *a loss of function that may effect localization of a regulator of cap initiation in* Acetabularia. http:// weber. washington. edu/~mandoli

73. Mandoli DF, Larsen T. 1993. Improved mating efficiency in *Acetabularia acetabulum*: recovery of 48–100% of the expected zygotes. *Protoplasma* 176:53–63

74. Mandoli DF, Wexler A, Teschmacher J, Zukowski A. 1995. Brief incubation of gametangia-bearing caps in antibiotics eliminates branching in progeny cell populations of *Acetabularia acetabulum* (Chlorophyta). *J. Phycol.* 31:844–48

75. Marin A, Ros JD. 1992. Dynamics of a peculiar plant-herbivore relationship: the photosynthetic ascoglossan *Elysia timida* and the chlorophycean *Acetabularia acetabulum*. *Mar. Bot.* 112:677–82

75a. Markussen F-H, Michon A-M, Breitwieser W, Ephrussi A. 1995. Translational control of *oskar* generates Short OSK, the isoform that induces pole plasm assembly. *Development* 121:3723–32

76. McDaniel CN. 1996. Developmental physiology of floral initiation in *Nicotiana tabacum* L. *J. Exp. Bot.* 47:465–75

76a. Deleted in proof

76b. Deleted in proof

76c. Deleted in proof

77. Menzel D. 1981. Development and fine structure of plugs in the cap rays of *Acetabularia acetabulum* (*mediterranea*) (L.) Silva (Dasycladales). *Phycologia* 20:56–64

78. Menzel D. 1994. Tansley Rev. No. 77. Cell differentiation and the cytoskeleton in *Acetabularia*. *New Phytol.* 128:369–93

79. Menzel D, Elsner-Menzel C. 1989. Induction of actin-based cytoplasmic contraction in the siphonous green alga *Acetabularia* (Chlorophyceae) by locally restricted calcium influx. *Bot. Acta* 102:164–71

80. Messmer J, Mandoli DF. 1998. *Functional significance of phase change in* A. acetabulum: *recapitulation of adult phase after loss of the shoot apex.* http:// weber.washington.edu/~mandoli

81. Meyerowitz EM. 1996. Plant Development: local control, global patterning. *Curr. Opin. Genet. Dev.* 6:475–79

82. Moose SP, Sisco PH. 1996. *Glossy15*, an APETALA2–like gene from maize that regulates leaf epidermal cell identity. *Genes Dev.* 10:3018–27

83. Moritani C, Ohhashi T, Kadowaki H, Tagaya M, Fukui T, et al. 1997. The primary structure of the Cl$^-$-translocating ATPase, b subunit of *Acetabularia acetabulum*, which belongs to the F-type ATPase family. *Arch. Biochem. Biophys.* 339:115–24

84. Naumova LP, Pressman EK, Sandakchiev LS. 1976. Gradient of RNA distribution in the cytoplasm of *Acetabularia mediterranea*. *Plant Sci. Lett.* 6:231–35

85. Neuhaus G, Neuhaus-Url G, de Groot EJ, Schweiger H-G. 1986. High yield and stable transformation of the unicellular green alga *Acetabularia* by microinjection of SV40 DNA and pSV2neo. *EMBO J.* 5:1437–44

86. Niggemeyer J, Mandoli DF. 1998. *Long*

rhizoid *of* Acetabularia *phenoconverts wild type to lor and changes apex identity, converting rhizoids into stalks, when in cell::cell grafts.* http://weber.washington.edu/~mandoli

87. Nishimura NJ, Mandoli DF. 1992. Vegetative growth of *Acetabularia acetabulum* (Chlorophyta): structural evidence for juvenile and adult phases in development. *J. Phycol.* 28:669–77

88. Nüsslein-Volhard C. 1996. Gradients that organize embryo development. *Sci. Am.* 275:54–61

89. Olszewska MJ, Brachet J. 1960. Incorporation de la DL-méthionine-^{35}S dans l'algue *Acetabularia mediterranea*. *Arch. Int. Physiol. Biochim.* 68:693–94

90. Olszewska MJ, Brachet J. 1961. Incorporation de la DL-méthionine ^{35}S dans les fragments nucléés et anucléés d'*Acetabularia mediterranea*. *Exp. Cell Res.* 22:370–80

91. Olszewska MJ, deVitry F, Brachet J. 1961. Influence d'irradiations localisées sur l'incorporation de l'adénine-8-^{14}C, de l'uridine-^{3}H et de la DL-méthionine-^{35}S dans l'algue *Acetabularia mediterranea*. *Exp. Cell Res.* 24:58–63

92. Puiseux-Dao S. 1962. Recherches biologiques et physiologiques sur quelques Dasycladacées, en particulier, le *Batophora oerstedii* J. Ag. et l'*Acetabularia mediterranea* (Lamouroux). *Rev. Gén. Bot.* 50:1–36

93. Puiseux-Dao S. 1963. Les Acétabulaires, matériel de laboratoire. Les résultats obtenus avec ces Chlorophycées. *L'Année Biol.* 2:99–154

93a. Rongo C, Gavis ER, Lederman R. 1995. Localization of *oskar* RNA regulates *oskar* translation and requires Oskar protein. *Development* 121:2737–46

94. Runft LL, Mandoli DF. 1996. Coordination of the cellular events that precede reproductive onset in *Acetabularia acetabulum:* evidence for a "loop" in development. *Development* 122:1187–94

95. Schmid R, Idziak E-M, Tunnerman M. 1987. Action spectrum for the blue-light-dependent morphogenesis of hair whorls in *Acetabularia mediterranea*. *Planta* 171:96–103

96. Schweiger E, Wallraff HG, Schweiger H-G. 1964. Endogenous circadian rhythm in cytoplasm of *Acetabularia*: influence of the nucleus. *Science* 146:658–59

97. Schweiger H-G, Dillard WL, Gibor A, Berger S. 1967. RNA synthesis in *Acetabularia*. I. RNA synthesis in enucleated cells. *Protoplasma* 64:1–12

98. Schweiger H-G, Werz G, Reuter W. 1969.

Tochtergenerationen von heterologen implantaten bei *Acetabularia*. *Protoplasma* 68:351–56

99. Shephard DC. 1965. Chloroplast multiplication and growth in the unicellular alga *Acetabularia mediterranea*. *Exp. Cell Res.* 37:93–110

100. Shephard DC. 1970. Axenic culture of *Acetabularia* in a synthetic medium. *Methods Cell Physiol.* 4:49–69

101. Shoeman RL, Neuhaus G, Schweiger H-G. 1983. Gene expression in *Acetabularia*. III. Comparison of stained cytosolic proteins and *in vivo* and *in vitro* translation products. *J. Cell. Sci.* 60:1–12

102. Shoeman RL, Schweiger H-G. 1982. Gene expression in *Acetabularia*. I. Calibration of wheat germ cell–free translation system proteins as internal references for two-dimensional electrophoresis. *J. Cell Sci.* 58:23–33

103. Shoeman RL, Schweiger H-G. 1982. Gene expression in *Acetabularia*. II. Analysis of *in vitro* translation products. *J. Cell Sci.* 58:35–48

104. Shoeman RL, Wasilewska L, Schweiger H-G. 1980. Translation of polyadenylated RNA from *Acetabularia*. *Protoplasma* 105:366

105. Spring H, Grierson D, Hemleben V, Stohr M, Krohne G, et al. 1978. DNA contents and numbers of nucleoli and pre-rRNA genes in nuclei of gametes and vegetative cells of *Acetabularia mediterranea*. *Exp. Cell Res.* 114:203–15

106. Spring H, Scheer U, Franke WW, Trendelenburg MF. 1975. Lampbrush-type chromosomes in the primary nucleus of the green alga *Acetabularia mediterranea*. *Chromosoma* 50:25–43

107. St. Johnston D, Beuchle D, Nüsslein-Volhard C. 1991. *staufen*, a gene required to localize maternal RNAs in the Drosophila egg. *Cell* 66:51–63

108. VanDen Driessche T, Petiau-de Vries GM, Guisset J-L. 1997. Tansley Rev. No. 91. Differentiation, growth and morphogenesis: *Acetabularia* as a model system. *New Phytol.* 135:1–20

109. van den Hock C, Mann AG, Johns HM. 1995. *Algae. An Introduction to Phycology*. Cambridge: Cambridge Univ. Press. 623 pp.

110. Van Langendonckt A, Vanden Driessche T. 1992. Changes in intracellular distribution of thiol groups during the development of *Acetabularia mediterranea*. *J. Exp. Bot.* 43:1643–50

111. Werz G. 1955. Kernphysiologische Untersuchungen an *Acetabularia*. *Planta* 46:113–53

112. Werz G. 1965. Determination and realization of morphogenesis in *Acetabularia. Brookhaven Symp. Biol.* 18:185–
203
113. Zeller A, Mandoli DF. 1993. Growth of
Acetabularia acetabulum on solid substrata at specific cell densities. *Phycologia*
32:136–42
114. Zetsche K. 1964. Der Einfluss von Actinomycin D auf die Abgabe morpho-
genetischer Substanzen aus dem Zellkern

von *Acetabularia mediterranea. Naturwissenschaften* 51:18–19
115. Zetsche K. 1966. Regulation der zeitli-
chen Aufeinanderfolge von Differenzie-
rungsvorgängen bei *Acetabularia. Z.
Naturforsch. TeilB* 21:375–79
116. Zetsche K, Grieninger GE, Anders J.
1970. Regulation of enzyme activity during morphogenesis of nucleate and anucleate cells of *Acetabularia.* See Ref. 16a,
pp. 87–110

Annu. Rev. Plant Physiol. Plant Mol. Biol. 1998. 49:199–222

ABSCISIC ACID SIGNAL TRANSDUCTION

Jeffrey Leung and Jérôme Giraudat
Institut des Sciences Végétales, Unité Propre de Recherche 40, Centre National de la
Recherche Scientifique, 1 Avenue de la Terrasse, 91190 Gif-sur-Yvette, France;
e-mail: jerome.giraudat@isv.cnrs-gif.fr

KEY WORDS: guard cell, mutants, second messenger, seed development, stress tolerance

ABSTRACT

The plant hormone abscisic acid (ABA) plays a major role in seed maturation
and germination, as well as in adaptation to abiotic environmental stresses. ABA
promotes stomatal closure by rapidly altering ion fluxes in guard cells. Other
ABA actions involve modifications of gene expression, and the analysis of ABA-
responsive promoters has revealed a diversity of potential *cis*-acting regulatory
elements. The nature of the ABA receptor(s) remains unknown. In contrast,
combined biophysical, genetic, and molecular approaches have led to considerable progress in the characterization of more downstream signaling elements.
In particular, substantial evidence points to the importance of reversible protein phosphorylation and modifications of cytosolic calcium levels and pH as
intermediates in ABA signal transduction. Exciting advances are being made in
reassembling individual components into minimal ABA signaling cascades at the
single-cell level.

CONTENTS

INTRODUCTION

The scientific origins of abscisic acid (ABA) have been traced to several independent investigations in the late 1940s, but it was only in the 1960s that ABA was isolated and identified (1). Mutants affected in ABA biosynthesis are known in a variety of plant species (39, 146). The characterization of these mutants, together with physicochemical studies, has enabled the pathway of ABA biosynthesis to be elucidated in higher plants (134, 135, 146, 166). Two ABA biosynthetic genes have been recently cloned (88, 135) and should provide insights into the regulation and the sites of ABA biosynthesis.

A large body of evidence indicates that ABA plays a major role in adaptation to abiotic environmental stresses, seed development, and germination. The present review focuses on our current knowledge concerning how the ABA signal could be faithfully transduced to mediate these well-characterized physiological and developmental processes. Physicochemical and molecular genetic approaches have already provided fundamental insights into the diversity of ABA perception sites and other downstream components that could couple ABA stimuli to particular responses. It should be emphasized that many concepts about the relative importance of these components in the ABA signaling network are far from firmly established. Nonetheless, it is clear that the present development of single-cell systems should allow some of these components to be assembled into minimal signaling pathways within a cellular context.

ABA SIGNAL TRANSDUCTION IN SEEDS

Endogenous ABA content peaks during roughly the last two thirds of seed development before returning to lower levels in the dry seed (121). ABA is thus thought to regulate several essential processes occurring during the developmental stages that follow pattern formation of the embryo. These processes include the induction of seed dormancy, the accumulation of nutritive reserves, and the acquisition of desiccation tolerance.

Seed Dormancy and Germination

Exogenous ABA can inhibit the precocious germination of immature embryos in culture (121). Embryos of the ABA-biosynthetic mutants from maize exhibit precocious germination while still attached to the mother plant or vivipary (91). Seeds of the ABA biosynthetic mutants from *Arabidopsis thaliana* (71, 77) and *Nicotiana plumbaginifolia* (88) fail to become dormant. These observations support that endogenous ABA inhibits precocious germination and promotes seed dormancy.

Thus far, our knowledge on the signaling elements that mediate the regulation of seed dormancy and germination by ABA is primarily derived from genetic analysis. As summarized in Table 1, mutations that alter the sensitivity to ABA

Table 1 Main characteristics of the various mutations known to affect ABA sensitivity

Species	Mutation	Dominance[a]	Phenotype	Gene product[b]	Ref.
Hordeum vulgare	*cool*		ABA insensitivity in guard cells		118
Zea mays	*vp1*	R	ABA insensitivity in seeds	Seed-specific transcription factor	92, 120
	rea	R	ABA insensitivity in seeds		144
Craterostigma plantagineum	*cdt-1*	D	Constitutive ABA response in callus	Regulatory RNA or short polypeptide	31
Arabidopsis thaliana	*abi1*	SD	ABA insensitivity	Protein phosphatase 2C	72, 78, 97
	abi2	SD	ABA insensitivity	Protein phosphatase 2C	72, 79
	abi3	R	ABA insensitivity in seeds	Seed-specific transcription factor	38, 72, 102, 108
	abi4/5	R	ABA insensitivity in seeds		27
	era1	R	ABA hypersensitivity[c]	β subunit of farnesyl transferase	24
	era2/3	R	ABA hypersensitivity[c]		24
	gca1/8		ABA insensitivity[d]		10
	axr2	D	Resistance to auxin, ethylene, and ABA[d]		163
	jar1	R	Resistance to MeJa and hypersensitivity to ABA[c]		143
	jin4	R	Resistance to MeJa and hyper-sensitivity to ABA[c]		11
	bri1	R	Resistance to brass-inosteroids and hyper-sensitivity to ABA[d]		22
	sax	R	Hypersensitivity to auxin and ABA[e]		[f]
Nicotiana plum-baginifolia	*iba1* (*aba1*)[g]	R	Resistance to auxin, cytokinin, and ABA[c]	Molybdenum cofactor biosynthesis	13, 80

[a]Dominance of the mutant alleles over wild-type; D, dominant; SD, semidominant; R, recessive.
[b]Molecular function of the product of the wild-type gene.
[c]Sensitivity of seed germination to exogenous ABA.
[d]Sensitivity of seedling (root) growth to exogenous ABA.
[e]Sensitivities of root growth and stomatal closing to exogenous ABA.
[f]M. Fellner and G. Ephritikhine, personal communication.
[g]The *IBA1* locus was renamed *ABA1* when it was found that *iba1* and other alleles are in fact ABA-deficient, as a result of a defect in the biosynthesis of a molybdenum cofactor that is required for multiple enzymatic activities including ABA aldehyde oxidase (the last step in ABA biosynthesis).

in seeds have been described in maize and in *A. thaliana*, and several of the corresponding genes have been cloned.

The maize *vp1* (*viviparous1*) (119) and *rea* (*red embryonic axis*) (144) mutations lead, like ABA biosynthetic mutations in this species, to vivipary. However, *vp1* (120) and *rea* (144) embryos do not have reduced ABA content but rather exhibit a reduced sensitivity to germination inhibition by exogenous ABA in culture. As is discussed below, the *VP1* gene encodes a seed-specific transcription factor (92).

The *A. thaliana* ABA-insensitive *ABI1* to *ABI5* loci were all identified by selecting for seeds capable of germinating in the presence of ABA concentrations (3–10 μM) that are inhibitory to the wild type (27, 72). The *abi1* (72), *abi2* (72), and *abi3* (72, 101, 108) mutants also exhibit, like *A. thaliana* ABA-deficient mutants, a marked reduction in seed dormancy. These three loci thus seem to mediate the inhibitory effects of endogenous ABA on seed germination. The *ABI1* (78, 97) and *ABI2* (79) genes encode homologous protein serine/threonine phosphatase 2C (PP2C) (12, 79). The *ABI3* gene is the ortholog of the maize *VP1* gene mentioned above (38). It is presently unclear whether the *ABI1*, *ABI2*, and *ABI3* loci act in the same or in partially distinct ABA signaling cascade in seeds (28, 110).

Mutations in the *ERA1* to *ERA3* (Enhanced Response to ABA) loci of *A. thaliana* were identified by a lack of seed germination in the presence of low concentrations of ABA (0.3 μM) that are not inhibitory to the wild type (24). The *era1* mutations also markedly increase seed dormancy. The *ERA1* gene encodes the β subunit of a farnesyl transferase, which may possibly function as a negative regulator of ABA signaling by modifying signal transduction proteins for membrane localization (24). The exact relationships between *ERA1* and the above-mentioned *ABI* loci are unknown.

Reserve Accumulation and Acquisition of Desiccation Tolerance

The accumulation of nutritive reserves and the acquisition of desiccation tolerance are associated with the expression of specific sets of mRNAs (60, 121). Transcripts encoding either storage proteins or late-embryogenesis-abundant (LEA) proteins thought to participate in desiccation tolerance can be precociously induced by exogenous ABA in cultured embryos (60, 121). The characterization of ABA-deficient mutants in *A. thaliana* (70, 77, 96, 101, 112) and in maize (91, 109) supports a contribution of endogenous ABA in the developmental expression of these genes in seeds.

The functional dissection of such ABA-responsive promoters, conducted primarily in transient expression systems, has identified several *cis*-acting elements involved in ABA-induced gene expression.

Table 2   *cis*-acting promoter elements involved in the activation of gene expression by ABA

Gene (Species)	Element	Sequence[a]	Ref.
Rab16	Motif I	GTACGTGGC	141
(*Oryza sativa*)	Motif III	GCCGCGTGGC	107
Em	Em1a	ACACGTGGC	44
(*Triticum aestivum*)	Em1b	ACACGTGCC	44
HVA22	ABRE3	GCCACGTACA	138
(*Hordeum vulgare*)	CE1	TGCCACCGG	138
HVA1	ABRE2	CCTACGTGGC	139
(*Hordeum vulgare*)	CE3	ACGCGTGTCCTC	139
C1	Sph	CGTGTCGTCCATGCAT	65
(*Zea mays*)			
CDeT27-45		AAGCCCAAATTTCACA-	104
(*Craterostigma plantagineum*)		GCCCGATAACCG	

[a]The ACGT core in G-box-type ABREs is underlined.

CIS-ACTING PROMOTER ELEMENTS   A first category of elements is exemplified by the related motif I of the *Rab16* LEA gene from rice (107, 141) and Em1a motif of the wheat *Em* LEA gene from wheat (44). These sequences share a G-box ACGT core motif (Table 2) and have been designated ABREs, for ABA Response Elements. ABRE-related sequence motifs are present in many other ABA-inducible genes, although their function in ABA signaling often remains speculative in the absence of experimental tests. For instance, only a subset of the multiple ABRE-like motifs present in certain promoters are indeed required for ABA regulation (20, 48, 117, 138, 139). A second type of *cis*-acting element has been identified in the maize *C1* gene, a regulator of anthocyanin biosynthesis in seeds. The ABRE-like motifs present in *C1* are not major determinants of the ABA responsiveness of this gene, whereas the distinct sequence motif designated as Sph element (Table 2) is essential for ABA induction of the *C1* promoter (65).

Multimerized copies of ABREs (107, 141, 155) or of the Sph element (65) can confer ABA responsivity to minimal promoters. Single copies of these elements were, however, not sufficient for ABA response, suggesting that multimerization substituted for other aspects of the native promoter contexts of these elements. In fact, the smallest promoter units designated ABRCs shown to be both necessary and sufficient for ABA induction of gene expression appear to consist of (at least) two essential *cis*-elements. ABRCs can be composed of two G-boxes as in the case of the wheat *Em* promoter, where both the Em1a and

Em1b G-box motifs (Table 2) contribute to the activity of the 75-bp Complex I (44, 155). In contrast, several other ABRCs consist of an ABRE and of a *cis*-element not related to G-boxes. The G-box-type motif I and the distinct motif III are essential for the ABA responsiveness of a 40-bp fragment of the rice *Rab16B* promoter (107). In the barley *HVA22* promoter, ABRC1 is a 49-bp fragment that comprises ABRE3 and the Coupling Element CE1 located 20-bp downstream (138). In the barley *HVA1* promoter, the 22-bp long ABRC3 consists of the coupling element CE3 directly upstream of ABRE2 (139). These results indicate that a diversity of ABRCs participate in ABA signaling in seeds.

TRANS-ACTING FACTORS   The nature of the transcription factors that mediate ABA regulation of seed genes via the various above-mentioned *cis*-elements remains largely unknown. Most proteins that bind DNA sequences with ACGT core motifs contain a basic region adjacent to a leucine-zipper motif (bZIP-proteins). Several plant bZIP-proteins that bind to ABREs in vitro and/or are inducible (at the mRNA level) by ABA have been cloned (44, 68, 74, 100, 103, 105). However, few of these genes are known to be expressed in seeds, and thus far none has been unambiguously shown to be involved in ABA signaling in vivo (83).

In contrast, genetic and molecular evidence supports that the maize VP1 and *A. thaliana* ABI3 proteins are homologous transcription factors essential for ABA action in seeds. The *vp1* (91, 109) and *abi3* (28, 101, 108, 112) mutants are similarly altered in sensitivity to ABA, and in various other aspects of seed maturation including the developmental expression of storage proteins and LEA genes. The maize *VP1* (92) and *A. thaliana ABI3* (38) genes, as well as the rice *OsVP1* (47) and *Phaseolus vulgaris PvALF* (19) genes, encode homologous proteins (Figure 1). These genes are all exclusively expressed in seeds (19, 47, 92, 112). However, when ectopically expressed in transgenic *A. thaliana* plants, *ABI3* rendered vegetative tissues hypersensitive to ABA and

Figure 1 Schematic diagram of the architecture of the VP1/ABI3-like proteins. The maize VP1 (92), *A. thaliana* ABI3 (38), rice OsVP1 (47), and *Phaseolus vulgaris* PvALF (19) proteins display a similar and novel structural organization. They contain, in particular, four domains of high amino acid sequence identity: the A domain located in the large acidic (−) N-terminal region and three basic domains designated as B1, B2, and B3 in order from the N terminus. The sizes of these proteins range from 691 (VP1) to 752 (PvALF) amino acids.

activated expression of several otherwise seed-specific genes in leaves when exogenous ABA was supplied (110, 112). Similarly, in transient expression studies, VP1 (65, 92, 155), OsVP1 (47, 48), and PvALF (18, 19) could *trans*-activate various seed-specific promoters. When investigated, these *trans*-activations were found to require the acidic N-terminal domain of the relevant VP1/ABI3-like protein (48, 92), and domain-swapping experiments showed that the acidic N-terminal regions of VP1 (92) and PvALF (19) are indeed functional transcriptional activation domains. Thus, taken together, the above-mentioned results indicate that VP1/ABI3-like proteins act as transcription factors.

The maize *vp1* (56, 91, 109) and *A. thaliana abi3* (70, 77, 101, 108, 112) mutants exhibit a larger spectrum of seed phenotypes than the ABA-deficient mutants in the corresponding species. A possible explanation would be that the physiological roles of VP1 and ABI3 are not strictly confined to ABA signaling and rather that these proteins integrate ABA and other seed developmental factors. This model is also consistent with the observation that VP1/ABI3-like proteins can substantially *trans*-activate seed promoters in the absence of (added) ABA (19, 48, 92, 110), and that, within a given promoter, the *cis*-elements involved in VP1 responsiveness are partially separable from those involved in ABA responsiveness (65, 155).

In the wheat *Em* promoter, the ABRE Em1a (Table 2) that is essential for ABA induction also mediates transactivation by VP1 and synergism between ABA and VP1 (155). The conserved domain B2 of VP1 is required for *Em* transactivation (54), but its exact role in this process remains unclear (54, 131). Furthermore, no specific binding of VP1 to G-box-like ABREs has been observed thus far (54, 92, 145), suggesting that VP1 is a coactivator protein that physically interacts with G-box-binding proteins (91, 155). Unlike for *Em*, the combined response of the maize *C1* promoter to ABA and VP1 is less than additive , and transactivation of the *C1* promoter by VP1 is mediated exclusively by the sequence CGTCCATGCAT located within the Sph promoter element (Table 2) (65). The conserved basic domain B3 of VP1 is essential for *C1* transactivation, and the isolated B3 domain binds specifically to the above-mentioned sequence element in vitro (145). VP1 is, however, likely to interact with other DNA-binding proteins involved in ABA regulation of *C1*, because the entire Sph element is required for responsiveness of *C1* to both ABA and VP1 (65). The VP1/ABI3-like proteins thus appear to be multifunctional transcription factors that integrate ABA and other regulatory signals of seed maturation, most likely by interacting with distinct *trans*-acting factors that remain to be identified.

UPSTREAM SIGNALING ELEMENTS   Genetic studies in *A. thaliana* have identified a few other intermediates in the ABA regulation of gene expression in seeds. The already mentioned *abi4* and *abi5* mutants share with *abi3* mutants a decreased ABA sensitivity in germinating seeds, as well as a reduced developmental expression of certain LEA genes. These three loci have thus been proposed to act in a common regulatory pathway of seed maturation (27). In addition, the recent analysis of *abi3 fus3* and *abi3 lec1* double mutants indicates that *ABI3* acts synergistically with the FUSCA3 (*FUS3*) and LEAFY COTYLEDON1 (*LEC1*) genes to control multiple elementary processes during seed maturation, including sensitivity to ABA and accumulation of storage protein mRNAs (111). The *ABI4*, *ABI5*, *FUS3*, and *LEC1* gene products are currently unknown.

Studies on barley aleurone protoplasts indicate that reversible protein phosphorylation is likely to be implicated in the regulation of gene expression by ABA in seeds. ABA rapidly stimulated the activity of a MAP kinase, and this stimulation appeared to be correlated with the induction of the *Rab16* mRNA (69). Okadaic acid (an inhibitor of the PP1 and PP2A classes of serine/threonine protein phosphatases) inhibited the induction of the *HVA1* (73) and *Rab16* (51) mRNAs by ABA, and phenylarsine oxide (an inhibitor of tyrosine protein phosphatases) blocked the induction of *Rab16* (51). The protein kinase and phosphatases revealed by these studies remain to be isolated.

Finally, suggestive evidence indicates that modifications of intracellular cytosolic free Ca^{2+} levels ($[Ca]_i$) and cytoplasmic pH (pH_i) may act as intracellular second messengers in the ABA regulation of gene expression in seeds. In barley aleurone protoplasts, ABA triggers an increase in pH_i (153) and a decrease in $[Ca]_i$ (35, 158). However, the exact contribution of these alterations in $[Ca]_i$ and pH_i to the regulation of gene expression by ABA could not be clearly established (152, 153).

ABA SIGNAL TRANSDUCTION IN STRESS RESPONSE

During vegetative growth, endogenous ABA levels increase upon conditions of water stress, and ABA is an essential mediator in triggering the plant responses to these adverse environmental stimuli (166). As is discussed below, substantial evidence supports that the increased ABA levels limit water loss through transpiration by reducing stomatal aperture. ABA is also involved in other aspects of stress adaptation. For instance, ABA-deficient mutants of *A. thaliana* are impaired in cold acclimation (87) and in a root morphogenetic response to drought (drought rhizogenesis) (154). The role of ABA in signaling stress conditions has also been extensively documented by molecular studies showing that ABA-deficient mutants are affected in the regulation of numerous genes by drought, salt, or cold (39, 60, 140). It should, however, be noted that the adaptation to these adverse environmental conditions also involves ABA-independent pathways (140). Finally, ABA is involved in the induction of gene expression by mechanical damage (116).

Regulation of Gene Expression in Response to Stress

PROMOTER STUDIES   The regulation of gene expression by ABA in vegetative tissues appears to involve several signaling pathways. The ABA induction of distinct genes exhibit differential requirements for protein synthesis (164). In addition, several types of *cis*-acting elements are involved in ABA-induced gene expression in vegetative tissues. Many of the LEA genes that are abundantly expressed in desiccating seeds are also responsive to drought stress and ABA in vegetative tissues (60, 121). For several of these genes, the G-box-type ABREs already described (Table 2) are required for ABA induction both in seeds and in vegetative tissues (20, 44, 48, 107, 117).

In contrast, ABRE-like motifs are not involved in the ABA regulation of other stress-inducible genes such as the *A. thaliana RD22* (63) and the *Craterostigma plantagineum CDeT27-45* genes (104). The distinct sequence motif shown in Table 2 is essential for ABA responsiveness of *CDeT27-45* (104). In this case, as for ABREs (see section on ABA Signal Transduction in Seeds), the corresponding physiological *trans*-acting factors are presently unknown. Genes that are induced by ABA and encode other types of potential transcription factors include the *A. thaliana* homeobox gene *ATHB-7* (142) and several *myb* homologues from *A. thaliana* (151) and *Craterostigma* (62). The role of these genes in ABA signaling has, however, not been tested experimentally.

POTENTIAL INTERMEDIATES   Studies performed on various model systems have identified several potential signaling intermediates in the ABA activation of gene expression in response to stress.

The resurrection plant *Craterostigma plantagineum* can tolerate extreme dehydration. However, in vitro propagated callus derived from this plant has a strict requirement for exogenously applied ABA in order to survive a severe dehydration (31). This property has been exploited for isolation of dominant mutants by activation tagging, in which high expression of resident genes activated by insertion of a foreign promoter would confer desiccation tolerance to the transformed cells without prior ABA treatments. One gene was identified (*CDT-1*), whose high expression did confer the expected phenotypes in calli and led to constitutive expression of several ABA- and dehydration-inducible genes (31). The function of the *CDT-1* gene is not immediately obvious, because it encodes a transcript with no large open reading frame. It is possible that the biologically active product of *CDT-1* is a regulatory RNA or a short polypeptide (31).

The ABA regulation of gene expression in vegetative tissues is likely to involve reversible protein phosphorylation events. Several stress- and ABA-inducible mRNAs that encode protein kinases have been identified (57–59). In epidermal peels of *Pisum sativum*, the ABA-induced accumulation of dehydrin mRNA was reduced by K-252a (an inhibitor of serine/threonine protein kinases) and also by okadaic acid and cyclosporin A (an inhibitor of the PP2B class of serine/threonine protein phosphatases) (53).

The involvement of particular protein phosphatases 2C was revealed by the cloning of the *ABI1* (78, 97) and *ABI2* (79) genes of *A. thaliana*. Their corresponding mutants show decreased sensitivities to the ABA inhibition of seedling growth, and defects in various morphological and molecular responses to applied ABA in vegetative tissues (28, 41, 72). Despite the highly similar architecture of the ABI1 and ABI2 proteins (79), their functions are unlikely to be completely redundant for all ABA actions. The *abi1-1* and *abi2-1* mutations lead to identical amino acid substitutions at equivalent positions in the ABI1 and ABI2 proteins, respectively (79), but have (at least quantitatively) different effects on some of the responses to exogenous ABA or water stress (25, 33, 41, 124, 142, 154).

The identification of stress- and ABA-inducible mRNAs that code for a Ca^{2+}-binding membrane protein in rice (30) and for a phosphatidylinositol-specific phospholipase C in *A. thaliana* (55), respectively, provides circumstantial evidence for the possible involvement of Ca^{2+} in ABA signaling in vegetative tissues. In addition, ABA has been shown to induce increases in $[Ca]_i$ and pH_i in cells of corn coleoptiles and of parsley hypocotyls and roots (32).

SINGLE-CELL SYSTEMS   Ambitious attempts are being made to reconstruct minimal stress and ABA signaling pathways by reassembling candidate components in single-cell systems.

In maize leaf protoplasts, the barley *HVA1* promoter fused to reporter genes could be activated by applied ABA or by various stress conditions (136). In the absence of these stimuli, expression of the reporter gene could be induced by treating the protoplasts with 1 mM Ca^{2+} plus Ca^{2+} ionophores, or by overexpressing constitutively active forms of the normally Ca^{2+}-dependent ATCDPK1 and ATCDPK1a protein kinases from *A. thaliana*. Overexpressing the catalytic domain of the wild-type *A. thaliana* ABI1 PP2C inhibited the activation of *HVA1* by ABA or by ATCDPK1 (136).

Microinjection experiments were conducted in hypocotyl cells of the tomato *aurea* mutant to investigate the signaling cascade that mediates ABA activation of the *A. thaliana rd29A* (165) and *kin2/cor6.6* (157) promoters fused to the GUS reporter gene. This work supports the existence of a minimal linear cascade in which (*a*) ABA triggers a transient accumulation of cyclic ADP-ribose (cADPR) (for information on cADPR, see 3), (*b*) cADPR induces a release of Ca^{2+} from internal stores, and (*c*) a K-252a-sensitive kinase acting downstream of Ca^{2+} is required for activation of *rd29A* and *kin2* (Y Wu & N-H Chua, personal communication). Microinjection of the mutant abi1-1 protein inhibited ABA-, cADPR-, and Ca^{2+}-induced gene expression, and these effects could be reversed by an excess of wild-type ABI1 protein (Y Wu, A Himmelbach, E Grill & N-H Chua, personal communication).

These two sets of results illustrate that combining genetics with single-cell analyses represents a promising approach for deciphering possible epistatic relationships between candidate components in ABA signaling cascades.

Regulation of Stomatal Aperture

The stomatal pore is defined by a pair of surrounding guard cells. The closing and opening of the pore result from osmotic shrinking and swelling of these guard cells, respectively. In conditions of water stress, the increase in cellular ABA (45) or in apoplastic ABA at the surface of guard cells (161) is thought to provoke a reduction in turgor pressure of the guard cells, which in turn leads to stomatal closure to limit transpirational water loss. Applied ABA inhibits the opening and promotes the closure of stomatal pores (8, 16, 159). The enhanced transpiration of ABA-biosynthetic mutants (71, 77, 88) and transgenic plants expressing an anti-ABA antibody (7) provides evidence for such a role of endogenous ABA.

Unlike most other cells in higher-plant tissues, guard cells at maturity lack plasmodesmata (162). This property has afforded a single-cell system, directly accessible in planta, to investigate how ABA is perceived and transduced to trigger integrated physiological responses.

ABA PERCEPTION   Available evidence suggests that guard cells possess at least two sites of ABA perception involved in the regulation of stomatal aperture, one located at the plasma membrane and a second located intracellularly.

Stomatal closure can be triggered by extracellular application of ABA. However, the protonated form of the weak acid ABA (ABAH) readily permeates the lipid bilayer of the cell membrane (64). Guard cells of *Commelina communis* appear to have, in addition, a significant carrier-mediated uptake of ABA (86, 133). The requirement for an extracellular ABA receptor is indicated by the failure of ABA to inhibit the stomatal opening when microinjected directly into the cytosol of *Commelina* guard cells (5). In this species, extracellular ABA was, however, less effective in regulating stomatal aperture at high than at low external pH (that favors passive uptake of ABAH), providing indirect evidence for an intracellular ABA receptor (5, 106, 113, 133). This conclusion is also supported by the reports that stomatal closure could be triggered in *Commelina* by ABA released inside guard cells from caged ABA microinjected into the cytosol (2), and by ABA microinjected into guard cells in the presence of 1 μM extracellular ABA (133).

ABA closes stomatal pores by inducing net efflux of both K^+ and Cl^- from the vacuole to the cytoplasm, and from the cytoplasm to the outside of guard cells (85, 86). Tracer flux studies, measuring rate of loss of ^{86}Rb$^+$ from isolated *Commelina* guard cells, also indicate multiple actions of ABA, both inside and outside the cell. It has been proposed that internal receptors regulate tonoplast ion channels for release of vacuolar ions, whereas external receptors mediate the stimulation of ion efflux at the plasma membrane (85, 86).

ION CHANNELS REGULATED BY ABA   Electrophysiological studies, either by whole cell impalement or by patch clamping of the plasma membrane of guard cell protoplasts or of isolated vacuoles, have identified a number of ionic channels present in these guard cell membranes. The nature of the tonoplast ion channels involved in the regulation of stomatal aperture by ABA is not yet clearly identified (159). Attention is thus focused here on ion transport mechanisms across the plasma membrane of guard cells.

ABA causes a depolarization of the plasma membrane (149). Two types of anion channels, which may reflect different states of a single channel protein (26, 132), have been identified in the plasma membrane of guard cells (50, 129). One of these types of anion channels (S-type) shows slow and sustained activation properties and is active over a wide range of voltage. Inhibitor studies (132) and the recent demonstration that ABA activates S-type anion channels (43, 114) indicate that these channels can account for the membrane depolarization and prolonged anion efflux required for ABA-mediated stomatal closure. Inactivation of the plasma membrane H$^+$-ATPase by ABA may, however, also contribute to the membrane depolarization during stomatal closure (40).

The guard cell plasma membrane contains two types of K$^+$ channels. The inward-rectifying K$^+$ channel is responsible for K$^+$ influx to the guard cell and is inhibited by ABA (76, 130, 149). The outward-rectifying K$^+$ channel is active only at membrane potentials positive to the equilibrium potential for K$^+$, and mediates K$^+$ efflux required for stomatal closure (14, 130, 149). The ABA-induced membrane depolarization mentioned above will thus activate outward-rectifying K$^+$ channels, and the magnitude of this outward K$^+$ current is further enhanced by ABA in a largely voltage-independent manner (14, 76).

SECOND MESSENGERS   One of the earliest responses of guard cells to ABA is an increase in the concentration of cytosolic free Ca^{2+} ([Ca]$_i$) (2, 34, 61, 88a, 89, 128). The origin of the Ca^{2+} required to elevate [Ca]$_i$ in response to ABA is unclear (90), but evidence favoring influx across the plasma membrane (128) or release from internal stores mediated by inositol 1,4,5-trisphosphate (17, 37, 75) and/or cyclic ADP-ribose (3, 90) has been presented. Experimental elevation of [Ca]$_i$ is sufficient to induce stomatal closure (37) and mimics several of the above-mentioned effects of ABA on ion channels in the plasma membrane of guard cells. Increased [Ca]$_i$ activates the S-type anion channel and inhibits the inward-rectifying K^+ channel (50, 127). These results thus provide strong evidence for the involvement of Ca^{2+}-coupled signal transduction cascade(s) in the regulation of guard cell turgor by ABA.

Several observations, however, indicate that Ca^{2+}-independent signaling pathways are also involved. The extent of ABA-induced increase in $[Ca]_i$ is variable, and stomata could even close in response to ABA without any detectable change in $[Ca]_i$ (2, 34, 89). It remains, however, unclear whether ABA always produces a local increase in $[Ca]_i$ that does not systematically translate into global cytoplasmic change (89), or whether ABA can produce stomatal closure using only Ca^{2+}-independent signaling pathways (2). In any case, the fact that the outward-rectifying K^+ channel is insensitive to increases in $[Ca]_i$ (76, 127) indicates that other second messengers must be involved in the regulation of plasma membrane ion channels by ABA.

In particular, ABA increases the cytoplasmic pH (pH_i) of guard cells (15, 61). Studies in which guard cells were loaded with pH buffers or pH_i was experimentally modified support that the ABA-induced cytoplasmic alkalinization can account for ABA activation of the outward-rectifying K^+ channel and also contribute to ABA inactivation of the inward-rectifying K^+ channel (15, 42, 76, 98). The ABA regulation of ion channels in guard cells thus appears to involve, at least, both Ca^{2+}-dependent and pH_i-dependent signaling pathways.

DOWNSTREAM SIGNALING ELEMENTS   How the ABA-induced Ca^{2+} and pH_i signals are then decoded and relayed by other cellular components in the transduction chain is not clear. The recent report that pH_i regulates the outward-rectifying K^+ channel in isolated membrane patches from *Vicia faba* guard cells suggests that the signaling elements acting downstream of the pH_i signal are membrane associated (98). The Ca^{2+} signal is possibly relayed by particular protein kinases and phosphatases. Both Ca^{2+}-dependent protein kinases (115) and phosphatases (4, 23, 84) have been inferred to be potential candidates in modulating the activity of various ion channels in guard cells, but their precise roles in ABA signaling have not been clearly established at present.

There is nonetheless incontrovertible evidence that protein phosphorylation is a critical component in the ABA signal transduction controlling stomatal closure. In *Vicia faba* and *Commelina*, stomatal closure induced by ABA was abolished by kinase inhibitors while enhanced by inhibitors of the protein phosphatases PP1 and/or PP2A (125). These protein phosphatases, whose actions are Ca^{2+}-independent, have been implicated in the regulation of the inward-rectifying K^+ channel (82, 148), outward-rectifying K^+ channel (148), and S-type anion channel (114, 125) in *Vicia faba* and *Commelina* guard cells.

Recent reports described two protein kinases whose activities are rapidly stimulated by ABA in protoplasts from guard cells but not mesophyll cells in

Vicia faba (81, 99). Although they have a similar apparent molecular weight, the two protein kinases are probably distinct because they differ in their ability to autophosphorylate and in their substrate specificity in vitro (81, 99). The cellular targets of these kinases are not yet known, but because their activities are detectable minutes after induction by physiological concentrations of ABA, these kinases may be involved in the modulation of ion channel activities that are essential for rapid stomatal responses.

Genetic dissection of ABA signaling in guard cells has been rather limited despite an initial and promising report of the barley mutant *cool*, which displayed excessive transpiration and ABA-insensitive guard cells (118). Mutants with specific impairment in ABA-induced stomatal closure have not yet been identified in *A. thaliana*.

However, among the *gca1-gca8* (Growth Control by ABA) mutants that were isolated based on their reduced sensitivities to the inhibition of seedling growth by exogenous ABA, the *gca1* and *gca2* mutants display a disturbed stomatal regulation (10). The *abi1-1* and *abi2-1* mutants mentioned earlier also have ABA-insensitive guard cells as part of their pleiotropic phenotypes (114, 122).

What are the roles of the Ca^{2+}-independent (12) ABI1 and ABI2 protein phosphatases 2C (PP2Cs) in stomatal regulation by ABA? Taking advantage of the fact that the *abi1-1* mutation is dominant (72, 78), the mutant *A. thaliana* gene was introduced into the diploid tobacco *Nicotiana benthamiana* to facilitate electrophysiological measurements (6). Voltage clamp studies found that the mutant *abi1-1* transgene had no detectable effect on anion channels, but decreased ABA sensitivity of both the inward- and outward-rectifying K^+ channels in the guard cell plasma membrane (6, 43). ABA sensitivity of these K^+ channels and ABA-induced stomatal closure could be partially reestablished in the *abi1-1* transgenic plants by adding protein kinase inhibitors (6). Recent and important advances made in single-cell techniques have permitted measurements of ion channel activities in *A. thaliana* guard cells, which had previously posed technical difficulties because of their small sizes (114, 123). In patch-clamp studies of *A. thaliana* guard cell protoplasts, the ABA response of S-type anion channels was impaired in the *abi1-1* mutant. However, consistent with the reversibility of the effects of the *abi1-1* transgene in tobacco (6), treatment with the protein kinase inhibitor K-252a could partially restore ABA regulation of the anion channel and ABA-induced stomatal closure in the *A. thaliana abi1-1* mutant (114). The *abi2-1* mutation also suppressed the ABA activation of S-type anion channels, but curiously, this inhibition could not be counterbalanced by K-252a treatments (114). In addition, the *abi2-1* mutation also diminished the background activity of inward-rectifying K^+ channels (114), an effect observed neither in the *abi1-1 A. thaliana* mutant (114) nor in the *abi1-1* transgenic tobacco plants (6). The differential effects of the *abi1* and *abi2* mutations on the background activity of inward-rectifying K^+ channels, and the differential sensitivity of these mutants to K-252a, thus indicate that, although the ABI1 and ABI2 proteins are homologous PP2Cs, the functions of these proteins in stomatal regulation, as in other physiological responses, are nonredundant.

The apparent discrepancies between the above-mentioned results on *abi1-1* transgenic tobacco and *abi1-1* mutant *A. thaliana* plants may be indicative of the functional flexibility of ABI1 in maintaining an integrated ABA regulation of both K^+ and anion channels, depending on the imposed (environmental or experimental) conditions. This would be analogous to the relative importance of the ABA-induced increase in $[Ca]_i$, which is subjected to both experimental (89) and environmental (2) influences. A more serious problem in terms of formulating a unified model of stomatal regulation, however, is how many species-dependent differences exist in ABA signaling. For instance, various pharmacological studies indicate that PP1/PP2A protein phosphatases (distinct from the ABI1 and ABI2 PP2Cs) act as positive regulators of anion currents in *A. thaliana* guard cells (114), but as negative regulators in *Vicia*, *Commelina*, and *N. benthamiana* (43, 125). Therefore, it seems that detailed comparative studies with different species would be necessary before a fuller understanding of the signaling cascades that mediate stomatal regulation by ABA would emerge.

ABA SIGNALING TO GUARD CELL NUCLEUS   As described above, stomatal guard cells have been used extensively as a cellular system to explore the transduction pathways that target the ABA signal to ion channels during stomatal closing. These responses are rapid and are thus thought to be operationally distinct from long-term ABA responses that require RNA and protein synthesis (166). However, hormone induction of gene expression at the transcriptional level can also take place within minutes, as has been observed in the case of auxin (93). The relevance of ABA in gene activation in guard cells has only been explored recently. Several promoters or mRNAs have been shown to be induced in guard cells by exogenous ABA treatment (53, 110, 137, 147, 157). These studies demonstrate that guard cells are competent to relay ABA signals to the nucleus. Initial attempts are being made to analyze whether identical or distinct signaling pathways mediate the ABA regulations of ion channels and gene expression in guard cells (53). It also remains to be established whether there are gene products directly implicated in controlling stomatal movements that are regulated by ABA transcriptionally.

CONCLUSION AND PERSPECTIVES

In recent years, the research into the molecular and physiological action of ABA has benefited tremendously from cross-fertilization of different expertise and ideas. Studies with optically pure ABA analogues revealed that the stereo-chemical requirements of ABA are not identical in all ABA responses (21, 156), indicating that higher plants may contain several types of ABA receptors. Evidence for ABA perception sites located both at the cell surface and intracellularly has come from analyses on the regulation of ion fluxes in guard cells. Whether this model is applicable to other cell types, and to more long-term responses involving changes in gene expression, are important questions to be explored (for instance, see 36). In addition, even though locations of these reception sites can be detected physiologically, we have no firm idea whether they are functionally distinct, and if so, what physiological relevance they may harbor. Cloning genes of ABA receptors and creating corresponding mutants would thus constitute a major step toward dissecting these initial events of ABA action.

Besides reception sites, we also need to improve further our understanding of the more downstream events in the transduction cascades triggered by ABA signals. Recent advances point to the central role of reversible protein phosphorylation in mediating several of the physiological responses to ABA. Much work, however, is clearly needed to identify the pertinent kinases and phosphatases, and cellular targets of their action.

The ectopic expression of the seed-specific *ABI3* gene in leaves resulted in activation of several otherwise seed-specific genes. This observation and the pleiotropic phenotypes of some ABA response mutants raise the question of the degree of similarity between signaling cascades in the different tissues (or cell types in the same tissue). It might be possible that many core components of the ABA signaling network exist in different tissues. A simple model would be that these central cascades are then regulated by cell-specific factors such as the VP1/ABI3-like proteins. In this light, ABA-stimulated protein kinases with apparent specificity to guard cells may play similar deterministic roles in activating a branch of the ABA pathway in this cell type.

Isolation of additional mutants by judicious genetic screens (taking into account the potential for redundancy in signaling components) will no doubt enrich our knowledge about regulatory points in ABA action, which might include modulation in the activity of synthetic enzymes and transport of the hormone (150). These two latter points have not been the subject of intense investigations. Furthermore, mutations that simultaneously alter the responsiveness to ABA and other hormones or developmental clues, will permit a fuller appreciation of cross-talk between ABA and other cellular signals. Several mutants of this type have already been isolated in *A. thaliana* (Table 1). The eventual cloning of the genes identified by mutational analyses will reveal their molecular identity and, in some cases, suggest possible functions based on homologies with known proteins. Genes with unexpected structural features (such as *CDT-1* in *C. plantagineum*) will divulge novel signal transduction mechanisms, to the delight of researchers with iconoclastic tendencies.

The isolation of genes responsive to ABA, coupled with promoter analyses, are vital to our understanding of how combinatorial and synergistic actions of *cis*-acting elements and *trans*-acting factors are involved in the versatile control of the output response. Other approaches, such as biochemical purification and those based on interaction cloning, will undoubtedly complement the more established technologies in unraveling additional signaling elements or even signaling complexes.

Investigators with a pioneering spirit have already attempted to reconstruct particular signaling cascades by reassembling individual components suspected to play the relevant roles. These are much welcome advances, since these cellular systems will complement the more established guard cell model. It is certain that continuation of such a multidisciplinary approach will yield ever deeper insight not only into ABA signal transduction itself but also into how ABA and other signaling molecules jointly coordinate physiological responses.

ACKNOWLEDGMENTS

Work in the authors' laboratory is supported by the Centre National de la Recherche Scientifique and grants from the International Human Frontier Science Program (RG-303/95), the European Community BIOTECH program (BIO4-CT96-0062), and the Groupement de Recherches et d'Etudes sur les Génomes (Décision 9-95).

> Visit the *Annual Reviews home page* at
> http://www.AnnualReviews.org.

Literature Cited

1. Addicott FT, Carns HR. 1983. History and introduction. In *Abscisic Acid*, ed. FT Addicott, pp. 1–21. New York: Praeger Sci.
2. Allan AC, Fricker MD, Ward JL, Beale MH, Trewavas AJ. 1994. Two transduction pathways mediate rapid effects of abscisic acid in *Commelina* guard cells. *Plant Cell* 6:1319–28
3. Allen GJ, Muir SR, Sanders D. 1995. Release of Ca^{2+} from individual plant vacuoles by both InsP$_3$ and cyclic ADP-ribose. *Science* 268:735–37
4. Allen GJ, Sanders D. 1995. Calcineurin, a type 2B protein phosphatase, modulates the Ca^{2+}-permeable slow vacuolar ion channel of stomatal guard cells. *Plant Cell* 7:1473–83
5. Anderson BE, Ward JM, Schroeder JI.

1994. Evidence for an extracellular reception site for abscisic acid in *Commelina* guard cells. *Plant Physiol.* 104:1177–83
6. Armstrong F, Leung J, Grabov A, Brearley J, Giraudat J, Blatt MR. 1995. Sensitivity to abscisic acid of guard cell K$^+$ channels is suppressed by *abi1-1*, a mutant *Arabidopsis* gene encoding a putative protein phosphatase. *Proc. Natl. Acad. Sci. USA* 92:9520–24
7. Artsaenko O, Peisker M, zur Nieden U, Fiedler U, Weiler EW, et al. 1995. Expression of a single chain Fv antibody against abscisic acid creates a wilty phenotype in transgenic tobacco. *Plant J.* 8:745–50
8. Assmann SM. 1993. Signal transduction in guard cells. *Annu. Rev. Cell Biol.* 9:345–75

9. Deleted in proof
10. Benning G, Ehrler T, Meyer K, Leube M, Rodriguez P, Grill E. 1996. Genetic analysis of ABA-mediated control of plant growth. *Proc. Workshop Abscisic Acid Signal Transduction Plants, Madrid*, pp. 34. Madrid: Juan March Found.
11. Berger S, Bell E, Mullet JE. 1996. Two methyl jasmonate-insensitive mutants show altered expression of *AtVsp* in response to methyl jasmonate and wounding. *Plant Physiol.* 111:525–31
12. Bertauche N, Leung J, Giraudat J. 1996. Protein phosphatase activity of abscisic acid insensitive 1 (ABI1) protein from *Arabidopsis thaliana. Eur. J. Biochem.* 241:193–200
13. Bitoun R, Rousselin P, Caboche M. 1990. A pleiotropic mutation results in cross-resistance to auxin, abscisic acid and paclobutrazol. *Mol. Gen. Genet.* 220:234–39
14. Blatt MR. 1990. Potassium channel currents in intact stomatal guard cells: rapid enhancement by abscisic acid. *Planta* 180:445–55
15. Blatt MR, Armstrong F. 1993. K$^+$ channels of stomatal guard cells: abscisic acid–evoked control of the outward rectifier mediated by cytoplasmic pH. *Planta* 191:330–41
16. Blatt MR, Thiel G. 1993. Hormonal control of ion channel gating. *Annu. Rev. Plant Physiol. Plant Mol. Biol.* 44:543–67
17. Blatt MR, Thiel G, Trentham DR. 1990. Reversible inactivation of K$^+$ channels of *Vicia* stomatal guard cells following the photolysis of caged inositol 1,4,5-trisphosphate. *Nature* 346:766–69
18. Bobb AJ, Chern M-S, Bustos MM. 1997. Conserved RY-repeats mediate transactivation of seed-specific promoters by the developmental regulator PvALF. *Nucleic Acids Res.* 25:641–47
19. Bobb AJ, Eiben HG, Bustos MM. 1995. PvAlf, an embryo-specific acidic transcriptional activator enhances gene expression from phaseolin and phytohemagglutinin promoters. *Plant J.* 8:331–43
20. Busk PK, Jensen AB, Pagès M. 1997. Regulatory elements in vivo in the promoter of the abscisic acid responsive gene *rab17* from maize. *Plant J.* 11:1285–95
21. Chandler JW, Abrams SR, Bartels D. 1997. The effect of ABA analogs on callus viability and gene expression in *Craterostigma plantagineum. Physiol. Plant.* 99:465–69
22. Clouse SD, Langford M, McMorris TC.

1996. A brassinosteroid-insensitive mutant in *Arabidopsis thaliana* exhibits multiple defects in growth and development. *Plant Physiol.* 111:671–78
23. Cousson A, Cotelle V, Vavasseur A. 1995. Induction of stomatal closure by vanadate or a light/dark transition involves Ca^{2+}-calmodulin-dependent protein phosphorylations. *Plant Physiol.* 109:491–97
24. Cutler S, Ghassemian M, Bonetta D, Cooney S, McCourt P. 1996. A protein farnesyl transferase involved in abscisic acid signal transduction in *Arabidopsis. Science* 273:1239–41
25. de Bruxelles GL, Peacock WJ, Dennis ES, Dolferus R. 1996. Abscisic acid induces the alcohol dehydrogenase gene in *Arabidopsis. Plant Physiol.* 111:381–91
26. Dietrich P, Hedrich R. 1994. Interconversion of fast and slow gating modes of GCAC1, a guard cell anion channel. *Planta* 195:301–4
27. Finkelstein RR. 1994. Mutations at two new *Arabidopsis* ABA response loci are similar to the *abi3* mutations. *Plant J.* 5:765–71
28. Finkelstein RR, Somerville CR. 1990. Three classes of abscisic acid (ABA)–insensitive mutations of *Arabidopsis* define genes that control overlapping subsets of ABA responses. *Plant Physiol.* 94:1172–79
29. Deleted in proof
30. Frandsen G, Müller-Uri F, Nielsen M, Mundy J, Skriver K. 1996. Novel plant Ca^{2+}-binding protein expressed in response to abscisic acid and osmotic stress. *J. Biol. Chem.* 271:343–48
31. Furini A, Koncz C, Salamini F, Bartels D. 1997. High level transcription of a member of a repeated gene family confers dehydration tolerance to callus tissue of *Craterostigma plantagineum. EMBO J.* 16:3599–608
32. Gehring CA, Irving HR, Parish RW. 1990. Effects of auxin and abscisic acid on cytosolic calcium and pH in plant cells. *Proc. Natl. Acad. Sci. USA* 87:9645–49
33. Gilmour SJ, Thomashow MF. 1991. Cold acclimation and cold-regulated gene expression in ABA mutants of *Arabidopsis thaliana. Plant Mol. Biol.* 17:1233–40
34. Gilroy S, Fricker MD, Read ND, Trewavas AJ. 1991. Role of calcium in signal transduction of *Commelina* guard cells. *Plant Cell* 3:333–44
35. Gilroy S, Jones RL. 1992. Gibberellic acid and abscisic acid coordinately regulate cytoplasmic calcium and secretory activity in barley aleurone protoplasts. *Proc. Natl. Acad. Sci. USA* 89:3591–95

36. Gilroy S, Jones RL. 1994. Perception of gibberellin and abscisic acid at the external face of the plasma membrane of barley (*Hordeum vulgare* L.) aleurone protoplasts. *Plant Physiol.* 104:1185–92

37. Gilroy S, Read ND, Trewavas AJ. 1990. Elevation of cytoplasmic calcium by caged calcium or caged inositol trisphosphate initiates stomatal closure. *Nature* 343:769–71

38. Giraudat J, Hauge BM, Valon C, Smalle J, Parcy F, Goodman HM. 1992. Isolation of the *Arabidopsis ABI3* gene by positional cloning. *Plant Cell* 4:1251–61

39. Giraudat J, Parcy F, Bertauche N, Gosti F, Leung J, et al. 1994. Current advances in abscisic acid action and signaling. *Plant Mol. Biol.* 26:1557–77

40. Goh C-H, Kinoshita T, Oku T, Shimazaki K-I. 1996. Inhibition of blue light–dependent H+ pumping by abscisic acid in *Vicia* guard-cell protoplasts. *Plant Physiol.* 111:433–40

41. Gosti F, Bertauche N, Vartanian N, Giraudat J. 1995. Abscisic acid–dependent and –independent regulation of gene expression by progressive drought in *Arabidopsis thaliana*. *Mol. Gen. Genet.* 246:10–18

42. Grabov A, Blatt MR. 1997. Parallel control of the inward-rectifier K+ channel by cytosolic free Ca2+ and pH in *Vicia* guard cells. *Planta* 201:84–95

43. Grabov A, Leung J, Giraudat J, Blatt MR. 1997. Alteration of anion channel kinetics in wild-type and *abi1-1* transgenic *Nicotiana benthamiana* guard cells by abscisic acid. *Plant J.* 12:203–13

44. Guiltinan MJ, Marcotte WR, Quatrano RS. 1990. A plant leucine zipper protein that recognizes an abscisic acid response element. *Science* 250:267–71

45. Harris MJ, Outlaw WH Jr. 1991. Rapid adjustment of guard-cell abscisic acid levels to current leaf-water status. *Plant Physiol.* 95:171–73

46. Deleted in proof

47. Hattori T, Terada T, Hamasuna S. 1994. Sequence and functional analyses of the rice gene homologous to the maize *Vp1*. *Plant Mol. Biol.* 24:805–10

48. Hattori T, Terada T, Hamasuna S. 1995. Regulation of the *Osem* gene by abscisic acid and the transcriptional activator VP1: analysis of *cis*-acting promoter elements required for regulation by abscisic acid and VP1. *Plant J.* 7:913–25

49. Deleted in proof

50. Hedrich R, Busch H, Raschke K. 1990. Ca2+ and nucleotide dependent regulation of voltage dependent anion channels

in the plasma membrane of guard cells. *EMBO J.* 9:3889–92

51. Heimovaara-Dijkstra S, Nieland TJF, van der Meulen RM, Wang M. 1996. Abscisic acid–induced gene-expression requires the activity of protein(s) sensitive to the protein-tyrosine phosphatase inhibitor phenylarsine oxide. *Plant Growth Regul.* 18:115–23

52. Deleted in proof

53. Hey SJ, Bacon A, Burnett E, Neill SJ. 1997. Abscisic acid signal transduction in epidermal cells of *Pisum sativum* L. *Argenteum*: Both dehydrin mRNA accumulation and stomatal responses require protein phosphorylation and dephosphorylation. *Planta* 202:85–92

54. Hill A, Nantel A, Rock CD, Quatrano RS. 1996. A conserved domain of the *Viviparous-1* gene product enhances the DNA binding activity of the bZIP protein EmBP-1 and other transcription factors. *J. Biol. Chem.* 271:3366–74

55. Hirayama T, Ohto C, Mizoguchi T, Shinozaki K. 1995. A gene encoding a phosphatidylinositol-specific phospholipase C is induced by dehydration and salt stress in *Arabidopsis thaliana*. *Proc. Natl. Acad. Sci. USA* 92:3903–7

56. Hoecker U, Vasil IK, McCarty DR. 1995. Integrated control of seed maturation and germination programs by activator and repressor functions of Viviparous-1 of maize. *Genes Dev.* 9:2459–69

57. Holappa LD, Walker-Simmons MK. 1995. The wheat abscisic acid-responsive protein kinase mRNA, PKABA1, is up-regulated by dehydration, cold temperature, and osmotic stress. *Plant Physiol.* 108:1203–10

58. Hong SW, Jon JH, Kwak JM, Nam HG. 1997. Identification of a receptor-like protein kinase gene rapidly induced by abscisic acid, dehydration, high salt, and cold treatments in *Arabidopsis thaliana*. *Plant Physiol.* 113:1203–12

59. Hwang IW, Goodman HM. 1995. An *Arabidopsis thaliana* root-specific kinase homolog is induced by dehydration, ABA, and NaCl. *Plant J.* 8:37–43

60. Ingram J, Bartels D. 1996. The molecular basis of dehydration tolerance in plants. *Annu. Rev. Plant Physiol. Plant Mol. Biol.* 47:377–403

61. Irving HR, Gehring CA, Parish RW. 1992. Changes in cytosolic pH and calcium of guard cells precede stomatal movements. *Proc. Natl. Acad. Sci. USA* 89:1790–94

62. Iturriaga G, Leyns L, Villegas A, Gharaibeh R, Salamini F, Bartels D. 1996.

A family of novel *myb*-related genes from the resurrection plant *Craterostigma plantagineum* are specifically expressed in callus and roots in response to ABA or desiccation. *Plant Mol. Biol.* 32:707–16

63. Iwasaki T, Yamaguchi-Shinozaki K, Shinozaki K. 1995. Identification of a *cis*-regulatory region of a gene in *Arabidopsis thaliana* whose induction by dehydration is mediated by abscisic acid and requires protein synthesis. *Mol. Gen. Genet.* 247:391–98

64. Kaiser WM, Hartung W. 1981. Uptake and release of abscisic acid by isolated photoautotrophic mesophyll cell, depending on pH gradients. *Plant Physiol.* 68:202–6

65. Kao C-Y, Cocciolone SM, Vasil IK, McCarty DR. 1996. Localization and interaction of the *cis*-acting elements for abscisic acid, VIVIPAROUS1, and light activation of the *C1* gene of maize. *Plant Cell* 8:1171–79

66. Deleted in proof

67. Deleted in proof

68. Kim SY, Chung H-J, Thomas TL. 1997. Isolation of a novel class of bZIP transcription factors that interact with ABA-responsive and embryo-specification elements in the *Dc3* promoter using a modified yeast one-hybrid system. *Plant J.* 11:1237–51

69. Knetsch MLW, Wang M, Snaar-Jagalska BE, Heimovaara-Dijkstra S. 1996. Abscisic acid induces mitogen-activated protein kinase activation in barley aleurone protoplasts. *Plant Cell* 8:1061–67

70. Koornneef M, Hanhart CJ, Hilhorst HWM, Karssen CM. 1989. In vivo inhibition of seed development and reserve protein accumulation in recombinants of abscisic acid biosynthesis and responsiveness mutants in *Arabidopsis thaliana*. *Plant Physiol.* 90:463–69

71. Koornneef M, Jorna ML, Brinkhorst-van der Swan DLC, Karssen CM. 1982. The isolation of abscisic acid (ABA) deficient mutants by selection of induced revertants in nongerminating gibberellin sensitive lines of *Arabidopsis thaliana* (L.) Heynh. *Theor. Appl. Genet.* 61:385–93

72. Koornneef M, Reuling G, Karssen CM. 1984. The isolation and characterization of abscisic acid-insensitive mutants of *Arabidopsis thaliana*. *Physiol. Plant.* 61:377–83

73. Kuo A, Cappelluti S, Cervantes-Cervantes M, Rodriguez M, Bush DS. 1996. Okadaic acid, a protein phosphatase inhibitor, blocks calcium changes, gene expression, and cell death induced by gib-

berellin in wheat aleurone cells. *Plant Cell* 8:259–69

74. Kusano T, Berberich T, Harada M, Suzuki N, Sugawara K. 1995. A maize DNA-binding factor with a bZIP motif is induced by low temperature. *Mol. Gen. Genet.* 248:507–17

75. Lee YS, Choi YB, Suh S, Lee J, Assmann SM, et al. 1996. Abscisic acid–induced phosphoinositide turnover in guard cell protoplasts of *Vicia faba*. *Plant Physiol.* 110:987–96

76. Lemtiri-Chlieh F, MacRobbie EAC. 1994. Role of calcium in the modulation of *Vicia* guard cell potassium channels by abscisic acid: a patch clamp study. *J. Membr. Biol.* 137:99–107

77. Léon-Kloosterziel KM, Gil MA, Ruijs GJ, Jacobsen SE, Olszewski NE, et al. 1996. Isolation and characterization of abscisic acid-deficient *Arabidopsis* mutants at two new loci. *Plant J.* 10:655–61

78. Leung J, Bouvier-Durand M, Morris P-C, Guerrier D, Chefdor F, Giraudat J. 1994. *Arabidopsis* ABA-response gene *ABI1*: features of a calcium-modulated protein phosphatase. *Science* 264:1448–52

79. Leung J, Merlot S, Giraudat J. 1997. The *Arabidopsis ABSCISIC ACID-INSENSITIVE 2* (*ABI2*) and *ABI1* genes encode redundant protein phosphatases 2C involved in abscisic acid signal transduction. *Plant Cell* 9:759–71

80. Leydecker M-T, Moureaux T, Kraepiel Y, Schnorr K, Caboche M. 1995. Molybdenum cofactor mutants, specifically impaired in xanthine dehydrogenase activity and abscisic acid biosynthesis, simultaneously overexpress nitrate reductase. *Plant Physiol.* 107:1427–31

81. Li JX, Assmann SM. 1996. An abscisic acid-activated and calcium-independent protein kinase from guard cells of Fava bean. *Plant Cell* 8:2359–68

82. Li WW, Luan S, Schreiber SL, Assmann SM. 1994. Evidence for protein phosphatase 1 and 2A regulation of K^+ channels of two types of leaf cells. *Plant Physiol.* 106:963–70

83. Lu GH, Paul A-L, McCarty DR, Ferl RJ. 1996. Transcription factor veracity: Is GBF3 responsible for ABA-regulated expression of Arabidopsis *Adh*? *Plant Cell* 8:847–57

84. Luan S, Li WW, Rusnak F, Assmann SM, Schreiber SL. 1993. Immunosuppressants implicate protein phosphatase regulation of K^+ channels in guard cells. *Proc. Natl. Acad. Sci. USA* 90:2202–6

85. MacRobbie EAC. 1995. Effects of ABA on $^{86}Rb^+$ fluxes at plasmalemma and

tonoplast of stomatal guard cells. *Plant J.* 7:835–43

86. MacRobbie EAC. 1995. ABA-induced ion efflux in stomatal guard cells: multiple actions of ABA inside and outside the cell. *Plant J.* 7:565–76

87. Mäntylä E, Lang V, Palva ET. 1995. Role of abscisic acid in drought-induced freezing tolerance, cold acclimation, and accumulation of LTI78 and RAB18 proteins in *Arabidopsis thaliana. Plant Physiol.* 107:141–48

88. Marin E, Nussaume L, Quesada A, Gonneau M, Sotta B, et al. 1996. Molecular identification of zeaxanthin epoxidase of *Nicotiana plumbaginifolia*, a gene involved in abscisic acid biosynthesis and corresponding to *ABA* locus of *Arabidopsis thaliana. EMBO J.* 15:2331–42

88a. McAinsh MR, Brownlee AM, Hetherington AM. 1990. Abscisic acid-induced elevation of guard cell cytosolic Ca^{2+} precedes stomatal closure. *Nature* 343:186–88

89. McAinsh MR, Brownlee C, Hetherington AM. 1992. Visualizing changes in cytosolic-free Ca^{2+} during the response of stomatal guard cells to abscisic acid. *Plant Cell* 4:1113–22

90. McAinsh MR, Brownlee C, Hetherington AM. 1997. Calcium ions as second messengers in guard cell signal transduction. *Physiol. Plant.* 100:16–29

91. McCarty DR. 1995. Genetic control and integration of maturation and germination pathways in seed development. *Annu. Rev. Plant Physiol. Plant Mol. Biol.* 46:71–93

92. McCarty DR, Hattori T, Carson CB, Vasil V, Lazar M, Vasil IK. 1991. The *viviparous-1* developmental gene of maize encodes a novel transcriptional activator. *Cell* 66:895–905

93. McClure BA, Hagen G, Brown CS, Gee MA, Guilfoyle TJ. 1989. Transcription, organization, and sequence of an auxin-regulated gene cluster in soybean. *Plant Cell* 1:229–39

94. Deleted in proof

95. Deleted in proof

96. Meurs C, Basra AS, Karssen CM, van Loon LC. 1992. Role of abscisic acid in the induction of desiccation tolerance in developing seeds of *Arabidopsis thaliana. Plant Physiol.* 98:1484–93

97. Meyer K, Leube MP, Grill E. 1994. A protein phosphatase 2C involved in ABA signal transduction in *Arabidopsis thaliana. Science* 264:1452–55

98. Miedema H, Assmann SM. 1996. A membrane-delimited effect of internal pH on the K^+ outward rectifier of *Vicia faba* guard cells. *J. Membr. Biol.* 154:227–37

99. Mori IC, Muto S. 1997. Abscisic acid activates a 48-kilodalton protein kinase in guard cell protoplasts. *Plant Physiol.* 113:833–39

100. Nakagawa H, Ohmiya K, Hattori T. 1996. A rice bZIP protein, designated OSBZ8, is rapidly induced by abscisic acid. *Plant J.* 9:217–27

101. Nambara E, Keith K, McCourt P, Naito S. 1995. A regulatory role for the *ABI3* gene in the establishment of embryo maturation in *Arabidopsis thaliana. Development* 121:629–36

102. Nambara E, Naito S, McCourt P. 1992. A mutant of *Arabidopsis* which is defective in seed development and storage protein accumulation is a new *abi3* allele. *Plant J.* 2:435–41

103. Nantel A, Quatrano RS. 1996. Characterization of three rice basic/leucine zipper factors, including two inhibitors of EmBP-1 DNA binding activity. *J. Biol. Chem.* 271:31296–305

104. Nelson D, Salamini F, Bartels D. 1994. Abscisic acid promotes novel DNA-binding activity to a desiccation-related promoter of *Craterostigma plantagineum. Plant J.* 5:451–58

105. Oeda K, Salinas J, Chua N-H. 1991. A tobacco bZip transcription activator (TAF-1) binds to a G-box-like motif conserved in plant genes. *EMBO J.* 10:1793–802

106. Ogunkami AB, Tucker DJ, Mansfield TA. 1973. An improved bioassay for abscisic acid and other antitranspirants. *New Phytol.* 72:277–78

107. Ono A, Izawa T, Chua N-H, Shimamoto K. 1996. The *rab16B* promoter of rice contains two distinct abscisic acid-responsive elements. *Plant Physiol.* 112:483–91

108. Ooms JJJ, Léon-Kloosterziel KM, Bartels D, Koornneef M, Karssen CM. 1993. Acquisition of desiccation tolerance and longevity in seeds of *Arabidopsis thaliana.* A comparative study using abscisic acid–insensitive *abi3* mutants. *Plant Physiol.* 102:1185–91

109. Paiva R, Kriz AL. 1994. Effect of abscisic acid on embryo-specific gene expression during normal and precocious germination in normal and *viviparous* maize (*Zea mays*) embryos. *Planta* 192:332–39

110. Parcy F, Giraudat J. 1997. Interactions between the *ABI1* and the ectopically expressed *ABI3* genes in controlling abscisic acid responses in *Arabidopsis* vegetative

tissues. *Plant J.* 11:693–702
111. Parcy F, Valon C, Kohara A, Miséra S, Giraudat J. 1997. The *ABSCISIC ACID-INSENSITIVE 3* (*ABI3*), *FUSCA 3* (*FUS3*) and *LEAFY COTYLEDON 1* (*LEC1*) loci act in concert to control multiple aspects of *Arabidopsis* seed development. *Plant Cell.* 9:1265–77
112. Parcy F, Valon C, Raynal M, Gaubier-Comella P, Delseny M, Giraudat J. 1994. Regulation of gene expression programs during *Arabidopsis* seed development: roles of the *ABI3* locus and of endogenous abscisic acid. *Plant Cell* 6:1567–82
113. Paterson NW, Weyers JDB, Brook RA. 1988. The effect of pH on stomatal sensitivity to abscisic acid. *Plant Cell Environ.* 11:83–89
114. Pei Z-M, Kuchitsu K, Ward JM, Schwarz M, Schroeder JI. 1997. Differential abscisic acid regulation of guard cell slow anion channels in *Arabidopsis* wild-type and *abi1* and *abi2* mutants. *Plant Cell* 9:409–23
115. Pei Z-M, Ward JM, Harper JF, Schroeder JI. 1996. A novel chloride channel in *Vicia faba* guard cell vacuoles activated by the serine/threonine kinase, CDKP. *EMBO J.* 15:6564–74
116. Pena-Cortes H, Fisahn J, Willmitzer L. 1995. Signals involved in wound-induced proteinase inhibitor II gene expression in tomato and potato plants. *Proc. Natl. Acad. Sci. USA* 92:4106–13
117. Pla M, Vilardell J, Guiltinan MJ, Marcotte WR, Niogret M-F, et al. 1993. The *cis*-regulatory element CCACGTGG is involved in ABA and water-stress responses of the maize gene *rab28*. *Plant Mol. Biol.* 21:259–66
118. Raskin I, Ladyman JAR. 1988. Isolation and characterization of a barley mutant with abscisic acid-insensitive stomata. *Planta* 173:73–78
119. Robertson DS. 1955. The genetics of vivipary in maize. *Genetics* 40:745–60
120. Robichaud C, Sussex IM. 1986. The response of *viviparous-1* and wild-type embryos of *Zea mays* to culture in the presence of abscisic acid. *J. Plant Physiol.* 126:235–42
121. Rock CD, Quatrano RS. 1995. The role of hormones during seed development. In *Plant Hormones*, ed. PJ Davies, pp. 671–97. Dordrecht: Kluwer
122. Roelfsema MRG, Prins HBA. 1995. Effect of abscisic acid on stomatal opening in isolated epidermal strips of *abi* mutants of *Arabidopsis thaliana*. *Physiol. Plant.* 95:373–78

123. Roelfsema MRG, Prins HBA. 1997. Ion channels in guard cells of *Arabidopsis thaliana*. *Planta* 202:18–27
124. Savouré A, Hua X-J, Bertauche N, Van Montagu M, Verbruggen N. 1997. Abscisic acid-independent and abscisic acid-dependent regulation of proline biosynthesis following cold and osmotic stresses in *Arabidopsis thaliana*. *Mol. Gen. Genet.* 254:104–9
125. Schmidt C, Schelle I, Liao Y-J, Schroeder JI. 1995. Strong regulation of slow anion channels and abscisic acid signaling in guard cells by phosphorylation and dephosphorylation events. *Proc. Natl. Acad. Sci. USA* 92:9535–39
126. Deleted in proof
127. Schroeder JI, Hagiwara S. 1989. Cytosolic calcium regulates ion channels in the plasma membrane of *Vicia faba* guard cells. *Nature* 338:427–30
128. Schroeder JI, Hagiwara S. 1990. Repetitive increases in cytosolic Ca^{2+} of guard cells by abscisic acid activation of non-selective Ca^{2+}-permeable channels. *Proc. Natl. Acad. Sci. USA* 87:9305–9
129. Schroeder JI, Keller BU. 1992. Two types of anion channel currents in guard cells with distinct voltage regulation. *Proc. Natl. Acad. Sci. USA* 89:5025–29
130. Schroeder JI, Raschke K, Neher E. 1987. Voltage dependence of K^+ channels in guard cells protoplasts. *Proc. Natl. Acad. Sci. USA* 84:4108–12
131. Schultz TF, Spiker S, Quatrano RS. 1996. Histone H1 enhances the DNA binding activity of the transcription factor EmBP-1. *J. Biol. Chem.* 271:25742–45
132. Schwartz A, Ilan N, Schwarz M, Scheaffer J, Assmann SM, Schroeder JI. 1995. Anion-channel blockers inhibit S-type anion channels and abscisic responses in guard cells. *Plant Physiol.* 109:651–58
133. Schwartz A, Wu W-H, Tucker EB, Assmann SM. 1994. Inhibition of inward K^+ channels and stomatal response by abscisic acid: an intracellular locus of phytohormone action. *Proc. Natl. Acad. Sci. USA* 91:4019–23
134. Schwartz SH, Léon-Kloosterziel KM, Koornneef M, Zeevaart JAD. 1997. Biochemical characterization of the *aba2* and *aba3* mutants in *Arabidopsis thaliana*. *Plant Physiol.* 114:161–66
135. Schwartz SH, Tan BC, Gage DA, Zeevaart JAD, McCarty DR. 1997. Specific oxidative cleavage of carotenoids by VP14 of maize. *Science* 276:1872–74
136. Sheen J. 1996. Ca^{2+}-dependent protein kinases and stress signal transduction in plants. *Science* 274:1900–2

137. Shen LM, Outlaw WH, Epstein LM. 1995. Expression of an mRNA with sequence similarity to pea dehydrin (Psdhn1) in guard cells of *Vicia faba* in response to exogenous abscisic acid. *Physiol. Plant.* 95:99–105

138. Shen QX, Ho T-HD. 1995. Functional dissection of an abscisic acid (ABA)-inducible gene reveals two independent ABA-responsive complexes each containing a G-box and a novel *cis*-acting element. *Plant Cell* 7:295–307

139. Shen QX, Zhang PH, Ho T-HD. 1996. Modular nature of abscisic acid (ABA) response complexes: composite promoter units that are necessary and sufficient for ABA induction of gene expression in barley. *Plant Cell* 8:1107–19

140. Shinozaki K, Yamaguchi-Shinozaki K. 1996. Molecular responses to drought and cold stress. *Curr. Opin. Biotechnol.* 7:161–67

141. Skriver K, Olsen FL, Rogers JC, Mundy J. 1991. *Cis*-acting DNA elements responsive to gibberellin and its antagonist abscisic acid. *Proc. Natl. Acad. Sci. USA* 88:7266–70

142. Söderman E, Mattson J, Engström P. 1996. The *Arabidopsis* homeobox gene *ATHB-7* is induced by water deficit and by abscisic acid. *Plant J.* 10:375–81

143. Staswick PE, Su WP, Howell SH. 1992. Methyl jasmonate inhibition of root growth and induction of a leaf protein are decreased in an *Arabidopsis thaliana* mutant. *Proc. Natl. Acad. Sci. USA* 89:6837–40

144. Sturaro M, Vernieri P, Castiglioni P, Binelli G, Gavazzi G. 1996. The *rea* (red embryonic axis) phenotype describes a new mutation affecting the response of maize embryos to abscisic acid and osmotic stress. *J. Exp. Bot.* 47:755–62

145. Suzuki M, Kao CY, McCarty DR. 1997. The conserved B3 domain of VIVIPAROUS1 has a cooperative DNA binding activity. *Plant Cell* 9:799–807

146. Taylor IB. 1991. Genetics of ABA synthesis. In *Abscisic Acid Physiology and Biochemistry*, ed. WJ Davies, HG Jones, pp. 23–37. Oxford: Bios Sci.

147. Taylor JE, Renwick KF, Webb AAR, McAinsh MR, Furini A, et al. 1995. ABA-regulated promoter activity in stomatal guard cells. *Plant J.* 7:129–34

148. Thiel G, Blatt MR. 1994. Phosphatase antagonist okadaic acid inhibits steady state K^+ currents in guard cells of *Vicia faba*. *Plant J.* 5:727–33

149. Thiel G, MacRobbie EAC, Blatt MR. 1992. Membrane transport in stomatal

150. Trejo CL, Clephan AL, Davies WJ. 1995. How do stomata read abscisic acid signals? *Plant Physiol.* 109:803–11

151. Urao T, Yamaguchi-Shinozaki K, Urao S, Shinozaki K. 1993. An *Arabidopsis myb* homolog is induced by dehydration stress and its gene product binds to the conserved MYB recognition sequence. *Plant Cell* 5:1529–39

152. Van der Meulen RM, Visser K, Wang M. 1996. Effects of modulation of calcium levels and calcium fluxes on ABA-induced gene expression in barley aleurone. *Plant Sci.* 117:75–82

153. van der Veen R, Heimovaara-Dijkstra S, Wang M. 1992. Cytosolic alkalinization mediated by abscisic acid is necessary, but not sufficient, for abscisic acid–induced gene expression in barley aleurone protoplasts. *Plant Physiol.* 100:699–705

154. Vartanian N, Marcotte L, Giraudat J. 1994. Drought rhizogenesis in *Arabidopsis thaliana*. Differential responses of hormonal mutants. *Plant Physiol.* 104:761–67

155. Vasil V, Marcotte WR Jr, Rosenkrans L, Cocciolone SM, Vasil IK, et al. 1995. Overlap of Viviparous1 (VP1) and abscisic acid response elements in the *Em* promoter: G-box elements are sufficient but not necessary for VP1 transactivation. *Plant Cell* 7:1511–18

156. Walker-Simmons MK, Anderberg RJ, Rose PA, Abrams SR. 1992. Optically pure abscisic acid analogs—tools for relating germination inhibition and gene expression in wheat embryos. *Plant Physiol.* 99:501–7

157. Wang H, Cutler AJ. 1995. Promoters from *kin1* and *cor6.6*, two *Arabidopsis thaliana* low-temperature- and ABA-inducible genes, direct strong β-glucuronidase expression in guard cells, pollen and young developing seeds. *Plant Mol. Biol.* 28:619–34

158. Wang M, Van Duijn B, Schram AW. 1991. Abscisic acid induces a cytosolic calcium decrease in barley aleurone protoplasts. *FEBS Lett.* 278:69–74

159. Ward JM, Pei Z-M, Schroeder JI. 1995. Roles of ion channels in initiation of signal transduction in higher plants. *Plant Cell* 7:833–44

160. Deleted in proof

161. Wilkinson S, Davies WJ. 1997. Xylem sap pH increase: a drought signal received at the apoplastic face of the guard cell that involves the suppression of saturable abscisic acid uptake by the epidermal sym-

plast. *Plant Physiol.* 113:559–73

162. Wille AC, Lucas WJ. 1984. Ultrastructural and histochemical studies on guard cells. *Planta* 160:129–42

163. Wilson AK, Pickett FB, Turner JC, Estelle M. 1990. A dominant mutation in *Arabidopsis* confers resistance to auxin, ethylene and abscisic acid. *Mol. Gen. Genet.* 222:377–83

164. Yamaguchi-Shinozaki K, Shinozaki K. 1993. The plant hormone abscisic acid mediates the drought-induced expression but not the seed-specific expression of

rd22, a gene responsive to dehydration stress in *Arabidopsis thaliana. Mol. Gen. Genet.* 238:17–25

165. Yamaguchi-Shinozaki K, Shinozaki K. 1993. Characterization of the expression of a desiccation-responsive *rd29* gene of *Arabidopsis thaliana* and analysis of its promoter in transgenic plants. *Mol. Gen. Genet.* 236:331–40

166. Zeevaart JAD, Creelman RA. 1988. Metabolism and physiology of abscisic acid. *Annu. Rev. Plant Physiol. Plant Mol. Biol.* 39:439–73

Annu. Rev. Plant Physiol. Plant Mol. Biol. 1998. 49:223–47

DNA METHYLATION IN PLANTS

E. J. Finnegan,[1] R. K. Genger,[1,2] W. J. Peacock,[1] and E. S. Dennis[1]

[1]Commonwealth Scientific and Industrial Research Organization, Plant Industry,
P.O. Box 1600, Canberra, ACT 2601, Australia, and Cooperative Research Centre
for Plant Science, P.O. Box 475, Canberra, ACT 2601, Australia;
e-mail: jean.finnegan@pican.pi.csiro.au

[2]Division of Biochemistry and Molecular Biology, Australian National University,
Canberra, ACT 0200, Australia

KEY WORDS: DNA methyltransferase, epigenetics, defense, gene silencing, development

ABSTRACT

Methylation of cytosine residues in DNA provides a mechanism of gene control. There are two classes of methyltransferase in Arabidopsis; one has a carboxy-terminal methyltransferase domain fused to an amino-terminal regulatory domain and is similar to mammalian methyltransferases. The second class apparently lacks an amino-terminal domain and is less well conserved. Methylcytosine can occur at any cytosine residue, but it is likely that clonal transmission of methylation patterns only occurs for cytosines in strand-symmetrical sequences CpG and CpNpG. In plants, as in mammals, DNA methylation has dual roles in defense against invading DNA and transposable elements and in gene regulation. Although originally reported as having no phenotypic consequence, reduced DNA methylation disrupts normal plant development.

CONTENTS

1040-2519/98/0601-0223$08.00

INTRODUCTION

The information content of the primary DNA sequence can be enhanced by addition of a methyl group to the ring structure of cytosine or adenine residues. Chemical modification of DNA affects protein-DNA interactions; in prokaryotes, modification of DNA by methyltransferases prevents cleavage by the cognate restriction endonucleases (reviewed in 102). In higher eukaryotes, cytosine methylation can inhibit binding of regulatory proteins (reviewed in 39), and methylation of promoter and coding sequences of genes can repress transcription, both in vitro and in vivo (reviewed in 64). Methylation of DNA has been implicated in the timing of DNA replication; in determination of chromatin structure; in increasing mutation frequency; as a causal agent for some human diseases; and as a basis for epigenetic phenomena (reviewed in 64). It has been suggested that DNA methylation in mammals reduces the background of nonspecific transcripts from complex genomes (25) and is a mechanism to maintain the many transposable elements present in the mammalian genome in a quiescent state (16, 18, 38).

Although the evidence that DNA methylation plays a role in the developmental regulation of gene expression is largely correlative, targeted mutations of the mouse DNA methyltransferase have demonstrated that DNA methylation is essential for normal mammalian development. Mouse embryos homozygous for a methyltransferase knockout mutation that reduces methylation to about 30% of normal abort spontaneously during gestation (81). The two parental genomes are not functionally equivalent in mammalian development, and a number of genes, known as imprinted genes, are differentially expressed depending on their parental origin. In methyltransferase-knockout embryos, imprinted genes from both maternal and paternal genomes are expressed equally (80), confirming that DNA methylation is essential for the maintenance of genomic imprinting. X chromosome inactivation is also perturbed in embryos with reduced DNA methylation because expression of *Xist*, a gene important for X chromosome inactivation, is aberrant (9, 111). In embryos with reduced levels of methylation, *Xist* is transcribed from each X chromosome, resulting in inappropriate X inactivation, which may contribute to embryo lethality (104).

Plants that have substantially reduced levels of DNA methylation also display a number of phenotypic abnormalities (46, 67, 119). In this review, we discuss the distribution of methylcytosine in plant genomes and the regulation of DNA methylation. We consider the role of DNA methylation in regulating gene expression, as a mechanism controlling transposable elements, and as a defense against foreign DNA.

DISTRIBUTION OF METHYLCYTOSINE IN PLANT GENOMES

Early papers reported that methylation in plant DNA occurs predominantly in cytosines of symmetrical sequences such as CpG and CpNpG (54, 91a). This distribution of methylcytosine was determined using methylation-sensitive restriction enzymes and nearest-neighbor analyses modified to allow determination of trinucleotide sequences. While these conclusions were to some extent dictated by the choice of restriction enzymes, nearest neighbor analyses indicated that methylation occurred more frequently in CpApG and CpTpG than in CpApT (54). The strand symmetry of CpG and CpNpG motifs provides an obvious mechanism for transmitting methylation patterns through cycles of cell division (Figure 1).

More recently, genomic sequencing (49) has shown that methylation of cytosines in nonsymmetrical sites often occurs (97, 103, 141). In 182 bp of the promoter of a silenced chimeric 35SA1 transgene in Petunia, 27 cytosine residues located in CpG or CpNpG (100%) and 40 cytosines in asymmetric sequences (69%) were either completely or partially methylated (97). Methylation in the promoter region of the tobacco T85 gene was observed in pollen DNA while the corresponding region was unmethylated in leaf DNA. In the region examined, 10 out of 49 cytosines were highly methylated but only three methylated cytosines were located in symmetric sites, suggesting that methylation at asymmetric sites may also be common in endogenous genes (103).

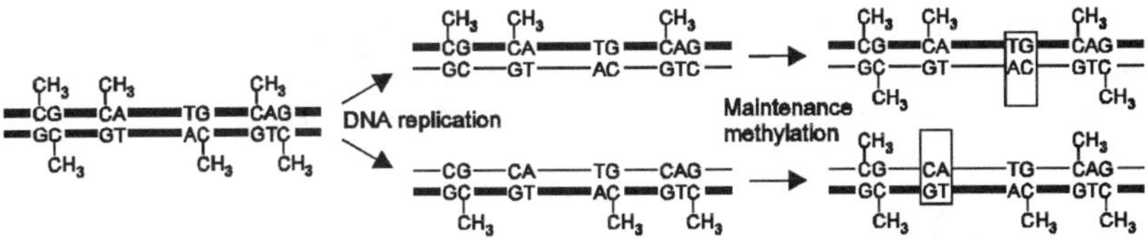

Figure 1 Transmission of methylation patterns through a cycle of DNA replication. Patterns of methylation based on cytosines located in symmetric sequences (CpG and CpNpG) are transmitted to both daughter strands following replication by the action of a methyltransferase, which preferentially methylates hemimethylated DNA at the replication fork, a process known as maintenance methylation. Methylation of cytosines in nonsymmetric sequences (CpXpX where X is any base other than G) is not transmitted to the newly synthesized daughter strand. Parental strand, *thick line*; newly synthesized strand, *thin line*. Loss of methylation at nonsymmetric sites on the daughter molecules is indicated by *box* surrounding these residues.

Methylation of symmetrically located cytosines may act as a nucleation center for the spread of methylation to adjacent nonsymmetric cytosines. The methylation pattern of individual DNA molecules was examined for the 35SA1 transgene (97). Methylation of cytosine residues in CpG and CpNpG occurred at 100% of sites in almost every molecule. In contrast, many nonsymmetric sites were methylated at low frequency, and the pattern of methylation at these sites was variable between molecules. Methylation at these sites was probably not transmitted through cycles of DNA replication (Figure 1).

In plants, DNA methylation is mainly restricted to the nuclear genome, suggesting that the smaller chloroplast and mitochondrial genomes do not require this additional level of gene control (reviewed in 43). Methylcytosine is not randomly distributed throughout the nuclear genome but is concentrated in repeated sequences. Repetitive DNA consists of long tandem arrays that are normally clustered around the centromere, at telomeres, or in the nucleolar organizer region and of middle to highly repetitive sequences made up of retroelements and derivatives. At least 50% of the maize genome is composed of different families of retroelements (122); in bean, there are more than one million copies of one *copia*-like retroelement that makes up about 25% of the genome (47). It is likely that most methylcytosine in plant genomes, as in mammalian genomes (151), is located within retroelements that are heavily methylated (11, 13, 89). Modification and condensation into heterochromatin render retroelements transcriptionally and recombinationally inactive (13).

Approximately 80% of cytosines in CpG dinucleotides in plant genomes are modified (54), but despite this plants also contain regions of unmethylated DNA with closely spaced, unmethylated *Hpa*II restriction enzyme sites (CCGG). *Hpa*II digestion released a prominent low-molecular-weight DNA fraction with fragments ranging in size from ~25 bp to ~250 bp (3) that resemble the *Hpa*II tiny fragments characteristic of CpG islands in vertebrate genomes (35). Islands are typically unmethylated in a wide range of tissues whether or not the associated gene is transcribed (24). Unmethylated CpG islands have been identified in the maize *A1* (dihydroflavonol 4-reductase), *Adh1* (alcohol dehydrogenase), and *Sh1* (sucrose synthase) genes (3, 101).

METHYLATION OF DNA

DNA methyltransferase catalyzes transfer of a methyl group from S-adenosylmethionine (S-adomet) to the pyrimidine ring of cytosine residues (22, 75) in newly replicated DNA (23) (maintenance methylation, Figure 1). Changes in methylation patterns occur by de novo methylation or by passive demethylation through failure of maintenance methylation. Active demethylation has been reported in chicken and mouse (63, 144) but not yet in plants.

Plant DNA Methyltransferases

Prokaryote cytosine 5 (5mC) methyltransferases belong to a single class of enzyme consisting of 10 conserved protein motifs (I–X), arranged sequentially and separated by nonconserved sequences of variable length (76, 113, 130). Motifs I and X form the binding site for the cofactor *S*-adomet (32, 60). The cysteine residue of a completely conserved proline-cysteine doublet in motif IV forms the active site involved in methyl transfer (31, 98, 147, 148). The variable region between motifs VIII and IX—the target recognition domain (5, 72, 135, 147)—directs the enzyme to the recognition sequence flanking the target cytosine.

Eukaryote cytosine methyltransferases (2, 14, 19, 44, 70, 132, 150, 152) also have a conserved structure including a carboxy-terminal methyltransferase domain containing 8 of the 10 prokaryote motifs. In most eukaryote enzymes, this domain is fused to a large regulatory amino-terminal domain (Figure 2) which in the mouse enzyme, Dnmt1, targets the protein to the replication fork in S phase nuclei (78) and causes preferential methylation of hemimethylated templates (17). Mouse ES cells, homozygous for a null mutation of the Dnmt1 methyltransferase, contain low but stable amounts of methylcytosine and can methylate incoming retroviral DNA, demonstrating that there is a second methyltransferase in mouse (77). A second putative methyltransferase gene, Dnmt2, which appears to lack an amino-terminal domain, has been cloned from both mouse and human (JA Yoder & TH Bestor, personal communication).

There are two classes of genes encoding putative DNA methyltransferases in Arabidopsis. One class, represented by *METI* and *METII*, encodes proteins that are similar to the mouse Dnmt1 methyltransferase (Figure 2). Homology between *METI* and *METII* is higher in the methyltransferase domain

→

Figure 2 Comparison of methyltransferase proteins from prokaryote, mammalian, and plant sources. (*a*) Schematic representation of the methyltransferase structure showing the location of the methyltransferase and amino terminal domains and the approximate molecular weights of the different proteins. Plant class 1 methyltransferases include METI and METII from Arabidopsis, the two carrot methyltransferases (14), and the pea methyltransferase (S Pradhan & R Adams, personal communication). Arabidopsis METIII is the sole representative of a Plant class 2 methyltransferase (S Henikoff & L Comai, personal communication). (*b*) Amino acid identity between the different methyltransferases. Comparisons for the methyltransferase domain shown in the *lower half* of the figure include amino acids from the beginning of conserved motif I (113) to the carboxy-terminus. The *upper half* of the figure shows the comparisons for the amino-terminal domain spanning the amino-terminus to the beginning of the lysine-glycine repeat separating the two domains (19, 44). Sequence data for this figure were taken from the following references: 14, 44, 150, 152; RK Genger, unpublished observation; S Henikoff & L Comai, personal communication; JA Yoder & TH Bestor, personal communication.

than in the amino-terminal domain; these proteins are more divergent than the human and mouse Dnmt1 enzymes (44; RK Genger, unpublished observation). A third putative methyltransferase gene (*METIII*), identified in a database search of genomic sequence (H Goodman, unpublished observation), does not cross hybridize with *METI* or *METII* and belongs to the second class of methyltransferases (Figure 2) (EJ Finnegan, unpublished observation). The

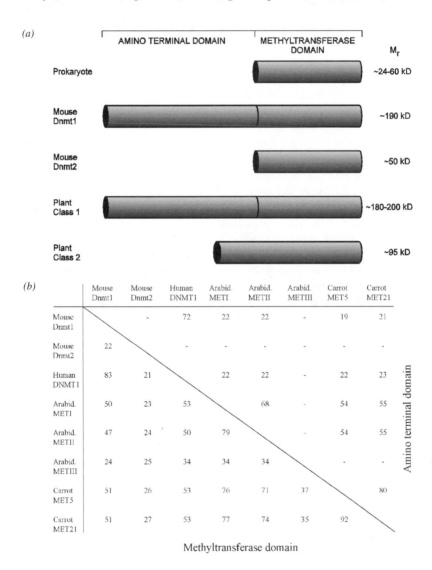

(a)

	AMINO TERMINAL DOMAIN	METHYLTRANSFERASE DOMAIN	M_r
Prokaryote			~24-60 kD
Mouse Dnmt1			~190 kD
Mouse Dnmt2			~50 kD
Plant Class 1			~180-200 kD
Plant Class 2			~95 kD

(b)

	Mouse Dnmt1	Mouse Dnmt2	Human DNMT1	Arabid. METI	Arabid. METII	Arabid. METIII	Carrot MET5	Carrot MET21
Mouse Dnmt1		-	72	22	22	-	19	21
Mouse Dnmt2	22		-	-	-	-	-	-
Human DNMT1	83	21		22	22	-	22	23
Arabid. METI	50	23	53		68	-	54	55
Arabid. METII	47	24	50	79		-	54	55
Arabid. METIII	24	25	34	34	34		-	-
Carrot MET5	51	26	53	76	71	37		80
Carrot MET21	51	27	53	77	74	35	92	

Amino terminal domain

Methyltransferase domain

predicted METIII protein lacks an amino-terminal domain and has a chromo-domain, a short motif found in chromatin-associated proteins, Heterochromatin 1 (Hp1) and Polycomb (Pc) from Drosophila, inserted between motifs II and IV (S Henikoff & L Comai, personal communication).

Two very closely related DNA methyltransferase genes encoding class 1 methyltransferases (Figure 2) have been cloned from carrot (14). The two carrot methyltransferases, which are probably the products of a recent duplication, have higher homology to Arabidopsis METI than METII in the methyltransferase domain, suggesting that both are METI homologues (Figure 2).

A single methyltransferase cDNA encoding a protein similar to METI has been cloned from pea; in vitro assays with the expressed protein ($M_r \sim 180\,\mathrm{kDa}$) suggest that it has the capacity to methylate both CpG and CpA/TpG (S Pradhan & R Adams, personal communication). This conflicts with earlier studies reporting partial purification of two methyltransferases from pea, one of which preferentially methylates CpG ($M_r \sim 150$ kDa) and the other CpA/TpG ($M_r \sim 110$ kDa) (114). Proteolytic processing to yield smaller proteins in vivo has been invoked to reconcile the observed differences in protein size. Smaller proteins ($M_r \sim 50$–85 kDa) with methyltransferase activity have also been partially purified from wheat and rice (50, 133, 136); these may be the products of proteolytic cleavage.

Crystallization of a ternary covalent complex between a synthetic oligonucleotide template, purified *Hha*I methyltransferase, and *S*-adenosyl homocysteine (the product of methyltransfer from *S*-adomet) showed that the DNA is bound in a cleft between the catalytic and target recognition domains of the enzyme (32, 71). Methyltransfer involves extrusion of the target cytosine from the DNA helix, leading to distortion of the DNA backbone. A second prokaryote methyltransferase, *Hae*III, also gains access to target cytosines by their extrusion from the helix (117). Modeling the 3D structure of the METI methyltransferase domain, based on the *Hha*I methyltransferase crystal structure (32, 108, 109, 110), shows that the tertiary structure is conserved between a bacterial and a plant methyltransferase (RK Genger, unpublished observation). It is likely that these enzymes use similar mechanisms to access their target cytosine.

The identification of multiple DNA methyltransferases in plants raises the question of whether the proteins recognize and methylate cytosines in different sequence contexts, whether the different enzymes catalyze maintenance or de novo methylation, or whether they are active in different tissues or stages of development. There is evidence supporting the notion that plant methyltransferases may differ in target specificity. Two purified pea methyltransferases, which may originate from a single gene, differ in target specificity (114). In tobacco, methylation of cytosines in CpG dinucleotides and in CpNpG sequences showed differential sensitivity to methylation inhibitors 5-azacytidine (5-azaC) and ethionine, suggesting that different enzymes may catalyze methylation of CpG and CpNpG sites, respectively (73). A *METI* antisense preferentially reduced methylation of cytosines in CpG and CpCpG sequences in transgenic Arabidopsis (46).

Most of the known plant methyltransferase transcripts are expressed ubiquitously in vegetative and reproductive tissues (14, 119; RK Genger, unpublished observation) but generally show higher expression in meristematic cells (14, 119). However, *METI* transcripts are at least 10,000-fold more abundant than those of *METII* (RK Genger, unpublished observation). *METIII* was expressed at low levels in both vegetative and reproductive tissues (EJ Finnegan, unpublished observation); expression was higher in floral tissue (S Henikoff & L Comai, personal communication), which is perhaps indicative of a role during gametogenesis. In some ecotypes of Arabidopsis, the protein is truncated by insertion of a retroelement upstream of some of the conserved motifs required for methyltransfer (S Henikoff & L Comai, personal communication). This shows that METIII is not essential for apparently normal development.

DNA Methylation Requires Proteins Other Than Methyltransferases

The first hint that proteins other than methyltransferases may be required for DNA methylation came from the identification of Arabidopsis mutants (*ddm*) at four loci that have decreased DNA methylation of the centromeric repeat (137). Mutations at *ddm1* and *ddm2* mutations are recessive, implying loss of function, but the remaining two mutations, *ddm0* and *ddmB*, are dominant or semidominant (EJ Richards, personal communication). *DDM2* maps to the previously described *METI* locus, but the other loci do not cosegregate with known methyltransferase genes (EJ Richards & J Jeddeloh, personal communication). *DDM1* does not encode a methyltransferase because *ddm1* mutants have normal levels of both CpG and CpNpG methyltransferase activity in vitro. The intracellular pools of *S*-adomet are also normal. DDM1 is only required for methylation in vivo and may have a role in chromatin structure, or in the interaction between methyltransferases and DNA (67).

Demethylation of both centromeric and ribosomal repeat sequence DNA was observed in M2 homozygous *ddm1* mutants; after repeated rounds of self-pollination, single copy sequences became demethylated in homozygous progeny (67). Demethylation of cytosines in both CpG and CpNpG sequences was observed in *ddm1* homozygotes (137); the effect on methylcytosine in nonsymmetric sequences has not been measured.

Regulation of De Novo Methylation

There are many unanswered questions about the regulation of DNA methylation; for example, what regulates de novo methylation of some but not all cytosines in CpG and CpNpG motifs? Does methylation at asymmetric cytosines depend on methylation at adjacent symmetric sites? Is methylation regulated by chromatin structure or by other DNA-binding proteins that compete with the methyltransferase for access to DNA?

Transgenes (and transposable elements) that have been inactivated and methylated de novo show a high density of methylcytosine, in CpG, in CpNpG, and in asymmetric sequences. In some cases, methylation does not extend beyond the ends of the foreign DNA or transposon (95, 141); in other examples, methylation extends, at lower density, into the flanking plant DNA (86, 123).

De novo methylation of endogenous genes has been studied in plants in which DNA methylation has been perturbed by either mutation, antisense technology, or treatment with 5-azaC. Homozygous *ddm1* mutant plants, which had about 30% of normal methylation, were outcrossed to the wild-type progenitor. The level of methylation in progeny was intermediate between the mutant and wild-type parent, suggesting that even though the *ddm1* mutation is recessive, sequences that were demethylated in the genome of the mutant parent remained hypomethylated in the progeny (137). Repeated backcrossing to a wild-type parent showed that the level of methylation in each generation was intermediate between the parental lines, consistent with dilution of the hypomethylated genome rather than replacement of methylcytosine by de novo methylation (137).

Arabidopsis plants carrying a *MET1* antisense transgene have reduced levels of DNA methylation (46, 119). When *MET1* antisense plants, hemizygous for the transgene, were outcrossed to wild-type plants the hypomethylation phenotype was transmitted to progeny that did not inherit the antisense transgene (119). In a different study, gradual recovery of methylation levels was observed in progeny of selfed hemizygous methyltransferase antisense plants that did not inherit the transgene (antisense-null) (46). Methylation in first generation antisense-null plants was significantly below normal but was higher than in sibling plants that had inherited the transgene. The level of methylation increased in progeny of antisense-null plants but was still below normal. Remethylation of repeated sequences located at the centromere and at least some single copy sequences, including Ta3, occurred in first generation antisense-null plants (EJ Finnegan & T Kakutani, unpublished observations). Remethylation of endogenous sequences in these plants presumably occurred by de novo methylation.

Treatment of tobacco seed with 5-azaC results in extensive demethylation of HRS60 subtelomeric repeat DNA. Hypomethylated, HRS60 repetitive DNA was transmitted to progeny arising from self-pollination or outcrossing (140). In contrast, when tobacco suspension cells were subjected to a brief 5-azaC treatment, remethylation of the majority of DNA sequences occurred during several months' culture in the absence of the drug (1).

Thus, the capacity for de novo methylation in plants may be limited, particularly when global methylation patterns have been substantially altered.

ROLE OF DNA METHYLATION IN PLANTS

More than twenty years ago, Riggs (118) and Holliday & Pugh (57) proposed that cytosine methylation influences gene expression, a hypothesis based on the frequency of methylcytosine and on the proposed transmission of methylation patterns through cycles of DNA replication. It has since been shown that patterns of methylation are clonally inherited (146), that DNA methylation does repress transcription, and that changes in DNA methylation can be correlated with changes in gene expression in a tissue-specific or developmentally regulated manner (reviewed in 64). These data are correlative, and it is still not clear whether DNA methylation plays a primary role in regulating gene expression during development.

Methylation Changes During Development

If DNA methylation is important for tissue-specific or developmentally regulated gene expression, then there must be a mechanism to reset methylation patterns between generations. In mammals, global demethylation early in embryogenesis is followed by remethylation around the time of implantation (reviewed in 116). Germ cells are set aside early in development and are therefore not affected by organ-specific changes in methylation. Plants probably do not undergo global demethylation followed by remethylation during embryo development (46, 67) because *ddm1* mutant or methyltransferase antisense Arabidopsis plants that have a substantial reduction in DNA methylation do not restore normal levels of methylation in progeny that have lost the mutation or antisense transgene by outcrossing. Because the germ line is not set aside early in plant development, pro-gametic cells inherit methylation changes that accumulate in the vegetative meristem during development.

How are methylation patterns erased and reset in plants? The activity of maize transposable elements, *Ac* and *Spm*, differed when inherited from male or female gamete (42, 125), implying that methylation can change during gametogenesis. *Spm* and *Mu* elements are more heavily methylated in leaves at the top of the plant than in the first emerging leaves, showing that DNA methylation of transposable elements increases during development (6, 12, 86, 87).

Methylation levels of DNA from young seedlings were approximately 20% lower than in mature leaves of both tomato and Arabidopsis (92; EJ Finnegan, unpublished observation). In DNA from young Arabidopsis seedlings, the centromeric repeat was undermethylated compared to DNA of mature leaves (EJ Finnegan, unpublished observation), but it is not known whether methylation of single-copy DNA also increases during development. Demethylation of these repetitive sequences probably occurred during gametogenesis or early in seed development because the centromeric repeat was also undermethylated in DNA from Arabidopsis seeds (EJ Finnegan, unpublished observation).

The level of DNA methylation was higher in tomato and Arabidopsis seeds than in mature leaves (92; EJ Finnegan, unpublished observation); this may be artefactual because the ratio of nuclear (methylated) to plastid (hypomethylated) genomes is probably higher in seed than leaf. If levels of methylation are increased in seed, then this may reflect increased methylation of single-copy sequences. Hypermethylation of single-copy DNA during embryogenesis, followed by demethylation of specific sequences during germination, would provide a mechanism for resetting methylation patterns in plants.

The data from Arabidopsis and maize indicate that methylation of transposable elements and other repeated sequences increases during development, perhaps to ensure that these sequences are packaged into heterochromatin before gamete formation. This could reduce meiotic recombination between repeated sequences (8, 13, 28, 33, 58, 134) at different chromosomal locations, which has the potential to cause major rearrangements leading to loss of gamete viability. Rearrangements resulting from recombination between nonallelic transposons is more common in yeast and Drosophila (74, 124), which lack DNA methylation, than in mammals (56) and plants (145).

A Role for DNA Methylation in Plant Development

Studies of the maize transposable elements, *Ac* and *Spm*, cycling between active and inactive states suggested that DNA methylation plays an important role in regulating the activity of these elements (7, 41, 126). The first indication that DNA methylation may regulate plant development came from work on the molecular basis of vernalization, that is, the promotion of flowering following prolonged exposure to low temperatures.

The vernalization signal is perceived by mitotically active cells that will form the inflorescence meristem; it is inherited through mitotic cell divisions but is not transmitted through meiosis (reviewed in 37). On the basis of these properties, Burn et al (29) proposed that vernalization is mediated by demethylation in the promoter of gene(s) whose expression is critical for the initiation of flowering and that subsequent expression of these genes triggers early flowering. Treatment with 5-azaC caused early flowering, mimicking the vernalization response and supporting the involvement of DNA methylation (27, 29). The early flowering response to chemical demethylation was specific to the vernalization-dependent flowering pathway because it was restricted to plants that respond to vernalization; there was no promotion of flowering in plants that are insensitive to vernalization (Figure 3) (27, 29, 37). Demethylation caused by a methyltransferase antisense construct also promoted flowering in the absence

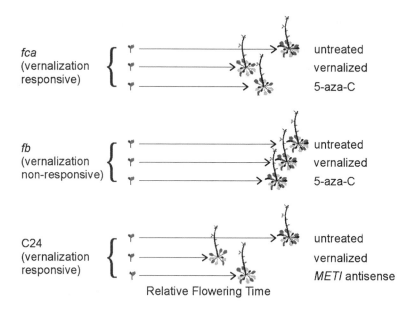

Figure 3   Changes in flowering time in response to vernalization; treatment with the demethylating agent, 5-azaC; or demethylation caused by a *METI* antisense construct. The length of the *line* separating the young seedling and flowering plant represents the flowering time in response to the treatment indicated, relative to the flowering time for the untreated control, which is taken as 100%. The late flowering mutants *fca* and *fb* are in ecotype Ler (data adapted from 29), and the *METI* antisense is in ecotype C24 (46; RK Genger, unpublished observation).

of a cold treatment, again suggesting that decreased DNA methylation is sufficient to cause early flowering (Figure 3) (37; RK Genger & EJ Finnegan, unpublished observations).

One prediction from this hypothesis is that exposure to low temperatures will decrease methylation, perhaps by uncoupling replication and maintenance methylation. DNA methylation of cultured tobacco cells decreased after incubation for 1 week at 8°C (29). Similarly, a 4- or 8-week cold treatment of germinating Arabidopsis seeds caused a transient, 15% decrease in DNA methylation compared with control plants harvested at the same stage of development (EJ Finnegan, unpublished observation).

The cloning of genes required for the vernalization response will help elucidate any role of methylation.

Loss of DNA Methylation Affects Plant Development

DNA methylation is essential for normal plant development. Arabidopsis with reduced levels of DNA methylation display a range of abnormalities including loss of apical dominance, reduced stature, altered leaf size and shape, reduced root length, homeotic transformation of floral organs, and reduced fertility (46, 67, 119). A subset of these morphological abnormalities was observed in independent plant lines in which methylation had been reduced either by mutagenesis or by introduction of a methyltransferase antisense gene (46, 67).

Decreased DNA methylation also altered flowering time in independent *METI* antisense and *ddm1* mutant plants; the effect depended on growth conditions and on the response of the wild-type progenitor to vernalization (65, 67, 119; RK Genger & K Kovac, unpublished observations). Other heterochronic changes in development occurred; some antisense plants formed aerial rosettes and produced more cauline leaves and secondary inflorescences on the primary bolt stem (119). Plants that had lost the antisense or *ddm1* mutation by segregation or outcrossing inherited both phenotypic abnormalities and reduced levels of DNA methylation, implicating demethylation as the cause of abnormal development (46, 66, 119).

Plants with the lowest methylation levels were most severely affected, and the abnormal phenotype became more severe in successive generations of progeny from self-pollinated plants (46). Although originally reported as having no phenotype, homozygous *ddm1* mutants developed phenotypic abnormalities after a number of generations of selfing; this correlated with demethylation of unique sequences (67). Arabidopsis plants in which methylation had been reduced by at least 70% became infertile after four to five generations of selfing (46). A comparable reduction in methylation is embryo lethal for mammals (81).

Melandrium album is a dioecious plant in which sex determination of the heterogametic male is based on the presence of a Y chromosome that plays a role in both male-determining and female-suppressing functions. Treatment with 5-azaC resulted in hermaphroditism in about 21% of male plants but did not affect sex determination of females (62). The karyotypically normal androhermaphrodites were sexual mosaics with both male and hermaphrodite flowers that showed differing degrees of gynoecium development and seed set. Although inheritance was incomplete, the bisexual trait was transmitted through two generations, but only when the androhermaphrodite plants were used as pollen donors. Janousek et al (62) proposed that female sex suppression in males depends on methylation of specific sequences either on the Y chromosome or on the autosomes. If an autosomal sequence is involved, then uniparental transmission suggests that this sequence must be imprinted following passage through the female gamete.

Imprinted Genes Are Important in Endosperm Development

The maternal and paternal genomes are not functionally equivalent in endosperm development (reviewed in 55, 69), a situation that is similar to genomic imprinting in mammals where the two parental genomes differ. The endosperm of many angiosperms is triploid, having two copies of the maternal genome and a single copy of the paternal genome (82). When this balance is perturbed, development is impaired, often leading to failure of seed maturation. Using translocations, it has been shown that some chromosomal arms must be derived from the male parent for normal endosperm development to occur (55, 69); these arms may carry imprinted genes that are essential for endosperm development. Nonessential genes, *r* and *dzr1*, are also subject to imprinting (30, 68); differential methylation, which correlated with differential expression, of maternal and paternal copies of alleles at the *r* locus has been observed (M Alleman, unpublished observation), suggesting that, as in mammals (80), DNA methylation may also be important for genomic imprinting in plants.

Imprinting in plants differs from mammalian genomic imprinting in two aspects. First, even though differential activity of transposable elements has been observed after transmission through male and female gametes (42, 125), imprinting has so far been demonstrated only in endosperm, which does not contribute to the next generation (55, 69). Second, not all alleles at imprinted loci show differential expression and/or methylation (30, 68; M Alleman, personal communication).

DNA Methylation Modulates Endogenous Gene Expression

The Arabidopsis *PAI* genes encoding phosphoribosylanthranilate isomerase, an enzyme of the tryptophan pathway, are examples of endogenous genes whose expression is modulated by methylation. The ecotype Ws has four copies of the gene at three unlinked loci. All four genes are methylated in the homologous coding sequences, but sufficient enzyme is made to prevent tryptophan deficiency (10). In a spontaneous tryptophan mutant, the two linked genes *PAI1* and *PAI4*, present as a tail-to-tail inverted repeat, were deleted by recombination between flanking, direct-repeat sequences. The mutant phenotype was unstable and occasionally yielded revertant progeny or somatic sectors. Reversion of the mutant phenotype was associated with increased expression and hypomethylation of the two remaining unlinked genes, *PAI2* and *PAI3* (10). When the *ddm1* mutation was introduced into the *pai1-pai4* deletion, hypomethylation and reactivation of *PAI2* and *PAI3* genes occurred at high frequency (J Jeddeloh, personal communication). When the complex, methylated Ws *PAI1-PAI4* locus was combined with unmethylated *PAI2* and *PAI3* genes by crossing to another ecotype, the unmethylated copies became methylated (J Bender, personal communication), a process resembling *trans*-inactivation of 9 transgene (91).

When independent homozygous *ddm1* mutant lines were established by five generations of selfing, their progeny displayed overlapping sets of phenotypic abnormalities, suggesting that a limited number of genes was affected (66). Five phenotypic traits were inherited, independent of the *ddm1* mutation, in F2 outcross progeny. These epimutations have been mapped to loci unlinked to *DDM1*, and attempts to clone the genes responsible are under way (65, 66).

Reduction in DNA methylation in *MET1* antisense plants is associated with aberrant expression of several floral regulatory genes. Ectopic expression of *APETALA3* and *AGAMOUS* (*AG*), which are normally expressed in floral tissue, was observed in leaves (46); expression was low and was probably not the cause of the floral abnormalities. Another gene, *SUPERMAN* (*SUP*), showed wild-type expression in the ovule but was ectopically expressed in the carpel wall (H Sakai & EJ Finnegan, unpublished observation). In wild-type flowers, *SUP* is also expressed in the floral bud (121); in situ hybridization showed that the genes, *SUP* and *AG*, were not transcribed in the developing floral buds of antisense plants that had *sup* or *sup ag* mutant flowers, respectively (EJ Finnegan & H Sakai, unpublished observation). Methylation of the *SUP* gene was examined in wild-type and in *MET1* antisense plants with *sup* flowers. Restriction enzyme analyses showed that sites some distance from the gene were methylated in wild-type but not in antisense plants (N Kishimoto & EJ Finnegan, unpublished observation). Genomic sequencing of the *SUP* gene in wild-type plants showed that there was no cytosine methylation in the coding sequence or within 1 kb upstream of the transcription start. Unexpectedly, the corresponding region was densely methylated in the *MET1* antisense plants (61; N Kishimoto & EJ Finnegan, unpublished observation). Hypermethylation of cytosines located in both symmetric and nonsymmetric sequences was associated with repression of *SUP* transcription in the developing flower bud and with a *sup*-like phenotype. Repression of *AG* was also associated with hypermethylation of this gene in *MET1* antisense plants (S Jacobsen, personal communication). It is likely that methyltransferases other than MET1 methylate these sites; transcripts of *MET1I* were present in *MET1* antisense plants (RK Genger, unpublished observation).

Hypermethylation at the *SUP* gene was also observed in seven independent lines that showed a weak, unstable *sup* phenotype, termed *clark kent* (*clk*). These *clk* mutants were isolated from plants that had undergone EMS, diepoxybutane, fast neutron, X irradiation, or insertional mutagenesis (61). The initiating event for hypermethylation of *SUP* in these plants is not known.

There are other examples of mutations in Arabidopsis that are associated with local changes in DNA methylation. Two independent mutations at the *fwa* locus, which confers late flowering, are associated with hypomethylation of at least 5 Mb, spanning *FWA*, as determined by methylation RFLPs (W Soppe & M Koornneef, personal communication). A semidominant late flowering mutant that was identified in *ddm1* mutants and that can be segregated from *ddm1* maps at *FWA* (65); perhaps the late flowering phenotype in this line is also associated with hypomethylation. Mutagenesis may result in perturbation of methylation patterns at unlinked sites because the *clk1* allele arose in the same mutagenesis as the *fwa1* mutant.

The observation that de novo methylation can be stimulated by global demethylation and by mutagenic agents suggests that plants use DNA methylation as a defense against factors that have the potential to perturb normal genome organization. The local hypermethylation observed at *SUP* and *AG* resembles that seen in silenced transgenes and inactive transposable elements. It is possible that changes in chromatin structure caused by demethylation, mutagenesis, or insertion of mobile elements or transgenes stimulates dense methylation. It is likely that genes involved in flowering are no more susceptible to changes in methylation than other genes, but are more easily scored.

DNA Methylation Is Associated with Transgene Silencing

Transgenes are a tool both to investigate gene function and to improve plants of agronomic importance by improving characters. One limitation to plant improvement is that transgenes are frequently inactivated, by at least two independent mechanisms (reviewed in 45, 48, 88, 131). In transcriptional silencing, RNA production from the introduced gene is repressed. In posttranscriptional silencing, transcripts do not accumulate in the cytoplasm even though transcription rates are comparable with or higher than those in cells where transcripts do accumulate (99).

Transcriptional silencing is associated with transgene methylation, particularly in the promoter (84). Methylated transgenes are packaged into condensed chromatin, as determined by reduced accessibility to nucleases (94, 149; R van Blockland & P Meyer, personal communication). Such silenced transgenes can be reactivated by treatment with 5-azaC (143) or by crossing into a *ddm1* mutant background (O Mittelsten Scheid, personal communication); in each case, transgene methylation decreased, suggesting that methylation plays an integral part in the maintenance of silencing. However, introduction of a *METI* antisense construct did not reactivate the same 35S*Hyg* transgene reactivated in *ddm1* homozygotes, even though some demethylation occurred (O Mittelsten Scheid, personal communication). The *METI* antisense does not remove methylation of cytosine in all sequence contexts.

DNA structure may be one factor that identifies foreign DNA as a target for methylation, as silencing occurs more frequently when multiple copies of the transgene are inserted (4, 84), perhaps because of pairing of transgenes at the same or different loci (88). Inverted transgene repeats may form a cruciform structure, a structure that is recognized and efficiently modified by mammalian methyltransferases (15, 129). Other proposed mechanisms for targeting foreign DNA include lack of homology between sequences flanking paired transgenes inserted at different loci (ER Signer, cited in 20), RNA-mediated de novo methylation (142), or differences in base composition between a transgene and flanking DNA (95). For example, the maize *A1* gene, which encodes dihydroflavonol-4-reductase (DFR), is frequently silenced in transgenic Petunia; silencing was associated with methylation of multiple copy inserts (40, 91a). When the Gerbera homologue (*gdfr*), which has a C+G content more compatible with the Petunia genome, was used in place of the *A1* gene, multiple copy insertions of the transgene remained unmethylated and flowers were intensely pigmented (40). Most reports of transcriptional silencing involve transgenes from heterologous sources, suggesting that DNA heterogeneity caused by insertion of foreign DNA may be a common theme in transgene inactivation.

The frequency of inactivation of a single copy of the chimeric 35S*A1* gene in transgenic Petunia was increased by a prolonged period of elevated temperatures and high light intensity. There was a parallel increase in methylation of the transgene (96). Methylation may be increased by heat and/or high light intensity, contrasting with demethylation, which appears to accompany vernalization.

Posttranscriptional silencing, which affects both transgenes and homologous endogenous genes, is also associated with transgene methylation, but within the coding sequence rather than the promoter (59, 128). It is not clear whether methylation is required for posttranscriptional silencing. Expression of a transgene that has homology to sequences present in the genome of an invading RNA virus can lead to degradation of the viral RNA and resistance to infection (83).

It is likely that both forms of transgene silencing reflect normal cellular defenses against invading or mobile DNAs (45). Posttranscriptional silencing is functionally equivalent to natural processes that allow tobacco plants to recover from, and develop immunity to, infection by nepoviruses (115). Transcriptional silencing is similar to inactivation of transposable elements, retroelements, and T-DNA. *Trans*-inactivation of an incoming transgene by an inactive, resident transgene (91), resembles paramutation (90), which has been defined as a heritable change in one allele (the paramutable allele) induced by a second, paramutagenic allele (26). Changes in DNA methylation are detected by restriction enzymes associated with paramutation at the maize *r* locus (M Alleman & J Kermicle, cited in 107). To date, no changes in methylation have been detected using restriction enzymes after paramutation of the *b* locus of maize (106, 107).

IS DNA METHYLATION A PRIMARY OR SECONDARY MECHANISM FOR REGULATING GENE EXPRESSION?

There are many examples in plants of genes that are both transcriptionally inactive and methylated within the promoter and/or coding sequences. It is not clear, however, whether DNA methylation is the primary cause of inactivation or a secondary consequence of some other process that has resulted in transcriptional inactivation.

It can be difficult to determine cause and effect, but the *Pl-Blotched (Pl-Bh)* allele of the maize purple plant (*Pl*) gene, encoding a transcription factor that regulates anthocyanin production, provides an example where changes in methylation are secondary to alterations in chromatin structure (O Hoekenga & K Cone, personal communication). The *Pl-Bh* allele causes variegated pigmentation throughout the plant, which ranges from heavy blotching at the base to sparse blotching at the top. In husk tissue, decreased pigmentation was associated with hypermethylation of *Pl-Bh* compared with *Pl-Rhoades*, an allele that conditions uniform intense pigmentation (34). Hypermethylation correlated with decreased sensitivity to DNase I and with reduced gene expression of *Pl-Bh*. *Pl-Bh* was also resistant to DNase I in juvenile tissue where methylation RFLP analyses showed that the gene was hypomethylated (O Hoekenga & K Cone, personal communication). In this case, the primary event appears to be compaction of chromatin around *Pl-Bh* followed by hypermethylation. Because the two alleles show only 10 nucleotide changes over 5.5 kb, including the coding region and flanking sequences, it is likely that differences in expression relate to chromatin structure. The patchy expression of *Pl-Bh* throughout the plant may result from chromatin condensation, with the density of blotches decreasing as this structure is stabilized by methylation.

Organisms lacking methylated DNA, such as Drosophila, regulate gene expression by establishing chromatin structures that are stably inherited through mitosis and that are compatible with either transcriptionally active or repressed states. Polycomb-group (Pc-G) proteins are found in chromatin associated with stably repressed genes and Trithorax-group (Trx-G) proteins with transcribed genes (reviewed in 105, 112). The protein encoded by the Arabidopsis *CURLY LEAF (CLF)* gene has sequence and functional homology to *E(z)*, a Pc-G protein of Drosophila (51). Long-term repression of the *AG* gene is dependent on CURLY LEAF; in *clf* mutants, *AG* is expressed ectopically in leaves and in flower buds at a late stage in development. If DNA methylation stabilizes chromatin structure in plants (46), then perturbation of DNA methylation could affect chromatin structure around *AG*. The observation that *AG* is expressed ectopically in leaves of methyltransferase antisense plants is consistent with this prediction (46).

METIII, a putative methyltransferase containing a chromodomain, provides a link between DNA methylation and chromatin structure, as chromodomain proteins Hp1 and Pc from Drosophila are associated with condensed chromatin (105). If METIII is associated with compacted chromatin via the chromodomain this could direct methylation of the underlying DNA. Alternatively, compaction of DNA could arise by association of other chromodomain proteins with METIII during methylation. Dnmt1, the mouse methyltransferase, has homology to a human homologue of the trithorax protein (Trx), again suggesting a link between DNA methylation and chromatin (21). Dnmt1-like plant methyltransferases lack homology to Trx.

CONCLUSIONS

In both prokaryotes and lower eukaryotes, DNA methylation provides a defense against invading or mobile DNA (52, 102, 127). Duplicated sequences in *Ascobolus immersus* and *Neurospora crassa* are methylated and inactivated just prior to cells entering the sexual phase of the life cycle (52, 127). Methylation prevents recombination between nonallelic repeat sequences and the spread of mobile elements in these fungi (53, 120). Mobile elements and other repeated sequences are methylated in plant genomes, supporting the idea that DNA methylation also plays an important role in genome defense in plants.

Reduction of methylation in Arabidopsis demonstrated that methylation is essential for normal development. In methyltransferase antisense plants with reduced levels of DNA methylation, genes involved in flower development showed both low-level ectopic expression and transcriptional repression associated with hypermethylation (46, 61). Flowering time was altered, consistent with DNA methylation having a role in vernalization. The dysregulation of gene expression in these plants implies that even if methylation of DNA does not play a primary role in regulating gene expression it provides the appropriate genomic context essential for programmed development.

DNA methylation has dual roles in defense and gene regulation. How did this arise? Plant genomes are punctuated with methylated retroelements and their remnants. Many of these elements have inserted into intergenic spacers, and some are located within promoter regions contributing sequences important for promoter function (138). Did an ancient role in genome defense evolve into one of gene regulation because methylated retroelements reside in promoters? Did a primary role in regulating gene expression in complex genomes expand to include genome defense? Or was methylation independently recruited to control gene expression and defend against invading DNA in complex genomes?

While it is unlikely that we will be able to answer these questions from studies of DNA methylation, we may gain insights into what enables a plant to activate defense mechanisms to silence invading DNA. Gene inactivation by hypermethylation in the *METI* antisense plants suggests that a defense mechanism has been activated against endogenous genes. An understanding of the factors that regulate this process may provide insights into what makes an endogenous gene appear as foreign DNA. This in turn may assist in the generation of transgenic plants that stably express transgenes.

ACKNOWLEDGMENTS

The authors thank the following people for allowing inclusion of results prior to publication: R Adams, M Alleman, J Bender, T Bestor, L Comai, K Cone, T Elmayan, S Henikoff, O Hoekenga, S Jacobsen, J Jeddeloh, T Kakutani, N Kishimoto, K Kovac, M Koornneef, M Matzke, A Matzke, P Meyer, O Mittelsten Scheid, J Paszkowski, S Pradhan, E Richards, H Sakai, W Soppe, R van Blockland, H Vaucheret, and J Yoder; and L Thorpe for assistance in preparation of the manuscript.

> **Visit the *Annual Reviews* home page** at
> **http://www.AnnualReviews.org.**

Literature Cited

1. Amasino RM, Powell ALT, Gordon MP. 1984. Changes in T-DNA methylation and expression are associated with phenotypic variation and plant regeneration in a crown gall tumor line. *Mol. Gen. Genet.* 197:437–46

2. Aniello F, Locascio A, Fucci L, Geraci G, Branno M. 1996. Isolation of cDNA clones encoding DNA methyltransferase of sea urchin *P. lividus*: expression during embryo development. *Gene* 178:57–61

3. Antequera F, Bird AP. 1988. Unmethylated CpG islands associated with genes in higher plant DNA. *EMBO J.* 7:2295–99

4. Assaad FF, Tucker KL, Signer ER. 1993. Epigenetic repeat-induced silencing (RIGS) in *Arabidopsis*. *Plant Mol. Biol.* 22:1067–85

5. Balganesh TS, Reiners L, Lauster R, Noyer-Weidner M, Wilke K, et al. 1987. Construction and use of chimeric SPR/f3T DNA methyltransferases in the definition of sequence recognizing enzyme regions. *EMBO J.* 6:3543–49

6. Banks JA, Fedoroff N. 1989. Patterns of developmental and heritable change in methylation of the *Suppressor-mutator* transposable element. *Dev. Genet.* 10:425–37

7. Banks JA, Masson P, Fedoroff N. 1988. Molecular mechanisms in the developmental regulation of the maize *Suppressor-mutator* transposable element. *Gen. Dev.* 2:1364–80

8. Barton DW. 1951. Localized chiasmata in the differentiated chromosomes of the tomato. *Genetics* 36:374–81

9. Beard C, Li E, Jaenisch R. 1995. Loss of methylation activates *Xist* in somatic but not embryonic cells. *Genet. Dev.* 9:2325–34

10. Bender J, Fink GR. 1995. Epigenetic control of an endogenous gene family is revealed by a novel blue fluorescent mutant of Arabidopsis. *Cell* 83:725–34

11. Bennetzen JL. 1996. The contributions of retroelements to plant genome organization, function and evolution. *Trends Microbiol.* 4:347–53

12. Bennetzen JL, Brown WE, Springer PS. 1988. The state of DNA modification within and flanking maize transposable elements. See Ref. 100, pp. 237–50

13. Bennetzen JL, Schrick K, Springer PS, Brown WE, SanMiguel P. 1994. Active maize genes are unmodified and flanked by diverse classes of modified, highly repetitive DNA. *Genome* 37:565–76

14. Bernacchia G, Primo A, Giorgetti L, Pitto L, Cella R. 1998. Carrot DNA-methyltransferase is encoded by two classes of genes with differing patterns of expression. *Plant J.* In press

15. Bestor T. 1987. Supercoiling-dependent sequence specificity of mammalian DNA methyltransferase. *Nucleic Acids Res.* 15:3835–43

16. Bestor T. 1990. DNA methylation: evolution of a bacterial immune function into a regulator of gene expression and genome structure in higher eukaryotes. *Philos. Trans. R. Soc. London Ser. B* 326:179–87

17. Bestor TH. 1992. Activation of mammalian DNA methyltransferase by cleavage of a Zn binding regulatory domain. *EMBO J.* 11:2611–17

18. Bestor TH, Coxon A. 1993. The pros and

cons of DNA methylation. *Curr. Biol.* 3:384–86

19. Bestor T, Laudano A, Mattaliano R, Ingram V. 1988. Cloning and sequencing of a cDNA encoding DNA methyltransferase of mouse cells. *J. Mol. Biol.* 203:971–83

20. Bestor TH, Tycko B. 1996. Creation of genomic methylation patterns. *Nat. Genet.* 12:363–67

21. Bestor TH, Verdine GL. 1994. DNA methyltransferases. *Curr. Opin. Cell Biol.* 6:380–89

22. Billen D. 1968. Methylation of the bacterial chromosome: an event at the replication point. *J. Mol. Biol.* 31:477–86

23. Bird AP. 1978. Use of restriction enzymes to study eukaryote DNA methylation. II. The symmetry of methylated sites supports semi-conservative copying of the methylation pattern. *J. Mol. Biol.* 118:49–60

24. Bird AP. 1986. CpG-rich islands and the function of DNA methylation. *Nature* 321:209–13

25. Bird AP. 1995. Gene number, noise reduction and biological complexity. *Trends Genet.* 11:94–100

26. Brink RA. 1973. Paramutation. *Annu. Rev. Genet.* 7:129–52

27. Brock RD, Davidson JL. 1994. 5-azacytidine and gamma rays partially substitute for cold treatment in vernalizing winter wheat. *Environ. Exp. Bot.* 34:195–99

28. Brown SW. 1966. Heterochromatin. *Science* 151:417–25

29. Burn JE, Bagnall DJ, Metzger JD, Dennis ES, Peacock WJ. 1993. DNA methylation, vernalization, and the initiation of flowering. *Proc. Natl. Acad. Sci. USA* 90:287–91

30. Chaudhuri S, Messing J. 1994. Allele-specific parental imprinting of dzr1, a posttranscriptional regulator of zein accumulation. *Proc. Natl. Acad. Sci. USA* 91:4867–71

31. Chen L, MacMillan AM, Chang W, Ezaz-Nikpay K, Lane WS, et al. 1991. Direct identification of the active-site nucleophile in a DNA (cytosine-5)-methyltransferase. *Biochemistry* 30:11018–25

32. Cheng XD, Kumar S, Posfai J, Pflugrath JW, Roberts RJ. 1993. Crystal structure of the HhaI DNA methyltransferase complexed with S-adenosyl-L-methionine. *Cell* 74:299–307

33. Civardi L, Xia YJ, Edwards KJ, Schnable PS, Nikolau BJ. 1994. The relationship between genetic and physical distances in the cloned *a1-sh2* interval of the *Zea mays* L. genome. *Proc. Natl. Acad. Sci. USA* 91:8268–72

34. Cocciolone SM, Cone KC. 1993. *Pl-Bh*, an anthocyanin regulatory gene of maize that leads to variegated pigmentation. *Genetics* 135:575–88

35. Cooper DN, Taggart MH, Bird AP. 1983. Unmethylated domains in vertebrate DNA. *Nucleic Acids Res.* 11:647–57

36. Deleted in proof

37. Dennis ES, Finnegan EJ, Bilodeau P, Chaudhury A, Genger R, et al. 1996. Vernalization and the initiation of flowering. *Semin. Cell Dev. Biol.* 7:441–48

38. Doerfler W. 1991. Patterns of DNA methylation—evolutionary vestiges of foreign DNA inactivation as a host defense mechanism. *Biol. Chem. Hoppe-Seyler* 372:557–64

39. Ehrlich M, Ehrlich KC. 1993. Effect of DNA methylation on the binding of vertebrate and plant proteins to DNA. See Ref. 64, pp. 145–68

40. Elomaa P, Helariutta Y, Griesbach RJ, Kotilainen M, Seppänen P, et al. 1995. Transgene inactivation in *Petunia hybrida* is influenced by the properties of the foreign gene. *Mol. Gen. Genet.* 248:649–56

41. Fedoroff N, Masson P, Banks JA. 1989. Mutations, epimutations, and the developmental programming of the maize *Suppressor-mutator* transposable element. *BioEssays* 10:139–44

42. Fedoroff NV, Banks JA. 1988. Is the *Suppressor-mutator* element controlled by a basic developmental regulatory mechanism? *Genetics* 120:559–77

43. Finnegan EJ, Brettell RIS, Dennis ES. 1993. The role of DNA methylation in the regulation of plant gene expression. See Ref. 64, pp. 218–61

44. Finnegan EJ, Dennis ES. 1993. Isolation and identification by sequence homology of a putative cytosine methyltransferase from *Arabidopsis thaliana*. *Nucleic Acids Res.* 21:2383–88

45. Finnegan EJ, McElroy D. 1994. Transgene inactivation: plants fight back! *Bio-Technology* 12:883–88

46. Finnegan EJ, Peacock WJ, Dennis ES. 1996. Reduced DNA methylation in *Arabidopsis thaliana* results in abnormal plant development. *Proc. Natl. Acad. Sci. USA* 93:8449–54

47. Flavell AJ, Pearce SR, Kumar A. 1995. Plant transposable elements and the genome. *Curr. Opin. Genet. Dev.* 4:838–44

48. Flavell RB. 1994. Inactivation of gene expression in plants as a consequence of specific sequence duplication. *Proc. Natl. Acad. Sci. USA* 91:3490–96

49. Frommer M, McDonald LE, Millar DS, Collis CM, Watt F, et al. 1992. A genomic sequencing protocol that yields a positive display of 5-methylcytosine residues in individual DNA strands. *Proc. Natl. Acad. Sci. USA* 89:1827–31

50. Giordano M, Mattachini ME, Cella R, Pedrali-Noy G. 1991. Purification and properties of a novel DNA methyltransferase from cultured rice cells. *Biochem. Biophys. Res. Commun.* 177:711–19

51. Goodrich J, Puangsomlee P, Martin M, Long D, Meyerowitz EM, et al. 1997. A polycomb-group gene regulates homeotic gene expression in *Arabidopsis*. *Nature* 386:44–51

52. Goyon C, Faugeron G. 1989. Targeted transformation of *Ascobolus immersus* and de novo methylation of the resulting duplicated DNA sequences. *Mol. Cell Biol.* 9:2818–27

53. Goyon C, Rossignol J-L, Faugeron G. 1996. Native DNA repeats and methylation in *Ascobolus*. *Nucleic Acids Res.* 24:3348–56

54. Gruenbaum Y, Naveh-Many T, Cedar H, Razin A. 1981. Sequence specificity of methylation in higher plant DNA. *Nature* 292:860–62

55. Haig D, Westoby M. 1991. Genomic imprinting in endosperm: effects on seed development in crosses between species, and between different ploidies of the same species and its implications for the evolution of apomixis. *Philos. Trans. R. Soc. London Ser. B* 333:1–13

56. Hartl DL. 1996. The most unkindest cut of all. *Nat. Genet.* 12:227–29

57. Holliday R, Pugh JE. 1975. DNA modification mechanisms and gene activity during development. *Science* 187:226–32

58. Hsieh C-L, Lieber MR. 1992. CpG methylated minichromosomes become inaccessible for V(D)J recombination after undergoing replication. *EMBO J.* 11:315–25

59. Ingelbrecht I, van Houdt H, van Montagu M, Depicker A. 1994. Posttranscriptional silencing of reporter transgenes in tobacco correlates with DNA methylation. *Proc. Natl. Acad. Sci. USA* 91:10502–6

60. Ingrosso D, Fowler AV, Bleibaum J, Clarke S. 1989. Sequence of the D-aspartyl/L-isoaspartyl protein methyltransferase from human erythrocytes. *J. Biol. Chem.* 264:20131–39

61. Jacobsen SE, Meyerowitz EM. 1997. Hypermethylated *SUPERMAN* epigenetic alleles in *Arabidopsis*. *Science* 277:1100–3

62. Janousek B, Siroky J, Vyskot B. 1996. Epigenetic control of sexual phenotype in a dioecious plant, *Melandrium album*. *Mol. Gen. Genet.* 250:483–90

63. Jost J-P. 1993. Nuclear extracts of chicken embryos promote an active demethylation of DNA by excision repair of 5-methyldeoxycytidine. *Proc. Natl. Acad. Sci. USA* 90:4684–88

64. Jost J-P, Saluz HP, eds. 1993. *DNA Methylation: Molecular Biology and Biological Significance*. Basel: Springer-Verlag

65. Kakutani T. 1998. Genetic characterization of late-flowering traits induced by DNA hypomethylation mutation in *Arabidopsis thaliana*. *Plant J.* In press

66. Kakutani T, Jeddeloh JA, Flowers SK, Munakata K, Richards EJ. 1996. Developmental abnormalities and epimutations associated with DNA hypomethylation mutations. *Proc. Natl. Acad. Sci. USA* 93:12406–11

67. Kakutani T, Jeddeloh J, Richards EJ. 1995. Characterization of an *Arabidopsis thaliana* DNA hypomethylation mutant. *Nucleic Acids Res.* 23:130–37

68. Kermicle JL. 1978. Imprinting of gene action in maize endosperm. In *Maize Breeding and Genetics*, ed. DB Walden, pp. 357–71. New York: Wiley

69. Kermicle JL, Alleman M. 1990. Gametic imprinting in maize in relation to the angiosperm life cycle. *Development Suppl.*, pp. 9–14

70. Kimura H, Ishihara G, Tajima S. 1996. Isolation and expression of a Xenopus laevis DNA methyltransferase cDNA. *J. Biochem.* 120:1182–89

71. Klimasauskas S, Kumar S, Roberts RJ, Cheng XD. 1994. HhaI methyltransferase flips its target base out of the DNA helix. *Cell* 76:357–69

72. Klimasauskas S, Nelson JL, Roberts RJ. 1991. The sequence specificity domain of cytosine-C5 methylases. *Nucleic Acids Res.* 19:6183–90

73. Kovarik A, Koukalova B, Holy A, Bezdek M. 1994. Sequence-specific hypomethylation of the tobacco genome induced with dihydroxypropyladenine, ethionine and 5-azacytidine. *FEBS Lett.* 353:309–11

74. Kupiec M, Petes TD. 1988. Allelic

and ectopic recombination between Ty elements in yeast. *Genetics* 119:549–59

75. Lark K. 1968. Studies on the *in vivo* methylation of DNA in *Escherichia coli* 15T. *J. Mol. Biol.* 31:389–99

76. Lauster R, Trautner TA, Noyer-Weidner M. 1989. Cytosine-specific type II DNA methyltransferases. A conserved enzyme core with variable target-recognizing domains. *J. Mol. Biol.* 206:305–12

77. Lei H, Oh SP, Okano M, Jutterman R, Goss KA, et al. 1996. De novo DNA cytosine methyltransferase activities in mouse embryonic stem cells. *Development* 122:3195–205

78. Leonhardt H, Page AW, Weier H-U, Bestor TH. 1992. A targeting sequence directs DNA methyltransferase to sites of DNA replication in mammalian nuclei. *Cell* 71:865–73

79. Deleted in proof

80. Li E, Beard C, Jaenisch R. 1993. Role for DNA methylation in genomic imprinting. *Nature* 366:362–65

81. Li E, Bestor TH, Jaenisch R. 1992. Targeted mutation of the DNA methyltransferase gene results in embryonic lethality. *Cell* 69:915–26

82. Lin B-Y. 1984. Ploidy barrier to endosperm development in maize. *Genetics* 107:103–15

83. Lindbo JA, Silva-Rosales L, Proebsting WM, Dougherty WG. 1993. Induction of a highly specific antiviral state in transgenic plants: implications for regulation of gene expression and virus resistance. *Plant Cell* 5:1749–69

84. Linn F, Heidmann I, Saedler H, Meyer P. 1990. Epigenetic changes in the expression of the maize A1 gene in *Petunia hybrida*: role of numbers of integrated gene copies and state of methylation. *Mol. Gen. Genet.* 222:329–36

85. Deleted in proof

86. Martienssen R, Barkan A, Taylor WC, Freeling M. 1990. Somatically heritable switches in the DNA modification of Mu transposable elements monitored with a suppressible mutant in maize. *Genet. Dev.* 4:331–43

87. Martienssen R, Baron A. 1994. Coordinate suppression of mutations caused by Robertson's *Mutator* transposons in maize. *Genetics* 136:1157–70

88. Matzke MA, Matzke AJM. 1995. How and why do plants inactivate homologous (*Trans*)genes. *Plant Physiol.* 107:679–85

89. Matzke MA, Matzke AJM. 1998. Gene

silencing in plants: relevance for genome evolution and the acquistion of genomic methylation patterns. In *Ciba Foundation Symposium 214. Epigenetics*, ed. G Cardew. In press

90. Matzke MA, Matzke AJM, Eggleston WB. 1996. Paramutation and transgene silencing: a common response to invasive DNA? *Trends Plant Sci.* 1:382–88

91. Matzke MA, Primig M, Trnovsky J, Matzke AJM. 1989. Reversible methylation and inactivation of marker genes in sequentially transformed tobacco plants. *EMBO J.* 8:643–49

91a. McClelland M. 1983. The frequency and distribution of methylatable DNA sequences in leguminous plant protein coding genes. *J. Mol. Evol.* 19:346–54

92. Messeguer R, Ganal MW, Steffens JC, Tanksley SD. 1991. Characterization of the level, target sites and inheritance of cytosine methylation in tomato nuclear DNA. *Plant Mol. Biol.* 16:753–70

93. Deleted in proof

94. Meyer P. 1995. DNA methylation and transgene silencing in *Petunia hybrida*. In *Current Topics in Microbiology and Immunology: Gene Silencing in Higher Plants and Related Phenonmena in Other Eukaryotes*, ed. P Meyer, pp. 15–28. Berlin: Springer-Verlag

95. Meyer P, Heidmann I. 1994. Epigenetic variants of a transgenic petunia line show hypermethylation in transgene DNA: an indication for specific recognition of foreign DNA in transgenic plants. *Mol. Gen. Genet.* 243:390–99

96. Meyer P, Linn F, Heidmann I, Meyer H, Niedenhof I, et al. 1992. Endogenous and environmental factors influence 35S promoter methylation of a maize A1 gene construct in transgenic petunia and its colour phenotype. *Mol. Gen. Genet.* 231:345–52

97. Meyer P, Niedenhof I, ten Lohuis M. 1994. Evidence for cytosine methylation of nonsymmetrical sequences in transgenic *Petunia hybrida*. *EMBO J.* 13:2084–88

98. Mi S, Roberts RJ. 1992. How M. *Msp*I and M.*Hpa*II decide which base to methylate. *Nucleic Acids Res.* 20:4811–16

99. Mol J, van Blokland R, Kooter J. 1991. More about co-suppression. *Trends Biotechnol.* 9:182–83

100. Nelson OE Jr, ed. 1988. *Plant Transposable Elements*. New York: Plenum

101. Nick H, Bowen B, Ferl RJ, Gilbert W. 1986. Detection of cytosine methylation

in the maize alcohol dehydrogenase gene by genomic sequencing. *Nature* 319:243–46
102. Noyer-Weidner M, Trautner TA. 1993. Methylation of DNA in prokaryotes. See Ref. 64, pp. 39–108
103. Oakeley EJ, Jost J-P. 1996. Nonsymmetrical cytosine methylation in tobacco pollen DNA. *Plant Mol. Biol.* 31:927–30
104. Panning B, Jaenisch R. 1996. DNA hypomethylation can activate *Xist* expression and silence X-linked genes. *Genet. Dev.* 10:1991–2002
105. Paro R, Harte PJ. 1996. The role of polycombe group and thrithorax group chromatin complexes in the maintenance of determined cell states. See Ref. 120a, pp. 507–28
106. Patterson GI, Chandler VL. 1995. Paramutation in maize and related allelic interactions. See Ref. 94, pp. 121–41
107. Patterson GI, Thorpe CJ, Chandler VL. 1993. Paramutation, an allelic interaction, is associated with a stable and heritable reduction of transcription of the maize *b* regulatory gene. *Genetics* 135:881–94
108. Peitsch MC. 1995. Protein modeling by e-mail. *Bio-Technology* 13:658–60
109. Peitsch MC. 1996. ProMod and Swissmodel: internet-based tools for automated comparative protein modelling. *Biochem. Soc. Trans.* 24:274–79
110. Peitsch MC, Jongeneel CV. 1993. A 3-D model for the CD40 ligand predicts that it is a compact trimer similar to the tumor necrosis factors. *Int. Immunol.* 5:233–38
111. Penny GD, Kay GF, Sheardown SA, Rastan S, Brockdorff N. 1996. Requirement for *Xist* in X chromosome inactivation. *Nature* 379:131–37
112. Pirrotta V. 1996. Stable chromatin states regulating homeotic genes in *Drosophila*. See Ref. 120a, pp. 489–505
113. Pósfai J, Bhagwat AS, Pósfai G, Roberts RJ. 1989. Predictive motifs derived from cytosine methyltransferases. *Nucleic Acids Res.* 17:2421–35
114. Pradhan S, Adams RLP. 1995. Distinct CG and CNG DNA methyltransferases in *Pisum sativum. Plant J.* 7:471–81
115. Ratcliff F, Harrison BD, Baulcombe DC. 1997. A similarity between viral defense and gene silencing in plants. *Science* 276:1558–60
116. Razin A, Cedar H. 1993. DNA methylation and embryogenesis. See Ref. 64, pp. 343–57
117. Reinisch KM, Chen L, Verdine GL, Lipscomb WN. 1995. The crystal structure

of HaeIII methyltransferase covalently complexed to DNA: an extrahelical cytosine and rearranged base pairing. *Cell* 82:143–53
118. Riggs AD. 1975. X-inactivation, differentiation and DNA methylation. *Cytogenet. Cell Genet.* 14:9–25
119. Ronemus MJ, Galbiati M, Ticknor C, Chen JC, Dellaporta SL. 1996. Demethylation-induced developmental pleiotropy in *Arabidopsis. Science* 273: 654–57
120. Rossignol J-L, Faugeron G. 1994. Gene inactivation triggered by recognition between DNA repeats. *Experientia* 50:307–17
120a. Russo VEA, Martienssen RA, Riggs AD, eds. 1996. *Epigenetic Mechanisms of Gene Regulation.* Cold Spring Harbor, NY: Cold Spring Harbor Lab. Press
121. Sakai H, Medrano LJ, Meyerowitz EM. 1995. Role of *SUPERMAN* in maintaining *Arabidopsis* floral whorl boundaries. *Nature* 378:199–203
122. SanMiguel P, Tikhonov A, Jin Y-K, Motchoulskaia N, Zakharov D, et al. 1996. Nested retrotransposons in the intergenic regions of the maize genome. *Science* 274:765–68
123. Schmülling T, Röhrig H. 1995. Gene silencing in transgenic tobacco hybrids: frequency of the event and the visualization of somatic inactivation pattern. *Mol. Gen. Genet.* 249:375–90
124. Schneuwly S, Kuroiwa A, Gehring WJ. 1987. Molecular analysis of the dominant homeotic *Antennapedia* phenotype. *EMBO J.* 6:201–6
125. Schwartz D. 1988. Comparison of methylation of the male- and female-derived *wx-m9 Ds-cy* allele in endosperm and sporophyte. See Ref. 100, pp. 351–54
126. Schwartz D, Dennis E. 1986. Transposase activity of the *Ac* controlling element in maize is regulated by its degree of methylation. *Mol. Gen. Genet.* 205:476–82
127. Selker EU, Cambareri EB, Jensen BC, Haack KR. 1987. Rearrangement of duplicated DNA in specialized cells of *Neurospora. Cell* 51:741–52
128. Smith HA, Swaney ST, Parks TD, Wernsman EA, Dougherty WG. 1994. Transgenic plant virus resistance mediated by untranslatable sense RNAs: expression, regulation, and fate of nonessential RNAs. *Plant Cell* 6:1441–53
129. Smith SS, Laayoun A, Lingeman RG, Baker DJ, Riley J. 1994. Hypermethyla-

tion of telomere-like foldbacks at codon 12 of the human c-Ha-*ras* gene and the trinucleotide repeat of the *FMR-1* gene of fragile X. *J. Mol. Biol.* 243:143–51

130. Som S, Bhagwat AS, Friedman S. 1987. Nucleotide sequence and expression of the gene encoding the *Eco*RII modification enzyme. *Nucleic Acids Res.* 15:313–31

131. Stam M, Mol JNM, Kooter JM. 1997. The silence of genes in transgenic plants. *Ann. Bot.* 79:3–12

132. Tajima S, Tsuda H, Wakabayashi N, Asano A, Mizuno S, et al. 1995. Isolation and expression of a chicken DNA methyltransferase cDNA. *J. Biochem.* 117:1050–57

133. Theiss G, Schleicher R, Schimpff-Weiland R, Follmann H. 1987. DNA methylation in wheat. *Eur. J. Biochem.* 167:89–96

134. Thuriaux P. 1977. Is recombination confined to structural genes on the eukaryotic genome? *Nature* 268:460–62

135. Trautner TA, Balganesh TS, Pawlek B. 1988. Chimeric multispecific DNA methyltransferases with novel combinations of target recognition. *Nucleic Acids Res.* 16:6649–58

136. Vlasova TI, Demidenko ZN, Kirnos MD, Vanyushin BF. 1995. In vitro DNA methylation by wheat nuclear cytosine DNA methyltransferase: effects of phytohormones. *Gene* 157:279–81

137. Vongs A, Kakutani T, Martienssen RA, Richards EJ. 1993. *Arabidopsis thaliana* DNA methylation deficient mutants. *Science* 260:1926–28

138. Voytas DF. 1996. Retroelements in genome organization. *Science* 274:737–38

139. Deleted in proof

140. Vyskot B, Koukalova B, Kovarik A, Sachambula L, Reynolds D, et al. 1995. Meiotic transmission of a hypomethylated repetitive DNA family in tobacco. *Theor. Appl. Genet.* 91:659–64

141. Wang LH, Heinlein M, Kunze R. 1996. Methylation pattern of *Activator* transposase binding sites in maize endosperm. *Plant Cell* 8:747–58

142. Wassenegger M, Heimes S, Riedel L, Sanger HL. 1994. RNA-directed de novo methylation of genomic sequences in plants. *Cell* 76:567–76

143. Weber H, Ziechmann C, Graessmann A. 1990. *In vitro* DNA methylation inhibits gene expression in transgenic tobacco. *EMBO J.* 9:4409–15

144. Weiss A, Keshet I, Razin A, Cedar H. 1996. DNA demethylation in vitro—involvement of RNA. *Cell* 86:709–18

145. White SE, Habera LF, Wessler SR. 1994. Retrotransposons in the flanking regions of normal plant genes: a role for copia-like elements in the evolution of gene structure and expression. *Proc. Natl. Acad. Sci. USA* 91:11792–96

146. Wigler M, Levy D, Perucho M. 1981. The somatic replication of DNA methylation. *Cell* 24:33–40

147. Wilke K, Rauhut E, Noyer-Weidner M, Lauster R, Pawlek B, et al. 1988. Sequential order of target-recognizing domains in multispecific DNA-methyltransferases. *EMBO J.* 7:2601–9

148. Wyszynski MW, Gabbara S, Bhagwat AS. 1992. Substitutions of a cysteine conserved among DNA cytosine methylases result in a variety of phenotypes. *Nucleic Acids Res.* 20:319–26

149. Ye F, Signer ER. 1996. RIGS (repeat-induced gene silencing) in *Arabidopsis* is transcriptional and alters chromatin configuration. *Proc. Natl. Acad. Sci. USA* 93:10881–86

150. Yen R-WC, Vertino PM, Nelkin BD, Yu JJ, El-Deiry W, et al. 1992. Isolation and characterization of the cDNA encoding human DNA methyltransferase. *Nucleic Acids Res.* 20:2287–91

151. Yoder JA, Walsh CP, Bestor TH. 1997. Cytosine methylation and the ecology of intragenomic parasites. *Trends Genet.* 13:335–40

152. Yoder JA, Yen R-WC, Vertino PM, Bestor TH, Baylin SB. 1996. New 5′ regions of the murine and human genes for DNA (cytosine-5)-methyltransferase. *J. Biol. Chem.* 271:31092–97

Annu. Rev. Plant Physiol. Plant Mol. Biol. 1998. 49:249–79
Copyright © 1998 by Annual Reviews. All rights reserved

ASCORBATE AND GLUTATHIONE:
Keeping Active Oxygen Under Control

Graham Noctor

Laboratoire du Métabolisme, Institut National de la Recherche Agronomique,
Route de Saint Cyr, 78026 Versailles cedex, France

Christine H. Foyer

Department of Environmental Biology, Institute of Grassland and Environmental
Research, Plas Gogerddan, Aberystwyth, Ceredigion SY23 3EB, United Kingdom

KEY WORDS: oxidative stress, antioxidative enzymes, photosynthesis, photorespiration, signal
               transduction

ABSTRACT

To cope with environmental fluctuations and to prevent invasion by pathogens,
plant metabolism must be flexible and dynamic. Active oxygen species, whose
formation is accelerated under stress conditions, must be rapidly processed if
oxidative damage is to be averted. The lifetime of active oxygen species within the
cellular environment is determined by the antioxidative system, which provides
crucial protection against oxidative damage. The antioxidative system comprises
numerous enzymes and compounds of low molecular weight. While research
into the former has benefited greatly from advances in molecular technology, the
pathways by which the latter are synthesized have received comparatively little
attention. The present review emphasizes the roles of ascorbate and glutathione in
plant metabolism and stress tolerance. We provide a detailed account of current
knowledge of the biosynthesis, compartmentation, and transport of these two
important antioxidants, with emphasis on the unique insights and advances gained
by molecular exploration.

CONTENTS

249

INTRODUCTION

Life in oxygen has led to the evolution of biochemical adaptations that exploit the reactivity of active oxygen species (AOS). The term AOS is generic, embracing not only free radicals such as superoxide (O_2^-) and hydroxyl radicals but also H_2O_2 and singlet oxygen. While it is generally assumed that the hydroxyl radical and singlet oxygen are so reactive that their production must be minimized (59, 91), O_2^- and H_2O_2 are synthesized at very high rates even under optimal conditions. They are involved in virtually all major areas of aerobic biochemistry (e.g. respiratory and photosynthetic electron transport; oxidation of glycolate, xanthine, and glucose) and are produced in copious quantities by several enzyme systems [e.g. plasmalemma-bound NADPH-dependent superoxide synthase (3) and superoxide dismutase (SOD) (21)]. The chief toxicity of O_2^- and H_2O_2 is thought to reside in their ability to initiate cascade reactions that result in the production of the hydroxyl radical and other destructive species such as lipid peroxides. These dangerous cascades are prevented by efficient operation of the cell's antioxidant defenses. In some circumstances, however, the destructive power and signaling potential of AOS are utilized as an effective means of defense (33, 62, 114).

The term antioxidant can be considered to describe any compound capable of quenching AOS without itself undergoing conversion to a destructive radical (149, 178). Antioxidant enzymes are considered as those that either catalyze such reactions or are involved in the direct processing of AOS. Hence, antioxidants and antioxidant enzymes function to interrupt the cascades of uncontrolled oxidation. Of the numerous enzymes and metabolites potentially covered by the above definitions, many remain uncharacterized (see, for example, 74, 201). The comparatively few classes of antioxidant enzymes that have been well characterized are listed in Table 1. These enzymes catalyze redox reactions, many of which rely on electrons supplied by reductants of low molecular weight. Among these low-molecular-weight antioxidants, ascorbate and glutathione are of paramount importance. They fulfill multiple roles in defense reactions and

Table 1   Major antioxidative enzymes

Enzyme	Abbreviation in text	EC number
Superoxide dismutase	SOD	1.15.1.1
Ascorbate peroxidase	APX	1.11.1.11
Monodehydroascorbate reductase	MDHAR	1.6.5.4
Dehydroascorbate reductase	DHAR	1.8.5.1
Glutathione reductase	GR	1.6.4.2
Catalase	CAT	1.11.1.6
Glutathione peroxidase	GPX	1.11.1.9
Guaiacol-type peroxidases	—	1.11.1.7
Glutathione S-transferases	GST	2.5.1.18

are major assimilate sinks, present in many tissues at millimolar concentrations (19, 57, 63, 67, 154, 173). In spite of this, and despite the repeated observation of increased levels of ascorbate and glutathione in stress conditions, very little consideration has been given to the regulation of the pathways that influence their cellular concentration. Consequently, while this review considers recent developments in various systems that keep active oxygen under control, we lend particular emphasis to the metabolism of ascorbate and glutathione.

THE ANTIOXIDANT ENZYMES

Efficient destruction of O_2^- and H_2O_2 requires the action of several antioxidant enzymes acting in synchrony. Superoxide produced in the different compartments of plant cells is rapidly converted to H_2O_2 by the action of SOD (21). In organelles such as chloroplasts, which contain high concentrations of ascorbate, direct reduction of O_2^- by ascorbate is also rapid (25). As has frequently been pointed out, dismutation of O_2^- simply serves to convert one destructive AOS to another. Since H_2O_2 is a strong oxidant that rapidly oxidizes thiol groups, it cannot be allowed to accumulate to excess in organelles such as chloroplasts, where photosynthesis depends on thiol-regulated enzymes (95).

Catalases (CAT) convert H_2O_2 to water and molecular oxygen (for a review, see 230). These enzymes have extremely high maximum catalytic rates but low substrate affinities, since the reaction requires the simultaneous access of two H_2O_2 molecules to the active site (230). Furthermore, the absence of CAT in the chloroplast precludes a role in protection of the thiol-regulated enzymes of the Calvin cycle. An alternative mode of H_2O_2 destruction is via peroxidases, which are found throughout the cell (93, 236) and which have a much higher affinity for H_2O_2 than CAT. Peroxidases, however, require a reductant, since they reduce H_2O_2 to H_2O. In animals, peroxidases that use reduced glutathione (GSH) are important in H_2O_2 detoxification (134). Other than glutathione S-transferases (GSTs) (125), enzymes capable of catalyzing GSH-dependent reduction of H_2O_2 have not been characterized in plants. Several plant genes have recently been isolated that show homology to mammalian phospholipid hydroperoxide GSH peroxidase (39, 84, 94). One of these genes encodes a protein that catalyzes GSH-dependent reduction of phospholipid hydroperoxides, albeit at much lower rates than the enzyme from pig heart (18).

In plant cells, the most important reducing substrate for H_2O_2 detoxification is ascorbate (132, 145). Ascorbate peroxidase (APX) uses two molecules of ascorbate to reduce H_2O_2 to water, with the concomitant generation of two molecules of monodehydroascorbate (MDHA; Figure 1). MDHA is a radical with a short lifetime that, if not rapidly reduced, disproportionates to ascorbate and dehydroascorbate (DHA; Figure 1). Within the cell [for example at the plasmalemma (86) or at the thylakoid membrane (12)], MDHA can be reduced directly to ascorbate. The electron donor for MDHA reduction may be b-type cytochrome (85), reduced ferredoxin (139), or NAD(P)H (12). The latter reaction is catalyzed by MDHA reductases, which are found in several cellular compartments (12). Despite the possibility of enzymic and nonenzymic regeneration of ascorbate directly from MDHA, rapid disproportionation of the MDHA radical means that some DHA is always produced when ascorbate is

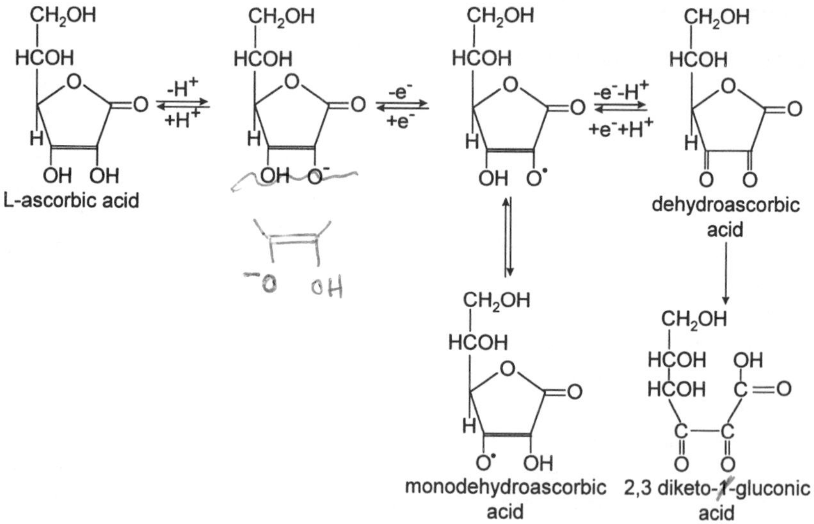

Figure 1   L-Ascorbic acid and its oxidation products (after Reference 226).

oxidized in leaves and other tissues. DHA is reduced to ascorbate by the action of DHA reductase, using GSH as the reducing substrate (57). This reaction generates glutathione disulphide (GSSG), which is in turn re-reduced to GSH by NADPH, a reaction catalyzed by glutathione reductase (GR). The removal of H_2O_2 through this series of reactions is known as the ascorbate-glutathione cycle (Figure 2). Ascorbate and glutathione are not consumed in this pathway but participate in a cyclic transfer of reducing equivalents, involving four

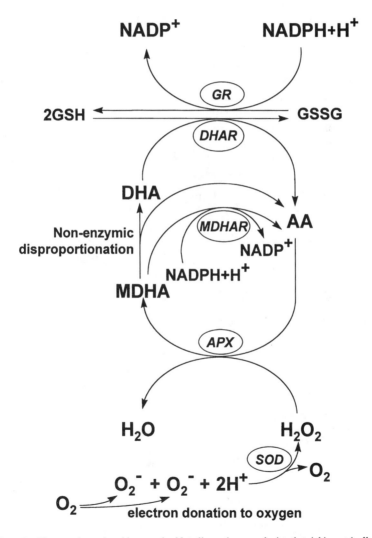

Figure 2   The ascorbate-glutathione cycle. Not all reactions are depicted stoichiometrically.

enzymes, which permits the reduction of H_2O_2 to H_2O using electrons derived from NAD(P)H (Figure 2).

Antioxidants and the Regulation of Photosynthesis

The production of O_2^- and other AOS is frequently considered to be a deleterious event since oxidant accumulation invariably leads to oxidative stress. During photosynthesis, however, AOS are produced and destroyed in a concerted manner that contributes to the regulation of electron transport (11, 12, 55). Many components of the thylakoid electron transport chain have electrochemical potentials commensurate with the reduction of molecular oxygen. Indeed, "leakage" of electrons to O_2 "poises" electron carriers, allowing them to function more efficiently. The highest rates of O_2 reduction probably occur on the reducing side of photosystem I (PSI) (Mehler reaction). Here, all the electron transport components, from the Fe-S centers to reduced thioredoxin, are auto-oxidizable (13). The sum of the Mehler reaction is electron transfer from H_2O to O_2, producing molecular oxygen (at PSII), O_2^- (at PSI), and a proton gradient across the thylakoid membrane. H_2O_2 resulting from O_2^- dismutation can be reduced by thylakoid-bound or stromal forms of APX (12). The reduction of O_2 at PSI, dismutation of O_2^-, reduction of H_2O_2, and regeneration of ascorbate together constitute the Mehler-peroxidase cycle (62). Operation of the complete pathway involves H_2O oxidation at PSII accompanied by the production of equimolar amounts of H_2O at PSI. ATP can therefore be formed without concomitant generation of reducing power. Hence, not only does the chloroplast ascorbate pool detoxify H_2O_2, thereby preventing enzyme inactivation and the generation of more dangerous radicals, it also allows flexibility in the production of photosynthetic assimilatory power (55, 56). Moreover, electron transfer to O_2 prevents overreduction of the electron transport chain, which reduces the risk of harmful back reactions within the photosystems. This minimizes direct energy exchange between the activated states of chlorophyll and ground state molecular oxygen and so avoids formation of activated singlet oxygen (59). In this way, the production of AOS is directly involved in the regulation (and protection) of photosynthetic electron transport. Ascorbate is also implicated in the regulation of photosynthetic light harvesting, since it supplies the electrons for violaxanthin de-epoxidase (VDE), a thylakoid-bound enzyme that synthesizes zeaxanthin in high light conditions (23). Zeaxanthin contributes to the processes of thermal energy dissipation affording photoprotection of PSII activity (87). The mechanism by which ascorbate is rereduced within the thylakoid lumen following oxidation by VDE is unknown.

Abiotic Stress and Signal Transduction

Recent years have witnessed a plethora of reports correlating increases in one or more of the antioxidant enzymes with either stress conditions or ameliorated stress resistance. Abiotic conditions that have been studied include, among others, low temperature (4, 43, 47, 51, 166, 225), high salinity (43, 223), herbicide challenge (122, 187), drought (137, 138, 200), wounding (69), ultraviolet irradiation (231), SO_2 fumigation (105, 121, 231), and ozone exposure (34, 105, 168, 192, 195, 215, 231). In work where several enzymes have been studied under the same stress conditions, differential responses have frequently been observed (e.g. 34, 215, 225, 231). The degree to which the activities of individual antioxidant enzymes are increased as a result of stress imposition is extremely variable and, in many cases, relatively minor. Moreover, determinations of total foliar activities may not adequately reflect the importance of compartment-specific changes. Each of the antioxidant enzymes comprises a family of isoforms, often with different characteristics. For example, the APX family consists of at least five different isoforms (11), including thylakoid and microsomal membrane-bound forms, as well as soluble stromal, cytosolic, and apoplastic isozymes (93, 139, 236). The genes encoding these APX isoforms respond differentially to metabolic and environmental signals (96, 105, 193, 217). Furthermore, although transcripts for specific isoforms can be quantified, increases in transcript abundance may not necessarily be accompanied by corresponding increases in enzyme activities (137, 138).

Plant transformation offers an alternative means of studying the significance of individual antioxidative enzymes. To date, such studies have been limited in both approach (because appropriate cDNAs have not been available) and scope (because the analysis of stress tolerance has been largely superficial). Only two enzymes, SOD and GR, have been studied intensively.

While the first plants overexpressing SOD did not demonstrate enhanced tolerance to stress (164, 216), subsequent studies have suggested that SOD overexpression can endow better stress tolerance (1, 20, 129, 130, 194, 198, 223, 224). Whether improved stress tolerance is observed probably depends on a multitude of factors, including nature of the stress imposed, isoenzyme overexpressed (223), intracellular targeting (223), strength of overexpression (193), leaf age (223), and growth conditions (198).

In transformed poplars overexpressing the first enzyme of GSH biosynthesis (see below), total glutathione levels are markedly enhanced, but GSH:GSSG ratios and GR activities are unchanged (9, 155). This implies that in untransformed plants, GR activities are more than high enough to maintain high foliar GSH:GSSG ratios. Despite this, overexpression of the *Escherichia coli* GR in the chloroplast, but not the cytosol, increased both the GSH:GSSG ratio and total foliar glutathione contents (61, 64, 142). In contrast, the foliar GSH:GSSG ratio in tobacco was not affected by overexpression of pea GR in either compartment (24). Enhanced chloroplastic GR activities were also associated with increases in foliar ascorbate contents (64), better protection of ascorbate and glutathione pools against paraquat stress (61, 64), decreased sensitivity to photoinhibition (64), and mitigated foliar damage during exposure to paraquat (6). However, enhanced glutathione content is not in itself sufficient to improve resistance to high rates of AOS production (151). While overexpression of GR in tobacco improved the foliar response of some transformed lines to ozone (24), no effect was observed in poplar (M Strohm, M Eiblmeier, C Langebartels, L Jouanin, A Polle, et al, submitted for publication).

The phenomenon of "cross tolerance" has emphasized the unity of diverse stress conditions and underlined the common feature of enhanced AOS production (21, 45, 121, 122, 231). Nevertheless, different stress conditions must involve distinct processes and, consequently, evoke disparate responses. Moreover, the literature undoubtedly exaggerates the importance of antioxidative enzymes in many stress situations, since positive results find their way into press easier than negative ones. Consequently, the absence of evidence for a massive, concerted up-regulation of antioxidant enzyme activities under stress conditions is all the more conspicuous. Perhaps the fault lies not in the plant cell but in our presuppositions. Rather than a system designed to annihilate AOS, the antioxidative system may have evolved primarily to allow adjustment of the cellular redox state and to enable redox signaling. Accumulation of H_2O_2, for example, is perceived by the plant as a signal of environmental change. As a diffusible signal-transducing molecule, H_2O_2 alerts metabolism to the presence of both biotic and abiotic threats (45). It is involved in well-characterized stress responses such as the hypersensitive response, systemic required resistance, (33, 114) and tolerance to chilling (167), as well as cross tolerance to a variety of stresses (21, 45). Although the absence of rapid, potent up-regulation of antioxidant enzymes during stress has frequently been interpreted as a failure of the plant to adapt to adverse conditions, such apparent shortcomings may have strategic relevance, particularly in the short term. Since ascorbate and glutathione are involved in the regulation of gene expression (48, 232), adjustment of the ratios of the reduced and oxidized forms of these antioxidants may also be of regulatory significance. Seen in this light, the role of antioxidative enzymes in stress situations would be not only to control AOS accumulation and limit oxidative damage but also to orchestrate gene expression through the generation of appropriate signals (H_2O_2, GSH, GSSG, ASC, DHA).

The Hypersensitive Response and Plant-Pathogen Interactions

During many incompatible plant-pathogen interactions, the recognition of an invading pathogen results in a coordinated activation of plant defense mechanisms, including programmed cell death (PCD) (136). The hypersensitive response (HR) is defined as rapid cell death associated with disease resistance (45, 114). Activation of PCD results in the formation of a zone of dead cells around the infection site that inhibits the proliferation of the invading pathogen.

Cell death is accompanied by an increase in the production of AOS and in lipid peroxidation (126, 131). Several studies have implicated H_2O_2 as a component of the signal transduction pathway that leads to the induction of pathogenesis-related proteins (22) and systemic resistance in uninfected parts of the plant (33). While the precise role of H_2O_2 in HR and systemic resistance remains a matter of debate (16), it is clear that the presence of foreign organisms, pathogens, and elicitors triggers an oxidative burst of O_2^- production on the apoplastic face of the plasmalemma (16). Apoplastic SODs rapidly convert the O_2^- to H_2O_2, which accumulates at the site of contact (218). Apoplastic O_2^- generation alone is sufficient to trigger cell death in a lesion mimic mutant (90).

Pathogen-induced H_2O_2 production has several effects. First, it hinders penetration of the pathogen by stimulating peroxidase activity and by cross-linking cell walls at the site of contact; second, it poses a stress on the pathogen as well as the host cell generating the oxidative burst; and third, it acts as a diffusible signal that leads to systemic acquired resistance (3).

Cell death will result only from H_2O_2 accumulation if the antioxidative defenses of the plant are overwhelmed and oxidative damage ensues. Although little information is available concerning the responses of foliar antioxidants to pathogen attack, glutathione appears to play a central role. Both GSH and GSSG may function as signal molecules in HR (62, 229). When transformed tobacco lacking the major catalase isoform (Cat 1) were transferred to high light, expression of pathogenesis-related genes was observed (31). These plants accumulated glutathione rather than H_2O_2 and showed a marked drop in the GSH:GSSG ratio (229). Since CAT is known to undergo rapid light-induced turnover, akin to the D1 protein of PSII (53, 83), these observations may be of considerable physiological relevance. Phenylalanine ammonia-lyase (PAL) and chalcone synthase transcripts may be induced by GSH (232), while GSH-responsive elements on the promoters of GST genes and in genes involved in the synthesis of phytoalexins have been identified (48, 114). Treatment with fungal elicitor induced GPX and GSTs (114) and led to increased GSH levels (50, 144). Inhibition of glutathione synthesis triggered phytoalexin accumulation, a response mimicked by addition of H_2O_2 (72). The cellular GSH:GSSG ratio may be more important in the regulation of defense-related gene expression than the absolute amounts of either form (131).

ASCORBIC ACID

All plants, and all animals except primates and guinea pigs (27), can synthesize ascorbic acid. In plants, ascorbate can accumulate to millimolar concentrations in both photosynthetic and nonphotosynthetic tissues (63). Leaves often contain more ascorbate than chlorophyll, with the ascorbate pool representing over

10% of the soluble carbohydrate. Of the many functions ascribed to ascorbic acid, relatively few are well characterized. It is clear, however, that ascorbate is a major primary antioxidant (148), reacting directly with hydroxyl radicals, superoxide, and singlet oxygen (25). In addition to its importance in photo-protection and the regulation of photosynthesis (54, 59), ascorbate plays an important role in preserving the activities of enzymes that contain prosthetic transition metal ions (163). Ascorbate is also a powerful secondary antioxi-dant, reducing the oxidized form of α-tocopherol, an important antioxidant in nonaqueous phases (163).

Ascorbate and Growth

Ascorbate is the reductant used for the hydroxylation of proline residues during extensin biosynthesis (115) and is implicated in root elongation, cell vacuolar-ization, and cell wall expansion (10, 37, 68, 146, 147, 212, 213). In the cell wall and apoplast, peroxidases are involved in the formation of both isodityrosine bonds between extensin glycoproteins and difrulate bridges between polysac-charide polymers (161). During lignification, a wide range of aromatic and phenolic compounds can be oxidized by either laccase or peroxidase using ex-tracellular H_2O_2 (40). Because phenoxyl radicals react much more readily with ascorbate than with each other, low concentrations of ascorbate can completely inhibit these oxidation reactions (212, 213). Apoplastic ascorbate is also con-sidered crucial in scavenging AOS, particularly those arising from exposure to atmospheric pollutants such as ozone (119, 165).

Ascorbate may also be involved in regulation of the cell cycle (98). When the ascorbate pool in meristems is diminished, cells arrest in G1-phase. Arrest during oxidative stress, which prevents replication of damaged DNA, appears to correlate with a decreased ratio of ascorbate to DHA. The details of the mech-anism by which this occurs are unknown but could involve histone hydroxy-lation and the transcription factor, ASF-1 (98). Hence, intracellular ascorbate may influence DNA replication, whereas apoplastic ascorbate may regulate cell wall expansion and cell elongation. Redox-related mechanisms for transport of ascorbate and DHA across the plasma membrane (see below) would be central to such coordinate control of intra- and extracellular functions.

Biosynthesis

Since ascorbate is an essential metabolite implicated in vital cell functions, it is surprising that the pathway of ascorbate synthesis in plants remains to be established (149, 199). In animals, ascorbate deficiency leads to "scurvy." In plants, however, the effects of ascorbate depletion are less easily discernible. A mutant with a putative lesion in the pathway of ascorbate biosynthesis has recently been described (36). Isolated via its increased sensitivity to ozone, this *Arabidopsis* mutant, *vtc1* (originally called *soz1*), contains about 30% of wild-type foliar levels of ascorbate (35, 36).

The hypothesis that ascorbate is synthesized from glucose is widely accepted, but measured rates of conversion of labeled glucose into ascorbate are very low (183). Furthermore, analysis of the *vtc1* mutant suggests that several pathways of ascorbate biosynthesis may operate simultaneously in plants (35, 199). It is possible that separate pathways of ascorbate biosynthesis occur at different sites in the plant cell and that redundant pathways operate in the same organelle (35). While the following discussion restricts itself to current concepts of the biosynthesis of ascorbate, it should be noted that no conclusive evidence exists in favor of the operation of either pathway in vivo.

The two putative routes of ascorbate biosynthesis are illustrated in Figure 3. The "inversion" pathway is so called because the carbon skeleton of glucose is inverted in the intermediates. This pathway has common features with the pathway of biosynthesis in animals (116, 124), which proceeds through D-glucuronate and D-gulono-1,4-lactone. Little evidence supports the operation of the inversion pathway in plants (80), apart from the presence of the mitochondrial enzyme L-galactono-1,4-lactone dehydrogenase (GLDH) in leaves and other tissues (15, 41, 123, 157). This enzyme, which has been purified from sweet potato mitochondria (158), readily converts L-galactono-1,4-lactone to ascorbate (75). While correlations exist between GLDH and ascorbate accumulation in *Cuscuta reflexa* and wounded potato tubers (157, 219), the steps leading to the formation of L-galactono-1,4-lactone (Figure 3) remain to be elucidated in plants.

In the "noninversion" pathway (Figure 3), ascorbate synthesis from glucose is proposed to occur via first D-glucosone then sorbosone (71, 117, 183). Evidence favoring the operation of this pathway comes mainly from isotope studies and is equivocal. Label from D-[^{14}C]glucosone but not from [^{14}C]-sorbosone was incorporated more efficiently into ascorbate than label from D-[^{14}C]glucose (183). Similarly, isotope dilution with unlabeled D-glucosone inhibited the conversion of D-[^{14}C]glucose to L-[^{14}C]ascorbate, but dilution with unlabeled sorbosone did not (183). Conklin et al (35) reported that neither D-glucosone nor L-sorbosone had positive effects on ascorbate accumulation in the *Arabidopsis vtc1* mutant. An enzyme catalyzing the conversion of L-sorbosone to ascorbate has been identified (117), but no enzymes have been found that catalyze the conversion of D-glucose to D-glucosone or D-glucosone to L-sorbosone.

The Importance of Dehydroascorbate Recycling

The ascorbate redox system consists of L-ascorbic acid, MDHA, and DHA (Figure 2). Although ascorbate oxidase has been assigned various biochemical functions, its influence in the redox balance of the ascorbate pool is unknown.

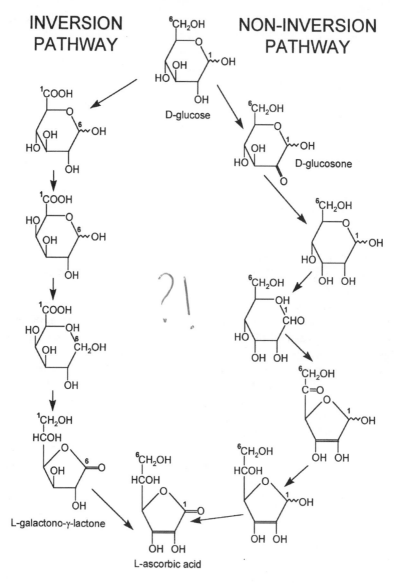

Figure 3 The putative pathways of ascorbate biosynthesis from glucose in higher plants.

Ascorbate oxidase activity may be particularly relevant at certain stages of development such as fruit ripening or senescence. Both oxidized forms of ascorbate are relatively unstable in aqueous environments. While DHA can be chemically reduced by GSH to ascorbate (57), enzymes that catalyze this reaction (DHAR) have been purified from rice, spinach, and potato (44, 58, 88, 97). In addition, DHAR activity has been attributed to glutaredoxins (thiol transferases), protein disulphide isomerases, and even a Kunitz-type trypsin inhibitor (220, 228). However, the amino acid sequence of the rice DHAR was quite distinct from these enzymes (97). Evidence to support the critical role of DHAR, GSH, and GR in maintaining the foliar ascorbate pool has been obtained in transformed plants overexpressing GR (64). These plants have higher foliar ascorbate contents and improved tolerance to oxidative stress (6, 61, 64). Conversely, depleted GR activity resulted in increased stress sensitivity (7).

If not rapidly rereduced to ascorbate, DHA is catabolized to two- and four-carbon products such as oxalate and tartrate, which can accumulate to relatively high levels (116, 226). While the reactions through which ascorbate is catabolized in plants remain obscure, degradation is considered to be an enzymic process involving 2,3-diketo-4-gulonic acid (Figure 1), as well as α-ketoaldehyde, and enediol intermediates (116). Cleavage at C_2/C_3 of the ascorbate molecule, which yields oxalic and L-threonic acids, appears to occur through both oxygenase and hydrolase reactions (184). In contrast, L-tartric acid results from hydrolytic cleavage at C4/C5 of 5-keto-D-gluconic acid (184).

Compartmentation and Transport

The site of ascorbate biosynthesis in plant cells is unknown (116). Since ascorbate is synthesized in both green and nongreen tissues, its formation is not directly dependent on photosynthesis. Ascorbate may be synthesized in the cytosol (116), and the location of GLDH suggests that mitochondria contribute to ascorbate biosynthesis. Although vacuolar ascorbate concentrations are relatively low (0.6 mM; 171), high values have been calculated for both chloroplastic and cytosolic compartments (20–50 mM; 60, 63, 67, 171).

Ascorbate transport into isolated intact chloroplasts across the chloroplast envelope occurs by a process of facilitated diffusion (5, 17, 60). In contrast, the thylakoid membranes have no ascorbate transport system (60). Regeneration of reduced ascorbate from its oxidized forms appears to occur in nearly all cell compartments, perhaps even in the apoplast, where glutathione is either absent or present at very low concentrations (HBJ Vanacker, TLW Carver & CH Foyer, submitted for publication). While apoplastic (and vacuolar) DHA may be transported into the cytosol for re-reduction, membrane-bound redox systems could also operate. For instance, the plasma membrane contains at least three different mechanisms capable of facilitating ascorbate-mediated transport of reducing equivalents between the cytosol and apoplast. These are (a) a highly specific b-type cytochrome transferring electrons from cytosolic ascorbate to extracellular acceptors, including MDHA; (b) a plasma membrane-localized MDHAR; and (c) ascorbate carriers selectively transporting ascorbate and DHA between the cytosol and apoplast (14, 85, 86).

GLUTATHIONE

The reduced form of glutathione, GSH, is a tripeptide (γ-glu-cys-gly) that exists interchangeably with the oxidized form, GSSG. Certain plants contain tripeptide homologs of GSH, in which the carboxy terminal gly is replaced by other amino acids. These are γ-glu-cys-β-ala (homoglutathione; 100), γ-glu-cys-ser (hydroxymethylglutathione; 101), and γ-glu-cys-glu (135). The oxidized forms of γ-glu-cys-β-ala and γ-glu-cys-ser can be reduced by yeast GR (100, 101), suggesting similar physiological and biochemical roles to the more widespread γ-glu-cys-gly.

In plants, the physiological significance of glutathione may be divided into two categories: sulfur metabolism and defense. GSH is the predominant non-protein thiol (173), and it regulates sulfur uptake at root level (82, 109). It is used by the GSH S-transferases in the detoxification of xenobiotics (108, 125) and is a precursor of the phytochelatins, which are crucial in controlling cellular heavy metal concentrations (70, 169, 186). In both plants and animals, GSH is important as an antioxidant and redox buffer (57, 106, 111, 134). In addition to its effects on expression of defense genes (48, 232), glutathione may also be involved in redox control of cell division (185, 196).

Biosynthesis of Glutathione

Unlike ascorbate synthesis, the pathway of glutathione synthesis has been elucidated and appears to be common to all organisms that contain GSH. Two ATP-dependent steps, catalyzed by γ-glutamylcysteine synthetase (γ-ECS) and glutathione synthetase (GS), lead to the sequential formation of γ-EC and GSH (Figure 4; 133). Both enzymes have been extracted from plant tissues and partially characterized (78, 79, 112, 120, 180, 190). Species that produce γ-glu-cys-β-ala have homoglutathione synthetase activity, whose affinity for β-ala is much higher than for gly (103, 120). Reports of the cloning of plant genes for γ-ECS (128) and GS (172, 221) have recently appeared.

Under some stress conditions, oxidation of GSH is accompanied by net glutathione degradation (64, 111, 235). Other studies have shown that glutathione accumulates in response to increased AOS generation (121, 127, 187, 192, 202, 203, 229) or is constitutively higher in plants adapted to exacting conditions (47, 51, 76, 119, 215). Differences in GSH contents may be wholly or partly

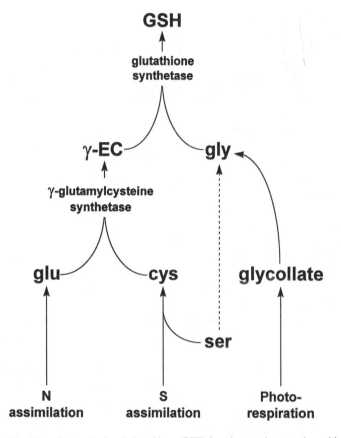

Figure 4   The biosynthesis of reduced glutathione (GSH) from its constituent amino acids.

due to modulated rates of GSH synthesis. The following discussion seeks to assess the in vivo significance of the different factors likely to control GSH biosynthesis.

Since glutathione is an important pool of reduced sulfur (173), integration of GSH synthesis with sulfate assimilation and cys formation might be expected. When cys contents were increased by supplying sulfate (42), by fumigation with H_2S (28–30), or by direct feeding of cys itself (52, 153, 155, 190, 210), higher GSH contents were observed. Since supplying glu has little effect on GSH contents (153, 155), cys-induced increases in GSH content presumably reflect a kinetic limitation over the rate of γ-EC synthesis. Most measurements of total tissue cys contents give values below or equal to the K_m value of γ-ECS (79, 190). It is unclear whether regulation of GSH synthesis by

cys concentration is of significance under conditions of enhanced AOS generation. Increased rates of glutathione accumulation provoked by H_2O_2 generation
were accompanied by changes in the rate of sulfate uptake into barley leaves
(204). Some evidence for concerted up-regulation of synthesis of cys and GSH
was obtained during fruit ripening (177). A more detailed study with Canola
roots indicated that interactions between GSH synthesis and sulfate assimilation are different from those between GSH synthesis and oxidative stress (110).
However, unlike studies in which glutathione was shown to accumulate (e.g.
202, 203, 229), exposure of roots to H_2O_2 caused glutathione levels to decrease
(110). Further work is required to establish whether sulfate assimilation and
cys synthesis are increased in response to oxidative stress, as they appear to be
upon exposure to Cd^+ (156).

Feedback inhibition of γ-ECS by GSH has often been considered a fundamental control over synthesis of GSH (2, 173, 202, 203). Inhibition of γ-ECS
by GSH was first described for the animal enzyme and shown to be competitive with respect to glutamate (176). Subsequently, in vitro studies with the
enzymes from tobacco and parsley cells showed that the plant γ-ECS was inhibited similarly (79, 190). It was proposed that alleviation of this inhibition was
responsible for increased rates of glutathione synthesis in response to increased
H_2O_2 levels (202, 203). Thus, under stress conditions, oxidation of GSH to
GSSG would decrease GSH levels and allow increased γ-ECS activity (203).
Some studies have shown initial decreases in GSH during glutathione accumulation in oxidizing conditions (192, 202). In other studies, such decreases were
less evident (121, 203). Moreover, the decreases were sometimes reversed well
before accumulation of total glutathione had ceased (e.g. 192).

Feedback inhibition may be important in controlling glutathione synthesis in
circumstances where the rate of glutathione synthesis is not closely related to
glutathione accumulation. One example is phytochelatin synthesis, in which
GSH is an intermediate rather than an end product (70). Several reports have
shown that phytochelatin accumulation, which entails increased flux through
the pathway of glutathione synthesis, is associated with decreased GSH contents (169, 170, 179, 186, 190). It remains debatable whether altered severity
of γ-ECS inhibition by GSH modulates glutathione levels under conditions of
oxidative stress.

A third mechanism through which tissue GSH contents might be modified
is altered de novo synthesis of γ-ECS and/or GS. Cadmium-resistant tomato
cells, containing increased levels of GSH, were shown to possess twofold higher
extractable γ-ECS activities than susceptible cells (32). Exposure of roots or
cultured cells to herbicide safeners or Cd provoked increases in the rate of
GSH synthesis and concomitant increases in γ-ECS and/or GS activities (of
50–200%; 52, 179, 181, 190).

An alternative approach to purely biochemical studies of the control of GSH synthesis was made possible by the cloning of genes for γ-ECS (227) and GS (73) from *Escherichia coli*. Introduction of these genes into yeast brought about constitutive increases in GSH (160). Transformation of poplar with the same genes showed that strong overexpression of GS in either the cytosol or the chloroplast had no effect on foliar GSH contents (64, 210; G Noctor, ACM Arisi, L Jouanin & C Foyerr, submitted for publication). Leaf discs from the transformants had a greater capacity for GSH synthesis from γ-EC but not from cys and glu, implying that GSH synthesis is restricted by formation of γ-EC (Figure 4; 210). This view was confirmed by the observation of pronounced constitutive increases in foliar contents of both γ-EC and GSH in poplars overexpressing γ-ECS in either cytosol (9, 155) or chloroplast (G Noctor, ACM Arisi, L Jouanin & C Foyer, submitted for publication), demonstrating that increases in the amount of the endogenous plant γ-ECS might be one way in which sustained increases in GSH synthesis could be achieved. Simultaneous up-regulation of GS could also be required for optimal GSH production, since this enzyme activity limits conversion of γ-EC to GSH when γ-ECS activity is augmented (151).

Similar increases in γ-EC and GSH have been obtained in tobacco plants overexpressing γ-ECS (38, 143). Unlike the poplars, however, the tobacco transformants suffer leaf necrosis and symptoms of severe oxidative stress (143). These effects, although of considerable interest, are difficult to evaluate. The absence of such effects in poplars overexpressing γ-ECS in the chloroplast, which have 100 times higher γ-ECS activity and 50 times more γ-EC than untransformed poplars (G Noctor, ACM Arisi, L Jouanin & C Foyer, submitted for publication), suggests that caution should be exercised in extrapolating observations in the tobacco transformants to glutathione metabolism in other plant species.

Increased contents of γ-EC and GSH engendered by overexpression of γ-ECS do not discount feedback control of GSH synthesis, since increases in enzyme amounts are expected to overcome competitive feedback inhibition (154). Thus, marked increases in the rate of γ-EC synthesis in poplars overexpressing γ-ECS do not imply that feedback control is insignificant in vivo. Indeed, the *E. coli* γ-ECS introduced into the poplars is sensitive to inhibition by GSH (208). Why would in vivo γ-ECS activities necessitate rigorous control by both low amounts of γ-ECS and sensitivity to GSH? One possibility is that the two controls allow temporal flexibility in the integration of cys synthesis with GSH synthesis. End-product inhibition may enable rapid homeostatic responses that avoid, among other things, depletion of the cys pool, while developmental or adaptive adjustments of GSH contents could be brought about by changes in the amount of γ-ECS.

Comparison of transgenic studies indicates that GSH synthesis is regulated in similar ways in animal and plant cells. Transfection of cDNAs encoding human γ-ECS into cancer cell lines increased cellular GSH contents by about twofold (107, 141). Moreover, high cellular GSH contents were shown to correlate with high extractable γ-ECS activities (140, 197). In addition, despite the apparent differences in the structure of the animal and plant γ-ECS (cf 79 and 89), both are inhibited by GSH (79, 176). It remains to be seen whether protein phosphorylation, recently reported for the animal enzyme (211), also acts to modify the kinetic properties of the plant enzyme.

The Role of Photorespiration

One of the differences between GSH synthesis in animals and plants is the light-dependent increase in glycine concentrations in plants of type C3 (e.g. 152, 233, 234). This is due to the process of photorespiration (for a review, see 113). Several studies have shown that foliar glutathione contents are higher in the light (19, 104, 152, 191). Attention was first brought to a possible role of photorespiration by Buwalda et al (29). When cys was increased by H_2S fumigation of spinach leaves, conversion of γ-EC to GSH was light-dependent (29). In the dark, high quantities of γ-EC accumulated (29). Analogous, though less pronounced, effects were observed in other species (28).

Poplars also exhibited light-dependent changes in γ-EC and GSH contents (152, 153; G Noctor, ACM Arisi, L Jouanin & C Foyer, submitted for publication). In darkened leaves of poplars overexpressing γ-ECS, γ-EC accumulated to very high levels (twice as high as the GSH contents of untransformed poplars), and GSH contents were halved. Similar effects were observed in illuminated leaves placed under nonphotorespiratory conditions (152). Under all conditions, an inverse correlation was observed between foliar gly contents and γ-EC levels (152, 153; G Noctor, ACM Arisi, L Jouanin & C Foyer, submitted for publication).

In both H_2S-fumigated spinach and poplars overexpressing γ-ECS, dark accumulation of γ-EC was prevented by supplying gly and other photorespiratory intermediates (30, 153; G Noctor, ACM Arisi, L Jouanin & C Foyer, submitted for publication). Importantly, the less-marked accumulation of γ-EC in darkened leaves of untransformed poplars was also prevented by exogenous glycine (153). It is clear, therefore, that the requirement of GSH synthesis for photorespiratory gly is not restricted to either H_2S-fumigated plants or plants overexpressing γ-ECS. Common to both cases, however, is an increased capacity for γ-EC synthesis. Thus, it may be inferred that photorespiration will become crucial in sustaining GSH production under any condition in which synthesis is up-regulated by increases in flux to γ-EC.

Glutathione synthesis is, to our knowledge, the first biosynthetic route unambiguously shown to utilize intermediates from the photorespiratory C_2 cycle. The possibility that photorespiration provides biosynthetic intermediates has often been ignored (e.g. 77, 162) or considered improbable (159). It is thought that most of the glycolate carbon must be recycled to 3-phosphoglycerate to support ongoing rates of CO_2 fixation (99, 159). However, even high rates of removal of glycine or serine would not represent a loss of reduced nitrogen (159). Leegood et al (113) considered that utilization of photorespiratory carbon in other pathways was "probably quite minor." No doubt the amounts of glycine used in GSH synthesis are, under most conditions, a small fraction of the huge quantities of carbon passing through the photorespiratory cycle. This may not be the case under all circumstances. One possible example is low temperature, where photorespiration is attenuated (99) and high GSH contents may be required (47, 51, 76, 119, 225).

Compartmentation, Transport, and Degradation

Glutathione can be synthesized in the cytosol and the chloroplast. Both γ-ECS and GS have been detected in chloroplastic and extrachloroplastic fractions (78, 79, 102, 103, 112, 180), and overexpression of γ-ECS in either compartment led to a threefold increase in foliar GSH contents (9; G Noctor, ACM Arisi, L Jouanin & C Foyer, submitted for publication). Sufficient substrates can therefore be made available in both locations. In spinach and barley leaves, chloroplastic and cytosolic concentrations of glu are similar (233, 234). Although less information is available concerning intracellular cys concentrations, the enzymes responsible for cys synthesis have been localized to the chloroplast, mitochondrion, and cytosol (118, 182, 214). Glycine is found in both cytosol and chloroplast, although cytosolic concentrations are probably higher (233, 234).

Estimations of chloroplastic GSH concentrations have yielded values of 1–4.5 mM (19, 57, 111, 173), but information is scarce concerning concentrations in other compartments. Accurate determination of organellar concentrations may be complicated by exchange of GSH between compartments during aqueous fractionation (102, 204), begging the question of the existence of intracellular GSH transporters. Glutathione transport at the plasmalemma has been investigated in tobacco cells (189) and broad bean protoplasts (92), but no studies of possible intracellular transporters have yet appeared. It remains unclear whether amino acid permeases such as those cloned recently (65) are likely to transport glutathione: The presence of glu, cys, or gly did not affect uptake of GSH and GSSG by bean protoplasts (92). Further investigation is required to establish whether glutathione associated with GR activity in the mitochondrion (49) and peroxisome (93) is produced in situ or must be imported. Apoplastic glutathione (HBJ Vanacker, TLW Carver & CH Foyer, submitted for publication) is probably supplied by export from the cytosol.

Glutathione is the major form in which reduced sulfur is exported from tobacco leaves (174). In poplars overexpressing γ-ECS, increases in foliar GSH contents were accompanied by markedly higher contents in the phloem (C Herschbach, L Jouanin, & H Rennenberg, submitted for publication) and roots (8). Since the introduced construct contained the non-tissue-specific 35S CaMV promoter (9, 155), some of the increase in root GSH was probably due to in situ synthesis. In the roots of untransformed plants, the relative rates of import and in situ synthesis are not clear. Both γ-ECS and GS activities have been detected in the roots of pea and maize (180, 181). Synthesis of GSH in the root probably varies according to developmental stage (170, 180) as well as external factors such as Cd (156, 179, 181) and xenobiotics (108). Little information is available concerning GSH synthesis in other tissues, although high concentrations of glutathione are found in seeds (100), and GS activity has been reported in fruits (177).

In maize, glutathione is more abundant in bundle sheath cells than in mesophyll cells (46). Interestingly, literature data indicate that sulfate assimilation in C_4 species is confined almost exclusively to the bundle sheath cells (26, 66, 188), although cys synthesis may occur in both types of cell (26).

Knowledge of one pathway of GSH degradation in plant tissues has come from work on tobacco cells (175, 205–207). This route is confined to the cytosol and, with the possible exception of the first step (206), proceeds via the same sequence as the γ-glutamyl cycle in animals (133). Further work is required to ascertain whether this pathway is of any significance in the degradation of GSH sometimes observed under oxidizing or stress conditions (64, 111, 235). Processing of GSH S-conjugates (108, 125) may also make a significant contribution to the degradation of GSH.

CONCLUSIONS

AOS have pleiotropic effects in plants as they do in animals. When they are produced in a controlled manner within specific compartments, AOS have key roles in plant metabolism and molecular biology. When they are produced in excess, the resultant uncontrolled oxidation leads to cellular damage and eventual death. To prevent damage, yet allow beneficial functions of AOS to continue, the antioxidant defenses must keep active oxygen under control. Technological advances in recent years have facilitated a deeper understanding of the key roles of enzymes such as SOD (ACM Arisi, G Cornic, L Jouanin, & CH Foyer, submitted for publication), APX (237), and catalase (229) in antioxidant defenses. While these individual enzymes have central roles in the antioxidant defense network, the exploration of the enzymes involved in the synthesis and metabolism of ascorbate and glutathione is, in our view, potentially of even greater interest and importance.

The past decade has witnessed rapid evolution of our knowledge of the biological roles of ascorbate and glutathione in plants. Molecular technologies hold the key to future development. Studies with transformed plants have helped to tease apart the network of interactions affecting GSH concentrations in plants and have also highlighted the multiplicity of regulatory controls. It cannot be assumed that the same regulatory factors operate under all conditions, but metabolic cross talk enables GSH biosynthesis to respond to different environmental and metabolic triggers. Further important information concerning the control of GSH levels is likely to come from under- or overexpression of the recently cloned plant genes for the enzymes of GSH biosynthesis as well as from the study of cadmium-sensitive mutants.

Molecular technologies are making an important contribution to the elucidation of biochemical problems such as the elucidation of the pathway of ascorbate biosynthesis, which, only a few years ago, seemed intractable. Many important issues that remain to be resolved include (*a*) the pathways of ascorbate biosynthesis and degradation in plants; (*b*) the influence of degradation over tissue levels of ascorbate and GSH; (*c*) the role of enzymes such as ascorbate oxidase; (*d*) the relationships between the reduction state of antioxidants and their pool sizes; (*e*) the importance of the Mehler-peroxidase and ascorbate-glutathione cycles in the regulation of photosynthesis; and (*f*) the role of these cycles in the adjustment of cellular redox potentials and in the regulation of gene expression. The exciting discovery of promoters that are sensitive to the reduced or oxidized forms of glutathione and ascorbate may herald the recognition of redox sensing as central to cellular chemistry and development. Such information, together with the provision of molecular tools, will allow exploitation of AOS and antioxidants to manipulate cellular redox status and identify traits that contribute to the vigor and sustainability of agricultural productivity for plant breeding programs.

Visit the *Annual Reviews home page* at
http://www.AnnualReviews.org.

Literature Cited

1. Allen RD. 1995. Dissection of oxidative stress tolerance using transgenic plants. *Plant Physiol.* 107:1049–54
2. Alscher RG. 1989. Biosynthesis and antioxidant function of glutathione in plants. *Physiol. Plant.* 77:457–64
3. Alvarez ME, Lamb C. 1997. Oxidative burst-mediated defense responses in plant disease resistance. See Ref. 185a, pp. 815–39
4. Anderson JV, Chevone BI, Hess JL. 1992. Seasonal variation in the antiox-

idant system of eastern white pine needles: evidence for thermal dependence. *Plant Physiol.* 98:501–8

5. Anderson JW, Foyer CH, Walker DA. 1983. Light-dependent reduction of dehydroascorbate and uptake of exogenous ascorbate by spinach chloroplasts. *Planta* 158:442–50

6. Aono M, Kubo A, Saji H, Tanaka K, Kondo N. 1993. Enhanced tolerance to photooxidative stress of transgenic *Nicotiana tabacum* with high chloroplastic glutathione reductase activity. *Plant Cell Physiol.* 34:129–35

7. Aono M, Saji H, Fujiyama K, Sugita M, Kondo N, Tanaka K. 1995. Decrease in activity of glutathione reductase enhances paraquat sensitivity in transgenic *Nicotiana tabacum. Plant Physiol.* 107:645–48

8. Arisi ACM. 1997. *Tolérance au stress de peupliers transformés surexprimant la superoxyde dismutase la Γ-glutamyl cystéine sythétase ou la glutathion synthétase.* PhD thesis. Univ. Paris XI, France

9. Arisi ACM, Noctor G, Foyer CH, Jouanin L. 1997. Modification of thiol contents in poplars (*Populus tremula x P. alba*) overexpressing enzymes involved in glutathione synthesis. *Planta.* 203:362–72

10. Arrigoni O. 1994. Ascorbate system in plant development. *J. Bioenerg. Biomembr.* 26:407–19

11. Asada K. 1992. Ascorbate peroxidase—a hydrogen peroxide-scavenging enzyme in plants. *Physiol. Plant.* 85:235–41

12. Asada K. 1997. The role of ascorbate peroxidase and monodehydroascorbate reductase in H_2O_2 scavenging in plants. See Ref. 185a, pp. 715–36

13. Asada K, Takahashi M. 1987. Production and scavenging of active oxygen in photosynthesis. In *Photoinhibition*, ed. DJ Kyle, CB Osmond, CJ Arntzen, pp. 227–87. Amsterdam: Elsevier

14. Asard H, Horemans N, Caubergs RJ. 1995. Involvement of ascorbic acid and a b-type cytochrome in plant plasma membrane redox reactions. *Protoplasma* 184:36–41

15. Baig MM, Kelly S, Loewus F. 1970. L-ascorbic acid biosynthesis in higher plants from L-gulono-1,4-lactone and L-galactono-1,4-lactone. *Plant Physiol.* 46:277–80

16. Baker CJ, Orlandi EW. 1995. Active oxygen in plant pathogenesis. *Annu. Rev. Phytopathol.* 33:299–321

17. Beck E, Burkert A, Hofmann M. 1983.

Uptake of L-ascorbate by intact chloroplasts. *Plant Physiol.* 73:41–45

18. Beeor-Tzahar T, Ben-Hayyim G, Holland D, Faltin Z, Eshdat Y. 1995. A stress-associated citrus protein is a distinct plant phospholipid hydroperoxide glutathione peroxidase. *FEBS Lett.* 366:151–55

19. Bielawski W, Joy KW. 1986. Reduced and oxidised glutathione and glutathione reductase activity in tissues of *Pisum sativum. Planta* 169:267–72

20. Bowler C, Slooten L, Vandenbranden S, De Rycke R, Botterman J, et al. 1991. Manganese superoxide dismutase can reduce cellular damage mediated by oxygen radicals in transgenic plants. *EMBO J.* 10:1723–32

21. Bowler C, Van Montagu M, Inzé D. 1992. Superoxide dismutase and stress tolerance. *Annu. Rev. Plant Physiol. Plant Mol. Biol.* 43:83–116

22. Bowles DJ. 1990. Defense-related proteins in higher plants. *Annu. Rev. Biochem.* 59:873–907

23. Bratt CE, Arvidsson P-O, Carlsson M, Akerlund H-E. 1995. Regulation of violaxanthin de-epoxidase activity by pH and ascorbate concentration. *Photosynth. Res.* 45:169–75

24. Broadbent P, Creissen GP, Kular B, Wellburn AR, Mullineaux P. 1995. Oxidative stress responses in transgenic tobacco containing altered levels of glutathione reductase activity. *Plant J.* 8:247–55

25. Buettner GR, Jurkiewicz BA. 1996. Chemistry and biochemistry of ascorbic acid. In *Handbook of Antioxidants*, ed. E Cadenas, L Packer, pp. 91–115. New York: Dekker

26. Burnell JN. 1984. Sulfate assimilation in C_4 plants. Intercellular and intracellular location of ATP sulfurylase, cysteine synthase and cystathionine b-lyase in maize leaves. *Plant Physiol.* 75:873–75

27. Burns JJ. 1957. Missing step in man, monkey and guinea pig required for the biosynthesis of L-ascorbic acid. *Nature* 180:553

28. Buwalda F, De Kok LJ, Stulen I. 1993. Effects of atmospheric H_2S on thiol composition of crop plants. *J. Plant Physiol.* 142:281–85

29. Buwalda F, De Kok LJ, Stulen I, Kuiper PJC. 1988. Cysteine, γ-glutamylcysteine and glutathione contents of spinach leaves as affected by darkness and application of excess sulfur. *Physiol. Plant.* 74:663–68

30. Buwalda F, Stulen I, De Kok LJ, Kuiper PJC. 1990. Cysteine, γ-glutamylcysteine and glutathione contents of spinach leaves as affected by darkness and application of excess sulfur. II. Glutathione accumulation in detached leaves exposed to H_2S in the absence of light is stimulated by the supply of glycine to the petiole. *Physiol. Plant.* 80:196–204

31. Chamnongpol S, Willekens H, Langebartels C, Van Montagu M, Inzé D, Van Camp W. 1996. Transgenic tobacco with a reduced catalase activity develops necrotic lesions and induces pathogenesis-related expression under high light. *Plant J.* 10:491–503

32. Chen JJ, Goldsbrough PB. 1994. Increased activity of γ-glutamylcysteine synthetase in tomato cells selected for cadmium tolerance. *Plant Physiol.* 106:233–39

33. Chen ZX, Silva H, Klessig DF. 1993. Active oxygen species in the induction of plant systemic acquired resistance by salicylic acid. *Science* 262:1883–86

34. Conklin PL, Last RL. 1995. Differential accumulation of antioxidant mRNAs in *Arabidopsis thaliana* exposed to ozone. *Plant Physiol.* 109:203–12

35. Conklin PL, Pallanca J, Last RL, Smirnoff N. 1997. L-ascorbic acid metabolism in the ascorbate deficient *Arabidopsis* mutant *vtc1*. *Plant Physiol.* 115:1277–85

36. Conklin PL, Williams EH, Last RL. 1996. Environmental stress sensitivity of an ascorbic acid–deficient *Arabidopsis* mutant. *Proc. Natl. Acad. Sci. USA* 93:9970–74

37. Cordoba-Pedregosa MC, Gonzalez-Reyes JA, Sanadillas MS, Navas P, Cordoba F. 1996. Role of apoplastic and cell-wall peroxidases on the stimulation of root elongation by ascorbate. *Plant Physiol.* 112:1119–25

38. Creissen G, Broadbent P, Stevens R, Wellburn AR, Mullineaux P. 1996. Manipulation of glutathione metabolism in transgenic plants. *Biochem. Soc. Trans.* 24:465–69

39. Criqui MC, Jamet E, Parmentier Y, Marbach J, Durr A, Fleck J. 1992. Isolation and characterization of a plant cDNA showing homology to animal glutathione peroxidases. *Plant Mol. Biol.* 18:623–27

40. Czaninski Y, Sachot RM, Catesson AM. 1993. Cytochemical localisation of hydrogen peroxide in lignifying cell walls. *Ann. Bot.* 72:547–50

41. De Gara L, Paciolla C, Tommasi F, Liso R, Arrigoni O. 1994. *In vivo* inhibition of galactono-γ-lactone conversion to ascorbate by lycorine. *J. Plant Physiol.* 144:649–53

42. De Kok LJ, Kuiper PJC. 1986. Effect of short-term dark incubation with sulfate, chloride and selenate on the glutathione content of spinach leaf discs. *Physiol. Plant.* 68:477–82

43. De Kok LJ, Oosterhuis FA. 1983. Effects of frost-hardening and salinity on glutathione and sulfhydryl levels and on glutathione reductase activity in spinach leaves. *Physiol. Plant.* 58:47–51

44. Dipierro S, Borranccino G. 1991. Dehydroascorbate reductase from potato tubers. *Phytochemistry* 30:427–29

45. Doke N. 1997. The oxidative burst: roles in signal transduction and plant stress. See Ref. 185a, pp. 785–813

46. Doulis AG, Debian N, Kingston-Smith AH, Foyer CH. 1997. Differential localization of antioxidants in maize leaves. *Plant Physiol.* 114:1031–37

47. Doulis AG, Hausladen A, Mondy B, Alscher RG, Chevone BI, et al. 1993. Antioxidant response and winter hardiness in red spruce (*Picea rubens* Sarg.). *New Phytol.* 123:365–74

48. Dron M, Clouse SD, Dixon RA, Lawton MA, Lamb CJ. 1988. Glutathione and fungal elicitor regulation of a plant defense promoter in electroporated protoplasts. *Proc. Natl. Acad. Sci. USA* 85:6738–42

49. Edwards EA, Rawsthorne S, Mullineaux PM. 1990. Subcellular distribution of multiple forms of glutathione reductase in leaves of pea (*Pisum sativum* L.). *Planta* 180:278–84

50. Edwards R, Blount JW, Dixon RA. 1991. Glutathione and elicitation of the phytoalexin response in legume cell cultures. *Planta* 184:403–9

51. Esterbauer H, Grill D. 1978. Seasonal variation of glutathione and glutathione reductase in needles of *Picea abies*. *Plant Physiol.* 61:119–21

52. Farago S, Brunold C. 1994. Regulation of thiol contents in maize roots by intermediates and effectors of glutathione synthesis. *J. Plant Physiol.* 144:433–37

53. Feierabend J, Dehne S. 1996. Fate of the porphyrin cofactors during the light-dependent turnover of catalase and of the photosystem II reaction centre protein D1 in mature rye leaves. *Planta* 198:413–22

54. Forti G, Elli G. 1995. The function of

ascorbic acid in photosynthetic phosphorylation. *Plant Physiol.* 109:1207–11

55. Foyer CH. 1997. Oxygen metabolism and electron transport in photosynthesis. See Ref. 185a, pp. 587–22

56. Foyer CH, Furbank RT, Harbinson J, Horton P. 1990. The mechanisms contributing to photosynthetic control of electron transport by carbon assimilation in leaves. *Photosynth. Res.* 25:83–100

57. Foyer CH, Halliwell B. 1976. The presence of glutathione and glutathione reductase in chloroplasts: a proposed role in ascorbic acid metabolism. *Planta* 133:21–25

58. Foyer CH, Halliwell B. 1977. Purification and properties of dehydroascorbate reductase from spinach leaves. *Phytochemistry* 16:1347–50

59. Foyer CH, Harbinson J. 1994. Oxygen metabolism and the regulation of photosynthetic electron transport. In *Causes of Photooxidative Stresses and Amelioration of Defense Systems in Plants*, ed. CH Foyer, P Mullineaux, pp. 1–42. Boca Raton, FL: CRC Press

60. Foyer CH, Lelandais M. 1996. A comparison of the relative rates of transport of ascorbate and glucose across the thylakoid, chloroplast and plasma membranes of pea leaf mesophyll cells. *J. Plant Physiol.* 148:391–98

61. Foyer CH, Lelandais M, Galap C, Kunert K-J. 1991. Effects of elevated cytosolic glutathione reductase activity on the cellular glutathione pool and photosynthesis in leaves under normal and stress conditions. *Plant Physiol.* 97:863–72

62. Foyer CH, Lopez-Delgado H, Dat JF, Scott IM. 1997. Hydrogen peroxide- and glutathione-associated mechanisms of acclimatory stress tolerance and signalling. *Physiol. Plant.* 100:241–54

63. Foyer CH, Rowell J, Walker D. 1983. Measurements of the ascorbate content of spinach leaf protoplasts and chloroplasts during illumination. *Planta* 157:239–44

64. Foyer CH, Souriau N, Perret S, Lelandais M, Kunert KJ, et al. 1995. Overexpression of glutathione reductase but not glutathione synthetase leads to increases in antioxidant capacity and resistance to photoinhibition in poplar trees. *Plant Physiol.* 109:1047–57

65. Frommer WB, Hummel S, Rentsch D. 1994. Cloning of an *Arabidopsis* transporting protein related to nitrate and peptide transporters. *FEBS Lett.* 347:185–89

66. Gerwick BC, Ku SB, Black CC. 1980. Initiation of sulfate activation: a variation in C_4 photosynthesis plants. *Science* 209:513–15

67. Gillham DJ, Dodge AD. 1986. Hydrogen peroxide scavenging systems within pea chloroplasts. *Planta* 167:246–51

68. Gonzalez-Reyes JA, Alcain FJ, Caler JA, Serrano A, Cordoba F, Navas P. 1994. Relationship between apoplastic ascorbate regeneration and the stimulation of root growth in *Allium apa* L. *Plant Sci.* 100:23–29

69. Grantz AA, Brummell DA, Bennett AB. 1995. Ascorbate free radical reductase mRNA levels are induced by wounding. *Plant Physiol.* 108:411–18

70. Grill E, Löffler S, Winnacker EL, Zenk MH. 1989. Phytochelatins, the heavy-metal binding peptides of plants, are synthesized from glutathione by a specific γ-glutamylcysteine dipeptidyl transpeptidase (phytochelatin synthase). *Proc. Natl. Acad. Sci. USA* 86:6838–42

71. Grun M, Renstrom B, Loewus FA. 1982. Loss of hydrogen from carbon 5 of D-glucose during conversion of D-[5-^{3}H,6-^{14}C]glucose to L-ascorbic acid in *Pelargonium crispum* (L.) L'Her. *Plant Physiol.* 70:1233

72. Guo Z-J, Nakagawara S, Sumitani K, Ohta Y. 1993. Effect of intracellular glutathione level on the production of 6-methoxymellein in cultured carrot (*Daucus carota*) cells. *Plant Physiol.* 102:45–51

73. Gushima H, Yasuda S, Soeda E, Yokota M, Kondo M, Kimura A. 1984. Complete nucleotide sequence of the *E.coli* glutathione synthetase *gsh-II*. *Nucleic Acids Res.* 12:9299–307

74. Halliwell B, Aeschbach R, Loliger I, Aruoma OI. 1995. The characterization of antioxidants. *Food Chem. Toxicol.* 333:601–17

75. Hausladen A, Kunert KJ. 1990. Effects of artificially enhanced levels of ascorbate and glutathione on the enzymes monodehydroascorbate reductase, dehydroascorbate reductase, and glutathione reductase in spinach (*Spinacia oleracea*). *Physiol. Plant.* 79:384–88

76. Hausladen A, Madamanchi NR, Fellows S, Alscher RG, Amundson RG. 1990. Seasonal changes in antioxidants in red spruce as affected by ozone. *New Phytol.* 115:447–58

77. Heber U, Krause H. 1980. What is the physiological role of photorespiration? *Trends Biochem. Sci.* 9:32–34

78. Hell R, Bergmann L. 1988. Glutathione synthetase in tobacco suspension cultures: catalytic properties and localization. *Physiol. Plant.* 72:70–76

79. Hell R, Bergmann L. 1990. γ-Glutamylcysteine synthetase in higher plants: catalytic properties and subcellular localization. *Planta* 180:603–12

80. Helsper JP, Kagan L, Hilby CL, Maynard TM, Loewus FA. 1982. L-ascorbic acid synthesis in *Ochromonas danica*. *Plant Physiol.* 69:465–68

81. Deleted in proof

82. Herschbach C, Rennenberg H. 1994. Influence of glutathione (GSH) on net uptake of sulphate and sulphate transport in tobacco plants. *J. Exp. Bot.* 45:1069–76

83. Hertwig B, Streb P, Feirerabend J. 1992. Light dependence of catalase synthesis and degradation in leaves and the influence of interfering stress conditions. *Plant Physiol.* 100:1547–53

84. Holland D, Ben-Hayyim G, Faltin Z, Camoin L, Strosberg AD, Eshdat Y. 1993. Molecular characterization of a salt-stress associated protein in citrus: protein and cDNA sequence homology to mammalian glutathione peroxidases. *Plant Mol. Biol.* 21:923–27

85. Horemans N. 1997. *Ascorbate-mediated functions at the plasma membrane of higher plants*. PhD thesis. Univ. Antwerpen, Belg.

86. Horemans N, Asard H, Caubergs RJ. 1994. The role of ascorbate free radical as an electron acceptor to cytochrome b-mediated trans-plasma membrane electron transport in higher plants. *Plant Physiol.* 104:1455–58

87. Horton P, Ruban AV, Walters RG. 1996. Regulation of light harvesting in green plants. *Annu. Rev. Plant. Physiol. Plant Mol. Biol.* 47:655–84

88. Hossain MA, Asada K. 1984. Purification of dehydroascorbate reductase from spinach and its characterisation as a thiol enzyme. *Plant Cell Physiol.* 25:85–92

89. Huang CS, Chang LS, Anderson ME, Meister A. 1993. Catalytic and regulatory properties of the heavy subunit of rat kidney γ-glutamylcysteine synthetase. *J. Biol. Chem.* 268:19675–80

90. Jabs T, Dietrich RA, Dangl JL. 1996. Initiation of runaway cell death in an *Arabidopsis* mutant by extracellular superoxide. *Science* 273:1853–55

91. Jakob B, Heber U. 1996. Photoproduction and detoxification of hydroxyl radicals in chloroplasts and leaves in relation to photoinactivation of photosystems I and II. *Plant Cell Physiol.* 37:629–35

92. Jamaï A, Tommasini R, Martinoia E, Delrot S. 1996. Characterization of glutathione uptake in broad bean leaf protoplasts. *Plant Physiol.* 111:1145–52

93. Jiménez A, Hernandez JA, del Rio LA, Sevilla F. 1997. Evidence for the presence of the ascorbate-glutathione cycle in mitochondria and peroxisomes of pea leaves. *Plant Physiol.* 114:275–84

94. Johnson RR, Cranston HJ, Chaverra ME, Dyer WE. 1995. Characterization of cDNA clones for differentially expressed genes in embryos of dormant and nondormant *Avena fatua* L. cayopses. *Plant Mol. Biol.* 28:113–22

95. Kaiser WM. 1979. Reversible inhibition of the Calvin cycle and activation of oxidative pentose phosphate cycle in isolated intact chloroplasts by hydrogen peroxide. *Planta* 145:377–82

96. Kaprinski S, Escobar C, Karprinska B, Creissen G, Mullineaux PM. 1997. Photosynthetic electron transport regulates the expression of cytosolic ascorbate peroxidase genes in *Arabidopsis* during excess light stress. *Plant Cell* 9:627–40

97. Kato Y, Urano J, Maki Y, Ushimaru T. 1997. Purification and characterisation of dehydroascorbate reductase from rice. *Plant Cell Physiol.* 38:173–78

98. Kerk NM, Feldman LJ. 1995. A biochemical model for initiation and maintenance of the quiescent center: implications for organisation of root meristems. *Plant Development* 121:2825–33

99. Keys AJ. 1980. Prospects for increasing photosynthesis by control of photorespiration. *Pestic. Sci.* 19:313–16

100. Klapheck S. 1988. Homoglutathione: isolation, quantification and occurrence in legumes. *Physiol. Plant.* 74:727–32

101. Klapheck S, Chrost B, Starke J, Zimmerman H. 1992. γ-Glutamylcysteinylserine—a new homologue of glutathione in plants of the family *Poaceae*. *Botanica Acta* 105:174–79

102. Klapheck S, Latus C, Bergmann L. 1987. Localization of glutathione synthetase and distribution of glutathione in leaf cells of *Pisum sativum* L. *J. Plant Physiol.* 131:123–31

103. Klapheck S, Zopes H, Levels HG, Bergmann L. 1988. Properties and localization of the homoglutathione synthetase from *Phaseolus coccineus* leaves. *Physiol. Plant.* 74:733–39

104. Koike S, Patterson BD. 1988. Diurnal variation of glutathione levels in tomato seedlings. *Hort. Sci.* 23:713–14

105. Kubo A, Saji H, Tanaka K, Kondo N. 1995. Expression of *Arabidopsis* cytoso-

lic ascorbic peroxidase gene in response to ozone or sulfur dioxide. *Plant Mol. Biol.* 29:479–86

106. Kunert KJ, Foyer C. 1993. Thiol/ disulfide exchange in plants. In *Sulfur Nutrition and Sulfur Assimilation in Higher Plants*, ed. LJ de Kok, I Stulen, H Rennenberg, C Brunold, WE Rauser, pp. 139–51. The Hague: SPB Acad.

107. Kurokawa H, Ishida T, Nishio K, Arioka H, Sata M, et al. 1995. γ-Glutamylcysteine synthetase gene overexpression results in increased activity of the ATP-dependent glutathione S-conjugate export pump and cisplatin resistance. *Biochem. Biophys. Res. Commun.* 216:258–64

108. Lamoureux GL, Rusness DG. 1993. Glutathione in the metabolism and detoxification of xenobiotics in plants. In *Sulfur Nutrition and Assimilation in Higher Plants*, ed. LJ de Kok, I Stulen, H Rennenberg, C Brunold, WE Rauser, pp. 221–37. The Hague: SPB Acad.

109. Lappartient AG, Touraine B. 1996. Demand-driven control of root ATP sulfurylase activity and sulphate uptake in intact Canola. *Plant Physiol.* 111:147–57

110. Lappartient AG, Touraine B. 1997. Glutathione-mediated regulation of ATP sulfurylase activity, sulfate uptake, and oxidative stress response in intact Canola roots. *Plant Physiol.* 114:177–83

111. Law MY, Charles SA, Halliwell B. 1983. Glutathione and ascorbic acid in spinach (*Spinacea oleracea*) chloroplasts. *Biochem. J.* 210:899–903

112. Law MY, Halliwell B. 1986. Purification and properties of glutathione synthetase from spinach (*Spinacia oleracea*) leaves. *Plant Sci.* 4:185–91

113. Leegood RC, Lea PJ, Adcock MD, Häusler RE. 1995. The regulation and control of photorespiration. *J. Exp. Bot.* 46:1397–414

114. Levine A, Tenhaken R, Dixon R, Lamb C. 1994. H_2O_2 from the oxidative burst orchestrates the plant hypersensitive disease resistance response. *Cell* 79:583–93

115. Liso R, De Gara L, Tommasi F, Arrigoni O. 1985. Ascorbic acid requirement for increased peroxidase activity during potato tuber slice aging. *FEBS Lett.* 187:141–45

116. Loewus FA. 1988. Ascorbic acid and its metabolic products. In *The Biochemistry of Plants*, ed. J Priess, 14:85–107. New York: Academic

117. Loewus MW, Bedgar DL, Saito K,

Loewus FA. 1990. Conversion of L-sorbosone to L-ascorbic acid by a NADP-dependent dehydrogenase in bean and spinach leaf. *Plant Physiol.* 94:1492–95

118. Lunn JE, Droux M, Martin J, Douce R. 1990. Localization of ATP sulphurylase and *O*-acetylserine(thiol)lyase in spinach leaves. *Plant Physiol.* 94:1345–52

119. Luwe M. 1996. Antioxidants in the apoplast and symplast of beech (*Fagus sylvatica* L.) leaves: seasonal variations and responses to changing ozone concentrations in air. *Plant Cell Environ.* 19:321–28

120. Macnicol PK. 1987. Homoglutathione and glutathione synthetases of legume seedlings: partial purification and substrate specificity. *Plant Sci.* 53:229–35

121. Madamanchi NR, Yu X, Doulis A, Alscher RG, Hatzios KK, Cramer CL. 1994. Acquired resistance to herbicides in pea cultivars through pretreatment with sulfur dioxide. *Pestic. Biochem. Physiol.* 48:31–40

122. Malan C, Greyling MM, Gressel J. 1990. Correlation between CuZn superoxide dismutase and glutathione reductase, and environmental and xenobiotic stress tolerance in maize inbreds. *Plant Sci.* 69:157–66

123. Mapson LW, Breslow E. 1958. Biological synthesis of ascorbic acid: L-galactono-γ-lactone dehydrogenase. *Biochem. J.* 68:359–406

124. Mapson LW, Isherwood FA. 1956. Biological synthesis of ascorbic acid: the conversion of derivatives of D-galacturonic acid into L-ascorbic acid by plant extracts. *Biochem. J.* 64:151–57

125. Marrs K. 1996. The functions and regulation of glutathione S-transferases in plants. *Annu. Rev. Plant Physiol. Plant Mol. Biol.* 47:127–58

126. May MJ, Hammond-Kosack KE, Jones DG. 1996. Involvement of reactive oxygen species, glutathione metabolism and lipid peroxidation in the *Cf*-gene-dependent defense response of tomato cotyledons induced by race-specific elicitors of *Cladosporium fulvum*. *Plant Physiol.* 110:1367–79

127. May MJ, Leaver CJ. 1993. Oxidative stimulation of glutathione synthesis in *Arabidopsis thaliana* suspension cultures. *Plant Physiol.* 103:621–27

128. May MJ, Leaver CJ. 1994. *Arabidopsis thaliana* γglutamylcysteine synthetase is structurally unrelated to mammalian, yeast, and *Escherichia coli* homologs.

Proc. Natl. Acad. Sci. USA 91:10059–63

129. McKersie BD, Bowley SR, Harjanto E, Lepreince O. 1996. Water-deficit tolerance and field performance of transgenic alfalfa overexpressing superoxide dismutase. Plant Physiol. 111:1171–81

130. McKersie BD, Chen YR, De Beus M, Bowley SR, Bowler C, et al. 1993. Superoxide dismutase enhances tolerance of freezing stress in transgenic alfalfa (Medicago sativa L.). Plant Physiol. 103:1155–63

131. Mehdy MC. 1994. Active oxygen species in plant defense against pathogens. Plant Physiol. 105:467–42

132. Mehlhorn H, Lelandais M, Korth HG, Foyer CH. 1996. Ascorbate is the natural substrate for plant peroxidases. FEBS Lett. 378:203–6

133. Meister A. 1988. Glutathione metabolism and its selective modification. J. Biol. Chem. 263:17205–8

134. Meister A. 1994. Glutathione-ascorbic acid antioxidant system in animals. J. Biol. Chem. 269:9397–400

135. Meuwly P, Thibault P, Rauser WE. 1993. γ-Glutamylcysteinylglutamic acid—a new homologue of glutathione in maize seedlings exposed to cadmium. FEBS Lett. 336:472–76

136. Mittler R, Lam E. 1996. Sacrifice in the face of foes: pathogen-induced programmed cell death in higher plants. Trends Microbiol. 4:10–15

137. Mittler R, Zilinskas BA. 1992. Molecular cloning and characterization of a gene encoding pea cytosolic ascorbate peroxidase. J. Biol. Chem. 267:21802–7

138. Mittler R, Zilinskas BA. 1994. Regulation of pea cytosolic ascorbate peroxidase and other antioxidant enzymes during the progression of drought stress and following recovery from drought. Plant J. 5:397–405

139. Miyake C, Asada K. 1992. Thylakoid-bound ascorbate peroxidase in spinach chloroplasts and photoregeneration of its primary oxidation product monodehydroascorbate radicals in thylakoids. Plant Cell Physiol. 33:541–53

140. Mulcahy RT, Bailey HH, Gipp JJ. 1994. Up-regulation of γ-glutamylcysteine synthetase activity in melphalan-resistant human multiple myeloma cells expressing increased glutathione levels. Cancer Chemother. Pharmacol. 34:67–71

141. Mulcahy RT, Bailey HH, Gipp JJ. 1995. Transfection of complementary DNAs for the heavy and light subunits of human γ-glutamylcysteine synthetase results in an elevation of intracellular glutathione and resistance to melphalan. Cancer Res. 55:4771–75

142. Mullineaux PM, Creissen G, Broadbent P, Reynolds H, Kular B, Wellburn A. 1994. Elucidation of the role of glutathione reductase using transgenic plants. Biochem. Soc. Trans. 22:931–36

143. Mullineaux PM, Wellburn AR, Baker NR, Creissen GP. 1997. Increased capacity for glutathione biosynthesis in the chloroplast paradoxically promotes oxidative stress in transgenic tobacco. In Sulphur Metabolism in Higher Plants. Molecular, Ecophysiological and Nutritional Aspects, ed. WJ Cram, LJ De Kok, I Stulen, C Brunold, H Rennenberg, pp. 269–70. Leiden: Backhuys

144. Nakagawara S, Nakamura N, Guo Z-J, Sumitani K, Katoh K, Ohta Y. 1993. Enhanced formation of a constitutive sesquiterpenoid in culture cells of liverwort, Calypogeia granulata Inoue during elicitation; effects of vanadate. Plant Cell Physiol. 34:421–29

145. Nakano Y, Asada K. 1987. Purification of ascorbate peroxidase in spinach chloroplasts: its inactivation in ascorbate-depleted medium and reactivation by monodehydroascorbate radical. Plant Cell Physiol. 28:131–40

146. Navas P, Gomez-Diaz C. 1995. Ascorbate free radical and its role in growth control. Protoplasma 184:8–13

147. Nemoto S, Otsuka M, Arakawa N. 1996. Inhibitory effect of ascorbate on cell growth: relation to catalase activity. J. Nutr. Sci. Vitaminol. 43:77–85

148. Nijs D, Kelley PM. 1991. Vitamins C and E donate single hydrogen atoms in vivo. FEBS Lett. 284:147–51

149. Nishikimi M, Yagi K. 1996. Biochemistry and molecular biology of ascorbic acid biosynthesis. In Subcellular Biochemistry Ascorbic Acid: Biochemistry and Biomedical Cell Biology, ed. J Harris, 25:17–39. New York: Plenum

150. Deleted in proof

151. Noctor G, Arisi ACM, Jouanin L, Kunert KJF, Rennenberg H, Foyer CH. 1997. Glutathione biosynthesis and metabolism explored in transformed poplars. J. Exp. Bot. In press

152. Noctor G, Arisi ACM, Jouanin L, Valadier MH, Roux Y, Foyer CH. 1997. Light-dependent modulation of foliar glutathione synthesis and associated amino acid metabolism in transformed poplar. Planta 202:357–69

153. Noctor G, Arisi ACM, Jouanin L, Valadier MH, Roux Y, Foyer CH. 1997. The role of glycine in determining the rate of glutathione synthesis in poplars. Possible implications for glutathione production during stress. *Physiol. Plant.* 100:255–63

154. Noctor G, Jouanin L, Foyer CH. 1997. The biosynthesis of glutathione explored in transgenic plants. In *Regulation of Enzymatic Systems Detoxifying Xenobiotics in Plants*, ed. K Hatzios, pp. 109–24. Dordrecht: Kluwer Academic

155. Noctor G, Strohm M, Jouanin L, Kunert KJ, Foyer CH, Rennenberg H. 1996. Synthesis of glutathione in leaves of transgenic poplar (*Populus tremula x P. alba*) overexpressing γ-glutamylcysteine synthetase. *Plant Physiol.* 112: 1071–78

156. Nussbaum S, Schmutz D, Brunold C. 1988. Regulation of assimilatory sulfate reduction by cadmium in *Zea mays* L. *Plant Physiol.* 88:1407–10

157. Oba K, Fukui M, Imai Y, Iriyama S, Nogami K. 1994. L-galactono-γ-lactone dehydrogenase: partial characterization, induction of activity and role in the synthesis of ascorbic acid in wounded white potato tuber tissue. *Plant Cell Physiol.* 35:473–78

158. Oba K, Ishikawa S, Nichikawa M, Mizuno H, Yamamoto T. 1995. Purification and properties of L-galactono-γ-lactone dehydrogenase, a key enzyme for ascorbic acid biosynthesis, from sweet potato roots. *J. Biochem.* 117:120–24

159. Ogren WL. 1984. Photorespiration: pathways, regulation and modification. *Annu. Rev. Plant Physiol.* 35:415–42

160. Ohtake Y, Kunihiko W, Tezuka H, Ogata T, Yabuuchi S, et al. 1989. Expression of the glutathione synthetase gene of *Escherichia coli* B in *Saccharomyces cervisiae*. *J. Ferment. Bioeng.* 68:390–94

161. Olson PD, Varner JE. 1993. Hydrogen peroxide and lignification. *Plant J.* 4:887–92

162. Osmond CB, Grace SC. 1995. Perspectives on photoinhibition and photorespiration in the field: quintessential inefficiencies of the light and dark reactions of photosynthesis? *J. Exp. Bot.* 46:1351–62

163. Padh H. 1990. Cellular functions of ascorbic acid. *Biochem. Cell Biol.* 68: 1166–73

164. Pitcher LH, Brennan E, Hurley A, Dunsmuir P, Tepperman JM, Zilinskas BA.

165. Polle A, Wieser G, Havranek WM. 1995. Quantification of ozone influx and apoplastic ascorbate content in needles of Norway spruce trees (*Picea abies* L. Karst.) at high altitudes. *Plant Cell Environ.* 18:681–88

166. Prasad TK. 1996. Mechanisms of chilling-induced oxidative stress injury and tolerance in developing maize seedlings: changes in antioxidant system, oxidation of proteins and lipids, and protease activities. *Plant J.* 10:1017–26

167. Prasad TK, Anderson MD, Martin BA, Stewart CR. 1994. Evidence for chilling-induced oxidative stress in maize seedlings and regulatory role of hydrogen peroxide. *Plant Cell* 6:65–74

168. Ranieri A, D'Urso G, Nali C, Lorenzini G, Soldatini GF. 1996. Ozone stimulates apoplastic antioxidant systems in pumpkin leaves. *Physiol. Plant.* 97:381–87

169. Rauser WE. 1987. Changes in glutathione content of maize seedlings exposed to cadmium. *Plant Sci.* 51:171–75

170. Rauser WE, Schupp R, Rennenberg H. 1991. Cysteine, γ-glutamylcysteine and glutathione levels in maize seedlings. *Plant Physiol.* 97:128–34

171. Rautenkranz AAF, Li J, Machler F, Martinoia E, Oertli JJ. 1994. Transport of ascorbic and dehydroascorbic acids across protoplast and vacuole membranes isolated from barley (*Hordeum vulgare* L. cv. Gerbel) leaves. *Plant Physiol.* 106:187–93

172. Rawlins MR, Leaver CJ, May MJ. 1995. Characterisation of an *Arabidopsis thaliana* cDNA encoding glutathione synthetase. *FEBS Lett.* 376:81–86

173. Rennenberg H. 1982. Glutathione metabolism and possible biological roles in higher plants. *Phytochemistry* 21: 2771–81

174. Rennenberg H, Schmitz K, Bergmann L. 1979. Long-distance transport of sulfur in *Nicotiana tabacum*. *Planta* 147:57–62

175. Rennenberg H, Steinkamp R, Kesselmeier J. 1981. 5-oxo-prolinase in *Nicotiana tabacum*: catalytic properties and subcellular localization. *Physiol. Plant.* 62:211–16

176. Richman PG, Meister A. 1975. Regulation of γ-glutamylcysteine synthetase by nonallosteric feedback inhibition by glutathione. *J. Biol. Chem.* 250:1422–26

177. Römer S, Harlingue A, Camara B,

1991. Overproduction of petunia copper/zinc superoxide dismutase does not confer ozone tolerance in transgenic tobacco. *Plant Physiol.* 97:452–55

Schantz R, Kuntz M. 1992. Cysteine synthase from *Capsicum annuum* chromoplasts. Characterization and cDNA cloning of an up-regulated enzyme during fruit development. *J. Biol. Chem.* 267:17966–70

178. Rose RC, Bode AM. 1993. Biology of free radical scavengers: an evaluation of ascorbate. *FASEB J.* 7:1135–42

179. Rüegsegger A, Brunold C. 1992. Effect of cadmium on γ-glutamylcysteine synthesis in maize seedlings. *Plant Physiol.* 99:428–33

180. Rüegsegger A, Brunold C. 1993. Localization of γ-glutamylcysteine synthetase and glutathione synthetase activity in maize seedlings. *Plant Physiol.* 101:561–66

181. Rüegsegger A, Schmutz D, Brunold C. 1990. Regulation of glutathione synthesis by cadmium in *Pisum sativum* L. *Plant Physiol.* 93:1579–84

182. Ruffet ML, Lebrun M, Droux M, Douce R. 1995. Subcellullar distribution of serine acetyltransferase from *Pisum sativum* and characterization of an *Arabidopsis thaliana* putative cytosolic isoform. *Eur. J. Biochem.* 227:500–9

183. Saito K, Nick JA, Loewus FA. 1990. D-glucosone and L-sorbosone, putative intermediates of L-ascorbate biosynthesis in detached bean and spinach leaves. *Plant Physiol.* 94:1496–500

✳ 184. Saito K, Ohmoto J, Kuriha N. 1997. Incorporation of ^{18}O into oxalic, L-threonic and L-tartaric acids during cleavage of L-ascorbic and 5-keto-D-gluconic acids in plants. *Phytochemistry* 44:805–9

185. Sanchez-Fernandez R, Fricker M, Corben LB, White NS, Sheard N, et al. 1997. Cell proliferation and hair tip growth in the *Arabidopsis* root are under mechanistically different forms of redox control. *Proc. Natl. Acad. Sci. USA* 94:2745–50

185a. Scandalios J, ed. 1997. *Oxidative Stress and the Molecular Biology of Antioxidant Defences*. Cold Spring Harbor, NY: Cold Spring Harbor Lab. Press

186. Scheller HV, Huang B, Hatch E, Goldsbrough PB. 1987. Phytochelatin synthesis and glutathione levels in response to heavy metals in tomato cells. *Plant Physiol.* 85:1031–35

187. Schmidt A, Kunert KJ. 1986. Lipid peroxidation in higher plants: the role of glutathione reductase. *Plant Physiol.* 82:700–2

188. Schmutz D, Brunold C. 1984. Intercel-lular localization of assimilatory sulfate reduction in leaves of *Zea mays* and *Triticum aestivum*. *Plant Physiol.* 74:866–70

189. Schneider A, Schatten T, Rennenberg H. 1992. Reduced glutathione (GSH) transport in cultured tobacco cells. *Plant Physiol. Biochem.* 30:29–38

190. Schneider S, Bergmann L. 1995. Regulation of glutathione synthesis in suspension cultures of parsley and tobacco. *Bot. Acta* 108:34–40

191. Schupp R, Rennenberg H. 1990. Diurnal changes in the thiol composition of spruce needles. In *Sulfur Nutrition and Sulfur Assimilation in Higher Plants*, ed. H Rennenberg, CH Brunold, LJ de Kok, I Stulen, pp. 89–96. The Hague: SPB Acad.

192. Sen Gupta A, Alscher RG, McCune D. 1991. Response of photosynthesis and cellular antioxidants to ozone in *Populus* leaves. *Plant Physiol.* 96:650–55

193. Sen Gupta A, Heinen JL, Holladay AS, Burke JJ, Allen RD. 1993. Increased resistance to oxidative stress in transgenic plants that overexpress chloroplastic Cu/Zn superoxide dismutase. *Proc. Natl. Acad. Sci. USA* 90:1629–33

194. Sen Gupta A, Webb RP, Holladay AS, Allen RD. 1993. Overexpression of superoxide dismutase protects plants from oxidative stress. Induction of ascorbate peroxidase in superoxide dismutase-overexpressing plants. *Plant Physiol.* 103:1067–73

195. Sharma YK, Davis KR. 1994. Ozone-induced expression of stress-related genes in *Arabidopsis thaliana*. *Plant Physiol.* 105:1089–96

196. Shaul O, Mironov V, Burssens S, Van Montagu M, Inzé D. 1996. Two *Arabidopsis* cyclin promoters mediate distinctive transcriptional oscillation on synchronised tobacco 3Y-2 cells. *Proc. Natl. Acad. Sci. USA* 93:4868–72

197. Shi MM, Kugelman A, Iwamoto T, Tian L, Forman HJ. 1994. Quinone-induced oxidative stress elevates glutathione and induces γ-glutamylcysteine synthetase activity in rat lung epithelial L2 cells. *J. Biol. Chem.* 42:26512–17

198. Slooten L, Capiau K, Van Camp W, Van Montagu M, Sybesma C, Inzé D. 1995. Factors affecting the enhancement of oxidative stress tolerance in transgenic tobacco overexpressing manganese superoxide dismutase in the chloroplasts. *Plant Physiol.* 107:737–50

199. Smirnoff N. 1996. The function and

metabolism of ascorbic acid in plants. *Ann. Bot.* 78:661–99

200. Smirnoff N, Colombe SV. 1988. Drought influences the activity of enzymes of the hydrogen peroxide scavenging system. *J. Exp. Bot.* 39:1097–108

201. Smirnoff N, Cumbes Q. 1989. Hydroxyl radical scavenging activity of compatible solutes. *Phytochemistry* 28:1957–60

202. Smith IK. 1985. Stimulation of glutathione synthesis in photorespiring plants by catalase inhibitors. *Plant Physiol.* 79:1044–47

203. Smith IK, Kendall AC, Keys AJ, Turner JC, Lea PJ. 1984. Increased levels of glutathione in a catalase-deficient mutant of barley (*Hordeum vulgare* L.). *Plant Sci. Lett.* 37:29–33

204. Smith IK, Kendall AC, Keys AJ, Turner JC, Lea PJ. 1985. The regulation of the biosynthesis of glutathione in leaves of barley (*Hordeum vulgare* L.). *Plant Sci.* 41:11–17

205. Steinkamp R, Rennenberg H. 1984. γ-Glutamyltranspeptidase in tobacco suspension cultures: catalytic properties and subcellular localization. *Physiol. Plant.* 61:251–56

206. Steinkamp R, Rennenberg H. 1985. Degradation of glutathione in plant cells: evidence against the participation of a γ-glutamyltranspeptidase. *Z. Naturforsch. Teil C* 40:29–33

207. Steinkamp R, Schweihofen B, Rennenberg H. 1987. γ-Glutamylcyclotransferase in tobacco suspension cultures: catalytic properties and subcellular localization. *Physiol. Plant.* 69:499–503

208. Strohm M. 1996. *Biochemische, physiologische und molekulare Grundlagen des Glutathion-Stoffwechsels in Pappeln (Populus tremula x P. alba).* PhD thesis. Univ. Freiburg, Ger.

209. Deleted in proof

210. Strohm M, Jouanin L, Kunert KJ, Pruvost C, Polle A, et al. 1995. Regulation of glutathione synthesis in leaves of transgenic poplar (*Populus tremula x P. alba*) overexpressing glutathione synthetase. *Plant J.* 7:141–45

211. Sun WM, Huang ZZ, Lu SC. 1996. Regulation of γ-glutamylcysteine synthetase by protein phosphorylation. *Biochem. J.* 320:321–28

212. Takahama U. 1993. Redox state of ascorbic acid in the apoplast of stems of *Kalanchoe daigremontiana. Physiol. Plant.* 89:791–98

213. Takahama U, Oniki T. 1994. Effects of ascorbate on the oxidation of derivatives of hydroxycinnamic acid and the mechanism of oxidation of sinapic acid by cell wall-bound peroxidases. *Plant Cell Physiol.* 35:593–600

214. Takahashi H, Saito K. 1996. Subcellular localization of spinach cysteine synthase isoforms and regulation of their gene expression by nitrogen and sulfur. *Plant Physiol.* 112:273–80

215. Tanaka K, Suda Y, Kondo N, Sugahara K. 1985. O_3 tolerance and the ascorbate-dependent H_2O_2 decomposing system in chloroplasts. *Plant Cell Physiol.* 26:1425–31

216. Teppermann JM, Dunsmuir P. 1990. Transformed plants with elevated levels of chloroplastic SOD are not more resistant to superoxide toxicity. *Plant Mol. Biol.* 14:501–11

217. Thomsen B, Drumm-Herrel H, Mohr H. 1992. Control of the appearance of ascorbate peroxidase (EC 1.11.1.11) in mustard seedling cotyledons by phytochrome and photooxidative treatments. *Planta* 186:600–8

218. Thordal-Christensen H, Zhang Z, Wei Y, Collinge DB. 1997. Subcellular localization of H_2O_2 in plants. H_2O_2 accumulation in papillae and hypersensitive response during the barley-powdery mildew interaction. *Plant J.* 11:1187–94

219. Tommasi FLD, Liso R, Arrigoni O. 1990. The ascorbic acid system in *Cuscuta reflexa* Roxb. *J. Plant Physiol.* 135:766–68

220. Trumper S, Follmann H, Haberlein I. 1994. A novel dehydroascorbate reductase from spinach chloroplasts homologous to plant trypsin inhibitor. *FEBS Lett.* 352:159–62

221. Ullman P, Gondet L, Potier S, Bach TJ. 1996. Cloning of *Arabidopsis thaliana* glutathione synthetase (*GSH2*) by functional complementation of a yeast *gsh2* mutant. *Eur. J. Biochem.* 236:662–69

222. Deleted in proof

223. Van Camp W, Capiau K, Van Montagu M, Inzé D, Slooten L. 1996. Enhancement of oxidative stress tolerance in transgenic tobacco plants overproducing Fe superoxide dismutase in chloroplasts. *Plant Physiol.* 112:1703–14

224. Van Camp W, Willekens H, Bowler C, Van Montagu M, Inzé D, et al. 1994. Elevated levels of superoxide dismutase protect transgenic plants against ozone damage. *BioTechnology* 12:165–68

225. Walker MA, McKersie BD. 1993. Role of the ascorbate-glutathione antioxidant

system in chilling resistance of tomato. *J. Plant Physiol.* 141:234–39

226. Washko PW, Welch RW, Dhariwal KR, Wang Y, Levine M. 1992. Ascorbic acid and dehydroascorbic acid analysis in biological samples. *Anal. Biochem.* 204:1–14

227. Watanabe K, Yamano Y, Murata K, Kimura A. 1986. The nucleotide sequence of the gene for γ-glutamylcysteine synthetase of *Escherichia coli*. *Nucleic Acids Res.* 14:4393–400

228. Wells WW, Xu DP, Yang Y, Rocque PA. 1990. Mammalian thioltransferase (glutaredoxin) and protein disulfide isomerase have dehydroascorbate reductase activity. *J. Biol. Chem.* 265:15361–64

229. Willekens H, Chamnongpol S, Davey M, Schraudner M, Langebartels C, et al. 1997. Catalase is a sink for H_2O_2 and is indispensable for stress defense in C_3 plants. *EMBO J.* 14:4806–16

230. Willekens H, Inzé D, Van Montagu M, Van Camp W. 1995. Catalase in plants. *Mol. Breed.* 1:207–28

231. Willekens H, Van Camp W, Van Montagu M, Inzé D, Langebartels C, Sandermann H. 1994. Ozone, sulfur dioxide, and ultraviolet B have similar effects on mRNA accumulation of antioxidant genes in *Nicotiana plumbaginifolia* L. *Plant Physiol.* 106:1007–14

232. Wingate VPM, Lawton MA, Lamb CJ. 1988. Glutathione causes a massive and selective induction of plant defense genes. *Plant Physiol.* 31:205–11

233. Winter H, Robinson G, Heldt HW. 1993. Subcellular volumes and metabolite concentrations in barley leaves. *Planta* 191:180–90

234. Winter H, Robinson G, Heldt HW. 1994. Subcellular volumes and metabolite concentrations in spinach leaves. *Planta* 193:530–35

235. Wise RR, Naylor AW. 1987. Chilling-enhanced photooxidation. Evidence for the role of singlet oxygen and superoxide in the breakdown of pigments and endogenous antioxidants. *Plant Physiol.* 83:278–82

236. Yamaguchi K, Mori H, Nishimura M. 1995. A novel isozyme of ascorbate peroxidase localized on glyoxysomal and leaf peroxisomal membranes in pumpkin. *Plant Cell Physiol.* 36:1157–62

237. Torsethaugen G, Pitcher LH, Zilinskas BA, Pell EJ. 1997. Overproduction of ascorbate peroxidase in the tobacco chloroplast does not provide protection against ozone. *Plant Physiol.* 114:529–37

Annu. Rev. Plant Physiol. Plant Mol. Biol. 1998. 49:281–309
Copyright © 1998 by Annual Reviews. All rights reserved

PLANT CELL WALL PROTEINS

Gladys I. Cassab
Department of Plant Molecular Biology, Institute of Biotechnology, National
University of Mexico, Apdo. 510-3 Cuernavaca, Morelia 62250, Mexico;
e-mail: gladys@ibt.unam.mx

KEY WORDS: extensins, extensin chimeras, arabinogalactan proteins, proline-rich proteins,
              glycine-rich proteins, hydroxyproline

ABSTRACT

The nature of cell wall proteins is as varied as the many functions of plant cell
walls. With the exception of glycine-rich proteins, all are glycosylated and contain
hydroxyproline (Hyp). Again excepting glycine-rich proteins, they also contain
highly repetitive sequences that can be shared between them. The majority of
cell wall proteins are cross-linked into the wall and probably have structural
functions, although they may also participate in morphogenesis. On the other
hand, arabinogalactan proteins are readily soluble and possibly play a major role
in cell-cell interactions during development. The interactions of these proteins
between themselves and with other wall components is still unknown, as is how
wall components are assembled. The possible functions of cell wall proteins are
suggested based on repetitive sequence, localization in the plant body, and the
general morphogenetic pattern in plants.

CONTENTS

281

1040-2519/98/0601-0281$08.00

INTRODUCTION

It took 85 million years for land plants to evolve from their green-algae ancestors. One of the most notable changes that occurred was the development of different kinds of cell walls. The vast ecological diversity of plants is directly associated with a great variety in size, shape, form, and function of individual cells. Since each cell type has its own distinctive wall, mutations of genes that control growth, size, and form of cells, via their action on cell walls, are major agents of evolutionary diversification (140). The cell wall is a complex molecular entity made of polysaccharides, lignin, suberin, waxes, proteins, enzymes, calcium, boron, and water that has the ability to self-assemble. This complexity is indispensable for the prosperous adaptation of plants to almost all microclimates on earth. Even the most superficial survey of plant anatomy indicates the adaptive and ecological importance of cell wall differentiation. The most prominent example is the development of a water and mineral conducting system (xylem) as well as a system for the conduction of photosynthates (phloem). The xylem also provide support by the incorporation of lignin in cell walls. Other examples are (*a*) formation of a waxy layer (cuticle) on aerial branch systems that prevents desiccation; (*b*) production of spores with cell walls impregnated with sporopollenin, a substance that prevents desiccation and is virtually indestructible by microorganisms; and (*c*) stems with rigid versus flexible branches, herbaceous cortex versus thin or thick bark produced by a specialized cork cambium, and specialized stem derivates such as tendrils, underground rhizomes, and tubers (57). Diversity of cell shape based on wall growth is equally great among leaves, for which it serves to provide maximum exposure to light; to protect from overheating, abscission, and dormancy under adverse conditions; and in recovery from insect damage and grazing by large animals.

During growth, development, environmental stresses, and infection, the cell wall is continuously modified by enzyme action (20). Moreover, morphogenesis in plants is the result of differential growth of the organs at the level of cell walls and not at the plane of cell division (74). Thus, it is important to study how cell walls are made more pliable locally to result in the differential growth process observed in plant morphogenesis. A plant cell is polyhedral in appearance and, therefore, is composed of discrete facets, edges, and vertices: Each of these three types of cell walls might be structurally and compositionally distinct from others and, hence, might be capable of changing to new states independently of other elements of a cell (83). In fact, different cell wall polysaccharide epitopes are not evenly distributed (or accessible) within the walls of a given cell, nor are these epitopes distributed uniformly across the two walls set by adjacent cells (52). Hence, it seems that the biosynthesis and differentiation of cell walls are accurately regulated in a temporal, spatial, and developmental manner. The cell wall superstructure also has association properties with the nucleo-cytoplasmic system via plasmodesmata, plasma membrane, and cytoskeleton. In fact, cytoplasm in living cells is present in cell walls in the form of plasmodesmata. It is through plasmodesmata that plant cells can communicate by symplastic continuity (99).

Proteins localized in cell walls are ubiquitous and relatively abundant in land plants and green algae. They are unusually rich in one or two amino acids, contain highly repetitive sequence domains, and are highly or poorly glycosylated. Among these are the hydroxyproline-rich glycoproteins (HRGPs) or extensins, the arabinogalactan proteins (AGPs), the glycine-rich proteins (GRPs), the proline-rich proteins (PRPs), and chimeric proteins that contain extensin-like domains. Nevertheless, recent characterization of new cell wall proteins show that the classifications proline-rich, hydroxyproline-rich, and glycine-rich may be more relevant to sequence domains within proteins than to the proteins themselves because there are proteins that have a mixture of these domains (17). These proteins exhibit different abundances in the walls of different cell types. Thus, each can be assumed to have functions specific to its particular cell type. However, there is little direct evidence as to what these functions might be. On the other hand, these are not the only cell wall proteins that are known. There are cysteine-rich thionins; a histidine-tryptophan–rich protein, a calmodulin-binding protein, and a leucine-rich repeat protein; and several cell wall enzymes (20, 73, 147, 155). Nevertheless, the above types are the most abundant and studied plant cell wall proteins. This review discusses the most common types of cell wall proteins present in green algae and plants, their cellular distribution in the plant body, and their possible function.

HYDROXYPROLINE-RICH GLYCOPROTEINS

Earliest HRGPs and Their Relation to Extensin Chimeras

How cell wall proteins were initially detected in algae and plants has been reviewed (20). Since the discovery in 1960 of hydroxyproline (Hyp) as a major amino acid constituent of hydrolysates of cell walls from tissue cultures (41, 88), hundreds of Hyp-rich proteins have been described in a wide variety of plants and algae. All Hyp-rich proteins exhibit repeating amino acid motifs, and the amino acid sequence of the repeat unit is generally used to designate a newly isolated cell wall protein to a particular class. Extensins, for example, include Ser(Pro)$_4$ repeats, the PRPs carry ProProProXYLys repeats, and the AGPs have Ala(Hyp)$_n$ repeats. Nevertheless, as the number of characterized HRGPs has increased, considerable variation has been found in the repeat units. Kieliszewski & Lamport (78) suggested that, instead, focus be placed on the similarity of the HRGPs because they appear to belong to a common superfamily. If this is the case, it should be feasible to trace the evolutionary origin of all Hyp-rich proteins to a small number of archetypal peptide motifs. To date, however, most of the characterized genes and proteins derive from dicots; a few derive from monocots, gymnosperms, and green algae. Important resources for this evolutionary approach are members of the green algae class Chlorophyceae, such as *Chlamydomonas* and *Volvox*, which also contain cell wall HRGPs (161). For some time it was assumed that some group of the class Chlorophyceae was the progenitor of land plants from which bryophytes and vascular plants later evolved. However, recent studies have pointed toward the green algae class Charophyceae (stoneworts), including *Chara* and *Nitella*, as being more likely candidates (119). This may be questionable because cell walls of *Chara* and *Nitella* lack Hyp (143). Therefore, the Chlorophyceae should be reexamined as the more probable progenitor of land plants, since the first step in the long evolutionary development of the upright system was the acquisition of specialized cell walls with Hyp. Unfortunately, even though specialized cell walls are the rule among land plants, phylogenetic trends involving them have not been fully described. Hence, conserved motifs identified in Chlorophyceae may represent archetypal peptide domains subsequently wielded by land plants.

An invariant feature of all of the volvocalean cell wall proteins that have been sequenced so far is their subdivision into distinct domains: Domains that are repetitive and Pro-rich, and domains that lack repeating motifs, have few prolines, frequently are Cys-rich, or contain motifs of cell adhesion molecules (45, 46, 70, 159–161). Extensin-like cell wall proteins in *Chlamydomonas reinhardtii* contain the $Ser(Pro)_x$ motif, but in *Volvox carteri* the $Ser(Pro)_{3-7}$ domain has only been observed. Nevertheless, electron microscope analysis of these proteins in both organisms shows that they contain two regions: a terminal knob, which corresponds to an N-terminal domain lacking numerous Pro or repeated units; and a rod coded by a C-terminal domain, including the $Ser(Pro)_x$- or $Ser(Pro)_{3-7}$-repeating motifs (45, 70, 161). Plant extensins, on the other hand, appear as simple rods and lack these knobs (139). However, there are several intra- and extracellular proteins that resemble the multidomain structure found in volvocalean wall proteins, such as cucumber-peel cupredoxin, which contains a Hyp-rich C-terminal domain (113); cysteine-rich extensin-like proteins (CELPs) in tobacco (164); solanaceous lectins from potato, tomato, and thorn apple that include a chitin-binding and an HRGP motif (40, 79, 111); a class of pollen-specific, extensin-like proteins in *Zea mays* that has a putative globular domain at the N terminus and an extensin-like domain at the C terminus (*Pex1*) (126); a histidine-rich HRGP from *Z. mays* and a style-specific HRGP from *Nicotiana alata* with properties of both extensins and AGPs (77, 94); and tobacco chitinase and an anther-specific gene from sunflower that contain a Hyp-rich linker domain (49, 141).

The sulphated, Hyp-rich glycoprotein ISG (inversion-specific glycoprotein) from *V. carteri* is composed of a globular and a rod-shaped extensin-like domain. ISG is synthesized for only a few minutes in inverting embryos during the 48-h life cycle of *Volvox*. ISG is perhaps the first extracellular matrix (ECM) component produced in the embryo, and thus, ISG is a candidate for involvement in this important function. In addition, a distinct feature at the C-terminal end of ISG is the occurrence of highly charged residues, specifically five negatively charged amino acids followed by five positively charged residues. This C-terminal region has been implicated in the emergence of the extracellular matrix during *Volvox* embryogenesis, since a synthetic decapeptide matching this sequence is able to desegregate the organism into individual cells (45). Interestingly, CELPs that are a family of tobacco genes that code for a set of proteins with an extensin-like Pro-rich motif and a Cys-rich domain also contain a highly charged C terminus (164). The extensin-like sequence contains the $X(Pro)_{3-7}$ motif where X may be Ser, Cys, or Trp. A remarkable difference between CELPs and extensins is the low Tyr content, the absence of His residues, and the abundance of Cys groups. A number of these Cys groups are found within the $X(Pro)_{3-7}$ motif. Some of these Cys residues may form disulfide bonds under the proper redox conditions and may participate in major inter- and intramolecular interactions involving the CELPs in the cell wall and the ECM. The presence of distinct Pro-rich and Cys-rich domains and their solubility properties are reminiscent of the solanaceous lectins, a class of Hyp-rich glycoproteins that are also rich in Ser and Cys residues (79, 164). In the pistil, CELP mRNAs gather in stylar cells delimiting the transmitting tissue, and in the ovary, they center in a narrow row of cells lining the placenta that are a continuation from the stylar transmitting tissue to which ovules are attached. The expression pattern of CELP mRNAs in these tissues suggests that these genes might be involved in pollination and fertilization processes. In addition, the fact that CELPs contain several distinct domains indicates that they may be multifunctional as *Volvox* ISG and algal-CAM (cell adhesion molecules) (45, 70). One possibility is that the extensin-like domain of these chimeric proteins is embedded in the wall and the other domains are either interacting with the membrane, the cytoplasm, the cytoskeleton, or the extracellular milieu.

A second extensin-like protein from *Volvox* was identified by using monoclonal antibodies that inhibit aggregation of embryonic cells (70). This protein, named algal-CAM, contains a three-domain structure: an N-terminal extensin-like domain with $Ser(Pro)_{3-7}$ sequences; and two repeats with homology to fasciclin I, a CAM involved in the neuronal development of *Drosophila*. Since algal-CAM represents the first plant homolog to animal CAM, its association with an extensin-like domain, a structural element unique to plants, has significant implications for the function of extensin chimeras. Therefore, extensin chimeras might play a major role in cell adhesion during plant development. For instance, one well-characterized cell-cell recognition process involving HRGPs is the mating system of *Chlamydomonas*. The sexual agglutinins of *Chlamydomonas* mediate the initial recognition of opposite mating types. Although the primary sequence of these agglutinins is not known, analysis of the sexual agglutinins at the EM level has shown that they have rod-shaped domains (similar to extensins) in addition to globular domains and, thus, are extensin chimeras. It has been suggested that one domain of the agglutinin is inserted in the flagellar membrane and the other is responsible for interaction with the agglutinin of the opposite mating type (1). *Chlamydomonas*, in addition, elaborates two biochemically and morphologically distinct cell walls during its life cycle: One surrounds the vegetative and gametic cell, and the other encompasses the zygote. There is evidence that *Chlamydomonas* appears to switch between two distinct sets of HRGP genes, some expressed by vegetative/gametic cells and the others by zygotic cells (159). Thus, since early times, distinct classes of HRGPs were present at different cell wall types or developmental times, as seen later in higher plants. There is a gene that is specifically expressed in the transmitting tissue of *Antirrhinum* flowers with high similarity to the HRGP from the zygote wall of *Chlamydomonas*. This gene, *ptl1*, is also an extensin chimera, since it contains a 49-residue C-terminal extensin motif attached to a 46-residue N-terminal module (5). In *Nicotiana tabacum*, seven genes that are specifically expressed in the pistil are also extensin chimeras with an N-terminal extensin motif with several repetitions of Ser-$(Pro)_4$ and a C-terminal tail (60). These genes encode for proteins with low Tyr content, as seen in CELPs (164) and in the 120-kDa glycoprotein from styles of *N. alata* (94). There are also some chimeric proteins whose extensin module is quite small, like the 9-residue Thr-Hyp linker between the lectin and chitinase modules of tobacco chitinase (141), or the 15-residue Pro-rich linker embedded in Lys/Cys-rich proteins specific to the sunflower anther epidermis (49).

Another example of chimeric extensin is *Pex1*, a maize pollen-specific gene with a hypothetical rod domain with $Ser(Pro)_4$ repeats at the C terminus, and a putative globular domain that is Leu-rich and Pro poor at the N terminus (126). Antibodies raised against a *Pex* fusion protein and a *Pex* synthetic peptide immunolabel the intine in mature pollen, and the inner, mainly callosic layer of the pollen tube wall. In maize, there are two closely related genes, *Pex1* and *Pex2*, that represent the first class of genes that encode proteins specifically targeted to the pollen tube wall (127). Localization to the pollen tube wall strikingly suggests that *Pex* proteins participate in pollen tube growth during pollination. The extensin-like domain of *Pex* may give structural support for the pollen tube crucial for its rapid growth. The maize pollen tube grows at the extraordinary rate of ~1 cm/h (6), and thus it must be able to support the mechanical force produced as it penetrates the cuticle of the stigma and crosses first the silk cortex and then the growth between the cells of the transmitting tissue during pollination. Formation of an HRGP-cellulose framework is associated with an increase in the tensile strength of cell walls. In addition, HRGPs are covalently cross-linked to the cell wall. *Pex* proteins could not be removed with high salt, sodium dodecyl sulfate, or chaotropic or reducing agents, suggesting a close association with the pollen tube wall. However, the extensin-like domain of *Pex1* has only 0.7% Tyr, and thus, it is not insolubilized by isodityrosine cross-links (27, 53). This is not surprising, since (*a*) soybean seed coat HRGP is also cross-linked into the wall, (*b*) it does not contain isodityrosine, and (*c*) it has 8.5% Tyr (20).

In maize, for instance, specific cell-cell interactions between pollen and pistil seem to be needed for the location of the transmitting tract by pollen tubes. Pollen of sorghum and pennisetum, as well as graminaceous monocots, can germinate on maize silks (equivalent to the stigma), but the pollen tubes lack directionality and are incapable of locating the transmitting tract (64). Pollen has the ability to support its own activity during germination and tube growth (156). The pistil, nevertheless, must also participate in this process, since pollen tubes grown in vitro protrude in random directions, their growth rates are notably lower, and they attain considerably shorter distances than pollen tubes grown in vivo. Tubes that emerge from pollen grains on the stigmatic surface are guided toward the ovary within this secretory matrix by as yet unknown mechanisms. It has been suggested that transmitting tissues may be a source of nutrients and/or chemotropic factors, or they may grant mechanical properties necessary for pollen tube extension toward the ovules (63, 98). This process clearly depends on species-specific cell-cell signaling between the pollen tube and the pistil. The directional cues may be mediated by a diverse array of glycoproteins, such as transmitting tissue-specific AGPs (TTS, see below), CELPs, and *Pexs* (38, 127, 163, 164), and also probably by Ca^{2+} and B (102, 124). These glycoproteins might function as, for instance, algal-CAMs, which play a significant role in *Volvox* development. CAMs are central morphoregulatory proteins (44). They play a major topobiological role in animal development, since disruption of their binding alters morphology and disruption of morphology alters regulation of CAM expression. Topobiology is the study of the place-dependent regulation of cells resulting from interaction of molecules at cell surfaces with those of other cells or substrates. Such place-dependent molecular interaction can regulate the primary processes of development and lead to changes in morphology by epigenetic means (44). Because plant cell wall proteins tend to be large multisubunit and multivalent structures, they also have the capability of cross-linking and "patching" mobile cell surface–putative receptors or of altering their association with the cytoskeleton. Morphogenesis and cell signaling in plants may depend on the interactions of cell wall proteins—particularly on interaction with the entire cell wall and ECM in a complex modulation network offering many combinatorial possibilities of protein binding. Plant cells can communicate through plasmodesmata, possibly signaling the formation of groups of coupled cells responding similarly to inductive signals. While the picture is incomplete, some evidence supports the view that all of these glycoproteins play mechanochemical as well as regulatory roles in development.

Another class of extensin chimera is the unusually rich in His extensin-AGP (HHRGP) (77). This glycoprotein shows repetitive motifs related to both extensins and AGPs, since it contains Ala-Hyp$_3$ and Ala-Hyp$_4$ repeats that may be related to the classical Ser-Hyp$_4$ extensin domain by the single T → G (Ser → Ala) base change. In addition, HHRGP also possesses AGP characteristics, particularly an elevated Ala content, near sequence identity with the known *Lolium* AGP peptide Ser-Hyp-Hyp-Ala-Pro-Ala-Pro, the putative presence of glucuronoarabinogalactan, and precipitation by Yariv antigen. The HHRGP was isolated from maize cell suspension cultures, so its precise location in the plant body is uncertain. Another HRGP with structural features of both extensins and AGPs is a 120-kDa style-specific HRGP isolated from *N. alata* (94). This glycoprotein is rich in Hyp, Pro, Lys, and Ser and poor in Tyr, His, and Cys. These characteristics are similar to the extensins and distinguish it from typical AGPs. The low level of Cys and the high level of Lys in this 120-kDa glycoprotein distinguishes it from the solanaceous lectins, which contain a Cys-rich domain that is essential for carbohydrate binding (3). The low Tyr content distinguishes the 120-kDa glycoprotein from the extensins. This is consistent with high solubility of the 120-kDa glycoprotein, since Tyr residues are thought to be involved in cross-linking extensin into the cell wall. The linkage composition of the monosaccharide residues on the 120-kDa glycoprotein showed 1,3-linked Gal, 1,6-linked Gal, and 1,3,6-linked Gal, which are characteristic of AGPs. The 120-kDa Hyp-rich glycoprotein is a major component in the ECM of the transmitting tissue of *N. alata* and accounts for approximately 8.8% (wt/wt) of the protein in the transmitting tract fluid. It is developmentally regulated in that it is barely detectable in extracts of green bud styles (which contain little transmitting tract ECM) but is very abundant in the buffer-soluble fraction of mature styles. This glycoprotein appears widespread in the Solanaceae and thus might play an important role in pistil function by providing adhesiveness in the stigma surface, might contribute to the gel structure of the ECM of the transmitting tissue, and/or might participate in gamete recognition.

One of the best-characterized HRGP chimeras is potato lectin (79). It is a developmentally regulated, chitin-binding glycoprotein (50% carbohydrate, wt/wt) consisting of at least two evolutionary autonomous domains: a lectin module fused to an HRGP module (3). The lectin module contains a β-turn secondary structure (151) and has a composition similar to other plant chitin-binding proteins in that it is rich in Cys and Gly, contains no Tyr and Trp, and has no sugar (3). The HRGP motif, on the other hand, is rich in Hyp, is highly glycosylated with abundant Ara and minor amounts of Gal (3), and has a polyproline-II secondary conformation (151). Peptide sequence information from the HRGP domain shows that it contains similar sequences to the P3-type extensin; the lectin domain comprises homologous sequences to the hevein lectin family. Therefore, potato lectin maybe a mosaic protein, which by definition evolved via the exchange of domains between genes of previously existing proteins (via symmetrical exons; 120). Potato lectin is prominently expressed in vascular tissue, particularly xylem and also phloem, although it is not present in the cambium. It is also present in the epidermis and in the cells lining secretory cavities. Extensin has a similar expression pattern, although it is also observed in the cambium (91).

It is feasible that certain chimeric extensins might be preferentially localized either in pits or in primary pit fields where they could function in cell-cell recognition and defense processes. Both primary and secondary walls are commonly characterized by the presence of depressions or cavities varying in depth, expanse, and detailed structure. Such cavities are termed pits in secondary walls and primary pit-fields in primary walls (47; Figure 1A,B). In primary pit-fields, the primary wall is particularly thin but continuous across the pit-field area. On the other hand, the distinguishing feature of a pit is that the secondary wall layers are completely interrupted at the pit; that is, the primary wall is not covered by secondary layers in the pit region. A pit consists of a pit cavity and a pit membrane. The pit cavity is open internally to the lumen to the cell and is closed by the pit membrane along the line of junction of two cells (Figure 1C). The pit membrane is common to both pits of a pair and consists of two primary walls and a lamella of intercellular substance. Primary pit-fields are present mainly in meristems and parenchyma and are colocalized with groups of plasmodesmata (PD). PD are regarded as cytoplasmic threads interconnecting the living protoplasts of the plant body into an organic whole. The relation of PD to primary pit-fields is characteristic: In two adjoining cells, cytoplasmic processes extend into the cavities of a pair of pit-fields, and the thin wall in the pit-field is traversed by very fine threads connecting the two small masses of cytoplasm filling the depressions of the pit-fields (47, Figure 1B). Thus, the presence of chimeric extensin in primary pit-fields of living cells could play a major role in cell-cell recognition and signaling in morphogenesis. Furthermore, they could also constitute an important barrier for pathogen movements through PD. PD allow a diverse array of molecules such as sucrose, virus, nucleic acids, proteins, etc, to pass selectively from cell to cell.

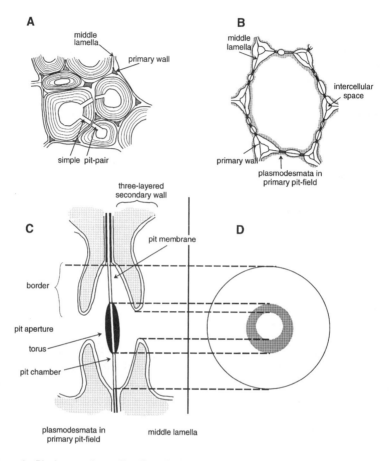

Figure 1 Pits in secondary cell walls and primary pit-fields in primary walls. (*A*) Phloem fiber cells with secondary walls and simple pits (x560). (*B*) Primary pit-field and plasmodesmata (PD) in parenchyma cells with primary walls (x325). Bordered pit-pair in sectional (*C*) and face (*D*) views. The pit membrane consists of two primary walls and the intercellular lamella. The torus is formed by thickening of the primary wall. (Adapted from Esau, Reference 47.)

Moreover, PD seem to play a major role in orchestrating plant development, via cell-to-cell transport of regulatory signals (99). It has been recently proposed that PD channels could be controlled by the binding of acetylcholine to its receptors (112). Consequently, the transport of hormones and substances could be regulated by the opening and/or closing of the conductive channels. Then, morphoregulatory molecules such as extensin chimeras and other cell wall and ECM proteins could also interact with PD regulators, and this cascade of interactions will be required for establishing morphology. In pits, however, chimeric extensins could hold additional functions according to the cell type where they are expressed. For instance, pits can be present in tracheids and vessel elements, which conduct water; in sclereids, which provide mechanical support; and in fibers that may retain their protoplasts and thus combine a vital function, such as starch storage, with the mechanical one of support.

Extensins

Extensin is certainly the most well-studied cell wall structural protein of plants. It is amply distributed in dicots and less common in graminaceous monocots. The difference in extensin distribution seems correlated with the fact that the chemical structure of the primary cell wall of grasses and their progenitors differ from those of all other flowering plant species (15). In addition, grasses have a low boron demand, and all other flowering species have a high boron requirement (59). B is localized primarily in the cell wall (96), and it is linked with the rhamnogalacturonan (RG) II fraction of pectins (69, 71, 82, 117). Loomis & Durst (96) proposed that borate esters with apiose (which is preferentially present in RG II) are responsible for cross-linking cell wall polymers and thus are necessary for cell wall stability. Thus, both the low demand of B and the distinct chemical structure of cell walls of grasses yield Hyp-poor walls, although there are some tissue exceptions, such as pollen (126) and pericarp (67).

From sequence determinations of salt-extractable precursors and/or cDNA or genomic clones, extensin consists of a repeating Ser-$(Hyp)_4$ motif that seems to be significant to structure. However, a chenopod extensin lacks this motif (93) but contains a repetitive motif in which the tetra-Hyp block, Ser-Hyp-Hyp-[X]-Hyp-Hyp-Thr-Hyp-Val-Tyr-Lys, is split by the insertion of [X] (Val-His-Glu/Lys-Tyr-Pro). Extensin has, in addition to the Ser-$(Hyp)_4$ motif, other repetitive sequences. These sequences vary from plant to plant and from dicots to monocots, and/or to gymnosperms and/or green algae, although there are some notable similarities among them (78, 134). However, all extensins are rich in Hyp and Ser and in some combination of the amino acids Val, Tyr, Lys, and His; most of the Hyp residues are glycosylated with one to four arabinosyl residues, and some of the Ser residues are glycosylated with a single Gal unit. Extensins are basic proteins with isoelectric points of ~ 10, owing to their high Lys content. They commonly assume a polyproline II helical structure in solution, and they have a rod-like appearance when viewed in the EM (139, 152). Hyp oligoarabinosides reinforce extensin structure because removal of the sugars results in loss of the rod-like appearance (139) and reduces the polyproline II structure in solution (152). The arrangement of the arabinosides possibly determines how these glycoproteins interact with other polymers of the wall; it also instructs them how to assemble into the wall.

It has been proposed that extensin is slowly insolubilized in the cell wall by a covalent link (27, 87). One proposed covalent link is isodityrosine formed between two Tyr residues from different extensin molecules (53). This proposition, however, seems to be vague, since the evidence is circumstantial. Isodityrosine is present in many but not all cell walls, and to date no intermolecular cross-link has been characterized. A novel amino acid, di-isodityrosine, has been isolated from hydrolysates of cell walls of tomato cell culture (12). This compound could form an interpolypeptide linkage between cell wall proteins such as extensin. Nonetheless, extensin is indeed cross-linked into the wall by an unknown mechanism, since in many cases it is recalcitrant to extraction. It has been recently found that, at least in cotton suspension cultures, there is a covalent linkage between pectin (RG I fraction) and most or all of the extensin that has been incorporated into the cell wall matrix (122). The exact nature of the linkage between pectin and extensin remains to be resolved, although the linkage could be via either a 3,6-linked galactan or a phenolic cross-link from a feruloylated sugar in the pectin to an amino acid in the protein (122). Furthermore, it has recently been reported that in B-deficient root nodules, the insolubilization of Hyp-rich glycoproteins is severely affected (9). As mentioned earlier, B is linked to the RG II fraction of pectins, and extensin is linked to the RG I fraction. Hence, it seems plausible that both fractions of pectins need to be present in a wall matrix for extensin to become insolubilized. On the other hand, complete deglycosylation of cell walls is not sufficient to solubilize the bulk of extensin (122). Thus, some form of protein-protein or protein-phenolic-protein cross-link is likely to exist between extensin polypeptides in the cell wall.

Another possible cross-link could be derived from the peroxidization of extensin (153). Proteins can be peroxidized by exposure to reactive oxygen species, such as hydroxyl free radicals produced from hydrogen peroxide and Fe^{2+} ions (55). The formation of these rather stable hydroperoxides raises the possibility that they can initiate further reactions in their vicinity. The three amino acids most readily peroxidized—Val, Pro, and Lys—are abundant in the motifs Val-Tyr-Lys and Val-Lys-Pro-Tyr-His-Pro, present in several extensins. Although His is not so readily peroxidized, it can be oxidized to open the imidazole ring and generate an aldehyde and an amine group. If this were to form a Schiff's base with the ϵ-amino group of Lys and be reduced, a stable protein-protein cross-link could be generated. It is interesting that Ser, Thr, and Hyp, the amino acid residues most involved in generating the polyproline II structure of extensins, are the least susceptible to peroxidation (153).

The significance of extensin cross-linking into the wall matrix is still controversial. Lamport (87) proposed that extensin weaves cellulose microfibrils of the primary wall and that breakage of the cross-linked extensin network could allow the microfibrils to slide apart; thus, extensin may be integral to extension growth. It is now thought that expansins (107) and xyloglucan endotransglyco-sylases (135) are more likely to be the primary determinants of wall elongation. The data on appearance of extensins and their relationship to growth do not sustain this view. Extensin precursors are seen early in cell wall formation (168), but a large increase in the amount of extensin in the cell wall is associated with cessation of growth. In some cells, a strong correlation is found between final cell length and amount of Hyp in the cell walls (19), whereas the correlation is not as obvious in other systems (168). Formation of an extensin-cellulose framework is associated with an increase in the tensile strength of the cell wall in sclerenchyma cells (19). One possibility is that incorporation of extensin into the cell wall prepares the cell to stop growing and, thus, fixed its final shape (16). This seems to be the case in sclerenchyma cells (19), and even in the vascular cambium, where extensin is highly abundant (168). The vascular cambium is the lateral meristem that forms the secondary vascular tissues (47). Interestingly, there is a general resemblance between the cambium cells and their derivatives, and the shape and arrangement of cells in the secondary xylem and the secondary phloem are foreshadowed in the shape and arrangement of the cambial cells. Hence, extensin may have already fixed the shape of the cambial cells before they give origin to the different vascular cell types.

The variability observed in the content of non-Hyp amino acids, in motifs other than Ser-(Hyp)$_4$, and in the arabinosylation profiles in the different extensins may be plant and tissue specific. In some cases, there are different extensins in the same tissue or cell type. This may mean that extensin can have different functions according to cell type, or even in the same cell type or tissue. Immunocytochemistry, tissue printing, and in situ hybridization have all been used to study extensin distribution in different plants and tissues (134). These studies revealed that extensin is preferentially localized in sclerenchyma and cambium cells, although it is also associated with secondary xylem and phloem tissues. Unfortunately, many of these experiments have been done by using nonspecific antibodies or nucleic acid probes. For example, polyclonal antibodies against extensin soybean seed coat have been utilized to study the localization of extensin in several tissues and plants (167, 168). However, these antibodies also cross-react with other extensin-related sequences found in PRPs and *ENOD2* (9; GI Cassab, unpublished results). Consequently, the results obtained may be inaccurate. Therefore, the immunolocalization experiments should be repeated with species-specific antibodies. Yet experiments with specific nucleic acid probes have been performed (146), as have analyses of extensin expression in heterologous systems (133). These studies have shown that *Brassica napus* extensin is expressed in the sclerified phloem fiber cells of transgenic tobacco stem nodes. These fibers have a major strengthening function in the plant, since they tolerate compression stresses resulting from the development of vascular bundles. In *Nicotiana plumbaginifolia*, the extensin gene is expressed in cortical cells of the stems in regions presumed to require strengthening (146). A similar pattern was observed in cortical cells associated with the developing vascular bundles of the axillary flowering stalk (133). In tobacco roots infected with the nematode *Meloidogyne javanica*, extensin transcript is accumulated in cortical cells of galls, presumably as a result of the mechanical pressure exerted by the developing gall after infection (114). It is also apparent that the extensin gene *extA* of *B. napus* is regulated by the degree of tensile stress experienced by cortical cells: As the level of stress increases, so does the intensity of expression in transgenic tobacco plants (133). This was also confirmed by hanging weights from young developing axillary flowering stalks of transgenic tobacco plants and examining the pattern of *extA* promoter-driven β-glucuronidase expression. In axillary stalks from which 10-g weights had been suspended, a tight, distinct band of expression extending through the parenchyma cells of the pith and cortex was seen in the node. This band was absent from axillary branch nodes of a similar age and size that were not weight loaded. Extensin is also present in carpel wall cells, which experience considerable tensile stress as pod tissues begin to desiccate prior to the explosive shedding of seeds that occurs when the dry pod wall ruptures (133). Although lignification is almost certainly the major factor in strengthening, it is likely that the accumulation of extensin in these cells in response to tensile stress is also significant. In fact, there appears to be no correlation between the strength of fibers and the degree of lignification of their walls (see 61). This possibly means that extensins contribute to the tensile strength of these cells.

Extensin seems to contribute to plant defense, helping to protect against pathogen attack, elicitation, or mechanical wounding (134). Immunochemical studies have shown that extensins accumulate in cell walls close to sites where microbial growth is restricted by the plant (48). There is some evidence that extensin may act as an impenetrable physical barrier or may immobilize the pathogens by binding to their surfaces (105). The latter probably results from positively charged extensin molecules interacting ionically with negatively charged surfaces of plant pathogens (108). Treatment of bean or soybean cells with fungal elicitor or glutathione causes a rapid (2–5 min) insolubilization of two (Hyp)Pro-rich structural proteins in the cell wall (11). In addition, it has been shown that a short elicitation of 30 min rendered cell walls more refractory to enzyme digestion, as assayed by the yield of protoplasts released. This effect could be ascribed to protein cross-linking because of its insensitivity to inhibitors of transcription and translation and its induction by exogenous H_2O_2 (13). Elicitors and suppressors of extensin accumulation are solubilized from plant cell walls by endopolygalacturonase purified from pathogens (10). Small galacturonides can trigger extensin gene expression and elicit a 40–70% Hyp increase in the cell wall. In contrast, pectic fragments of higher molecular mass, composed of galacturonic acid, with minor RG II constituents, had the ability to suppress Hyp deposition in the cell wall and extensin gene expression. Besides defense, signaling via pectic fragments is likely to play important roles during plant development in situations where the cell needs to be loosened or reinforced. Pollen germination, fruit ripening, cell elongation, and wound-healing responses are such examples. In addition, the regulation of extensin by endogenous suppressors at certain stages of cell growth and development might explain the long reported inverse relationship between the level of extensin in the cell wall and the extent of cell elongation. Thus, cell wall carries its own information, i.e. signals that regulate its own biogenesis. Recently, a novel soybean extensin gene was shown to require sucrose for its wound-inducible expression in transgenic tobacco plants (2). Sucrose may mediate wound signal transduction in phloem tissues and cambium cells by binding to a specific receptor(s) in their cell membranes.

Although the structural and regulatory properties as well as the localization of extensin may indicate its function, direct functional evidence is needed. To date, extensins have been proposed to be structural cell wall proteins that may also play a significant role in development, wound healing, and plant defense. Studies that putatively could unravel the function of extensin, such as (anti)sense gene technology, have demonstrated that tobacco transgenic plants expressing sense or antisense extensin gene constructs tolerate a large variation in total Hyp concentration and soluble HRGP content without apparent effects on their phenotype and cell wall structure (109). The absence of a phenotype could be because only one cell wall component (extensin) was manipulated and other wall proteins (such as PRPs or GRPs) and/or polysaccharides may have compensated for its absence. Therefore, we are challenged to design suitable systems to test extensin function.

PROLINE-RICH PROTEINS

PRPs and nodulins contain essentially one repetitive motif: variations of (Pro-Hyp-Val-Tyr-Lys)$_n$, notably lacking Ser and lightly glycosylated (33, 39, 66, 95, 101, 145). This repetitive pentamer occurs in more complex extensins (tomato P1) as a pentameric variation, combined with the Ser-Hyp$_4$ motif (78). Recently, the first nodulin PRP has been isolated and partially characterized (9). Apparently, this nodulin corresponds to *ENOD2*, since its amino acid composition is highly similar to the derived amino acid composition of the *ENOD2* cDNA clone (51). *ENOD2* contains approximately equimolar quantities of Pro and Hyp, low levels of Val in contrast to other PRPs, and high levels of Glu.

Antibodies against extensin of soybean seed coat recognize the purified *ENOD2* protein and only one polypeptide in total nodule extracts by Western blot analysis (9). It is presumed that the anti-extensin antibodies may recognize the sequence Pro-Pro-Pro-Val-Tyr, which is found twice in *ENOD2* from soybean (51), and which could also be present in some extensins, such as in the cDNA for HRGP-3 from soybean hypocotyl (65). In addition, because there is no sequence data from the soybean seed coat extensin, it could also be speculated that this glycoprotein may contain sequences similar to the sugarbeet extensin 15-mer core: Hyp-Hyp-[Val-His-Glu-Tyr-Pro]-Hyp-Hyp (93). This sequence is similar to the repetitive motif Pro-Pro-[His-Glu-Lys-Pro]-Pro-Pro present in soybean *ENOD2* (51). *ENOD2* mRNA and protein are localized in the nodule parenchyma (9, 148). The nodule parenchyma appears to be an important tissue in the *Rhizobium*-legume symbiosis, since it forms a barrier to the diffusion of gaseous O_2, as demonstrated by O_2 microelectrode measurements (158). It has been suggested that *ENOD2* contributes to the diffusion barrier by modifying cell walls (148). The occlusion of the small intercellular spaces by ECM proteins and water would theoretically provide 104 times the resistance to gas diffusion as would continuous airways (158). Interestingly, the morphology of the nodule parenchyma is severely affected in B-deficient root nodules, since it contains expanded intercellular spaces (9). Further, in B-deficient root nodules, *ENOD2* could not be immunolocalized in cell walls of nodule parenchyma, and little Hyp was covalently bound to the nodule cell walls. The lack of *ENOD2* deposition in cell walls of nodule parenchyma may be related to the decrease in their capability to fix N_2. In the nodule parenchyma, *ENOD2* may contribute to their occlusion, perhaps by functioning as a gel plug or an adhesive. In fact, there is a strong sequence similarity between PRPs and an adhesive protein from mussels (78). B is an essential micronutrient for higher plants (136); however, its mechanism of action is not clearly understood. The fact that Hyp/Pro-rich proteins are not covalently bound to the walls of B-deficient root nodules suggests that B also has a role in the assembly of some wall protein components (9).

Another PRP that has recently been isolated and structurally characterized is the galactose-rich PRP (*GaRSGP*) from styles of *N. alata* (137). This is a basic glycoprotein that is a component of the walls of the transmitting-tract cells. It corresponds to the cDNA previously described and designated *NaPRP4* (31). The glycoprotein is rich in the amino acids Lys, Pro, and Hyp and in the monosaccharides Gal and Ara and is composed of approximately 25% (wt/wt) protein and 75% (wt/wt) carbohydrate. This novel Gal-rich glycosylation pattern has not been described previously for this group of proteins. Unlike the extensin, *GaRSGP* is not insolubilized in the cell wall. This may be related with its low content of Tyr, which is believed to form intra- and possibly intermolecular isodityrosine cross-links in extensin (134). How *GaRSGP* relates to the

TTS proteins from *N. tabacum* described by Wang et al (157) remains uncertain. *NaPRP4* cDNA is 96.5% identical at the nucleotide level with the TTS-1 cDNA from *N. tabacum* isolated by Cheung et al (37). However, *GaRSGP* is localized in the inner cell wall of the transmitting tract cells and not in the intercellular areas of the extracellular matrix where the TTS proteins were localized (157). Since *GaRSGP* is located in the inner cell wall, a part of the cell wall that normally does not come into contact with the pollen tubes, a direct function in pollen tube growth seems unlikely. *GaRSGP* could thus be a structural cell wall glycoprotein, the function of which is related to the specialized secretory nature of the transmitting tract cells. The protein backbone of *GaRSGP* has similarities to the agglutinins from potato and tobacco (90, 108).

Several studies indicate that PRPs are implicated in various aspects of development, ranging from germination, to pod formation, to the early stages of nodulation (134). The expression of PRP genes is also influenced by wounding, endogenous and fungal elicitors, ethylene, cell culturing, drought, and light (101, 110, 132, 145). Proline-rich proteins display tissue- and cell-specific patterns of expression. They commonly are present in the xylem, fibers, sclereids, epidermis, aleurone, and nodule parenchyma (148, 165, 167). The localization pattern of PRPs is similar to that of most dicot glycine-rich proteins (GRPs, see below), although some PRPs' expression occurs in the same cell types as extensins, e.g. in sclereids.

As extensins, PRPs are presumably insolubilized in the cell wall matrix. In fact, indirect immunological evidence argues for PRP cross-linking in the wall (9, 11, 101), although there is no direct evidence. The insolubilization process seems to occur rapidly in response to fungal elicitors and, as mentioned previously, may be mediated by the release of hydrogen peroxide and catalyzed by a wall peroxidase (11). Nevertheless, the mechanism that provokes the insolubilization of PRPs is also an enigma. That PRPs are mainly present in the xylem may imply that these proteins are involved either in xylem differentiation or in lignification. The possible influence of PRPs and HRGPs on the process of lignification has been proposed in the light of the observed affinity of phenolic substrates for prolyl residues in protein structures (100). These glycoproteins may provide sites for selective complexation with its phenolic precursors (the *p*-hydroxycinnamyl alcohols) and, therefore, points of growth for the polymer lignin. Furthermore, if there are preferred modes of complexation of the *p*-hydroxycinnamyl alcohols with the Hyp residues, then the relative disposition of these groups in the network may direct the polymerization along particular pathways rather than in the random manner present theories suggest (100). Interestingly, plant polyphenols (tannins) repel predators and act as general feeding deterrents in herbivores. Herbivorous mammals are able to adapt to the adverse effects of the dietary tannin by producing unique PRPs (up to

45% Pro) in the salivary glands, which have a very high affinity for polyphenols (14). The induction of these PRPs constitutes the first line of defense against the presence of polyphenols in the digestive tract. Without these proteins, the nutritional consequences of the consumption of tannins would be much more severe. Therefore, if PRPs from mammalian salivary glands make strong complexes with plant polyphenols (100), plant PRPs-HRGPs may have the same affinity for these cell wall compounds.

The PRPs seem to participate in several aspects of development, such as xylem, nodule, and pod differentiation (28, 97, 148) and ovary, embryo, and microspore development (72, 125), but their specific functions are unknown. In the case of PRP nodulins, *ENOD2* seem to be involved in nodule morphogenesis, particularly in the production of a cell wall oxygen barrier for the oxygen-sensitive nitrogenase (9, 148). In contrast, *ENOD12*, a PRP gene, is not required for symbiotic N_2 fixation, since *ENOD12* alfalfa-deficient plants were similar to their wild-type parents in viability, nodule development and structure, and N_2 fixation efficiency (30). However, it cannot be excluded that the function of the *ENOD12* gene product required for nodulation is taken over by another cell wall protein.

ARABINOGALACTAN PROTEINS

AGPs are found in higher plants and in liverworts (85). The AGPs have been located on cell membranes, in the ECM, and in gum exudates. One of the best-known AGPs is gum arabic that is secreted by *Acacia senegal* upon wounding. Gum arabic has been used since ancient times, and nowadays gums are still being used in the food industry as additives for their aggregating and gelling capacity. AGPs are not covalently linked to the cell wall and therefore do not have a structural function. Upon wounding, AGPs are secreted in large amounts and they might act as a physical barrier by producing a gel plug. However, AGPs are present in many different tissues and are not exclusively produced upon wounding, and thus they may display other functions. A role of AGPs in plant differentiation has been proposed (81, 84, 131).

The majority of the AGPs characterized so far have a protein content of less than 10% (wt/wt) and contain more than 90% (wt/wt) carbohydrate. The protein moiety is rich in Hyp, Ala, Thr, Gly, and Ser (22), but the content of amino acids may vary among AGPs of different species and tissues (23). The carbohydrate side chains are primarily linked by *O*-glycosylation to the OH group of Ser and Hyp. These side chains may consist of more than 50 carbohydrate residues (75). In principle, the variation in type of branching of the side chains can be unlimited. As a consequence, there are few data on the precise structure of the side chains. The heterogeneity of the polysaccharide side chains of AGPs is not yet understood. The protein core of AGPs often contains several Ala-Hyp repeats or closely related repeats (58). Recently, the first cDNAs coding for the protein backbone of AGPs have been cloned (32, 42, 43, 56, 92, 104, 121). The analysis of the derived amino acid composition of these clones revealed that there is little homology between them. A characteristic feature of AGPs is that they can be specifically precipitated by the β-glucosyl Yariv reagent (166). In fact, this provides an operational definition of AGPs. The β-glucosyl Yariv reagent can be used for isolating AGPs and in tissue localization studies. The interactions between AGPs and this reagent indicates that AGPs are also β-lectins with a broad binding specificity directed toward β-D-glycopyranosyl linkages (134).

AGPs are chemically stable, being resistant to high-temperature and cold-alkali treatment (85), as well as to proteolysis in its native state. Pulse-chase experiments have shown that the half-life of AGPs of cell suspension cells is about 10–15 min, indicating a rapid removal of the labeled molecules (150). The brief presence of AGPs in tissues is presumably due to an active system of degradation. The oligosaccharins, presently considered signal molecules in different processes (54), have been shown to be internalized by means of receptor-mediated endocytosis to multivesicular bodies and are deposited into the vacuole (68). AGPs have also been found on the internal membranes of multivesicular bodies and in the vacuole (62, 131). The multivesicular bodies can fuse with the tonoplast, or in styles they can fuse with the plasmalemma-releasing vesicles containing AGPs into the extracellular space, both eventually leading to degradation of the AGPs. That AGPs are often localized on multi-vesicular bodies may illustrate an active system of degradation and recycling and explain the high turnover of AGPs.

Electrophoretic separation of AGPs in the presence of Yariv's antigen shows that AGPs are expressed in a tissue-specific manner and that a given plant tissue can contain more than one kind of AGP (18, 149). Furthermore, in a determined tissue, the composition of AGPs changed during development. The heterogeneity observed by this method seems to be caused by the sugar moiety, since there is a limited number of protein cores reported and AGPs do not belong to a gene family. Carbohydrate side chains can be subjected to tissue-specific partial degradation, or the production of side chains can be regulated.

A powerful tool to characterize and solve the function of AGPs are poly-clonal or monoclonal antibodies. All antibodies have been directed against the sugar moiety of the AGPs (81). Monoclonal antibodies have been used to localized AGP epitopes in different tissues. The JIM4 antibody was obtained after immunization with carrot protoplasts (80). This antibody recognized AGP epitopes from the medium of suspension-cultured carrot cells and gum arabic. It was shown that the epitope was located at a specific set of cells during the development of the carrot root apex. The expression of the antigen occurred well before pattern formation was visible and was restricted to two small groups of cells in the future vascular bundle. The same epitope has been found in carrot somatic embryos (138). Here, the epitope was restricted to the protoderm of early somatic embryos and to the provascular tissue of the root apex and the cotyledons. Thus, the JIM4 epitope seems to accompany the differentiation of the vascular tissue. This and studies with other AGP antibodies (81, 118, 131) show that specific epitopes are present only on a limited number of cells or cell types during their differentiation and that some epitopes may show only transient expression. The AGP epitopes may reflect a tissue pattern determined by cell position. This raises the question of whether the different AGP epitopes are merely a result of differentiation or if they can cause these events, or possibly both. This question of what came first, the AGP epitope or the differentiation event, may be hard to answer.

AGPs seem to have a morphoregulatory role in bryophytes where they participate in the correlative control of leaf and branch development (7, 8). In particular, AGPs suppress cell division and growth at specific places on developing hepaticae leaves, leading to a species-specific leaf and branching morphology. In higher plants, AGPs are essential for somatic embryogenesis of carrot (84). Plant cells often differentiate according to position rather than to lineage (142). This implies that positional information may be communicated by cell wall components during development and differentiation. Since AGPs are soluble and diffusible, components of the ECM and of the plasmalemma are good candidates to act as cell positional markers or as messengers in cell-cell interactions during differentiation (50).

Because of their high sugar content and stickiness, AGPs have been ascribed important functions in pollination, such as in pollen recognition and adhesion on the stigma, and in serving as nutrient molecules and surface adhesives for pollen-tube growth (50, 98). One of these glycoproteins, the TTS proteins from the tobacco transmitting tissue, has been shown to be crucial for pollen-tube growth (38, 163). In vitro, TTS protein stimulates pollen-tube growth and, when provided at a distance in a semi–in vivo culture system, attracts pollen tubes. TTS proteins also adhere to the pollen-tube surface and tips, suggesting that they may serve as adhesive substrates for promoting pollen-tube growth. Moreover, TTS proteins are incorporated in vivo into pollen tubes, implying that they may provide nutrients to this process. Pollinated transmitting tissue accumulates an underglycosylated TTS protein species, which is not present in the prepollinated styles, suggesting that pollen tubes may also deglycosylate TTS proteins in vivo. Interestingly, TTS proteins added to pollen-tube growth cultures in vitro are deglycosylated by the pollen tubes, and these proteins are incorporated into the walls of these tubes. These observations are consistent with a role for TTS proteins as a nutrient resource for pollen-tube growth. In the tobacco transmitting tissue, TTS proteins display a gradient of increasing glycosylation coincident with the direction of pollen-tube growth. Together with its pollen-tube growth-stimulating and -attracting activity, this TTS protein-bound sugar gradient may be a directional cue for pollen-tube elongation in the style (36). Numerous other AGPs are present in the ECM along the pollen-tube growth pathway (e.g. in *N. alata*; 43, 94) and the extensin-like proteins in tobacco (60, 164). It seems possible that multiple cell wall glycoproteins play overlapping roles to sustain successful pollen-tube growth. The transmitting tissue ECM components must be interacting with some pollen-tube tip surface components, such as the tube-wall–associated extensins (126, 127). Interestingly, B is essential for pollen germination and tube growth (96). Pollen tubes extend by tip growth, i.e. by stretching and deposition of new cell wall material at the growing point rather than by overall cell wall stretching. In the absence of B, pollen tubes burst explosively, always at the outermost tip, indicating weakness of the newly formed cell wall. In addition, there seems to be a chemotropic response of pollen tubes of *Petunia* and *Agapanthus* to boron (124). That the deposition of some cell wall proteins seems to be regulated by the presence of boron in the wall (9) suggests that B may not only be a chemotropic agent but also an important player in the assembly of structural proteins in the pollen tube wall matrix.

The complex structure of AGPs has a possibility of generating many different types of molecules, all of which can have diverse functions. AGPs in the cell wall may be involved in directing planes of growth and development and participating in cell shape. Transport of AGPs or small components of AGPs through cell walls from one cell type to another may modify wall composition and ultimately add a new developmental path. Intercellular signaling between cells is crucial for a correct differentiation process. Since the addition of nanomolar quantities of specific AGPs to cell cultures changes the developmental fate of cells (84), it is presumed that AGPs have the ability to change a cell's fate.

GLYCINE-RICH PROTEINS

GRPs are another class of structural wall proteins. However, the isolation and characterization of GRPs have been done partially on almond seeds (162), strawberries (123), and pumpkin seed coats (154). The purification and characterization of a 30-kDa GRP from cell walls of the aleurone layer of soybean seeds have been reported (103). The protein is 68% Gly and 12% Ser and is slightly glycosylated. The N-terminal amino acid sequence of this GRP contains a novel poly-Gly structure that includes at least 20 Gly-repeated sequences. GRP gene sequences are characterized by their repetitive primary structure, which contains up to 70% Gly arranged in short amino acid repeat units. They usually contain an amino terminal signal peptide and, thus, are believed to be cell wall proteins. The derived amino acid sequence of the GRP gene isolated from *Petunia* contains 67% Gly and is arranged primarily in Gly-X repeat units, where X is most frequently Gly but can also be Ala or Ser (26). The predicted proteins of the bean GRP genes contain 63% and 58% Gly, and these are organized predominantly in repeating Gly-X units, similar to those predicted by the petunia GRP gene (76). It was predicted that the protein structure of *Petunia* GRP was capable of forming a β-pleated sheet composed of eight antiparallel strands (26). It was then expected that the aleurone GRP might form a β-pleated sheet because of its poly-Gly structure (103). However, circular dichroism (CD) measurements could not detect the β-pleated sheet in this glycoprotein. Perhaps the sugar chains interfere with the formation of this particular secondary structure (103). Several other groups have also isolated and characterized GRP cDNAs or genes from diverse plant species (134). Interestingly, there also are nodulins that are GRPs, such as *ENOD-GRP3* from *Vicia faba* (86). The encoded early nodulin *GRP3* is characterized by the presence of an N-terminal signal peptide and a C-terminal domain displaying a Gly content of 31% that is arranged primarily in Gly-X repeats, where X is usually Gly, Glu, or Gln. The *ENOD-GRP3* transcript is present in the interzone II-III region of broad bean root nodules. It is generally assumed that GRPs are insolubilized in the wall (153). Some GRPs have Tyr residues, and thus these might easily become linked to the aromatic residues of lignin. Tyrosine residues enhance cross-linking of synthetic proteins (polylysine/tyrosine) into lignin-like dehydrogenation products (106). The mechanism by which the phenolic Tyr residues enhance polymerization is not yet known, but they may act as templates or nucleation sites for the initiation of free radical reactions. However, not all GRPs have Tyr residues, and thus His, Val, Glu, and Gln could probably participate in the kinds of derivatization and linkage already mentioned for extensins. Transglutamylation is also a possible source of cross-linking, for GRPs as well as for nodulin PRPs, since both proteins contain high levels of Glu. However, the extraction of GRP from soybean aleurone layers with hot water suggests that GRP is associated with cell wall polysaccharides by nonionic bonds (103).

GRPs seem to play important roles in the development of vascular tissues, nodules, and flowers and during wound healing and freezing tolerance (21, 25, 86, 116, 128, 129). Ultrastructural studies reveal that GRP 1.8 protein is mainly localized in the modified primary walls of protoxylem cells (129), which according to light microscopic cytochemistry and ultrastructural observations are not lignified (115). Bean GRP 1.8 accumulates relatively late in dying or already dead xylem cells and is synthesized at least in part by adjacent xylem parenchyma cells. Thus, GRP 1.8 is secreted into the wall of neighboring xylem elements and probably repairs the walls of these dead cells that are subjected to intense passive stretching (128, 129). Xylem parenchyma has long been known to export certain substances, such as gums and resins, during the formation of heartwood (47), and thus presumably these cells also secrete proteins. Recently, it was reported that GRPs seem to mediate a repair process in the development of protoxylem cells (129). It is proposed that the modified primary walls of protoxylem cells are not just breakdown products, as has been believed, but that they are reinforced with GRP 1.8 to produce primary walls with an unusually high protein content and specialized chemical and physical properties.

CONCLUDING REMARKS

The diversity of cell wall proteins is perplexing, as is the functional multiplicity of plant cells. This complexity is probably essential for the developmental events, morphogenetic processes, and ecological adaptations that take place in plants. Some evidence supports the view that all these glycoproteins play mechanochemical and regulatory roles in development. As we begin to understand these remarkable gene products (none of which is necessary to the survival of a cell proper), and the sequence of primary processes in development that they help to regulate, we will be in a position to ask how we can account for genetically determined plant forms. We then must develop new techniques for studying the function and assembly of all these glycoproteins. That, together with knowledge of how the other nonprotein components are assembled in the wall, will help us finally comprehend plant development and morphogenesis.

ACKNOWLEDGMENTS

This work was supported by the following grants: DeGAPA IN204 496, IN206 694, and CONACyT 0268P-N9506. I thank Raúl Noguez for expert assistance with the preparation of the figures.

> **Visit the *Annual Reviews home page* at
> http://www.AnnualReviews.org.**

Literature Cited

1. Adair WS, Hwang C, Goodenough UW. 1983. Identification and visualization of the sexual agglutinin from the mating-type plus flagellar membrane of Chlamydomonas. *Cell* 33:183–93
2. Ahn JH, Choi Y, Kwon YM, Kim S-G, Choi YD, Lee JS. 1996. A novel extensin gene encoding a hydroxyproline-rich glycoprotein requires sucrose for its wound-inducible expression in transgenic plants. *Plant Cell* 8:1477–90
3. Allen AK, Desai NN, Neuberger A, Creeth JM. 1978. Properties of potato lectin and the nature of its glycopeptide linkages. *Biochem. J.* 171:665–74
4. Deleted in proof

5. Baldwin TC, Coen ES, Dickinson HG. 1992. The *ptl1* gene expressed in the transmitting tissue of *Antirrhinum* encodes an extensin-like protein. *Plant J.* 2:733–39

6. Barnabas B, Fridvalszky L. 1984. Adhesion and germination of differently treated maize pollen grains on the stigma. *Acta Bot. Hung.* 30:329–32

7. Basile DV. 1980. A possible mode of action for morphoregulatory hydroxyproline-proteins. *Bull. Torrey Bot. Club* 107:325–38

8. Basile DV, Basile MR. 1984. Probing the evolutionary history of bryophytes experimentally. *J. Hattori Bot. Lab.* 55:173–85

9. Bonilla I, Mergold-Villaseñor C, Campos M, Sánchez N, Pérez H, et al. 1997. The aberrant cell walls of boron deficient bean root nodules have no covalently-bound hydroxyproline-proline-rich proteins. *Plant Physiol.* 115:1329–40

10. Boudart G, Dechamp-Guillaume G, Lafitte C, Ricart G, Barthe J-P, et al. 1995. Elicitors and suppressors of hydroxyproline-rich glycoprotein accumulation are solubilized from plant cell walls by endopolygalacturonase. *Eur. J. Biochem.* 232:449–57

11. Bradley DJ, Kjellboom P, Lamb CJ. 1992. Elicitor- and wound induced oxidative cross-linking of a proline-rich plant cell wall proteins: a novel rapid defense response. *Cell* 70:21–30

12. Brady JD, Sadler IH, Fry SC. 1996. Diisodityrosine a novel tetrameric derivative of tyrosine in plant cell wall proteins: a new potential cross-link. *Biochem. J.* 315:323–27

13. Brisson LF, Tenhaken R, Lamb CJ. 1994. Function of oxidative cross-linking of cell wall structural protein in plant disease resistance. *Plant Cell* 6:1703–12

14. Butler LG. 1989. *Toxicants of Plant Origin*, pp. 95–121. Boca Raton, FL: CRC

15. Carpita NC. 1996. Structure and biogenesis of the cell wall of grasses. *Annu. Rev. Plant Physiol. Plant Mol. Biol.* 47:445–76

16. Carpita NC, Gibeaut DM. 1993. Structural models of primary cell walls in flowering plants: consistency of molecular structure with the physical properties of the walls during growth. *Plant J.* 3:1–30

17. Carpita NC, McCann M, Griffing LR. 1996. The plant extracellular matrix: news from the cell's frontier. *Plant Cell* 8:1451–63

18. Cassab GI. 1986. Arabinogalactan-proteins during the development of soybean root nodules. *Planta* 168:441–46

19. Cassab GI, Varner JE. 1987. Immunolocalization of extensin in developing soybean seed coats by immunogold-silver staining and tissue printing on nitrocellulose paper. *J. Cell Biol.* 105:2581–88

20. Cassab GI, Varner JE. 1988. Cell wall proteins. *Annu. Rev. Plant Physiol. Plant Mol. Biol.* 39:321–53

21. Castonguay Y, Nadeau P, Laberge S. 1993. Freezing tolerance and alteration of translatable mRNAs in alfalfa (*Medicago sativa* L.) hardened at subzero temperatures. *Plant Cell Physiol.* 34:31–38

22. Clarke AE, Anderson RL, Stone BA. 1979. Form and function of arabinogalactans and arabinogalactan-proteins. *Phytochemistry* 18:521–40

23. Clarke AE, Gleeson PA, Jermyn MA, Know RB. 1978. Characterization and localization of β-lectins in lower and higher plants. *Aust. J. Plant Physiol.* 5:707–22

24. Deleted in proof

25. Condit CM, McLean M, Meagher RB. 1990. Characterization of the expression of the petunia glycine-rich protein-1 gene product. *Plant Physiol.* 93:596–602

26. Condit CM, Meagher RB. 1986. A gene encoding a novel glycine-rich structural protein of petunia. *Nature* 323:178–81

27. Cooper JB, Varner JE. 1983. Insolubilization of hydroxyproline-rich glycoprotein in aerated carrot root slices. *Biochem. Biophys. Res. Commun.* 112:161–67

28. Coupe SA, Taylor JE, Isaac PG, Roberts JA. 1993. Identification and characterization of a proline-rich mRNA that accumulates during pod development in oilseed rape (*Brassica napus* L.). *Plant Mol. Biol.* 23:1223–32

29. Deleted in proof

30. Csanádi G, Szécsi J, Kaló P, Kiss P, Endre G, et al. 1994. *ENOD12*, an early nodulin gene is not required for nodule formation and efficient nitrogen fixation in alfalfa. *Plant Cell* 6:201–13

31. Chen C-G, Mau S-L, Clarke AE. 1993. Nucleotide sequence and style-specific expression of a novel proline-rich protein gene from *Nicotiana alata*. *Plant Mol. Biol.* 21:391–95

32. Chen C-G, Pu Z-Y, Mortiz RH, Simpson RJ, Bacic A, et al. 1994. Molecular cloning of a gene encoding an arabinogalactan-protein from pear (*Pyrus communis*) cell suspension culture. *Proc. Natl. Acad. Sci. USA* 91:10305–9

33. Chen J, Varner JE. 1985. Isolation and characterization of cDNA clones for carrot extensin and a proline-rich 33-kDa protein. *Proc. Natl. Acad. Sci. USA* 82:4399–403

34. Deleted in proof
35. Deleted in proof
36. Cheung AY. 1996. Pollen-pistil interaction during pollen-tube growth. *Trends Plant Sci.* 1:45–51
37. Cheung AY, May B, Kawata EE, Qing G, Wu H-M. 1993. Characterization of cDNAs for stylar transmitting tissue-specific proline-rich proteins in tobacco. *Plant J.* 3:151–60
38. Cheung AY, Wang H, Wu H-M. 1995. A floral transmitting tissue-specific glycoprotein attracts pollen tubes and stimulates their growth. *Cell* 82:383–89
39. Datta K, Schmidt A, Marcus A. 1989. Characterization of two soybean repetitive proline-rich proteins and a cognate cDNA from germinated axes. *Plant Cell* 1:945–52
40. Desai NN, Allen AK, Neuberger A. 1981. Some properties of the lectin from *Datura stamonium* (thorn-apple) and the nature of its glycoprotein linkages. *Biochem. J.* 197:345–53
41. Dougall DK, Shimbayashi K. 1960. Factors affecting growth of tobacco callus tissue and its incorporation of tyrosine. *Plant Physiol.* 35:396–404
42. Du H, Simpson RJ, Clark AE, Bacic A. 1996. Molecular characterization of a stigma-specific gene encoding an arabinogalactan-protein (AGP) from *Nicotiana alata*. *Plant J.* 9:313–23
43. Du H, Simpson RJ, Moritz RL, Clarke AE, Bacic A. 1994. Isolation of the protein backbone of an arabinogalactan-protein from the styles of *Nicotiana alata* and characterization of a corresponding cDNA. *Plant Cell* 6:1643–53
44. Edelman GM. 1988. *Topobiology.* New York: Basic Books. 212 pp.
45. Ertl H, Hallmann A, Wenzl S, Sumper M. 1992. A novel extensin that may organize extracellular matrix biogenesis in *Volvox carteri*. *EMBO J.* 11:2055–62
46. Ertl H, Mengele R, Wenzl S, Engel J, Sumper M. 1989. The extracellular matrix of *Volvox carteri*: molecular structure of cellular compartment. *J. Cell Biol.* 109:3493–501
47. Esau K. 1965. *Plant Anatomy.* New York: Wiley. 767 pp.
48. Esquerré-Tugayé M-T, Mazau D, Pélissier B, Roby D, Rumeau D, Toppan A. 1985. Induction by elicitors and ethylene of proteins associated to the defense of plants. In *Cellular Biology of Plant Stress*, ed. JL Key, T Kosuge, pp. 459–73. New York: Liss
49. Evrad J-L, Jako C, Saint-Guily A, Weil J-H, Kuntz M. 1991. Anther-specific developmentally regulated expression of genes encoding a new class of proline-rich proteins in sunflower. *Plant Mol. Biol.* 16:271–81
50. Fincher GB, Stone BA, Clarke AE. 1983. Arabinogalactan-proteins: structure biosynthesis and function. *Annu. Rev. Plant Physiol.* 34:47–70
51. Franssen HJ, Nap J-P, Gloudemans T, Stiekema W, van Dam H, et al. 1987. Characterization of cDNA for nodulin-75 of soybean: a gene product involved in early stages of root nodule development. *Proc. Natl. Acad. Sci. USA* 84:4495–99
52. Freshour G, Clay RP, Fuller MS, Albersheim P, Darvil AG, Hahn MG. 1996. Developmental and tissue-specific structural alterations of the cell-wall polysaccharides of *Arabidopsis thaliana* roots. *Plant Physiol.* 110:1413–29
53. Fry SC. 1986. Cross-linking of matrix polymers in the growing cell walls of angiosperms. *Annu. Rev. Plant Physiol.* 37:165–86
54. Fry SC, Aldington S, Hetherington PR, Aitken J. 1993. Oligosaccharides as signals and susbstrates in the plant cell wall. *Plant Physiol.* 103:1–5
55. Gebicki S, Gebicki JM. 1993. Formation of peroxides in amino acids and proteins exposed to oxygen free radicals. *Biochem. J.* 289:743–49
56. Gerster J, Allard S, Robert LS. 1996. Molecular characterization of two *Brassica napus* pollen-expressed genes encoding putative arabinogalactan proteins. *Plant Physiol.* 110:1231–37
57. Gifford EM, Foster AS. 1989. *Morphology and Evolution of Vascular Plants.* New York: Freeman. 626 pp. 3rd ed.
58. Gleeson PA, McNamara M, Wettenhall EH, Stone BA, Fincher GB. 1989. Characterization of the hydroxyproline rich protein core of an arabinogalactan-protein secreted from suspension cultured *Lolium multiflorum* (Italian ryegrass) endosperm cells. *Biochem. J.* 264:857–62
59. Goldbach HE, Blaser-Grill J, Lindemann N, Porzelt M, Hörrmann C, et al. 1991. Influence of boron on net proton release and its relation to other metabolic processes. In *Current Topics in Plant Biochemistry and Physiology*, ed. DD Randall, DG Blevins, CD Miles, 10:195–220. Columbia, MO: Univ. Columbia. 320 pp.
60. Goldman MHS, Pezzotti M, Seurinck J, Mariani C. 1992. Developmental expression of tobacco pistil-specific genes encoding novel extensin-like proteins. *Plant Cell* 4:1041–51

61. Haberlandt G. 1914. *Physiological Plant Anatomy*. New Delhi: Today & Tomorrow. 775 pp.

62. Herman EM, Lamb CJ. 1992. Arabinogalactan-rich glycoproteins are localized on the cell surface and in intravacuolar multivesicular bodies. *Plant Physiol.* 98:264–72

63. Heslop-Harrison J. 1987. Pollen germination and pollen tube growth. *Int. Rev. Cytol.* 107:1–78

64. Heslop-Harrison Y, Reger BJ, Heslop-Harrison J. 1984. The pollen-stigma interaction in the grasses. 6. The stigma ("silk") of *Zea mays* L. as host to pollen of *Sorghum bicolor* (L.) Moench and *Pennisetum americanum* (L.) Leeke. *Acta Bot. Neerl.* 33:205–27

65. Hong JC, Cheon YW, Nagao RT, Bahk JD, Cho MJ, Key JL. 1994. Isolation and characterization of three soybean extensin cDNAs. *Plant Physiol.* 104:793–96

66. Hong JC, Nagao RT, Key JL. 1987. Characterization and sequence analysis of a developmentally regulated putative cell wall protein gene isolated from soybean. *J. Biol. Chem.* 262:8367–76

67. Hood EE, Shen Q-X, Varner JE. 1988. A developmentally regulated hydroxyproline-rich glycoprotein in maize pericarp cell walls. *Plant Physiol.* 87:138–42

68. Horn MA, Heistein PF, Low PS. 1989. Receptor-mediated endocytosis in plant cells. *Plant Cell* 1:1003–9

69. Hu HI, Brown PH. 1994. Localization of boron in cell walls of squash and tobacco and its association with pectin. *Plant Physiol.* 105:681–89

70. Huber O, Sumper M. 1994. Algal-CAMs: isoforms of a cell adhesion molecule in embryos of the alga *Volvox* with homology to *Drosophila* fasciclin-I. *EMBO J.* 13:4212–22

71. Ishi T, Matsunaga T. 1996. Isolation and characterization of boron rhamnogalacturonan II complex from sugar beet pulp. *Carbohydr. Res.* 284:1–9

72. Josè-Estanyol M, Puigdomènech P. 1998. Developmental and hormonal regulation of genes coding for proline-rich proteins in maize embryo and ovary. *Plant Physiol.* In press

73. Jun T, Shupin W, Daye A. 1996. Extracellular calmodulin-binding proteins in plants: purification of a 21-kDa calmodulin-binding protein. *Planta* 198:510–16

74. Kaplan DR, Hagemann W. 1991. The relationship of cell and organism in vascular plants. *BioScience* 41:693–703

75. Keegstra K, Talmadge KW, Bauer WD, Albersheim P. 1973. The structure of plant cell walls. III. A model of the walls of suspension cultured sycamore cells based on the interconnections of the macromolecular components. *Plant Physiol.* 51:188–96

76. Keller B, Sauer N, Lamb CJJ. 1988. Glycine-rich cell wall proteins in beans: gene structure and association of the proteins with the vascular system. *EMBO J.* 7:3625–33

77. Kieliszewski MJ, Kamyab A, Leykam JF, Lamport DTA. 1992. A histidine-rich extensin from *Zea mays* is an arabinogalactan protein. *Plant Physiol.* 99:538–47

78. Kieliszewski MJ, Lamport DTA. 1994. Extensin: repetitive motifs functional sites post-translational codes and phylogeny. *Plant J.* 5:157–72

79. Kieliszewski MJ, Showalter AM, Leykam JF. 1994. Potato lectin: a modular protein sharing sequence similarities with the extensin family, the hevein lectin family and snake venom disintegrins (platelet aggregation inhibitors). *Plant J.* 5:849–61

80. Knox JP, Day S, Roberts K. 1989. A set of surface glycoproteins forms an early marker of cell position but not cell type in the root apical meristem of *Daucus carota* L. *Development* 106:47–56

81. Knox JP, Lindstead PJ, Peart J, Cooper C, Roberts K. 1991. Developmentally regulated epitopes of cell surface arabinogalactan proteins and their relation to root tissue pattern formation. *Plant J.* 1:317–26

82. Kobayashi M, Matoh T, Azuma J-C. 1996. Two chains of rhamnogalacturonan II are cross-linked by borate-diol ester bonds in higher plant cell walls. *Plant Physiol.* 110:1017–20

83. Korn RW. 1982. Positional specificity within plant cells. *J. Theor. Biol.* 95:543–68

84. Kreuger M, van Holst G-J. 1993. Arabinogalactan proteins are essential in somatic embryogenesis of *Daucus carota* L. *Planta* 189:243–48

85. Kreuger M, van Holst G-J. 1996. Arabinogalactan proteins and plant differentiation. *Plant Mol. Biol.* 30:1077–86

86. Küster H, Schröeder G, Frühling M, Pich U, Rieping M, et al. 1995. The nodule-specific *VfENOD-GRP3* gene encoding a glycine-rich early nodulin is located on chromosome I of *Vicia faba* L. and is predominantly expressed in the interzone II-III of root nodules. *Plant Mol. Biol.* 28:405–21

87. Lamport DTA. 1986. Roles for peroxidases in cell wall genesis. In *Molecular*

and Physiological Aspects of Plant Per-oxidases, ed. H Greppin, C Penel, T Gaspar, pp. 199–208. Geneva: Univ. Geneva Press

88. Lamport DTA, Northcote DH. 1960. Hydroxyproline in primary cell walls of higher plants. *Nature* 118:665–66
89. Deleted in proof
90. Leach JE, Cantrell MA, Sequeira L. 1982. Hydroxyproline-rich bacterial agglutinin from potato. *Plant Physiol.* 70:1353–58
91. Li S-X, Showalter AM. 1996. Immunolocalization of extensin and potato tuber lectin in carrot tomato and potato. *Physiol. Plant.* 97:708–18
92. Li S-X, Showalter AM. 1996. Cloning and developmental/stress-regulated expression of a gene encoding a tomato arabinogalactan protein. *Plant Mol. Biol.* 32:641–52
93. Li X-B, Kieliszewski MJ, Lamport DTA. 1990. A chenopod extensin lacks repetitive tetrahydroxyproline blocks. *Plant Physiol.* 92:327–33
94. Lind JL, Bacic A, Clark AE, Anderson MA. 1994. A style-specific hydroxyproline-rich glycoprotein with properties of both extensins and arabinogalactan proteins. *Plant J.* 6:491–502
95. Lindstrom JT, Vodkin LO. 1991. A soybean cell wall protein is affected by seed color genotype. *Plant Cell* 3:561–71
96. Loomis WD, Durst RW. 1992. Chemistry and biology of boron. *BioFactors* 3:229–39
97. Loopstra CA, Sederoff RR. 1995. Xylem-specific gene expression in loblolly pine. *Plant Mol. Biol.* 27:277–91
98. Lord EM, Sanders LC. 1992. Roles for the extracellular matrix in plant development and pollination: a special case of cell movement in plants. *Dev. Biol.* 153:16–28
99. Lucas WJ. 1995. Plasmodesmata: intercellular channels for macromolecular transport in plants. *Curr. Opin. Cell Biol.* 7:673–80
100. Luck G, Liao H, Murray NJ, Grimmer HR, Warminski EE, et al. 1994. Polyphenols astringency and proline-rich proteins. *Phytochemistry* 37:357–71
101. Marcus A, Greenberg J, Averyhart-Fullard V. 1991. Repetitive proline-rich proteins in the extracellular matrix of the plant cell. *Physiol. Plant.* 81:273–79
102. Mascarenhas JP, Machlis L. 1962. Chemotropic response of *Antirrhinum majus* pollen to calcium. *Nature* 196:292–93
103. Matsui M, Toyosawa I, Fukuda M. 1995. Purification and characterization of a glycine-rich protein from the aleurone layer of soybean seeds. *Biosci. Biotechnol. Biochem.* 59:2231–34

104. Mau S-L, Chen C-G, Pu Z-Y, Mortiz RL, Simpson RJ, et al. 1995. Molecular cloning of cDNAs encoding the protein backbones of arabinogalactan-proteins from the filtrate of suspension-cultured cells of *Pyrus communis* and *Nicotiana alata. Plant J.* 8:269–81
105. Mazau D, Rumeau D, Esquerré-Tugayé M-T. 1987. Molecular approaches to understanding cell surface interactions between plants and fungal pathogens. *Plant Physiol. Biochem.* 25:337–43
106. McDougall G-J, Stewart D, Morrison IM. 1996. Tyrosine residues enhance cross-linking of synthetic proteins into lignin-like dehydrogenation products. *Phytochemistry* 41:43–47
107. McQueen-Mason SJ, Durachko DM, Cosgrove DJ. 1992. Two endogenous proteins that induce cell wall extension in plants. *Plant Cell* 4:1425–33
108. Mellon JE, Helgeson JP. 1982. Interaction of a hydroxyproline-rich glycoprotein from tobacco callus with potential pathogens. *Plant Physiol.* 70:401–5
109. Memelink J, Swords KMM, de Kam RJ, Schilperoot RA, Hoge JHC, Staehelin LA. 1993. Structure and regulation of tobacco extensin. *Plant J.* 4:1011–22
110. Mergold-Villaseñor CA, Castrejón L, Cassab GI. 1996. The role of turgor pressure in the cross-linking of plant cell wall proteins. *Plant Physiol.* 111:670 (Abstr.)
111. Merkle RK, Cumming RD. 1987. Tomato lectin is located predominantly in the locular fluid of ripe tomatoes. *Plant Sci.* 48:71–78
112. Momonoki YS. 1997. Asymmetric distribution of acetylcholinesterase in gravi-stimulated maize seedlings. *Plant Physiol.* 114:47–53
113. Nersissian AM, Mehrabian ZB, Nalbandyan RM, Hart PJ, Fraczkiewicz G, et al. 1996. Cloning expression and spectroscopic characterization of *Cucumis sativus* stellacyanin in its nonglycosylated form. *Protein Sci.* 5:2184–92
114. Niebel A, Engler JA, de Tiré C, Engler G, Van Montagu M, Gheysen G. 1993. Induction patterns of an extensin gene in tobacco upon nematode infection. *Plant Cell* 5:1697–710
115. O'Brian TP. 1981. The primary xylem. In *Xylem Cell Development*, ed. JR Barnett, pp. 14–46. Turnbridge Wells, Kent, UK: Castle House Publ.
116. Oliveira DD, de Franco L, Simoens C, Seurinck J, Coppiters J, et al. 1993.

Inflorescence-specific genes from *Arabidopsis thaliana* encoding glycine-rich proteins. *Plant J.* 3:495–507

117. O'Neill MA, Warrenfeltz D, Kates K, Pellerin P, Doco T, et al. 1996. Rhamnogalacturonan-II, a pectic polysaccharide in the walls of growing plant cells form a dimer that is covalently cross-linked by a borate ester. *J. Biol. Chem.* 271:22923–30

118. Pennel RI, Janniche L, Scofield GN, Booij H, Vries SC, de Roberts K. 1992. Developmental regulation of a plasma membrane arabinogalactan protein epitope in oilseed rape flowers. *Plant Cell* 3:1317–26

119. Pickett-Heaps J. 1976. Cell division in eucaryotic algae. *BioScience* 26:445–50

120. Platthy L. 1988. Detecting distant homologies of mosaic proteins: analysis of the sequences of thrombomodulin thrombospondin complement components C9, C8a, C8b, vitronectin and plasma cell membrane glycoproteins PC-1. *J. Mol. Biol.* 202:689–96

121. Pogson BJ, Davies C. 1995. Characterization of a cDNA encoding the protein moiety of a putative arabinogalactan protein from *Lycopersicon esculentum*. *Plant Mol. Biol.* 28:347–52

122. Qi XY, Beherens BX, West PR, Mort AJ. 1995. Solubilization and partial characterization of extensin fragments from cell walls of cotton suspension cultures. *Plant Physiol.* 108:1691–701

123. Reddy ASN, Poovaiah BW. 1987. Accumulation of a glycine-rich protein in auxin-deprived strawberry fruits. *Biochem. Biophys. Res. Commun.* 147:885–91

124. Robbertse PJ, Lock JJ, Stoffberg E, Coetzer LA. 1990. Effect of boron on directionality of pollen tube growth in *Petunia* and *Agapanthus*. *S. Afr. J. Bot.* 56:487–92

125. Roberts MR, Foster GD, Blundell RP, Robinson SW, Kumar A, et al. 1993. Gametophytic and sporophytic expression of an anther-specific *Arabidopsis thaliana* gene. *Plant J.* 3:111–20

126. Rubinstein AL, Broadwater AH, Lowrey KB, Bedinger PA. 1995. *Pex1*, a pollen-specific gene with an extensin-like domain. *Proc. Natl. Acad. Sci. USA* 92:3086–90

127. Rubinstein AL, Márquez J, Suárez-Cervera M, Bedinger PA. 1995. Extensin-like glycoproteins in the maize pollen tube wall. *Plant Cell* 7:2211–25

128. Ryser U, Keller B. 1992. Ultrastructural localization of a bean glycine-rich protein in unlignified primary walls of protoxylem cells. *Plant Cell* 4:773–83

129. Ryser U, Schorderet M, Zhao G-F, Studer D, Ruel K, et al. 1997. Structural cell-wall proteins in protoxylem development: evidence for a repair process mediated by a glycine-rich protein. *Plant J.* 12:97–111

130. Deleted in proof

131. Schindler T, Bergfeld R, Schopfer P. 1995. Arabinogalactan proteins in maize coleoptiles: developmental relationship to cell death during xylem differentiation but not to extension growth. *Plant J.* 7:25–36

132. Sheng J, D'Ovidio R, Mehdy MC. 1991. Negative and positive regulation of a novel proline-rich protein mRNA by fungal elicitor and wounding. *Plant J.* 1:345–54

133. Shirsat AH, Bell A, Spence J, Harris JN. 1996. The *Brassica napus extA* extensin gene is expressed in regions of the plant subject to tensile stresses. *Planta* 199:618–24

134. Showalter AM. 1993. Structure and function of plant cell wall proteins. *Plant Cell* 5:9–23

135. Smith RC, Fry SC. 1991. Endotransglycosylation of xyloglucans in plant cell suspension cultures. *Biochem. J.* 279:529–35

136. Sommer RAC, Lipman SC. 1926. Evidence of the indispensable nature of zinc and boron for higher green plants. *Plant Physiol.* 1:231–49

137. Sommer-Knudsen J, Clark AE, Bacic A. 1996. A galactose-rich cell-wall glycoprotein from styles of *Nicotiana alata*. *Plant J.* 9:71–83

138. Stacey NJ, Roberts K, Knox JP. 1990. Patterns of expression of the JIM4 arabinogalactan-protein epitopes in cell cultures and during somatic embryogenesis in *Daucus carota* L. *Planta* 180:285–92

139. Stafstrom JP, Staehlin LA. 1986. Cross-linking patterns in salt-extractable extensin from carrot cell walls. *Plant Physiol.* 81:234–41

140. Stebbins GL. 1992. Comparative aspects of plant morphogenesis: a cellular molecular and evolutionary approach. *Am. J. Bot.* 79:589–98

141. Sticher L, Hofsteenge J, Milani A, Neuhaus J-M, Meins F Jr. 1992. Vacuolar chitinase of tobacco: a new class of hydroxyproline-containing proteins. *Science* 257:655–57

142. Sussex IM. 1989. Developmental programming of the shoot meristem. *Cell* 56:225–29

143. Thompson EW, Preston RD. 1967. Proteins in the cell wall of some green algae. *Nature* 213:684–85

144. Deleted in proof

145. Tierney ML, Wiechert J, Pluymers D. 1988. Analysis of the expression of extensin and p33-related cell wall proteins in carrot and soybean. *Mol. Gen. Genet.* 211:393–99

146. Tiré C, Rycke R, de Loose M, Inzé D, Van Montagu M, Engler G. 1994. Extensin gene expression is induced by mechanical stimuli leading to local cell wall strengthening in *Nicotiana plumbaginifolia. Planta* 195:175–81

147. Tornero P, Mayda E, Gómez MD, Cañas L, Conejero V, Vera P. 1996. Characterization of LRP, a leucine-rich repeat (LRR) protein from tomato plants that is processed during pathogenesis. *Plant J.* 10:315–30

148. van de Wiel C, Scheres B, Franssen H, van Lierop M-J, van Lammeren A, et al. 1990. The early nodulin transcript ENOD2 is located in the nodule parenchyma (inner cortex) of pea and soybean root nodules. *EMBO J.* 9:1–7

149. van Holst G-J, Clarke AE. 1986. Organ specific arabinogalactan-proteins of *Lycopersicum peruvianum* (Mill.) demonstrated by crossed electrophoresis. *Plant Physiol.* 80:786–89

150. van Holst G-J, Klis FM, Wildt PJM, Hazenberg CAM, Buijs J, Stegwee D. 1981. Arabinogalactan-protein from a crude well organelle fraction of *Phaseolus vulgaris* L. *Plant Physiol.* 68:910–13

151. van Holst G-J, Martin SR, Allen AK, Ashford D, Desai NN, Neuberger A. 1986. Protein conformation of potato (*Solanum tuberosum*) lectin determined by circular dichroism. *Biochem. J.* 233:731–36

152. van Holst G-J, Varner JE. 1984. Reinforces polyproline II conformation in a hydroxyproline-rich cell wall glycoprotein from carrot root. *Plant Physiol.* 74:247–51

153. Varner JE. 1994. Cell wall structural proteins—can we deduce function from sequence? *Biochem. Soc. Symp.* 60:1–4

154. Varner JE, Cassab GI. 1986. A new protein in petunia. *Nature* 323:110

155. Varner JE, Lin S-L. 1989. Plant cell wall architecture. *Cell* 56:231–39

156. Vasil IK. 1987. Physiology and culture of pollen. *Int. Rev. Cytol.* 107:127–75

157. Wang H, Wu H-M, Cheung AY. 1993. Development and pollination regulated accumulation and glycosylation of a stylar transmitting tissue specific proline-rich protein. *Plant Cell* 5:1639–50

158. Witty JF, Minchin FR, Skot L, Sheehy JE. 1986. Nitrogen fixation and oxygen in legume root nodules. *Oxford Surv. Plant Mol. Cell. Biol.* 3:275–314

159. Woessner JP, Goodenough UW. 1989. Molecular characterization of a zygote wall protein: an extensin-like molecule in *Chlamydomonas reinhardtii. Plant Cell* 1:901–11

160. Woessner JP, Goodenough UW. 1994. Volvocine cell walls and their constituent glycoproteins: an evolutionary perspective. *Protoplasma* 181:245–58

161. Woessner JP, Molendijk AL, van Egmond P, Klis FM, Goodenough UW, Haring MA. 1994. Domain conservation in several volvocalean cell wall proteins. *Plant Mol. Biol.* 26:947–60

162. Wolf WJ. 1995. Gel electrophoresis and amino analysis of the nonprotein nitrogen fractions of defatted soybean and almond meals. *Cereal Chem.* 72:115–21

163. Wu H-M, Wang H, Cheung AY. 1995. A pollen tube growth stimulatory glycoprotein is deglycosylated by pollen tubes and displays a glycosylation gradient in the flower. *Cell* 82:395–403

164. Wu H-M, Zou JT, May B, Gu Q, Cheung AY. 1993. A tobacco gene family for flower cell wall proteins with a proline-rich domain and a cysteine-rich domain. *Proc. Natl. Acad. Sci. USA* 90:6829–33

165. Wyatt RE, Nagao RT, Key JL. 1992. Patterns of soybean proline-rich protein gene expression. *Plant Cell* 4:99–110

166. Yariv J, Rapport MM, Graf L. 1962. The interaction of glycosides and saccharides with antibody to the corresponding phenylazo glycoside. *Biochem. J.* 85:383–88

167. Ye Z-H, Song Y-R, Marcus A, Varner JE. 1991. Comparative localization of three classes of cell wall proteins. *Plant J.* 1:175–83

168. Ye Z-H, Varner JE. 1991. Tissue-specific expression of cell wall proteins in developing soybean tissues. *Plant Cell* 3:23–37

Annu. Rev. Plant Physiol. Plant Mol. Biol. 1998. 49:311–43

MOLECULAR-GENETIC ANALYSIS OF PLANT CYTOCHROME P450-DEPENDENT MONOOXYGENASES

Clint Chapple

Department of Biochemistry, Purdue University, West Lafayette, Indiana 47907-1153;
e-mail: chapple@biochem.purdue.edu

KEY WORDS: cloning, hydroxylase, metabolism, mutants

ABSTRACT

Cytochrome P450-dependent monooxygenases are a large group of heme-containing enzymes, most of which catalyze NADPH- and O_2-dependent hydroxylation reactions. The cloning of plant P450s has been hampered because these membrane-localized proteins are typically present in low abundance and are often unstable to purification. Since the cloning of the first plant P450 gene in 1990, there has been an explosion in the rate at which genes encoding plant P450s have been identified. These successes have largely been the result of advances in purification techniques, as well as the application of alternative methods such as mutant- and PCR-based cloning strategies. The availability of these cloned genes has made possible the analysis of P450 gene regulation and may soon reveal aspects of the evolution of P450s in plants. This new knowledge will significantly improve our understanding of many metabolic pathways and may permit their manipulation in the near future.

CONTENTS

INTRODUCTION

It has been forty years since the first report of the presence of a carbon monoxide–binding pigment in rat and pig liver microsomes treated with sodium dithionite (52, 76). Under these conditions, this pigment was found to strongly absorb 450 nm light. Six years later, Omura & Sato reported that this unique pigment had characteristics of a heme-containing protein and was distinct from cytochrome b_5, and named the substance "P-450" (P for pigment) (107). At about the same time, research in a number of labs was focused on oxygen- and NADPH-dependent microsomal enzymes that catalyzed a variety of hydroxylation, deamination, oxidation, and N-dealkylation reactions involved in the metabolism of steroids (40) and xenobiotics (96, reviewed in 18). It was found that these enzymes could be inhibited by carbon monoxide and that this inhibition could be reversed by irradiation with 450 nm light, thus linking the enigmatic "P-450" with enzymatic function. It soon became clear that these enzymes represented a major class of proteins that we now know as cytochrome P450-dependent monooxygenases (P450s). Very soon thereafter, an enzyme of the phenylpropanoid pathway in plants, cinnamate-4-hydroxylase, was identified in the microsomal fraction of plant tissue extracts (118) and was shown to have the classic P450 characteristic of light-reversible inhibition of enzymatic activity by carbon monoxide (117). Although the microsomes in which the enzyme was assayed also showed an absorbance peak at 450 nm in the presence of carbon monoxide and sodium dithionite (113), it was a further twenty years before this characteristic was shown to be intrinsic to the purified protein (97).

Since the time of their discovery, P450s have been found in bacteria, insects, fish, mammals, plants, and fungi. We now know that these enzymes represent a superfamily of heme-containing proteins, most of which catalyze NADPH- and O_2-dependent hydroxylation reactions. Most of these enzymes do not use NADPH directly but instead interact with a flavoprotein known as a P450 reductase that transfers electrons to the P450 from the nicotinamide cofactor. A great deal is now known about the roles, sequences, and catalytic mechanisms of animal and bacterial P450s, particularly due to the study of those P450s that function in the detoxification of xenobiotics. In several notable examples of bacterial origin, the three dimensional structures of these enzymes have also been solved, and this work has contributed greatly to the understanding of the role of specific amino acids in catalytic function (114).

Plant P450s participate in myriad biochemical pathways, including those devoted to the synthesis of plant products such as phenylpropanoids, alkaloids, terpenoids, lipids, cyanogenic glycosides, and glucosinolates, and plant growth regulators such as gibberellins, jasmonic acid, and brassinosteroids. Many comprehensive reviews have recently been published concerning the breadth of reactions known to be catalyzed by plant P450s (14, 39, 121), the role of P450s in the biosynthesis and degradation of compounds important in plant insect interactions (122), the complex redox and radical chemistry catalyzed by P450s (61), and the importance of plant P450s in the metabolism of herbicides (5, 48). The primary purpose of this chapter is to review the advances in our understanding of plant P450s at the genetic and molecular level that have come about since the cloning of the first plant P450 gene at the beginning of this decade (16).

P450 PRIMARY SEQUENCE MOTIFS

Most eukaryotic P450s are 50- to 60-kDa proteins associated with the endoplasmic reticulum (ER) and are cotranslationally inserted in a signal recognition particle-dependent fashion (119). They remain anchored to the ER membrane by a hydrophobic helix near the N-terminus with most of the protein residing on the cytosolic face of the membrane (120, 153). The transmembrane helix is usually followed by a series of basic amino acid residues that may interact with the negatively charged head groups of ER membrane lipids; however, it has been shown that these residues are not required to stop the continued transfer of the nascent polypeptide into the ER lumen (120). Additional ER targeting signals have been identified in the cytoplasmic domain of P450s, suggesting that the N-terminal anchor may be the major but not the sole mechanism for ER retention (137).

A proline-rich region immediately after the N-terminal hydrophobic helix typically obeys the consensus (P/I)PGPx(G/P)xP. Considering that both proline and glycine are known to destabilize α-helices, this region is thought to act as a "hinge" that is required for optimal orientation of the enzyme with regard to the membrane (158). Deletion of this motif has been shown to result in a complete loss of enzyme activity (138), and mutation of the proline residues to alanine can disrupt protein structure sufficiently to eliminate heme incorporation (158).

Although these changes resulted in an increase in protease susceptibility, in neither case was there an impact on the mutant proteins' targeting to, or retention in, the ER.

Analysis of the crystal structure of prokaryotic P450s identified a series of residues that form a threonine-containing binding pocket for the oxygen molecule required in catalysis. Sequences obeying the consensus (A/G)Gx(D/E)T(T/S) have now been found in eukaryotic P450s where they are thought to perform a similar function (38, 153). It is of interest that plant and animal P450s which no longer use molecular oxygen in catalysis often have amino acid residues in this region of their primary sequence that deviate from this consensus (see sections below on allene oxide synthase and fatty acid hydroperoxide lyase).

Finally, P450s are heme proteins. The presence of heme is essential for catalysis, and gives P450s the carbon monoxide-binding ability that is often important in their characterization. The heme-binding domain contains the signature sequence FxxGxxxCxG where the conserved cysteine serves as the fifth ligand to the heme iron (153).

On the basis of overall identity and sequence conservation within these domains, phylogenetic analysis of plant P450s has led to the identification of "group A" and "non–group A" proteins (38). The non–group A P450s are a disparate set of families that cluster closer to animal and fungal enzymes than do the group A proteins. Group A P450s are a tighter group in the phylogenetic sense and appear to have evolved since the divergence of plants from the common ancestor they share with animals and fungi.

P450 NOMENCLATURE

The diversity of P450 sequence, function, and distribution has required the development of a systematic nomenclature that assigns each protein to a specific family and subfamily (Table 1) based upon its primary sequence (103). All P450 systematic gene names include the designation *CYP* for **cy**tochrome **P**450, (CYP when refering to mRNA, cDNA and protein) although it has recently been suggested that the term "heme-thiolate protein" is a preferable name for P450s since they are not actually cytochromes (103). If the term cytochrome P450 is used, "P450" is preferred over "P-450." P450 family names are assigned chronologically following the determination of their primary sequence. Relationships between P450s are determined on the basis of amino acid sequence identity. If the amino acid sequence of a newly discovered P450 has more than 40% identity to one or more previously identified P450s, the proteins are said to be members of the same P450 family. If the new sequence has 40% or less identity with known sequences, it defines a new family. The numbering of plant

P450 gene families begins with *CYP71* through *CYP99*. These family designations have now been exhausted and continue from *CYP701*, since *CYP101* begins a block reserved for prokaryotic P450s. If two P450s are in the same family but are less than 55% identical, they are determined to be members of two subfamilies, such as *CYP74A* and *CYP74B* (82, 89, 109, 131). Proteins that are more than 55% identical are designated as members of the same subfamily, such as the CYP71C subfamily of maize, which includes CYP71C1 through CYP71C5, each of which catalyzes a different reaction in DIMBOA biosynthesis (49, 50). Similarly, genes encoding the same protein from different species are often members of the same subfamily, such as in the case of the genes encoding cinnamate-4-hydroxylase (*CYP73A1* through *CYP73A16*). P450s that are 97% identical or greater are assumed to be allelic variants of the same gene unless otherwise demonstrated (103).

Although the "40% rule" was an arbitrary standard set when relatively few P450 sequences were available, identity values are usually significantly lower or higher than 40%, making the *CYP* family classifications relatively unambiguous. Some exceptions do exist, as in the placement of the plant obtusifoliol 14α-demethylase in the CYP51 family along with the fungal and mammalian lanosterol 14α-demethylases (see section on obtusifoliol 14α-demethylase below). In other cases, additional criteria such as intron/exon boundaries of P450 genes are considered in the assignment of P450 family designations.

The rapid identification of new P450s made it essential that a committee be struck to insure the uniform use of P450 nomenclature in the literature. Authors are encouraged to submit new P450 sequences before publication to David Nelson (dnelson@utmem1.utmem.edu) at the University of Tennessee, Memphis, for consideration and assignment of *CYP* designation.

MOLECULAR ANALYSIS OF PLANT P450s

The identification of P450 genes and the determination of their biochemical function are now at the center of plant P450 research. The biochemical irony of these investigations is that most plant P450 genes have not been cloned, yet most of those that have been cloned have not been associated with an in vivo function (Table 1). The cloning of plant P450 genes via protein purification is often difficult because these enzymes are present at low levels in plant cells and are difficult to purify. Nevertheless, standard biochemical approaches have been successful in a number of cases. Conversely, while it is relatively straightforward to identify P450 genes using PCR or through the use of heterologous probes, the sequence of a newly identified gene generally provides little or no information about its in vivo function. Recently, the limitations to both of these approaches have been circumvented by using mutants defective in P450-catalyzed reactions

Table 1   Cloned genes encoding plant cytochrome P450–dependent monooxygenases[a]

P450 family/ subfamily	Enzyme name	Species	Genbank accession #	Reference
CYP51	Obtusifoliol 14α-demethylase	*Sorghum bicolor*	U74319	3
CYP71A1	Unknown	*Persea americana*	M32885	16
CYP71A2	Unknown	*Solanum melongena*	X71654	149
CYP71A3	Unknown	*Solanum melongena*	X70982	149
CYP71A4	Unknown	*Solanum melongena*	X70981	149
CYP71A5	Unknown	*Nepeta racemosa*	Y09423	25
CYP71A6	Unknown	*Nepeta racemosa*	Y09424	25
CYP71B1	Unknown	*Thlaspi arvense*	L24438	148
CYP71B7	Unknown	*Arabidopsis thaliana*	X97864	90
CYP71C1	HBOA synthase[b]	*Zea mays*	X81827, X81828	49, 50
CYP71C2	Indolin-2-one hydroxylase[b]	*Zea mays*	X81829, Y11404	49, 50
CYP71C3	HBOA-*N*-hydroxylase[b]	*Zea mays*	X81830, Y11403	49, 50
CYP71C4	Indolin-2-one synthase[b]	*Zea mays*	X81831, Y11368	49, 50
CYP71D6	Unknown	*Solanum chacoense*	U48434	73
CYP71D7	Unknown	*Solanum chacoense*	U48435	73
CYP72A1	Unknown	*Catharanthus roseus*	L10081, L19074, L19075	88, 151
CYP73A1	Cinnamate-4-hydroxylase	*Helianthus tuberosus*	Z17369	143
CYP73A2	Cinnamate-4-hydroxylase	*Phaseolus aureus*	L07634	99
CYP73A3	Cinnamate-4-hydroxylase	*Medicago sativa*	L11046	93
CYP73A4	Cinnamate-4-hydroxylase	*Catharanthus roseus*	X69788, Z32563	72, 92
CYP73A5	Cinnamate-4-hydroxylase	*Arabidopsis thaliana*	U71080, U71081 D78596, D78597	8, 98
CYP73A9	Cinnamate-4-hydroxylase	*Pisum sativum*	U29243	46
CYP73A10	Cinnamate-4-hydroxylase	*Petroselinum crispum*	L38898	86
CYP73A12	Cinnamate-4-hydroxylase	*Zinnia elegans*	U19922	159
CYP73A13	Cinnamate-4-hydroxylase	*Populus tremuloides*	U47293	53
CYP73A16	Cinnamate-4-hydroxylase	*Populus kitakamiensis*	D82815, D82812	75
CYP73A?	Cinnamate-4-hydroxylase	*Populus kitakamiensis*	D82813	75
CYP73A?	Cinnamate-4-hydroxylase	*Populus kitakamiensis*	D82814	75
CYP74A1	Allene oxide synthase	*Linum usitatissimum*	U00428	131
CYP74A2	Rubber particle protein	*Parthenium argentatum*	X78166	109
CYP74A3	Allene oxide synthase	*Arabidopsis thaliana*	X92510	82

(*Continued*)

Table 1   (*Continued*)

P450 family/ subfamily	Enzyme name	Species	Genbank accession #	Reference
CYP74B	Fatty acid hydroperoxide lyase	*Capsicum annuum*	U51674	89
CYP75A1	Flavonoid-3′,5′-hydroxylase	*Petunia hybrida*	Z22545, X71130	71, 144
CYP75A2	Flavonoid-3′,5′-hydroxylase	*Solanum melongena*	X70824	146
CYP75A3	Flavonoid-3′,5′-hydroxylase	*Petunia hybrida*	Z22544	71
CYP75A4	Flavonoid-3′,5′-hydroxylase	*Gentiana triflora*	D85184	141
CYP76A1	Unknown	*Solanum melongena*	X71658	145
CYP76A2	Unknown	*Solanum melongena*	X71657	145
CYP77A1	Unknown	*Solanum melongena*	X71655	147
CYP77A2	Unknown	*Solanum melongena*	X71656	147
CYP78A1	Unknown	*Zea mays*	L23209	80
CYP78A2	Unknown	*Phalaenopsis* sp.	U34744	102
CYP79	Tyrosine-*N*-hydroxylase	*Sorghum bicolor*	U32624	77
CYP80	Berbamunine synthase	*Berberis stolonifera*	U09610	78
CYP82	Unknown	*Pisum sativum*	U29333	46
CYP83	Unknown	*Arabidopsis thaliana*	U18929	22
CYP84	Ferulate-5-hydroxylase	*Arabidopsis thaliana*	U38416	94
CYP85	Unknown (brassinosteroid synthesis?)	*Lycopersicon esculentum*	U54770	13
CYP86	Unknown	*Arabidopsis thaliana*	X90458	11
CYP88	Unknown (gibberelin synthesis)[c]	*Zea mays*	U32579	157
CYP89A2	Unknown	*Arabidopsis thaliana*	U61231	29
CYP90	Cathasterone-23-hydroxylase[c]	*Arabidopsis thaliana*	X87367, X87368	139
CYP92A2	Unknown	*Nicotiana tabacum*	X95342	31
CYP92A3	Unknown	*Nicotiana tabacum*	X96784	32
CYP93A1	Unknown	*Glycine max*	D83968	135
CYP93A2	Unknown	*Glycine max*	D86531	136
CYP97B1	Unknown	*Pisum sativum*	Z49263	4

[a]Unpublished sequences, PCR products, and ESTs do not appear in this table. Many sequences of this type are in the database, and although some can be associated with limited certainty to specific P450 families, it may be premature to do so before the entire open reading frame is sequenced. For a list of these sequences, please refer to the P450 WWW page <http://drnelson.utmem.edu/homepage.html>. Gaps in *CYP* family designations or the absence of certain subfamilies are the result of confidential submissions to the Nomenclature Committee. Abbreviations: HBOA, 2-hydroxy-1,4-benzoxazin-3-one; GA, gibberellic acid.

[b]Provisional name.

[c]Substrate specificity uncertain.

(Table 2), enabling the cloning of these loci and the unambiguous demonstration of gene function through mutant complementation.

P450s of Unknown Function

The first P450 to be purified to homogeneity from a plant extract was isolated from ripening avocado mesocarp (106). Although the purified protein catalyzed the demethylation of the model substrate p-chloro-N-methylaniline, its endogenous substrate was unknown. The gene encoding the avocado P450 was cloned by differential screening of libraries constructed from ripe and unripe fruit and was subsequently used to demonstrate that the accumulation of its corresponding mRNA is highly up-regulated during the climacteric period of fruit ripening (16). The amino acid sequence derived from the avocado gene exactly matched the N-terminal sequence determined from the purified protein (106), provided the first data indicating that the primary structure of plant P450s is similar to those found in other organisms, and defined the first plant P450 family, $CYP71$. In an effort to demonstrate the in vivo function of the protein, CYP71 was expressed in yeast. The heterologously expressed protein catalyzed p-chloro-N-methylaniline demethylation (15) and was later shown to convert nerol and geraniol to their 2,3- and 6,7-epoxy derivatives (66). Despite the demonstration of these activities, neither the terpenoid substrates nor their reaction products was found in extracts of avocado mesocarp. Thus, although CYP71 (now CYP71A1) was the first P450 characterized at the molecular level in plants, its function is still unknown.

Table 2 Mutants defective in P450-catalyzed reactions

Mutant	Species	Defective enzyme	Gene cloned?	Reference
b	Matthiola incana	flavonoid-3'-hydroxylase	No	44
bx3	Zea mays	indolin-2-one hydroxylase	Yes	49
cpd	Arabidopsis thaliana	cathasterone-23-hydroxylase[a]	Yes	139
dwarf	Lycopersicon esculentum	unknown	Yes	13
dwarf3	Zea mays	GA$_{12}$-13-hydroxylase[a]	Yes	157
eos	Antirrhinum majus	flavonoid-3'-hydroxylase	No	45
fah1	Arabidopsis thaliana	ferulate-5-hydroxylase	Yes	23, 94
ga3	Arabidopsis thaliana	kaurenol hydroxylase	No	43
hf1, hf2	Petunia hybrida	flavonoid-3',5'-hydroxylase	Yes	71
ht1, ht2	Petunia hybrida	flavonoid-3'-hydroxylase	No	35, 140
p	Verbena hybrida	flavonoid-3',5'-hydroxylase	Yes (see hf1)	7
tt7	Arabidopsis thaliana	flavonoid-3'-hydroxylase	No	127
643	Mentha × gracilis	(-)-limonene-6-hydroxylase	No	30

[a]Substrate specificity uncertain.

Following the cloning of CYP71A1 from avocado, many P450 cDNAs have been cloned using PCR (4, 25, 29, 46, 73, 92, 135, 148) and low-stringency hybridization with previously identified P450 probes (32, 88, 144, 146, 149). Many of these genes identified new P450 families and subfamilies. Unfortunately, virtually all of these efforts have resulted in the identification of genes of unknown function (Table 1). These approaches also depend upon a priori assumptions concerning the primary sequence of P450 genes and the proteins that they encode. In at least a few cases, there is reason to believe that atypical P450s exist in plants. Among these is benzoic acid 2-hydroxylase (BA2H), which catalyzes the hydroxylation of benzoic acid to salicylic acid, a mediator of plant responses to pathogens and wounding. BA2H is unusual in that it is a 160-kDa P450 that uses NADPH directly and is thus catalytically self-sufficient (83, 84). It seems unlikely that a PCR-based screening approach would identify a P450 with these atypical characteristics. The gene encoding BA2H may only be isolated following purification of the protein to homogeneity, or by the identification of a salicylic acid-deficient mutant that is defective in the gene encoding BA2H.

In some cases, P450 genes have been identified by differential screening or differential display (31, 80, 102, 135, 136, 151). These procedures were used to identify methyljasmonate-inducible genes from soybean suspension cultures, one of which encodes a P450 of the CYP93 family (135). This clone was subsequently used to identify another member of the same subfamily, CYP93A2, which is 80% identical to CYP93A1 but is completely unresponsive to methyljasmonate, suggesting a distinct in vivo function (136). In another example, a differential screen for genes specifically expressed during ovule development in *Phalaenopsis* identified a novel P450 clone (CYP78A2) (102). This gene was specifically expressed five to seven weeks after pollination (ovule development and pollination in orchids is a very protracted process compared with most angiosperms), and in situ hybridization with a CYP78A2 probe revealed that this P450 is expressed specifically in pollen tubes. Interestingly, the other member of the CYP78 family (CYP78A1), with which the orchid P450 shares 54% identity, is a P450 of unknown function that was isolated by differential screening of a maize tassel cDNA library (80). It was suggested that CYP78A2 may play a role in the biosynthesis of a plant growth regulator required for pollen tube development; however, its function at this time remains enigmatic.

Other P450s have been identified among expressed sequence tag (EST) libraries (28, 104, 123). For example, a new member of the *CYP71B* subfamily was characterized and shown to be highly expressed in Arabidopsis rosette leaves (90). Despite its successful expression in yeast, and the observation that the heterologously expressed enzyme was capable of catalyzing the cumene hydroperoxide-mediated dealkylation of 7-ethoxycoumarin, the endogenous substrate for this P450 remains unknown.

Cinnamate-4-Hydroxylase

Cinnamate-4-hydroxylase (C4H) catalyzes the *para*-hydroxylation of *trans*-cinnamic acid (Figure 1), which is derived from phenylalanine by the action of phenylalanine ammonia-lyase. C4H was the first plant P450 to be characterized (117, 118). In the years since its discovery, C4H has been extensively studied because of its involvement in the phenylpropanoid pathway, which gives rise to a wide array of important metabolites including lignin, flavonoids, hydroxycinnamic acid esters, lignans, stilbenes, and a host of other secondary metabolites. The isolation of its corresponding cDNA identified the first plant P450 gene that could be associated with an enzymatic function (42, 99, 143).

The isolation of the C4H protein was facilitated by the development of a phase-partitioning technique in which P450s from solubilized microsomal preparations are extracted into a detergent-rich phase in the presence of high concentrations of glycerol (155). Using this technique, C4H was purified from Jerusalem artichoke (51) and mung bean (97), and the corresponding cDNAs were isolated (99, 143) using peptide sequence and antibodies prepared against the purified protein (154). In parallel, the C4H gene from alfalfa was cloned (42) using antibodies that had been generated against the avocado *CYP71A1* (106). C4H cDNAs and genomic clones have since been isolated from many species (Table 1). C4H defines the CYP73 family and is a typical group A P450.

C4H activity and gene expression are modulated by a wide array of factors. C4H enzyme activity is induced by wounding (12) and exposure to metal ions (115), effects that appear to be regulated at the transcriptional level (6, 8, 46, 98). C4H mRNA accumulation is also induced by pathogen attack (19). In addition, the expression of C4H is regulated in a temporal and spatial manner consistent

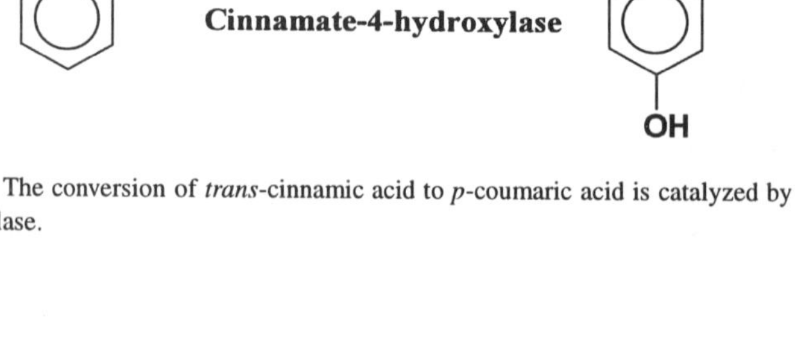

Figure 1 The conversion of *trans*-cinnamic acid to *p*-coumaric acid is catalyzed by cinnamate-4-hydroxylase.

with the role of this enzyme in lignification. The Arabidopsis C4H promoter directs GUS expression to lignifying cells of the Arabidopsis rachis (8), and the induction of tracheary element formation from isolated *Zinnia* mesophyll cells leads to a dramatic increase in the steady state levels of C4H mRNA (159).

C4H has been expressed in yeast and subjected to detailed spectroscopic and kinetic analyses. These studies revealed that the enzyme has a very high substrate specificity (111), a low spectral dissociation constant (8 μM) and K_M for cinnamate (4 μM) and an extremely high turnover number of over 400 min^{-1} (150). C4H also exhibited a high level of coupling between NADPH oxidation and substrate hydroxylation (1.05:1). The enzyme did not catalyze the hydroxylation of a wide range of plant products that have been shown to be metabolized by P450s in crude extracts or semipurified preparations (111), although it did catalyze the dealkylation or hydroxylation of several xenobiotics with low efficiency. These data strongly suggest that, unlike xenobiotic-metabolizing P450s, C4H is a highly efficient and specific enzyme that has evolved under the evolutionary constraints dictated by its anabolic role (150).

Allene Oxide Synthase

It had been reported as early as 1966 (160) that plant tissue extracts would metabolize 13-hydroperoxy-linoleic acid, the product of lipoxygenase action on linoleic acid, to yield a keto-hydroxy fatty acid derivative. It has since been shown that this substance is not the direct product of enzymatic activity but the result of the spontaneous degradation ($t_{1/2} \approx 9$ s) (130) of an epoxy fatty acid known as allene oxide (17). Allene oxide is a very important lipid intermediate in plants because it is converted to the plant signaling molecule jasmonic acid (JA) by the combined activity of allene oxide cyclase and subsequent steps of reduction and β-oxidation. The enzyme that catalyzes the conversion of 13-hydroperoxy-linoleic acid to allene oxide is now known as allene oxide synthase (AOS) (Figure 2).

AOS was demonstrated to be a P450 by its absorbance spectrum in the presence of dithionite and carbon monoxide following purification of the enzyme from an acetone powder of flax seed (130). The enzyme exhibited a number of unusual characteristics, including the lack of requirement for oxygen, NADPH, and P450 reductase and an extremely high turnover number of 70–80,000 min^{-1}. Using peptide sequence derived from the purified protein, the gene encoding allene oxide synthase was cloned (131), and heterologous probing using the flax sequence has permitted the cloning of an Arabidopsis orthologue (82). Compared to the high degree of sequence conservation within the heme-binding domain of most P450s, AOS has a number of notable substitutions. As mentioned above, the heme-binding domain of most P450s has the signature

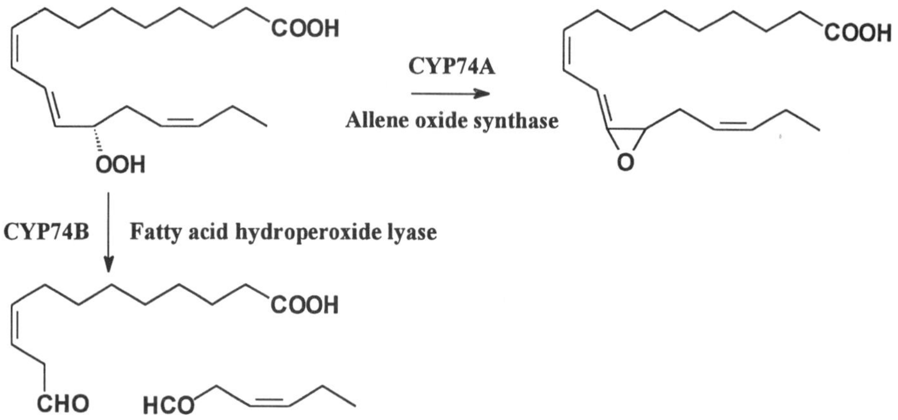

Figure 2 Two reactions that use 13-hydroperoxy-linoleic acid as a substrate are catalyzed by members of the *CYP74* family. Allene oxide synthase catalyzes its conversion to the corresponding epoxy derivative, allene oxide. Fatty acid hydroperoxide lyase catalyzes the cleavage of the same substrate to generate hex-3-enal and dodec-9-enoic acid-12-al.

sequence FxxGxxxCxG. This domain of the flaxseed AOS is **PSVANKQCAG**, where the proline and alanine residues shown in bold deviate from the consensus. The Arabidopsis sequence only carries a F6P substitution. Similarly, the region that, by analogy to prokaryotic P450s, forms the oxygen-binding pocket in eukaryotic P450s, is also divergent in both sequences. Specifically, the conserved threonine that is thought to be involved in oxygen activation (153) is replaced with an isoleucine. These and other sequence differences placed AOS in its own P450 family, CYP74. Interestingly, a P450 that catalyzes a similar reaction in prostaglandin synthesis in animals, thromboxane synthase, has an identical isoleucine substitution at the otherwise conserved threonine (131). The substitutions seen in AOS, its oxygen and NADPH independence, and its high turnover number probably reflect the fact that the enzyme no longer binds molecular oxygen and catalyzes a peroxidative dehydration instead.

The sequence of AOS cDNAs from flax and Arabidopsis has revealed that the protein has an N-terminus whose amino acid sequence resembles a plastid or mitochondrial transit peptide (82, 131). This is consistent with previous observations that most of the AOS activity in spinach is associated with the chloroplast membranes (152). Subsequent experiments in transgenic potato have indicated that the flax protein is directed to the chloroplast (67). This work also demonstrated that AOS overexpression under the control of the cauliflower mosaic virus 35S promoter leads to a six- to twelve-fold increase in JA accumulation. These data suggest that linolenic acid availability and lipoxygenase activity do not limit JA biosynthesis in unwounded plants and that AOS catalyzes the rate-limiting step in JA biosynthesis. In keeping with these observations, AOS mRNA was found to be highly wound-inducible in Arabidopsis, paralleling the increase in JA levels within one hour of wounding (82). These experiments also indicated that at least a portion of the JA biosynthetic pathway takes place in the chloroplast. Consistent with these data is the recent report of rice and tomato lipoxygenases that are targeted to the chloroplast (68) or that carry putative plastid transit peptides (110).

AOS is the major P450 in many plant organs such as tulip bulbs (81). Recently, the rubber particle protein (RPP) found in the desert shrub guayule (*Parthenium argentatum*) was demonstrated to be AOS (109). This finding was unexpected since RPP makes up approximately 50% of the protein associated with the rubber particles and had been thought to be involved in rubber synthesis. When the protein was purified and its corresponding gene was cloned, it was found to be 65% identical to the flax AOS and was subsequently shown to have AOS activity. Like the flax AOS, RPP has a T $\rightarrow$ I substitution in the region corresponding to the P450 oxygen-binding pocket; however, it lacks a plastid transit sequence. The absence of this targeting peptide may permit the association of the protein with the rubber particles, although its function there is still unknown.

Fatty Acid Hydroperoxide Lyase

The fatty acid hydroperoxide lyase (HPO lyase) is another enzyme that, like AOS, uses lipoxygenase-derived hydroperoxides of unsaturated fatty acids as its substrate. In contrast to AOS, HPO lyase cleaves the carbon-carbon bond adjacent to the hydroperoxy functional group, thus producing a pair of shorter chain aldehydes (Figure 2). These aldehydes and their reduced alcohol counterparts have been implicated in plant pathogen interactions and are important components of fruit and vegetable volatiles. Shibata et al (125) developed a scheme that allowed the purification of two HPO lyase isoforms from detergent-treated green pepper fruit membrane preparations. The enzyme was later shown to be a heme protein on the basis of its absorption spectrum (124). The gene encoding HPO lyase has since been cloned, providing definitive proof that the enzyme is a P450 (89). Like AOS, HPO lyase has an isoleucine in place of the conserved oxygen-binding threonine, and the sequence of its heme-binding domain is divergent from that of the P450 consensus. The enzyme is 40% identical to AOS, and although this value lies at the cutoff for defining a new P450 family, it has been classified as the first member of a new CYP74 subfamily, CYP74B. HPO lyase is included in the same family as AOS because they share unusual amino acid substitutions, their activities are dependent on neither oxygen nor NADPH, and they use the same hydroperoxy fatty acids as substrates.

Tyrosine-N-Hydroxylase

Two groups of plant natural products, cyanogenic glycosides and glucosinolates, are derived from protein and nonprotein amino acids with their corresponding aldoximes acting as key intermediates in their biosynthesis. Dhurrin is the major cyanogenic glycoside accumulated by *Sorghum bicolor*, and its synthesis can be readily studied using microsomes from etiolated seedlings (62). In radiotracer feeding experiments, these microsomes rapidly convert ^{14}C-labeled tyrosine to labeled *p*-hydroxybenzaldehyde, the spontaneous breakdown product of the last intermediate in the pathway, the cyanohydrin *p*-hydroxymandelonitrile (100, 126). These interconversions are inhibited by carbon monoxide and require NADPH, implicating the participation of at least two P450s in the pathway. Inhibition experiments using anti-P450 reductase antibodies confirmed the involvement of P450s and also indicated that the initial hydroxylation of tyrosine is the rate-limiting step in the pathway (63).

Although this pathway contains a number of intermediates, only radiolabeled *p*-hydroxyphenylacetaldoxime could be recovered in substantial quantities in these feeding experiments (63, 100, 126). In addition, tyrosine-derived intermediates produced by the microsomal system were metabolized preferentially over exogenously supplied compounds (65). On the basis of these biochemical data, Møller & Conn (100) concluded that the cyanogenic glycoside biosynthetic pathway is channeled by "two bifunctional systems, the first channeling the flow of tyrosine into *p*-hydroxyphenylacetaldoxime via *N*-hydroxytyrosine and the second channeling the flow of *p*-hydroxyphenylacetaldoxime into *p*-hydroxymandelonitrile via *p*-hydroxyphenylacetonitrile."

The P450 catalyzing the *N*-hydroxylation of tyrosine (tyrosine-*N*-hydroxylase or $P450_{TYR}$) was purified to homogeneity (128), and antibodies generated against the polypeptide were used to clone the corresponding gene (77). The protein defines the first member of the CYP79 family of P450s and is unusual in that it has several amino acid substitutions in conserved regions including an asparagine residue in place of the threonine usually found in the oxygen-binding pocket. This asparagine substitution, and the unusual presence of an adjacent proline residue, are also found in CYP56, a tyrosine-metabolizing enzyme of fungal origin (77), suggesting that these residues may be related to enzyme substrate specificity.

The explanation for the observations of pathway channeling became apparent following the purification of tyrosine-*N*-hydroxylase and the expression of the cloned gene in *Escherichia coli* (64). The purified protein exhibited a type I binding spectrum that is indicative of substrate binding with both tyrosine and *N*-hydroxytyrosine (129). Further, when the purified protein was reconstituted into micelles containing P450 reductase, the protein catalyzed all steps of the conversion of tyrosine to *p*-hydroxyphenylacetaldoxime (Figure 3) (129), thus

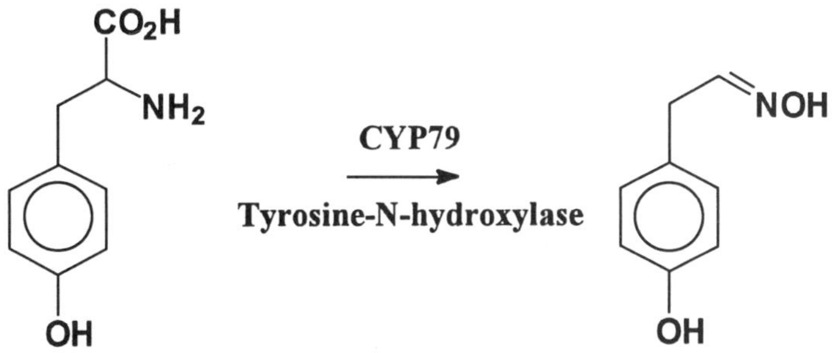

Figure 3   Tyrosine *N*-hydroxylase (P450$_{TYR}$) catalyzes a multistep reaction that converts tyrosine to *p*-hydroxyphenylacetaldoxime.

confirming the hypothesis of Møller & Conn (100). Thus, while the first half of the pathway leading to cyanogenic glycosides might have been envisioned to require a handful of enzymes, it is instead catalyzed by a single multifunctional P450. The involvement of similar aldoxime-producing activities in the synthesis of glucosinolates (36, 37) suggests a possible evolutionary relationship to cyanogenic glycoside biosynthesis, although there is evidence for alternative routes to aldoximes and glucosinolates in some plants (9).

Obtusifoliol 14α-Demethylase

Membrane sterols lack the 14α-methyl group found in their biosynthetic precursors. The demethylases that act on these sterol precursors, lanosterol in animals and fungi and obtusifoliol in plants, are P450s of the CYP51 family. Due to the essential nature of the demethylated products, azole inhibitors of the demethylases are effective antifungal and herbicidal agents (59).

By analogy to its animal counterpart, the plant obtusifoliol 14α-demethylase catalyzes three consecutive oxidations to introduce a double bond into the sterol nucleus with the concomitant release of the methyl carbon as formic acid (Figure 4) (108). When assayed in maize microsomal preparations, the enzyme showed no activity toward lanosterol and many other closely related compounds, indicating that the enzyme has high substrate specificity (142). Obtusifoliol 14α-demethylase was recently purified from *Sorghum bicolor* seedlings (74), and peptide amino acid sequence data were used to design degenerate PCR primers with which the cDNA was cloned (3). Low-stringency hybridization using CYP81 as a probe has also permitted the isolation of CYP51 from wheat (20). Recombinant obtusifoliol 14α-demethylase expressed in *E. coli* gives a type I binding spectrum with obtusifoliol but not with lanosterol, consistent with previous reports of the enzyme's high substrate specificity (3).

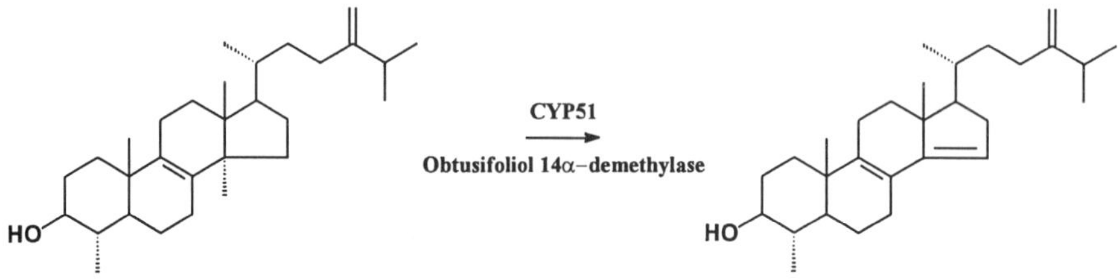

Figure 4   The conversion of obtusifoliol to 4α-methyl-5α-ergosta-8,14,24(28)-trien-3β-ol is catalyzed by the plant CYP51, obtusifoliol 14α-demethylase.

Obtusifoliol 14α-demethylase is a non-group A P450 and is the first plant P450 to have been identified that has a nonplant enzyme as its closest homologue. The assignment of obtusifoliol 14α-demethylase to the CYP51 family is an example of an exception made in the P450 nomenclature rules. The yeast and mammalian versions of CYP51 are 39–42% identical; however, these overall identity values underestimate the sequence conservation within domains that dictate substrate binding and catalysis (1, 56). Similarly, when the obtusifoliol 14α-demethylase gene from sorghum was cloned, the deduced amino acid sequence of the enzyme was found to be 32% and 36% identical to the lanosterol 14α-demethylases from yeast and rat, respectively (3). According to the 40% rule, these enzymes would be placed in three P450 families. Considering that their identity is close to this threshold value, and that they catalyze identical reactions on related substrates, the lanosterol 14α-demethylases and the obtusifoliol 14α-demethylases have been placed into one P450 family, CYP51.

Berbamunine Synthase

Many reports have documented the role of P450s in alkaloid biosynthesis, but in many cases, these compounds and the catalysts that function in their synthesis are found only in low levels and often only in specific cell types. The study of berbamunine synthase which catalyzes the oxidative coupling of two molecules of *N*-methylcoclaurine to form the dimeric alkaloid berbamunine (Figure 5) was facilitated by the isolation of a cell culture of *B. stolonifera* that accumulated up to 1% of its dry mass as bisbenzylisoquinoline alkaloids (21). When the purified enzyme was reconstituted with P450 reductase, the enzyme exhibited high regio- and stereo-specificity but some ability to accommodate alternative substrates. When incubated with equimolar amounts of its two substrates, (*R*)- and (*S*)-*N*-methylcoclaurine, berbamunine synthase produced almost exclusively (*R,S*)-berbamunine; however, it was also capable of synthesizing the berbamunine diastereomer guattegaumerine or the related alkaloid

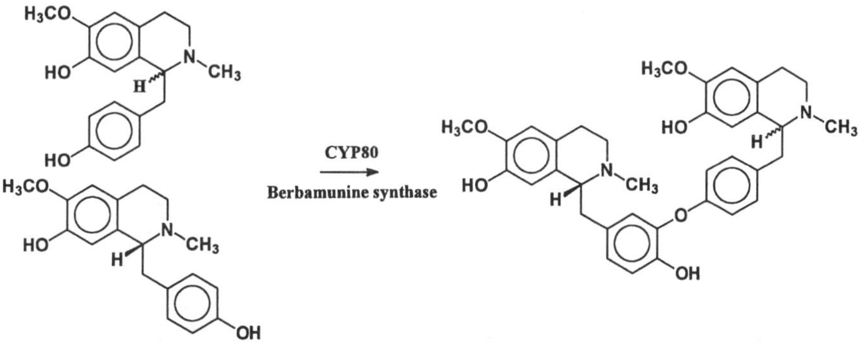

Figure 5 Berbamunine synthase catalyzes the coupling of two molecules of *N*-methylcoclaurine to generate berbamunine or guattegaumerine. The name of the dimeric product depends upon the stereochemistry of the monomers condensed.

2′-norberbamunine when presented with appropriate monomeric substrates (132). These data indicate that the relative abundance of the various dimeric alkaloids found in vivo may depend upon the molar ratios of the monomers available for coupling. Interestingly, no synthesis of the more oxidized dimer aromoline was seen when the enzyme was presented with berbamunine, despite the fact that aromoline is also found in *Berberis stolonifera* cell cultures, suggesting that yet another P450 is required for its synthesis.

The cDNA encoding berbamunine synthase was cloned following microsequencing of the purified protein, and its identity was verified by expression in a baculovirus system (78) that permitted the production and purification of almost 20,000-fold more enzyme per liter than in the original *Berberis stolonifera* cell cultures. The heterologously expressed enzyme showed a turnover rate of 34–62 min^{-1} (calculated in terms of dimers produced), in comparison with 50 min^{-1} for the purified enzyme from its natural source, but marked differences were seen in the type of dimeric alkaloids produced, with guattegaumerine being a major product even in the presence of an equimolar mixture of (*R*)- and (*S*)-*N*-methylcoclaurine. Although the factors that were responsible for these differences in activity were not fully elucidated, the source of P450 reductase employed in the assay had a substantial impact, suggesting that subtle aspects of protein folding or protein-protein interactions required for electron transfer may impact the ratio of dimers produced. Southern analysis using the berbamunine synthase cDNA as a probe against *B. stolonifera* genomic DNA indicated the presence of strongly hybridizing bands corresponding to the berbamunine synthase genomic sequence as well as two cross-hybridizing bands, suggesting the presence of related sequences in the *Berberis* genome, possibly encoding those P450s responsible for the further oxidation of berbamunine to aromoline and obamegine.

The reaction catalyzed by berbamunine synthase is of note because it requires the dimerization of two monomers by a mechanism that is suggested to involve biradical coupling (132). If this hypothesis is correct, it would require an enzyme active site that can accommodate two alkaloid monomers. It will be interesting to see whether comprehensive alignment of plant P450 sequences, or future structural studies, will verify the presence of these two substrate binding sites.

Flavonoid-3',5'-Hydroxylase

The gene encoding the flavonoid-3',5'-hydroxylase (F3'5'H) from *Petunia hybrida* was the first plant P450 gene to be cloned using a combined biochemical and genetic approach (71). F3'5'H catalyzes the stepwise conversion of the flavanone naringenin to eriodictyol and pentahydroxyflavanone, as well as the parallel hydroxylation of the dihydroflavonol dihydrokaempferol to dihydroquercetin and dihydromyricetin (Figure 6) (93, 133). The products of the F3'5'H reaction are intermediates in the biosynthesis of anthocyanins. The glucosides of the di- and tri-hydroxylated derivatives, cyanidin and delphinidin, are common purple-blue flower pigments in *Petunia* and many other species. The potential for the manipulation of flower color using F3'5'H provided the driving force for the isolation of its corresponding gene (70).

In *Petunia*, the *Hf1* and *Hf2* loci encode F3'5'H, and *hf1/hf2* double mutants are altered in color because they lack the dihydromyricetin-derived anthocyanins found in wild type (Table 2) (133). To clone the genes encoded by the *Hf* loci, a pool of PCR products was generated using primers directed against the conserved region of the P450 heme-binding domain (71). Two clones were identified that failed to recognize transcripts in RNA prepared from *hf1/hf2* double mutant corollas. These two cDNAs were 93% identical to one another at the nucleotide level and identified a new P450 family, CYP75. RFLP analysis demonstrated that these clones were tightly linked to the *Hf1* and *Hf2* loci, and heterologous expression in yeast showed that the protein encoded by each clone catalyzed the F3'5'H reaction. Complementation of the *Petunia* mutants was also used to demonstrate the identity of the isolated clones, and the complemented mutants accumulated higher levels of delphinidin-based anthocyanins, suggesting that the expression of F3'5'H is rate-limiting for anthocyanin biosynthesis.

The F3'5'H gene from *Petunia* was independently cloned using a CYP75 cDNA isolated from UV-irradiated eggplant seedlings (144) as a heterologous probe. Although the function of the eggplant gene was unknown at the time, RNA blot analysis indicated that the corresponding gene from *Petunia* was expressed in a manner consistent with a role in flavonoid biosynthesis. The

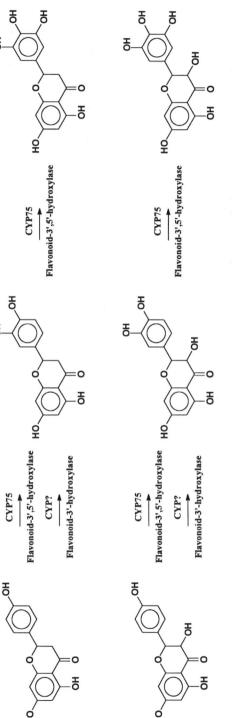

Figure 6 The conversions of naringenin (*upper*) and dihydrokaempferol (*lower*) to eriodictyol and dihydroquercetin, respectively, are catalyzed by flavonoid-3′,5′-hydroxylase (*CYP75*) and flavonoid-3′-hydroxylase. The products of these reactions can be further hydroxylated by flavonoid-3′,5′-hydroxylase to generate pentahydroxyflavanone and dihydromyricetin.

absence of *CYP75* expression in the *Petunia hf1/hf2* mutants again provided strong evidence that the *CYP75* gene encodes F3'5'H.

It should be noted that F3'5'H is biochemically distinct from the flavonoid-3'-hydroxylase (F3'H) that converts naringenin and dihydrokaempferol to eriodictyol and dihydroquercetin, respectively (Figure 6) (93). Although the two meta-positions (C3 and C5) of the flavonoid B ring are chemically equivalent due to free rotation about the C-C bond that connects the B and C rings, once hydroxylated the two positions become distinguishable. It would appear that F3'5'H is capable of accommodating a dihydroxylated B ring at its active site, while F3'H cannot. The availability of *Petunia* and Arabidopsis mutants (Table 2) defective in flavonoid-3'-hydroxylase activity (35, 127, 140) may soon permit the isolation of F3'H genes following mutant-based cloning strategies. It will be of great interest to see how the primary sequences of F3'5'H and F3'H compare, and whether any similarities exist between the sequences of F3'H and ferulate-5-hydroxylase, another P450 catalyzing phenolic ring *meta*-hydroxylation (94).

Ferulate-5-Hydroxylase

Only one paper has appeared in the literature demonstrating the enzymatic activity of ferulate-5-hydroxylase (F5H) in vitro (58). Although this study demonstrated that the conversion of ferulate to 5-hydroxyferulate (Figure 7) was catalyzed by a P450, very little additional characterization was possible because the enzyme was very labile even when stored at 4°C, and detergent solubilization resulted in a rapid loss of activity.

Recently, an F5H mutant was identified in Arabidopsis (Table 2). The *fah1* mutant was identified by its inability to synthesize sinapoylmalate, one of the major soluble phenylpropanoid secondary metabolites in Arabidopsis. In wild type, sinapoylmalate is accumulated in the adaxial leaf epidermis where it plays a role in the resistance of Arabidopsis to UV-B (79). The distribution of this

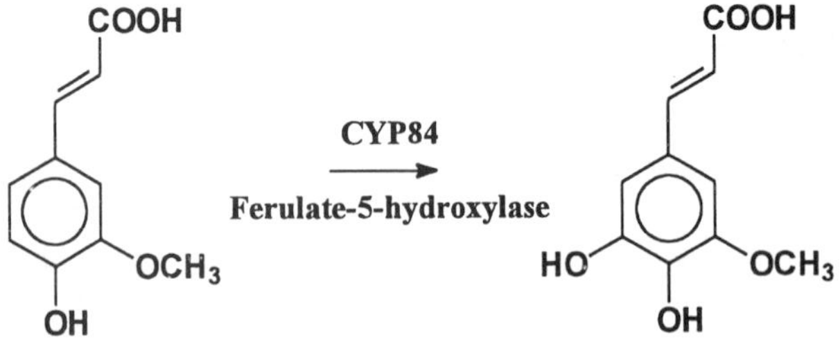

Figure 7 The conversion of ferulic acid to 5-hydroxyferulic acid is catalyzed by ferulate-5-hydroxylase.

blue-fluorescent secondary metabolite can be exploited as a rapid method for isolating *fah1* mutants. Because their leaves lack sinapoylmalate, *fah1* mutants can be readily identified by their dark red chlorophyll fluorescence under UV light among a population of blue fluorescent wild-type plants (23). Using this visual screen, a T-DNA-tagged allele of *fah1* was identified and was used to clone the gene encoding F5H. The identity of the cloned gene was verified by complementation of the mutant phenotype using a wild-type version of the cloned gene (94).

The lesion in the F5H gene and its position in the lignin biosynthetic pathway prevents the *fah1* mutants from accumulating sinapic acid-derived (syringyl) lignin (23). In wild-type Arabidopsis, syringyl units comprise only 20% of the polymer and are accumulated only in the sclerified parenchyma of the rachis. Expression of the F5H gene under the direction of the cauliflower mosaic virus 35S promoter abolished the tissue specificity of lignin monomer deposition and resulted in a 50% increase in syringyl content, thus demonstrating that F5H expression is responsible for quantitative and developmental regulation of lignin biosynthesis in Arabidopsis (K Meyer, JC Cusumano, DA Bell-Lelong & C Chapple, unpublished observations). Surprisingly, expression of the F5H gene under the control of the Arabidopsis C4H promoter was far more efficacious, generating a lignin with a syringyl content of over 90%, demonstrating that both P450 promoters and their coding sequences may be useful in biotechnological applications.

Gibberellin Biosynthetic P450s

The gibberellins are important plant growth regulators, and soon after the discovery of P450s it was shown that enzymes of this type mediated the three-step conversion of kaurene to kaurenoic acid (34, 101). In this set of reactions, the C-19 methyl group of kaurene is hydroxylated to yield kaurenal. The same carbon is presumably again hydroxylated to a dihydroxymethenyl function that would spontaneously convert to the corresponding aldehyde. In the last of the three steps, kaurenal is further oxidized to the corresponding carboxylic acid, kaurenoic acid. It has since been shown that a number of the subsequent steps in gibberellin biosynthesis are also P450-catalyzed (57, 67a).

The dwarf phenotype of GA-deficient plants has been of great utility in the cloning of "non-P450" GA biosynthetic loci using genetic approaches (24, 134); however, to date, only one P450-encoding gene from this pathway has been isolated. Biochemical complementation experiments and the analysis of GA intermediate pool sizes suggested that the maize mutant *dwarf 3* (*d3*) (Table 2) is defective in the P450 that converts GA_{12} to GA_{53} (157); however, this has not been demonstrated unequivocally (67a). A transposon tagging approach was used to clone the gene that is defective in the mutant (157). The *D3* gene product defined a new P450 family (*CYP88*), which is of the non–group A type (38). Its sequence is typical of a P450 other than the unusual presence of an imperfect 27–amino acid repeat at its amino terminus, the function of which is unknown.

Five loci affecting GA biosynthesis (*ga1-ga5*) have been identified in Arabidopsis by screening for mutants that fail to germinate in the absence of exogenous GA. The *ga3* mutant may be defective in a specific kaurenol oxidase (43), or a multifunctional kaurene oxidase that converts kaurene to kaurenoic acid (75a). Confirmation of these data must await cloning of the *GA3* gene, which may be achieved in the near future using a genetic approach.

Brassinosteroid Biosynthetic P450s

Although the biological activity of brassinosteroids was demonstrated several decades ago (60, 95), particular attention was drawn to these phytohormones recently by the identification of Arabidopsis mutants that resemble light-grown plants when grown under etiolated conditions (26). Two of these mutants identified brassinosteroid biosynthetic genes, one of which encodes a P450 (85, 139). The *cpd* mutant (Table 2) was isolated from a T-DNA-tagged population of plants, and cloning of the corresponding wild-type gene revealed that the CPD locus encodes a non–group A P450 that has been designated the first member of the CYP90 family. Exogenous application of the 23-hydroxylated brassinolide precursor teasterone was capable of complementing the *cpd* phenotype, while precursors earlier in the pathway were not. On the basis of these biochemical complementation experiments, CYP90 appears to catalyze the 23-hydroxylation of cathasterone to teasterone (Figure 8). Analysis of the *CYP90* genomic sequence revealed the presence of eight introns, two of which are in the same position as introns of the human *CYP21A2* gene that encodes progesterone-21-hydroxylase (69, 156). This finding suggests that these plant and human steroid-hydroxylating P450s may share a more recent evolutionary ancestor than most plant and animal P450s.

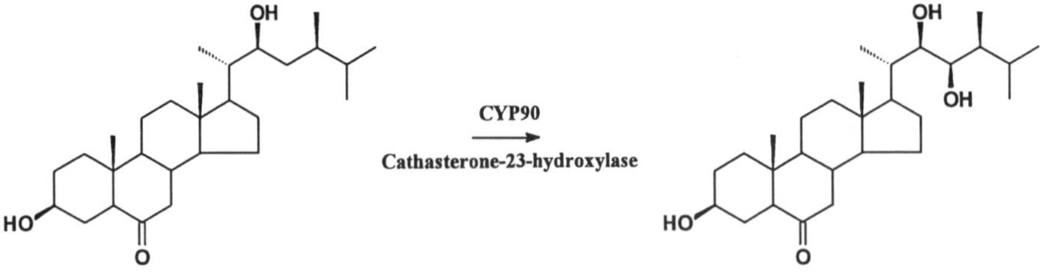

Figure 8 Genetic evidence indicates that the conversion of cathasterone to teasterone is catalyzed by the Arabidopsis *CPD* gene that encodes CYP90.

Another non–group A P450 that was recently identified using a mutant approach may represent an additional P450 involved in brassinosteroid biosynthesis (13). The tomato *Dwarf* gene (Table 2) was cloned by transposon tagging using the maize Activator element. The *Dwarf* gene product has been designated CYP85. Although it is not clear that this P450 is involved in brassinosteroid biosynthesis, the fact that the dwarf phenotype of the mutant is not reversed by exogenous GA is consistent with this hypothesis. Further, in the transposon-tagged mutant, somatic excision of the Activator element generated distinct revertant sectors, suggesting that the *Dwarf* gene is involved in the biosynthesis of a substance that is not readily transported or diffusible throughout the plant, possibly a hydrophobic molecule such as a brassinosteroid (13).

DIMBOA Biosynthetic P450s

Members of the Gramineae accumulate 1,4-benzoxazin-3-one derivatives including DIMBOA and DIBOA that play an important role in plant pathogen and insect resistance (105). The synthesis of DIMBOA has been shown to involve a P450 that catalyzes the N-hydroxylation of 2-hydroxy-1,4-benzoxazin-3-one (HBOA) to its dihydroxy derivative, DIBOA (2). Indirect evidence for the involvement of additional P450s comes from the finding that the other oxygen atoms of DIMBOA are also derived from molecular oxygen (54). The identification of a DIMBOA-deficient maize mutant (*bx1* for *benzoxazineless*) prompted efforts to clone the genes of this pathway and to evaluate whether the *Bx1* locus encodes a P450 involved in DIMBOA biosynthesis (50). Differential cDNA screening using high- and low-DIMBOA-producing maize lines identified a P450 cDNA of the CYP71 family, which in turn was used to identify three additional *CYP71* clones. All of these clones mapped to within four map units of one another and close to the *Bx1* locus.

Recently, the involvement of the four CYP71 clones (CYP71C1-CYP71C4) in DIMBOA biosynthesis has been unequivocally demonstrated. First, a reverse genetic approach (10) was used to identify a maize mutant that is defective in DIMBOA synthesis due to a *Mu* transposon insertion in the *CYP71C2* gene (49). Second, by expression of each of the cDNAs in yeast, it was possible to demonstrate that each of the cloned P450s catalyzed one of the oxidation steps between indole and DIBOA (Figure 9). *CYP71C3* was found to encode the *N*-hydroxylase that converts HBOA to DIBOA (2). These experiments employed the same yeast system that was used to characterize cinnamate-4-hydroxylase, but also used the WAT11 yeast strains in which the yeast P450 reductase has been replaced by one of the two P450 reductase genes of Arabidopsis by homologous recombination (112). Sequencing of *CYP71C* genomic clones revealed that several of the genes have introns in common locations. The *Bx1* locus was also cloned as part of this work, and was demonstrated to encode

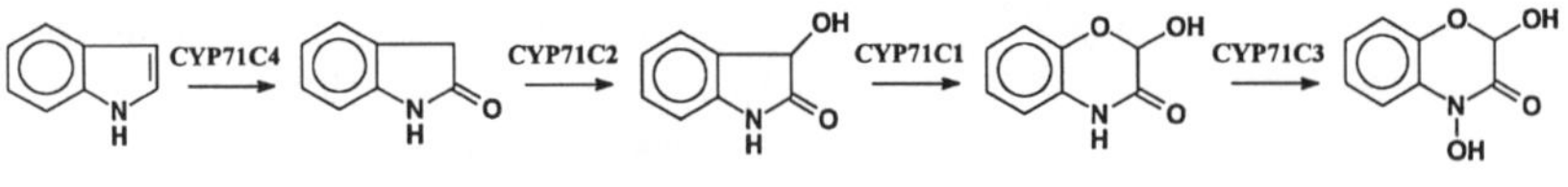

Figure 9   The four-step conversion of indole to DIBOA is catalyzed by four members of the maize CYP71 family.

a tryptophan synthase α homologue that converts indole-3-glycerol phosphate to indole for DIMBOA biosynthesis (49). The *Bx1* gene is separated from the *CYP71C4* gene by less than 2.5 kb. In the context of the evolution of plant secondary metabolic pathways, it is interesting that the entire DIBOA biosynthetic pathway is encoded in a relatively small region of chromosome 4 in maize, and that at least four of its genes appear to have a common evolutionary ancestry.

NEW P450 RESEARCH RESOURCES

Genomics

The sequencing of plant genomes will soon have a dramatic impact on P450 research. Genomic sequencing projects have already identified new plant P450 genes (for one example, see <http://pgec-genome.pw.usda.gov/F5I14.anno. html>). This research will also provide a comprehensive view of the genomic organization and evolution of plant P450s and will indicate whether clustering of P450 genes is a general phenomenon or whether the clustering of the maize *CYP71C* subfamily is unique (49, 50). The EST databases (28, 104, 123) have already been used by a number of researchers to isolate new plant P450s (11, 90), and orthologues of previously identified genes (8). These cDNA sequencing efforts have also provided the first estimates of P450 gene diversity in plant genomes. With over 50% of Arabidopsis genes represented by ESTs (116), the number of P450s in Arabidopsis alone may be as high as 200. ESTs have also been used to select at least eight T-DNA insertional mutants in P450 genes (KA Feldmann, personal communication) using the PCR-based approach previously applied to the isolation of Arabidopsis actin mutants (91). If novel phenotypes can be associated with these insertional mutants, this research will identify functions for P450s that are otherwise defined only by their nucleotide and inferred amino acid sequences.

Internet Resources

Several World Wide Web sites track P450 research progress and provide investigators with up-to-date catalogs of cloned genes (33, 41). The Directory of

P450-Containing Systems (DPS) server was established in 1994 and provides information on animal, fungal, plant, and bacterial P450s, as well as related proteins such as P450 reductases and cytochrome b_5. The DPS server can be accessed at <http://www.icegeb.trieste.it/p450/>. The site documents the total number of genes identified, the organisms from which they have been cloned, and their database accession numbers, as well as particular information regarding the substrates and products of steroid-metabolizing P450s. ESTs are not included in this database. A similar database is maintained by David Nelson at the Department of Biochemistry, University of Tennessee, and can be found at <http://drnelson.utmem.edu/homepage.html>. This database also catalogs the full list of P450s cloned from both eukaryotes and prokaryotes, as well as a current phylogenetic tree that depicts P450 gene divergence.

CONCLUDING REMARKS

The cloning of P450 genes is contributing significantly to our understanding of these linchpin enzymes of plant metabolism. The analysis of tyrosine-N-hydroxylase has demonstrated that it is a multifunctional enzyme, and future investigations may reveal a general role for P450s in the channeling of metabolic pathways. In addition, the overexpression of three P450s, AOS, F3'5'H, and F5H has demonstrated that these enzymes catalyze rate-limiting steps in the biochemical pathways in which they are involved. If this proves to be a general characteristic of P450s, the overexpression of their corresponding genes may permit the modification of flux through pathways that give rise to economically important products such as alkaloids, or to substances that affect plant growth and development such as GA, brassinosteroids, and JA. Similarly, the introduction of foreign P450 genes into alternative host plants, yeast, or bacteria may allow for the engineering of novel biochemical pathways and the synthesis of so-called designer phytochemicals. It will also be very interesting to determine in what ways plant P450s differ from their animal counterparts, and how these differences impact upon the reactions that they catalyze. Considering the number of P450s that remain to be discovered and analyzed at the molecular level, it is clear that we still have much to learn about these enzymes.

ACKNOWLEDGMENTS

I would like to thank Joanne C Cusumano for her help in the preparation of this manuscript. This is journal paper number of 15564 the Purdue University Agricultural Experiment Station. This work was supported by a grant from the Division of Energy Biosciences, United States Department of Energy DE-FG02-94ER20138.

Literature Cited

1. Aoyama Y, Noshiro M, Gotoh O, Imaoka S, Funae Y, et al. 1996. Sterol 14-demethylase P450 (P45014DM*) is one of the most ancient and conserved P450 species. *J. Biochem.* 119:926–33

2. Bailey BA, Larson RL. 1991. Maize microsomal benzoxazinone N-monooxygenase. *Plant Physiol.* 95:792–96

3. Bak S, Kahn RA, Olsen CE, Halkier BA. 1997. Cloning and expression in *Escherichia coli* of the obtusifoliol 14α-demethylase of *Sorghum bicolor* (L) Moench, a cytochrome P450 orthologous to the sterol 14α-demethylases (CYP51) from fungi and mammals. *Plant J.* 11:191–201

4. Baltrusch M, Fulda M, Wolter F-P, Heinz E. 1997. Cloning and sequencing of a cytochrome P450 from *Pisum sativum* L. (Accession No. Z49263). *Plant Physiol.* 114:1568

5. Barrett M. 1995. Metabolism of herbicides by cytochrome P450 in corn. *Drug Metab. Drug Interact.* 12:299–315

6. Batard Y, Schalk M, Pierrel M-A, Zimmerlin A, Durst F, Werck-Reichhart D. 1997. Regulation of the cinnamate 4-hydroxylase (CYP73A1) in Jerusalem artichoke tubers in response to wounding and chemical treatments. *Plant Physiol.* 113:951–59

7. Beale GH, Price JR, Scott-Moncrieff R. 1940. The genetics of *Verbena* II. Chemistry of the flower colour variations. *J. Genet.* 41:65–74

8. Bell-Lelong DA, Cusumano JC, Meyer K, Chapple C. 1997. Cinnamate-4-hydroxylase expression in Arabidopsis. Regulation in response to development and the environment. *Plant Physiol.* 113:729–38

9. Bennett RN, Hick AJ, Dawson GW, Wallsgrove RM. 1995. Glucosinolate biosynthesis. Further characterization of the aldoxime-forming microsomal monooxygenases in oilseed rape leaves. *Plant Physiol.* 109:299–305

10. Bensen RJ, Johal GS, Crane VC, Tossberg JT, Schnable PS, et al. 1995. Cloning and characterization of the maize *An1* gene. *Plant Cell* 7:75–84

11. Benveniste I, Durst F. 1995. Cloning, sequencing and expression of CYP86, a new cytochrome P450 from *Arabidopsis thaliana* (Accession No. X90458) (PGR95–074). *Plant Physiol.* 109:722

12. Benveniste I, Salaun J-P, Durst F. 1977. Wounding-induced cinnamic acid hydroxylase in Jerusalem artichoke tuber. *Phytochemistry* 16:69–73

13. Bishop GJ, Harrison K, Jones JDG. 1996. The tomato *dwarf* gene isolated by heterologous transposon tagging encodes the first member of a new cytochrome P450 family. *Plant Cell* 8:959–69

14. Bolwell GP, Bozak K, Zimmerlin A. 1994. Plant cytochrome P450. *Phytochemistry* 37:1492–506

15. Bozak KR, O'Keefe DP, Christoffersen RE. 1992. Expression of a ripening-related avocado (*Persea americana*) cytochrome P450 in yeast. *Plant Physiol.* 100:1976–81

16. Bozak KR, Yu H, Sirevåg R, Christoffersen RE. 1990. Sequence analysis of ripening-related cytochrome P-450 cDNAs from avocado fruit. *Proc. Natl. Acad. Sci. USA* 87:3904–8

17. Brash AR, Song W-C. 1995. Structure-function features of flaxseed allene oxide synthase. *J. Lipid Mediators Cell Signal.* 12:275–82

18. Brodie BB, Gillette JR, La Du BN. 1958. Enzymatic metabolism of drugs and other foreign compounds. *Annu. Rev. Biochem.* 27:427–54

19. Buell CR, Somerville SC. 1995. Expression of defense-related and putative signaling genes during tolerant and susceptible interactions of *Arabidopsis* with *Xanthomonas campestris* pv. *campestris*. *Mol. Plant-Microbe Interact.* 8:435–43

20. Cabello-Hurtado F, Zimmerlin A, Rahier A, Taton M, DeRose R, et al. 1997. Cloning and functional expression in yeast of a cDNA coding for an obtusifoliol 14α-demethylase (CYP51) in wheat. *Biochem. Biophys. Res. Commun.* 230:381–85

21. Cassels BK, Breitmaier E, Zenk MH. 1987. Bisbenzylisoquinoline alkaloids in *Berberis* cell cultures. *Phytochemistry* 26:1005–8

22. Chapple CCS. 1995. A cDNA encoding a novel cytochrome P450 dependent

monooxygenase from *Arabidopsis thaliana. Plant Physiol.* 108:875–76

23. Chapple CCS, Vogt T, Ellis BE, Somerville CR. 1992. An Arabidopsis mutant defective in the general phenylpropanoid pathway. *Plant Cell* 4:1413–24

24. Chiang H-H, Hwang J, Goodman HM. 1995. Isolation of the Arabidopsis *GA4* locus. *Plant Cell* 7:195–201

25. Clark IM, Forde BG, Hallahan DL. 1997. Spatially distinct expression of two new cytochrome P450s in leaves of *Nepeta racemosa*: identification of a trichome-specific isoform. *Plant Mol. Biol.* 33:875–85

26. Clouse SD. 1996. Molecular genetic studies confirm the role of brassinosteroids in plant growth and development. *Plant J.* 10:1–8

27. Deleted in proof

28. Cooke R, Raynal M, Laudié M, Grellet F, Delseny M, et al. 1996. Further progress towards a catalogue of all *Arabidopsis* genes: analysis of a set of 5000 nonredundant ESTs. *Plant J.* 9:101–24

29. Courtney KJ, Percival FW, Hallahan DL, Christoffersen RE. 1996. Cloning and sequencing of a cytochrome P450, CYP89, from *Arabidopsis thaliana* (Accession No. U61231) (PGR96–061). *Plant Physiol.* 112:445

30. Croteau R, Karp F, Wagschal KC, Satterwhite DM, Hyatt DC, Skotland CB. 1991. Biochemical characterization of a spearmint mutant that resembles peppermint in monoterpene content. *Plant Physiol.* 96:744–52

31. Czernic P, Huang HC, Marco Y. 1996. Characterization of *hsr201* and *hsr515*, two tobacco genes preferentially expressed during the hypersensitive reaction provoked by phytopathogenic bacteria. *Plant Mol. Biol.* 31:255–65

32. Czernic P, Moliere F, Marco Y. 1996. Structural organization of a genomic clone encoding a cytochrome P450-dependent monooxygenase from *Nicotiana tabacum* (Accession No. X96784) (PGR96–034). *Plant Physiol.* 111:652

33. Degtyarenko KN, Fábián P. 1996. The directory of P450-containing systems on WorldWide Web. *Comput. Appl. Biosci.* 12:237–40

34. Dennis DT, West CA. 1967. Biosynthesis of gibberellins III. The conversion of (-)-kaurene to (-)-kauren-19-oic acid in endosperm of *Echinocystis macrocarpa* Greene. *J. Biol. Chem.* 242:3293–300

35. Doodeman M, Tabak AJH, Schram AW, Bennink GJH. 1982. Hydroxylation of cinnamic acids and flavonoids during

biosynthesis of anthocyanins in *Petunia hybrida. Horticult. Planta* 154:546–49

36. Du L, Halkier BA. 1996. Isolation of a microsomal enzyme system involved in glucosinolate biosynthesis from seedlings of *Tropaeolum majus* L. *Plant Physiol.* 111:83–37

37. Du L, Lykkesfeldt J, Olsens CE, Halkier BA. 1995. Involvement of cytochrome P450 in oxime production in glucosinolate biosynthesis as demonstrated by an in vitro microsomal enzyme system isolated from jasmonic acid-induced seedlings of *Sinapis alba* L. *Proc. Natl. Acad. Sci. USA* 92:12505–9

38. Durst F, Nelson DR. 1995. Diversity and evolution of plant P450s and P450-reductases. *Drug Metab. Drug Interact.* 12:189–206

39. Durst F, O'Keefe DP. 1995. Plant cytochromes P450: an overview. *Drug Metab. Drug Interact.* 12:171–87

40. Estabrook RW, Cooper DY, Rosenthal O. 1963. The light reversible carbon monoxide inhibition of the steroid C21-hydroxylase system of the adrenal cortex. *Biochem. Z.* 338:741–55

41. Fábián P, Degtyarenko KN. 1997. The directory of P450-containing systems in 1996. *Nucleic Acids Res.* 25:274–77

42. Fahrendorf T, Dixon RA. 1993. Stress responses in alfalfa (*Medicago sativa* L.) XVIII: molecular cloning and expression of the elicitor-inducible cinnamic acid 4-hydroxylase cytochrome P450. *Arch. Biochem. Biophys.* 305:509–15

43. Finkelstein RR, Zeevaart JAD. 1994. Gibberellin and abscisic acid biosynthesis and response. In *Arabidopsis*, ed. EM Meyerowitz, CR Somerville, pp. 523–53. Cold Spring Harbor, NY: Cold Spring Harbor Press

44. Forkmann G, Heller W, Grisebach H. 1980. Anthocyanin biosynthesis in flowers of *Matthiola incana*. Flavanone 3- and flavonoid 3'-hydroxylases. *Z. Naturforsch. Teil* 35:691–95

45. Forkmann G, Stotz G. 1981. Genetic control of flavanone 3-hydroxylase activity and flavonoid 3'-hydroxylase activity in *Antirrhinum majus* (snapdragon). *Z. Naturforsch. Teil C* 36:411–16

46. Frank MR, Deyneka JM, Schuler MA. 1996. Cloning of wound-induced cytochrome P450 monooxygenases expressed in pea. *Plant Physiol.* 110:1035–46

47. Deleted in proof

48. Frear DS. 1995. Wheat microsomal

cytochrome P450 monooxygenases: characterization and importance in the metabolic detoxification and selectivity of wheat herbicides. *Drug Metab. Drug Interact.* 12:329–57

49. Frey M, Chomet P, Glawischnig E, Stettner C, Grün S, et al. 1997. Analysis of a chemical plant defense mechanism in grasses. *Science* 277:696–99

50. Frey M, Kliem R, Saedler H, Gierl A. 1995. Expression of a cytochrome P450 gene family in maize. *Mol. Gen. Genet.* 246:100–9

51. Gabriac B, Werck-Reichhart D, Teutsch H, Durst F. 1991. Purification and immunocharacterization of a plant cytochrome P450: the cinnamic acid 4-hydroxylase. *Arch. Biochem. Biophys.* 288:302–9

52. Garfinkel D. 1958. Studies on pig liver microsomes. I. Enzymic and pigment composition of different microsomal fractions. *Arch. Biochem. Biophys.* 77: 493–509

53. Ge L, Chiang VL. 1996. A full length cDNA encoding trans-cinnamate 4-hydroxylase from developing xylem of *Populus tremuloides* (Accession No. U47293) (PGR96–075). *Plant Physiol.* 112:861

54. Glawischnig E, Eisenreich W, Bacher A, Frey M, Gierl A. 1997. Biosynthetic origin of oxygen atoms in DIMBOA from maize: NMR studies with $^{18}O_2$. *Phytochemistry* 45:715–18

55. Deleted in proof

56. Gotoh O. 1992. Substrate recognition sites in cytochrome P450 family 2 (CYP2) proteins inferred from comparative analyses of amino acid and coding nucleotide sequences. *J. Biol. Chem.* 267:83–90

57. Graebe JE. 1987. Gibberellin biosynthesis and control. *Annu. Rev. Plant Physiol.* 38:419–65

58. Grand C. 1984. Ferulic acid 5-hydroxylase: a new cytochrome P-450-dependent enzyme from higher plant microsomes involved in lignin synthesis. *FEBS Lett.* 169:7–11

59. Grausem B, Chaubet N, Gigot C, Loper JC, Benveniste P. 1995. Functional expression of a *Saccharomyces cerevisiae CYP51A1* encoding lanosterol-14-demethylase in tobacco results in bypass of endogenous sterol biosynthetic pathway and resistance to an obtusifoliol-14-demethylase herbicide inhibitor. *Plant J.* 7:761–70

60. Grove MD, Spencer GF, Rohwedder WK, Mandava N, Worley JF, et al. 1979. Brassinolide, a plant growth-promoting steroid isolated from *Brassica napus* pollen. *Nature* 281:216–17

61. Halkier BA. 1996. Catalytic reactivities and structure/function relationships of cytochrome P450 enzymes. *Phytochemistry* 43:1–21

62. Halkier BA, Møller BL. 1989. Biosynthesis of the cyanogenic glucoside dhurrin in seedlings of *Sorghum bicolor* (L.) Moench and partial purification of the enzyme system involved. *Plant Physiol.* 90:1552–59

63. Halkier BA, Møller BL. 1991. Involvement of cytochrome P-450 in the biosynthesis of dhurrin in *Sorghum bicolor* (L.) Moench. *Plant Physiol.* 96:10–17

64. Halkier BA, Nielsen HL, Koch B, Møller BL. 1995. Purification and characterization of recombinant cytochrome $P450_{TYR}$ expressed at high levels in *Escherichia coli*. *Arch. Biochem. Biophys.* 322:369–77

65. Halkier BA, Olsen CE, Møller BL. 1989. The biosynthesis of cyanogenic glycosides in higher plants. The (E)- and (Z)-isomers of p-hydroxyphenylacetaldehyde oxime as intermediates in the biosynthesis of dhurrin in *Sorghum bicolor* (L.) Moench. *J. Biol. Chem.* 264:19487–94

66. Hallahan DL, Lau S-MC, Harder PA, Smiley DWM, Dawson GW, et al. 1994. Cytochrome *P-450*-catalyzed monoterpenoid oxidation in catmint (*Nepeta racemosa*) and avocado (*Persea americana*); evidence for related enzymes with different activities. *Biochim. Biophys. Acta* 1201:94–100

67. Harms K, Atzorn R, Brash A, Kühn H, Wasternack C, et al. 1995. Expression of a flax allene oxide synthase cDNA leads to increased endogenous jasmonic acid (JA) levels in transgenic potato plants but not to a corresponding activation of JA-responding genes. *Plant Cell* 7:1645–54

67a. Hedden P, Kamiya Y. 1997. Gibberellin biosythesis: genes and their regulation. *Annu. Rev. Plant Physiol.* 48:431–60

68. Heitz T, Bergey DR, Ryan CA. 1997. A gene encoding a chloroplast-targeted lipoxygenase in tomato leaves is transiently induced by wounding, systemin, and methyl jasmonate. *Plant Physiol.* 114:1085–93

69. Higashi Y, Yoshioka H, Yamane M, Gotoh O, Fujii-Kuriyama Y. 1986. Complete nucleotide sequence of two steroid 21-hydroxylase genes tandemly arranged in human chromosome: a

pseudogene and a genuine gene. *Proc. Natl. Acad. Sci. USA* 83:2841–45

70. Holton TA. 1995. Modification of flower colour via manipulation of P450 gene expression in transgenic plants. *Metab. Drug Interact.* 12:359–68

71. Holton TA, Brugliera F, Lester DR, Tanaka Y, Hyland CD, et al. 1993. Cloning and expression of cytochrome P450 genes controlling flower colour. *Nature* 366:276–79

72. Hotze M, Schroeder G, Schroeder J. 1995. Cinnamate 4-hydroxylase from *Catharanthus roseus*, and a strategy for the functional expression of plant cytochrome P450 proteins as translational fusions with P450 reductase in *Escherichia coli. FEBS Lett.* 374:345–50

73. Hutvágner G, Barta E, Bánfalvi Z. 1997. Isolation and sequence analysis of a cDNA and a related gene for cytochrome P450 proteins from *Solanum chacoense. Gene* 188:247–52

74. Kahn RA, Bak S, Olsen CE, Svendsen I, Møller BL. 1996. Isolation and reconstitution of the heme-thiolate protein obtusifoliol 14α-demethylase from *Sorghum bicolor* (L.) Moench. *J. Biol. Chem.* 271:32944–50

75. Kawai S, Mori A, Shiokawa T, Kajita S, Katayama Y, Morohoshi N. 1996. Isolation and analysis of cinnamic acid 4-hydroxylase homologous genes from a hybrid aspen, *Populus kitakamiensis. Biosci. Biotechnol. Biochem.* 60:1586–97

75a. Kende H, Zeevaart JAD. 1997. The five "classical" plant hormones. *Plant Cell* 9:1197–1210

76. Klingenberg M. 1958. Pigments of rat liver microsomes. *Arch. Biochem. Biophys.* 75:376–86

77. Koch BM, Sibbesen O, Halkier BA, Svendsen I, Møller BL. 1995. The primary sequence of cytochrome $P450_{TYR}$, the multifunctional N-hydroxylase catalyzing the conversion of L-tyrosine to *p*-hydroxyphenylacetaldehyde oxime in the biosynthesis of the cyanogenic glucoside dhurrin in *Sorghum bicolor* (L.) Moench. *Arch. Biochem. Biophys.* 323:177–86

78. Kraus PFX, Kutchan TM. 1995. Molecular cloning and heterologous expression of a cDNA encoding berbamunine synthase, a C-O phenol-coupling cytochrome P450 from the higher plant *Berberis stolonifera. Proc. Natl. Acad. Sci. USA* 92:2071–75

79. Landry LG, Chapple CCS, Last R. 1995.

Arabidopsis mutants lacking phenolic sunscreens exhibit enhanced ultraviolet-B injury and oxidative damage. *Plant Physiol.* 109:1159–66

80. Larkin JC. 1994. Isolation of a cytochrome P450 homologue preferentially expressed in developing inflorescences of *Zea mays. Plant Mol. Biol.* 25:343–53

81. Lau S-MC, Harder PA, O'Keefe DP. 1993. Low carbon monoxide affinity allene oxide synthase is the predominant cytochrome P450 in many plant tissues. *Biochemistry* 32:1945–50

82. Laudert D, Pfannschmidt U, Lottspeich F, Holländer-Czytko H, Weiler EW. 1996. Cloning, molecular and functional characterization of *Arabidopsis thaliana* allene oxide synthase (CYP 74), the first enzyme of the octadecanoid pathway to jasmonates. *Plant Mol. Biol.* 31:323–35

83. León J, Shulaev V, Yalpani N, Lawton MA, Raskin I. 1995. Benzoic acid 2-hydroxylase, a soluble oxygenase from tobacco, catalyzes salicylic acid biosynthesis. *Proc. Natl. Acad. Sci. USA* 92: 10413–17

84. León J, Yalpani N, Raskin I, Lawton MA. 1993. Induction of benzoic acid 2-hydroxylase in virus-inoculated tobacco. *Plant Physiol.* 103:323–28

85. Li J, Nagpal P, Vitart V, McMorris TC, Chory J. 1996. A role for brassinosteroids in light-dependent development of Arabidopsis. *Science* 272:398–401

86. Logemann E, Parniske M, Hahlbrock K. 1995. Modes of expression and common structural features of the complete phenylalanine ammonia-lyase family in parsley. *Proc. Natl. Acad. Sci. USA* 92: 5905–9

87. Deleted in proof

88. Mangold U, Eichel J, Batschauer A, Lanz T, Kaiser T, et al. 1994. Gene and cDNA for plant cytochrome P450 proteins (CYP72 family) from *Catharanthus roseus*, and transgenic expression of the gene and a cDNA in tobacco and *Arabidopsis thaliana. Plant Sci.* 96:129–36

89. Matsui K, Shibutani M, Hase T, Kajiwara T. 1996. Bell pepper fruit fatty acid hydroperoxide lyase is a cytochrome P450 (CYP74B). *FEBS Lett.* 394:21–24

90. Maughan JA, Nugent JHA, Hallahan DL. 1997. Expression of CYP71B7, a cytochrome P450 expressed sequence tag from *Arabidopsis thaliana. Arch. Biochem. Biophys.* 341:104–11

91. McKinney EC, Ali N, Traut A, Feldmann KA, Belostotsky DA, et al.

1995. Sequence-based identification of T-DNA insertion mutations in *Arabidopsis*: actin mutants *act2-1* and *act4-1*. *Plant J.* 8:613–22

92. Meijer AH, Souer E, Verpoorte R, Hoge JHC. 1993. Isolation of cytochrome P-450 cDNA clones from the higher plant *Catharanthus roseus* by a PCR strategy. *Plant Mol. Biol.* 22:379–83

93. Menting JGT, Scopes RK, Stevenson TW. 1994. Characterization of flavonoid 3′,5′-hydroxylase in microsomal membrane fraction of *Petunia hybrida* flowers. *Plant Physiol.* 106:633–42

94. Meyer K, Cusumano JC, Somerville C, Chapple CCS. 1996. Ferulate-5-hydroxylase from *Arabidopsis thaliana* defines a new family of cytochrome P450-dependent monooxygenases. *Proc. Natl. Acad. Sci. USA* 93:6869–74

95. Mitchell JW, Mandava N, Worley JF, Plimmer JR, Smith MV. 1970. Brassins—a new family of plant hormones from rape pollen. *Nature* 225:1065–66

96. Mitoma C, Posner HS, Reitz HC, Udenfriend S. 1956. Enzymatic hydroxylation of aromatic compounds. *Arch. Biochem. Biophys.* 61:431–41

97. Mizutani M, Ohta D, Sato R. 1993. Purification and characterization of a cytochrome P450 (*trans*-cinnamic acid 4-hydroxylase) from etiolated mung bean seedlings. *Plant Cell Physiol.* 34:481–88

98. Mizutani M, Ohta D, Sato R. 1997. Isolation of a cDNA and a genomic clone encoding cinnamate 4-hydroxylase from Arabidopsis and its expression manner in planta. *Plant Physiol.* 113:755–63

99. Mizutani M, Ward E, DiMaio J, Ohta D, Ryals J, Sato R. 1993. Molecular cloning and sequencing of a cDNA encoding mung bean cytochrome P450 (P450$_{C4H}$) possessing cinnamate 4-hydroxylase activity. *Biochem. Biophys. Res. Commun.* 190:875–80

100. Møller BL, Conn EE. 1980. The biosynthesis of cyanogenic glucosides in higher plants. Channeling of intermediates in dhurrin biosynthesis by a microsomal system from *Sorghum bicolor* (Linn) Moench. *J. Biol. Chem.* 255:3049–56

101. Murphy PJ, West CA. 1969. The role of mixed function oxidases in kaurene metabolism in *Echinocystis macrocarpa* Greene endosperm. *Arch. Biochem. Biophys.* 133:395–407

102. Nadeau JA, Zhang XS, Li J, O'Neill SD. 1996. Ovule development: identification of stage-specific and tissue-specific cDNAs. *Plant Cell* 8:213–39

103. Nelson DR, Koymans L, Kamataki T, Stegeman JJ, Feyereisen R, et al. 1996. P450 superfamily: update on new sequences, gene mapping, accession numbers and nomenclature. *Pharmacogenetics* 6:1–42

104. Newman T, de Bruijn FJ, Green P, Keegstra K, Kende H, et al. 1994. Genes galore: a summary of methods for accessing results from large-scale partial sequencing of anonymous *Arabidopsis* cDNA clones. *Plant Physiol.* 106:1241–55

105. Niemeyer HM. 1988. Hydroxamic acids (4-hydroxy-1,4-benzoxazin-3-ones), defence chemicals in the Gramineae. *Phytochemistry* 27:3349–58

106. O'Keefe DP, Leto KJ. 1989. Cytochrome P-450 from the mesocarp of avocado (*Persea americana*). *Plant Physiol.* 89:1141–49

107. Omura T, Sato R. 1964. The carbon monoxide-binding pigment of liver microsomes. I. Evidence for its hemoprotein nature. *J. Biol. Chem.* 239:2370–78

108. Ortiz de Montellano PR. 1995. Oxygen activation and reactivity. See Ref. 108a, pp. 245–303

108a. Ortiz de Montellano PR, ed. 1995. *Cytochrome P450: Structure, Mechanism, and Biochemistry*, Vol. 2. New York: Plenum

109. Pan Z, Durst F, Werck-Reichhart D, Gardner HW, Camara B, et al. 1995. The major protein of guayule rubber particles is a cytochrome P450. Characterization based on cDNA cloning and spectroscopic analysis of the solubilized enzyme and its reaction products. *J. Biol. Chem.* 270:8487–94

110. Peng Y-L, Shirano Y, Ohta H, Hibino T, Tanaka K, Shibata D. 1994. A novel lipoxygenase from rice. Primary structure and specific expression upon incompatible infection with rice blast fungus. *J. Biol. Chem.* 269:3755–61

111. Pierrel MA, Batard Y, Kazmaier M, Mignotte-Vieux C, Durst F, Werck-Reichhart D. 1994. Catalytic properties of the plant cytochrome P450 CYP73 expressed in yeast. Substrate specificity of a cinnamate hydroxylase. *Eur. J. Biochem.* 224:835–44

112. Pompon D, Louerat B, Bronine A, Urban P. 1996. Yeast expression of animal and plant P450s in optimized redox environments. *Methods Enzymol.* 272:51–64

113. Potts JRM, Weklych R, Conn EE. 1974. The 4-hydroxylation of cinnamic acid by

sorghum microsomes and the requirement for cytochrome P-450. *J. Biol.
Chem.* 249:5019–26

114. Poulos TL, Cupp-Vickery J, Li H. 1995.
Structural studies on prokaryotic cytochromes P450. See Ref. 108a, pp.
125–80

115. Reichhart D, Salaun J-P, Benveniste I,
Durst F. 1979. Induction by manganese,
ethanol, phenobarbital, and herbicides
of microsomal cytochrome *P-450* in
higher plant tissues. *Arch. Biochem. Biophys.* 196:301–3

116. Rounsley SD, Glodek A, Sutton G,
Adams MD, Somerville CR, et al. 1996.
The construction of Arabidopsis ex-
pressed sequence tag assemblies. A new
resource to facilitate gene identification.
Plant Physiol. 112:1177–83

117. Russell DW. 1971. The metabolism of
aromatic compounds in higher plants.
X. Properties of the cinnamic acid 4-
hydoxylase of pea seedlings and some
aspects of its metabolic and developmental control. *J. Biol. Chem.* 246:
3870–78

118. Russell DW, Conn EE. 1967. The
cinnamic acid 4-hydroxylase of pea
seedlings. *Arch. Biochem. Biophys.* 122:
256–58

119. Sakaguchi M, Mihara K, Sato R. 1984.
Signal recognition particle is required
for co-translational insertion of cytochrome P-450 into microsomal membranes. *Proc. Natl. Acad. Sci. USA* 81:
3361–64

120. Sakaguchi M, Mihara K, Sato R. 1987.
A short amino-terminal segment of microsomal cytochrome P-450 functions as
an insertion signal and as a stop-transfer
sequence. *EMBO J.* 6:2425–31

121. Schuler MA. 1996. Plant cytochrome
P450 monooxygenases. *Crit. Rev. Plant
Sci.* 15:235–84

122. Schuler MA. 1996. The role of cytochrome P450 monooxygenases in plant-
insect interactions. *Plant Physiol.* 112:
1411–19

123. Shen B, Carneiro N, Torres-Jerez I,
Stevenson B, McCreery T, Helentjaris
T, et al. 1994. Partial sequencing and
mapping of clones from two maize
cDNA libraries. *Plant Mol. Biol.* 26:
1085–101

124. Shibata Y, Matsui K, Kajiwara T,
Hatanaka A. 1995. Fatty acid hydroper-
oxide lyase is a heme protein. *Biochem.
Biophys. Res. Commun.* 207:438–43

125. Shibata Y, Matsui K, Kajiwara T,
Hatanaka A. 1995. Purification and
properties of fatty acid hydroperoxide

lyase from green bell pepper fruits. *Plant
Cell Physiol.* 36:147–56

126. Shimada M, Conn EE. 1977. The
enzymatic conversion of *p*-hydroxy-
phenlacetaldoxime to *p*-hydroxyman-
delonitrile. *Arch. Biochem. Biophys.*
180:199–207

127. Shirley BW, Kubasek WL, Storz G,
Bruggemann E, Koornneef M, et al.
1995. Analysis of *Arabidopsis* mutants
deficient in flavonoid biosynthesis. *Plant
J.* 8:659–71

128. Sibbesen O, Koch B, Halkier BA, Møller
BL. 1994. Isolation of the heme-thiolate
enzyme cytochrome P-450$_{TYR}$, which
catalyzes the committed step in the
biosynthesis of the cyanogenic glucoside dhurrin in *Sorghum bicolor* (L.)
Moench. *Proc. Natl. Acad. Sci. USA* 91:
9740–44

129. Sibbesen O, Koch B, Halkier BA, Møller
BL. 1995. Cytochrome P-450$_{TYR}$ is a
multifunctional heme-thiolate enzyme
catalyzing the conversion of L-tyrosine
to *p*-hydroxyphenylacetaldehyde oxime
in the biosynthesis of the cyanogenic
glucoside dhurrin in *Sorghum bicolor*
(L.) Moench. *J. Biol. Chem.* 270:3506–
11

130. Song W-C, Brash AR. 1991. Purification
of allene oxide synthase and identification of the enzyme as a cytochrome P-
450. *Science* 253:781–84

131. Song W-C, Funk CD, Brash AR. 1993.
Molecular cloning of an allene oxide
synthase: a cytochrome P450 specialized for the metabolism of fatty acid hydroperoxides. *Proc. Natl. Acad. Sci. USA*
90:8519–23

132. Stadler R, Zenk MH. 1993. The purification and characterization of a unique cytochrome P-450 enzyme from *Berberis
stolonifera* plant cell cultures. *J. Biol.
Chem.* 268:823–31

133. Stotz G, Forkmann G. 1981. Hydroxylation of the B-ring of flavonoids in
the 3'- and 5'-position with enzyme extracts from flowers of *Verbena hybrida*.
Z. Naturforsch. Teil C 37:19–23

134. Sun T-P, Goodman HM, Ausubel FM.
1992. Cloning the Arabidopsis *GA1* locus by genomic subtraction. *Plant Cell*
4:119–28

135. Suzuki G, Ohta H, Kato T, Igarashi
T, Sakai F, et al. 1996. Induction of a
novel cytochrome P450 (CYP93 family) by methyl jasmonate in soybean
suspension-cultured cells. *FEBS Lett.*
383:83–86

136. Suzuki G, Ohta H, Kato T, Shibata
D, Takano A, et al. 1997. Molecular

cloning of a cDNA encoding cytochrome P450 CYP93A2 (Accession No. D86531) from soybean suspension-cultured cells (PGR97-087). *Plant Physiol.* 114:748

137. Szczesna-Skorupa E, Ahn K, Chen C-D, Doray B, Kemper B. 1995. The cytoplasmic and N-terminal transmembrane domains of cytochrome P450 contain independent signals for retention in the endoplasmic reticulum. *J. Biol. Chem.* 270:24327–33

138. Szczesna-Skorupa E, Straub P, Kemper B. 1993. Deletion of a conserved tetrapeptide, PPGP, in P450 2C2 results in loss of enzymatic activity without a change in its cellular location. *Arch. Biochem. Biophys.* 304:170–75

139. Szekeres M, Németh K, Koncz-Kálmán Z, Mathur J, Kauschmann A, et al. 1996. Brassinosteroids rescue the deficiency of CYP90, a cytochrome P450, controlling cell elongation and de-etiolation in Arabidopsis. *Cell* 85:171–82

140. Tabak AJH, Meyer H, Bennink GJH. 1978. Modification of the B-ring during flavonoid synthesis in *Petunia hybrida*: introduction of the 3'-hydroxyl group regulated by the gene Ht1. *Planta* 139:67–71

141. Tanaka Y, Yonekura K, Fukuchi-Mizutani M, Fukui Y, Fujiwara H, et al. 1996. Molecular and biochemical characterization of three anthocyanin synthetic enzymes from *Gentiana triflora*. *Plant Cell Physiol.* 37:711–16

142. Taton M, Rahier A. 1991. Properties and structural requirements for substrate specificity of cytochrome P-450-dependent obtusifoliol 14α demethylase from maize (*Zea mays*) seedlings. *Biochem. J.* 277:483–92

143. Teutsch HG, Hasenfratz MP, Lesot A, Stoltz C, Garnier J-M, et al. 1993. Isolation and sequence of a cDNA encoding the Jerusalem artichoke cinnamate 4-hydroxylase, a major plant cytochrome P450 involved in the general phenylpropanoid pathway. *Proc. Natl. Acad. Sci. USA* 90:4102–6

144. Toguri T, Azuma M, Ohtani T. 1993. The cloning and characterization of a cDNA encoding a cytochrome P450 from the flowers of *Petunia hybrida*. *Plant Sci.* 94:119–26

145. Toguri T, Kobayashi O, Umemoto N. 1993. The cloning of eggplant seedling cDNAs encoding proteins from a novel cytochrome P-450 family (CYP76). *Biochim. Biophys. Acta* 1216:165–69

146. Toguri T, Tokugawa K. 1994. Cloning of eggplant hypocotyl cDNAs encoding cytochromes P450 belonging to a novel family (CYP77). *FEBS Lett.* 338:290–94

147. Toguri T, Umemoto N, Kobayashi O, Ohtani T. 1993. Activation of anthocyanin synthesis genes by white light in eggplant hypocotyl tissues, and identification of an inducible P-450 cDNA. *Plant Mol. Biol.* 23:933–46

148. Udvardi MK, Metzger JD, Krishnapillai V, Peacock WJ, Dennis ES. 1994. Cloning and sequencing of a full-length cDNA from *Thlaspi arvense* L. that encodes a cytochrome P-450. *Plant Physiol.* 105:755–56

149. Umemoto N, Kobayashi O, Ishizaki-Nishizawa O, Toguri T. 1993. cDNAs sequences encoding cytohrome P450 (CYP71 family) from eggplant seedlings. *FEBS Lett.* 330:169–73

150. Urban P, Werck-Reichhart D, Teutsch H, Durst F, Regnier S, et al. 1994. Characterization of recombinant cinnamate 4-hydroxylase produced in yeast. Kinetic and spectral properties of the major plant P450 of the phenylpropanoid pathway. *Eur. J. Biochem.* 222:843–50

151. Vetter H-P, Mangold U, Schröder G, Marner F-J, Werck-Reichhart D, Schröder J. 1992. Molecular analysis and heterologous expression of an inducible cytochrome P-450 protein from periwinkle (*Catharanthus roseus* L.). *Plant Physiol.* 100:998–1007

152. Vick BA, Zimmerman DC. 1987. Pathways of fatty acid hydroperoxide metabolism in spinach leaf chloroplasts. *Plant Physiol.* 85:1073–78

153. Von Wachenfeldt C, Johnson EF. 1995. Structures of eukaryotic cytochrome P450 enzymes. See Ref. 108a, pp. 183–223

154. Werck-Reichhart D, Batard Y, Kochs G, Lesot A, Durst F. 1993. Monospecific polyclonal antibodies directed against purified cinnamate 4-hydroxylase from *Helianthus tuberosus*. *Plant Physiol.* 102:1291–98

155. Werck-Reichhart D, Benveniste I, Teutsch H, Durst F, Gabriac B. 1991. Glycerol allows low-temperature phase separation of membrane proteins solubilized in Triton X-114: application to the purification of plant cytochromes P-450 and b_5. *Anal. Biochem.* 197:125–31

156. White PC, New MI, Dupont B. 1986. Structure of human steroid 21-hydroxylase genes. *Proc. Natl. Acad. Sci. USA* 83:5111–15

157. Winkler RG, Helentjaris T. 1995. The maize *dwarf3* gene encodes a cytochrome P450-mediated early step in gibberellin biosynthesis. *Plant Cell* 7:1307–17

158. Yamazaki S, Sato K, Suhara K, Sakaguchi M, Mihara K, Omura T. 1993. Importance of the proline-rich region following signal-anchor sequence in the formation of correct conformation of microsomal cytochrome P-450s. *J. Biochem.* 114:652–57

159. Ye A-H. 1996. Expression of the cinnamic acid 4-hydroxylase gene during lignification in *Zinnia elegans*. *Plant Sci.* 121:133–41

160. Zimmerman DC. 1966. A new product of linoleic acid oxidation by a flaxseed enzyme. *Biochem. Biophys. Res. Commun.* 23:398–402

Annu. Rev. Plant Physiol. Plant Mol. Biol. 1998. 49:345–70

GENETIC CONTROL OF FLOWERING TIME IN ARABIDOPSIS

Maarten Koornneef, Carlos Alonso-Blanco, Anton J. M. Peeters, and Wim Soppe

Department of Genetics, Wageningen Agricultural University, Dreijenlaan 2, NL-6703 HA Wageningen, The Netherlands

KEY WORDS: *Arabidopsis thaliana*, photoperiod, circadian rhythm, gibberellins, vernalization

ABSTRACT

The timing of the transition from vegetative to reproductive development is of great fundamental and applied interest but is still poorly understood. Recently, molecular-genetic approaches have been used to dissect this process in Arabidopsis. The genetic variation present among a large number of mutants with an early- or late-flowering phenotype, affecting the control of both environmental and endogenous factors that influence the transition to flowering, is described. The genetic, molecular, and physiological analyses have led to identification of different components involved, such as elements of photoperception and the circadian rhythm. Furthermore, elements involved in the signal transduction pathways to flowering have been identified by the cloning of some floral induction genes and their target genes.

CONTENTS

345

1040-2519/98/0601-0345$08.00

INTRODUCTION

In order to achieve successful sexual reproduction, plants must be able to flower under favorable environmental conditions, and the proper timing of flowering is, therefore, supposed to have an important adaptive value for plants. The transition from vegetative to reproductive development is controlled by both environmental and endogenous factors. Plant physiologists have studied this important process by changing environmental factors and analyzing the subsequent morphological, physiological, and biochemical consequences of these treatments. More recently, genetics has been used to study the mechanism of flower initiation by analysis of genetic variation in species, such as pea and Arabidopsis. Especially in Arabidopsis, the possibility to pursue the genetic analysis down to the molecular level is attractive. This topic and aspects of it have been reviewed (4, 33, 51, 72, 91, 105, 138). In the present review, we summarize the current progress made in the analysis of the transition to flowering using the genetic and molecular approaches as they have been applied to Arabidopsis.

The Transition to Flowering—Meristem Fate Changes

Arabidopsis thaliana has a distinct vegetative phase during which the apical meristem produces lateral meristems developing into leaves subtending an axillary bud. The nodes do not elongate, resulting in the formation of a rosette. Floral transition is marked by the establishment of a floral fate in these meristems and by the suppression of leaf production.

A bi-directional development has been shown in this transition, with flowers being initiated acropetally and leaf primordia being suppressed basipetally (53). After floral initiation and following this basipetal direction, the axillary buds of the leaf primordia mostly develop into a secondary shoot (or paraclades or coflorescences). In specific genotypes, they replicate the fate of the initial meristem by forming axillary rosettes. Following the fate change of these lateral meristems, internode elongation takes place (bolting). The elongated stem or inflorescence bears cauline leaves and flowers that are not subtended by leaves at higher internodes. The part of the inflorescence with leaves, which was called early inflorescence by Haughn et al (51), should be considered as part of the vegetative phase. As a consequence of this, total leaf number together with time to flowering are the best quantitative parameters to monitor flowering initiation. Although the appearance of flowers is the final and most dramatic result of the phase change, other changes occur earlier. These are somewhat gradual and can be observed in leaf morphology (93) and in the gradual appearance of trichomes at the abaxial side of the leaves and their gradual disappearance at the adaxial side (27, 131). It has been proposed that phase changes involve a decrease of a floral repressor (130), called a *c*ontroller *of p*hase *s*witching (COPS), which at critical low levels leads to the activation of the *fl*oral *i*nitiation *p*rocess (FLIP)

(124). The latter is controlled by the so-called Floral Meristem Identity or FLIP genes, such as *LEAFY (LFY)*, *APETALA1* and *2 (AP1, AP2)*, *AULIFLOWER (CAL)*, and *UNUSUAL FLORAL ORGANS (UFO)* (51).

Environmental and Endogenous Control of Flowering

Arabidopsis is a facultative long-day (LD) plant, which means that plants flower earlier under LDs than under short days (SDs). When plants of the common early laboratory genotypes are of sufficient age, indicating a certain competence for flowering, one LD is sufficient to induce flowering (32, 55, 99). This treatment has been used to monitor the morphological (54) and molecular changes (55) involved.

The photoperiodic control of flowering is thought to be mediated by the interaction of photoreceptors, such as phytochrome and cryptochrome, and a clock mechanism or circadian rhythm. Photoreceptors play a role to set the phase of the circadian rhythm, but they can also affect flowering directly, thereby involving light quality in the control of this process. Blue (B) light and far-red (FR) light are known to be more effective to promote flowering than red (R) light (16, 41). Besides, the sensitivity of plants to light quality itself depends on a circadian rhythm (20). The importance of light quality in flowering is determined by the mechanism of light perception, since the ratio red:far-red (R:FR) determines the phytochrome status in the plant. Nevertheless, light is not a prerequisite for flowering, since flowering occurs rapidly in complete darkness when sufficient carbohydrates are provided to the growing shoot meristem (88, 115). A higher light intensity also promotes flowering probably by its effect on carbohydrate supply (7, 67).

Another important treatment promoting flowering is vernalization, which is a transient exposure to low temperatures. The effectiveness of vernalization depends on the stage of the plant, the length of the treatment, and the temperature employed (100, 102). Furthermore, the growing temperature also affects flowering as measured not only by flowering time but also by leaf number (6), which should correct for differences in temperature effects on growth.

In Arabidopsis, the effect of (sensitivity for) the environmental factors strongly depends on the genotype (see below). These environmental factors are thought to modulate certain endogenous components, thus affecting and controlling flowering. Many chemical treatments have been shown to promote flowering (91), of which the application of gibberellins (GAs) (7, 143) and base analogues (91, 114) has attracted most attention, because of their relatively large effects.

GENES AFFECTING FLOWERING TIME

The genetic differences present among ecotypes and, mainly, the genetic variation induced by mutagenic treatments are very important for the analysis of flowering time in Arabidopsis. Many mutants with an early- or late-flowering phenotype have been described that affect genes controlling both environmental and endogenous factors that influence the transition to flowering. Besides, some cloned genes of unknown function are involved in flowering through their constitutive expression in transgenic plants. Furthermore, the regulation of gene expression through DNA methylation changes has been suggested to play a role in this process.

Natural Variation

Genetic variation for flowering time has been described within and among Arabidopsis natural populations (ecotypes) since the earliest research (76, 77, 102, 114). Arabidopsis has a wide range of distribution in the Northern hemisphere (114), and the differences found when growing different ecotypes under the same laboratory conditions are supposed to reflect particular adaptations to different natural environments. To illustrate this genetic variation, Karlsson et al (66) analyzed 32 ecotypes under SD and LD light conditions, with and without a vernalization treatment. Interactions between the three parameters—ecotype, photoperiod, and vernalization—were found. The first genetic analyses of Arabidopsis flowering time made use of this natural variation to establish the minimum number of genes involved in particular crosses. These early studies often showed the segregation of one or two major genes (65, 102, 132). However, because different parental combinations were analyzed it is not clear whether the same genes were segregating in those populations. Furthermore, segregation of genes with relatively small effects (minor genes) escaped detection in such studies. Napp-Zinn (100, 102) studied in detail the flowering time differences and vernalization requirement between the late ecotype Stockholm and the early Limburg-5 and isolated genotypes with single major flowering time gene differences. This analysis showed that at least four genes were involved and that alleles at the loci with larger effect were more or less epistatic to the alleles with smaller effect. At the locus *FRIGIDA* (*FRI*), the dominant allele produced a large delay in flowering time, and at the *KRYOPHILA* (*KRY*) and the *JUVENALIS* (*JUV*) loci the recessive alleles did so with a smaller effect. Vernalization overcame most of the effects of these late alleles (101).

The advent of molecular markers and the development of genetic maps has facilitated the localization in the genome and the characterization of some of the major loci controlling flowering time differences between very late and very early ecotypes. Napp-Zinn's *FRI* gene has been mapped on top of chromosome 4 (30). It has been shown that the extreme lateness present in several ecotypes is due to dominant alleles at a locus mapping at a similar position, which is probably *FRI* (18, 46, 81, 103, 123). The late-flowering phenotype of *FRI* is very much suppressed under LDs by the Landsberg *erecta* (L*er*) allele at locus *Flowering Locus C* (*FLC*) mapping on top of chromosome 5 (69, 82), likely at

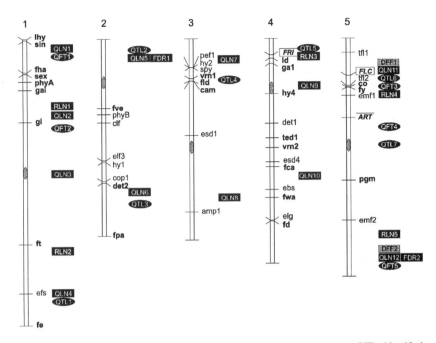

Figure 1 Arabidopsis genetic map showing the mutant loci and polymorphic QTLs identified affecting flowering time. Loci in *bold* correspond to genes with late-flowering mutant phenotype; otherwise the mutant is early. *FLC, FRI,* and *ART* loci, identified from natural populations, are indicated with *white boxes*. *Black* and *gray boxes* correspond to the approximate position of putative quantitative trait loci (QTLs) identified in different crosses; DFF1-2, QTLs in a Hannover/Münden F2 population (73); RLN1-5, QTLs in a Ler × H51 F2/F3 population (31); QLN1-12 in Ler × Col RIL population (60); FDR1-2 in the same Ler × Col RIL population (97); QTL1-7 in a backcross to Limburg-5, with selective genotyping, from F1 Limburg-5 × Naantali (74); QFT 1-5 in a Ler × Cape Verde Island RIL population (3).

a different position than any of the known flowering mutant loci (see Figure 1). Therefore, the flowering time differences between late and early ecotypes are largely determined by these two loci, each one by itself having a small effect and requiring dominant alleles at both to produce extreme lateness. So far, only the laboratory strains Ler and C24 (69, 120) have been found to contain early *FLC* alleles. The late-flowering phenotype of *FRI* and *FLC*, present under both LD and SD conditions, is reduced by FR-enriched light and eliminated by vernalization; saturation of vernalization abolishes a further effect of FR light (79). The Ler early *FLC* alleles also suppress the lateness of mutant alleles at several loci (see below) such as *ld* (69, 82) and *fld* (121), which were isolated in Col background but not in Ler.

A third locus, *Aerial RosetTe (ART)*, located on chromosome 5, has been identified by analyzing another very late ecotype, Skye (46). The dominant

ART allele in combination with dominant alleles at another gene located on chromosome 4, probably *FRI*, delays the transition from vegetative to reproductive in the axillary meristems, giving rise to aerial rosettes under LDs. *ART* alone seemed to produce lateness, but taking into account the close location to *FLC*, it is unclear how much of the *ART* late phenotype comes from *FLC* and whether late *FLC* alleles are also necessary to produce the aerial phenotype. Epistatic analysis shows that aerial rosettes are produced by combining *ART* not only with *FRI* but also with the late-flowering mutants *fca, fve, fpa, ld, fwa, co,* and *gi* (see below) (47). Thus, *ART* might act downstream in the flowering pathways, and in a late-flowering background it would produce a prolonged insensitivity to the floral evocation signals in the axillary meristems.

To find other natural alleles of smaller effect has required the combination of molecular genetic maps with statistical methods to map quantitative trait loci (QTLs) (59). QTL analyses have been performed using crosses between late and early ecotypes (31, 74) and between early ones (3, 60, 73, 97) (Figure 1). Multiple QTLs have been found in all the crosses and therefore differences in behavior of flowering mutant alleles in different genetic backgrounds cannot be directly attributed to a single gene differing between ecotypes. Further analyses are needed to detect the interacting genes in each particular case. The spectrum of natural variation is different from the spectrum of flowering-time variants obtained by mutational analyses. This is at least due to the limitations of the reduced number of ecotypes used to generate mutants, and to the possible deleterious pleiotropic effects of some of the induced mutations. For example, no mutant allele has been identified for the *FRI* locus. Some dominant late-flowering mutants such as McKelvie's *florens* (*F*) mutant (95) and the M73, L4, L5, and L6 mutants (133) were reported allelic to *FRI*, but it was unclear whether they were mutants or contaminant natural variants (69). Some of the putative QTLs locate at mutant gene positions, and therefore it is expected that part of the natural variants will correspond to alleles of mutant flowering genes. However, there are known mutant flowering genes scattered all over the genome (see Figure 1), and complex situations such as very closely linked QTLs might be expected. As an example, several analyses have detected QTLs on top of chromosome 5, a region enriched for mutant flowering genes, and at least some of these QTLs are likely to correspond to a different locus than *FLC* (3, 74). Therefore, the identification of the individual alleles controlling this variation is necessary.

Late-Flowering Mutants

Late-flowering mutants with a strong effect but with no other obvious pleiotropic effects were described for the first time by Rédei (113). He isolated the *constans* (*co*), *gigantea* (*gi*), and *luminidependens* (*ld*) mutants in Col background. Later on, more mutant alleles at these and 11 other loci in L*er* were isolated and described by Koornneef et al (70) and in Wassilewskija (Ws) by Lee et al (80). Thus, the loci *LD, CO, GI, FE, FT, FD, FY, FCA, FHA, FPA, FVE,* and *FWA* have been considered the classical late-flowering genes (Figure 1). They have been physiologically characterized, and epistatic relationships have been examined in relation to early, late, and meristem identity genes (50, 68, 119). Koornneef et al (68) constructed 42 double mutants among 10 of these loci. The epistatic interactions proved to be complex, but groups of loci similar to the ones established on the basis of their physiological behavior were identified. A major epistatic group could be identified corresponding to the group of mutants *co, fd, fe, fha, ft, fwa,* and *gi.* These mutants are late mainly under LD conditions, i.e. they show little or no response to daylength, and they have a low response to FR supplementary light and to vernalization treatments. In contrast, the epistatic behavior of the mutants that are much more responsive to these environmental factors (*fca, fpa, ld, fve,* and *fy*) is more complex. Combining the *FLC*-Col allele with late-flowering mutants in L*er* background, Sanda & Amasino (122) showed that the mutants *fca, fpa,* and *fve,* of the same group, all have very enhanced late phenotypes like those of *ld, FRI,* and *fld. Flowering locus D* (*fld*) is another late-flowering mutant without apparent pleiotropic effects (121). This mutant retains its response to photoperiod, and its flowering time can be reduced by cold treatment and low R:FR ratio.

Five of these late-flowering genes, *LD, FCA, CO, FT,* and *FHA,* have been cloned. *LD* encodes a glutamine-rich nuclear protein containing a possible homeodomain (80). Its function remains unknown.

FCA encodes a protein containing two RNA-binding domains and a WW protein interaction domain, suggesting that it is functioning in the posttranscriptional regulation of transcripts involved in flowering (87). An interesting characteristic of this gene is that the transcript is alternatively spliced; four different *FCA* transcripts have been found, the full-length transcript being only one third of the total amount. The WW domain, which is only present in the full-length transcript, seems essential for the flowering time effect.

The *CO* gene was found to encode a protein with similarity to GATA-1–type transcription factors (111). Constitutive expression of *CO* leads to earliness (127), thereby confirming that this gene has flowering promoting properties. Besides, transgenic plants with extra copies of *CO* flower earlier than wild type, suggesting that *CO* activity is limiting flowering time (111). The *CO* mRNA appears more abundant in plants grown under LDs than under SDs, in agreement with the role of this gene in promotion of flowering under LDs. It is interesting to note that two homologues of the *CO* gene, *CONSTANS LIKE 1* (*COL1*) and *COL2,* have been described recently (78, 110), and although quite similar in structure, their role in flowering has not yet been demonstrated.

The *FT* and *FHA* genes have been recently cloned. The *FT* gene shows strong homology to the *TERMINAL FLOWER 1* (*TFL1*) gene (5; D Weigel, personal communication). The *FHA* gene encodes the CRY2 protein (85; C Lin, personal communication) and is thought to be involved in blue-light perception.

Late mutants have been identified that are involved in light perception or transduction. The mutants *long hypocotyl 4* (*hy4*) and *phytochrome A* (*phyA*) correspond to the blue-light photoreceptor CRY1 (1) and phytochrome A (phyA) (142), respectively. An elongated hypocotyl is also shown by the *late hypocotyl* (*lhy*) mutant (126). This mutant is daylength insensitive and lacks circadian rhythms for leaf movement (R Schaffer, I Carré & G Coupland, personal communication). It is suggested that this Myb-like transcription factor (R Schaffer & G Coupland, personal communication) might be a component of the circadian clock.

Mutants deficient in GA biosynthesis, like *ga1*, or action, like *gibberellin insensitive* (*gai*), show a late phenotype under SD conditions (143).

Late-flowering mutants have also been identified as defective in starch metabolism, such as *phosphoglucomutase* (*pgm*) (22) and *ADP glucose pyrophosphorylase 1* (*adg1*) (86), which lack leaf starch and flower late, mainly under SD conditions. In contrast, *starch excess 1* (*sex1*) (23) and *carbohydrate accumulation mutant 1* (*cam1*) (40), which also flower late, have increased starch content in leaves. This characteristic was also observed in the late mutant *gi* (6, 40). In the *pgm* and *sex1* mutants, the late-flowering phenotype could be suppressed by a vernalization treatment (11). The late-flowering phenotype observed in these mutants is not due to the defect in starch accumulation and the slow growth but more to the inability to mobilize the stored carbohydrates (11, 40). Nevertheless, it remains unclear how carbohydrate metabolism affects flowering time in Arabidopsis.

Additional mutants that show lateness either under specific conditions and/or with more pronounced pleiotropic effects are *de-etiolated 2* (*det2*), *ted1* (a suppressor of *det1*) (108), *ethylene insensitive* (*ein*) (39), *ethylene responsive* (*etr1*) (14), *short integument* (*sin*) (112), and *vernalization* (*vrn*) (25). Several of these genes have been cloned and are known to encode steps in brassinosteroid biosynthesis (*DET2*) (84) and ethylene action (*EIN, ETR1*) (39).

Early-Flowering Mutants

Early-flowering mutants were described later than the late ones, probably due to the use of early ecotypes growing in LD conditions, which makes the effects of early mutants less pronounced.

The early-flowering mutants with the most dramatic phenotypes are *embryonic flower 1* and *2* (*emf1* and *emf2*). The *emf* mutants do not produce a normal rosette after germination, but they make only a few cauline leaves followed by floral buds. In addition, their flowers are usually abnormal and incomplete (130). The phenotype indicates that most of the normal vegetative phase is bypassed, and *EMF* genes are therefore likely to play a central role in the COPS mechanism (51, 145). Double-mutant analyses indicated that the *emf* is epistatic to both early- and late-flowering mutants (145), although differences have been found among double mutants of *emf* with several late-flowering mutants (146). Interactions between *EMF* and genes regulating inflorescence meristem development and floral organ identity were revealed in the analysis of double mutants between *emf* and *terminal flower* (*tfl*) and *agamous* (*ag*). It has been proposed that the *EMF* genes play a role during the different phase transitions of the plant by a gradual reduction in its activity (145).

Several early-flowering mutants are involved in light perception and light signal transduction pathways. Among these, *long hypocotyl 1 and 2* (*hy1* and *hy2*), which are defective in phytochrome chromophore biosynthesis (106), and *phytochrome B* (*hy3* = *phyB*), deficient in phytochrome B (128), are daylength sensitive (45). Overexpression of phytochrome B also leads to early flowering (9), suggesting that the balance between different phytochromes is important for the proper timing of transition to flowering. Furthermore, phytochromes A and B are not the only phytochromes influencing this transition because *phyA phyB* double mutants still respond to increases in the proportion of FR light by flowering early (37).

The *phytochrome-signaling early-flowering* (*pef1*) mutant shows a similar phenotype to *hy1* and *hy2* but cannot be rescued by the chromophore precursor biliverdin, which can rescue *hy1* and *hy2*. This suggests that *pef1* has a mutation in a signaling intermediate interacting with all the phytochrome family members (2). The *pef2* and *pef3* mutants more closely resemble *phyB* mutants. Therefore, they may have lesions early in the signaling pathway primarily mediated by phyB and/or some of the other phytochrome gene family members (phyC, D, E) (2).

The *sucrose-uncoupled 2* (*sun2*) mutant has an early-flowering phenotype, at least under LD conditions, and shows a long hypocotyl and reduced fertility (38). This mutant was initially isolated as showing reduced repression by sucrose of a transgenic plastocyanin promotor. These phenotypes suggest an interaction between carbohydrate metabolism repression and light signaling in the flowering process.

Some of the mutants influence the circadian rhythm. The *early-flowering 3* (*elf3*) mutant lacks rhythmicity in circadian-regulated processes under constant light conditions (56), while the *cop1* and *det1* show shorter circadian period lengths in constant darkness (96). The *elf3* mutant flowers early under both LDs and SDs, is photoperiod insensitive, and has a long hypocotyl (most noticeably in blue and green light). Double mutant analysis with *hy4* and *hy2* indicates that *ELF3* is involved in blue light–regulated photomorphogenesis (147). In contrast, the *cop1* and *det1* mutants are early flowering in SDs and also have a constitutive photomorphogenic phenotype. *DET1* encodes a novel nuclear-localized protein, suggesting that it controls cell type–specific expression of light-regulated promotors (109). *COP1* encodes a protein with both a zinc-binding motif and a G_β homologous domain (35). Double mutant analysis with *hy1* and *hy4* suggests that *COP1*, together with other *COP* and *DET* genes, acts downstream of phytochrome and the blue-light photoreceptor (28, 75). The DET/COP protein complex formed in darkness negatively regulates transcription of certain genes involved in photomorphogenesis (134). It is thought that light signals mediated by multiple photoreceptors can be transduced to inactivate the pleiotropic COP/DET regulators and thus release the repression of seedling photomorphogenesis. Nevertheless, since the *cop/det* mutants also have a clear phenotype in light-grown plants, these genes may also function in other pathways (94) that are not directly related to photomorphogenesis.

Cytokinins, applied to wild-type plants, result in a phenocopy of *det1* mutants (29). Consistent with this the *altered meristem program 1* (*amp1* = *pt* = *hpt* = *cop2*) mutant, which has high levels of cytokinin, shows a constitutive photomorphogenic phenotype, flowers early, and is daylength insensitive, like the *det1* mutant (26). This suggests a role for cytokinins in the light signal transduction. Nevertheless, this mutant shows a strongly altered growth and leaf formation rate rather than altered flowering time. Other mutants, like *spy*, show the role of GAs in the transition to flowering. The *spy* mutant has the phenotype of wild-type plants treated with GAs and is therefore early flowering. The *SPY* gene is probably involved in the GA signal transduction pathway (58).

The early-flowering *elongated* (*elg*) mutant shows a pleiotropic phenotype that suggests a disruption of phytochrome and/or GA function. However, it has been shown that *ELG* acts independently of phytochrome and GA action (49).

Another group of mutants involves genes whose function in the transition to flowering has not yet been determined. Two of these mutants, *early-flowering 1* and *2* (*elf1* and *elf2*), do not show clear pleiotropic phenotypes and have a daylength response (148). In contrast, *early in short days 1* (*esd1*) (JM Martínez-Zapater, C Gómez-Mena, L Ruiz-García & J Salinas, personal communication) and *4* (*esd4*) (126; G Murtas, P Reeves & G Coupland, personal communication) *early bolting in short days* (*ebs* = *speedy*) (JM Martínez-Zapater, C Gómez-Mena, M Pineiro & G Coupland, personal communication), and *early flowering in short days* (*efs*) (129) have a reduced daylength response and show pleiotropic phenotypes such as reduced fertility and/or plant size. Double mutant analysis indicated that these mutants might interact with some of the late-flowering mutants (126, 129). The *ESD4* gene has recently been cloned but did not show homology to other genes of known function, although related sequences were found in a range of other organisms (117).

A number of early-flowering mutants are involved in the later stages of floral transition. These genes are regulating the expression of floral meristem identity genes like *AP1*, *LFY*, and *AG* (*AGAMOUS*). Mutations in *TFL1* result in early flowering, replacement of coflorescences by flowers, and determinated growth of the apical meristem, which develops into a flower (125). The *tfl* mutation shows ectopic expression of *LFY* and *AP1* in the apical meristem (13, 48), agreeing with overexpression in transgenic plants of *LFY* and *AP1*, giving a phenotype reminiscent of *tfl1* (13, 89, 139). Therefore, it appears that the *tfl1* mutant fails in negatively regulating *LFY* and *AP1*, thereby promoting early flowering with the formation of a terminal flower (125). The *TFL1* gene has been cloned and encodes a putative phosphatidylethanolamine-binding and nucleotide-binding protein (15, 104).

Mutations in the *curly leaf* (*CLF*) gene cause a very similar phenotype to the one conferred by constitutive expression of the meristem identity gene *AG*, showing narrow and upwardly curled leaves as well as early flowering in SDs (98). *CLF* has been cloned and encodes a protein with homology to polycomb-group genes. *CLF* is required to repress *AG* transcription in leaves, inflorescence stems, and flowers (44).

Flowering Time Genes Identified by Constitutive Expression in Transgenic Plants

Constitutive expression of cloned genes is commonly used as a tool to confirm and further analyze the role of genes cloned on the basis of a mutant phenotype. Furthermore, when no mutants are available, the function of cloned genes can be inferred also by analyzing transgenic plants that constitutively express these genes.

For a number of genes of unknown function transgenic plants suggested their role in promoting flowering, although no late mutants were available. The *FPF1* gene is expressed immediately after photoperiodic induction. Constitutive expression of this gene leads to early flowering under LDs and SDs and to other associated changes that mimic the effect of GA applications (64). In a search for genes whose products bind to the promotor of the meristem identity gene *AP1* (and its *Antirrhinum* ortholog *SQUAMOSA*), the *SPL3* gene was isolated. Its constitutive expression leads to earliness (19). Although overexpression phenotypes show the sufficiency of these genes to promote flowering, they do not prove that these genes are necessary for the timing of the transition. Therefore, late mutants at these loci may not be found. This is because the function of these genes may be redundant or they may be involved in other related processes. This is illustrated with the meristem identity genes *AP1* (89), *LFY* (139), and the meristem-organ identity gene *AG* (98), for which mutants are available without an obvious flowering-time phenotype. However, transgenic plants expressing these genes constitutively do flower early.

Another way by which overexpression may indicate the function of a gene is by providing the endogenous gene with constitutive promotors or enhancers. A transposable element with outward-directing 35S promotor has generated the dominant mutant *lhy* (34, 126), described above, which constitutively expresses this gene. A phenocopy of the *lhy* mutant was obtained in transgenic plants with constitutive expression of a related Myb-type gene called *CCA1* (136, 137). The protein encoded by this gene binds to phytochrome-regulated promoters, indicating a link with the phytochrome-related long hypocotyl phenotype.

Methylation and Epigenetics

During the past few years, it has become clear that DNA methylation plays an important role during development of eukaryotes. DNA (de)methylation is involved in the control of gene expression during development and differentiation, by either negative or positive regulation (12, 90). An important observation is that mice carrying a homozygous mutation of the DNA methyltransferase gene do not develop beyond a certain embryonic stage (83). Furthermore, there is evidence that DNA methylation is one of the mechanisms to silence foreign DNA in eukaryotes (93).

There is some indication that DNA methylation might be involved in the vernalization response. Arabidopsis plants either cold treated or treated with the demethylating compound 5-azacytidine show reduced amounts of 5-methylcytosine in their DNA. Among the late-flowering mutants there are some, like *fca* and *fy*, responsive to vernalization (70), and others such as *gi*, *fd* and *ft* that show little response to this treatment. After treating these mutants with 5-azacytidine (17), earliness was observed in the responsive genotypes but not in the nonresponsive ones, thus imitating the effect of vernalization.

DNA methylation has also been reduced in transgenic plants. Transgenic C24 plants were constructed in which methylation was suppressed by the antisense methyltransferase cDNA *MET1* from *Arabidopsis thaliana* (36, 42, 43). This resulted in a reduction of total genomic cytosine methylation, which induced several developmental effects, and a correlation was found between demethylation and reduction in flowering time. This was particularly clear under SDs where C24 shows a pronounced vernalization response (EJ Finnegan and ES Dennis, personal communication). The authors suggest that demethylation is involved in the process of flower promotion by vernalization. Surprisingly, using the same antisense approach, Ronemus et al (118) found a late-flowering phenotype in Col genetic background.

In addition to effects on flowering time, reduced methylation led to abnormal flowers due to ectopic expression of genes such as *AG* and *APETALA3* (*AP3*), probably caused by changes in chromatin structure (43). These phenotypes are in some aspects similar to the phenotype of the early-flowering mutant *clf*, defective in a gene encoding for a polycomb-like protein, which is known to affect chromatin structure (44).

Changes in the DNA methylation state, either by antisense methyltransferase genes or by mutation, also lead to heritable mutant phenotypes. Among these, flowering time phenotypes are relatively abundant. A mutant, designated *ddm1* (*decrease in DNA methylation*) affected in DNA methylation but not exhibiting a flowering time phenotype, has been isolated in Arabidopsis (135). The *ddm1* mutation causes hypomethylation up to 70% of the total genomic 5-methylcytosine levels, although these plants exhibit normal methyltransferase activity. The *ddm1* mutation induces other heritable mutations after repeated self-pollination (63). Among them, there is a late-flowering mutant designated *fts* mapped on chromosome 4 at a similar position as *fwa* (62). The latter late-flowering mutant was described by Koornneef et al (70), and both alleles, *fwa-1* and *fwa-2*, show strong hypomethylation in at least a 5-Mbase region where the gene has been mapped (129). Recently, Jacobsen & Meyerowitz (57) showed that a *superman* (*sup*) mutant epi-allele found in antisense cytosine methyltransferase lines is due to highly localized hypermethylation in the *SUP* gene. The regulation of transcription of certain genes that are involved in the flowering initiation process is apparently either under control or may be influenced by DNA methylation as a component of cell memory.

DISCUSSION: A WORKING MODEL FOR THE CONTROL OF FLOWERING TIME

The complex multigenic control of flowering as revealed by genetic analysis in Arabidopsis (91, 107, 138) and pea (141) indicates that the process is complex and influenced by many factors. This observation supports physiological evidence for a multifactorial control of the transition to flowering (10). Recently, it has been proposed that the transition to flowering is the developmental default state (51, 91, 115, 130, 138). This hypothesis is mainly based on two observations. First, Arabidopsis can flower with very few leaves in complete darkness when sufficient sucrose is provided to the shoot meristem (88, 115). Under these conditions the late mutants, as far as tested, are as early as wild type with the exception of *fwa* and *ft* (88). Second, no mutants without flower-like structures have been described, but in contrast, the *emf1* and *emf2* mutants with hardly any vegetative development have been isolated (130). This default state is then thought to be suppressed by a floral repressor or a promotor of vegetative development, which may be encoded by the *EMF* genes (51, 91, 130, 138). During development, the effect of this repressor decreases in accordance with gradients observed for the various parameters related to phase changes (51, 130). The genes identified by the late and early mutants are assumed to play a promotive or repressive role, respectively. Environmental factors such as daylength and vernalization may regulate the flowering time genes.

The genetic and physiological classification of several late mutants has led to group these genes into two different general modifying promotion pathways (Figure 2). The late-flowering genes, *FCA*, *FY*, *FPA*, *FVE*, *LD*, and *FLD*, are assumed to promote flowering constitutively, under LD and SD, and are therefore involved in the so-called constitutive or autonomous promotion pathway. These mutants are highly daylength sensitive, presumably because when this pathway is defective the transition to flowering becomes very dependent on another pathway that is largely regulated by photoperiod. This second pathway has been called the LD promotion pathway, involving the late-flowering genes, *CO*, *FD*, *FE*, *FHA*, *FT*, *FWA* and *GI*, which are believed to promote flowering mainly under photoperiodically inductive conditions, i.e. LDs. Nevertheless, since the mRNA level of *CO*, a gene that promotes flowering, is reduced in SD, the effect of LD might be the removal of a hypothetical SD repressor, and therefore this pathway could also be referred to as SD repression.

The reduced responsiveness to vernalization of these LD promotion mutants does not imply that these genes are involved in sensing the cold signal, because long vernalization treatments are effective in these mutants (24) and the parental genotype L*er* also has a limited vernalization responsiveness compared with mutants such as *fca*, even when it flowers late under SD (24, 25). Furthermore, double mutants involving representative genes of the two pathways are sensitive to vernalization, although the absence of the LD promotion cannot be replaced by the vernalization treatment (68). In contrast, the stronger vernalization sensitivity of the constitutive promotion mutants suggests that this pathway and a third one, the vernalization promotion pathway, might converge downstream and are able to replace each other. The candidate genes affecting the sensing or transduction of the cold signal are the *VRN* genes isolated on the basis of their lack of a vernalization response in an *fca* mutant background (25).

Analysis of double mutants places the LD promotion mutations with similar phenotypes in the same epistatic group. This study also indicated that the situation for the constitutive pathway mutants is more complex, suggesting parallel subpathways within this group. In particular, the *fpa* mutant shows a complex behavior and might play a role in both pathways (68).

To place other flowering genes, including those for which the recessive (probably loss of function) phenotype, is earliness, in relation to these two general pathways will be attempted. However, since detailed genetic analyses of double

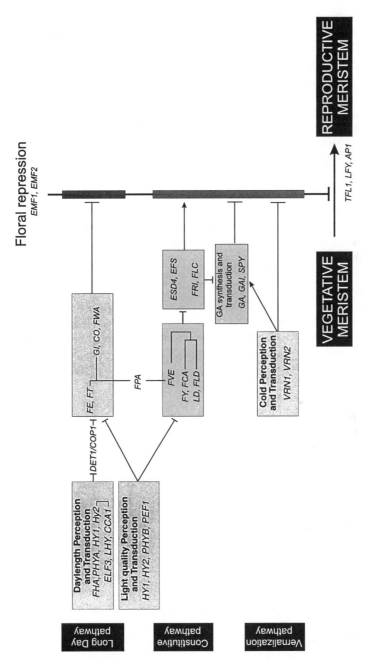

Figure 2   A model describing the interactions of flowering time genes in Arabidopsis. Different groups of genes, established according to their genetic and physiological behavior, are shown in *boxes*. *Lines within boxes* indicate subgroups. Promotive effects are shown by "⟶"; repressive effects by "⊣".

mutants are lacking in most cases, this can only be done in a provisional manner. Furthermore, not knowing whether mutants are true null alleles complicates the interpretation of any double mutant analysis (68, 138). In the case of the *FRI* and *FLC* genes, which in combination lead to the extreme late-flowering phenotype and vernalization responsiveness observed in many late ecotypes, it is thought that they also affect the constitutive pathway in an inhibitory way. "Double mutant" analyses between the early *FLC-Ler* allele and the late mutations *fld*, *ld*, *fca*, *fve*, and *fpa* flower relatively early in comparison with the late-flowering phenotype observed in these late mutants in an *FLC-Col* background (122). This suggests that these late genes antagonize inhibitors. A vernalization treatment might have the same effect.

The phenotype of double mutants between late and early mutants has already allowed the placement of some early genes in one of the pathways. The *esd4* and *efs* mutants are epistatic to *fve*, suggesting that *FVE* might interact with the inhibiting gene products of these early genes (126, 129).

The analysis of mutants originally not isolated for their flowering phenotype, but having an effect on it mainly either under SDs or LDs, has given indications about the mode of action of GAs and light in the floral initiation process. The extreme lateness of GA-deficient or GA-insensitive mutants, only in SD, suggests that this hormone is required in these conditions, acting through the constitutive pathway and compensating for the absence of the LD promotion pathways. The synergistic effect of the *co ga1* and *co gai* double mutants in LD is in agreement with this (111).

It has been suggested that the outcome of the constitutive pathway is similar to that of vernalization and GAs. In agreement with this, a detailed morphogenetic analysis of *fve* mutants indicated that they show some symptoms of reduced GA levels or reduced GA action, although these are far less extreme than in *ga1* and *gai* mutants (92). Besides, the implication of GA synthesis in vernalization has been strongly suggested, not only by the work in *Thlaspi arvense* (52) but also in Arabidopsis by the finding that the *ga1-3* mutant does not respond to vernalization in SDs. However, the observation that the *fca ga1-3* double mutant responds well to vernalization under continuous light argues against the hypothesis that vernalization acts through GA biosynthesis or through the *FCA* gene product (25). Nevertheless, GAs have been shown to be crucial for a number of processes associated with flowering, such as internode elongation and the suppression of adaxial trichomes, which indicates that there is a higher GA activity after the transition to flowering, which might be partially due to the promotive effect of LDs on the GA 20-oxidase encoded by the *GA5* locus (144). Therefore, the actual sequence in the interaction among the constitutive promotion pathway, GAs, and vernalization remains to be solved, and further research in this area is necessary. Besides, it has been suggested that the vernalization promotion involves modulation of gene expression through changes in methylation, which needs further confirmation by the study of the target genes. It is possible that GAs, vernalization, and the constitutive pathway have a similar target that leads to floral induction. Therefore, their functions may overlap, and different environmental conditions may modulate the three promotion pathways in a different way. Whether this putative common target is at the level of *FRI/FLC* or downstream is unknown, but a candidate target gene that probably is specific for GAs is *FPF1* (64).

The chromophore and *phyB* mutations cause early flowering, indicating that this phytochrome has an inhibitory role in flowering, which seems independent from the daylength sensing mechanism. The earliness conferred by the *hy* mutants to the *co*, *gi*, and *fwa* mutant backgrounds under both LDs and SDs (71) further indicates that early flowering caused by the *hy* mutations does not act exclusively through these flowering time genes. However, the *hy* mutants in the *fca* mutant background are late under SDs, suggesting that phyB, apparently, mainly represses the *FCA* gene pathway under SD conditions (71). In contrast, under LDs, *hy* mutants in the *fca* background are early, suggesting that under these conditions another promotion pathway is repressed by phyB. Therefore, the phyB and other light-stable phytochromes might repress both the constitutive and the LD promotion pathways. Reed et al (116) have shown that phyB decreases responsiveness to GAs, which suggests that this phytochrome might repress flowering through this mechanism.

The effect of the light labile phytochrome A is very different and more or less opposite to that of the light-stable phytochromes. Phytochrome A promotes flowering, since overexpression of this gene leads to earliness (9), and the mutant is late when SDs are extended by 8 h of light with a low R:FR ratio (61). Under LDs provided by "normal" fluorescent lamps, no lateness is observed, probably because other photoreceptors can compensate for the lack of phyA. In pea, PHYA-deficient mutants have a much more pronounced late phenotype under LDs and are photoperiod insensitive (140). In this species, SDs lead to the production of a graft transmissible inhibitor that is under control of the pea genes *Sn*, *Dne*, and *Ppp*. Based on grafting studies and the analysis of double mutants, it was concluded that phyA reduces the level of this inhibitor under LD conditions (140).

In addition to phytochromes, blue-light receptors, called cryptochromes, play a role in flowering. As in the case of phytochrome, the different members of this family of photoreceptor seem to have distinct roles in the transition to flowering. The promotive role of the cryptochrome I encoded by the *HY4* gene seems minor since the flowering time effect of this mutant is limited (8). The effect of the cryptochrome II (CRY2) appears more important in LDs, since these mutants (*fha*) are clearly late. The similarity in phenotype of these mutants with the LD promotion pathway mutants strongly suggests that CRY2 and phyA are the photoreceptors for this pathway.

To measure the length of the photoperiod, apart from photoreceptors, a time measurement mechanism is required, which is probably provided by a circadian rhythm. The relation between daylength and a circadian rhythm mechanism affecting leaf movement and *CAB2* gene expression was studied in the Arabidopsis *elf3* mutant, which is early and daylength insensitive (56, 147). The *elf3* mutant lacks these circadian rhythms in continuous light but not in light/dark cycles and continuous darkness, suggesting that *ELF3* is involved in circadian regulation, especially in the transduction of light signals to a component of the clock (20, 56). Two other genes that may affect directly the clock are *LHY* and *CCA1*, both encoding Myb-related proteins. In wild type the *LHY* mRNA is expressed rhythmically (R Schaffer & G Coupland, personal communication). In the presence of the overexpressed copy of *LHY*, transcription from the endogenous *LHY* promotor is repressed, indicating that LHY is part of a transcriptional feed-back loop. Both the phenotypic effects and molecular properties of this gene are expected for circadian clock components (21).

In relation to this, the early-flowering phenotype under SD of mutants such as *det1* (108) and *cop1* (138), suggest that the DET1/COP1 proteins suppress flowering under SD, which may be done by repressing floral promotors such as *CO*. The simplest hypothesis to explain this SD inhibition would be through repression by DET1/COP1 in the absence of the LD signal, and this would predict that photoreceptor-deficient mutants, which would not be able to remove the suppression of flowering by DET1/COP1, should be late in LD. Although this might be the case for phyA and blue-light receptor mutants (8, 61), this is not the case for mutants affecting phyB (*phyB = hy3*) and the chromophore (*hy1* and *hy2*), which are relatively early in SD (45) due to the inhibiting effect of phyB discussed above. Nevertheless, analyses of double mutants involving these genes are still needed in order to understand the role of *DET1/COP1* in this process.

Based on grafting studies, daylength is perceived by the leaves, and the signal is then transported to the apical meristem (10). It is not clear whether the crucial target is the apical shoot meristem or the lateral leaf/flower primordia itself. The latter is suggested by the chimeric structures observed by Hempel & Feldman (54) after the transfer of plants from SD to LD. In Arabidopsis, the shoot apical or inflorescence meristem remains undetermined, and to maintain this state the *TFL1* and *TFL2* genes are required. The *TFL1* gene is strongly expressed in a group of cells just below the apical dome of the inflorescence in accordance with a role in this meristem (15). Bradley et al (15) suggested that *TFL1* delays the commitment to flowering during vegetative phase where it is also weakly expressed. In contrast, its *Antirrhinum* ortholog *CEN* is not expressed during vegetative development, and *cen* mutants are not early (15). Double-mutant analyses between *tfl* and the late-flowering *fpa*, *fve*, *fwa*, and *co* indicates that to repress flower initiation *TFL* requires the function of the late-flowering loci tested (119).

The floral meristem identity genes *LFY*, *AGL8*, and *AP1* are crucial early targets of the floral promotion process and thereby of the flowering time genes as evidenced from mutant phenotypes, from studies using transgenic plants with constitutive expression, and from expression analyses after flower induction. Elegant studies in which the *CO* function was regulated by the ligand-binding domain of the rat glucocorticoid receptor showed that *LFY* expression increased within 24 h after the activation of *CO* (127) and that *AP1* is expressed later. This sequence of gene expression was also observed after the shift from SDs to LDs (55). Both LFY and AP1 can convert shoot meristems into floral meristems as shown by the early flowering of transgenic plants that constitutively express these genes. However, expression of these genes may only trigger floral development after the main shoot has acquired competence to respond to its activity, since constitutive expression of *LFY* still allows the formation of some leaves (139).

Two lines of evidence indicate that *FT* and *FWA* also have effects in the floral induction process. Double mutants of *ft* and *fwa* with *lfy* virtually lack floral initiation and do not show *AP1* mRNA in the inflorescence apex, indicating the importance of these genes for the initiation of *LFY* and *AP1* expression (119). Furthermore, in contrast to other late-flowering mutants, *ft* and *fwa* still are late in continuous darkness (88). This indicates that their role is not restricted to modifying the level or effect of the light-induced floral repressor only, but instead these genes may work at the meristem level and may be required (also) for the flower initiation process itself. The normal flowers of these mutants show that genetic redundancy exists for the flower initiation program as well as for the control of flowering time (119). The cloning of *FT* (5) revealed strong homology with *TFL1*. The opposite effect of mutations in these genes points to a different role, and the two genes might have in common their interaction with *LFY* and *AP1*.

In what way the promoting flowering environmental signals interact with the flowering genes, how these genes interact, and how they activate their targets is still unknown. The phenotypic and epistatic analyses indicate a complex network and suggest various redundant pathways. Since some of the promotive flowering time genes may act as transcription factors (*LD* and *CO*) or may affect RNA stability (*FCA*), a sequence of gene activation events is a likely mechanism. The combined genetic, physiological, and molecular analyses will provide answers to this just-started and evolving picture of the network.

CONCLUDING REMARKS

Recent genetic, molecular, and physiological analysis of flowering initiation in Arabidopsis has led to the identification of components in this important developmental process. Molecular elements involved in some of the initial steps such as photoreceptors and components of the circadian clock, in intermediate steps such as some of the cloned flowering genes, and in the target genes of floral induction, are now known. However, many questions remain: how do these elements interact and transmit the signals? Intriguing questions are, for example, how light and clock signals are integrated and how these interact with the flowering genes. The effect of vernalization at the molecular level is not yet understood. Furthermore, although a role for GAs in flowering is strongly indicated, its function remains unclear, as does the role of other hormones such as cytokinins, and factors such a carbohydrates. Besides, the sequence of events and redundancy suggested by the genetics and physiology is not yet understood at the molecular level. However, the molecular and genetic tools are available in Arabidopsis and will further refine and modify the model presented in this review. It will be important to relate and complement these studies in Arabidopsis with those in other plants to identify both the differences and common aspects, as it has been done for flower development between *Antirrhinum* and Arabidopsis. For flowering timing, pea is particularly important because of its similarity with Arabidopsis in the physiological responses and its suitability for grafting studies (140). This may aid in identifying the nature of the floral repressor, deduced thus far only from genetic and physiological studies, and in determining whether any of the flowering time genes encode the elusive graft-transmissible florigen.

ACKNOWLEDGMENTS

Our work on flowering was supported by the BRIDGE and TMR programs of the European Union (contract BIOT-CT90-0207, BIOT-CT92-0529, and BIO4–CTg6-5008).

We thank our colleagues who provided us with unpublished information and their discussions on this complex topic.

> Visit the *Annual Reviews home page* at
> http://www.AnnualReviews.org.

Literature Cited

1. Ahmad M, Cashmore AR. 1993. *HY4* gene of *Arabidopsis thaliana* encodes a protein with characteristics of a blue-light photoreceptor. *Nature* 366:162–66
2. Ahmad M, Cashmore AR. 1996. The *PEF* mutants of *Arabidopsis thaliana* define lesions early in the phytochrome signaling pathway. *Plant J.* 10:1103–10
3. Alonso-Blanco C, Peeters AJM, Koornneef M. 1997. Genetic analysis of developmental traits based on the Ler/Cvi natu-ral variation. *Int. Conf. Arabidopsis Res., 8th, Madison, Wis.* 12-1 (Abstr.)
4. Amasino RM. 1996. Control of flowering time in plants. *Curr. Opin. Genet. Dev.* 6: 480–87
5. Araki T, Kaya H, Kobayashi Y, Nakatani M, Iwabuchi M. 1997. Flowering time gene *FT* encodes a *TFL1* homolog. *Int. Conf. Arabidopsis Res., 8th, Madison, Wis.* 5-1 (Abstr.)
6. Araki T, Komeda Y. 1993. Analysis of the

role of the late-flowering locus, *GI* in the flowering of *Arabidopsis thaliana*. *Plant J.* 3:231–39

7. Bagnall DJ. 1992. The control of flowering of *Arabidopsis thaliana* by light, vernalisation and gibberellins. *Aust. J. Plant Physiol.* 19:401–9

8. Bagnall DJ, King RW, Hangarter RP. 1996. Blue-light promotion of flowering is absent in *hy4* mutants of *Arabidopsis*. *Planta* 200:278–80

9. Bagnall DJ, King RW, Whitelam GC, Boylan MT, Wagner D, Quail PH. 1995. Flowering responses to altered expression of phytochrome in mutants and transgenic lines of *Arabidopsis thaliana* (L.) Heynh. *Plant Physiol.* 108:1495–503

10. Bernier G. 1988. The control of floral evocation and morphogenesis. *Annu. Rev. Plant Physiol.* 39:175–219

11. Bernier G, Havelange A, Houssa C, Petitjean A, LeJeune P. 1993. Physiological signals that induce flowering. *Plant Cell* 5:1147–55

12. Bird A. 1992. The essentials of DNA methylation. *Cell* 70:5–8

13. Blázquez MA, Soowal LN, Lee I, Weigel D. 1997. *LEAFY* expression and flower initiation in *Arabidopsis*. *Development.* 24:3835–44

14. Bleecker AB, Estelle MA, Somerville C, Kende H. 1988. Insensitivity to ethylene conferred by a dominant mutation in *Arabidopsis thaliana*. *Science* 241:1086–89

15. Bradley D, Ratcliffe O, Vincent C, Carpenter R, Coen E. 1997. Inflorescence commitment and architecture in *Arabidopsis*. *Science* 275:80–83

16. Brown JAM, Klein WH. 1971. Photomorphogenesis in *Arabidopsis thaliana* (L.) Heynh. Threshhold intensities and blue-far-red synergism in floral induction. *Plant Physiol.* 47:393–99

17. Burn JE, Bagnall DJ, Metzger JD, Dennis ES, Peacock WJ. 1993. DNA methylation, vernalization, and the initiation of flowering. *Proc. Natl. Acad. Sci. USA* 90:287–91

18. Burn JE, Smyth DR, Peacock WJ, Dennis ES. 1993. Genes conferring late flowering in *Arabidopsis thaliana*. *Genetica* 90:147–55

19. Cardon GH, Höhmann S, Nettesheim K, Saedler H, Huijser P. 1997. Functional analysis of the *Arabidopsis thaliana* SBP-box gene *SPL3*: a novel gene involved in the floral transition. *Plant J.* 12:367–77

20. Carré IA. 1996. Genetic analysis of the photoperiodic clock in *Arabidopsis*. *Flower. Newsl.* 22:20–24

21. Carré IA, Ramsay N, Schaffer R, Coupland G. 1997. Characterisation of a putative circadian clock gene in *Arabidopsis*. *Eur. Symp. Photomorphogen., Leicester, UK*. (Abstr.)

22. Caspar T, Huber SC, Somerville CR. 1985. Alterations in growth, photosynthesis, and respiration in a starchless mutant of *Arabidopsis thaliana* (L.) deficient in chloroplast phosphoglucamutase activity. *Plant Physiol.* 79:11–17

23. Caspar T, Lin T-P, Kakefuda G, Benbow L, Preiss J, Somerville CR. 1991. Mutants of *Arabidopsis* with altered regulation of starch degradation. *Plant Physiol.* 95:1181–88

24. Chandler J, Dean C. 1994. Factors influencing the vernalization response and flowering time of late flowering mutants of *Arabidopsis thaliana* (L.) Heynh. *J. Exp. Bot.* 45:1279–88

25. Chandler J, Wilson A, Dean C. 1996. *Arabidopsis* mutants showing an altered response to vernalization. *Plant J.* 10:637–44

26. Chaudhury AM, Letham S, Craig S, Dennis ES. 1993. *amp1*—A mutant with high cytokinin levels and altered embryonic pattern, faster vegetative growth, constitutive photomorphogenesis and precocious flowering. *Plant J.* 4:907–16

27. Chien JC, Sussex IM. 1996. Differential regulation of trichome formation on the adaxial and abaxial leaf surfaces by gibberellins and photoperiod in *Arabidopsis thaliana*. *Plant Physiol.* 111:1321–28

28. Chory J. 1992. A genetic model for light-regulated seedling development in *Arabidopsis*. *Development* 115:337–54

29. Chory J, Reinecke D, Sim S, Washburn T, Brenner M. 1994. A role for cytokinins in de-etiolation in *Arabidopsis*. *Plant Physiol.* 104:339–47

30. Clarke JH, Dean C. 1994. Mapping *FRI*, a locus controlling flowering time and vernalization response in *Arabidopsis thaliana*. *Mol. Gen. Genet.* 242:81–89

31. Clarke JH, Mithen R, Brown JKM, Dean C. 1995. QTL analysis of flowering time in *Arabidopsis thaliana*. *Mol. Gen. Genet.* 248:555–64

32. Corbesier L, Gadisseur I, Silvestre G, Jacqmard A, Bernier G. 1996. Design in *Arabidopsis thaliana* of a synchronous system of floral induction by one long day. *Plant J.* 9:947–52

33. Coupland G. 1995. Genetic and environmental control of flowering time in *Arabidopsis*. *Trends Genet.* 11:393–97

34. Coupland G. 1997. Regulation of flowering by photoperiod in *Arabidopsis*. *Plant Cell Environ.* 20:785–89

35. Deng X-W, Matsui M, Wei N, Wagner D, Chu AM, et al. 1992. *COP1*, an *Arabidopsis* regulatory gene, encodes a protein with both a zinc-binding motif and a G_b homologous domain. *Cell* 71:791–801

36. Dennis ES, Finnegan EJ, Bilodeau P, Chaudhury A, Genger R, et al. 1996. Vernalization and the initiation of flowering. *Semin. Cell Dev. Biol.* 7:441–48

37. Devlin PF, Halliday KJ, Harberd NP, Whitelam GC. 1996. The rosette habit of *Arabidopsis thaliana* is dependent upon phytochrome action: novel phytochromes control internode elongation and flowering time. *Plant J.* 10:1127–34

38. Dijkwel PP, Huijser C, Weisbeek PJ, Chua NH, Smeekens SCM. 1997. Sucrose control of phytochrome signalling in *Arabidopsis*. *Plant Cell* 9:583–95

39. Ecker JR. 1995. The ethylene signal transduction pathway in plants. *Science* 268:667–75

40. Eimert K, Wang S-M, Lue W-L, Chen JC. 1995. Monogenic recessive mutations causing both late floral initiation and excess starch accumulation in *Arabidopsis*. *Plant Cell* 7:1703–12

41. Eskins K. 1992. Light-quality effects on *Arabidopsis* development. Red, blue and far-red regulation of flowering and morphology. *Physiol. Plant.* 86:439–44

42. Finnegan EJ, Dennis ES. 1993. Isolation and identification by sequence homology of a putative cytosine methyl transferase from *Arabidopsis thaliana*. *Nucleic Acids Res.* 21:2383–88

43. Finnegan EJ, Peacock WJ, Dennis ES. 1996. Reduced DNA methylation in Arabidopsis thaliana results in abnormal plant development. *Proc. Natl. Acad. Sci. USA* 93:8449–54

44. Goodrich J, Puangsomlee P, Martin M, Long D, Meyerowitz EM, Coupland G. 1997. A polycomb group gene regulates homeotic gene expression in *Arabidopsis*. *Nature* 386:44–51

45. Goto N, Kumagai T, Koornneef M. 1991. Flowering responses to light-breaks in photomorphogenic mutants of *Arabidopsis thaliana*, a long-day plant. *Physiol. Plant.* 83:209–15

46. Grbic V, Bleecker AB. 1996. An altered body plan is conferred on *Arabidopsis* plants carrying dominant alleles of two genes. *Development* 122:2395–403

47. Grbic V, Gray J. 1997. Aerial rosette 1, ART1, is a new late flowering gene of Arabidopsis thaliana. *Int. Conf. Arabidopsis Res., 8th, Madison, Wis.* 5-9 (Abstr.)

48. Gustafson-Brown C, Savidge B, Yanofsky M. 1994. Regulation of the *Arabidopsis* floral homeotic gene *APETALA1*. *Cell* 76:131–43

49. Halliday KJ, Devlin PF, Whitelam GC, Hanhart CJ, Koornneef M. 1996. The *ELONGATED* gene of *Arabidopsis* acts independently of light and gibberellins in the control of elongation growth. *Plant J.* 9:305–12

50. Halliday KJ, Koornneef M, Whitelam GC. 1994. Phytochrome B and at least one other phytochrome mediate the accelerated flowering response of *Arabidopsis thaliana* L. to low red/far-red ratio. *Plant Physiol.* 104:1311–15

51. Haughn GW, Schultz EA, Martínez-Zapater JM. 1995. The regulation of flowering in *Arabidopsis thaliana*: meristems, morphogenesis, and mutants. *Can. J. Bot.* 73:959–81

52. Hazebroek JP, Metzger JD, Mansager ER. 1993. Thermoinductive regulation of gibberellin metabolism in *Thlaspi arvense* L. II. Cold induction of enzymes in gibberellin synthesis. *Plant Physiol.* 102:547–52

53. Hempel FD, Feldman LJ. 1994. Bidirectional inflorescence development in *Arabidopsis thaliana*: acropetal initiation of flowers and basipetal initiation of paraclades. *Planta* 192:276–86

54. Hempel FD, Feldman LJ. 1995. Specification of chimeric shoots in wild-type *Arabidopsis*. *Plant J.* 8:725–31

55. Hempel FD, Weigel D, Mandel MA, Ditta G, Zambryski PC, et al. 1997. Floral determination and expression of floral regulatory genes in *Arabidopsis*. *Development* 124:3845–53

56. Hicks KA, Millar AJ, Carré IA, Somers DE, Straume M, et al. 1996. Conditional circadian dysfunction of the Arabidopsis early-flowering 3 mutant. *Science* 274:790–92

57. Jacobsen SE, Meyerowitz EM. 1997. Hypermethylated *SUPERMAN* epigenetic alleles in *Arabidopsis*. *Science* 277:1100–3

58. Jacobsen SE, Olszewski NE. 1993. Mutations at the *SPINDLY* locus of *Arabidopsis* alter gibberellin signal transduction. *Plant Cell* 5:887–96

59. Jansen RC. 1996. Complex plant traits: time for polygenic analysis. *Trends Plant Sci.* 1:89–94

60. Jansen RC, Van Ooijen JW, Stam P, Lister C, Dean C. 1995. Genotype by

environment interaction in genetic mapping of multiple quantitative trait loci. *Theor. Appl. Genet.* 91:33–37

61. Johnson E, Bradley M, Harberd NP, Whitelam GC. 1994. Photoresponses of light-grown phyA mutants of *Arabidopsis*. Phytochrome A is required for the perception of daylength extensions. *Plant Physiol.* 105:141–49

62. Kakutani T. 1998. Genetic dissection of developmental abnormalities induced by DNA hypomethylation mutation in *Arabidopsis thaliana. Plant J.* In press

63. Kakutani T, Jeddeloh JA, Flowers SK, Munakata K, Richards EJ. 1996. Developmental abnormalities and epimutations associated with DNA hypomethylation mutations. *Proc. Natl. Acad. Sci. USA* 93: 12406–11

64. Kania T, Russenberger D, Peng S, Apel K, Melzer S. 1997. *FPF1* promotes flowering in *Arabidopsis. Plant Cell.* 9:1327–38

65. Karlovska V. 1974. Genotypic control of the speed of development in *Arabidopsis thaliana* (L.) Heynh. Lines obtained from natural populations. *Biol. Plant.* 16:107–17

66. Karlsson BH, Sills GR, Nienhuis J. 1993. Effects of photoperiod and vernalization on the number of leaves at flowering in 32 *Arabidopsis thaliana* (*Brassicaceae*) ecotypes. *Am. J. Bot.* 80:646–48

67. King RW, Bagnall DJ. 1996. Photoreceptors and the photoperiodic response controlling flowering time in *Arabidopsis. Semin. Cell Dev. Biol.* 7:449–54

68. Koornneef M, Alonso-Blanco C, Blankestijn-de Vries H, Hanhart CJ, Peeters AJM. 1998. Genetic interactions among late flowering mutants of *Arabidopsis. Genetics.* In press

69. Koornneef M, Blankestijn-de Vries H, Hanhart CJ, Soppe W, Peeters AJM. 1994. The phenotype of some late-flowering mutants is enhanced by a locus on chromosome 5 that is not effective in the Landsberg *erecta* wild-type. *Plant J.* 6: 911–19

70. Koornneef M, Hanhart CJ, Van der Veen JH. 1991. A genetic and physiological analysis of late flowering mutants in *Arabidopsis thaliana. Mol. Gen. Genet.* 229:57–66

71. Koornneef M, Hanhart CJ, Van Loenen-Martinet P, Blankestijn-de Vries H. 1995. The effect of daylength on the transition to flowering in phytochrome deficient, late-flowering and double mutants of *Arabidopsis thaliana. Physiol. Plant.* 95:260–66

72. Koornneef M, Peeters AJM. 1997. Floral transition mutants in *Arabidopsis. Plant Cell Environ.* 20:779–84

73. Kowalski SP, Lan TH, Feldmann KA, Paterson AH. 1994. QTL mapping of naturally-occurring variation in flowering time of *Arabidopsis thaliana. Mol. Gen. Genet.* 245:548–55

74. Kuittinen H, Sillanpaa MJ, Savolainen O. 1997. Genetic basis of adaptation: flowering time in *Arabidopsis thaliana. Theor. Appl. Genet.* 95:573–83

75. Kwok SF, Piekos B, Misera S, Deng X-W. 1996. A complement of ten essential and pleiotropic Arabidopsis *cop/det/fus* genes is necessary for repression of photomorphogenesis in darkness. *Plant Physiol.* 110:731–42

76. Laibach F. 1951. Über Sommer und Winterannuelle Rasse von *Arabidopsis thaliana* (L.) Heynh. Ein Beitrag zur Ätiologie der Blütenbildung. *Beitr. Biol. Pflanz.* 28:173–210

77. Lawrence MJ. 1976. Variations in natural populations of *Arabidopsis thaliana* (L.) Heynh. In *The Biology and Chemistry of the Cruciferae*, ed. JG Vaughan, AJ Macleod, BMG Jones, pp. 167–90. New York: Academic

78. Ledger SE, Dare AP, Putterill J. 1996. *COL2* is a homologue of the *Arabidopsis* flowering time gene *CONSTANS* (PGR96-081). *Plant Physiol.* 112:862

79. Lee I, Amasino RM. 1995. Effect of vernalization, photoperiod, and light quality on the flowering phenotype of *Arabidopsis* plants containing the *FRIGIDA* gene. *Plant Physiol.* 108:157–62

80. Lee I, Aukerman MJ, Gore SL, Lohman KN, Michaels SD, et al. 1994. Isolation of *LUMINIDEPENDENS*: a gene involved in the control of flowering time in *Arabidopsis. Plant Cell* 6:75–83

81. Lee I, Bleecker A, Amasino R. 1993. Analysis of naturally occurring late flowering in *Arabidopsis thaliana. Mol. Gen. Genet.* 237:171–76

82. Lee I, Michaels SD, Masshardt AS, Amasino RM. 1994. The late-flowering phenotype of *FRIGIDA* and mutations in *LUMINIDEPENDENS* is suppressed in the Landsberg *erecta* strain of *Arabidopsis. Plant J.* 6:903–9

83. Li E, Bestor TH, Jaenisch R. 1992. Targeted mutation of the DNA methyltransferase gene results in embryonic lethality. *Cell* 69:915–26

84. Li J, Nagpal P, Vitart V, McMorris TC, Chory J. 1996. A role for brassinosteroids in light-dependent development in *Arabidopsis. Science* 272:398–401

85. Lin C, Ahmad M, Chan AR, Cashmore AR. 1996. *CRY2*: a second member of the *Arabidopsis* cryptochrome gene family. *Plant Physiol.* 110:1047

86. Lin T-P, Caspar T, Somerville C, Preiss J. 1988. Isolation and characterization of a starchless mutant of *Arabidopsis thaliana* (L.) Heynh lacking ADPglucose pyrophosphorylase. *Plant Physiol.* 86:1131–35

87. MacKnight R, Bancroft I, Page T, Lister C, Schmidt R, et al. 1997. *FCA*, a gene controlling flowering time in *Arabidopsis*, encodes a protein containing RNA-binding domains. *Cell* 89:737–45

88. Madueño F, Ruiz-García L, Salinas J, Martínez-Zapater JM. 1996. Genetic interactions that promote the floral transition in Arabidopsis. *Semin. Cell Dev. Biol.* 7:401–7

89. Mandel MA, Yanofsky MF. 1995. A gene triggering flower formation in *Arabidopsis*. *Nature* 377:522–24

90. Martienssen RA, Richards EJ. 1995. DNA methylation in eukaryotes. *Curr. Opin. Genet. Dev.* 5:234–42

91. Martínez-Zapater JM, Coupland G, Dean C, Koornneef M. 1994. The transition to flowering in *Arabidopsis*. In *Arabidopsis*, ed. EM Meyerowitz, CR Somerville, pp. 403–34. Cold Spring Harbor, NY: Cold Spring Harbor Lab. Press

92. Martínez-Zapater JM, Jarillo JA, Cruz-Alvarez M, Roldán M, Salinas J. 1995. *Arabidopsis* late-flowering *fve* mutants are affected in both vegetative and reproductive development. *Plant J.* 7:543–51

93. Matzke MA, Matzke AJM, Eggleston WB. 1996. Paramutation and transgene silencing: a common response to invasive DNA? *Trends Plant Sci.* 1:382–88

94. Mayer R, Raventos D, Chua NH. 1996. *Det1*, *cop1*, and *cop9* mutations cause inappropriate expression of several gene sets. *Plant Cell* 8:1951–59

95. McKelvie AD. 1962. A list of mutant genes in *Arabidopsis thaliana* (L.) Heynh. *Radiat. Bot.* 1:233–41

96. Millar AJ, Straume M, Chory J, Chua NH, Kay SA. 1995. The regulation of circadian period by phototransduction pathways in *Arabidopsis*. *Science* 267:1163–66

97. Mitchell-Olds T. 1996. Genetic constraints of life-history evolution: quantitative-trait loci influencing growth and flowering in *Arabidopsis thaliana*. *Evolution* 50:140–45

98. Mizukami Y, Ma H. 1997. Determination of *Arabidopsis* floral meristem identity by *AGAMOUS*. *Plant Cell* 9:393–408

99. Mozley D, Thomas B. 1995. Developmental and photobiological factors affecting photoperiodic induction in *Arabidopsis thaliana* Heynh. Landsberg *erecta*. *J. Exp. Bot.* 46:173–79

100. Napp-Zinn K. 1957. Untersuchungen zur Genetik des Kältebedürfnisses bei *Arabidopsis thaliana* (L.) Heynh. *Z. Indukt. Abstamm. Vererbungsl.* 88:253–85

101. Napp-Zinn K. 1962. Über die genetischen Grundlagen des Vernalisationsbedurgnisses bei *Arabidopsis thaliana*. I. Die Zahl der beteiligten Faktoren. *Z. Vererbungsl.* 93:154

102. Napp-Zinn K. 1969. *Arabidopsis thaliana* (L.) Heynh. In *The Induction of Flowering: Some Case Histories*, ed. LT Evans, pp. 291–304. Melbourne: Macmillan

103. Napp-Zinn K. 1987. Vernalization. Environmental and genetic regulation. In *Manipulation of Flowering*, ed. JG Atherton, pp. 123–32. London: Butterworths

104. Ohshima S, Murata M, Sakamoto W, Ogura Y, Motoyoshi F. 1997. Cloning and molecular analysis of the *Arabidopsis* gene *terminal flower 1*. *Mol. Gen. Genet.* 254:186–94

105. Okada K, Shimura Y. 1994. Genetic analyses of signalling in flower development using *Arabidopsis*. *Plant Mol. Biol.* 26:1357–77

106. Parks BM, Quail PH. 1991. Phytochrome-deficient *hy1* and *hy2* long hypocotyl mutants of *Arabidopsis* are defective in phytochrome chromophore biosynthesis. *Plant Cell* 3:1177–86

107. Peeters AJM, Koornneef M. 1996. Genetic variation in flowering time in *Arabidopsis thaliana*. *Semin. Cell Dev. Biol.* 7:381–89

108. Pepper AE, Chory J. 1997. Extragenic suppressors of the *Arabidopsis det1* mutant identify elements of flowering time and light response regulatory pathways. *Genetics* 145:1125–37

109. Pepper AE, Delaney T, Washburn T, Poole D, Chory J. 1994. *DET1*, a negative regulator of light-mediated development and gene expression in *Arabidopsis*, encodes a novel nuclear-localized protein. *Cell* 78:109–16

110. Putterill J, Ledger SE, Lee K, Robson F, Murphy G, Coupland G. 1997. The flowering time gene *CONSTANS* and homologue *CONSTANS LIKE 1* exist as a tandem repeat on chromosome 5 of *Arabidopsis* (PGR97–077). *Plant Physiol.* 114:396

111. Putterill J, Robson F, Lee K, Simon R, Coupland G. 1995. The *CONSTANS* gene of Arabidopsis promotes flowering and

encodes a protein showing similarities to zinc finger transcription factors. *Cell* 80:847–57

112. Ray A, Lang JD, Golden T, Ray S. 1996. *SHORT INTEGUMENT (SINI)*, a gene required for ovule development in *Arabidopsis*, also controls flowering time. *Development* 122:2631–38

113. Rédei GP. 1962. Supervital mutants of *Arabidopsis*. *Genetics* 47:443–60

114. Rédei GP. 1970. *Arabidopsis thaliana* (L.) Heynh. A review of the genetics and biology. *Bibliogr. Genet.* 20:1–151

115. Rédei GP, Acedo G, Gavazzi G. 1974. Flower differentiation in *Arabidopsis*. *Stadler Genet. Symp.* 6:135–68

116. Reed JW, Foster KR, Morgan PW, Chory J. 1996. Phytochrome B affects responsiveness to gibberellins in *Arabidopsis*. *Plant Physiol.* 112:337–42

117. Reeves P, Murtas G, Bancroft I, Dean C, Dash S, Coupland G. 1997. Cloning of *ESD4*, a gene controlling flowering time in *Arabidopsis*. *Int. Conf. Arabidopsis Res., 8th, Madison, Wis.* 5-24 (Abstr.)

118. Ronemus MJ, Galbiati M, Ticknor C, Chen JC, Dellaporta SL. 1996. Demethylation-induced developmental pleiotropy in *Arabidopsis*. *Science* 273:654–57

119. Ruiz-García L, Madueño F, Wilkinson M, Haughn GW, Salinas J, Martínez-Zapater JM. 1997. Different roles of flowering time genes in the activation of floral initiation genes in Arabidopsis. *Plant Cell* 9:1921–34

120. Sanda SL, Amasino RM. 1995. Genetic and physiological analysis of flowering time in the C24 line of *Arabidopsis thaliana*. *Weeds World* 2:2–8

121. Sanda SL, Amasino RM. 1996. Ecotype-specific gene expression of a flowering mutant phenotype in *Arabidopsis thaliana*. *Plant Physiol.* 111:641–44

122. Sanda SL, Amasino RM. 1996. Interaction of *FLC* and late-flowering mutations in *Arabidopsis thaliana*. *Mol. Gen. Genet.* 251:69–74

123. Sanda SL, John M, Amasino RM. 1997. Analysis of flowering time in ecotypes of *Arabidopsis thaliana*. *J. Hered.* 88:69–72

124. Schultz EA, Haughn GW. 1993. Genetic analysis of the floral initiation process (Flip) in *Arabidopsis*. *Development* 119:745–65

125. Shannon S, Meeks-Wagner DR. 1991. A mutation in the *Arabidopsis TFL1* gene affects inflorescence meristem development. *Plant Cell* 3:877–92

126. Simon R, Coupland G. 1996. *Arabidopsis* genes that regulate flowering time in response to day-length. *Semin. Cell Dev. Biol.* 7:419–25

127. Simon R, Igeño MI, Coupland G. 1996. Activation of floral meristem identity genes in Arabidopsis. *Nature* 384:59–62

128. Somers DE, Sharrock RA, Tepperman JM, Quail PH. 1991. The *hy3* long hypocotyl mutant of *Arabidopsis* is deficient in phytochrome B. *Plant Cell* 3:1263–74

129. Soppe W, Alonso-Blanco C, Koornneef M, Peeters AJM. 1997. Cloning of *FWA* and characterisation of *EFS*, two genes involved in floral induction in Arabidopsis. *Int. Conf. Arabidopsis Res., 8th, Madison, Wis.* 5-33 (Abstr.)

130. Sung ZR, Belachew A, Shunong B, Bertrand-García R. 1992. *EMF*, an *Arabidopsis* gene required for vegetative shoot development. *Science* 258:1645–47

131. Telfer A, Bollman KM, Poethig RS. 1997. Phase change and the regulation of trichome distribution in *Arabidopsis thaliana*. *Development* 124:645–54

132. Van der Veen JH. 1965. Genes for late flowering in *Arabidopsis thaliana*. In *Arabidopsis Research, Proceedings of the Göttingen Symposium*, ed. G Röbbelen, pp. 62–71. Wasmund, Gelschenkirchen, Germany

133. Vetrilova M. 1973. Genetic and physiological analysis of induced late mutants of *Arabidopsis thaliana* (L.) Heynh. *Biol. Plant.* 15:391–97

134. Von Armin AG, Osterlund MT, Kwok SF, Deng X-W. 1997. Genetic and developmental control of nuclear accumulation of cop1, a repressor of photomorphogenesis in *Arabidopsis*. *Plant Physiol.* 114:779–88

135. Vongs A, Kakutani T, Martienssen RA, Richards EJ. 1993. *Arabidopsis thaliana* DNA methylation mutants. *Science* 260:1926–28

136. Wang Z-Y, Green R, Tobin E. 1997. The functions of the *CCA1* gene in the regulation of gene expression and plant development by light and the circadian clock. *Int. Conf. Arabidopsis Res., 8th, Madison, Wis.* 9-31 (Abstr.)

137. Wang Z-Y, Kenigsbuch D, Sun L, Harel E, Ong MS, Tobin EM. 1997. A Myb-related transcription factor is involved in the phytochrome regulation of an *Arabidopsis lhcb* gene. *Plant Cell* 9:491–507

138. Weigel D. 1995. The genetics of flower development: from floral induction to ovule morphogenesis. *Annu. Rev. Genet.* 29:19–39

139. Weigel D, Nilsson O. 1995. A developmental switch sufficient for flower initiation in diverse plants. *Nature* 377:495–500

140. Weller JL, Murfet IC, Reid JB. 1997. Pea mutants with reduced sensitivity to far-red light define an important role for phytochrome A in daylength detection. *Plant Physiol.* 114:1225–36

141. Weller JL, Reid JB, Taylor SA, Murfet IC. 1997. The genetic control of flowering in pea. *Trends Plant Sci.* 2:412–188

142. Whitelam GC, Harberd NP. 1994. Action and function of phytochrome family members revealed through the study of mutant and transgenic plants. *Plant Cell Environ.* 17:615–25

143. Wilson RN, Heckman JW, Somerville CR. 1992. Gibberellin is required for flowering but not for senescence in *Arabidopsis thaliana* under short days. *Plant Physiol.* 100:403–8

144. Xu Y-L, Gage DA, Zeevaart JAD. 1997. Gibberellins and stem growth in *Arabidopsis thaliana*: effect of photoperiod on expression of *GA4* and *GA5* loci. *Plant Physiol.* 114:1471–76

145. Yang CH, Chen LJ, Sung ZR. 1995. Genetic regulation of shoot development in *Arabidopsis*: role of the *EMF* genes. *Dev. Biol.* 169:421–35

146. Yang CH, Haung MD. 1997. *EMF* genes interact with late-flowering mutant genes to regulate *Arabidopsis* floral competence. *Int. Conf. Arabidopsis Res., 8th, Madison, Wis.* 5-31 (Abstr.)

147. Zagotta MT, Hicks KA, Jacobs C, Young JC, Hangarter RP, Meeks-Wagner DR. 1996. The *Arabidopsis ELF3* gene regulates vegetative photomorphogenesis and the photoperiodic induction of flowering. *Plant J.* 10:691–702

148. Zagotta MT, Shannon S, Jacobs C, Meeks-Wagner DR. 1992. Early-flowering mutants of *Arabidopsis thaliana*. *Aust. J. Plant Physiol.* 19:411–18

Annu. Rev. Plant Physiol. Plant Mol. Biol. 1998. 49:371–95

MEIOTIC CHROMOSOME ORGANIZATION AND SEGREGATION IN PLANTS

R. Kelly Dawe

Department of Botany and Department of Genetics, University of Georgia, Athens, Georgia 30602

KEY WORDS: meiosis, chromosome pairing, chromosome segregation, recombination nodule, kinetochore

ABSTRACT

During meiosis, homologous chromosomes are brought together to be recombined and segregated into separate haploid gametes. This requires two cell divisions, an elaborate prophase with five substages, and specialized mechanisms that regulate the association of sister chromatids. This review focuses on plant chromosomes and chromosome-associated structures, such as recombination nodules and kinetochores, that ensure accurate meiotic chromosome segregation.

CONTENTS

OVERVIEW

Meiosis in most plants can be summarized as shown in Figure 1. After DNA replication in the premeiotic interphase, meiotic chromosomes are first

371

1040-2519/98/0601-0371$08.00

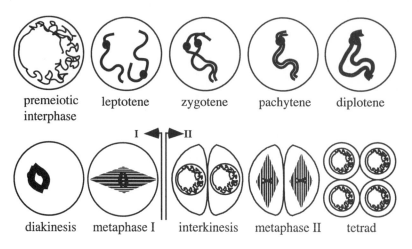

premeiotic   leptotene   zygotene   pachytene   diplotene
interphase

diakinesis   metaphase I   interkinesis   metaphase II   tetrad

Figure 1   Meiosis in plants. The process of microsporogenesis beginning with the premeiotic interphase is illustrated in a plant that forms tetragonal tetrads (in other species the four daughters form a tetrahedron).

identifiable in leptotene as long threads with the sister chromatids tightly pressed together (e.g. 38, 68, 103). The two sister chromatids of each leptotene chromosome are bound to a common protein core known as an axial element, which appears to hold the meiotic chromatin in a looped configuration (at least in animals and fungi; 97). It is not known whether the axial element attachment regions correspond to the scaffold attachment regions of mitotic chromosomes (16). During zygotene, genetic recombination is probably initiated (see below), the chromosomes begin to coil (68, 103), and for a brief period the sister chromatids become visibly distinct (27). These specialized condensation patterns may be regulated in part by prophase I-specific chromatin proteins such as meiotin-1 (106). The homologous zygotene chromosomes begin to synapse along their length via a ribbon-like structure called the synaptonemal complex (SC). The axial elements become the lateral elements of the SC, which are joined together by transverse elements and a central element. Recent three-dimensional reconstructions indicate that the maize SC has the structure illustrated in Figure 2. For a review of plant SCs, consult Gillies (38), and for a more general discussion of SCs consult Heyting (48).

At pachytene, the chromosomes are fully synapsed and often dispersed in the nucleus so that they can be easily identified. Pachytene chromosomes are much longer than mitotic prophase chromosomes and have been used in several species to make rough cytological maps (e.g. 31, 100). In diplotene, the homologous chromosomes separate but remain associated by chiasmata

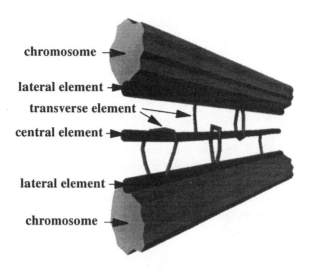

chromosome →

lateral element →

transverse element →

central element →

lateral element →

chromosome →

Figure 2   The synaptonemal complex in maize. The structure is an interpretation of data obtained using high voltage electron microscopy followed by computerized axial tomography (J Fung, J Sedat, D Agard, unpublished data).

(a result of crossovers). In diakinesis, the chromosomes contract lengthwise by a spiraling process (102, 124), and by prometaphase I (immediately before metaphase I) they are thickened and highly condensed. The spindle is formed in prometaphase-metaphase, and in anaphase I the chiasmata are released and sister chromatids segregate to the same pole (108).[1] During the interphase between meiosis I and II (called interkinesis) there is no DNA replication. The chromosomes again become visible at prophase II, and after a mitotic-like division in meiosis II, the sister chromatids disjoin to form four haploid daughter cells.

The focus in this review is on the major events that distinguish meiosis from mitosis: the pairing and recombination of chromosomes and the unique attributes of sister chromatids in meiosis that allow them to first segregate together in anaphase I and then away from each other in anaphase II.

CHROMOSOME PAIRING AND RECOMBINATION

Gross Chromosome Alignment

In an excellent review on the initiation of chromosome pairing, Loidl (72) identified three possible mechanisms for the early stages of chromosome alignment:

---

[1] A few plants, such as *Luzula echinata*, have nonlocalized kinetochores that extend all along the chromosomes (14). In *Luzula*, the sister kinetochores disjoin from each other in meiosis I in a process known as inverted meiosis (57, 108).

premeiotic associations, specific interactions at prophase (cross-talk over long distances), and random contacts. The first two, premeiotic associations and specific interactions at prophase, have received little support. In several studies premeiotic associations in plants (either the premeiotic mitosis or premeiotic interphase) were proposed, but these studies have been questioned on several grounds (72). The most serious argument against the role of premeiotic associations is that if they do exist premeiotically, they are not apparent in the stages that precede synapsis (2, 27, 51, 54, 68). Specific interactions at prophase have been discussed in the form of unsubstantiated "elastic connectors" (75) and with reference to an observed fibrillar material in cereal meiocytes (12). While the fibrillar material may have a function in chromosome pairing (discussed below), there is no evidence that it connects homologous chromosomes. Recent studies tend to support the third model in which random contacts initiate synapsis (9, 27).

THE BOUQUET STAGE   It is generally thought that if homology is identified by a trial and error process, there must a mechanism(s) in the early prophase cell that increases the number or efficiency of random contacts. One such mechanism could be a widespread phenomenon known as the bouquet stage: the clustering of telomeres to a small region of the nuclear envelope during zygotene (29). The bouquet stage has been observed in every plant species where three-dimensional reconstructions have been performed (9, 38). Bouquet formation is an active process in plants (9, 27, 136). Using three-dimensional light microscopy in maize, it was shown that telomeres are randomly distributed in the premeiotic interphase and early leptotene and then transported to a small region of the nuclear envelope in prezygotene (9, 27). The clustering of telomeres preceded the alignment of nontelomeric loci (27), suggesting that the bouquet is one of the first steps in the pairing process.

Role of the bouquet in pairing   Two views on the role of the bouquet have been presented (72). The first emphasizes the importance of telomeres in initiating synapsis, and the second emphasizes a general stirring process that brings chromosomes into close proximity. The central feature of the first proposal is that the complex three-dimensional problem of pairing could be reduced to two dimensions on the inner surface of the nuclear envelope. Homologs could be identified at the telomeres, the process of synapsis initiated, and synapsis completed by a zippering process (72, 108). Supporting this model is the observation that synapsis, as measured by SC formation (2, 38, 44) or cytogenetic analysis (17), is often initiated in telomeric regions. There are, however, several observations that conflict with the idea that synapsis must be initiated at telomeres (78). Ring chromosomes in maize, which lack telomeres, have been shown to pair normally with a homologous ring or rod chromosome (90, 114);

newly broken chromosomes deficient for telomeres pair normally in maize (91); and synapsis proceeds to completion in rye even though large heterochromatic regions at the ends of the chromosomes interfere with the end-to-end associations (40). While these data make it unlikely that telomeres are required for homology identification, it nevertheless remains possible that the efficiency of homology identification is improved by the close proximity of telomeres.

The alternative proposal is that the telomere cluster generates a general stirring process that brings otherwise distant chromosomes into close proximity and thereby increases the likelihood that homologous contacts will occur (27, 72). The extent to which telomere movement alone can affect the movement of chromosomes within the nucleus is not known, but early data from living plant meiocytes indicate that prophase I movements within the nucleus are quite fast (49). It was not possible to identify individual chromosomes in these early studies, but the nucleolus could be clearly resolved (attached to at least one chromosome). Nucleolar movement began in leptotene and became most rapid in zygotene and early pachytene, achieving rates as high as 3.9 μm $\cdot$ min^{-1} in *Acacia*, and 8.0 μm $\cdot$ min^{-1} in *Salvinia*. These rates are considerably higher than the rate of anaphase chromosome movement in plants (e.g. 145) and could conceivably be generated by forces within the nucleus. In several cereal species, a meiosis-specific intranuclear network of fibrillar material has been identified that might mediate such intranuclear movement (12). Chromosome movement may also be facilitated by the structure of the chromosomes and nucleus during the bouquet stage. Coincident with telomere migration in maize, the chromosomes undergo a global chromatin reorganization, involving a separation of chromatids, an elongation of heterochromatic knobs, and a 50% increase in chromosome and nuclear volume (27). Similar chromatin and nuclear changes have been documented in other plant species immediately before synapsis (11, 27, 68, 103, 108).

Mechanism of bouquet formation   A better understanding of the role of the bouquet may be accomplished through further studies of how and when the bouquet is formed. One possibility for the mechanism of bouquet formation is that telomeres move by *trans*-nuclear envelope interactions with the microtubule cytoskeleton (29, 115). In plants, it has long been known that the microtubule-destabilizing drug colchicine interferes with chromosome pairing (72, 115), but the mode of action for colchicine is unknown. In rye and some wheat studies, pairing was only disrupted when colchicine was applied well before the bouquet stage, during the end of the preceding mitosis and early premeiotic interphase (72). In *Lilium* and *Allium*, however, colchicine reduced pairing when applied during bouquet formation (72, 131). A colchicine-binding protein was identified in the nuclear envelope of *Lilium*, suggesting that at least in this species, colchicine affects chromosome/nuclear envelope interactions or a prerequisite step (131).

An important link between the telomeres and microtubule-based motility in plants has recently been demonstrated by Schmit et al (111, 112). Using a monoclonal antibody to calf centrosomes (6C6), they demonstrated cross-reactivity with kinetochores, the nuclear surface, and synaptonemal complex of several plant species. In early prophase I, the immunostaining was distributed over the entire nuclear surface; at zygotene, the ends of the chromosomes were also stained; and at pachytene, the nuclear surface lost its staining and the chromosomes became stained throughout their length. The facts that animal centrosomes interact with microtubules, and that the nuclear envelope functions as a microtubule-organizing center (MTOC) in plants, suggests that the 6C6 immunostaining identifies an epitope specific to plant MTOCs (69, 111). If so, the switch from nuclear envelope to chromosome-end staining in zygotene could provide the telomeres with the MTOC activity that enables them to interact with microtubules on the nuclear surface. A different antiserum (CREST EK, see below) reacts with both the kinetochores and synaptonemal complex in maize, providing a further link between an organelle that interacts with microtubules (the kinetochore) and the SC (Figure 3). In some species, distinctly staining structures have been observed at the telomeres (44, 51, 115, 125), which may have a role in mediating the interactions between telomeres and the cytoskeleton.

Although the results with 6C6 immunostaining provide a plausible explanation for the mechanism of telomere movement, it is not clear how the telomeres cluster in a defined region of the plant nuclear envelope. In animals and fungi, the telomere cluster interacts with the centrosomes or spindle pole bodies, which are nuclear envelope–associated microtubule-organizing organelles (29). This is especially pronounced in Mantids; during pachytene the telomere cluster is divided in two as the centrosomes separate in preparation for spindle formation (53). In contrast to the localized MTOCs in fungi and animals, plant MTOC activity is distributed over the entire surface of the nucleus (69, 112). In the absence of any polarity on the nuclear envelope it is difficult to envisage what could serve as the focus for the telomere cluster. Differential nuclear pore densities have been observed in the vicinity of the telomeres (38), but it is not clear whether the changes in pore density are a cause or an effect of the bouquet. Clear answers to questions about the mechanism of bouquet formation, as well as the function of the bouquet, will require either mutants or effective drug treatments that can be used to inhibit the process of telomere migration.

Homology Recognition and Recombination Nodules

Once chromosomes are brought into close contact, homology must be identified at the molecular level so that recombination can occur. Historically, it was

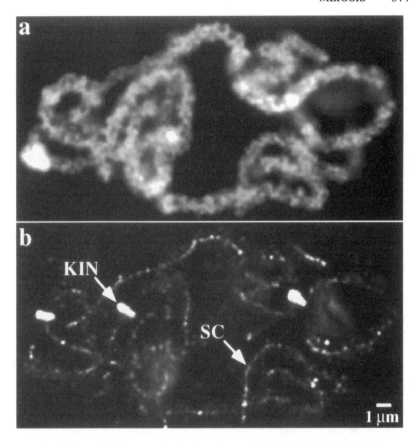

Figure 3   Labeling of maize kinetochores and synaptonemal complex by CREST EK serum. See text for description of the CREST serum. (*a*) DAPI (4,6-diamidine-2-phenylindole dihydrochloride)-stained chromosomes at pachytene. Only three chromosomes are shown. (*b*) Immunolocalization of CREST EK serum to the same chromosomes in A. Both kinetochores (KIN) and synaptonemal complex (SC) are labeled (RK Dawe & WZ Cande, unpublished data).

proposed that the chromosomes synapsed before recombination was initiated; however, it is now thought that the early events of recombination may precede synapsis (46). Three significant observations in yeast are most responsible for this reappraisal. First, a gene (*RAD50*) known to encode an enzyme that repairs double-strand breaks (which initiate recombination in yeast) is also required for SC formation. The simplest interpretation of this result is that recombination must be initiated before the SC can be installed. Second, recombination intermediates in yeast are first observed well in advance of SC formation. Finally, two yeast mutations that abolish SC formation (*zip1* and *mer1*) reduce but do not abolish recombination (109). Early observations by Maguire indicate that a similar sequence of events occurs in plants (74, 82). Using a chromosomal inversion stock that included a ~19 map unit region, she demonstrated a nearly 1:1 correlation between the frequency of synapsis and the frequency of recombination in the inverted region. Because conventional wisdom would have predicted that only a fraction of the successful pairing events would lead to recombination (19/50 in this case), Maguire effectively argued that recombination is associated with the initiation of synapsis. As will been seen below, recent cytological studies further support the contention that the early events of recombination precede SC formation.

PRESYNAPTIC ALIGNMENT   The first visual evidence of homology identification in many plants is a phenomenon known as presynaptic alignment (72). In species where it occurs regularly, presynaptic alignment is a discrete phase in the pairing process that results in a remarkably uniform and apparently accurate alignment of axial elements over distances that greatly exceed the width of the synaptonemal complex (up to 2.5 μm). Such long-distance alignment has been observed consistently in *Allium* species. In diploid and tetraploid *Allium*, presynaptic alignment can be detected in late leptotene, but it is most prevalent at zygotene between regions that have already completed synapsis (2, 127). The alignment is also pronounced in triploid *Allium*, where at any given position two zygotene chromosomes are completely synapsed while the third remains aligned at a distance (73). In late pachytene, the unsynapsed chromosome loses its long-distance alignment, indicating that the mechanism leading to presynaptic alignment is specific to zygotene-early pachytene. Similar evidence of presynaptic alignment was obtained in both tomato and maize (77, 125). Events akin to presynaptic alignment were also observed in maize homozygous for the *asynaptic* (*as1*) mutation (85). In *as1* plants, synapsis ceases in the early stages, leaving the centromeric regions with only fragments of synaptonemal complex (84). In some meiocytes, the interrupted pairing conditioned by *as1* revealed a long-distance interaction of apparently homologous axial elements (85). The authors speculate that the defect in *as1* confers weak or otherwise defective transverse elements that span the distance between homologs. An alternative interpretation is that the mutant plants cannot convert the long-distance interactions into close-range synapsis.

EARLY RECOMBINATION NODULES   How homology can be identified over distances typical of presynaptic alignment (2.5 μm) is not known, but recent evidence suggests that it is a function carried out by recombination nodules. Recombination nodules (RNs) are small proteinaceous particles that associate with the SC (1, 2, 6, 38, 50, 125–128). In zygotene, the RNs are abundant and

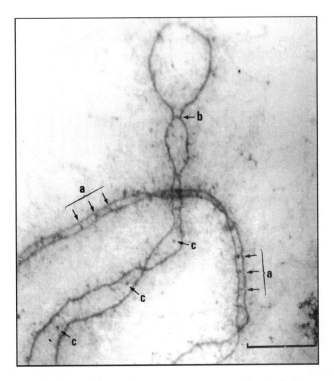

Figure 4 Recombination nodules at zygotene in *Allium cepa*. Early nodules (*a, b, c*) and axial elements (*long threads*) were stained with phosphotungstic acid. Bar is 2.5 μm. [From Albini & Jones (2) with permission from Springer-Verlag.]

referred to as early RNs (128). The idea that early RNs have a role in the homology search is supported by their location on unpaired chromosomes (2, 125) and by their occasional presence in regions of nonhomologous synapsis at zygotene (50, 128). Albini & Jones (2) observed that early RNs were associated with onion chromosomes at several different states of pairing (Figure 4). In some cases, two nodules were found at matching positions on presynaptically aligned (not synapsed) chromosomes. In other cases, single RNs were suspended between two aligned lateral elements (Figure 4c), and in still other cases an RN appeared to be centered in a region where two lateral elements had converged (Figure 4b). In regions of extended synapsis, RNs were observed at a density of greater than one every 2 μm (Figure 4a). Remarkably similar results were obtained in tomato and in the lower vascular plant *Psilotum nudum* (6, 128). In these species, RNs were not only observed suspended between converging lateral elements, but fibers were also frequently observed connecting the two chromosomes and the RN. The data from *Allium*, tomato, and *Psilotum* collectively suggest that early RNs (and perhaps their associated fibers) have a contractile role by pulling chromosomes from presynaptically aligned distances (2–3 μm) to the proximity required to form the synaptonemal complex (0.3 μm) (128).

INITIATION OF SYNAPSIS   An estimate of the number and location of successful homology identification events per chromosome can be obtained by direct observation of SC formation. With few exceptions, SC initiation sites are correlated with the presence of at least one early RN (128). Observations are consistent with the idea that telomeres frequently initiate pairing: SC is first formed in the subterminal regions of the chromosomes (2, 38, 39, 44, 125). Internal sites of pairing initiation are also very frequent (38). In maize, there are about 4 initiation sites per bivalent (36); in rye, there are 9 to 20 per bivalent (1); in lily, there are 5 to 36 per bivalent (51); in *Trandescantia*, up to an average of ~9 per bivalent (44); and in *Allium*, from 1 to 9 in per bivalent (2). The analysis of triploids and trisomics demonstrates that the observed initiation sites are converted into stable associations in pachytene. In triploid/trisomic plants, the individual chromosomes of a trivalent alternate between synapsis and asynapsis; the number of partner switches serves as a minimum estimate of the number of pairing initiation sites. An analysis of trisomic *Crepis capillaris* indicated an average of ~7 initiation sites per chromosome (139). Similarly, the number of partner switches in triploid *Allium* suggested that there were ~6.1 initiation sites per chromosome (73). The number of SC initiation sites invariably exceeds the number of chiasmata typical of the species, indicating that SC initiation sites, and their associated early RNs, do not necessarily correspond to sites of reciprocal recombination.

LATE RECOMBINATION NODULES   During the transition between zygotene and pachytene, the majority of early RNs are either degraded or dissociate from the SC. In tomato, the number of RNs in pachytene is 15 times fewer than in zygotene (125): in *Allium*, an even greater discrepancy was reported (2). The RNs in pachytene are referred to as late RNs (128). With rare exceptions (e.g. 37), the number of late RNs very closely matches the number of chiasmata, suggesting that late RNs mediate reciprocal recombination or become associated with the sites where recombination occurs (1, 3, 117, 126). For instance, in *Lilium*, the average number of RNs was 55.1, and the average number of chiasmata was 54.8 (126). Most notable are the studies that have correlated changes in chiasmata frequency or localization with changes in late RN frequency. In one such study, Albini & Jones (3) compared the localization of late RNs in two *Allium* species that differ with respect to the localization of chiasmata. In

A. fistulosum, the chiasmata are localized almost exclusively in centromere-proximal regions, while in *A. cepa*, the chiasmata are found in more distal regions of the chromosome arms. When the localization of late RNs was determined in these species, a pattern remarkably similar to chiasma localization was observed (3). In another study, the frequency of late RNs was determined in tomato reciprocal translocation heterozygotes (47). The translocations were found to reduce the frequency of both chiasmata and late RNs, strongly supporting a one-to-one correspondence between late RNs and crossovers.

COMPOSITION OF RECOMBINATION NODULES   The available data are consistent with a role of early RNs in the homology search and the initiation of synapsis (2, 128), whereas late RNs appear to be involved in completing and/or stabilizing crossover events (2, 125, 142). It has also been suggested by several authors that a subset of the early RNs (homologous contacts) are converted into late RNs (recombination events) (2, 18, 116, 128). A clear prediction from the cytological data is that RNs contain enzymes involved in recombination. Two of the genes that are thought to mediate recombination in yeast are *RAD51* and *DMC1* (109). Homologs for both *RAD51* and *DMC1* (called *LIM15*) were recently identified in *Lilium* by Teresawa et al (135). Immunolocalization of the Rad51 and Lim15 proteins revealed foci of staining on both zygotene and pachytene chromosomes. Whereas in zygotene the Rad51 and Lim15 co-localized to the same foci, in pachytene the number of Rad51-stained foci decreased, and Lim15 was no longer detected. The spot-like nature of the staining and the reduced number of foci in pachytene are consistent with antisera reaction with RNs and further suggested that Lim15 has a function that is specific to early RNs (135). Confirmation that the Rad51 and Lim15 proteins are a component of early RNs was recently provided by Anderson and coworkers (5) using electron microscopic immunogold localization and an antibody that identifies both proteins. The localization data of Anderson and coworkers not only indicate that RNs contain enzymes that mediate meiotic recombination but provide strong support for the idea that the early events of recombination precede synapsis.

The next important step will be to identify mutations in plant genes with the functions of yeast *RAD51* and *DMC1/LIM15*. A *DMC1* homolog has been identified in *Arabidopsis*, but the mutant phenotype is unknown (65). In maize, the genes for at least two homologs of *RAD51* and two homologs of *LIM15/DMC1* have been cloned, and mutations have been generated at each locus using a reverse genetics procedure (B Bowen and S Tabata, personal communication; and see 92 for reverse genetics in maize). None of the single *rad51* mutations alone have definite meiotic phenotypes; it is likely that the individual *Rad51*-like genes have overlapping functions and that double mutant analysis will be required for functional analysis. Mutations that affect RN function may also exist among the large collection of plant mutants that cause premature desynapsis in late pachytene-diplotene (43, 66).

Role of the Synaptonemal Complex

Although the SC normally forms only between homologous chromosomes at pachytene, intimate pairing and SC formation can occur between nonhomologous chromosomes. Nonhomologous associations were first observed by McClintock (89) in monosomics, trisomics, and a variety chromosomal rearrangements in maize. Nonhomologous pairing, mediated by apparently normal SC, has now been documented in a number of plants (e.g. 37, 44, 50, 51, 73, 118, 138). In addition, mutations in both maize and onion that cause failures in homologous pairing are associated with indiscriminate synapsis and fold-back pairing (55, 86). Late RNs are not observed on nonhomologously synapsed chromosomes (SM Stack, personal communication), and nonhomologous associations generally do not lead to recombination. The one exception is in haploids, where very limited recombination (rarely more than one event per cell) has been detected in specific, presumably homologous, regions (99, 142). Because nonhomologous pairing can occur during zygotene as well as pachytene (44, 50, 51, 118), homology is probably not a prerequisite for SC formation at any stage of meiosis (118).

It would be appropriate to suggest that there is no causal relationship between the SC and genetic recombination in plants if not for the *asynaptic* mutation of wheat (var. Aziziah). No recombination is observed in the wheat *asynaptic* mutation (87). Lateral elements are present, but a mature synaptonemal complex is not formed (67). If the mutation was simply recombination-defective, nonhomologous pairing and SC formation would be expected (as observed in other synaptic mutants; 55, 86). The data suggest that *asynaptic* is defective for a component of the SC, and that either the protein encoded by *asynaptic* or the mature SC is required to complete recombination. The effect of the SC on recombination may be indirect. For instance, the SC may be required as a scaffold to house and/or stabilize recombination nodules during the final stages of recombination.

INTERFERENCE   An important role for the SC may be in the regulation of recombination frequency. A consistent feature of meiosis is that the distribution of mature recombination events leading to chiasmata are not evenly distributed. In *Hyacinthus amethystinus*, for example, there is a 20-fold variation in chromosome length, but the frequency of chiasmata on the smallest chromosomes is only slightly less than half the frequency on the largest chromosomes (23). The uneven distribution of recombination can be explained by genetic interference, which refers to the observation that a single recombination event inhibits additional recombination in nearby regions (60). How the existence of a crossover is detected and then communicated to nearby regions of a chromosome is not known, but recent authors have argued that it is a function of the SC (33, 60). The observation that several fungi lack both SC and genetic interference has been used to support this argument (33). Perhaps the strongest evidence that the SC mediates interference was provided by an analysis of the yeast *Zip1* gene, which encodes a structural component of the SC. In loss-of-function *zip1* mutants, recombination levels are slightly reduced, but interference is abolished (132, 134).

The idea that the SC mediates interference is consistent with the available data from plants. In an elegant study, Parker (105) determined chiasmata frequency in two regions that flanked a centromere in *Hypochoeris radicata*. When one chromosome of the bivalent was broken at the centromere such that the region available for synapsis was separated into two parts, interference between the two regions was significantly reduced. Single gene mutations are also available in both tomato and maize that provide a direct connection between the SC and interference. The as_1, as_4, and as_b mutations of tomato show incomplete synapsis at pachytene and reduced number of chiasmata per cell. By measuring the effect on recombination and interference over a 36-map unit interval, it was shown that each mutation caused a significant reduction in interference (96). The as_b mutation had the most severe effect, reducing interference by 80%. A limited ultrastructural analysis indicated that apparently normal SC as well as unpaired axial elements were present in the tomato *asynaptic* mutants. Similarly, the *as1* mutation of maize shows a variable degree of desynapsis (93) and a reduction of interference. Dempsey (28) measured recombination over two intervals on chromosomes 2 and 9 in *as1* plants, finding that recombination increased overall and that interference dropped by ~30% in both regions. The maize *as1* mutation is not a null allele but a hypomorph (8), which may explain the incomplete synaptonemal complex formation (84, 85) and limited effect on interference.

How the synaptonemal complex could promote interference is a matter of debate (33, 45, 64). Because the number of SC initiation sites greatly exceeds the number of chiasmata, interference presumably occurs after the SC has assembled. Genetic evidence in yeast suggests that the role of the SC might be to "transmit stress" along the chromosome (64, 132). Presumably, the stress occurs in the form of DNA coiling but could involve the interaction of the DNA with the lateral element of the SC (64). According to this model, crossover events are promoted by stress, and noncrossover-events are the default pathway (132). When a crossover occurs, stress is relieved in surrounding regions and all flanking homologous contacts are resolved into noncrossover events. An alternative model suggests that RNs generate interference, and that the SC simply serves as scaffold for RN movement along the chromosome (45). While either model would accommodate the observations in plants, they both remain highly speculative.

CHROMOSOME SEGREGATION

A unique feature of meiosis I in most plants is that the homologous chromosomes, each containing two sister chromatids, segregate away from each other. As shown in Figure 5, two conditions must be met for this to happen: At least one chiasma must be present (the result of pairing and recombination), and the sister kinetochores must orient together to the same spindle pole. When a chiasma is not present, the resulting univalents are free to segregate randomly and may arrive at the same pole (Figure 5b). When the sister kinetochores fail to orient together, the chromatids may disjoin (Figure 5c). Disjoined sister chromatids are not only subject to random segregation in meiosis I but are incapable of regular disjunction in meiosis II. The orientation of sister chromatids typical of meiosis I is referred to as a *co*-orientation, and the behavior of sister chromatids typical meiosis II is referred to as *auto*-orientation (108).

Chiasmata

CHIASMA MAINTENANCE   Chiasmata are the result of recombination between at least two nonsister chromatids in a bivalent (Figure 5; 81, 108). A variety

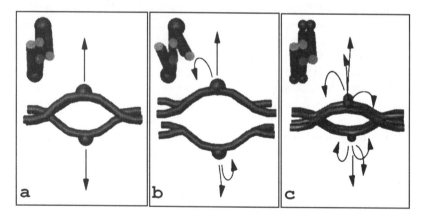

Figure 5 Metaphase I chromosome structure. In each panel, a side view (*upper left*) and frontal view are shown. (*a*) In normal cells, the chiasmata hold the chromosomes together, and the two sister kinetochores are observed as a single unit. (*b*) If chiasmata do not form (as in the maize *dy* mutant), the homologous chromosomes can segregate in either direction and may arrive at the same pole. (*c*) If the sister kinetochores are separated (as in the tomato *pc* mutant), they can interact with different spindle poles and randomly segregate.

of data indicate that chiasmata are stable structures (59, 60, 71) that do not "terminalize" as once thought (24). Maguire (80, 81) has recently emphasized that recombination is insufficient to hold the chiasmata in place; additional factors, located either at the chiasmata or between sister chromatids, are required to maintain chiasmata. In several fungi, remnants of the SC have been detected at crossover sites, indicating that the SC may be partly responsible for holding chiasmata in place (140). Remnants of SC at diplotene have also been detected in plants, but the fragments that remain generally do not coincide with the location of chiasmata (36, 52, 125, 126). It is possible that as yet unidentified proteins (or protein complexes) bind specifically to chiasmata and hold them in place.

An alternative view is that the association of sister chromatids distal to the crossover provides the glue that holds chiasmata in place (57, 81, 95). Evidence that sister chromatid cohesiveness is responsible for chiasma maintenance in plants was provided by Maguire (76) working with the desynaptic (*dy*) mutation of maize. The *dy* mutation is representative of a large group of mutations typified by univalent formation at diakinesis (66). Most desynaptic mutations are thought to be defects in recombination. In *dy* plants, however, it was possible to determine that at least some of the univalents had undergone recombination of cytological markers, suggesting a defect in chiasma maintenance. The *dy* mutation also conferred a general tendency of the univalents to prematurely dissociate, suggesting that the plants were defective for sister chromatid cohesiveness (76). More recent studies demonstrating that *dy* plants have defective SC have led to the suggestion that the substance binding sister chromatids is derived from the SC (83, 86). This idea is supported by the recent observation that a component of the animal SC (Cor1) is present along chromosome arms at metaphase 1 (30). Whether components of the SC remain associated with plant chromosomes during metaphase I is unclear (34, 124).

CHIASMATA IN CELL CYCLE CONTROL AND SPINDLE FORMATION   In insect spermatocytes, chiasmata formation is required for the cell to proceed from metaphase I to anaphase I (101). The formation of a chiasma provides a connection between the homologs that restrains poleward movement and causes tension at the kinetochores. In the absence of a chiasma, the resulting univalent kinetochores do not sense tension and produce a signal that causes the cell to delay anaphase. Among the factors that are thought to sense tension at the kinetochores in *Drosophila melanogaster* is a protein called ZW10 (4, 143). An Arabidopsis homolog of the *zw10* gene has recently been identified, suggesting that the tension-sensing mechanisms similar to those described in animals could operate in plants (129).

An additional role for chiasmata in plants is to ensure proper meiotic spindle assembly. In synaptic mutations where recombination or chiasma maintenance are disrupted (resulting in univalents), a general failure in bipolar spindle formation is observed (66). The effects of univalents in meiosis I can range from small extra minispindles to tripolar, quadripolar, and multiple spindles in the same cell (e.g. 10). Similar observations have been made in triploid and haploid plants (88, 137), indicating that spindle aberrations are a secondary effect caused by the presence of unpaired chromosomes. The simplest explanation for these findings is that unpaired chromosomes disrupt the spindle because they have only a single functional kinetochore (25). The idea that chromosomes can affect spindle structure is well supported by data from both plant mitotic cells and animal meiotic cells (107, 122).

The Meiotic Kinetochore

The kinetochore is known to regulate chromosome movement in plant meiosis (145), and its structure predicts its behavior: When the two sister kinetochores compose a single structure in meiosis I they usually segregate together (15, 70, 144); when the kinetochores are visibly separated from each other they usually disjoin (14, 70, 123). More direct data on the importance of the kinetochore in meiosis I segregation are available from yeast. In strains of yeast that undergo a phenomenon known as single-division meiosis, some chromosomes preferentially undergo a meiosis I–type disjunction (co-orient), and some undergo a meiosis II–type segregation (auto-orient). By moving centromeric DNA sequences from one chromosome to another it was shown that the segregation behavior of a chromosome was encoded by the centromere itself (119). Limited data suggest that this is also true in plants. An analysis of maize trisomic strains indicates that each chromosome has its own propensity to co-orient or auto-orient in meiosis I, and a minichromosome with a reduced centromeric region has been identified that auto-orients nearly 100% of the time (79).

COMPOSITION OF THE MEIOTIC KINETOCHORE   Plant meiotic kinetochores are most often described as amorphous and ball-shaped with a granular substructure (e.g. 15, 35, 144). Little is known of the composition of kinetochores in plants, but significant progress has been made in understanding the structure of animal kinetochores (32). The most valuable tool in animal kinetochore research was discovered in patients with the CREST variant of systemic sclerosis. Sera from the majority of CREST patients recognizes antigens in the centromeric region of mammalian mitotic and meiotic chromosomes (32). Several proteins recognized by CREST sera have been characterized and studied extensively (32). One CREST serum (EK) identifies an 80-kDa protein that is required to complete meiosis in mice (120). The 80-kDa protein identified by CREST EK is presumably CENP-B, a centromere-binding protein. The same CREST serum was shown to identify mitotic kinetochores in *Haemanthus* (98) and *Tradescantia* (104) and identifies the meiotic kinetochores in pachytene-diplotene chromosomes of maize (Figure 3). These data suggest that CENP-B-like proteins may be present in plant meiotic kinetochores; however, the function of CENP-B is not well understood even in animals (32).

In addition to protein, DNA is found in animal kinetochores (21). The observation that some plant kinetochores contain aceto-orcein-stained chromomeres that stretch towards the poles in anaphase (Figure 6) suggests that there may be DNA in plant kinetochores as well. With the availability of several plant centromere sequences (7, 56, 61), it should be possible to test for the presence of DNA in plant kinetochores.

MEIOSIS I KINETOCHORE CO-ORIENTATION   An important first step toward understanding the mechanism of co-orientation in meiosis I was taken by Stern & Hotta (130). They discovered that in *Lilium*, meiocytes from the leptotene stage or later could be removed and cultured in an artificial medium where they completed a normal meiotic division. In contrast, when meiocytes in S

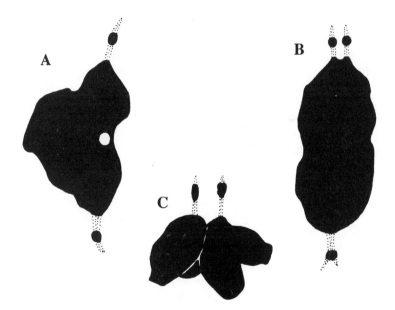

Figure 6   Kinetochore maturation in *Tradescantia*. Chromosomes were stained with aceto-orcein. (*a*) Early metaphase in *T. virginiana*. (*b*) Late metaphase in *T. virginiana*. (*c*) Anaphase in *T. bracteata*. Note the centromeres appear to be pulled poleward and each contains a conspicuous chromomere (*darkly-stained region of the chromosome*). [From Lima-de-Faria (70) with permission from *Hereditas*.]

phase were removed, mitosis occurred instead of meiosis. Interestingly, meiocytes removed immediately after S phase underwent a normal mitotic prophase followed by failed anaphase, apparently because the centromeres could not divide (130). The simplest interpretation of the data is that the orientation of the kinetochores is one of the first events to be determined as a cell initiates meiosis.

The contention that kinetochore co-orientation occurs very early in meiosis is supported by studies of the maize *absence of first division* (*afd1*) mutation (42). In *afd1* meiocytes, all the chromosomes auto-orient in meiosis I, and the sister chromatids segregate away from each other. Meiosis II is ineffective (only single chromatids are present) but appears to occur on schedule, indicating that *afd1* does not convert the meiotic program to a mitotic one. Analysis of the early prophase stages indicates that the chromosomes do not condense into the long thin threads typical of leptotene but appear thick and short as if they preceded directly to diakinesis without the intervening pairing stages (42). Electron micrographs of *afd1* meiocytes indicate that SC formation ceases quickly after it is initiated and produces only ~12% of the SC found in wild-type plants (41). Much of the SC that is installed in *afd1* plants appears defective, lacking either the lateral or the central elements (41). After their analysis of *afd1*, Golubovskaya & Mashnenkov (42) argued that if the cell passes through leptotene, the kinetochores become committed to co-orientation. The *afd1* phenotype is similar to a *Drosophila* mutation called *orientation disrupter* (*ord*) (94). In *ord* mutants, recombination is reduced to 10–13% of the wild-type levels, and sister kinetochore separation regularly occurs during prometaphase, metaphase, and anaphase of meiosis I. The recombination defect in *ord* may be associated with a defect in the SC, but the appropriate studies have not yet been carried out. The *ORD* gene encodes a novel protein with characteristics that suggest it is regulated by proteolysis (13).

Studies of synaptic mutants further support the idea that sister kinetochore co-orientation is an early meiotic event. As a rule, the phenotypes of synaptic mutations are first observed late in meiotic propase and are classed together by their failure to form bivalents. Surveys of published synaptic mutants, identified in over 90 plant species, indicate that there is no correlation between synaptic defects and the segregation behavior of the univalents at anaphase I (66). For example, no univalents divide equationally (in half) in *Vicia faba* (121), whereas in *Oenothera* (19), all univalents divide equationally. In *Brassica campestris* (133) there is a positive correlation between the degree of asynapsis and the frequency of equational division, while in rice there is no such correlation (63). The data are broadly consistent with the interpretation that meiosis I kinetochore orientation is determined at or around leptotene. The fact that univalents sometimes do divide in meiosis I is best interpreted as the result of kinetochore maturation during late metaphase/anaphase, as discussed in the next section.

Although largely circumstantial, the available data suggest that the orientation of kinetochores in meiosis I is determined at or around leptotene, either directly or indirectly by *Afd1* or similar genes. Because *afd1* plants are deficient for axial elements, it is possible that these elements promote co-orientation simply by holding the sister chromatids together during kinetochore formation. Other factors promoting or stabilizing the close apposition of kinetochores could be the inherent "stickiness" of the prophase I kinetochores (e.g. 38, 125) and the tendency for the microtubules that are attached to the kinetochores to bundle together (e.g. 22).

KINETOCHORE MATURATION AND MEIOSIS II   Whereas the onset of anaphase I is marked by the dissolution of chiasmata, the onset of anaphase II is marked by the separation of sister kinetochores. Therefore, in the period between anaphase I and anaphase II, the kinetochores must lose their co-orientation and adopt an auto-orientation. An important clue to the timing of kinetochore separation can be inferred from the behavior of univalents in synaptic mutants. The great majority of synaptic mutants have phenotypes that for one reason or another produce both univalents and bivalents at meiosis I (43, 66). It has been reported in all cases that the univalents divide after the bivalents. In most species, the univalent division occurs very late, if at all, during anaphase. A notable exception occurs in *asynaptic Oenothera*, where the univalents divide after bivalents but still early enough to be segregated properly to telophase nuclei. In such plants, Catcheside (19) noted that when the univalents divide, they divide at roughly the same time regardless of whether they are located in the polar region or plate region of the meiosis I spindle (19). The overall synchrony of equational segregation at late anaphase indicates that sister kinetochores are not physically torn apart by their interactions with the spindle but rather that they are separated in a regulated manner.

The idea that sister kinetochores begin their separation during late metaphase and anaphase is supported by light microscopic observations of the kinetochores themselves (70, 113). As shown in Figure 6, Lima-de-Faria (70) demonstrated that the sister kinetochores of *Tradescantia* appear as a single unit in early metaphase, whereas in late metaphase and anaphase the kinetochores become visibly distinct. Essentially, the same results were obtained by silver-staining of *Allium cepa* and *Rhoeo discolor* kinetochores (123). Similarly in maize, it was demonstrated using a cloned cereal centromere repeat (56) that the late metaphase I kinetochore consists of two units (145; EN Hiatt & RK Dawe, unpublished observations). Two distinct sister kinetochores at metaphase I were also demonstrated in wheat univalents at the electron microscope level (141). The univalents first adopted positions close to a pole, presumably reflecting the close apposition of the two kinetochores. Later in metaphase, even though the sister kinetochores remain closely apposed, the wheat univalents moved to the spindle midzone. Only after anaphase I was in progress (as judged by the segregation of bivalents) did the sister chromatids separate, allowing the chromatids to segregate equationally (140).

At least one mutation has been identified that alters the timing of sister kinetochore separation. The tomato *precocious centromere division* (*pc*) mutation causes the kinetochores to separate prematurely during anaphase I and interkinesis such that at prophase II only single chromatids are observed (20). During anaphase I, it appeared that a few univalents became auto-oriented and divided. A similar phenotype was described in *Alopecurus myosuroides*, although the genetic basis of this defect is not known (58). The kinetochore separation phenotypes in these plants is almost identical to the phenotype of the *Drosophila mei-S332* mutation. *Mei-S332* has been cloned, and the encoded protein localizes to kinetochores in meiosis I but disappears at the onset of anaphase II (62). The localization pattern of the MEI-S332 protein suggests that it is not only responsible for maintaining the association of sister chromatids during meiosis I but may also regulate the disjunction of chromosomes in meiosis II.

FUTURE PROSPECTS

In recent years, significant progress has been made in understanding chromosome pairing and recombination nodules (RNs) in plants (e.g. 5, 9, 112, 128). In other areas the most important research progress has been made in yeast and *Drosophila* (e.g. 62, 109), and it is only possible to speculate on the similarities to plants. Future studies will likely make use of randomly sequenced cDNAs (ESTs) and genome-sequencing projects to identify plant homologs to important proteins from other species. With the availability of plant homologs to fungal and animal genes it will be possible to use immunolocalization and newly established reverse genetic strategies (92) to determine whether there are functional similarities across organismal boundaries. This approach has already been employed in recombination nodules studies, where it has been established that plants have homologs to yeast *RAD51* and *DMC1* (5, 135). In addition, it will be important to pursue forward genetic approaches in combination with high-resolution cytological analysis (e.g. 26) to identify genes that have roles unique to plant meiosis. Genetics and cytology can be combined in maize, tomato, and even Arabidopsis, which despite its small genome size can be analyzed cytologically at all stages of meiosis (110).

ACKNOWLEDGMENTS

I thank L. Arthur, M. Brickman, C. Hasenkampf, and S. Stack for helpful discussions, as well as B. Bowen, S. Tabata, J. Fung, D. Agard, and J. Sedat for sharing unpublished data. This work was supported by a grant from the National Science Foundation.

Visit the *Annual Reviews home page* at
http://www.AnnualReviews.org.

Literature Cited

1. Abirached-Darmency M, Zickler D, Cauderon Y. 1983. Synaptonemal complex and recombination nodules in rye (*Secale cereale*). *Chromosoma* 88:299–306

2. Albini SM, Jones GH. 1987. Synaptonemal complex spreading in *Allium cepa* and *Allium fistulosum*. I. The initiation and sequence of pairing. *Chromosoma* 95:324–38

3. Albini SM, Jones GH. 1988. Synaptonemal complex spreading in *Allium cepa* and *Allium fistulosum*. II. Pachytene observations: the SC karyotype and the correspondence of late recombination nodules and chiasmata. *Genome* 30:339–410

4. Allshire RC. 1997. Centromeres, checkpoints, and chromatid cohesion. *Curr. Opin. Genet. Dev.* 7:264–73

5. Anderson LK, Offenberg HH, Verkuijlen WMHC, Heyting C. 1997. RecA-like proteins are components of early meiotic nodules in lily. *Proc. Natl. Acad. Sci. USA* 94:6868–73

6. Anderson LK, Stack SM. 1988. Nodules associated with axial cores and synaptonemal complexes during zygotene in *Psilotum nudum*. *Chromosoma* 97:96–100

7. Aragon-Alcaide L, Miller T, Schwarzacher T, Reader S, Moore G. 1996. A cereal centromeric sequence. *Chromosoma* 105:261–68

8. Baker RL, Morgan DT. 1969. Control of pairing in maize and meiotic interchromosomal effects of deficiencies in chromosome 1. *Genetics* 61:91–106

9. Bass HW, Marshall WF, Sedat JW, Agard DA, Cande WZ. 1997. Telomeres cluster de novo before the initiation of synapsis: a three dimensional spatial analysis of telomere positions before and during meiotic prophase. *J. Cell Biol.* 137:5–18

10. Beadle GW. 1930. Genetical and cytological studies of Mendelian asynapsis in *Zea mays. Cornell Agric. Exp. Stn. Mem.* 129:1–23

11. Beasley JO. 1938. Nuclear size in relation to meiosis. *Bot. Gaz.* 99:865–71

12. Bennett MD, Smith JB, Simpson S, Wells B. 1979. Intranuclear fibrillar material in cereal pollen mother cells. *Chromosoma* 71:289–332

13. Bickel SE, Wyman DW, Miyazaki Wy, Moore DP, Orr-Weaver TL. 1996. Identification of ORD, a Drosophila protein essential for sister chromatid cohesion. *EMBO J.* 15:1451–59

14. Braselton JP. 1981. The ultrastructure of meiotic kinetochores in *Luzula*. *Chromosoma* 82:143–51

15. Braselton JP, Bowen CC. 1971. The ultrastructure of the kinetochores of *Lilium longiflorum* during the first meiotic division. *Caryologia* 24:49–58

16. Breyne P, Van Montagu M, Gheysen G. 1994. The role of scaffold attachment regions in the structural and functional organization of plant chromatin. *Transgenic Res.* 3:195–202

17. Burnham CR, Stout JT, Weinheimer WH, Kowles RV, Phillips RL. 1972. Chromosome pairing in maize. *Genetics* 71:111–26

18. Carpenter ATC. 1987. Gene conversion, recombination nodules, and the initiation of meiotic synapsis. *BioEssays* 6:232–36

19. Catcheside DG. 1939. An asynaptic *Oenethera*. *New Phytol.* 38:323–34

20. Clayberg CD. 1959. Cytogenetic studies of precocious meiotic centromere division in *Lycopersicon esculentum* Mill. *Genetics* 44:1335–45

21. Cooke CA, Bazett-Jones DP, Earnshaw WC, Rattner JB. 1993. Mapping DNA within the mammalian kinetochore. *J. Cell Biol.* 120:1083–91

22. Cyr RJ, Palevitz BA. 1989. Microtubule-binding proteins from carrot. I. Initial characterization and microtubule bundling. *Planta* 177:245–60

23. Darlington CD. 1932. The origin and behavior of chiasmata VI. *Hyacynthus amethystinus*. *Biol. Bull.* 63:368–71

24. Darlington CD. 1932. *Recent Advances in Cytology*. London: Churchill

25. Dawe RK, Cande WZ. 1995. The role of chromosomes in maize meiotic spindle

morphogenesis. *J. Cell. Biochem. Suppl.* 21A:438

26. Dawe RK, Cande WZ. 1996. Induction of centromeric activity in maize by *Suppressor of meiotic drive 1. Proc. Natl. Acad. Sci. USA* 93:8512–17

27. Dawe RK, Sedat JW, Agard DA, Cande WZ. 1994. Meiotic chromosome pairing in maize is associated with a novel chromatin organization. *Cell* 76:901–12

28. Dempsey E. 1958. Analysis of crossing over in haploid gametes of asynaptic plants. *Maize Genet. Coop. Newslett.* 33:54–55

29. Dernburg AF, Sedat JW, Cande WZ, Bass HW. 1995. Cytology of telomeres. In *Telomeres*, ed. EH Backburn, CW Greider, pp. 295–338. Cold Spring Harbor, NY: Cold Spring Harbor Lab. Press

30. Dobson MJ, Pearlman RE, Karaiskakis A, Spyropoulos B, Moens PB. 1994. Synaptonemal complex proteins: occurrence, epitope mapping and chromosome disjunction. *J. Cell Sci.* 107:2749–60

31. Dundas IS, Britten EJ, Byth DE. 1983. Pachytene chromosome identification by a key based on chromomeres in the pigeonpea. *J. Hered.* 74:461–64

32. Earnshaw WC. 1994. Structure and molecular biology of the kinetochore. In *Microtubules*, ed. JS Hyams, CW Lloyd, pp. 393–412. New York: Wiley-Liss

33. Egel R. 1995. The synaptonemal complex and the distribution of meiotic recombination events. *Trends Genet.* 11:206–8

34. Fedotova YS, Kolomiets OL, Bogdanov YF. 1989. Synaptonemal complex transformations in rye microsporocytes at the diplotene stage of meiosis. *Genome* 32:816–23

35. Gillies CB. 1973. Ultrastructural analysis of maize pachytene karyotypes by three dimensional reconstruction of synaptonemal complexes. *Chromosoma* 43:145–76

36. Gillies CB. 1975. An ultrastructural analysis of chromosomal pairing in maize. *C. R. Trav. Lab. Carlsberg* 40:135–61

37. Gillies CB. 1983. Ultrastructural studies of the association of homologous and nonhomologous parts of chromosomes in the mid-prophase of meiosis in *Zea mays. Maydica* 28:265–87

38. Gillies CB. 1984. The synaptonemal complex in higher plants. *Crit. Rev. Plant Sci.* 2:81–116

39. Gillies CB. 1985. An electron microscope study of synaptonemal complex formation at zygotene in rye. *Chromosoma* 92:165–75

40. Gillies CB, Lukaszewski AJ. 1989. Synaptonemal complex formation in rye (*Secale cereale*) heterozygous for telomeric C-bands. *Genome* 32:901–7

41. Golubovskaya I, Khristolyubova NB. 1985. The cytogenetic evidence of the gene control of meiosis. In *Plant Genetics*, ed. M Freeling, pp. 723–38. New York: Liss

42. Golubovskaya IN, Mashnenkov AS. 1975. Genetic control of meiosis. I. meiotic mutation in corn (*Zea mays* L.) afd, causing the elimination of the first meiotic division. *Genetika* 11:11–17

43. Gottschalk W, Kaul MLH. 1980. Asynapsis and desynapsis in flowering plants. II. Desynapsis. *Nucleus* 23:97–120

44. Hasenkampf CA. 1984. Synaptonemal complex formation in pollen mother cells of *Tradescantia. Chromosoma* 90:275–84

45. Hasenkampf CA. 1996. The synaptonemal complex—the chaperone of crossing over. *Chromosome Res.* 4:133–40

46. Hawley RS, Arbel T. 1993. Yeast meiosis and the fall of the classical view of meiosis. *Cell* 72:301–3

47. Herickhoff L, Stack S, Sherman J. 1993. The relationsip between synapsis, recombination nodules and chiasmata in translocation heterozygotes. *Heredity* 71:373–85

48. Heyting C. 1996. Synaptonemal complexes: structure and function. *Curr. Opin. Cell Biol.* 8:389–96

49. Hiraoka T. 1952. Observational and experimental studies of meiosis with special reference to the bouquet stage. XI. Locomotory movement of the nucleolus in the bouquet stage. *Cytologia* 17:201–9

50. Holbolth P. 1981. Chromosome pairing in allohexaploid wheat var. Chinese Spring. Transformation of multivalents into bivalents, a mechanism for exclusive bivalent formation. *Carlsberg Res. Commun.* 46:129–73

51. Holm PB. 1977. Three-dimensional reconstruction of chromosome pairing during the zygotene stage of meiosis in *Lilium longiflorum* (Thunb.). *Carlsberg Res. Commun.* 42:103–51

52. Holm PB. 1986. Chromosome pairing and chiasma formation in allohexaploid wheat, *Triticum aestivum* analyzed by spreading of meiotic nuclei. *Carlsberg Res. Commun.* 51:239–84

53. Hughes-Schrader S. 1943. Polarization, kinetochore movements, and bivalent structure in the meiosis of male mantids. *Biol. Bull.* 85:265–300

54. Jenkins G. 1983. Chromosome pairing in *Triticum aestivum* cv. Chinese Spring. *Carlsberg Res. Commun.* 48:255–83

55. Jenkins G, Okomus A. 1992. Indiscriminate synapsis in achiasmate *Allium fistulosum* L. (Liliaceae). *J. Cell Sci.* 103:415–22

56. Jiang JM, Nasuda S, Dong FG, Scherrer CW, Woo SS, et al. 1996. A conserved repetitive DNA element located in the centromeres of cereal chromosomes. *Proc. Natl. Acad. Sci. USA* 93:14210–13

57. John B. 1990. *Meiosis.* Cambridge: Cambridge Univ. Press

58. Johnsson H. 1944. Meiotic aberrations and sterility in *Alopecurus myosuroides* Huds. *Hereditas* 30:469–566

59. Jones GH. 1978. Giesma C-banding of rye meiotic chromosomes and the nature of "terminal" chiasmata. *Chromosoma* 66:45–57

60. Jones GH. 1984. The control of chiasma distribution. In *Controlling Events in Meiosis*, ed. CW Evans, HG Dickinson, pp. 293–320. Cambridge: Company Biol.

61. Kaszas E, Birchler JA. 1996. Misdivision analysis of centromere structure in maize. *EMBO J.* 15:5246–55

62. Kerrebrock AW, Moore DP, Wu JS, Orr-Weaver TL. 1995. Mei-S332, a *Drosophila* protein required for sister-chromatid cohesion, can localize to meiotic centromere regions. *Cell* 83:247–56

63. Kitada K, Omura T. 1984. Genetic control of meiosis in rice, *Oryza sativa* L. IV. Cytogenetical analyses of asynaptic mutants. *Genome* 26:264–71

64. Kleckner N. 1996. Meiosis: How could it work? *Proc. Natl. Acad. Sci. USA* 93:8167–74

65. Klimyuk VI, Jones JDG. 1997. AtDMC1, the Arabidopsis homologue of the yeast DMC1 gene: characterization, transposon-induced allelic variation and meiosis-associated expression. *Plant J.* 11:1–14

66. Koduru PRK, Rao MK. 1981. Cytogenetics of synaptic mutants in higher plants. *Theor. Appl. Genet.* 59:197–214

67. La Cour LF, Wells B. 1970. Meiotic prophase in anthers of asynaptic wheat. *Chromosoma* 29:419–27

68. La Cour LF, Wells B. 1971. The chromomeres of prepachytene chromosomes. *Cytologia* 36:111–20

69. Lambert A-M. 1993. Microtubule organizing centers in higher plants. *Curr. Opin. Cell Biol.* 5:116–22

70. Lima-de-Faria A. 1956. The role of the kinetochore in chromosome organization. *Hereditas* 42:85–160

71. Loidl J. 1979. C-band proximity of chiasmata and absence of terminalization in *Allium flavum* (Liliaceae). *Chromosoma* 73:45–51

72. Loidl J. 1990. The initiation of meiotic chromosome pairing: the cytological view. *Genome* 33:759–78

73. Loidl J, Jones GH. 1986. Synaptonemal complex spreading in *Allium* I. Triploid *A. sphaerocephalon*. *Chromosoma* 93:420–28

74. Maguire MP. 1972. The temporal sequence of synaptic initiation, crossing over and synaptic completion. *Genetics* 70:353–70

75. Maguire MP. 1977. Homologous chromosome pairing. *Philos. Trans. R. Soc. London Ser. B* 277:245–58

76. Maguire MP. 1978. Evidence for separate genetic control of crossing over and chiasma maintenance in maize. *Chromosoma* 65:173–83

77. Maguire MP. 1983. Homologue pairing and synaptic behavior at zygotene in maize. *Cytologia* 48:811–18

78. Maguire MP. 1984. The mechanism of meiotic homologue pairing. *J. Theor. Biol.* 106:605–15

79. Maguire MP. 1987. Meiotic behavior of a tiny fragment chromosome that carries a transposed centromere. *Genome* 29:744–47

80. Maguire MP. 1993. Sister chromatid association at meiosis. *Maydica* 38:93–106

81. Maguire MP. 1995. Is the synaptonemal complex a disjunction machine? *J. Hered.* 86:330–40

82. Maguire MP. 1995. The relationship of homologous synapsis and crossing over in a maize inversion. *Genetics* 137:281–88

83. Maguire MP, Paredes AM, Riess RW. 1991. The desynaptic mutant of maize as a combined defect of synaptonemal complex and chiasma maintenance. *Genome* 34:879–87

84. Maguire MP, Riess RW. 1991. Synaptonemal complex behavior in asynaptic maize. *Genome* 34:163–68

85. Maguire MP, Riess RW. 1996. Synaptic defects of asynaptic homozygotes in maize at the electron microscope level. *Genome* 39:1194–98

86. Maguire MP, Riess RW, Paredes AM. 1993. Evidence from a maize desynaptic mutant points to a probable role of synaptonemal complex central region components in provision for subsequent chiasma maintenance. *Genome* 36:797–807

87. Martini G, Bozzini A. 1966. Radiation induced asynaptic mutations in durum

wheat (*Triticum durum* Desf.). *Chromosoma* 20:251–66

88. McClintock B. 1929. A cytological and genetical study of triploid maize. *Genetics* 14:180–222

89. McClintock B. 1933. The association of nonhomologous parts of chromosomes in the mid-prophase of *Zea mays*. *Z. Zellforsch. Mikrosk Anat.* 19:191–237

90. McClintock B. 1941. The association of mutants with homozygous deficiencies in *Zea mays*. *Genetics* 25:542–71

91. McClintock B. 1951. Chromosome organization and genic expression. *Cold Spring Harbor Symp. Quant. Biol.* 16:13–47

92. Mena M, Ambrose BA, Meeley RB, Briggs SP, Yanofsky MF, Schmidt RJ. 1996. Diversification of C-function activity in maize flower development. *Science* 274:1537–40

93. Miller OL. 1963. Cytological studies of asynaptic maize. *Genetics* 48:1445–66

94. Miyazaki WY, Orr-Weaver TL. 1992. Sister-chromatid misbehavior in Drosophila *ord* mutants. *Genetics* 342:1047–61

95. Miyazaki WY, Orr-Weaver TL. 1994. Sister chromatid cohesiveness in mitosis and meiosis. *Annu. Rev. Genet.* 28:167–87

96. Moens PB. 1969. Genetic and cytological effects of three desynaptic genes in the tomato. *Can. J. Genet. Cytol.* 11:857–69

97. Moens PB, Pearlman RE. 1988. Chromatin organization at meiosis. *BioEssays* 9:151–53

98. Mole-Bajer J, Bajer AS, Zinkowski RP, Balczon RD, Brinkley BR. 1990. Autoantibodies from a patient with scleroderma CREST recognized kinetochores of the higher plant *Haemanthus*. *Proc. Natl. Acad. Sci. USA* 87:3599–603

99. Neijzing MG. 1982. Chiasma formation in duplicated segments of the haploid rye genome. *Chromosoma* 85:287–98

100. Neuffer MG, Coe EH, Wessler SR. 1997. *Mutants of Maize*. Cold Spring Harbor, NY: Cold Spring Harbor Lab. Press

101. Nicklas RB. 1997. How cells get the right chromosomes. *Science* 275:632–37

102. Nokkala S, Nokkala C. 1985. Spiral structures of meiotic chromosomes in plants. *Hereditas* 103:187–94

103. Oehlkers F, Eberle P. 1957. Spiralen und chromomeren in der meiosis von *Bellevalia romana*. *Chromosoma* 8:351–63

104. Palevitz BA. 1990. Kinetochore behavior during generative cell division in *Tradescantia virginiana*. *Protoplasma* 157:120–27

105. Parker JS. 1987. Increased chiasma frequency as a result of chromosome rearrangement. *Heredity* 58:87–94

106. Qureshi M, Hasenkampf C. 1995. DNA, histone H1 and meiotin-1 immunostaining patterns along whole-mount preparations of *Lilium longiflorum* pachytene chromosomes. *Chromosome Res.* 3:214–20

107. Reider CL, Ault J, Eichenlaub-Ritter U, Sluder G. 1993. Morphogenesis of the mitotic and meiotic spindle: conclusions from one system are not necessarily applicable to the other. In *Chromosome Segregation and Aneuploidy*, ed. BK Vig, pp. 183–97. Berlin: Springer-Verlag

108. Rhoades MM. 1961. Meiosis. In *The Cell*, ed. J Brachet, AE Mirsky, pp. 3:1–75. New York: Academic

109. Roeder GS. 1995. Sex and the single cell: meiosis in yeast. *Proc. Natl. Acad. Sci. USA* 92:10450–56

110. Ross KJ, Fransz GH, Jones GH. 1996. A light microscopic atlas of meiosis in *Arabidopsis thaliana*. *Chromosome Res.* 4:507–16

111. Schmit A-C, Endle M-C, Lambert A-M. 1996. The perinuclear microtubule-organizing center and the synaptonemal complex of higher plants share a common antigen: its putative transfer and role in meiotic chromosomal ordering. *Chromosoma* 104:405–13

112. Schmit A-C, Stoppin V, Chevrier V, Job D, Lambert A-M. 1994. Cell cycle dependent distribution of centrosomal antigen at the perinuclear MTOC or at the kinetochores of higher plant cells. *Chromosoma* 103:343–51

113. Schrader F. 1939. The structure of the kinetochore at meiosis. *Chromosoma* 1:230–37

114. Schwartz D. 1953. The behavior of an X-ray-induced ring chromosome in maize. *Am. Nat.* 87:19–28

115. Sheldon J, Willson C, Dickinson HG. 1988. Interaction between the nucleus and cytoskeleton during the pairing stages of male meiosis in flowering plants. *Kew Chromosome Conf., 3rd*, pp. 27–35. London: HMSO

116. Sherman JD, Herickhoff LA, Stack SM. 1992. Silver staining two types of meiotic nodules. *Genome* 35:907–15

117. Sherman JD, Stack SM. 1995. Two-dimensional spreads of synaptonemal complexes from solanaceous plants. VI. High-resolution recombination nodule map for tomato. *Genetics* 141:683–708

118. Sherman JD, Stack SM, Anderson LK. 1989. Two-dimensional spreads of

synaptonemal complexes from solanaceous plants. IV. Synaptic irregularities. *Genome* 32:743–53

119. Simchen G, Hugerat Y. 1993. What determines whether chromosomes segregate reductionally or equationally in meiosis? *BioEssays* 15:1–8

120. Simerly C, Balczon R, Brinkley BR, Schatten G. 1990. Microinjected kinetochore antibodies interfere with chromosome movement in meiotic and mitotic mouse embryos. *J. Cell Biol.* 111:1491–504

121. Sjodin J. 1970. Induced asynaptic mutations in *Vicia faba* L. *Hereditas* 66:215–32

122. Smirnova EA, Bajer AS. 1992. Spindle poles in higher plant mitosis. *Cell Motil. Cytoskelet.* 23:1–7

123. Stack SM. 1975. Differential giemsa staining of kinetochores in meiotic chromosomes of two higher plants. *Chromosoma* 51:357–63

124. Stack SM. 1991. Staining plant cells with silver. II. Chromosome cores. *Genome* 34:900–8

125. Stack SM, Anderson LK. 1986. Two-dimensional spreads of synaptonemal complexes from Solanaceous plants II. Synapsis in *Lycopersicum esculentum* (tomato). *Am. J. Bot.* 73:264–81

126. Stack SM, Anderson LK, Sherman JD. 1989. Chiasmata and recombination nodules in *Lilium Longiflorum*. *Genome* 32:486–98

127. Stack SM, Roelofs D. 1996. Localized chiasmata and meiotic nodules in the tetraploid onion *Allium porrum*. *Genome* 39:770–83

128. Stack SM, Sherman JD, Anderson LK, Herickhoff LS. 1993. Meiotic nodules in vascular plants. *Chromosomes Today* 11:301–11

129. Starr DA, Williams BC, Li Z, Etemad-Moghadam B, Dawe RK, Goldberg M. 1997. Conservation of the centromere/kinetochore protein ZW10. *J. Cell Biol.* 138:1289–1301

130. Stern H, Hotta Y. 1968. Biochemical studies of male gametogenesis in liliaceous plants. In *Current Topics of Developmental Biology*, ed. AA Moscona, A Monroy, pp. 37–63. New York: Academic

131. Stern H, Hotta Y. 1973. Biochemical controls of meiosis. *Annu. Rev. Genet.* 7:37–66

132. Storlazzi A, Xu LZ, Schwacha A, Kleckner N. 1996. Synaptonemal compex (SC)

component Zip1 plays a role on meiotic recombination independent of SC polymerization along the chromosomes. *Proc. Natl. Acad. Sci. USA* 93:9043–48

133. Stringam GR. 1970. A cytogenetic analysis of three asynaptic mutants in *Brassica campestris* L. *Can. J. Genet. Cytol.* 12:743–49

134. Sym M, Roeder S. 1994. Crossover interference is abolished in the absence of a synaptonemal complex protein. *Cell* 79:283–92

135. Teresawa M, Shinohara A, Hotta Y, Ogawa H, Ogawa T. 1995. Localization of RecA-like recombination proteins on chromosomes of the lily at various meiotic stages. *Genes Dev.* 6:925–34

136. Thomas JB, Kaltiskes PJ. 1976. A bouquet-like attachment plate for telomeres in leptotene of rye revealed by heterochromatin staining. *Heredity* 36:155–62

137. Ting YC. 1966. Duplication and meiotic behavior of chromosomes in haploid maize. *Int. J. Cytol.* 31:324–29

138. Ting YC. 1973. Synaptonemal complex and crossing over in haploid maize. *Chromosomes Today* 4:161–67

139. Vincent JE, Jones GH. 1993. Meiosis in autopolyploid *Crepis capillaris*. I. Triploids and trisomics; implications for models of chromosome pairing. *Chromosoma* 102:195–206

140. von Wettstein D, Rasmussen SW, Holm PB. 1984. The synaptonemal complex in genetic segregation. *Annu. Rev. Genet.* 18:331–413

141. Wagenaar EB, Bray DF. 1973. The ultrastructure of kinetochores of unpaired chromosomes in a wheat hybrid. *Can. J. Genet. Cytol.* 15:801–6

142. Weber DF, Alexander DE. 1972. Redundant segments in *Zea mays* detected by translocations of monoploid origin. *Chromosoma* 39:27–42

143. Williams BC. 1996. Bipolar spindle attachments affect redistributions of ZW10, a Drosophila centromere/kinetochore component required for accurate chromosome segregation. *J. Cell Biol.* 134:1127–40

144. Wilson HJ. 1968. The fine structure of the kinetochore in meiotic cells of *Tradescantia*. *Planta* 78:379–85

145. Yu H-G, Hiatt E, Chan A, Sweeney M, Dawe RK. 1997. Neocentromere-mediated chromosome movement in maize. *J. Cell Biol.* 139:831–40

Annu. Rev. Plant Physiol. Plant Mol. Biol. 1998. 49:397–425

PHOTOSYNTHETIC CYTOCHROMES c IN CYANOBACTERIA, ALGAE, AND PLANTS

Cheryl A. Kerfeld
219 Molecular Biology Institute, University of California at Los Angeles, Box 951570, Los Angeles, California 90095-1570; e-mail: kerfeld@ewald.mbi.ucla.edu

David W. Krogmann
Biochemistry Department, Purdue University, West Lafayette, Indiana 47907-1153; e-mail: krogmann@biochem.purdue.edu

KEY WORDS: cytochrome c_6, cytochrome f, cytochrome M, LP cytochrome c_{549}, photosynthesis

ABSTRACT

The cytochromes that function in photosynthesis in cyanobacteria, algae, and higher plants have, like the other photosynthetic catalysts, been largely conserved in their structure and function during evolution. Cyanobacteria and algae contain cytochrome c_6, which is not found in higher plants and which may enhance survival in their planktonic mode of life. Cyanobacteria and algae contain another cytochrome, low-potential c_{549}, which is not found in higher plants. This cytochrome has a structural role in PSII and may contribute to anaerobic survival. There is a third unique cytochrome, cytochrome M, in the planktonic photosynthesizers, and its function is unknown. New evidence is appearing to indicate evolution of cytochrome interaction mechanisms during the evolution of photosynthesis. The ease of cytochrome gene manipulation in cyanobacteria and in *Chlamydomonas reinhardtii* now provides great advantages in understanding of photosynthesis. The solution of tertiary and quaternary structures of cytochromes and cytochrome complexes will provide structural and functional detail at atomic resolution.

397

1040-2519/98/0601-0397$08.00

CONTENTS

Introduction[1]

The history of cytochromes is relatively brief but certainly voluminous. A rural physician, CA McMunn, practicing medicine in Wolverhampton, England, at the end of the past century, devoted his spare time to "medical spectroscopy" by examining transparent and translucent tissues of insects with a microspectroscope. He identified absorption bands due to cytochromes *a*, a_3, *b*, and *c*. By manipulating their oxidation and reduction, McMunn realized they were participants in respiration. He reported his observations at a meeting of the Physiological Society in London in 1881 but failed to draw any notice to his work from the authorities of those times. In 1925, David Keilin, who had found McMunn's work and extended it to other organisms, published a paper entitled *On Cytochrome, A Respiratory Pigment Common to Animals, Yeast and Higher Plants.* Keilin established the nearly universal significance of the cytochromes and observed cytochromes in fine slices of shallot, garlic, leeks, beans, and pollen—all of the colorless plant tissues he examined. In 1930, Robin Hill published with Keilin a paper entitled *The Porphyrin Component of Cytochrome*, and so began the chemical characterization of these molecules. Keilin wrote a charming history of the cytochromes (48). Hill was much devoted to botany and told, with a mischievous twinkle in his eye, of the remark of FG Hopkins, the Professor of Biochemistry at Cambridge in Hill's youth. "Plants are disgusting. They don't excrete." Undaunted by this egregious animal chauvinism, Hill made chloroplast cytochromes his life's work, and his student, Derek Bendall, continues in that tradition at Cambridge. Thus a span of four careers has carried cytochromes from in vivo observation to site-directed mutagenesis, modern protein dynamics, and atomic resolution structure determination.

[1]Abbreviations: CBC, methyl carbon adjacent to the heme carbon that is covalently linked to second Cys in CXXCH; cyt b_6, cytochrome b_6; cyt $b_6 f$, cytochrome $b_6 f$ complex; cyt c_6, cytochrome c_6; cyt f, cytochrome f; cyt M, cytochrome M; E*m*, midpoint oxidation reduction potential; LP cyt c, low potential cytochrome c; OA, propionate oxygen of the A ring; OD, propionate oxygen of the D ring; P700, photo oxidized electron acceptor of PSI; PC, plastocyanin, pI, isoelectric point; PS, photosystem.

Two earlier reviews on photosynthetic cytochromes have been published. (11, 12). The two volumes by Pettigrew & Moore on cytochrome c are excellent resources (76, 77). A new article will review the application of molecular biology techniques to photosynthesis research (62). The papers of Meyer et al (67) and Kerfeld et al (52) are rich in comparative data on bacterial and photosynthetic cytochromes.

A Speculative Overview

While most efforts to understand the process of photosynthesis have centered on higher plants, there is reason to examine this process in simpler forms of life. The origin, evolution, and influence of life on this planet are interesting. Current speculation posits a period of abiogenesis of organic compounds, the assembly of which led to the formation of living cells. The generous abundance of these compounds would sustain heterotrophic life. Geochemical data indicate an absence of oxygen in the earliest times, so these first heterotrophs are supposed to have been anaerobes that would use fermentative electron transfer to supply their energy needs. There is fossil evidence of filamentous cyanobacteria of 3.4×10^9 years of age. The Earth's age is 4.4–4.8×10^9 years. The cyanobacteria and the early eukaryotic algae produced oxygen and, over several billion years, brought the atmosphere to its present condition in which aerobic life predominates. The anaerobic skills of these planktonic photosynthesizers have persisted to good effect in allowing survival of these small organisms through prolonged winter burial in sediments in the temperate zone. Hydrogen-producing fermentation is characteristic of the planktonic photosynthesizers. Perhaps the low-potential cytochrome c (LP cyt c) of cyanobacteria and algae is involved in this or some other low-potential electron transfer reaction associated with anaerobic survival. Planktonic lifestyle may have another consequence, manifested in the persistence of cytochrome c_6 (cyt c_6) in cyanobacteria and algae. Cyt c_6 is a high-potential cytochrome similar to and related to mitochondrial respiratory cytochrome c. Cyt c_6 catalyzes the transfer of electrons from cytochrome f (cyt f) to P700 and is functionally interchangeable with the blue copper protein plastocyanin (PC). In many species, cyt c_6 replaces PC in response to copper deficiency. That planktonic organisms frequently experience copper deficiency is evidenced by the high concentrations of cyt c_6 in cyanobacteria collected from natural blooms. Cyt c_6 may also have a role in the respiration of cyanobacteria. Cyt c_6 has not been detected in higher plants. Perhaps the colonizing of a more stable terrestrial environment allowed efficient uptake of insoluble copper and obviated the need for cyt c_6. Cyt M is a c type cytochrome recognized recently from its gene sequence found in cyanobacteria. It is likely to be the cytochrome c_{552} of *Anacystis nidulans* described by Holton & Myers (38, 39) and noted as present in eukaryotic algae (45, 78). The function of cytochrome M is unknown.

In addition to the unusual LP cyt c and cyt c_6, cyanobacteria and algae contain cyt f, cyt b_6, and cyt b_{559}—the universal participants in all oxygenic photosynthesis. In fact, most of the molecules participating as catalysts or structural elements in oxygenic photosynthesis are very similar in cyanobacteria and higher plants. This review begins with an examination of the cytochromes c unique to cyanobacteria and algae and then summarizes recent work on cyt f.

Cytochrome c_6

This cytochrome is similar in size, pI, and Em, to PC. Cyt c_6 alternates with PC as an electron carrier between cyt f and P700 in the photosynthetic electron transport chain. Cyt c_6 has a single heme attached to a polypeptide of 83 to 90 amino acids. The iron atom in the heme has a histidine and methionine as its fifth and sixth ligands (Figure 1). The Em of cyt c_6 varies from 335 to 390 mV. The isoelectric point is 9.3 in *Aphanizomenon flos-aquae* and *Anabaena variabilis*, which are complex filamentous cyanobacteria in group IV of the Stanier classification (83). In contrast, the pI varies from 3.8 to 5.2 in the unicellular and some of the simple filamentous cyanobacteria in groups I through III and in the eukaryotic algae. Within a given species that synthesizes both cyt c_6 and PC, the isoelectric points of the two proteins are similar. The radical alterations of the pIs of cyt c_6 and PC in cyanobacteria have been accompanied by alterations in charged residues in the docking region for these proteins on cyt f. Perhaps there are similar alterations in the cyt c_6 and PC docking regions of PSI. The parallel evolution of these proteins is intriguing.

Figure 2 shows 12 cyanobacterial cyt c_6 amino acid sequences: 2 from red algae, 3 from brown algae, 4 from green algae, and 2 from *Euglena* (10, 53). There are crystal structures for cyt c_6 from two cyanobacteria—*Anacystis nidulans* (61) and *Arthrospira maxima* (formerly *Spirulina*), and from two green algae—*Chlamydomonas reinhardtii* (53) and *Monoraphidium braunii* (19).

The first crystal structure of a cyt c_6, the 3.0-Å structure of *A. nidulans* cyt c_6, established its similarity in fold to the other small (S-type) respiratory cytochromes such as *Pseudomonas* c_{551} (61). The 1.9-Å structure of *C. reinhardtii* cytochrome c_6 (53) and the 1.2-Å structure of *M. braunii* cyt c_6 (19) verified that the overall fold of cyt c_6, which encloses about 95% of the two-stranded antiparallel beta sheet in the vicinity of the methionine axial ligand to the heme surface, is predominantly alpha helical. In addition, there is a short, two-stranded beta sheet that connects to a segment of poly-proline helix in the vicinity of methionine axial ligand (Figure 2). The primary structure of this region of the molecule (residues 53–68, *C. reinhardtii*, beginning at the arrow in Figure 2), is the most highly conserved segment of primary structure in cytochrome c_6. Electron transfer is likely through exposed atoms of the heme: the CBC (the methyl

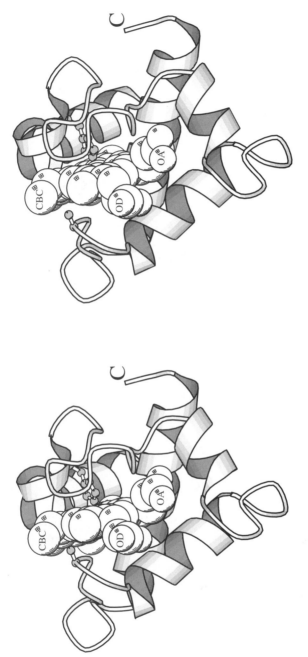

Figure 1   Stereo ribbon diagram of *Arthrospira maxima* cytochrome c_6. The histidine and methionine axial ligands are in ball-and-stick representation. The heme is shown in space-filling representation with the three most solvent-exposed atoms shown in white and labeled. This figure and Figures 3 and 5 were prepared with MOLSCRIPT (54).

```
euggr  .....GGADVFADNCSTCHVNGNVISAGKVLSKTAIEEYLD....GGYTKEAIEYQVRNGKGPMPAWEGVLSEDEIVAVTDYVTQAGGAWANVS.
eugvi  .....SGAEVFGNNCSSCHVNGNIIIPGHVLSQSAMEEYLD....GGYTKEAIEYQVRNGKGPMPAWEGVLDESEIKEVTDYVYSQASGPWANAS.
chlre  .ADLALGAQVFNGNCAACHMGGRNSVMPEKTLDKAALEQYLD..GGFKVE..SIIYQVENGKGAMPAWADRLSEEEIQAVAEYVFKQATDAAWKY.
monbr  EADLALGKAVFDGNCAACHAGGGNNVIPDHTLQKAAIEQFLD..GGFNIE..AIVYQIENGKGAMPAWDGRLDEDEIAGVAAYVYDQAAGNKW...
bryma  GGDLEIGADVFTGNCAACHAGGANSVEPLKTLNKEDVTKYLD..GGLSIE..AITSQVRNGKGAMPAWSDRLDEEIDGVAYVFKNINEGW....
plebo  .ADAAAGGKVFNANCAACHASGGGQINGAKTLKKNALTAN...GKDTVE..AIVAQVTNGNGAMPAFKGRLSDDQIQSVALYVLDKAEKGW...
syny3  .ADLAHGKAIFAGNCAACHNGGLNAINPSKTLKMADLEAN...GKNSVA..AIVAQITNGNGAMPGFKGRISDSDMEDVAAYVLDQAEKGW..
synli  .ADIANGAKVFSGNCAACHMGGGNVVMANKTLKKEALEQF...GMNSED..AIIYQVQHGKNAMPAFAGRLTDEQIQDVAAYVLDQAAKGWAG.
synsp  .ADIADGAKVFSANCAACHMGGGNVVMANKTLKKEALEQF...GMNSAD..AIMYQVQNGKNAMPAFGGRLSEAQIENVAAYVLDQSSNKWAG.
alaes  .IDINNGENIFTANCSACHAGGNNVIMPEKTLKKDALADN...KMVSVN..AITYQVTNGKNAMPAFGSRLAETDIEDVANFVLTQSDKGWD..
petfa  .VDINNGESVFTANCSACHAGGNNVIMPEKTLKKDALEEN...EMNNIK..STTYQVTNGKNAMPAFGGRLSETDIEDVANFVISQSQKGW..
porpu  .ADLDNGEKVFSANCAACHAGGNNAIMPDKTLKKDVLEAN...SMNGID..AITYQVTNGKNAMPAFGGRLVDEDIEDAANYVLSQSEKGW..
porte  .ADLDNGEKVFSANCAACHAGGNNAIMPDKTLKKDVLEAN...SMNTID..AITYQVQNGKNAMPAFGGRLVDEDIEDAANYVLSQSEKGW..
bumfi  .ADIENGERIFTANCAACHAGGNNVIMPDKTLKKDALEAN...GMNAVS..AITTQVTNGKGAMPAFGGRLSDSDIEDVANYVLSQSEQGWD.
anani  .ADLAHGGQVFSANCASCHLGGRNVVNPAKTLEKADLDEY...GMASIE..AITTQVTNGKGAMPAFGAKLSADDIEGVASYALDQSGKEW..
synp7  .ADLAHGGQVFSANCAACHLGGRNVVNPAKTLQKADLDQY...GMASIE..AITTQVTNGKGAMPAFGSKLSADDIADVASYVLDQSEKGWQG.
micae  ...DGASIFSANCASCHMGGKNVVNAAKTLKKEDLVKY...GKDSVE..AIVTQVTKGMGAMPAFKGRLKPEQIEDVAAYVLGKADADWK.
anasp  .ADSVNGAKIFSANCASCHAGGKNLVQAQKTLKKADLEKY...GMYSAE..AIIAQVTNGKNAMPAFKGRLKPEQIEDVAAYVLGQADSWK.
anasq  .ADVANGAKIFSANCASCHAGGKNLVQAQKTLKKEDLEKF...GMYSAE..AIIAQVTNGKNAMPAFKGRLKPDQIEDVAAYVLGQADKSWK.
anava  .ADSVNGAKIFSANCASCHAGGKNLGVAQKTLKKADLEKY...GAYSAM..AIGAQVTNGKNAMPAFKGRLKPEEIZBVAAYVLGKAEAEWK.
aphfl  .ADTVSGAALFKANCAQCHVGGGNLVNRAKTLKKEALEKY...NMYSAK..AIIAQVTHGKGAMPAFGIRLKAEQIENVAAYVLEQADNGWKK.
spima  .GDVAAGASVFSANCAACHMGGRNVIVANKTLSKSDLAKYLKGFDDDAVA..AVAYQVTNGKNAMPGFNGRLSPKQIENVAAYVVEQADNGWKK.
monlu  .GDIANGEQVFTGNCAACHS.......VZZZKTLELSSLWKAKSYLANFNGDESAIVYQVTNGKNAMPAFGGRLEDDEIANVASYVLSKAG......
```

Key:- ◁▷ Helix   ⇧ Beta strand   ── Random coil
Accessibility shading: Black=buried, White=accessible

carbon adjacent to the heme carbon that is covalently linked to the second Cys in the CXXCH motif) and the propionate oxygen atoms (Figures 1 and 3).

C. reinhardtii cyt c_6 crystallized in two unique crystal forms. In both, an oligomerization about the heme crevice is observed resulting in a dimer (Form 2) or a trimer (Form 1). Similar oligomerization is observed in the crystals of *A. nidulans* cyt c_6 (dimer) and in *M. braunii* cyt c_6 (trimer). Likewise, two isoforms of cytochrome c_6 from *A. maxima* have been isolated and crystallized. The structure of one form has been solved, and again a trimeric organization about the heme crevice is apparent in the crystal. The second crystal form of cytochrome c_6 exhibits unusually high symmetry (the space group is $i4_132$ with approximately 12 molecules in the asymmetric unit, possibly in a complex related by tetrahedral symmetry). Solution studies support the hypothesis that cyt c_6 may form oligomers that are functionally relevant and that the oligomeric state may be related to the level of posttranslational modification (see below). Oligomerization of cyt c_6 may be advantageous for efficient transfer of electrons from the dimeric b_6f complex to the trimeric PSI reaction center.

The 1.2-Å crystal structure of cyt c_6 of *M. braunii* is the highest-resolution structure of a heme protein presently determined (19). Comparison of the surface electrostatic potential of *M. braunii* cyt c_6 and PC indicates that there are similar regions of negative potential on the two molecules (19). Docking studies of *M. braunii* cyt c_6 to cyt *f* suggest that the conformation of cytochrome *f* observed in the turnip crystal structure is not poised for efficient electron transfer. As in other interprotein electron transfers, rearrangements involving

←―――

Figure 2   Primary structures of the cytochromes c_6 as reported in the SWISS Protein Data Base. The beginning of the highly conserved region that forms a short, double-stranded beta-sheet is *underlined*. The methionine axial ligand is marked with a *plus* and the CXXC heme binding site is denoted by a *curved line*. The secondary structure assignment shown is based on the *Arthrospira maxima* cyt c_6 structure. The secondary structure panel in this figure and in Figure 4 was generated by PROCHECK (57).

KEY (each species name is followed by its data base accession number): cyc$_6$_euggr, *Euglena gracilis* (P00119); cyc$_6$_eugvi, *Euglena viridis* (P22343); cyc$_6$_chlre, *Chlamydomonas reinhardtii* (P08197); cyc$_6$_monbr, *Monophoridium braunii* (Q09099); cyc$_6$_bryma, *Bryopsis maxima* (P11448); cyc$_6$_plebo, *Plectonema borynaum* (P00117); cyc$_6$_syny3, *Synechocystis* sp. 6803 (P46445); cyc$_6$_synli, *Synechococcus lividus* (P00114); cyc$_6$_synsp, *Synechococcus* sp. (P00115); cyc$_6$_alaes, *Alaria esculenta* (P00109); cyc$_6$_petfa, *Petalonia fascia* (P00108); cyc$_6$_porpu, *Porphyrea purpurea* (P51200); cyc$_6$_porte, *Porphyra tenera* (P00111); cyc$_6$_bumi, *Bumilleropsis filiformis* (P00110); cyc$_6$_anani, *Anacystis nidulans* (P07497); cyc$_6$_synp7, *Synechococcus* sp PCC 7942 (P25935); cyc$_6$_micae, *Microcystis aeruginosa* (P00112); cyc$_6$_anasp, *Anabena* sp. 7120 (P28596); cyc$_6$_anasq, *Anabena* sp. 7937 (P28597); cyc$_6$_anava, *Anabena variabilis* (P00113); cyc$_6$_aphfl, *Aphanizomenon flos-aquae* (P00116); cyc$_6$_spima, *Arthrospira (Spirulina) maxima* (P00118); cyc$_6$_monlu, *Monochrysis lutheri* (P00107).

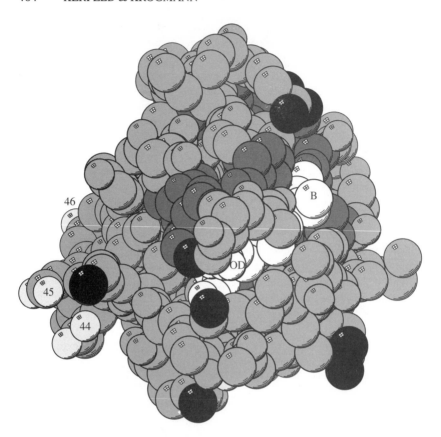

Figure 3 Space-filling model of *Arthrospira maxima* cytochrome c_6. The heme is in white. The propionate oxygen of the D ring is labeled *OD*, and the methyl group adjacent to heme carbon linked to the second Cys in CXXCH is labeled *B*. Side-chain oxygen atoms of Asp and Glu residues are in light gray; side-chain nitrogen atoms of Lys and Arg are black, and invariant residues of the cyt c_6 primary structure are dark gray.

factors other than electrostatics may drive the complex into a configuration optimal for electron transfer. The structure of cyt c_6 of *M. braunii* in solution has been thoroughly characterized with NMR, EPR, and Mossbauer spectroscopy in a fine paper by Campos et al (6). Here evidence is presented for multiple forms of the methionine-histidine heme iron ligand field.

Purified cyt c_6 is required not only for the structural studies mentioned above, but also for analyses of function by reaction reconstruction and kinetic analysis to be reviewed below. Thus far, it has proved difficult to produce cyt c_6 by overexpression of the gene in a heterotrophic bacterium. Diaz et al expressed the gene for cyt c_6 from *Synechocystis* PCC6803 in *Escherichia coli,* but the amount of cytochrome produced was small (15). The preparation of large amounts of cyt c_6 from cyanobacteria has been described by Gómez Lojero & Krogmann (22), and Ho (33) has reviewed the purification methodology. Zhang et al (103) and Ho & Tan (35) have added excellent final steps to this method, which lead to a very pure product.

Isoforms of a protein are often detected during protein purification. The isoforms may suggest multiple genes that express subtly different proteins for different purposes. Thus far the photosynthetic cytochromes appear to originate from single-copy genes. Detection of isoforms of cyt c_6 raises the possibility of posttranslational modifications that may have functional relevance. Cyt c_6 from a variety of cyanobacteria fractionates on ion exchange chromatography in several distinctly eluting fractions or bands (35). These multiple bands were not found in cyt c_6 isolated from the red alga *Porphyridium cruentum* (17). Recently, the techniques of protein mass spectrometry were used by Krogmann to further investigate this phenomenon with cyt c_6 of *A. maxima.* Each chromatographic band contained a family of five or six proteins, the smallest and most abundant corresponding to the calculated mass of the cytochrome. The additional mass isoforms were 14 to 16 Daltons, or multiples of these increments, heavier. A few of the plus–16 Dalton peaks could be eliminated if the protein was treated with sodium hydrosulfite. The loss of 16 Daltons on reduction suggests the presence of methionine sulfoxide residues in the sample. The occurrence of methionine sulfoxide in cyt c_6 from *C. reinhardtii* was suggested by the results of the structural studies of Kerfeld et al (52). The two crystal forms of the *C. reinhardtii* cyt c_6 were distinguished as a dimer and a trimer. Comparison of the molecular masses of the proteins in the two crystal forms revealed an interesting difference. The dimer form showed modest amounts of cyt c_6 with excess masses of 16 and 32 Daltons. The trimer form showed larger amounts of these heavy species. The mass changes suggest oxygen atom addition, and the likely site is a methionine that can be oxidized to the sulfoxide and the sulfone. Met26 showed excess electron density around the sulfur atom that was more prominent in the trimeric protein. Met26 is not the iron ligand. It is solvent exposed in the monomer but is part of the dimerization interface. Methionine is susceptible to oxidation by singlet oxygen or by partially reduced oxygen radicals in superoxide anion, peroxides, or hydroxyl radicals that may be generated in photosynthesis. The two crystal forms of *A. maxima* cytochrome c_6 noted above also differ in the level of posttranslational modification of the protein preparations from which they were grown. *Synechococcus elongatus* cyt c_6 also has been purified as a mixture of mass forms (M Sutter, personal communication). Other nonreducible mass gaps of 16 (or 32) Daltons may be due to further oxidation of the methionine sulfoxide to methionine sulfone. Smaller mass increases of 14 Daltons may be due to methylation. A methyl lysine in the cyt c_6 of the alga *Monochrysis lutheri* is clearly in evidence (59). One suspects that the different chromatographic species are the result of formation of large oligomers with different surface charges. Not all separable isoforms of cyanobacterial cyt c_6 are due to oligomerization. Ho & Krogmann (34) reported isoforms from *Oscillatoria princeps* separated on a reverse-phase HPLC column in the presence of trifluoroacetic acid, which would dissociate all oligomers.

There is an improved assay for light-driven electron transport through PSI particles from cyanobacteria (21, 41) using cyt c_6 as the electron donor. High rates of electron throughput are achieved and are rate limited by the interaction of cyt c_6 with its receptor site. Thus the assay is most suitable for measuring differences in catalytic activity of the cytochrome from different sources. *A. maxima* cyt *c* activity in this assay was doubled by a treatment of the cytochrome with low-potential reducing agents. This suggests that a rather small percentage of the cyt c_6 molecules have a reducible methionine sulfoxide in the vicinity of the heme electron transfer site that inhibits electron transfer.

Prince et al describe the isolation and characterization of a single form of cyt c_6 from *Porphyra umbilicalis* and two isoforms of c_6 from *Chondrus crispus* (79). These forms are present in an approximate ratio of four to one, and the major form has an Em of +330 mV, while that of the minor form is +285 mV. The green alga *Ulva pertusa* yielded two forms of cyt c_6 with different absorption spectra, but one of these may be LP cyt *c* (91).

Cytochrome c_6 and Plastocyanin

In many cyanobacterial and algal species, cyt c_6 is replaced by the copper protein PC when copper is available. The two proteins are functionally interchangeable in cell-free assays of photosynthetic electron transport. Cyt c_6 has not been detected in higher plants; instead, PC seems solely responsible for the transport of electrons from cyt *f* to P700. As yet, no clear, functional advantage of PC over cyt c_6 has emerged from studies of these molecules. Readers may find details of PC structure and function in three reviews (25, 47, 81).

Wood (96) first realized that PC and cyt c_6 were interchangeable proteins in *C. reinhardtii*. Laudenbach et al (58) isolated and sequenced the gene (*petJ*) from *Synechococcus* PCC7942 and then deleted it. Contrary to expectation, the deletion mutant showed no diminution in photosynthesis, since this organism was thought to lack the gene for PC. Later the PC gene and its product were found in *Synechococcus* PCC 7942. Zhang et al (101) readdressed this problem of elimination of both PC and cyt c_6 with *Synechocystis* PCC 6803. They showed that copper deficiency allows cyt c_6 synthesis, and copper sufficiency causes exclusive synthesis of PC. Ghassemian et al (20) characterized the genes for PC and cyt c_6 in *Anabaena sp.* PCC7120 and demonstrated a similar regulation of their expression by copper. Next, Zhang et al (101) isolated the *petJ* gene for cyt c_6 and constructed a deletion mutant for it. The suppression of synthesis of PC by copper depletion and cyt c_6 by gene deletion still had no effect on photosynthesis. This is a wonderful dilemma. The movement of electrons between cyt f and P700 is so important that cyanobacteria and algae maintain and regulate two separate genes to supply catalysts for this step. When both of the genes are inactivated, the process that these catalysts serve goes on unabated. A third alternative electron carrier for the transfer of electrons from cytochrome f to P700 is a possible solution to this conundrum.

The widely held assumption that PC is not synthesized by *Synechococcus* PCC7942 has been disproved. An unusually large divergence in amino acid sequence of the acidic PC confounded conclusions based on antibody cross reactivity. Early warning might have been taken from the publications of Stewart & Kaethner (90) and of Tan & Ho (93), which documented the lability and the special difficulties in the isolation of the acidic PCs from unicellular and simple, filamentous cyanobacteria. Clarke & Campbell (8) found the gene (*petE*) for PC in *Synechococcus* PCC7942. Unlike *Synechocystis* PCC6803, other cyanobacteria, and eukaryotic algae, expression of this gene in *Synechococcus* PCC7942, as measured by transcript levels, is not influenced by copper availability. Inactivation of the *petE* gene does raise the level of mRNA transcripts for cyt c_6, indicating that some control is exercised to prevent excess production of both PC and cyt c_6. Inactivation of *petE* likewise alters the photosynthetic capacity of the cells, reducing the maximum rate of O_2 evolution by 20%. The *petE* mutant showed a dramatic increase in susceptibility to inactivation of O_2 evolution when the temperature was lowered from 37°C to 25°C. This cold shift in wild-type cells decreases electron transport, which slows the removal of electrons from PSII, and this effect is intensified in the absence of PC. Clarke & Campbell (8) argue that PC is the preferred electron donor to PSI in *Synechococcus* PCC7942 since its absence gives a slightly lower growth rate and a lower net O_2 evolution. There is a relatively high level of the PC message expression and greater message stability. PC provides better protection against photoinhibition at suboptimal growth temperatures.

Cyt c_6 Interaction with P700

In higher plants and eukaryotic algae, electrons are thought to be delivered to Tyr83 or His87 of PC from cyt f. This Tyr sits in an "acid patch" on the east face of the molecule (81). Cyanobacterial PCs have a Tyr in the analogous position but have either no acid patch (8) or a remnant of one (94). The reduced PC would then donate the electron to P700 via His87, which is surrounded by a hydrophobic patch in the PC molecule from higher plants and algae (81). The hydrophobic patch is said to be different (81) or absent (8) in cyanobacterial

PC. In contrast, cyt c_6 is thought to both receive and donate electrons via a common pathway through the exposed heme edge at the surface of the molecule (Figure 3). The point is that PC has the same catalytic role as cyt c_6, but their interaction with their membrane-bound redox partners is different.

The site of interaction of cyt c_6 with PSI may soon be revealed when a crystal structure of PSI is completed. A first approximation was attempted by chemical cross linking. Wynn & Malkin (97) found that spinach PC was cross linked to spinach PSI particles through a polypeptide identified as the product of the *psaF* gene. They repeated this experiment with PSI particles and cyt c_6 from *Synechococcus* PCC6301 and confirmed cross linking to the *psaF* gene product (98). The clarity of the observations was elegant. The conclusion that cyt c_6 must bind to the product of the *psaF* gene was temporarily confounded by several observations. Hatanaka et al (27) prepared PSI particles from *Synechococcus elongatus* and, by detergent treatments, selectively removed some of the polypeptides associated with these particles. Measurement of the reduction kinetics of flash-oxidized P700 revealed that cyt c_6 rapidly reduced P700 in particles that retained the *PsaC* and *PsaD* polypeptides, but removal of the *PsaF* polypeptide had no influence on the reaction. Xu et al (99) reached a similar conclusion by gene deletion experiments with *Synechocystis* PCC6803. Cells of the *psaF* deletion mutant and wild type had similar rates of photosynthetic electron transfer and P700 reduction. Similar results were obtained in cell-free reduction of P700 in PSI particles by cyt c_6. The cross-linking experiment establishes proximity of *psaF* to the site of electron transfer, but the gene deletion shows that the presence of the gene product is not essential for electron transfer to PSI in cyanobacteria.

Measurements of Bottin et al on oxidation of PC by PSI using preparations from a higher plant are of interest (4). Flash kinetics experiments with chloroplast PSI indicated two distinct phases of PC oxidation—one rapidly oxidized form "near" P700 and another "distant" slowly oxidized form that replaced or reduced the first oxidized PC. A long series of experiments by Hervas, DeLaRosa, and others at Sevilla; Bottin at Gif-sur-Yvette; and Tollin at Tucson have documented the evolution of the dual sites for oxidation of the donors to P700 in higher plants and algae from a single site of slower oxidation in cyanobacteria (14, 28, 31, 65). The overall conclusions from this work are summarized in a paper describing the evidence for the evolution of the reaction mechanism for P700 reduction (29). The results obtained with PC, cyt c_6, and PSI reaction centers from four organisms at distinct stages of evolution are compared. These organisms are *Anabaena* PCC 7119, *Synechocystis* PCC 6803, *M. braunii*, and spinach. Three different kinetic models are proposed. The first and simplest is an electrostatically oriented collision mechanism observed with *Anabaena* PSI particles oxidizing *Anabaena* or spinach PC. Mechanism I also occurs with spinach PSI particles oxidizing the acidic pI donors *Synechocystis* PC and cyt c_6. At the next level of complexity, there is a minimal two-step reaction mechanism involving complex formation between P700 and its reductant followed by intracomplex electron transfer (Mechanism II). This is seen when spinach PSI particles react with *Anabaena* or spinach PC. The most elaborate mechanism is complex formation followed by rearrangement of the reaction partners within the complex before electron transfer takes place (Mechanism III). This is seen in the reaction of *Anabaena* PSI particles with *Anabaena* cyt c_6—which is suggested as the first stage of evolution. Mechanism III is also seen in the reactions of the alga *M. braunii* PSI particles with *M. braunii* PC or cyt c_6 or spinach PC. The fast phase kinetics of reduction are hardly visible with PC from either source but are more evident with *M. braunii* cyt c_6. This suggests that the algae evolved the fast phase mechanism first using cyt c_6 and the higher plants adapted it to PC. Mechanism III with fast phase kinetics is found in spinach PSI particles reacting with spinach PC or either the *M. braunii* PC or cyt c_6. A recent paper by Hippler et al (32) shows efficient electron transfer from both PC and cyt c_6 to PSI in *C. reinhardtii*. The PSI complex of *C. reinhardtii* is unique among photosynthetic organisms in that both PC and cyt c_6 reduce P700 with first-order kinetics and a half-time of three microseconds—the fast phase mechanism of algae and higher plants.

Reduction of Cyt c_6

Recently, Navarro et al have used flavin-photosensitized oxidation and reduction of PC and cyt c_6 from two different cyanobacteria to study the path of electrons in and out of these molecules (72). *Anabaena* PCC7119 and *Synechocystis* PCC6803 cyts c_6 show steric and electrostatically influenced responses to oxidation and reduction by flavins that indicate that the same sites are being used for electron insertion and removal. *Anabaena* PC may use different sites—reduction at the location equivalent to the acid patch on the east face and oxidation at the location of the hydrophobic north pole. Hervas et al have added a thermodynamic analysis of reactions between PSI and both PC and cyt c_6 from various species (30). Long-range electrostatic interactions appear to be attractive in *Anabaena* PCC7119 but repulsive in *Synechocystis* PCC6803 and spinach. These electrostatic interactions are needed in vitro, but they may not be rate limiting in vivo according to Soriano et al (89). Short distance forces seem to have increased in importance as the interaction evolved through algae and higher plants.

Low-Potential Cytochrome c

Like cyt c_6, LP cyt c is found in cyanobacteria and algae but not in higher plants. Amino acid and gene sequences have been done for the molecules from several cyanobacteria (9, 46, 73, 88). Navarro et al (71) have published an updated purification and characterization of LP cyt c that confirms the unusual bis-histidine coordination at the fifth and sixth ligand positions. LP cyt c has been identified in red algae (16, 17, 82), in the green alga *Bryopsis maxima* (45), and in the diatom *Navicula pelliculosa* (100). This cytochrome has a molecular weight of 15.6 to 15.8 kDa, a pI of 3.9 to 4.5, and an Em of -250 to -300 mV. Regions of LP cyt c are related by sequence similarity to regions of cyt c_6 (46). Crystals of *A. maxima* LP cyt c diffract X rays to 2.3 Å, and the structure determination is under way in Kerfeld's laboratory.

LP cyt c is essential to the structure of Photosystem II. Shen, Inoue, and their colleagues have published a series of excellent papers documenting this. Their first step was to isolate a PSII core complex from *Synechococcus vulcanis* using gentle detergents that allowed the retention of PSII extrinsic proteins (85). Three extrinsic proteins were released from the particles by washing with solutions containing high concentrations of salts. The residual core complex contained a set of proteins easily identified with those found in higher plant PSII core complexes. The released proteins included a 33-kDa protein found in higher plants but also two unusual proteins of 17 and 12 kDa that replaced the 23- and 17-kDa extrinsic proteins of higher plants. There is no sequence similarity between the two smaller extrinsic proteins of cyanobacteria and the two from higher plants. The 17-kDa protein is LP cyt c. On studying the rebinding of these proteins to the stripped PSII core, LP cyt c bound appreciably in the absence of the 33- and 12-kDa proteins but their presence facilitated full rebinding (86). LP cyt c alone did not restore O_2 evolving activity, but if the 33-kDa protein was present, partial restoration occurred, and with both the 33- and 12-kDa proteins, nearly complete restoration occurred. Measurement of the light intensity dependence of oxygen evolution and the thermoluminescence of the particles indicated that LP cyt c regulates the efficiency of some S state transitions of O_2 production and that the 12-kDa protein modulates a step in the dark reactions. Shen & Inoue next demonstrated that, in their preparations of *S. vulcanis*, all of the LP cyt c was tightly bound to PSII and sonication would not dislodge it unless 1 M $CaCl_2$ was present (87). Han et al (26) studied the chemical cross linking of these extrinsic proteins. LP cyt c formed cross links to both the 12-kDa and the 33-kDa manganese stabilizing protein and to both the 12-kDa and the D2 protein that holds elements of the reaction center. The cytochrome is certainly close to crucial electron transfer catalysts in PSII, but there is no evidence that it undergoes reduction and oxidation in this location. Shen et al (88) switched to *Synechocystis* PCC6803 to isolate the gene (*psbV*) and used it to construct insertion and deletion mutant lines of cells. The mutant cells grew photoautotrophically but at a reduced rate and contained only half the amount of PSII found in wild type. LP cyt c is thus thought to contribute to optional functional stability of PSII as a lumenal-side extrinsic protein enhancing oxygen production. Next, a mutant lacking both the *psbV* and *psbO* genes for LP cyt *c* and the 33-kDa protein was constructed (84). This double deletion mutant could not grow photoautotrophically and showed less than 10% of wild-type oxygen-evolving activity. Thermoluminescence measurements showed the mutant incapable of S-state transitions. Enami et al (16) have repeated the critical biochemical experiments with a PSII complex from the red alga *Cyanidium calderium* and have found it to possess LP cyt *c* and the 12-kDa protein like the cyanobacteria.

Nishiyama et al (73) found that extraction of PSII particles from the thermophilic *Synechococcus* sp. PCC7002 with the detergent Triton X-100 diminished the stability of oxygen evolution to heating. This destabilization could be reversed by the addition of the Triton extracted material. The specific restoring agent was identified as LP cyt *c*.

These data for a structural role of LP cyt *c* in PSII are convincing and of impeccable technical quality. Cytochromes usually function in electron transfer reactions but, as yet, there is no evidence for redox turnover of the LP cyt *c* in vivo.

Kinzel & Peschek (53) demonstrated with cell-free reactions that LP cyt *c* could serve as redox cofactor for cyclic photophosphorylation. Kang et al (46) noted a reduction of LP cyt *c* with ferredoxin II in a reaction with unusual specificity toward this electron donor. Morand et al (69) suggested a possible role of this cytochrome in hydrogen metabolism. Shen et al (84, 85) have clearly established that the cytochrome is a constitutively expressed component of the PSII complex that can be released by high-salt treatment. Similar results were obtained by Hoganson et al (36). Nishiyama et al (73), Bowes et al (5), and Krinner et al (55) used strong detergents to solubilize LP cyt *c*. However, others (22, 38, 53, 71) have obtained this cytochrome without the use of high salt or detergent. These latter observations suggest that the LP cyt *c* may be expressed and may function in a soluble form. Perhaps this cytochrome has an electron transfer function in addition to its structural role in the PSII reaction center. There is clearly only one copy of the gene for this cytochrome, and there is, as yet, no evidence for posttranslational modification beyond removal of the leader sequence. It seems plausible that this protein with its heme prosthetic group would be retained over such a long history because it provides an advantage to the cell through a redox function.

Cytochrome M, Cytochrome c_{552}

Malakhov et al (63) described a gene found in *Synechocystis* PCC6803 that encoded the sequence-CXXCH-, which is the heme-binding site signature of *c* type cytochromes. The gene product is called cytochrome *M*. A transit peptide or leader sequence precedes a putative protein of 76 amino acids in length.

In addition to its small size, the putative mature protein has equal numbers of positively and negatively charged residues. The predicted pI is 7.3, and the protein is hydrophilic. The gene was inactivated by cartridge mutagenesis, and the mutant cells showed a photoautotrophic growth rate and rates of oxygen evolution and respiration that were identical to those of wild-type cells. Ho has purified a cytochrome from *Synechocystis* PCC6803 that has mass and purification characteristics one would expect for cytochrome M (KK Ho, personal communication). The behavior of this cytochrome through purification suggests its identity to cytochrome c_{552}, described by Holton & Myers (39). It is possible that the high-potential cytochrome c_{549} of the green alga *Bryopsis maxima* described by Kamimura et al (45) and, perhaps, the cytochrome c_{549} of *Scenedesmus obliquus* D_3 described by Powls et al (78) are also cyts M. This cytochrome is reduced by ascorbate, indicating that its Em is close to that of cyt c_6. It might thus qualify as the third alternative to PC or cyt c_6. Thus far, its abundance in cyanobacteria appears to be very low. The concentration of cyt M is closer to that of cytochrome oxidase than to the concentrations of photosynthetic catalysts.

The Kazusa Institute sequence of the *Synechocystis* 6803 genome may yield more cytochromes. Whitmarsh et al (personal communication), using the CXXCH signature sequence for heme-binding sites, have found a gene for a large polypeptide with three closely grouped sites of this type.

Cytochrome f

Cyt f is an intrinsic membrane protein and has been difficult to purify. The intact protein forms octamers that may interfere with measurements of its properties. The amino acid sequences of cyt f from 18 sources—plants, algae, and cyanobacteria—are available, and these proteins are 284 to 289 amino acids in length (Figure 4). The single heme is bound close to the N terminus at Cys 21 and Cys 24. The sixth ligand to the iron in the heme is unusual in that it is a tyrosine at or perhaps near the N-terminus. The protein is anchored to the photosynthetic membrane by a stretch of hydrophobic amino acids from residue 261 to 270. The Em of these proteins varies from 345 to 395 mV, which is close to the Em of PC and cyt c_6. Metzger et al (66) have recently revised the value of the extinction coefficient of cyt f. Their measurements indicate a 20 to 30% increase in the value of the extinction coefficient of cyt f and perhaps of cyt b_6 is warranted. This would impose a slight downward revision of many estimates of cyt f concentration and of rates of its oxidation and reduction.

The amino acid sequences of higher plant cytochromes f are remarkably similar, showing 80% identity (Figure 4). Of the 19 sequences available from all sources at the time of this writing, 34% of the amino acid residues are invariant; many of these invariant residues are visible in the view shown in

Figure 5A. The pI varies from 4.4 in cyanobacteria to above 7 in some higher plants to 8.6 in *Porphyra purpura* and 8.8 in *Cyanophora paradoxa*.

Research on cyt f has been greatly facilitated by the observation that, during purification of this protein from cruciferous plants, organic solvent initiates protease cleavage at residue 250, releasing the cytochrome from the membrane, and results in a protein that remains in the monomeric form (23, 24). It was this monomeric protein without the hydrophobic membrane anchor that Martinez et al (64) purified from turnip, crystallized, and used to solve the structure. In the 2.3-Å model of turnip cyt f, the heme atoms with the most surface exposure are the CBC and the proprionate oxygen atoms, similar to the exposed heme edge in cyt c_6. In this respect, both cyt c_6 and cyt f differ from the bacterial cytochromes c_2 and the respiratory cytochromes c in which the heme propionate oxygen atoms are buried within the protein interior. In cyt f, the exposed edge of the heme is flanked by two bands of invariant residues (Figure 5A). The amino acid sidechains within 4 Å of the exposed CBC and propionate oxygen atoms are likewise highly conserved in cyt f amino acid sequences. Glutamine 59 and Asn 70 border the propionate oxygen atoms of the A ring (pyrrole ring IV of the heme). Tyr 160 and Arg 156 border the oxygen atoms of the D ring (pyrrole ring III of the heme). Cys 24, Pro 161, and an aromatic amino acid at position 4 border the exposed CBC atom (Figure 5A). The structure of the turnip cyt f refined to 1.9 Å resolution has recently been described. Refinement at higher resolution revealed a chain of five water molecules that are buried in the cytochrome and that form hydrogen bonds to residues that are conserved in all 18 of the presently available f sequences. This chain stretches from the iron ligand His 25 to the vicinity of Lys 66 (which lies just behind Lys 65 in Figure 5A), which has been suggested as part of the docking site for PC. The authors propose that this chain might serve as a redox linked proton wire. The $b_6 f$ complex can transport one to two protons into the thylakoid lumen for each electron transported from plastoquinol to PC. Might this wire of water molecules transport the protons from the site of proton generation at the heme to the lumen? Recently, Berry (E Berry, personal communication) has determined the crystal structure of cyt f from *C. reinhardtii*. In this case, the gene for the cytochrome was shortened by removing the membrane anchor sequence-encoding region and expressed in *C. reinhardtii* (56). This cyt f model is very similar to turnip cytochrome f. Ho has found that cyt f prepared from the cyanobacterium *Arthrospira maxima* shows a solvent-activated protease cleavage of the membrane anchor peptide (KK Ho, personal communication). This protein was purified and crystallized, and the structure has been solved. Since Kerfeld (49) has also obtained the structure of cyt c_6 from *A. maxima*, there is a rare opportunity to study the interaction of two proteins from the organism in which they interact in vivo.

(a)

```
                              1                                                  50                                                100
906   YPFWAQQTYPETPREPTGRIVCANCHLAAKPTEVEVPQSVLPDTVFKAVV KIPYDTSAQQVGADGSKVGLNVGAVLMLPEGFKIAPEDRISEELQEEIGD
Pho   YPFWAQQNYA.NPREATGRIVCANCHLAAKPAEIEVPQSVLPDTVFKAVV KIPYDHSVQQVQADGSKGPLNVGAVLMLPEGFKIAPEDRIPEEMKEVGP
Art   YPFWAQETAPETPREATGRIVCANCHLAAKPIEVEVPQSVLPDTVFKAVV KIPYDHSVQQVLGDGSKGALNVGAVLMLPEGFKIAPADRIPEEIAEVGG
803   YPFWAQQTAPETPREATGRIVCANCHLAAKAAEVEIPQAVLPDTVFEAVV KIPYDLDSQQVLGDGSKGGLNVGAVLMLPEGFKIAPPDRLSEGLKEKVGG
002   YPFWAQQTAPETPREATGRIVCANCHLAAKBAEVEIPQSVLPDQVFEAVV KIPYDHSQQQVLGDGSKGGLNVGAVLMLPDGFKIAPADRLSDELKEKTEG

Cya   FPIYAQQAY.QIPREATGRIVCANCHLGKKPVEIEVPQAVLPNTVFEA.V KIPIDKGAQQIQANGQKGPLNVGAVLMLPEGFKLKPAERLSEELKAKTAG
Por   FPIYAQQAY.ESPREATGRIVCANCHLAQKPVEIEAPQAVLPNTVFETVV KIPYDNNAKQILGNGSKGGLNVGAVILPEGFKLIAPANRLSPELKEKTKN
Chl   YPVFAQQNY.ANPREANGRIVCANCHLAQKAVEIEVPQAVLPDTVFEAVI ELPYDKQVKQVLANGKKGDLNVGMVILPEGFKVIAPDRDVPAEIKEVGN

Liv   FPIYAQQGY.ENPREATGRIVCANCHLAKKPVDIEVPQSVLPNTVFEAVV KIPYDMOIKQVLANGKKGSLNVGAVLILPEGFELAPSDRIPPEMKEIGN
Pea   YPIFAQQGY.ENPREATGRIVCANCHLANKPVDIEVPQAVLPDTVFEAVV RIPYDMOVKQVLANGKKGALNVGAVLILPEGFELAPPRLSPQIKEKIGN
Bea   YPIFAQQGY.ENPREATGRIVCANCHLANKPVDIEVPQAILPDTVFEAVV RIPYDMOVKQVLANGKKGALNVGAVLILPEGFELAPPDRLSPEIKEKIGN
Ric   YPIFAQQGY.ENPREATGRIVCANCHLANKPVDIEVPQAVLPDTVFEAVL RIPYDMLKQVLANGKKGGLNVGAVLILPEGFELAPPDRISPELKEKIGN
Mai   YPIFAQQGY.ENPREATGRIVCANCHLANKPVDIEVPQAVLPDTVFEAVL RIPYDMLKQVLANGKKGGLNVGAVLILPEGFELAPPDRISPELKEKIGN
Whe   YPIFAQQGY.ENPREATGRIVCANCHLASKPVDIEVPQAVLPDTVFEAVL RIPYDMLKQVLANGKKGGLNVGAVLILPEGFELAPPDRISPELKEKIGN
Tob   YPIFAQQGY.ENPREATGRIVCANCHLANKPVEIEVPQAVLPDTVFEAVV RIPYDMLKQVLANGKKGGLNVGAVLILPEGFELAPPDRISPEMKEKIGN
Hoo   YPIFAQQGY.ENPREATGRIVCANCHLANKPVDIEVPQAVLPDTVFEAVV RIPYDMLKQVLANGKKGGLNVGAVLILPEGFELAPPARISPEMKERIGN
Spi   YPIFAQQGY.ENPREATGRIVCANCHLANKPVDIEVPQAVLPDTVFEAVV RIPYDMLKQVLANGKKGGLNVGAVLILPEGFELAPPDRISPEMKEMGN
Tur   YPIFAQQNY.ENPREATGRIVCANCHLASKPVDIEVPQAVLPDTVFEAVV RIPYDMLKQVLANGKKGALNVGAVLILPEGFELAPPDRISPEMKEIGN
Pin   YPIFAQQGY.ENPREATGRIVCANCHLAKKPVDIEVPQSVLPNTVFEAVV KIPYDMQMKQVLANGKKGALNVGAVLILPEGFELAPPDRISPEIRQKTGN
```

(b)

```
                              101                                                150                                               200
906   TY.FQPYSEDKENIVIVGPLPGEQYQEIVFPVLSNPATDKNIHFGKYSV HVGGNRGRGQVYPTGEKSNNNLYNASATGTIAKIAKEDEDGNVKYQVNI
Pho   SY.FQPYADDKONIVAGGPAAGEPVEIVFPALSENPATDKSIHFGKYA1 LDGANRGRGQIYPAGNASNNTVYKASVGGTITEITQEEVG.YQVRVKI
Art   LY.FQPYSADMENVVIVGPIPGQYOEIVFPVLSPDPATNKSINYGKFAVHLGANRGRGQLYPAGNASNNTVYKASVGGTITEITQEEVG.YQVLIL
803   LY.FKQSYARDQENVVIVGPISGDQYEEIVFPVLSPDPAKDKSINYGKFAVHLGANRGRGQVYPTGLLSNNNAKRAPNAGTINIAVNEAAG...?DITI

Cya   LY.FQPYSADKENIVVIGPIPGDKNQEIIFPILSPNEETNKNVKYLKYQLHVGGNRGRGQVSPTGEKTNNTIYNASVNGRISEITKL..ENGGYEITI.T
Por   LY.IYOPYSTKQSNIVIVGPIPGDKNREIIFPILSPDPAKDKQAHFFKYPI VVGGNRGRGQIYPTGDKSNNNLISASASGKINKIEAL..EKGGFIVHI.T
Chl   LY.FQSYSPEQKNIVIVGPPVPGKKYSEMVPILSPDPAKNKNVSYLKYPI YFGGNRGRGQVYPTGDGKSNNTIYNASAAGKIVAITALSEKKGGFEVSI.E

Liv   LF.FQPYSNDKKNIVIVGPVPGKKYSEMVFPILSPDPATNKEAHFLKYPI YVGGNRGRGQIYPDGSKSNNTVVNASITGKVSKI.FRKEKGGYEITI.D
Pea   LS.FQSYRPTKKNIIVIGPVPGKKYSETFPILSPDPATKRDVFLKYPI YVGGNRGRGQIYPDGSKSNNNVSNATATGVVKQI.IRKEKGGYEITI.V
Bea   LS.FQSYRPTKKNIIVIGPPVPGKKYSETFPILSPDPATKRDYFLKYPI YVGGTRGRGQIYPDGSKSNNNVVNATATGVVNKK.IRKEKGGYEITI.V
Ric   LS.FQSYRPNKKNILVIGPVPGKKYSEIVFPILSPDPAMKKDVHFLKYPI YVGGNRGRGQIYPDGSKSNNTVVNATSTGVVRKI.LRKEKGGVEISI.V
Mai   LA.FQSYRPDKKNILVIGPPVPGKKYSEIVFPILSPDPATKKDAHFLKYPI YVGGNRGRGQIYPDGSKSNNTVVNATSTGIVKKI.LRKEKGGVEISI.V
Whe   LA.FQSYRPDKKNILVIGPPVPGKKYSEIVFPILSPDPATKKDAHFLKYPI YVGGNRGRGQIYPDGSKSNNTVVNATSTGIVRKI.LRKEKGGVEISI.V
Tob   LS.FQSYRPNKKNILVIGPVPGQKYSETFPILSPDPATNKDVHFLKYPI YVGGNRGRGQIYPDGSKSNNTVVNATAAGIVSKI.IRKEKGGYEITI.T
Hoo   PS.FQSYRPTKKNILVIGPVPGQKYSETFPILADPDATNKKDVHFLKYPI YVGGNRGRGQIYPDGSKSNNTVVNATAAGIVSKI.IRKEKGGYEITI.A
Spi   LS.FQSYRPNKQNILVIGPVPGQKYSETFPILADPDATNKDVHFLKYPI YVGGNRGRGQIYPDGSKSNNTVVNATAAGIVSKI.IRKEKGGYEITI.A
Tur   LS.FQSYRPNKKNILVIGPVPGQKYSETFPILSDPDATNKDVHFLKYPI YVGGNRGRGQIYPDGSKSNNTVVNATAGGIISKI.LRKEKGGYEITI.V
Pin   LY.FQNRPNKKNIIVIGPVPGQKYSELVFPILSDVSTDKEAHFLKYPI IYGGNRGRGQIYPDGSKSNNTVVSASATGRVSKI.LRKEKGGYEITI.D
```

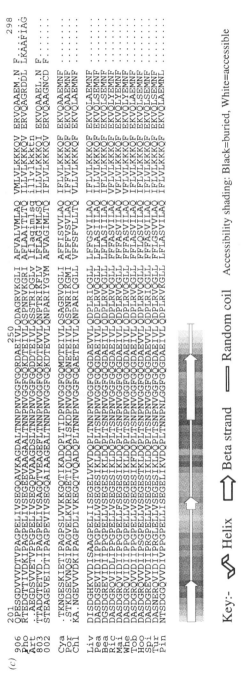

Key:- ⋙ Helix ⇧ Beta strand ══ Random coil   Accessibility shading: Black=buried, White=accessible

Figure 4 Primary structures of the cytochromes *f* aligned with a secondary structure assignment based on the 2.3 Å turnip cyt *f* structure (Protein Data Bank code 1CTM). Amino acids in lower case have yet to be verified.

KEY (each species name is followed by its data base accession number): 803, *Synechocystis* 6803 (S16572); 002, *Synechococcus* 7002 (M74514); 906, *Nostoc* 6903 (B35580); Pho, *Phoridium laminosum* (Y09612); Pea, garden pea (S19109); Bea, fava bean (S26576); Ric, rice (CFRZ); Mai, maize (S58564); Whe, wheat (S07296); Tob, tobacco (A00149); Hoo, Hooker's evening primrose (S00431); Spi, spinach (S00430); Tur, turnip (S45661); Liv, liverwort (P06246); Pin, Pinus (D17510); Chl, *Chlamydomonas* (S16916); Cya, *Cyanophora* (U30821) Por, *Porphyra* (S73186).

(A)

Figure 5   (*A*) Space-filling model of turnip cytochrome *f*. Side-chain oxygen atoms of Asp and Glu residues are in light gray, side-chain nitrogen atoms of Lys and Arg are black, and invariant residues of the cyt *f* primary structure are dark gray. Lys 66 lies just behind Lys 65. (*B*) Space-filling representation of a hypothetical model of *Nostoc* cytochrome *f* based on the turnip cyt *f* structure (generated by Swiss-Model; 74). Charged side chains are colored as in (*A*).

(B)

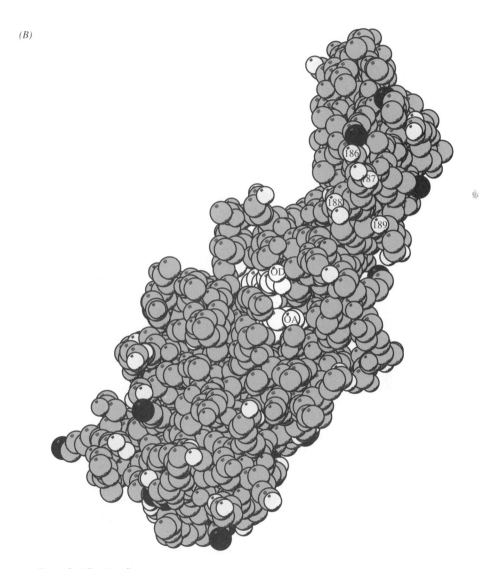

Figure 5   (Continued)

Site-directed mutagenesis has yielded some interesting insights into the structure-function relations of cyt f. Zhou et al (104) constructed mutations in the first three positions at the N terminus. Only one mutation, Y1P, which removes the primary N-terminal amine group that serves as the sixth ligand to the heme iron, failed to assemble cyt f and consequently failed in photoautotrophic growth. In three of the currently available 18 sequences of cyt f, the Tyr is replaced by Phe. The N-terminal aromatic residue is followed by an invariant Pro. When this Pro was replaced by Val, there was an appreciable drop in O_2 evolution and photoautotrophic growth. The rate of stigmatellin-sensitive re-reduction of the cytochrome was diminished by a factor of 10 in this mutant. This suggests a very interesting difference in the routes of arrival and departure of electrons in cyt f.

Soriano et al (89) mutagenized the cytochrome f of *C. reinhardtii* to test the role of selected Lys residues in the docking of plastocyanin to cyt f. A docking site was suggested by the crystal structure of a turnip cyt f, and by earlier cross-linking experiments of Morand et al (70). Three lysine residues at 58, 65, and 66 (Figure 5) constitute a large domain of possible interaction. Lys 187 (in the small domain that is uppermost in the view shown in Figure 5) cross links to PC. Equivalent residues in the *C. reinhardtii* large and small domains of cytochrome were altered to neutralize the positive charges. Mutants were constructed with neutralized large domain, neutralized small domain, and with both domains neutralized. All the mutants grew photoautotrophically with a 20–30% increase in generation time and a comparable decrease in O_2 evolution. There was a small increase in the half-time of the cyt f oxidation and up to a doubling of the half-time of cyt f reduction. This is a smaller effect of electrostatic interaction between cyt f and its oxidant than was expected from in vitro experiments. This might be the result of the high ionic strength and small diffusional space in the lumen. Support for the idea of interaction at these sites has come from the work of Wagner et al, who studied cyt f from the cyanobacterium *Phormidium luridum* (95). The *P. luridum* cyt f has acid or neutral residues in most of the positions equivalent to basic residue positions in turnip cyt f. Of these basic residues, only Lys 66 is conserved in the two structures. The higher plant Lys 187 is replaced by an Asp 188 in the *P. luridum* protein. Higher plant plastocyanins have an acid patch on their eastern face, which is thought to interact with the positive domains on higher plant cyt f. Many of the corresponding residues on the eastern face of *P. laminosum* PC are basic or neutral. Thus the charges on the interacting partners in *P. laminosum* are reversed in these partners from turnip. Wagner et al measured the rate of electron transfer in vitro of *P. laminosum* b_6f complex to *P. laminosum* PC at varying ionic strengths. There was an optimum rate in response to ionic strength, as is the case in the interaction of higher plant cyt f with higher plant PC. This is reasonably interpreted as a long range electrostatic interaction that initiates docking followed by a rearrangement of the two proteins to allow electron transfer. *Nostoc* cyt c_6 and PC have a basic isoelectric point. A hypothetical model of cyt f from *Nostoc* is shown in Figure 5*B*. Amino acid substitution in *Nostoc* cyt f (186–189) gives the region thought to be involved in interaction with cyt c_6 or PC an overall negative charge in contrast to the positive electrostatic potential observed in cyt f (Figure 5*A*), corroborating the results of Wagner et al (95).

Little is known about the reduction of cyt f by the Rieske protein, but this may soon change. Zhang et al have reported on the characterization and crystallization of the lumen side domain of the chloroplast Rieske iron-sulfur protein (102). As with turnip cyt f, the proteolytic removal of the membrane anchor region of the spinach Rieske protein produced a soluble peptide of substantial size (139 of the 179 residues). This peptide was purified and crystallized, and its structure has been solved to 1.85 Å resolution (7). This opens the way to modeling the interactions of the Rieske protein with cyt f, to site-directed mutagenesis of both partners, and to in vitro kinetic measurements of electron transfer between these partners.

The Cytochrome b_6f Complex

This multimeric, integral membrane protein complex is well characterized from higher plants and is attracting investigations on the lower plant forms as well. The complex consists of four large subunits—cyt f, cyt b_6, Rieske iron sulfur protein, and subunit IV lacking redox prosthetic groups—and three small peptides. The review by Cramer et al (11) gives a fine picture of its structure and function. Krinner et al (55) were the first to report an isolation of cyt b_6f complex from a cyanobacterium, *A. variabilis*, and they pointed out that the apparent absence of a cytochrome bc_1 respiratory complex in cyanobacteria might indicate a dual function for cyt b_6f in respiration and photosynthesis, as had been suggested by Peschek & Schmetterer (75). This complex reduced the basic cyt c_6 and PC from *A. variabilis* at a substantially higher rate than it reduced acidic cyt c_6 and PC. Next, Rayas-Vera et al described the isolation of b_6f complex from *A. maxima* using an affinity column of horse heart cytochrome c bound to Sepharose (80). Minami et al (68) isolated a similar complex from *Spirulina* sp. and used a purification step of a DEAE-Toyopearl 650 M column. The purified complex contained small amounts of carotenoid and chlorophyll. *Spirulina* (now called *Arthrospira*) could provide an advantage in experiments where large quantities of the complex are required since this cyanobacterium is available in large amounts from commercial sources (22). Minami et al (68) found that the b_6f complex could be dissociated into its individual protein components by treatment with 15% 2-mercaptoethanol, and the components were resolved by chromatography. Most of the cyt b_6 heme was dissociated from its 23-kDa polypeptide in the DEAE Toyopearl column. Some of the cyt b_6

was not absorbed and passed directly through the column. These observations may provide new access to purification of the individual components of the complex.

In 1988, Kallas et al identified and sequenced the genes for cyt f and the Rieske iron sulfur protein in *Nostoc* PCC7906 (43) and then characterized the two operons encoding the complex (44). Kallas has written an excellent review of the molecular biology of the cyt $b_6 f$ complex in cyanobacteria (42). Recently, Boronowski et al have cloned and sequenced the genes for cyt b_6 and subunit IV of *Synechocystis* PCC6803 and have overexpressed the subunit IV gene in *E. coli* (3).

C. reinhardtii has become an important organism in photosynthesis research since elegant techniques for site-directed mutagenesis have been developed for it. Lemaire et al (60) isolated and characterized a cyt $b_6 f$ particle from *C. reinhardtii*. Recent work has resulted in the isolation, sequencing, and deletion of the genes *petG*, *petL*, and *petX* (2, 13, 92) for the small peptides of the $b_6 f$ complex from *C. reinhardtii*. Deletion of *petG* eliminated photosynthetic growth and gave reduced levels of the four large subunits of the $b_6 f$ complex, indicating a role in either its assembly or stability. Deletion of *petL* diminished both photoautotrophic growth and the amount of $b_6 f$ complex so that electron transfer through the complex is reduced.

Finazzi et al (18) have published a mutagenesis experiment on the *C. reinhardtii* $b_6 f$ complex that seems to presage a wave of similar efforts to understand the structure and function of this unit of photosynthetic electron transport. These authors characterized a mutant that contains a 36–base pair duplication in the chloroplast *petB* gene that results in the duplication of a 12–amino acid sequence in the cd loop of cyt b_6. This mutant has a fluorescence phenotype typical of mutants blocked in electron transfer through the cyt $b_6 f$ complex. The mutant cyt $b_6 f$ protein complexes are present in wild-type amounts. The $b_6 f$ complex shows loosened attachment of the Rieske protein and is more vulnerable to degradation. Electron transfer through the complex is greatly diminished. There is evidence for a 100-fold decrease in plastoquinol affinity to the mutant protein.

A very hopeful sign has come from the Cramer laboratory (40), where the cytochrome $b_6 f$ particle from the thermophilic cyanobacterium *Mastigocladus laminosus* has been crystallized. The crystals diffract to 14 Å and intense effort is under way to improve this. A high-resolution crystal structure will be a great aid to understanding the cytochrome $b_6 f$ complex.

ACKNOWLEDGMENTS

The authors are indebted to WA Cramer, KK Ho, and J Whitmarsh for their helpful discussions and careful reading of this manuscript. R Knutson and KK

Ho gave valuable assistance in the preparation of Figure 2. We are most grateful to KK Ho, who provided the amino acid sequence of *A. maxima* cyt *f*.

Visit the *Annual Reviews home page* at
http://www.AnnualReviews.org.

Literature Cited

1. Deleted in proof
2. Berthold DA, Schmidt CL, Malkin R. 1996. The deletion of *petG* in *Chlamydomonas reinhardtii* disrupts the cytochrome b/f complex. See Ref. 64a, pp. 571–74
3. Boronowsky U, Kruip J, Rogner M. 1996. Cloning and sequencing of *petB* and *petD* genes: overexpression and characterization of subunit IV from a cyanobacterial b₆/f complex. See Ref. 64a, pp. 583–86
4. Bottin H, Sétif P, Mathis P. 1987. Study of the photosystem I acceptor side by double and triple flash experiments. *Biochim. Biophys. Acta* 894:39–48
5. Bowes JM, Stewart AC, Bendall D. 1983. Purification of photosystem II particles from *Phormidium laminosum* using the detergent dodecyl-β-D-maltoside. *Biochim. Biophys. Acta* 725:210–19
5a. Bryant DA, ed. 1994. *The Molecular Biology of Cyanobacteria.* Dordrecht: Kluwer
6. Campos AP, Aguiar AP, Hervas M, Regalla M, Navarro JA, et al. 1993. Cytochrome c_6 from *Monoraphidium braunii*, a cytochrome with an unusual heme axial coordination. *Eur. J. Biochem.* 216: 329–41
7. Carrell CJ, Zhang H, Cramer WA, Smith JL. 1997. Conductive pathways for electron transfer are suggested by the structure of the lumen-side domain of the chloroplast Rieske protein. *Structure.* In press
8. Clarke AK, Campbell D. 1996. Inactivation of the *petE* gene for plastocyanin lowers photosynthesis capacity and exacerbates chilling-induced photoinhibition in the cyanobacterium *Synechococcus. Plant Physiol.* 112:1551–61
9. Cohen CL, Sprinkle JR, Alam J, Hermodson M, Meyer T, Krogmann DW. 1989. The amino acid sequence of low potential cytochrome c_{550} from the cyanobacterium *Microcystis aeruginosa. Arch. Biochem. Biophys.* 270:227–35
10. Cohn CL, Hermodson M, Krogmann DW. 1989. The amino acid sequence of cytochrome c_{553} from *Microcystis aeruginosa. Arch. Biochem. Biophys.* 270:219–26
11. Cramer WA, Soriano GM, Ponomarev M, Huang D, Zhang H, et al. 1996. Some new aspects and old controversies concerning the cytochrome b₆/f complex of oxygenic photosynthesis. *Annu. Rev. Plant Physiol. Plant Mol. Biol.* 47:477–508
12. Cramer WA, Whitmarsh J. 1977. Photosynthetic cytochromes. *Annu. Rev. Plant Physiol.* 28:133–72
13. De Vitry C, Breyton C, Pierre Y, Popot J-L. 1996. The 4-kDa chloroplast polypeptide of cytochrome b₆/f complex encoded by the nuclear *petX* gene: nucleic acid and protein sequences, targeting signals, and membrane topology. See Ref. 64a, pp. 595–98
14. Diaz A, Hervas M, Navarro JA, DeLaRosa MA, Tollin G. 1994. A thermodynamic study of laser-flash photolysis of plastocyanin and cytochrome c_6 oxidation by photosystem I from the green alga *Monoraphidium braunii. Eur. J. Biochem.* 222:1001–7
15. Diaz A, Navarro F, Hervas M, Navarro JA, Chavez S, et al. 1994. Cloning and correct expression in *E. coli* of the *petJ* gene encoding cytochrome c_6 from *Synechocystis* 6803. *FEBS Lett.* 347:173–77
16. Enami I, Murayama H, Ohta H, Kamo M, Nakazato K, et al. 1995. Isolation and characterization of a photosystem II complex from the red alga *Cyanidium caldarium*: association of cytochrome c_{550} and a 12-kDa protein with the complex. *Biochim. Biophys. Acta* 1232:208–16
17. Evans PK, Krogmann DW. 1983. Three c-type cytochromes from the red alga *Porphyridium cruentum. Arch. Biochem. Biophys.* 277:494–510
18. Finazzi G, Büschlen S, de Vitry C, Rappaport F, Joliot P, et al. 1997. Function directed mutagenesis of the cytochrome b₆/f complex in *Chlamydomonas reinhardtii*: involvement of the cd loop of cytochrome b₆ in quinol binding to the Qo site. *Biochemistry* 36:2867–74

19. Frazao C, Soares CM, Carrondo MA, Pohl E, Dauter Z, et al. 1995. *Ab initio* determination of the crystal structure of cytochrome c_6 and comparison with plastocyanin. *Structure* 3:1159–69

20. Ghassemian M, Wong B, Ferreira F, Markley JL, Straus NA. 1994. Cloning, sequencing and transcriptional studies of the genes for cytochrome c_{553} and plastocyanin from *Anabaena* Sp. PCC 7120. *Microbiology* 140:1151–59

21. Golbeck JH. 1998. A comparison of *in vitro* and *in vivo* mutants of Photosystem I: protocols for mutagenesis and techniques for analysis. *Methods Enzymol.* In press

22. Gómez Lojero C, Krogmann DW. 1996. Large scale preparations of photosynthetic catalysts from cyanobacteria. *Photosynth. Res.* 47:293–99

23. Gray JC. 1992. Cytochrome f: structure, function and biogenesis. *Photosynth. Res.* 34:359–74

24. Gray JC, Rochford RJ, Packman LC. 1994. Proteolytic removal of the C-terminal transmembrane region of cytochrome f during extraction from turnip and charlock leaves generates a water soluble monomeric form of the protein. *Eur. J. Biochem.* 223:481–88

25. Gross EL. 1993. Plastocyanin: structure and function. *Photosynth. Res.* 37:103–16

26. Han K-C, Shen J-R, Ikeuchi M, Inoue Y. 1994. Chemical crosslinking studies of extrinsic proteins in cyanobacterial photosystem II. *FEBS Lett.* 355:121–24

27. Hatanaka H, Senoike K, Hirano M, Katoh S. 1993. Small subunits of photosystem I reaction center complex from *Synechococcus elongatus*: 1. Is the *psaF* gene product required for oxidation of cytochrome c_{553}. *Biochim. Biophys. Acta* 1141:45–51

28. Hervas M, DeLaRosa MA, Tollin G. 1992. A comparative laser flash absorption spectroscopy study of algal plastocyanin and cytochrome c_{552} photooxidation by photosystem I particles from spinach. *Eur. J. Biochem.* 203:115–20

29. Hervas M, Navarro JA, Diaz A, Bottin H, DeLaRosa MA. 1995. Laser-flash kinetic analysis of the fast electron transfer from plastocyanin and cytochrome c_6 to photosystem I. Experimental evidence on the evolution of the reaction mechanism. *Biochemistry* 34:11321–26

30. Hervas M, Navarro JA, Diaz A, DeLaRosa MA. 1996. Comparative thermodynamic analysis by laser-flash absorption spectroscopy of photosystem I reduction by plastocyanin and cytochrome c_6 in *An-*

abaena PCC 7119, *Synechocystis* PCC 6803 and spinach. *Biochemistry* 35:2693–98

31. Hervas M, Ortega JM, Navarro JA, DeLaRosa MA, Bottin H. 1994. Laser flash kinetic analysis of *Synechocystis* PCC 6803 cytochrome c_6 and plastocyanin oxidation by photosystem I. *Biochim. Biophys. Acta* 1184:235–46

32. Hippler M, Drepper F, Farah J, Rochaix J-D. 1997. Fast electron transfer from cytochrome c_6 and plastocyanin to photosystem I of *Chlamydomonas reinhardtii* requires *psaF*. *Biochemistry* 36:6343–49

33. Ho KK. 1997. Isolation and characterization of soluble electron-transfer proteins of cyanobacteria. In *Handbook of Photosynthesis*, ed. M Pessarakli, pp. 513–23. New York: Marcel Decker. 1027 pp.

34. Ho KK, Krogmann DW. 1984. Electron donors to P700 in cyanobacteria and algae an instance of unusual genetic-variability. *Biochim. Biophys. Acta* 766:310–16

35. Ho KK, Tan S. 1994. Use of adsorption chromatography on Sephacryl S–500 for improved separation of isoforms of soluble photosynthetic catalysts from cyanobacteria. *J. Liquid Chromatogr.* 17: 833–45

36. Hoganson W, Lagenfelt G, Andreasson L-E. 1990. EPR and redox potentiometric studies of cytochrome c_{549} of *Anacystis nidulans*. *Biochim. Biophys. Acta* 1016:203–6

37. Deleted in proof

38. Holton RW, Myers J. 1967. Water soluble cytochromes from a blue-green alga. I. Extraction, purification and spectral properties of cytochromes c (549, 522 and 554, *Anacystis nidulans*). *Biochim. Biophys. Acta* 131:362–74

39. Holton RW, Myers J. 1963. Cytochromes of blue-green algae: extraction of a c type with a strongly negative redox potential. *Science* 142:234–35

40. Huang D, Zang H, Krahn JM, Carrell CJ, Soriano GM, et al. 1996. Crystallization of *Mastigocladus laminosus* cytochrome b_6f complex, and of the p-side domain of the chloroplast Rieske protein. *Biophys. J.* 72:A248

41. Jung YS, Yu L, Golbeck JH. 1995. Reconstitution of iron-sulfur center F_B results in complete restoration of $NADP^+$ photoreduction in Hg-treated photosystem I complexes from *Synechococcus* sp. PCC 6301. *Photosynth. Res.* 46:249–55

42. Kallas T. 1994. The cytochrome b_6/f complex. See Ref. 5a, pp. 259–71

43. Kallas T, Spiller S, Malkin R. 1988.

Characterization of two operons encoding the cytochrome b_6/f complex of the cyanobacterium *Nostoc* PCC 7906. *J. Biol. Chem.* 263:14334–42

44. Kallas T, Spiller S, Malkin R. 1988. Primary structure of cotranscribed genes encoding the Rieske Fe-S and cytochrome f proteins of the cyanobacterium *Nostoc* PPC 7906. *Proc. Natl. Acad. Sci. USA* 85:5794–98

45. Kamimura Y, Yamasaki T, Matsuzaki E. 1977. Cytochrome components of green alga, *Bryopsis maxima. Plant Cell Physiol.* 18:317–24

46. Kang C, Chitnis PR, Smith S, Krogmann DW. 1994. Cloning and sequence analysis of the gene encoding the low potential cytochrome c of *Synechocystis* PCC 6803. *FEBS Lett.* 344:5–9

47. Katoh S. 1995. The discovery and function of plastocyanin: a personal account. *Photosynth. Res.* 43:177–89

48. Keilin D. 1966. *The History of Cell Respiration and Cytochrome.* Cambridge: Cambridge Univ. Press. 416 pp.

49. Kerfeld CA. 1997. Structural comparison of c_6 and c_2. *Photosyn. Res.* 54:81–98

50. Deleted in proof

51. Deleted in proof

52. Kerfeld CA, Anwar HP, Interrante R, Merchant S, Yates TO. 1995. The structure of chloroplast c_6 at 1.9Å resolution: evidence for functional oligomerization. *J. Mol. Biol.* 250:627–47

53. Kinzel PF, Peschek GA. 1983. Cytochrome c_{549}—an endogenous cofactor of cyclic photophosphorylation in the cyanobacterium *Anacystis nidulans? FEBS Lett.* 162:76–80

54. Kraulis PJ. 1991. MOLSCRIPT: a program to produce both detailed and schematic plots of protein structure. *J. Appl. Crystallogr.* 24:946–50

55. Krinner M, Hauska G, Hurt E, Lockau W. 1982. A cytochrome f/b_6 complex with plastoquinol cytochrome c oxidoreductase activity from *Anabaena variabilis. Biochim. Biophys. Acta* 681:110–17

56. Kuras R, Wollman FA, Joliot P. 1995. Conversion of cytochrome f to a soluble form in vivo in *Chlamydomonas reinhardtii. Biochemistry* 34:7468–75

57. Laskowski RA, MacArthur MW, Moss DS, Thornton JM. 1993. PROCHECK: a program to check the stereochemical quality of protein structures. *J. Appl. Crystallogr.* 26:283–91

58. Laudenbach DE, Herbert SK, McDowell C, Fork DC, Grossman AR, et al. 1990. Cytochrome c_{553} is not required for photosynthetic activity in the cyanobac-

terium *Synechococcus. Plant Cell* 2:913–24

59. Laycock MU. 1972. The amino acid sequence of cytochrome c_{553} from the chrysophycean alga *Monochrysis lutheri. Can. J. Biochem.* 50:1311–25

60. Lemaire C, Girard-Bascou J, Wollman FA, Bennoun P. 1986. Studies on the cytochrome b_6/f complex. I. Characterization of the complex subunits in *Chlamydomonas reinhardtii. Biochim. Biophys. Acta* 851:229–38

61. Ludwig ML, Pattridge KA, Powers TB, Dickerson RE, Takano T. 1983. Structure analysis of ferricytochrome c from the cyanobacterium *Anacystis nidulans.* In *Electron Transport* and *Oxygen Utilization*, ed. C Ho, pp. 27–32. Amsterdam: Elsevier

62. MacIntosh L. 1998. *Photosynthesis: The Molecular Biology of Energy Capture.* San Diego: Academic. In press

63. Malakhov MP, Wada H, Los DA, Semenenko VE, Murata N. 1994. A new type of cytochrome c from *Synechocystis* PCC 6803. *J. Plant Physiol.* 144:259–64

64. Martinez SE, Huang D, Ponomarev M, Cramer WA, Smith JL. 1996. The heme redox center of chloroplast cytochrome f is linked to a buried five-water chain. *Protein Sci.* 5:1081–92

64a. Mathis P, ed. 1996. *Photosynthesis: From Light to Biosphere.* Dordrecht: Kluwer. 2nd ed.

65. Medina M, Diaz A, Hervas M, Navarro JA, Gomez-Moreno C, et al. 1993. A comparative laser flash absorption spectroscopy study of *Anabaena* PCC 7119 plastocyanin and cytochrome c_6 photooxidation by photosystem I particles. *Eur. J. Biochem.* 213:1133–38

66. Metzger SU, Cramer WA, Whitmarsh J. 1997. Critical analysis of the extinction coefficient of chloroplast cytochrome f. *Biochim. Biophys. Acta* 1319:233–41

67. Meyer TE, Tollin G, Cusanovich MA. 1994. Protein interaction sites obtained via sequence homology. The site of complexation of electron transfer partners of cytochrome c revealed by mapping amino acid substitutions onto three-dimensional protein surfaces. *Biochimie* 76:480–88

68. Minami Y, Wada K, Matsubara H. 1989. The isolation and characterization of a cytochrome b_6f complex from the cyanobacterium *Spirulina* sp. *Plant Cell Physiol.* 30:91–98

69. Morand LZ, Cheng RH, Ho KK, Krogmann DW. 1994. Soluble electron transfer catalysts in cyanobacteria. See Ref. 5a, pp. 381–407

70. Morand LZ, Frame MK, Colvert KK, Johnson DA, Krogmann DW, Davis DJ. 1989. Plastocyanin cytochrome f interaction. *Biochemistry* 28:8039–47
71. Navarro JA, Hervas M, DeLaCedra B, DeLaRosa MA. 1995. Purification and physicochemical properties of the low-potential cytochrome c_{549} from the cyanobacterium *Synechocystis* sp. PCC 6803. *Arch. Biochem. Biophys.* 318:46–52
72. Navarro JA, Hervas M, Gutierrez-Merino C, DeLaRosa MA. 1996. A comparative kinetic analysis of the flavin-photosensitized oxidation and reduction of plastocyanin and cytochrome c_6 from different organisms. *Photochem. Photobiol.* 63:86–91
73. Nishiyama Y, Hayashi H, Watanabe T, Murata N. 1994. Photosynthetic oxygen evolution is stabilized by cytochrome c_{550} against heat inactivation in *Synechococcus* sp. PCC 7002. *Plant Physiol.* 105:1313–19
74. Peitsch MC. 1996. ProMod and Swiss-Model: Internet-based tools for automated comparative protein modeling. *Biochem. Soc. Trans.* 24:274–79
75. Peschek GA, Schmetterer G. 1982. Evidence for plastoquinol-cytochrome f/b_6 reductase as a common electron donor to P700 and cytochrome oxidase in cyanobacteria. *Biochem. Biophys. Res. Commun.* 108:1188–95
76. Pettigrew GW, Moore GR. 1987. *Cytochromes c: Biological Aspects.* Berlin: Springer-Verlag. 282 pp.
77. Pettigrew GW, Moore GR. 1990. *Cytochromes c: Evolutionary, Structural and Physicochemical Aspects.* Berlin: Springer-Verlag. 478 pp.
78. Powls R, Wong J, Bishop NI. 1969. Electron transfer components of wild-type and photosynthetic mutants of *Scenedesmus obliquus* D3. *Biochim. Biophys. Acta* 180:490–99
79. Prince NT, Smith AJ, Sykes AG, Rogers LU. 1992. Cytochrome c_{553} from two species of macroalgae. *Phytochemistry* 30:2843–48
80. Rayas-Vera A, Gonzalez-Halphen D, Gómez-Lojero C. 1983. Oxido-reduction reactions of different membrane preparations from the cyanobacterium *Spirulina maxima*. In *Photosynthetic Prokaryotes, Cell Differentiation and Function*, ed. GC Papageorgiou, L Packer, pp. 185–97. Amsterdam: Elsevier
81. Redinbo MR, Yeats RO, Merchant S. 1994. Plastocyanin: structural and functional analysis. *J. Bioenerg. Biomembr.* 26:49–66

82. Reith M, Mulholland J. 1993. A high resolution gene map of the chloroplast genome of the red alga *Porphyra purpurea*. *Plant Cell* 5:465–75
83. Rippka R, Deruelles J, Waterbury JB, Herdman M, Stanier RY. 1979. Generic assignments, strain histories and properties of pure cultures of cyanobacteria. *J. Gen. Microbiol.* 111:1–61
84. Shen J-R, Burnap RL, Inoue Y. 1995. An independent role of cytochrome c_{550} in cyanobacterial photosystem II as revealed by double-deletion mutagenesis of the psbO and psbV genes in *Synechocystis* sp. PCC 6803. *Biochemistry* 34:12661–68
85. Shen J-R, Ikeuchi M, Inoue Y. 1992. Stoichiometric association of extrinsic cytochrome c_{550} and 12-kDa protein with a highly purified oxygen-evolving photosystem II core complex from *Synechococcus vulcanus*. *FEBS Lett.* 301:145–49
86. Shen J-R, Inoue Y. 1993. Binding and functional properties of two new extrinsic components, cytochrome c_{550} and a 12-kDa protein, in cyanobacterial photosystem II. *Biochemistry* 32:1825–32
87. Shen J-R, Inuoe Y. 1993. Cellular localization of cytochrome c_{550}. *J. Biol. Chem.* 268:20408–13
88. Shen J-R, Vermaas W, Inoue Y. 1995. The role of cytochrome c_{550} as studied through reverse genetics and mutant characterization in *Synechocystis* sp. PCC 6803. *J. Biol. Chem.* 270:6901–7
89. Soriano GM, Ponamareu MV, Tae GS, Cramer WA. 1996. Effect of the interdomain basic region of cytochrome f on its redox reactions in vivo. *Biochemistry* 35:14590–98
90. Stewart AC, Kaethner TM. 1983. Extraction and partial purification of an acidic plastocyanin from a blue-green alga. *Photobiochem. Photobiophys.* 6:67–73
91. Sugimura Y, Toda F, Murata T, Yakushiji T. 1968. Studies on algal cytochromes. In *Structure and Function of Cytochromes*, ed. K Okunuki, MB Kamem, I Sukuzu, pp. 452–58. Tokyo: Univ. Tokyo Press
92. Takahashi Y, Rahire M, Breyton C, Popot JL, Joliot P, et al. 1996. The chloroplast ycf7 (*petL*) open reading frame of *Chlamydomonas reinhardtii* encodes a small functionally important subunit of the cytochrome b_6/f complex. *EMBO J.* 15:3498–506
93. Tan S, Ho KK. 1989. Purification of an acidic plastocyanin from *Microcystis aeruginosa*. *Biochim. Biophys. Acta* 973:111–17
94. Varley JPA, Moehrle JJ, Manasse RS, Bendall DS, Howe CJ. 1995. Characteri-

zation of plastocyanin from the cyanobacterium *Phormidium laminosum*: copper inducible expression and SecA-dependent targeting in *Escherichia coli*. *Plant Mol. Biol.* 27:179–90

95. Wagner MJ, Packer JCL, How CJ, Bendall DS. 1996. Some characteristics of cytochrome f in the cyanobacterium *Phormidum laminosum*: its sequence and charge properties in the reaction with plastocyanin. *Biochim. Biophys. Acta* 1276:246–52

96. Wood PM. 1978. Interchangeable copper and iron proteins in algal photosynthesis. Studies on plastocyanin and cytochrome c_{552} in *Chlamydomonas*. *Eur. J. Biochem.* 87:9–19

97. Wynn RM, Malkin R. 1988. Interaction of plastocyanin with photosystem I: a chemical cross-linking study of the polypeptide that binds plastocyanin. *Biochemistry* 27:5863–69

98. Wynn RM, Omaha J, Malkin R. 1989. Structural and functional properties of the cyanobacterial photosystem I complex. *Biochemistry* 28:5554–60

99. Xu Q, Yu L, Chitnis VP, Chitnis PR. 1994. Function and organization of photosystem I in a cyanobacterial mutant strain that lacks PsaF and PsaJ subunits. *J. Biol. Chem.* 269:3205–11

100. Yamanaka T, DeKlerk H, Komen MD. 1967. Highly purified cytochromes c derived from the diatom, *Navicula pelliculosa*. *Biochim. Biophys. Acta* 143:416–24

101. Zhang H, Pakrasi HB, Whitmarsh J. 1994. Photoautotrophic growth of the cyanobacterium *Synechocystis* sp. PCC 6803 in the absence of cytochrome c_{553} and plastocyanin. *J. Biol. Chem.* 269:5036–42

102. Zhang H, Carrell CJ, Huang D, Sled V, Onishi T, et al. 1996. Characterization and crystallization of the lumen side domain of the chloroplast Rieske iron-sulfur protein. *J. Biol. Chem.* 271:31360–66

103. Zhang L, McSpadden B, Pakrasi HB, Whitmarsh J. 1992. Copper-mediated regulation of cytochrome c_6 and plastocyanin in the cyanobacterium *Synechocystis* 6803. *J. Biol. Chem.* 267:19054–59

104. Zhou JH, Fernandez-Velasco JG, Malkin R. 1996. N-terminal mutants of chloroplast cytochrome f: effect on redox reactions and growth in *Chlamydomonas reinhardtii*. *J. Biol. Chem.* 271:6225–32

Annu. Rev. Plant Physiol. Plant Mol. Biol. 1998. 49:427–51

BRASSINOSTEROIDS: Essential Regulators of Plant Growth and Development

Steven D. Clouse

Department of Horticultural Science, North Carolina State University, Raleigh, North Carolina 27695

Jenneth M. Sasse

School of Forestry and Resource Conservation, University of Melbourne, Parkville, Victoria 3052, Australia

KEY WORDS: steroid biosynthesis, cell elongation, signal transduction, hormone-insensitive mutant, hormone-deficient mutant

ABSTRACT

Brassinosteroids (BRs) are growth-promoting natural products found at low levels in pollen, seeds, and young vegetative tissues throughout the plant kingdom. Detailed studies of BR biosynthesis and metabolism, coupled with the recent identification of BR-insensitive and BR-deficient mutants, has greatly expanded our view of steroids as signals controlling plant growth and development. This review examines the microchemical and molecular genetic analyses that have provided convincing evidence for an essential role of BRs in diverse developmental programs, including cell expansion, vascular differentiation, etiolation, and reproductive development. Recent advances relevant to the molecular mechanisms of BR-regulated gene expression and BR signal transduction are also discussed.

CONTENTS

427

INTRODUCTION

Brassinosteroid (BR) research began nearly thirty years ago when Mitchell et al (73) reported that organic extracts of *Brassica napus* pollen promoted stem elongation and cell division in plants. The announcement that the active component of these extracts was a unique steroid (46) stimulated international research interest on the chemistry and physiology of these very potent plant growth regulators, and by 1988 (when the first BR review appeared in this series; 66), over 100 publications on BR-related topics were available. Survey of this literature (66) led Mandava to conclude that "BRs may thus be regarded as a new group of plant hormones with a regulatory function in cell elongation and cell division...." Although the case for BRs as endogenous plant growth regulators was strong (95), there was not widespread acceptance of BRs as hormones and little or no attention was paid to these steroids in botany textbooks or general reviews of plant development.

Recent application of molecular genetics to the analysis of BR biosynthesis and signal transduction (24, 56, 63, 64, 79, 107) has confirmed predictions of chemists and physiologists that BRs are essential for normal plant growth and must be considered along with auxins, cytokinins, gibberellins, abscisic acid, and ethylene in any model of plant development. A proliferation of short reviews and reports has drawn attention to these developments and prompted renewed general interest in plant steroids as signaling molecules (21, 22, 39, 52, 87, 132). BR research is moving rapidly along three convergent lines of analysis. Microchemical techniques are revealing the details of biosynthesis, distribution, and metabolism; the analysis of BR-deficient and BR-insensitive mutants is helping to clarify the physiological roles of BRs in growth and development; and the cloning of BR-regulated genes is providing insight into the molecular mechanism of plant steroid hormone action. This review focuses on recent advances in these three areas.

OCCURRENCE AND LOCALIZATION OF BRs

Distribution and Endogenous Levels

The first BR, brassinolide, $(22R,23R,24S)$-$2\alpha,3\alpha,22,23$-tetrahydroxy-24-methyl-B-homo-7-oxa-5α-cholestan-6-one (Figure 1), was isolated from pollen of *Brassica napus* (46), and over forty related compounds have now been identified in plant extracts. All are hydroxylated derivatives of cholestane and, given the possibilities of combinations of substructures in rings A and B and the side chain, the family probably has many more members. The compounds can be classified as C_{27}, C_{28}, or C_{29} BRs, depending on the alkyl-substitution pattern of the side chain (132). Their individual structures have been illustrated in several reviews, and Fujioka & Sakurai (39) include an extensive list of the families, genera, species, and plant organs from which they have been rigorously identified, and their occurrence in another five families has been summarized (96). BRs have been identified in an alga and a pteridophyte, and three families of gymnosperms; in angiosperms, they have been shown to occur in 16 families of dicots, and 5 of monocots. So BRs are probably ubiquitous in the plant kingdom, and they certainly occur in shoots and seeds of the important experimental plant *Arabidopsis thaliana* (36, 101).

Levels of endogenous BRs vary among plant tissues, and so do the suites of congeners accompanying the most active members of the family. Seeds of *Phaseolus vulgaris* contain a wide array of BRs (134), as does pollen from *Cupressus arizonica* (45) where concentrations of congeners can be ~6000-fold greater than the concentration of brassinolide. BRs occur endogenously

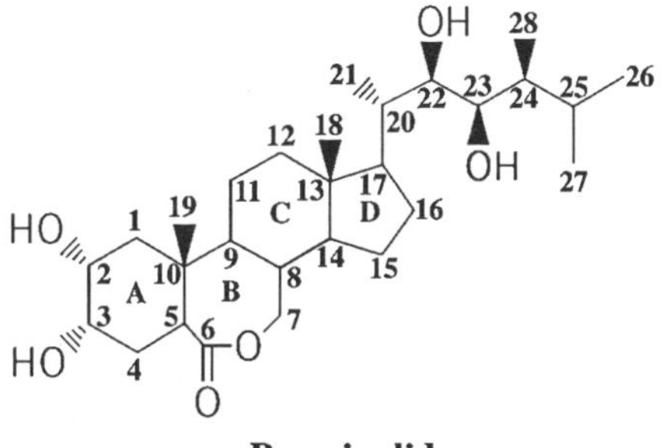

Brassinolide

Figure 1 The structure of brassinolide, a commonly occurring BR with high biological activity, showing *numbered positions* mentioned in the text. In natural BRs, hydroxylation can occur in *ring A* at positions 3-, and/or 2-, and/or 1-; also found are epoxidation at 2,3-, or a 3-oxo-group. In *ring B*, alternatives are 6-oxo- and 6-deoxo- forms. In the side chain methyl-, ethyl-, methylene-, ethylidene-, or *nor-* alkyl groups can occur at 24-, and the 25-methyl- series is also represented.

at quite low levels. Pollen, the original source of brassinolide, and immature seeds are the richest sources with ranges of 1–100 ng $\cdot$ g^{-1} fw, while shoots and leaves usually have lower amounts of 0.01–0.1 ng $\cdot$ g^{-1} fw (110, 137). Cultured crown gall cells of *Catharanthus roseus* have levels of brassinolide and castasterone (30 ng $\cdot$ g^{-1} fw) equivalent to that of pollen (81). In general, young growing tissues contain higher levels of BRs than mature tissues (110), which is not surprising considering that BRs show greater physiological effect on immature vs older tissue (66). While bioassay results suggest that root tissue contains BRs, and that etiolated stem tissue is also comparatively rich in them, they have yet to be identified and quantified rigorously (JM Sasse, unpublished data).

Methods of Analysis

The extraction and purification of BRs depend on solvent partitioning and subsequent chromatographic separations. The choice and sequence of these may differ, with later fractionation often guided by bioassay. The most frequently used is the rice lamina inclination test (119). Analysis of purified fractions is mostly done by GC-MS and GC-MS-SIM (reviewed in 4, 110). FAB-MS is successful with pure BRs (13, 99) but not with partially purified extracts (99). LC-MS methods are also being explored (41–43) and were found useful for the detection of epimerization (77) and identification of teasterone esters (8). Direct analysis of BRs with LC and electrospray techniques of MS is promising (60) and (PG Griffiths, JM Sasse, DW Cameron & G Currie, unpublished data).

Assay by liquid chromatography, with UV, fluorescence, or electrochemical detection of derivatives, is also a sensitive method (reviewed in 43). Fluorescence detection at long wavelengths is particularly useful for checking fractions from small samples of plant tissue. However, derivatization can be difficult in very dilute solutions, and conditions must be adjusted when very small-scale extractions and fractionations are attempted (PG Griffiths, JM Sasse & DW Cameron, unpublished data). For accurate quantitation and calculation of losses, internal standards include appropriate d-labeled BRs in MS, and selected "spikes" in LC analysis.

Identification of the conformations of BRs may be assisted by crystallographic (62) and NMR studies (reviewed in 4), and such knowledge will be useful in receptor studies, and in understanding structure/activity relationships. These relationships have been examined by several groups, and while some differences were noted between bioassays, the 7-oxalactone and 6-keto forms were generally the most active, with distinct effects of the hydroxylation patterns in ring A and the side chain, and of the alkylation pattern of the side chain (66, 84, 123, 124, 135, 140). A computer analysis of interatomic distances in energy-minimized structures of various BR structures showed that the distances between C_{16} on the ring and the C_{22}, C_{23}, C_{24}, and C_{28} carbons as well as the O_{22} and O_{23} oxygens, were critical for optimal activity, suggesting that the overall dimensions of the side chain may be as important as the configuration at the individual chiral carbons (70).

A quantitative structure/activity study has begun (12), and the finding of high activity with inversion of the 2- and 3-hydroxyl groups together with a *cis* A/B ring junction is interesting—this is the arrangement in ecdysones. However, 20-hydroxyecdysone, alone or together with brassinolide, does not affect the rice lamina bending assay (JM Sasse, unpublished data) confirming early work with ecdysones in other assays (26, 125) and suggesting the lamina inclination assay can discriminate between patterns of substitution in the side chain, even if conformations in the ring A region permit activity.

BIOSYNTHESIS AND METABOLISM

Early and Late C_6 Oxidation Pathways

A detailed understanding of how endogenous BR levels are regulated via synthesis, breakdown, and conjugation is an essential component of a molecular model of BR action. A coordinated effort by several Japanese groups has led to rapid progress in the elucidation of the biosynthetic pathways leading to BRs in plant cell cultures and seedlings (reviewed in 38, 39, 132). Campesterol was predicted to be the plant sterol progenitor of brassinolide based on side chain structure; and the relative biological activities, co-occurrence, and molecular structure of teasterone, typhasterol, and castasterone suggested that brassinolide was synthesized from campesterol through these intermediates (136).

The reduction of campesterol to campestanol and the oxidation of campestanol to 6-oxocampestanol (Figure 2) has been demonstrated by feeding experiments (39), and the hydroxylation of 6-oxocampestanol to cathasterone is presumed, but direct demonstration of this step by feeding experiments has not been accomplished, possibly because the endogenous pool of cathasterone is 500-fold less than that of 6-oxocampestanol (37). The large difference in pool sizes suggests that this conversion may be the rate-limiting step in brassinolide biosynthesis. Conversion of cathasterone to brassinolide via the intermediates shown in Figure 3 has been demonstrated by feeding experiments (39). In the final step, some differences were seen in seedlings; castasterone was converted to brassinolide in *Catharanthus roseus*, but not in tobacco and rice (106). However, since brassinolide and castasterone co-occur as endogenous BRs in rice seedlings, it is likely the full pathway is operational in this plant as well and that the exogenous labeled castasterone used in the feeding experiments may not have reached the site of brassinolide synthesis in the rice seedlings (106). The conversion of teasterone to brassinolide did occur in lily cells (8), and

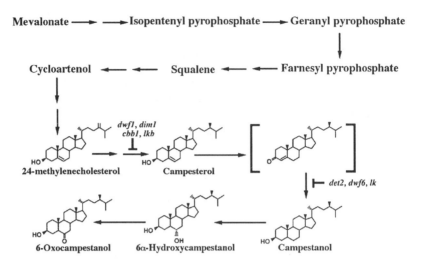

Figure 2 Biosynthesis of early members of the BR biosynthetic pathway via mevalonate and the isoprenoid pathway. Campesterol is a bulk sterol, also found in membranes, while campestanol and later derivatives are considered committed to BR biosynthesis. Proposed blocks in BR biosynthesis in Arabidopsis (*dwf*1, *dim*1, *cbb*1; *det*2, *dwf*6) and pea (*lkb, lk*) mutants are indicated. The structure in *brackets* is a probable intermediate based on molecular genetic and biochemical studies (62a).

the co-occurrence of teasterone, typhasterol, castasterone, and brassinolide in at least four other species (*Phaseolus vulgaris, Lilium elegans, Citrus unshiu,* and *Thea sinensis*) suggests that the complete pathway is widespread in plants (106).

C_6-oxidation is a very early step in this pathway (Figure 2). However, appreciable quantities of 6-deoxocastasterone have frequently been identified in extracts of BRs, and when other 6-deoxo-congeners were discovered, these compounds were also proposed as precursors to BRs (45, 90), and this has now been confirmed (17, 18). Thus, "early C_6-oxidation" and "late C_6-oxidation" pathways for the biosynthesis of BRs coexist in cultured cells of *Catharanthus roseus,* and in seedlings of tobacco and rice. Representatives from both pathways co-occur in many plants, e.g. Arabidopsis (36), so both could be widespread in the plant kingdom. The multiplicity of substitution patterns in the 2- and 3-positions of BRs, and in the alkyl substituents of the side chain, imply even more pathways and interconnections, and these remain to be elucidated, as does the detailed enzymology. Determining the specificity of the enzymes involved will assist our understanding, and the contribution of BR biosynthetic mutants has already been significant.

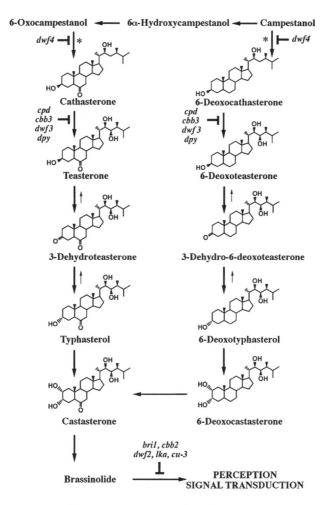

Figure 3 Biosynthesis of brassinolide via the early C_6 oxidation pathway (*left*) and the late C_6 oxidation pathway (*right*). Both pathways can occur together in the same plant. Steps marked by * have not been confirmed by feeding experiments. Putative assignments of biosynthetic and insensitive mutants in Arabidopsis (*dwf4*; *cpd*, *cbb3*, *dwf3*; *bri1*, *cbb2*, *dwf2*), pea (*lka*), and tomato (*dpy*, *cu-3*) are indicated.

BR Biosynthetic Mutants

In the past two years, several dwarf mutants, isolated during screens for physiological processes apparently unrelated to BRs, have been shown in fact to be lesions in genes encoding BR biosynthetic enzymes. Perhaps the best characterized of these is *det2*, a de-etiolated Arabidopsis mutant that was originally proposed to be a negative regulator of photomorphogenesis since it shows characteristics of light-grown plants even when grown in the dark (20). Sequence analysis of the *DET2* gene showed considerable identity (38–42% overall, 80% in conserved regions) with mammalian steroid 5α-reductases that catalyze the reduction of 3-oxo,$\Delta^{4,5}$ steroids such as testosterone (64). A similar reduction in BR biosynthesis occurs between campesterol and campestanol, and examination of endogenous levels showed that campestanol was much reduced in the *det2* mutant when compared with wild type (91).

Evidence that *DET2* encodes a biosynthetic enzyme comes from the fact that BRs, but not other growth regulators, rescued the Det2 phenotype to wild type, as did the transformation of the mutant with human 5α-reductases driven by the CaMV 35S promoter (62a, 64). Rescue by the human genes did not require BR application and was prevented by specific inhibitors of the mammalian 5α-reductases (62a). Moreover, recombinant DET2 protein expressed in human embryonic kidney cells was able to reduce several 3-oxo,$\Delta^{4,5}$ mammalian steroids, including testosterone and progesterone, but failed to reduce 3β-hydroxy,$\Delta^{5,6}$ steroids such as cholesterol (62a). These experiments provide convincing evidence that the DET2 enzyme performs the same function as human steroid 5α-reductases. However, campesterol is a 3β-hydroxy,$\Delta^{5,6}$ steroid, implying that plants possess an additional enzyme that converts campesterol to its 3-oxo,$\Delta^{4,5,}$ isomer before reduction by the DET2 5α-reductase. In mammals, a membrane-bound 3β-hydroxysteroid dehydrogenase/$\Delta^{5,6}$-$\Delta^{4,5}$ isomerase performs this function (62a) and use of this gene as a probe may allow cloning of the plant homolog.

Mutants for an earlier step in the BR pathway involving a reductase have been putatively assigned. The Arabidopsis *dwf*1 (33) and its alleles *dim*1 (109) and *cbb*1 (56) are rescued to wild type by BR treatment (K Feldmann, personal communication; 56, 107). Recent examination of BR levels in *dim*1 has shown that 24-methylenecholesterol levels are elevated while campestanol and campesterol are reduced, suggesting that *dim*1 (*dwf*1, *cbb*1) is blocked in the conversion of 24-methylenecholesterol to campesterol (U Khlare, personal communication). The gene has been cloned and shows some homology to FAD-dependent oxidases (74). Since the conversion of 24-methylenecholesterol to campesterol involves a reduction, the oxidase homology is either artefactual or the oxidase functions in the reverse reaction as a reductase. Experiments with recombinant protein and labeled substrates, as described for DET2 above, would help to resolve this question. The *lkb* mutant of pea is a BR-deficient dwarf that shows normalization of internode growth upon application of a range of BRs (79). Recent studies showed that 24-methylenecholesterol accumulated in *lkb* mutants while campesterol and campestanol were dramatically reduced, suggesting that *lkb*, like *dwf*1 in Arabidopsis, is a lesion in the gene encoding the biosynthetic enzyme responsible for 24-methylenecholesterol to campesterol conversion (T Yokota, personal communication).

Putative steroid hydroxylases have also been cloned from Arabidopsis BR-deficient dwarf mutants. Brassinolide, castasterone, typhasterol, 3-dehydroteasterone, and teasterone all rescued the *cpd* mutant to wild type phenotype in the light and the dark, while cathasterone and its precursors had no effect, implicating CPD in the C_{23} hydroxylation of cathasterone to teasterone (107). Sequence analysis of the cloned gene supports this view; CPD shows 24% identity to rat testosterone-16α-hydroxylase and 19% identity to human progesterone-21α-hydroxylase, including conserved amino acids in the steroid substrate-binding domains (107). Moreover, there is 50–90% identity with conserved domains of microsomal cytochrome P450s, and CPD has been classified as a CYP90 P450 monooxygenase (107). CPD must also serve to hydroxylate 6-deoxocathasterone to 6-deoxoteasterone, since the involvement of a different enzyme in the late C_6 oxidation pathway would allow synthesis of brassinolide in the *cpd* mutant and a dwarf phenotype should not be observed. Some direct evidence that the same hydroxylase catalyzes both early and late C_6 oxidation steps comes from the *dpy* mutant of tomato, an intermediate dwarf with severely altered leaf morphology, including the downward curling and dark-green color typical of the Arabidopsis BR mutants. We have found that teasterone and 6-deoxoteasterone, along with all subsequent intermediates, rescue *dpy* to wild type while cathasterone, 6-deoxocathasterone, and all of its precursors fail to do so (E Cerny, S Fujioka & S Clouse, unpublished data). This suggests that *DPY* is the tomato homolog of *CPD*, and cloning by transposon tagging is underway to determine the extent of sequence homology between the tomato and Arabidopsis genes (E Cerny, G Bishop & S Clouse, unpublished data). Another tomato gene, *DWARF*, was cloned by transposon tagging and shown to encode a cytochrome P450 (CYP85) with 38% identity to CPD (11). Preliminary feeding experiments suggest that DWARF is also a BR biosynthetic enzyme, but later in the pathway than CPD (E Cerny, G Bishop, S Fujioka & S Clouse, unpublished data).

The *DWF4* gene, encoding the enzyme responsible for the C_{22} hydroxylation of 6-oxocampestanol to cathasterone (the potential rate-limiting step of BR biosynthesis), has also been cloned. Cathasterone, 6-deoxocathasterone, and all subsequent intermediates in both the early and late C_6 oxidation pathway rescued the Arabidopsis *dwf4* to wild type, but campestanol and 6-oxocampestanol failed to do so. Sequence analysis showed that DWF4 was another cytochrome P450 with 42% sequence identity to CPD (S-W Choe & K Feldmann, personal communication).

Conjugation and Metabolism

As discussed recently (4), conjugates of plant hormones are considered to be transport, storage, or inactivated forms. Two examples of 23-glucosyl conjugates of BRs were known (134), and reversible acyl conjugation at the 3-position has now been reported (2, 7, 8). Glucosyl, sulfate, and acyl conjugates were also detected after BRs were supplied to explants and cell cultures. The importance of the 3-β-epimerization for acylation (and also for glucosidation) has been confirmed, and reduction at C_6-, hydroxylation, and glucosidation or degradation of the side chain were also observed (4, 5, 47, 48, 77, 89, 103). Diglycosidation of exogenous 24-*epi*easterone occurred in cell cultures of tomato, *Lycopersicon esculentum* (60), and in the same system, hydroxylation at the 25- and 26-positions of 24-*epi*brassinolide required two separate hydroxylases, with only the 25-hydroxylase sensitive to cytochrome P450 inhibitors (127). How many of these transformations occur in vivo as part of normal biosynthesis and turnover of BRs remains to be seen, but the possibility of very low levels of even more active members of the family, like those produced by fungal metabolism (48, 117, 118) is tantalizing.

PHYSIOLOGICAL RESPONSES TO BRs

Early work explored the range of effects of BRs in various bioassays and their interactions with inhibitors and promoters. The responses include effects on elongation, bending, cell division, reproductive and vascular development, membrane polarization and proton pumping, source/sink relationships, and modulation of stress. BRs also interact with environmental signals and can affect insect and fungal development (for reviews, see 3, 6, 19, 66, 67, 72, 83, 89, 94, 96). The sites for BR synthesis in planta are not yet elucidated. It may be that all tissues produce them, since BR biosynthetic and signal transduction genes are expressed in a wide range of plant organs (63), and short-distance effects (as seen in pollen, seeds, and cell cultures) can be assumed. Studies on the distribution of labeled BRs supplied exogenously suggest long-distance transport is predominately acropetal (e.g. 78, 133), but it is not yet known whether their long-distance transport is important in normal plant growth and development.

Examination of the phenotype of BR-deficient and insensitive mutants provides independent confirmation that many of the effects observed by exogenous application of BRs to bioassay systems do in fact occur in planta. In Arabidopsis, BR mutants show extreme dwarfism (as small as one thirtieth the size of wild type) which is rescued only by BR application in the deficient but not the insensitive mutants, and microscopic examination of cell files in various organs shows that mutant cells are shorter than the corresponding wild-type cells, confirming the role of BRs in elongation (24, 56, 64, 107). In Arabidopsis and tomato, BR mutants show a de-etiolated phenotype in which dark-grown seedlings exhibit the short hypocotyl and open cotyledons characteristic of light-grown plants. In the light, these same mutants exhibit extremely altered leaf morphology (22). Thus, an important role of BRs in photomorphogenesis and leaf morphogenesis can be assumed. BR mutants generally have reduced fertility or male sterility, delayed senescence, and altered vascular development, implicating BRs in all these developmental processes (21).

Cell Expansion

BR application at nM to μM levels causes pronounced elongation of hypocotyls, epicotyls, and peduncles of dicots, as well as coleoptiles and mesocotyls of monocots (21, 66, 94). Young vegetative tissue is particularly responsive to BRs, and, if endogenous BRs are directly involved in the control of cell expansion, they must be present in such tissue. Approaches to establishing this include the analysis of levels in a BR-sensitive zone of pea stem (99) and localization of an exogenously supplied ^{125}I-BR, which accumulated in the elongating zone of mung bean epicotyls and the apex of cucumber seedlings (129).

While both BR and auxin promote elongation, their kinetics are quite different. Auxin generally shows a very short lag time of 10 to 15 min between application and the onset of elongation, with maximum rates of elongation reached within 30 to 45 min (108). In contrast, BR has lag times of at least 45 min with elongation rates continuing to increase for several hours (26, 55, 68, 140). This difference in kinetics is also seen at the level of gene expression in Arabidopsis, where auxin induces the *TCH4* gene much more rapidly than BR (130). Other differences in the effect of auxin and BR on elongation have been observed in physiological (93, 115) and molecular studies (23, 24, 26, 140). However, synergisms between BR and auxins occur in many systems (66). Additive effects on elongation were often seen with gibberellins, and enhancement of lateral enlargement induced by cytokinins, and inhibitory effects of cytokinins, abscisic acid, and ethylene on BR-induced elongation have been described for stem tissue (55, 93 and references therein).

BR-induced expansion is accompanied by proton extrusion and hyperpolarization of cell membranes and these effects have also been observed in the asymmetric expansion of the joint pulvini of rice laminae (14) and in an alga (9) where BRs at concentrations from 10^{-15}–10^{-8} M markedly stimulated and accelerated the growth cycle. However, some authors found little effect in vivo of 24-*epi*brassinolide on plasmalemma ATPases of wheat roots (128) and decreased activity in diploid and tetraploid buckwheat (28). Indirect modulation of ATPase activity had also been invoked to explain BR-induced effects on sucrose transport (3, 82).

Inhibitory effects of BRs on expansion have been widely reported in root tissue. In general, exogenous application of BRs inhibits primary root extension and lateral root formation, with occasional promotions of elongation or adventitious rooting seen with <pM concentrations (23, 24, 85, 97). However, there is some evidence for involvement of endogenous BRs in the control of lateral root initiation; uniconazole treatment produced many stunted lateral roots in *Lotus*, but concomitant brassinolide treatment reduced the number to the control value (57). Inhibitory effects, particularly on expansion, are often mediated via the induction of ethylene biosynthesis, and treatment with exogenous BRs increases the production of ethylene in stem tissue (summarized in 6). However, recent work on the inhibitory effects of a brief brassinolide treatment of cress seeds showed ethylene levels were not increased in the germinants, suggesting an independent inhibitory action of BR (54). It is not clear what role endogenous BRs play in the early stages of germination, but changes in levels of castasterone and brassinolide were seen after radish seeds germinated (102).

Cell Division

Reports of promotive effects on cell division in whole plants (66) proved hard to confirm in model systems, with elongation (92) or inhibition (44, 86, 126) noted instead. Furthermore, microscopic examination of BR-deficient and BR-insensitive mutants in Arabidopsis showed that the dwarf phenotype was due to reduced cell size, not cell number (56). However, in cultured parenchyma cells of *Helianthus tuberosus*, application of nanomolar concentrations of BR-stimulated cell division by at least 50% in the presence of auxin and cytokinin (25). In Chinese cabbage protoplasts, 24-*epi*brassinolide, when applied with 2,4-D and kinetin, promoted cell division in a dose-dependent manner and enhanced cluster and colony formation. The data suggested that dedifferentiation of the protoplasts was enhanced and that BR promoted or accelerated the necessary regeneration of the cell wall before division (75). Brassinolide also accelerated the rate of cell division in Petunia protoplasts in the presence of auxin and cytokinin but could not take the place of either hormone (M-H Oh & S Clouse, unpublished data). The contradictory results make the role of BRs in cell division unclear, and much further work is required to resolve this issue, including studies of the effect of BR on genes controlling cell division.

Vascular Differentiation

Auxin and cytokinin are required for the initiation of xylem development both in vivo and in vitro (40). However, evidence continues to mount that BRs may also play a significant role in vascular differentiation. In *H. tuberosus* explants, one of the major in vitro systems for studying xylem differentiation, Clouse & Zurek (25) found that nM concentrations of exogenous brassinolide increased differentiation of tracheary elements 10-fold after only 24 h. Normally, tracheary element differentiation requires 72 h in this system. In isolated mesophyll cells of *Zinnia elegans*, a second widely used model system for xylem differentiation, tracheary element formation has been divided into three stages (40). In Stage I, the mesophyll cells dedifferentiate after induction by auxin, cytokinin, and wounding, and specific transcripts are induced including the phenylpropanoid pathway members phenylalanine ammonia-lyase and cinnamate hydroxylase. During Stage II, phenylpropanoid pathway gene expression abates and three-dimensional networks of actin filaments form. In Stage III, phenylalanine ammonia-lyase and cinnamate hydroxylase gene expression again increases, the highly lignified secondary wall is formed, and programmed cell death ensues. Current evidence suggests that endogenous BRs are required for entry into Stage III (40). Uniconazole, an inhibitor of both gibberellin and BR biosynthesis, prevented differentiation of *Z. elegans* mesophyll cells into tracheary elements, and this inhibition was overcome by BR but not by gibberellin application (53). Uniconazole also inhibited Stage III–specific genes but not those specific to Stages I or II. Moreover, expression of phenylalanine ammonia-lyase and cinnamate hydroxylase was inhibited by uniconazole during Stage III but not Stage I, and this inhibition was overcome by brassinolide and several BR biosynthetic intermediates (40).

The spatial expression of *BRU*1, a BR-regulated gene encoding a xyloglucan endotransglycosylase (XET) in soybean (139), also points to a role of BRs in xylem differentiation. XETs are thought to be involved in processes requiring cell wall modification, including expansion (see below), vascular differentiation, and fruit ripening (35). In cross-sections of elongating soybean epicotyls, *BRU*1 expression was most intense in paratracheary parenchyma cells surrounding vessel elements (80) suggesting a role for BRs and XETs in xylem formation. The modification of cambial division seen in a BR-deficient mutant (56, 107) also suggests involvement of endogenous BRs in xylem differentiation in vivo, and it is relevant that BRs have been identified in cambial scrapings of *Pinus silvestris* (58) and a *Eucalyptus* species (T Yokota, unpublished work).

Pollen and Reproductive Biology

Pollen is a rich source of endogenous BRs and in vitro studies have suggested that pollen tube elongation could depend in part on BRs (51). Male sterility of BR-insensitive mutants would support this (24, 56, 63), but the failure of the filament to elongate such that the pollen, although viable, cannot reach the stigma was suggested as an alternative mechanism of male sterility for the BR-deficient *dwf4* mutant (S-W Choe & K Feldmann, personal communication). However, the *cpd* mutant was reported to be male sterile because the pollen itself failed to elongate during germination (107). In addition, pollination is often the initial step for the genesis of haploid plants, and in *Arabidopsis thaliana* and *Brassica juncea*, treatment with brassinolide induced the formation of haploid seeds that developed into stable plants (59). Subcellular localization of BRs was explored in pollen of *Brassica napus* and *Lolium temulentum*, using polyclonal antibodies generated against castasterone, and the data suggested BRs could be stored (or trapped) in developing starch granules and be released on imbibition (99, 114). The relative distribution of BRs in maturing pollen has also been explored chemically (8), and conjugated teasterone was present at the microspore stage. Its level decreased as the pollen developed, and levels of free BRs increased. Taken together, these data suggest that BRs have important physiological roles in the fertilization of plants.

With respect to the general effect of BRs on sex differentiation in plants, Suge (105) found that direct application of brassinolide to the staminate influo-rescence of *Luffa cylindrica* induced bisexual and pistillate flowers. Numerous model systems of sexual morphogenesis in plants are currently available, and application of BRs to these systems could be a profitable exercise.

Senescence

Senescence of leaf and cotyledon tissue has often been shown to be retarded in vitro by administration of cytokinins; in contrast, 24-*epi*brassinolide acceler-ated senescence in such systems (29, 50, 138). Altered activities of peroxidase, superoxide dismutase, and catalase and a marked increase in the level of mal-ondialdehyde were observed, and the authors suggested BRs might regulate these effects via "activated oxygen." Delayed senescence in Arabidopsis BR mutants would tend to support the role of BRs in accelerating senescence in normal plants (24, 56, 64, 107). However, work concerned with lipid perox-idation suggests 24-epibrassinolide inhibits oxidative degradation, decreases malondialdehyde levels (31), and acts as a membrane protectant, thus delay-ing senescence. Examination of the effect of BR application on senescence-associated mutants of Arabidopsis and study of the expression of senescence-associated genes in the BR mutants will be necessary to help clarify the role of BRs in this process.

Modulation of Stress Responses

Since the early reports of the ameliorative effects of BR-treatment of stressed plants (49), most work has focused on chilling stress (summarized in 55, 112,

126). In rice, 24-*epi*brassinolide treatment reduced electrolyte leakage during chilling at 1–5°, reduced malondialdehyde content and slowed the decrease in activity of superoxide dismutase; while levels of ATP and, initially, proline, were enhanced. The enhanced resistance was attributed to BR-induced effects on membrane stability and osmoregulation (121). In a model system that develops both cold- and thermotolerance, treatment with 24-*epi*brassinolide enhanced both tolerances. While abscisic acid (at higher concentration) was a more effective regulator in both cases, there were interesting differences in gene expression resulting from the application of BR or abscisic acid (126). Changes in the spectrum of heat-shock proteins after BR administration, and promotion of heat shock granule formation, had also been reported in heat-stressed wheat (61).

Mild drought stress in sugar beet is ameliorated by treatment with a synthetic BR (100), and effects in other crop plants have been explored (88, 104). Improvement in salt tolerance in BR-treated rice has been confirmed (111, 112), and a protective effect on barley leaf ultrastructure after a 24-h exposure to 500 mM sodium chloride described (61). A promotive effect of 10 μM 24-*epi*brassinolide was noted on both germination rate and percentage of *Eucalyptus camaldulensis* seeds in 150 mM sodium chloride (98). BR treatment was also claimed to protect against various pathogens (49), but this may not be general, because very low concentrations of two BRs induced susceptibility of potato to *Phytophora infestans*, and the effect was long lasting (116). However, it is interesting to note that in the BR-deficient *cpd* mutant, the expression of pathogenesis related genes is greatly reduced (107). Little further work has been done on the effects of BRs as herbicide safeners (49, 112, 113), but interest in the interactions of BRs with insects (16, 83) continues, with the object of using BR-analogues as antiecdysones (65).

MOLECULAR MECHANISMS OF ACTION

Aspects of BR-Promoted Cell Elongation

The plant cell wall forms a highly cross-linked, rigid matrix that opposes cell expansion and differentiation. In order for elongation and other morphogenetic processes to occur, the cell wall must be modified, i.e. by wall relaxation or loosening and by incorporation of new polymers into the extending wall to maintain wall integrity. Several proteins with possible roles in cell wall modification processes have been identified, including glucanases, xyloglucan endotransglycosylases (XETs), and expansins (27). While the complexity of the wall has made a definitive molecular model of wall extension elusive, a plausible scenario has been presented by Cosgrove (27) in which expansins are primarily responsible for wall relaxation, but glucanases and XETs affect the extent of expansin activity by altering the viscosity of the hemi-cellulose matrix. Furthermore,

XETs may function to incorporate new xyloglucan into the growing wall, and cellulose biosynthesis would also be expected to occur. It is certain, however, that BRs alter the biophysical properties of plant cell walls (115, 122, 140) and also increase the abundance of mRNA transcripts for wall-modifying proteins such as XETs (15, 130, 139).

In elongating soybean epicotyls, BR application resulted in increased plastic extensibility of the walls within 2 h with a concomitant increase in the mRNA level of a gene named *BRU1*, which showed significant homology to numerous XETs (139). Subsequent enzyme assays with recombinant protein showed that BRU1 was indeed a functional XET (80). Moreover, increasing concentrations of applied BR during early stages of elongation lead to a linear increase in extractable XET activity in the epicotyls (80). The *BRU1* gene was regulated specifically by BRs during early stages of elongation, and increased expression was not simply the consequence of enhanced elongation (139). Therefore, BRU1 is likely to play an important role in BR-promoted epicotyl elongation in soybean. The role of expansins in this system has yet to be examined.

BR-regulated XETs have also been identified in Arabidopsis. The *TCH4* gene was shown to encode an XET whose expression was increased within 30 min of BR treatment, with a maximum at 2 h (130). In contrast to the soybean *BRU1*, *TCH4* was also rapidly induced by auxin treatment (130). Environmental stimuli such as touch, darkness, and temperature also affected *TCH4* expression. The expression of *TCH4* was restricted to expanding tissue and organs that undergo cell wall modification such as vascular elements (130). *TCH4* expression was greatly reduced in BR-deficient and insensitive mutants, again suggesting a role for XETs in BR-promoted elongation (56).

Cell expansion also depends on an adequate supply of wall components, and recent work (68) confirmed the inhibition of BR-induced elongation in stem tissue by inhibitors of cellulose biosynthesis or microtubule orientation (93, 94). Brassinolide, alone or in combination with auxin, enhanced the percentage of transversely oriented cortical microtubules (68). The orientation of cortical microtubules, which generally correlates with the orientation of microfibrils, follows a cyclic pattern, and phosphorylation of proteins, possibly those linking the microtubules to the plasmalemma, is an essential component in the maintenance of transversely oriented microtubules. BR-induced elongation also depends on such phosphorylation (69).

Levels of Gene Regulation by BRs

Transcriptional regulation of gene expression by BRs has been demonstrated with the *TCH4* gene of Arabidopsis (130), using *TCH4* promoter:*GUS* fusions lacking any 5' or 3' *TCH4* transcribed sequences (W Xu & J Braam, personal communication). Subsequent work has localized the promoter region responsible for BR-regulation to within 100 bp, and linker-scanning mutagenesis is under way to identify the specific BR response element for this gene (B Torisky, J Braam & S Clouse, unpublished data). While transcriptional regulation of gene expression by animal steroid hormones is more commonly discussed (32), posttranscriptional regulation has also been observed (76). The *BRU1* gene of soybean represents an example of such posttranscriptional gene regulation by a plant steroid (139). The posttranscriptional regulation, which apparently involves *BRU1* mRNA stability, is maintained in a BR-dependent manner in soybean cell suspension cultures, thus providing an excellent model system for dissecting *cis*-acting sequences responsible for this regulation (J Jiang & S Clouse, unpublished data).

BR Signal Transduction

The value of hormone-insensitive mutants in unraveling signal transduction pathways in plants has been demonstrated for ethylene (30) and abscisic acid (34), and such an approach has proven successful for BRs as well. Clouse et al (24) identified a BR-insensitive mutant in Arabidopsis by the ability of mutant plants to elongate roots in the presence of inhibitory concentrations of BR with respect to wild type. The mutant, named *bri1*, showed severe pleiotropic effects on development including dwarfism, de-etiolation, male sterility, and altered leaf morphology, which suggested that the BRI1 protein product played an important role in BR signal perception or transduction. Several alleles of *BRI1* with identical phenotype have been isolated in independent screens (K Feldmann, personal communication; 56, 63)

In animals, there are two major paradigms of signal transduction. The first involves a membrane-bound receptor with an extracellular ligand-binding domain and an intracellular domain responsible for transducing the signal to the next member of the pathway, often a kinase or G protein. Amplification and proliferation of the signal proceeds through cascades of phosphorylation and dephosphorylation and involves second messengers such as calcium, cyclic AMP, and diacyl glycerol (131). The second pathway involves intracellular receptors that recognize steroid or steroid-like ligands to directly affect the transcription of specific genes by binding directly to the promoter of hormone-responsive genes (10). Because of the structural similarity of animal and plant steroid signaling molecules, it is reasonable to assume that plants might have members of the intracellular superfamily of steroid receptors. The *BRI1* gene has now been cloned, and perhaps somewhat surprisingly, shows strong sequence homology not to steroid receptors but to leucine-rich receptor kinases that function at the cell surface to transduce extracellular signals (63).

The *BRI1* gene shares homology with plant and animal receptor kinases in all conserved domains including ligand binding domain, membrane domain, and cytoplasmic kinase domain. Moreover, sequence analysis of five mutant alleles confirms that the putative ligand binding and kinase domains are essential for in vivo function (63). On the basis of the dramatic pleiotropic phenotype of the *bri*1 mutants and the sequence homology of BRI1 to important signal transduction molecules, it is obvious that BRI1 is a critical component of the BR signal transduction pathway. However, its role as the BR receptor has not been confirmed by direct binding studies. No ligands have been identified for plant receptor kinases, and in animals, all known ligands for such receptors are polypeptides or glycoproteins (120). It is possible that the BR receptor is a distinct polypeptide that binds to BRI1 in the presence of BR, or there may be an unknown ligand that is required for BR activity. Even if BR binds directly to BRI1, it does not exclude the possibility that there are also intracellular BR receptors, since it is now known that in animals both intracellular and extracellular steroid receptors co-occur, with the intracellular receptor mediating gene expression and the extracellular receptor modulating nongenomic responses such as calcium ion flux and phosphorylation status of a variety of proteins (71).

We have recently identified a gene, *TCH4-BF*1, that may be involved in the terminal end of BR signal transduction (M-H Oh, T Altmann & SD Clouse, unpublished data). Recombinant TCH4-BF1 binds specifically to the *TCH4* promoter in the region thought to contain the BR response element (see above), and sequence analysis shows homology to a type of zinc finger protein called the PHD finger, found in plant and Drosophila transcriptional regulators of homeodomain genes and in mammalian transcription factors (1). A hypothetical scheme that incorporates both BRI1 and TCH4-BF1 in BR signal transduction is shown in Figure 4.

CONCLUSION AND FUTURE PERSPECTIVES

Recent research on the chemistry, physiology, and molecular biology of BRs provides a convincing body of evidence that these plant steroids are essential regulators of plant growth and development. The pace of BR research is accelerating rapidly, and with the proliferation of cloned genes and advances in microchemical techniques, the range of experimental approaches to understanding BR action continues to expand. Much remains to be done, however. Determination of whether or not BRI1 is the BR receptor by measuring the binding of high-specific activity radiolabeled BRs to recombinant BRI1 protein is of top priority. Studies on the effect of phosphorylation/dephosphorylation states on BR signal transduction using available inhibitors of kinases and phosphatases will be informative; e.g. okadaic acid, an inhibitor of type I phosphatases, blocks BR-promoted elongation in soybean (S Clouse, unpublished data), suggesting a role for phosphatases in addition to the BRI1 kinases in BR action. The availability of BRI1 and TCH4-BF1 as probes will allow application of interactive cloning techniques to identify new components of the BR signal transduction

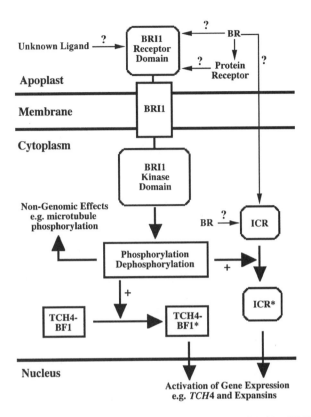

Figure 4   A hypothetical scheme for BR signal transduction in Arabidopsis. BRI1, a putative leucine-rich receptor kinase, may bind BR directly (or BR bound to a distinct protein receptor that functions at the cell surface), which causes activation of the kinase domain and subsequent phosphorylation of additional kinases and/or phosphatases. Signal transduction results in activation (*shown with + or **) of transcription factors (such as TCH4-BF1 and other unknown proteins), which then bind to promoters of BR-responsive genes such as *TCH4*. It is also possible the BRI1 binds an unknown ligand, while BR binds to an intracellular receptor (ICR, similar to animal steroid receptors), which requires activation by the kinase domain of BRI1. There is currently no evidence for an intracellular receptor. Nongenomic activation by BR via BRI1-mediated phosphorylation of microtubules or metabolic enzymes, etc, is also a possibility.

pathway. The study of *BRI*1, *DET*2, *DWF*4, and *CPD* expression by in situ hybridization in different organs at different developmental stages, coupled with direct measurements of endogenous BR levels, will help verify the role of BRs in different physiological processes. Epistasis studies using BR mutants crossed with other hormone and developmental mutants should help to place BRs in the overall pattern of development, as will studies of the ectopic expression of *BRI*1 and biosynthetic genes in a variety of different transgenic plants.

By the time the next review on BRs appears in this series, our knowledge of the molecular mechanisms of BR action will have expanded dramatically, and the degree of evolutionary conservation between plant and animal steroid signaling pathways will be determined in much more detail.

ACKNOWLEDGMENTS

Thanks to G Adam, A Bennett, J Chory, S Fujioka, K Feldmann, K Uhle, and T Yokota for providing updates before publication. Research in the laboratory of SDC is supported by the National Science Foundation, the USDA NRICGP, and the North Carolina Agricultural Research Service.

Visit the *Annual Reviews home page* at
http://www.AnnualReviews.org.

Literature Cited

1. Aasland R, Gibson T, Stewart A. 1995. The PHD finger: implications for chromatin-mediated transcriptional regulation. *Trends Biol. Sci.* 20:56–59
2. Abe H, Asakawa S, Natsume M. 1996. Interconvertible metabolism between teasterone and its conjugate with fatty acid in cultured cells of lily. *Proc. Plant Growth Regul. Soc. Am.* 23:9
3. Adam G. 1994. Brassinosteroide-eine neue Phytohormon Gruppe. *Naturwissenschaften* 81:210–17
4. Adam G, Porzel A, Schmidt J, Schneider B, Voigt B. 1996. New developments in brassinosteroid research. In *Studies in Natural Products Chemistry*, ed. ST Attaur-Ranman, pp. 495–549. Amsterdam: Elsevier
5. Adam G, Schneider B, Kolbe A, Hai T, Porzel A, et al. 1996. Progress in brassinosteroid metabolism. *Proc. Plant Growth Regul. Soc. Am.* 23:7
6. Arteca RN. 1995. Brassinosteroids. In *Plant Hormones Physiology, Biochemistry and Molecular Biology*, ed. PJ Davies, pp. 206–13. Dordrecht: Kluwer. 2nd ed.
7. Asakawa S, Abe H, Kyokawa Y, Nakamura S, Natsume M. 1994. Teasterone-3-myristate: a new type of brassinosteroid derivative in *Lilium longiflorum* anthers. *Biosci. Biotechnol. Biochem.* 58:219–20
8. Asakawa S, Abe H, Nishikawa N, Natsume M, Koshioka M. 1996. Purification and identification of new acyl-conjugated teasterones in lily pollen. *Biosci. Biotechnol. Biochem.* 60:1416–20
9. Bajguz A, Czerpak R. 1996. Effect of brassinosteroids on growth and proton extrusion in the alga *Chlorella vulgaris* Beijerinck (Chlorophyceae). *J. Plant Growth Regul.* 15:153–56
10. Beato M, Herrlich P, Schutz G. 1995. Steroid hormone receptors: many actors in search of a plot. *Cell* 83:851–57
11. Bishop G, Harrison K, Jones J. 1996. The tomato *Dwarf* gene isolated by heterologous transposon tagging encodes the first member of a new cytochrome P450 family. *Plant Cell* 8:959–69
12. Brosa C. 1997. Biological effects of brassinosteroids. In *Biochemistry and Function of Sterols*, ed. EJ Parish, WD Nes, pp. 201–20. Boca Raton, FL: CRC Press
13. Caballero G, Centurion OT, Galagovsky L, Gros E. 1996. FAB mass spectrometry of 5-α-H and 5-α-OH derivatives of 28-homocastasterone and 28-homoteasterone. *Proc. Plant Growth Regul. Soc. Am.* 23:50–55
14. Cao HP, Chen SK. 1995. Brassinosteroid-induced rice lamina joint inclination and its relation to indole-3-acetic acid and ethylene. *Plant Growth Regul.* 16:189–96
15. Catala C, Rose J, Bennett A. 1997. Auxin-regulation and spatial localization of an endo-1,4-β-D-glucanase and a xyloglucan endotransglycosylase in expanding tomato hypocotyls. *Plant J.* 12:417–26
16. Charrois GJR, Mao H, Kaufman WR. 1996. Impact on salivary gland degeneration by putative ecdysteroid antagonists

and agonists in the ixodid tick *Amblyomma hebraeum*. *Pestic. Biochem. Physiol.* 55:140–49

17. Choi Y-H, Fujioka S, Harada A, Yokota T, Takatsuto S, et al. 1996. A new brassinolide biosynthetic pathway via 6-deoxocastasterone. *Phytochemistry* 43: 593–96

18. Choi Y-H, Fujioka S, Nomura T, Harada AYT, Takatsuto S, et al. 1997. An alternative brassinolide biosynthetic pathway via late C-6 oxidation. *Phytochemistry* 44:609–13

19. Chory J, Chatterjee M, Cook RK, Elich T, Fankhauser C, et al. 1996. From seed germination to flowering, light controls plant development via the pigment phytochrome. *Proc. Natl. Acad. Sci. USA* 93:12066–71

20. Chory J, Nagpal P, Peto CA. 1991. Phenotypic and genetic analysis of *det2*, a new mutant that affects light-regulated seedling development in *Arabidopsis*. *Plant Cell* 3:445–59

21. Clouse SD. 1996. Molecular genetic studies confirm the role of brassinosteroids in plant growth and development. *Plant J.* 10:1–8

22. Clouse SD. 1997. Molecular genetic analysis of brassinosteroid action. *Physiol. Plant.* 100:702–9

23. Clouse SD, Hall AF, Langford M, McMorris TC, Baker ME. 1993. Physiological and molecular effects of brassinosteroids on *Arabidopsis thaliana*. *J. Plant Growth Regul.* 12:61–66

24. Clouse SD, Langford M, McMorris TC. 1996. A brassinosteroid-insensitive mutant in *Arabidopsis thaliana* exhibits multiple defects in growth and development. *Plant Physiol.* 111:671–78

25. Clouse SD, Zurek D. 1991. Molecular analysis of brassinolide action in plant growth and development. See Ref. 27a, pp. 122–40

26. Clouse SD, Zurek DM, McMorris TC, Baker ME. 1992. Effect of brassinolide on gene expression in elongating soybean epicotyls. *Plant Physiol.* 100:1377–83

27. Cosgrove D. 1997. Relaxation in a high-stress environment: the molecular basis of extensible cell walls and enlargement. *Plant Cell* 9:1031–41

27a. Cutler HG, Yokota T, Adam G, eds. 1991. *Brassinosteroids Chemistry, Bioactivity, and Applications*. Washington, DC: Am. Chem. Soc.

28. Deeva VP, Pavlova IP, Khripach VA. 1996. The effect of 24-*epi*brassinolide on ATPase activity of plasmalemma and cytoplasmatic components of different buckwheat genotypes. *Proc. Plant Growth Regul. Soc. Am.* 23:32–35

29. Ding W-M, Zhao Y-J. 1995. Effect of *epi*-BR on activity of peroxidase and soluble protein content of cucumber cotyledon. *Acta Phytophysiol. Sin.* 21:259–64

30. Ecker JR. 1995. The ethylene signal transduction pathway in plants. *Science* 268:667–75

31. Ershova A, Khripach V. 1996. Effect of epibrassinolide on lipid peroxidation in *Pisum sativum* at normal aeration and under oxygen deficiency. *Russ. J. Plant Physiol.* 43:750–52

32. Evans RM. 1988. The steroid and thyroid hormone receptor superfamily. *Science* 240:889–95

33. Feldmann KA, Marks MD, Christianson ML, Quatrano RS. 1989. A dwarf mutant of *Arabidopsis* generated by T-DNA insertion mutagenesis. *Science* 243:1351–54

34. Finkelstein R, Zeevart J. 1994. Gibberellin and abscisic acid biosynthesis and response. In *Arabidopsis*, ed. E Meyerowitz, C Somerville, pp. 523–54. Cold Spring Harbor, NY: Cold Spring Harbor Lab. Press

35. Fry SC, Smith RC, Renwick KF, Martin DJ, Hodge SK, et al. 1992. Xyloglucan endotransglycosylase, a new wall-loosening enzyme activity from plants. *Biochem. J.* 282:821–28

36. Fujioka S, Choi Y-H, Takatsuto S, Yokota T, Li JM, et al. 1996. Identification of castasterone, 6-deoxocastasterone, typhasterol and 6-deoxotyphasterol from the shoots of *Arabidopsis thaliana*. *Plant Cell Physiol.* 37:1201–3

37. Fujioka S, Inoue T, Takatsuto S, Yanagisawa T, Yokota T, et al. 1995. Identification of a new brassinosteroid, cathasterone, in cultured cells of *Catharanthus roseus* as a biosynthetic precursor of teasterone. *Biosci. Biotechnol. Biochem.* 59:1543–47

38. Fujioka S, Sakurai A. 1997. Biosynthesis and metabolism of brassinosteroids. *Physiol. Plant.* 100:710–15

39. Fujioka S, Sakurai A. 1997. Brassinosteroids. *Nat. Prod. Rep.* 14:1–10

40. Fukuda H. 1997. Tracheary element differentiation. *Plant Cell* 9:1147–56

41. Gamoh K, Abe H, Shimada K, Takatsuto S. 1996. Liquid chromatography/mass spectrometry with atmospheric pressure chemical ionization of free brassinosteroids. *Rapid Commun. Mass Spectrom.* 10:903–6

42. Gamoh K, Prescott MC, Goad LJ, Takatsuto S. 1996. Analysis of brassino-

steroids by liquid chromatography/mass spectrometry. *Bunseki Kagaku* 45:523–27

43. Gamoh K, Takatsuto S. 1994. Liquid chromatographic assay of brassinosteroids in plants. *J. Chromatogr. A* 658:17–25

44. Gaudinova A, Sussenbekova H, Vojtechnova M, Kaminek M, Eder J, et al. 1995. Different effects of two brassinosteroids on growth, auxin and cytokinin concentrations in tobacco callus tissue. *Plant Growth Regul.* 17:121–26

45. Griffiths PG, Sasse JM, Yokota T, Cameron DW. 1995. 6-Deoxotyphasterol and 3-Dehydro-6-deoxoteasterone, possible precursors to brassinosteroids in the pollen of *Cupressus arizonica*. *Biosci. Biotechnol. Biochem.* 59:956–59

46. Grove MD, Spencer GF, Rohwedder WK, Mandava NB, Worley JF, et al. 1979. Brassinolide, a plant growth-promoting steroid isolated from *Brassica napus* pollen. *Nature* 281:216–17

47. Hai T, Schneider B, Adam G. 1995. Metabolic conversion of 24-*epi*brassinolide into pentahydroxylated brassinosteroid glucosides in tomato cell cultures. *Phytochemistry* 40:443–48

48. Hai T, Schneider B, Porzel A, Adam G. 1996. Metabolism of 24-*epi*castasterone in cell suspension culture of *Lycopersicon esculentum*. *Phytochemistry* 41:197–201

49. Hamada K. 1986. Brassinolide in crop cultivation. In *Plant Growth Regulators in Agriculture*, ed. P Macgregor, pp. 190–96. Taiwan: Food Fertil. Technol. Cent. Asian Pac. Reg.

50. He Y-J, Xu R-J, Zhao Y-J. 1996. Enhancement of senescence by *epi*brassinolide in leaves of mung bean seedling. *Acta Phytophysiol. Sin.* 22:58–62

51. Hewitt FR, Hough T, O'Neill P, Sasse JM, Williams EG, et al. 1985. Effect of brassinolide and other growth regulators on the germination and growth of pollen tubes of *Prunus avium* using a multiple hanging drop assay. *Aust. J. Plant Physiol.* 12:201–11

52. Hooley R. 1996. Plant steroid hormones emerge from the dark. *Trends Genet.* 12:281–83

53. Iwasaki T, Shibaoka H. 1991. Brassinosteroids act as regulators of tracheary-element differentiation in isolated *Zinnia* mesophyll cells. *Plant Cell Physiol.* 32:1007–14

54. Jones-Held S, VanDoren M, Lockwood T. 1996. Brassinolide application to *Lepidum sativum* seeds and the effects on seedling growth. *J. Plant Growth Regul.* 15:63–67

54a. Karssen CM, Loon LCV, Vreugdenhil D, eds. 1992. *Progress in Plant Growth Regulation*. Dordrecht: Kluwer

55. Katsumi M. 1991. Physiological modes of brassinolide action in cucumber hypocotyl growth. See Ref. 27a, pp. 246–54

56. Kauschmann A, Jessop A, Koncz C, Szekeres M, Willmitzer L, Altmann T. 1996. Genetic evidence for an essential role of brassinosteroids in plant development. *Plant J.* 9:701–13

57. Kawaguchi M, Imaizumi-Anraku H, Fukai S, Syono K. 1996. Unusual branching in the seedlings of *Lotus japonicus*—gibberellins reveal the nitrogen-sensitive cell divisions within the pericycle on roots. *Plant Cell Physiol.* 37:461–70

58. Kim S-K, Abe H, Little CHA, Pharis RP. 1990. Identification of two brassinosteroids from the cambial region of Scots pine (*Pinus silvestris*) by gas chromatography–mass spectrometry, after detection using a dwarf rice lamina inclination bioassay. *Plant Physiol.* 94:1709–13

59. Kitani Y. 1994. Induction of parthenogenetic haploid plants with brassinolide. *Jpn. J. Genet.* 69:35–39

60. Kolbe A, Porzel A, Schneider B, Adam G. 1997. Diglycosidic metabolites of 24-*epi*teasterone in cell suspension cultures of *Lycopersicon esculentum* L. *Phytochemistry*. In press

61. Kulaeva ON, Burkhanova EA, Fedina AB, Khokhlova VA, Bokebayeva GA, et al. 1991. Effect of brassinosteroids on protein synthesis and plant-cell ultrastructure under stress conditions. See Ref. 27a, pp. 141–55

62. Kutschabsky L, Adam G, Vorbrodt H-M. 1990. Molekul- und Kristallstruktur von (22S,23S)-Homobrassinolid. *Z. Chem.* 30:136–37

62a. Li JM, Biswas MG, Chao A, Russell DW, Chory J. 1997. Conservation of function between mammalian and plant steroid 5α-reductases. *Proc. Natl. Acad. Sci. USA* 94:3534–39

63. Li JM, Chory J. 1997. A putative leucine-rich repeat receptor kinase involved in brassinosteroid signal transduction. *Cell* 90:929–38

64. Li JM, Nagpal P, Vitart V, McMorris TC, Chory J. 1996. A role for brassinosteroids in light-dependent development of *Arabidopsis*. *Science* 272:398–401

65. Luu B, Werner F. 1996. Sterols that modify moulting in insects. *Pestic. Sci.* 46:49–53

66. Mandava NB. 1988. Plant growth-promoting brassinosteroids. *Annu. Rev. Plant Physiol. Plant Mol. Biol.* 39:23–52

67. Marquardt V, Adam G. 1991. Recent advances in brassinosteroid research. In *Chemistry of Plant Protection, Herbicide Resistance—Brassinosteroids, Gibberellins, Plant Growth Regulators*, ed. H Boerner, D Martin, V Sjut, HJ Stan, J Stetter, 7:103–39. Berlin: Springer-Verlag

68. Mayumi K, Shibaoka H. 1995. A possible double role for brassinolide in the reorientation of cortical microtubules in the epidermal cells of Azuki bean epicotyls. *Plant Cell Physiol.* 36:173–81

69. Mayumi K, Shibaoka H. 1996. The cyclic reorientation of cortical microtubules on walls with a crossed polylamellate structure: effects of plant hormones and an inhibitor of protein kinases on the progression of the cycle. *Protoplasma* 195:112–22

70. McMorris TC, Patil PA, Chavez RG, Baker ME, Clouse SD. 1994. Synthesis and biological activity of 28-homobrassinolide and analogs. *Phytochemistry* 36:585–89

71. Mendoza C, Soler A, Tesarik J. 1995. Nongenomic steroid action: independent targeting of a plasma membrane calcium channel and a tyrosine kinase. *Biochem. Biophys. Res. Commun.* 210:518–23

72. Meudt WJ. 1987. Chemical and biological aspects of brassinolide. In *Ecology and Metabolism of Plant Lipids. ACS Symp. Ser.*, ed. G Fuller, WD Nes, 325:53–75. Washington, DC: Am. Chem. Soc.

73. Mitchell JW, Mandava NB, Worley JF, Plimmer JR, Smith MV. 1970. Brassins: a new family of plant hormones from rape pollen. *Nature* 225:1065–66

74. Mushegian A, Koonin E. 1995. A putative FAD-binding domain in a distinct group of oxidases including a protein involved in plant development. *Protein Sci.* 4:1243–44

75. Nakajima N, Shida A, Toyama S. 1996. Effects of brassinosteroid on cell division and colony formation of Chinese cabbage mesophyll protoplasts. *Jpn. J. Crop Sci.* 65:114–18

76. Nielsen D, Shapiro D. 1990. Estradiol and estrogen receptor-dependent stabilization of a minivitellogenin mRNA lacking 5,100 nucleotides of coding sequence. *Mol. Cell Biol.* 10:371–76

77. Nishikawa N, Abe H, Natsume M, Shida A, Toyama S. 1995. Epimerization and conjugation of ^{14}C-labeled epibrassino-lide in cucumber seedlings. *J. Plant Physiol.* 147:294–300

78. Nishikawa N, Toyama S, Shida A, Futatsuya F. 1994. The uptake and transport of ^{14}C-labeled epibrassinolide in intact seedlings of cucumber and wheat. *J. Plant Res.* 107:125–30

79. Nomura T, Nakayama M, Reid JB, Takeuchi Y, Yokota T. 1997. Blockage of brassinosteroid synthesis and sensitivity causes dwarfism in *Pisum sativum. Plant Physiol.* 113:31–37

80. Oh M-H, Romanow W, Smith R, Zamski E, Sasse J, Clouse S. 1998. Soybean *BRU1* encodes a functional xyloglucan endo-transglycosylase that is highly expressed in inner epicotyl tissues during brassinosteroid-promoted elongation. *Plant Cell Physiol.* In press

81. Park KH, Saimoto H, Nakagawa S, Sakurai A, Yokota T, et al. 1989. Occurrence of brassinolide and castasterone in crown gall cells of *Catharanthus roseus. Agric. Biol. Chem.* 53:805–11

82. Petzold U, Peschel S, Dahse I, Adam G. 1992. Stimulation of source-applied ^{14}C-sucrose export in *Vicia faba* plants by brassinosteroids, GA_3 and IAA. *Acta Bot. Neerl.* 41:469–79

83. Richter K, Koolman J. 1991. Antiecdysteroid effects of brassinosteroids in insects. See Ref. 27a, pp. 265–78

84. Roddick JG. 1994. Comparative root growth inhibitory activity of four brassinosteroids. *Phytochemistry* 37:1277–81

85. Roddick JG, Guan M. 1991. Brassinosteroids and root development. See Ref. 27a, pp. 231–45

86. Roth PS, Bach TJ, Thompson MJ. 1989. Brassinosteroids: potent inhibitors of transformed tobacco callus cultures. *Plant Sci.* 59:63–70

87. Russell DW. 1996. Green light for steroid hormones. *Science* 272:370–71

88. Sairam RK. 1994. Effects of homobrassinolide application on plant metabolism and grain yield under irrigated and moisture-stress conditions of two wheat varieties. *Plant Growth Regul.* 14:173–81

89. Sakurai A, Fujioka S. 1993. The current status of physiology and biochemistry of brassinosteroids. *Plant Growth Regul.* 13:147–59

90. Sakurai A, Fujioka S. 1996. Biosynthetic pathways of brassinosteroids. *Proc. Plant Growth Regul. Soc. Am.* 23:8

91. Sakurai A, Fujioka S. 1997. Studies on biosynthesis of brassinosteroids. *Biosci. Biotechnol. Biochem.* 61:757–62

92. Sala C, Sala F. 1985. Effect of brassino-

steroid on cell division and enlargement in cultured carrot (*Daucus carota* L.) cells. *Plant Cell Rep.* 4:144–47

93. Sasse JM. 1990. Brassinolide-induced elongation and auxin. *Physiol. Plant.* 80: 401–8

94. Sasse JM. 1991. Brassinolide-induced elongation. See Ref. 27a, pp. 255–64

95. Sasse JM. 1991. The case for brassinosteroids as endogenous plant hormones. See Ref. 27a, pp. 158–66

96. Sasse JM. 1997. Recent progress in brassinosteroid research. *Physiol. Plant.* 100:696–701

97. Sasse JM, Sasse JM. 1994. Brassinosteroids and roots. *Proc. Plant Growth Regul. Soc. Am.* 21:228–32

98. Sasse JM, Smith R, Hudson I. 1995. Effect of 24-*epi*brassinolide on germination of seeds of *Eucalyptus camaldulensis* in saline conditions. *Proc. Plant Growth Regul. Soc. Am.* 22:136–41

99. Sasse JM, Yokota T, Taylor PE, Griffiths PG, Porter QN, et al. 1992. Brassinolide-induced elongation. See Ref. 54a, pp. 319–25

100. Schilling G, Schiller C, Otto S. 1991. Influence of brassinosteroids on organ relations and enzyme activities of sugar-beet plants. See Ref. 27a, pp. 208–19

101. Schmidt J, Altmann T, Adam G. 1997. Brassinosteroids from seeds of *Arabidopsis thaliana*. *Phytochemistry* 45:1325–27

102. Schmidt J, Yokota T, Adam G, Takahashi N. 1990. Castasterone and brassinolide in *Raphanus sativus* seeds. *Phytochemistry* 30:364–65

103. Schneider B, Kolbe A, Hai T, Porzel A, Adam G. 1996. Metabolism of brassinosteroids in plant cell cultures. *Proc. Plant Growth Regul. Soc. Am.* 23:43

104. Singh J, Nakamura S, Ota Y. 1993. Effect of *epi*brassinolide on gram (*Cicer arietinum*) plants grown under water stress in the juvenile stage. *Indian J. Agric. Sci.* 63:395–97

105. Suge H. 1986. Reproductive development of higher plants as influenced by brassinolide. *Plant Cell Physiol.* 27:199–205

106. Suzuki H, Fujioka S, Takatsuto S, Yokota T, Murofushi N, et al. 1995. Biosynthesis of brassinosteroids in seedlings of *Catharanthus roseus*, *Nicotiana tabacum*, and *Oryza sativa*. *Biosci. Biotechnol. Biochem.* 59:168–72

107. Szekeres M, Nemeth K, Koncz-Kalman Z, Mathur J, Kauschmann A, et al. 1996. Brassinosteroids rescue the deficiency of CYP90, a cytochrome P450, controlling cell elongation and de-etiolation in *Arabidopsis*. *Cell* 85:171–82

108. Taiz L. 1984. Plant cell expansion: regulation of cell wall mechanical properties. *Annu. Rev. Plant. Physiol.* 35:585–657

109. Takahashi T, Gasch A, Nishizawa NH, Chua NH. 1995. The *DIMINUTO* gene of *Arabidopsis* is involved in regulating cell elongation. *Genes Dev.* 9:97–107

110. Takatsuto S. 1994. Brassinosteroids: distribution in plants, bioassays and microanalysis by gas chromatography-mass spectrometry. *J. Chromatogr.* 658:3–15

111. Takematsu T, Takeuchi Y, Choi CD. 1986. Overcoming effects of brassinosteroids on growth inhibition of rice caused by unfavourable growth conditions. *Shokucho* 20:2–12

112. Takeuchi Y. 1992. Studies on physiology and applications of brassinosteroids. *Shokubutsu no Kogaku Chosetsu* 27:1–10

113. Takeuchi Y, Ogasawara M, Konnai M, Takematsu T. 1992. Application of brassinosteroids in agriculture in Japan. *Proc. Plant Growth Regul. Soc. Am.* 19:343–52

114. Taylor PE, Spuck K, Smith PM, Sasse JM, Yokota T, et al. 1993. Detection of brassinosteroids in pollen of *Lolium perenne* L. by immunocytochemistry. *Planta* 189:91–100

115. Tominaga R, Sakurai N, Kuraishi S. 1994. Brassinolide-induced elongation of inner tissues of segments of squash (*Cucurbita maxima* Duch.) hypocotyls. *Plant Cell Physiol.* 35:1103–6

116. Vasyukova NJ, Chalenko GI, Kaneva IM, Khripach VA. 1994. Brassinosteroids and potato blight. *Prikl. Biokhim. Mikrobiol.* 30:464–70

117. Voigt B, Porzel A, Naumann H, Horhold-Schubert C, Adam G. 1993. Hydroxylation of the native brassinosteroids 24-*epi*castasterone and 24-*epi*brassinolide by the fungus *Cunninghamella echinulata*. *Steroids* 58:320–23

118. Voigt B, Porzel A, Undisz K, Horhold-Schubert C, Adam G. 1993. Microbial hydroxylation of 24-*epi*castasterone by the fungus *Cochliobolus lunatus*. *Nat. Prod. Lett.* 3:123–29

119. Wada K, Marumo S, Abe H, Morishita T, Nakamura K, et al. 1984. A rice lamina inclination test—a micro-quantitative bioassay for brassinosteroids. *Agric. Biol. Chem.* 48:719–26

120. Walker J, Stone J, Collinge M, Horn M, Braun D, et al. 1996. Receptor-like protein kinases. In *Protein Phosphorylation in Plants*, ed. P Shewry, N Halford, R Hooley, pp. 227–38. Oxford: Oxford Univ. Press

121. Wang B, Zeng G. 1993. Effect of epi-brassinolide on the resistance of rice seedlings to chilling injury. *Zhiwu Shengli Xuebao* 19:53–60
122. Wang T-W, Cosgrove DJ, Arteca RN. 1993. Brassinosteroid stimulation of hypocotyl elongation and wall relaxation in pakchoi (*Brassica chinensis* cv Leichoi). *Plant Physiol.* 101:965–68
123. Wang Y-Q, Luo W-H, Xu R-J, Zhao Y-J, Zhou W-S, et al. 1994. Biological activity of brassinosteroids and relationship of structure to plant growth promoting effects. *Chin. Sci. Bull.* 39:1573–77
124. Wang Y-Q, Zhao Y-J. 1989. Relationship between structure and biological acitivity of brassinolide analogues. *Acta Phytophysiol. Sin.* 15:18–23
125. West CA. 1980. New growth factors—summary of session. In *Plant Growth Substances 1979*, ed. F Skoog, pp. 289–90. Berlin: Springer-Verlag
126. Wilen RW, Sacco M, Gusta LV, Krishna P. 1995. Effects of 24-*epi*brassinolide on freezing and thermotolerance of bromegrass (*Bromus inermis*) cell cultures. *Physiol. Plant.* 95:195–202
127. Winter J, Schneider B, Strack D, Adam G. 1997. Role of a cytochrome P450-dependent monooxygenase in the hydroxylation of 24-*epi*brassinolide. *Phytochemistry* 45:233–37
128. Xu R, He Y, Wu D, Zhao Y-J. 1995. Effect of 24-*epi*brassinolide on the distribution of cAMP and activity of plasma membrane ATPase in plant tissues. *Shiwu Shengli Xuebao* 21:143–48
129. Xu R-J, He Y-J, Wang Y-Q, Zhao Y-J. 1994. Preliminary study of brassinosterone binding sites from mung bean epicotyls. *Acta Phytophysiol. Sin.* 20:298–302
130. Xu W, Purugganan MM, Polisenksy DH, Antosiewicz DM, Fry SC, et al. 1995. Arabidopsis *TCH4*, regulated by hormones and the environment, encodes a xyloglucan endotransglycosylase. *Plant Cell* 7:1555–67
131. Yang Z. 1996. Signal transducing proteins in plants: an overview. In *Signal Transduction in Plant Growth and Development*, ed. D Verma, pp. 1–37. New York: Springer
132. Yokota T. 1997. The structure, biosynthesis and function of brassinosteroids. *Trends Plant Sci.* 2:137–43
133. Yokota T, Higuchi K, Kosaka Y, Takahashi N. 1992. Transport and metabolism of brassinosteroids in rice. See Ref. 54a, pp. 298–305
134. Yokota T, Koba S, Kim SK, Takatsuto S, Ikekawa N, et al. 1987. Diverse structural variations of the brassinosteroids in *Phaseolus vulgaris* seed. *Agric. Biol. Chem.* 51:1625–31
135. Yokota T, Mori K. 1992. Molecular structure and biological activity of brassinolide and related brassinosteroids. In *Molecular Structure And Biological Activity of Steroids*, ed. M Bohl, WL Duax, pp. 317–40. Boca Raton: CRC Press
136. Yokota T, Ogino Y, Suzuki H, Takahashi N, Saimoto H, et al. 1991. Metabolism and biosynthesis of brassinosteroids. See Ref. 27a, pp. 86–96
137. Yokota T, Takahashi N. 1986. Chemistry, physiology and agricultural application of brassinolide and related steroids. In *Plant Growth Substances 1985*, ed. M Bopp, pp. 129–38. Berlin: Springer-Verlag
138. Zhao Y-J, Xu R-J, Luo W-H. 1990. Inhibitory effects of abscisic acid on *epi*brassinolide-induced senescence of detached cotyledons in cucumber seedlings. *Chin. Sci. Bull.* 35:928–31
139. Zurek DM, Clouse SD. 1994. Molecular cloning and characterization of a brassinosteroid-regulated gene from elongating soybean (*Glycine max* L.) epicotyls. *Plant Physiol.* 104:161–70
140. Zurek DM, Rayle DL, McMorris TC, Clouse SD. 1994. Investigation of gene expression, growth kinetics, and wall extensibility during brassinosteroid-regulated stem elongation. *Plant Physiol.* 104:505–13

Annu. Rev. Plant Physiol. Plant Mol. Biol. 1998. 49:453–80
Copyright © 1998 by Annual Reviews. All rights reserved

NUCLEAR CONTROL OF PLASTID AND MITOCHONDRIAL DEVELOPMENT IN HIGHER PLANTS

P. Leon and A. Arroyo

Departamento de Biologia Molecular de Plantas, Instituto de Biotecnología UNAM, Cuernavaca, Morelos 62250 Mexico; e-mail: patricia@ibt.unam.mx

S. Mackenzie

Department of Agronomy, Lilly Hall of Life Sciences, Purdue University, West Lafayette, Indiana 47907

KEY WORDS: organelle biogenesis, transcriptional regulation, posttranscriptional regulation, chloroplast mutants, mitochondrial mutants

ABSTRACT

The nucleus must coordinate organelle biogenesis and function on a cell and tissue-specific basis throughout plant development. The vast majority of plastid and mitochondrial proteins and components involved in organelle biogenesis are encoded by nuclear genes. Molecular characterization of nuclear mutants has illuminated chloroplast development and function. Fewer mutants exist that affect mitochondria, but molecular and biochemical approaches have contributed to a greater understanding of this organelle. Similarities between organelles and prokaryotic regulatory molecules have been found, supporting the prokaryotic origin of chloroplasts and mitochondria. A striking characteristic for both mitochondria and chloroplast is that most regulation is posttranscriptional.

CONTENTS

INTRODUCTION

It is generally believed that higher plant evolution occurred through two differ-ent endosymbiotic events, both involving prokaryotic organisms. Eukaryotic cells initially acquired mitochondria through endosymbiosis of a bacterium. Later, the chloroplast was derived from a cyanobacteria-type organism (149). Mitochondria and plastids each contain an autonomous genome, yet the nu-cleus now plays a major role in determining organelle properties, because it encodes the majority of the genes required for function and maintenance. The presence of organelle genes in two different cellular compartments creates an obvious regulatory problem. The cell must coordinate the expression of nuclear-encoded genes, present in only one or a few copies per cell, with the expression of organellar-encoded genes present in several hundred or even thousands of copies. Coordinate control is a necessity for mitochondrial and chloroplast function and likely requires new regulatory pathways not present in the original endosymbionts (48).

Many of the structural genes required for the photosynthetic and respiratory reactions have been cloned and sequenced. We have much less information about the many nuclear genes required in development. This is a broad subject for study, as higher plants contain a variety of differentiated plastids types, re-flecting cell and tissue type (85). This review concentrates primarily on recent advances in organelle development and function in vascular plants. Recent pub-lished reviews cover other photosynthetic organisms such as *Chlamydomonas* and *Euglena* (45, 54, 142, 160).

A DEVELOPMENTAL CONSIDERATION

Plastids exhibit a very clear developmental program. All plastids are derived from proplastids present in meristematic cells (85). Upon cell differentiation, proplastids also differentiate. Plastids can also redifferentiate in response to external environmental signals (85). The best characterized differentiation process results in a chloroplast.

Almost every step of plastid development depends on the direct action of nuclear-encoded molecules. Molecules required in the initial stages of organelle differentiation could affect all plastid types, whereas molecules necessary at later stages should affect only one plastid type.

Chloroplast differentiation appears to start very early during plant development. The leaf primordia that develop from the apical meristem contain mesophyll tissue from which most of the chloroplasts will be derived. Mesophyll chloroplasts can then undergo further differentiation, as is well-documented for differentiation of mesophyll and bundle sheath chloroplasts of C4 plants (124).

Microscopic studies have provided a sequential picture of the events during plastid differentiation (95). Further refinement of these events at the molecular level has been difficult. It is possible that many events during chloroplast differentiation are concurrent or occur in such a brief period that they cannot be dissected with the available techniques. In monocot leaves initial differentiation events include DNA replication and production of the chloroplast decoding apparatus (11, 123). Production of thylakoid membranes and accumulation of specific photosynthetic complexes are detected in later stages (85, 123).

The study of the participation of nuclear genes in plastid development has benefited from the isolation and characterization of an enormous number of mutants, many of which will be described in this review. In contrast to the well-defined developmental program of the chloroplast, the mitochondrion does not have an obvious developmental program. However, it is clear that mitochondria play essential roles throughout cellular differentiation and take on a pronounced role at particular points in higher plant development.

NUCLEAR REGULATION OF THE EARLY PHASES OF CHLOROPLAST DEVELOPMENT

Initial Steps

Transcription is a central regulatory point during the early stages of chloroplast differentiation (54, 123). Transcription is low in the proplastid, and is activated in the immature chloroplast (11). In parallel, the synthesis of both the transcription and translation apparatus takes place (63). Several lines of evidence suggest a nuclear origin for the enzyme responsible for initial transcriptional activity in the chloroplast (69, 70, 74, 122). For example, in the parasitic plant *Epifagus virginiana*, which lacks the chloroplast-encoded subunits of RNA polymerase, and in the *albostrians* barley mutant, which lacks ribosomes, transcription of particular plastid genes has been detected (70, 122).

Allison et al (2) have shown that a subset of the chloroplast-encoded genes are transcribed by a second RNA polymerase in which at least some subunits are nuclear-encoded. The role of this nuclear-encoded RNA polymerase (NEP polymerase) seems to be primarily, but not exclusively, the expression of housekeeping genes during the early phases of chloroplast development (54, 123, 160). Among the genes transcribed by this enzyme are the rRNA genes (2), ribosomal proteins (70), and the *rpoB* operon (72), which encodes the subunits of the plastid-encoded RNA polymerase. Candidates for the NEP enzyme have been purified from spinach (97). NEP is probably encoded by the *RpoPt* gene recently isolated from *Arabidopsis* (68); it shares important similarities with phage-type RNA polymerases.

Concomitant with transcription of the rRNA genes, nuclear-encoded ribosomal proteins are synthesized (63) and imported into the chloroplast to form ribosomes, in preparation for translation in later stages of development. The current view is that the nucleus initiates chloroplast differentiation and provides key components of the transcriptional and translational machinery required for later stages of development (123).

MUTANTS THAT AFFECT EARLY CHLOROPLAST DEVELOPMENT

Early Developmental Mutants

Although many chloroplast developmental mutants have been isolated, few of them affect initial events in organelle biogenesis. Because the plastid produces many essential compounds such as hormones, lipids, and cofactors, mutations that interrupt these pathways will probably disrupt both plastid biogenesis and plant function. The *amidophosphoribosyl-transferase deficient* (*atd*) mutant from *Arabidopsis* is an example; chloroplast development is arrested at an early stage in *atd* lines because one of the genes encoding the ATase enzyme, (*ATase2* gene) is disrupted; a key enzyme of purine biosynthesis. This nuclear gene provides the only ATase activity in photosynthetic tissue, but a second gene might permit purine biosynthesis in other tissues (170).

Similar examples of fundamental mutants include *dcl* from tomato (80), *dag* from *Antirrhinum* (18), and the *Arabidopsis* albino T-DNA tagged *cla1-1* (110). Plastids present in these plants are very small, with almost no thylakoid membrane, and thus resemble proplastids. mRNA analyses showed that *CLA1*, *DAG*, and *DCL* genes are normally expressed not only in photosynthetic but also in nonphotosynthetic tissue, suggesting that these genes are probably required for the development of different plastid types. In both *cla1-1* (J Estevez, A Arroyo, A Cantero & P Leon, unpublished results) and *dag* (18) plants, chloroplasts and etioplasts are abnormal. Each of these genes encodes a novel plastid-localized protein of unknown function. Null mutations of the *DCL* gene cause embryo abortion (JS Keddie & W Gruissem, unpublished data). Proteins homologous to CLA1 are found in photosynthetic and nonphotosynthetic prokaryotes; recent data indicate that this protein may be involved in the biosynthesis of a novel isoprenoid pathway conserved in evolution (J Estevez, A Arroyo, A Cantero & P Leon, unpublished results). At the morphological level, *dag* and *dcl* plants apparently lack palisade cells (18, 80).

Similar alterations occur in plastids after disruption of the *PALE CRESS* (*PAC*), *IMMUTANS* (*IM*), and *ATD* genes of *Arabidopsis* (139, 170, 178). Because it seems unlikely that all these genes directly regulate leaf differentiation, a novel signaling system during early plastid development has been hypothesized that would affect the final divisions of mesophyll cells in the prepalisade stage (18). Compelling evidence indicates that a factor originating in the chloroplast signals its developmental stage to the nucleus, coordinating the expression of many nuclear-encoded genes (reviewed in 162, 165). Alterations in this signal transduction pathway do not have any obvious impact on mesophyll morphology. These results suggest that multiple signals might exist to ensure proper coordination between cell and organelle during plastid development.

Genes involved in early plastid development have also been described in C4 maize plants. Phenotypic characteristics of the *bsd3* (R Roth & J Langlade, unpublished data) mutant resemble those described above, with no thylakoid formation and alterations in different plastid types. However, in the *bsd3* mutant, neither the morphology of the photosynthetic tissue nor accumulation of the C4 nuclear-encoded photosynthetic genes is altered. These results suggest that, in contrast to what it is found in C3 plants, leaf development in C4 plants such as maize might not rely on chloroplast signals for final differentiation.

Pigment Mutants

Direct participation of pigment biosynthesis in chloroplast development has been difficult to establish because of pleiotropy. Several mutants with altered chlorophyll accumulation have been described (158). In all cases, chloroplast development was arrested at initial membrane assembly, suggesting that chlorophyll synthesis and chloroplast development may be interdependent (174). Because these kinds of mutations can disrupt different metabolic activities, loss of chlorophyll along with other plastid components might represent secondary effects (62). Recent data that support the hypothesis that chlorophyll production has the potential to influence organelle development come from work with mutants altered in chlorophyll *a/b* ratio (40) and the *olive* (*oli*) mutant from *Antirrhinum* (73). The *OLI* gene seems to encode a key enzyme of the chlorophyll biosynthetic pathway, the porphyrin IX Mg-chelatase. *oli* conditionally affects chlorophyll synthesis and organelle development, and the chloroplasts in this mutant contain thylakoids but no grana. This phenotype seems to be the direct result of a block in chlorophyll biosynthesis, not a result of photo-oxidation (73). Chlorophyll *b*–deficient mutants usually result in smaller photosynthetic unit sizes (115), possibly because of LHC destabilization and an abnormal thylakoid membrane system (39). Chlorophyll appears to be necessary for either translation or stability of the nuclear-encoded apoproteins of the light-harvesting complex and the formation of grana-deficient thylakoid membranes (64, 88). Gene disruptions at other steps of chlorophyll biosynthesis display alterations in plastid development even in the dark, where photo-oxidative damage does not exist, supporting the hypothesis that chlorophyll is involved directly (39, 64, 146, 158).

NUCLEAR REGULATION DURING CHLOROPLAST DEVELOPMENT

Division

Organelle division is a fundamental process in chloroplast and mitochondrial biogenesis. With the recent isolation of several mutants from *Arabidopsis thaliana*, Pyke and coworkers have contributed to our understanding of the molecular basis of control of chloroplast division (135, 136). These mutants, named *arc* for "accumulation and replication of the chloroplast," demonstrate that several nuclear genes are required. Interestingly, all mutants analyzed so far show a reduction in chloroplast number per mesophyll cell, but no obvious plant morphological alterations (134a, 135). It is likely that the total amount of chloroplast material within a cell is the important parameter for cell function rather than chloroplast number.

Physiological analysis of most *arc* mutants shows plastid division to be altered only in chloroplast (134a). An exception is the *arc6* mutant, the most extreme mutation isolated, which may affect division at the proplastid state (134a) the gene affected in the *arc6* mutant appears to be a homolog of the *FtsZ* protein, implicated in bacterial cell division (12, 31, 131). *FtsZ*, which polymerizes similarly to eukaryotic cytoskeletal elements, may be a progenitor of tubulin

(38), which raises the intriguing possibility that the mechanism of organelle division may be, to some degree, conserved between prokaryotes and plants.

Transcription

Although transcriptional control is the key element setting expression of nuclear-encoded plastid-localized gene products (90), the relative transcription rates of many chloroplast-encoded genes are constant in different tissues and at different developmental stages. This suggests that posttranscriptional events may predominate in chloroplast gene regulation (29). Evidence suggests, however, that stage-specific transcriptional regulation occurs during leaf maturation. For example, some plastid-encoded genes fluctuate in transcription levels in response to factors such as light and plastid type (103, 114, 123, 138). This is the case for the light-activated genes such as *psbA* in barley and maize (86), *psbD-psbC* in barley (150), and *petG* from maize (59). Similarly, the transcription rates of *rbcL*, *psbA*, and *atpB/E* genes from *Arabidopsis* are specifically reduced in amyloplasts (77). All these genes are preferentially transcribed by the better-characterized plastid-encoded RNA polymerase (74). It is tempting to attribute the selectivity of the transcriptional response during chloroplast development to the interaction of the core RNA polymerase with specific regulatory molecules (103). These factors might be available only under certain conditions and in this way might regulate gene transcription during development.

Two types of factors seem to play a role in the regulation of chloroplast transcription: factors that bind DNA only in the presence of the RNA polymerase (sigma-like factors) and sequence-specific proteins acting as activators or repressors. Several sigma-like factors have been recognized (102, 167). For example, a 90-kDa sigma-like factor from mustard is present in young but not in mature spinach leaves (96). In contrast, the transcriptional activity of chloroplasts and etioplasts in mustard appears to be modulated by differential phosphorylation of three sigma factors in response to light (166). In this way only dephosphorylated factors permit efficient transcription initiation of promoters like that of *psbA*. The most likely candidate responsible for phosphorylation of these sigma-like factors is a serine/threonine kinase found to associate with the active transcription complex (103).

Evidence has emerged recently for a number of sequence-specific binding factors that might also modulate expression of photosynthetic genes under specific conditions. A nuclear factor involved in differential light-mediated transcription of the *psbD* gene has been isolated (84). This element (AGF) interacts at a specific site within the promoter region of the *psbD* gene and shares similarity with a specific DNA binding factor, CDF2 (for chloroplast DNA binding factor 2), isolated recently (76). Both factors have similar properties, suggesting that they may belong to a larger family of DNA-binding proteins, but their specific function remains to be seen.

Posttranscriptional Processing

Posttranscriptional and translational events play a major role in regulating the differential accumulation of many plastid-encoded mRNAs during chloroplast development (for extensive reviews, see 53, 54, 114, 142). Most chloroplast-encoded genes are transcribed as polycistronic units (53). These precursors undergo a series of complex maturation events that include processing of the mature mRNAs, although only in *Chlamydomonas* (98, 141, 142) the importance of posttranscriptional processing understood. One of the few exceptions in higher plants is the *CRP1* gene from maize, which seems to be required for the efficient translation of two chloroplast mRNAs (*pet A* and *pet D*) (9). This gene affects processing of the *petD* mRNA in such a way that the monocistronic transcript is absent, implying a mechanistic link between RNA processing and translation. It is hypothesized that in this mutant the petD protein is no longer synthesized, because the ribosome cannot initiate within the polycistronic mRNA, due to masking of the translation initiation region. The *CRP1* gene has been cloned, but the encoded protein shows no significant similarity to any known protein (M Walker & A Barkan, unpublished data).

In contrast to processing, mRNA stability seems to be important for the accumulation of several photosynthetic mRNAs required during the transition from proplastid to chloroplast. For example, as measured by run-on experiments, steady state levels of the *psbA* and *atpB* transcripts increase during greening of etiolated spinach cotyledons without changes in the transcription rate (29). In addition, the stability of several chloroplast-encoded transcripts seems to be dramatically increased during chloroplast development in barley (83) and spinach (51). Stabilization seems to occur by the correct processing of the polycistronic mRNAs in the differentiated chloroplasts, whereas in proplastids these mRNAs appear to be rapidly degraded (53). It has recently been established that nuclear-encoded enzymes are required for correct mRNA processing (66). Both exo- and endoribonuclease activities are found in a high-molecular-weight complex that binds near stem and loop sequences found in the 3' region of chloroplast transcripts (53, 114, 142). Although this complex has not been fully characterized, similarities have been postulated with a ribonucleolytic complex ("degradosome") in *Escherichia coli* (66). In addition to the action of the chloroplast high-molecular-weight complex, stabilization of transcripts requires additional factors that inhibit rapid degradation (53, 66). One of these factors is a nuclear-encoded 28-kDa protein (28RNP) (53) that interacts with both the transcript and the high-molecular-weight complex (66). This protein is differentially expressed during plant development, is modified by phosphorylation, and is likely necessary to direct the correct 3′-end processing of plastid transcripts under particular developmental conditions (66).

Unexpectedly, polyadenylation is the signal for degradation of specific chloroplast transcripts. Following the endonucleolytic cleavage events in the *petD* mRNA by components of the high-molecular-weight complex, the products are then polyadenylated (89). By increasing susceptibility to degradation in response to conditions such as light, polyadenylation seems to play a critical role in regulating chloroplast transcript stability. For example, in dark-adapted plants, the extent of polyadenylated *petD*-specific transcripts is higher than in plants grown in light (89). This unexpected process is probably not restricted to a particular transcript and will most likely require the action of additional nuclear regulatory factors.

The modulation of transcript stability during the differentiation of chloroplasts in C4 plants has also been documented. In the dimorphic cells of the C4 plant, the photosynthetic genes are expressed in a cell-specific manner. The characterization of plants altered in bundle sheath cell differentiation in maize has led to identification of a nuclear gene (*Bsd2*) that influences the regulation of transcript stability and/or translation of *rbcL* mRNA. In this mutant, chloroplast development is prevented in bundle sheath cells; this is likely the result of photo-oxidative damage in the absence of the RuBPCase protein in these cells (145). It is conceivable that the product of the *Bsd2* gene might interact in a sequence-specific manner with the *rbcL* mRNA in bundle sheath cells to stabilize this transcript.

A few other mutants have been described that alter stability of specific RNA transcripts, including the maize *hcf2* and *hcf6* mutations (118). Still others have a more general effect on the stability of multiple transcripts such as *hcf38* from maize (7, 118) and the *hcf109* and *hcf5* mutants from *Arabidopsis* (35, 116).

In both mitochondria and chloroplasts unspliced transcripts can accumulate to high levels (177). The role of splicing in the regulation of chloroplast development in higher plants has not been extensively studied, but recent studies in maize show that the relative abundance of spliced and unspliced transcripts differs among plastid types. Compared to chloroplasts, unspliced forms predominate in both amyloplasts and proplastids (6). RNA splicing could exert control in plastid development by regulating the abundance of plastid-encoded proteins.

Probably the best evidence for participation of splicing in chloroplast development comes from the work with the *crs* (chloroplast RNA splicing) mutants in maize. Chloroplast genes in higher plants contain multiple group II introns, from both the A and B subgroups (117). These introns are capable of self-splicing in vitro; genetic evidence indicates that *trans*-acting factors are required for efficient splicing in vivo (148). Recently, two nuclear genes, *crs1* and *crs2*, were reported to be required for the proper splicing of group II introns in maize chloroplasts (79). Mutation at *crs1* specifically blocks splicing of the *atpF* type IIA intron that encodes subunit I of the chloroplast ATP synthase, whereas mutation at the recently cloned *crs2* gene blocks the splicing of many group IIB introns. Disruption of chloroplast development in these mutants provides important biological evidence for the structural division of type II introns because different splicing factors are required to process IIA and IIB introns. Previously described mutants might also reflect splicing deficiencies. For example, in plants with the nuclear gene *iojap* (*ij*) splicing of most IIA introns is defective (16). Whether there is a direct relationship between *ij* and *crs1* remains to be deduced.

Translational and posttranslational regulation has been documented in *Chlamydomonas*, *Euglena*, and higher plants (45, 54, 142, 160). This control could underlie the rapid changes in protein accumulation prominent in several photosynthetic factors during chloroplast development; for example, the product of the *psbA* transcripts does not accumulate during senescence of bean leaves (10), in etiolated leaves (159), in roots, and in cultured cells (71). Translational control is possible at several steps, but the best analyzed so far is during ribosome binding to the mRNA. Both 5′ and 3′ untranslated leader (UTR) sequences play important roles during translation in eukaryotes, and several *cis*-acting sequences have been identified (45). Nuclear-encoded proteins have been identified that interact with specific stem-loop sequences in the 5′ UTR of chloroplast transcripts in several organisms (27, 65, 71). These 5′ UTR binding complexes seem to modulate ribosome binding, acting as translational activators or repressors (28, 114). The nuclear-encoded factors identified to date are mostly from *Chlamydomonas*; these factors can regulate the translation of either individual or a group of transcripts (45, 142). In higher plants, a specific protein factor(s) was recently demonstrated to be required for the efficient translation of *psbA* mRNA in an in vitro translation system from tobacco chloroplasts. This factor interacts with specific elements of the 5′ UTR of the *psbA* mRNA (71). This approach has an enormous potential to analyze the molecular mechanisms of translational regulation in vascular plants.

Coordinate protein accumulation for different photosynthetic protein complexes depends on translational control. It is known that in order to maintain the stoichiometric levels of different plastid complexes, their subunits are rapidly degraded when prevented from assembling (142). Until now, the molecular mechanism has been little studied, but evidence currently exists to suggest it as an important point of translational regulation. For example, in *Chlamydomonas* mutants, the accumulation of cytochrome *f* decreases in the absence of other genes of the cytochrome b_6f complex (*petB* or *petD*), although its transcript level remains unchanged. The most likely interpretation of this phenomenon is the direct autoregulation of the cytochrome f so that when unassembled, its translation is attenuated (160).

A similar mechanism appears to be present in higher plants. It has been shown that the translation of the large subunit protein (LS) of the Rubisco enzyme is controlled by levels of the small subunit (SS). The levels of the *rbcL* transcript remain unchanged, again suggesting translational regulation. Though the mechanism underlying this mode of regulation is still unclear, one might postulate that proteins such as SS may act as activators during translation of the *rbcL* transcript (144).

Posttranslational proteolysis is important during plastid differentiation. For example, during chloroplast to chromoplast differentiation, massive degradation of the photosynthetic complexes and the accumulation of new sets of proteins must occur (111). Likewise, photosystem D1 protein is specifically degraded after damage by oxygen radicals during electron transport (92), and this degradation is mediated by a specific protease (75). Several nuclear proteins involved in proteolytic degradation in chloroplasts have been identified. The gene homologous to ClpA, a subunit of the ATP-dependent serine protease (Clp) (47), is nuclear-encoded and transported into the chloroplast (81), where it may be involved in the degradation of soluble and thylakoidal proteins (151). This offers a means by which the nucleus may regulate protein accumulation during specific stages in the biogenesis of the organelle. This subject, though not yet extensively studied, likely represents a very important point in plastid developmental regulation.

Concomitant with the accumulation of the photosynthetic complexes, extensive proliferation of thylakoid membranes and the formation of granal stacks occurs in maturing chloroplast. It has been suggested that incorporation of the light-harvesting complex (LHCII) into the thylakoid membrane plays an important role in thylakoid stacking in *Chlamydomonas* and higher plants (35, 158). Only recently has participation of other nuclear genes been analyzed (36). One interesting example is development of transgenic *Arabidopsis* plants with low levels of a dynamin-like gene (*ADL1*), which show a greatly reduced thylakoid membranes and an increase in lipid granules. Because dynamin-like proteins in other systems are involved in the formation of membrane vesicles (180), it is proposed that the nuclear-encoded *ADL1* gene participates in thylakoid biogenesis in higher plants (JM Park, JH Cho, SG Kang, HJ Jang, KT Pih, et al, unpublished data).

Nuclear Regulation of Targeting

The proper allocation of the nuclear-encoded subunits is required for organelle functionality and development. Plastids have developed an elaborate sorting system to ensure proper targeting of multimeric protein complexes. The presence of three distinct membranes requires that proteins are incorporated into the chloroplast and later reallocated within the organelle (21). General aspects of chloroplast protein import have been known for some time, but the mechanisms for sorting and import are just beginning to be elucidated. Most genes involved in this process are nuclear-encoded, and their proper expression regulates correct targeting during chloroplast development (21).

A fairly detailed picture has emerged recently for the integration of proteins into the thylakoid membrane. Mutations in four nuclear genes of maize interfere with plastid biogenesis by disrupting the translocation of thylakoid proteins (8). Mutations in the *tha1* and *tha5* genes affect a SecA-type pathway, whereas mutations in the *hcf106* and *tha4* genes disrupt a DpH pathway that depends on pH (112, 172). These nuclear mutations are recessive and result in pale-green, nonphotosynthetic seedlings that die after endosperm depletion. The *tha1* mutant affects chloroplast development specifically in the mesophyll cells in which giant grana are found, but does not apparently affect bundle sheath chloroplasts that look normal (173). In the *hcf106* mutant, both plastid types are abnormal (112). *tha1* encodes a chloroplast-localized SecA protein homologous to the SecA protein from *E. coli*. This suggests that at least one of the plastid targeting mechanisms is similar to the prokaryotic *secA/Y/E* pathway (173). Recently the *tha4* gene has also been cloned; it is closely related to an open reading frame of unknown function of *Synechocystis*. In *tha4* plants the Dph pathway is disrupted; this pathway may have evolved from a cyanobacterial progenitor (M Walker & A Barkan, unpublished data). It is unknown why two distinct translocation pathways exist for the tylakoidal proteins and what the differences are between both mechanisms. Given the recent identification of the proteins involved in tylakoidal protein import, important progress in the future is expected.

NUCLEAR SIGNALS FOR CHLOROPLAST DEVELOPMENT

Light Regulation

Organelle development and plant development respond to external signals, particularly light. Light regulation is mediated by the photoreceptors: a phytochromes, blue/UV-A cryptochromes, and the UV-A and -B receptors. Several nuclear genes are part of the signal transduction system coupling light to plastid and plant development (19, 37, 137). Mutants with a de-etiolated phenotype—*det, fus*, and *cop* classes—have a morphology when grown in the dark similar to light-grown wild-type seedlings (91, 137, 173a); these mutants have altered expression of several nuclear- and chloroplast-encoded photosynthetic genes (119, 137). The plastids in *det1*, *cop1*, and *cop9* develop a mature thylakoid system when grown in the dark, even though they do not accumulate chlorophyll. These mutants together with several *fus* mutants also affect the amyloplast, by promoting differentiation of a chloroplast, resulting in a constitutive default photomorphogenic developmental pathway (91, 119). In contrast, other de-etiolated mutants such as *det2* show no impact in chloroplast development (20). Several of the genes affected in these mutants have been cloned (30, 100, 132, 176), and the characteristics and localization of their gene products suggest that they may be nuclear regulators (17). The COP1 protein is probably one of the best studied light regulators; it is a negative transcriptional regulator capable of direct interaction with upstream DNA sequences of its target genes (137). Other genes such as COP8, COP9, and COP11, seemed to act in the same pathway (176a).

The *cue* mutants, were selected to identify elements that play a positive role during de-etiolation in *Arabidopsis* (99, 105). Most *cue* genes influence chloroplast development in the light and appear to modulate expression levels of particular nuclear-encoded photosynthetic genes. *CUE1* encodes a phosphate/phosphoenolpyruvate translocator (PPT) of the chloroplast inner envelope membrane (42, 161). PPT is likely required early in chloroplast development; loss of this function probably directly influences signals from the chloroplast that affect nuclear expression of light-regulated genes.

MUTATIONAL ANALYSIS OF PLANT MITOCHONDRIA

Unlike the chloroplast case, there are few mutations in mitochondrial development. Most loss of function mutation in mitochondria are lethal unless maintained in a heteroplasmic state (125). Some dominant mitochondrial mutations, resulting from expression of novel open reading frames in mtDNA, survive in higher plants. For the most part, these mutations lead to a similar phenotype of pollen inviability known as cytoplasmic male sterility (*cms*) (61). This maternally inherited trait has been reported in over 150 plant species, and in all cases examined in detail results from expression of novel mitochondrial polypeptides. Usually, these novel polypeptides contain at least one hydrophobic domain, likely facilitating membrane association. Each *cms*-associated polypeptide identified has been unique, and their modes of action are unclear.

Cms mutations have proven particularly useful in defining nuclear regulation of mitochondrial functions. Nuclear suppressors of the *cms* phenotype are readily identified as nuclear fertility restorer (*Rf*) genes. In the majority of cases, *Rf* genes act as single, dominant loci, though examples exist of polygenic restorer systems. Nuclear-directed mitochondrial functions identified via the analysis of fertility restoration mechanisms include mitochondrial transcript processing (154, 164, 182), posttranscriptional functions (157, 185), possible modes of biochemical detoxification (26), and alteration of mitochondrial genome organization (108).

NUCLEAR REGULATION OF MITOCHONDRIAL GENE EXPRESSION

Regulation of Transcription

A particularly unusual feature of plant mitochondria is the variable and complex pattern of transcripts arising from a given mitochondrial gene. Variation in transcript size can arise from use of multiple transcription start sites and termination sites (50). Complexity is compounded by posttranscriptional processing, described below.

Transcriptional regulation of mitochondrial gene expression does occur, although it does not appear to be the predominant means of gene modulation. For example, tissue-specific differences in transcript accumulation have been detected in various tissues of the maize seedling using in situ hybridization (101). It is unknown whether these differences reflect cell-specific changes in transcription rate, as opposed to posttranscriptional processes; it it clear that cell-to-cell modulation of mitochondrial expression can occur. Plant mitochondrial transcription appears to be mediated by at least one nuclear-encoded RNA polymerase bearing striking similarity to the RNA polymerases of bacteriophages T7, T3, and SP6 (68).

Data to date suggest that variation in promoter strength is likely a primary influence on relative transcription rates (15, 41, 50). Nuclear factors regulating transcription rate or promoter selection are difficult to detect. Perhaps the most compelling demonstration of nuclear influence on mitochondrial transcription is described in a Zea mays/Zea perennis alloplasmic line. A single nuclear gene, designated Mct, influences promoter selection in the cytochrome oxidase subunit II (coxII) gene in maize (22, 126). Transcriptional initiation at position −907 produces the predominant coxII transcript of ∼1900 nucleotides. The dominant Mct allele apparently directs transcriptional initiation at a second site (−347) upstream to the coxII locus. Interestingly, this alternate initiation site, detected as a shorter coxII transcript, does not conform to the consensus promoter sequence described for maize. This provides the first evidence that specific cofactors may influence promoter selection in plant mitochondria, unlike yeasts and mammalian systems.

Certain nuclear-directed changes in mitochondrial transcript levels appear to be tissue-specific. This has been suggested by in situ hybridization studies of maize seedling tissues, where several mitochondrial transcripts appeared to be present at different levels depending on tissue type (101). Similar studies in developing anthers of sunflower demonstrated a marked accumulation of *atpA*, *atp9*, *cob*, and *rrn26* transcripts in young meiotic cells, with a concomitant increase in their respective protein products (157). Moreover, *orf522* transcripts decrease in *cms* sunflower; the encoded 15-kDa protein is observed in anthers when the nuclear *Rf* gene, likely acting posttranscriptionally, is present (121). These results imply that nuclear-directed modulation of mitochondrial gene expression may occur in a cell-specific pattern. Further support comes from observations in *cms* Petunia where restoration of fertility by a nuclear *Rf* gene is associated with the tissue-specific reduction in a particular transcript derived from the *pcf* mitochondrial region (185).

Transcript Processing

Nuclear background influences mitochondrial transcription patterns in particular genomic regions. For example, nuclear background alters transcription of *atpA* in the Ogura cytoplasm of radish (109) and the *cms* gene (*T-urf 13*) in the T cytoplasm of maize. The *T-urf13* sterility-associated sequence is co-transcribed with the *orf 221* open reading frame. Specific lines of maize revealed a marked influence of nuclear background on its pattern of transcription (82). Rocheford & Pring (1994) demonstrated that the transcript changes were not only a function of the mitochondrial genomic environment encompassing the *T-urf13/orf221* region but also of dominant nuclear gene action influencing the pattern of posttranscriptional processing these transcripts undergo (143).

Transcript processing is a widespread phenomenon in plant mitochondria though not yet well understood mechanistically. It was first deduced by the observation that unusually complex transcript in certain regions of the plant mitochondrial genome were produced not only by multiple transcription start and stop sites, but also by internal transcript processing (reviewed in 50). Although the role of processing in gene regulation is not yet clear, nuclear regulation of transcript processing could be an important means of mitochondrial gene suppression. This conclusion is based on the observation that three different nuclear *Rf* loci directly influence transcript processing within the respective mitochondrial *cms*-associated regions.

In *cms*-T maize, perhaps the most well-investigated example of cytoplasmic male sterility, fertility restoration requires dominant allels of two unlinked nuclear loci, *Rf1* and *Rf2* (94). Compelling evidence demonstrates that the product of *Rf1*, essential though not sufficient to restore fertility, directly influences transcript processing of the *T-urf13* mitochondrial region (182). Processing of *T-urf13* transcripts appears to be directly associated with a marked reduction in the expression of the encoded 13-kDa T-URF13 polypeptide (43). The action of *Rf2* will be discussed in a later section. In sorghum line IS1112C, *cms* is associated with expression of the *orf107* open reading frame (164). Again, fertility restoration is associated with altered processing of *orf107* transcripts, and the concomitant reduction in the accumulation of a 12-kDa polypeptide presumed to be the gene product (164). Of particular interest is the observation that the transcript processing sites described in both *cms*-T maize and *cms* sorghum share sequence features (34), implying that sequence motifs exist within plant mitochondrial genes that may act as targets for nuclear-directed gene modulation.

cms in the *Brassica napus* (oilseed rape) Polima cytoplasm is associated with aberrant expression of a region encoding the *atp6* gene cotranscribed with a downstream chimeric sequence, *orf224* (152). Fertility restoration, conditioned by either of two dominant single nuclear loci, results in a transcript processing event that generates predominantly monocistronic *atp6* transcripts (152, 153). Generation of monocistronic *atp6* transcripts cosegregates with a single dominant fertility restorer locus, *Rfp1* (154). Most intriguing, however, is the observation that an alternate recessive allele at this locus, (*rfp1*) or a second locus tightly linked to *rfp1*, influences transcript processing of two other mitochondrial genes not associated with sterility, *nad4* and a *ccl1*-like gene that may be involved in cytochrome *c* biogenesis. At all four processing sites under *Rfp1* or *rfp1* control, there is UUGUGG or UUGUUG, a sequence motif, located very near the processing site.

Transcript Editing

One of the more surprising features distinguishing plant mitochondrial gene expression from that of yeast and mammals is the extensive editing of plant mitochondrial transcripts. The biochemistry and pattern of transcript editing has been reviewed recently (49, 60). The mechanisms regulating the rate of editing, as well as the local sequence features determining sites to be edited, are not yet understood. Evidence to date indicates that local features of the editing site are important (24, 55, 56) and that sites that alter codons may be preferentially edited (181). Nuclear genotype (181) as well as tissue type, developmental stage, and growth conditions may influence the extent of transcript editing (52). Perhaps the most definitive evidence of nuclear influence on transcript editing involves differential *nad3* editing in a particular Petunia mitochondrial genotype when combined with different nuclear backgrounds. The extent of *nad3* editing changes dramatically in response to different nuclear backgrounds, with a high extent of editing segregating as a single dominant nuclear locus. Interestingly, this genetic variation for extent of transcript editing pertained exclusively to *nad3*; this implies specificity in nuclear control of editing and predicts that additional nuclear factors exist.

Translational/Posttranslational Regulation

Relatively little evidence currently exists for nuclear-directed, translational regulation in plant mitochondria. In contrast, several nuclear factors influence translation of specific mitochondrial genes in yeast (44). In plants, sequence conservation exist upstream of some mitochondrial start codons, but it is not clear whether these sequences actually function as specific binding sites for translational components (134). Furthermore, although particular mitochondrial polypeptide differences have been associated with changes in plant nuclear background (23), it is not yet determined whether the protein differences result from differential transcription or translation.

Perhaps more provocative at this stage is the accumulating evidence suggesting surprisingly limited regulation of translation in mitochondria, particularly in light of extensive translational control in plastids. One unexpected observation of transcript editing is that both edited and unedited transcripts are translationally competent. Both edited and partially edited transcripts appear to be represented in association with polysomes (57, 106). Moreover, little discrimination by the ribosome occurs with regard to untranslated leader sequences or transcript splicing status (184).

It may be argued that polysome association does not, in itself, demonstrate nonselective translation. Using antibodies against the predicted polypeptide product of the *rps12* unedited transcript, compelling evidence demonstrates that polypeptides are produced from partially edited or unedited transcripts, as well as edited forms (107, 133). Not unexpectedly, these aberrant polypeptides are not incorporated into functional ribosomes (133).

These observations raise the obvious general question about the fate of aberrant translation products, and why they are not more commonly detected within plant mitochondria. One explanation may be differences in half-life for products of edited vs unedited transcripts. Although protein import and the concomitant processing events associated with import have been well described (reviewed in 46, 179), little is yet known about posttranslational proteolysis of mitochondrial gene products. Mitochondrial proteolysis has been best detailed in yeast, where several proteases, both matrix and membrane-localized, have been characterized (reviewed in 140). In plants, no proteolytic activity has been detected in the matrix of mitochondria, but limited activity is detected within the inner membrane; and there is some indication that this activity may be involved in the proteolysis of unassembled, imported proteins (87).

In common bean, posttranslational regulation of the *cms*-associated mito-chondrial protein appears to occur in vegetative tissues. In bean, *cms* is associ-ated with the expression of a 239–amino acid polypeptide, (ORF239) (1, 67) that accumulates only in reproductive tissues, with no detectable ORF239 in vegeta-tive (seedling, leaf, or root) tissues (1). More extensive investigation shows that the ORF239 protein is subject to proteolysis in vegetative tissues, dependent in part on a mitochondrial protease related to the *lon* homologs of yeast (163, 171) and human (175; R Sarria, A Lyznik, E Vallejos & S Mackenzie, submitted manuscript). Whether this newly identified plant protease is involved in the degradation of other aberrant mitochondrial gene products remains an active area of inquiry.

Detoxification of metabolic poisons is a quite unexpected means of influenc-ing mitochondrial functions posttranslationally. The *Rf2* fertility restorer gene in *cms-T* maize encodes a putative aldehyde dehydrogenase (26). Although the function of this gene in fertility restoration is not yet clear, recent biochemical studies have confirmed its enzymatic activity identity (P Schnable, personal communication). It has been speculated that the *Rf2* gene product may play a role in the detoxification of a pollen-specific product that interacts with the T-URF13 protein to cause premature tapetal breakdown.

NUCLEAR INFLUENCE ON THE MITOCHONDRIAL GENOME

The plant mitochondrial genome, now fully sequenced in *Arabidopsis* (169) and *Marcantia* (128), is organized in a much more complex and variable struc-ture than is observed in other higher eukaryotes (reviewed in 183). In most higher plants, this organization is defined by the presence of recombinationally active repeated sequences that allow for high- and low-frequency inter- and intramolecular recombination events to occur (3). The physical organization of the mitochondrial genome in plants has been difficult to define. Although most genomes map as circular molecules defined by overlapping clones, direct physical observation by pulsed field gel electrophoresis, electron microscopy, and other procedures has indicated that the genome may consist of both lin-ear and circular forms, with molecules much larger than the multiple circles constructed by clone analysis (3, 14, 130).

The nucleus definitely affects mitochondrial genome organization. One value of the remarkable mitochondrial DNA variation existing within plant families is the information it provides regarding evolution of this unusual genome and the cellular forces molding current organization. Examination of variation within the legume family (Fabaceae) provides striking evidence for the ongoing evolutionary transfer of functional genetic information from the mitochondrion to the nucleus via RNA intermediates (25, 127). This evolutionary transfer of mitochondrial genes to the nucleus has presented some intriguing problems, most notably the requirement to move tRNA into mitochondria (33). There is evidence that this transfer likely requires association with the appropriate aminoacyl tRNA synthase to mediate transmembrane import (32).

A multipartite genome organization exists in the mitochondrion of most plant species, with each molecule containing only a portion of the genetic information. Overall structure is further complicated by the variable stoichiometry of specific subgenomic regions. Remarkably, in several plant species, a subset of mitochondrial DNA molecules, atypical genomic organizations termed "sublimons," can be retained indefinitely in nearly undetectable levels (155, 156). The relative copy number of the various mitochondrial DNA forms and their recombinational activity appears to be under nuclear control. One of the most pronounced examples is the observed loss of a mitochondrial genomic molecule in response to a single nuclear gene in common bean. The *cms*-associated mitochondrial mutation in common bean, *pvs-orf239*, appears to be maintained on a single 210-kb molecule within a tripartite mitochondrial genome organization (78). Introduction of a single dominant nuclear factor, *Fr*, results in a genomic shift of the *pvs-orf239*–containing molecule to substoichiometric levels within the genome, thus restoring pollen fertility (H Janska, R Sarria, M Woloszynska, M Arrieta-Montiel & S Mackenzie, manuscript submitted).

The development of alloplasmic lines, derived by recurrent backcrossing strategies or protoplast fusion to combine different mitochondrial and nuclear genotypes, routinely gives rise to changes in relative stoichiometries and mitochondrial genomic rearrangements in *Nicotiana* (5, 13, 58), *Brassica* (93, 104), and *Triticum* (120, 129, 168). In some cases, it has been possible to identify specific nuclear loci essential to establishing compatibility in individual nuclear-cytoplasmic combinations (4). Moreover, particular nuclear-cytoplasmic genetic combinations in maize can be predicted to give rise to a high frequency of specific mitochondrial mutations. These mutations, referred to collectively as nonchromosomal stripe mutations, generally result in loss of mitochondrial gene function, leaf striping, severe growth impairment, and infertility; *ncs* mutations are maintained in a heteroplasmic state with wild-type, functional mitochondria (125). Several *ncs* mutations affect in distinct loci and have arisen by what appear to be different molecular events.

In *Arabidopsis*, the appearance of mitochondrial mutations is likewise associated with modification of the nuclear genotype. In the case of *Arabidopsis*, mutation at a single dominant nuclear gene, *CHM*, yields mitochondrial genomic rearrangements (113). Notably, the mutant mitochondrial forms arising upon *CHM* mutation are already present in the wild-type lines at substoichiometric levels (147), implying that the role of the *CHM* locus, like that of the *Fr* locus in bean, may be to suppress copy number of mutant mitochondrial forms. Efforts are ongoing to determine whether the *Fr* locus in *cms* bean involve homologous functions (B Li & S Mackenzie, personal communication).

PERSPECTIVES

Chloroplast and mitochondrial development are regulated by complex intracellular interactions. In the case of the plastids, it is likely that the discrete nuclear factors set the program that determines to a large degree the stage of organelle development. In contrast, plastid signaling to the nucleus may be continual throughout development. In the next few years, we will likely see important advances in the characterization of chloroplast signals that trigger changes in nuclear expression.

In the past years, we have seen an important number of breakthroughs in several aspects of the chloroplast field as protein import, transcription, and posttranscriptional events. Now a clearer picture is emerging of the regulatory events that take place in each of these processes. Collectively, in the near future, it will be possible to understand the surprising variety of regulatory mechanisms that in concert permit chloroplast development. In the years to come, the implementation of new approaches will contribute to the identification of molecules required for alternative plastid development pathways. Novel classes of genetic loci in *Arabidopsis* define unsolved steps of early chloroplast differentiation. Detailed study of these genes during plant development will certainly help us understand the mechanisms operating in early chloroplast development.

In mitochondrial research, several laboratories are now quite close to cloning additional fertility restorers involved in transcript processing, as well as the *CHM* and *Fr* loci involved in differential amplification of the genome. A surprising picture emerges as we investigate mitochondrial regulation; namely, that the mitochondrial genome is permitted to expend what would appear to be tremendous energy in the execution of transcription and, in most cases, translation of products destined for rapid turnover. These observations imply that, unlike prokaryotes, factors other than energy conservation may be most influential in plant mitochondria. We predict that future investigations will discover the impetus for retaining functions as perplexing as substoichiometric genomic forms and prolific editing functions. Until such time, the plant mitochondrial genome remains an enigma.

ACKNOWLEDGMENTS

We thank everyone who provided us with reprints, preprints, and unpublished results of their recent work, and especially K Pyke and D Stern for sharing their expertise on selected topics. In addition, we thank Helena Porta, Kenneth Luehrsen, and Stewart Gillmor for critical reading of the manuscript. The work in our laboratories was supported by a grant of the DGAPA program IN206294 and the Pew Charitable Trust to PL and from the USDA, National Science Foundation, and National Institutes of Health-GMS to SM.

Visit the *Annual Reviews home page* at
http://www.AnnualReviews.org.

Literature Cited

1. Abad A, Mehrtens B, Mackenzie S. 1995. Specific expression in reproductive tissues and fate of a mitochondrial sterility-associated protein in cytoplasmic male sterile bean. *Plant Cell* 7:271–85

2. Allison L, Simon L, Maliga P. 1996. Deletion of *rpoB* reveals a second distinct transcription system in plastids of higher plants. *EMBO J.* 15:2802–9

3. Andre C, Levy A, Walbot V. 1992. Small repeated sequences and the structure of plant mitochondrial genomes. *Trends Genet.* 8:128–32

4. Asakura N, Nakamura C, Ohtsuka I. 1997. RAPD markers linked to the nuclear gene from *Triticum timopheevii* that confers compatibility with *Aegilops squarrosa* cytoplasm on alloplasmic durum wheat. *Genome* 40:201–10

5. Aviv D, Galun E. 1987. Chondriome analysis in sexual progenies of Nicotiana cybrids. *Theor. Appl. Genet.* 73:821–26

6. Barkan A. 1989. Tissue-dependent plastid RNA splicing in maize: transcripts from four plastid genes are predominantly unspliced in leaf meristems and roots. *Plant Cell* 1:437–45

7. Barkan A, Miles D, Taylor W. 1986. Chloroplast gene expression in nuclear, photosyntetic mutants of maize. *EMBO J.* 5:1421–27

8. Barkan A, Voelker R, Mendel-Hartvig J, Johnson D, Walker M. 1995. Genetic analysis of chloroplast biogenesis in higher plants. *Physiol. Plant.* 93:163–70

9. Barkan A, Walker M, Nolasco M, Johnson D. 1994. A nuclear mutation in maize blocks the processing and translation of several chloroplast mRNAs and provides evidence for the differential translation of alternative mRNA forms. *EMBO J.* 13:3170–81

10. Bate NJ, Straus NA, Thompson JE.

1990. Expression of chloroplast photosynthetic genes during leaf senescence. *Plant Physiol.* 80:217–25

11. Baumgartner BJ, Rapp JC, Mullet JE. 1989. Plastid transcription activity and DNA copy number increase early in barley chloroplast development. *Plant Physiol.* 89:1011–18

12. Begg K, Donachie WD. 1985. Cell shape and division in *Escherichia coli:* experiments with shape and division mutants. *J. Bacteriol.* 163:615–22

13. Belliard G, Vedel F, Pelletier G. 1979. Mitochondrial recombination in cytoplasmic hybrids of *Nicotiana tabacum* by protoplast fusion. *Nature* 281:401–3

14. Bendich AJ. 1993. Reaching for the ring: the study of mitochondrial genome structure. *Curr. Genet.* 24:279–90

15. Binder S, Marchfelder A, Brennicke A. 1996. Regulation of gene expression in plant mitochondria. *Plant Mol. Biol.* 32:303–14

16. Byrne M, Taylor WC. 1996. Analysis of *Mutator*-induced mutations in the *Iojap* gene of maize. *Mol. Gen. Genet.* 252:216–20

17. Chamovitz DA, Wei N, Osterlund MT, von Arnim AG, Staub JM, et al. 1996. The COP9 complex, a novel multisubunit nuclear regulator involved in light control of a plant developmental switch. *Cell* 86:115–21

18. Chatterjee M, Sparvoli S, Edmunds C, Garosi P, Findlay K, Martin C. 1996. *DAG*, a gene required for chloroplast differentiation and palisade development in *Antirrhinum majus. EMBO J.* 15:4194–207

19. Chory J. 1993. Out of darkness: mutants reveal pathways controlling light-regulated development in plants. *Trends Genet.* 9:167–72

20. Chory J, Nagpal P, Peto CA. 1991. Phenotypic and genetic analysis of *det2*, a

new mutant that affects light-regulated seedling development in *Arabidopsis. Plant Cell* 3:445–59

21. Cline K, Henry R. 1996. Import and routing of nucleus-encoded chloroplast proteins. *Annu. Rev. Cell Dev. Biol.* 12:1–26

22. Cooper P, Butler E, Newton KJ. 1990. Identification of a maize nuclear gene which influences the size and number of *cox2* transcripts in mitochondria of perennial teosintes. *Genetics* 126:461–67

23. Cooper P, Newton KJ. 1989. Maize nuclear background regulates the synthesis of a 22-kDa polypeptide in *Zea luxurians* mitochondria. *Proc. Natl. Acad. Sci. USA* 86:7423–26

24. Covello PS, Gray MW. 1990. Differences in editing at homologous sites in messenger RNAs from angiosperm mitochondria. *Nucleic Acids Res.* 18:5189–96

25. Covello PS, Gray MW. 1992. Silent mitochondrial and active nuclear genes for subunit 2 of cytochrome c oxidase (*cox2*) in soybean: evidence for RNA-mediated gene transfer. *EMBO J.* 11:3815–20

26. Cui XQ, Wise RP, Schnable PS. 1996. The *rf2* nuclear restorer gene of male sterile T-cytoplasm maize. *Science* 272:1334–36

27. Danon A, Mayfield SP. 1991. Light regulated translational activators: identification of chloroplast gene specific mRNA binding proteins. *EMBO J.* 10:3993–4001

28. Danon A, Mayfield SP. 1994. ADP-dependent phosphorylation regulates RNA-binding in vitro: implications in light-modulated translation. *EMBO J.* 13:2227–35

29. Deng X-W, Gruissem W. 1987. Control of plastid gene expression during development: the limited role of transcriptional regulation. *Cell* 49:379–87

30. Deng X-W, Matsui M, Wei N, Wagner D, Chu AM, et al. 1992. *COP1*, an Arabidopsis regulatory gene, encodes a protein with both a zinc-binding motif and a G_β homologous domain. *Cell* 27:791–801

31. de Souza D, Osteryoung K, Pyke KA. 1997. Molecular genetics of chloroplast development in *Arabidopsis. Int. Conf. Arabidopsis Res., 8th, Madison, Wis.*

32. Dietrich A, Marechaldorouard L, Carneiro V, Cosset A, Small I. 1996. A single base change prevents import of cytosolic tRNA (ala) into mitochondria in transgenic plants. *Plant J.* 10:913–18

33. Dietrich A, Small I, Cosset A, Weil JH, Marechaldrouard L. 1996. Editing and import—strategies for providing plant mitochondria with a complete set of functional transfer RNAs. *Biochimie* 78:518–29

34. Dill CL, Wise RP, Schnable PS. 1998. *Rf8* and *Rf** mediate unique *T-urf13*-transcript accumulation, revealing a mitochondrial consensus sequence associated with RNA processing and restoration of pollen fertility in T-cytoplasm maize. *Genetics*. In press

35. Dinkins RD, Bandaranayake H, Baeza L, Griffiths AJF, Green BR. 1997. *hcf5*, a nuclear photosynthetic electron transport mutant of *Arabidopsis thaliana* with a pleiotropic effect on chloroplast gene expression. *Plant Physiol.* 113:1023–31

36. Dörmann P, Hoffmann-Benning S, Balbo I, Benning C. 1995. Isolation and characterization of an Arabidopsis mutant deficient in the thylakoid lipid digalactosyl diacylglycerol. *Plant Cell* 7:1801–10

37. Duckett CM, Gray JC. 1995. Illuminating plant development. *BioEssays* 17:101–3

38. Erickson HP. 1995. FtsZ, a prokaryotic homolog of tubulin? *Cell* 80:367–70

39. Falbel TG, Meehl JB, Staehelin LA. 1996. Severity of mutant phenotype in a series of chlorophyll-deficient wheat mutants depends on light intensity and the severity of the block in chlorophyll synthesis. *Plant Physiol.* 112:821–32

40. Falbel TG, Staehelin LA. 1994. Characterization of a family of chlorophyll-deficient wheat and barley mutants with defects in the Mg-insertion step on chlorophyll biosynthesis. *Plant Physiol.* 104:639–48

41. Finnegan PM, Brown GG. 1990. Transcriptional and post-transcriptional regulation of RNA levels in maize mitochondria. *Plant Cell* 2:71–83

42. Fisher K, Kammerer B, Gutensohn M, Arbirger B, Weber A, et al. 1997. A new class of plastidic phosphate translocators: a putative link between primary and secondary metabolism by the phosphoenolpyruvate/phosphate antiporter. *Plant Cell* 9:453–62

43. Forde BG, Oliver RJC, Leaver CJ. 1978. Variation in mitochondrial translation products associated with male-sterile cytoplasms in maize. *Proc. Natl. Acad. Sci. USA* 75:3841–45

44. Fox TD. 1996. Genetics of mitochondrial translation. In *Translational Control*, ed. J Hershey, M Mathews,

N Sonenberg, pp. 733–58. Cold Spring Harbor, NY: Cold Spring Harbor Lab. Press

45. Gillham NW, Boynton JE, Hauser CR. 1994. Translational regulation of gene expression in chloroplast and mitochondria. *Annu. Rev. Genet.* 28:71–93

46. Glaser E, Ljoling S, Szigyarto C, Eriksson AC. 1996. Plant mitochondrial protein import: precursor processing is catalysed by the integrated mitochondrial processing peptidase (MPP)/bc1 complex and degradation by the ATP-dependent proteinase. *Biochem. Biophys. Acta* 1275:33–37

47. Gottesman S, Wickner S, Maurizi MR. 1997. Protein quality control: triage by chaperones and proteases. *Genes Dev.* 11:815–23

48. Gray MW. 1992. The endosymbiont hypothesis revisited. *Int. Rev. Cytol.* 141:233–357

49. Gray MW. 1996. RNA editing in plant organelles—a fertile field. *Proc. Natl. Acad. Sci. USA* 96:8157–59

50. Gray MW, Hanic-Joyce PJ, Covello PS. 1992. Transcription, processing and editing in plant mitochondria. *Annu. Rev. Plant Physiol. Plant Mol. Biol.* 43:145–75

51. Green CD, Hollingsworth MJ. 1992. Expression of the large ATP synthase gene cluster in spinach plastids during light-induced development. *Plant Physiol.* 100:1164–70

52. Grosskopf D, Mulligan RM. 1996. Developmental and tissue-specificity of RNA editing in mitochondria of suspension-cultured maize cells and seedlings. *Curr. Genet.* 29:556–63

53. Gruissem W, Schuster G. 1993. Control of mRNA degradation in organelles. In *Control of Messenger RNA Stability*, ed. J Belasco, G Brawerman, pp. 329–65. New York: Academic

54. Gruissem W, Tonkyn JC. 1993. Control mechanisms of plastid gene expression. *Crit. Rev. Plant Sci.* 12:19–55

55. Gualberto JM, Bonnard G, Lamattina L, Grienenberger JM. 1991. Expression of the wheat mitochondrial *nad3-rps12* transcription unit: correlation between editing and mRNA maturation. *Plant Cell* 3:1109–20

56. Gualberto JM, Weil JH, Grienenberger JM. 1990. Editing of the wheat *coxIII* transcript: evidence for twelve C to U and one U to C conversions and for sequence similarities around editing sites. *Nucleic Acids Res.* 18:3771–76

57. Gualberto JM, Wintz H, Weil JH, Grie-nenberger JM. 1988. The genes coding for subunit 3 of NADH dehydrogenase and for ribosomal protein S12 are present in the wheat and maize mitochondrial genome and are co-transcribed. *Mol. Gen. Genet.* 215:118–27

58. Hakansson G, Glimelius K. 1991. Extensive nuclear influence on mitochondrial transcription and genome structure in male-fertile and male-sterile alloplasmic *Nicotiana* materials. *Mol. Gen. Genet.* 229:380–88

59. Haley J, Bogorad L. 1990. Alternative promoters are used for genes within maize chloroplast polycistronic transcription units. *Plant Cell* 2:323–33

60. Hanson M, Sutton C, Lu B. 1996. Plant organelle gene expression-altered by RNA editing. *Trends Plant Sci.* 1:57–64

61. Hanson MR. 1991. Plant mitochondrial mutations and male sterility. *Annu. Rev. Genet.* 25:461–86

62. Harpster MH, Mayfield SP, Taylor WC. 1984. Effects of pigment-deficient mutants on the accumulation of photosynthetic proteins in maize. *Plant Mol. Biol.* 3:59–71

63. Harrak H, Langrange T, Bisanz-Seyer C, Lerbs-Mache S, Mache R. 1995. The expression of nuclear encoding plastid ribosomal proteins precedes the expression of chloroplast gene during early phases of chloroplast development. *Plant Physiol.* 108:685–92

64. Harrison MA, Nemson JA, Melis A. 1993. Assembly and composition of the chlorophyll *a-b* light harvesting complex of barley: immunochemical analysis of the chlorophyll *b*-less and chlorophyll b-deficient mutants. *Photosynth. Res.* 38:141–51

65. Hauser CR, Gillham NW, Boynton JE. 1996. Translational regulation of chloroplast genes. Proteins binding to the 5′-untranslated regions of chloroplast mRNAs in *Chlamydomonas reinhardtii. J. Biol. Chem.* 271:1486–97

66. Hayes R, Kudla J, Schuster G, Gabay L, Maliga P, Gruissem W. 1996. Chloroplast mRNA 3′-end processing by a high molecular weight protein complex is regulated by nuclear encoded RNA binding proteins. *EMBO J.* 15:1132–41

67. He S, Abad AR, Gelvin SB, Mackenzie SA. 1996. A cytoplasmic male sterility-associated mitochondrial protein causes pollen disruption in transgenic tobacco. *Proc. Natl. Acad. Sci. USA* 93:11763–68

68. Hedtke B, Borner T, Weihe A. 1997. Mitochondrial and chloroplast phage-type

RNA polymerases in *Arabidopsis*. *Science* 277:809–11

69. Hess WR, Muller A, Nagy F, Borner T. 1994. Ribosome-deficient plastids affect transcription of light-induced nuclear genes: genetic evidence for a plastid-derived signal. *Mol. Gen. Genet.* 242:305–12

70. Hess WR, Prombona A, Fieder B, Subramanian AR, Borner T. 1993. Chloroplast *rps*15 and the *rpo*B/C1/C2 gene cluster are strongly transcribed in ribosome-deficient plastids: evidence for a functioning nonchloroplast encoded RNA polymerase. *EMBO J.* 12:563–71

71. Hirose T, Sugiura M. 1996. *Cis*-acting elements and *trans*-acting factors for accurate translation of chloroplast *psbA* mRNAs: development of an in vitro translation system from tobacco chloroplasts. *EMBO J.* 15:1687–95

72. Hu J, Bogorad L. 1990. Maize chloroplast RNA polymerase: the 180-, 120-, and 38-kilodalton polypeptides are encoded in chloroplast genes. *Proc. Natl. Acad. Sci. USA* 87:1531–35

73. Hudson A, Carpenter R, Doyle S, Coen ES. 1993. *Olive*: a key gene required for chlorophyll biosynthesis in *Antirrhinum majus. EMBO J.* 12:3711–19

74. Igloi GL, Kössel H. 1992. The transcriptional apparatus of chloroplast. *Crit. Rev. Plant Sci.* 10:525–58

75. Inagaki N, Fujita S, Satoh K. 1989. Solubilization and partial purification of a thylakoidal enzyme of spinach involved in the processing of D1 protein. *FEBS Lett.* 246:218–22

76. Iratni R, Baeza R, Andreeva A, Mache R, Lerbs-Mache S. 1994. Regulation of the rDNA transcription in chloroplasts: promoter exclusion by a constitutive repression. *Genes Dev.* 8:2928–38

77. Isono K, Niwa Y, Satoh K, Kobayashi H. 1997. Evidence for transcriptional regulation of plastid photosynthesis genes in *Arabidopsis thaliana* roots. *Plant Physiol.* 114:623–30

78. Janska H, Mackenzie SA. 1993. Unusual mitochondrial genome organization in cytoplasmic male sterile common bean and the nature of cytoplasmic reversion to fertility. *Genetics* 135:869–79

79. Jenkins BD, Kulhanek DJ, Barkan A. 1997. Nuclear mutations that block group II RNA splicing in maize chloroplast reveal several intron classes with distinct requirements for splicing factors. *Plant Cell* 9:283–96

80. Keddie JS, Carroll B, Jones JDG, Gruissem W. 1996. The *DCL* gene of tomato

81. Keegstra K, Olsen LJ, Theg SM. 1989. Chloroplastic precursors and their transport across the envelope membranes. *Annu. Rev. Plant Physiol. Plant Mol. Biol.* 40:471–501

82. Kennell JC, Wise RP, Pring DR. 1987. Influence of nuclear background on transcription of a maize mitochondrial region associated with Texas male sterile cytoplasm. *Mol. Gen. Genet.* 210:399–406

83. Kim MY, Christopher DA, Mullet JE. 1993. Direct evidence for selective modulation of *psbA*, *rpoA*, *rbcL* and 16S RNA stability during barley chloroplast development. *Plant Mol. Biol.* 22:447–63

84. Kim MY, Mullet JE. 1995. Identification of a sequence-specific DNA binding factor required for transcription of the barley chloroplast blue light–responsive *psbD-psbC* promoter. *Plant Cell* 7:1445–57

85. Kirk JTO, Tilney-Bassett RAE. 1978. *The Plastids: Their Chemistry, Structure, Growth and Inheritance.* Amsterdam: Elsevier/North Holland. 960 pp.

86. Klein RR, Mullet JE. 1990. Light-induced transcription of chloroplast genes. *psbA* transcription is differentially enhanced in illuminated barely. *J. Biol. Chem.* 265:1895–902

87. Knorpp C, Szigyarto C, Glaser E. 1996. Evidence for a novel ATP-dependent membrane-associated protease in spinach leaf mitochondria. *Biochem. J.* 310:527–31

88. Król M, Spangfort MD, Huner NPA, Öquist G, Gustafsson P, Jansson S. 1995. Chlorophyll *a/b* binding proteins, pigment conversions and early light induced proteins in a chlorophyll *b*-less barley mutant. *Plant Physiol.* 107:873–83

89. Kudla J, Hayes R, Gruissem W. 1996. Polyadenylation accelerates degradation of chloroplast mRNA. *EMBO J.* 15:7137–46

90. Kuhlemeier C. 1992. Transcriptional and post-transcriptional regulation of gene expression in plants. *Plant Mol. Biol.* 19:1–14

91. Kwok SF, Piekos B, Miséra S, Deng X-W. 1996. A complement of ten essential and pleiotropic Arabidopsis *COP/DET/FUS* genes is necessary for repression of photomorphogenesis in darkness. *Plant Physiol.* 110:731–42

92. Kyle DJ, Ohad I, Arntzen CJ. 1984.

Membrane protein damage and repair: selective loss of a quinone-protein function in chloroplast membranes. *Proc. Natl. Acad. Sci. USA* 81:4070–74

93. Landgren M, Zetterstrand M, Sundberg, Glimelius K. 1996. Alloplasmic male-sterile *Brassica* lines containing *B. tournefortii* mitochondria express an ORF 3' of the *atp*6 gene and a 32 kDa protein. *Plant Mol. Biol.* 32:879–90

94. Laughnan JR, Gabay-Laughnan S. 1983. Cytoplasmic male sterility in maize. *Annu. Rev. Genet.* 17:27–48

95. Leech RM. 1986. Stability and plasticity during chloroplast development. In *Plasticity in Plants*, ed. D Jennings, A Trewavas, pp. 121–53. Cambridge: Soc. Exp. Biol. Symp., 40

96. Lerbs S, Brautigam E, Mache R. 1988. DNA-dependent RNA polymerase of spinash chloroplast: Characterization of α a-like and σ-like polypeptides. *Mol. Gen. Genet.* 211:459–64

97. Lerbs-Mache S. 1993. The 110-kDa polypeptide of spinach plastid DNA-dependent RNA polymerase: single-subunit enzyme or catalytic core of multimeric enzyme complexes? *Proc. Natl. Acad. Sci. USA* 90:5509–13

97a. Levings CS, Vasil I, eds. 1995. *The Molecular Biology of Plant Mitochondria*. Boston: Kluwer

98. Levy H, Kindle KL, Stern D. 1997. A nuclear mutation that affects the 3' processing of several mRNAs in Chlamydomonas chloroplasts. *Plant Cell* 9:825–36

99. Li H-M, Culligan K, Dixon RA, Chory J. 1995. *CUE1*: a mesophyll cell-specific positive regulator of light-controlled gene expression in Arabidopsis. *Plant Cell* 7:1599–610

100. Li JM, Nagpal P, Vitart V, McMorris TC, Chory J. 1996. A role for brassinosteroids in light-dependent development of *Arabidopsis*. *Science* 272:398–401

101. Li X-Q, Zhang M, Brown GG. 1996. Cell-specific expression of mitochondrial transcripts in maize seedlings. *Plant Cell* 8:1961–75

102. Link G. 1994. Plastid differentiation: organelle promoters and transcription factors. In *Plant Promoters and Transcription Factors*, ed. L Nover, pp. 63–83. Heidelberg: Springer-Verlag

103. Link G. 1996. Green life: control of chloroplast gene transcription. *BioEssays* 18:465–71

104. Liu JH, Landgren M, Glimelius K. 1996. Transfer of the *B. tournefortii* cytoplasm to *B. napus* for the production of cytoplasmic male sterile *B. napus*. *Physiol. Plant* 96:123–29

105. Lopez-Juez E, Stretfield S, Chory J. 1996. Light signals and autoregulated chloroplast development. In *Regulation of Plant Growth and Development by Light*, ed. WR Briggs, RL Heath, EM Tobin, pp. 144–52. Rockville, MD: Am. Soc. Plant Physiol.

106. Lu BW, Hanson MR. 1996. Fully edited and partially edited *nad9* transcripts differ in size and both are associated with polysomes in potato mitochondria. *Nucleic Acids Res.* 24:1369–74

107. Lu BW, Wilson RK, Phreaner CG, Mulligan RM, Hanson MR. 1996. Protein polymorphism generated by differential RNA editing of a plant mitochondrial *rps12* gene. *Mol. Cell Biol.* 16:1543–49

108. Mackenzie SA, Chase CD. 1990. Fertility restoration is associated with loss of a portion of the mitochondrial genome in cytoplasmic male sterile common bean. *Plant Cell* 2:905–12

109. Makaroff CA, Apel IJ, Palmer JD. 1990. Characterization of radish mitochondrial *atpA*: influence of nuclear background on transcription of *atpA*-associated sequences and relationship with male sterility. *Plant Mol. Biol.* 15:735–46

110. Mandel MA, Feldmann KA, Herrera-Estrella L, Rocha-Sosa M, Leon P. 1996. *CLA1*, a novel gene required for chloroplast development, is highly conserved in evolution. *Plant J.* 9:649–58

111. Marano M, Serra E, Orellano E, Carrillo N. 1993. The path of chromoplast development in fruits and flowers. *Plant Sci.* 94:1–17

112. Martienssen RA, Barkan A, Freeling M, Taylor WC. 1989. Molecular cloning of a maize gene involved in photosynthetic membrane organization that is regulated by Robertson's *Mutator*. *EMBO J.* 8:1633–39

113. Martínez-Zapater JM, Gil P, Capel J, Somerville CR. 1992. Mutations at the Arabidopsis *CHM* locus promote rearrangements of the mitochondrial genome. *Plant Cell* 4:889–99

114. Mayfield SP, Yohn CB, Cohen A, Danon A. 1995. Regulation of chloroplast gene expression. *Annu. Rev. Plant Physiol. Plant Mol. Biol.* 46:147–66

115. Melis A. 1991. Dynamics of photosynthetic membrane composition and function. *Biochim. Biophys. Acta* 1058:87–106

116. Meurer J, Meierhoff K, Westhoff P.

1996. Isolation of high chlorophyll-fluorescence mutants of *Arabidopsis thaliana* and their characterisation by spectroscopy, immunoblotting and northern hybridization. *Planta* 198:385–96

117. Michel F, Umesono K, Ozeki H. 1989. Comparative and functional anatomy of group II catalytic introns—A review. *Gene* 82:5–30

118. Miles D. 1994. The role of high chlorophyll fluroscence photosynthesis mutants in the analysis of chloroplast thylakoid membrane assembly and function. *Maydica* 39:35–45

119. Millar AJ, McGrath RB, Chua N-H. 1994. Phytochrome phototransduction pathways. *Annu. Rev. Genet.* 28:325–49

120. Mohr S, Schulte-Kappert E, Odenbach W, Oettler G, Kuck U. 1993. Mitochondrial DNA of cytoplasmic male-sterile *Triticum timopheevi*: rearrangement of upstream sequences of the *atp6* and *orf25* genes. *Theor. Appl. Genet.* 86:259–68

121. Moneger F, Smart CJ, Leaver CJ. 1994. Nuclear restoration of cytoplasmic male sterility in sunflower is associated with the tissue-specific regulation of a novel mitochondrial gene. *EMBO J.* 13:8–17

122. Morden CW, Wolfe KH, de Pamphilis CW, Palmer JD. 1991. Plastid translation and transcription genes in a nonphotosynthetic plant: intact, missing and pseudo genes. *EMBO J.* 10:3281–88

123. Mullet JE. 1993. Dynamic regulation of chloroplast transcription. *Plant Physiol.* 103:309–13

124. Nelson T, Langdale JA. 1989. Patterns of leaf development in C4 plants. *Plant Cell* 1:3–13

125. Newton KJ. 1995. Aberrant growth phenotypes associated with mitochondrial genome rearrangements in higher plants. See Ref. 97a, pp. 585–96

126. Newton KJ, Winberg B, Yamato K, Lupold S, Stern DB. 1995. Evidence for a novel mitochondrial promoter preceding the *cox2* gene of perennial teosintes. *EMBO J.* 14:585–93

127. Nugent JM, Palmer JD. 1991. RNA-mediated transfer of the gene *coxII* from the mitochondrion to the nucleus during flowering plant evolution. *Cell* 66:473–81

128. Oda K, Yamato K, Ohta E, Nakamura Y, Takemura M, et al. 1992. Gene organization deduced from the complete sequence of liverwort *Marchantia polymorpha* mitochondrial DNA: a primitive form of plant mitochondrial genome. *J. Mol. Biol.* 223:1–7

129. Ogihara Y, Futami K, Tsuji K, Murai K. 1997. Alloplasmic wheats with *Aegilops crassa* cytoplasm which express photoperiod-sensitive homeotic transformations of anthers, show alterations in mitochondrial DNA structure and transcription. *Mol. Gen. Genet.* 255:45–53

130. Oldenburg DJ, Bendich AJ. 1996. Size and structure of replicating mitochondrial DNA in cultured tobacco cells. *Plant Cell* 8:447–61

131. Osteryoung KW, Vierling E. 1995. Conserved cell and organelle division. *Nature* 376:473–74

132. Pepper A, Delaney T, Washburn T, Poole D, Chory J. 1994. *DET1*, a negative regulator of light-mediated development and gene expression in Arabidopsis, encodes a novel nuclear-localized protein. *Cell* 78:109–16

133. Phreaner CG, Williams MA, Mulligan RM. 1996. Incomplete editing of *rps12* transcripts results in the synthesis of polymorphic polypeptides in plant mitochondria. *Plant Cell* 8:107–17

134. Pring DR, Mullen JA, Kempken F. 1992. Conserved sequence blocks 5' to start codons of plant mitochondrial genes. *Plant Mol. Biol.* 19:313–17

134a. Pyke KA. 1997. The genetic control of plastid division in higher plants. *Am. J. Bot.* 84:1017–27

135. Pyke KA, Leech RM. 1992. Chloroplast division and expansion is radically altered by nuclear mutations in *Arabidopsis thaliana*. *Plant Physiol.* 99:1005–8

136. Pyke KA, Leech RM. 1994. A genetic analysis of chloroplast division and expansion in *Arabidopsis thaliana*. *Plant Physiol.* 104:201–7

137. Quail PH, Boylan MT, Parks BM, Short TW, Xu Y, Wagner D. 1995. Phytochromes: photosensory perception and signal transduction. *Science* 268:675–80

138. Rapp JC, Baumgartner BJ, Mullet J. 1992. Quantitative analysis of transcription and RNA levels of 15 barley chloroplast genes. Transcription rates and mRNA levels vary over 300-fold; predicted mRNA stabilities vary 30-fold. *J. Biol. Chem.* 267:21404–11

139. Reiter RS, Coomber SA, Bourett TM, Bartley GE, Scolnik PA. 1994. Control of leaf and chloroplast development by the Arabidopsis gene *pale cress*. *Plant Cell* 6:1253–64

140. Rep M, Grivell LA. 1996. The role

of protein degradation in mitochondrial function and biogenesis. *Curr. Genet.* 30:367–80

141. Rochaix JD. 1992. Post-transcriptional steps in the expression of chloroplast genes. *Annu. Rev. Cell Biol.* 8:1–28

142. Rochaix JD. 1996. Post-transcriptional regulation of chloroplast gene expression in *Chlamydomonas reinhardtii*. *Plant Mol. Biol.* 32:327–41

143. Rocheford TR, Kennell JC, Pring DR. 1992. Genetic analysis of nuclear control of *T-urf13/orf221* transcription in T cytoplasm maize. *Theor. Appl. Genet.* 84:891–98

144. Rodermel S, Haley J, Jiang C-Z, Tsai C-H, Bogorad L. 1996. A mechanism for intergenomic integration: abundance of ribulose biphosphate carboxilase small-subunit protein influences the translation of the large-subunit mRNA. *Proc. Natl. Acad. Sci. USA* 93:3881–85

145. Roth R, Hall LN, Brutnell TP, Langdale JA. 1996. *bundle sheath defective 2*, a mutation that disrupts the coordinate development of bundle sheath and mesophyll cells in the maize leaf. *Plant Cell* 8:915–27

146. Runge S, van Cleve B, Lebedev N, Armstrong G, Apel K. 1995. Isolation and classification of chlorophyll-deficient *xantha* mutants of *Arabidopsis thaliana*. *Planta* 197:490–500

147. Sakamoto W, Kondo H, Murata M, Motoyoshi F. 1996. Altered mitochondrial genome expression in a maternal distorted leaf mutant of Arabidopsis induced by chloroplast mutator. *Plant Cell* 8:1377–90

148. Saldanha R, Mohr G, Belfort M, Lambowitz AM. 1993. Group I and group II introns. *FASEB J.* 7:15–24

149. Schwartz RM, Dayhoff MO. 1978. Origins of prokaryotes, eukaryotes, mitochondria, and chloroplast. *Science* 199:395–403

150. Sexton TB, Christopher DA, Mullet JE. 1990. Light-induced swich in barley *psbD-psbC* promoter utilization: a novel mechanism regulating chloroplast gene expression. *EMBO J.* 9:4485–94

151. Shanklin J, DeWitt ND, Flanagan JM. 1995. The stroma of higher plant plastids contain ClpP and ClpC, functional homologs of *Escherichia coli* ClpP and ClpA: an archetypal two-component ATP-dependent protease. *Plant Cell* 7:1713–22

152. Singh M, Brown GG. 1991. Suppression of cytoplasmic male sterility by nuclear genes alters expression of a novel mitochondrial gene region. *Plant Cell* 3:1349–62

153. Singh M, Brown GG. 1993. Characterization of expression of a mitochondrial gene region associated with the Brassica "Polima" CMS: developmental influences. *Curr. Genet.* 24:316–22

154. Singh M, Hamel N, Menassa R, Li X-Q, Young B, et al. 1996. Nuclear genes associated with a single Brassica CMS restorer locus influence transcripts of three different mitochondrial gene regions. *Genetics* 143:505–16

155. Small ID, Isaac PG, Leaver CJ. 1987. Stoichiometric differences in DNA molecules containing the *atpA* gene suggest mechanisms for the generation of mitochondrial diversity in maize. *EMBO J.* 6:865–69

156. Small ID, Suffolk R, Leaver CJ. 1989. Evolution of plant mitochondrial genomes via sub-stoichiometric intermediates. *Cell* 58:69–76

157. Smart CJ, Moneger F, Leaver CJ. 1994. Cell-specific regulation of gene expression in mitochondria during anther development in sunflower. *Plant Cell* 6:811–25

158. Somerville CR. 1986. Analysis of photosynthesis with mutants of higher plants and algae. *Annu. Rev. Plant Physiol.* 37:467–507

159. Staub JM, Maliga P. 1993. Accumulation of the D1 polypeptide in tobacco plastids is regulated via the untranslated region of the *psbA* mRNA. *EMBO J.* 12:601–6

160. Stern DB, Higgs DC, Yang J. 1997. Transcription and translation in chloroplasts. *Trends Plant Sci.* 2:308–15

161. Streatfield SJ, Post-Beittenmiller D, Chory J. 1997. Analysis of the *cuel* (cab underexpressed) mutant of *Arabidopsis thaliana*. *Int. Conf. Arabidopsis Res., 8th, Madison, Wis.*

162. Susek R, Chory J. 1992. A tale of two genomes: role of a chloroplast signal in coordinating nuclear and plastid genome expression. *Aust. J. Plant Physiol.* 19:387–99

163. Suzuki CK, Suda K, Wang N, Schatz G. 1994. Requirement for the yeast gene *LON* in intramitochondrial proteolysis and maintenance of respiration. *Science* 264:273–76

164. Tang HV, Pring DR, Shaw LC, Salazar RA, Muza FR, et al. 1996. Transcript processing internal to a mitochondrial open reading frame is correlated with fertility restoration in male-sterile sorghum. *Plant J.* 10:123–33

165. Taylor WC. 1989. Regulatory interactions between nuclear and plastid genomes. *Annu. Rev. Plant Physiol. Plant Mol. Biol.* 40:211–33

166. Tiller K, Link G. 1993. Phosphorylation and dephosphorylation affect functional characteristics of chloroplast and etioplast transcription systems from mustard (*Sinapis alba* L.). *EMBO J.* 12:1745–53

167. Troxler RF, Zhang F, Hu J, Bogorad L. 1994. Evidence that σ factors are components of chloroplast RNA polymerase. *Plant Physiol.* 104:753–59

168. Tsunewaki K. 1993. Genome-plasmon interactions in wheat. *Jpn. J. Genet.* 68:1–34

169. Unseld M, Marienfeld JR, Brandt P, Brennicke A. 1997. The mitochondrial genome of *Arabidopsis thaliana* contains 57 genes in 366,924 nucleotides. *Nat. Genet.* 15:57–61

170. van der Graaff E. 1997. *Developmental mutants of Arabidopsis thaliana obtained after T-DNA transformation.* PhD thesis. Leiden Univ. 179 pp.

171. Van Dyck L, Pearce DA, Sherman F. 1994. *PIM1* encodes a mitochondrial ATP-dependent protease that is required for mitochondrial function in the yeast *Saccharomyces cerevisiae. J. Biol. Chem.* 269:238–42

172. Voelker R, Barkan A. 1995. Two nuclear mutations disrupt distinct pathways for targeting proteins to the chloroplast thylakoid. *EMBO J.* 14:3905–14

173. Voelker R, Mendel-Hartvig J, Barkan A. 1997. Transposon-disruption of a maize nuclear gene, *thaI*, encoding a chloroplast SecA homologue: in vivo role of cp-SecA in thylakoid protein targeting. *Genetics* 145:467–78

173a. von Arnim A, Deng X-W. 1996. Light control of seedling development. *Annu. Rev. Plant Physiol. Plant. Mol. Biol.* 47:215–43

174. von Wettstein D, Henningsen KW, Boynton JE, Kannangara GC, Nielsen OF. 1971. The genetic control of chloroplast development in barley. In *Autonomy and Biogenesis of Mitochondria and Chloroplast*, ed. N Boardman, A Linnae, R Smillie, pp. 205–23. Amsterdam: North-Holland

175. Wang N, Gottesman S, Willingham M, Gottesman MM, Maurizi MR. 1993. A human mitochondrial ATP-dependent protease that is highly homologous to bacterial *Lon* protease. *Proc. Natl. Acad. Sci. USA* 90:11247–51

176. Wei N, Chamovitz DA, Deng X-W. 1994. Arabidopsis COP9 is a component of a novel signaling complex mediating light control of development. *Cell* 78:117–24

176a. Wei N, Deng X-W. 1996. The role of the COP/DET/FUS genes in light control of Arabidopsis seedling development. *Plant Physiol.* 112:871–78

177. Westhoff P, Hermann RG. 1988. Complex RNA maturation in chloroplasts. *Eur. J. Biochem.* 171:551–64

178. Wetzel CM, Jiang C, Meehan LJ, Voytas DF, Rodermel SR. 1994. Nuclear-organelle interactions: the *immutans* variegation mutant of *Arabidopsis* is plastid autonomous and impaired in carotenoid biosynthesis. *Plant J.* 6:161–75

179. Whelan J, Glaser E. 1997. Protein import into plant mitochondria. *Plant Mol. Biol.* 33:771–89

180. Wilsbach K, Payne GS. 1993. Vps1p, a member of the dynamin GTPase family, is necessary for Golgi membrane protein retention in *Saccharomyces cerevisiae. EMBO J.* 12:3049–59

181. Wilson RK, Hanson MR. 1996. Preferential RNA editing at specific sites within transcripts of two plant mitochondrial genes does not depend on transcriptional context or nuclear genotype. *Curr. Genet.* 30:502–8

182. Wise RP, Dill CL, Schnable PS. 1996. *Mutator*-induced mutations of the *rf1* nuclear fertility restorer of T-cytoplasm maize alter the accumulation of *T-urf13* mitochondrial transcripts. *Genetics* 143:1383–94

183. Wolstenholme DR, Fauron CM-R. 1995. Mitochondrial genome organization. See Ref. 97a, pp. 1–60

184. Yang AJ, Mulligan RM. 1993. Distribution of maize mitochondrial transcripts in polysomal RNA: evidence for nonselectivity in recruitment of mRNAs. *Curr. Genet.* 23:532–36

185. Young EG, Hanson MR. 1987. A fused mitochondrial gene associated with cytoplasmic male sterility is developmentally regulated. *Cell* 50:41–49

Annu. Rev. Plant Physiol. Plant Mol. Biol. 1998. 49:481–500

BORON IN PLANT STRUCTURE AND FUNCTION

Dale G. Blevins and Krystyna M. Lukaszewski

Interdisciplinary Plant Group, University of Missouri, Columbia, Missouri 65211;
blevins@psu.missouri.edu

KEY WORDS: boron deficiency, root growth, cell walls, membranes, ascorbate

ABSTRACT

New and exciting developments in boron research in the past few years greatly contributed to better understanding of the role of boron in plants. Purification and identification of the first boron-polyol transport molecules resolved much of the controversy about boron phloem mobility. Isolation and characterization of the boron-polysaccharide complex from cell walls provided the first direct evidence for boron crosslinking of pectin polymers. Inhibition and recovery of proton release upon boron withdrawal and restitution in plant culture medium demonstrated boron involvement in membrane processes. Rapid boron-induced changes in membrane function could be attributed to boron-complexing membrane constituents. Boron may affect metabolic pathways by binding apoplastic proteins to *cis*-hydroxyl groups of cell walls and membranes, and by interfering with manganese-dependent enzymatic reactions. In addition, boron has been implicated in counteracting toxic effects of aluminum on root growth of dicotyledonous plants. Molecular investigations of boron nutrition have been initiated by the discovery of a novel mutant of *Arabidopsis thaliana* with an altered requirement for boron.

CONTENTS

481

1040-2519/98/0601-0481$08.00

Introduction

World-wide, boron deficiency is more extensive than deficiency of any other plant micronutrient (39, 103). It is particularly prevalent in light textured soils, where water-soluble boron readily leaches down the soil profile and becomes unavailable for plants (131). Adequate boron nutrition is critical not only for high yields but also for high quality of crops. Boron deficiency causes many anatomical, physiological, and biochemical changes, most of which represent secondary effects. Because of the rapidity and the wide variety of symptoms that follow boron deprivation, determining the primary function of boron in plants has been one of the greatest challenges in plant nutrition.

There are excellent reviews summarizing boron research from agricultural and physiological perspectives (20a, 21, 22, 31, 32, 39, 67, 70, 73, 77, 96, 98, 100, 115, 121, 122). One comprehensive review of the early work (32) was divided into 15 separate sections, each describing a different possible site of boron action. Some later reviews proposed unifying hypotheses that pulled together many of the effects of boron on plant growth and developmental processes (67, 73). In the absence of conclusive evidence, however, the actual role of boron in plants remained speculative.

In recent years, research has progressed to the point where it is possible to demonstrate boron involvement in three main aspects of plant physiology. Thus, this chapter features an in-depth look at a structural role for boron in cell walls, a role for boron in membrane function, and boron involvement in metabolic activities. Analysis of these aspects of boron nutrition provides explanations for several of the controversial areas in the literature. In addition, following the recent breakthrough in isolation, purification, and identification of boron-containing compounds from plant tissues, this review presents and discusses the first direct evidence for boron-bound molecules in plants.

History

In 1910, Agulhon (1) reported that several diverse plant species contained boron but did not claim that boron was essential (70). Subsequently, Mazé (82, 83) claimed that boron, aluminum, fluoride, and iodine were essential, but his experimental techniques have been questioned. Therefore, Warington (132) is credited with the first definitive proof of boron requirement for a higher plant. A key observation by Warington was that plants required a continuous supply of boron, an important concept for today's understanding of boron function in growth. Following Warington (132), Sommer & Lipman (123) established the boron requirement for six nonleguminous dicots and for one graminaceous plant, barley.

Shortly after boron was introduced as an essential element for higher plants, structural damage described as cracked stem of celery, stalk rot of cauliflower, heart rot and internal black spot of beets, top rot of tobacco, internal cork of apples, and yellows of alfalfa, was attributed to boron deficiency (32). Application of boron fertilizer became a common practice for production of several horticultural crops, sugar beets, and alfalfa. This led to an observation that the boron requirement among species was highly variable, and that the optimum quantity for one species could be either toxic or insufficient for other species. Based on their boron requirement, plants could be divided into three groups: graminaceous plants, which have the lowest demand for boron; the remaining monocots and most dicots with an intermediate requirement; and latex-forming plants, with the highest boron requirement among plant species (84). Another important classification was made by Shkolnik (121), who subdivided dicots and monocots based on stage of growth and localization of boron deficiency symptoms. In some boron-deficient dicots (sunflower, tomato, squash, alfalfa), inhibition of root growth and degeneration of meristematic regions appear quickly and simultaneously; in other dicots (pea, soybean, lupine), degeneration of growing points is delayed. Some monocots (maize, sorghum, millet, spiderwort, onion) are able to maintain normal root growth and vegetative growth in boron-free conditions much longer then dicots. The small grains and grasses (rye, oat, wheat, timothy) have normal vegetative growth and show boron deficiency symptoms only during formation of reproductive organs. High demand for boron during reproductive growth is a common feature among plant species and is discussed in a different section (see "Reproduction, Pollen Tube Growth, and Pollen Germination").

Other than vascular plants, boron is required by diatoms, some species of marine algal flagellates, and *Cyanobacteria* dependent on heterocysts for nitrogen fixation (9, 10, 32, 78). As for humans, Nielsen et al (91, 92) presented evidence that boron may be beneficial, especially for calcium retention by older women; however, a strict requirement has not yet been established.

Boron Phloem Mobility and Transport Molecules

In vascular plants, boron moves from the roots with the transpiration stream and accumulates in growing points of leaves and stems. It has been suggested that these local concentrations in apical tissues led to the evolutionary development of dependency on boron for some aspects of metabolism in plant meristems (72). Once in the leaves, boron retranslocation is restricted and it becomes fixed in the apoplast. Based on this pattern, boron is generally considered phloem immobile. However, tracer studies with stable isotope ^{10}B demonstrated that in some fruit trees, foliar application of boron in the fall temporarily increased boron concentration of leaves, but during late fall and winter boron moved to the bark. In the spring, the boron moved from the bark into flowers and resulted in increased fruit set. This movement of boron required phloem transport (40–42).

Besides fruit trees, phloem transport of boron was reported for some *Brassicas*, radish, cauliflower, and rutabaga (114, 116–119). Subsequently, Hu & Brown (49) evaluated boron mobility in some species within the genera of *Pyrus, Malus*, and *Prunus* and connected their phloem boron transport with the key fact that these species transported carbon as polyols. Since the beginning of the twentieth century, chemists have used polyols, such as glycerol or mannitol, to enhance the acidity of borate solutions. The basis for these reactions is the ability of borate to form cyclic diesters with some diols and polyols. Brown and associates (51) isolated and characterized soluble sorbitol-boron-sorbitol complexes from the floral nectar of peach, and mannitol-boron-mannitol complexes from phloem sap of celery. This was the first isolation and identification of boron transport molecules in plants. Brown's group also obtained evidence for phloem boron movement in species transporting dulcitol (51). These results explain much of the confusion about boron phloem mobility in plants. We can now conclude that phloem movement of boron depends on the sugar or polyol transport molecules used by the particular plant. It could be of interest that a major crop in the United States, soybean, contains large quantities of the polyol pinitol and shows some response to foliar applications of boron (38, 101, 112). In the future, perhaps phloem sap from soybean will be analyzed for pinitol-B-pinitol complexes.

Role for Boron in Cell Wall Structure

The primary cell wall of higher plants is an important factor determining cell size and shape during plant development. The mechanical properties of growing cell walls can be modified by crosslinks between their major components, cellulosic polymers, and matrix polymers such as hemicellulosic and pectic polysaccharides (15). Over the years, researchers have observed a close relationship between the primary cell walls and boron nutrition. Up to 90% of the cellular boron has been localized in the cell wall fraction (70). The first symptoms of boron deprivation include abnormalities in cell wall and middle lamella organization (48, 70, 79). Recently, formation of borate esters with hydroxyl groups of cell wall carbohydrates and/or glycoproteins has been proposed as a mechanism for crosslinking cell wall polymers (70). Borate bridging could explain many of the characteristics of boron-deficient and boron-toxic plants. This type of bonding could account for brittle leaves of boron-deficient plants, while plants grown with supraoptimal levels of boron produce leaves that are plastic or elastic in their response to bending (32, 48, 70). In addition, the slipping and sliding properties of "slime" (17, 109), permitted by the H-bonding of hydroxyl groups on borate molecules and the hydroxyl groups of the polyvinyl alcohol, could explain the properties of primary cell walls at early stages of development (15).

The specific plant molecules that could participate in borate bridging of cell wall polymers were discussed at length by Loomis & Durst (70). Borate forms the most stable diesters with *cis*-diols on a furanoid ring. The compounds in plants that have this configuration are limited to ribose and apiose. According to Loomis & Durst (70), apiose, found in cell walls of a variety of plant species within both dicots and monocots (16), can be the key sugar moiety for borate-crosslinking cell wall polymers, while ribose, present in abundance in ribonucleotides, is likely involved in the chemistry of boron toxicity (70). Another possible candidate for borate cell wall bridging is fucose. Diatoms and certain algae that require boron contain fucose in walls (5, 70). Higher plants, other than the gramineae, have cell wall xyloglucans and rhamnogalacturonans with terminal fucose moieties (5, 16, 44). In addition, many of the glycoproteins secreted into the apoplast are fucosylated (56). It is noteworthy that mutants of *Arabidopsis thaliana*, which lack fucose, have brittle leaves (104), a condition found in boron-deficient plants (70).

It was an early observation that plant boron content was closely correlated with pectin (32). In 1961, Ginsburg (33) showed that a strong chelator, EDTA, mixed with a weak chelator (e.g. IAA), was effective in causing cell separation by removing the pectin/protein matrix, but borate buffer kept the matrix intact longer than any other buffer. Clarkson & Hanson (19) proposed that by forming crosslinks in pectin, boron protects Ca in the cell wall. Results that supported this idea came from Yamanouchi (135) and Yamauchi et al (136), who found that boron-deficient cell walls of tomato contained less calcium. We could hypothesize that the hydroxyl H-bonding and borate ester formation may pull carboxylate groups of polymers into close proximity and allow calcium or magnesium binding by the polymers. In this theme, based on a sophisticated growth analysis of pine cell cultures supplied with various concentrations of boron, calcium, and magnesium, Teasdale & Richards (127) proposed that plant cell wall acceptor molecules efficiently bind calcium after loosely binding boron. The acceptor molecules also bind magnesium competitively with calcium. Teasdale & Richards' work (127) supports the idea that borate esters, formed with hydroxyls of sugars (like apiose or fucose) on pectin or glycoprotein polymers, provide areas for the chelation of calcium or magnesium in cell walls.

Shortly after Loomis & Durst (70) presented their model for boron bridging cell wall polymers, with apiose being the key sugar moiety binding borate, Matoh et al (80) isolated the first boron-polysaccharide complex from driselase-treated radish root cell walls. The complex had a molecular weight of 7.5 kDa and contained 0.232% boron, 52.3% uronic acid, 17.4% arabinose, 9.8% rhamnose, 4.9% galactose, and 0.3% xylose. Up to 80% of the cell wall boron was localized in this complex, and ^{11}B-NMR analysis suggested that the boron was present as a two-ligand borate-diol, BL2. Matoh's work was complemented by Brown and his group (48), who reported a correlation between pectin fraction and boron in cell walls of squash and tobacco, and by their survey of 14 plant species (50) showing very close correlation between the uronic fraction, pectin sugars, and boron content of the plant.

Kobayashi et al (59) purified the pectin fraction from radish root cell walls and isolated the first boron-containing pectic polysaccharide complex from plants, boron-rhamnogalacturonan-II (RG-II-B). This group also produced the first direct evidence for borate crosslinking two RG-II monomers, by demonstrating that the removal of boron from the RG-II-B complex reduced the molecular weight of the complex by half. Subsequently, Matoh et al (81) found the RG-II-B complex in cell walls of 22 other plant species, including two of *Brassicaceae*, three *Cucurbitaceae*, four *Leguminosae*, two *Apiaceae*, two *Chenopodiaceae*, two *Solanaceae*, two *Asteraceae*, one *Liliaceae*, one *Araceae*, two *Amaryllidaceae*, and three *Gramineae*. On the basis of this research, Matoh et al (81) proposed that RG-II may be the exclusive polysaccharide-binding boron in cell walls.

Using cultured sycamore cells, etiolated pea stems, and red wine as sources of RG-II-B complex, O'Neill et al (95) found that the borate ester was located on C-2 and C-3 of two of the four 3'-linked apiosyl residues of dimeric RG-II. The authors postulated that the dimeric RG-II-B covalently crosslinks the cell wall pectic matrix in dicots, nongraminaceous monocots, and graminaceous plants, though the pectin content of the grasses is much lower than that of the other species (95). Ishii & Matsunaga (52) and Kaneko et al (58) isolated and characterized RG-II-B complexes from sugar beet (a dicot) pulp and bamboo (a monocot) shoot, respectively.

Recently, research on RG-II-B complexes of the wall has moved into a new phase, addressing the formation of dimers from monomers and borate. According to O'Neill et al (95), in vitro dimer formation was increased by addition of strontium, lead, and cadmium, but calcium and magnesium were ineffective. The authors suggest that a catalyst, for example an enzyme, may be required for a rapid dimer formation by boric acid at physiological pH.

Other candidates for borate crosslinking in primary cell walls are hydroxyproline-rich glycoproteins and proline-rich proteins, e.g. extensin. It has been observed that cell walls of boron-deficient bean root nodules contain very low levels of hydroxyproline-rich proteins, compared with those of boron-sufficient controls (11). Interestingly, the mRNA of one of these proteins (an early nodulin called ENOD2) was present in the nodules, but evidently the proteins were not assembled into the wall structure. This is consistent with the work of

Jackson (53) on *Petunia* pollen tube growth, where without boron, proteins were secreted but not assembled into tube walls and therefore "lost" to the medium.

Membranes and Membrane-Associated Reactions

The evidence provided by recent cell wall studies explains many problems caused by boron deficiency. However, there are some aspects of plant boron nutrition that go beyond cell wall structure. These include rapid changes in membrane function induced by addition of boron to boron-deficient tissues.

Boron was first localized in maize root membranes by Pollard et al (99) and was later found in membrane fractions from protoplasts of mung bean by Tanada (125). Although the quantities of boron in membranes were not large, especially compared with those in cell wall fractions, they were significant for ion uptake. Uptake of rubidium (which is a potassium analog) by roots of *Vicia faba* was inhibited in boron-free solutions, and the problem was localized in the terminal 10-mm section of the root (107). Uptake of phosphorus, which was slow in boron-deficient roots of *Vicia faba*, was restored to normal levels when the roots were pretreated with boron for 1 h before the absorption experiment (107). Phosphorus, chloride, and rubidium uptake by boron-deficient roots of maize and *Vicia faba* was restored to 40% of normal within 20 min after boron was added to the rooting medium (99). Both uptake and efflux of phosphorus were decreased in boron-deficient sunflower roots, and both were restored within 1 hr after addition of boron (35). Goldbach et al (36) demonstrated that boron deficiency in suspension cultures of carrot and tomato caused a 50% reduction in the ferricyanide-induced net proton release. The inhibition was reversed within 60 min after addition of boron. This boron effect occurred only in the presence of auxin, so the authors concluded that boron was required for the auxin stimulation of ferricyanide-induced proton release. Vanadate suppressed the proton release, indicating that the plasmalemma proton pump was key to the process (36). In fact, Ferrol and coworkers (27, 28, 108) showed that boron deficiency inhibited the vanadate-sensitive H^+-ATPase in microsomes isolated from sunflower roots. Although immunoblotting showed that the quantity of the enzyme was not affected by boron deficiency, fluorescence anisotropy showed a difference between the membrane preparations from boron-deficient and boron-sufficient roots (26). This difference was interpreted as an increase in rigidity of boron-deficient membranes. In other studies, Lawrence et al (63) showed lower ATPase activity in plasmalemma-enriched vesicles from boron-deficient chickpea roots than in vesicles from control roots, and Obermeyer et al (94) reported boron stimulation of the plasmalemma ATPase from ungerminated pollen grains of lily.

Barr & Crane (3, 4) reported a boron effect on plasma membrane electron transport reactions and showed that the auxin-sensitive plasma membrane NADH oxidase was inhibited in boron-deficient carrot cell cultures. The authors demonstrated that proton secretion associated with the H^+-ATPase was also decreased by boron deficiency, but not as severely as the ferricyanide-stimulated proton release. This means that ferricyanide activates transmembrane electron transport that is coupled to proton release only when boron is present. In the same work, Barr & Crane (3, 4) showed that addition of exogenous boric acid (with or without 2,4-D) to low boron cells caused an instantaneous stimulation of the plasma membrane NADH oxidase. This was the fastest boron response reported. An earlier report by Schon et al (113) indicated significant hyperpolarization of sunflower membranes within 3 min following the addition of boron.

In 1986, Morré et al (88) presented evidence that identified the activity of auxin-sensitive plasma membrane NADH oxidase as ascorbate free radical (AFR) oxidoreductase. This finding supported an earlier hypothesis of JC Brown (13) that boron nutrition was important in maintaining the "reducing atmosphere" in the apoplast to support ion uptake. By stimulating NADH oxidase, boron could be involved in keeping ascorbate reduced at the cell wall/membrane interface. It is noteworthy that both NADH oxidase activity and ascorbate have been linked with plant growth processes (37, 45, 68, 87, 88).

In summary, boron treatment of low-boron plants leads to hyperpolarization of root membranes, and stimulation of ferricyanide-dependent H^+ release, ATPase activity, NADH oxidase activity, and ion transport (3, 26, 28, 36, 71, 108, 113). Though these changes are associated with membrane function, several researchers have speculated that boron may be affecting physical properties of membrane proteins. In 1977, Pollard et al (99) suggested that rapid restoration of ATPase activity and potassium uptake by boron-deficient roots following supplementation with boron could be explained by boron complexing with polyhydroxy groups of membrane components. Marschner's group (14) demonstrated leakage of potassium, sucrose, phenolics, and amino acids in boron deficient sunflower leaves and discussed a role for boron in maintaining the integrity of plasma membranes. They proposed that boron stabilized the structure of the plasma membrane by complexing membrane constituents. Either H-bonding or ester formation with glycolipids or glycoproteins could easily keep enzymes or channels in an optimum conformation and anchored in the membrane. In agreement with this hypothesis, less phospholipid and galactolipid were found in membranes of boron deficient plants (121).

In addition, Shkolnik (121) observed that several enzymes, normally bound to membranes or walls in a latent form, become active when released under boron deficient conditions. These enzymes include ribonuclease, glucose-6-phosphate dehydrogenase, phenylalanine ammonia lyase, β-glucosidase and polyphenoloxidase. Release of these enzymes under boron-insufficient conditions could severely alter plant metabolism, deplete RNA, and increase phenolic synthesis. Many of the phenolics are potent growth inhibitors (66), the same phenolics also inhibit ion uptake and thus retard membrane function (34).

The work of Shkolnik (121) indicates that although this response to boron deficiency is common in dicots, it has not been observed in graminaceous monocots. This raises a question of whether glycosylation of apoplastic proteins is different in graminaceous plants, or whether these proteins are less abundant in membranes and cell walls of grasses.

Reproduction, Pollen Tube Growth, and Pollen Germination

Based on the latest research, cell wall composition may be of primary importance in determining the quantity of boron required for growth. However, it has been observed that in most plant species the boron requirement for reproductive growth is much higher than for vegetative growth (32, 70). This is especially true for gramineaceous plants, which have the lowest cell wall pectin content and the lowest boron requirement to maintain normal vegetative growth, but need as much boron as other species at the reproductive stage. The physiological basis for the high boron demand for plant reproduction is not fully understood.

Boron requirement for reproductive growth in plants has long been recognized. Gauch & Dugger (32) cited over 70 references that reported boron effects on pollen germination, or on flowering and fruiting of plants. Boron deficiency caused sterility in maize and flower malformations in a wide variety of both monocots and dicots (32). Piland et al (97) found that boron treatment increased seed yield of alfalfa by 600%, while hay yield was increased by only 3%. Schmucker (110) found that pollen from a tropical species of *Nymphaea* failed to germinate in 1% glucose but germinated readily in a stigma extract. Later, he determined that the stigma extract contained borate and found that addition of boric acid to 1% glucose made it a very effective germination medium. Schmucker (111) proposed that boric acid was bound to hydroxyl-rich organic molecules, like sugars, and was involved in pollen tube wall formation.

Pollen grains of most species are naturally low in boron, but in the styles, stigma, and ovaries, boron concentrations are generally high (32). Visser (130) showed that a continuous and ample supply of boron was required for pollen tube growth, and speculated that the boron was complexing with cellular materials during the tube elongation process. Along this line, Johri & Vasil (57) demonstrated that boron was more critical for pollen tube elongation than for pollen germination.

Rapid growth of pollen tube depends on constant fusion of vesicles forming the plasmalemma, and continuous secretion of cell wall material. Jackson (54) proposed that the "capture" of secreted pollen proteins for membrane and wall building, proceeds through borate complexes with sugar residues.

Pollen of *Petunia* contains many glycoproteins, and the oligosaccharides of plant glycoproteins contain significant amounts of mannose and fucose, both known to form stable esters with borate. Jackson (54) studied protein secretion during pollen tube growth of *Petunia* at different temperatures. He observed that the phase change patterns of lipids in membranes was completely different when boron was present in the medium. This could be related to the fact that in the presence of boron, a greater proportion of the protein was assembled into the membrane and wall matrices. Jackson (54) also noted that pollen tube germination was completely inhibited at temperatures over 21°C unless boron was present. This could explain the importance of boron in reproductive growth of warm season crops, like maize (129).

Robbertse et al (106) found a boron gradient from the stigma through the style to the ovary and showed that pollen tubes of *Petunia* grew toward higher boron concentrations. Perhaps boron is a chemotactic agent for pollen tube growth through reproductive tissues. This idea is consistent with the high boron concentration generally found in female flower parts (32).

The results of pollen growth studies are consistent with boron-complexing cell wall polymers, while the lipid thermostability results show that boron is important in membrane structure and function. Whatever the mechanism, the role of boron in reproductive growth is particularly striking. The uniformly high boron requirement for reproductive growth across the plant kingdom is intriguing and indicates similarities between reproductive structures, so unlike cell walls, perhaps the composition of the pollen tube wall is similar across plant species. Gauch & Dugger (32) quoted Lohnis (69), who said that "it is quite conceivable it will be the study of pollen which may elucidate the very fundamental part boron plays in the biochemical processes."

Nitrogen Fixation

Boron was found essential for nitrogen fixation by heterocysts of *Anabaena* ACC 7119 (78). Loss of nitrogenase activity under boron-deficient conditions was explained by the destruction of nitrogenase by O_2. Altered O_2 status of the boron-deficient heterocysts was also implied in the increased activity of super-oxide dismutase, catalase, and peroxidase in boron-deficient heterocysts (29). Heterocysts are morphologically and functionally distinct from vegetative cells and are capable of nitrogen fixation because they maintain a reducing (low O_2) environment (105). The low O_2 status in heterocysts is possible because of a thick envelope comprised of an inner layer of specific glycolipids (133). These glycolipids are absent from vegetative cells and therefore are predicted to provide the O_2 diffusion barrier in heterocysts (62). Mutants of *Cyanobacteria* deficient in these envelope glycolipids fail to fix nitrogen when assayed aerobically (43). Extraction and quantification of heterocyst envelope glycolipids showed that within 6 h following boron removal, glycolipid content dropped by 33% and within 24 h was less than 1% of that found in boron-sufficient cells (30). It was suggested that boron stabilizes the inner glycolipid layer of heterocyst envelopes and thus retards O_2 diffusion.

In early work, Brenchley & Thornton (12) showed a major reduction in nodule number and in nitrogen fixation by inoculated boron-deficient fava bean. Vascular connections to the nodule were reduced, and so was the number of bacteria that changed into bacteroid. The authors speculated that under boron-deficient conditions, the symbionts may become parasitic. Results of this early study are consistent with the recent work by Bolaños et al (8). In boron-deficient pea nodules, the number of infected host cells was much lower then in sufficient controls. Host cells in boron-deficient plants developed enlarged and abnormally shaped infection threads, which frequently burst. Binding of the plant matrix glycoprotein to the cell surface of *Rhizobium leguminosarum* was inhibited by the presence of borate in the incubation buffer. The authors proposed that binding of matrix glycoprotein in the absence of boron may block the interaction between bacterial cell surface and the plant membrane glycocalyx. Developing soybean root nodules were more sensitive to low boron nutrition than large fully developed nodules (134). Both development and nitrogen fixation of young nodules were retarded after boron removal, while acetylene reduction rates remained unchanged in large nodules.

In the most recent study on bean root nodules (11), the ratio of hydroxyproline to cell wall dry weight was fivefold lower in boron-deficient nodules than in boron-sufficient controls. The levels of hydroxyproline-rich covalently bound ENOD2 protein were extremely low in walls of boron-deficient nodule parenchyma cells, although the Northern blot analysis showed that the mRNA was present in both boron-sufficient and -deficient nodules. The absence of the ENOD2 protein in the wall correlated with an irregular wall structure. The researchers concluded that a failure to incorporate the ENOD2 protein in the absence of boron could lead to wall abnormalities that prevent proper formation of the O_2 barrier, which protects the dinitrogenase complex and allows symbiotic nitrogen fixation.

Sites of Boron Action in Plant Metabolism

Primary cell wall structure and membrane function are now closely linked to boron nutrition. In contrast, boron role in plant metabolism is still a subject of considerable debate. Focusing on the diversity of early responses to boron deficiency, Lovatt & Dugger (73) postulated that boron can be involved in a number of metabolic pathways and can act in regulation of metabolic processes similarly to plant hormones. However, due to a lack of suitable information, boron function in metabolic events has never been properly evaluated.

There is substantial evidence supporting the association of boron with ascorbate metabolism. Among earlier studies, two reports on vegetable crops are particularly interesting. Chandler & Miller (18) found that rutabaga treated with boron had more ascorbate than untreated controls, and that following dehydration and storage, the boron-treated plants maintained about twice as much ascorbate. Mondy & Munshi (86) found that foliar application of borax on potatoes at 10 and 13 weeks after planting resulted in significantly greater quantities of ascorbate in tubers harvested 18 weeks after planting. Following storage for 6 months, discoloration of tubers harvested from boron-treated plants was decreased, and the decrease was attributed to lower phenolic concentrations and higher ascorbate concentrations in tubers from boron-treated plants.

One way boron could increase ascorbate concentration is through its effect on plasma membrane electron transport reactions. Barr and associates (3, 4) showed that boron instantaneously stimulated the auxin-sensitive plasmalemma NADH oxidase. This enzyme, also called ascorbate free radical oxidoreductase (88), catalyzes the transfer of electrons to ascorbate free radical. Inhibition of this process in the absence of boron could result in deprivation of reduced ascorbate. In our studies (74), inhibition of squash root elongation caused by inadequate boron nutrition was correlated with a decline in ascorbate concentration in root apices. However, no boron- or growth-related variation was observed in the concentration of ascorbate free radical and dehydroascorbate, the oxidized forms of ascorbate. This indicates that the decline in ascorbate concentration induced by boron deficiency cannot be ascribed, at least not in full, to boron interference with the redox cycle, but represents a decrease in the total pool of ascorbate in root meristems. It should be noted that ascorbate added to hydroponic medium promoted root elongation in the absence of boron and under low boron conditions. This shows that supplemental ascorbate can, to some degree, compensate for boron in root growth processes (74).

Another site of boron action that is not connected with a structural role in cell walls or membranes is auxin metabolism. Boron interaction with auxin has long been postulated, and although the issue remains controversial, it may be central to our understanding of the role of boron in plants. In the past, boron deprivation has been reported to cause IAA accumulation and toxicity (20, 55, 76, 85, 89) as well as IAA depletion and deficiency (23, 24, 120). The mechanisms of the responses have not been explained. In 1977, Bohnsack & Albert (7) demonstrated a 20-fold increase in IAA oxidation rate in root apices 24 h after boron was withheld from the nutrient medium. The authors attributed the increase to stimulation of the activity by high levels of IAA accumulated in boron-deficient tissues. However, Hirsch & Torrey (47) demonstrated that the ultrastructural changes caused by boron deficit were different from those resulting from IAA toxicity. Furthermore, boron deficiency symptoms were observed with unchanged or reduced IAA levels in apical tissues (25, 46).

Studies in our laboratory confirmed the sizable increase in IAA oxidation rate in boron-deprived squash root tips and ascribed it to boron interaction with the cofactors of IAA oxidase, manganese, and p-coumaric acid. It has been reported that foliar fertilization with boron resulted in lower leaf Mn concentrations in mint (124) and soybean (102). Previously, we observed boron inhibition of a manganese-dependent enzyme in soybean leaves, allantoate amidohydrolase (75). Boric acid, applied foliarly on field-grown nodulated soybeans, caused up to a 10-fold increase in allantoate concentration in treated leaf tissue. Accumulation of allantoate in response to boron was either eliminated or greatly reduced in plants presprayed with manganese. In vitro, the activity of partially purified allantoate amidohydrolase reflected the boron/manganese ratio in the incubation solution.

Recently, we observed that the activity of IAA oxidase in squash root apices depended on boron nutrition in a manner very similar to soybean allantoate amidohydrolase. IAA oxidation rate was low in root tips of boron-sufficient plants and high in plants grown without boron. Supplemental manganese had little effect on IAA oxidase activity in boron-deficient root tips but resulted in a substantial increase in the activity in boron-sufficient plants, showing that in the presence of boron more manganese was required for stimulation of the enzyme. The activity of IAA oxidase in excised root tips was modified by a change in the boron/manganese ratio during incubation (90; KM Lukaszewski & DG Blevins, unpublished observations).

IAA oxidation rate was also negatively correlated with root tip ascorbate content, which has been shown to depend on boron nutrition (74; KM Lukaszewski & DG Blevins, unpublished observations). Ascorbate added to the medium suppressed IAA oxidase activity in boron-deficient squash root meristems. Inhibition of IAA oxidase by ascorbate coincided with a decrease in the tissue concentration of p-coumaric acid, the monophenolic cofactor of the activity. Although the regulation of endogenous levels of IAA is not fully understood and the importance of IAA oxidase in IAA catabolism has been challenged by experiments with transgenic plants (2, 61, 93), in our studies we found a close negative correlation between root growth and oxidative degradation of IAA, both greatly affected by boron nutrition. In the future, it may be important to determine IAA levels in specific tissues and in the apoplast of cells in the elongation zone of boron-deficient and sufficient plants.

Amelioration of Aluminum-Induced Root Growth Inhibition

Aluminum is an important factor that impairs plant growth in acid soils by causing structural and functional damage to the roots. The mechanisms of aluminum toxicity are complex and not fully understood (60). The toxic form of aluminum in soil solution or nutrient medium is Al^{3+}, and this form is abundant at pH 4.0–4.5. Once aluminum is inside the plant, it is likely to be in the form of aluminate $Al(OH)_3$, which is structurally similar to boric acid $B(OH)_3$ (6, 60).

Based on the similarities of the molecules and of the symptoms characteristic for aluminum-stressed and boron-deficient plants, most dealing with cell walls, membrane function, and root growth, it was proposed that aluminum may exert its toxic effect by inducing boron deficiency (6). The results from our laboratory supported this hypothesis. We showed that supplemental boron incorporated into acidic, high-aluminum subsoil promoted root penetration into this soil (64). In hydroponic culture, supraoptimal boron prevented aluminum inhibition of root growth at all criteria examined: root length, cell elongation, cell production rate, tissue organization and cell structure, root morphology, and maturation (65). This reaction may be limited to nongraminaceous plants, since Taylor & MacFie (126) showed that boron did not alleviate aluminum toxicity symptoms in wheat. However, as stated earlier, cell wall structure and boron requirement of graminaceous plants are different from dicots, and boron deficiency does not affect root growth of wheat.

Aluminum added to squash rooting medium caused a decline in root tip ascorbate concentration that was very similar to that caused by boron deprivation. The reduction in ascorbate content in root apices by aluminum was parallel to the inhibition of root elongation (74). Supplemental boron in aluminum-toxic medium produced root apices with higher ascorbate concentrations. IAA oxidase activity in the root tips of aluminum-stressed plants decreased in a manner reversely correlated with boron concentration in nutrient solutions (KM Lukaszewski & DG Blevins, unpublished observations). In summary, our findings support the hypothesis that toxic levels of aluminum induce boron deficiency in plants. The results also indicate that root growth inhibition under both boron-deficient or aluminum-toxic conditions may be a consequence of a disrupted ascorbate metabolism (74).

Molecular Approach

Molecular genetics can be a powerful tool in studying plant nutritional disorders. Recently, a mutant of *Arabidopsis thaliana* (*bor1-1*) with increased requirement for boron was identified and characterized by Noguchi et al (92a). In contrast to the wild type, *bor1-1* plants grown in "normal" concentrations of boron showed severe boron deficiency symptoms (reduced expansion of rosette leaves, reduced fertility, and loss of apical dominance) that were reversed by "excess" boron. Tracer experiments with [10]B suggested a defect in boron uptake and/or translocation. Genetic mapping showed a linkage of the *bor1* mutation with the simple sequence length polymorphism (SSLP) markers on chromosome 2. This approach may provide much needed information on the genetic mechanisms controlling boron metabolism in plants.

Concluding Remarks

Boron is essential for plant growth and development, and adequate boron nutrition of cultivated plants can be of great economic importance. Boron affects the yield of fruits, vegetables, nuts, and grains as well as the quality of harvested crops. Increased boron applications may promote root elongation in acidic, high-aluminum soils.

In one of his last articles, Joe Varner listed the boron requirement as one of the important unknowns in plant biology (128). Although recent progress in the isolation and characterization of plant boron-polyol transport molecules and pectin RG-II-B complexes greatly improved our understanding of boron mobility and boron chemistry in plant cell walls, it also highlighted the need to learn more about boron complexes with glycolipids and/or glycoproteins in membranes. The concept of boron-binding apoplastic proteins, as well as the effect of boron on manganese-activated enzymes, may be of importance in many metabolic processes. Molecular investigations of boron requirement in plants open new possibilities for improving boron deficiency/toxicity stress tolerance of crops. Elucidation of these aspects of boron nutrition will be a challenging goal for future research.

> Visit the *Annual Reviews home page* at
> http://www.AnnualReviews.org.

Literature Cited

1. Agulhon H. 1910. Présence et utilité du bore chez les végétaux. *Ann. Inst. Pasteur* 24:321–29
2. Bandurski RS, Cohen JD, Slovin JP, Reinecke DM. 1995. Auxin biosynthesis and metabolism. In *Plant Hormones*, ed. PJ Davies, pp. 39–65. Dordrecht: Kluwer
3. Barr R, Böttger M, Crane FL. 1993. The effect of boron on plasma membrane electron transport and associated proton secretion by cultured carrot cells. *Biochem. Mol. Biol. Int.* 31:31–39
4. Barr R, Crane FL. 1991. Boron stimulates NADH oxidase activity of cultured carrot cells. See Ref. 99a, p. 290
5. Bidwell RGS. 1979. *Plant Physiology*. New York: Macmillan
6. Blevins DG. 1987. Future developments in plant nutrition research. In *Future Developments in Soil Science Research*, ed. LL Boersma, DE Elrick, RB Corey, HH Cheng, TC Tucker, et al, pp. 445–48. Madison, Wis: Soil Sci. Soc. Amer.
7. Bohnsack CW, Albert LS. 1977. Early effects of boron deficiency on indoleacetic

acid oxidase levels of squash root tips. *Plant Physiol.* 59:1047–50
8. Bolaños L, Brewin NJ, Bonilla I. 1996. Effects of boron on Rhizobium-Legume cell-surface interactions and nodule development. *Plant Physiol.* 110:1249–56
9. Bolaños L, Mateo P, Bonilla I. 1993. Calcium-mediated recovery of boron deficient *Anabaena* sp. PCC 7119 grown under nitrogen fixing conditions. *J. Plant Physiol.* 142:513–17
10. Bonilla I, Bolaños L, Mateo P. 1995. Interaction of boron and calcium in the cyanobacteria *Anabaena* and *Synechococcus*. *Physiol. Plant.* 94:31–36
11. Bonilla I, Mergold-Villaseñor C, Campos ME, Sánches N, Pérez H, et al. 1997. The aberrant cell walls of boron deficient bean root nodules have no covalently-bound hydroxyproline/proline-rich proteins. *Plant Physiol.* In press
12. Brenchley WE, Thornton BA. 1925. The relation between the development, structure and functioning of the nodules on

Vicia faba, as influenced by the presence or absence of boron in the nutrient medium. *Proc. R. Soc. London Ser. B Biol. Sci.* 98:373–98

13. Brown JC. 1979. Effects of boron stress on copper enzyme activity in tomato. *J. Plant Nutr.* 1:39–53

14. Cakmak I, Kurz H, Marschner H. 1995. Short-term effects of boron, germanium and high light intensity on membrane permeability in boron deficient leaves of sunflower. *Physiol. Plant.* 95:11–18

15. Carpita NC. 1987. The biochemistry of the 'growing' plant wall. In *Physiology of Cell Expansion During Plant Growth*, ed. DJ Cosgrove, DP Knievel, pp. 28–45. Rockville, MD: Am. Soc. Plant Physiol.

16. Carpita NC, Gibeaut DM. 1993. Structural models of primary cell walls in flowering plants: consistency of molecular structure with the physical porperties of the walls during growth. *Plant J.* 3:1–30

17. Casassa EZ, Sarquis AM, Van Dyke CH. 1986. The gelation of polyvinyl alcohol with borax. A novel class participation experiment involving the preparation and properties of a "slime." *J. Chem. Educ.* 63:57–60

18. Chandler FB, Miller MC. 1946. Effect of boron on the vitamin C content of rutabagas. *Proc. Am. Soc. Hortic Sci.* 47:331–34

19. Clarkson DT, Hanson JB. 1980. The mineral nutrition of higher plants. *Annu. Rev. Plant Physiol. Plant Mol. Biol.* 31:239–98

20. Coke L, Whittington WJ. 1967. The role of boron in plant growth. IV. Interrelations between boron and indol-3yl-acetic acid in the metabolism of bean radicles. *J. Exp. Bot.* 19:295–308

20a. Dell B, Brown PH, Bell RW, eds. 1997. *Boron in Soils and Plants: Reviews.* Dordrecht, Kluwer. 219 pp.

21. Dugger WM. 1973. Functional aspects of boron in plants. *Adv. Chem. Ser.* 123:112–29

22. Dugger WM. 1983. Boron in plant metabolism. In *Encyclopedia of Plant Physiology*, ed. A Lauchli, RL Bieleski, 15B:626–50. Berlin: Springer-Verlag

23. Dyar JJ, Webb KL. 1961. A relationship between boron and auxin in C^{14} translocation in bean plants. *Plant Physiol.* 36:672–76

24. Eaton FM. 1940. Interrelation in the effect of boron and indoleacetic acid on plant growth. *Bot. Gaz.* 101:700–805

25. Fackler U, Goldbach H, Weiler EW,

Amberger A. 1985. Influence of boron-deficiency on indol-3yl-acetic acid and abscisic acid levels in root and shoot tips. *J. Plant Physiol.* 119:295–99

26. Ferrol N, Belver A, Roldán M, Rodriguez-Rosales MP, Donaire JP. 1993. Effects of boron on proton transport and membrane properties of sunflower (*Helianthus annuus* L.) cell microsomes. *Plant Physiol.* 103:763–69

27. Ferrol N, Donaire JP. 1992. Effect of boron on plasma membrane proton extrusion and redox activity in sunflower cells. *Plant Sci.* 86:41–47

28. Ferrol N, Rodriguez-Rosales MP, Roldán M, Belver A, Donaire JP. 1992. Characterization of proton extrusion in sunflower cell cultures. *Rev. Esp. Fisiol.* 48:22–27

29. Garcia-Gonzáles M, Mateo P, Bonilla I. 1988. Boron protection for O_2 diffusion in heterocysts of *Anabaena* PCC 7119. *Plant Phiol.* 87:785–89

30. Garcia-Gonzáles M, Mateo P, Bonilla I. 1991. Boron requirement for envelope structure and function in *Anabaena* PCC 7119 heterocysts. *J. Exp. Bot.* 42:925–29

31. Gauch HG. 1972. Roles of micronutrients in higher plants. In *Inorganic Plant Nutrition*, pp. 239–80. Stroudsburg, PA: Dowden, Hutchinson & Ross

32. Gauch HG, Dugger WM Jr. 1954. *The Physiological Action of Boron in Higher Plants: A Review and Interpretation.* College Park: Univ. Md., Agric. Exp. Stn.

33. Ginsburg BZ. 1961. Evidence for a protein gel structure crosslinked by metal cations in the intercellular cement of plant tissue. *J. Exp. Bot.* 12:85–107

34. Glass ADM, Dunlop J. 1974. Influence of phenolic acids on ion uptake. IV. Depolarization of membrane potentials. *Plant Physiol.* 54:855–58

35. Goldbach HE. 1984. Influence of boron nutrition on net uptake and efflux of ^{32}P and ^{14}C-glucose in *Helianthus annuus* roots and cell cultures of *Daucus carota*. *J. Plant Physiol.* 118:431–38

36. Goldbach HE, Hartmann D, Rötzer T. 1990. Boron is required for the ferricyanide-induced proton release by auxins in suspension-cultured cells of *Daucus carota* and *Lycopersicon esculentum*. *Physiol. Plant.* 80:114–18

37. Gonzales-Reyes JA, Alcain FJ, Caler JA, Serrano A, Cordoba F, Navas P. 1994. Relationship between apoplastic ascorbate regeneration and the stimulation of root growth in *Allium cepa* L. *Plant Sci.* 100:23–29

38. Guertal EA, Abaye AO, Lippert BM, Miner GS, Gascho GJ. 1996. Sources of boron for foliar fertilization of cotton and soybean. *Commun. Soil Sci. Plant Anal.* 27:2815–28

39. Gupta UC. 1979. Boron nutrition of crops. *Adv. Agron.* 31:273–307

40. Hanson EJ. 1991. Movement of boron out of tree fruit leaves. *HortScience* 26:271–73

41. Hanson EJ, Breen PJ. 1985. Effects of fall boron sprays and environmental factors on fruit set and boron accumulation in "Italian" prune flowers. *J. Am. Hortic Sci.* 110:566–70

42. Hanson EJ, Breen PJ. 1985. Xylem differentiation and boron accumulation in 'italian' prune flower buds. *J. Am. Hortic Sci.* 110:389–92

43. Haury JF, Wolk CP. 1978. Classes of *Anabaena variabilis* mutants with oxygensensitive nitrogenase activity. *J. Bacteriol.* 136:688–92

44. Hayashi T. 1989. Xyloglucans in the primary cell wall. *Annu. Rev. Plant Physiol. Plant Mol. Biol.* 40:139–68

45. Hidalgo A, Garcia-Herdugo G, Gonzáles-Reyes JA, Morré DJ, Navas P. 1991. Ascorbate free radical stimulates onion root growth by increasing cell elongation. *Bot. Gaz.* 152:282–88

46. Hirsch AM, Pengelly WL, Torrey JG. 1982. Endogenous IAA levels in boron-deficient and control root tips of sunflower. *Bot. Gaz.* 143:15–19

47. Hirsch AM, Torrey JG. 1980. Ultrastructural changes in sunflower root cells in relation to boron deficiency and added auxin. *Can. J. Bot.* 58:856–66

48. Hu H, Brown PH. 1994. Localization of boron in cell walls of squash and tobacco and its association with pectin. *Plant Physiol.* 105:681–89

49. Hu H, Brown PH. 1996. Phloem mobility of boron is species dependent: evidence for boron mobility in sorbitol-rich species. *Ann. Bot.* 77:497–505

50. Hu H, Brown PH, Labavitch JM. 1996. Species variability in boron requirement is correlated with cell wall pectin. *J. Exp. Bot.* 47:227–32

51. Hu H, Penn SG, Lebrilla CB, Brown PH. 1997. Isolation and characterization of soluble boron complexes in higher plants. *Plant Physiol.* 113:649–55

52. Ishii T, Matsunaga T. 1996. Isolation and characterization of a boron-rhamnogalacturonan-II complex from cell walls of sugar beet pulp. *Carbohydr. Res.* 284:1–9

53. Jackson JF. 1989. Borate control of protein secretion from *Petunia* pollen exhibits critical temperature discontinuities. *Sex. Plant Reprod.* 2:11–14

54. Jackson JF. 1991. Borate control of energy-driven protein secretion from pollen and interaction of borate with auxin or herbicide—a possible role for boron in membrane events. See Ref. 99a, pp. 221–29

55. Jaweed MM, Scott EG. 1967. Effect of boron on ribonucleic acid and indoleacetic acid metabolism in the apical meristem of sunflower plants. *Proc. W. Va. Acad. Sci.* 39:186–93

56. Johnson KD, Chrispeels MJ. 1987. Substrate specificities in N-acetylglucosaminyl-, fucosyl-, and xylosyl-transferases that modify glycoproteins in the Golgi apparatus of bean cotyledons. *Plant Physiol.* 84:1301–8

57. Johri BM, Vasil IK. 1961. Physiology of pollen. *Bot. Rev.* 27:325–81

58. Kaneko S, Ishii T, Matsunaga T. 1997. A boron-rhamnogalacturonan-II complex from bamboo shoot cell walls. *Phytochemistry* 44:243–48

59. Kobayashi M, Matoh T, Azuma J. 1996. Two chains of rhamnogalacturonan II are cross-linked by borate-diol ester bonds in higher plant cell walls. *Plant Physiol.* 110:1017–20

60. Kochian LV. 1995. Cellular mechanism of aluminum toxicity and resistance in plants. *Annu. Rev. Plant Physiol. Plant Mol. Biol.* 46:237–60

61. Lagrimini LM. 1991. Peroxidase, IAA oxidase and auxin metabolism in transformed tobacco plants. *Plant Physiol.* 96:S–77

62. Lambein F, Wolk CP. 1973. Structural studies on the glycolipids from the envelope of the heterocyst of *Anabaena cylindrica*. *Biochemistry* 12:791–98

63. Lawrence K, Bhalla P, Misra PC. 1995. Changes in (NADP)H-dependent redox activities in plasmalemma-enriched vesicles isolated from boron- and zinc-deficient chick pea roots. *J. Plant Physiol.* 146:652–57

64. LeNoble ME, Blevins DG, Miles RJ. 1996. Prevention of aluminum toxicity with supplemental boron. II. Stimulation of root growth in acidic, high aluminum subsoil. *Plant Cell Environ.* 19:1143–48

65. LeNoble ME, Blevins DG, Sharp RE, Cumbie BG. 1996. Prevention of aluminum toxicity with supplemental boron. I. Maintenance of root elongation and cellular structure. *Plant Cell Environ.* 19:1132–42

66. Leopold AC, Kreidemann PE. 1975. Auxins. In *Plant Growth and Development*, ed. WJ Willey, A Stryker-Rodda, C First, pp. 109–38. New York: McGraw-Hill
67. Lewis DH. 1980. Boron, lignification and the origin of vascular plants. *New Phytol.* 84:209–29
68. Lin LS, Varner JE. 1991. Expression of ascorbic acid oxidase in zucchini squash (*Cucurbita pepo* L.). *Plant Physiol.* 96:159–65
69. Lohnis MP. 1937. Plant development in the absence of boron. *Overgedrukt uit Mededeelingen van de Landbouwhoogeschool* 41:3–36
70. Loomis WD, Durst RW. 1992. Chemistry and biology of boron. *BioFactors* 3:229–39
71. Loughman B, White P. 1984. The role of minor nutrients in the control of ion movement across membranes. In *Membrane Transport in Plants*, ed. WJ Cram, K Janacek, R Rybova, K Sigler, pp. 501–2. New York: Wiley
72. Lovatt CJ. 1985. Evolution of xylem resulted in a requirement for boron in the apical meristems of vascular plants. *New Phytol.* 99:509–22
73. Lovatt CJ, Dugger WM. 1984. Boron. In *The Biochemistry of the Essential Ultra Trace Elements*, ed. E Frieden, pp. 389–421. New York: Plenum
74. Lukaszewski KM, Blevins DG. 1996. Root growth inhibition in boron-deficient or aluminum-stressed squash plants may be a result of impaired ascorbate metabolism. *Plant Physiol.* 112:1–6
75. Lukaszewski KM, Blevins DG, Randall DD. 1992. Asparagine and boric acid cause allantoate accumulation in soybean leaves by inhibiting manganese-dependent allantoate amidohydrolase. *Plant Physiol.* 99:1670–76
76. MacVicar R, Tottingham WE. 1947. A further investigation of the replacement of boron by indoleacetic acid. *Plant Physiol.* 22:598–602
77. Marschner H. 1995. *Mineral Nutrition of Higher Plants*. New York: Academic
78. Mateo P, Bonilla I, Fernandez-Valiente E, Sanchez-Maseo E. 1986. Essentiality of boron for dinitrogen fixation in *Anabaena* sp PCC 7119. *Plant Physiol.* 81:430–33
79. Matoh T, Ishigaki K, Mizutani M, Matsunaga, Takabe K. 1992. Boron nutrition of cultured tobacco B Y-2 cells. I. Requirement for and intracellular localization of boron and selection of cells that tolerate low levels of boron. *Plant Cell Physiol.* 33:1135–41
80. Matoh T, Ishigaki K, Ohno K, Azuma J. 1993. Isolation and characterization of a boron-polysaccharide complex from radish roots. *Plant Cell Physiol.* 34:639–42
81. Matoh T, Kawaguchi S, Kobayashi M. 1996. Ubiquity of a borate-rhamnogalacturonan II complex in the cell walls of higher plants. *Plant Cell Physiol.* 37:636–40
82. Mazé P. 1915. Determination des elements mineraux rares necessaires as development du mais. *Compt. Rend.* 160:211–14
83. Mazé P. 1919. Recherche d'une solution purement minerale capable d'assurer l'evolution complete du mais cultive a l'abri des microbes. *Ann. Inst. Pasteur* 33:139–73
84. Mengel K, Kirkby EA. 1987. *Principles of Plant Nutrition*. Bern: Int. Potash Inst.
85. Moinat AD. 1943. Nutritional relationships of boron and indolacetic acid on head lettuce. *Plant Physiol.* 18:517–23
86. Mondy NI, Munshi CB. 1993. Effect of boron on enzymatic discoloration and phenolic and ascorbic acid content of potatoes. *J. Agri. Food Chem.* 41:554–56
87. Morré DJ, Crane FL, Sun IL, Navas PC. 1987. The role of ascorbate in biomembrane energetics. *Ann. NY Acad. Sci.* 498:153–71
88. Morré DJ, Navas P, Penel C, Castillo FJ. 1986. Auxin-stimulated NADH oxidase (semidehydroascorbate reductase) of soybean plasma membrane: role in acidification of cytoplasm? *Protoplasma* 133:195–97
89. Neales TF. 1960. Some aspects of boron in root growth. *Aust. J. Biol. Sci.* 13:232–48
90. Nguyen MN, Lukaszewski KM, Blevins DG 1993. IAA oxidase activity in squash roots may be regulated by boron and manganese interaction. *Plant Physiol.* 102:S.7
91. Nielson FH. 1991. The saga of boron in food: from a banished food preservative to a beneficial nutrient for humans. See Ref. 99a, pp. 274–86
92. Nielson FH, Hunt CF, Mullen LM, Hunt JR. 1987. Effect of dietary boron on mineral, estrogen, and testosterone metabolism in postmenopausal women. *FASEB J.* 1:394–97
92a. Noguchi K, Yasumori M, Imai T, Naito S, Matsunaga T, et al. 1997. *bor1-1*, an *Arabidopsis thaliana* mutant that requires a high level of boron. *Plant Physiol.* 115:901–6
93. Normanly J, Slovin JP, Cohen JD. 1995. Rethinking auxin biosynthesis and

metabolism. *Plant Physiol.* 107:323–29

94. Obermeyer G, Kriechbaumer R, Strasser D, Maschessnig A, Bentrup FW. 1996. Boric acid stimulates the plasma membrane H^+-ATPase of ungerminated lily pollen grains. *Physiol. Plant.* 98:281–90

95. O'Neill MA, Warrenfeltz D, Kates K, Pellerin P, Doco T, et al. 1996. Rhamnogalacturonan-II, a pectic polysaccharide in the walls of growing plant cell, forms a dimer that is covalently cross-linked by a borate ester. *J. Biol. Chem.* 271:22923–30

96. Parr AJ, Loughman BC. 1983. Boron in membrane functions in plants. In *Metals and Micronutrients: Uptake and Utilization by Plants*, ed. DA Robb, WS Pierpoint, pp. 87–107. *Annu. Proc. Phytochem. Soc. Eur. 21.* London: Academic

97. Piland JR, Ireland CF, Reisenauer HM. 1944. The importance of borax to legume seed production in the south. *Soil Sci.* 57:75–84

98. Pilbeam DJ, Kirkby EA. 1983. The physiological role of boron in plants. *J. Plant Nutr.* 6:563–82

99. Pollard AS, Parr AJ, Loughman BC. 1977. Boron in relation to membrane function in higher plants. *J. Exp. Bot.* 28:831–41

99a. Randall DD, Blevins DG, Miles CD, eds. 1991. *Current Topics in Plant Biochemistry and Physiology*, Vol. 10. Columbia: Univ. Mo. Press

100. Reed HS. 1947. A physiological study of boron deficiency in plants. *Hilgardia* 17:377–411

101. Reinbott TM, Blevins DG. 1995. Response of soybean to foliar-applied boron and magnesium and soil-applied boron. *J. Plant Nutr.* 18:179–200

102. Reinbott TM, Blevins DG, Schon MK. 1997. Content of boron and other elements in main stem and branch leaves and seed of soybean. *J. Plant Nutr.* 20:831–43

103. Reisenauer HM, Walsh LM, Hoeft RG. 1973. Testing soils for sulfur, molybdenum and chlorine. In *Soil Testing and Plant Analysis*, ed. LM Walsh, JD Beaton, pp. 173–200. Madison, Wis: Soil Sci. Soc. Amer.

104. Reiter WD, Chapple CCS, Somerville CR. 1993. Altered growth and cell walls in a fucose-deficient mutant of *Arabidopsis. Science* 261:1032–35

105. Rippka R, Stanier RY. 1979. The effects of anaerobiosis on nitrogenase synthesis by cyanobacteria. *Can. J. Microbiol.* 105:83–94

106. Robbertse PJ, Lock JJ, Stoffberg E, Coetzer LA. 1990. Effect of boron on directionality of pollen tube growth in *Petunia* and *Agapanthus. S. Afr. J. Bot.* 56:487–92

107. Robertson GA, Loghman BC. 1974. Reversible effects of boron on the absorption and incorporation of phosphate in *Vicia faba* L. *New Phytol.* 73:291–98

108. Roldán M, Belver A, Rodriguez-Rosales MP, Ferrol N, Donaire JP. 1992. In vivo and in vitro effects of boron on the plasma membrane proton pump of sunflower roots. *Physiol. Plant.* 84:49–54

109. Sarquis AM. 1986. Dramatization of polymeric bonding using slime. *J. Chem. Educ.* 63:60–61

110. Schmucker T. 1933. Zur Blutenbiologie tropischer *Nymphaea* Arten. II. Bor, als entscheidener Faktor. *Planta* 18:641–50

111. Schmucker T. 1935. Uber den Einfluss von Borsaure und Pflanzen, insbesondere keimende Pollekorner. *Planta* 23:264–83

112. Schon MK, Blevins DG. 1990. Foliar boron applications increase the final number of branches and pods on branches of field-grown soybean. *Plant Physiol.* 92:602–7

113. Schon MK, Novacky A, Blevins DG. 1990. Boron induces hyperpolarization of sunflower root cell membranes and increases membrane permeability to K^+. *Plant Physiol.* 93:566–71

114. Shelp BJ. 1987. Boron mobility and nutrition in broccoli (*Brassica oleracea* var. *italica*). *Ann. Bot.* 61:83–91

115. Shelp BJ. 1993. Physiology and biochemistry of boron in plants. In *Boron and Its Role in Crop Production*, ed. UC Gupta, pp. 53–85. Boca Raton, FL: CRC Press

116. Shelp BJ, Marentes E, Kitheka AM, Vivekanandan P. 1995. Boron mobility in plants. *Physiol. Plant.* 94:356–61

117. Shelp BJ, Shattuck VI. 1987. Boron nutrition and mobility, and its relation to hollow stem and the elemental composition of greenhouse grown cauliflower. *J. Plant Nutr.* 10:143–62

118. Shelp BJ, Shattuck VI. 1987. Boron nutrition and mobility, and its relation to the elemental composition of greenhouse grown root crops. I. Rutabaga. *Soil Sci. Plant Anal.* 18:187–201

119. Shelp BJ, Shattuck VI. 1987. Boron nutrition and mobility, and its relation to the elemental composition of greenhouse grown root crops. II. Radish. *Soil Sci. Plant Anal.* 18:203–19

120. Shkolnik MY, Krupnikova TA, Dmitrieva NN. 1964. Influence of boron deficiency on some aspects of auxin metabolism in the sunflower and corn. *Sov. Plant Physiol.* 11:164–69

121. Shkolnik MY. 1984. *Trace Elements in Plants*. New York: Elsevier

122. Skok J, McIlrath WJ. 1958. Distribution of boron in cells of dicotyledonous plants in relation to growth. *Plant Physiol.* 33:428–31

123. Sommer AL, Lipman CB. 1926. Evidence on the indispensaable nature of zinc and boron for higher green plants. *Plant Physiol.* 1:231–49

124. Srivastava NK, Luthra R. 1992. Influence of boron nutrition on essential oil biogenesis, glandular scales, CO_2 assimilation and growth in *Mentha arvensis* L. *Photosynthetica* 26:405–13

125. Tanada T. 1983. Localization of boron in membranes. *J. Plant Nutr.* 6:743–49

126. Taylor GJ, MacFie SM. 1994. Modeling the potential for boron amelioration of aluminum toxicity using the Weibull function. *Can. J. Bot.* 72:1187–96

127. Teasdale RD, Richards DK. 1990. Boron deficiency in cultured pine cells. *Plant Physiol.* 93:1071–77

128. Varner JE. 1995. Foreword: 101 reasons to learn more plant biochemistry. *Plant Cell* 7:795–96

129. Vaughan AKF. 1977. The relation between the concentration of boron in the reproductive and vegetative organs of maize plants and their development. *Rhod. J. Agric.* 15:163–70

130. Visser T. 1955. Germination and storage of pollen. *Meded. Landb. Hoogesch.* 55:1–68

131. Walsh T, Golden JD. 1953. The boron status of Irish soils in relation to the occurrence of boron deficiency in some crops in acid and alkaline soils. *Int. Soc. Soil Trans.* II:167–71

132. Warington K. 1923. The effect of boric acid and borax on the broad bean and certain other plants. *Ann. Bot.* 37:629–72

133. Winkenbach F, Wolk CP, Jost M. 1972. Lipids of membranes and of the cell envelope in heterocysts of a blue-green alga. *Planta* 107:69–80

134. Yamagishi M, Yamamoto Y. 1994. Effects of boron on nodule development and symbiotic nitrogen fixation in soybean plants. *Soil Sci. Plant Nutr.* 40:265–74

135. Yamanouchi M. 1971. The role of boron in higher plants. I. The relations between boron and calcium or the pectic substances in plants. *J. Sci. Soil Manure* 42:207–13

136. Yamauchi T, Hara T, Sonoda Y. 1986. Distribution of calcium and boron in the pectin fraction of tomato leaf cell wall. *Plant Cell Physiol.* 27:729–32

Annu. Rev. Plant Physiol. Plant Mol. Biol. 1998. 49:501–23

HORMONE-INDUCED SIGNALING DURING MOSS DEVELOPMENT

Karen S. Schumaker and Margaret A. Dietrich
Department of Plant Sciences, University of Arizona, Tucson, Arizona 85721;
e-mail: schumake@ag.arizona.edu

KEY WORDS: asymmetric division, cell fate, cell differentiation, calcium signaling, cytokinin

ABSTRACT

Understanding how a cell responds to hormonal signals with a new program of cellular differentiation and organization is an important focus of research in developmental biology. In *Funaria hygrometrica* and *Physcomitrella patens*, two related species of moss, cytokinin induces the development of a bud during the transition from filamentous to meristematic growth. Within hours of cytokinin perception, a single-celled initial responds with changes in patterns of cell expansion, elongation, and division to begin the process of bud assembly. Bud assembly in moss provides an excellent model for the study of hormone-induced organogenesis because it is a relatively simple, well-defined process. Since buds form in a nonrandom pattern on cells that are not embedded in other tissues, it is possible to predict which cells will respond and where the ensuing changes will take place. In addition, bud assembly is amenable to biochemical, cellular, and molecular biological analyses. This review examines our current understanding of cytokinin-induced bud assembly and the potential underlying mechanisms, reviews the state of genetic analyses in moss, and sets goals for future research with this organism.

CONTENTS

INTRODUCTION

The study of nonflowering plants has contributed important information about the nature of changes in form and function that occur during the development of both flowering and nonflowering plants. While even these simple plants grow and develop via complex processes and interactions, their less complicated morphology makes the study of certain aspects of development more feasible than is possible in higher plants. In *Funaria hygrometrica* and *Physcomitrella patens*, two related species of moss, assembly of a bud from an initial cell involves hormone-induced organogenesis beginning in a single cell that is not embedded in other tissues. In this review, we describe what is known about early events in bud assembly, examine the advantages and limitations of studying moss to understand the underlying elements of eukaryotic development, and identify areas of future investigation and the tools that will be critical for these studies. Finally, we set some immediate goals for the study of development in moss.

MOSS GAMETOPHYTE DEVELOPMENT

The earliest stage of vegetative development in *Funaria* and *Physcomitrella* is characterized by cellular differentiation during filament growth. The cellular dimensions and timing of the events described here for *Funaria* grown in culture have been recently described in detail (56). Spore germination (Figure 1*A* and *B*) leads to the formation of a filament that consists of a tip (apical) cell and a linear array of subapical cells produced by successive divisions of the tip cell. These cells, the chloronema (Figure 1*C*), are filled with disc-shaped chloroplasts and have cross walls that are perpendicular to the filament axis. As is characteristic of the subapical cells of tip-growing organisms (36), no further growth occurs in the subapical chloronema cells. The tip cell elongates, reaches a maximum length, and divides to produce a new subapical cell to extend the filament. Chloronema filament growth continues until, in response to increases in light (18) and auxin, the appearance of the chloronema tip cell begins to change. This cellular differentiation leads to formation of the second filament

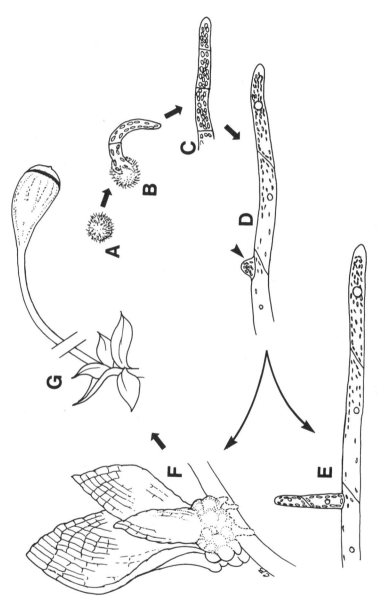

Figure 1   Stages of moss development.   Haploid spores (*A*) germinate to form a filament consisting of chloronema cells (*B* and *C*). Subsequently, light and auxin induce changes in the tip cell to give rise to caulonema cells (*D*). A single-celled initial (*D, arrowhead*) forms on the second subapical cell of the caulonema filament.   This initial cell has two potential fates.   In the absence of cytokinin, the initial cell will continue to grow by tip growth to form a new lateral filament (*E*). In the presence of cytokinin, the initial cell takes on the morphology associated with the assembly of a bud to form the leafy shoot (*F* and *G*) that eventually bears the gametangia (not shown). Following fertilization, a diploid capsule (*G*) forms on the leafy shoot.   Ultimately, meiosis occurs within the capsule to produce haploid spores.

cell type, the caulonema (Figure 1*D*). In comparison to a chloronema tip cell, a fully developed caulonema tip cell elongates dramatically, exhibits decreased time for tip cell division, and has smaller, elongated, flattened chloroplasts that contain less chlorophyll. During the transition from chloronema to caulonema, the newly formed cells appear intermediate in character between the two cell types, but after five to six days, caulonema cells are long, nearly clear, and have cross walls that are oblique to the filament axis.

Once caulonema cell differentiation has begun, a new axis of cellular polarity is set up during the formation of initial cells. Very shortly after division of a caulonema tip cell, a small swelling appears in the second subapical cell (the third cell of the filament). This outgrowth, which will give rise to an initial cell (Figure 1*D*, *arrowhead*), appears at the apical end of the cell near the apical-most end of the oblique cross wall. The outgrowth continues to expand, and the division that will produce the fully formed initial cell occurs five to six hours after visible evidence of initial cell formation is first seen. Before this division, the nucleus migrates from midway in the filament cell to the initial cell site, where it divides. One daughter nucleus moves into the forming initial cell, and the second moves back to the middle of the filament cell. A cell wall, oriented parallel to the longitudinal axis of the filament cell, separates the initial cell from the filament to produce the fully formed initial cell (Figure 1*D*).

The next stages of development in *Funaria* are characterized by hormone-induced organogenesis as a bud is assembled from the caulonema initial cell and the leafy gametophyte develops from the bud. The caulonema initial cell has two potential fates that are developmentally distinct. In the absence of cytokinin, the initial cell will continue to grow by tip growth to produce a new lateral filament (side branch) (Figure 1*E*), thus maintaining the filamentous growth habit. However, in the presence of cytokinin, the initial cell takes on a distinct morphology associated with the assembly of a bud in transition from filamentous to meristematic growth.

In culture, bud assembly can occur in both the presence and absence of exogenous hormone; however, treatment with cytokinin leads to the production of significantly more buds. Tissue can respond to added cytokinin for a period of time after caulonema initial cells begin to form, but prior to the appearance of buds in untreated tissue (8, 13). The very small number of buds that form in the absence of added cytokinin presumably arise in response to endogenous hormone.

Early changes during bud assembly include an altered pattern of cell expansion and elongation of the initial cell to produce the single-celled bud. Later changes involve divisions within the bud to give rise to a simple meristem that produces a leafy shoot (Figure 1*F* and *G*) that eventually bears the gametangia (not shown). Following the production of gametes and fertilization, the zygote develops into a sporophyte (Figure 1*G*) with a stalk 3–4 cm long that will bear a single capsule containing hundreds of thousands of haploid spores.

In the context of this description of moss development, we turn our discussion to what is known at the cellular and subcellular levels about how the caulonema initial cell is assembled into a bud.

CELLULAR CHANGES DURING BUD ASSEMBLY

Cellular changes are apparent in the initial cell within two to three hours after the addition of cytokinin. At the time of the division that separates the initial cell from the filament cell, the initial cell (Figure 2*A*) is approximately 20 μm long and contains many large chloroplasts. The first visible indication of cytokinin-induced bud assembly is the dramatic swelling of the initial cell resulting from a lack of further tip growth and a change in the pattern of cell expansion and elongation. The apical area of the initial cell becomes dome-shaped as the deposition of new wall material moves from the very tip region to the sides of the cell, forming a rounded single-celled bud with an elongating stalk (Figure 2*B*) (10, 16). Other early cellular changes in the single-celled bud include a reduction in chloroplast size and an alteration in chloroplast shape. The first division of the bud occurs when it is approximately 65 μm long. This division is asymmetric and, therefore, produces daughter cells of different developmental fates. The bud divides transversely with respect to its long axis, producing a large, highly vacuolate stalk cell and a small, densely cytoplasmic apical cell (Figure 2*C*). The apical cell then divides longitudinally, resulting in two densely cytoplasmic cells (Figure 2*D*) (16). Subsequent unequal cell divisions give rise to a tetrahedral apical cell (meristem) that continues to divide in three planes to form the relatively simple multicellular bud (Figure 2*E*). The subapical cells of this bud then divide more frequently than the apical cell to give rise to a larger, more complex bud (Figure 2*F*) (24). Subsequently, leaf primordia arise as projections from the side of the bud (Figure 2*F*, *arrowhead*). One of the derivatives of each apical cell division gives rise to one primordium; the leaflets (Figure 2*G*) derived from it are composed of files of cells that grow by general expansion.

SUBCELLULAR CHANGES UNDERLYING BUD ASSEMBLY

During initial cell formation there is a change in the cellular organization at the presumptive initial cell site; directed growth at this site takes place via stratification of organelles (16). TEM (transmission electron microscopy) studies have shown that the apex of the outgrowth contains Golgi bodies, associated

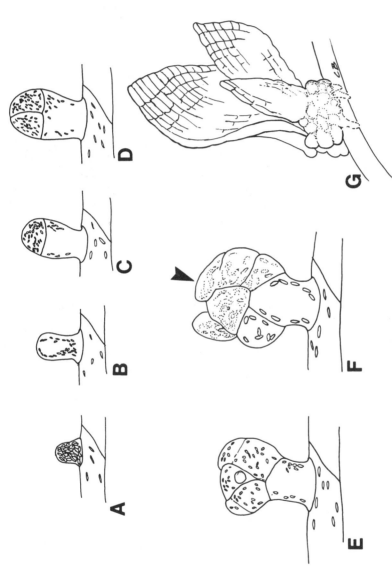

Figure 2   Developmental transition from filamentous to meristematic growth. Changes are apparent in the initial cell (*A*) within two to three hours of cytokinin addition. The first visible indication of bud assembly is a dramatic swelling of the initial cell (compare *A* and *B*). This is followed by an asymmetric division to produce a large, highly vacuolate stalk cell and a small, densely cytoplasmic apical cell (*C*). The apical cell divides longitudinally, resulting in two densely cytoplasmic cells (*D*). Subsequent unequal divisions give rise to a tetrahedral apical cell that continues to divide in three planes to form the relatively simple multicellular bud (*E*). The subapical cells of the bud divide more frequently than the apical cell to give rise to a larger, more complex bud (*F*). Subsequently, the leaf primordia (*F, arrowhead*), each of which will develop into a leaflet of the leafy shoot (*G*), arise as projections from the side of the bud.

vesicles, and cortical (immediately adjacent to the plasma membrane) endoplasmic reticulum (ER). The outgrowth is filled with cytoplasm but contains only a few vacuoles and chloroplasts that are positioned in the cell cortex. While tip-growing cells are distinguished by perpetual stratification of organelles (35), the moss initial cell loses this organization after the division separating it from the filament occurs (16).

A fully formed initial cell that has not been stimulated to form a bud becomes a side branch, and the cellular organization resumes the tip cell pattern of organellar stratification. If perception of cytokinin occurs, organelle distribution remains random during the dramatic swelling that follows, but there is a qualitative and quantitative change in internal membranes (41, 42). The structure, quantity, and distribution of the ER during bud assembly have been studied using both the fluorescent, lipophilic carbocyanine dye, 3,3'-dihexyloxacarbocyanine iodide, and rapid freeze-fixation/freeze-substitution. These studies have shown that while the cortex of the bud contains the same cellular components as side branches, during bud assembly there is an increase in ER membrane density and cortical ER volume. The ER network becomes "tighter" and forms a gradient within the developing single-celled bud as the stalk region becomes delineated. As vacuolation increases in the stalk region, its ER network becomes more open. In contrast, the apex of the one- or two-celled bud has closely packed ER. This ER is associated with ribosomes and forms a shell in the periphery of the bud apex with close apposition of the outermost ER and plasma membrane throughout the bud cortex. The new configuration and quantity of ER has been found to be the most significant subcellular change observed during bud development, and this bud-like pattern has never been observed in side branches. The ER continues to be closely spaced in the apical region of the bud as it develops into a multicellular structure and forms the tetrahedral apical cell. It is not clear how the change in cortical ER density during bud assembly is accomplished.

McCauley & Hepler (42) suggested that the cortical ER may be a general indicator of the metabolic status of a cell during bud assembly. Cytokinin-induced bud assembly may be mediated through release of calcium, and the increased quantity of cortical ER in buds may represent the mechanism to change calcium sequestration capabilities and intracellular calcium levels.

Doonan et al (24) demonstrated that reorganization of the microtubule cytoskeleton is also correlated with changes in the pattern of cell growth during bud assembly. In a newly formed caulonema initial cell, there is a meshwork of microtubules that have a random orientation and do not focus to any particular site in the cell. As an initial cell that has not been stimulated to form a bud resumes tip growth, the microtubules associated with the nucleus become aligned along the axis of cell elongation and focus to the surface of the tip apex. In cytokinin-treated initial cells, the nucleus-to-cortex microtubules are oriented randomly and do not focus to the tip, which is consistent with the more diffuse pattern of growth in the bud. Evidence from this study suggests that cytokinin may be specifically affecting the cytoplasmic microtubules in the developing bud. While cytokinin has no apparent effect on either microtubules in nonbud-forming tissue or on spindle and phragmoplast microtubules within the bud, the cytoplasmic microtubules in assembling buds appear to be poorly preserved. Based on this lack of microtubule preservation, it has been suggested that the cytokinin-induced changes in cytoplasmic microtubule organization may account for the loss or prevention of tip-directed growth resulting in a swelled initial cell.

SIGNALS INITIATING BUD ASSEMBLY

Experimental evidence supports the existence of two distinct developmental stages in bud formation: (*a*) caulonema initial cell formation and (*b*) assembly of the bud from the initial cell (11). In many of the studies outlined below, these two processes have not been distinguished from one another. Where possible, we have tried to separate them and to focus our discussion on bud assembly.

Light

Light has a marked influence on a number of processes in moss development such as spore germination, growth of chloronemal and caulonemal side branches, and bud assembly (3, 4, 18, 19, 60). Evidence suggesting a light requirement in bud formation includes the absence of buds in dark-grown plants and the induction of buds in dark-grown plants exposed to light (3, 4, 19, 60). Because the starting tissue for these studies was not defined in most cases, we cannot conclude whether initial cell formation, bud assembly, or both are light-sensitive.

Some of the characteristics of the light requirement for bud formation have been determined. Production of buds is dependent on the intensity of red light, and weak white light delays further development to the leafy gametophyte (3, 44, 60). In low-intensity continuous light, buds do not assemble from initial cells in response to cytokinin (18). However, as white light intensities are raised, bud formation increases steadily (3). Large numbers of buds can also be induced with exposure of tissue to red light with maximum bud production occurring at >16 μmol quanta m$^{-2} \cdot$ s^{-1} (3). It has been shown in *Physcomitrium turbinatum* that there is a relatively large cumulative light dose required for bud formation, suggesting that light energy may be used for the synthesis of a product that must accumulate before buds form (44). Further support for the accumulation of such a product comes from experiments in which moss tissue "remembers" exposure to light. In *Physcomitrella*, some buds are formed when dark-grown cultures are exposed to light for several hours and then simultaneously treated with cytokinin and returned to darkness (18).

Hormones

In assessing the role of regulatory molecules in caulonema filament and initial cell formation, we have previously suggested criteria that would need to be satisfied to show the involvement of a molecule in a physiological process (56). For example, this effector should be present at the appropriate times and in the correct concentrations to elicit the response. This requires mechanisms to alter the amounts of the effector or the sensitivity of the target cell to the effector. It should also be possible to show that altering the level of the effector in wild-type or mutant plants changes the response. We will use these criteria to provide a framework with which to evaluate the evidence implicating auxin and cytokinin in bud assembly.

The genetic nomenclature that follows uses the respective authors' strain designations. Italicized lower-case letters represent mutant alleles that have been shown in crosses to segregate in a Mendelian manner. Italicized upper-case letters indicate strain designations based on phenotypic analyses; it has not been determined that these strains are the result of single mutations (5, 26).

AUXIN   To our knowledge, auxin levels have not been measured during bud formation. However, evidence of a role for auxin in bud formation comes from experiments with cytokinin-resistant (benzyladenine-resistant, BAR) mutants of Physcomitrella. BAR mutants do not produce buds even in the presence of exogenous cytokinin (3, 5). When grown in white light on medium lacking hormones, one class of BAR mutants produces a normal amount of caulonema but forms few or no leafy gametophytes even though these mutants are producing initial cells. With the addition of low levels of auxin, however, these mutants show normal development through the formation of leafy gametophytes. These results suggest that in addition to cytokinin, bud assembly requires auxin, but presumably at higher levels than is needed to produce caulonema cells (3, 56).

CYTOKININ   At the time of the last comprehensive review in this series on gametophyte development in ferns and bryophytes (12), it had been shown that addition of cytokinin to moss cultures stimulated bud formation (see 13 and references therein). Since that time, it has been shown that cytokinin specifically affects both initial cell formation and the subsequent assembly of a bud. Bopp & Jacob (11) have shown that in Funaria picomolar levels of cytokinin induce a caulonema filament to produce initial cells, whereas nanomolar to micromolar concentrations are required for the assembly of a bud from an initial cell.

All synthetic and natural substances with the characteristics of a cytokinin (adenine derivatives with an N^6-substituted side chain of five or more carbon atoms) evaluated to date can cause a change from filamentous growth to at least the early stages of bud formation (9). Both cytokinin bases and ribosides are active; however, the ribosides are less so (67). Additional studies have shown the concentration dependence of cytokinin action over a range of 50 nM to 1 μM (13, 67).

Few measurements of tissue-derived cytokinin levels have been made for wild-type *Funaria* or *Physcomitrella*, and to our knowledge, no measurements have been made to relate cytokinin levels to different stages of development. In a hybrid generated from a cross between *Funaria* and *Physcomitrium piriforme*, Beutelmann & Bauer (7) characterized the endogenous hormone and showed that its chromatographic behavior was identical to that of N^6-(Δ^2-isopentenyl)adenine (i^6Ade) and that it was found in both tissue and culture medium at ~1 μM. Apart from this report, most of the information about the nature of the endogenous cytokinin responsible for bud formation has come from studies of mutants of *Physcomitrella*. Ashton et al (5) isolated and characterized a number of mutants that responded abnormally to auxin and cytokinin and have shown that a relationship exists between the presence of the hormones and bud formation. Based on their responses to added hormones, the mutants were divided into two categories, those that have altered endogenous auxin or cytokinin levels (possibly due to altered levels of synthesis, increased production of molecules that modulate hormone activity, or changes in the degradation of endogenous hormone) and those altered in their response to one or both of the hormones.

The characterization of one group of mutants that appears to have altered endogenous cytokinin levels, the bud *over*producing (*ove*) mutants, has provided evidence for the involvement of cytokinin in bud formation. 1. These mutants produce more buds than wild type on media lacking hormones, thus resembling wild-type plants treated with cytokinin (4). 2. When grown with wild-type tissue, members of one *ove* group can induce bud production in neighboring wild-type cells (4). 3. Under conditions in which the medium is continuously replaced, *ove* mutants do not produce buds; the colonies show a morphology identical to that of wild-type tissue grown under the same conditions (4). 4. Subsequent studies measured the cytokinin content of the medium from cultures of the wild type and several *ove* mutants. It was shown that the i^6Ade concentrations in the culture medium from the *ove* mutants reached a maximum of 100 nM and zeatin concentrations reached 5 nM (25, 63, 65). Cytokinin levels in the medium from the wild-type culture were approximately 1% of what was found for the *ove* mutants (65).

Very little is known about mechanisms of synthesis or catabolism of cytokinin during moss development. Evidence for cytokinin biosynthesis comes from experiments in which cultures of *ove* mutants were fed with radiolabeled adenine (62). Within hours, radiolabeled cytokinin was found in the culture medium and in the tissue. Gerhäuser & Bopp (30) provided preliminary evidence for a cytokinin degradation pathway. They showed that *Funaria* cultures convert radiolabeled cytokinin to adenine and its derivatives.

CALCIUM AS AN INTRACELLULAR MESSENGER IN BUD ASSEMBLY

In a series of papers, Saunders & Hepler (51–53), Saunders (50), and Conrad & Hepler (15) reported experiments with *Funaria* addressing the role of calcium in bud formation. The authors concluded that calcium is involved; however, they did not determine in which process(es) it plays a role. Analysis of the data presented in these papers suggests that the studies have most often evaluated the role of calcium in the formation of the caulonema initial cell. Two observations lead to this conclusion. First, the authors often started with tissue that had not yet produced initial cells, the target cells for cytokinin-induced bud assembly. Because they added cytokinin to caulonema tissue that was not producing initial cells, this addition first induced the formation of those targets. Second, they measured the levels and distribution of intracellular calcium immediately after the addition of cytokinin, during the period of initial cell formation, before targets of bud assembly were present.

There is, however, indirect evidence suggesting that calcium in the external medium is required for bud assembly. Saunders & Hepler (52) determined the effect of artificially increasing intracellular calcium levels on initial cell formation. They found that in the absence of cytokinin, but in the presence of calcium, the calcium ionophore A23187 induced initial cell formation. When treatment with the ionophore was prolonged (under the same conditions), in some instances initial cells continued to divide and form buds with typical tetrahedral apical cells. Moreover, Markmann-Mulisch & Bopp (40) attempted to induce buds in *Funaria* in cytokinin-containing medium in which the effective concentration of calcium had been reduced using cobalt. Fewer buds formed, and those that did form were unable to undergo cell division remaining round and unicellular.

Saunders & Hepler (51) provided further evidence implicating calcium in bud assembly. They measured membrane-associated calcium at various stages of development after cytokinin addition using the fluorescent calcium-chelating probe chlorotetracycline (CTC). They found that fluorescence was four times greater in the single-celled bud than in its subtending caulonema cell, suggesting calcium levels increased during early stages of bud assembly. As bud assembly progressed, the stalk cell became highly vacuolate and less fluorescent while the dividing cells of the bud continued to display bright fluorescence at least through the formation of the tetrahedral apical cell. Using the fluorescent membrane probe N-phenyl-1-naphthylamine (NPN) as a measure of the amount of total membrane present, some NPN fluorescence was observed in the single-celled buds, but at lower levels than observed with CTC in cells at the same stage of development. The authors concluded that the relative amount of calcium per quantity of membrane, which had increased during initial cell formation, was maintained during bud assembly. They suggested that the increases in membrane-associated calcium reflect a localized cytokinin-induced increase in intracellular free calcium.

Evidence suggesting that calcium may not be required for early events in cytokinin-induced bud assembly comes from experiments in which *Funaria* was grown in calcium-free medium (40). In this medium, initial cells swelled even in the absence of apical calcium-CTC fluorescence. The buds remained round and unicellular, suggesting that although calcium may not be required for the early events in bud assembly, it appears to be required for the subsequent cell divisions that lead to the formation of a multicellular bud.

PROSPECTS FOR THE ANALYSIS OF BUD ASSEMBLY

Early Events

While it is clear that cytokinin can stimulate the assembly of a bud from a caulonema initial cell, many questions remain about the processes involved in the cytokinin-induced signaling. For example, where is endogenous cytokinin made and where does the perception that leads to bud assembly take place? Is cytokinin made in filament cells and perceived intracellularly by the initial cell or at its plasma membrane? Do endogenous cytokinin levels change during development, or is the sensitivity of the initial cell altered in a developmentally programmed manner?

To localize endogenous cytokinin and identify sites of perception during normal development, it will be necessary to develop methods that allow detection of low levels of hormone and quantitative measurement of cytokinin in situ. In addition, information is needed about the transport of endogenous cytokinin and mechanisms of cytokinin synthesis, catabolism, or differential activation during bud assembly.

Several approaches could be used to provide information about the site of perception of exogenous cytokinin. For example, Brandes & Kende (13) monitored the distribution of radiolabeled cytokinin in *Funaria*. They treated cells with radiolabeled cytokinin and saw significant localization of radioactivity at the single-celled bud. While some radioactivity was associated with the subtending caulonema cells, little or none was localized to caulonema cells that were not or had not produced initial cells. Based on these studies, Brandes

(12) suggested that the presence of binding sites for cytokinin may be the biochemical basis for the difference between cells that form buds and those that do not. Additional experiments will be required to determine whether radiolabeled cytokinin binds to the plasma membrane or whether it is taken up into the cells. Another approach that should provide information about the site of perception of exogenous cytokinin would involve the comparison of early events in bud assembly in cells treated with impermeant cytokinin with those events in cells that have been injected with the hormone.

An initial cell that does not encounter cytokinin will form a side branch by resuming tip growth. One characteristic feature of tip growth is the presence of an oscillating gradient of calcium focused at the apex of the tip cell (43, 46). These calcium gradients may be due in part to regulated influx of calcium at the growing tip. If these gradients are disrupted (e.g. with calcium and a calcium ionophore), normal tip growth is altered (28, 39). Could a cytokinin-induced increase in calcium early in bud assembly disrupt the normal tip-focused gradient of calcium? A delocalized calcium influx might then lead to an altered distribution of organelles and vesicles as part of the change from tip growth to the pattern of expansion associated with the developing bud. An early difference between a branch or a bud then may be due to this change in calcium, and cytokinin may induce and/or maintain delocalized calcium entry.

Experiments from our laboratory have shown that a calcium transport mechanism is present on the plasma membrane in *Physcomitrella* (57–59). This transport is sensitive to 1,4-dihydropyridines (DHPs), molecules that are known to modulate calcium entry through voltage-dependent calcium channels in animal cells (14). In our studies, we have provided evidence for DHP modulation of calcium influx into moss protoplasts (57). Influx was stimulated by DHP agonists and inhibited by DHP antagonists. Calcium accumulation increased dramatically within 15 s of addition of cytokinin to protoplasts, suggesting a potential interaction of the hormone and the transporter. As has been shown for DHP-sensitive calcium transport in animal cells, this influx into moss cells was stimulated by a depolarization of the plasma membrane and was affected by numerous classes of calcium channel blockers. We have also shown that there are abundant sites for DHP binding in the *Physcomitrella* plasma membrane; DHPs bind with high affinity and specificity (58). This ligand/receptor interaction was stimulated by cytokinin at low concentrations and by heterotrimeric GTP-binding proteins (58, 59).

We are presently examining the distribution, activity, and regulation of the DHP-sensitive calcium transport activity during different stages of development. With this information, we will be able to determine whether (*a*) the transporter plays a role in initial cell formation (56), (*b*) the transporter regulates calcium oscillations during tip growth in initial cells that have not been stimulated by cytokinin to form a bud, and (c) cytokinin regulation of this transporter disrupts the normal tip-focused gradient of calcium.

Determining the role of calcium in early events in cytokinin-induced bud assembly will be complicated by the apparent diversity of calcium-dependent stages during moss development. At a minimum, it will be necessary to separate the changes taking place in bud assembly immediately after cytokinin addition from processes involved in initial cell formation and from those involved in the cell divisions during later stages of bud assembly. In addition, it will be important to avoid experiments in which calcium is removed from the medium as resulting changes in bud assembly may be due to calcium's effect on other critical developmental processes. It should be possible, however, to determine whether calcium is involved in early events in cytokinin-induced bud assembly using single cell assays, which we believe represent one of the major advantages of using moss to study development. In these assays, molecules that may influence a specific process are delivered via microinjection to a particular cell at a specific stage of development. Using this approach, it will be possible to monitor cellular calcium changes immediately after the addition of cytokinin and to alter intracellular calcium levels in the initial cell in the absence of cytokinin. If calcium is involved, it will be possible to determine the timing of its involvement since bud assembly occurs progressively along cells of a single filament.

If calcium is not found to mediate these very early events in bud assembly, it will be necessary to determine how cytokinin is regulating the altered pattern of growth that takes place during the assembly of the bud. For example, what molecules are synthesized or activated in response to cytokinin? How do they lead to the changes in cell expansion and elongation seen early in bud assembly or to the cell divisions and subsequent morphological changes that result in formation of the multicellular bud? Is cytokinin causing structural changes by regulating the rate of wall synthesis and degree of wall extensibility as the initial cell swells? Does it alter the stability or biochemical composition of the cytoskeleton to allow for changes in deposition of wall and membrane material during initial cell swelling or the orientation of the mitotic apparatus during subsequent cell divisions?

Later Events

The processes that result in daughter cells with different fates are fundamental to the generation of cell diversity during development (37). During cytokinin-induced bud assembly, the initial cell reaches a specific size and then undergoes an asymmetric division to produce two cells of different sizes with very different fates. This asymmetric division must require the coordination of two events: establishment of cytoplasmic polarity and orientation of the mitotic apparatus along the axis of polarity. A number of questions need to be answered if we are to understand how these processes take place. For example, what cell fate determinants are distributed differentially to the daughter cells? Do polarized components of the cytoskeleton provide a structural basis for localizing these determinants? Which genes play a role in the regulation of the asymmetric division? How are these genes regulated and how do the resulting gene products act?

Asymmetric divisions are common in many organisms, and genes that are responsible for these divisions, and the subsequent cell fate, have been identified in plants, nematodes, insects, yeast, and bacteria (1, 20–22, 31, 34, 49). Recognizing that multiple mechanisms may be responsible for the generation of asymmetry, careful comparative studies of asymmetric division in numerous organisms may suggest candidate molecules that may underlie this process in moss. Once these molecules are identified, it should be possible to determine whether homologs exist in moss. For example, conserved sequences found in genes in these other organisms could be used as hybridization probes or primers for polymerase chain reaction. Methods that enable isolation of cDNAs from moss that functionally complement asymmetry defects in other organisms also offer a potentially powerful route for identifying homologs of molecules important in this process. In vivo assays altering the levels and distribution of putative regulatory molecules before, during, and after asymmetric division of the initial cell should provide important insights into the role and regulation of these molecules during bud assembly.

Studies have shown that cytokinin is not just a trigger for bud assembly; its presence is required for several hours to prevent reversion to normal filament growth (13). In experiments using a photolabile cytokinin, Sussman & Kende (61) showed that reversion can be induced by exposing the tissue to UV light to destroy the cytokinin, as well as by washing the tissue (13). Since it is not clear if the caulonema filaments used as starting tissue for these studies were producing initial cells, it is not possible to determine the duration of cytokinin exposure required for bud assembly alone. However, as was shown by the formation of a filament from a multicellular bud after cytokinin removal, clearly some prolonged exposure to cytokinin is required to commit growth to bud assembly. A number of questions arise concerning this prolonged requirement for cytokinin. Is continued exposure to cytokinin required to ensure production of sufficient levels of a product that is required for commitment to bud assembly? Is cytokinin required at multiple steps in the pathway of a cytokinin-induced cascade?

GENETIC ANALYSES OF BUD ASSEMBLY

The studies outlined above have begun to provide insight into the events that take place during cytokinin-induced bud assembly and some of the mechanisms involved. Continued progress toward understanding the underlying biochemical and molecular mechanisms will be facilitated by thorough characterization of previously identified developmental mutants. Progress will also be aided by the identification of additional relevant mutations, such as those leading to altered perception and transduction of the cytokinin signal or altered levels of downstream interacting components. Potentially, mutants will allow the identification of genes important in these processes. In addition, simple mutations can be used in conjunction with the cellular and molecular tools currently available to characterize the underlying defects. In the following section, we describe the current status of genetic analyses in moss and identify technological advances that will be required for genetics to further contribute to our understanding of mechanisms underlying development in this organism.

Most genetic studies have been performed with *Physcomitrella*. Recent estimates suggest that in the wild type $n = 27$, and the DNA content is 0.6 pg per haploid genome corresponding to 600 megabase pairs (48). Sporophytes can arise as a result of either self- or cross-fertilization via union of gametes from gametophytes produced from spores of the same or different sporophytes. Wild-type strains are normally self-fertile. In culture, *Physcomitrella* has a short generation time of approximately 12 weeks.

Conditions have been described for the isolation and characterization of morphological and amino acid and purine analog-resistant mutants (2, 4, 5). Mutations have been induced in haploid spores or filament cells with N-methyl-N'-nitro-N-nitrosoguanidine (NTG), ethylmethane sulphonate (EMS), or ultraviolet light. During somatic mutagenesis, filament tissue is treated with a mutagen, and protoplasts are isolated (64). Regeneration of protoplasts leads to the formation of tissue that can be screened for mutant phenotypes. Following mutagenesis of either spores or filament cells, mutant strains have been isolated most often using nonselective isolation (47). With this approach, the mutagenized protoplasts or spores are cultured initially on medium that will promote filamentous growth. After 1–2 weeks of growth, the tissue is transferred to conditions that select for the desired mutant phenotype; putative mutants can usually be identified within 2–3 weeks. Male and female gametes borne on the same individual gametophyte are genetically identical. Thus, sporophytes arising after fusion of these gametes are homozygous for the induced mutation, and all resulting spores contain the mutation. If strains are fertile, classical techniques can be used for complementation analysis, dominance testing in nonhaploid tissue, and linkage studies.

It is easy to identify characteristics of moss that make it well suited for genetic analyses. The production of single-celled spores and subsequent development of the prolonged gametophytic stage allow genetic studies at the haploid level. As described above, it is possible to mutagenize spores and identify gametophytic mutations. In addition, self-fertilization of gametophytes results in completely homozygous sporophytes. Because moss can be propagated vegetatively, mutants whose terminal phenotypes are expressed at a later stage in development can often be maintained at an earlier stage (23).

Some of the same characteristics that make moss well suited for genetic analyses also present limitations. The prolonged and complex, multicellular haploid stage means that for a greater portion of the moss life cycle, lethal mutations cannot be masked by a wild-type allele, as would be possible in a diploid. In addition, genetic analyses have been hampered by the discovery that certain mutant strains possess alleles that are dominant or incompletely dominant to their respective wild-type alleles (5).

One of the greatest limitations to genetic studies in moss has been the fact that many of the developmentally abnormal mutants are sterile (32). Somatic hybridization following protoplast fusion has been used to circumvent the sterility that is a consequence of mutants whose development is blocked prior to gamete production (5, 26, 32, 33). In this procedure, protoplasts made from filamentous tissue of one (mutant) strain are mixed with protoplasts prepared from another strain. Cellular and nuclear fusion are induced using chemical (5, 26, 32, 33) or electrical (66) methods. The protoplasts then regenerate into filamentous gametophytes under conditions that select for hybrids. While karyotypic analyses have not been reported, segregation ratios of progeny resulting from self-fertilization of such hybrids suggest that most of the hybrids are diploid (17, 32). Possibly due to the change in gene dosage, the morphologies of the somatic hybrids are variable and are unlike the morphologies of the parental haploid strains.

Somatic hybrids have generally been found to have low rates of reversion to the wild-type phenotype, and the hybrid phenotypes are stable after numerous subcultures. These hybrids can be used for more detailed genetic analysis if sporophyte production occurs. Since many of the hybrids have been found to produce fewer sporophytes than wild type or to be sterile, they have been used most often in complementation analyses (26, 33). For example, using somatic hybridization, it has been shown that one group of *ove* mutants occurs relatively frequently, is recessive to wild type, and is associated with at least three complementation groups (26).

IMMEDIATE GOALS FOR MOSS RESEARCH

Hormone-induced bud assembly in moss provides a unique opportunity to study differentiation, morphogenesis, and organogenesis in vivo in a relatively simple organism. The advantages of using moss for these studies include the following: 1. The process of bud assembly is well defined and normal development can be manipulated experimentally (13). 2. The moss has only a few cell types, so it is possible to isolate the events in bud assembly from other developmental events. 3. Because bud assembly occurs progressively along a filament, it should be possible to determine when and where molecules involved in this process are important. 4. The multicellular gametophyte enables the study of more complex development than is possible with most other haploid organisms. 5. The cells involved are large and accessible and, as a result, are amenable to testing the in vivo function of molecules using microinjection technology. 6. Using liquid cultures of *Funaria* that are enriched for specific stages of development (38), it is possible to produce stage-specific tissue for biochemical, cellular, and molecular biological analyses. 7. Moss cells have been successfully transformed (55; K Schumaker, unpublished results); this should provide an additional approach with which to study gene function during development. For example, this can be done in transformation experiments in which the levels of a gene product are altered in the wild type. Alternatively, a mutant phenotype may suggest candidate molecules that can rescue the mutant; transformation will provide a functional assay for such molecules. Successful transformations have used polyethylene glycol-mediated uptake of DNA into protoplasts. Protoplast regeneration takes approximately one week and always leads to the production of chloronema filaments first, indicating that the developmental program appears to reset at this stage. As a result, it will be necessary to isolate stage-specific promoters in order to study expression and function of cloned genes at particular stages during development. Preliminary experiments examining the differential patterns of mRNA expression during development suggest that isolation of stage-specific promoters will be feasible (K Schumaker & M Dietrich, unpublished data). 8. A recent report describes homologous recombination in *Physcomitrella* (54). The authors have provided evidence for tandem insertions at several independent, targeted sites; this will be extremely useful for producing null mutations. Further refinements of this technique, resulting in gene replacement, will make this an invaluable tool for moss research.

From whole plant, cellular, and subcellular studies over the past 20 years, we know that bud formation in moss involves in sequence: differentiation of a chloronema tip cell into a caulonema tip cell, production of caulonema filament cells, caulonema initial cell formation, and cytokinin-induced assembly of the bud from the caulonema initial cell. In order to understand the mechanisms underlying these processes, two overriding goals will need to be met. It is critical to experimentally isolate the specific process under study. Studies of bud formation that do not distinguish initial cell formation and bud assembly will not permit an understanding of the underlying mechanisms of either process and will produce results that are difficult to interpret. Of equal importance is the need to standardize the protocols used for culturing moss tissue. This will allow research using the most appropriate tissue for the specific developmental event and permit meaningful evaluation of the data among laboratories.

Once these major goals have been met, where should the focus of research with moss be directed? While we anticipate that significant progress will be made in answering many of the questions outlined above, we believe that the identification and characterization of additional developmental mutants should be an immediate focus of research. Generation of these mutants will ultimately allow identification of novel molecules critical for the processes underlying bud assembly.

Generation of Additional Developmental Mutants

CONDITIONAL MUTANTS   It is critical that a strategy be developed to identify mutants in which expression of the mutation can be controlled, for example, by isolation of developmentally abnormal mutants that are temperature-, pH-, or calcium-sensitive (45, 47). Conditional mutants would be an extremely powerful tool for the study of protein function in vivo, as they provide a reversible mechanism with which to lower the level of a specific gene product at any stage during growth simply by changing the growth conditions. Isolation of conditional mutants will be beneficial because many of the moss developmental mutants are sterile. In addition, such an approach might enable the identification and characterization of mutations that would otherwise be lethal. If the mutations are extreme and if moss genomes do not include redundant or alternate gene products to assume the function of the altered gene, nonconditional mutations in genes essential for development might otherwise never be identified.

Since there is currently no general method to predict which mutations in a protein will give rise to a conditional phenotype, mutants must be generated by random mutagenesis followed by screening for conditional phenotypes. This approach should work well with moss due to its relatively simple morphology, the ease of screening tissue derived from a large number of spores or protoplasts, and its short generation time in culture. Isolation of temperature-sensitive mutations should be especially feasible in *Funaria* based on the relatively wide range of temperatures in which normal wild-type growth occurs (10°C to greater than 25°C). There is one report in the literature describing the isolation of a temperature-sensitive developmental mutant in *Physcomitrella* (29). This mutant, *ove* 409, produces leafy gametophytes with wild-type morphology when maintained at low temperature. As the temperature is increased, the phenotype changes to that of a bud-overproducing mutant, suggesting that the mutant may be temperature-sensitive for cytokinin production. However, further analysis of media from wild type, a nontemperature-sensitive *ove* mutant, and *ove* 409 showed that all contain more cytokinin at elevated temperatures. The authors suggested that although the mutant is temperature-sensitive for bud production, the allele might not encode a temperature-sensitive gene product. Rather, the *ove* phenotype at higher temperatures might be due to the production of cytokinin at levels high enough to increase bud formation. Effort in isolating conditional mutants will be fundamental to the identification of critical defects in essential functions that, at the same time, allow maintenance of the mutant strain.

INSERTIONAL MUTANTS   Insertional mutagenesis has become an important approach for the identification of developmental mutants in plants (6, 27). This general strategy allows insertion of known sequences at random sites in the genome to disrupt the function of unknown genes and create a molecular tag for subsequent gene isolation. Isolation, cloning, sequencing, and further analysis of the developmental gene affected are then feasible. Success with this approach in moss will require the identification of endogenous transposable elements or introduction of foreign elements through transformation.

CONCLUDING REMARKS

The goals of this review were to describe what is known about early events in bud assembly and to evaluate the advantages and limitations of studying moss to understand hormone-induced changes that take place during development. It is clear that moss has great potential as an experimental organism for understanding the underlying physiological processes using biochemical, cellular, and molecular biological approaches. We have identified several objectives that, once achieved, will enable moss to reach its full potential as a model for developmental studies. Some of these objectives, such as defining the stages of development under study and standardizing the methods of tissue growth, should be relatively easy to meet. Others, such as developing the tools required for routine genetic analyses, will require a more significant commitment of time and resources. The effort will be worthwhile, since many of the developmental changes taking place during cytokinin-induced bud assembly in moss appear to be common to other more complex organisms. Therefore, results from studies with moss should provide important information about mechanisms underlying hormone-induced development in general.

ACKNOWLEDGMENTS

The authors gratefully acknowledge support for their work from the Department of Energy (Energy Biosciences Program). We thank Rachel Pfister and Drs. Robert Dietrich, Whitney Hable, Bruce McClure, Steve Smith, Frans Tax, and Mary Alice Webb for helpful discussions and comments on the manuscript.

Literature Cited

1. Amon A. 1996. Mother and daughter are doing fine: Asymmetric cell division in yeast. *Cell* 84:651–54
2. Ashton NW, Cove DJ. 1977. The isolation and preliminary characterisation of auxotrophic and analogue resistant mutants of the moss, *Physcomitrella patens. Mol. Gen. Genet.* 154:87–95
3. Ashton NW, Cove DJ. 1990. Mutants as tools for the analytical dissection of cell differentiation in *Physcomitrella patens* gametophytes. In *Bryophyte Development: Physiology and Biochemistry*, ed. RN Chopra, pp. 17–31. Boca Raton, FL: CRC Press
4. Ashton NW, Cove DJ, Featherstone DR. 1979. The isolation and physiological analysis of mutants of the moss *Physcomitrella patens*, which over-produce gametophores. *Planta* 144:437–42
5. Ashton NW, Grimsley NH, Cove DJ. 1979. Analysis of gametophytic development in the moss, *Physcomitrella patens*, using auxin and cytokinin resistant mutants. *Planta* 144:427–35
6. Bancroft I, Bhatt AM, Sjodin C, Scofield S, Jones JDG, Dean C. 1992. Development of an efficient two-element transposon tagging system in *Arabidopsis thaliana. Mol. Gen. Genet.* 233:449–61
7. Beutelmann P, Bauer L. 1977. Purification and identification of a cytokinin from moss callus cells. *Planta* 133:215–17
8. Bopp M. 1961. Morphogenese der laubmoose. *Biol. Rev.* 36:237–80
9. Bopp M. 1983. Developmental physiology of bryophytes. See Ref. 59a, pp. 276–324
10. Bopp M. 1984. The hormonal regulation of protonema development in mosses. II. The first steps of cytokinin action. *Z. Pflanzenphysiol. Bd.* 113:435–44
11. Bopp M, Jacob HJ. 1986. Cytokinin effect on branching and bud formation in *Funaria. Planta* 169:462–64
12. Brandes H. 1973. Gametophyte development in ferns and bryophytes. *Annu. Rev. Plant Physiol.* 24:115–28
13. Brandes H, Kende H. 1968. Studies on cytokinin-controlled bud formation in moss protonemata. *Plant Physiol.* 43: 827–37
14. Catterall WA. 1995. Structure and function of voltage-gated ion channels. *Annu. Rev. Biochem.* 64:493–531
15. Conrad PA, Hepler PK. 1988. The effect of 1,4-dihydropyridines on the initiation and development of gametophore buds in the moss *Funaria. Plant Physiol.* 86:684–87
16. Conrad PA, Steucek GL, Hepler PK. 1986. Bud formation in *Funaria*: organelle redistribution following cytokinin treatment. *Protoplasma* 131:211–23
17. Cove DJ. 1983. Genetics of Bryophyta. See Ref. 59a, pp. 222–31
18. Cove DJ, Ashton NW. 1988. Growth regulation and development in *Physcomitrella patens*: an insight into growth regulation and development of bryophytes. *Bot. J. Linn. Soc.* 98:247–52
19. Cove DJ, Schild A, Ashton NW, Hartmann E. 1978. Genetic and physiological studies of the effect of light on the development of the moss *Physcomitrella patens. Photochem. Photobiol.* 27:249–54
20. Di Laurenzio L, Wysocka-Diller J, Malamy JE, Pysh L, Helariutta Y, et al. 1996. The *SCARECROW* gene regulates an asymmetric cell division that is essential for generating the radial organization of the Arabidopsis root. *Cell* 86:423–33
21. Doe CQ. 1996. Spindle orientation and asymmetric localization in Drosophila: both inscuteable? *Cell* 86:695–97
22. Dolan L. 1997. *SCARECROW*: specifying asymmetric cell divisions throughout development. *Trends Plant Sci.* 2:1–2
23. Doonan JH. 1991. The cytoskeleton and moss morphogenesis. In *The Cytoskeletal Basis of Plant Growth and Form*, ed. CW Lloyd, pp. 289–301. London: Academic
24. Doonan JH, Cove DJ, Corke FMK, Lloyd CW. 1987. Pre-prophase band of microtubules, absent from tip-growing moss filaments, arises in leafy shoots during transition to intercalary growth. *Cell Motil. Cytoskelet.* 7:138–53
25. Eberle J, Wang TL, Cook S, Wells B, Weiler EW. 1987. Immunoassay and ultrastructural localization of isopentenyladenine and related cytokinins using monoclonal antibodies. *Planta* 172:289–97
26. Featherstone DR, Cove DJ, Ashton NW. 1990. Genetic analysis by somatic hybridization of cytokinin overproducing developmental mutants of the moss, *Physcomitrella patens. Mol. Gen. Genet.* 222:217–24
27. Feldmann K. 1991. T-DNA insertion mutagenesis in *Arabidopsis*: Mutational spectrum. *Plant J.* 1:71–82

28. Franklin-Tong VE, Drøbak BK, Allan AC, Watkins PAC, Trewavas AJ. 1996. Growth of pollen tubes of *Papaver rhoeas* is regulated by a slow-moving calcium wave propagated by inositol 1,4,5-trisphosphate. *Plant Cell* 8:1305–21

29. Futers TS, Wang TL, Cove DJ. 1986. Characterisation of a temperature-sensitive gametophore over-producing mutant of the moss, *Physcomitrella patens. Mol. Gen. Genet.* 203:529–32

30. Gerhäuser D, Bopp M. 1990. Cytokinin oxidases in mosses. 1. Metabolism of kinetin and benzyladenine *in vivo. J. Plant Physiol.* 135:680–85

31. Gönczy P, Hyman AA. 1996. Cortical domains and the mechanisms of asymmetric cell division. *Trends Cell Biol.* 6:382–87

32. Grimsley NH, Ashton NW, Cove DJ. 1977. The production of somatic hybrids by protoplast fusion in the moss, *Physcomitrella patens. Mol. Gen. Genet.* 154:97–100

33. Grimsley NH, Ashton NW, Cove DJ. 1977. Complementation analysis of auxotrophic mutants of the moss, *Physcomitrella patens*, using protoplast fusion. *Mol. Gen. Genet.* 155:103–7

34. Guo S, Kemphues KJ. 1996. Molecular genetics of asymmetric cleavage in the early *Caenorhabditis elegans* embryo. *Curr. Opin. Genet. Dev.* 6:408–15

35. Harold FM. 1990. To shape a cell: an inquiry into the causes of morphogenesis of microorganisms. *Microbiol. Rev.* 54:381–431

36. Harold RL, Harold FM. 1986. Ionophores and cytochalasins modulate branching in *Achyla bisexualis. J. Gen. Microbiol.* 132:213–19

37. Horvitz HR, Herskowitz I. 1992. Mechanisms of asymmetric cell division: Two Bs or not two Bs, that is the question. *Cell* 68:237–55

38. Johri MM. 1974. Differentiation of caulonema cells by auxins in suspension cultures of *Funaria hygrometrica*. In *Plant Growth Substances*, ed. NG Kaigi, pp. 925–33. Tokyo: Hirokawa Publishing

39. Malhó R, Trewavas AJ. 1996. Localized apical increases of cytosolic free calcium control pollen tube orientation. *Plant Cell* 8:1935–49

40. Markmann-Mulisch U, Bopp M. 1987. The hormonal regulation of protonema development in mosses. IV. The role of Ca^{2+} as cytokinin effector. *J. Plant. Physiol.* 129:155–68

41. McCauley MM, Hepler PK. 1990. Visualization of the endoplasmic reticulum in living buds and branches of the moss *Funaria hygrometrica* by confocal laser scanning microscopy. *Development* 109:753–64

42. McCauley MM, Hepler PK. 1992. Cortical ultrastructure of freeze-substituted protonemata of the moss *Funaria hygrometrica. Protoplasma* 169:168–78

43. Miller DD, Callaham DA, Gross DJ, Hepler PK. 1992. Free Ca^{2+} gradient in growing pollen tubes of *Lilium. J. Cell Sci.* 101:7–12

44. Nebel BJ, Naylor AW. 1968. Light, temperature and carbohydrate requirements for shoot-bud initiation from protonemata in the moss *Physcomitrium turbinatum. Am. J. Bot.* 55:38–44

45. Ohya Y, Miyamoto S, Oshumi Y, Anraku Y. 1986. Calcium-sensitive *cls4* mutant of *Saccharomyces cerevisiae* with a defect in bud formation. *J. Bacteriol.* 165:28–33

46. Pierson ES, Miller DD, Callaham DA, van Aken J, Hackett G, Hepler PK. 1996. Tip-localized calcium entry fluctuates during pollen tube growth. *Dev. Biol.* 174:160–73

47. Pringle JR. 1975. Induction, selection, and experimental uses of temperature-sensitive and other conditional mutants of yeast. *Methods Cell Biol.* 12:233–72

48. Reski R, Faust M, Wang X-H, Wehe M, Abel WO. 1994. Genome analysis of the moss *Physcomitrella patens* (Hedw.) B.S.G. *Mol. Gen. Genet.* 224:352–59

49. Rothfield LI, Zhao C-R. 1996. How do bacteria decide where to divide? *Cell* 84:183–86

50. Saunders MJ. 1986. Cytokinin activation and redistribution of plasma-membrane ion channels in *Funaria*. A vibrating-microelectrode and cytoskeleton-inhibitor study. *Planta* 167:402–9

51. Saunders MJ, Hepler PK. 1981. Localization of membrane-associated calcium following cytokinin treatment in *Funaria* using chlorotetracycline. *Planta* 152:272–81

52. Saunders MJ, Hepler PK. 1982. Calcium ionophore A23187 stimulates cytokinin-like mitosis in *Funaria. Science* 217:943–45

53. Saunders MJ, Hepler PK. 1983. Calcium antagonists and calmodulin inhibitors block cytokinin-induced bud formation in *Funaria. Dev. Biol.* 99:41–49

54. Schaefer D, Zryd J-P, Knight CD, Cove DJ. 1991. Stable transformation of the moss *Physcomitrella patens. Mol. Gen. Genet.* 226:418–24

55. Schaefer DG, Zryd J-P. 1997. Efficient

gene targeting in the moss *Physcomitrella patens*. *Plant J.* 11:1195–206

56. Schumaker KS, Dietrich MA. 1997. Programmed changes in form during moss development. *Plant Cell* 9:1099–107

57. Schumaker KS, Gizinski MJ. 1993. Cytokinin stimulates dihydropyridine-sensitive calcium uptake in moss protoplasts. *Proc. Natl. Acad. Sci. USA* 90: 10937–41

58. Schumaker KS, Gizinski MJ. 1995. 1,4-dihydropyridine binding sites in moss plasma membranes: properties of receptors for a calcium channel antagonist. *J. Biol. Chem.* 270:23461–67

59. Schumaker KS, Gizinski MJ. 1996. G proteins regulate dihydropyridine binding to moss plasma membranes. *J. Biol. Chem.* 271:21292–96

59a. Schuster RM, ed. 1983. *New Manual of Bryology*. Nichinan, Miyazaki: Hattori Bot. Lab.

60. Simon PE, Naef JB. 1981. Light dependency of the cytokinin-induced bud initiation in protonema of the moss *Funaria hygrometrica*. *Physiol. Plant.* 53:13–18

61. Sussman MR, Kende H. 1977. The synthesis and biological properties of 8-azi-do-N^6-benzyladenine, a potential photoaffinity reagent for cytokinin. *Planta* 137:91–96

62. Wang TL, Beutelmann P, Cove DJ. 1981. Cytokinin biosynthesis in mutants of the moss *Physcomitrella patens*. *Plant Physiol.* 68:739–44

63. Wang TL, Cove DJ, Beutelmann P, Hartmann E. 1980. Isopentenyladenine from mutants of the moss, *Physcomitrella patens*. *Phytochemistry* 19:1103–5

64. Wang TL, Futers TS, McGeary F, Cove DJ. 1984. Moss mutants and the analysis of cytokinin metabolism. In *The Biosynthesis and Metabolism of Plant Hormones*, ed. A Crozier, JR Hillman, pp. 135–64. Cambridge: Cambridge Univ. Press

65. Wang TL, Horgan R, Cove D. 1981. Cytokinins from the moss *Physcomitrella patens*. *Plant Physiol.* 68:735–38

66. Watts JW, Doonan JH, Cove DJ, King JM. 1985. Production of somatic hybrids of moss by electrofusion. *Mol. Gen. Genet.* 199:349–51

67. Whitaker BD, Kende H. 1974. Bud formation in *Funaria hygrometrica*: A comparison of the activities of three cytokinins with their ribosides. *Planta* 121:93–96

Annu. Rev. Plant Physiol. Plant Mol. Biol. 1998. 49:525–55

EVOLUTION OF LIGHT-REGULATED PLANT PROMOTERS

Gerardo Argüello-Astorga and Luis Herrera-Estrella

Departamento de Ingeniería Genética de Plantas, Centro de Investigación y de Estudios Avanzados del IPN, Apartado Postal 629, 36500 Irapuato, Guanajuato, México

KEY WORDS: gene evolution, plant evolution, plastid-nuclear interactions, photoreceptors

ABSTRACT

In this review, we address the phylogenetic and structural relationships between light-responsive promoter regions from a range of plant genes, that could explain both their common dependence on specific photoreceptor-associated transduction pathways and their functional versatility. The well-known multipartite light-responsive elements (LREs) of flowering plants share sequences very similar to motifs in the promoters of orthologous genes from conifers, ferns, and mosses, whose genes are expressed in absence of light. Therefore, composite LREs have apparently evolved from *cis*-regulatory units involved in other promoter functions, a notion with significant implications to our understanding of the structural and functional organization of angiosperm LREs.

CONTENTS

1040-2519/98/0601-0525$08.00

INTRODUCTION

Photosynthesis is an ancient biochemical process that probably evolved 3500 million years ago (111). The nuclear genome of higher plants encodes proteins homologous to cyanobacterial proteins of photosynthesis; because these bacteria are the living descendants of the original photosynthetic organisms, plant genes encoding such proteins are among the oldest genes of eukaryotes.

To date, molecular evolution studies have concentrated on the coding sequences of gene families. The evolution of regulatory sequences, which determine where, when, and the level at which genes are transcribed, has been largely neglected. In the case of the photosynthesis-associated nuclear genes (PhANGs) from higher plants, interesting evolutionary aspects of the molecular mechanisms by which transcription is activated by light receptors (e.g. phytochrome) could be addressed through the comparative analysis of promoter sequences. For instance, why does light profoundly affect transcription of PhANGs in monocotyledonous and dicotyledonous plants, while PhANG promoters in conifers, ferns, and mosses are either light insensitive or, at most, weakly photoresponsive (4, 71, 89, 99, 129). The systematic comparisons of angiosperm and nonflowering plant PhANGs promoter sequences provide a unique opportunity to explore how a new regulatory function, light responsiveness, was incorporated into the promoters of a wide range of genes whose expression is coordinately regulated.

Besides its obvious relevance for evolutionary studies, a comparative analysis of the structure of photoregulated promoters can be useful to address other important issues in plant gene expression. Comparative analysis of PhANG upstream sequences may contribute to reduce the apparent diversity of light-responsive elements (LREs) by revealing concealed phylogenetic and structural relationships between dissimilar promoter regions with analogous functions. Other important issues concerning the composition and functional organization of LREs in plant genes could also be uncovered by their analysis from an evolutionary perspective.

The purpose of this review is to address the phylogenetic and structural relationships between light-responsive promoter regions from orthologous and paralogous plant genes. We review LREs from a wide range of genes, explore their common dependence on specific phototransduction pathways, and analyze correlations between the composition of multipartite LREs and their overall functional properties.

LIGHT REGULATION OF GENE TRANSCRIPTION: A BRIEF OVERVIEW

Control of Gene Expression by Photoreceptors

The responses of plants to light are complex: seed germination, de-etiolation of seedlings, chloroplast development, stem growth, pigment biosynthesis, flowering, and senescence (67). Most of these responses require changes in both chloroplast and nuclear gene expression, which are mediated by three major classes of photoreceptors: phytochromes, blue/UV-A light receptors, and UV-B light receptor(s) (2, 61, 100). Light-regulated genes may respond to more than one photoreceptor, thus allowing a finely tuned control of their expression when stimulated by light (123).

The best characterized light receptor is phytochrome (PHY), which exists in two photochemically interconvertible forms, Pr and Pfr, and is encoded by a small family of genes in angiosperms (42, 99, 100). PHY controls the expression of diverse genes at the transcriptional, posttranscriptional, and translational levels (43, 112, 123). Gene expression is regulated by at least three different signal transduction pathways activated by PHY: one depends on cyclic GMP (cGMP), which regulates genes such as those involved in anthocyanin biosynthesis in some species; a second pathway depends on calcium/calmodulin, which activates a subset of chloroplast-associated nuclear genes (17, 85); and a third signal pathway, which requires both calcium and cGMP, activates a subset of genes necessary for chloroplast development (e.g. the gene encoding ferredoxin NADP+ oxidoreductase) (18), and represses transcription of *phy-A* and the gene encoding asparagine synthetase (91).

Light-Responsive Promoters

In flowering plants, light regulation of nuclear genes occurs mainly at the transcriptional level, as demonstrated by nuclear run-on transcription assays and promoter-reporter gene fusions (112, 122, 126). The most extensively studied light-responsive genes are those encoding the small subunit of ribulose-1,5-bisphosphate carboxylase/oxygenase (*rbcS*) and the chlorophyll *a/b* binding proteins (*Lhc*, formerly called *Cab*), both of which are a paradigm for PhANGs controlled by the calcium/calmodulin phototransduction pathway (17). Chalcone synthase (*chs*) genes, on the other hand, have been the model for genes involved in the anthocyanin biosynthetic pathway in some dicots and, hence, for genes targeted by the phytochrome signal pathway dependent on cGMP (17, 85). More recently, PHY-A genes have become a model for the diverse genes whose transcription is down-regulated by light and which are controlled by the third phytochrome signaling pathway, dependent on calcium and cGMP (91).

A plethora of *cis*-acting elements and protein factors presumably involved in transcriptional light responses have been identified (13, 16, 122, 126); however, conclusive evidence for an essential role in light responsiveness has been obtained for only a few of them (5, 41, 66, 136). Several general conclusions are possible: 1. No single conserved sequence element is found in all light-responsive promoters. 2. The smallest native promoter sequences, sufficient to confer light inducibility on heterologous minimal promoters, are multipartite regulatory elements that contain different combinations of *cis*-acting sequences. 3. Even the smallest known photoresponsive promoter regions, when examined in a heterologous context, display fairly complex responses, often retaining the dependence on light wavelength, developmental stage of chloroplasts, and tissue specificity of the native promoter. 4. Numerous protein factors bind sequences within photoregulated promoters, although their actual contribution to light-activated transcription remains uncertain (13, 122, 126, 128).

The Phylogenetic-Structural Sequence Analysis

A commonly used approach to identify regulatory *cis*-elements is to search for conserved DNA motifs within the promoter region of orthologous genes. Such analyses have been successful in identifying *cis*-acting elements involved in the light responsiveness of PhANGs, such as the G-box and I-box elements from *rbcS* genes (48) and the GATA motifs of *Lhcb1* genes (44). However, regulatory elements showing sequence degeneracy are not easily recognized in comparative analyses.

Sequence heterogeneity of regulatory elements may be functionally overcome if multiprotein regulatory complexes facilitate binding to imperfect target sites (86, 135). Because conventional computer programs for DNA sequence comparisons can fail to detect evolutionarily related but structurally variable promoter regions with analogous functions, alternative approaches have been developed, such as the "phylogenetic-structural method" of sequence analysis (8, 10). This method is based on the search of "homologous" (rather than "similar") DNA sequences of a functionally characterized promoter. Two sequences are homologous when they share common ancestry, regardless of the degree of similarity between them (37). In this sense, the G-box elements of *rbcS*, *chs*, and *Lhcb1* genes are not homologous but only similar because they have different evolutionary origins, whereas the *rbcS* I-G unit of solanaceous species is probably homologous to the *rbcS* X-Y promoter region of *Lemna*, an aquatic monocotyledonous plant, despite their rather low overall sequence similarities (10).

The individual elements found within a multipartite *cis*-regulatory region are termed phylogenetic footprints (PFs); they share high conservation over a segment of 6 contiguous base pairs in alignments of orthologous upstream sequences and represent potential binding sites for transcription factors (53). A cluster of PFs whose arrangement (combination, spacing, and relative orientation) is conserved through a phylogenetic series of homologous promoter regions is termed a conserved modular arrangement (CMA). If a given CMA consistently correlates with experimentally defined light-responsive promoter regions, then it is called an LRE-associated CMA. In this method, sequence comparisons are made in a phylogenetically ordered fashion, because the overall structure of multipartite regulatory units tends to diverge in evolution. Thus, homologous genes from species belonging to the same plant taxon should be compared first. Comparisons with other taxons follow. This procedure allows us to discern how an ancestral multipartite regulatory module has changed in an evolutionary context and can establish phylogenetic relationships between promoter regions that are apparently unrelated by structural criteria. An example of this analysis applied to *Lhcb1* light-responsive regions is presented in Figure 1a.

EVOLUTION OF LREs IN *Lhcb* PROMOTERS

The Proximal LRE of Lhcb 1 Genes

Three types of chlorophyll *a/b* binding proteins are found in the major light-harvesting complex of photosystem II, encoded by *Lhcb1*, *Lhcb2*, and *Lhcb3* (52, 60). Most *Lhcb* promoters that have been functionally analyzed are from the *Lhcb1* gene family and typically lack introns. In *Lhcb1* genes from dicotyledons, an LRE is located in the proximal promoter region of these genes. This LRE is characterized by three conserved GATA motifs spaced by 2 and 6 bp, respectively, which are located between the CCAAT and TATA boxes (44, 87). In Arabidopsis *Lhcb1*1* (formerly *Cab2*), the −111 to −33 region is sufficient to confer a pattern of expression dependent on both PHY and a circadian clock, to a heterologous minimal promoter (6). A nuclear factor that specifically interacts with the triple GATA repeats found in this LRE was identified and named

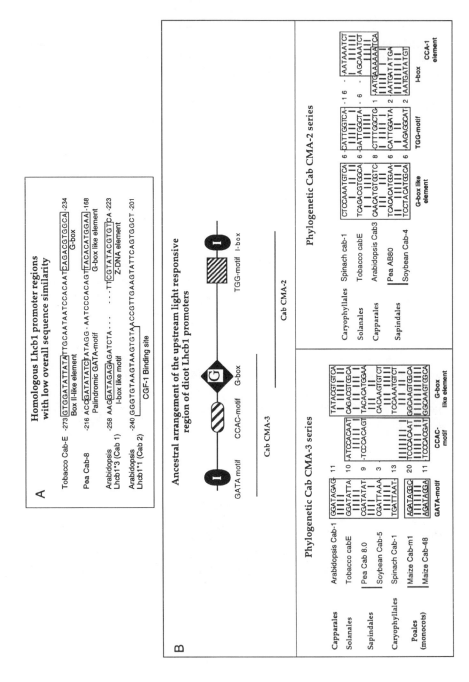

A

Homologous Lhcb1 promoter regions with low overall sequence similarity

Tobacco Cab-E -273 GTGGAATATTATA TTGCAATAATCCACAAT CAGACGTGGCA -234
                                    Box II-like element   G-box

Pea Cab-8 -216 ACCGATATATC ATAGG - AATCCCACAGT TACACATGGAA -168
                        Palindromic GATA-motif   G-box like element

Arabidopsis -258 AACGATAGAG AGATCTA - - - · · · TT CGTATACGTGT CA -223
Lhcb1*3 (Cab 1)   I-box like motif   Z-DNA element

Arabidopsis -240 GGGTGTAAGTAAGTGTA ACCGTTGAAGTATTCAGTGGCT -201
Lhcb1*1 (Cab 2)   CGF-1 Binding site

B   **Ancestral arrangement of the upstream light responsive region of dicot Lhcb1 promoters**

GATA motif   CCAC-motif   G-box   TGG-motif I-box

Cab CMA-3   Cab CMA-2

Phylogenetic Cab CMA-3 series

		GATA-motif	CCAC-motif	G-box like element
Capparales	Arabidopsis Cab-1	GGATAGAG 11		
Solanales	Tobacco cabE	GGATATTA 10	ATCCACAAT TATACGTGTCA	
	Pea Cab 8.0	CGATATAT 9	-TCCCACAGT CAGACGTGGCA	
Sapindales	Soybean Cab-5	CGATTAAA 3	TACACATGGAA	
	Spinach Cab-1	TGATTAAT 13	CACAAGTGTCT	
Caryophyllales			TCCAAATGTCT	
Poales (monocots)	Maize Cab-m1	AGATAGGC 20	TCCCACAAT GGCAAGTGGCA	
	Maize Cab-48	AGATAGGA 11	TCCACGAGT GGCAAGTGGCA	

Phylogenetic Cab CMA-2 series

		G-box like element	TGG-motif	I-box	CCA-1 element
Caryophyllales	Spinach cab-1	CTCCAAATGTCA 6	CATTGGTCA -1 6	AATAAATCT	
Solanales	Tobacco cabE	TCAGACGTGGCA 6	GATTGGCTA -6	AGCAAATCT	
Capparales	Arabidopsis Cab3	CAACATGTGTC 8	CTTTGGCTG 1	AATGAAAAATCA	
Sapindales	Pea AB80	TCACACATGGAA 6	CATTGGATA 2	AATGATATGA	
	Soybean Cab-4	TCCTACATGGCA 6	AAGAGGCAT 2	AATGATATGT	

CGF-1 (6). If the three GATA motifs are altered by site-directed mutagenesis, the interaction of CGF-1 is disrupted and responsiveness to PHY and part of the light-induced circadian oscillation of *Cab2* expression are comprised (5). Mutational analysis of the Arabidopsis *Lhcb1*3* (*Cab1*) gene promoter identified additional *cis*-acting motifs, that could be involved in the functionality of this LRE. Mutations in a 27-bp region upstream the CCAAT box, which is bound by the CA-1 factor (75a), drastically reduced the overall promoter activity and also eliminated PHY responsiveness (68). More recently, a Myb-related transcription factor that specifically interacts with a conserved sequence motif [AA(C/A)AAATCT] within the CA-1 region was cloned (131). Transgenic *Arabidopsis* expressing an antisense mRNA for this factor (called CCA-1) showed reduced PHY induction of the endogenous *Lhcb1*3* gene, whereas expression of other PHY-regulated genes, such as *rbcS*, was not affected; thus CCA-1 has a specific role in *Lhcb1* photoregulation (131).

Evolution of the Proximal Region of Lhcb1 Promoters

At least five PFs are found within the first 110–130 bp of the *Lhcb1* promoter sequences from dicotyledons. All these PFs bind defined proteins. The most distal PF is the CUF-1 binding site, a G-box-related sequence which functions as a general activating element (6). The second PF is the CCA-1 binding site (131). The third is the CCAAT-box, which is part of the binding site of Tac, a protein factor proposed to be involved in the regulation of *Arabidopsis cab2* by the circadian clock (22). The fourth and fifth PFs are located in the CGF-1 binding region, comprising the three GATA motifs located upstream to the TATA box. Comparative analyses indicate that these GATA elements are distinct from both the phylogenetic and the structural point of view. 1. GATA I is part of a PF whose consensus is GATAAGR (the I-box motif) (48), whereas GATA II/III form a single PF with a GATANNGATA consensus in dicotyledons. 2. Apparently GATA I is a more ancient element than GATA II/III, inferred from the fact that the former, but not the latter, element is present in *Lhcb* promoters of gymnosperms (11, 71).

Figure 1   (a) *Lhcb1* promoter regions that are homologous but dissimilar. The indicated CGF-1 binding site was defined by Teakle & Kay (121). The Z-DNA element of the *Arabidopsis Lhcb1*3* promoter has been functionally characterized by Ha & An (54) and by Puente et al (98). The illustrated region corresponds to the cabCMA3 of the genes indicated in the figure. (b) Hypothetical arrangement of the upstream LRE of *Lhcb1* promoters in the common ancestor of Dicotyledons. Two putative LREs derived from the ancestor, which have been previously identified as cabCMA-3 and cabCMA-2 (10), are shown. GATA-containing sequences of maize genes that are inverted with respect to dicot homologous motifs are *underlined*. Notice that in some cases, a CCA-1 motif is the evolutionary counterpart of an I-box element.

Within dicots only the CCAAT and GATA motifs are invariably present in the TATA-proximal region. The CUF-1 and CCA-1 binding sites are found in some but not all *Lhcb1* genes. Only one tomato gene (*Lhcb1*5*) contains both CUF-1 and CCA-1 binding sites, four (*Lhcb1*1*, **2*, **3*, and **4*) contain the CUF-1 element but lack the CCA-1 site, and two (*Lhcb1*6* and **7*) lack a discernible CUF-1 site but contain an inverted CCA-1 binding site. Although it is probable that these structural differences determine qualitative differences in the regulation of these paralogous promoters, transcript abundance in tomato leaves is not correlated with the composition of their proximal region (97).

Most of the known *Lhcb1* genes of monocotyledons have promoters clearly distinct from those of dicots, with two exceptions: the *Lemna Lhcb1*1 (AB30)* and the maize *Lhcb1*2 (cab-1)* genes (69, 118), both containing a (CUF-1 site)-(CCAAT-box)-(GATA II)-(I-box) arrangement homologous to that of dicot promoters (10). This arrangement in *Lhcb1* promoters may either predate the divergence of monocots and dicots or result from convergent evolution. Support for an ancient origin comes from an orthologous gene of a conifer (*Pinus contorta*) that has a promoter (CCAAT-box)-(I-box)-(TATA-box) arrangement that is similar to that found in angiosperms (see Figure 6). The pine gene lacks the GATA III, CCA-1, and CUF-1 elements, suggesting that these elements probably evolved after the divergence of the lineages leading to modern conifers and angiosperms.

LREs in the Central Region of Lhcb1 Promoters

Lhcb1 genes contain additional LREs, besides those found in the TATA-proximal region. Castresana et al (23) demonstrated by gain-of-function experiments the photoresponsiveness of the −396 to −186 region of the tobacco *Cab-E* gene. A light-responsive region has also been mapped to the −347 to −100 promoter region of the pea *Lhcb1*2 (AB80)* gene. This 247-bp fragment was shown to function as a light-responsive, tissue-specific enhancer in gain-of-function experiments (113). Two nuclear factors binding this regulatory region were identified, one of which (ABF-1) is found only in green tissues (7). By systematic deletion the *AB80* enhancer light-responsive core is located in the −200 to −100 region (Argüello-Astorga & Herrera-Estrella, manuscript in preparation). Evidence for LREs in the −240 to −100 region of the *Arabidopsis Cab2* promoter has been also reported (121).

Evolution of the Central LRE of Lhcb1 Promoters

The comparative analysis of the tobacco *CabE* and pea AB80 upstream LREs, with the corresponding promoter segments from other *Lhcb1* genes from dicots, suggests a structure of this promoter region in the common ancestor of dicotyledons. An extensive ancestral CMA encompassing at least five PFs is inferred (Figure 1*b*). This ancestral organization is still conserved in some genes of leguminous species [e.g. pea *Cab-8* (3)]. The central PF of this CMA is a sequence similar, but not identical, to the G-box from *rbcS* genes. In tobacco *CabE* this G box–like element (G*box) is associated with a Box II–like element corresponding to a PF with a GATA core motif (23). Homologous G box–like and 5′ associated GATA-motif elements (I*box) exist in members of the *Lhcb1* gene family in plants belonging to four orders of dicotyledons. A third PF (YCCACART) is found immediately upstream of the G-box element in several *Lhcb1* genes of legumes and the tobacco *Cab-E* promoter but not in other dicot genes. Interestingly, this PF is also found in the homologous region of two maize *Lhcb1* genes (Figure 1*b*), suggesting a very ancient origin of this PF. A fourth PF with the consensus CATTGGCTA closely precedes a PF encompassing an I-box motif located 15–20 bp downstream of the G-box element. The arrangement of this *Lhcb1* region (cabCMA2) is highly analogous to that of the G-, I-box region of dicot *rbcS* promoters, with a very similar PF associated 5′ to the I-box motif. The resemblance in some cases is so striking that a *Lhcb1* promoter (e.g. pea *AB80*) can display more similarity in this region with the analogous *rbcS* promoter segment (e.g. pea *rbcS 3A*) than with certain homologous *Lhcb1* regions (10). The analogies between the overall structural organization of the I*-G*-I promoter region of *Lhcb1* genes and the I-G-I unit of *rbcS* promoters are intriguing and suggest either a common ancestry or convergent evolution of these regulatory promoter modules.

Delimitation of a LRE in Lhcb2 Promoters

Although comparatively few members of the *Lhcb2* gene family have been studied, the promoter of one, the *Lhcb2*1* (formerly AB19) gene from *Lemna gibba*, has been characterized in great detail. Deletion analysis, linker scanning, and site-directed mutagenesis identified two 10-bp elements in the −134 to −105 region that are critical for light responsiveness. One of them contains the I box–related GATAGGG motif and the other a CCAAT motif. Mutation of the latter element led to high levels of expression in the dark, suggesting that it binds a repressor in the absence of light. Mutations in the I-box motif led to complete loss of red-light responsiveness, suggesting that this region is involved in PHY- mediated light activation of transcription (66). More detailed site-directed mutagenesis of those 10-bp regions allowed the identification of two shorter sequences (REα = AACCAA and REβ = CGGATA) that are critical for *AB19* light regulation (31).

Evolution of Lhcb2 Promoters

Genomic clones of *Lhcb2* genes have been isolated from the moss *Physcomitrella patens* (79), the fern *Polystichum munitum* (96), the conifer *Pinus thunbergii* (71), the monocot *Lemna gibba* (64), and three dicotyledonous plants (pea, *Petunia*, and cotton) (39, 103, 115). These promoters contain a number of PFs, two of which correspond to the *Lemna* REα and REβ elements (Figure 2). Only a few of the identified PFs are present in all species. Among the PFs common to most if not all the known *Lhcb2* promoters are the *Lemna* REα element, generally recognized as a conserved CCAAT box, and a G(G/A)AAATCT motif, which is similar to the CCA-1 binding site of the *Lhcb1* genes.

An interesting observation is that the gene harboring sequences with the highest similarity to the *Lemna* REα and REβ light-responsive elements is that of *P. thunbergii* (see Figure 2), whose promoter directs light-independent gene expression in both its native context (71) and in transgenic angiosperms (70, 138). This paradoxical observation is discussed below. Another relevant observation is that *Lhcb2* genes from dicotyledons lack a REβ element, instead displaying three lineage-specific PFs, two of which include inverted GATA motifs (Figure 2). It would be interesting to determine whether those elements are functionally equivalent to REβ.

Upstream sequences of *Lhcb2* and *Lhcb1* genes are probably derived from a common ancestral promoter, because they display a similar overall structural organization, with a (G Box/CUF-1 site)-(CCA-1 element)-(CCAAT-box)-(I-box)-(TATA box) basic arrangement. Spacing between these elements is, however, very different in the two gene families.

EVOLUTION OF LREs IN *rbcS* PROMOTERS

The Light-Responsive Box II-III Region

The pea *rbcS-3A* gene has been used as a paradigm for the study of light-regulated gene expression in dicotyledons. Analysis of the *rbcS-3A* promoter has uncovered three independent regions that contain an LRE (47). The LRE located in the −166 to −50 region has been characterized in most detail (45). This region includes two elements called Box II and Box III, both of which are binding sites for a nuclear factor named GT-1 (49). This factor binds in vitro to sequences related to the degenerate consensus (A/T) GTGPu (T/A) AA (T/A) (50). A synthetic tetramer of the pea Box II element (GTGTGGTTAATAATG) conferred light-responsive transcriptional activity to the −90 CaMV 35S promoter (75, 98) but was unable to enhance transcription when fused to either the −46 CaMV 35S or the −50 *rbcS-3A* minimal promoters (28). This element appears necessary, but not sufficient, for light-regulated transcriptional activation. *Arabidopsis* and tobacco cDNAs encoding GT-1, a Box II DNA binding proteins, have been cloned and partially characterized (46, 58, 95). Interestingly, GT-1 is closely related to the GATA-binding nuclear factors CGF-1 and IBF-2b and can bind to similar cognate DNA sequences (121). Two other

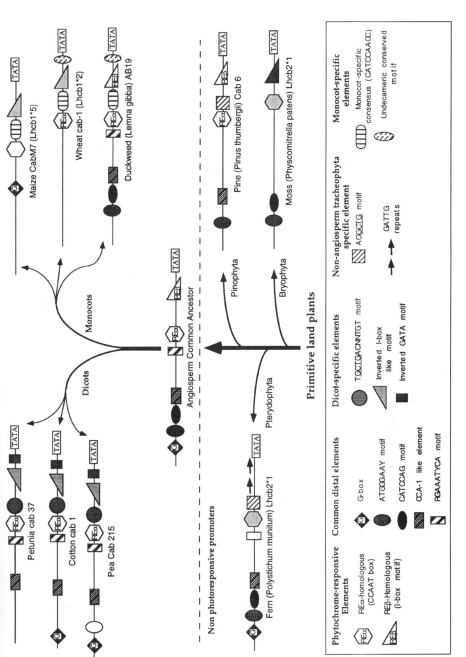

Figure 2 Sequence changes in *Lhcb2* gene promoters from land plants. DNA motifs represented by identical symbols but shaded differently indicate related but not identical sequences.

proteins called 3AF-3 and 3AF-5 probably act together with GT-1 to confer *rbcS-3A* light-regulated expression; they bind inverted GATA motifs located at each end of Box III. Mutations in these GATA sequences severely reduced promoter activity (104).

Evolution of the Box II-III Homologous Regions

Upstream sequences of many angiosperm *rbcS* genes are known. However, only two *rbcS* promoters are known from nonflowering plants, the conifer *Larix laricina* (59) and the homosporous fern *Pteris vittata* (56). Comparison of the *Pteris* and *Larix* proximal promoter regions ($-130/-1$) reveals only four PFs, two of which are inverted I box–related motifs: one overlapping the putative TATA-box of the *Larix* gene, and the second, with the sequence GTTATCC, found ∼70 bp upstream. In the *Larix* promoter, this PF is found as an imperfect direct repeat, flanked by two additional PFs, one of which encompasses a CCAAT motif. The relative position of this CMA in the conifer promoter (i.e. ∼25 bp downstream to the I-box element) is practically identical to that of the Box II–3AF3 region in dicot *rbcS* promoters. Comparison of these CMAs uncovers several interesting characteristics:

1. The Box II element seems to be evolutionarily derived from two separate *Pteris/Larix* PFs (Figure 3). One of them is the most 5' GTTATCC motif in *Larix*, which is homologous to the 3' half of Box II. This relationship is especially clear in *rbcS* promoters of the Brassicaceae (see Figure 6). Therefore, Box II could be a composite element bound in vivo by two protein factors, one of them being a GATA-binding factor.

2. The second repeat of the GTTATCC motif of *Larix* is homologous to a conserved sequence immediately downstream to Box II, which in Arabidopsis and *Brassica* genes is almost identical to the so-called LAMP binding site (51), an inverted I-box element. In nonbrassicaceous dicots the sequences immediately downstream of Box II do not resemble the LAMP motif or the *Pteris/Larix* PF, but their structural relationship is easily recognizable in a phylogenetic series (not shown, but see 82).

3. The *Pteris/Larix* CCAAT-box is apparently homologous to the pea 3AF3 binding site (104). In the conifer promoter the LAMP-like motif and the CCAAT box are close, but in dicots the 3AF3 element is separated from the LAMP-related motif by 10–24 bp. Sequences in this intermediate DNA are functionally relevant, encompassing in pea *rbcS-3A* the Box III and the 3AF5 elements (104) (Figure 3).

The Box II–containing CMA is absent in all known *rbcS* genes of monocotyledons (∼10), and in the orthologous genes of a dicot species, the common

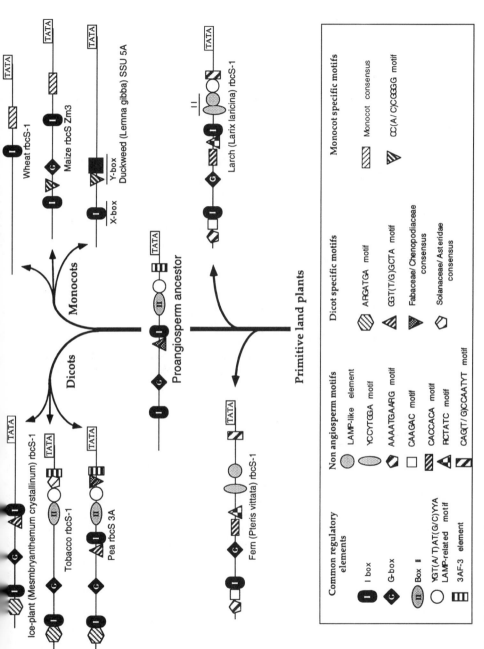

Figure 3   Sequences of *rbcS* gene promoters in vascular plants.

ice plant (Azoideae) (35). Our interpretation is that these groups have lost Box II, which we consider an ancestral feature of *rbcS* genes.

The Light-Responsive I-G Unit

rbcS promoters usually contain two closely associated elements, the I and G boxes. Mutations of either the G-box or the two flanking I-box elements of the Arabidopsis *rbcS-1A* promoter almost abolished its activity (36). A similar drastic drop of transcriptional activity was observed in the spinach *rbcS-1* promoter when the G-box element was mutated (80). In spite of its functional relevance in dicots, deletion of a G-box in the *rbcSZm1* gene of maize had no significant effects on promoter activity. Deletion of the associated I-box motif reduced expression in light 2.5-fold (106).

Lemna rbcS promoters lack a typical G-box element but contain a canonical I-box motif within the so-called X-box element, which is part of the binding site of LRF-1, a light-regulated nuclear factor (21). This I-box motif is included in a 30-bp region of the *Lemna rbcS SSU5B* promoter necessary for PHY regulation (31).

Gain-of-function experiments showed that the region of the *rbcS* promoters encompassing the I- and G-box elements functions as a composite LRE, able to direct a tissue-specific pattern of expression almost identical to the native *rbcS* promoter, including the dependence on the developmental stage of plastids (9). Mutation of either the I-box or the G-box eliminated detectable transcription (9). The I-G region functions as a complex regulatory unit, similar to that of the light-responsive Unit 1 of the parsley *chs* gene (108).

Evolution of the I-G Region

The two I-box elements of the I-G-I CMA display a different sequence consensus and are flanked by different conserved sequence motifs. The more upstream I-box has the consensus GATAAGAT (A/T) and is adjacent 3' to a PF with the consensus (A/T) ARGATGA; the second I-box has the consensus ATGATAAGG and is 5' flanked by a PF with the consensus TGGTGGCTA (Figure 3). The I-box-associated PFs are conserved in genes of plants belonging to five orders of dicotyledons. A homologous I-G-I arrangement is found in some maize *rbcS* genes but not in other genes from monocotyledons. However, CMAs that are probably derived from an ancestral I-G-I structure are present in all these promoters (Figure 3).

The finding that the I-G (-I) arrangement is also found in the homologous promoters of a conifer and a fern (Figure 3) is unexpected because these promoters are presumably light-insensitive, whereas the I-G unit is apparently involved in responses to light signals in angiosperms. Based on the persistence of the I-G-I arrangement since at least the divergence of ferns and seed plants, 395 mya (116), we propose that it has played, and probably still does, one or several important functional roles, different to light-regulation, in the control of *rbcS* gene transcription.

EVOLUTION OF THE LIGHT-REPRESSED PHYTOCHROME A PROMOTER

Functional Organization of phyA Promoters

Phytochrome (PHY) is encoded by small gene families in angiosperms; in *Arabidopsis* the PHY apoprotein is encoded by five genes (99). Each PHY is proposed to have a different physiological role (100). PHYs have been classified into two types. The type I, or "etiolated-tissue" PHY, is most abundant in dark-grown plants, and its Pfr form is rapidly degraded in light by an ubiquitin-mediated proteolytic process. Type II or "green-tissue" PHY are present in much lower levels, but their Pfr form is stable in light (99, 100). The only known type I PHY is that encoded by the *phyA* gene, whose mRNA abundance also decreases in light. This inhibition of *phyA* gene activity is autoregulatory (PHY-dependent) and operates at the transcriptional level (19, 65, 77).

The *phyA* promoters of two monocotyledons, oat and rice, have been functionally characterized. A combination of deletion analysis and linker-scan mutagenesis identified in oat three *cis*-regulatory elements designated PE1 (positive element-1), PE3 (positive element-3), and RE1 (repressor element-1) (19, 20). PE1 and PE3 act synergistically to support maximal expression under derepressed conditions (i.e. low Pfr levels); mutation of either element decreases expression to basal levels. In contrast, mutation of the RE1 element results in maximal transcription under all conditions, suggesting that Pfr represses *phyA* transcription through this negatively acting element (19).

The rice *phyA* promoter has no element similar to PE1. Instead, this promoter contains a triplet of GT-elements that have been shown, in transient expression assays, to be functionally equivalent to the oat PE1 element (32, 99). These elements, related in sequence to the GT-1 binding sites of *rbcS* promoters (65), are bound by a transcription factor, named GT-2 (32, 34).

In addition to *phyA*, several other genes are down-regulated by light (88, 92, 123), including the genes encoding asparagine synthetase (127). Recently a 17-bp element was identified that is both necessary and sufficient for the PHY-mediated repression of the pea asparagine synthetase gene. This sequence is very similar to the *phyA* RE1 element and is the target for a highly conserved PHY-generated repressor, whose activity is regulated by both calcium and cGMP (91).

Evolution of phyA Promoters

Phytochrome genes have been found in phototropic eukaryotic organisms ranging from algae to angiosperms (42, 99), and genomic clones of *phy* genes have been isolated from angiosperms and several lower plants. The latter include species representative of lineages dated to the Silurian and Devonian, and which diverged more than 400 million years ago (27, 110, 116).

The cereal PE3-RE1 region is conserved in the *phyA* promoters of plant species belonging to three different orders of dicotyledons (1, 33, 105), but no obvious counterpart of the monocot GT-boxes region was detected. The conservation of the PE3-RE1 arrangement in dicots suggests a conserved regulatory activity of these elements (33). The *phyA* upstream sequences of bryophytes (the mosses *Physcomitrella* and *Ceratodon*; 107, 124), a lycopodiophyta (*Selaginella*; 57), and a psilotophyta (*Psilotum*; 107) contain a number of PFs. Two clusters of these PFs or CMAs are common to all reported lower plant *phyA* promoters (Figure 4). The CMA nearest the start codon of the mosses and lower vascular plant genes includes sequences similar to the PE3 core element and other flanking sequence elements. The central, most conserved DNA motif of the lower plant CMA is nearly identical to a sequence element in the *Arabidopsis* PE3–RE1 region, immediately upstream of the RE1 motif. No canonical RE1 element is found in lower plant *phy* promoters.

Interestingly, the second, more distal lower plant *phyA*-CMA seems to be the evolutionary counterpart of the region encompassing the GT boxes in monocot *phyA* promoters (Figure 4). Thus, the *cis*-regulatory elements of monocotyledon *phyA* genes seem to have evolved from *cis*-acting sequences already present in orthologous genes of the common ancestor of land plants. What the regulatory function was of such elements in the primitive land plants is an intriguing question that could be partially solved by the functional characterization of the identified *phyA*CMAs in mosses and vascular cryptogams.

Evolution of chs Promoters

In some species, induction of *chs* gene expression by light is required for flavonoid accumulation, which provides a protective shield against potentially harmful UV irradiation (55).

The promoters of parsley and mustard *chs* genes have been functionally analyzed, and a 50-bp light-responsive conserved region, named light-regulatory unit 1 (LRU1), has been studied in great detail (41, 62, 63). LRU1 is sufficient to confer light- responsiveness to heterologous minimal promoters (63, 102, 132) and consists of at least two distinct *cis*-acting elements, ACEchs and MREchs (formerly Box II and Box I, respectively) (108, 109). ACEchs contains a G-box element that interacts in vivo and in vitro with bZIP regulatory factors (40, 132). MREchs was originally identified as a 17-bp in vivo DNA footprint (109) with a conserved sequence motif called the H-box core (78), which is recognized by PcMYB1, a Myb-related transcription factor of parsley (41). Recently, it was shown that *chs* LRU1 activity is controlled by PHY through the cGMP-dependent transduction pathway (136). Comparative analysis of

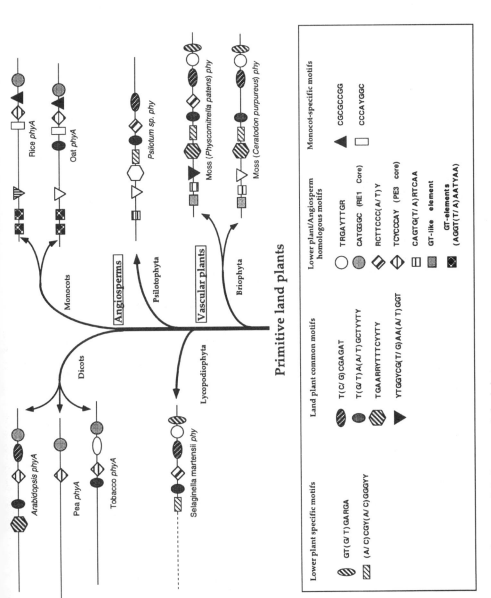

Figure 4   Sequences of *phyA* promoters in land plants.

orthologous *chs* promoters showed that LRU1 homologous sequences are present only in genes of cereals and brassicaceous plants (10). Homologous genes of legumes, snapdragon, and carrot contain, in the same relative position as LRU1, a CMA comprising G-box and H-box elements as in LRU1, but spaced differently (6–8 bp). Several regulatory functions have been assigned to this G-H box CMA, including transcriptional activation by p-coumaric acid (78) and tissue-specific transcription (38). Although this CMA is found within light-responsive regions of some *chs* genes (e.g. *Anthirrinum chs*; 76), there is no direct evidence of a role in light regulation. Because of both their proximal position to the TATA box and their PF composition, LRU1 and the G-H CMA are clearly homologous *cis*-regulatory regions. Nonetheless, evolutionary divergence of *chs* genes harboring LRU1 and G-H CMA, respectively, date back to an era before the divergence of lines that gave rise to monocotyledons and dicotyledons, as inferred from the phylogenetical distribution of LRU1 (10, 26).

EVOLUTION OF LIGHT-REGULATED PARALOGOUS GENE PROMOTERS

Differential Activity of Paralogous Gene Promoters

Several light-regulated genes such as *Lhcb*, *rbcS*, and *chs* are found in multiple copies in most genomes. Members of these gene families frequently display quantitative and/or qualitative differences in expression reflecting in part transcriptional regulation (29a, 30, 82, 117, 133). Some of these functional differences correlate with differences in promoter architecture. For example, among the eight *rbcS* genes of Petunia, only the two most highly expressed (*SSU301* and *SS611*) contain the I-G-box arrangement (29a). Insertion of an 89-bp fragment of the SSU301 promoter containing the I-G unit into the equivalent region of the weakly expressed *SSU911* gene increased its expression 25-fold (29).

Tomato has five *rbcS* genes (117). Three have promoters with the I-G unit. The mRNAs from all five tomato *rbcS* genes accumulate to similarly high levels in leaves and light-grown cotyledons; however, only the genes containing the I-G unit are coordinately expressed in dark-grown cotyledons, water-stressed leaves, and developing fruits (12). Interestingly, the spacer DNA sequence between the I-box and the G-box elements is highly divergent in tomato *rbcS* genes but conserved in orthologous genes from other plant species. A fruit-specific factor (FBF) specifically interacts with the I-G spacer DNA sequence of the tomato *rbcS-3A* promoter, which correlates with its reduced activity in developing tomato fruit (84). The tomato *rbcS-2* and tobacco *rbcS 8.0* display a very similar sequence to the pea *rbcS-3A* I-G spacer element, which is bound in vitro by GT-1 (50). It is conceivable that these "paralogous gene-specific motifs" modulate environmental or tissue-specific effects on the structurally invariant I-G regulatory unit.

Do Paralogous Promoters Evolve by Nonrandom Processes?

Paralogous gene promoters often diverge in discrete segments, displaying non-random patterns of structural variation. For example, in the *rbcS* family of *Arabidopsis* the paralogous 1B, 2B, and 3B gene promoters clearly differ from that of *rbcS-1A* by a 60-bp internal deletion. This molecular event juxtaposed the photoresponsive Box II–LAMP element with downstream conserved motifs, creating a new combination of *cis*-regulatory elements. In the *rbcS-1B* promoter an additional deletion event removed a 44-bp region encompassing the I-G-I Unit, juxtaposing the two I boxes originally flanking the G-box sequence (72). This gene is the only member of the *Arabidopsis rbcS* gene family unable to respond to light pulses (30).

The evolution of discrete, short regulatory elements interspersed along the promoters of paralogous genes also seems to occur, and this is exemplified by the maize *rbcSZm1* and *rbcZm3* genes. Their promoter sequences are very similar but differ in the presence of small insertions, which are distributed in a discontinuous pattern (Figure 5*a*). Some of these paralogous gene-specific motifs are *cis*-regulatory elements (106).

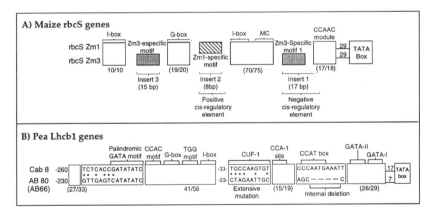

Figure 5 Paralogous gene promoters. (*a*) Schematic comparisons of two maize *rbcS* promoters mainly differing by small insertion/deletions events; some changes create *cis*-acting elements [based on data from Schäffner & Sheen (106)]. (*b*) Schematic comparison of two functionally distinct *Lhcb1* promoters of pea. Segments of high sequence divergence coinciding with relevant *cis*-regulatory elements are shown. *Clear rectangles* represent blocks of conserved sequences, whose similarity is indicated as the ratio between the number of identical nucleotides and the overall longitude (in bp) of the compared promoter segments.

A probable example of evolution of differential function by loss of specific *cis*-acting elements is available with the *Lhcb1* genes of pea. Two of these genes (*Cab-8* and *AB96*) showed significant transcript accumulation after a red light pulse, whereas the other three *Lhcb1* genes (*AB80, AB66,* and *Cab-9*) require continuous red light for significant expression (133, 134). The pea *Cab-8* gene encodes a mature protein 99.5% similar to that of the AB80 and AB66 genes (3, 125) and it displays significant similarity in the 350-bp proximal part of the promoter. *Cab-8* has a consensus dicot *Lhcb1* promoter, with a distal I*-G*-I structural unit and a proximal arrangement of elements (CUF-1)-(CCA-1)-(CCAAT-box)-(CGF-1 site). In the AB80 and AB66 promoters the original CCAAT box sequence seems to have been eliminated by an 8 nt internal deletion, the CUF-1 binding site is lost by multiple nucleotide substitutions, and the conserved GATA motif, upstream to the G-box, is mutated in the G residue and in a few additional upstream nucleotides (Figure 5*b*). Because the Cab-8 promoter has the organization of the hypothetical, ancestral dicot *Lhcb1* promoter, the architecture of the AB80 and AB66 promoters could be considered as derived, by mutation and deletion of specific *cis*-regulatory elements, from a *Cab-8*-like promoter. The cause of the mutations and the selective forces that fixed them in the species remain unknown.

LRE-ASSOCIATED CMAs

CMAs in Additional Plant Genes

Photoresponsive regions have been defined in other genes by diverse approaches (13, 122). Some *cis*-regulatory elements different to those identified in *rbcS*, *chs*, and *Lhcb* promoters have been proposed for light-regulated transcription (14, 81, 93). Discrete arrays of sequence motifs in light-responsive regions of those genes are conserved between phylogenetically distant plant species (10). Some of these LRE-associated CMAs are present in orthologous genes from both monocots and dicots, indicating a very remote evolutionary origin. These ancient CMAs are found in genes encoding ferredoxin, and the pyruvate–orthophosphate dikinase, sedoheptulose-bisphophatase, and the A subunit of chloroplast glyceraldehyde 3-phosphate dehydrogenase. LRE-associated CMAs were also identified in dicot genes encoding plastocyanin, subunits of the chloroplast ATPase, a 10-kDa protein of photosystem II, 4-coumarate:CoA ligase, and phenylammonia-lyase (10).

Structural Analogies Between LRE-Associated CMAs

Most of the ∼30 identified CMAs (10) can be grouped in a small set of structural and phylogenetic types.

1. The (I-box)-(G-box)-(I-box) arrangement. Included are the I-G and G-I units of *rbcS* promoters and their evolutionary variants, including the (X-box)-(Y-box) region of *Lemna* genes (Figure 3); the I*-G*-I ancestral arrangement of *Lhcb1* genes found in legumes, and their evolutionary derivatives such as the (GATA-motif)-(Z-DNA) region of *Arabidopsis* Lhcb1*3, the (Box II)-(G-box) arrangement of tobacco *CabE*, and the (G-box)-(CCA-1 element) region of some spinach, tobacco, and *Arabidopsis* genes (Figure 1); also in this CMA group are included the I*-G-(I) array of *Fed* promoters; and the inverted I-G-I unit of dicot and monocot *sbp* genes.

2. The (GT-1)-(LAMP-site)-(GT-1) arrangement, observed in both the Box II-3AF3 region of *rbcS* genes (*rbcS* CMA-3) and in the CMA-1 of *LS* genes

3. The (CCAAT-motif)-(GATA/I-box) combination, found in several light-responsive promoters, including the REα-REβ unit of *Lemna Lhcb2*1* (i.e. *cab*CMA4), the (CCAAT)-(GATA I-III) arrangement from *Lhcb1* promoters (i.e. *cab*CMA1), the (CCAAT box)-(motif 15) from solanaceous *rbcS* genes (i.e. *rbcS* CMA-2), and the CMA containing the PC2 region from plastocyanin gene promoters (i.e. *Pc*CMA-2).

4 The (LAMP-site)-(TATA-box) arrangement, characteristic of all of the solanaceous *rbcS* promoters (82) and found in a modified form in the −50/+15 light-responsive region of pea *rbcS-3A* (73). This arrangement is also observed in the LRE-associated CMA of *atpC*, in plastocyanin gene promoters (i.e. *Pc*CMA-1), and in *gapA* CMA-1.

5. The (G-box)-(H-box) arrangement, found in the three identified CMAs of *chs* genes, including the photoresponsive units 1 and 2 of parsley *chs* gene (108, 109).

Common Structural Features of Composite LREs

Do all, or most, of the genes whose transcription is dependent on PHY harbor a common *cis*-regulatory element in their promoters? This appealing idea has not been confirmed by the sequence data from the dozens of PHY-dependent genes. The identification of LRE-associated CMAs provides a new opportunity to assess whether the regulatory regions of those genes share structural features that could explain their common dependence on such a photoreceptor. The comparative analysis of ∼30 of these natural combinations of sequence motifs led to two important findings. 1. All of the LRE-associated CMAs present in PhANGs include at least one sequence identical, or related, to either the I-box core motif or its inverted version, the LAMP-site. 2. All of the CMAs found in genes encoding enzymes involved in the metabolism of phenylpropanoids

(PhEMAGs) share a conserved module related to the *chs* H-box core motif, ACCTA(A/C) C (A/C) (10).

Because PHY regulates expression of its target genes by three different transduction pathways (85), the superfamily-specific conserved motifs could be binding sites for transcription factors targeted by specific PHY-signaling pathways (10). Based on the knowledge of the genes that are activated or repressed by these phototransduction pathways, it has been proposed that the I-box/GATA-binding factors are direct or indirect targets of the Ca^{2+}/calmodulin-dependent transduction pathway, whereas transcription factors binding at the H-box motifs in PheMAG CMAs would be affected by the cGMP-dependent phototransduction pathway (10).

A general model of LRE function proposes that LREs are multipartite *cis*-regulatory elements with two general components: "light-specific" elements and "coupling elements." The former are bound by transcription factors targeted by the light-signal transduction pathways (i.e. I-box/GATA-binding factors and HBFs), which confer photoresponsiveness. Coupling elements are bound by either cell-specific factors or regulatory proteins targeted by other signaling systems; consequently, the light stimulus to transcription is coupled to other endogenous and exogenous signals.

Using microinjection into single cells of the tomato *aurea* mutant, Wu et al (136) established that constructs containing either 11 copies of the *rbcS* Box II element or 4 copies of *chs* Unit 1 are activated by different PHY signaling pathways. Box II is affected by the calcium-dependent pathway and the *chs* Unit 1 is activated by the cGMP pathway. Taking into account that GT-1, the factor that binds Box II, is highly related to the nuclear factors CGF-1 and IBF-2b, both of which bind I-box/GATA motifs (121) the work by Wu et al (136) supports the hypothesis that factors interacting with I-box-related sequences are targets for the Ca2+/calmodulin PHY activated pathway. Moreover, Feldbrügge et al (41) recently determined that the light-responsive core of parsley *chs* LRU1 is indeed the H-box, making it the most likely target of the cGMP pathway.

Evolution of LREs: The Chloroplast Connection

The finding that LREs from angiosperm gene promoters are very similar to putative regulatory units present in promoters from conifers and lower plants (Figure 6) is unexpected, because it is generally assumed that such promoters are either light insensitive or, at most, weakly photoresponsive. Such physiology is

---→

Figure 6 Similarities of light-responsive elements of angiosperm with homologous promoter sequences in nonflowering plants. Notice that Petunia *Cab 22R* and *P. thunbergii* Cab-6 belong to different gene families (i.e. *Lhcb1* and *Lhcb2*, respectively) and that *rbcS* Box II contains two PFs. Asterisks denote mismatches between aligned sequences.

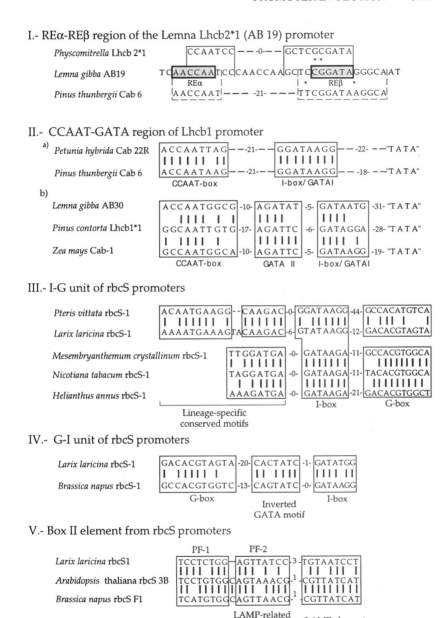

I.- REα-REβ region of the Lemna Lhcb2*1 (AB 19) promoter

Physcomitrella Lhcb 2*1

Lemna gibba AB19

Pinus thunbergii Cab 6

II.- CCAAT-GATA region of Lhcb1 promoter

a) *Petunia hybrida* Cab 22R

Pinus thunbergii Cab 6

b) *Lemna gibba* AB30

Pinus contorta Lhcb1*1

Zea mays Cab-1

III.- I-G unit of rbcS promoters

Pteris vittata rbcS-1

Larix laricina rbcS-1

Mesembryanthemum crystallinum rbcS-1

Nicotiana tabacum rbcS-1

Helianthus annus rbcS-1

IV.- G-I unit of rbcS promoters

Larix laricina rbcS-1

Brassica napus rbcS-1

V.- Box II element from rbcS promoters

Larix laricina rbcS1

Arabidopsis thaliana rbcS 3B

Brassica napus rbcS F1

well established in conifers (4, 71, 89, 94, 137), although to our knowledge, no detailed molecular studies have been carried out in pteridophytas and bryophytas. Conifers and other nonflowering plants (*Gingko biloba* being an exception; 24, 25) develop chloroplasts when grown in darkness (83, 129, 130), indicating that their PhANG promoters are active in the dark. A central question emerges: did light-responsiveness evolve by changes in *cis*-regulatory elements or transacting factors, or by both.

Because I-box/LAMP elements, which are critical for PHY responsiveness in angiosperms, are also components of ancestral, presumably non-photoresponsive, regulatory units of *Lhcb* and *rbcS* genes, it is probable that factors binding these motifs became direct or indirect targets of light-signaling pathways in organisms preceding flowering plants. This notion is in agreement with the available experimental evidence, including the recent demonstration that synthetic pairwise combinations of I-box sequences with diverse conserved elements function as complex LREs (98). However, the presence of an I-box in combination with other conserved sequence motifs is not necessarily sufficient for light regulation. The *Pinus thunbergii cab-6* promoter containing Reα and Reβ (an I-box motif) elements identical to those of the *Lemna AB19* gene (Figure 6) directs a light-independent and tissue-specific expression of a reporter gene in both dicots and monocots transgenic plants (70, 138). Therefore, other *cis*-acting signals in addition to the I-box core seem to be necessary for proper light control in flowering plants.

Our data suggest that composite LREs evolved from regulatory units that performed functions other than light-regulation. What could these functions have been? In the case of PhANGs, we hypothesize that these functions were related to nuclear gene regulation by chloroplast-derived signals. This possibility is supported by the finding that even the smallest light-responsive PhANG promoter segments display a tissue-specific and chloroplast-dependent pattern of expression similar to that of entire promoters; to date no evidence has been obtained that these two functions can be separated (9, 14, 15, 74, 98, 114, 120). Because coordination of gene expression between nuclear and chloroplast genomes should have evolved a long time before terrestrial plants (101), it is plausible that PhANG *cis*-acting promoter elements targeted by plastid signal transduction pathways evolved before LREs. Therefore, it is possible that photoreceptor-mediated transcriptional regulation was produced during evolution by targeting, either directly or through new regulators (i.e. via protein-protein interactions), the same transcription factors and *cis*-regulatory elements that mediated the influence of chloroplasts on PhANGs transcription. This possibility is attractive because it suggests a simple mechanism by which different gene families whose expression is coordinate could simultaneously acquire a new, coordinated pattern of regulation.

CONCLUDING REMARKS

Much remains to be learned about the structure, mode of action, and evolution of LREs. It is clear that LREs are complex, composed of at least two *cis*-acting elements, that can be targeted by different photoreceptor-activated signal transduction pathways. The composite and variable structure of LREs could explain the specific properties of individual light-responsive promoters. Phylogenetic analysis of LREs clearly indicates that they have evolved from ancient regulatory elements, whose original, primary function was probably not light regulation. Transformation of mosses and other nonflowering plants in which the expression of photosynthesis-associated genes is not regulated by light could help in answering some questions concerning the evolution of LREs and the different signal transduction pathways that activate them. It will be of great interest to explore how a new mode of regulation, affecting many gene families involved in photosynthesis, arose during evolution.

ACKNOWLEDGMENTS

We are grateful to June Simpson for critical reading of this manuscript. This work was supported by a grant from the HHMI to LH-E (75 197-526902).

> Visit the *Annual Reviews home page* at
> http://www.AnnualReviews.org.

Literature Cited

1. Adam E, Deak M, Kay S, Chua N-H, Nagy F. 1993. Sequence of a tobacco (*Nicotiana tabacum*) gene coding for type A phytochrome. *Plant Physiol.* 101:1047–48
2. Ahmad M, Cashmore AR. 1996. Seeing blue: the discovery of chryptochrome. *Plant Mol. Biol.* 30:851–61
3. Alexander L, Falconet D, Fristensky BW, White MJ, Watson JC, et al. 1991. Nucleotide sequence of *Cab-8*, a new type I gene encoding a chrorophyll a/b-binding protein in LHC II in *Pisum. Plant Mol. Biol.* 17:523–26
4. Alosi MC, Neale DB, Kinlaw CS. 1990. Expression of *cab* genes in Douglas-fir is not strongly regulated by light. *Plant Physiol.* 93:829–32
5. Anderson SL, Kay SA. 1995. Functional dissection of circadian clock- and phytochrome-regulated transcription of the *Arabidopsis CAB2* gene. *Proc. Natl. Acad. Sci. USA* 92:1500–4
6. Anderson SL, Teakle GR, Martino-Catt SJ, Kay SA. 1994. Circadian clock- and

phytochrome-regulated transcription is conferred by a 78 bp *cis*-acting domain of the *Arabidopsis CAB2* promoter. *Plant J.* 6:457–70
7. Argüello G, García-Hernández E, Sánchez M, Gariglio P, Herrera-Estrella LR, Simpson J. 1992. Characterization of DNA sequences that mediate nuclear protein binding to the regulatory region of the *Pisum sativum* (pea) chlorophyll a/b binding protein gene AB80: identification of a repeated heptamer motif. *Plant J.* 2:301–9
8. Argüello-Astorga GR, Guevara-González RG, Herrera-Estrella LR, Rivera-Bustamante RF. 1994. Geminivirus replication origins have a group-specific organization of iterative elements: a model for replication. *Virology* 203:90–100
9. Argüello-Astorga GR, Herrera-Estrella LR. 1995. Theoretical and experimental delimitation of Minimal Photoresponsive Elements in *Cab* and *rbcS* genes. In *Current Issues in Plant Molecular and Cellular Biology*, ed. M Terzi, R Cella, A Falavigna, pp. 501–11. Dordrecht: Kluwer

10. Argüello-Astorga GR, Herrera-Estrella LR. 1996. Ancestral multipartite units in light-responsive plant promoters have structural features correlating with specific phototransduction pathways. *Plant Physiol.* 112:1151–66

11. Barrett JW, Beech RN, Dancik BP, Strobeck C. 1994. A genomic clone of a type I *cab* gene encoding a light harvesting chlorophyll a/b binding protein of photosystem II identified from lodgepole pine. *Genome* 37:166–72

12. Bartholomew DM, Bartley GE, Scolnik PA. 1991. Abscisic acid control of *rbcS* and *cab* transcription in tomato leaves. *Plant Physiol.* 96:291–96

13. Batschauer A, Gilmartin PM, Nagy F, Schäfer E. 1994. The molecular biology of photoregulated genes. See Ref. 67, pp. 559–99

14. Bolle C, Kusnetsov VV, Herrmann RG, Oelmüller R. 1996. The spinach *AtpC* and *AtpD* genes contain elements for light-regulated, plastid-dependent and organ-specific expression in the vicinity of the transcription start sites. *Plant J.* 9:21–30

15. Bolle C, Sopory S, Lübberstedt T, Klösgen RB, Herrmann RG, Oelmüller R. 1994. The role of plastids in the expression of nuclear genes for thylakoid proteins studied with chimeric β- glucuronidase gene fusions. *Plant Physiol.* 105:1355–64

16. Borello U, Ceccarelli E, Guiliano G. 1993. Constitutive, light-responsive and circadian clock-responsive factors compete for the different I-box elements in plant light-regulated promoters. *Plant J.* 4:611–19

17. Bowler C, Neuhaus G, Yamagata H, Chua N-H. 1994. Cyclic GMP and calcium mediate phytochrome phototransduction. *Cell* 77:73–81

18. Bowler C, Yamagata H, Neuhaus G, Chua N-H. 1994. Phytochrome signal transduction pathways are regulated by reciprocal control mechanisms. *Genes Dev.* 8:2188–202

19. Bruce WB, Deng X-W, Quail PH. 1991. A negatively acting DNA sequence element mediates phytochrome-directed repression of *phyA* gene transcription. *EMBO J.* 10:3015–24

20. Bruce WB, Quail PH. 1990. Cis-acting elements involved in photoregulation of an oat phytochrome promoter in rice. *Plant Cell* 2:1081–89

21. Buzby JS, Yamada T, Tobin EM. 1990. A light-regulated DNA-binding activity interacts with a conserved region of a *Lemna gibba rbcS* promoter. *Plant Cell* 2:805–14

22. Carré I, Kay SA. 1995. Multiple DNA-protein complexes at a circadian-regulated promoter element. *Plant Cell* 7:2039–51

23. Castresana C, Garcia-Luque I, Alonso E, Malick VS, Cashmore AR. 1988. Both positive and negative regulatory elements mediate expression of a photoregulated CAB gene from *Nicotiana plumbaginifolia*. *EMBO J.* 7:1929–36

24. Chinn E, Silverthorne J. 1993. Light-dependent chloroplast development and expression of a light-harvesting chlorophyll a/b binding protein gene in the gymnosperm *Gingko biloba*. *Plant Physiol.* 103:727–32

25. Chinn E, Silverthorne J, Hohtola A. 1995. Light-regulated and organ specific expression of types 1, 2, and 3 light-harvesting complex b mRNAs in *Ginkgo biloba*. *Plant Physiol.* 107:727–32

26. Clegg MT, Cummings MP, Durbin ML. 1997. The evolution of plant nuclear genes. *Proc. Natl. Acad. Sci. USA* 94: 7791–98

27. Crane PR, Friis EM, Pedersen KR. 1995. The origin and early diversification of angiosperms. *Nature* 374:27–33

28. Davis MC, Yong M-H, Gilmartin PM, Goyvaerts E, Kuhlemeier C, Chua N-H. 1990. Minimal sequence requirements for the regulated expression of rbcS-3A from *Pisum sativum* in transgenic tobacco plants. *Photochem. Photobiol.* 52:43–50

29. Dean C, Favreau M, Bedbrook J, Dunsmuir P. 1989. Sequences 5′ to translation start regulate expression of petunia *rbcS* genes. *Plant Cell* 1:209–15

29a. Dean C, Pichersky E, Dunsmuir P. 1989. Structure, evolution, and regulation of RbcS genes in higher plants. *Annu. Rev. Plant Physiol. Plant Mol. Biol.* 40:415–39

30. Dedonder A, Rethy R, Fredericq H, Van Montagu M, Krebbers E. 1993. Arabidopsis rbcS genes are differentially regulated by light. *Plant Physiol.* 101:801–8

31. Degenhardt J, Tobin E. 1996. A DNA binding activity for one of two closely defined phytochrome regulatory elements in an Lhcb promoter is more abundant in etiolated than in green plants. *Plant Cell* 8:31–41

32. Dehesh K, Bruce WB, Quail PH. 1990. A trans-acting factor that binds to a GT-motif in a phytochrome gene promoter. *Science* 259:1397–99

33. Dehesh K, Franci C, Sharrock RA, Somers DE, Welsh JA, Quail PH. 1994. The *Arabidopsis* phytochrome A gene has multiple transcription start sites and a promoter sequence motif homologous to

the repressor element of monocot phytochrome A genes. *Photochem. Photobiol.* 59:379–84

34. Dehesh K, Hung H, Tepperman JM, Quail PH. 1992. GT2: a transcription factor with twin autonomous DNA-binding domains of closely related but different target sequence specificity. *EMBO J.* 11:4131–44

35. DeRocher EJ, Quigley F, Mache R, Bohnert HJ. 1993. The six genes of the rubisco small subunit multigene family from *Mesembryanthemum crystallinum*, a facultative CAM plant. *Mol. Gen. Genet.* 239:450–62

36. Donald RGK, Cashmore AR. 1990. Mutation in either G box or I box sequences profoundly affects expression from the *Arabidopsisthaliana* rbcS-1A promoter. *EMBO J.* 9:1717–26

37. Doolittle RF. 1987. *Of URFs and ORFs. A Primer on How to Analyze Derived Amino Acid Sequences.* Mill Valley, CA: Univ. Sci. Books

38. Faktor O, Loake G, Dixon RA, Lamb CJ. 1997. The G-box and H-box in a 39-bp region of a french bean chalcone synthase promoter constitute a tissue-specific regulatory element. *Plant J.* 11:1105–13

39. Falconet D, White MJ, Fristensky BW, Dobres MS, Thompson WF. 1991. Nucleotide sequence of Cab-215, a Type II gene encoding a Photosystem II chlorophyll a/b binding protein in *Pisum*. *Plant Mol. Biol.* 17:135–39

40. Feldbrügge M, Sprenger M, Dinkelbach M, Yazaki K, Harter K, Weisshaar B. 1994. Functional analysis of a light-responsive plant bZIP transcriptional regulator. *Plant Cell* 6:1607–21

41. Feldbrügge M, Sprenger M, Hahlbrock K, Weisshaar B. 1997. PcMYB1, a novel plant protein containing a DNA-binding domain with one MYB repeat, interacts in vivo with a light-regulatory promoter unit. *Plant J.* 11:1079–93

42. Furuya M. 1993. Phytochromes: their molecular species, gene families and functions. *Annu. Rev. Plant Physiol. Plant Mol. Biol.* 44:617–45

43. Gallie DR. 1993. Posttranscriptional regulation of gene expression in plants. *Annu. Rev. Plant Physiol. Plant Mol. Biol.* 44:77–105

44. Gidoni D, Brosio P, Bond-Nutter D, Bedbrook J, Dunsmuir P. 1989. Novel cis-acting elements in petunia Cab gene promoters. *Mol. Gen. Genet.* 215:337–44

45. Gilmartin PM, Chua N-H. 1990. Localization of a phytochrome-responsive element within the upstream region of pea rbcS-3A. *Mol. Cell. Biol.* 10:5565–68

46. Gilmartin PM, Memelink J, Hiratsuka K, Kay SA, Chua N-H. 1992. Characterization of a gene encoding a DNA binding protein with specificity for a light-responsive element. *Plant Cell* 4:839–49

47. Gilmartin PM, Sarokin L, Memelink J, Chua N-H. 1990. Molecular light-switches for plant genes. *Plant Cell* 2:369–78

48. Giuliano G, Pichersky E, Malik VS, Timko MP, Scolnik PA, Cashmore AR. 1988. An evolutionarily conserved protein binding sequence upstream of a plant light-regulated gene. *Proc. Natl. Acad. Sci. USA* 85:7089–93

49. Green PJ, Kay SA, Chua N-H. 1987. Sequence-specific interactions of a pea nuclear factor with light-responsive elements upstream of the rbcS-3A gene. *EMBO J.* 6:2543–49

50. Green PJ, Yong M-H, Cuozzo M, Kano-Mukarami Y, Silverstein P, Chua N-H. 1988. Binding site requirements for pea nuclear protein GT-1 correlate with sequences required for light-dependent transcriptional activation of the rbcS-3A gene. *EMBO J.* 7:4035–44

51. Grob U, Stuber K. 1987. Discrimination of phytochrome-dependent light-inducible from nonlight-inducible plant genes. Prediction of a common light-responsive element (LRE) in phytochrome dependent light-inducible genes. *Nucleic Acids Res.* 15:9957–72

52. Grossman AR, Bhaya D, Apt KE, Kehoe DM. 1995. Light-harvesting complexes in oxygenic photosynthesis: diversity, control, and evolution. *Annu. Rev. Genet.* 29:231–88

53. Gumucio DL, Shelton DA, Bailey WJ, Slightom JL, Goodman M. 1993. Phylogenetic footprinting reveals unexpected complexity in trans factor binding upstream from the b-globin gene. *Proc. Natl. Acad. Sci. USA* 90:6018–22

54. Ha S-B, An G. 1988. Identification of upstream regulatory elements involved in the developmental expression of the *Arabidopsis thaliana* cab1 gene. *Proc. Natl. Acad. Sci. USA* 85:8017–21

55. Hahlbrock K, Scheel D. 1989. Physiology and molecular biology of phenylpropanoid metabolism. *Annu. Rev. Plant Physiol. Plant. Mol. Biol.* 40:347–69

56. Hanania U, Zilberstein A. 1994. Ribulose-1,5-biphosphate carboxilase/oxygenase small subunit gene from the fern *Pteris vittata*. *Plant Physiol.* 106:1685–86

57. Hanelt S, Braun B, Marx S, Schneider-Poetsch HA. 1992. Phytochrome evolution: a phylogenetic tree with the first complete sequence of phytochrome from a cryptogamic plant (*Selaginella martensii* Spring). *Photochem. Photobiol.* 56:751–58

58. Hiratsuka K, Wu XD, Fukuzawa H, Chua N-H. 1994. Molecular dissection of GT-1 from Arabidopsis. *Plant Cell* 6:1805–13

59. Hutchinson KW, Harvie PD, Singer PV, Brunner AF, Greenwood MS. 1990. Nucleotide sequence of the small subunit of ribulose-1,5-biphosphate carboxylase from the conifer *Larix laricina*. *Plant Mol. Biol.* 14:281–84

60. Jansson S, Pichersky E, Bassi R, Green BR, Ikeuchi M, et al. 1992. A nomenclature for the genes encoding the chlorophyll a/b- binding proteins of higher plants. *Plant Mol. Biol. Rep.* 10:242–53

61. Jordan BR. 1996. The effects of ultraviolet-B radiation on plants: a molecular perspective. *Adv. Bot. Res.* 22:97–162

62. Kaiser T, Batschauer A. 1995. Cis-acting elements of the CHS1 gene from white mustard controlling promoter activity and spatial patterns of expression. *Plant Mol. Biol.* 28:231–43

63. Kaiser T, Emmler K, Kretsch T, Weisshaar B, Schäfer E, Batschauer A. 1995. Promoter elements of the mustard CHS1 gene are sufficient for light regulation in transgenic plants. *Plant. Mol. Biol.* 28:219–29

64. Karlin-Neumann GA, Kohorn BD, Thornber JP, Tobin EM. 1985. A chlorophyll a/b-protein encoded by a gene containing an intron with characteristics of a transposable element. *J. Mol. Appl. Genet.* 3:45–61

65. Kay SA, Keith B, Shinozaki K, Chye ML, Chua N-H. 1989. The rice phytochrome gene: structure, autoregulated expression, and binding of GT-1 to a conserved site in the 5' upstream region. *Plant Cell* 1:351–60

66. Kehoe DM, Degenhardt J, Winicov I, Tobin EM. 1994. Two 10-bp regions are critical for phytochrome regulation of a *Lemna gibba* Lhcb gene promoter. *Plant Cell* 6:1123–34

67. Kendrick RE, Kronenberg GHM, eds. 1994. *Photomorphogenesis in Plants.* Dordrecht: Kluwer. 2nd ed.

68. Kenigsbuch D, Tobin EM. 1995. A region of the Arabidopsis *Lhcb1*3 promoter that binds to CA-1 activity is essential for high expression and phytochrome regulation. *Plant Physiol.* 108:1023–27

69. Kohorn BD, Harel D, Chitnis PR, Thornber JP, Tobin EM. 1986. Functional and mutational analysis of the light-harvesting chlorophyll a/b protein of thylakoid membranes. *J. Cell Biol.* 102:972–81

70. Kojima K, Sasaki S, Yamamoto N. 1994. Light-independent and tissue-specific expression of a reporter gene mediated by the pine cab-6 promoter in transgenic tobacco. *Plant J.* 6:591–96

71. Kojima K, Yamamoto N, Sasaki S. 1992. Structure of the pine (*Pinus thunbergii*) chlorophyll a/b binding protein gene expressed in the absence of light. *Plant Mol. Biol.* 19:405–410

72. Krebbers E, Seurinck J, Herdies L, Cashmore AR, Timko MP. 1988. Four genes in two diverged subfamilies encode the ribulose-1,5-bisphosphate carboxylase small subunit polypeptides of *Arabidopsis thaliana*. *Plant Mol. Biol.* 11:745–59

73. Kuhlemeier C, Strittmatter G, Ward K, Chua N-H. 1989. The pea *rbcS-3A* promoter mediates light-responsiveness but not organ specificity. *Plant Cell* 1:471–78

74. Kusnetsov V, Bolle C, Lübberstedt T, Sopory S, Herrmann RG, Oelmüller R. 1996. Evidence that the plastid signal and light operate via the same *cis*-acting elements in the promoters of nuclear genes for plastid proteins. *Mol. Gen. Genet.* 252:631–39

75. Lam E, Chua N-H. 1990. GT-1 binding site confers light-responsive expression in transgenic tobacco. *Science* 248:471–74

75a. Sun L, Doxsee RA, Harel E, Tobin EM. 1993. CA-1, a novel phosphoprotein, interacts with the promoter of the cab 140 gene in Arabidopsis and is undetectable in det1 mutant seedlings. *Plant Cell* 5:109–21

76. Lipphardt S, Brettschneider R, Kreuzaler F, Schell J, Dangl JL. 1988. UV-inducible transient expression in parsley protoplasts identifies regulatory cis-elements of a chimeric *Antirrhinum majus* chalcone synthase gene. *EMBO J.* 7:4027–33

77. Lissemore JL, Quail PH. 1988. Rapid transcriptional regulation by phytochrome of the genes for phytochrome and chlorophyll a/b-binding protein in *Avena sativa*. *Mol. Cell. Biol.* 8:4840–50

78. Loake GJ, Faktor O, Lamb CJ, Dixon RA. 1992. Combination of H-box [CC-TACC(N)7CT] and G-box (CACGTG) cis-elements is necessary for feed-forward stimulation of a chacone synthase promoter by the phenylpropanoid-pathway intermediate p-coumaric acid.

Proc. Natl. Acad. Sci. USA 89:9230–34
79. Long Z, Wang S-Y, Nelson N. 1989. Cloning and nucleotide sequence analysis of genes coding for the major chlorophyll-binding protein of the moss *Physcomitrella patens* and thehalotolerant alga *Dunaliella salina. Gene* 76:299–312

80. Lubberstedt T, Bolle CEH, Sopory S, Flieger K, Herrmann RG, Oelmüller R. 1994. Promoters from genes for plastid proteins possess regions with different sensitivities toward red and blue light. *Plant Physiol.* 104:997–1006

81. Lubberstedt T, Oelmüller R, Wanner G, Herrmann RG. 1994. Interacting cis elements in the plastocyanin promoter from spinach ensure regulated high-level expression. *Mol. Gen. Genet.* 242:602–13

82. Manzara T, Gruissem W. 1988. Organization and expression of the genes encoding ribulose-1,5-bisphosphate carboxylase in higher plants. *Photosynth. Res.* 16:117–39

83. McNellis TW, Deng X-W. 1995. Light control of seedling morphogenetic pattern. *Plant Cell* 7:1749–61

84. Meier I, Callan KL, Fleming AJ, Gruissem W. 1995. Organ-specific differential regulation of a promoter subfamily for the ribulose-1,5-bisphosphate carboxylase/oxygenase small subunit genes in tomato. *Plant Physiol.* 107:1105–18

85. Millar AJ, McGrath B, Chua N-H. 1994. Phytochrome phototransduction pathways. *Annu. Rev. Genet.* 28:325–49

86. Miner JN, Yamamoto KR. 1991. Regulatory crosstalk at composite response elements. *Trends Biochem. Sci.* 16:423–26

87. Mitra A, Choi HK, An G. 1989. Structural and functional analyses of *Arabidopsis thaliana* chlorophyll a/b-binding protein (cab) promoters. *Plant Mol. Biol.* 12:169–79

88. Mösinger E, Batschauer A, Schäfer E, Apel K. 1985. Phytochrome control of in vitro transcription of specific genes in isolated nuclei from barley (*Horedum vulgare*). *Eur. J. Biochem.* 147:137–42

89. Mukai Y, Tazaki K, Fujii T, Yamamoto N. 1992. Light-independence expression of three photosynthetic genes, cab, rbcS, and rbcL, in coniferous plants. *Plant Cell Physiol.* 33:859–66

90. Deleted in proof

91. Neuhaus G, Bowler Ch, Hiratsuka K, Yamagata H, Chua N-H. 1997. Phytochrome regulated repression of gene expression requires calcium and cGMP. *EMBO J.* 16:2554–64

92. Okubara PA, Williams SA, Doxsee RA,

Tobin EM. 1993. Analysis of genes negatively regulated by phytochrome action in *Lemna gibba* and identification of a promoter region required for phytochrome responsiveness. *Plant Physiol.* 101:915–24

93. Park S-C, Kwon H-B, Shih MC. 1996. Cis-acting elements essential for light regulation of the nuclear gene encoding the A subunit of chloroplast glyceraldehyde 3-phosphate dehydrogenase in *Arabidopsis thaliana. Plant Physiol.* 112:1563–71

94. Peer W, Silverthorne J, Peters JL. 1996. Developmental and light-regulated expression of individual members of the light-harvesting complex b gene family in *Pinus palustris. Plant Physiol.* 111:627–34

95. Perisic O, Lam E. 1992. A tobacco DNA binding protein that interacts with a light responsive box II element. *Plant Cell* 4:831–38

96. Pichersky E, Soltis D, Soltis P. 1990. Defective chlorophyll a/b-binding protein genes in the genome of a homosporous fern. *Proc. Natl. Acad. Sci. USA* 87:195–99

97. Piechulla B, Kellmann JW, Pichersky E, Schwartz E, Forster HH. 1991. Determination of steady-state mRNA levels of individual chlorophyll a/b binding protein genes of the tomato cab gene family. *Mol. Gen. Genet.* 230:413–22

98. Puente P, Wei N, Deng XW. 1996. Combinatorial interplay of promoter elements constitutes the minimal determinants for light and developmental control of gene expression in Arabidopsis. *EMBO J.* 15:3732–43

99. Quail PH. 1994. Phytochrome genes and their expression. See Ref. 67, pp. 71–104

100. Quail PH, Boylan MT, Parks BM, Short TW, Xu Y, Wagner D. 1995. Phytochromes: photosensory perception and signal transduction. *Science* 268:675–80

101. Reith M. 1995. Molecular biology of rhodophyte and chromophyte plastids. *Annu. Rev. Plant Physiol. Plant Mol. Biol.* 46:549–75

102. Rocholl M, Talke-Messerer C, Kaiser T, Batschauer A. 1994. Unit 1 of the mustard chalcone synthase promoter is sufficient to mediate light responses from different photoreceptors. *Plant Sci.* 97:189–98

103. Sagliocco F, Kapazoglou A, Dure L. 1992. Sequence of cab-151, a gene encoding a photosystem II type II chlorophyll a/b-binding protein in cotton. *Plant Mol. Biol.* 18:841–42

104. Sarokin LP, Chua N-H. 1992. Binding sites for two novel phosphoproteins,

3AF5 and 3AF3, are required for rbcS-3A expression. *Plant Cell* 4:473–83

105. Sato N. 1988. Nucleotide sequence and expression of the phytochrome gene in *Pisum sativum*: differential regulation by light of multiple transcripts. *Plant Mol. Biol.* 11:697–710

106. Schäffner AR, Sheen J. 1991. Maize rbcS promoter activity depends on sequence elements not found in dicot rbcS promoters. *Plant Cell* 3:997–1012

107. Schneider-Poetsch HAW, Marx S, Koluklisaoglu HU, Hanelt S, Braun B. 1994. Phytochrome evolution: phytochrome genes in ferns and mosses. *Physiol. Plant.* 91:241–50

108. Schulze-Lefert P, Becker-André M, Schulz W, Hahlbrock K, Dangl JL. 1989. Functional architecture of the light-responsive chalcone synthase promoter from parsley. *Plant Cell* 1:707–14

109. Schulze-Lefert P, Dangl JL, Becker-André M, Hahlbrock K. 1989. Inducible in vivo DNA footprints define sequences necessary for UV light activation of the parsley chalcone synthase gene. *EMBO J.* 8:651–56

110. Shear WA. 1991. The early development of terrestrial ecosystems. *Nature* 351:283–89

111. Shopf JW. 1993. Microfossils of the early archean apex chert: new evidence of the antiquity of life. *Science* 260:640–46

112. Silverthorne J, Tobin E. 1984. Demonstration of transcriptional regulation of specific genes by phytochrome action. *Proc. Natl. Acad. Sci. USA* 81:1112–16

113. Simpson J, Schell J, Van Montagu M, Herrera-Estrella L. 1986. Light-inducible and tissue-specific pea lhcp gene expression involves an upstream element combining enhancer and silencer-like properties. *Nature* 323:551–54

114. Simpson J, Van Montagu M, Herrera-Estrella L. 1986. Photosynthesis-associated gene families: differences in response to tissue-specific and environmental factors. *Science* 233:34–38

115. Stayton MM, Black M, Bedbrook J, Dunsmuir P. 1986. A novel chlorophyll a/b binding (Cab) protein gene from petunia which encodes the lower molecular weight Cab precursor proteins. *Nucleic Acids Res.* 14:9781–96

116. Stewart WN, Rothwell GW. 1993. *Paleobotany and the Evolution of Plants.* Cambridge: Cambridge Univ. Press. 2nd ed.

117. Sugita M, Manzara T, Pichersky E, Cashmore AR, Gruissem W. 1987. Genomic organization, sequence analysis

and expression of all five genes encoding the small subunit of ribulose-1,5-bisphosphate carboxylase/oxygenase from tomato. *Mol. Gen. Genet.* 209:247–56

118. Sullivan TD, Christensen AH, Quail PH. 1989. Isolation and characterization of a maize chlorophyll a/b binding protein gene that produces high levels of mRNA in the dark. *Mol. Gen. Genet.* 215:431–40

119. Deleted in proof

120. Taylor WC. 1989. Regulatory interactions between nuclear and plastid genomes. *Annu. Rev. Plant Physiol. Plant Mol. Biol.* 40:211–33

121. Teakle GR, Kay SA. 1995. The GATA-binding protein CGF-1 is closely related to GT-1. *Plant Mol. Biol.* 29:1253–66

122. Terzaghi WB, Cashmore AR. 1995. Light-regulated transcription. *Annu. Rev. Plant Physiol. Plant Mol. Biol.* 46:445–74

123. Thompson WF, White MJ. 1991. Physiological and molecular studies of light-regulated nuclear genes in higher plants. *Annu. Rev. Plant Physiol. Plant Mol. Biol.* 42:423–66

124. Thümmler F, Dufner M, Kreisl P, Dittrich P. 1992. Molecular cloning of a novel phytochrome gene of the moss *Ceratodon purpureus* which encodes a putative light regulated protein kinase. *Plant Mol. Biol.* 20:1003–17

125. Timko MP, Kaush AP, Hand JM, Cashmore AR, Herrera-Estrella LR, et al. 1985. Structure and expression of nuclear genes encoding polypeptides of the photosynthetic apparatus. In *Molecular Biology of the Photosynthetic Apparatus*, pp. 381–95. Cold Spring Harbor, NY: Cold Spring Harbor Lab. Press

126. Tobin EM, Kehoe DM. 1994. Phytochrome regulated gene expression. *Semin. Cell Biol.* 5:335–46

127. Tsai F-Y, Coruzzi G. 1991. Light represses transcription of asparagine synthetase in photosynthetic and nonphotosynthetic organs of plants. *Mol. Cell. Biol.* 11:4966–72

128. von Arnim A, Deng XW. 1996. A role for transcriptional repression during light control of plant development. *BioEssays* 18:905–10

129. Wada M, Kadota A. 1989. Photomorphogenesis in lower green plants. *Annu. Rev. Plant Physiol. Plant Mol. Biol.* 40:169–91

130. Wada M, Sugai M. 1994. Photobiology of ferns. See Ref. 67, pp. 783–802

131. Wang ZY, Kenigsbuch D, Sun L, Harel E, Ong MS, Tobin EM. 1997. A myb-related transcription factor is involved in the phytochrome regulation of an Arabidopsis *Lhcb* gene. *Plant Cell* 9:491–507

132. Weisshaar B, Armstrong GA, Block A, de Costa e Silva O, Hahlbrock K. 1991. Light-inducible and constitutively expressed DNA-binding proteins recognizing a plant promoter element with functional relevance in light responsiveness. *EMBO J.* 10:1777–86

133. White MJ, Fristensky BW, Falconet D, Childs LC, Watson JC. 1992. Expression of the chlorophyll-a/b-protein multigene family in pea (*Pisum sativum* L.). *Planta* 188:190–98

134. White MJ, Kaufman LS, Horwitz BA, Briggs WR, Thompson WF. 1995. Individual members of the Cab gene family differ widely in fluence response. *Plant Physiol.* 107:161–65

135. Wright WE, Funk WD. 1993. Casting for multicomponent DNA-binding complexes. *Trends Biochem. Sci.* 18:77–80

136. Wu Y, Hiratsuka K, Neuhaus G, Chua N-H. 1996. Calcium and cGMP target distinct phytochrome-responsive elements. *Plant J.* 10:1149–54

137. Yamamoto N, Kojima K, Matsuoka M. 1993. The presence of two types of gene that encode the chlorophyll a/b-binding protein (LHCPII) and their light-independent expression in pine (*Pinus thunbergii*). *Plant Cell Physiol.* 34:457–63

138. Yamamoto N, Tada Y, Fujimura T. 1994. The promoter of a pine photosynthetic gene allows expression of a β-glucuronidase reporter gene in transgenic rice plants in a light-independent but tissue-specific manner. *Plant Cell Physiol.* 35:773–78

Annu. Rev. Plant Physiol. Plant Mol. Biol. 1998. 49:557–83

GENES AND ENZYMES OF CAROTENOID BIOSYNTHESIS IN PLANTS

F. X. Cunningham, Jr. and E. Gantt

Department of Microbiology, University of Maryland, College Park, MD 20742;
e-mail: fc18@umail.umd.edu; eg37@umail.umd.edu

KEY WORDS: carotene, chloroplast, isoprenoid, metabolism, xanthophyll

ABSTRACT

Carotenoids are integral and essential components of the photosynthetic membranes in all plants. Within the past few years, genes encoding nearly all of the enzymes required for the biosynthesis of these indispensable pigments have been identified. This review focuses on recent findings as to the structure and function of these genes and the enzymes they encode. Three topics of current interest are also discussed: the source of isopentenyl pyrophosphate for carotenoid biosynthesis, the progress and possibilities of metabolic engineering of plants to alter carotenoid content and composition, and the compartmentation and association of the carotenogenic enzymes. A speculative schematic model of carotenogenic enzyme complexes is presented to help frame and provoke insightful questions leading to future experimentation.

CONTENTS

1040-2519/98/0601-0557$08.00

INTRODUCTION

Indispensable and Other Roles of Carotenoids in Plants

The carotenoid pigments are essential components of the photosynthetic membranes in all plants, algae, and cyanobacteria and serve an extraordinary variety of functions in plants (18, 28, 32, 38, 111). A defining characteristic of these organisms is the use of readily available water as the reductant for photosynthesis. This ability has been of central importance to the evolutionary success of plants and algae. However, the production of molecular oxygen that accompanies the oxidation of water near the reaction centers of photosystem II is problematic. Excited triplet state chlorophyll, formed in the reaction centers and light-harvesting antenna upon illumination, can interact with oxygen to form the reactive and damaging singlet oxygen. Carotenoids associated with the reaction centers and antenna complexes react with and efficiently quench triplet chlorophyll, singlet oxygen, and also superoxide anion radicals (18, 28, 32). These protective functions are so critical that an inability to form cyclic carotenoids is eventually lethal in oxygen-evolving photosynthetic organisms (84).

Carotenoids in plants also dissipate excess light energy absorbed by the antenna pigments (28, 32, 111), harvest light for photosynthesis (18, 28, 32, 111), serve as precursors for biosynthesis of the plant growth regulator abscisic acid (108, 112), and are exploited as coloring agents in flowers and fruits to attract pollinators and agents of seed dispersal (38). The yellow, orange, and red colors provided by these pigments are of important agronomic value in many crop and ornamental plants. Carotenoids are precursors of vitamin A in human and animal diets (52), play other roles in human nutrition (74), and are of interest as potential anticancer agents (64). A number of carotenoids are valuable poultry and fish feed additives (61), or important colorants in the cosmetics and food industries (49). There is currently considerable interest in the manipulation of carotenoid content and composition in plants to improve the agronomic and nutritional value for human and animal consumption.

Historical Context and Scope of This Review

The sequence of the biochemical reactions that constitute the pathway of carotenoid biosynthesis in plants was reasonably well established by the early to
mid-1960s from perceptive analyses of the intermediates that accumulated in
naturally occurring mutants, under certain environmental conditions, or in the
presence of herbicides and inhibitors (101). In subsequent years, the characterization of many of the reactions in cell-free systems and in vivo using radiolabeled precursors lent further support and filled in many of the details. However,
some early successes in the 1970s in the isolation of enriched fractions catalyzing the early, soluble reactions of the pathway in tomato (see 101) were
not brought to fruition until the relatively recent isolation of the three soluble
enzymes that begin the pathway in pepper (29, 30). The membrane-associated
enzymes that catalyze subsequent reactions of the pathway have proven even
more recalcitrant. Both the substrates and products are lipid-soluble compounds, and the radiolabeled substrates needed for enzyme assays in vitro were
not (and are not) available commercially. These membrane-associated enzymes
are typically present in relatively low amounts, rapidly lose activity after detergent solubilization, and may require additional plastid components for activity.
The many attempts to purify individual enzymes of the membrane-associated
portion of the pathway were almost invariably without success (15) until a few
quite recent and notable examples that are mentioned in the appropriate sections
below.

In the past few years, genes and cDNAs encoding nearly all the enzymes
required for carotenoid biosynthesis in green plants have been identified and
sequenced, and their products characterized. In part, these recent accomplish-
ments are the fruits of long-term research programs that have employed a biochemical approach to study carotenogenesis in well-chosen experimental systems (e.g. pepper and daffodil chromoplasts). Important also has been the use
of cyanobacteria, with the many advantages that these procaryotic organisms
provide, as models for carotenogenesis in plants. Advances in plant genetics,
in particular the availability of tagged mutants, and the information resources
of the *Arabidopsis thaliana* EST data base and the genome sequence of the
cyanobacterium *Synechocystis PCC6803* have also contributed. However, the
recent and rapid progress in the study of the plant pathway can perhaps be
attributed most of all to the pioneering work on carotenogenesis in bacterial
systems. In a landmark study that built upon the earlier work of Marrs and
coworkers (60), Armstrong et al (3, 5) genetically defined the pathway and determined the sequences of genes encoding the enzymes of carotenoid biosynthesis in the photosynthetic bacterium *Rhodobacter capsulatus*. Soon thereafter,
Misawa et al (68) reported the sequences and functions of the products of

the carotenogenic genes in the bacterium *Erwinia uredovora*, and the pathway in *Erwinia herbicola* was also genetically dissected (45, 86). These bacterial genes have proven to be indispensable tools in the elucidation of the pathway in plants. Their importance lies not in their use as heterologous probes; the sequence identity is insufficient for this application in most cases. Rather, the ability to engineer strains of *Escherichia coli* that will accumulate a variety of carotenoids and carotenoid precursors has provided a simple and powerful in vivo system for assay of enzyme function and substrate specificity without the need for radiolabeled precursors. Furthermore, in a procedure we refer to as color complementation (26), the different colors displayed by colonies of *E. coli* that accumulate carotenoids (Figure 1, see color plate) have been exploited to visually screen cDNA and genomic libraries, enabling the identification of a number of previously unidentified plant, algal, and cyanobacterial genes and cDNAs that encode enzymes of carotenoid biosynthesis.

This review is concerned primarily with enzymes of the pathway of carotenoid[1] biosynthesis and the genes[2] encoding these enzymes in plants, with an emphasis on advances in knowledge since the 1994 review in this series (7). The enzymes are discussed sequentially in their order within the pathway, giving specific details for one example of each and using the *A. thaliana* (hereafter referred to as Arabidopsis) gene/enzyme as the archetype whenever possible. The extensive literature on carotenoid biosynthesis in bacteria, photosynthetic bacteria, and fungi (reviewed in 2, 2a, 4, 83) is considered and discussed only insofar as it contributes to a better understanding of the plant pathway.

Carotenoids in the Context of Plant Metabolism

The carotenoid pigments are synthesized in the plastids of plants. In chloroplasts they accumulate primarily in the photosynthetic membranes in association with the light-harvesting and reaction center complexes. In the chromoplasts of ripening fruits and flower petals and in the chloroplasts of senescing leaves the carotenoids may be found in membranes or in oil bodies or other structures within the stroma.

The lipid-soluble carotenoid pigments are but one example of the plethora of chemical compounds that are produced by what are collectively known as the pathways of isoprenoid biosynthesis. The one feature in common to the many isoprenoids (more than 20,000 in plants; 6, 24, 65) is their biosynthesis from the central metabolite and building block for all isoprenoid compounds:

---

[1] The common names of the carotenoids are used in this review. At first use, and where needed for clarity, the more descriptive systematic or semisystematic names (109) are also given.

[2] Gene designations given in the cited publications are used in this review in order to minimize confusion when consulting the original materials. A recommended nomenclature for genes of carotenoid biosynthesis may be found elsewhere (94a).

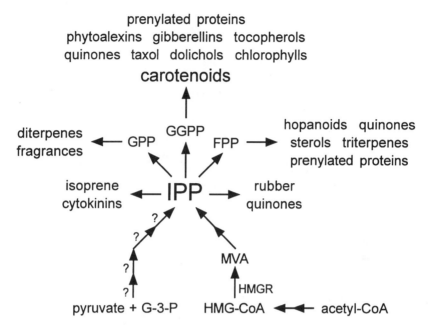

Figure 2   Isopentenyl pyrophosphate (IPP) serves as the central metabolite leading to an immense variety of different isoprenoid compounds in plants. A biosynthetic route to IPP from acetyl-CoA in the cytosol is well characterized. A second route from pyruvate and glyceraldehyde-3-phosphate (G-3-P) is thought to operate in chloroplasts (71). FPP, farnesyl pyrophosphate; GPP; geranyl pyrophosphate; GGPP, geranylgeranyl pyrophosphate; HMGR, hydroxymethylglutaryl-CoA reductase.

the 5-carbon compound isopentenyl pyrophosphate (IPP; Figure 2). A modular assembly process that produces compounds of 5, 10, 15, 20, or more carbons (in multiples of 5) allows the biosynthesis of the basic skeletons for the many and various isoprenoids with a relatively small number of basic reaction steps (65). The C_{40} skeleton of plant carotenoid pigments, for instance, is assembled from two molecules of a C_{20} compound, geranylgeranyl pyrophosphate (GGPP), that is itself assembled from four units of IPP and that also serves as a precursor for many other branches of the isoprenoid pathway in plants (Figure 2).

ASSEMBLY OF THE C_{40} BACKBONE

IPP Isomerase

A description of the carotenoid biosynthetic pathway quite often commences with the first committed step of the pathway, the formation of phytoene from

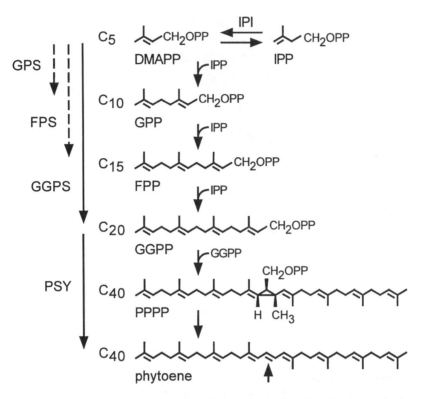

Figure 3 The C_{40} carotenoid phytoene is derived by a head-to-head condensation of two molecules of the C_{20} geranylgeranyl pyrophosphate (GGPP), which itself is assembled from three molecules of the C_5 isopentenyl pyrophosphate (IPP) and one molecule of its isomer, dimethylallyl pyrophosphate (DMAPP). FPP, farnesyl pyrophosphate; GPP, geranyl pyrophosphate; PPPP, prephytoene pyrophosphate.

two molecules of GGPP (Figure 3). However, it is both useful and appropriate to begin discussion of the pathway with the isomerization of IPP to its allylic isomer, dimethylallyl pyrophosphate (DMAPP; Figure 3). DMAPP is the initial, activated substrate for formation of long chain polyisoprenoid compounds such as GGPP (Figure 3). The formation of DMAPP from IPP is a reversible reaction that is catalyzed by the enzyme IPP isomerase (EC 5.3.3.2). This soluble enzyme has been isolated from pepper (29) and daffodil (58), but more detailed information on enzyme structure, cofactors, and reaction mechanisms is available from studies of the related yeast and mammalian enzymes (40, 79, 102). A review of the older work on plant enzymes may be found in Gray (39).

A plant cDNA for IPP isomerase was first identified in *Clarkia brewerii* (*Ipi1*; 10). A second *C. brewerii* gene has also been reported (*Ipi2*; 9), and two distinct cDNAs for this enzyme in Arabidopsis (FM Hahn & CD Poulter, direct submission to GenBank), lettuce (FX Cunningham Jr, G Jiang & E Gantt, unpublished data), and *Haematococcus pluvialis* (46, 104, 105) have been identified. The Arabidopsis *Ipp2* (GenBank accession number U49259) predicts a polypeptide of 284 amino acids (32,607 mol wt) with an N-terminal extension, relative to the mammalian and bacterial enzymes, that has been suggested to target this enzyme to the chloroplast (10). The Arabidopsis *Ipp1* (U47324) predicts a polypeptide of 233 amino acids (27,110 mol wt) and lacks the N-terminal extension of *Ipp2*, thereby suggesting a cytosolic location.

Both plastid and cytosolic locations for IPP isomerase are amply supported by evidence (6, 29, 50). A peroxisomal location for a human enzyme has been reported (75), and a mitochondrial location for a plant one inferred (57). A localization in several different cell compartments ordinarily implies the existence of specific genes or multiple transcripts to produce polypeptides targeted to these compartments. As yet, no more than two different cDNAs or genes have been identified for any plant.

A reversible isomerization reaction as catalyzed by the IPP isomerase would seem to be an unlikely candidate for a controlling or regulatory step in isoprenoid biosynthesis. However, it has been found that the activity of this enzyme in *E. coli* is limiting for isoprenoid production as indicated by the accumulation of carotenoids in strains engineered to produce these pigments. Introduction of any of a number of different plant, algal, or yeast IPP isomerase cDNAs, or of additional copies of the *E. coli* gene for this enzyme, enhances the accumulation of carotenoid pigments by several-fold (46, 104, 105). These observations raise the possibility that IPP isomerase activity might also limit biosynthesis of carotenoids and other isoprenoids in plants.

GGPP Synthase

The 20-carbon GGPP, which serves as the immediate precursor for carotenoid biosynthesis, is formed by the sequential and linear addition of three molecules of IPP to one molecule of DMAPP (Figure 3). The enzyme that catalyzes these reactions, the GGPP synthase (GGPS; EC 2.5.1.29), is one member of a closely related family of prenyl transferase enzymes that are distinguished by the length of the final product (73). A molecular understanding of the basis for the determination of chain length by these prenyl transferases has emerged from the construction and functional analysis of site-directed mutants (106a) and from the selection and analysis, after random mutagenesis, of enzymes with altered function (90).

A GGPP synthase has been isolated as a soluble and functional homodimer from the chromoplasts of pepper (29), and the corresponding cDNA has been identified and sequenced (54). Immunolocalization experiments confirmed a predominant localization in the chromoplast for GGPP synthase in pepper fruits (54). The enzyme was not, however, distributed homogeneously throughout the stroma; rather, it appeared to be concentrated in discrete locations, in particular at the developing stroma globuli where carotenoid accumulation is thought to occur (25).

In Arabidopsis, five different cDNA or genomic clones that predict polypeptides with substantial sequence similarity to the pepper GGPP synthase have been identified (91, 93, 95, 113). A GGPP synthase activity has been demonstrated for the products of two of these Arabidopsis cDNAs, *Ggps1* (L25813; FX Cunningham Jr, unpublished data) and *Ggps5* (D85029; 113) by functional complementation in *E. coli*. *Ggps1* encodes a 371–amino acid polypeptide (40,206 mol wt) with an N-terminal extension of 76 amino acids, relative to bacterial enzymes, that has been suggested to target this enzyme to the chloroplast (91). However, the specific roles and subcellular locations of the various Arabidopsis GGPP synthases have not been ascertained. Whether there might be a specific isoform of GGPP synthase in the plastid that is dedicated to carotenogenesis is also unknown. The presence of GGPP synthase genes in bacterial carotenogenic gene clusters (45, 68) underscores the importance of this enzyme in carotenoid biosynthesis.

Phytoene Synthase

The formation of the symmetrical 40-carbon phytoene (7,8,11,12,7',8',11',12'-octahydro-ψ,ψ-carotene) from two molecules of GGPP (Figure 3) is the first reaction specific to the pathway of carotenoid biosynthesis. The biosynthesis of phytoene from GGPP is a two-step reaction catalyzed by the enzyme phytoene synthase (PSY; EC 2.5.1.32). The amino acid sequences of plant, algal, and cyanobacterial PSY resemble those of the analogous bacterial and fungal phytoene synthase enzymes (CRTB; 2, 83) and share an extensive prenyl transferase domain with squalene synthase enzymes (FDFT; Figure 4). The condensation of two molecules of farnesyl pyrophosphate (FPP; Figure 3) to produce squalene, the C_{30} precursor of sterols, very much resembles the formation of phytoene from GGPP (103).

The PSY and CRTB enzymes share an additional conserved sequence domain not found in squalene synthases, and the known plant PSY enzymes (including representatives from both monocot and dicot species) have a third conserved sequence region that is not found in the algal, cyanobacterial, or bacterial phytoene synthases. The role of this third region and whether it remains in the mature polypeptide are not known. The Arabidopsis PSY (92) predicts a

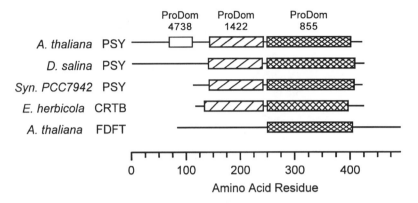

Figure 4 Plant, algal, and cyanobacterial phytoene synthase enzymes (PSY) share conserved amino acid sequence regions with bacterial phytoene synthases (CRTB) and with plant and mammalian (not shown) squalene synthases [FDFT (farnesyl diphosphate: farnesyl transferase)]. Species and SWISS-PROT accession numbers: *Arabidopsis thaliana* PSY, P37271; *Dunaliella salina* PSY, GenBank U91900; *Synechococcus PCC7942* PSY, P37269; *Erwinia herbicola* CRTB, P22872; *A. thaliana* FDFT, P53799. ProDom numbers refer to conserved domains identified by ProDom release 34.1 (100).

polypeptide of 423 amino acids with a mol wt of 47,611, but the mature size is probably about 40 kDa (48). The PSY of pepper chromoplasts was purified as a soluble, monomeric polypeptide (30); however, it is thought that this enzyme may normally be loosely (8) if not tightly (85) associated with chloroplast or chromoplast membranes. A specific requirement for galactolipid was demonstrated for catalytic activity of the PSY of *Narcissus pseudonarcissus* (daffodil) chromoplasts (85). A membrane association of PSY is expected because of the need to deliver the lipid-soluble phytoene to the membranes of the chloroplast where phytoene and subsequent intermediates and end products of the pathway are localized.

DESATURATION AND CYCLIZATION

The Desaturases

Phytoene undergoes a series of four desaturation reactions (Figure 5) that result in the formation of first phytofluene (7,8,11,12,7′,8′-hexahydro-ψ,ψ-carotene) and then, in turn, ζ-carotene (7,8,7′,8′-tetrahydro-ψ,ψ-carotene), neurosporene (7,8-dihydro-ψ,ψ-carotene), and lycopene (ψ,ψ-carotene). These desaturation reactions serve to lengthen the conjugated series of carbon-carbon double bonds that constitutes the chromophore in carotenoid pigments, and thereby transform the colorless phytoene into the pink-colored lycopene (Figure 1, see

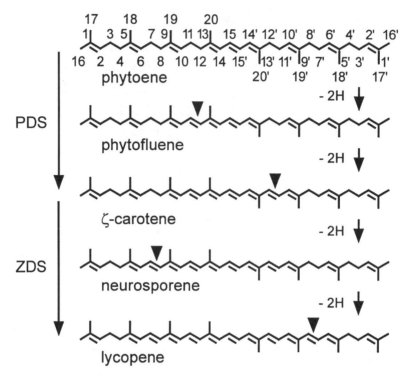

Figure 5 A series of four consecutive desaturation reactions at the 11-12, 11'-12', 7-8, and 7'-8' positions extend the conjugated series of double bonds that constitutes the chromophore in carotenoid pigments. Double bonds introduced by the desaturation reactions are indicated by *inverted triangles.* Conventional numbering of the carbon atoms is shown for phytoene.

color plate). The four sequential desaturations undergone by phytoene are catalyzed by two related enzymes in plants: phytoene desaturase (PDS) and ζ-carotene desaturase (ZDS). In contrast, bacteria and fungi achieve the same result with a single gene product (CRTI; reviewed in 2, 83).

Plant and cyanobacterial PDS are unusually well conserved in amino acid sequence. An alignment of the enzymes of *Synechococcus PCC7942* (23) and Arabidopsis (90), for instance, yields 65% identity and 79% similarity for the aligned residues. The Arabidopsis *Pds* (90) specifies a polypeptide of 566 amino acids (61,964 mol wt). A transit peptide cleavage site has not been identified. Sequence conservation in plant PDS in a region extending about 40 amino acids upstream of the start of the *Synechococcus* enzyme (474 amino acids; 23) would seem to indicate cleavage before amino acid 54, resulting in a mol wt of 57,000 or more for the mature Arabidopsis enzyme.

A cDNA encoding a ZDS was recently identified by functional analysis in *E. coli* of a pepper cDNA that predicts a polypeptide distantly resembling the known plant PDS (1). The pepper ZDS is about equidistant from plant and cyanobacterial PDS in predicted amino acid sequence comparisons (33–35% identities; 1). An Arabidopsis homologue of the pepper *Zds* has also been described (94). The putative Arabidopsis *Zds* encodes a polypeptide of 558 amino acids (60,532 mol wt). Cleavage sites suggested for the pepper and Arabidopsis ZDS (1, 94) would leave a well-conserved N-terminal extension of about 30 amino acids relative to the homologous *Synechococcus* PDS, and produce a mature Arabidopsis ZDS with about 59,000 mol wt.

PDS and ZDS both have amino acid sequence signatures that are conserved in pyridine nucleotide-disulphide oxidoreductases, and an isolated pepper chromoplast PDS was shown to bind added FAD (44). Such sequence motifs are also observed in the bacterial phytoene desaturase enzymes (CRTI; 2). Even so, the relatively low sequence similarities for plant PDS and bacterial CRTI gave rise to the suggestion that *Pds* and *crtI* are not related but rather arose independently (76).

The desaturases are demonstrably membrane-associated in plants (15), although the predicted amino acid sequences are not particularly hydrophobic overall. From a titration of the amount of detergent required to release PDS from daffodil chromoplast membranes, it was concluded that PDS is not an integral membrane protein (85). Also for the daffodil PDS, it was observed that FAD must be bound before or at the time of membrane integration, otherwise the membrane-associated enzyme will not be active (85). Enzyme assays of the flavinylated membrane-associated enzyme do not require additional FAD, supporting a role for tightly bound FAD as cofactor and implicating a membrane-associated electron acceptor (85). The involvement of quinones as electron acceptors for the desaturase reactions has experimental support (63, 71), most recently with a report that genetic lesions that impair the quinone biosynthetic pathway in Arabidopsis are accompanied by an accumulation of phytoene (72).

The Cyclases

The carotenoids in the photosynthetic apparatus of plants are bicyclic compounds, most commonly with two β or modified β rings (Figure 6). A single gene product, the lycopene β-cyclase (LCYB), catalyzes the formation of the bicyclic β-carotene from the linear, symmetrical lycopene in plants and cyanobacteria (26, 27, 43, 77; Figure 6). Because ζ-carotene (7,8,7',8'-tetrahydro-ψ,ψ-carotene) (Figure 5) was not a substrate for the tomato (77), Arabidopsis (26), and *Synechococcus PCC7942* (27) enzymes in *E. coli*, it was initially thought that desaturation of the 7-8 carbon-carbon bond (for numbering, see Figures 5 and 6) was a prerequisite for the cyclization reaction (26, 27, 77). However,

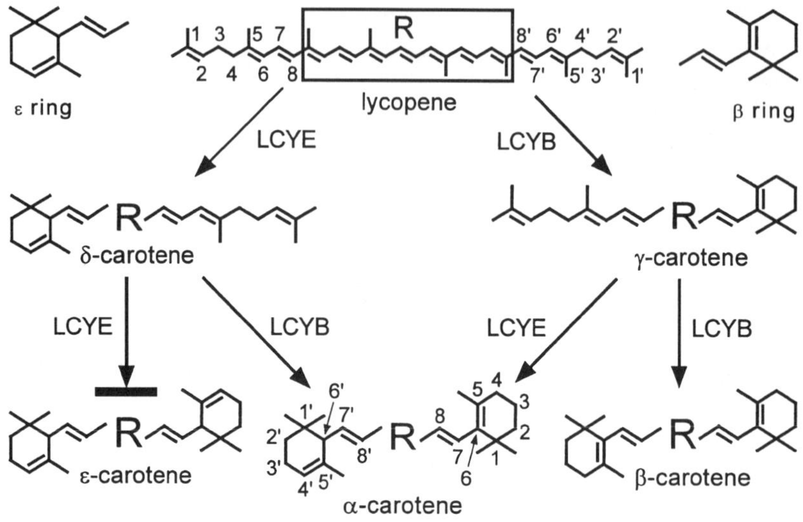

Figure 6 Cyclization of lycopene is a branch point in carotenoid biosynthesis. β-Carotene, with two β rings, is an essential end product and serves as the precursor for several other carotenoids that are commonly found in the photosynthetic apparatus of plants. α-Carotene, with one β and one ε ring, is the immediate precursor of lutein, the predominant carotenoid pigment in the photosynthetic membranes of many green plants. Carotenoids with two ε rings are not commonly found in plants.

in a subsequent report on a pepper LCYB it was found that the bicyclic compound 7,8-dihydro-β,β-carotene was produced in *E. coli* when neurosporene (7,8,-dihydro-ψ,ψ-carotene) was provided as the substrate (106). Therefore, it appears that desaturation at the 7-8 position, while not an absolute requirement for cyclization, may play a role in substrate recognition and/or binding. An association of the cyclases in the form of a homodimer could explain cyclization at both ends of neurosporene and the lack of cyclization for ζ-carotene. Neurosporene, recognized and bound at one end by virtue of desaturation of the 7-8 carbons at that end, would, at the same time, be constrained at the other end of the molecule in close proximity to and in the proper orientation for cyclization by a second cyclase subunit.

A second type of cyclic end group, the ε ring, is also common in plants and in certain algal classes. Lutein, the predominant carotenoid in the photosynthetic tissues of many plants and algae, has one β ring and one ε ring. Carotenoids with two ε rings are uncommon in plants and algae (38). The ε ring differs from the β ring only in the position of the double bond within the cyclohexene ring (Figure 6). A cDNA encoding the enzyme that catalyzes formation of the ε ring in Arabidopsis was identified by making use of the pink-to-yellow color change that accompanies lycopene cyclization in *E. coli* (Figure 1, see color plate; 26). The lycopene ε-cyclase (LCYE) of Arabidopsis is a homologue of the β-cyclase, and both enzymes are encoded by related, single-copy genes. However, unlike the β-cyclase, the ε-cyclase adds only one ring to the symmetrical lycopene, forming the monocyclic δ-carotene (ε,ψ-carotene). When combined, the β- and ε-cyclases convert lycopene to α-carotene (β,ε-carotene), a carotenoid with one β and one ε ring that serves as the precursor for formation of lutein (26). Whether the route to α-carotene (Figure 6) is only via δ-carotene (ε ring first) or can also proceed via γ-carotene (β ring first) has not been determined.

The inability of the ε-cyclase of Arabidopsis to add more than one ε ring to lycopene has been suggested to provide a mechanism for control of cyclic carotenoid formation (26). The apportioning of substrate into the pathways leading to β,ε-carotenoids (e.g. the abundant lutein) and to β,β-carotenoids (e.g. β-carotene, zeaxanthin, and violaxanthin) could be determined quite simply by the relative amounts and/or activities of the ε- and β-cyclase enzymes. Evidence that this control mechanism operates in vivo is provided by the results of Pogson et al (78), wherein a heterozygote of Arabidopsis carrying only one wild-type copy of the *lut2* allele (the mapped location of the ε-cyclase gene) accumulated only about 85% of the β,ε-carotenoids present in the wild type. The total amount of cyclic carotenoids did not differ significantly in the heterozygote and wild-type plants.

Lettuce is one of the rare examples of plants known to accumulate substantial amounts of a carotenoid with two ε rings: lactucaxanthin (ε,ε-carotene-3,3'-diol; 99). The ε-cyclase of romaine lettuce has been found to be a close homologue of the Arabidopsis ε-cyclase (ca 80% identity for the amino acid sequences). Yet this enzyme will, indeed, produce the bicyclic ε-carotene (ε,ε-carotene) rather than δ-carotene when lycopene is provided as the substrate in *E. coli* (FX Cunningham Jr, G Jiang & E Gantt, unpublished data).

The signature pigments of pepper, capsanthin and capsorubin, contain an unusual cyclopentane ring (the κ ring) that is formed from the 3-hydroxy-5,6-epoxy-β rings found in violaxanthin and antheraxanthin (Figure 7). The capsanthin-capsorubin synthase enzyme (CCS) has been purified, and a cDNA encoding it has been identified and sequenced (13). The pepper CCS (498 amino acids and 56,659 mol wt) closely resembles β-cyclases in its predicted amino acid sequence (13), and the gene product has more recently been shown to also possess a β-cyclase activity (43, 77). The pepper CCS has been purified from detergent-solubilized chromoplast membranes to yield a pale-yellow, intensely fluorescent polypeptide of approximately 60 kDa, as estimated by gel filtration (13).

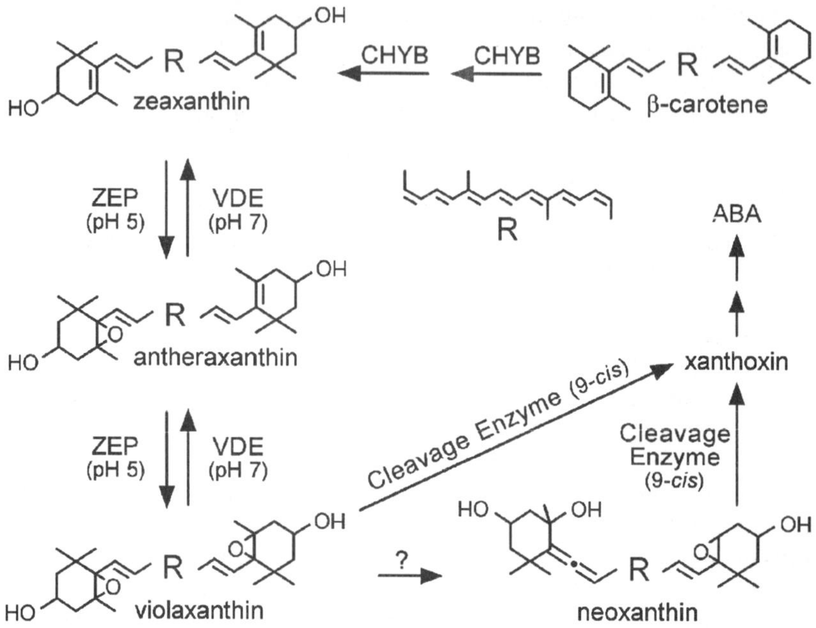

Figure 7 Formation of some common xanthophylls from β-carotene, and cleavage of 9-*cis* epoxy-carotenoids to produce xanthoxin, the precursor of abscisic acid (ABA).

The ε- and β-cyclases and CCS all have an amino acid sequence signature that is conserved in enzymes that bind dinucleotides such as FAD and NADP (26), but otherwise the cyclases bear little resemblance to other sequences in the data bases. Sequence similarity to the known bacterial β-cyclases is remote overall (20–25% amino acid sequence identity; 26) but, in addition to the dinucleotide-binding sequence signature, there are a few limited regions of conserved amino acid sequence for all cyclases (4, 13, 26).

The predicted amino acid sequences of the Arabidopsis β-cyclase (501 amino acids and 56,174 mol wt) and ε-cyclase (524 amino acids; 58,512 mol wt) are as much as 100 amino acids longer at the N terminus than the homologous *Synechococcus* enzyme (411 amino acids and 46,085 mol wt) when the sequences are aligned (26). Highly conserved but distinctly different sequence regions in the β- and ε-cyclases (26; FX Cunningham Jr, G Jiang & E Gantt, unpublished data) suggest that the mature plant ε- and β-cyclases may retain 30–50 amino acids of N-terminal sequence upstream of the position coincident with the start of the cyanobacterial β-cyclase. This would yield polypeptides of slightly greater than 50,000 in mol wt. The tomato ε-cyclase has been mapped

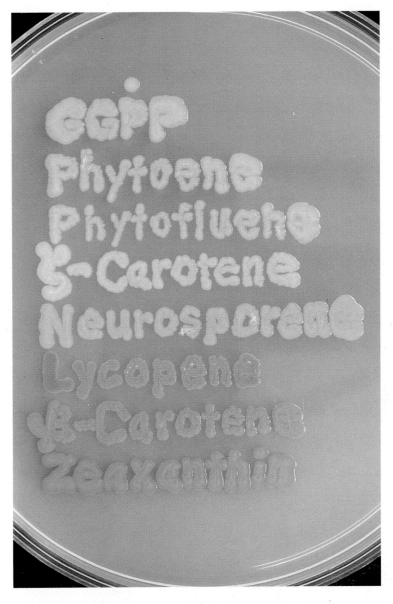

Figure 1   The pathway of carotenoid biosynthesis in living color. *Escherichia coli* strain TOP10 was genetically engineered to accumulate the carotenoid indicated. Details may be found elsewhere (26, 27, 105).

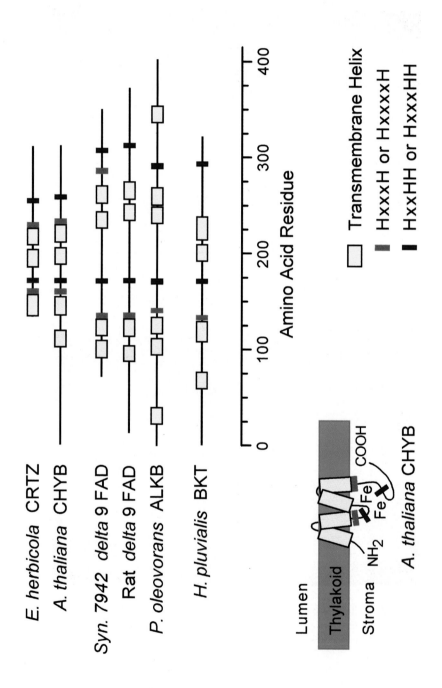

Figure 8   Plant and bacterial β ring hydroxylases (CHYB and CRTZ) and β-C-4-oxygenases (BKT and CRTW) are structurally similar to cyanobacterial and mammalian *delta*-9 fatty acid desaturases (*delta* 9 FAD), a bacterial alkane hydroxylase (ALKB), and many other integral membrane, oxygen-dependent, di-iron hydroxylases, desaturases, and oxygenases (not shown). A model for the predicted orientation of the *Arabidopsis* β ring hydroxylase (CHYB) in thylakoid membranes is displayed at the *lower left*. Species names and GenBank accession numbers: *Erwinia herbicola* Eho10 CRTZ, Q01332; *Arabidopsis thaliana* CHYB, U58919; *Synechococcus PCC7942 delta*-9 FAD, U36390; *Rattus norvegicus delta*-9 FAD, P07308; *Pseudomonas oleovorans* ALKB, P12691; *Haematocuccus pluvialis* BKT, S65078. Transmembrane helices were predicted as described elsewhere (82).

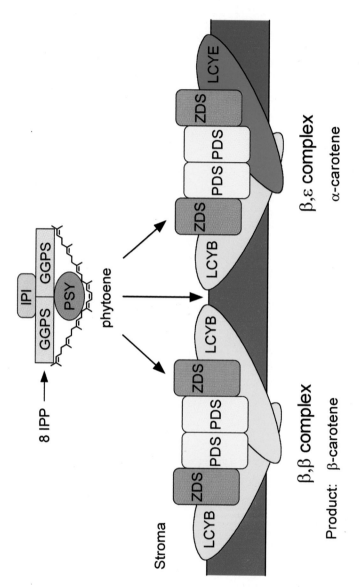

Figure 9 Schematic illustration and model of hypothetical multienzyme carotenogenic complexes in the thylakoid membranes and stroma of plants and algae. GGPS, geranylgeranyl pyrophosphate synthase; IPI, isopentenyl pyrophosphate isomerase; LCYB, lycopene β-cyclase; LCYE, lycopene ε-cyclase; PDS, phytoene desaturase; PSY, phytoene synthase; ZDS, ζ-carotene desaturase.

(J Hirschberg, personal communication) to the *del* locus defined long ago (107) in tomato plants with fruits that accumulate δ-carotene rather than lycopene.

XANTHOPHYLL FORMATION

The Hydroxylases

Xanthophylls or oxygenated carotenoids comprise most of the carotenoid pigment in the thylakoid membranes of plants. Hydroxylation at the number three carbon of each ring of the hydrocarbons β-carotene and α-carotene (Figure 6) will produce the well-known xanthophyll pigments zeaxanthin (β,β-carotene-3,3'-diol; Figure 7) and lutein (β,ε-carotene-3,3'-diol), respectively. Because the chirality of the hydroxyl on the ε ring is opposite that of the hydroxyl on the β ring of lutein (17), it is thought that different enzymes catalyze these reactions. Genetic evidence (78) and functional analysis of an Arabidopsis β-hydroxylase enzyme (105) support the existence of separate hydroxylases specific for the β and ε rings.

A cDNA (*Chyb*) encoding the β ring hydroxylase of Arabidopsis was recently identified by functional complementation in *E. coli* (105). The enzyme, in *E. coli*, was shown to hydroxylate both β rings of β-carotene, as well as the single β ring in α-carotene and β-zeacarotene (7',8'-dihydro-β,ψ-carotene). The ε rings of α-carotene and δ-carotene were found to be poor substrates. The Arabidopsis β-hydroxylase cDNA predicts a polypeptide of 310 amino acids (34,423 mol wt) that resembles the five known bacterial β ring hydroxylase enzymes (*crtZ* genes; 31–37% identities; 105), but has an N-terminal extension of more than 130 amino acids when aligned with the bacterial enzymes (105) (Figure 8, see color plate). Truncation of the gene to produce a polypeptide of a length comparable to the bacterial enzymes resulted in the accumulation of the monohydroxy carotenoid β-cryptoxanthin (β,β-caroten-3-ol) when β-carotene was the substrate in *E. coli*. The ratio, 11 to 1, of β-cryptoxanthin (one ring available for hydroxylation) to β-carotene (two rings available for hydroxylation) indicated a marked preference of the truncated enzyme for a β ring of β-carotene. It was speculated that a portion of the cleaved N-terminal region might be involved in formation of enzyme homodimers (105), though other plausible explanations (e.g. accessibility of the second ring) could certainly be offered.

Several regions in the predicted amino acid sequences of the plant and bacterial β-hydroxylases are strongly predicted to form transmembrane helices (Figure 8, see color plate; 105), suggesting a membrane-integral location in vivo. An oxygen requirement and stimulation by iron for two of the bacterial β ring hydroxylases was recently reported (34), and the authors noted the resemblance of a series of conserved histidine motifs (all of which are also present in the Arabidopsis enzyme) to those of enzymes containing nonheme iron. In fact, the arrangement of the histidine motifs and their position with respect to the predicted transmembrane helices are notable in their resemblance to the arrangement and positioning of similar motifs found in a structurally related group of oxygen-dependent di-iron-containing, membrane-integral enzymes (Figure 8, see color plate; 96–98). Members of this large and diverse group of di-iron oxygenases, which includes the membrane-associated fatty acid desaturases and various hydroxylases and oxidases such as the β-C-4-oxygenase or ketolase described in the following section, share an ability to attack unactivated carbons (96). They thereby provide an alternative class of enzymes with an ability to carry out the type of reactions usually associated with cytochromes P-450 (31a). Progress in understanding the structure, function, cofactor requirements, and catalytic mechanisms of others in this class of enzymes should provide valuable insights into the function of the plant and bacterial β ring hydroxylases and oxygenases.

β-C-4-Oxygenase

Carotenoids with keto groups in the 4 position of the rings are widely distributed in nature (38). Addition of a keto group at the 4 position of one or both rings of the yellow β-carotene will produce the reddish-orange to red pigments echinenone (β,β-caroten-4-one) and canthaxanthin (β,β-carotene-4,4'-dione). These differences in color have been exploited to screen libraries of the green alga *Haematococcus pluvialis* and the green plant *Adonis palaestina* for cDNAs that would produce a functional ketolase or "β-C-4-oxygenase" enzyme in a β-carotene-accumulating strain of *E. coli* (47, 56). The *crtO* gene product of Lotan & Hirschberg (56) and the *Bkt* gene product of Kajiwara et al (47), isolated from two different strains of *H. pluvialis*, are very much alike (greater than 80% identity for the predicted amino acid sequences) and resemble the two known bacterial ketolase enzymes (*crtW* gene products of *Agrobacterium aurantiacum* and *Alcaligenes PC-1*; 66, 69). In contrast, a cDNA encoding a ketolase enzyme in *A. palaestina* predicts a polypeptide that is homologous to the Arabidopsis (more than 60% identity) and bacterial (30–34% identities) β-carotene hydroxylases (FX Cunningham Jr, unpublished data). A gene (also named *crtO*) from *Synechocystis* sp. *PCC6803* that predicts a polypeptide similar in sequence to bacterial phytoene desaturase enzymes (*crtI* gene products) provides an example of a third type of ketolase enzyme (31b). This cyanobacterial enzyme is also unusual in that the product formed from the symmetrical β-carotene is the monoketo echinenone rather than the diketo canthaxanthin.

The *H. pluvialis* and bacterial ketolases (but not the *Synechocystis* ketolase) are, like the Arabidopsis and bacterial β ring hydroxylases (and the *Adonis* ketolase), members of a large class of membrane-integral, di-iron oxygenase enzymes (Figure 8, see color plate; 96–98). A requirement for molecular oxygen and a stimulatory effect of iron has been reported for the *H. pluvialis Bkt* gene product and two bacterial *crtW* gene products (16, 34). The 320–amino acid polypeptide (35,989 mol wt) predicted by the *Bkt* of *H. pluvialis* has an N-terminal extension of 65 amino acids when aligned with the two bacterial CRTW enzymes (both 242 amino acids and ca 27,000 mol wt). The CRTO enzyme likewise has such an N-terminal extension. Truncation of the *Bkt* cDNA to yield a polypeptide lacking the first 32 amino acids did not eliminate enzyme activity in *E. coli*, but removal of 60 amino acids did result in loss of activity (47). A portion of the N-terminal extension might comprise a plastid transit sequence. The location of the CRTO and BKT enzymes in *H. pluvialis* has not been determined but is of particular interest because the ketocarotenoids in this green alga accumulate outside of the chloroplast. Because the addition of the keto groups likely precedes hydroxylation (see below), an extraplastidic site for the ketolase would imply a location outside of the plastid, also, for the hydroxylase enzyme.

The end product ketocarotenoid in *H. pluvialis* and in many bacteria and fungi is the diketo, dihydroxy compound astaxanthin (3,3′-dihydroxy-β,β-carotene-4,4′-dione). The initial report for the *crtO* gene product of *H. pluvialis* (56) indicated that the enzyme, expressed in *E. coli*, was unable to utilize the dihydroxy carotenoid zeaxanthin as a substrate. Therefore it was concluded that the route to astaxanthin proceeds with addition of the keto groups before hydroxylation (56). The *Bkt* gene product, also expressed in *E. coli*, exhibited a similar preference for β rings lacking a 3-hydroxyl (16). However, in vitro assay of the *H. pluvialis* BKT found only a moderate (two fold) preference for β-carotene over zeaxanthin (34).

Epoxidase and De-epoxidase

The epoxidation of zeaxanthin to form violaxanthin (5,6,5′,6′-diepoxy-5,6,5′,6′-tetrahydro-β,β-carotene-3,3′-diol) via antheraxanthin (5,6-epoxy-5,6-dihydro-β,β-carotene-3,3′-diol) and de-epoxidation of violaxanthin to regenerate zeaxanthin comprise what is variously referred to as the xanthophyll, violaxanthin, or epoxide cycle (28, 110, 111). A cDNA encoding the zeaxanthin epoxidase (ZEP) or "β-cyclohexenyl xanthophyll epoxidase" in *Nicotiana plumbaginifolia* was identified by transposon tagging (59). The ABA-deficient phenotype of this mutant (*aba2*) and an ABA-deficient Arabidopsis mutant (*aba*) were complemented by the cDNA in transgenic plants. In vitro assay of a pepper ZEP was found to require ferredoxin and ferredoxin:NADP oxidoreductase in addition to NADPH, thereby implicating reduced ferredoxin as the reductant for the epoxidase reaction (12). The 3-hydroxy-β rings of zeaxanthin and antheraxanthin (Figure 7) were found to be substrates for the pepper enzyme but, surprisingly, the 3-hydroxy-β ring of lutein was not epoxidated in vitro (12).

The predicted amino acid sequences of the pepper and *N. plumbaginifolia* epoxidases contain a dinucleotide-binding motif, presumably for binding FAD, much like those in the carotenoid desaturases and cyclases (see above; 12, 59). The amino acid sequences of the pepper and tobacco ZEP are similar to that of a flavoprotein monooxygenase (salicylate-1-monooxygenase) and those of a number of other bacterial hydroxylases. The tobacco epoxidase is predicted to encode a polypeptide of 663 amino acids (72,524 mol wt). Import experiments using pea chloroplasts indicated that the mature size of the enzyme is about 5–6 kDa smaller than the initial product of translation in vitro (59).

A violaxanthin de-epoxidase cDNA (*Vde*) was obtained from romaine lettuce (19) using degenerate PCR primers based on N-terminal and internal amino acid sequences obtained from the purified enzyme (80). Tobacco and Arabidopsis cDNAs for the VDE have also been sequenced (RC Bugos & HY Yamamoto, direct submission to GenBank). N-terminal sequencing of the purified lettuce enzyme (80) indicates a cleavage site for the transit peptide of the lettuce enzyme immediately after residue 125 to yield a mature polypeptide of 348 amino acids (39,929 mol wt). The predicted amino acid sequences of the lettuce, tobacco, and Arabidopsis de-epoxidase enzymes do not significantly resemble any proteins in GenBank. A specific association of the purified enzyme with monogalactosyldiacylglyceride at pH 5.2, which is the approximate optimum for enzymatic activity, was noted (80). This observation very nicely explains the association of the de-epoxidase with the lumenal face of thylakoid membranes at low pH but solubility in the lumen at neutral pH (111). The properties and cofactor requirements of violaxanthin de-epoxidase have been reviewed (80, 110).

Epoxycarotenoid Cleavage Enzyme

Carotenoids, and in particular the epoxy carotenoids violaxanthin (Figure 7) and neoxanthin (5′,6′-epoxy-6,7-didehydro-5,6,5′,6′-tetrahydro-β,β-carotene-3,5,3′-triol), have long been thought to be precursors for biosynthesis of the plant growth regulator ABA (112). A maize enzyme that carries out the first step in the conversion of epoxycarotenoids to ABA was recently characterized (88). The maize *vp14* gene encodes a protein that quite specifically cleaves only the 9-*cis* geometrical isomers of violaxanthin and neoxanthin to yield *cis*-xanthoxin (Figure 7). The enzyme also was shown to cleave 9-*cis* zeaxanthin, a carotenoid that has no epoxide group. Enzyme activity in vitro required oxygen, iron, and a detergent (88).

The cDNA encoding the maize VP14 enzyme predicts a polypeptide of 604 amino acids (64,587 mol wt). Whether this includes a plastid targeting sequence that is subsequently cleaved is not yet known. The location of this enzyme in plants is an interesting question. The predominance of violaxanthin and neoxanthin in the envelope membranes of chloroplasts (11) is suggestive. The amino acid sequence of the maize cleavage enzyme resembles that of another oxidative cleavage enzyme, a bacterial lignostilbene dioxygenase. Most intriguing, Schwartz et al (88) noted a resemblance to microsomal membrane-associated polypeptides in human retinal pigment epithelia (41). The long sought and elusive cleavage enzyme that forms retinal from β-carotene in the human liver and intestine (52, 74) might well resemble these retinal proteins and the VP14 enzyme.

TOPICS OF CURRENT AND FUTURE INTEREST

Source of IPP/DMAPP for Carotenoid Biosynthesis

Isoprenoids are formed from IPP in at least three different compartments of plant cells: the cytosol/endoplasmic reticulum, the mitochondria (and/or Golgi apparatus), and the plastids (6, 65). The source of IPP for isoprenoid biosynthesis in these different compartments has long been a matter of some controversy and debate (6, 39, 50, 65). The well-known "classical" or acetate/mevalonate route to IPP proceeds from acetyl-CoA via 3-hydroxy-3-methylglutaryl-CoA (HMG-CoA) and mevalonic acid (MVA; Figure 2). The critical, rate-determining, and irreversible step in this route in animals is the reduction of HMG-CoA to produce MVA, catalyzed by the enzyme HMG-CoA reductase (HMGR; EC 1.1.1.34; reviewed in 37). HMGR has been localized in the cytosol of plants, associated with membranes of the endoplasmic reticulum (31). There is no convincing experimental evidence that this key enzyme also resides in the plastids or mitochondria of plants (6, 31, 39). However, the demonstrated ability of isolated chloroplasts of *Acetabularia* (70) and spinach (87) to synthesize carotenoids and other isoprenoids from $^{14}CO_2$ necessarily implies a chloroplast pathway for IPP biosynthesis. An explanation for this apparent contradiction is provided by the recent discovery of an "alternative" or nonmevalonate pathway for biosynthesis of IPP in the chloroplasts of plants (55) and in algae (89). Plausible routes to IPP from pyruvate and glyceraldehyde-3-phosphate have been suggested (6, 81, 89), but the precise nature and sequence of biochemical reactions in the nonmevalonate pathway (or pathways) remain to be determined. The elucidation of this pathway is a topic of intense current interest and import for carotenoid and isoprenoid biosynthesis.

Metabolic Engineering

The unraveling of the pathway of carotenoid biosynthesis in plants provides the requisite tools (if not yet the requisite insight and understanding of the regulatory mechanisms) for genetic modification of carotenoid content and composition. Because PSY is the branching enzyme that directs substrate irreversibly to carotenoids, it has been the target in several studies. Expression of antisense RNA to the *Psy1* gene of tomato was found to reduce the accumulation of carotenoids in the fruits of this plant by as much as 97% (14, 35) without noticeable effect on carotenoids in leaf tissue. However, the levels of gibberellins (GA) and other isoprenoids were also perturbed in these plants (33).

Approaches designed to increase the level of phytoene synthase lead to substantial increases in carotenoid accumulation and/or result in carotenoid accumulation in plant tissues that do not normally produce carotenoid pigments in tobacco (53), tomato (36), and rice (20). An unexpected result in the tomato experiments was a diminution in plant stature, a consequence attributed to the 30-fold reduction in the level of GA that is thought to result from a competition of the two pathways for GGPP (36). An accumulation of phytoene, albeit in relatively low concentration, was obtained in rice endosperm by expression of a daffodil *Psy* in this tissue (20). The objective of this work is to eventually engineer the production of β-carotene in rice to provide a much needed source of provitamin A.

The phytoene desaturase enzyme has also been the target of genetic manipulation in plants. Expression in tobacco of a *crtI* gene, the bacterial counterpart of *Pds*, produced transgenic plants resistant to a number of herbicides that are known to inhibit PDS but not CRTI (67). An unexpected and inexplicable reduction in carotenoids with ε rings (e.g. lutein) relative to those with only β rings was observed, but the total amount of carotenoids was unaffected.

Expression of a *H. pluvialis* β-C-4-oxygenase in *Synechococcus PCC7942* (42) and in tobacco (J Hirschberg, personal communication) illustrates the possibilities for production of the valuable ketocarotenoid astaxanthin in organisms in which this pigment is not normally made. In tobacco constitutively expressing the β-C-4-oxygenase, there was very little ketocarotenoid in the photosynthetic tissues. However, the ring of chromoplast-containing tissue at the base of the flowers (the nectary), normally yellow in color, became red from a substantial accumulation (ca 85% of the total carotenoid) of ketocarotenoids.

Carotenogenic Complexes

The site of biosynthesis of the carotenoids in green plants, the plastids, is the site of synthesis for a number of other compounds that are derived or assembled from IPP (Figure 2; 24, 65). An enduring question has been what determines the partitioning of this precursor into the various branches of the isoprenoid pathway. Regulation of carotenoid biosynthesis also entails more than simply a determination of flux through the pathway, because this pathway itself branches to produce the typically five or six different carotenoids with roles in the photosynthetic apparatus of green plants.

A long-held and attractive hypothesis holds that carotenogenesis occurs in multisubunit enzyme aggregates associated with the plastid membranes (51). The three "soluble" enzymes of the pathway that catalyze the formation of phytoene have been isolated together in multienzyme complexes (21, 29, 62). The evidence for complexes containing the membrane-associated desaturases, cyclases, and hydroxylases is less direct (15, 22, 50, 51, 101).

A schematic model of our conception of these hypothetical carotenogenic complexes is shown in Figure 9 (see color plate). Because GGPP synthase is a dimer (29), the soluble phytoene synthase complex is shown containing a single copy of IPP isomerase, two copies of the GGPP synthase, and one of PSY. The end product of these reactions, the central *cis*-isomer of phytoene, is a lipid-soluble compound that must somehow reach the desaturases and/or the membrane. The observed PSY-membrane association in daffodil (85) would suggest that such a complex is actually associated with the membranes.

The postulated membrane-associated desaturase/cyclase enzyme aggregates are shown to contain two copies of each desaturase and two cyclase subunits. There may well be other polypeptide components that serve in a regulatory capacity, act as cofactors, or otherwise interact with the catalytic subunits. Two types of the desaturase/cyclase complexes are displayed in order to account for the formation of cyclic carotenoids with two β rings and those with one β and one ε ring. To explain the absence in most plants of carotenoids with two ε rings, we hypothesize that complexes with two ε-cyclases are not formed.

Concluding Comments

The pursuit of genes encoding enzymes of the pathway leading to carotenoids that function in the photosynthetic apparatus has been a major endeavor in many laboratories. This hunt for genes is near an end, although pathways to such abundant algal carotenoids as peridinin and fucoxanthin, and enzymes that produce the many species-specific carotenoids and the secondary carotenoids found in nonphotosynthetic tissues, await investigation.

The progress achieved in the last few years has brought us to the point where models such as that in Figure 9 (see color plate) are now testable. Many of the obstacles that hindered earlier attempts at purification and characterization of carotenogenic enzymes and assemblies are rendered irrelevant by the availability of specific antisera. The ability to produce large quantities of the enzymes in *E. coli* and yeast has made possible new genetic and biochemical approaches for identifying amino acid residues or domains involved in catalysis, membrane association, subunit interactions, substrate binding, interaction with herbicides, association with cofactors, and regulation of carotenoid biosynthesis. There is much to be done.

ACKNOWLEDGMENTS

We thank J Hirschberg for communicating unpublished results. Financial support for the authors' research has been provided by NSF and USDA.

Visit the *Annual Reviews home page* at
http://www.AnnualReviews.org.

Literature Cited

1. Albrecht M, Klein A, Hugueney P, Sandmann G, Kuntz M. 1995. Molecular cloning and functional expression in *E. coli* of a novel plant enzyme mediating *zeta*-carotene desaturation. *FEBS Lett.* 372:199–202
2. Armstrong GA. 1994. Eubacteria show their true colors: genetics of carotenoid pigment biosynthesis from microbes to plants. *J. Bacteriol.* 176:4795–802
2a. Armstrong GA. 1997. Genetics of eubacterial carotenoid biosynthesis: a colorful tale. *Annu. Rev. Microbiol.* 51:629–59
3. Armstrong GA, Alberti M, Leach F, Hearst JE. 1989. Nucleotide sequence, organization, and nature of the protein products of the carotenoid biosynthesis gene cluster of *Rhodobacter capsulatus*. *Mol. Gen. Genet.* 216:254–68
4. Armstrong GA, Hearst JE. 1996. Carotenoids 2: genetics and molecular biology of carotenoid pigment biosynthesis. *FASEB J.* 10:228–37
5. Armstrong GA, Schmidt A, Sandmann G, Hearst JE. 1990. Genetic and biochemical characterization of carotenoid biosynthesis mutants of *Rhodobacter capsulatus*. *J. Biol. Chem.* 265:8329–38
6. Bach TJ. 1995. Some new aspects of isoprenoid biosynthesis in plants—a review. *Lipids* 30:191–202
7. Bartley GE, Scolnik PA, Giuliano G. 1994. Molecular biology of carotenoid biosynthesis in plants. *Annu. Rev. Plant. Physiol. Plant. Mol. Biol.* 45:287–301
8. Bartley GE, Viitanen PV, Bacot KO, Scolnik PA. 1992. A tomato gene expressed during fruit ripening encodes an enzyme of the carotenoid biosynthesis pathway. *J. Biol. Chem.* 267:5036–39
8a. Bauernfeind JC, ed. 1981. *Carotenoids as Colorants and Vitamin A Precursors*. London: Academic
9. Blanc VM, Mullin K, Pichersky E. 1996. Nucleotide sequences of *Ipi* genes from *Arabidopsis* and *Clarkia*. (Accession Nos. U48961, U48962, U48963)

(PGR96–036). *Plant Physiol.* 111:652
10. Blanc VM, Pichersky E. 1995. Nucleotide sequence of a *Clarkia breweri* cDNA clone of *Ipi1*, a gene encoding isopentenyl pyrophosphate isomerase. *Plant Physiol.* 108:855–56
11. Block MA, Dorne A-J, Joyard J, Douce R. 1983. Preparation and characterization of membrane fractions enriched in outer and inner envelope membranes of spinach chloroplasts. *J. Biol. Chem.* 258:13281–86
12. Bouvier F, d'Harlingue A, Hugueney P, Marin E, Marion-Poll A, Camara B. 1996. Xanthophyll biosynthesis: cloning, expression, functional reconstitution, and regulation of *beta*-cyclohexenyl carotenoid epoxidase from pepper (*Capsicum annuum*). *J. Biol. Chem.* 271:28861–67
13. Bouvier F, Hugueney P, d'Harlingue A, Kuntz M, Camara B. 1994. Xanthophyll biosynthesis in chromoplasts: isolation and molecular cloning of an enzyme catalyzing the conversion of 5,6-epoxycarotenoids into ketocarotenoid. *Plant J.* 6:45–54.
14. Bramley P, Teulieres C, Blain I, Bird C, Schuch W. 1992. Biochemical characterization of transgenic tomato plants in which carotenoid synthesis has been inhibited through the expression of antisense RNA to pTOM5. *Plant J.* 2:343–49
15. Bramley PM. 1985. The *in vitro* biosynthesis of carotenoids. *Adv. Lipid Res.* 21:243–79
16. Breitenbach J, Misawa N, Kajiwara S, Sandmann G. 1996. Expression in *Escherichia coli* and properties of the carotene ketolase from *Haematococcus pluvialis*. *FEMS Microbiol. Lett.* 140:241–46
17. Britton G. 1990. Carotenoid biosynthesis—an overview. In *Carotenoids: Chemistry and Biology*, ed. NI Krinsky, M Mathews-Roth, RF Taylor, pp. 167–84. New York: Plenum

18. Britton G. 1995. Structure and properties of carotenoids in relation to function. *FASEB J.* 9:1551–58

19. Bugos RC, Yamamoto HY. 1996. Molecular cloning of violaxanthin deepoxidase from romaine lettuce and expression in *Escherichia coli. Proc. Natl. Acad. Sci. USA* 93:6320–25

20. Burkhardt PK, Beyer P, Wunn J, Kloti A, Armstrong GA, et al. 1997. Transgenic rice (*Oryza sativa*) endosperm expressing daffodil (*Narcissus pseudonarcissus*) phytoene synthase accumulates phytoene, a key intermediate of provitamin A biosynthesis. *Plant J.* 11: 1071–78

21. Camara B. 1993. Plant phytoene synthase complex: component enzymes, immunology, and biogenesis. *Methods Enzymol.* 214:352–65

22. Candau R, Bejarano ER, Cerdá-Olmedo E. 1991. *In vivo* channeling of substrates in an enzyme aggregate for beta-carotene biosynthesis. *Proc. Natl. Acad. Sci. USA* 88:4936–40

23. Chamovitz D, Pecker I, Hirschberg J. 1991. The molecular basis of resistance to the herbicide norflurazon. *Plant Mol. Biol.* 16:967–74

24. Chappell J. 1995. Biochemistry and molecular biology of the isoprenoid biosynthetic pathway in plants. *Annu. Rev. Plant. Physiol. Plant. Mol. Biol.* 46: 521–47

25. Cheniclet C, Rafia F, Saint-Guily A, Verna A, Carde J-P. 1992. Localization of the enzyme geranylgeranylpyrophosphate synthase in Capsicum fruits by immunogold cytochemistry after conventional chemical fixation or quick-freezing followed by freeze-substitution. Labelling evolution during fruit ripening. *Biol. Cell* 75:145–54

26. Cunningham FX Jr, Pogson B, Sun ZR, McDonald KA, DellaPenna D, Gantt E. 1996. Functional analysis of the β and ε lycopene cyclase enzymes of Arabidopsis reveals a mechanism for control of cyclic carotenoid formation. *Plant Cell* 8:1613–26

27. Cunningham FX Jr, Sun ZR, Chamovitz D, Hirschberg J, Gantt E. 1994. Molecular structure and enzymatic function of lycopene cyclase from the cyanobacterium *Synechococcus* sp. strain PCC7942. *Plant Cell* 6:1107–21

28. Demmig-Adams B, Gilmore AM, Adams WW III. 1996. Carotenoids 3: *in vivo* function of carotenoids in higher plants. *FASEB J.* 10:403–12

29. Dogbo O, Camara B. 1987. Purification of isopentenyl pyrophosphate isomerase and geranylgeranyl pyrophosphate synthase from *Capsicum* chromoplasts by affinity chromatography. *Biochim. Biophys. Acta* 920:140–48

30. Dogbo O, Laferrière A, D'Harlingue A, Camara B. 1988. Carotenoid biosynthesis: isolation and characterization of a bifunctional enzyme catalyzing the synthesis of phytoene. *Proc. Natl. Acad. Sci. USA* 85:7054–58

31. Enjuto M, Balcells L, Campos N, Caelles C, Arro M, Boronat A. 1994. *Arabidopsis thaliana* contains two differentially expressed 3-hydroxy-3-methylglutaryl-CoA reductase genes, which encode microsomal forms of the enzyme. *Proc. Natl. Acad. Sci. USA* 91: 927–31

31a. Estabrook RW. 1996. The remarkable P450s: a historical overview of these versatile hemeprotein catalysts. *FASEB J.* 10:202–4

31b. Fernandez-Gonzalez B, Sandmann G, Vioque A. 1997. A new type of asymmetrically acting *beta*-carotene ketolase is required for the synthesis of echinenone in the cyanobacterium *Synechocystis* sp. PCC 6803. *J. Biol. Chem.* 272:9728–33

32. Frank HA, Cogdell RJ. 1996. Carotenoids in photosynthesis. *Photochem. Photobiol.* 63:257–64

33. Fraser PD, Hedden P, Cooke DT, Bird CR, Schuch W, Bramley PM. 1995. The effect of reduced activity of phytoene synthase on isoprenoid levels in tomato pericarp during fruit development and ripening. *Planta* 196:321–26

34. Fraser PD, Miura Y, Misawa N. 1997. *In vitro* characterization of astaxanthin biosynthetic enzymes. *J. Biol. Chem.* 272:6128–35

35. Fray RG, Grierson D. 1993. Identification and genetic analysis of normal and mutant phytoene synthase genes of tomato by sequencing, complementation and co-suppression. *Plant Mol. Biol.* 22:589–602

36. Fray RG, Wallace A, Fraser PD, Valero D, Hedden P, et al. 1995. Constitutive expression of a fruit phytoene synthase gene in transgenic tomatoes causes dwarfism by redirecting metabolites from the gibberellin pathway. *Plant J.* 8:693–701

37. Goldstein JL, Brown MS. 1990. Regulation of the mevalonate pathway. *Nature* 343:425–30

38. Goodwin TW. 1980. *The Biochemistry of the Carotenoids*, Vol. 1. London:

Chapman & Hall. 377 pp. 2nd ed.

39. Gray JC. 1987. Control of isoprenoid biosynthesis in higher plants. *Adv. Bot. Res.* 14:25–91

40. Hahn FM, Xuan JW, Chambers AF, Poulter CD. 1996. Human isopentenyl diphosphate: dimethylallyl diphosphate isomerase: overproduction, purification, and characterization. *Arch. Biochem. Biophys.* 332:30–34

41. Hamel CP, Tsilou E, Pfeffer BA, Hooks JJ, Detrick B, Redmond TM. 1993. Molecular cloning and expression of RPE65, a novel retinal pigment epithelium-specific microsomal protein that is post-transcriptionally regulated *in vitro. J. Biol. Chem.* 268:15751–57

42. Harker M, Hirschberg J. 1997. Biosynthesis of ketocarotenoids in transgenic cyanobacteria expressing the algal gene for β-C-4-oxygenase, *crtO. FEBS Lett.* 404:129–34

43. Hugueney P, Badillo A, Chen HC, Klein A, Hirschberg J, et al. 1995. Metabolism of cyclic carotenoids: a model for the alteration of this biosynthetic pathway in *Capsicum annuum* chromoplasts. *Plant J.* 8:417–24

44. Hugueney P, Romer S, Kuntz M, Camara B. 1992. Characterization and molecular cloning of a flavoprotein catalyzing the synthesis of phytofluene and *zeta*-carotene in *Capsicum* chromoplasts. *Eur. J. Biochem.* 209:399–407

45. Hundle B, Alberti M, Nievelstein V, Beyer P, Kleinig H, et al. 1994. Functional assignment of *Erwinia herbicola* Eho10 carotenoid genes expressed in *Escherichia coli. Mol. Gen. Genet.* 245: 406–16

46. Kajiwara S, Fraser PD, Kondo K, Misawa N. 1997. Expression of an exogenous isopentenyl diphosphate isomerase gene enhances isoprenoid biosynthesis in *Escherichia coli. Biochem. J.* 324:421–26

47. Kajiwara S, Kakizono T, Saito T, Kondo K, Ohtani T, et al. 1995. Isolation and functional identification of a novel cDNA for astaxanthin biosynthesis from *Haematococcus pluvialis*, and astaxanthin synthesis in *Escherichia coli. Plant. Mol. Biol.* 29:343–52

48. Karvouni Z, John I, Taylor JE, Watson CF, Turner AJ, Grierson D. 1995. Isolation and characterisation of a melon cDNA clone encoding phytoene synthase. *Plant. Mol. Biol.* 27:1153–62

49. Klaui H, Bauernfeind JC. 1981. Carotenoids as food color. See Ref. 8a, pp. 47–317

50. Kleinig H. 1989. The role of plastids in isoprenoid biosynthesis. *Annu. Rev. Plant. Physiol. Plant. Mol. Biol.* 40:39–59

51. Kleinig H, Britton G. 1982. Carotenoid biosynthesis in higher plants. *Physiol. Veg.* 20:735–55

52. Krinsky NI, Wang X-D, Tang T, Russell RM. 1994. Cleavage of β-carotene to retinoids. In *Retinoids: Basic Science and Clinical Applications*, ed. MA Livrea, G Vidali, pp. 21–28. Basel: Birkhaeuser

53. Kumagai MH, Donson J, Della-Cioppa G, Harvey D, Hanley K, Grill LK. 1995. Cytoplasmic inhibition of carotenoid biosynthesis with virus-derived RNA. *Proc. Natl. Acad. Sci. USA* 92:1679–83

54. Kuntz M, Romer S, Suire C, Hugueney P, Weil JH, et al. 1992. Identification of a cDNA for the plastid-located geranylgeranyl pyrophosphate synthase from *Capsicum annuum*: correlative increase in enzyme activity and transcript level during fruit ripening. *Plant J.* 2:25–34

55. Lichtenthaler HK, Schwender J, Disch A, Rohmer M. 1997. Biosynthesis of isoprenoids in higher plant chloroplasts proceeds *via* a mevalonate-independent pathway. *FEBS Lett.* 400:271–74

56. Lotan T, Hirschberg J. 1995. Cloning and expression in *Escherichia coli* of the gene encoding β-C-4-oxygenase, that converts β-carotene to the ketocarotenoid canthaxanthin in *Haematococcus pluvialis. FEBS Lett.* 364:125–28

57. Lütke-Brinkhaus F, Liedvogel B, Kleinig H. 1984. On the biosynthesis of ubiquinones in plant mitochondria. *Eur. J. Biochem.* 141:537–41

58. Lützow M, Beyer P. 1988. The isopentenyl-diphosphate Δ-isomerase and its relation to the phytoene synthase complex in daffodil chromoplasts. *Biochim. Biophys. Acta* 959:118–26

59. Marin E, Nussaume L, Quesada A, Gonneau M, Sotta B, et al. 1996. Molecular identification of zeaxanthin epoxidase of *Nicotiana plumbaginifolia*, a gene involved in abscisic acid biosynthesis and corresponding to the ABA locus of *Arabidopsis thaliana. EMBO J.* 15:2331–42

60. Marrs B. 1981. Mobilization of the genes for photosynthesis from *Rhodopseudomonas capsulata* by a promiscuous plasmid. *J. Bactiol.* 146:1003–12

61. Marusich WL, Bauernfeind JC. 1981.

Oxycarotenoids in poultry feed. See Ref. 8a, pp. 319–462

62. Maudinas B, Bucholtz ML, Papastephanou C, Katiyar SS, Briedis AV, Porter JW. 1977. The partial purification and properties of a phytoene synthesizing enzyme system. *Arch. Biochem. Biophys.* 180:354–62

63. Mayer MP, Beyer P, Kleinig H. 1990. Quinone compounds are able to replace molecular oxygen as terminal electron acceptor in phytoene desaturation in chromoplasts of *Narcissus pseudonarcissus* L. *Eur. J. Biochem.* 191:359–63

64. Mayne ST. 1996. *Beta*-carotene, carotenoids and disease prevention in humans. *FASEB J.* 10:690–701

65. McGarvey DJ, Croteau R. 1995. Terpenoid metabolism. *Plant Cell* 7:1015–26

66. Misawa N, Kajiwara S, Kondo K, Yokoyama A, Satomi Y, et al. 1995. Canthaxanthin biosynthesis by the conversion of methylene to ketogroups in a hydrocarbon β-carotene by a single gene. *Biochem. Biophys. Res. Commun.* 209:867–76

67. Misawa N, Masamoto K, Hori T, Ohtani T, Böger P, Sandmann G. 1994. Expression of an *Erwinia* phytoene desaturase gene not only confers multiple resistance to herbicides interfering with carotenoid biosynthesis but also alters xanthophyll metabolism in transgenic plants. *Plant J.* 8:481–89

68. Misawa N, Nakagawa M, Kobayashi K, Yamano S, Izawa Y, et al. 1990. Elucidation of the *Erwinia uredovora* carotenoid biosynthetic pathway by functional analysis of gene products expressed in *Escherichia coli. J. Bacteriol.* 172:6704–12

69. Misawa N, Satomi Y, Kondo K, Yokoyama A, Kajiwara S, et al. 1995. Structure and functional analysis of a marine bacterial carotenoid biosynthesis gene cluster and astaxanthin biosynthetic pathway proposed at the gene level. *J. Bacteriol.* 177:6575–84

70. Moore FD, Shephard DC. 1977. Biosynthesis in isolated *Acetabularia* chloroplasts. II. Plastid pigments. *Protoplasma* 92:167–75

71. Nievelstein V, Vandekerchove J, Tadros MH, Lintig JV, Nitschke W, Beyer P. 1995. Carotene desaturation is linked to a respiratory redox pathway in *Narcissus pseudonarcissus* chromoplast membranes. Involvement of a 23-kDa oxygen-evolving-complex-like protein. *Eur. J. Biochem.* 233:864–72

72. Norris SR, Barrette TR, DellaPenna D. 1995. Genetic dissection of carotenoid synthesis in Arabidopsis defines plastoquinone as an essential component of phytoene desaturation. *Plant Cell* 7: 2139–49

73. Ogura K, Koyama T, Sagami H. 1997. Polyprenyl diphosphate synthases. *Subcell. Biochem.* 28:57–87

73a. Ohnuma S, Nakazawa T, Hemmi H, Hallberg AM, Koyama T, Ogura K, Nishino T. 1996. Conversion from farnesyl diphosphate synthase to geranylgeranyl diphosphate synthase by random chemical mutagenesis. *J. Biol. Chem.* 271:10087–95

74. Parker RS. 1996. Absorption, metabolism, and transport of carotenoids. *FASEB J.* 10:542–51

75. Paton VG, Shackelford JE, Krisans SK. 1997. Cloning and subcellular localization of hamster and rat isopentenyl diphosphate dimethylallyl diphosphate isomerase. A PTS1 motif targets the enzyme to peroxisomes. *J. Biol. Chem.* 272:18945–50

76. Pecker I, Chamovitz D, Linden H, Sandmann G, Hirschberg J. 1992. A single polypeptide catalyzing the conversion of phytoene to *zeta*-carotene is transcriptionally regulated during tomato fruit ripening. *Proc. Natl. Acad. Sci. USA* 89: 4962–66

77. Pecker I, Gabbay R, Cunningham FX Jr, Hirschberg J. 1996. Cloning and characterization of the cDNA for lycopene β-cyclase from tomato reveals decrease in its expression during fruit ripening. *Plant. Mol. Biol.* 30:807–19

78. Pogson B, McDonald KA, Truong M, Britton G, DellaPenna D. 1996. Arabidopsis carotenoid mutants demonstrate that lutein is not essential for photosynthesis in higher plants. *Plant Cell* 8:1627–39

79. Reardon JE, Abeles RH. 1986. Mechanism of action of isopentenyl pyrophosphate isomerase: evidence for a carbonium ion intermediate. *Biochemistry* 25:5609–16

80. Rockholm DC, Yamamoto HY. 1996. Violaxanthin de-epoxidase. *Plant Physiol.* 110:697–703

81. Rohmer M, Seemann M, Horbach S, Bringer-Meyer S, Sahm H. 1996. Glyceraldehyde-3-phosphate and pyruvate as precursors of isoprenic units in an alternative nonmevalonate pathway for terpenoid biosynthesis. *J. Am. Chem. Soc.* 118:2564–66

82. Rost B, Casadio R, Fariselli P, Sander

C. 1995. Prediction of helical transmembrane segments at 95% accuracy. *Protein Sci.* 4:521–33

83. Sandmann G. 1994. Carotenoid biosynthesis in microorganisms and plants. *Eur. J. Biochem.* 223:7–24

84. Sandmann G, Böger P. 1989. Inhibition of carotenoid biosynthesis by herbicides. In *Target Sites of Herbicide Action*, ed. P Böger, G Sandmann, pp. 25–44. Boca Raton, FL: CRC

85. Schledz M, al-Babili S, von Lintig J, Haubruck H, Rabbani S, et al. 1996. Phytoene synthase from *Narcissus pseudonarcissus*: functional expression, galactolipid requirement, topological distribution in chromoplasts and induction during flowering. *Plant J.* 10:781–92

86. Schnurr G, Schmidt A, Sandmann G. 1991. Mapping of a carotenogenic gene cluster from *Erwinia herbicola* and functional identification of six genes. *FEMS Microbiol. Lett.* 62:157–61

87. Schulze-Seibert D, Schulz G. 1987. β-Carotene synthesis in isolated spinach chloroplasts. Its tight linkage to photosynthetic carbon metabolism. *Plant Physiol.* 84:1233–37

88. Schwartz SH, Tan BC, Gage DA, Zeevaart JA, McCarty DR. 1997. Specific oxidative cleavage of carotenoids by VP14 of maize. *Science* 276:1872–74

89. Schwender J, Seemann M, Lichtenthaler HK, Rohmer M. 1996. Biosynthesis of isoprenoids (carotenoids, sterols, prenyl side-chains of chlorophylls and plastoquinone) *via* a novel pyruvate/glyceraldehyde 3-phosphate nonmevalonate pathway in the green alga *Scenedesmus obliquus. Biochem. J.* 316:73–80

90. Scolnik PA, Bartley GE. 1993. Phytoene desaturase from Arabidopsis. *Plant Physiol.* 103:1475

91. Scolnik PA, Bartley GE. 1994. Nucleotide sequence of an Arabidopsis cDNA for geranylgeranyl pyrophosphate synthase. *Plant Physiol.* 104:1469–70

92. Scolnik PA, Bartley GE. 1994. Nucleotide sequence of an Arabidopsis cDNA for phytoene synthase. *Plant Physiol.* 104:1471–72

93. Scolnik PA, Bartley GE. 1995. Nucleotide sequence of a putative geranylgeranyl pyrophosphate synthase (GenBank L40577) from Arabidopsis. *Plant Physiol.* 108:1342

94. Scolnik PA, Bartley GE. 1995. Nucleotide sequence of *zeta*-carotene desaturase (Accession No. U38550) from

Arabidopsis (PGR95-111). *Plant Physiol.* 109:1499

94a. Scolnik PA, Bartley GE. 1996. A table of some cloned plant genes involved in isoprenoid biosynthesis. *Plant Mol. Biol. Report.* 14:305–19

95. Scolnik PA, Bartley GE. 1996. Two more members of an Arabidopsis geranylgeranyl pyrophosphate synthase gene family (Accession Nos. U44876 and U44877) (PGR96–014). *Plant Physiol.* 110:1435

96. Shanklin J, Cahoon EB. 1998. Desaturation and related modifications of fatty acids. *Annu. Rev. Plant. Physiol. Plant. Mol. Biol.* 49:611–41

97. Shanklin J, Achim C, Schmidt H, Fox BG, Münck E. 1997. Mössbauer studies of alkane-hydroxylase: evidence for a diiron cluster in an integral-membrane enzyme. *Proc. Natl. Acad. Sci. USA* 94:2981–86

98. Shanklin J, Whittle E, Fox BG. 1994. Eight histidine residues are catalytically essential in a membrane-associated iron enzyme, stearoyl-CoA desaturase, and are conserved in alkane hydroxylase and xylene monooxygenase. *Biochemistry* 33:12787–94

99. Siefermann-Harms D, Hertzberg S, Borch G, Liaaen-Jensen S. 1981. Lactucaxanthin, an ε,ε-carotene-3,3′-diol from *Lactuca sativa. Phytochemistry* 20:85–88

100. Sonnhammer ELL, Kahn D. 1994. Molecular arrangement of proteins as inferred from analysis of homology. *Protein Sci.* 3:482–92

101. Spurgeon SL, Porter JW. 1981. Biosynthesis of carotenoids. In *Biochemistry of Isoprenoid Compounds*, ed. JW Porter, SL Spurgeon, pp. 1–122. New York: Wiley

102. Street IP, Coffman HR, Baker JA, Poulter CD. 1994. Identification of Cys139 and Glu207 as catalytically important groups in the active site of isopentenyl diphosphate: dimethylallyl diphosphate isomerase. *Biochemistry* 33:4212–17

103. Summers C, Karst F, Charles AD. 1993. Cloning, expression and characterization of the cDNA encoding human hepatic squalene synthase, and its relationship to phytoene synthase. *Gene* 136:185–92

104. Sun ZR, Cunningham FX Jr, Gantt E. 1998. Differential expression of two isopentenyl pyrophosphate isomerases and enhanced carotenoid accumulation in a unicellular chlorophyte. *Proc. Natl. Acad. Sci. USA.* In press

105. Sun ZR, Gantt E, Cunningham FX Jr. 1996. Cloning and functional analysis of the β-carotene hydroxylase of *Arabidopsis thaliana*. *J. Biol. Chem.* 271:24349–52

106. Takaichi S, Sandmann G, Schnurr G, Satomi Y, Suzuki A, Misawa N. 1996. The carotenoid 7,8-dihydro-*psi* end group can be cyclized by the lycopene cyclases from the bacterium *Erwinia uredovora* and the higher plant *Capsicum annuum*. *Eur. J. Biochem.* 241:291–96

106a. Tarshis LC, Proteau PJ, Kellogg BA, Sacchettini JC, Poulter CD. 1996. Regulation of product chain length by isoprenyl diphosphate synthases. *Proc. Natl. Acad. Sci. USA* 93:15018–23

107. Tomes ML. 1967. The competitive effect of the *beta*- and *delta*-carotene genes on *alpha*- or *beta*-ionone ring formation in the tomato. *Genetics* 56:227–32

108. Walton DC, Li Y. 1995. Abscisic acid biosynthesis and metabolism. In *Plant Hormones*, ed. PJ Davies, pp. 140–57.

Dordrecht: Kluwer. 2nd ed.

109. Weedon BCL, Moss GP. 1995. Structure and nomenclature. In *Carotenoids, Vol. IB: Spectroscopy*, ed. G Britton, S Liaaen-Jensen, HP Pfander, pp. 27–70. Basel: Birkhäuser Verlag

110. Yamamoto HY. 1979. Biochemistry of the violaxanthin cycle in higher plants. *Pure Appl. Chem.* 51:639–48

111. Yamamoto HY, Bassi R. 1996. Carotenoids: localization and function. In *Oxygenic Photosynthesis: The Light Reactions*, ed. DR Ort, CF Yocum, pp. 539–63. The Netherlands: Kluwer

112. Zeevaart JAD, Creelman RA. 1988. Metabolism and physiology of abscisic acid. *Annu. Rev. Plant. Physiol. Plant. Mol. Biol.* 39:439–73

113. Zhu XF, Suzuki K, Okada K, Tanaka K, Nakagawa T, et al. 1997. Cloning and functional expression of a novel geranylgeranyl pyrophosphate synthase gene from *Arabidopsis thaliana* in *Escherichia coli*. *Plant Cell. Physiol.* 38: 357–61

Annu. Rev. Plant Physiol. Plant Mol. Biol. 1998. 49:585–609

RECENT ADVANCES IN UNDERSTANDING LIGNIN BIOSYNTHESIS

Ross W. Whetten,[1] *John J. MacKay,*[2] *and Ronald R. Sederoff*[1]

[1]Forest Biotechnology Group, Department of Forestry, North Carolina State University, Raleigh, North Carolina 27695-8008; [2]Institute of Paper Science and Technology, 500 10th Street N.W., Atlanta, Georgia 30318

KEY WORDS: monolignol biosynthesis, lignin composition, pulp manufacturing, forage quality, metabolic plasticity

ABSTRACT

After a long period of little change, the basic concepts of lignin biosynthesis have been challenged by new results from genetic modification of lignin content and composition. New techniques for making directed genetic changes in plants, as well as improvements in the analytical techniques used to determine lignin content and composition in plant cell walls, have been used in experimental tests of the accepted lignin biosynthetic pathway. The lignins obtained from genetically modified plants have shown unexpected properties, and these findings have extended the known range of variation in lignin content and composition. These results argue that the accepted lignin biosynthetic pathway is either incomplete or incorrect, or both; and also suggest that plants may have a high level of metabolic plasticity in the formation of lignins. If this is so, the properties of novel lignins could be of significant scientific and practical interest.

CONTENTS

1040-2519/98/0601-0585$08.00

INTRODUCTION

Lignins are phenolic polymers found in some plant cell walls. These complex polymers are believed to contribute compressive strength, resistance to degradation by microbial attack, and water impermeability to the polysaccharide-protein matrix of the cell wall. The basic outlines of the lignin monomer biosynthetic pathway have been known for about thirty years, but there are still ambiguities and uncertainties about how lignin monomers, or monolignols, are made, where they are stored, how they are transported to the cell wall, and how they are polymerized into lignins. Recent advances in structural studies of lignins have given new insights into the nature of the subunits in these complex and ill-defined polymers, while genetic engineering and analysis of mutations affecting monolignol biosynthesis have yielded additional information about monomer biosynthesis. The monomer biosynthetic pathway has been the subject of several reviews in the past few years, and so the pathway is only outlined here for the purposes of comparison. Readers seeking additional background information on lignin monomer biosynthesis are referred to the reviews by Boudet et al (11), Campbell & Sederoff (13), Douglas (23), and Whetten & Sederoff (103).

Although the structure and biosynthesis of lignins have been subjects of attention for more than a century [reviewed by Chen (16)], many fundamental questions remain unanswered. The structure of lignins varies between plant species, between cell types within a single plant, and between different parts of the wall of a single cell, but in no case has a complete structure for any lignin been defined. Lignins are unlike many biopolymers in that they are not simple linear polymers composed of subunits linked together by similar bonds at similar locations. There are thought to be about 20 types of intersubunit linkages in lignins, and several sites on each subunit can participate in intersubunit bonding (16). In this respect, lignins may be compared to complex polysaccharides. Some highly branched carbohydrate polymers also consist of several different types of subunits linked at various positions on the subunits, although the diversity of bond types is less than that found in lignins. Such complex carbohydrates are believed to be assembled by complexes of enzymes that join particular subunits in a stereospecific manner to form oligomers, which are then transported to the cell wall and assembled into highly complex yet ordered polymers [reviewed by Robinson (84)]. It is not yet clear whether polymerization of monolignols is organized to a similar extent, or occurs more randomly in the cell wall by a free-radical mechanism (16, 19).

Many uncertainties also exist about the pathway of precursor biosynthesis. Some enzymes carry out reactions in vitro but have not been demonstrated to act in lignin biosynthesis in vivo, while others are believed to act in vivo but do not efficiently catalyze the same reaction in vitro. Analysis of genetically modified plants has revealed the presence of subunits that cannot be explained by the traditional precursor biosynthetic pathway. Conflicting results from similar experiments carried out in different plant species by different laboratories leave open the question of whether a single monolignol biosynthetic pathway exists in higher plants, or whether different plant genera may use different strategies to synthesize and polymerize lignin monomers.

Great variation in lignin structure exists, but we have no information about the relationship of lignin structure and function. Broad questions of the role of stereospecific precursors, the mechanisms controlling lignin polymerization, and the potential plasticity of lignin structure are only beginning to be addressed. Recent results suggest that there is unexpected complexity and flexibility in the biosynthesis of lignins. The review summarizes recent results and discusses the implications for our understanding of lignin biosynthesis.

LIGNIN STRUCTURE, ANALYSIS, AND VARIATION

Lignins are traditionally considered to be polymers of three alcohol monomers, or monolignols: p-coumaryl alcohol, coniferyl alcohol, and sinapyl alcohol (32; see Figure 1). Each type of precursor may form several types of bonds with other precursors in forming a lignin polymer, and in some cases such bonds can obscure later analysis of the fractional composition of the three subunits. Monolignols can also form bonds to other cell wall polymers in addition to lignin, cross-linking polysaccharides and proteins with lignin in a very complex three-dimensional network (16).

Traditional methods for analyzing lignin monomer composition have relied on chemical degradation of lignin to release substituted aromatic residues with hydroxylation/methoxylation patterns similar to those of the monolignols (Figure 1). A variety of chemical degradation methods have been developed over the years, and new methods continue to emerge (11, 12, 61, 63). No single chemical degradation method of analysis yields complete information on the content and composition of lignins in plant samples, because some types of bonds are very difficult to break (11, 16, 61).

Nuclear magnetic resonance (NMR) spectroscopy has become a powerful tool for the determination of lignin structure (16, 55, 59), particularly when used in combination with chemical degradation analysis. NMR is carried out

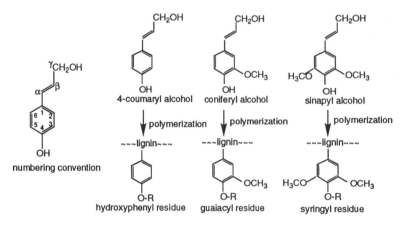

Figure 1 Structures of the three monolignols and the residues derived from them. The numbering convention used is shown on the *left*. H, G, and S residues are depicted with an R group bonded to the oxygen at the 4-position. This may be a proton, in the case of a free phenolic hydroxyl, or another monolignol residue.

by inducing transitions between different energy levels of atomic nuclei in a static magnetic field, resulting in signals that are highly diagnostic of chemical structure. This technique may be carried out on small amounts of material with essentially no destruction or modification. Pulsed two-dimensional NMR correlation analysis has been a critical tool for inference of molecular composition and organization of lignins and has been used to confirm basic aspects of lignin structure that had been predicted earlier by chemical analysis of degradation products (29).

Lignin content and composition have long been known to vary between the major groups of higher plants and between species (27, 88). The two major classes are (*a*) gymnosperm lignins, which primarily contain guaiacyl subunits (also called G units), polymerized from coniferyl alcohol, and a small proportion of *p*-hydroxyphenyl units (H units) polymerized from *p*-coumaryl alcohol; and (*b*) angiosperm lignins, which contain both syringyl units (S units), polymerized from sinapyl alcohol, and G units, with a small proportion of H units (Figure 1).

Exceptions to this basic classification are found in both gymnosperms and angiosperms. *Podocarpus* species, among the gymnosperms, and certain species within the Gnetales contain significant amounts of syringyl lignin; for example, *Ephedra trifurca* (Gnetales) is reported to contain lignin with a 1:1 S:G ratio similar to that of angiosperms (88). *Erythrina cristagalli* is an angiosperm that produces guaiacyl lignin similar to that characteristic of gymnosperms (48).

A few angiosperm families include species that produce lignins with low syringyl content, including the Magnoliaceae, which are sometimes referred to as "primitive" angiosperms. In the other extreme, some angiosperms are reported to have very high S:G ratios, up to 8:1 (70). Variation in the proportion of these conventional precursors is found within the cell wall, between cell types, and in response to biotic (e.g. pests and pathogens) and abiotic (e.g. mechanical) stresses. The existence of more than one type of lignin and the precise regulation of monomer composition during development and stress responses strongly implies that alternative lignins have different adaptive or functional properties. Detailed description of lignin variation has been the subject of many reviews (11, 13, 58, 69).

Natural variation of lignins extends beyond variation in the fractional composition of the three conventional monolignol precursors. Several new subunits and structures have recently been described. The refinement of lignin characterization has advanced greatly owing to application of NMR spectroscopy. Results from such studies have led to a more detailed view of natural variation of lignin structure, implicating new structures as lignin subunits. In addition, as mutational analysis and the use of transgenic plants are applied to the study of lignins, the limits of compositional variation have been pushed even further.

MONOLIGNOL BIOSYNTHESIS

Monolignol biosynthesis has been extensively reviewed in the past few years, and it is summarized here only to point out areas in which recent progress has been made and areas in which questions still remain. The monolignol biosynthetic pathway converts phenylalanine or tyrosine to one of three monolignols. The first step is deamination of phenylalanine or tyrosine by phenylalanine ammonia-lyase (PAL) or tyrosine ammonia-lyase (TAL), respectively. TAL activity has only been reported in grasses, and recent work has shown that a single maize cDNA expressed in *Escherichia coli* encodes both PAL and TAL activity (86). The product of PAL activity on phenylalanine is *trans*-cinnamic acid, which is the substrate for a P-450 hydroxylase, cinnamate 4-hydroxylase (C4H). C4H hydroxylates the 4 position of the aromatic ring of cinnamate to produce *para*-coumaric acid. The result of the combined activity of PAL and C4H on L-phenylalanine, or the direct product of TAL activity on tyrosine, is *para*-coumaric acid (Figure 2). Genetic engineering of plants with reduced PAL activity has been used to show that carbon flux into lignin is affected only in plants that have a severe reduction in PAL activity, although other branches of phenylpropanoid metabolism respond more readily to alterations in levels of PAL (5). Recent analysis of lignin composition in transgenic plants with

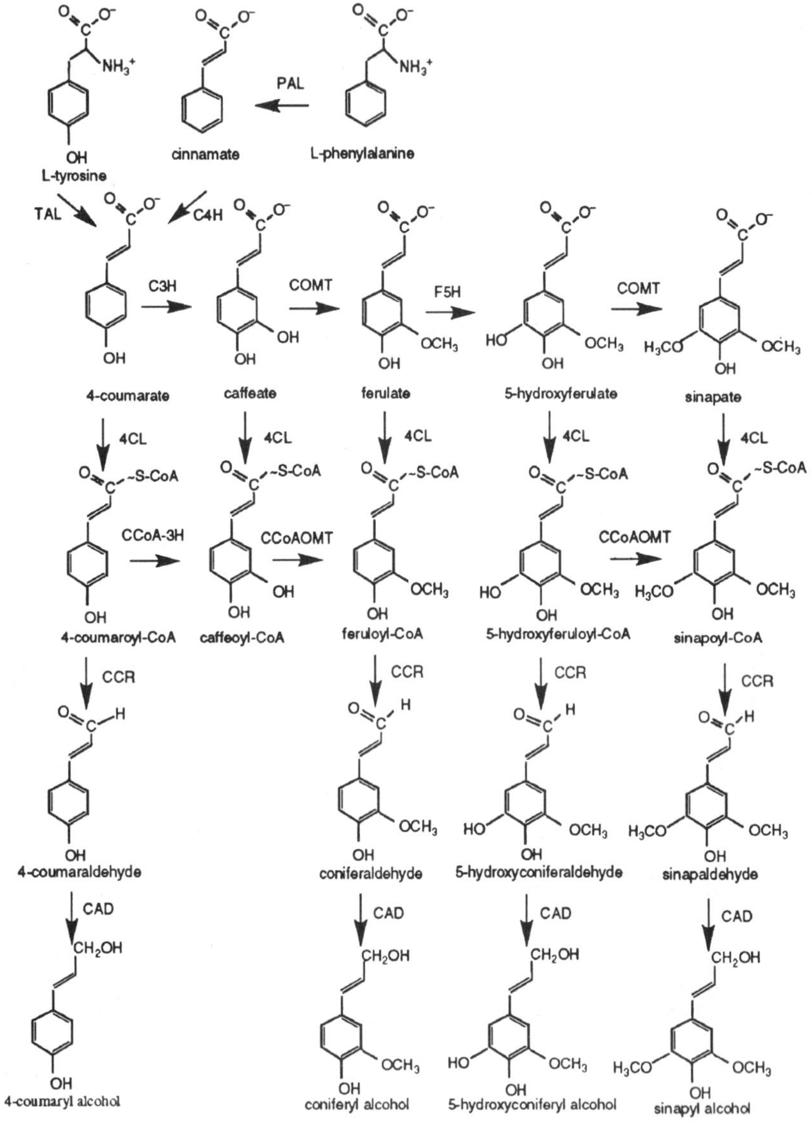
PAL
L-tyrosine
cinnamate
L-phenylalanine
TAL
C4H
C3H
COMT
F5H
COMT
4-coumarate
caffeate
ferulate
5-hydroxyferulate
sinapate
4CL
CCoA-3H
CCoAOMT
CCoAOMT
4-coumaroyl-CoA
caffeoyl-CoA
feruloyl-CoA
5-hydroxyferuloyl-CoA
sinapoyl-CoA
CCR
4-coumaraldehyde
coniferaldehyde
5-hydroxyconiferaldehyde
sinapaldehyde
CAD
4-coumaryl alcohol
coniferyl alcohol
5-hydroxyconiferyl alcohol
sinapyl alcohol

reduced levels of PAL or C4H activity suggests that changes in the amounts of these enzyme activities can affect the S:G ratios in ways that are not consistent with the accepted pathway of monolignol biosynthesis (92). Further analysis of such plants will be necessary to confirm these results, and to explore the authors' hypothesis that multienzyme complexes may partition metabolites into G-specific and S-specific pathways very early in monolignol biosynthesis (92).

The next step is ambiguous. The traditional view of the pathway has been that *p*-coumaric acid is hydroxylated at the 3-position by *p*-coumarate 3-hydroxylase (C3H). Enzymes that can carry out this hydroxylation reaction in vitro have been purified, but no corresponding evidence exists to support a physiological role for these enzymes in monolignol biosynthesis in vivo [reviewed by Whetten & Sederoff (103)]. Caffeic acid, the presumed product of hydroxylation of *p*-coumarate, is known to occur in plants, but the mechanism by which it is synthesized is not yet known. Whether caffeic acid is an obligatory intermediate in the synthesis of monolignols (as opposed to synthesis of other low-molecular weight phenolic compounds) has not been rigorously tested. An alternative hypothesis is that *p*-coumaric acid is converted to *p*-coumaroyl-CoA by 4-coumarate:CoA ligase (see next section) and hydroxylated at the level of the CoA thioester, to produce caffeoyl-CoA. It is also formally possible that *p*-coumarate is converted to some compound other than caffeic acid or its CoA derivative, either directly to ferulic acid or to another intermediate. No experimental evidence exists to support this alternative, and this area is an important gap in our understanding of monolignol synthesis.

The traditional pathway holds that caffeic acid is methylated on the 3-hydroxyl to produce ferulic acid by caffeic acid O-methyltransferase (COMT). More recently, a caffeoyl-CoA O-methyltransferase (CCoAOMT) enzyme has been hypothesized to play a role in monolignol biosynthesis (106). Recent work leading to isolation of a cDNA clone encoding pine COMT has revealed

---

Figure 2 An outline of the monolignol synthetic pathway. The general flow of the pathway is from L-phenylalanine or L-tyrosine at the *upper left*, through increasing degrees of substitution of the aromatic ring across the *top*, then through a two-stage reduction of the side chain acid group to the alcohols at the *bottom*. The enzymes that catalyze conversion of intermediates are shown next to the *arrows*. As described in the text, the ability of an enzyme to catalyze a reaction in vitro is not proof that the enzyme catalyzes the reaction in vivo, and some conversions that apparently occur in vivo cannot yet be carried out in vitro. Abbreviations: PAL, phenylalanine ammonia-lyase; TAL, tyrosine ammonia-lyase; C4H, cinnamate 4-hydroxylase; C3H, coumarate 3-hydroxylase; COMT, caffeate O-methyltransferase; CCoAOMT, caffeoyl-Coenzyme A O-methyltransferase; F5H, ferulate 5-hydroxylase; 4CL, 4-coumarate:Coenzyme A ligase; CCR, cinnamoyl-Coenzyme A reductase; CAD, cinnamyl alcohol dehydrogenase.

that the pine enzyme called AEOMT will methylate both free acids (A) and CoA esters (E) (60). Pine thus appears to differ from *Zinnia*, because assays of the *Zinnia* COMT expressed from a cDNA clone showed the enzyme lacks activity on CoA esters of hydroxycinnamic acids (107). This portion of the monolignol biosynthetic pathway appears to be more a network or grid than a linear pathway, but the relative contributions of these various enzyme activities in different plant species is far from clear.

Ferulic acid can be activated by 4-coumarate:CoA ligase (4CL) through conjugation to Coenzyme A, as can *p*-coumarate, and these CoA thioesters are presumed intermediates in the synthesis of coniferyl alcohol and *p*-coumaryl alcohol, respectively. An alternative fate for ferulic acid is hydroxylation at the 5-position by another P-450 hydroxylase, ferulate 5-hydroxylase (F5H). 5-Hydroxyferulic acid can be methylated by COMT to produce sinapic acid. In most cases, however, sinapic acid has been shown to be a poor substrate for purified 4CL enzymes in vitro (37, 49, 64, 102).

There have long been questions about how syringyl residues are produced in lignin if sinapic acid cannot be efficiently activated to the CoA thioester by 4CL. Higuchi (42) suggested an alternative route through 5-hydroxy-coniferaldehyde, followed by methylation by an OMT and reduction by a CAD, and Ye & Varner (107) suggested that 5-hydroxyferuloyl-CoA might be methylated by CCoAOMT to produce sinapoyl-CoA. Available evidence supports a role for COMT in formation of sinapyl alcohol, but the level at which methylation occurs (acid, thioester, aldehyde, or alcohol) cannot yet be resolved. Transgenic plants with altered COMT activity, as well as maize plants homozygous for mutations in the structural gene encoding COMT at the *brown-midrib* 3 (*bm3*) locus, show reductions in the levels of syringyl residues in lignin, as well as accumulation of a novel 5-hydroxyconiferyl alcohol residue (3, 54, 101). The appearance of the novel residue is consistent with the hypothesis that COMT acts on 5-hydroxyferulic acid or 5-hydroxyconiferaldehyde, and that reductions in COMT enzyme activity result in accumulation of the substrate, which is subsequently reduced to the alcohol and incorporated into lignin. In most cases, alterations in COMT activity in planta result in greater reductions in syringyl content than in guaiacyl content (see Table 1). These results are consistent with the hypothesis that CCoAOMT is unable to act on 5-hydroxylferuloyl-CoA in vivo to complement the reduction in COMT activity. This inability to act may be due to lack of access to substrate rather than intrinsic properties of the enzyme, as purified CCoAOMT is active with both caffeoyl-CoA and 5-hydroxyferuloyl-CoA (107).

Recent transgenic plant experiments testing the role of 4CL have yielded conflicting results in different species, suggesting that this step may vary between plant taxa. Transgenic Arabidopsis with reduced 4CL activity show decreased

Table 1   Modification of lignin in transgenic plants

Transgenic method	Species	Enzyme effect[1]	Lignin content[2]	Lignin composition[3]	Reference
Sense suppression of *pal* (heterologous *pal*)	Tobacco	PAL 5–30%	n.d. Decreased 10–80%		25 5
Sense suppression of *pal*	Tobacco	PAL decreased	Decreased	Increased S:G	92
Antisense C4H	Tobacco	C4H decreased	Decreased	Decreased S:G	92
Antisense *omt*	Tobacco	OMT 5%	No effect	Decreased S:G	3
Sense suppression *omt*	Tobacco	OMT 180%	No effect	No effect	3
Antisense *omt*	Poplar	OMT 2%	No effect	Decreased S:G 5-OH-G subunits	101
Antisense and sense suppression *omt*	Poplar	OMT decreased	No effect	Decreased S only	99
Antisense 4CL	Tobacco	4CL 8%	Decreased to 50%	Decreased G only	56
Antisense 4CL and sense suppression	Tobacco			Decreased G & S	44, 45
Antisense *ccr*	Tobacco	CCR 25%	Decreased to 75%	Increased S:G	10
Sense suppression *ccr*	Tobacco	CCR 2%	n.d.	Decreased S:G	74
Antisense *cad*	Tobacco	CAD 7%	No effect	Increased aldehyde	38
Antisense *cad*	Tobacco	CAD 50%	No effect	Increased aldehyde	40
Antisense *cad*	Poplar	CAD 30–50%	No effect	Increased aldehyde	6
Antisense *pod*	Tobacco	POD decreased	No effect		51
Overexpression *pod*	Tobacco	POD increased	Increased to ~130%	n.d. slower growth	50
Antisense *lac*	Tulip poplar	LAC 10% (preliminary)	n.d.		21
Constitutive *f5h*	Arabidopsis	F5H ectopic	No effect	Ectopic S lignin	67
Overexpression *f5h*-C4H prom fusion	Arabidopsis			Almost only S lignin	67
Introduction of *tdc* (tryptophan decarboxylase)	Potato	TDC active	Decreased 60%	Decrease S:G	105

[1]Enzyme effect: level of enzyme activity relative to wild type (%). POD, peroxidase; LAC, laccase.
[2]Effect on lignin content, % of wild type content. n.d., not determined.
[3]Effect on lignin composition relative to wild type.

levels of guaiacyl residues in lignin but little change in the levels of syringyl residues, even when 4CL activity is reduced to a few percent of normal levels (56). Transgenic tobacco with reduced 4CL activity, however, shows greater reduction of syringyl content than guaiacyl content in some cases, although both types of residues are affected (44, 45). Further comparisons between transgene constructs and between species are necessary to verify these results and to further test the hypothesis that monolignol biosynthesis differs between Arabidopsis and tobacco.

Evidence confirming that F5H functions in monolignol biosynthesis is available from genetic studies in Arabidopsis. The *fah*1 mutant of Arabidopsis lacks F5H activity, and the lignin of this mutant contains only guaiacyl residues (15). Overexpression of the F5H gene from the CaMV 35S promoter, or from the promoter of the Arabidopsis C4H gene, results in an increase in syringyl content and depletion in guaiacyl content of the lignin relative to wild-type control plants (C Chapple, personal communication). Taken together, the results of the Chapple and Douglas groups suggest that sinapic acid, or at least 5-hydroxyferulic acid, is an intermediate in the synthesis of monolignols, but that 4CL activity may not be required for production of sinapyl alcohol in Arabidopsis. It would be interesting to make a double mutant in Arabidopsis that lacks both COMT and 4CL activity, then examine the lignin to see whether 5-hydroxyconiferyl alcohol residues are detectable. This experiment would test the involvement of 4CL in activation of 5-hydroxyferulic acid in planta, as 4CL in Arabidopsis extracts is active on 5-hydroxyferulic acid in vitro (56). Such double mutant plants would also provide information about the potential role of CCoAOMT as an alternative methylation pathway in synthesis of sinapyl alcohol.

The CoA-thioesters of *p*-coumarate, ferulate, and sinapate are considered to be substrates for cinnamoyl-CoA reductase (CCR), which reduces these thioesters to yield the corresponding aldehydes. The aldehydes, in turn, are substrates for cinnamyl alcohol dehydrogenase (CAD), which reduces the aldehyde end group to an alcohol to yield the three monolignols. Transgenic plant experiments have recently shown that reduction of CCR activity by genetic engineering reduces lignin content (10), while a number of studies of both transgenic plants and plants homozygous for mutations reducing CAD activity have shown that cinnamaldehydes are incorporated into lignins. Both of these results are consistent with the roles CCR and CAD have been traditionally assigned in the pathway, but a few surprises have emerged from these studies as well.

The reduction of cinnamaldehydes by CAD is usually considered to be the last enzymatic reaction in the formation of all three cinnamyl alcohol lignin precursors. A long-standing question has been whether this is accomplished by single or several CAD enzymes. A number of CAD enzymes and *cad* genes have been investigated in diverse plant species, and the biochemistry and molecular biology of CAD have been the object of recent reviews (e.g. 11). In addition to *cad* cDNA clones isolated on the basis of information obtained from purified CAD enzymes, other cDNA clones have been isolated that have a lower level of similarity to this first group (approximately 50% identity). The functions of these gene products have been less well characterized. These include the ELI3 gene, an elicitor inducible gene isolated from parsley and present in several other plant species (93, 98). Expression of an Arabidopsis ELI3 protein in bacteria has recently demonstrated that ELI3 catalyzes the reduction of aromatic aldehydes, including 2-methoxybenzaldehyde (the preferred substrate) and the aldehyde forms of monolignols (94). These findings suggest that the unraveling of CAD enzyme diversity and specificities may yet be at its early stages.

Analyses of mutants and transgenic plants in which the main CAD enzyme associated with lignification has been down-regulated have helped confirm the function and role of CAD but have not clearly established whether a single enzyme is responsible for the formation of the three cinnamyl alcohols. Accumulation of CAD substrates in alcohol soluble fractions (65) or in alkaline hydrolysates (6) indicated that decreased CAD activity can result in an accumulation in all three of its cinnamaldehyde substrates. The *cad-n1* mutant of loblolly pine accumulated free coniferaldehyde to a high level and had a small accumulation of *p*-coumarylaldehyde, although levels of *p*-hydroxyphenyl units in lignin were unchanged (82). Transgenic poplar with reduced CAD activity yielded a significantly increased amount of vanillin (derived from coniferaldehyde) and syringaldehyde upon extractions with 1 N NaOH (6).

A key observation is that even severe reduction of CAD activity has only a small effect on total lignin content in several plant species (6, 38, 65). A few reports have documented the increased incorporation of certain cinnamaldehydes into the lignin polymer, accompanied by a decrease in the corresponding alcohol-derived subunit (38, 65, 82). Halpin et al (38) reported that in transgenic tobacco with reduced CAD activity, the shift in the aldehyde-to-alcohol ratio was greater for sinapyl than coniferyl moieties, perhaps reflecting preferential reduction of coniferaldehyde when CAD activity was limiting. Finally, the milled wood lignin (MWL) of the pine *cad-n1* mutant showed no differences in either *p*-coumaryl aldehyde– or *p*-coumaryl alcohol–derived subunits compared with the wild type (82). These findings are consistent with a single CAD enzyme being sufficient for the formation of lignin precursors during normal development but suggest that other enzymes may also contribute. The degree to which other enzymes contribute may increase in plants with reduced primary CAD activity, but this is a very difficult hypothesis to test.

CAD has been considered essential for lignin biosynthesis, but this hypothesis may need to be reexamined in view of the small effect of reduced CAD activity on lignin content and the incorporation of cinnamaldehyde into lignin of transgenic and mutant plants of several species. The polymer produced in these transgenic and mutant plants differs from the traditional view of lignin as a polymer of hydroxycinnamyl alcohols, but it seems functionally equivalent to traditional lignins in terms of plant support and water transport (6, 38). Incorporation of cinnamaldehyde into lignin is associated not only with a reduction in CAD activity but also with a reduction in COMT activity (99) or both CAD and COMT (76). It is often linked with a brown coloration of the vascular tissue, although a similar color is also found in grasses in association with brown-midrib mutations unrelated to CAD and in the absence of increased cinnamaldehyde. The results from mutants and transgenics do not yet offer a consistent explanation of the relationship between the level of CAD activity, the incorporation of cinnamaldehyde, and the brown coloration of the vascular tissue. For example, Hibino et al (40) found that tobacco plants harboring an antisense *cad* gene, in which CAD activity was about 50% of wild-type levels and vascular tissues were not brown, incorporated two times the wild-type level of cinnamaldehyde. The same increase in cinnamaldehydes (two times the wild-type level) was found in MWL of loblolly pine homozygous for the *cad-n1* mutation, although these plants have only about 1% of wild-type CAD activity, have brown wood, and accumulate free cinnamaldehyde to high levels (65, 82). This twofold increase of cinnamaldehyde incorporated into lignin, observed in plants with very different levels of CAD activity, may indicate the upper limit that can be achieved in plants. Wild-type lignins are estimated to contain 3–5% of cinnamaldehyde, but isolated lignins such as MWL can contain 10–15% aromatic aldehydes owing to oxidation during isolation (12, 108). An alternative explanation for the fact that increase in aldehydes does not seem to be correlated with the level of CAD activity is experimental inconsistencies because different approaches were used to estimate the amount of aldehyde. The fact that unconventional subunits are formed and incorporated into lignin in plants with severe reduction in CAD activity and concomitant limiting supply of conventional alcohol precursors (see next section; 82) offers additional support for the hypothesis that plants may limit the amount of cinnamaldehydes incorporated into lignin.

NONCONVENTIONAL LIGNIN SUBUNITS

Grasses

Within grasses (angiosperm monocots), recent studies have highlighted significant departures from the conventional definition of the lignin subunit. *Para*-coumaric acid is found esterified to hydroxycinnamyl subunits in lignins from many grasses (reviewed by 7). Ralph et al (81) used NMR to show that *p*-coumarate esterification occurs exclusively at the terminal (gamma) position of the side chain of the monolignol (Figure 3), and further analysis by thioacidolysis showed that about 90% of the *p*-coumarate is esterified to syringyl

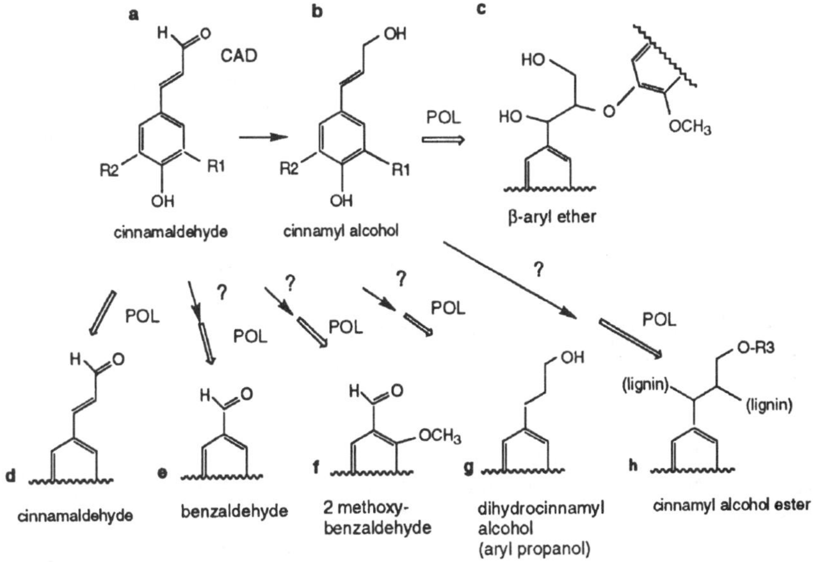

Figure 3 Novel lignin subunits. (*a*) A general structure for substituted cinnamaldehydes is shown. (*b*) The usual pathway of reduction by CAD to form a substituted cinnamyl alcohol and subsequent polymerization (POL) is shown to the *right*. The different cinnamaldehydes have the following substituents on the aromatic ring: p-coumaraldehyde, $R_1 = H$ and $R_2 = H$; coniferaldehyde, $R_1 = OCH_3$ and $R_2 = H$; 5-hydroxyconiferaldehyde, $R_1 = OCH_3$ and $R_2 = OH$; and sinapaldehyde, $R_1 = OCH_3$ and $R_2 = OCH_3$. (*c*) The β-O-4 ether bond between subunits is shown because it is common in normal lignins (16). *Below* the normal pathway are shown novel residues recently detected in lignins (see text). The *question marks* next to the *arrows* leading to these residues indicate that the biochemical pathways leading to these molecules are not yet known. Carbons 3, 4, and 5 of the aromatic ring are not shown but may be involved in a variety of linkages to other moieties in the lignin polymer. (*d*) Unmodified cinnamaldehyde side chains, (*e*) benzaldehyde, (*f*) 2-methoxybenzaldehyde, and (*g*) dihydrocinnamyl alcohol residues have been detected in lignins obtained from mutant or transgenic plants (see text). (*h*) Cinnamyl alcohol esters, in which R_3 may be either an acetate group or a p-coumarate group, have also been reported. The α and β carbons of the cinnamyl alcohol ester (*h*) side chain can be, but are not necessarily, involved in links to other residues in the polymer, shown here as lignin.

subunits (35). This regiospecificity implies that the monolignol-p-coumarate esters are formed before export from the cytoplasm to the cell wall. If this is true, both monolignol-p-coumarate esters and free monolignols are substrates for lignin formation in grasses (81).

Kenaf

An unusual lignin has been found in kenaf (*Hibiscus cannabinus*) that further extends the range of natural subunit variation. Early work reported association of acetate groups with lignin in kenaf, and more recent analyses have clarified the nature of the association (17, 78). NMR spectra revealed that about 50% of the subunits in kenaf lignin are acetylated, almost entirely at the gamma position of the side chain. Small amounts of acetylated subunits had previously been described in association with hardwood lignins (87). Acetylation of subunits is associated with a relatively high syringyl to guaiacyl ratio in kenaf lignin (6:1, compared with 1:1 in most angiosperms), and this association has led to the speculation that acetylation might play a role in formation of high syringyl-content lignins (78). There is as yet no conclusive evidence of formation of acetylated monolignols before export from the cytoplasm, but it is reasonable to speculate that similar mechanisms may operate in formation of monolignol acetates in kenaf and monolignol *p*-coumarates in grasses (78). This hypothesis deserves experimental testing, as enzymes capable of forming such monolignol esters should exist in several plant species.

Pine

A mutation (*cad-n1*) resulting in loss of virtually all CAD activity has recently been described in loblolly pine (82). NMR characterization of lignin from a tree homozygous for this mutation revealed two structural units not normally associated with the lignin biosynthetic pathway (82). The major subunit in this lignin is dihydroconiferyl alcohol (Figure 3). This compound has been reported in diverse plants, including pines, but has not been implicated as a lignin subunit. The formation of such a structure with an alcohol end group is unexpected in plants that are deficient in CAD activity. The saturated side chain of this subunit restricts its participation in β-O-4 linkages and α-O-4 linkages and increases the proportion of C-C bonds, as revealed by such structures as bibenzodioxicin (82). The MWL of the *cad-n1* mutant also contains elevated levels of vanillin, which can be derived from coniferaldehyde by an aldol reaction, and a 2-methoxybenzaldehyde derivative, not previously described as a lignin subunit (82). These two structures contribute to the aldehyde content of the lignin and to the reduction in the number of beta linkages (Figure 3). The mixture of unusual subunits and increased fractions of normally rare subunits (coniferaldehyde and vanillin) found in this lignin does not allow determination of the effect of each subunit individually.

Tobacco

It was recently shown that transgenic tobacco in which CAD activity has been reduced by antisense (10) incorporates the same 2-methoxybenzaldehyde derivative as the pine *cad-n1* mutant but does not incorporate dihydroconiferyl alcohol (79). This finding may provide some important clues regarding the compensatory mechanism that may be activated in CAD-deficient plants. The transgenic tobacco has significantly greater residual CAD activity than does the pine mutant. Differences in lignin monomer composition may reflect different compensatory mechanisms that are activated as a function of the level of residual CAD activity. Mutants or transgenics with varied levels of reduction in CAD activity can be expected to provide complementary information on modification of lignin structure and chemical behavior.

The characterization of subunits not considered as conventional intermediates or products of the lignin biosynthetic pathway is of great interest. These new subunits could give new insights into lignin biosynthesis and its regulation. The brown midrib mutants *bm3* of maize and transgenic tobacco and poplar plants in which COMT activity has been reduced all synthesize 5-hydroxyguaiacyl lignin subunits (Figure 1) (3, 54, 101). The incorporation of this novel subunit may provide considerable insight into the organization of pathways leading to syringyl and guaiacyl subunits and hence to the potential mechanisms regulating their relative amounts.

The formation of 5-hydroxyconiferyl alcohol can be predicted from the conventional lignin pathway but would require reduction of 5-hydroxyferuloyl-CoA to 5-hydroxyconiferaldehyde and reduction of 5-hydroxyconiferaldehyde to the alcohol, enzymatic reactions not normally attributed to CCR and CAD. The effect of this unusual subunit on lignin structure and properties has not yet been described in detail. Poplar trees containing this novel subunit were less efficiently delignified during pulping, but this could be due, at least in part, to their severe reduction in syringyl subunits (9).

CHEMISTRY OF MODIFIED LIGNIN

Modification of lignin composition could lead to differences in structure and properties that modify the behavior of lignin or cell walls in traditional methods of lignin determination or characterization. The behavior of lignins enriched in cinnamaldehydes when analyzed by thioacidolysis supports this idea. Thioacidolysis did not detect the increase in cinnamaldehyde subunits in transgenic tobacco (38) and poplar (6) lignins containing an increased proportion of these subunits. One explanation for this result is that cinnamaldehyde subunits were infrequently coupled through β-O-4 ether linkages in these samples, and thioacidolysis failed to detect the increased cinnamaldehyde content because it is rather selective for this linkage type (11). This hypothesis is supported by the NMR characterization of MWL from the pine *cad-n1* mutant, indicating a low proportion of β-O-4 ether linkages (82).

Modification of the type of coupling between lignin subunits could affect lignin removal during chemical pulping. Transgenic tobacco in which CCR activity was decreased by an antisense transgene contained a much greater fraction of the lignin that was removed by mild alkaline extraction than control plants (10). Furthermore, increased cinnamaldehyde was associated with lower residual lignin in pulp derived from low-CAD transgenic poplar (6) and mutant pine (J MacKay, T Presnell, H Taneda, unpublished data) wood samples. Increases in cinnamaldehyde levels have also been associated with increased delignification in mild alkaline extractions in transgenic tobacco (36). Mild alkaline extraction of wood from the pine *cad-n1* mutant (J MacKay, unpublished data) and the *cad* antisense poplars yielded large amounts of vanillin and syringaldehyde, respectively, further supporting the hypothesis that the incorporation of aromatic aldehydes facilitates lignin removal. Although the chemical bases of these modified properties have not been determined, lower molecular weight and/or a higher proportion of phenolic hydroxyl groups might be expected if there are fewer β-O-4 linkages. Alternatively, β-O-4 coupling involving the beta carbon of coniferaldehyde may prevent cross-linking to cell wall components at the alpha carbon. Cross-linking normally occurs with quinone methide intermediates after β-O-4 coupling (91); however, during β-O-4 coupling of coniferaldehyde, the quinone methide could form an enone structure preventing cross-linking at the alpha position (36).

IMPLICATIONS OF NONCONVENTIONAL SUBUNITS

Together, the incorporation of these diverse subunits indicated that considerable metabolic plasticity exists not only between taxonomic groups but also within an individual species (11, 82). The amount of lignin in a plant or a given tissue can be maintained at a relatively constant level, even when an enzyme normally involved in the synthesis of lignin precursors is suppressed or impaired, and subunits not found in normal lignins are frequently present in such cases. This result suggests that compensatory mechanisms, involving uncharacterized enzymes or at least new enzyme functions, must allow plants to synthesize nonconventional lignin subunits and thus form dramatically altered lignins. This raises an important question. How does altering lignin composition affect structure and function of lignin, and of the plant as a whole? It should be noted that monolignols are thought to be precursors for phenolic products other than lignin, particularly including phytoalexins and compounds with growth-regulating effects, and that suppression of key steps in the lignin pathway could have additional biological effects.

The formation of nonconventional lignin could provide new insights into the structure-function relationship of lignins. Lignin composition and structure may be under certain constraints to maintain the essential functions of lignin. Compensatory mechanisms may allow plants to synthesize nonconventional lignins that still maintain essential functions. In some transgenic plants, variation in lignin composition had no obvious effect on the physiology of the plant (11). However, the effect of reduced or altered lignin composition may not become obvious unless plants are tested under extreme conditions, such as drought, mechanical stress, or intense pathogen attack.

MONOLIGNOL TRANSPORT AND POLYMERIZATION

The 4-O-β-D-glucosides of monolignols that accumulate in the cambial sap of gymnosperms and some angiosperms (30, 89, 97) have long been considered to be transport or storage forms of monolignols. These glucosides are formed by transfer of a glucose residue from UDP-glucose to the phenolic hydroxyl of the monolignol, catalyzed by a specific UDPG-glucosyltransferase (43). Coniferin, the glucoside of coniferyl alcohol, accumulates to levels of a few percent of wet tissue weight in differentiating xylem of conifers (89, 97). Hydrolysis of coniferin by a beta-glucosidase is considered essential before polymerization in the cell wall may proceed. A coniferin-beta-glucosidase (CBG) has been purified and characterized from pines (22, 57). CBG has been implicated in lignification by its enzymatic specificity, cellular location, and time of expression in development in both gymnosperms and angiosperms (22; B Ellis, personal communication). The expression of the enzyme is tightly associated with initiation of lignin deposition in lignifying cells during xylem differentiation.

Nucleation of Lignification

The question of how plant cells regulate the deposition of lignin in the cell wall has been important since early workers identified differences in lignin subunit composition and cross-linking patterns between different parts of the cell wall (reviewed in 42). One mechanism that has been suggested is the nucleation of lignin deposition at specific sites by other cell wall components such as polysaccharide-phenolic esters or cell wall proteins.

The structural similarity of ferulic acid to coniferyl alcohol suggests that feruloyl esters of polysaccharides could participate in lignin-carbohydrate cross-linking reactions. Ferulate can be esterified to arabinose or arabinoxylans before transport to the cell wall (34, 68, 71) and has been shown to participate in cross-linking between the lignin polymer and cell wall polysaccharides (52, 53, 80). Recent studies in grasses indicate that ferulate may function as a site to which lignin monomers attach and from which lignin polymers grow (71, 80). This hypothesis can be experimentally tested as soon as more information is available about the enzymes that form ferulate esters during polysaccharide export to the cell wall. It may well prove that the distribution and amount of feruloylated polysaccharides in plant cell walls can be genetically manipulated, with desirable effects on patterns of lignin deposition and cross-linking to other wall components.

Proteins rich in aromatic amino acid residues, particularly tyrosine, have also been suggested as potential sites for nucleation of lignin deposition. Early work by Whitmore (104) suggested that linkages are formed between cell wall proteins and lignin. More recently, Keller et al (47) suggested that the tyrosine residues of glycine-rich cell wall proteins may be involved in linkages to lignin. To test this, McDougall et al (66) studied the incorporation of synthetic proteins into dehydrogenation polymers synthesized in vitro from the peroxidase-catalyzed polymerization of coniferyl alcohol. Synthetic proteins composed of a random co-polymer of lysine and tyrosine increased the yield of synthetic lignin polymer compared with a polymer of lysine alone. The authors suggested that the phenolic tyrosines may enhance polymerization as templates or as nucleation sites for free-radical polymerization during the formation of lignin in vivo.

Oxidases and Polymerization of Lignin

Erdtmann (26, 33) proposed a dehydrogenative mechanism for the production of free-radical intermediates that could polymerize into lignin. Since that time, many workers have looked for enzymes that could create such intermediates or had an alternative role in the formation of the lignin polymer. In early work, Freudenberg & Richtzenhain (33) used fungal enzymes, essentially crude laccase preparations, to polymerize monolignols in vitro. Later experimenters were able to polymerize lignin with enzyme-containing extracts from higher plants. Both peroxidase and laccase, identified in lignin-forming tissues, were candidates for carrying out the formation of higher-order polymers from monolignols (28, 31, 41). This is still the case, although laccase was for some period of time excluded from consideration as a potential participant in lignin polymerization.

For many years, peroxidase was favored over laccase for this potential role in lignification (39, 72), in part owing to the difficulties in solubilizing activities of enzymes strongly associated with the cell wall. The question of laccase involvement in lignification was revived in 1983 by the identification of a laccase in cell cultures of *Acer pseudoplatanus* (8). Improved procedures resulted in the characterization and purification of oxidase activities from several woody plants (4, 24, 95).

Both laccases and peroxidase activities are associated with differentiating xylem, although their activities on monolignols in vitro may differ (96). Lignifying tobacco xylem contains a number of enzyme activities capable of oxidizing coniferyl alcohol, including both hydrogen peroxide–dependent (peroxidase) and hydrogen peroxide–independent ("laccase-like") activities (83). The most abundant activity is a laccase-like protein similar in physical properties and substrate specificity to previously characterized laccases purified from tree species (4, 24, 95). Lignification in *Zinnia* stems also shows a strong correlation with hydrogen peroxide–independent oxidase activity (62). Discriminating between different types of hydrogen peroxide–independent oxidases is not a trivial problem, and it is difficult on the basis of histochemical staining to distinguish between catechol oxidases (also known as polyphenol oxidases) and laccases (also known as diphenol oxidases) (reviewed in 75). All of these classes of enzymes may be involved in lignification. Savidge & Udagama-Randeniya (90) identified a coniferyl alcohol oxidase associated with lignification in pines and later reported that the purified enzyme resembles catechol oxidase in substrate specificity (100). A polyphenol oxidase has been identified in mungbean hypocotyls that is capable of oxidizing coniferyl alcohol (14). Davin et al (18) characterized activities for both stereospecific and nonspecific enzymes from stems of *Forsythia* and later characterized the nature of the proteins involved in the stereospecific synthesis (see below).

Attempts to rule out laccase as a participant in lignification have frequently relied on histochemical staining of lignifying cells in the presence and absence of hydrogen peroxide (39, 85). Failure to detect H_2O_2-independent oxidase activity in such studies cannot be taken as definitive proof that laccases are absent in the tissue studied. The activity of either peroxidase or laccase on a particular chromogenic substrate need not be correlated with the activity of the same enzyme on monolignols, and assay conditions may differ for the two types of enzymes. Harkin & Obst (39) were unable to detect laccase activity in green ash stems after staining with syringaldazine and concluded that laccase was not present. Later workers were able to detect laccase activity in green ash by taking samples at a biologically relevant time of year (from growing trees in late spring, rather than dormant trees in early spring) and using a different substrate (75). A more convincing test of the participation of a particular enzyme in lignification is analysis of lignin from transgenic plants in which the activity of the enzyme is reduced. Such analyses have been conducted on transgenic tobacco deficient in anionic peroxidase, and no significant effect on lignin content or composition was detectable (51). This result shows that the anionic peroxidase is not the sole catalyst of lignin polymerization in tobacco but does not exclude participation of other peroxidase isozymes or of laccases. Analysis of the formation of an extracellular lignin-like precipitate, produced by cultured cells of loblolly pine (*Pinus taeda*) in the presence of 8% sucrose, suggests that hydrogen peroxide is necessary for polymerization of monolignols in this system (73).

ORDER AND DISORDER IN LIGNIN STRUCTURE

Does lignin have a highly ordered structure? Although lignin is widely thought of as a relatively disordered and amorphous polymer, there is evidence for significant order. Raman microprobe studies provided evidence indicating nonrandom orientation of the aromatic rings in lignin (1, 2). More recently, Radotic et al (77) studied the formation of synthetic polymers of coniferyl alcohol over time, using scanning tunneling microscopy (STM). The structure of these polymers shows long range order. After two days of polymerization, STM shows highly ordered structures on a graphite surface. Features of the polymers are less ordered on a gold surface but show the same structural motifs. They propose that polymerization products formed in vitro are organized in a lattice-like structure of connected "supermodules," each containing about 500 monomers. The importance of this finding is not necessarily in the suggestion that lignins in plant cell walls may have a similar structure but in the demonstration that order can arise spontaneously in a controlled system without obvious organizing factors. The appearance of order from apparent chaos is not an unusual phenomenon; multiple examples exist of systems in which self-assembly of macromolecular structures exploits the structural characteristics of the components to guide their interactions in forming large and complex structures (46).

Lewis and colleagues have focused attention on bimolecular phenoxy radical coupling reactions in the formation of lignans and lignin (18, 19). They have demonstrated the stereospecific coupling of E-coniferyl alcohol through a putative enzyme bound intermediate to form (+) pinoresinol, a lignan with a beta-beta linkage. This reaction was recently shown to require an oxidase or an oxidant and a 78-kDa nonenzymatic (dirigent) protein to effect stereospecific coupling (19). The traditional view of lignin polymerization is that one or more nonspecific oxidases produce lignin polymers in vivo through racemic coupling, in a manner analogous to the reactions that form dehydrogenation polymers in vitro (reviewed in 20, 42, 75). Davin et al (19) proposed that the polymerization of lignin is accomplished by stereospecific reactions analogous to that demonstrated for pinoresinol synthesis, arguing that "[i]t is inconceivable, however, that lignin formation would be left to the vagaries of such a wide range of enzymes, or be realized in a haphazard manner."

There is an apparent contrast between the biochemical stereospecificity evident in the synthesis of (+)-pinoresinol in *Forsythia* and the flexibility of lignin polymerization evident in the subunit composition of genetically modified plants. Incorporation of novel monomers into lignin in genetically modified plants is not, however, entirely inconsistent with a high level of regio- and stereospecificity in normal plants. The importance of lignin to vascular function and mechanical support in land plants lends credence to the argument that redundant systems may exist for the production of phenolic polymers in cell walls. A complex set of enzymes may exist for the production of highly ordered lignins in normal plants, but a plant with a genetic change that blocks the function of such an elaborate pathway is more likely to survive if it makes a random polymer of any available phenolic subunits than if it makes no lignin at all. The exact nature of the processes that produce lignins in normal and mutant plants is likely to be the basis of much discussion until the structure of lignin is known, the mode of polymerization is defined, and the role of specific, well-characterized oxidases is elucidated.

> **Visit the *Annual Reviews home page* at**
> **http://www.AnnualReviews.org.**

Literature Cited

1. Agarwal UP, Atalla RH. 1986. In-situ microprobe studies of plant cell walls: macromolecular organization and compositional variability in the secondary wall of *Picea mariana* (Mill.) B.S.P. *Planta* 169:325–32

2. Atalla RH, Agarwal UP. 1985. Raman microprobe evidence for lignin orientation in the cell walls of native woody tissue. *Science* 227:636–38

3. Atanassova R, Favet N, Martz F, Chabbert B, Tollier M-T, et al. 1995. Altered lignin composition in transgenic tobacco expressing O-methyltransferase sequences in sense and antisense orientation. *Plant J.* 8:465–77

4. Bao W, O'Malley DM, Whetten R, Sederoff RR. 1993. A laccase associated with lignification in loblolly pine xylem. *Science* 260:672–74

5. Bate N, Orr J, Ni W, Meromi A, Nadler-Hassar T, et al. 1994. Quantitative relationship between phenylalanine ammonia-lyase levels and phenylpropanoid accumulation in transgenic tobacco identifies a rate-determining step in natural product synthesis. *Proc. Natl. Acad. Sci. USA* 91:7608–12

6. Baucher M, Chabbert B, Pilate G, van Doorsselaere J, Tollier M-T, et al. 1996. Red xylem and higher lignin extractability by down-regulating cinnamyl alcohol dehydrogenase in poplar (*Populus tremula* and *Populus alba*). *Plant Physiol.* 112:1479–90

7. Besle JM, Cornu A, Jouany JP. 1994. Roles of structural phenylpropanoids in forage cell wall digestion. *J. Sci. Food Agric.* 64:171–90

8. Bligny R, Douce R. 1983. Excretion of laccase by sycamore (*Acer pseudoplatanus* L.) cells. *Biochem. J.* 209:489–96

9. Boerjan W, Meyermans H, Chen C, Leplé J-C, Christensen JH, et al. 1996. Genetic engineering of lignin biosynthesis in poplar. In *Somatic Cell Genetics and Molecular Genetics of Trees*, ed. MR Ahuja, W Boerjan, DB Neale, pp. 81–88. Dordrecht: Kluwer

10. Boudet AM, Grima-Pettenati J. 1996. Lignification genes: structure, function and regulation. *Proc. Keystone Symp. Extracell. Matrix Plants: Mol. Cell. Dev. Biol., Tamarron, Colo.*, p. 7. New York: Wiley-Liss

11. Boudet AM, Lapierre C, Grima-Pettenati J. 1995. Biochemistry and molecular biology of lignification. *New Phytol.* 129:203–36

12. Browning BL. 1967. *Methods of Wood Chemistry.* Appleton, Wis: Inst. Pap. Chem. 384 pp.

13. Campbell MM, Sederoff RR. 1996. Variation in lignin content and composition. *Plant Physiol.* 110:3–13

14. Chabanet A, Goldberg R, Catesson AM, Quinet-Szely M, Delauney AM, Faye L. 1994. *Plant Physiol.* 106:1095

15. Chapple CC, Vogt T, Ellis BE, Somerville CR. 1992. An *Arabidopsis* mutant defective in the general phenylpropanoid pathway. *Plant Cell* 4:1413–24

16. Chen C-L. 1991. Lignin: occurrence in woody tissues, isolation, reactions, and structure. See Ref. 57a, pp. 183–261

17. Das NN, Das SC, Sarkar AK, Mukherjee AK. 1984. Lignin-xylan ester linkage in

mesta fiber (*Hibiscus cannabinus*). *Carbohydr. Res.* 129:197–207

18. Davin LB, Bedgar DL, Katayama T, Lewis NG. 1992. On the stereoselective synthesis of (+)-pinoresinol in *Forsythia suspensa* from its achiral precursor, coniferyl alcohol. *Phytochemistry* 31:3869–74

19. Davin LB, Wang H-B, Crowell AL, Bedgar DL, Martin DM, et al. 1997. Stereoselective bimolecular phenoxy radical coupling by an auxiliary (dirigent) protein without an active center. *Science* 275:362–66

20. Dean JFD, Eriksson KEL. 1994. Laccase and the deposition of lignin in vascular plants. *Holzforschung* 48:21–33

21. Dean JFD, LaFayette PR, Rugh CL, Merkle SA, Eriksson KEL. 1996. The role of laccase in lignin biosynthesis. In *Proc. Keystone Symp. Extracell. Matrix Plants: Mol. Cell. Dev. Biol., Tamarron, Colo.*, p. 6. New York: Wiley-Liss

22. Dharmawardhana DP, Ellis BE, Carlson JE. 1995. A β-glucosidase from lodgepole pine xylem specific for the lignin precursor coniferin. *Plant Physiol.* 107:331–39

23. Douglas CJ. 1996. Phenylpropanoid metabolism and lignin biosynthesis: from weeds to trees. *Trends Plant Sci.* 1:171–78

24. Driouch A, Lainé AC, Vian B, Faye L. 1992. Characterization and localization of laccase forms in stem and cell cultures of sycamore. *Plant J.* 2:13–24

25. Elkind Y, Edwards R, Mavandad M, Hedrick SA, Ribak O, et al. 1990. Abnormal plant development and downregulation of phenylpropanoid biosynthesis in transgenic tobacco containing a heterologous phenylalanine ammonialyase gene. *Proc. Natl. Acad. Sci. USA* 87:9057–61

26. Erdtman H. 1933. Dehydrierungen in der Coniferylreihe. (1) Dehydrodieugenol und dehydrodiisoeugenol. *Z. Biochem.* 258:172–80

27. Freudenberg K. 1959. Biosynthesis and constitution of lignin. *Nature* 183:1152–55

28. Freudenberg K. 1965. Lignin: its constitution and formation from *p*-hydroxycinnamyl alcohols. *Science* 148:595–600

29. Freudenberg K. 1968. The constitution and biosynthesis of lignin. In *Constitution and Biosynthesis of Lignin*, ed. K Freudenberg, AC Neish, pp. 47–116. New York: Springer-Verlag

30. Freudenberg K, Harkin JM. 1963. The glucosides of cambial sap of spruce. *Phytochemistry* 2:189–93

31. Freudenberg K, Harkin JM, Reichert M, Fukuzumi T. 1958. Die an der Verholzung beteiligten Enzyme. Die Dehydrierung des Sinapinalkohols. *Chem. Ber.* 91:581–90

32. Freudenberg K, Neish AC. 1968. Constitution and Biosynthesis of Lignin. In *Molecular Biology, Biochemistry and Biophysics*, Vol. 2, ed. A Kleinzeller, GF Springer, HG Wittman. New York: Springer-Verlag. 129 pp.

33. Freudenberg K, Richtzenhain H. 1943. Enzymatische Versuche zur Entstehung des Lignins. *Ber. Dtsch. Chem. Ges. (Chem. Ber.)* 76:997–1006

34. Fry SC. 1987. Intracellular feruloylation of pectic polysaccharides. *Planta* 171:205–11

35. Grabber JH, Quideau S, Ralph J. 1996. *p*-coumaroylated syringyl units in maize lignin: implications for β-ether cleavage by thioacidolysis. *Phytochemistry* 43:1189–94

36. Grabber JH, Ralph J, Hatfield RD. 1998. Severe inhibition of cell-wall degradation by lignins formed with coniferaldehyde. *J. Sci. Food Agric.* In press

37. Gross GG, Mansell RL, Zenk MH. 1975. Hydroxycinnamate:Coenzyme A ligase from lignifying tissue of higher plants. *Biochem. Physiol. Pflanz.* 168:S41–S51

38. Halpin C, Knight ME, Foxon GA, Campbell MM, Boudet AM, et al. 1994. Manipulation of lignin quality by downregulation of cinnamyl alcohol dehydrogenase. *Plant J.* 6:339–50

39. Harkin JM, Obst TR. 1973. Lignification in trees: indication of exclusive peroxidase participation. *Science* 180:296–97

40. Hibino T, Takabe K, Kawazu T, Shibata D, Higuchi T. 1995. Increase of cinnamaldehyde groups in lignin of transgenic tobacco plants carrying an antisense gene for cinnamyl alcohol dehydrogenase. *Biosci. Biotechnol. Biochem.* 59:929–31

41. Higuchi T. 1959. Studies on the biosynthesis of lignin. In *Biochemistry of Wood*, ed. K Kratzl, G Billek, pp. 161–88. New York: Pergamon

42. Higuchi T. 1985. Biosynthesis of lignin. In *Biosynthesis and Biodegradation of Wood Components*, ed. T Higuchi, pp. 141–60. New York: Academic

43. Ibrahim R, Grisebach H. 1976. Purification and properties of UDP-glucose: coniferyl alcohol glucosyltransferase from suspension cultures of Paul's Scarlet rose. *Arch. Biochem. Biophys.* 176:700–8

44. Kajita S, Hishiyama S, Tomimura Y, Katayama Y, Omori S. 1997. Structural characterization of modified lignin in transgenic tobacco plants in which the activity of 4-coumarate: Coenzyme A ligase is depressed. *Plant Physiol.* 114:871–79

45. Kajita S, Katayama Y, Omoi S. 1996. Alterations in the biosynthesis of lignin in transgenic plants with chimeric genes for 4-coumarate:CoA ligase. *Plant Cell Physiol.* 37:957–65

46. Kauffman SA. 1993. *The Origins of Order: Self Organization and Selection in Evolution.* New York: Oxford Univ. Press. 709 pp.

47. Keller B, Sauer N, Lamb CJ. 1988. Glycine-rich proteins in bean: gene structure and assocation of the protein with the vascular system. *EMBO J.* 7:3625–34

48. Kutsuki H, Higuchi T. 1978. The formation of lignin in *Erythrina cristagalli.* *Mokuzai Gakkaishi* 24:625–31

49. Kutsuki H, Shimada M, Higuchi T. 1982. Distribution and roles of p-hydroxycinnamate:CoA ligase in lignin biosynthesis. *Phytochemistry* 21:267–71

50. Lagrimini LM. 1991. Wound-induced deposition of polyphenols in transgenic plants overexpressing peroxidase. *Plant Physiol.* 96:577–83

51. Lagrimini LM, Gingas V, Finger F, Rothstein S, Liu TTY. 1997. Characterization of antisense transformed plants deficient in the tobacco anionic peroxidase. *Plant Physiol.* 114:1187–96

52. Lam TBT, Iiyama K, Stone BA. 1992. Cinnamic acid bridges between cell wall polymers in wheat and phalaris internodes. *Phytochemistry* 31:1179–83

53. Lam TBT, Iiyama K, Stone BA. 1994. An approach to the estimation of ferulic acid bridges in unfractioned cell walls of wheat internodes. *Phytochemistry* 37:327–33

54. Lapierre C. 1993. Application of new methods for the investigation of lignin structure. In *Forage Cell Wall Structure and Digestibility,* ed. HG Jung, DR Buxton, RD Hatfield, J Ralph, pp. 133–66. Madison, Wis: ASA-CSSA-SSSA

55. Lapierre C, Monties B, Guittet E, Lallemand J-Y. 1987. RMN [13]C Bidimensionelle des lignines du peuplier: étude des corrélations entre atomes de carbone et réexamen par la methode INADEQUATE des attributions des signaux du spectre. *Holzforschung* 41:51–58

56. Lee D, Meyer K, Chapple C, Douglas CJ. 1997. Down-regulation of 4-coumarate:CoA ligase (4CL) in *Arabidopsis*: effect on lignin composition and implications for the control of monolignol biosynthesis. *Plant Cell.* 9:1985–98

57. Leinhos V, Udagama-Randeniya PV, Savidge RA. 1994. Purification of an acidic coniferin-hydrolysing β-glucosidase from developing xylem of *Pinus banksiana.* *Phytochemistry* 37:311–15

57a. Lewin M, Goldstein IS, eds. 1991. *Wood Structure and Composition.* New York: Dekker

58. Lewis NG, Yamamoto E. 1990. Lignin: occurrence, biogenesis and biodegradation. *Annu. Rev. Plant Physiol. Plant Mol. Biol.* 41:455–96

59. Lewis NG, Yamamoto E, Wooten JB, Just G, Ohoshi H, Towers GHN. 1987. Monitoring biosynthesis of wheat cellwall phenylpropanoids *in situ.* *Science* 237:1344–46

60. Li LG, Popko JL, Zhang XH, Osakabe K, Tsai C-J, et al. 1997. A novel multifunctional O-methyltransferase implicated in a dual methylation pathway associated with lignin biosynthesis in loblolly pine. *Proc. Natl. Acad. Sci. USA* 94:5461–66

61. Lin SY, Dence CW. 1992. *Methods in Lignin Chemistry.* Berlin: Springer-Verlag. 578 pp.

62. Liu L, Dean JFD, Friedman WE, Eriksson KE. 1994. A laccase-like phenoloxidase is correlated with lignin biosynthesis in *Zinnia elegans* stem tissue. *Plant J.* 6:213–24

63. Lu F, Ralph J. 1997. Derivatization followed by reductive cleavage (DFRC method), a new method for lignin analysis: protocol for analysis of DFRC monomers *J. Agric. Food Chem.* 45:2590–92

64. Lüderitz T, Schatz G, Grisebach H. 1982. Enzymic synthesis of lignin precursors–purification and properties of 4-coumarate:CoA ligase from cambial sap of spruce (*Picea abies* L). *Eur. J. Biochem.* 123:583–86

65. MacKay JJ, O'Malley DM, Presnell T, Fitzgerald LB, Campbell MM, et al. 1997. Inheritance, gene expression, and lignin characterization in a mutant pine deficient in cinnamyl alcohol dehydrogenase. *Proc. Natl. Acad. Sci. USA* 94:8255–60

66. McDougall GJ, Stewart D, Morrison IM. 1996. Tyrosine residues enhance crosslinking of synthetic proteins into lignin-like dehydrogenation products. *Phytochemistry* 41:43–47

67. Meyer K, Cusumano JC, Ruegger M, Bell-Lelong DA, Chapple CCS. 1997. *Regulation and manipulation of lignin monomer composition by expression and overexpression of ferulate 5-hydroxylase,*

a cytochrome P450-dependent monooxygenase required for syringyl lignin biosynthesis. Presented at Int. Wood Biotechnol. Symp., 2nd, Canberra, Pap. 9

68. Meyer K, Kohler A, Kauss H. 1991. Biosynthesis of ferulic acid esters of plant cell wall polysaccharides in endomembranes from parsley cells. FEBS Lett. 290:209–12

69. Monties B. 1989. Lignins. In Plant Phenolics, ed. JB Harbourne, 1:113–57, Methods in Plant Biochemistry, ed. PM Dey, JB Harbourne. New York: Academic

70. Musha Y, Goring DAI. 1975. Distribution of syringyl and guaiacyl moieties in hardwoods as indicated by ultraviolet microscopy. Wood Sci. Technol. 9:45–58

71. Myton KE, Fry SC. 1994. Intraprotoplasmic feruloylation of arabinoxylans in Festuca arundinacea cell cultures. Planta 193:326–30

72. Nakamura W. 1967. Studies on the biosynthesis of lignin. 1. Disproof against the catalytic activity of laccase in the oxidation of coniferyl alcohol. J. Biochem. Jpn. 62:54–60

73. Nose M, Bernards MA, Furlan M, Zajicek J, Eberhardt TL, Lewis NG. 1995. Towards the specification of consecutive steps in macromolecular lignin assembly. Phytochemistry 39:71–79

74. O'Connell AP, Holt K, Halpin C, Schuch W. 1996. Manipulation of lignin biosynthesis in transgenic plants. Proc. Keystone Symp. Extracell. Matrix Plants: Mol. Cell. Dev. Biol., Tamarron, Colo., p. 19. New York: Wiley-Liss

75. O'Malley DM, Whetten R, Bao W, Chen C-L, Sederoff RR. 1993. The role of laccase in lignification. Plant J. 4:751–57

76. Pillonel C, Mulder MM, Boon JJ, Forster B, Binder A. 1991. Involvement of cinnamyl-alcohol dehydrogenase in the control of lignin formation in Sorghum bicolor L. Moench. Planta 185:538–44

77. Radotic K, Simic-Krstic J, Jeremic M, Trifunovic M. 1994. A study of lignin formation at the molecular level by scanning tunneling microscopy. Biophys. J. 66:1763–67

78. Ralph J. 1996. An unusual lignin from kenaf. J. Nat. Prod. 59:341–42

79. Ralph J. 1997. Recent advances in characterizing "nontraditional" lignins. Presented at Int. Symp. Wood Pap. Chem., PL2-1–PL2-7, Montreal

80. Ralph J, Grabber JH, Hatfield RD. 1995. Lignin-ferulate crosslinks in grasses: active incorporation of ferulate polysaccha-

ride esters into ryegrass lignins. Carbohydr. Res. 275:167–78

81. Ralph J, Hatfield RD, Quideau S, Helm RF, Grabber JH, Jung H-JG. 1994. Pathway of p-coumaric acid incorporation into maize lignin as revealed by NMR. J. Am. Chem. Soc. 116:9448–56

82. Ralph J, MacKay JJ, Hatfield RD, O'Malley DM, Whetten RW, Sederoff RR. 1997. Abnormal lignin in a loblolly pine mutant. Science 277:235–39

83. Richardson A, McDougall GJ. 1997. A laccase-type polyphenol oxidase from lignifying xylem of tobacco. Phytochemistry 44:229–35

84. Robinson DG. 1985. Synthesis and Secretion of Extracellular Macromolecules in Plant Membranes. New York: Wiley

85. Ros Barcelo A. 1995. Peroxidase and not laccase is the enzyme responsible for cell wall lignification in the secondary thickening of xylem vessels in Lupinus. Protoplasma 186:41–44

86. Rösler J, Krekel F, Amrhein N, Schmid J. 1997. Maize phenylalanine ammonialyase has tyrosine ammonia-lyase activity. Plant Physiol. 113:175–79

87. Sarkanen KV, Chang H-M, Allan GG. 1967. Species variation in lignins. III. Hardwood lignins. Tappi J. 50:587–90

88. Sarkanen KV, Hergert HL. 1971. Classification and distribution. In Lignins, Occurrence, Formation, Structure and Reactions, ed. KV Sarkanen, CH Ludwig, pp. 43–94. New York: Wiley

89. Savidge RA. 1989. Coniferin, a biochemical indicator of commitment to tracheid differentiation in conifers. Can. J. Bot. 67:2663–68

90. Savidge RA, Udagama-Randeniya P. 1992. Cell wall–bound coniferyl alcohol oxidase associated with lignification in conifers. Phytochemistry 31:2959–66

91. Sederoff R, Chang HM. 1991. Lignin biosynthesis. See Ref. 57a, pp. 263–85

92. Sewalt VJH, Ni W, Blount JW, Jung HG, Masoud SA, et al. 1997. Reduced lignin content and altered lignin composition in transgenic tobacco down-regulated in expression of L-phenylalanine ammonialyase or cinnamate 4-hydroxylase. Plant Physiol. 115:41–50

93. Somssich IE, Bollmann J, Hahlbrock K, Kombrink E, Schulz W. 1989. Differential early activation of defense-related genes in elicitor-treated parsley cells. Plant Mol. Biol. 12:227–34

94. Somssich IE, Wernert P, Kiedrowski S, Hahlbrock K. 1996. Arabidopsis thaliana defense-related protein ELI3 is an aromatic alcohol:NADP+ oxidoreductase.

Proc. Natl. Acad. Sci. USA 93:14199–203

95. Sterjiades R, Dean JFD, Eriksson K-EL. 1992. Laccase from sycamore maple (*Acer pseudoplatanus*) polymerizes monolignols. *Plant Physiol.* 99:1162–68

96. Sterjiades R, Dean JFD, Gamble G, Himmelsbach DS, Eriksson K-EL. 1993. Extracellular laccases and peroxidases from sycamore maple (*Acer pseudoplatanus*) cell-suspension cultures—reactions with monolignols and lignin model compounds. *Planta* 190:75–87

97. Terazawa M, Okuyama H, Miyake M. 1984. Seasonal variations of phenolic glycosides in the cambial sap of wood. *Mokuzai Gakkaishi* 30:329–34

98. Trezzini GF, Horrichs A, Somssich IE. 1993. Isolation of putative defense-related genes from *Arabidopsis thaliana* and expression in fungal elicitor-treated cells. *Plant Mol. Biol.* 21:385–89

99. Tsai C-J, Mielke M, Popko J, Monossol J, Hu W, et al. 1997. *Altered lignin composition in quaking aspen through manipulation of caffeic acid/5-hydroxy ferulic acid O-methyltransferase gene expression.* Presented at Int. Wood Biotechnol. Symp., 2nd, Canberra, Pap. 12

100. Udagama-Randeniya P, Savidge R. 1995. Coniferyl alcohol oxidase—a catechol oxidase? *Trees* 10:102–7

101. Van Doorsselaere J, Baucher M, Chognot E, Chabbert B, Tollier M-T, et al. 1995. A novel lignin in poplar trees with reduced caffeic acid/5-hydroxyferulic acid O-methyltransferase activity. *Plant J.* 8:855–64

102. Voo KS, Whetten RW, O'Malley DM, Sederoff RR. 1995. 4-coumarate:CoA ligase in xylem of loblolly pine. *Plant Physiol.* 108:85–97

103. Whetten R, Sederoff R. 1995. Lignin biosynthesis. *Plant Cell* 7:1001–13

104. Whitmore FW. 1978. Lignin-carbohydrate complex formed in isolated cell walls of callus. *Phytochemistry* 17:421–25

105. Yao K, De Luca V, Brisson N. 1995. Creation of a metabolic sink for tryptophan alters the phenylpropanoid pathway and the susceptibility of potato to *Phytophthora infestans*. *Plant Cell* 7:1787–99

106. Ye Z-H, Kneusel RE, Matern U, Varner JE. 1994. An alternative methylation pathway in lignin biosynthesis in *Zinnia*. *Plant Cell* 6:1427–39

107. Ye Z-H, Varner JE. 1995. Differential expression of two O-methyltransferases in lignin biosynthesis in *Zinnia elegans*. *Plant Physiol.* 108:459–67

108. Zakis GF. 1994. *Functional Analysis of Lignins and Their Derivatives.* Atlanta: TAPPI. 94 pp.

Annu. Rev. Plant Physiol. Plant Mol. Biol. 1998. 49:611–41

DESATURATION AND RELATED MODIFICATIONS OF FATTY ACIDS[1]

John Shanklin and Edgar B. Cahoon

Department of Biology, Brookhaven National Laboratory, Upton, New York 11973;
e-mail: shanklin@bnl.gov

KEY WORDS: unsaturated fatty acid, protein engineering, binuclear iron, nonheme, oxygenase

ABSTRACT

Desaturation of a fatty acid first involves the enzymatic removal of a hydrogen from a methylene group in an acyl chain, a highly energy-demanding step that requires an activated oxygen intermediate. Two types of desaturases have been identified, one soluble and the other membrane-bound, that have different consensus motifs. Database searching for these motifs reveals that these enzymes belong to two distinct multifunctional classes, each of which includes desaturases, hydroxylases, and epoxidases that act on fatty acids or other substrates. The soluble class has a consensus motif consisting of carboxylates and histidines that coordinate an active site diiron cluster. The integral membrane class contains a different consensus motif composed of histidines. Biochemical and structural similarities between the integral membrane enzymes suggest that this class also uses a diiron cluster for catalysis. Soluble and membrane enzymes have been successfully re-engineered for substrate specificity and reaction outcome. It is anticipated that rational design of these enzymes will result in new and desired activities that may form the basis for improved oil crops.

CONTENTS

INTRODUCTION

Unsaturated fatty acids contain one or more double bonds, each of which lacks two hydrogen atoms relative to its saturated counterpart. Double bonds in fatty acids are predominantly of the *cis* (or Z) configuration. The number and position of double bonds in fatty acids profoundly affects their physical and therefore their physiological properties (50, 83).

Various mechanisms have evolved for the introduction of double bonds into fatty acids. Many prokaryotes, including *Escherichia coli*, introduce double bonds into fatty acids anaerobically (12). The advent of an aerobic environment several billion years ago allowed eukaryotes, cyanobacteria, and some bacilli to desaturate the methylene groups of long-chain fatty acids using enzymes called fatty acid desaturases (13, 44). Oxidative desaturation is more energy demanding than the anaerobic introduction of double bonds into fatty acids. However, the transition from anaerobic fermentation to aerobic respiration yielded greater than an order of magnitude of energy efficiency, leaving a surplus available for processes such as desaturation. In addition, the ability to regulate membrane fluidity by controlling the number of double bonds in fatty acids within the membrane in response to changing temperature likely conveyed a selective advantage to organisms capable of aerobic desaturation. Desaturases can be divided into two evolutionarily unrelated classes, one soluble, the other integral membrane enzymes (12). The membrane class is more widespread in nature. The fact that they are found in cyanobacteria and yeast suggests that they likely arose first, and that the soluble plastid desaturase is a more recent addition. In addition to unsaturated fatty acids, many plants also contain "unusual" fatty acids that have a wide variety of functional substituents (149). Many of these fatty acids arise by enzymatic modification of conventional fatty acids by enzymes closely related to fatty acid desaturases (see below).

For a historical perspective to the biochemistry of lipid modifications, the reader is directed to insightful reviews by Bloch (12) and Fulco (44). In recent years, several excellent reviews have been written on related topics: biosynthesis of polyunsaturated fatty acids (50), isolation of desaturases (132) and other genes involved in lipid biosynthesis (94), physiology and cell biology of desaturases and the effects of desaturation status on the function of membranes (83, 131), the distribution and basis for unusual fatty acid production (149), and uses of desaturases and related enzymes in biotechnology (64, 84, 93, 155). The focus of this review is first to summarize how recent investigations of these enzymes have contributed to our understanding of the principles that govern desaturation and how these principles might relate to other fatty acid modifications such as hydroxylation, acetylenic bond formation, and epoxidation. In this way, we develop a general context within which to view modifications of fatty acids. Second, we describe recent advances that have contributed to the identification of specific residues within the enzymes that control the specificity of desaturases both with respect to substrate selectivity and also both position and chemical nature of the introduced functional group. These advances point toward the rational design of a new generation of lipid modification enzymes.

COMMON BIOCHEMICAL CHARACTERISTICS OF DESATURATION SYSTEMS

Fatty acid desaturation involves an enzymatic reaction in which a double bond is introduced into an acyl chain and a molecule of dioxygen is completely reduced to water. The necessity for molecular oxygen for the growth of yeast was first reported by Pasteur (97) and subsequently shown to be overcome by unsaturated fatty acid and cholesterol supplementation of their growth media (3, 4). In the classic study of desaturation in yeast extracts, Bloomfield & Bloch showed that conversion of saturated to monounsaturated fatty acids required molecular oxygen and concluded that desaturation is an oxidative process in eukaryotes (13). This somewhat surprising observation, that oxygen is required for a reaction in which two hydrogens are removed and no oxygen is introduced into the product, is discussed in greater detail in the section dealing with potential reaction mechanisms. In contrast, the anaerobic biosynthesis of unsaturated fatty acids was shown to involve dehydration of a hydroxy substrate followed by *trans-cis* isomerization of the double bond (117).

Two classes of fatty acid desaturase enzymes capable of converting saturated to monounsaturated fatty acids were initially identified: a soluble form found in plants and a particulate or integral membrane form found in yeast and mammals (13, 73). Free fatty acids are not thought to be desaturated in vivo, rather they are esterified to acyl carrier protein (ACP) for the soluble plastid desaturases, or to coenzyme A (CoA) or lipids for the integral membrane desaturases. Interestingly, *Euglena gracilis* contains both soluble and membrane classes of desaturases and switches between the two systems in response to environmental conditions. Photoauxotrophic growth conditions result in the accumulation of a soluble acyl-ACP desaturase, whereas heterotrophic growth favors the accumulation of a particulate acyl-CoA desaturase (85). Despite their differences in cellular localization, both classes of desaturase share several common characteristics. They perform stereospecific Δ^9-desaturation of an 18:0 substrate with the removal of the 9-D and 10-D hydrogens (12, 122). As discussed above, they require molecular oxygen (13, 81) in addition to a short electron transport chain (44). Both reactions are inhibited by cyanide, but are unaffected by CO, suggesting that they involve nonheme metalloenzymes (61, 87, 96, 137).

For desaturation, two electrons are required for the formation of each double bond. Two different electron transport systems with functionally equivalent components supply these reducing equivalents, one in the plastid and the other in the endoplasmic reticulum (Figure 1). Interestingly, the systems are specific to the subcellular compartment, i.e. plastid versus endoplasmic reticulum, rather than to the class of the desaturase, i.e. soluble or membrane bound. In these systems, pairs of electrons arising from NADPH or NADH are simultaneously transferred to a flavoprotein that acts as a "step down transformer" that releases them one at a time to a carrier protein capable of carrying only a single electron. Two reduced carrier proteins then transfer electrons sequentially to the desaturase to effect the two-electron reduction required for catalysis.

For the soluble and integral membrane plastid desaturases, the source of reducing equivalents is NADPH, the flavoprotein is ferredoxin-NADP$^+$ oxidoreductase, and the electron carrier is the 2Fe-2S protein ferredoxin (86, 120, 152). This is likely the predominant pathway in nonphotosynthetic tissue such as castor seed and during desaturation in the dark in photosynthetic tissues. However, in photosynthetic tissues in the light, reducing equivalents arise from photosystem I and are directly transferred to ferredoxin, which in turn supplies the desaturase independently of ferredoxin-NADP$^+$ oxidoreductase (60). In the

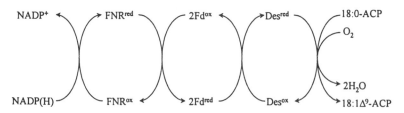

Figure 1 Δ^9-18:0-ACP desaturase electron transport chain under dark conditions or in nonphotosynthetic tissues. During photosynthesis, the ferredoxin can be directly reduced by photosystem I, as described in the text.

endoplasmic reticulum, the reducing equivalents are supplied by NADH, the flavoprotein is cytochrome b_5 reductase, and the electron carrier is the heme protein cytochrome b_5 (31,47,133). A major difference between these two systems is in the electrochemical potentials of the electron carriers ferredoxin and cytochrome b_5 [E°s of -420 mV for ferredoxin (146) and -24 mV for cytochrome b_5 (129)]. However, in the endoplasmic reticulum electron transport system, the frequency of interaction is enhanced because movement of the components is restricted to two dimensions rather than the three dimensions available to the soluble components (109). A detailed investigation into the relationship of specific electron transport systems for specific desaturases is lacking, though there is some evidence for a cyanobacterial (plastid type) Δ^6-desaturase (Δ^x where x represents the number of carbon atoms with respect to the carboxyl end of the fatty acid) functioning in the endoplasmic reticulum (105).

For the soluble desaturase under steady state conditions, the partially (one-electron) reduced state has yet to be observed (42). This implies that the first reduction is rate limiting and that the second occurs rapidly. This could occur in several different ways. The partially reduced desaturase could change conformation to favor a second interaction with ferredoxin. Alternatively, the second electron could be derived from a second partially reduced desaturase by an intra- or intermolecular disproportionation mechanism, as has been reported for ribonucleotide reductase (RR) (34). Sequencing of several desaturase-like genes yielded unexpected insight into the relationship of electron donors to desaturases. A desaturase-like gene of unknown function was isolated from sunflower that contained an N-terminal cytochrome b_5 domain (134). A similar gene encoding the borage microsomal Δ^6-desaturase was also isolated (116). In contrast, the yeast Δ^9-18:0-CoA desaturase was recently shown to contain a C-terminal cytochrome b_5 extension that was necessary for activity (80). These fusion proteins clearly support the immunological evidence that cytochrome b_5 is the in vivo electron donor for desaturation in the endoplasmic reticulum (63).

SOLUBLE FATTY ACID DESATURASES

Characterization of the Diiron Active Site of the Soluble acyl-ACP Desaturases

The C-H bond of a methylene group in a fatty acid is one of the most stable bonds in living systems at approximately 98 kcal $\cdot$ mol^{-1}. This energy is beyond the range of reactions mediated by amino acids alone. Thus a metal cofactor is necessary to harness the oxidative power of dioxygen to break this bond and initiate fatty acid modification. The diversity of metal cofactors identified in proteins and enzymes has recently been comprehensively reviewed (38).

Physiological experiments on the soluble acyl-ACP desaturase from *Euglena* showed that it was sensitive to metal chelators and required molecular oxygen, suggesting the presence of a metal cofactor (87). The sensitivity of the soluble class of desaturases to CN and NaN_3, together with their insensitivity to CO, suggested that the metal cofactor was not a heme group. UV-visible spectroscopy of purified safflower desaturase also revealed a lack of distinctive absorption features attributable to known metal centers (78). The isolation of a cDNA encoding the Δ^9-18:0-ACP desaturase enabled it to be overexpressed under the control of T7 polymerase in *E. coli* and for quantities of enzyme to be purified sufficient for biophysical characterization (125, 145). In contrast to the enzyme from natural sources, which was low in abundance and proved difficult to purify, the highly abundant recombinant protein was easily purified by conventional ion exchange and size exclusion chromatography (42, 54). Interestingly, the purified protein, when concentrated to millimolar levels, was straw yellow in color. Quantitative analysis revealed the presence of iron and the absence of other metals. It was also shown that the desaturase contains two moles of iron per mole of desaturase (42). UV-visible spectroscopy of the highly-concentrated protein revealed the presence of ligand-to-metal charge transfer absorption between 300 and 700 nm, consistent with its straw yellow color. There is a weak absorption feature ε_{340} of 8000 $M^{-1} \cdot cm^{-1}$ in the isolated oxidized protein, that closely resembled those seen in the diiron-oxo proteins methemerythrin and R2, the stable tyrosine radical-containing subunit of RR (14, 39). The 2:1 stoichiometry of iron:protein was independently confirmed by reductive titration of the 340-nm absorption.

Further and more definitive characterization employed the use of Mössbauer spectroscopy in which the nuclear shielding effect of the stable isotope ^{57}Fe was recorded with the use of gamma-ray absorption spectroscopy (82). This is a particularly powerful technique because, unlike techniques such as electron paramagnetic resonance, it does not rely on the existence of a mixed valent species for detection. However, it does depend on the production of high (millimolar) concentrations of isotopically enriched sample. Data collection involves recording spectra of the sample in different redox states, under different applied magnetic fields and at different temperatures. These spectra are compared with those of other well-characterized systems (either proteins or model complexes) and information regarding the structure of the metal center and its coordinating ligands is determined (103). Data on the purified desaturase showed that it indeed contained a diiron cluster with similar spectroscopic parameters to those previously identified in MMOH, the hydroxylase component of MMO, and R2 (42). Diiron clusters of this type involve two iron ions in a coordination sphere composed of nitrogen and oxygen ligands and typically contain an oxo or hydroxo bridge (91). The coordinating ligands are bidentate carboxylates glutamic and aspartic acid for oxygen, and the δN of histidines for nitrogen. The observation of a strong coupling constant of $J > 60 \text{ cm}^{-1}$ suggested the presence of an oxo bridge. In addition, a catalytic role was implied for the desaturase diiron cluster because the reduced cluster became re-oxidized upon the addition of substrate (42). This was particularly significant in that it suggested a formal link between several enzymes with diverse function including desaturases and hydroxylases that use activated oxygen to effect catalysis. Further support for the inclusion of desaturase in a class with MMOH and R2 came from resonance Raman spectroscopic experiments in which there were symmetric and asymmetric vibrational modes typical of oxo-bridged diiron clusters (41). Based on ^{18}O-labeling experiments, an Fe-O-Fe bond angle of 123 degrees was calculated (41).

A comparison of primary sequence alignment revealed a lack of discernible similarity between the desaturase and MMOH and R2. However, the coordination sphere for R2 had been determined by X-ray crystallography, and a search of the primary sequences of other diiron enzymes capable of interaction with O_2 revealed a consensus-binding motif of $[(D/E) X_2H]_2$ and general organization shown in Figure 2. A search of GenBank for other proteins containing the motif revealed several additional hydroxylases from *Pseudomonas* (41, 42) (Table 1). Of these, phenol hydroxylase was previously shown to require the addition of Fe^{2+} for activity (102), and toluene-4-monooxygenase has been

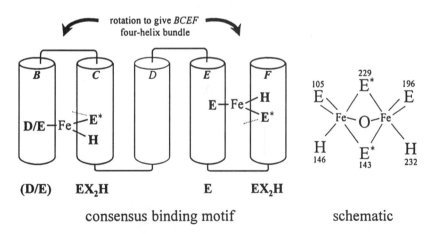

Figure 2 Organization of the diiron clusters in the large-helix bundle class (i.e. acyl-ACP desaturases, MMOH, and R2). General organization with respect to helices (*left panel*); schematic of the crystallographically defined castor Δ^9-18:0-ACP desaturase ligation sphere (*right panel*). Glutamates that coordinate both iron ions are indicated by *.

Table 1 Proteins containing motifs common to soluble desaturases and other diiron proteins: $(D/E\ X_2H)_2$

Enzyme	Organism	Substrate	Product	Reference
Δ^9 Stearoyl-ACP desaturase (sad)	*Ricinus/ Carthamus*	18:1-ACP	18:0-ACP	125, 145
Δ^4 Palmitoyl-ACP desaturase (TII)	*Coriandrum*	16:0-ACP	16:1-ACP	28
Δ^6 Palmitoyl-ACP desaturase (TAD4)	*Thunbergia*	16:0-ACP	16:1-ACP	25
Δ^9 Palmitoyl-ACP desaturase (milkweed)	*Asclepias*	16:0-ACP	16:1-ACP	24
Δ^9 Myristoyl-ACP desaturase (PHX-B)	*Pelargonium*	14:0-ACP	14:1-ACP	123
Methane monooxygenase	*Methylococcus*	Methane	Methanol	136
Ribonucleotide reductase (R2)	*Escherichia*	Ribonucleotides	Deoxyribonucleotides	89
Toluene-4-monooxygenase (T4MOH)	*Pseudomonas*	Toluene	P-cresol	154
Phenol hydroxylase	*Pseudomonas*	Phenol	Catechol	90
Alkene monooxygenase (amoC/AMO)	*Nocardia/ Mycobacterium*	Alkenes	Epoxides	49, 112
B-cell antigen (Des)	*Mycobacterium*	Unknown	Unknown	59

shown to contain a diiron cluster with properties similar to those of MMOH and the soluble desaturase (100). In addition to desaturases and hydroxylases, an alkene epoxidase with 30% homology to MMOH was identified (112). Proteins with identity greater than 25% are very likely to have the same fold (53), and because the MMOH diiron center coordination ligands are conserved, it has been proposed that the diiron active site is responsible for the oxidation of the alkene (112). This suggests that the diiron centers in soluble proteins are capable of at least three distinct activities: desaturation, hydroxylation, and epoxidation (Table 1).

There are four classes of soluble diiron-containing proteins: (*a*) large helix-bundle proteins; (*b*) simple helix-bundled proteins, with overhead connections; (*c*) simple helix bundles without overhead connections; and (*d*) α/β sandwich structures (91). R2, MMOH, desaturase, and likely the epoxidase comprise the large helix-bundle class.

Crystal Structure of the Desaturase

The determination of a crystal structure provides a unique global view of an enzyme, in contrast with spectroscopic investigation, which provides information about specific details of the active site. However, both techniques contribute unique and complementary information that, when integrated, can provide an unparalleled understanding of molecular catalysis. In this context, a more complete understanding of the soluble desaturases has been obtained from the crystal structure of the castor Δ^9-18:0-ACP desaturase (24, 70, 121). See Figure 3 for

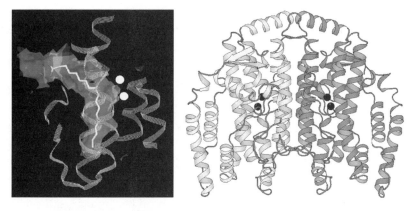

Figure 3   Structural representation of the castor Δ^9-18:0-ACP desaturase. Desaturase dimer (*right panel*); the two subunits are represented by *light or dark shading*, respectively; iron ions are *black spheres*. Expanded view showing the relationship of helices to the substrate-binding pocket shown in gray (*left panel*). A substrate molecule, stearic acid, is modeled into the pocket; iron ions are shown in *white*. Note: The detail view is rotated slightly from the overall view for optimal visualization of the cavity.

a schematic representation of the structure; more detail can be obtained via the internet at http://www.pdb.bnl.gov (accession 1AFR).

The Δ^9-18:0-ACP desaturase is a homodimer, as predicted from size exclusion chromatography experiments (78; Figure 3, *right panel*). The secondary structure of the desaturase monomer is almost exclusively α-helical with the exception of a short β-hairpin at the C terminus. The 41.6-kDa monomer comprises 11 helices, 9 of which form a core bundle with the remaining 2 helices capping each end (Figure 3). Of the 9 α-helices, 4 are involved in the coordination of two iron ions that constitute the active site diiron cluster. The diiron clusters are distantly spaced (>23 Å) within the dimer, suggesting that each monomer functions independently (70). As is the case in related enzymes, the diiron clusters are buried in the middle of the structure, resulting in an isolated local environment suitable for reactive oxygen chemistry (91). The fact that the diiron center is remote from the surface of the desaturase eliminates the possibility of direct transfer of electrons from the ferredoxin to the diiron center. Thus, electrons must be transferred from the ferredoxin-docking surface to the internal diiron cluster in order to affect its reduction. The crystal structure reveals potential routes for the movement of electrons through the side chains of aromatic and charged residues that are closely spaced from the surface to the interior of the desaturase (70).

The crystal structure confirmed the coordinating ligands proposed by sequence alignment, with the addition of a previously unidentified glutamic acid ligand (70; Figure 2). One of the iron ions interacts with the side chains of glutamic acid 196 and histidine 232, and the second iron ion interacts with the side chains of glutamic acid 105 and histidine 146. In addition, the side chains of the glutamic acid residues 143 and 229 participate in the coordination of both iron ions. The iron-iron distance for the diferric enzyme from the crystal structure was 4.2 Å, longer than the 3.2 Å determined by extended X-ray absorption fine structure (EXAFS) (2). In addition, the bridging oxo species detected by resonance Raman spectroscopy was absent from the crystal structure. Together these data suggest that the oxidized desaturase was photoreduced by the X-ray source during data collection, and thus the structure most likely represents the diferrous form of the enzyme (70).

Also identified in the crystal structure of the Δ^9-18:0-ACP desaturase is a hydrophobic channel that likely represents the substrate-binding pocket (Figure 3, *left panel*). Consistent with this, the channel extends from the surface to the deep interior of each subunit, and the path of this channel would place the fatty acid portion of the substrate in close proximity to the diiron center. If a substrate molecule stearate is modeled into the cavity with the fatty acid's methyl end at its base, the carboxyl end matches the annulus (Figure 3, *left panel*). In addition, the channel bends at the diiron cluster at a point corresponding to the insertion position of the double bond between carbons 9 and 10. As predicted by Bloch (12), a bend would place constraints on the free rotation of the substrate during catalysis and thus explain the observed stereochemistry.

As noted above, MMOH, R2, and the soluble desaturase share a consensus-binding motif for the coordination of the diiron center (42). This implies a possible common evolutionary origin for the enzymes, but the lack of sequence similarity outside the coordination sites would argue against such a scenario because proteins that have less than 25% homology rarely share the same fold (53, 115). Despite this lack of homology, superposition of the structures of the desaturase, R2 (92), and MMOH from *Methylococcus capsulatus* (110) with the diiron site for alignment showed that they are strikingly similar in fold (70). The desaturase had RMS fits of 1.90 Å for 144 Cα atoms for R2 and 1.98 Å for 117 Cα atoms for MMOH. It is interesting to note that this combination of fold/diiron cluster can be used to accomplish 2e$^-$ chemistry (e.g. desaturation and hydroxylation) and 1e$^-$ chemistry (e.g. ribonucleotide reduction)

Catalytic Mechanism

The topic of catalysis by nonheme-iron proteins is currently under intensive study (38). This is because these enzymes catalyze reactions that are of fundamental importance for life itself, such as facilitating the use of methane as a carbon source and catalyzing a key reaction in the metabolism of nucleic acids. In addition, they catalyze a diverse set of reactions including hydroxylation, desaturation, and epoxidation, and there is much interest in understanding the fundamental principles that allow biochemical reactions to be modulated to achieve different outcomes. In addition to the biological interest, bio-inorganic chemistry is amenable to investigation by an array of powerful spectroscopic techniques (55), in addition to molecular biology and X-ray crystallography. Several excellent reviews have recently appeared on the topic of diiron proteins and their chemistry (32, 68, 103, 127, 153).

MMO AS A MODEL FOR DESATURASE CHEMISTRY   Details of the reaction chemistry of MMO are currently better understood than those of acyl-ACP desaturase or other diiron proteins (128, 153). The proposal has recently been made that perhaps MMO, RR, and the desaturase share a common activated diiron-oxygen intermediate (103, 128, 153). For this reason, we first summarize current understanding of the MMO reaction cycle and then describe how the desaturase might function in this context.

MMO is the hydroxylase that initiates oxidation of methane to produce methanol (71). The enzyme has three protein components: MMOH, a hydroxylase containing the diiron center and substrate-binding site; MMOR, a reductase; and MMOB, an activator that binds to MMOH and increases its catalytic rate. Seven distinct states of the enzyme have been identified as reviewed by Wallar & Lipscomb (153); the five states pertinent to this discussion are shown in Figure 4 (39, 40, 43, 72, 111). In the resting form of MMOH, the diiron center is in the oxidized (diferric or Fe^{III}-Fe^{III}) form. Activation is initiated by 2-electron reduction from NADH via MMOR to produce the reduced (diferrous or Fe^{II}-Fe^{II}) form. After reduction, molecular oxygen binds to the iron center resulting in a peroxo form P. Scission of the O-O bond gives rise to compound Q, the key oxidizing intermediate responsible for hydrogen abstraction (128). At some point during oxygen activation a molecule of water is lost, but the precise timing of this step remains to be defined; consequently it is not shown in Figure 4. Application of rapid freeze quench experiments monitored by Mössbauer and EXAFS spectroscopies, showed that Q is an $Fe_2^{IV}O_2$ diamond core structure (128). According to current models, this oxidizing species then abstracts a hydrogen forming a caged hydroxyl intermediate R and a radical $H_3C\cdot$. This in turn undergoes oxygen rebound in a fashion described for cytochrome P450 hydroxylases (46) to yield the hydroxylated product methanol.

IMPLICATIONS FOR DESATURASE MECHANISM   It is envisaged that the acyl-ACP desaturase is activated in a similar manner as MMO because it shares a common fold, has a diiron cluster with similar spectral properties and ligands, and is dependent on oxygen and reductant for activity (153). In addition, several of the components of the catalytic cycle such as the oxidized and reduced forms

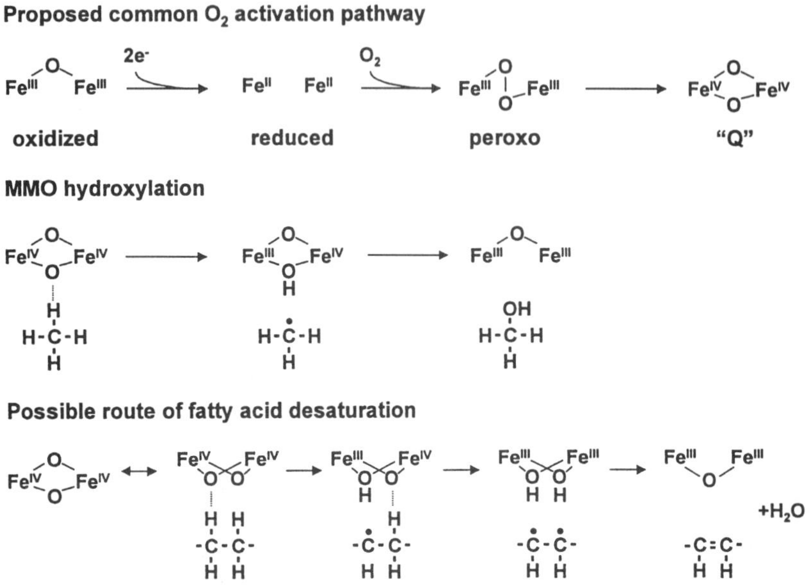

Proposed common O₂ activation pathway

MMO hydroxylation

Possible route of fatty acid desaturation

Figure 4 Proposed mechanism of action of MMO and acyl-ACP desaturase. *Top panel*, proposed common oxygen activation pathway; *middle panel*, MMO proposed reaction path; *lower panel*, possible route of fatty acid desaturation.

of the desaturase observed by Mössbauer and resonance Raman spectroscopy appear quite similar to those of the other diiron proteins (41, 42, 153). The use of azide as a probe of the active site in conjunction with UV-visible and resonance Raman spectroscopy identified a μ-1,3 azide complex that supports the proposal that oxygen forms a μ-1,2 diferric peroxo intermediate (2). Several workers have proposed that a bis (-oxo) diferryl cluster is the common activated intermediate for this class of enzymes (2, 153). Under this hypothesis, Figure 4 shows a possible route for fatty acid desaturation. Compound Q would first abstract a hydrogen as for MMOH, but instead of oxygen rebound, a second hydrogen would be abstracted. The removal of the first hydrogen would result in the formation of a radical intermediate, and removal of a second hydrogen would result in the formation of a transient diradical that would spontaneously recombine to form the olefinic double bond. The dihydroxy bridging species on the diiron cluster could decompose to form an oxo bridge and a second water molecule. Alternative schemes are also possible because the anaerobic mechanism of double bond introduction involves dehydration followed by *trans-cis* isomerization. A similar mechanism was proposed for aerobic desaturation (87). Under this hypothesis, the fatty acid first would be hydroxylated, then dehydrated, and the double bond isomerized; as described for the anaerobic mechanism. However, no conversion of hydroxy to unsaturated fatty acids was observed, suggesting that a hydroxylated substrate is likely not an intermediate in the desaturation reaction (87). Convincing evidence in support of any mechanism remains to be reported. Thus it should be stressed that the hydrogen abstraction scheme described above represents only one of several plausible desaturation schemes that are the subject of ongoing investigation.

Variant Acyl-ACP Desaturases

In addition to the nearly ubiquitous occurrence of the Δ^9-18:0-ACP desaturase in the plant kingdom (78, 86, 125, 145), a number of other acyl-ACP desaturases have been identified in specific tissues of certain plant species (24, 25, 27, 28, 123). These include a Δ^4-palmitoyl (16:0)-ACP desaturase from *Umbelliferae* seed (27, 28), a Δ^6-16:0-ACP desaturase from *Thunbergia alata* seed (25), and a Δ^9-myristoyl (14:0)-ACP desaturase from *Pelargonium xhortorum* trichomes (123). As indicated by their names, these enzymes display differences in substrate specificity and in the positioning of double bond insertion (or regiospecificity). As a result, the activities of these variant acyl-ACP desaturases give rise to unusual monounsaturated fatty acids, which accumulate primarily in seed oils (149) or, in the case of the Δ^9-14:0-ACP desaturase, are precursors for the production of pest-resistant compounds known as anacardic acids (123). Variant acyl-ACP desaturases characterized to date share $\geq 70\%$ amino acid sequence similarity with Δ^9-18:0-ACP desaturases (24, 25, 28, 123, 125, 145). The close relation of their primary structures suggests that variant acyl-ACP desaturases have likely evolved from the Δ^9-18:0-ACP desaturase. Of mechanistic significance, the natural occurrence of variant forms of the Δ^9-18:0-ACP desaturase with different activities has provided both tools and the impetus for protein engineering studies aimed at redesigning the substrate and regiospecificities of acyl-ACP desaturases (26).

Substrate and Regiospecificities of Acyl-ACP Desaturases: Properties and Redesign

An additional aspect of the mechanism of soluble desaturases is their ability to recognize the numbers of carbon atoms and the position of double bond placement in acyl-ACP substrates. In this regard, naturally occurring acyl-ACP desaturases typically display distinct specificities for the chain lengths of their fatty acid substrates (24–27, 45, 78, 123). For example, the Δ^9-18:0-ACP desaturase is approximately 100-fold more active with the 18:0-ACP than with 16:0-ACP (45, 78). Similarly, the relative activity of the Δ^6-16:0-ACP desaturase with 16:0-ACP is about six- to sevenfold greater than that detected with 18:0-ACP (26). In the case of the Δ^9-18:0-ACP desaturase, differences in activity with 16:0- and 18:0-ACP apparently do not result from differences in substrate binding, as the enzyme has similar K_ms for both acyl-ACP moieties (45, 78). Overall, the narrow substrate specificity profile of acyl-ACP desaturases is in contrast to that of the Δ^9-18:0-CoA desaturase from yeast and rat, which is nearly equally active with 16:0- and 18:0-CoA (13, 36).

In addition, acyl-ACP desaturases position the placement of double bonds relative to the carboxyl end of the fatty acids (27, 45), a property that is also observed with the yeast and rat Δ^9-18:0-CoA desaturase (13, 36). For example, it has been demonstrated that the activity of the soybean Δ^9-18:0-ACP desaturase with 16:0- or 18:0-ACP gives rise to Δ^9 monounsaturated products with either substrate (45). The "carboxyl-counting" nature of acyl-ACP desaturases suggests that a fixed distance exists in the active sites of these enzymes between the catalytic iron atoms and the ACP portion of bound substrates.

Information from the crystal structure of the castor Δ^9-18:0-ACP desaturase provides a partial explanation for the substrate and regiospecificities of acyl-ACP desaturases (70). As described above, a hydrophobic channel was identified in the crystal structure of the castor Δ^9-18:0-ACP desaturase that likely represents its substrate binding pocket. This channel extends from the surface to the deep interior of each monomer and passes in close proximity to the diiron-oxo cluster. An acyl chain modeled into the pocket would assume a bent conformation, with the Δ^9 carbon of the fatty acid (regardless of chain length) within catalytic distance of one of the iron ions of the diiron cluster. In this model, the ACP portion of the substrate would likely interact with residues at the surface of the desaturase, and the fatty acid, bound at its carboxyl end to ACP, would extend into the hydrophobic channel. The size of the channel is sufficient to accommodate an 18-carbon fatty acid, as well as fatty acids containing fewer carbon atoms (e.g. 14:0 and 16:0). This is consistent with the observation that the Δ^9-18:0-ACP desaturase binds 16:0- and 18:0-ACP with nearly equal affinity (45, 78). However, it is not obvious from the crystal structure why the Δ^9-18:0-ACP desaturase is more active with 18:0-ACP than with 16:0-ACP. One possibility is that interactions of the methyl end of the fatty acid substrate with the bottom of the binding pocket enhance rates of catalysis. With regard to regiospecificity, the diiron-oxo cluster in each subunit is at a fixed position relative to the hydrophobic channel. As such, differences in regiospecificities between the Δ^9-18:0-ACP desaturase and variant enzymes such as the Δ^6-16:0-ACP desaturase may result, in part, from differences in the length of the substrate binding channel from the surface of the enzyme to the diiron cluster. Surface interactions between ACP and the desaturase may also influence regiospecificity. However, a precise understanding of the determinants of regiospecificity awaits crystallographic data from an acyl-ACP desaturase with bound substrate.

Residues that line the lower portion of the hydrophobic channel likely limit the length of the fatty acid chain that can be accommodated by an acyl-ACP desaturase. This becomes apparent when these residues in the Δ^9-18:0-ACP desaturase are replaced in modeling studies with equivalent residues from variant acyl-ACP desaturases (26). For example, the deep portion of the modeled binding pocket of the Δ^9-14:0-ACP desaturase contains amino acids with bulkier side chains relative to those found in the Δ^9-18:0-ACP desaturase. As a result, the binding pocket of the Δ^9-14:0-ACP desaturase is only able to accommodate fatty acids with fewer carbon atoms beyond the point of double insertion relative to the Δ^9-18:0-ACP desaturase.

Such modeling studies have provided the basis for the redesign of substrate specificities of acyl-ACP desaturases (26). As a demonstration of this, the replacement of leucine 118 and proline 179 at the bottom of the binding pocket of castor Δ^9-18:0-ACP desaturase with the bulkier residues phenylalanine and isoleucine, respectively, converted this enzyme into one that functions primarily as a Δ^9-16:0-ACP desaturase. Conversely, the replacement of alanine 188 and tyrosine 189 in the Δ^6-16:0-ACP desaturase to the smaller glycine and phenylalanine, respectively, yielded an enzyme that was equally active with 16:0- and 18:0-ACP. Based on the active site model, these substitutions provide additional space at the lower portion of the binding pocket such that the Δ^6-16:0-ACP desaturase can accommodate the longer fatty acid chain of 18:0-ACP. Such studies demonstrate the ability to produce acyl-ACP desaturases with novel substrate specificities through rational design based on crystallographic and primary structural data.

Rational design of regiospecificities, however, is currently limited by the lack of detailed crystallographic data from a desaturase with bound acyl-ACP substrate. In spite of this, it has been shown that by replacement of five amino acid residues, a Δ^6-16:0-ACP desaturase can be converted into an enzyme that functions as a Δ^9-18:0-ACP desaturase (26). In this study, enzyme redesign was based solely on amino acid sequence alignments of Δ^6-16:0- and Δ^9-18:0-ACP desaturases. The implication of this work is that only a small subset of residues of the $\sim$360 amino acids in each subunit of an acyl-ACP desaturase act as determinants of regiospecificity.

INTEGRAL MEMBRANE DESATURASES

Sequences of the Integral Membrane Desaturases

Because of the technical difficulties in obtaining large quantities of purified membrane proteins, progress in understanding the membrane class of desaturases has lagged behind that of the soluble class. While large quantities of membrane desaturases have yet to be obtained, the cloning of the first desaturase gene, the Δ^9-18:0-CoA desaturase from rat liver, was made possible by heroic purification efforts of Strittmatter's group (137). Genetic approaches have generally proven superior to the biochemical approach for isolating desaturase genes from yeast (140), cyanobacteria (151), and higher plants (5). However, the plastid n-6 desaturase from spinach also yielded to an elegant biochemical purification, which provided sufficient protein for amino acid sequence determination (118). A full cDNA was obtained following PCR amplification with the use of a degenerate oligonucleotide based on the protein sequence. These approaches in addition to heterologous probing have resulted in the isolation of a large family of desaturase genes from diverse organisms (Table 2). For instance, the genes for at least eight different desaturase activities have been cloned from *Arabidopsis*, while at least two remain to be identified (132). Thus, while direct investigation of the enzymes has lagged behind that of the soluble desaturase, a wealth of sequence information has accumulated.

IDENTIFICATION OF AN INTEGRAL MEMBRANE DESATURASE MOTIF   Investigation of the deduced amino acid sequences corresponding to these genes showed that the iron-binding motif $[(D/E) X_2 H]_2$ of the soluble desaturases is not found in the integral membrane desaturases. However, several conserved histidines in equivalent positions with respect to potential membrane-spanning domains were identified by comparison of the deduced amino acid sequences of the yeast and rat Δ^9-18:0-CoA desaturases (140). This group of eight conserved histidines comprises a tripartite motif H $X_{(3-4)}$ H $X_{(7-41)}$ H $X_{(2-3)}$ HH $X_{(61-189)}$ H $X_{(2-3)}$ HH, which has now been identified in almost all membrane desaturases. The two exceptions to this pattern are the Δ^6-desaturases from *Anabaena* (104) and borage (116). In these enzymes, the first histidine of the third element is absent and a glutamine residue is found in its place, which also has a nitrogen-containing side chain. Thus, the motif has been amended to H $X_{(3-4)}$ H $X_{(7-41)}$ H $X_{(2-3)}$ HH $X_{(61-189)}$ (H/Q) $X_{(2-3)}$ HH to reflect this diversity. Site-directed mutagenesis experiments in which each of the eight conserved histidines was individually converted to alanines in the rat Δ^9-18:0-desaturase established that all the histidines are essential for catalysis (126). This result was supported by similar experiments on the cyanobacterial desA desaturase (7). An equivalent group of eight conserved histidines was also identified in two *Pseudomonas* monooxygenases, AlkB and XylM, enzymes that mediate the hydroxylation of alkanes and xylenes, respectively (126, 142). The identification of a conserved motif in different enzymes implies that they are evolutionarily related provided that the order and relative spacing of the motif is comparable (115). Indeed, not only are elements of the motif approximately equally spaced, they are also placed in equivalent positions with respect to (potential, or empirically defined) membrane-spanning domains (126, 140, 147). In addition, for each of

Table 2 Proteins containing motifs common to integral membrane desaturases and related proteins:
$H X_{(3-4)} H X_{(7-41)} H X_{(2-3)} HH X_{(61-189)} (H/Q) X_{(2-3)}$

Enzyme	Organism	Substrate	Product	Reference
18:0-CoA desaturase (scd/ole1)	Rattus/Saccharomyces	Stearoyl-CoA	Oleoyl-CoA	140, 143
Plastid-type Δ^9-desaturase (desC)	Synechocystis	Plastid glycerolipid-18:0	Plastid glycerolipid-18:1	114
Plastid-type Δ^6-desaturase (desD)	Anabaena	Plastid glycerolipid-18:2(3)	Plastid glycerolipid-18:3(4)	104
Plant microsomal desaturase (fad2)	Arabidopsis	Phospholipid-18:1	Phospholipid-18:2	95
Plant microsomal desaturase (fad3)	Arabidopsis	Phospholipid-18:2	Phospholipid-18:3	5
Plastid-type desaturase (desA/fad6)	Synechocystis/Arabidopsis	Plastid glycerolipid-18:1	Plastid glycerolipid-18:2	37, 52, 118, 151
Plastid-type desaturase (fad7/8 desB)	Spinacial/Arabidopsis Arabidopsis/Synechocystis	Plastid glycerolipid-18:2	Plastid glycerolipid-18:3	58, 113
Plant microsomal Δ^6-desaturase (BOD6)	Borago	Phospholipid-18:2	Phospholipid-18:3	116
Cytochrome b_5-desaturase fusion	Helianthus	Unknown	Unknown	134
Animal ω-3 desaturase (fat-1)	Caenorhabditis	Phospholipid-18(20):2	Phospholipid-18(20):3	135
Animal fatty acid desaturase (MLD)	Homo	Unknown	Unknown	23
Sterol-C-5 desaturase (Erg3)	Saccharomyces	Episterol	Ergosta-5,7,24(28)-trienol	6
Fatty acid hydroxylase (FAH12)	Ricinus	Phospholipid-oleate	Phospholipid-ricinoleic acid	148
Fatty acid acetylenase	Crepis	Phospholipid-18:2	Phospholipid-crepenynic acid	69
Fatty acid epoxygenase	Crepis/Vernonia	Phospholipid-18:2	Phospholipid-vernolic acid	S Stymne, A Kinney, unpublished data
Sphingolipid α-hydroxylase (Fah1P)	Saccharomyces	Sphingolipid-26:0	Sphingolipid-α-OH 26:0	80a
Sphingolipid 4-hydroxylase (Syr2)	Saccharomyces	Dihydroceramide	Phytoceramide	J Takemoto, unpublished data
Alkane ω-hydroxylase/epoxidase (AlkB)	Pseudomonas	Alkanes/alkenes	Primary alcohols/epoxides	67, 74
Xylene monooxygenase (Xmo)	Pseudomonas	Toluene/xylenes	(methyl)benzaldehydes	142
p-cymene monooxygenase (cymAa)	Pseudomonas	p-cymene	p-cumic alcohol	31a
β-carotene hydroxylase (crtZ)	Arabidopsis	β-carotene	Zeaxanthin	141
β-carotene oxygenase/ketolase (crtW)	Haematococcus	β-carotene	Canthaxanthin	62
C-4 sterol methyl oxidase (Erg25)	Saccharomyces	4,4-dimethylzymosterol	Zymosterol	10
Aldeyhyde decarbonylase (Cer1)	Arabidopsis	Long chain aldehydes	Alkanes	1

the desaturase-cytochrome b_5 fusion proteins, the motif is predicted to occur on the same face of the membrane as the cytochrome b_5 (80, 116, 134). Because the histidine-containing motif is critical to desaturase function and is located on the cytoplasmic face of the membrane, it has been proposed that it is involved in coordinating an active site metal center (7, 118, 126).

The identification of a motif common to integral-membrane desaturases, hydroxylases, and epoxidases parallels the identification of a soluble desaturase and hydroxylase consensus motif that coordinates an active site diiron cluster (Table 1 and Table 2). Under the hypothesis that the histidine motif is associated with a metal center that mediates the observed oxygen-dependent desaturation, hydroxylation, and epoxidation, we would expect that such a motif would be restricted to enzymes either with these or closely related chemistries. Results of database searches for proteins containing the histidine motif support this hypothesis because it showed that the motif, is indeed specific to membrane proteins that mediate reactions that either are, or are consistent with being, oxygen dependent (Table 2; 124, 126). In addition to the desaturase, hydroxylase, and epoxidase activities that parallel the soluble class of enzymes, the integral-membrane enzymes also include acetylenase, methyl oxidase, ketolase, and decarbonylase activities. Thus enzymes containing this motif can be found in both prokaryotes and eukaryotes, and they use a wide diversity of substrates.

Specificity of the Membrane Desaturases

The presence and spacing of the histidine motifs and the equivalent spacing of hydrophobic domains are consistent with the notion that they are evolutionarily related and share a common overall fold. However, in contrast to the soluble acyl-ACP desaturases in which regiospecificity is determined relative to the carboxyl end of the fatty acid (27, 45), different integral membrane desaturases have evolved at least three distinct methods of positioning double bonds (13, 51, 52, 119). Heinz has already presented a detailed summary of substrate specificity (50), so we restrict the following comments to counting mechanisms for the sake of contrast with the soluble enzymes. Several membrane enzymes, such as the rat and yeast Δ^9-18:0-CoA desaturases, position the double bond by counting from the carboxyl end of the molecule (13) in a fashion similar to that described for the soluble acyl-ACP desaturases described above. This mode is also seen for the *Limnanthes* Δ^5-desaturase (101) and the cyanobacterial and plant Δ^6-desaturases (104, 116), and for the algal Δ^7- and Δ^9-desaturases (57). In contrast, the so-called Δ^{15} cyanobacterial enzyme is an ω-3 desaturase counting three carbons from the methyl end of the fatty acid (51). A third class of desaturases, the plant Fad6s (51), sometimes called the ω-6 desaturases, uses neither end of the fatty acid as a counting reference point; rather they appear to count three carbons toward the methyl end from an existing double bond in the monoene (52). This enzyme might be described as a $\Delta^x + 3$, where the Δ^x refers to the position of the existing double bond. For efficient insertion of the Δ^{12} or Δ^{15} double bonds, the substrate must already have Δ^9 or $\Delta^{9,12}$ double bonds, respectively (79). Interestingly, both hydroxy and epoxy groups at the 9 or 12 position can function in place of these double bonds in substrate recognition for the subsequent desaturation (35).

Characterization of Histidine-Motif-Containing Enzymes

Successful characterization of membrane proteins requires (*a*) that a source be identified which can produce large quantities of protein, (*b*) that the protein yields to purification, and (*c*) that the purified protein can be sufficiently concentrated for analysis. Under certain dietary conditions, rat liver constitutes a rich source of the 18:0-CoA desaturase (144). Using such conditions, Strittmatter's group was able to isolate several milligrams of enzyme that appeared as the predominant band by gel electrophoresis (137). It was shown that the purified enzyme contained one mole of iron per mole of desaturase and that the iron was necessary for activity. UV-visible spectra of this purified desaturase showed low extinction in the 380–450-nm region precluding the involvement of heme. There was significant extinction in the 380–430-nm region compared with that seen for R2 (29). These features have subsequently been attributed to the ligand-to-metal charge transfer bands of the diiron cluster of R2 (14). The UV-visible spectrum is also similar to that seen for another soluble diiron enzyme, the castor Δ^9-18:0-ACP desaturase (42). Efforts to overexpress rat, cyanobacterial, or plant integral membrane desaturases in *E. coli* or yeast have thus far resulted in the production of only small quantities of active protein that are insufficient for spectroscopic investigation (30, 138, 150). However, overexpression in *E. coli* of the *Pseudomonas oleovorans* alkane ω-hydroxylase AlkB, a hydroxylase that contains the histidine motif, was successfully demonstrated (33, 88).

Alkane ω-Hydroxylase from Pseudomonas oleovorans

AlkB AS A MODEL FOR THE INTEGRAL MEMBRANE HISTIDINE-MOTIF-CONTAINING ENZYMES   AlkB is responsible for the oxygen- and rubredoxin-dependent oxidation of the methyl group of an alkane to produce the corresponding alcohol in a reaction that closely parallels desaturation (77, 99). When chemically induced in *E. coli*, AlkB accumulated to 10–15% of total protein (88), and similar accumulation was observed with the T7 expression system (124, 139). In both systems, the enzyme accumulates in a distinct cytoplasmic membrane fraction (88) composed of spherical lipoprotein vesicles containing approximately 80 monomers (124). The enzyme has been successfully solubilized in several detergents (98), though, as for many membrane proteins, the solubilized protein becomes somewhat labile (124). Large quantities of protein can thus be synthesized and the unique AlkB-enriched protein vesicles isolated by differential ultracentrifugation (88). AlkB has an unusual property of self-assembly into membrane vesicles that permits purification by ion-exchange chromatography without the prerequisite of detergent solubilization normally required for the purification of membrane proteins (124). With this method, extremely high concentrations 50–100 mg · ml^{-1} (i.e. 1–2 millimolar) of highly active enzyme could be isolated and stored for long periods with almost no loss of activity (124).

AlkB from the natural source *P. oleovorans* had been previously purified to near homogeneity and had been shown to require iron for activity (108). Its color was straw yellow, a color now associated with the presence of a diiron center, and its UV-visible spectrum was similar to that of the purified rat and recombinant castor desaturases (42, 137). The iron stoichiometry had been estimated to be one iron per protein, the same as that determined for the purified rat desaturase (107, 137). However, the purified recombinant AlkB stoichiometry was determined to be three irons per protein (124).

IDENTIFICATION OF A DIIRON ACTIVE SITE IN AlkB   Investigation of ^{57}Fe-enriched AlkB with Mössbauer spectroscopy revealed the presence of an exchange-coupled dinuclear iron cluster of the type found in soluble diiron proteins such as R2, MMOH, and Δ^9-18:0-ACP desaturase. Single turnover experiments monitored by Mössbauer spectroscopy showed that the dithionite-reduced enzyme was stable to oxygen, but in the presence of oxygen and the substrate octane became reoxidized. This experiment formally linked the redox state of the diiron center with catalytic turnover of the enzyme. In addition, the Mössbauer parameters of the cluster were consistent with a coordination environment rich in nitrogen-containing ligands. The identification of a diiron active site accounts for two of the three iron ions. What then is the role of the extra mole of iron? Detailed analysis of the Mössbauer spectra during the single turnover experiments showed that the extra mole of iron occurs as heterogeneous species that do not reoxidize in the presence of substrate, precluding a role in catalysis. It is envisaged that the lower stoichiometry of one iron per protein previously reported for AlkB reflects a loss of iron during the protein purification (107). It should be noted that the investigation of AlkB is at an early stage, and many experiments, including other forms of spectroscopy and the determination of its three-dimensional structure, remain to be reported.

The above data show that AlkB has a diiron active site with properties similar to those of the soluble desaturase and MMO. AlkB has many similarities with other histidine-motif-containing enzymes, suggesting that they belong to a superfamily of mechanistically related enzymes (Table 2). Biochemical similarities include oxidation chemistry involving two electrons, dependence on nonheme iron, oxygen, and a short electron transport chain for activity; in addition to similar responses to chemical inhibitors. Structural similarities include molecular size, histidine motif elements found in equivalent hydrophilic domains separated by equivalent hydrophobic domains, in addition to local sequence homology in the proximity of the histidine motif, and positioning of the histidine motif on the cytoplasmic face of the membrane. Taken together, these similarities suggest that the integral-membrane histidine-motif-containing enzymes are members of a class of diiron proteins that is evolutionarily distinct from the soluble class of diiron proteins to which the soluble desaturase belongs. If correct, this prediction implies a greatly expanded role for diiron proteins in biology, one that may eventually rival the functional diversity of the cytochrome P450s (48).

Relationship Among Desaturation, Hydroxylation, and Other Functionality: Implications for Mechanism

HYDROXYLATION   The *Ricinus* 12-hydroxylase is an enzyme that catalyzes the production of ricinoleic acid ($18:1\Delta^9$, 12-OH) from oleic acid esterified to the *sn*-2 position of phosphatidylcholine (8). It was shown to have properties similar to the desaturase, including the same electron transport chain and the same sensitivity to inhibitors (130). Its gene was identified by mass sequencing based on the hypothesis that the hydroxylase would be a close relative of the desaturase (148). Formal proof of the identity of the 12-hydroxylase came from the accumulation of ricinoleic and other hydroxy fatty acids upon its expression in transgenic tobacco and *Arabidopsis* (17, 148). This link between membrane desaturases and hydroxylases has implications for the reaction mechanism of both enzymes.

The same potential mechanisms can be envisaged for the membrane desaturases as for the soluble desaturases, including hydroxylation followed by dehydration, or direct hydrogen abstractions and formation of radical intermediates, as shown in Figure 4. Because the soluble acyl-ACP desaturases are evolutionarily unrelated to the membrane desaturases, the possibility exists that they have different mechanisms of desaturation. Thus while the hydroxylation/dehydration hypothesis had been rejected for soluble enzymes, it remained to be formally tested for the integral-membrane enzymes (87). Recent experiments show that the desaturation/hydroxylation mechanism is also inconsistent with the labeling patterns derived from desaturation of fluorinated fatty acids by the yeast Δ^9-CoA membrane desaturase (21). However, support for direct hydrogen abstraction was obtained for the yeast Δ^9-CoA desaturase by the demonstration of a kinetic isotope effect for the reaction (20). A maximal isotope effect was observed for the 9-position, but no isotope effect was seen for the 10-position, suggesting a sequential hydrogen abstraction mechanism initiated at the 9-position. This result is also in agreement with previous oxygen-trapping experiments in which sulfur was used as a methylene isostere (19, 22). Similar deuterium isotope effect experiments with the plant oleate-12 desaturase showed a maximal effect for the 12-position, with no detectable effect for the 13-position (PH Buist, unpublished results). These data are therefore consistent with the scheme for both hydroxylation and desaturation shown in Figure 4.

Other species including *Lesquerella fendleri* also possess a 12-hydroxylase activity (106). The gene encoding this 12-hydroxylase was recently isolated from this organism and shown to have high homology to both the *Ricinus* 12-hydroxylase and to the various 12-desaturases (15). When assayed, the *Lesquerella* 12-hydroxylase was found to be a bifunctional enzyme, such that it is partially active as a desaturase and partially as a hydroxylase. Both sequence similarity of the desaturases and hydroxylases and the bifunctionality again suggest a close mechanistic link between desaturation and hydroxylation. As in the investigation of the specificity factors in thioesterases (156) and soluble acyl-ACP desaturases (26), site-directed mutants of the desaturase and hydroxylase were made to identify the determinants of reaction outcome in terms of desaturation and hydroxylation. A yeast expression system (30) was used to assess the effects of such changes. By means of evaluation of amino acid sequence comparisons between two hydroxylases and five desaturases, seven amino acid positions were identified that could potentially be important determinants of the reaction outcome (15). Substitution of the equivalent residues from the desaturase into the hydroxylase shifted the ratio of desaturation:hydroxylation activity in the direction of hydroxylation (16). In reciprocal experiments, substitution of the equivalent residues from the hydroxylase into the desaturase shifted the ratio of hydroxylation:desaturation activity in the direction of desaturation. Further experiments suggested that multiple residues were responsible for the changes and that the effects of individual residues are additive (16). While none of the seven residues is in the histidine motif, four of the seven residues are immediately adjacent to the histidines of the motif, one is close by, and the other two are remote from the motif in the linear sequence. This proximity of residues to the histidines identified as necessary for catalysis supports the idea that they are part of the active site (124, 126). The working hypothesis is that the reaction proceeds along a common route via intermediate Q to hydrogen abstraction (Figure 4; 103, 128, 153). At this point, subtle changes in the active site geometry would favor either oxygen rebound or a second hydrogen abstraction leading to either hydroxylation or desaturation, respectively. For instance, the carbon that undergoes initial hydrogen abstraction could be positioned closer to the iron center favoring oxygen rebound, or the second hydrogen could be positioned more distant and hence out of range of the activated oxygen species.

OTHER ACTIVITIES   Recently, the 12-acetylenase with high homology to the 12-desaturase was cloned from *Crepis* (69). This reaction can be formally described as a second desaturation occurring on linoleoyl-phosphatidylcholine. Many plants contain epoxy fatty acids. However, biochemical evidence on the *Euphorbia lagascae* epoxygenase shows that it is inhibited by carbon monoxide, suggesting it is a cytochrome P450 enzyme (9, 11). However, the soluble class of diiron proteins contains at least one epoxidase enzyme (49, 112), and the only characterized membrane diiron protein AlkB is equally active as a hydroxylase and an epoxidase (74). This enzyme is unable to discriminate between alkanes and alkenes, presumably because it attacks the terminal carbon of the substrate and is unable to detect a bend in the molecule resulting from the double bond in the alkene. Thus, it is also possible that a desaturase-like diiron protein with similarity to integral membrane desaturases would also be capable of epoxidizing unsaturated fatty acids. Indeed, at the time of going to press two groups have isolated histidine-motif-encoding genes homologous to 12-desaturases and 12-hydroxylases that encode epoxygenases. The corresponding enzymes introduce epoxy groups at the 12-position of 18:2 (A Kinney & S Stymne, unpublished results).

In terms of mechanism, perhaps two general classifications of active site can be envisaged, one that favors desaturation-like reactions, and the other that favors hydroxylation-like reactions. When the enzyme is presented with a substrate saturated at the target position, desaturation or hydroxylation occurs. However, when presented with a double bond at the target position, the desaturase-type enzyme introduces a triple bond, whereas the hydroxylase type is unable to hydroxylate, and instead an epoxy group is formed (Figure 5). Under this hypothesis, a desaturase or hydroxylase would accomodate a straight (saturated) target region for substrate binding, whereas an acetylenase or an epoxidase would accomodate a bend, i.e. a *cis*-double bond [or perhaps a hydroxyl group (35)] for substrate binding. Because, as described above, it is possible to interconvert desaturase and hydroxylase functionality (16), it is likely that a common activated intermediate is steered toward different functional outcomes by alteration of the active site geometry. It is easy to imagine how a hydroxylase enzyme capable of binding an alkene would insert an oxygen into the π-system of the double bond. More difficult to understand is how the acetylenase could abstract two more hydrogens to introduce a triple bond without a similar oxygen insertion into the π-system of the double bond. This area will surely be fertile ground for future structure-function studies.

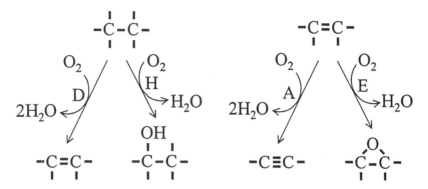

Figure 5 Proposed general scheme for fatty acid modification. Modification enzymes are indicated by letters: D, desaturase; H, hydroxylase; A, acetylenase; E, epoxidase.

FUTURE PERSPECTIVES

Advances in molecular biological techniques, heterologous expression, and purification of both soluble and integral-membrane enzymes have facilitated a substantial increase in our understanding of the biochemistry of lipid modification enzymes. We have identified commonalties both in the active sites of these soluble and membrane enzymes, and also in the range of substrate specificities and reaction outcomes. With our new understanding of the organization of the active sites of these enzymes and the availability of the atomic structure of one of the soluble class of desaturases, the stage is set for a rapid and detailed understanding of the reaction mechanism. Such experiments will involve rapid freeze quenched stop-flow spectroscopy with various substrates and kinetic isotope experiments. Determination of a crystal structure of one of the members of the integral-membrane class of enzymes will surely bring similar advances to our understanding of their function.

Of equal significance, we are starting to understand the factors within the proteins that determine substrate specificity and reaction outcome. For the soluble desaturases, we are no longer constrained by the availability of naturally-occurring enzyme activities. In this regard, we have already started to re-engineer substrate specificity rationally and have also engineered regiospecificity based on primary structural alignments (26).

The factors that determine reaction outcome are also starting to become understood. The ability to change the reaction outcome by site-directed mutagenesis is indeed a new paradigm (16). While only the interconversion of desaturase and hydroxylase function have been demonstrated to date, in the future it should be possible to change functionality at will between desaturase, hydroxylase, acetylenase, and epoxygenase functions. With further investigation, it should be possible to design lipid modification enzymes with desired substrate specificity, regiospecificity, and functional outcome. Plants are already being used to accumulate "designer" products of interest (64–66, 84, 93, 155). Perhaps rationally designed fatty acid modification enzymes will form the basis for a new generation of oilcrops (18).

ACKNOWLEDGMENTS

We thank the Office of Basic Energy Sciences of the US Department of Energy for support. We also thank Dr. J Ohlrogge and Dr. L Que for their critical reading of the manuscript and helpful discussion. We are grateful to Dr. P Buist, Dr. A Kinney, Dr. S Stymne, and Dr. J Takemoto for sharing data before publication.

Visit the *Annual Reviews home page* at
http://www.AnnualReviews.org.

Literature Cited

1. Aarts MG, Keijzer CJ, Stiekema WJ, Pereira A. 1995. Molecular characterization of the CER1 gene of *Arabidopsis* involved in epicuticular wax biosynthesis and pollen fertility. *Plant Cell* 7:2115–27

2. Ai J, Broadwater JA, Loehr TM, Sanders-Loehr J, Fox BG. 1997. Azide adducts of stearoyl-ACP desaturase: model for μ-1,2 bridging by diioxygen in the binuclear iron active site. *J. Biol. Inorgan. Chem.* 2:37–45

3. Andreasen AA, Stier TJB. 1953. Anaerobic nutrition of *Saccharomyces cerevisiae*. I. Ergosterol requirement for growth in a defined medium. *J. Cell Comp. Phys.* 41:23–36

4. Andreasen AA, Stier TJB. 1954. Anaerobic nutrition of *Saccharomyces cerevisiae*. II. Unsaturated fatty acid requirement for growth in a defined medium. *J. Cell Comp. Phys.* 43:271–81

5. Arondel V, Lemieux B, Hwang I, Gibson S, Goodman HM, Somerville CR. 1992. Map-based cloning of a gene controlling omega-3 fatty acid desaturation in *Arabidopsis*. *Science* 258:1353–55

6. Arthington BA, Bennett LG, Skatrud PL, Guynn CJ, Barbuch RJ, et al. 1991. Cloning, disruption and sequence of the gene encoding yeast C-5 sterol desaturase. *Gene* 102:39–44

7. Avelange-Macherel MH, Macherel D, Wada H, Murata N. 1995. Site-directed mutagenesis of histidine residues in the Δ^{12} acyl-lipid desaturase of *Synechocystis*. *FEBS Lett.* 361:111–14

8. Bafor M, Smith MA, Jonsson L, Stobart K, Stymne S. 1991. Ricinoleic acid biosynthesis and triacylglycerol assembly in microsomal preparations from developing castor-bean (*Ricinus communis*) endosperm. *Biochem. J.* 280:507–14

9. Bafor M, Smith MA, Jonsson L, Stobart K, Stymne S. 1993. Biosynthesis of vernoleate (*cis*-12-epoxyoctadeca-cis-9-enoate) in microsomal preparations from developing endosperm of *Euphorbia lagascae*. *Arch. Biochem. Biophys.* 303:145–51

10. Bard M, Bruner DA, Pierson CA, Lees ND, Biermann B, et al. 1996. Cloning and characterization of ERG25, the *Saccharomyces cerevisiae* gene encoding C-4 sterol methyl oxidase. *Proc. Natl. Acad. Sci. USA* 93:186–90

11. Blee E, Stahl U, Schuber F, Stymne S. 1993. Regio- and stereoselectivity of cytochrome P-450 and peroxygenase-dependent formation of *cis*-12,13-epoxy-9(Z)-octadecenoic acid (vernolic acid) in *Euphorbia lagascae*. *Biochem. Biophys. Res. Commun.* 197:778–84

12. Bloch K. 1969. Enzymatic synthesis of monounsaturated fatty acids. *Acc. Chem. Res.* 2:193–202

13. Bloomfield DK, Bloch K. 1960. Formation of Δ^9-unsaturated fatty acids. *J. Biol. Chem.* 235:337–45

14. Bollinger JM Jr, Edmondson DE, Huynh BH, Filley J, Norton JR, Stubbe J. 1991. Mechanism of assembly of the tyrosyl radical-dinuclear iron cluster cofactor of ribonucleotide reductase. *Science* 253:292–98

15. Broun P, Boddupalli S, Somerville C. 1998. A bifunctional oleate 12-hydroxylase:desaturase from *Lesquerella fendleri*. *Plant J.* 13:201–10

16. Broun P, Shanklin J, Whittle E, Somerville C. 1997. Switching between desaturation and hydroxylation of oleated by manipulating the key residues of a hydroxylase and a desaturase. *Abstr. Biochem. Mol. Biol. Plant Fatty Acids Glycerolipids Symp., Lake Tahoe, Calif., Jun. 4–8*, p. 16

17. Broun P, Somerville C. 1997. Accumulation of ricinoleic, lesquerolic, and densipolic acids in seeds of transgenic *Arabidopsis* plants that express a fatty acyl hydroxylase cDNA from castor bean. *Plant Physiol.* 113:933–42

18. Browse JA. 1996. Towards the rational engineering of plant oils: crystal structure of the 18:0–ACP desaturase. *Trends Plant Sci.* 1:403–4

19. Buist PH. 1993. Use of aromatic thia fatty acids as active site mapping agents for a yeast Δ^9-desaturase. *Can. J. Chem.* 72:176–81

20. Buist PH, Behrouzian B. 1996. Use of deuterium kinetic isotope effects to probe the cryptoregiochemistry of Δ^9–desaturation. *J. Am. Chem. Soc.* 118: 6295–96

21. Buist PH, Behrouzian B, Alexopoulos KA, Dawson B, Black B. 1996. Fluorinated fatty acids: new mechanistic probes for desaturases. *Chem. Commun.*, pp. 2671–72

22. Buist PH, Marecak DM. 1992. Stereochemical analysis of sulfoxides obtained by diverted desaturation. *J. Am. Chem. Soc.* 114:5073–80

23. Cadena DL, Kurten RC, Gill GN. 1997. The product of the MLD gene is a member of the membrane fatty acid desaturase family: overexpression of MLD inhibits EGF receptor biosynthesis. *Biochemistry* 36:6960–67

24. Cahoon EB, Coughlan S, Shanklin J. 1997. Characterization of a structurally and functionally diverged acyl-acyl carrier protein desaturase from milkweed seed. *Plant Mol. Biol.* 33:1105–10

25. Cahoon EB, Cranmer AM, Shanklin J, Ohlrogge JB. 1994. Δ^6 Hexadecenoic acid is synthesized by the activity of a soluble Δ^6-palmitoyl-acyl carrier protein desaturase in *Thunbergia alata* endosperm. *J. Biol. Chem.* 269:27519–26

26. Cahoon EB, Lindqvist Y, Schneider G, Shanklin J, 1997. Redesign of soluble fatty acid desaturases from plants for altered substrate specificity and double bond position. *Proc. Natl. Acad. Sci. USA* 94:4872–77

27. Cahoon EB, Ohlrogge JB. 1994. Metabolic evidence for the involvement of a Δ^4-palmitoyl-acyl carrier protein desaturase in the synthesis of petroselinic acid in coriander endosperm and transgenic tobacco cells. *Plant Physiol.* 104:827–38

28. Cahoon EB, Shanklin J, Ohlrogge JB. 1992. Expression of a coriander desaturase results in petroselinic acid production in transgenic tobacco. *Proc. Natl. Acad. Sci. USA* 89:11184–88

29. Capaldi RA, Vanderkooi G. 1972. The low polarity of many membrane proteins. *Proc. Natl. Acad. Sci. USA* 69:930–32

30. Covello PS, Reed DW. 1996. Functional expression of the extraplastidial *Arabidopsis thaliana* oleate desaturase gene (FAD2) in *Saccharomyces cerevisiae*. *Plant Physiol.* 111:223–26

31. Dailey HA, Strittmatter P. 1979. Modification and identification of cytochrome b_5 carboxyl groups involved in protein-protein interaction with cytochrome b_5 reductase. *J. Biol. Chem.* 254:5388–96

31a. Eaton RW. 1997. *p*-Cymene catabolic pathway in *Pseudomonas putida* F1: cloning and characterization of DNA encoding conversion of *p*-cymene to *p*-cumate. *J. Bacteriol.* 179:3171–80

32. Edmondson DE, Juynh BH. 1996. Diiron cluster intermediates in biological oxygen activation reactions. *Inorgan. Chim. Acta* 252:399–404

33. Eggink G, Lageveen RG, Altenburg B, Witholt B. 1987. Controlled and functional expression of the *Pseudomonas oleovorans* alkane utilizing system in *Pseudomonas putida* and *Escherichia coli*. *J. Biol. Chem.* 262:17712–18

34. Elgren TE, Lynch JB, Juarez-Garcia C, Münck E, Sjoberg BM, Que L Jr. 1991. Electron transfer associated with oxygen activation in the B2 protein of ribonucleotide reductase from *Escherichia coli*. *J. Biol. Chem.* 266:19265–68

35. Engeseth N, Stymne S. 1996. Desaturation of oxygenated fatty acids in *Lesquerella* and other oil seeds. *Planta* 198: 238–45

36. Enoch HG, Catala A, Strittmatter P.

1976. Mechanism of rat liver microsomal stearyl-CoA desaturase. Studies of the substrate specificity, enzyme-substrate interactions, and the function of lipid. *J. Biol. Chem.* 251:5095–103

37. Falcone DL, Gibson S, Lemieux B, Somerville C. 1994. Identification of a gene that complements an *Arabidopsis* mutant deficient in chloroplast ω 6 desaturase activity. *Plant Physiol* 106:1453–59

38. Fox BG. 1997. Catalysis by none-heme iron. In *Comprehensive Biological Catalysis*, ed. M Sinott, pp. 261–348. London: Academic

39. Fox BG, Froland WA, Dege JE, Lipscomb JD. 1989. Methane monooxygenase from *Methylosinus trichosporium* OB3b. Purification and properties of a three-component system with high specific activity from a type II methanotroph. *J. Biol. Chem.* 264:10023–33

40. Fox BG, Froland WA, Jollie DR, Lipscomb JD. 1990. Methane monooxygenase from *Methylosinus trichosporium* OB3b. *Methods Enzymol.* 188:191–202

41. Fox BG, Shanklin J, Ai JY, Loehr TM, Sanders-Loehr J. 1994. Resonance Raman evidence for an Fe-O-Fe center in stearoyl-ACP desaturase. Primary sequence identity with other diiron-oxo proteins. *Biochemistry* 33:12776–86

42. Fox BG, Shanklin J, Somerville C, Münck E. 1993. Stearoyl-acyl carrier protein Δ^9 desaturase from *Ricinus communis* is a diiron-oxo protein. *Proc. Natl. Acad. Sci. USA* 90:2486–90

43. Froland WA, Andersson KK, Lee SK, Liu Y, Lipscomb JD. 1992. Methane monooxygenase component B and reductase alter the regioselectivity of the hydroxylase component-catalyzed reactions. A novel role for protein-protein interactions in an oxygenase mechanism. *J. Biol. Chem.* 267:17588–97

44. Fulco AJ. 1974. Metabolic alterations of fatty acids. *Annu. Rev. Biochem.* 43:215–40

45. Gibson KJ. 1993. Palmitoleate formation by soybean stearoyl-acyl carrier protein desaturase. *Biochim. Biophys. Acta* 1169:231–35

46. Groves JT, McClusky GA. 1976. Aliphatic hydroxylation via oxygen rebound: oxygen transfer catalyzed by iron. *J. Am. Chem. Soc.* 98:859–61

47. Hackett CS, Strittmatter P. 1984. Covalent cross-linking of the active sites of vesicle-bound cytochrome b_5 and NADH-cytochrome b_5 reductase. *J. Biol. Chem.* 259:3275–82

48. Halkier BA. 1996. Catalytic reactivities and structure/function relationships of cytochrome P450 enzymes. *Phytochemistry* 43:1–21

49. Hartmans S, Weber FJ, Somhorst DP, de Bont JA. 1991. Alkene monooxygenase from *Mycobacterium*: a multicomponent enzyme. *J. Gen. Microbiol.* 137:2555–60

50. Heinz E. 1993. Biosynthesis of polyunsaturated fatty acids. In *Lipid Metabolism in Plants*, ed. TS Moore, pp. 34–89. Boca Raton, FL: CRC Press

51. Higashi S, Murata N. 1993. An in vivo study of substrate specificities of acyl lipid desaturases and acyltransferases in lipid synthesis in *Synechocystis* PCC6803. *Plant Physiol.* 102:1275–78

52. Hitz WD, Carlson TJ, Booth JR Jr, Kinney AJ, Stecca KL, Yadav NS. 1994. Cloning of a higher-plant plastid ω-6 fatty acid desaturase cDNA and its expression in a cyanobacterium. *Plant Physiol.* 105:635–41

53. Hobohm U, Sander C. 1995. A sequence property approach to searching protein databases. *J. Mol. Biol.* 251:390–99

54. Hoffman BJ, Broadwater JA, Johnson P, Harper J, Fox BG, Kenealy WR. 1995. Lactose fed-batch overexpression of recombinant metalloproteins in *Escherichia coli* BL21 (DE3): process control yielding high levels of metal-incorporated, soluble protein. *Protein Exp. Purif.* 6:646–54

55. Holm RH, Kennepohl P, Solomon EI. 1996. Structural and functional aspects of metal sites in biology. *Chem. Rev.* 96:2239–314

56. Deleted in proof

57. Howling D, Morris LJ, James AT. 1968. The influence of chain length on the dehydrogenation of saturated fatty acids. *Biochem. Biophys. Acta* 152:224–26

58. Iba K, Gibson S, Nishiuchi T, Fuse T, Nishimura M, et al. 1993. A gene encoding a chloroplast omega-3 fatty acid desaturase complements alterations in fatty acid desaturation and chloroplast copy number of the *fad7* mutant of *Arabidopsis thaliana*. *J. Biol. Chem.* 268:24099–105

59. Jackson M, Portnoi D, Catheline D, Dumail L, Rauzier J, et al. 1997. *Mycobacterium tuberculosis* Des protein: an immunodominant target for the humoral response of tuberculous patients. *Infect. Immun.* 65:2883–89

60. Jacobson BS, Jaworski JG, Stumpf PK. 1974. Fat metabolism in plants. LXII. Stearoyl-acyl carrier protein desaturase from spinach chloroplasts. *Plant Physiol.* 54:484–86

61. Jaworski JG, Stumpf PK. 1974. Fat

metabolism in higher plants: properties of a soluble stearoyl-acyl carrier protein desaturase from maturing *Carthamus tinctorius*. *Arch. Biochem. Biophys.* 162:158–65

62. Kajiwara S, Kakizono T, Saito T, Kondo K, Ohtani T, et al. 1995. Isolation and functional identification of a novel cDNA for astaxanthin biosynthesis from *Haematococcus pluvialis*, and astaxanthin synthesis in *Escherichia coli*. *Plant Mol. Biol.* 29:343–52

63. Kearns EV, Hugly S, Somerville CR. 1991. The role of cytochrome b_5 in Δ^{12} desaturation of oleic acid by microsomes of safflower (*Carthamus tinctorius* L.). *Arch. Biochem. Biophys.* 284:431–36

64. Kinney AJ. 1997. *Genetic Engineering of Oilseeds for Desired Traits*. In *Genetic Engineering*, ed. JK Seklow, pp. 149–66. New York: Plenum

65. Kishore GM, Somerville CR. 1993. Genetic engineering of commercially useful biosynthetic pathways in transgenic plants. *Curr. Opin. Biotechnol.* 4:152–58

66. Knauf VC. 1995. Genetic approaches for obtaining new products from plants. *Curr. Opin. Biotechnol.* 6:165–70

67. Kok M, Oldenhuis R, van der Linden MP, Raatjes P, Kingma J, et al. 1989. The *Pseudomonas oleovorans* alkane hydroxylase gene. Sequence and expression. *J. Biol. Chem.* 264:5435–41

68. Kurtz DM. 1997. Structural similarity and functional diversity in diiron-oxo proteins. *J. Biol. Inorgan. Chem.* 2:159–67

69. Lee AH, Lenman M, Banas A, Bafor M, Sjodahl S, et al. 1997. Cloning of a cDNA encoding an acetylenic acid forming enzyme and its functional expression in yeast. *Abstr. Biochem. Mol. Biol. Plant Fatty Acids Glycerolipids Symp., Lake Tahoe, Calif., Jun. 4–8*, p. A7

70. Lindqvist Y, Huang WJ, Schneider G, Shanklin J. 1996. Crystal structure of a Δ^9 stearoyl-acyl carrier protein desaturase from castor seed and its relationship to other diiron proteins. *EMBO J.* 15:4081–92

71. Liu KE, Lippard SJ. 1995. Studies of the soluble methane monooxygenase protein system: structure, component interactions, and hydroxylation mechanism. *Adv. Inorgan. Chem.* 42:263–89

72. Liu Y, Nesheim JC, Lee SK, Lipscomb JD. 1995. Gating effects of component B on oxygen activation by the methane monooxygenase hydroxylase component. *J. Biol. Chem.* 270:24662–65

73. Marsh JB, James AT. 1962. The con-

version of stearic to oleic acid by liver and yeast preparations. *Biochim. Biophys. Acta* 60:320–28

74. May SW, Abbott BJ. 1973. Enzymatic epoxidation. II. Comparison between the epoxidation and hydroxylation reactions catalyzed by the ω hydroxylation system of *Pseudomonas oleovorans*. *J. Biol. Chem.* 248:1725–30

75. Deleted in proof

76. Deleted in proof

77. McKenna EJ, Coon MJ. 1970. Enzymatic ω-oxidation. IV. Purification and properties of the ω-hydroxylase of *Pseudomonas oleovorans*. *J. Biol. Chem.* 245:3882–89

78. McKeon TA, Stumpf PK. 1982. Purification and characterization of the stearoyl-acyl carrier protein desaturase and the acyl-acyl carrier protein thioesterase from maturing seeds of safflower. *J. Biol. Chem.* 257:12141–47

79. Miquel M, Browse J. 1992. *Arabidopsis* mutants deficient in polyunsaturated fatty acid synthesis. Biochemical and genetic characterization of a plant oleoyl-phosphatidylcholine desaturase. *J. Biol. Chem.* 267:1502–9

80. Mitchell AG, Martin CE. 1995. A novel cytochrome b_5–like domain is linked to the carboxyl terminus of the *Saccharomyces cerevisiae* Δ^9 fatty acid desaturase. *J. Biol. Chem.* 270:29766–72

80a. Mitchell AG, Martin CE. 1997. Fah1p, a *Saccharomyces cerevisiae* cytochrome b_5 fusion protein, and its *Arabidopsis thaliana* homolog that lacks the cytochrome b_5 domain both function in the alpha hydroxylation of sphingolipid-associated very long chain fatty acids. *J. Biol. Chem.* 272:28281–88

81. Mudd JB, Stumpf PK. 1961. Fat metabolism in plants. XIV. Factors affecting the synthesis of oleic acid by particulate preparations from avocado mesocarp. *J. Biol. Chem.* 236:2602–9

82. Münck E. 1978. Mössbauer spectroscopy of proteins. *Methods Enzymol.* 54:346–79

83. Murata N, Wada H. 1995. Acyl-lipid desaturases and their importance in the tolerance and acclimatization to cold of cyanobacteria. *Biochem. J.* 308:1–8

84. Murphy DJ. 1994. Manipulation of lipid metabolism in transgenic plants: biotechnological goals and biochemical realities. *Biochem. Soc. Trans.* 22:926–31

85. Nagai J, Bloch K. 1965. Synthesis of oleic acid by *Euglena gracilis*. *J. Biol. Chem.* 240:3702–3

86. Nagai J, Bloch K. 1966. Enzymatic desaturation of stearoyl-acyl carrier protein. *J. Biol. Chem.* 241:1925–27

87. Nagai J, Bloch K. 1968. Enzymatic desaturation of stearoyl acyl carrier protein. *J. Biol. Chem.* 243:4626–33

88. Nieboer M, Kingma J, Witholt B. 1993. The alkane oxidation system of *Pseudomonas oleovorans*: induction of the *alk* genes in *Escherichia coli* W3110 (pGEc47) affects membrane biogenesis and results in overexpression of alkane hydroxylase in a distinct cytoplasmic membrane subfraction. *Mol. Microbiol.* 8:1039–51

89. Nilsson O, Aberg A, Lundqvist T, Sjoberg BM. 1988. Nucleotide sequence of the gene coding for the large subunit of ribonucleotide reductase of *Escherichia coli*. *Nucleic Acids Res.* 16(9):4174

90. Nordlund I, Powlowski J, Shingler V. 1990. Complete nucleotide sequence and polypeptide analysis of multicomponent phenol hydroxylase from *Pseudomonas* sp. strain CF600. *J. Bacteriol.* 172:6826–33

91. Nordlund P, Eklund H. 1995. Di-iron-carboxylate proteins. *Curr. Opin. Struct. Biol.* 5:758–66

92. Nordlund P, Sjoberg BM, Eklund H. 1990. Three-dimensional structure of the free radical protein of ribonucleotide reductase. *Nature* 345:593–98

93. Ohlrogge JB. 1994. Design of new plant products: engineering of fatty acid metabolism. *Plant Physiol.* 104:821–26

94. Ohlrogge J, Browse J. 1995. Lipid biosynthesis. *Plant Cell* 7:957–70

95. Okuley J, Lightner J, Feldmann K, Yadav N, Lark E, Browse J. 1994. *Arabidopsis* FAD2 gene encodes the enzyme that is essential for polyunsaturated lipid synthesis. *Plant Cell* 6:147–58

96. Oshino N, Imai Y, Sato R. 1966. Electron-transfer mechanism associated with fatty acid desaturation catalyzed. *Biochim. Biophys. Acta* 128:13–28

97. Pasteur L. 1879. *Studies on Fermentation.* London: Macmillan

98. Peters J, Witholt B. 1994. Solubilization of the overexpressed integral membrane protein alkane monooxygenase of the recombinant *Escherichia coli* W3110 [pGEc47]. *Biochim. Biophys. Acta* 1196: 145–53

99. Peterson JA, Kusunose M, Kusunose E, Coon MJ. 1967. Enzymatic ω-oxidation. II. Function of rubredoxin as the electron carrier in ω-hydroxylation. *J. Biol. Chem.* 242:4334–40

100. Pikus JD, Studts JM, Achim C, Kauffmann KE, Münck E, et al. 1996. Recombinant toluene-4-monooxygenase: catalytic and Mössbauer studies of the purified diiron and rieske components of a four-protein complex. *Biochemistry* 35:9106–19

101. Pollard MR, Stumpf PK. 1980. Biosynthesis of C20 and C22 fatty acids by developing seeds of *Limnanthes alba*. Chain elongation and Δ^5-desaturation. *Plant Physiol.* 66:649–55

102. Powlowski J, Shingler V. 1990. In vitro analysis of polypeptide requirements of multicomponent phenol hydroxylase from *Pseudomonas* sp. strain CF600. *J. Bacteriol.* 172:6834–40

103. Que L, Dong Y. 1996. Modeling the oxygen activation chemistry of methane monooxygenase and ribonucleotide reductase. *Acc. Chem. Res.* 29:190–96

104. Reddy AS, Nuccio ML, Gross LM, Thomas TL. 1993. Isolation of a Δ^6-desaturase gene from the cyanobacterium *Synechocystis* sp. strain PCC 6803 by gain-of-function expression in *Anabaena* sp. strain PCC 7120. *Plant Mol. Biol.* 22:293–300

105. Reddy AS, Thomas TL. 1996. Expression of a cyanobacterial Δ^6-desaturase gene results in gamma-linolenic acid production in transgenic plants. *Nat. Biotechnol.* 14:639–42

106. Reed DW, Taylor DC, Covello PS. 1997. Metabolism of hydroxy fatty acids in developing seeds in the genera *Lesquerella* (Brassicaceae) and *Linum* (Linaceae). *Plant Physiol.* 114:63–68

107. Reuttinger RT, Griffith GR, Coon MJ. 1977. Characterization of the ω-hydroxylase of *Pseudomonas oleovorans* as a nonheme iron protein. *Arch. Biochem. Biophys.* 183:528–37

108. Reuttinger RT, Olsen ST, Boyer RF, Coon MJ. 1974. Identification of the ω-hydroxylase of *Pseudomonas oleovorans* as a nonherme iron protein requiring phospholipid for catalytic activity. *Biochem. Biophys. Res. Commun.* 57:1011–17

109. Rogers MJ, Strittmatter P. 1974. Evidence for random distribution and translational movement of cytochrome b_5 in endoplasmic reticulum. *J. Biol. Chem.* 249:895–900

110. Rosenzweig AC, Frederick CA, Lippard SJ, Nordlund P. 1993. Crystal structure of a bacterial nonhaem iron hydroxylase that catalyses the biological oxidation of methane. *Nature* 366:537–43

111. Rosenzweig AC, Nordlund P, Takahara PM, Frederick CA, Lippard SJ. 1995. Geometry of the soluble methane monooxy-

genase catalytic diiron center in two oxidation states. *Chem. Biol.* 2:409–18

112. Saeki H, Furuhashi K. 1994. Cloning and characterization of a *Nocardia corallina* B-276 gene cluster encoding alkene monooxygenase. *J. Ferment. Bioeng.* 78:399–406

113. Sakamoto T, Los DA, Higashi S, Wada H, Nishida I, et al. 1994. Cloning of ω 3 desaturase from cyanobacteria and its use in altering the degree of membrane-lipid unsaturation. *Plant Mol. Biol.* 6:249–63

114. Sakamoto T, Wada H, Nishida I, Ohmori M, Murata N. 1994. Δ^9 acyl-lipid desaturases of cyanobacteria. Molecular cloning and substrate specificities in terms of fatty acids, *sn*-positions, and polar head groups. *J. Biol. Chem.* 269:25576–80

115. Sander C, Schneider R. 1991. Database of homology-derived protein structures and the structural meaning of sequence alignment. *Proteins* 9:56–68

116. Sayanova O, Smith MA, Lapinskas P, Stobart AK, Dobson G, et al. 1997. Expression of a borage desaturase cDNA containing an N-terminal cytochrome b_5 domain results in the accumulation of high levels of Δ^6-desaturated fatty acids in transgenic tobacco. *Proc. Natl. Acad. Sci. USA* 94:4211–16

117. Scheuerbrandt G, Goldfine H, Baronowsky PE, Bloch K. 1961. A novel mechanism for the biosynthesis of unsaturated fatty acids. *J. Biol. Chem.* 236: 2596–601

118. Schmidt H, Dresselhaus T, Buck F, Heinz E. 1994. Purification and PCR-based cDNA cloning of a plastidial n-6 desaturase. *Plant Mol. Biol.* 26:631–42

119. Schmidt H, Heinz E. 1993. Direct desaturation of intact galactolipids by a desaturase solubilized from spinach (*Spinacia oleracea*) chloroplast envelopes. *Biochem. J.* 289:777–82

120. Schmidt H, Heinz E. 1990. Involvement of ferredoxin in desaturation of lipid-bound oleate in chloroplasts. *Plant Physiol.* 94:214–20

121. Schneider G, Lindqvist Y, Shanklin J, Somerville C. 1992. Preliminary crystallographic data for stearoyl-acyl carrier protein desaturase from castor seed. *J. Mol. Biol.* 225:561–64

122. Schroepfer GJ, Bloch K. 1965. The stereospecific conversion of stearic acid to oleic acid. *J. Biol. Chem.* 240:54–63

123. Schultz DJ, Cahoon EB, Shanklin J, Craig R, Cox-Foster DL, et al. 1996. Expression of a Δ^9-14:0-acyl carrier protein fatty acid desaturase gene is necessary for the production of omega 5 anacardic acids found

in pest-resistant geranium (*Pelargonium xhortorum*). *Proc. Natl. Acad. Sci. USA* 93:8771–75

124. Shanklin J, Achim C, Schmidt H, Fox BG, Münck E. 1997. Mössbauer studies of alkane ω-hydroxylase: evidence for a diiron cluster in an integral-membrane enzyme. *Proc. Natl. Acad. Sci. USA* 94:2981–86

125. Shanklin J, Somerville C. 1991. Stearoyl-acyl-carrier-protein desaturase from higher plants is structurally unrelated to the animal and fungal homologs. *Proc. Natl. Acad. Sci. USA* 88:2510–14

126. Shanklin J, Whittle E, Fox BG. 1994. Eight histidine residues are catalytically essential in a membrane-associated iron enzyme, stearoyl-CoA desaturase, and are conserved in alkane hydroxylase and xylene monooxygenase. *Biochemistry* 33:12787–94

127. Shteinman AA. 1995. The mechanism of methane and dioxygen activation in the catalytic cycle of methane monooxygenase. *FEBS Lett.* 362:5–9

128. Shu LJ, Nesheim JC, Kauffmann K, Münck E, Lipscomb JD, Que L. 1997. An $Fe_2^{IV}O_2$ diamond core structure for the key intermediate Q of methane monooxygenase. *Science* 275:515–18

129. Smith MA, Cross AR, Jones OTG, Griffiths WT, Stymne S, Stobart K. 1990. Electron-transport components of the 1-acyl-l-2-oleoyl-*sn*-glycero-3-phosphocholine Δ^{12}-desaturase (Δ^{12}-desaturase) in microsomal preparations from developing safflower (*Carthamus tinctorius* L.) cotyledons. *Biochem. J.* 272:23–29

130. Smith MA, Jonsson L, Stymne S, Stobart K. 1992. Evidence for cytochrome b_5 as an electron donor in ricinoleic acid biosynthesis in microsomal preparations from developing castor bean (*Ricinus communis* L.). *Biochem. J.* 287:141–44

131. Somerville C. 1995. Direct tests of the role of membrane lipid composition in low-temperature-induced photoinhibition and chilling sensitivity in plants and cyanobacteria. *Proc. Natl. Acad. Sci. USA* 92:6215–18

132. Somerville CR, Browse JA. 1996. Dissecting desaturation; plants prove advantageous. *Trends Cell Biol.* 6:148–53

133. Spatz L, Strittmatter P. 1971. A form of cytochrome b_5 that contains an additional hydrophobic sequence of 40 amino acid residues. *Proc. Natl. Acad. Sci. USA* 68:1042–46

134. Sperling P, Schmidt H, Heinz E. 1995. A cytochrome-b_5-containing fusion protein

similar to plant acyl lipid desaturases. *Eur. J. Biochem.* 232:798–805

135. Spychalla JP, Kinney AJ, Browse J. 1997. Identification of an animal omega-3 fatty acid desaturase by heterologous expression in *Arabidopsis. Proc. Natl. Acad. Sci. USA* 94:1142–47

136. Stainthorpe AC, Lees V, Salmond GP, Dalton H, Murrell JC. 1990. The methane monooxygenase gene cluster of *Methylococcus capsulatus* (Bath). *Gene* 91:27–34

137. Strittmatter P, Spatz L, Corcoran D, Rogers MJ, Setlow B, Redline R. 1974. Purification and properties of rat liver microsomal stearoyl coenzyme A desaturase. *Proc. Natl. Acad. Sci. USA* 71: 4565–69

138. Strittmatter P, Thiede MA, Hackett CS, Ozols J. 1988. Bacterial synthesis of active rat stearoyl-CoA desaturase lacking the 26-residue amino-terminal amino acid sequence. *J. Biol. Chem.* 263:2532–35

139. Studier FW, Rosenberg AH, Dunn JJ, Dubendorff JW. 1990. Use of T7 RNA polymerase to direct expression of cloned genes. *Methods Enzymol.* 185:60–89

140. Stukey JE, McDonough VM, Martin CE. 1990. The OLE1 gene of *Saccharomyces cerevisiae* encodes the Δ^9 fatty acid desaturase and can be functionally replaced by the rat stearoyl-CoA desaturase gene. *J. Biol. Chem.* 265:20144–49

141. Sun ZR, Gantt E, Cunningham FX Jr. 1996. Cloning and functional analysis of the beta-carotene hydroxylase of *Arabidopsis thaliana. J. Biol. Chem.* 271: 24349–52

142. Suzuki M, Hayakawa T, Shaw JP, Rekik M, Harayama S. 1991. Primary structure of xylene monooxygenase: similarities to and differences from the alkane hydroxylation system. *J. Bacteriol.* 173:1690–95

143. Thiede MA, Ozols J, Strittmatter P. 1986. Construction and sequence of cDNA for rat liver stearoyl coenzyme A desaturase. *J. Biol. Chem.* 261:13230–35

144. Thiede MA, Strittmatter P. 1985. The induction and characterization of rat liver stearoyl-CoA desaturase mRNA. *J. Biol. Chem.* 260:14459–63

145. Thompson GA, Scherer DE, Foxall-Van Aken S, Kenny JW, Young HL, et al. 1991 Primary structures of the precursor and mature forms of stearoyl-acyl carrier protein desaturase from safflower embryos and requirement of ferredoxin for enzyme activity. *Proc. Natl. Acad. Sci. USA* 88:2578–82

146. Togawa K, Arnon DI. 1962. Ferredoxins as electron carriers in photosynthesis and in the biological production and consumption of hydrogen gas. *Nature* 195:537–43

147. van Beilen JB, Penninga D, Witholt B. 1992. Topology of the membrane-bound alkane hydroxylase of *Pseudomonas oleovorans. J. Biol. Chem.* 267:9194–201

148. van de Loo FJ, Broun P, Turner S, Somerville C. 1995. An oleate 12-hydroxylase from *Ricinus communis L.* is a fatty acyl desaturase homolog. *Proc. Natl. Acad. Sci. USA* 92:6743–47

149. van de Loo FJ, Fox BG, Somerville C. 1993. Unusual fatty acids. In *Lipid Metabolism in Plants*, ed. TS Moore, pp. 91–126. Boca Raton, FL: CRC Press

150. Wada H, Avelange-Macherel MH, Murata N. 1993. The desA gene of the cyanobacterium *Synechocystis* sp. strain PCC6803 is the structural gene for Δ^{12}-desaturase. *J. Bacteriol.* 175:6056–58

151. Wada H, Gombos Z, Murata N. 1990. Enhancement of chilling tolerance of a cyanobacterium by genetic manipulation of fatty acid desaturation. *Nature* 347:200–3

152. Wada H, Schmidt H, Heinz E, Murata N. 1993. In vitro ferredoxin-dependent desaturation of fatty acids in cyanobacterial thylakoid membranes. *J. Bacteriol.* 175:544–47

153. Wallar BJ, Lipscomb JD. 1996. Dioxygen activation by enzymes containing binuclear nonheme iron clusters. *Chem. Rev.* 96:2625–57

154. Yen KM, Karl MR, Blatt LM, Simon MJ, Winter RB, et al. 1991. Cloning and characterization of a *Pseudomonas mendocina* KR1 gene cluster encoding toluene-4-monooxygenase. *J. Bacteriol.* 173:5315–27

155. Yuan L, Knauf VC. 1997. Modification of plant components. *Curr. Opin. Biotechnol.* 8:227–33

156. Yuan L, Voelker TA, Hawkins DJ. 1995. Modification of the substrate specificity of an acyl-acyl carrier protein thioesterase by protein engineering. *Proc. Natl. Acad. Sci. USA* 92:10639–43

Annu. Rev. Plant Physiol. Plant Mol. Biol. 49:643–68

PHYTOREMEDIATION

D. E. Salt[1], R. D. Smith[2], and I. Raskin

AgBiotech Center, Rutgers University, New Brunswick, New Jersey 08903-0231;
[1]Present address: Chemistry Department, Northern Arizona University, Flagstaff,
Arizona 86071-5698; [2]Present address: DeKalb Genetics Corporation,
62 Maritime Drive, Mystic, Connecticut 06355-1958;
e-mail: raskin@aesop.rutgers.edu

KEY WORDS: decontamination, hyperaccumulator, phytoextraction, phytodegradation, heavy metals

ABSTRACT

Contaminated soils and waters pose a major environmental and human health problem, which may be partially solved by the emerging phytoremediation technology. This cost-effective plant-based approach to remediation takes advantage of the remarkable ability of plants to concentrate elements and compounds from the environment and to metabolize various molecules in their tissues. Toxic heavy metals and organic pollutants are the major targets for phytoremediation. In recent years, knowledge of the physiological and molecular mechanisms of phytoremediation began to emerge together with biological and engineering strategies designed to optimize and improve phytoremediation. In addition, several field trials confirmed the feasibility of using plants for environmental cleanup. This review concentrates on the most developed subsets of phytoremediation technology and on the biological mechanisms that make phytoremediation work.

CONTENTS

INTRODUCTION

Phytoremediation is defined as the use of green plants to remove pollutants from the environment or to render them harmless (42, 125). Several comprehensive reviews have been written on this subject, summarizing many important aspects of this novel plant-based technology (39, 41, 43, 44, 126, 136). The basic idea that plants can be used for environmental remediation is very old and cannot be traced to any particular source. However, a series of fascinating scientific discoveries combined with an interdisciplinary research approach have allowed the development of this idea into a promising, cost-effective, and environmentally friendly technology. Phytoremediation can be applied to both organic and inorganic pollutants, present in solid substrates (e.g. soil), liquid substrates (e.g. water), and the air. Phytoremediation is currently divided into the following areas:

- phytoextraction: the use of pollutant-accumulating plants to remove metals or organics from soil by concentrating them in the harvestable parts;

- phytodegradation: the use of plants and associated microorganisms to degrade organic pollutants;

- rhizofiltration: the use of plant roots to absorb and adsorb pollutants, mainly metals, from water and aqueous waste streams;

- phytostabilization: the use of plants to reduce the bioavailability of pollutants in the environment;

- phytovolatilization: the use of plants to volatilize pollutants; and

- the use of plants to remove pollutants from air.

Most of this review focuses on the phytoremediation of the metallic pollutants in soil, particularly the area of metal phytoextraction, which, arguably, is the area of major scientific and technological progress in the past years. This can be partially explained by the relative ease of detecting metals in various materials. Phytoremediation of metals is being developed as a potential cost-effective remediation solution for thousands of contaminated sites in the United States and abroad. Its development is driven by the prohibitively high cost of the available soil remediation methods, which mainly involve soil removal and burial at a price of about $1 million per acre. The metals of greatest importance as environmental pollutants and some of their regulatory limits are listed in Table 1. Elements in each category are ranked by authors according to their importance as environmental pollutants in the United States. This review focuses on phytoremediation technologies for removing toxins from the environment. We do not discuss rhizofiltration, which has been extensively reviewed (126).

Table 1 Currently found concentration ranges and regulatory guidelines for important metal and radionuclide contaminants in the order of relative importance

Element	Concentration range	Regulatory limit
Metals	$(\mu g \ kg^{-1})^a$	$(mg \ kg^{-1})^b$
Lead	1000–6,900,000	600
Cadmium	100–345,000	100
Arsenic	100–102,000	20
Chromium	5.1–3,950,000	100
Mercury	0.1–1,800,000	270
Copper	30–550,000	600
Zinc	150–5,000,000	1500
Radionuclides	Units (see below)	$pCi \ g^{-1}$
Uranium	0.2–16,000[c]	—
	0.06–18,700[d]	250[f]
Strontium	0.03–540,000[e]	—
Cesium	0.02–46,900[e]	—
Plutonium	0.00011–3,500,000[e]	—

[a]Riley et al (130).
[b]Nonresidential direct contact soil cleanup criteria. In *Cleanup Standards for Contaminated Sites*, New Jersey Department of Environmental Protection (1996).
[c]Micrograms per gram ($\mu g \ g^{-1}$).
[d]Picocuries per gram ($pCi \ g^{-1}$).
[e]Picocuries per kilogram ($pCi \ kg^{-1}$).
[f]Stern et al (149).

PHYTOEXTRACTION OF METALS

A review of the phytoremediation literature reveals that, at present, there are two basic strategies of phytoextraction being developed: chelate-assisted phytoextraction (Figure 1), which we term *induced phytoextraction*; and long-term *continuous phytoextraction* (Figure 2). Of the two processes, chelate-assisted phytoextraction is the more developed and is presently being implemented commercially. Continuous phytoextraction is also being studied by several groups for the removal of metals such as zinc, cadmium, and nickel and oxianionic metals such as selenium, arsenic, and chromium. Field trials have been performed using both phytoextraction strategies. The results, though encouraging, suggest that further development of these technologies is needed (Table 2).

Induced Phytoextraction

THE CONCEPT OF CHELATE-ASSISTED PHYTOEXTRACTION   There are no reliable reports of plants capable of naturally accumulating the most environmentally important toxic metals such as lead, cadmium, arsenic, and radionuclides.

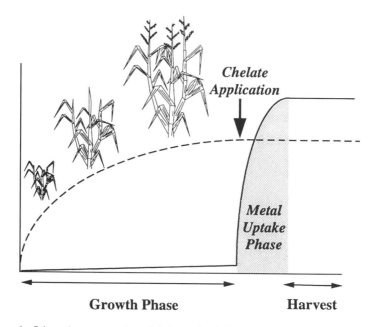

Figure 1   Schematic representation of chelate-assisted phytoremediation. *Solid line* represents metal concentration in shoot biomass; *dashed line* represents shoot biomass.

For example, vegetation growing on heavily lead-contaminated soil or solutions has been reported to contain only 0.01–0.06% of shoot dry biomass as lead (74, 81), levels well below that required for efficient phytoextraction. Early studies by Jøgensen (80) showed that application of synthetic metal chelates such as ethylenediaminetetraacetic acid (EDTA) to soils enhances lead accumulation by plants. Huang et al (74, 75) and Blaylock et al (18) were able to achieve rapid accumulation of lead in shoots to greater than 1% of shoot dry biomass. These discoveries paved the way to successful phytoremediation of lead and to defining strategies for the development of phytoextraction of other toxic metals using appropriate chelates.

The total amount of metal removed from a site is a product of metal concentration in the harvested plant material and the total harvested biomass. The observation that high biomass crop plants including Indian mustard, corn, and sunflower could be "induced" to accumulate high concentrations of lead (18, 74, 75) was another advance in the development of chelate-assisted phytoextraction.

The concept of chelate-assisted phytoextraction is applicable to other metals in addition to lead (18). The authors demonstrated the simultaneous accumulation of lead, cadmium, copper, nickel, and zinc in Indian mustard plants after

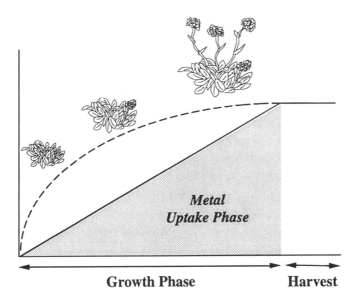

Growth Phase   **Harvest**

Figure 2   Schematic representation of continuous phytoextraction. *Solid line* represents metal concentration in shoot biomass; *dashed line* represents shoot biomass.

application of EDTA to soil contaminated with various heavy metals. Metal accumulation efficiency in these experiments was directly related to the affinity of the applied chelate for the metal. This suggests that for efficient phytoextraction synthetic chelates having a high affinity for the metal of interest should be used; for example, EDTA for lead, EGTA for cadmium (18), and possibly citrate for uranium.

Based on the above information, a hypothetical protocol for the chelate-assisted phytoextraction of a contaminated site can be outlined (Figure 1). 1. The site is evaluated and the appropriate chelate/crop combination is determined. 2. The site is prepared and planted, and the crop is cultivated. 3. Once optimal biomass is produced, the appropriate metal chelate is applied. 4. After a short metal-accumulation phase (several days or weeks), the crop is harvested. Depending on the crop and the season, the site could be replanted for further phytoextraction. Estimates suggest that plants can remove between 180 and 530 kg ha^{-1} of lead per year (18, 75), making remediation of sites contaminated with up to 2500 mg kg^{-1} lead possible in under 10 years. Following harvest, the weight and volume of contaminated material can be further reduced by ashing or composting. Metal-enriched plant residue can be disposed of as hazardous material or, if economically feasible, used for metal recovery.

Table 2   Examples of field trials for the phytoremediation of metals

Metal	Plant	Location	Method[a]	Comments	Reference
Pb	*Brassica juncea*	Trenton, N.J.	PE-CA	EDTA-enhanced uptake over one cropping season resulted in a 28% reduction in the Pb contamination area	
Cd Zn	*Thlaspi caerulescens* *Silene vulgaris*	Beltsville, Md.	PE-C	Phytoextraction of sludge-amended soils. Cd accumulation was similar in all three species. Zn accumulation in *T. caerulescens* was 10-fold higher then in other plants	26
Zn Cd Ni Cu Pb Cr	*Brassica oleracea* *Raphanus sativus* *Thlaspi caerulescens* *Alyssum lesbiacum* *Alyssum murale* *Arabidopsis thaliana*	Rothamstead, U.K.	PE-C	Sludge-amended soil	12
Se B	*Brassica juncea* *Festuca arundinacea* *Hibiscus cannibus* *Lotus corniculatus*	Los Baños, Calif.	PE-C PV	Water-extractable B was reduced between 24–52% and total Se reduced between 13–48% by all species	13
U	*Helianthus annus*	Asthabula, Ohio	RF	Removal of U from ground water[b]	

[a]Method of phytoremediation: PE, phytoextraction; PV, phytovolatilization; RF, rhizofiltration; CA, chelate-assisted phytoextraction; C, continuous phytoextraction.
[b]Phytotech Inc., personal communication.

DEVELOPMENT OF CHELATE-ASSISTED PHYTOEXTRACTION   Discovery of chelate-assisted metal uptake by plants is very recent, with only four publications appearing in the past four years. Chelate-assisted phytoextraction consists of two basic processes—release of bound metals into soil solution combined with transport of metals to the harvestable shoot. The role of chelates in increasing the soluble metal concentration in the soil solution can be explained using well-established equilibrium principles. However, the mechanisms involved in metal-chelate induced plant uptake and translocation of metals are not well understood.

Following EDTA application, lead accumulation in shoots is directly correlated with an accumulation of EDTA (A Vassel & D Salt, unpublished data). Thus, it is likely that lead is transported within the plant as a Pb-EDTA complex.

The presence of high levels of EDTA in plant tissues should increase soluble lead concentrations within the plant by formation of soluble Pb-EDTA, allowing its movement from roots to shoots where lead would likely accumulate as Pb-EDTA.

Clearly, transport of metal-chelate complexes within plants plays a pivotal role in chelate-assisted metal accumulation in plants. What are the mechanisms involved in transport of metal-chelate complexes in plants? A good place to start looking for answers to this question is the mineral nutrition literature. In the 1950s, Fe^{3+} chelates were introduced as a way to correct iron deficiency in plants. Since that time, the mechanism by which plant roots use iron from stable Fe^{3+} chelates has been debated. It appears that the roots of dicotyledonous plants acquire iron from Fe^{3+} chelate complexes either as Fe^{2+} after chelate splitting by a root Fe^{3+} chelate reductase (38) or as an intact Fe^{3+} chelate complex (72, 86, 133). The uptake mechanism depends on the iron nutritional status of the plant, with Fe^{2+} uptake predominating in iron-deficient plants (38). Thus, the highly stable Pb-EDTA complex, which cannot be split by the root Fe^{3+} chelate reductase, may be acquired in the same way as Fe^{3+} chelates. However, optimal "induction" of metal uptake occurs at chelate concentrations at least two orders of magnitude higher than those used in hydroponic nutrient solution (18). Is the mechanism of chelate uptake different at these elevated concentrations? An intriguing report by Jeffreys & Wallace (79) suggests that it is. Using the red iron chelate Fe-EDDHA, these authors showed that a threshold chelate concentration exists above which accumulation of iron chelate in shoots is induced and below which only low levels of iron chelate accumulate. This report predates the first observation of chelate-assisted metal accumulation by 25 years and suggests that there are at least two mechanisms involved in metal chelate uptake functioning at low and high chelate concentrations. Induction of metal chelate uptake by plants is correlated with severe plant stress and ultimately plant death; however, it is not clear if stress is necessary for induction or simply reflects the accumulation of high concentrations of synthetic chelate in the plant. More recently the biphasic nature of chelate uptake has been confirmed by the direct measurement of plant movement and distribution of ^{14}C-labeled EDTA and Pb-EDTA (A Vassel & D Salt, unpublished data).

Chelate-assisted transport of metal to shoots appears to occur in the xylem (74) via the transpiration stream (18). The metal appears to move to shoots as a metal-chelate complex (A Vassel & D Salt, unpublished) where water evaporates and the metal-chelate complex remains. In this way, after chelate-assisted induction the plant becomes a wick, which drives chelated metal from the soil solution into the leaves. The operation of the wick relies on a high-surface-area collection system provided by the roots and by the efficient capillary plumbing system inside the plant. Although it may be possible to design and build a similar system using engineering approaches, nature provides a more cost-effective and evolutionarily perfected approach.

Continuous Phytoextraction

An alternative approach to chelate-assisted metal accumulation is the reliance on the specialized physiological processes that allow plants to accumulate metals over the complete growth cycle. This type of metal uptake is epitomized by hyperaccumulating plants that grow on soils rich in heavy metals (9). These plants are naturally able to accumulate >1% of shoot dry biomass as Zn, Ni, Mn, or Se. It was the existence of this hyperaccumulation phenomenon that inspired Chaney in 1983 to formulate the concept of phytoextraction (37). Unlike induced metal uptake, continuous phytoextraction is based on the genetic and physiological capacity of specialized plants to accumulate, translocate, and resist high amounts of metals. Major disadvantages of using naturally occurring metal hyperaccumulators for continuous phytoextraction are their relatively low biomass, slow growth rates, and the lack of any hyperaccumulators for the most environmentally important metallic pollutants (e.g. lead, cadmium, arsenic, and uranium). However, understanding the biological mechanisms of hyperaccumulation may help in the development of superior plants for the phytoremediation of metals.

THE HYPERACCUMULATION CONCEPT   As early as 1885, A Baumann, a German botanist working near the border of Germany and Belgium, had observed that leaves of certain plant species growing on soils naturally enriched in zinc contained extraordinarily high levels of this element (14). Two species of particular note were the violet *Viola calaminaria* and the mustard *Thlaspi calaminare*, more recently classified as *Thlaspi caerulescens* (77), which contained about 1% and 1.7% zinc in dry leaves, respectively. This can be compared with zinc levels between 0.001% and 0.02% in dried leaves of plants growing on unmineralized soils. Fifty years later, studies in the United States implicated selenium as the plant component responsible for alkali disease in range animals in South Dakota. This observation led to the discovery of plants, notably of the genus *Astragalus*, capable of accumulating up to 0.6% selenium in dry shoot biomass (31, 32). Shortly thereafter, two Italian botanists (107) discovered plants that accumulate nickel. They observed that dried leaves of *Alyssum bertolonii* growing on nickel-enriched serpentenitic soils near Florence, Italy, contained about 1% nickel, over 100–1000 times higher than other plants growing nearby.

Since these early observations, plants that accumulate elevated levels of cobalt, copper, manganese, and possibly lead have also been described (9). However, the existence of hyperaccumulators for metals other than Ni, Zn, and

Se has been continuously questioned and requires further substantiation. The first hyperaccumulators characterized were members of the Brassicaceae and Fabaceae families. Presently, at least 45 plant families are known to contain metal-accumulating species. The number of metal-accumulating taxa identified to date has now grown to 397. This number is likely to change in the future. As more of the metal-enriched environments are investigated, new hyperaccumulators will be identified, and plants initially classified as hyperaccumulators from herbarium and field specimens may be reclassified as nonaccumulators after closer scrutiny.

The ecological role of metal hyperaccumulation is still not entirely clear. It has been suggested that metal accumulation provides protection against fungal and insect attack (21, 129). Recent evidence has confirmed the protective function of nickel hyperaccumulation against fungal and bacterial pathogens in *Streptanthus polygaloides* (23), and insect herbivory in *S. polygaloides* and *T. montanum* (22, 105). The antiherbivory effect of zinc has been also demonstrated in the zinc hyperaccumulator *T. caerulescens* (122).

DEVELOPMENT OF CONTINUOUS PHYTOEXTRACTION   The unique capacity of hyperaccumulators to accumulate high foliar metal concentrations makes these plants suitable for the development of phytoremediation crops for continuous phytoextraction. This idea was first introduced by Chaney (37) and Baker and coworkers (10). The ideal plant for continuous phytoextraction should grow on metal-polluted soils to high biomass and accumulate and resist high concentrations of metal in shoots. The first reported field trials of continuous phytoextraction were performed in 1991 with moderate success (12). Most known hyperaccumulator plants have low biomass and/or slow growth rates, whereas rapidly growing high-biomass crop plants are sensitive to metals and accumulate only low concentrations in shoots. To overcome these limitations, a two-component long-term strategy needs to be developed for continuous phytoextraction to succeed.

First, attempts to improve existing lines of phytoextracting plants should be continued (48, 85, 91) as well as the search for new high-biomass metal hyperaccumulators. The usefulness of this search was recently demonstrated by the identification of *Berkheya coddii* (Asteraceae), a tall, high-biomass plant from the northeastern Transvaal, South Africa, capable of accumulating up to 3.7% nickel in its shoot dry biomass (109). This (and related) species may have significant phytoremediation potential, due to strong hyperaccumulation, relatively high biomass production, and the ability to grow in dense stands. Biotechnological approaches to the production of high-biomass metal hyperaccumulators should also be considered. Modern genetics can be used to transfer hyperaccumulating genes to nonaccumulating plants.

Second, we need to understand and exploit the biological processes involved in metal acquisition, transport, and shoot accumulation in both hyperaccumulating and nonaccumulating plants. The plant mineral nutrition literature is rich in information relating to metal tolerance, metal ion uptake, transport, and accumulation (for recent reviews, see 86, 103, 169), and we will therefore only review here areas that have particular relevance to phytoextraction.

Metal Resistance Mechanisms

Continuous phytoextraction relies on the ability of plants to accumulate metals in their shoots, over extended periods. To achieve this, plants must possess efficient mechanisms for the detoxification of the accumulated metal. The recent observation that nickel resistance in *Thlaspi goesingense* is a primary determinant of nickel hyperaccumulation when plants are grown hydroponically (89) supports this conclusion. Therefore, the ability to manipulate metal tolerance in plants will be key to the development of efficient phytoremediation crops. As an elegant demonstration of this principle, Hg^{2+}-resistant *Arabidopsis thaliana* overexpressing bacterial mercury reductase was recently shown to remove Hg^{2+} efficiently from solution (134).

In order to develop hypertolerant plants capable of accumulating high concentrations of metals it will be vital to understand the existing molecular and biochemical strategies plants adopt to resist metal toxicity. The processes involved in intracellular detoxification of heavy metals have been extensively reviewed (56, 78, 152). Thus, in the interests of brevity, we only cover those processes that could potentially be manipulated to improve the metal resistance of phytoextraction crops. These mechanisms include chelation, compartmentalization, biotransformation, and cellular repair mechanisms.

CHELATION   Chelation of metal ions by specific high-affinity ligands reduces the solution concentration of free metal ions, thereby reducing their phytotoxicity. Two major classes of heavy metal chelating peptides are known to exist in plants—metallothioneins and phytochelatins. Metallothioneins are gene-encoded, low-molecular-weight, cysteine-rich polypeptides (131). Plant metallothioneins are induced by Cu and have high affinity for this metal (111, 178). Recent investigations of metallothionein (MT) expression levels in *A. thaliana* demonstrated that expression levels of MT2 mRNA strongly correlated with Cu resistance (112), suggesting that metallothioneins are involved in Cu resistance. Phytochelatins are low molecular weight, enzymatically synthesized cysteine-rich peptides known to bind cadmium and copper in plants (127, 128, 146). These peptides are essential for cadmium detoxification in *A. thaliana* (73). Although not strictly defined as chelation, precipitation of zinc as Zn-phytate has also been suggested as a zinc detoxification mechanism (156–158).

It is also likely that intra- and extracellular precipitation of lead as carbonates, sulfates, and phosphates plays a role in the detoxification of this metal in plant tissues.

COMPARTMENTALIZATION   Within cells, cadmium and phytochelatins accumulate in the vacuole (162), and this accumulation appears to be driven by a Cd/H antiport and an ATP-dependent PC-transporter (138, 139). A similar system of cadmium detoxification also exists in the fission yeast, *Schizosaccharomyces pombe*. Mutants lacking the ability to accumulate Cd-PC complex in the vacuole are Cd-sensitive and have a defect in *hmt1*, a gene encoding an ATP-binding cassette-type transport protein (117). The *hmt1* gene product is responsible for transporting Cd-PC complex into the vacuole (118). Once inside the vacuole, sulfide is added to the Cd-PC complex, forming a more stable high-molecular-weight Cd-PC-sulfide complex that may be essential for Cd resistance in the yeast (117, 145).

Intact vacuoles isolated from tobacco and barley exposed to Zn have also been shown to accumulate this metal (27, 90). Vacuolar Zn accumulation has been confirmed in roots and shoots of the Zn hyperaccumulator *Thlaspi caerulescens* (160, 161). Zinc accumulation within the vacuole, as a Zn detoxification mechanism, is also supported by the observation that the vacuolar volume fraction of meristematic cells of *Festuca rubra* increases during Zn exposure (46). Leaf trichomes also appear to provide a site for the sequestration of Cd (137), Mn (16), and Pb (104).

BIOTRANSFORMATION   The toxicity of such metals and metalloids as chromium, selenium, and arsenic can be reduced in plants by chemical reduction of the element and/or by its incorporation into organic compounds. Excess selenium is toxic to most plants because it is metabolized to selenocysteine and selenomethionine, which replace cysteine and methionine residues in proteins. By funneling selenium into the nonprotein amino acids methylselenocysteine and selenocystathionine, selenium accumulator species of *Astragalus* are able to reduce the amount of selenium incorporated into proteins, thereby tolerating elevated concentrations of selenium in shoots (93). Recently the enzyme responsible for the methylation of selenocysteine in the selenium accumulator *Astragalus bisculatus* has been isolated and characterized, a first step in determining the molecular basis of selenium resistance in plants (113). It also appears that several selenium-accumulating species are able to selectively exclude selenium from the methionine biosynthetic pathway, thereby avoiding the synthesis of selenomethionine, a toxic seleno-derivative of methionine (30). Selenium is also volatilized by plants by as yet uncharacterized mechanisms (see section on "Phytovolatilization of Metals").

Arsenic is toxic to plants, as demonstrated by the use of organoarsenical as herbicides, though little is known about arsenic detoxification in terrestrial plants. However, in marine macroalgae, arsenic is incorporated into various dimethyarsinylriboside and certain lipids (63), and it is likely that terrestrial plants also biotransform arsenic. Chromium is also toxic to plants, and there is limited evidence that plants, like certain bacteria and animals, can reduce Cr(VI) to Cr(III) as part of a detoxification mechanism (53).

CELLULAR REPAIR MECHANISMS   A primary component of cellular resistance to elevated Cu concentrations appears to be enhanced plasma membrane resistance to, or repair of, Cu-induced membrane damage (50, 111, 150). The intriguing observation that plant metallothioneins may be prenylated and targeted to the plasma membrane (111) suggests a possible mechanism whereby metallothioneins may be involved in plasma membrane repair. The involvement of membrane repair mechanisms in Cu resistance is also strongly supported by the recent observation that an acyl carrier protein (ACP) and an AcylCoA binding protein (ACBP), two proteins known to be involved in lipid metabolism, are induced in Cu-exposed *A. thaliana* (A Murphy & L Taiz, personal communication). These authors also showed that antisense down-regulation of ACBP expression caused increased sensitivity to Cu, supporting the role of membrane repair in Cu resistance (A Murphy & L Taiz, personal communication).

Metal resistance will clearly be an important characteristic of a phytoremediation crop. However, metal resistance alone may not be sufficient to allow plants to accumulate high concentrations of metals. Metal bioavailability, root uptake, and translocation are also essential for successful phytoextraction.

Metal Bioavailability, Root Uptake, and Shoot Accumulation

The enhancement of metal ion bioavailability in soil by addition of metal chelates is an essential component of chelate-assisted phytoextraction and may also be important for continuous phytoextraction. This is illustrated by the mechanism(s) involved in the acquisition of iron and other micronutrients by plants. Because of the high binding capacity for metallic micronutrients by soil particles, plants have evolved several strategies for increasing their soil bioavailability. These strategies include the production of metal-chelating compounds (phytosiderophores) such as mugenic and avenic acids (84), which are synthesized in response to iron (69, 70, 83) and possibly zinc (33, 34) deficiencies. In the rhizosphere, phytosiderophores chelate and mobilize Fe, Cu, Zn, and Mn (132). Once chelated to phytosiderophores, metal ions can be transported across the plasma membrane as a metal-phytosiderophore complex via specialized transporters (163–165). By reducing chelated Fe(III) with a root ferric chelate reductase (108), plants are also able to release soluble Fe(II) for root uptake (174). There is also some evidence that this ferric chelate reductase may play a more general role in Cu and Mn uptake (168). Plants can also solubilize iron and other metals by exuding protons from roots to acidify the rhizosphere (40). It may, therefore, be possible to enhance the bioavailability of metal pollutants by manipulating these root processes. Reliance on plant-produced chelating agents should also reduce the need for addition of synthetic chelates, thus reducing the cost of phytoextraction.

Possibly with the exception of Fe, little is known about the molecular mechanisms of metal entry into root cells. However, recently putative plasma membrane copper (COPT1) and iron (II) (IRT1) transporters have been cloned from *A. thaliana* using functional complementation in yeast (55, 82). Several genes have been also recently isolated from *A. thaliana* that appear to encode plasma membrane zinc transporters (65). Using a metal uptake screen in yeast, a wheat root gene has been identified that enhances both Cd and Pb uptake in transgenic yeast expressing the gene (7). It was suggested that this gene may encode a putative plasma membrane metal transporter. These data provide important molecular insight into plasma membrane metal ion transport in plants and suggest that it may soon be possible to manipulate metal ion transport systems in order to promote phytoextraction of toxic metals.

Once metal ions have entered the roots, they can either be stored or exported to the shoot. Metal transport to the shoot primarily takes place through the xylem. Cadmium loading into the xylem sap of *Brassica juncea* displays biphasic saturation kinetics (137), suggesting that xylem loading of metal ions is facilitated by specialized membrane transport processes. Recent evidence from work with Ni hyperaccumulators from the genus *Alyssum* suggests that xylem loading of Ni may be facilitated by the binding of Ni to free histidine (88). Movement of metal ions, particularly Cd, in xylem vessels appears to be mainly dependant on transpiration-driven mass flow (137).

Because xylem cell walls have a high cation exchange capacity, they are expected to retard severely the upward movement of metal cations. Therefore, noncationic metal-chelate complexes, such as Cd-citrate, should be transported more efficiently in the transpiration stream (143). Theoretical studies have predicted that the majority of the Fe(II) and Zn(II) in xylem sap should be chelated by citrate, whereas Cu(II) should be chelated by various amino acids including histidine and asparagine (170). Isolation of a citratonickelate (II) complex from the latex of the Ni hyperaccumulator *Sebertia acuminata* supports the role of organic acids in metal transport (94). X-ray absorbance fine structure (EXAFS) analysis showed that Cd in the xylem sap of *B. juncea* was chelated by oxygen or nitrogen atoms, suggesting the involvement of organic acids in Cd translocation (137). EXAFS analysis produced no evidence for sulfur coordination of Cd, confirming that phytochelatins and other thiol-containing ligands play no direct role in Cd transport in the xylem. X-ray spectroscopy also demonstrated that a portion of the Ni and Zn transported to the shoots of the Ni hyperaccumulator *T. goesingense* and the Zn hyperaccumulator *T. caerulescens* is coordinated with organic acids (D Salt, I Pickering & R Prince, unpublished data). However, this analysis also revealed that substantial amounts of Ni and Zn are transported in the xylem sap as hydrated cations. A similar speciation of Ni in the xylem sap of the Ni hyperaccumulator *Alyssum lesbiacum* was established by mathematical modeling (88).

Other chelating compounds may also play a role in metal ion mobility in plants. The nonproteinaceous amino acid nicotianamine is ubiquitous among plants and has the ability to form complexes with various divalent metal ions including Cu, Ni, Co, Zn, Fe, and Mn (147, 148). Investigations of the tomato mutant *chloronerva*, which lacks the ability to synthesize nicotianamine (71), demonstrated that nicotianamine is possibly involved in distributing Fe(II), Zn, and Mn in young growing tissues via the phloem (148) and in Cu(II) transport within the xylem (121). Recent evidence also suggests that in *A. thaliana* cellular Cu is transported chelated to a functional analogue of the yeast low-molecular-weight Cu-binding protein (ATX1) (R Amasino, personal communication). In addition, metals may be transported in the phloem chelated to other low-molecular-weight metabolites or proteins (102).

Enhanced rates of metal ion translocation from roots to shoots appear to be important for zinc hyperaccumulation in *T. caerulescens* (92), suggesting that modifications in the transport processes described above may allow development of plants with enhanced root to shoot transport of pollutant metal ions, an important development in the creation of effective phytoextraction crops.

PHYTOVOLATILIZATION OF METALS

Volatilization of selenium from plant tissues may provide a mechanism of selenium detoxification. As early as 1894, Hofmeister proposed that selenium in animals is detoxified by releasing volatile dimethyl selenide from the lungs. He based this proposal on the fact that the odor of dimethyl telluride was detected in the breath of dogs injected with sodium tellurite (referenced in 95). Using the same logic, it was suggested that the garlicky odor of plants that accumulate selenium may indicate the release of volatile selenium compounds. Lewis (97) was the first to show that both selenium nonaccumulator and accumulator species volatilize selenium. This was later confirmed by other authors (52, 57, 176, 177, 179). The volatile selenium compound released from the selenium accumulator *Astragalus racemosus* was identified as dimethyl diselenide (57). Selenium released from alfalfa, a selenium nonaccumulator, was different from the accumulator species and was identified as dimethyl selenide (96).

However, it is not clear whether plants are able to take inorganic selenium (as selenate or selenite) and reduce and methylate it to the volatile methyl forms. Recent work by Zayed & Terry (177) demonstrated that addition of the antibiotic penicillin to hydroponically grown Indian mustard (*Brassica juncea*) inhibited selenium volatilization by approximately 90% when selenium was provided as selenate. However, plants may still volatilize selenium in the absence of rhizobacteria when it is supplied as selenomethionine (151). This suggests that root-associated bacteria play an important role in reducing and assimilating selenium into organic forms. However, more work is needed to clarify the role of microbial and plant biochemical processes in selenium volatilization by plants.

Volatilization of arsenic as dimethylarsenite has also been postulated as a resistance mechanism in marine algae. However, it is not known whether terrestrial plants also volatilize arsenic in significant quantities. Studies on arsenic uptake and distribution in higher plants indicate that arsenic predominantly accumulates in roots and that only small quantities are transported to shoots. However, plants may enhance the biotransformation of arsenic by rhizospheric bacteria, thus increasing rates of volatilization.

More recently, a modified bacterial mercuric ion reductase has been introduced into transgenic *A. thaliana*, which converts Hg^{2+} into elemental mercury ($Hg°$). In addition to being more tolerant, these transgenic plants are very effective at volatilizing mercury (134). Phytovolatilization of metals may have unique advantages over phytoextraction, because it bypasses harvesting and disposal of metal-rich biomass. However, the environmental implications of metal volatilization have to be considered before this approach becomes accepted by regulators and the public.

PHYTOREMEDIATION OF ORGANICS

The use of plants to cleanse waters contaminated with organic and inorganic pollutants dates back hundreds of years and has been the basis for the present use of constructed wetlands in treating municipal and industrial waste streams (67). The concept of using plants to remediate soils contaminated with organic pollutants is a more recent development, based on observations that disappearance of organic chemicals is accelerated in vegetated soils compared with surrounding nonvegetated bulk soils (8, 29, 42, 45, 59, 68, 166). Subsequent metabolic studies have established the ability of plants to take up and metabolize a range of environmentally problematic organic pollutants, including ammunition wastes (e.g. TNT and GTN), polychlorinated phenols (PCBs), and trichloroethylene (TCE) (64, 76, 114, 141).

In addition to the direct uptake and metabolism of organics, plants release exudates from their roots that enhance microbial bioremediation in the rhizosphere, which has been termed *phytoremediation ex planta* (3, 41). A brief review of these two basic phytoremediation strategies is presented below. Additional reviews on this subject have been published recently (3, 42, 44, 141), including a timely review by Cunningham et al (41) that provides an in-depth discussion of the technical, logistical, and economic considerations and strategies for the phytoremediation of soils contaminated with organic pollutants.

Direct Uptake and Metabolism of Organics

BIOAVAILABILITY AND UPTAKE   By analogy with the phytoextraction of metals, direct uptake of organic contaminants is primarily limited by the availability of the target compound and uptake mechanisms. Plants can take up chemicals from three distinct soil phases: vapor, liquid, and solid (41). With a few notable exceptions [i.e. uptake of some polyaromatic hydrocarbons (PAHs) and herbicides from the vapor phase], movement of organics into plants occurs via the liquid phase, which has been extensively investigated in plants for uptake of pesticides and herbicides (25, 120, 153). A major criterion in assessing the probability that a target chemical will be taken up by plants is its lipophilicity. This governs its movement across plant membranes as well as its solubility in the water phase. Chemicals most likely to be taken up are moderately hydrophobic compounds with octanol-water partition coefficients ranging from 0.5 to 3.0 (25, 135).

In addition to the physicochemical properties of the target compound, other factors including soil conditions (e.g. pH, pKa, organic and water content, texture; 2, 19, 54, 135, 153) and plant physiology (100, 101) influence solubility and uptake of target compounds. Differences in uptake of organics among plant species and varieties are well recognized (66, 98) and should be a primary consideration in the development of effective phytoremediation strategies. Only limited screening of plants for uptake and metabolism of priority organic pollutants has been carried out by a few research groups (141). More exhaustive screening in the future may yield novel species or varieties that have enhanced phytoremediation capabilities for environmental pollutants, in the same way that screening for metal accumulators identified members of the Brassicaceae as superior phytoextractors (91).

Current strategies for remediating organics rely on long-term continuous phytoextraction approaches analogous to continuous phytoextraction of heavy metals (described above). Bioavailability has been found to be a major limiting step in the phytoextraction of metals. Similarly, availability of organics in soils appears to be a primary restriction for effective phytoremediation of organic pollutants (41, 141). While the application of soil amendments (e.g. EDTA) is considered a major breakthrough in the development of induced phytoextraction strategies, similar attempts to identify soil amendments that can induce the uptake and accumulation of organics in plants have not been made. The use of synthetic (e.g. triton X-100, SDS) and naturally produced biosurfactants (e.g. rhamnolipids) to enhance the apparent water solubility and bacterial degradation of organic contaminants is well documented (24, 49, 123, 155, 175). A recent study has also shown the beneficial use of cyclodextrins to increase the solubilities of both organics and heavy metals (28). Whether these or other chemical agents could be applied as amendments to enhance the availability and uptake of organic pollutants remains to be investigated. Potential advantages of using biosurfactants or cyclodextrins—aside from increasing the bioavailability of organics—include their rapid biodegradation in the environment, the capability of synthesizing these compounds in engineered plants (116) and rhizospheric microorganisms, and the ability of these compounds to solubilize both organics and metals (106, 115), which could be instrumental in remediation soils with mixed contaminants.

BIOTRANSFORMATION AND COMPARTMENTALIZATION   Following uptake, organic compounds may have multiple fates: They may be translocated to other plant tissues (36, 58, 142) and subsequently volatilized, they may undergo partial or complete degradation (64, 114, 141, 172), or they may be transformed to less toxic compounds and bound in plant tissues in nonavailable forms (60). Biotransformation and sequestration of herbicides and pesticides, in particular, have been extensively investigated in plants (reviewed by 60). More recently, metabolism of nonagricultural xenobiotics such as TCE, TNT, and nitroglycerin (GTN) has been studied using axenic cell cultures and whole plants (64, 114, 141). In general, most organics appear to undergo some degree of transformation in plant cells before being sequestered in vacuoles or bound to insoluble cellular structures, such as lignin. Metabolism of chloroacetanilide herbicides, for instance, results in the production of reduced and oxidized sulfur-containing compounds following conjugation to glutathione (60). The nitrate ester, GTN, is degraded to glycerol dinitrate and glycerol mononitrate in sugar beet cell cultures (64), and TCE metabolism in poplars generates trichloroethanol and di- and trichloroacetic acid (114). However, few chemicals appear to be fully mineralized by plants to water and CO_2, and where this does occur, it only represents a small percentage of the total parent compound (114). This property puts plants at a relative disadvantage compared with bacteria in degrading organic pollutants. In addition, the possibility that plant metabolites of pollutants may be more toxic than the original pollutants creates a difficult regulatory environment for phytoremediation of organics.

An important consideration in developing phytoremediation strategies for organics is the short- and long-term fate and potential toxicity of the metabolic end products of biodegradation. For classes of chemicals such as pesticides and herbicides, these questions have, in most cases, been thoroughly addressed, and the evidence indicates that most compounds are bound irreversibly with plant materials (18a, 60, 87, 154). Information on the metabolism of priority organic pollutants is limited, however, and will require further investigation given results from recent studies. For instance, the majority of the TNT transformation products in *Myriophyllum spicatum* could not be identified, and a significant fraction of these products were either released into the culture medium or associated with water-extractable cellular fractions (76). Studies by other groups, however, suggest that TNT degradation products in plants are not available (119, 141, 172). Plant differences in the partitioning of organics between roots and shoots as well as differences in metabolism of organic pollutants should also be considered when choosing specific plant species for phytoremediation.

Phytoremediation ex Planta

Plants may secrete 10–20% of their photosynthate in root exudates, which support the growth and metabolic activities of diverse fungal and bacterial communities in the rhizosphere (4, 99, 110, 144). Densities of rhizospheric bacteria can be as much as two to four orders of magnitude greater than populations in the surrounding bulk soils and display a greater range of metabolic capabilities, including the ability to degrade a number of recalcitrant xenobiotics (4, 167). It is not surprising, therefore, to find accelerated rates of biodegradation of organic pollutants in vegetated soils compared with nonvegetated soils (5, 29, 45, 59, 68). Some organic compounds in root exudates (i.e. phenolics, organic acids, alcohols, proteins) may serve as carbon and nitrogen sources for the growth and long-term survival of microorganisms that are capable of degrading organic pollutants. For instance, plant phenolics such as catechin and coumarin may serve as co-metabolites for PCB-degrading bacteria (15, 51, 61, 68).

The chemical composition of root exudates and rates of exudation differ considerably among plant species (124). This has led some research groups to screen for plant species that exude phenols capable of supporting PCB-degrading bacteria (61). Although studies directed at understanding mechanisms of plant-enhanced microbial degradation of organics are only beginning to emerge, several field and pilot studies have examined the use of specific plants for rhizospheric degradation of organic pollutants. Soils planted with crested wheat grass (*Agropyron desertorum*) showed enhanced mineralization of PCBs (59), while accelerated removal of PAHs was achieved using prairie grasses (5, 62). Similar studies have examined the degradation of TCE (5, 62) and TNT (172).

Rhizospheric microorganisms may also accelerate remediation processes by volatilizing organics such as PAHs or by increasing the humification of organic pollutants (41, 47). In particular, the release of oxidoreductase enzymes (e.g. peroxidase) by microbes, as well as by plant roots, can catalyze the polymerization of contaminants onto the soil humic fraction and root surfaces (1, 47). *Armoracia rusticana* (horseradish) has received particular attention with regard to the production of root peroxidases and its potential use for the remediation of polluted soils and water streams (47).

PLANT-DERIVED ENZYMES   In addition to secreting organic compounds that support the growth and activities of rhizospheric microorganisms, plants also release enzymes capable of degrading organic contaminants in soils. Soil enzymes derived from plant sources, based on immunological assays, include laccases, dehalogenases, nitroreductases, nitrilases, and peroxidases (20, 35, 141). Degradation of ammunition wastes (e.g. TNT, dinitromono-aminotoluene, and mononitrodiaminotoluene) and triaminotoluene is catalyzed by nitroreductases and laccases, respectively (141, 172). Whether release of plant enzymes from abscized plant tissue, root exudates, or guttation fluid can provide a cost-effective phytoremediation strategy for organic contaminants remains to be determined. The presence of plant-derived enzymes capable of degrading environmentally problematic xenobiotics (e.g. TNT and TCE) will no doubt be exploited for the development of future phytoremediation strategies.

CONCLUSIONS: FROM THE LABORATORY TO THE FIELD

At present, phytoremediation of metals and organics may be approaching commercialization. Additional, short-term advances in phytoremediation are likely to come from the selection of more efficient plant varieties and soil amendments and from optimizing agronomic practices used for plant cultivation. Major long-term improvements in phytoremediation should come when scientists isolate genes from various plant, bacterial, and animal sources, which can enhance the metal accumulation or degradation of organics. In addition, manipulating rhizospheric bacteria to enhance their role in phytoremediation can increase the efficiency of the future phytoremediation efforts.

However, biology alone cannot make phytoremediation work. The highly integrated nature of phytoremediation requires synergy with many other disciplines. For example, parallel developments in environmental and agricultural engineering should have a major impact on the efficiency of plant cultivation, amendment application, and disposal of metal-enriched biomass.

Only the future can tell whether phytoremediation will become a widely accepted technology. However, it is clear that the utilization of the remarkable potential of green plants to accumulate elements and compounds from the environment and to perform biochemical transformations is becoming a new frontier of plant biology.

ACKNOWLEDGMENTS

We thank the US Department of Agriculture (grant #96-35102-3838 to DES), US Department of Energy (grant #DE-FG07-96ER20251 to DES and RDS), Phytotech Inc., New Jersey Commission on Science and Technology, and Peter Day for comments.

Visit the *Annual Reviews home page* at
http://www.AnnualReviews.org.

Literature Cited

1. Adler PR, Arora R, El Ghaouth A, Glenn DM, Solar JM. 1994. Bioremediation of phenolic compounds from water with plant root surface peroxidases. *J. Environ. Qual.* 23:1113–17

2. Anderson TA. 1992. *Comparative plant uptake and microbial degradation of trichloroethylene in the rhizosphere of five plant species—implications for bioremediation of contaminated surface soils.* PhD thesis. Univ. Tenn., Knoxville

3. Anderson TA, Guthrie EA, Walton BT. 1993. Bioremediation. *Environ. Sci. Technol.* 27:2630–36

4. Anderson TA, Kruger EL, Coats JR. 1994. Enhanced degradation of a mixture of three herbicides in the rhizosphere of a herbicide-tolerant plant. *Chemosphere* 28:1551–57

5. Anderson TA, Walton BT. 1995. Comparative fate of ^{14}C-trichloroethylene in the root zone of plants from a former solvent disposal site. *Environ. Toxicol. Chem.* 14:2041–47

6. Deleted in proof

7. Antosiewicz D, Schachtman D, Clemens S, Schroeder JI. 1996. Plant root cDNA that enhances heavy metal uptake into yeast. *Plant Physiol.* 111:S130

8. Aprill W, Sims RC. 1990. Evaluation of the use of prairie grasses for stimulating polycyclic aromatic hydrocarbon treatment in soil. *Chemosphere* 20:253–65

9. Baker AJM, Brooks RR. 1989. Terrestrial higher plants which hyperaccumulate metallic elements—a review of their distribution, ecology and phytochemistry. *Biorecovery* 1:81–126

10. Baker AJM, Brooks RR, Reeves R. 1988. Growing for gold... and copper... and zinc. *New Sci.* 117:44–48

11. Deleted in proof

11a. Baker AJM, Proctor J, Reeves RD, eds. 1992. *The Vegetation of Ultramafic (Serpentine) Soils.* Andover, UK: Intercept. 509 pp.

12. Baker AJM, Reeves RD, McGrath SP. 1991. *In situ* decontamination of heavy metal polluted soils using crops of metal-accumulating plants—a feasibility study. In *In Situ Bioreclamation*, ed. RE Hinchee, RF Olfenbuttel, pp. 539–44. Stoneham, MA: Butterworth-Heinemann

13. Bañuelos GS, Cardon G, Mackey B, Ben-Asher J, Wu L, et al. 1993. Plant and environment interactions—boron and selenium removal in boron-laden soils by four springler irrigated plant species. *J. Environ. Qual.* 22:786–92

14. Baumann A. 1885. Das verhalten von zinksalzen gegen pflanzen und im boden. *Landwirtscha. Verss.* 31:1–53

15. Bedard DL, Wagner RE, Brennen MJ, Haberl ML, Brown JF Jr. 1987. Extensive degradation of aroclors and environmentally transformed polychlorinated biphenyls by *Alcaligenes eutrophus* H850. *Appl. Environ. Microbiol.* 53: 1094–102

16. Blamey FPC, Joyce DC, Edwards DG, Asher CJ. 1986. Role of trichomes in sunflower tolerance to manganese toxicity. *Plant Soil* 91:171–80

17. Deleted in proof

18. Blaylock MJ, Salt DE, Dushenkov S, Zakharova O, Gussman C, et al. 1997. Enhanced accumulation of Pb in Indian mustard by soil-applied chelating agents. *Environ. Sci. Technol.* 31:860–65

18a. Bockers M, Rivero CH, Thiede B, Jankowski T, Schmidt B. 1994. Uptake, translocation and metabolism of 3,4-dichloroaniline in soybean and wheat plants. *Z. Naturforsch. Teil C* 49:719–26

19. Bollag J-M, Myers C, Pal S, Huang PM. 1995. The role of abiotic and biotic catalysts in the transformation of phenolic compounds. In *Environmental Impacts*

of Soil Component Interactions, ed. PM Huang, J Berthelin, J-M Bollag, WB McGill, AL Page, 299–310. Boca Raton: Lewis. pp. 409

20. Boyajian GE, Carreira LH. 1997. Phytoremediation: a clean transition from laboratory to marketplace? *Nat. Biotechnol.* 15:127–28

21. Boyd RS, Martens SN. 1992. The *raison d'être* for metal hyperaccumulation by plants. See Ref. 11a, pp. 279–89

22. Boyd RS, Martens SN. 1994. Nickel hyperaccumulated by *Thlaspi montanum* var. *montanum* is acutely toxic to an insect herbivore. *OIKOS* 70:21–25

23. Boyd RS, Shaw JJ, Martens SN. 1994. Nickel hyperaccumulation defends *Streptanthus polygaloides* (Brassicaceae) against pathogens. *Am. J. Bot.* 81:294–300

24. Bragg JR, Prince RC, Harner EJ, Atlas RM. 1994. Effectiveness of bioremediation for the Exxon Valdez oil spill. *Nature* 368:413–18

25. Briggs GG, Bromilow RH, Evans AA. 1982. Relationship between lipophilicity and root uptake and translocation of nonionized chemicals by barley. *Pestic. Sci.* 13:495–504

26. Brown SL, Chaney RL, Angle JS, Baker AJM. 1995. Zinc and cadmium uptake by hyperaccumulator *Thlaspi caerulescens* and metal tolerant *Silene vulgaris* grown on sludge-amended soils. *Environ. Sci. Technol.* 29:1581–85

27. Brune A, Urbach W, Dietz K-J. 1994. Compartmentation and transport of zinc in barley primary leaves as basic mechanisms involved in zinc tolerance. *Plant Cell Environ.* 17:153–62

28. Brusseau ML, Wang XJ, Wang W-Z. 1997. Simultaneous elution of heavy metals and organic compounds from soil by cyclodextrin. *Environ. Sci. Technol.* 31:1087–92

29. Burken JG, Schnoor JL. 1996. Phytoremediation: plant uptake of atrazine and role of root exudates. *J. Environ. Eng.* 122:958–63

30. Burnell JN. 1981. Selenium metabolism in *Neptunia amplexicaulis*. *Plant Physiol.* 67:316–24

31. Byers HG. 1935. Selenium occurrence in certain soils in the United States, with a discussion of related topics. *US Dep. Agric. Technol. Bull.* 482:1–47

32. Byers HG. 1936. Selenium occurrence in certain soils in the United States, with a discussion of related topics. Second report. *US Dep. Agric. Technol. Bull.* 530:1–78

33. Cakmak I, Ozturk L, Karanlik S, Marschner H, Ekiz H. 1996. Zinc-efficient wild grasses enhance release of phytosiderophores under zinc deficiency. *J. Plant Nutr.* 19:551–63

34. Cakmak I, Sari N, Marschner H, Ekiz H, Kalayci M, et al. 1996. Phytosiderophore release in bread and duram wheat genotypes differing in zinc efficiency. *Plant Soil* 180:183–89

35. Carreira LH, Wolfe NL. 1995. *Isolation of a sediment nitroreductase, antibody production, and identification of possible plant sources.* Presented at IBC Int. Symp. Phytoremed., Arlington, VA, May 1996

36. Cataldo DA, Bean RM, Fellows RJ. 1987. Uptake and fate of phenol, aniline and quinoline in terrestrial plants. *Hanford Life Sci. Symp. Health Environ. Res. Complex Organic Mixtures*, ed. RH Gray, EK Chess, PJ Mellinger, RG Riley, DL Springer, 24:631–40. Richland, WA: Pacific Northwest Lab.

37. Chaney RL. 1983. Plant uptake of inorganic waste. In *Land Treatment of Hazardous Wastes*, ed. JE Parr, PB Marsh, JM Kla, pp. 50–76. Park Ridge, IL: Noyes Data Corp.

38. Chaney RL, Brown JC, Tiffin LO. 1972. Obligatory reduction of ferric chelates in iron uptake by soybeans. *Plant Physiol.* 50:208–13

39. Chaney RL, Mallik M, Li YM, Brown SL, Brewer EP, et al. 1997. Phytoremediation of soil metals. *Curr. Opin. Biotechnol.* 8:279–84

40. Crowley DE, Wang YC, Reid CPP, Szaniszlo PJ. 1991. Mechanisms of iron acquisition from siderophores by microorganisms and plants. *Plant Soil* 130:179–98

41. Cunningham SD, Anderson TA, Schwab AP, Hsu FC. 1996. Phytoremediation of soils contaminated with organic pollutants. *Adv. Agron.* 56:55–114

42. Cunningham SD, Berti WR. 1993. Remediation of contaminated soils with green plants: an overview. *In Vitro Cell. Dev. Biol.* 29:207–12

43. Cunningham SD, Berti WR, Huang JWW. 1995. Phytoremediation of contaminated soils. *Trends Biotechnol.* 13:393–97

44. Cunningham SD, Ow DW. 1996. Promises and prospects of phytoremediation. *Plant Physiol.* 110:715–19

45. Curl EA, Truelove B. 1986. *The Rhizosphere.* Berlin: Springer-Verlag. 288 pp.

46. Davies KL, Davies MS, Francis D. 1991. Zinc-induced vacuolation in root meristematic cells of *Festuca rubra L. Plant Cell Environ.* 14:399–406

47. Dec J, Bollag J-M. 1994. Use of plant

material for the decontamination of water polluted with phenols. *Biotechnol. Bioeng.* 44:1132–39

48. Delhaize E. 1996. A metal-accumulator mutant of *Arabidopsis thaliana. Plant Physiol.* 111:849–55

49. Desai JD, Banat IM. 1997. Microbial production of surfactants and their commercial potential. *Microb. Mol. Biol. Rev.* 61:47–64

50. De Vos CHR, Schat H, Vooijs R, Ernst WHO. 1989. Copper-induced damage to the permeability barrier in roots of *Silene cucubalus. J. Plant Physiol.* 135:164–69

51. Donnelly PK, Hegde RS, Fletcher JS. 1994. Growth of PCB-degrading bacteria on compounds from photosynthetic plants. *Chemosphere* 28:981–88

52. Duckart EC, Waldron LJ, Donner HE. 1992. Selenium uptake and volatilization from plants growing in soil. *Soil Sci.* 53:94–99

53. Dushenkov V, Kumar PBAN, Motto H, Raskin I. 1995. Rhizofiltration—the use of plants to remove heavy metals from aqueous streams. *Environ. Sci. Technol.* 29:1239–45

54. Edwards NT. 1986. Uptake, translocation and metabolism of anthracene in bush bean (*Phaseolus vulgaris* L.). *Environ. Toxicol. Chem.* 5:659–65

55. Eide D, Broderius M, Fett J, Guerinot ML. 1996. A novel iron-regulated metal transporter from plants identified by functional expression in yeast. *Proc. Natl. Acad. Sci. USA* 93:5624–28

56. Ernst WHO, Verkleij JAC, Schat H. 1992. Metal tolerance in plants. *Acta Bot. Neerl.* 41:229–48

57. Evans CS, Asher CJ, Johnson CM. 1968. Isolation of dimethyl diselenide and other volatile selenium compounds from *Astragalus racemosus* (Pursh.). *Aust. J. Biol. Sci.* 21:13–20

58. Fellows RJ, Harvey SD, Ainsworth CC, Cataldo DA. 1996. Biotic and abiotic transformation of munitions materials (TNT, RDX) by plants and soils. Potentials for attenuation and remediation of contaminants. *IBC Int. Conf. Phytoremed.,* Arlington, VA, May 1996

59. Ferro AM, Sims RC, Bugbee B. 1994. Hycrest crested wheatgrass accelerates the degradation of pentachlorophenol in soil. *J. Environ. Qual.* 23:272–79

60. Field JA, Thurman EM. 1996. Glutathione conjugation and contaminant transformation. *Environ. Sci. Technol.* 30:1413–18

61. Fletcher JS, Hegde RS. 1995. Release of phenols by perennial plant roots and their potential importance in bioremediation. *Chemosphere* 31:3009–16

62. Foth HD. 1990. *Fundamentals of Soil Science.* New York: Wiley. 8th ed.

63. Francesconi KA, Edmonds JS. 1994. Biotransformation of arsenic in the marine environment. In *Arsenic in the Environment, Part 1, Cycling and Characterization,* ed. JO Nriagu, 10:221–61. New York: Wiley. 430 pp.

64. Goel A, Kumar G, Payne GF, Dube SK. 1997. Plant cell biodegradation of a xenobiotic nitrate ester, nitroglycerin. *Nat. Biotechnol.* 15:174–77

65. Guerinot ML. 1997. Metal uptake in *Arabidopsis thaliana. J. Exp. Bot.* S48: 96

66. Harris CR, Sans WW. 1967. Absorption of organochlorine insecticide residues from agricultural soils by root crops. *J. Agric. Food Chem.* 15:861–63

67. Hartman WJ Jr. 1975. An evaluation of land treatment of municipal wastewater and physical siting of facility installations. Washington, DC: US Dep. Army

68. Hedge RS, Fletcher JS. 1996. Influence of plant growth stage and season on the release of root phenolics by mulberry as related to the development of phytoremediation technology. *Chemosphere* 23: 2471–79

69. Higuchi K, Kanazawa K, Nishizawa NK, Chino M, Mori S. 1994. Purification and characterization of nicotianamine synthase from Fe-deficient barley roots. *Plant Soil* 165:173–79

70. Higuchi K, Kanazawa K, Nishizawa NK, Mori S. 1996. The role of nicotianamine synthase in response to Fe nutrition status in Gramineae. *Plant Soil* 178:171–77

71. Higuchi K, Nishizawa N, Romheld V, Marschner H, Mori S. 1996. Absence of nicotianamine synthase activity in the tomato mutant Chloronerva. *J. Plant Nutr.* 19:1235–39

72. Hill-Cottingham DG, Lloyd-Jones CP. 1965. The behavior of iron chelating agents with plants. *J. Exp. Bot.* 16:233–42

73. Howden R, Goldsbrough PB, Andersen CR, Cobbett CS. 1995. Cadmium-sensitive, *cad1* mutants of *Arabidopsis thaliana* are phytochelatin deficient. *Plant Physiol.* 107:1059–66

74. Huang JWW, Chen JJ, Berti WR, Cunningham SD. 1997. Phytoremediation of lead-contaminated soils: role of synthetic chelates in lead phytoextraction. *Environ. Sci. Technol.* 31:800–5

75. Huang JW, Cunningham SD. 1996. Lead phytoextraction: species variation in lead

uptake and translocation. *New Phytol.* 134:75–84

76. Hughes JB, Shanks J, Vanderford M, Lauritzen J, Bhadra R. 1997. Transformation of TNT by aquatic plants and plant tissue cultures. *Environ. Sci. Technol.* 31:266–71

77. Ingrouille MJ, Smirnoff N. 1986. *Thlaspi caerulescenes* J. & C. Presl. (*T. alpestre* L.) in Britain. *New Phytol.* 102:219–33

78. Jackson PJ, Unkefer PJ, Delhaize E, Robinson NJ. 1990. Mechanisms of trace metal tolerance in plants. In *Environmental Injury to Plants*, ed. F Katterman, 10:231–55. San Diego: Academic

79. Jeffreys RA, Wallace A. 1968. Detection of iron ethylenediamine di (o-hydroxyphenylacetate) in plant tissues. *Agron. J.* 60:613–16

80. Jørgensen SE. 1993. Removal of heavy metals from compost and soil by ecotechnological methods. *Ecol. Eng.* 2:89–100

81. Kabata-Pendias A, Pendias H. 1989. *Trace Elements in Soils and Plants.* Boca Raton, FL: CRC Press

82. Kampfenkel K, Kushnir S, Babiychuk E, Inze D, Van Montagu M. 1995. Molecular characterization of a putative *Arabidopsis thaliana* copper transporter and its yeast homologue. *J. Biol. Chem.* 270:28479–86

83. Kanazawa K, Higuchi K, Nishizawa NK, Fushiya S, Chino M, Mori S. 1994. Nicotianamine aminotransferase activities are correlated to the phytosiderophore secretion under Fe-deficient conditions in Gramineae. *J. Exp. Bot.* 45:1903–6

84. Kinnersely AM. 1993. The role of phytochelates in plant growth and productivity. *Plant Growth Regul.* 12:207–17

85. Kneen BE, LaRue TA, Welch RM, Weeden NF. 1990. A mutation in *Pisum sativum* (L.) cv "Sparkle" conditioning decreased nodulation and increased iron uptake and leaf necrosis. *Plant Physiol.* 93:717–22

86. Kochian LV. 1991. Mechanisms of micronutrient uptake and translocation. In *Micronutrients in Agriculture*, ed. JJ Mortvedt, FR Cox, LM Shuman, WI Welch, 8:229–96. Madison, WI: Soil Sci. Soc. Am.

87. Komossa D, Langbartels C, Sandermann HJ. 1995. Metabolic processes for organic chemicals in plants. In *Plant Contamination: Modeling and Simulation of Organic Chemical Processes*, ed. S Trapp, JC McFarlane, pp. 69–106. Boca Raton, FL: CRC Press. 254 pp.

88. Krämer U, Cotter-Howells JD, Charnock JM, Baker AJM, Smith AC. 1996. Free histidine as a metal chelator in plants that accumulate nickel. *Nature* 379:635–38

89. Krämer U, Smith RD, Wenzel W, Raskin I, Salt DE. 1997. The role of nickel transport and tolerance in nickel hyperaccumulation by *Thlaspi goesingense* Hálácsy. *Plant Physiol.* 115:1641–50

90. Krotz RM, Evangelou BP, Wagner GJ. 1989. Relationship between cadmium, zinc, Cd-peptide, and organic acid in tobacco suspension cells. *Plant Physiol.* 91:780–87

91. Kumar PBAN, Dushenkov V, Motto H, Raskin I. 1995. Phytoextraction: the use of plants to remove heavy metals from soils. *Environ. Sci. Technol.* 29:1232–38

92. Lasat MM, Baker AJM, Kochian LV. 1996. Physiological characterization of root Zn^{2+} absorption and translocation to shoots in Zn hyperaccumulator and nonaccumulator species of *Thlaspi. Plant Physiol.* 112:1715–22

93. Läuchli A. 1993. Selenium in plants: uptake, functions, and environmental toxicity. *Bot. Acta* 106:455–68

94. Lee J, Reeves RD, Brooks RR, Jaffré T. 1977. Isolation and identification of a citrato-complex of nickel from nickel-accumulating plants. *Phytochemistry* 16:1503–5

95. Lewis B-AG. 1976. Selenium in biological systems, and pathways for its volatilization in higher plants. In *Environmental Biogeochemistry,* Vol. 1, *Carbon, Nitrogen, Phosphorus, Sulfur and Selenium Cycles*, ed. JO Nriagu, pp. 389–409. Ann Arbor, MI: Ann Arbor Sci. 423 pp.

96. Lewis BG, Johnson CM, Broyer TC. 1974. Volatile selenium in higher plants. The production of dimethyl selenide in cabbage leaves by enzymic cleavage of Se-methyl selenomethionine selenonium salt. *Plant Soil* 40:107–18

97. Lewis BG, Johnson CM, Delwiche CC. 1966. Release of volatile selenium compounds by plants. Collection procedures and preliminary observations. *J. Agric. Food Chem.* 14:638–40

98. Lichtenstein EP, Schultz KR. 1965. Residues of aldrin and heptachlor in soils and their translocation into various crops. *J. Agric. Food Chem.* 13:57–63

99. Lynch JM. 1982. Interactions between bacteria and plants in the root environment. In *Bacteria and Plants. Soc. Appli. Bacteriol. Symp. Ser.*, ed. ME Rhodes-Roberts, FA Skinner, 10:1–23. London: Academic

100. MacFarlane JC, Pfleeger T, Fletcher J. 1987. Transpiration effect on the uptake and distribution of bromacil,

nitrobenzene, and phenol in soybean plants. *J. Environ. Qual.* 16:372–76

101. MacFarlane JC, Pfleeger T, Fletcher J. 1990. Effect, uptake and distribution of nitrobenze in several terrestrial plants. *Environ. Toxicol. Chem.* 9:513–20

102. Marentes E, Stephens BW, Grusak MA. 1996. Identification of an iron translocator/putative signal molecule in the phloem of higher plants. *Plant Physiol.* S111:302 (Abstr.)

103. Marschner H. 1995. *Mineral Nutrition of Higher Plants.* San Diego: Academic. 889 pp.

104. Martell EA. 1974. Radioactivity of tobacco trichomes and insoluble cigarette smoke particles. *Nature* 249:215–17

105. Martens SN, Boyd RS. 1994. The ecological significance of nickel hyperaccumulation: a plant chemical defense. *Oecologia* 98:379–84

106. Miller RM. 1995. Biosurfactant-facilitated remediation of metal contaminated soils. *Environ. Health Perspect.* 103:59–62

107. Minguzzi C, Vergnano O. 1948. Il contenuto di nichel nelle ceneri di *Alyssum bertolonii Desv. Atti della Societa Toscana di Scienze Naturali, Mem. Ser. A* 55:49–77

108. Moog PR, Bruggemann W. 1994. Iron reductase systems on the plant plasma membrane—a review. *Plant Soil* 165:241–60

109. Morrey DR, Balkwill K, Balkwill M-J, Williamson S. 1992. A review of some studies of the serpentine flora of Southern Africa. See Ref. 11a, 12:147–57

110. Moser M, Haselwandter. 1983. In *Physiological Plant Ecology Vol. 3. Responses to the Chemical and Biological Environment,* ed. OL Lange, pp. 391–411. Berlin: Springer-Verlag

111. Murphy A, Zhou JM, Goldsbrough PB, Taiz L. 1997. Purification and immunological identification of metallothioneins 1 and 2 from *Arabidopsis thaliana. Plant Physiol.* 113:1293–301

112. Murphy AS, Taiz L. 1995. Comparison of metallothionein gene expression and nonprotein thiols in ten *Arabidopsis* ecotypes. *Plant Physiol.* 109:1–10

113. Neuhierl B, Böck A. 1996. On the mechanism of selenium tolerance in selenium-accumulating plants. Purification and characterization of a specific selenocysteine methyltransferase from cultured cells of *Astragalus bisculatus. Eur. J. Biochem.* 239:235–38

114. Newman LA, Strand SE, Choe N, Duffy J, Ekuan G, et al. 1997. Uptake and biotransformation of trichloroethylene by hybrid poplars. *Environ. Sci. Technol.* 31:1062–67

115. Nivas BT, Sabatini DA, Shiau B-J, Harwell JH. 1996. Surfactant enhanced remediation of subsurface chromium contamination. *Water Res.* 30:511–20

116. Oakes JV, Shewmaker CK, Stalker DM. 1991. Production of cyclodextrins, a novel carbohydrate, in the tubers of transgenic potato plants. *Bio-Technology* 9:982–86

117. Ortiz DF, Kreppel L, Speiser DM, Scheel G, McDonald G, Ow DW. 1992. Heavy metal tolerance in the fission yeast requires an ATP-binding cassette-type vacuolar membrane transporter. *EMBO J.* 11:3491–99

118. Ortiz DF, Ruscitti T, McCue KF, Ow DW. 1995. Transport of metal-binding peptides by HMT1, a fission yeast ABC-type B vacuolar membrane protein. *J. Biol. Chem.* 270:4721–28

119. Palazzo AJ, Leggett DC. 1986. Effect and disposition of TNT in a terrestrial plant. *J. Environ. Qual.* 15:49–52

120. Paterson S, Mackay D, Tam D, Shiu WY. 1990. Uptake of organic chemicals by plants: a review of processes, correlations and models. *Chemosphere* 21:297–331

121. Pich A, Scholz G. 1996. Translocation of copper and other micronutrients in tomato plants (*Lycopersicon esculentum* Mill)—nicotianamine-stimulated copper transport in the xylem. *J. Exp. Bot.* 47:41–47

122. Pollard JA, Baker AJM. 1997. Deterrence of herbivory by zinc hyperaccumulation in *Thlaspi caerulescens* (Brassicaceae). *New Phytol.* 135:655–58

123. Providenti MA, Flemming CA, Lee H, Trevore JT. 1995. Effect of addition of rhamnolipid biosurfactants or rhamnolipid-producing Pseudomonas aeruginosa on phenanthrene mineralization in soil slurries. *FEMS Microbiol. Ecol.* 17:15–26

124. Rao AS. 1990. Root flavonoids. *Bot. Rev.* 56:1–84

125. Raskin I, Kumar PBAN, Dushenkov S, Salt DE. 1994. Bioconcentration of heavy metals by plants. *Curr. Opin. Biotechnol.* 5:285–90

126. Raskin I, Smith RD, Salt DE. 1997. Phytoremediation of metals: using plants to remove pollutants from the environment. *Curr. Opin. Biotechnol.* 8:221–26

127. Rauser WE. 1990. Phytochelatins. *Annu. Rev. Biochem.* 59:61–86

128. Rauser WE. 1995. Phytochelatins and related peptides. Structure, biosynthesis, and function. *Plant Physiol.* 109:1141–49

129. Reeves RD, Brooks RR, Macfarlane RM. 1981. Nickel uptake by Californian Streptanthus and Caulanthus with particular reference to the hyperaccumulator *S. polygaloides* Gray (Brassicaceae). *Am. J. Bot.* 68:708–12

130. Riley RG, Zachara JM, Wobber FJ. 1992. *Chemical Contaminants on Doe Lands and Selection of Contaminant Mixtures for Subsurface Science Research.* Washington, DC: US Dep. Energy Off. Energy Res., Subsur. Sci. Prog.

131. Robinson NJ, Tommey AM, Kuske C, Jackson PJ. 1993. Plant metallothioneins. *Biochem. J.* 295:1–10

132. Römheld V. 1991. The role of phytosiderophores in acquisition of iron and other micronutrients in graminaceous species: an ecological approach. *Plant Soil* 130:127–34

133. Römheld V, Marschner H. 1981. Effect of Fe stress on utilization of Fe chelates by efficient and inefficient plant stress. *J. Plant Nutr.* 3:551–60

134. Rugh CL, Wilde HD, Stack NM, Thompson DM, Summers AO, Meagher RB. 1996. Mercuric ion reduction and resistance in transgenic *Arabidopsis thaliana* plants expressing a modified bacterial merA gene. *Proc. Natl. Acad. Sci. USA* 93:3182–87

135. Ryan JA, Bell RM, Davidson JM, O'Connor GA. 1988. Plant uptake of nonionic chemicals from soils. *Chemosphere* 17:2299–323

136. Salt DE, Blaylock M, Kumar NPBA, Viatcheslav D, Ensley BD, et al. 1995. Phytoremediation: a novel strategy for the removal of toxic metals from the environment using plants. *Bio-Technology* 13:468–74

137. Salt DE, Prince RC, Pickering IJ, Raskin I. 1995. Mechanisms of cadmium mobility and accumulation in Indian mustard. *Plant Physiol.* 109:427–33

138. Salt DE, Rauser WE. 1995. MgATP-dependent transport of phytochelatins across the tonoplast of oat roots. *Plant Physiol.* 107:1293–301

139. Salt DE, Wagner GJ. 1993. Cadmium transport across tonoplast of vesicles from oat roots. Evidence for a Cd^{+2}/H^+ antiport activity. *J. Biol. Chem.* 268:12297–302

140. Deleted in proof

141. Schnoor JL, Licht LA, McCutcheon SC, Wolfe NL, Carreira LH. 1995. Phytoremediation of organic and nutrient contaminants. *Environ. Sci. Technol.* 29:318A-23

142. Schroll R, Bierling B, Cao G, Dörfler U, Lahaniati M, et al. 1994. Uptake pathways of organic chemicals from soil by agricultural plants. *Chemosphere* 28:297–303

143. Senden MHMN, Van Paassen FJM, Van Der Meer AJGM, Wolterbeek HTh. 1990. Cadmium-citric acid-xylem cell wall interactions in tomato plants. *Plant Cell Environ.* 15:71–79

144. Shimp JF, Tracy JC, Davis LC, Lee E, Huang W, et al. 1993. Beneficial effects of plants in the remediation of soil and groundwater contaminated with organic materials. *Environ. Sci. Technol.* 23:41–77

145. Speiser DM, Ortiz DF, Kreppel L, Scheel G, McDonald G, Ow DW. 1992. Purine biosynthetic genes are required for cadmium tolerance in *Schizosaccharomyces pombe. Mol. Cell. Biol.* 12:5301–10

146. Steffens JC. 1990. The heavy metal-binding peptides of plants. *Annu. Rev. Plant Physiol. Mol. Biol.* 41:553–75

147. Stephan UW, Schmidke I, Stephan VW, Scholz G. 1996. The nicotianamin molecule is made-to-measure for complexation of metal micronutrients in plants. *Biometals* 9:84–90

148. Stephan UW, Scholz G. 1993. Nicotianamine: mediator of transport of iron and heavy metals in the phloem? *Physiol. Plant.* 88:522–29

149. Stern RJ, Moon J, Key T, Amidon T, Sickels F, et al. 1996. A pathway analysis approach for determining generic cleanup standards for radioactive materials. Draft Rep. Comment. Trenton, NJ: NJ Dep. Environ. Protect., Bur. Environ. Radiat.

150. Strange J, Macnair MR. 1991. Evidence for a role for the cell membrane in copper tolerence of *Mimulus guttatus* Fisher ex DC. *New Phytol.* 119:383–88

151. Terry N, Zayed A. 1997. Remediation of selenium-contaminated soils and waters by phytovolatilization. In *Proc. Extended Abstr. 4th Int. Conf. Biogeochemistry of Trace Elements*, ed. IK Iskandar, SE Hardy, AC Chang, GM Pierzynski, pp. 651–52, Washington, DC: US Gov. Print. Off.

152. Tomsett AB, Thurman DA. 1988. Molecular biology of metal tolerances of plants. *Plant Cell Environ.* 11:383–94

153. Topp E, Scheunert I, Attar A, Korte F. 1986. Factors affecting the uptake of 14C-labeled organic chemicals by plants from soil. *Ecotoxicol. Environ. Saf.* 11:219–28

154. Trapp S, MacFarlane JC. 1995. Plant contamination: modeling and simulation of organic chemical process. Boca Raton, FL: Lewis. 254 pp.

155. Van Dyke MI, Gulley SL, Lee H, Trevors JT. 1993. Evaluation of microbial surfactants for recovery of hydrophobic

pollutants from soil. *J. Ind. Microbiol.* 11:163–70

156. Van Steveninck RFM, Van Steveninck ME, Fernando DR. 1992. Heavy-metal (Zn, Cd) tolerance in selected clones of duck weed (*Lemna minor*). *Plant Soil* 146:271–80

157. Van Steveninck RFM, Van Steveninck ME, Fernando DR, Horst WJ, Marschner H. 1987. Deposition of zinc phytate in globular bodies in roots of *Deschampsia caespitosa* ecotypes; a detoxification mechanism? *J. Plant Physiol.* 131:247–57

158. Van Steveninck RFM, Van Steveninck ME, Wells AJ, Fernando DR. 1990. Zinc tolerance and the binding of zinc as zinc phytate in *Lemna minor*. X-ray microanalytical evidence. *J. Plant Physiol.* 137:140–46

159. Deleted in proof

160. Vazquez MD, Barcelo J, Poschenrieder CH, Madico J, Hatton P, et al. 1992. Localization of zinc and cadmium in *Thlaspi caerulescens* (Brassicaceae), a metallophyte that can hyperaccumulate both metals. *J. Plant Physiol.* 140:350–55

161. Vazquez MD, Poschenrieder CH, Barcelo J, Baker AJM, Hatton P, Cope GH. 1994. Compartmentation of zinc in roots and leaves of the zinc hyperaccumulator *Thlaspi caerulescens Bot. Acta* 107:243–50

162. Vogeli-Lange R, Wagner GJ. 1990. Subcellular localization of cadmium-binding peptides in tobacco leaves. Implications of a transport function for cadmium-binding peptides. *Plant Physiol.* 92:1086–93

163. Vonwiren N, Marschner H, Römheld V. 1995. Uptake kinetics of iron-phytosiderophores in two maize genotypes differing in iron efficiency. *Physiol. Plant.* 93:611–16

164. Vonwiren N, Marschner H, Römheld V. 1996. Roots of iron-efficient maize also absord phytosiderophore-chelated zinc. *Plant Physiol.* 111:1119–25

165. Vonwiren N, Mori S, Marschner H, Römheld V. 1994. Iron inefficiency in maize ys1 (*Zea mays* L. cv Yellow-Strip) is caused by a defect in uptake of iron phytosiderophores. *Plant Physiol.* 106:71–77

166. Walton BT, Anderson TA. 1992. Plant-microbe treatment systems for toxic waste. *Curr. Opin. Biotechnol.* 3:267–70

167. Walton BT, Hoylman AM, Perez MM, Anderson TA, Johnson TR, et al. 1994. Rhizosphere microbial community as a plant defense against toxic substances in soils. In *Bioremediation Through Rhizosphere Technology*, ed. TA Anderson, JR Coats, 563:82–92 Washington, DC: Am. Chem. Soc.

168. Welch RM, Norvell WA, Schaefer SC, Shaff JE, Kochian LV. 1993. Induction of iron(III) and copper(II) reduction in pea (*Pisum sativum* L.) roots by Fe and Cu status: does the root-cell plasmalemma Fe(III)-chelate reductase perform a general role in regulated cation uptake. *Planta* 190:555–61

169. Welch RM. 1995. Micronutrient nutrition of plants. *Crit. Rev. Plant Sci.* 14:49–82

170. White CW, Baker FD, Chaney RL, Decker AM. 1981. Metal complexation in xylem fluid II. Theoretical equilibrium model and computational computer program. *Plant Physiol.* 67:301–10

171. Deleted in proof

172. Wolfe NL, Ou T-Y, Carreira L. 1993. Biochemical remediation of TNT contaminated soils. Rep. US Army Corps Eng.

173. Deleted in proof

174. Yi Y, Guerinot ML. 1996. Genetic evidence that induction of root Fe(II) chelate reductase activity is necessary for iron uptake under iron deficiency. *Plant J.* 10:835–44

175. Zajic JE, Panchel CJ. 1976. Bio-emulsifiers. *CRC Crit. Rev. Microbiol.* 5:39–66

176. Zayed AM, Terry N. 1992. Selenium volatilization in Broccoli as influenced by sulfate suppy. *J. Plant Physiol.* 140:646–52

177. Zayed AM, Terry N. 1994. Selenium volatilization in roots and shoots: effects of shoot removal and sulfate level. *J. Plant Physiol.* 143:8–14

178. Zhou JM, Goldsbrough PB. 1995. Structure, organization and expression of the metallothionein gene family in *Arabidopsis. Mol. Gen. Genet.* 248:318–28

179. Zieve R, Peterson PJ. 1984. Volatilization of selenium from plants and soils. *Sci. Total Environ.* 32:197–202

Annu. Rev. Plant Physiol. Plant Mol. Biol. 1998. 49:669–96

MOLECULAR BIOLOGY OF CATION TRANSPORT IN PLANTS

Tama Christine Fox and Mary Lou Guerinot

6044 Gilman, Department of Biological Sciences, Dartmouth College, Hanover, New Hampshire 03755; e-mail: mary.lou.guerinot@dartmouth.edu

KEY WORDS: calcium, copper, iron, manganese, potassium, zinc

ABSTRACT

This review summarizes current knowledge about genes whose products function in the transport of various cationic macronutrients (K, Ca) and micronutrients (Cu, Fe, Mn, and Zn) in plants. Such genes have been identified on the basis of function, via complementation of yeast mutants, or on the basis of sequence similarity, via database analysis, degenerate PCR, or low stringency hybridization. Not surprisingly, many of these genes belong to previously described transporter families, including those encoding Shaker-type K^+ channels, P-type ATPases, and Nramp proteins. ZIP, a novel cation transporter family first identified in plants, also seems to be ubiquitous; members of this family are found in protozoa, yeast, nematodes, and humans. Emerging information on where in the plant each transporter functions and how each is controlled in response to nutrient availability may allow creation of food crops with enhanced mineral content as well as crops that bioaccumulate or exclude toxic metals.

CONTENTS

669

1040-2519/98/0601-0669$08.00

INTRODUCTION

Ions need to be transported from the soil solution into the root and then distributed throughout the plant, crossing both cellular and organellar membranes. Transport across the plant plasma membrane is driven by an electrochemical gradient of protons generated by plasma membrane H^+-ATPases (76, 106). These primary transporters pump protons out of the cell, thereby creating pH and electrical potential differences across the plasma membrane. Secondary transport systems then utilize these gradients for many functions, including nutrient uptake, phloem loading, and stomatal opening. Since the first reports on the cloning of H^+-ATPase genes from plants in 1989 (40, 88), a number of genes have been identified that encode various types of primary and secondary transporters (as reviewed in 3, 66). This review focuses on a group of plant cation transporters that has been identified either functionally or via sequence similarity. We highlight transporters for two macronutrient cations, K and Ca, and four micronutrient cations, Cu, Fe, Mn, and Zn. These are the mineral cations for which we currently have the most information at the molecular level.

When we think about cation transport, it is important to keep in mind that many cations, although essential, can also be toxic when present in excess (e.g. 39). Thus, plants may regulate uptake and efflux systems in order to control their intracellular cation concentrations. With cloned genes in hand, it now becomes possible not only to carry out structure-function studies on how various transporters move cations but also to investigate how transporters respond to changing cation levels. For most of the transporters we discuss, definitive proof of where they function within the plant has not yet been obtained. Nonetheless, what is becoming increasingly clear is that plant transporters belong to a number of well-described families, thus narrowing the perceived gap between the biology of animals and that of plants and allowing us to apply knowledge gained in one system directly to another.

Throughout this review, the term transporter is used inclusively to refer to both channels and carriers. Ion channels are distinguished from carriers by their several orders of magnitude higher flux rates and by their mediation of the passive flux of ions across membranes. The electrochemical gradient across the membrane in which the channel resides determines whether, and in which direction, ions move. Carriers may move ions either with or against their substrate concentration gradients and may function as uniporters or as cotransporters.

POTASSIUM

K^+ is the most abundant cation in higher plant cells. K^+ plays central roles in plant growth and development, including maintenance of turgor pressure, leaf and stomatal movement, and cell elongation (68, 69, 102, 111). Whereas K^+ levels in soil solution range from 1 μM to 10 mM, with many soils falling in the range of 0.3 to 5.0 mM, intracellular K^+ levels are maintained at 100–200 mM (56). Thus, uptake of K^+ from the soil must move K^+ against its concentration gradient. Recent articles have provided excellent overviews of K^+ transport in plants (10, 68, 69, 98, 102). We discuss genes identified to date that encode plant K^+ transporters. These transporters can be grouped into four categories: (a) high-affinity cotransporters, (b) inwardly rectifying K^+ channel α subunits, (c) K^+ channel β subunits, and (d) an outwardly rectifying K^+ channel (Table 1).

High-Affinity K^+ Transport

High-affinity K^+ transport occurs when external K^+ concentrations range from approximately 1–200 μM (56). The first high-affinity transporter, HKT1, was identified by complementation of a mutant yeast strain defective for K^+ uptake with a wheat (*Triticum aestivum*) root cDNA expression library. *HKT1* restores the ability of the yeast *trk1trk2* mutant to grow on low K^+-containing medium (96). In situ hybridization experiments show that *HKT1* mRNA localizes to the root cortex as well as to leaf and stem vascular tissue. *HKT1* confers high-affinity K^+ transport when expressed in yeast and in *Xenopus* oocytes (96).

Although originally thought to be a H^+-K^+ symporter, further investigations suggest that HKT1 may function as a K^+-Na^+ symporter (34, 92). Nontoxic (μM) Na^+ concentrations stimulate K^+ transport when *HKT1* is heterologously expressed in yeast or in *Xenopus* oocytes (92). Even more striking, HKT1 can mediate low-affinity Na^+ influx into oocytes. This Na^+ influx can be stimulated by external K^+. Furthermore, high (10–100 mM) external Na^+ concentrations inhibit K^+ influx into oocytes, probably via competition between Na^+ and K^+ for binding to the K^+ coupling site of HKT1 (92). The dual activities expressed

Table 1   Plant K^+ transporters and associated subunits

Gene	Species	Cloning approach	Direction of transport	Proposed function	Reference
High-Affinity K^+ Carriers					
HKT1	T. aestivum	Complementation of a yeast trk1 trk2 mutant	Inward	K^+-Na^+(H^+)-symporter	96
KEA1	A. thaliana	PCR	Inward	Putative K^+-H^+ or Na^+-H^+ antiporter	115
Low-Affinity K^+ Channels					
LCT1	T. aestivum	Complementation of a yeast trk1 trk2 mutant	Inward	K^+, Na^+ transport	95
KAT1	A. thaliana	Complementation of a yeast trk1 trk2 mutant	Inward	Channel	97
KST1	S. tuberosum	DNA hybridization based on KAT1 sequences	Inward	Channel	80
AKT1	A. thaliana	Complementation of a yeast trk1 trk2 mutant	Inward	Channel	99
AKT2/3	A. thaliana	DNA hybridization based on KAT1 or Shaker sequences	Inward	Channel	16, 52
SKT2	S. tuberosum	Yeast two-hybrid screen using KST1 as bait	Inward	Channel	28
SKT3	S. tuberosum	Yeast two-hybrid screen using KST1 as bait	Inward	Channel	28
KCO1	A. thaliana	EST database searched for P-domain sequences	Outward	Channel	27
K^+ Channel β Subunits					
KAB1	A. thaliana	EST database searched for animal β subunits sequences		Modulates channels	109

by HKT1—beneficial high-affinity K^+-Na^+ uptake under non-Na^+ stress conditions and detrimental low-affinity Na^+ uptake under Na^+ stress conditions—point out the importance of dissecting the selectivity of transporters. Na^+ stress conditions are frequently found in irrigated soils, and HKT1 may provide one of the pathways for Na^+ uptake, making it a potential target for engineering Na^+ tolerance in plants. Indeed, genetic selection in yeast has allowed the isolation of mutations in HKT1 that confer reduced Na^+ uptake and improved salt tolerance relative to wild-type HKT1. The next step is to test such mutant forms in plants. It is crucial that the observations obtained via heterologous expression in yeast and oocytes be verified experimentally in plants. To date, physiological studies of high affinity K^+ uptake in intact roots have not supported the importance of Na^+-coupled K^+ transport in roots of terrestrial plants (69a).

Another gene, *LCT1*, was identified in the same screen that identified *HKT1* (95). LCT1 mediates low-affinity Na^+, Rb^+, and possibly Ca^{2+} transport in yeast. The physiological role of LCT1 in plants is uncertain, but the competitive inhibition of cation uptake by Ca^{2+} fits well with studies that have shown the importance of Ca^{2+} in reducing Na^+ uptake and ameliorating Na^+ toxicity. Thus, LCT1, like HKT1, may be a good target for lowering Na^+ influx in plants to increase salinity tolerance (95). *LCT1* is expressed at a low level in roots and leaves. LCT1 is predicted to have 6 or 7 transmembrane domains and a hydrophilic N-terminus with two PEST sequences. There are currently no related proteins in the databases.

Finally, a cDNA—*KEA1*—has been cloned from Arabidopsis that shares 22% identity with bacterial K^+-H^+ and Na^+-H^+ transporters and mediates K^+-dependent inward currents in oocytes (115). It is unknown whether KEA1 also mediates Na^+ transport.

Low-Affinity K^+ Transport

Low-affinity K^+ transport occurs at mM external K^+ concentrations and is nonsaturating at physiologically relevant K^+ concentrations. Low-affinity K^+ transport is carried out by voltage-gated channels that allow the passive flow of K^+ down its electrochemical gradient (10, 69, 102). All of the known plant K^+ inward channels show amino acid similarity to the animal Shaker K^+-channel family (58). Family members have six conserved transmembrane domains (called S1–S6). An amphipathic S4 domain is involved in voltage sensing, and a hydrophobic hairpin region between S4 and S5, called the H5 or P(pore) domain, is proposed to form the channel pore and contain ion binding sites (Figure 1). Shaker channels are composed of four α-subunits, with the P domain from each subunit lining the channel pore (13). Some Shaker channels have additional β subunits as well (discussed below). All the plant K^+ inward channels, and many animal Shaker channels, contain a cyclic nucleotide binding

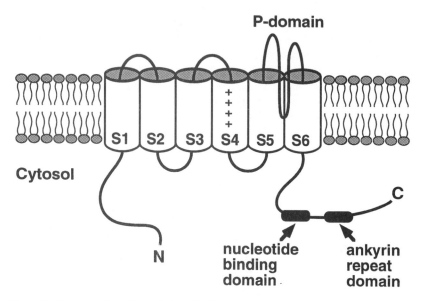

Figure 1   Plant inward-rectifying K channels all contain six transmembrane domains, a voltage sensing S4 domain, a nucleotide binding domain and a P domain. Some also have an ankyrin-repeat domain at the C terminus.

domain (1, 16, 52, 99) and some of the plant channels have an ankyrin repeat domain (16, 99).

Structural similarity does not predict direction of ion flux. Animal Shaker channels are all outward-rectifying. Outward-rectifying channels open upon depolarization of the membrane potential, thereby allowing K^+ efflux. The only plant outward-rectifying K^+ channel identified so far, *KCO1*, has four transmembrane domains and is more similar to human and yeast inward-rectifying channels than to the 6-transmembrane Shaker channels (27). In contrast, the plant 6-transmembrane Shaker channels are all inward-rectifying and thus activate K^+ influx upon membrane potential hyperpolarization. Animal inward-rectifying K^+ channels have only two transmembrane domains (11). Further details on specific plant K^+ channels are presented below.

KAT1/KST1   The Arabidopsis *KAT1* gene was identified by its ability to restore the growth of a yeast *trk1trk2* mutant strain on low K^+ media. KAT1 transports K^+ when expressed in *Xenopus* oocytes (97) and insect Sf9 cells (35, 72). KAT1-GUS constructs are expressed in guard cells and the vascular tissue of the stem and root of Arabidopsis (82). Thus, KAT1 is likely to have a role in stomatal opening and the transport of K^+ into vascular cells rather than the direct uptake of soil K^+. Further evidence that KAT1 functions in vivo as predicted comes from studies on transgenic plants expressing a mutant form of KAT1 that is resistant to Cs^+ blockage (48). These plants showed inward K^+ currents that were less sensitive to blockage by Cs^+ than controls. In addition, the Cs^+ inhibition of light-induced stomatal opening observed in wild-type plants was not seen in the transgenic plants; stomata in these plants opened in the presence of external Cs^+. This study clearly provides molecular biological evidence that K^+ channels constitute an important pathway for physiological K^+ uptake during stomatal opening.

The tripeptide $G^{262}Y^{263}G^{264}$ is conserved in the P domain of KAT1 and in animal inward-rectifying K^+ channels. This sequence may create selectivity by occupying the narrowest region of an hour glass–shaped pore (58). Normally, ion permeability through KAT1 in oocytes is in the order $K^+ > NH4^+ > Rb^+ \gg Na^+ \sim Li^+ \sim Cs^+$ (97). Altering the GYG tripeptide region or a T proposed to interact with GYG changed selectivity and/or converted acid activation to inactivation (81, 112). In addition, mutations were found that increased the sensitivity of yeast to $NH4^+$ or Na^+. KAT1 has an additional 14 amino acids in the P domain as compared to Drosophila Shaker channels. Mutation analyses suggest that the additional KAT1 amino acids are involved in pore selectivity and inhibition by Cs^+ and Ca^{2+} (6). Both plant inward- and animal outward-rectifying K^+ channels appear to have the same orientation in membranes, with the P domain on the extracellular side (Figure 1) (49). Evidence for conserved membrane orientation includes (a) chimeric channels containing portions of KAT1 and the Xenopus K^+ channel, Xsha2, are capable of transporting K^+ into Xenopus oocytes, suggesting that the two channels have the same orientation in the membrane (15), and (b) the KAT1 P domain mutants (H^{267} to T, and E^{269} to V) are less sensitive to Cs^+ and TEA blockage (without changing selectivity), suggesting that the P domain is accessible to the external side of the membrane as in animal Shaker channels (49). If inward- and outward-rectifying channels have the same orientation, then what determines whether a channel is inward or outward rectifying? The S4 domain appears to be involved; altering three amino acids in the S4 domain converts the Shaker B channel from outward to inward rectification (77).

A gene encoding a guard cell K^+ channel, KST1, was also cloned from a Solanum tuberosum cDNA library based on its similarity to KAT1 (80). While both KAT1 and KST1 mRNA are both predominantly expressed in guard cells, KST1 mRNA is present at low levels in flowers, whereas KAT1 mRNA is present in the vascular tissue of stems and roots (82).

The yeast two-hybrid system was used to screen for proteins that interact with KST1 (28). Because transmembrane domains cannot be used efficiently in the two-hybrid system, C-terminal regions of KST1 predicted to be cytosolic were used as bait. However, instead of identifying genes encoding proteins that might be involved in channel regulation or localization, two putative K^+ channel genes, *SKT2* and *SKT3*, were isolated. SKT2 and SKT3 are 88% similar to each other and both show similarity to Arabidopsis AKT2 (81% amino acid similarity for SKT2 and 77% similarity for SKT3; see next section for discussion of AKT2). A comparison of KST1, SKT2, and SKT3 with other K^+ channels revealed two conserved regions at the C-terminus that are found exclusively in all of the plant inward-rectifying K^+ channels. These regions, termed K_{HA} domains, are enriched for hydrophobic and acidic amino acid residues (28). Using insect cells expressing a GFP-KST1 fusion, clusters of K^+ channels could be visualized within the plasma membrane. Deletion of the K_{HA} domain resulted in an even distribution of the GFP-KST1ΔK_{HA} fluorescence pattern, suggesting that the K_{HA} domain is required for the clustering of KST1. KST1 lacking the K_{HA} domain is electrophysiologically active in insect cells, indicating that the K_{HA} domain is not required for assembly into α-subunit tetramers (28). All the potato inward-rectifying K^+ channels interact with each other indiscriminately in vitro, suggesting that the tissue-specific expression of these transporters likely influences which proteins interact to form channels.

AKT1   The *Arabidopsis AKT1* gene was cloned by complementing the yeast *trk1trk2* mutant (99). *AKT1* transports K^+ into transfected insect cells but not into *Xenopus* oocytes (35). Northern blot analyses indicate that *AKT1* is expressed mainly in roots (5, 59). Transgenic Arabidopsis expressing an *AKT1* promoter-GUS fusion exhibit GUS activity in the peripheral cell layers of mature root regions as well as in leaf primordia and hydathodes (59). *AKT1* expression in roots is consistent with a role in K^+ influx from the soil solution. This was confirmed using a T-DNA insertional mutant of *AKT1* called *akt1-1* (44). In *akt1-1* root cells, no inward K^+ currents could be detected. These *akt1-1* plants grew poorly on low K^+ medium and exhibited reduced high- and low-affinity Rb^+ uptake.

AKT2, which is approximately 60% identical to *AKT1* and *KAT1*, was identified using a *KAT1* PCR product to probe an Arabidopsis cDNA library (16). The same gene was also identified by using a degenerate oligonucleotide made from the conserved P domain of Shaker K^+ channels to probe an Arabidopsis genomic library (52). Although *AKT2* did not complement the yeast *trk1trk2* mutant (16), *AKT2* can function as an inward-rectifying K^+ channel in oocytes (52). *AKT2* was found by Northern blot analysis to be highly expressed in leaves (16, 52). This contrasts with the root expression seen for *AKT1* (5, 59). Both *AKT1* and *AKT2* genes contain similar C-terminal ankyrin-binding sequences. Ankyrin-binding sites facilitate the binding of proteins to the cytoskeleton, either to localize proteins to specific locations on the PM, or to promote protein-protein interactions (reviewed in 52). This cytoskeletal interaction is different from that proposed for KAT1, which has a C-terminal microtubule-binding site rather than ankyrin-binding sites. Colchicine, which destabilizes microtubules, decreases the current amplitude in oocytes expressing KAT1 but has not been tested on AKT1 or AKT2.

KCO1 The first plant outward-rectifying K^+ channel, KCO1 (for K^+ channel, Ca^{2+} activated, *outward-rectifying*), was found by searching the Arabidopsis EST database for sequences containing the conserved P domain of Shaker channels (27). KCO1 belongs to a new class of K^+ channels, recently described in yeast and in humans, that contain not one but two P domains and have four transmembrane segments. Sequences with similarity to KCO1 have also been identified in other plant species (79). When expressed in insect cells, KCO1 directs outward-rectifying K^+ currents that are strongly dependent on the presence of nanomolar concentrations of cytosolic free Ca^{2+}. Indeed, two EF-hand motifs (Ca^{2+}-binding sites) are present at the C-terminal end of KCO1. KCO1 is activated by depolarization of the plasma membrane. This depolarization is affected by external K^+ concentration. *KCO1* expression can be detected in seedlings, as well as in the leaves, and flowers of older plants, albeit at a low level, requiring the use of RT-PCR for detection. This level of localization offers few clues about KCO1 function. However, physiological studies have identified a number of outwardly rectifying K^+ channels and it is hoped that we will soon be in a position to correlate KCO1 expression with one of these known K^+ channel activities.

KAB1 An Arabidopsis EST was found to encode a protein with 49% identity to rat and bovine K^+ channel β subunits (74, 109). Co-expression of certain β subunits with Shaker α subunits can convert these channels from noninactivating to fast-inactivating. The rat β subunit, while hydrophilic, co-localizes with native α subunits isolated from brain membranes, suggesting that α and β subunits may interact (reviewed in 58). β subunits have been proposed to modulate K^+ channels by (*a*) blocking the pore sterically, causing channels to be fast inactivating (58); (*b*) reducing α subunits, based on the similarity of β subunits to NAD(P)H oxidoreductases (22, 74); and/or (*c*) modulating the maturation and direction of α/β complexes to membranes (100).

KAB1 RNA is highly expressed in Arabidopsis leaf (especially guard) cells and to a lesser extent in root cells, as shown by in situ hybridization (109). Western blot analyses demonstrated the presence of KAB1 in both soluble and membrane protein fractions of leaves, flowers, and roots (110). Immunocytochemical analyses show that KAB1 is present in the plasma membrane, tonoplast, chloroplast inner envelope, and mitochondrial inner membrane. The staining pattern suggests that the native polypeptide is present in regularly spaced individual protein complexes (H Tang, AC Vasconcelos & GA Berkowitz, submitted manuscript). KAB1 physically associates with KAT1 α subunits; in vitro translated ^{35}S-KAT1 binds specifically to an affinity column made from immobilized affinity-purified KAB1 and polyclonal antibodies directed against KAB1 specifically immunoprecipitate both KAB1 and ^{35}S-KAT1. The widespread expression of KAB1 mRNA in various tissues suggests that KAB1 may interact with multiple K$^+$ channels rather than exclusively with shoot-localized KAT1 (110).

CALCIUM

Ca is an abundant element in soils and is usually present in sufficient amounts to meet plant needs. Ca constitutes 0.1–2.0% dry weight of plants and structurally stabilizes membranes and cell walls (71). Ca is also an important signaling molecule: stimulation by red light, gravity, touch, cold shock, fungal elicitors, hormones, or salt stress induces the opening of gated Ca^{2+} channels (reviewed in 14). Ca^{2+} is transiently transported from organelles and the apoplasm into the cytosol, thereby activating signal transduction pathways. In resting plant cells, Ca^{2+} is actively exported by high-affinity Ca^{2+}-ATPases and Ca^{2+}/H$^+$ antiporters out of the cell or into organelles to maintain cytosolic Ca$^{2+} \leq 0.1$ μM. This prevents the activation of signal transduction pathways until a stimulus is sensed (reviewed in 14).

P-type Ca-ATPases

High-affinity Ca^{2+}-ATPases localize to the plant plasma membrane (2, 31) and endomembranes (J Harper, personal communication; 91), including the ER (63, 65), the Golgi Apparatus (65), the tonoplast (2, 20, 30, 31, 70), and the plastid envelope (47). These transporters remove cytosolic Ca^{2+} into organelles or the apoplasm. All the plant Ca^{2+}-ATPases cloned are P-type ATPases that are characterized by a phosphorylated intermediate, inhibition by vanadate, a requirement for Mg^{2+}, and a high substrate specificity for MgATP (47). Alignment of eukaryotic Ca^{2+}-ATPases shows that the plant and animal genes are related (70) and can be divided into "PM-type" pumps that are calmodulin-stimulated and "ER-type" pumps that are not (78).

CALMODULIN-STIMULATED CA^{2+}-ATPASES   At present, three genes encoding calmodulin-stimulated Ca^{2+}-ATPases have been cloned from plants (Table 2). Although similar to animal PM-type pumps, none localizes to the PM. *PEA1* [(47), later called *ACA1* for Arabidopsis Ca^{2+}-ATPase; (70)], was cloned from Arabidopsis using antibodies generated against spinach chloroplast envelope proteins (47). ACA1 is 40–44% identical to mammalian PM-type Ca^{2+}-ATPases

Table 2 Ca^{2+} transporters in plants

Gene	Species	Cloning approach	Location	Reference
Calmodulin-Stimulated Ca^{2+}-ATPases				
PEA1/ACA1	A. thaliana	Antibody to chloroplast protein	Chloroplast inner envelope	47
ACA2	A. thaliana	PCR using degenerate primers	Endomembranes	40a
BCA1	B. oleracea	Calmodulin-binding protein micro-sequencing	Tonoplast	70
Calmodulin-Insensitive Ca^{2+}-ATPases				
LCA1	L. esculentum	PCR using degenerate primers	Tonoplast and plasma membrane	113
pH27	N. tabacum	PCR using degenerate primers	Endomembrane	90
DCBA1	D. bioculata	PCR using degenerate primers	Endomembrane	91
ACA3/ECA1	A. thaliana	PCR using degenerate primers (ACA3) and DNA hybridization using pH27 as a DNA probe (ECA1)	Endoplasmic reticulum	63
OsCa-atpase	O. sativum	Differential display +/− gibberellin	Endomembrane	21
Ca^{2+}/H^+-Antiporters				
CAX1	A. thaliana	Complementation of a yeast vcx1pmc1 mutant	Tonoplast	45
CAX2	A. thaliana	Complementation of a yeast vcx1pmc1 mutant	Tonoplast	45

but lacks the C-terminal calmodulin-binding domain. ACA1 has been proposed to export Ca^{2+} from the plastids into the cytosol because (a) ATP-dependent Ca^{2+} import could not be shown in spinach leaf chloroplasts (albeit ACA1 expression is lower in leaves than in roots); (b) the evolutionary relationship of plastids to cyanobacteria, which export Ca^{2+} out of the cytoplasmic membrane; and (c) the low activity of Ca^{2+} in the stroma as compared with total plastid Ca^{2+} concentration (47).

A vacuolar calmodulin-affinity purified Ca^{2+}-ATPase from *Brassica oleracea* was microsequenced, and the gene, *BCA1*, was cloned (70). BCA1 shares 62% identity with ACA1 and >80% identity with proteins specified by three Arabidopsis EST sequences. BCA1 contains a potential calmodulin-binding, amphipathic helix at the N-terminus (instead of at the C-terminus as in animal PM-type ATPases). Immunoblotting confirmed that BCA1 is in the tonoplast, and isolated vacuoles were shown to have calmodulin-stimulated, ATP-dependent Ca^{2+} influx (2).

An Arabidopsis cDNA, *ACA2*, was identified using primers designed against sequences conserved among P-type ATPases (40a). ACA2 is 78% and 62% identical to ACA1 and BCA1, respectively, at the amino acid level. Biochemical assays confirm that ACA2 encodes a calmodulin-regulated Ca^{2+}-ATPase and provide evidence that the N-terminal domain of ACA2 functions as a calmodulin-regulated autoinhibitor. Removal of this autoinhibitory domain results in constitutively high ACA2 activity. Only the truncated pump allows a yeast mutant defective in calcium homeostasis (lacking both tonoplast and Golgi apparatus Ca^{2+}-ATPases as well as calcineurin) to grow on Ca^{2+}-depleted medium. Anti-ACA2 antibodies recognize a protein that is most abundant in root and flower endomembranes.

CALMODULIN-INSENSITIVE Ca^{2+}-ATPASES   *LCA1*, a *Lycopersicon esculentum* Ca^{2+}-ATPase, was identified by using a degenerate sequence of the conserved ATP-binding domain of P-type ATPases to probe a root cDNA library (113). LCA1 shares over 60% amino acid identity with regions of animal ER-type Ca^{2+}-ATPases. Northern blots of tomato identified three *LCA1* transcripts in roots and one in leaves. *LCA1* expression was induced by 50 mM NaCl in both leaves and roots (113), consistent with a model in which salt stress perturbs intracellular Ca^{2+} levels (62). A similar Ca^{2+}-ATPase (encoded by a cDNA called pH27) was identified in *Nicotiana tabacum* (90). Expression of the *N. tabacum* gene was highest in cultured cells and stems, intermediate in roots, and lowest in leaves. High NaCl induces expression of this Ca^{2+}-ATPase in cultured cells. Both LCA1 and the protein encoded by pH27 were initially proposed to localize to the ER (90, 113). However, antibodies raised against LCA1 (31) reacted instead with tonoplast and PM vesicle proteins. ATP-dependent $^{45}Ca^{2+}$ transport occurred only in those vesicles containing LCA1 and was inhibited by the anti-LCA1 antibodies (31).

A Ca^{2+}-ATPase, DCBA1, with 54% identity to mammalian ER-type Ca^{2+}-ATPases was identified from the highly salt-tolerant alga *Dunaliella bioculata* using a conserved region of the *Dunaliella* P-type H^+-ATPase gene, *DBPMA1*, to probe a cDNA library (91). *DCBA1* is expressed at very low levels as shown by Northern blots.

An Arabidopsis Ca^{2+}-ATPase gene, *ACA3/ECA1* (for ER-type Ca^{2+}-ATPase), was cloned using the same screen that identified *ACA2* (Y Wang, B Hong &

JF Harper, submitted manuscript) and by probing a cDNA library with *pH27* (63). ECA1 is 69% identical to OsCa-atpase (see below), 64% identical to LCA1 and 42% identical to a rabbit ER-type Ca^{2+}-ATPase. Western blots showed highest expression of ECA1 in roots and flowers versus shoots and siliques. ECA1 is located mainly in the ER (63). ECA1 restores growth to both a *pmr1* yeast mutant (defective in a Golgi Apparatus Ca^{2+}-pump), and a *pmr1pmc1cnb1* mutant (lacking also the vacuolar Ca^{2+}-pump and calcineurin) on low Ca^{2+}-containing medium (63). ECA1 may also transport Mn^{2+} (discussed in the Mn section; 63).

A differential display approach was used to compare gene expression in rice aleurone cells treated ± gibberellin for 1 h (21). A gibberellin-inducible Ca^{2+}-ATPase, OsCa-atpase, was identified that, when transiently expressed in aleurone cells, bypasses the gibberellin requirement for stimulating α-amylase. Thus, OsCa-atpase may be an early downstream component of the gibberellin signal transduction pathway.

Calcium/Proton Antiporters

Previously, high-affinity vacuolar Ca^{2+}-ATPases were thought to lower Ca^{2+} activities in the cytosol of resting cells, whereas Ca^{2+}/H^{+}-antiporters were thought to pump Ca^{2+} from the cytosol only when cytosolic Ca^{2+} levels were elevated (30). However, genes encoding two high- and low-affinity $Ca^{2+}-H^{+}$-antiporters, *CAX1* and *CAX2*, have now been cloned from Arabidopsis (45) via their ability to suppress the Ca^{2+} hypersensitivity of a *Saccharomyces cerevisiae vcx1pmc1* double mutant that lacks both the vacuolar P-type Ca^{2+}-ATPase and the Ca^{2+}/H^{+}-antiporter (25, 26; Table 2). CAX1 and CAX2 share a central acidic motif, and 11 proposed membrane-spanning domains with Ca^{2+}/H^{+}-antiporters identified in *S. cerevisiae* and *E. coli*. CAX1 and CAX2 catalyze ΔpH-dependent Ca^{2+} transport. CAX1-mediated Ca^{2+} transport is concentration dependent and exhibits Michaelis-Menton kinetics with a K_m of 13 μM. Thus CAX1, along with Ca^{2+}-ATPases, is proposed to keep cytosolic Ca^{2+} activities below 1 μM in resting plant cells (45). However, Cd, Hg, and La strongly inhibit ^{45}Ca uptake, suggesting that CAX1 may transport other cations (121). In support of this, CAX1 mediates $Na^{+}-H^{+}$ antiport as well as Cd transport (121). CAX2-mediated Ca^{2+} transport was also concentration dependent but exhibits a lower affinity ($Km > 100$ μM) for Ca^{2+} (45). The high K_m suggests that CAX2 does not normally function in Ca^{2+} transport. Rather, preliminary experiments suggest that CAX2 may function as a high-affinity H^{+}/heavy metal cation antiporter with a role in metal homeostasis rather than Ca^{2+} signaling (K Hirschi, personal communication). *CAX1* and *CAX2* RNA are both expressed in Arabidopsis roots, leaves, stems, flowers and siliques.

COPPER

Uptake

Cu is an essential redox component required for a wide variety of processes, including the electron transfer reactions of respiration (cytochrome c oxidase, alternate oxidase) and photosynthesis (plastocyanin), the detoxification of superoxide radicals (Cu-Zn superoxide dismutase) and lignification of plant cell walls (laccase). Cu levels in soils range from 10^{-4} to 10^{-9} M, with up to 98% of the Cu in soil solution complexed to low molecular weight organic compounds (71). Plants require from 5 to 20 μg gDW^{-1} Cu, depending on the species. Whether Cu needs to be reduced before transport, as has been shown to be the case for yeast (41), is still an open question (see section below on Fe).

A putative Cu transporter from Arabidopsis, encoded by the *COPT1* gene (Table 3), can suppress the growth defects of a yeast *ctr1-3* strain that lacks high-affinity Cu uptake (51). *COPT1* is expressed in flowers, stems, and leaves but is undetectable in roots. The lack of root expression may indicate that COPT1 is not responsible for Cu uptake from soil; however, it is not known whether Cu deficiency causes an increase in the expression of *COPT1* in roots (or in any other tissue). Although originally reported to be present as a single copy in the genome, several ESTs in the database encode proteins with similarity to COPT1, suggesting that *COPT1* may be part of a small gene family in Arabidopsis.

COPT1 is similar to Ctr1p (49% similarity as predicted by the GAP program) and to a newly described human copper transporter, hCTR1 (56% similarity as predicted by the GAP program), that was also identified by its ability to rescue a yeast *ctr1* mutant (122). COPT1 (169 aa) and hCTR1 (190 aa) are significantly smaller than Ctr1p (406 aa), mainly due to truncations at the N- and C-termini. However, COPT1, like Ctr1p and hCTR1, has three potential transmembrane domains, and all three proteins contain an N-terminal putative metal-binding domain rich in methionine and serine residues. This metal-binding domain is predicted to lie on the extracellular surface and is similar to those found in several bacterial copper-binding proteins, including the *Enterococcus hirae* copper ATPase (CopB) and the CopA and CopB proteins from *Pseudomonas syringae* (101). Two other genes in yeast (*CTR2* and *CTR3*; see 51, 53) and one other gene in humans (h*CTR2*; see 122) are similar to *COPT1*. Ctr2p and hCTR2 were originally identified as being similar to COPT1 and to hCTR1, respectively, in database searches (51). Neither *CTR2* nor h*CTR2* can rescue a *ctr1-3* mutant, and neither has a recognizable metal-binding motif (51, 122). However, *CTR2* overexpression confers sensitivity to Cu, and *ctr2* mutants are more resistant to Cu, suggesting that Ctr2p functions as a low-affinity Cu transporter. Ctr3p is a small, integral membrane protein identified by its ability to restore high-affinity Cu uptake to a *ctr1-3*–deficient yeast mutant (53). Mutations in

Table 3 Genes involved in micronutrient transport in plants

Gene	Species	Cloning approach	Proposed function	Reference
COPT1	*A. thaliana*	Complementation of a yeast *ctr1* mutant	Cu transport	51
PAA1	*A. thaliana*	PCR using degenerate primers	Cu efflux from cytosol	107
AMA1	*A. thaliana*	PCR using degenerate primers	Cu efflux from cytosol	J Harper, personal communication
OsNramp1	*O. sativa*	BLAST search for Nramp ESTs and DNA probing of a cDNA library	Mn transport, roots	8
OsNramp2	*O. sativa*	BLAST search for Nramp ESTs and DNA probing of a cDNA library	Mn transport, leaves	8
OsNramp3	*O. sativa*	BLAST search for Nramp ESTs and DNA probing of a cDNA library	Mn transport, roots and leaves	8
frohA, frohB, frohC, frohD	*A. thaliana*	PCR using primers based on yeast ferric reductases	*b*-type cytochrome; Fe(III)-chelate reductase	91a
IRT1	*A. thaliana*	Complementation of a yeast *fet3fet4* mutant	Fe(II) transport Mn(II) transport	29; H Pakrasi, personal communication
ZIP1, ZIP2, ZIP3	*A. thaliana*	Complementation of a yeast *zrt1zrt2* mutant	Zn transport	ML Guerinot, unpublished data
NA synthase	*H. vulgare*	Protein purification; cDNA has not been identified	Synthesis of phytosiderophores	42
NAAT	*H. vulgare*	PCR with degenerate primers based on microsequencing data	Synthesis of phytosiderophores	108
IDS2, IDS3	*H. vulgare*	Subtractive hybridization of +/− Fe roots	Synthesis of phytosiderophores	83, 85

both Ctr1p and Ctr3p are required to completely eliminate high-affinity copper uptake. Interestingly, in many laboratory strains, *CTR3* is interrupted by a transposable element (53). The exact role that Ctr3p plays in Cu transport is not entirely clear. Ctr3p is thought to function in an endocytic copper transport pathway, based on its localization pattern (53).

Intracellular Transport

It is likely that multiple Cu-trafficking pathways come into play after transport of Cu across the plasma membrane. Cu must be delivered to the mitochondria, to a compartment needed for activation of cytosolic Cu proteins and to the secretory pathway. In *S. cerevisiae*, *ATX1* encodes a 73 amino acid polypeptide that is believed to act in the intracellular transport of Cu to the secretory system (64). Atx1p contains the highly conserved metal-binding motif MTCXXC that functions as the mercury binding site in the *Escherichia coli* MerP and MerA proteins and as the putative copper-binding site for the Cu-ATPases from *E. hirae* (CopA), *S. cerevisiae* (Ccc2p), and humans (the Wilson and Menkes proteins). Two plant ESTs, from Arabidopsis and rice, encode peptides having greater than 60% identity to Atx1p; each contains a presumptive metal-binding motif, MXCXXC. Although the role of these Atx1p-like proteins in metal homeostasis in plants has not yet been established, their similarity to Atx1p suggests a similar function and makes these genes interesting targets for further investigation.

Efflux

P-type ATPases belong to a large superfamily of ATP-driven pumps involved in the transmembrane transport of a variety of cations across cell membranes. A recent analysis of 159 P-type ATPases demonstrates that these proteins can be organized in a phylogenetic tree with five major branches according to substrate specificity and not according to the evolutionary relationship of parent species. This indicates that abrupt changes in the rate of sequence evolution are accompanied by the acquisition of new substrate specificities (4). Analysis of the Arabidopsis EST database reveals that there are at least 18 different P-type ATPases represented.

Using degenerate oligonucleotides based on residues conserved among metal-transporting P-type ATPases, an Arabidopsis cDNA has been identified encoding a protein, PAA1, that is 42.6% identical to the PacS Cu-ATPase from *Synechococcus* and 37.9% identical to the Wilson Cu-ATPase (107). PAA1 contains all four motifs commonly found in P-type ATPases, a phosphatase region (TGES), an ion transduction region (xPC), a phosphorylation site (DKTGT), and an ATP binding domain (GDGxNDxP). It also contains the highly conserved MTCXXC metal-binding motif in its N terminal domain. *PAA1* transcripts are not very abundant in either roots or shoots and are not copper inducible. Another P-type ATPase, AMA1, has also been identified from Arabidopsis using degenerate PCR. There is also an Arabidopsis gene, recently sequenced as part of the genome effort, that has up to 80% similarity to Ccc2p, Menkes protein, and Wilson's protein (ML Guerinot, unpublished data). All three of these proteins probably function in the export of Cu from the cytosol into an extracytosolic compartment (118).

MANGANESE

Uptake

Mn is required for a number of essential processes in plants, including oxygen evolution in photosynthesis (water-splitting enzyme S in the Hill reaction), detoxification of oxygen-free radicals (Mn-superoxide dismutase), and $CO2$ fixation in C4 and CAM plants (PEP carboxylase). Mn is taken into plants mainly as the free Mn^{2+} ion. Concentrations of Mn can vary greatly in the soil, ranging from less than 0.1 μM in well-aerated alkaline soils to greater than 400 μM in submerged soils (54). The Mn concentration required by plants also spans a wide range, from 0.01 to 50 μM. Although there are a large number of reports linking Fe and Mn nutrition, almost no work has been done at the molecular level on the transport of Mn in plants. A number of proteins similar to Nramp may function as Mn transporters (see below). In addition, *IRT1*, identified originally by its ability to complement a yeast mutant with a defect in Fe uptake (see section on Fe), has been shown to rescue a *smf1* mutant of yeast that has a defect in high-affinity Mn uptake (H Pakrasi, personal communication). There is accumulating evidence from both yeast and Arabidopsis that Ca and Mn may be substrates for the same intracellular transporter (60, 63).

Members of the Nramp (*N*atural *r*esistance *a*ssociated *m*acrophage *p*rotein) family have now been implicated in Mn transport in *S. cerevisiae* (104) and in Fe transport in mammals (32, 38). Two ESTs from *O. sativa* and one from Arabidopsis show strong similarity (60% as determined by a BLAST search) to mammalian Nramp (7). Using the ESTs as hybridization probes, three different cDNAs have been isolated from rice (8). Northern blots indicate that *OsNramp1* is expressed primarily in roots, whereas *OsNramp2* is primarily expressed in leaves. *OsNramp3* is expressed in both tissues. As yet, there is no functional proof of what substrate(s) the plant Nramp proteins are transporting. We would also like to know whether any of the plant Nramp proteins respond to a deficiency of any particular metal. For example, dietary Fe deficiency upregulates the expression of *DCT1*, the rat isoform of human Nramp2, possibly via an iron responsive element found in the 3′ UTR of *DCT1* (38). DCT1 has an unusually broad substrate range that includes Fe, Mn, Co, Cd, Cu, Ni, and Pb. It is not known whether other dietary deficiencies can also upregulate the expression of this transporter.

The Nramp family has been highly conserved throughout evolution with representatives found in yeast, birds, Drosophila, *Caenorhabditis elegans* and bacteria in addition to those found in mammals and plants (17, 18). Nramp proteins have 10 conserved transmembrane domains, a glycosylated loop (between TM6 and TM7), and a sequence signature (TMT[X]4G[D/Q[X]4GF in the TM8 to TM9 interval) that shares similarity to the permeation pore of the K^+ channel family (17). Originally, Nramp1 was identified in mice because a mutation in this gene leads to susceptibility to intracellular pathogens such as *Mycobacterium*, *Salmonella*, and *Listeria*. If all members of the Nramp family function as metal transporters, this could nicely explain the original observations on Nramp1 (for models, see 38, 105). Briefly, Nramp is proposed to transport certain divalent cations, possibly Mn and/or Fe, from the extracellular milieu into the cytoplasm of a macrophage, and after the generation of a phagosome removes these divalent cations from the organelle, thus depriving the invading microorganism of divalent cations needed for production of defense enzymes such as superoxide dismutase (Mn) and catalase (Fe). It is also possible that Fe^{2+} uptake by macrophages may allow production of toxic hydroxyl radicals via the Fenton reaction, killing pathogens in the phagosome as part of the defense mechanism.

Intracellular Transport

ECA1 encodes a Ca^{2+}-ATPase in Arabidopsis that localizes mainly to the ER (63). In addition to being able to rescue the Ca pumping defect of a *pmr1* mutant of *S. cerevisiae*, ECA1 could also restore the growth of the *pmr1* mutant in Mn-containing medium, suggesting that this Ca pump may also catalyze Mn transport. Mn has been shown to stimulate the formation of a phosphorylated intermediate of ECA1 in microsomes prepared from yeast expressing ECA1. Mn is likely to be sequestered in endomembrane compartments; ECA1 may be responsible for transport of both Ca and Mn into the lumen of the ER or the Golgi. Indeed, Pmr1p has been implicated in supplying both Mn and Ca to the Golgi in yeast (60). An Arabidopsis line with a T-DNA insertion in *ACA3* (identical to *ECA1*) can now be used to examine the role of this pump in plants (J Harper, personal communication). For example, one would predict that such a line would be more sensitive than wild type to high levels of Mn.

IRON

Fe is found in the soil mainly as insoluble oxyhydroxide polymers of the general composition FeOOH. In an aerobic, aqueous environment at neutral pH, free Fe^{3+} is limited to an equilibrium concentration of approximately 10^{-17} M, a value far below that required for the optimal growth of plants (10^{-6} M). Thus, before plants can take up Fe for transport into the root, they must somehow solubilize these Fe(III) oxides. A recent summary of plant mutants affected in iron uptake (12) will provide additional information to that covered here on components of iron uptake systems now identified at the molecular level.

Strategy I

All plants except the grasses use a strategy (termed Strategy I) to acquire Fe that is similar to the system used by *S. cerevisiae* (37). The initial reduction of Fe(III), carried out by a plasma membrane–bound Fe(III) chelate reductase, is followed by transport of Fe(II) across the root epidermal cell membrane. Both the Fe(III) chelate reductase (117) and the Fe(II) transport activities

(33) are enhanced under Fe deficiency. Several candidate genes (*frohA, frohB, frohC,* and *frohD*) that may encode Fe(III) reductases have been identified in *Arabidopsis* using degenerate PCR with primers designed against motifs common to the yeast Fe(III) reductase proteins, Fre1p, Fre2p, and Frp1 (91a, 93). Each of these *froh* genes encodes a *b*-type cytochrome belonging to a larger family whose other members include the respiratory burst oxidase of mammalian neutrophils (gp91-phox). Arabidopsis mutants—*frd1* and *frd3*—that exhibit defects in Fe(III) reduction have also been identified (116, 117). We are now testing to see whether the *frohA* or *frohB* genes can complement *frd1* mutants, because these two genes map to the same location as *FRD1* (Q Groom, C Procter, E Connolly, ML Guerinot & N Robinson, unpublished data). Whereas there is good evidence that Fe is required in its reduced form for uptake (19, 117), it is still not clear whether Cu needs to be reduced before transport and whether Fe(III) chelate reductases also reduce Cu(II) chelates. Because Arabidopsis has a Cu transporter (COPT1) similar to the yeast Cu transporter Ctr1p that uses Cu(I) as a substrate, it may indeed turn out that plants also reduce Cu before transport. Cu deficiency has been shown to induce Fe(III) chelate reductase activity in pea plants; deficiencies of other cations (K, Mg, Ca, Mn, and Zn) do not elicit a similar response (23). Furthermore, *frd1* mutants that do not show induction of Fe(III) chelate reductase activity under iron-deficient growth conditions have also lost the ability to reduce Cu(II) chelates (117).

Proton release is also enhanced under Fe deficiency, which lowers rhizosphere pH and thereby increases the solubility of Fe(III). There is a large family of H^+-ATPase genes in Arabidopsis. One of these, *AHA2*, is upregulated in response to Fe deficiency (BA Parry & ML Guerinot, unpublished data).

The Arabidopsis *IRT1* gene (for *I*ron *R*egulated *T*ransporter) was isolated because its expression in yeast could restore iron-limited growth to a yeast *fet3fet4* mutant defective in iron uptake (29). Consistent with its proposed role as a metal ion transporter, yeast expressing *IRT1* possess a novel iron uptake system that is specific for Fe(II) over Fe(III). Moreover, IRT1 is specific for iron over other potential substrates; Fe(II) uptake was not greatly inhibited by high concentrations of other physiologically relevant metal ions such as Cu(I), Cu(II), Mn(II), and Zn(II). Most interestingly, Cd has been shown to inhibit iron uptake by IRT1. This suggests that Cd may serve as a substrate for this transporter. This will need to be tested directly using radiolabeled Cd. In Arabidopsis, *IRT1* is expressed in roots and is induced by iron-deficient growth conditions. Furthermore, its expression is altered in mutant strains with defects in regulation of the root Fe(III)-chelate reductase. Based on these results, we proposed that IRT1 is an Fe(II) transporter that takes up iron from the soil.

The significance of IRT1 in the field of metal uptake research is twofold. *IRT1* is the first Fe transporter gene to be isolated from plants, and it provides a useful handle on the mechanism and regulation of Fe uptake in plants. Second, IRT1 has led to the discovery of a family of transporters involved in metal ion uptake. We have named this group the "ZIP" gene family (for ZRT/IRT-related proteins) for the first three members to be isolated and characterized, ZRT1, ZRT2, and IRT1. *ZRT1* and *ZRT2* encode Zn transporters in *S. cerevisiae* (119, 120). *ZIP* family genes are found in a diverse array of eukaryotic organisms (eight genes in Arabidopsis, one in rice, one gene in trypanosomes, two genes in *S. cerevisiae*, four genes in nematodes, and two in humans). Based on our studies of *IRT1* and the two *ZRT* genes, we propose that the other genes in this family also function as metal transporters.

All ZIP proteins are predicted to have eight transmembrane domains. These proteins range from 309 to 476 amino acids in length; this difference is largely due to the length between transmembrane domains III and IV, designated the "variable region." This region is particularly intriguing because in all but two members this domain contains a histidine-rich motif that may serve as a metal-binding site. Similar domains have also been found in the Zn efflux transporters ZRC1 (50), COT1 (24), ZnT-1 (87), and ZnT-2 (86).

Strategy II

Strategy II plants, the grasses, release phytosiderophores, low molecular weight Fe(III)-specific ligands, in response to iron deficiency. These molecules are nonproteinogenic amino acids synthesized from methionine via nicotianamine (NA) to give mugineic acids that efficiently chelate Fe(III) with their amino and carboxyl groups (67). The Fe(III)–mugineic acid complexes are then believed to be internalized by specific transport systems that have yet to be characterized at the molecular level. A similar chelating strategy is used by a wide variety of bacteria and fungi (36).

The first step in phytosiderophore synthesis is the combination of three molecules of *S*-adenosylmethionine to form NA via NA synthase. Genes encoding *S*-adenosylmethionine synthetase have been cloned from a number of species, including Arabidopsis. NA synthase has been purified from iron-deficient barley roots, and efforts are now under way to identify cDNAs encoding this enzyme (42). Besides serving as a precursor for mugineic biosynthesis, NA is also thought to play a role in long-distance metal transport in Strategy I plants (103). Based on electrophoretic and potentiometric techniques, NA is predicted to transport Cu and Zn in the xylem and Cu, Zn, and Fe(II) in the phloem (43).

After formation of NA, NA aminotransferase (NAAT) is then thought to transfer an amino group to produce an unstable intermediate that is rapidly reduced to form deoxymugineic acid. cDNAs from barley-encoding NAAT have been identified; NAAT is strongly induced under iron deficiency (108). Using a subtractive hybridization approach, two other barley genes—*ids2* and *ids3*—that may function in synthesis of mugineic acid have been identified (83, 85). Each encodes a protein with some similarity to 2-oxoglutarate dioxygenases, making them good candidates for the conversion of deoxymugineic acid and mugineic acid to epihydroxymugineic acid via hydroxylation.

In order to identify the Fe(III)–mugineic acid transporter, a yeast *ctr1* mutant that is unable to grow on iron-deficient media was transformed with a barley cDNA expression library, and clones that could use Fe(III)–mugineic acid as an iron source were isolated. One clone, designated *SFD1* (*S*uppressor of *F*errous uptake *D*efect), can restore the ability of a *ctr1* mutant of yeast to grow on iron-deficient media when either Fe(III) mugineic acid or Fe(III) citrate are provided (114). SFD1 has no similarity to any protein of known function, and the exact mechanism by which the growth arrest is bypassed remains to be determined.

ZINC

Uptake

Zn is taken up from the soil solution as a divalent cation (71). Once taken up, Zn is neither oxidized nor reduced; thus, the role of Zn in cells is based on its behavior as a divalent cation that has a strong tendency to form tetrahedral complexes (for a review, see 9). Zn is an essential catalytic component of over 300 enzymes, including alkaline phosphatase, alcohol dehydrogenase, Cu-Zn superoxide dismutase, and carbonic anhydrase. Zn also plays a critical structural role in many proteins. For example, several motifs found in transcriptional regulatory proteins are stabilized by Zn, including the Zn finger, Zn cluster, and RING finger domains. Proteins containing these domains are very common; it has been estimated that as many as 2% of all yeast gene products contain Zn-binding domains. Despite the importance of Zn as an essential micronutrient for plant growth, relatively few studies have examined the mechanisms and regulation of Zn absorption by roots. Currently, there is little agreement on whether Zn enters via ion channels or via a divalent cation carrier and whether there is a link between uptake and metabolic energy transduction (55). Attention has mainly been focused on hyperaccumulators, i.e. plants that can grow in soils containing high levels of Zn and accumulate high concentrations of Zn in their shoots. Certain populations of *Thlaspi caerulescens* can tolerate up to 40,000 μg Zn g^{-1} tissue in their shoots; for most plants, optimal Zn concentration is between 20 and 100 μg g^{-1} tissue. Radiotracer studies with *T. caerulescens* and a nonhyperaccumulating related species, *T. arvense*, have shown that the Vmax for the uptake of Zn was 4.5-fold greater for *T. caerulescens* than for the nonhyperaccumulator *T. arvense*, while their Km values were not significantly different (61). This suggests that Zn uptake is controlled by regulating the number of transporters in the membrane (61). Once in the shoot, Zn is believed to be stored in the vacuoles of leaf cells, preventing the buildup of toxic levels in the cytoplasm.

Using a method similar to the one used to isolate *IRT1*, we isolated the *ZIP1*, *ZIP2*, and *ZIP3* genes of Arabidopsis by functional expression cloning in a *zrt1zrt2* mutant yeast strain; expression of these genes in yeast restored Zn-limited growth to this strain (N Grotz, T Fox, E Connolly, ML Guerinot, W Park & D Eide, submitted manuscript). Biochemical analysis of metal uptake has demonstrated that these genes encode Zn transporters. Yeast expressing *ZIP1*, *ZIP2*, and *ZIP3* each have a different time-, temperature-, and concentration-dependent Zn uptake activity with apparent Km values between 10 and 100 nM Zn(II). These values are similar to the levels of free Zn available in the rhizo-sphere (84). Moreover, no Fe uptake activity has been detected with any of these proteins in uptake experiments using [55]Fe. We propose that each of these three genes plays a role in Zn transport in the plant. These represent the first Zn transporter genes to be cloned from any plant species.

CONCLUSIONS

DNA-based strategies (as opposed to biochemical approaches) have success-fully identified a number of genes involved in cation transport in plants. Despite this success, we must keep in mind that such approaches may not identify all the relevant transporters. For example, yeast complementation may not allow us to identify multimeric transporters if yeast cannot provide the appropriate partner proteins. And although plant genes are well represented in several distinctive families of cation transporters such as P-type ATPases, there are currently no members of the RND (resistance/nodulation/cell division) (94), the CDF (cation diffusion facilitator) (89), or ABC (ATP-binding cassette) transporter families that have been implicated in cation transport in plants. Searching for family members will undoubtedly continue; this search will surely be facilitated by ongoing efforts to complete the sequence of Arabidopsis and other plant (rice, maize) genomes. Space limitations did not allow us to cover another possible avenue for identifying transporter genes, namely searching for mutants with transporter defects.

With many cloned genes already in hand, the obvious challenge now is to decipher the role of each of the transporters encoded by these genes. For most transporters, physiologically important information such as expression pattern and mechanism of regulation is still lacking. Various molecular approaches ultimately can tell not only in what tissue and cell types certain transporters are expressed but where within a cell each is expressed. They can also tell whether gene expression is directly influenced by changes in cation concentra-tions. We are also now in a position to identify plant mutants carrying insertions in particular transporter genes (57, 75); this will greatly help in assigning functions. Having cloned genes is also allowing us to undertake structure-function studies on the encoded proteins themselves. Many transporter activities have been well characterized at the electrophysiological level. Our ability to now combine such information with structural information about the proteins will hopefully lead to an understanding of the molecular mechanisms of transport. Finally, moving beyond how any one transporter functions, we need to keep in mind that ultimately we want to understand cation transport at the whole plant level and to use such knowledge to create plants with enhanced mineral content as well as plants that bioaccumulate or exclude toxic cations such as cadmium and lead.

ACKNOWLEDGMENTS

We thank Erin Connolly, David Eide, Natasha Grotz, and Rob McClung for critical reading of the text. We also thank numerous colleagues for sharing unpublished or prepublication materials. Research in the Guerinot lab is supported by grants from the National Science Foundation, the United States Department of Agriculture, and the Department of Energy.

> Visit the *Annual Reviews home page* at
> http://www.AnnualReviews.org.

Literature Cited

1. Anderson JA, Huprikar SS, Kochian LV, Lucas WJ, Gaber RF. 1992. Functional expression of a probable *Arabidopsis thaliana* potassium channel in *Saccharomyces cerevisiae*. *Proc. Natl. Acad. Sci. USA* 89:3736–40
2. Askerlund P. 1997. Calmodulin-stimulated Ca^{2+}-ATPases in the vacuolar and plasma membranes in cauliflower. *Plant Physiol.* 114:999–1007
3. Assmann SM, Haubrick LL. 1996. Transport proteins of the plant plasma membrane. *Curr. Opin. Cell Biol.* 8:458–67
4. Axelsen KB, Palmgren MG. 1998. Evolution of substrate specificities in the P-type ATPase superfamily. *J. Mol. Evol.* 46:84–101
5. Basset M, Conejero G, Lepetit M, Fourcroy P, Sentenac H. 1995. Organization and expression of the gene coding for the potassium transport system AKT1 of *Arabidopsis thaliana*. *Plant Mol. Biol.* 29:947–58
6. Becker D, Dreyer I, Hoth S, Reid JD, Busch H, et al. 1996. Changes in voltage activation, Cs^+ sensitivity, and ion permeability in H5 mutants of the plant K^+ channel KAT1. *Proc. Natl. Acad. Sci. USA* 93:8123–28
7. Belouchi A, Cellier M, Kwan T, Saini HS, Leroux G, Gros P. 1995. The macrophage specific membrane protein Nramp controlling natural resistance in mice has homologues expressed in the root systems of plants. *Plant Mol. Biol.* 29:1181–96
8. Belouchi A, Kwan T, Gros P. 1997. Cloning and characterization of the *Os-Nramp* family from *Oryza sativa*, a new family of membrane proteins possibly implicated in the transport of metal ions. *Plant Mol. Biol.* 33:1085–92
9. Berg JM, Shi Y. 1996. The galvanization of biology: a growing appreciation for the roles of zinc. *Science* 271:1081–85
10. Berkowitz GA, Ma J, Tang H, Zhang X, Fang Z, Vasconcelos AC. 1998. Molecular characterization of plant potassium channels. In *Frontiers in Potassium Nutrition: New Perspectives on the Effects of Potassium on Crop Plant Physiology*, ed. D Oostechnis, GA Berkowitz. Madison, Wis: Am. Soc. Agron.

11. Blatt MR. 1997. Plant potassium channels double up. *Trends Plant Sci.* 2:244–46

12. Briat J-F, Lobréaux S. 1997. Iron transport and storage in plants. *Trends Plant Sci.* 2:187–93

13. Brown AM. 1993. Functional bases for interpreting amino acid sequences of voltage-dependent K$^+$ channels. *Annu. Rev. Biophys. Biomol. Struct.* 22:173–98

14. Bush DS. 1995. Calcium regulation in plant cells and its role in signaling. *Annu. Rev. Plant Physiol. Plant Mol. Biol.* 46:95–122

15. Cao Y, Crawford NM, Schroeder JI. 1995. Amino terminus and the first four membrane-spanning segments of the *Arabidopsis* K$^+$ channel KAT1 confer inward-rectification property of plant-animal chimeric channels. *J. Biol. Chem.* 270:17697–701

16. Cao Y, Ward JM, Kelly WB, Ichida AM, Gaber RF, et al. 1995. Multiple genes, tissue specificity, and expression-dependent modulation contribute to the functional diversity of potassium channels in *Arabidopsis thaliana. Plant Physiol.* 109:1093–106

17. Cellier M, Belouchi A, Gros P. 1996. Resistance to intracellular infections: comparative genomic analysis of *Nramp. Trends Genet.* 12:201–4

18. Cellier M, Privé G, Belouchi A, Kwan T, Rodrigues V, et al. 1995. Nramp defines a family of membrane proteins. *Proc. Natl. Acad. Sci. USA* 92:10089–93

19. Chaney RL, Brown JC, Tiffin LO. 1972. Obligatory reduction of ferric chelates in iron uptake by soybeans. *Plant Physiol.* 50:208–13

20. Chanson A. 1993. Active transport of proton and calcium in higher plants. *Plant Physiol. Biochem.* 31:943–55

21. Chen XF, Chang MC, Wang BY, Wu R. 1997. Cloning of a Ca^{2+}-ATPase gene and the role of cytosolic Ca^{2+} in the gibberellin-dependent signaling pathway in aleurone cells. *Plant J.* 11:363–71

22. Chouinard SW, Wilson GF, Schlimgen AK, Ganetzky B. 1995. A potassium channel β subunit related to the aldo-keto reductase superfamily is encoded by the *Drosophila* hyperkinetic locus. *Proc. Natl. Acad. Sci. USA* 92:6763–67

23. Cohen CK, Norvell WA, Kochian LV. 1997. Induction of the root cell plasma membrane ferric reductase. *Plant Physiol.* 114:1061–69

24. Conklin DS, McMaster JA, Culbertson MR, Kung C. 1992. COT1, a gene involved in cobalt accumulation in *Saccharomyces cerevisiae. Mol. Cell. Biol.* 12:3678–88

25. Cunningham KW, Fink GR. 1994. Calcineurin dependent growth control in *Saccharomyces cerevisiae* mutants lacking *PMC1*, a homolog of plasma membrane Ca^{2+} ATPases. *J. Cell Biol.* 124:351–63

26. Cunningham KW, Fink GR. 1996. Calcineurin inhibits *VCX1*-dependent H$^+$/Ca^{2+} and induces Ca^{2+} ATPases in *Saccharomyces cerevisiae. Mol. Cell. Biol.* 16:2226–37

27. Czempinski K, Zimmermann S, Ehrhardt T, Müller-Röber B. 1997. New structure and function in plant K$^+$ channels: KCO1, an outward rectifier with a steep Ca^{2+} dependency. *EMBO J.* 16:2565–75

28. Ehrhardt T, Zimmermann S, Müller-Röber B. 1997. Association of plant K$^+$ in channels is mediated by conserved C-termini and does not affect subunit assembly. *FEBS Lett.* 409:166–70

29. Eide D, Broderius M, Fett J, Guerinot ML. 1996. A novel iron-regulated metal transporter from plants identified by functional expression in yeast. *Proc. Natl. Acad. Sci. USA* 93:5624–28

30. Evans DE, Briars S-A, Williams LE, Chanson A. 1991. Active transport of proton and calcium in higher plant cells. *J. Exp. Bot.* 42:285–303

31. Ferrol N, Bennett AB. 1996. A single gene may encode differentially localized Ca^{2+}-ATPases in tomato. *Plant Cell* 8:1159–69

32. Fleming MD, Trenor CC, Su MA, Foernzler D, Beier DR, et al. 1997. Microcytic anaemia mice have a mutation in *Nramp2*, a candidate iron transporter gene. *Nat. Genet.* 16:383–86

33. Fox TC, Shaff JE, Grusak MA, Norvell WA, Chen Y, et al. 1996. Direct measurement of 59 labeled Fe^{2+} influx in roots of *Pisum sativum* using a chelator buffer system to control free Fe^{2+} in solution. *Plant Physiol.* 111:93–100

34. Gassmann W, Rubio F, Schroeder JI. 1996. Alkali cation selectivity of the wheat root high-affinity potassium transporter HKT1. *Plant J.* 10:869–82

35. Gaymard F, Cerutti M, Horeau C, Lemaillet G, Urbach S, et al. 1996. The baculovirus/insect cell system as an alternative to *Xenopus* oocytes; first characterization of the AKT1 K$^+$ channel from *Arabidopsis thaliana. J. Biol. Chem.* 271:22863–870

36. Guerinot ML. 1994. Microbial iron transport. *Annu. Rev. Microbiol.* 48:743–72

37. Guerinot ML, Yi Y. 1994. Iron: nutritious,

noxious, and not readily available. *Plant Physiol.* 104:815–20

38. Gunshin H, Mackenzie B, Berger UV, Gunshin Y, Romero MF, et al. 1997. Cloning and characterization of a mammalian proton-coupled metal-ion transporter. *Nature* 388:482–88

39. Halliwell B, Gutteridge JMC. 1992. Biologically relevant metal ion-dependent hydroxyl radical generation. *FEBS Lett.* 307:108–12

40. Harper JF, Surowy TK, Sussman MR. 1989. Molecular cloning and sequence of cDNA encoding the membrane proton pump (H⁺-ATPase) of *Arabidopsis thaliana. Proc. Natl. Acad. Sci. USA* 86: 1234–38

40a. Harper JF, Hong B, Hwang I, Guo HQ, Stoddard R, et al. 1998. A novel calmodulin-regulated Ca²⁺-ATPase (*ACA2*) from *Arabidopsis* with an N-terminal autoinhibitory domain. *J. Biol. Chem.* 273:1099–1106

41. Hassett R, Kosman DJ. 1995. Evidence for Cu(II) reduction as a component of copper uptake by *Saccharomyces cerevisiae. J. Biol. Chem.* 270:128–34

42. Herbik A, Giritch A, Bäumlein H, Stephan UW, Mock H-P. 1997. *Characterization and purification of nicotianamine synthase.* Presented at Int. Symp. Iron Nutr. Interact. Plants, 9th, Stuttgart, Ger.

43. Hider RC, von Wiren N, Leigh R, Bansal S, Briat JF. 1997. *Physicochemical characterization of Fe-nicotianamine complexes and their physiological implications.* Presented at Int. Symp. Iron Nutr. Interact. Plants, 9th, Stuttgart, Ger.

44. Hirsch RE, Lewis B, Spalding E, Sussman MR. 1997. *Characterization of an Arabidopsis potassium channel mutant.* Presented at Int. Conf. Arabidopsis Res., 8th, Madison, Wis.

45. Hirschi KD, Zhen R-G, Cunningham KW, Rea PA, Fink GR. 1996. CAX1, an H⁺/Ca²⁺ antiporter from *Arabidopsis. Proc. Natl. Acad. Sci. USA* 93:8782–86

46. Deleted in proof

47. Huang LQ, Berkelman T, Franklin AE, Hoffman NE. 1993. Characterization of a gene encoding a Ca²⁺- ATPase-like protein in the plastid envelope. *Proc. Natl. Acad. Sci. USA* 90:10066–70

48. Ichida AM, Pei Z-M, Turner KJ, Schroeder JI. 1997. Expression of a Cs⁺ resistant guard cell K⁺ channel confers Cs⁺ resistant light-induced stomatal opening in transgenic *Arabidopsis. Plant Cell* 9(10):1843–57

49. Ichida AM, Schroeder JI. 1996. Increased resistance to extracellular cation block

by mutation of the pore domain of the *Arabidopsis* inward-rectifying K⁺ channel KAT1. *J. Membr. Biol.* 151:53–62

50. Kamizono A, Nishizawa M, Teranishi Y, Murata K, Kimura A. 1989. Identification of a gene conferring resistance to zinc and cadmium ions in the yeast *Saccharomyces cerevisiae. Mol. Gen. Genet.* 219:161–67

51. Kampfenkel K, Kushnir S, Babiychuk E, Inzé D, Van Montagu M. 1995. Molecular characterization of a putative *Arabidopsis thaliana* copper transporter and its yeast homologue. *J. Biol. Chem.* 270:28479–86

52. Ketchum KA, Slayman CW. 1996. Isolation of an ion channel gene from *Arabidopsis thaliana* using the H5 signature sequence from voltage-dependent K⁺ channels. *FEBS Lett.* 378:19–26

53. Knight SAB, Labbé S, Kwon LF, Kosman DJ, Thiele DJ. 1996. A widespread transposable element masks expression of a yeast copper transport gene. *Genes Dev.* 10:1917–29

54. Kochian LV. 1991. Mechanisms of micronutrient uptake and translocation in plants. In *Micronutrients in Agriculture*, ed. JJ Mortvedt, FR Cox, LM Shuman, RM Welch, pp. 229–96. Madison, Wis: Soil Soc. Am. 2nd ed.

55. Kochian LV. 1993. Zinc absorption from hydroponic solutions by plant roots. In *Zinc in Soils and Plants*, ed. AD Robson, pp. 45–57. Boston/Dordrecht: Kluwer

56. Kochian LV, Lucas WJ. 1988. Potassium transport in roots. *Adv. Bot. Res.* 15:93–178

57. Krysan PJ, Young JC, Tax F, Sussman MR. 1996. Identification of transferred DNA insertions within *Arabidopsis* genes involved in signal transduction and ion transport. *Proc. Natl. Acad. Sci. USA* 93: 8145–50

58. Kukuljan M, Labarca P, Latorre R. 1995. Molecular determinants of ion conduction and inactivation in K⁺ channels. *Am. J. Physiol.* 268:C535–56

59. Lagarde D, Basset M, Lepetit M, Conejero G, Gaymard F, et al. 1996. Tissue-specific expression of *Arabidopsis AKT1* gene is consistent with a role in K⁺ nutrition. *Plant J.* 9:195–203

60. Lapinskas PJ, Cunningham KW, Liu XF, Fink GR, Culotta VC. 1995. Mutations in *PMR1* suppress oxidative damage in yeast cells lacking superoxide dismutase. *Mol. Cell Biol.* 15:1382–88

61. Lasat MM, Baker AJM, Kochian LV. 1996. Physiological characterization of root Zn²⁺ absorption and translocation

to shoots in Zn hyperaccumulator and nonaccumulator species of *Thlaspi*. *Plant Physiol.* 112:1715–22

62. Lauchli A. 1990. Calcium, salinity and the plasma membrane. In *Calcium in Plant Growth and Development*, ed. RT Leonard, PK Hepler, pp. 26-35. Rockville: Am. Soc. Plant Physiol.

63. Liang F, Cunningham KW, Harper JF, Sze H. 1997. *ECA1* complements yeast mutants defective in Ca^{2+} pumps and encodes an endoplasmic reticulum-type Ca^{2+}-ATPase in *Arabidopsis thaliana*. *Proc. Natl. Acad. Sci. USA* 94:8579–84

64. Lin S-J, Pufahl RA, Dancis A, O'Halloran TV, Culotta VC. 1997. A role for the *Saccharomyces cerevisiae ATX1* gene in copper trafficking and iron transport. *J. Biol. Chem.* 272:9215–20

65. Logan DC, Venis MA. 1995. Characterisation and immunological identification of a calmodulin-stimulated Ca^{2+}- ATPase from maize shoots. *J. Plant Physiol.* 145:702–10

66. Logan H, Basset M, Véry A-A, Sentenac H. 1997. Plasma membrane transport systems in higher plants: from black boxes to molecular physiology. *Physiol. Plant.* 100:1–15

67. Ma JF, Nomoto K. 1996. Effective regulation of iron acquisition in graminaceous plants. The role of mugineic acids as phytosiderophores. *Physiol. Plant.* 97:609–17

68. Maathuis FJM, Ichida AM, Sanders D, Schroeder JI. 1997. Roles of higher plant K^+ channels. *Plant Physiol.* 114:1141–49

69. Maathuis FJM, Sanders D. 1996. Mechanisms of potassium absorption by higher plant roots. *Physiol. Plant.* 96:158–68

69a. Maathuis FJM, Verlin D, Smith FA, Sanders D, Fernandez JA, Walker NA. 1996. The physiological relevance of Na^+-coupled K^+ transport. *Plant Physiol.* 112: 1609–16

70. Malmström S, Askerlund P, Palmgren MG. 1997. A calmodulin-stimulated Ca^{2+}-ATPase from plant vacuolar membranes with a putative regulatory domain at its N-terminus. *FEBS Lett.* 400:324–28

71. Marschner H. 1995. *Mineral Nutrition of Higher Plants*. Boston: Academic. 2nd ed.

72. Marten I, Gaymard F, Lemaillet G, Thibaud J-B, Sentenac H, Hedrich R. 1996. Functional expression of the plant K^+ channel KAT1 in insect cells. *FEBS Lett.* 380:229–32

73. Deleted in proof

74. McCormack T, McCormack K. 1994. *Shaker* K^+ channel β subunits belong to an NAD(P)H-dependent oxidoreductase superfamily. *Cell* 79:1133–35

75. McKinney EC, Ali N, Traut A, Feldmann KA, Belostotsky DA, et al. 1995. Sequence-based identification of T-DNA insertion mutations in *Arabidopsis*: actin mutants *act2-1* and *act4-1*. *Plant J.* 8: 613–22

76. Michelet B, Boutry M. 1995. The plasma membrane H^+-ATPase. *Plant Physiol.* 108:1–6

77. Miller AG, Aldrich RW. 1996. Conversion of a delayed rectifier K^+ channel to a voltage-gated inward rectifier K^+ channel by three amino acid substitutions. *Neuron* 16:853–58

78. Møller JV, Juul B, le Maire M. 1996. Structural organization, ion transport and energy transduction of P-type ATPases. *Biochim. Biophys. Acta* 1286:1–51

79. Müller-Röber B, Czempinski K, Ehrhardt T, Zimmermann S. 1997. *Properties of a cloned Arabidopsis outward rectifying K^+ channel, KCO1, indicate a role in Ca^{2+}- mediated signal transduction*. Presented at Int. Conf. Arabidopsis Res., 8th, Madison, Wis.

80. Müller-Röber B, Ellenberg J, Provart N, Willmitzer L, Busch H, et al. 1995. Cloning and electrophysiological analysis of KST1, an inward rectifying K^+ channel expressed in potato guard cells. *EMBO J.* 14:2409–16

81. Nakamura RL, Anderson JA, Gaber RF. 1997. Determination of key structural requirements of a K^+ channel pore. *J. Biol. Chem.* 272:1011–18

82. Nakamura RL, McKendree WL, Hirsch RE, Sedbrook JC, Gaber RF, Sussman MR. 1995. Expression of an *Arabidopsis* potassium channel gene in guard cells. *Plant Physiol.* 109:371–74

83. Nakanishi H, Okumura N, Umehara Y, Nishizawa N-K, Chino M, Mori S. 1993. Expression of a gene specific for iron deficiency (*Ids3*) in the roots of *Hordeum vulgare*. *Plant Cell Physiol.* 34:401–10

84. Norvell WA, Welch RM. 1993. Growth and nutrient uptake by barley (*Hordeum vulgare* L. cv. Herta): studies using an N-(2-hydroxyethyl)ethylenedinitrilotriacetic acid-buffered nutrient solution technique. I. Zinc ion requirements. *Plant Physiol.* 101:619–25

85. Okumura N, Nishizawa N-K, Umehara Y, Ohata T, Nakanishi H, et al. 1994. A dioxygenase gene (*Ids2*) expressed under iron deficiency conditions in the roots of

Hordeum vulgare. Plant Mol. Biol. 25: 705–19

86. Palmiter RD, Cole TB, Findley SD. 1996. ZnT-2, a mammalian protein that confers resistance to zinc by facilitating vesicular sequestration. *EMBO J.* 15:1784–91

87. Palmiter RD, Findley SD. 1995. Cloning and functional characterization of a mammalian zinc transporter that confers resistance to zinc. *EMBO J.* 14:639–49

88. Pardo JM, Serrano R. 1989. Structure of a plasma membrane H^+-ATPase gene from the plant *Arabidopsis thaliana. J. Biol. Chem.* 264:8557–62

89. Paulsen LT, Saier MH. 1997. A novel family of ubiquitous heavy metal ion transport proteins. *J. Membr. Biol.* 156:99–103

90. Perez-Prat E, Narasimhan ML, Binzel ML, Botella MA, Chen Z, et al. 1992. Induction of a putative Ca^{2+}-ATPase mRNA in NaCl-adapted cells. *Plant Physiol.* 100:1471–78

91. Raschke BC, Wolf AH. 1996. Molecular cloning of a P-type Ca^{2+}-ATPase from the halotolerant alga *Dunaliella biocu-lata. Planta* 200:78–84

91a. Robinson NJ, Sadjuga MR, Groom QJ. 1997. The *froh* gene family from *Arabidopsis thaliana*: putative iron-chelate reductases. In *Plant Nutrition- for sustainable food production and environment*, ed. T Ando, 191–94. Kluwer Academic Pub.

92. Rubio F, Gassmann W, Schroeder JI. 1995. Sodium-driven potassium uptake by the plant potassium transporter HKT1 and mutations conferring salt tolerance. *Science* 270:1660–63

93. Sadjuga MR, Procter CM, Robinson NJ, Groom QJ. 1997. Isolation and analysis of genes encoding novel NAD(P)H oxidases from *Arabidopsis thaliana*: their putative role(s) as ferric/cupric reductases. *J. Exp. Bot.* 48:101 (Suppl.)

94. Saier MH, Tam R, Reizer A, Reizer J. 1994. Two novel families of bacterial membrane proteins concerned with nodulation, cell division and transport. *Mol. Microbiol.* 11:841–47

95. Schachtman DP, Kumar R, Schroeder JI, Marsh EL. 1997. Molecular and functional characterization of a novel low-affinity cation transporter (LCT1) in higher plants. *Proc. Natl. Acad. Sci. USA* 94:11079–84

96. Schachtman DP, Schroeder JI. 1994. Structure and transport mechanism of a high-affinity potassium uptake transporter from higher plants. *Nature* 370:655–58

97. Schachtman DP, Schroeder JI, Lucas WJ,

Anderson JA, Gaber RF. 1992. Expression of an inward-rectifying potassium channel by the *Arabidopsis KAT1* cDNA. *Science* 258:1654–58

98. Schroeder JI, Ward JM, Gassmann W. 1994. Perspectives on the physiology and structure of inward-rectifying K^+ channels in higher plants. *Annu. Rev. Biophys. Biomol. Struct.* 23:441–71

99. Sentenac H, Bonneaud N, Minet M, Lacroute F, Salmon J-M, et al. 1992. Cloning and expression in yeast of a plant potassium ion transport system. *Science* 256:663–65

100. Shi G, Nakahira K, Hammond S, Rhodes KJ, Schechter LE, Trimmer JS. 1996. β subunits promote K^+ channel surface expression through effects early in biosynthesis. *Neuron* 16:843–52

101. Silver S, Phung LT. 1996. Bacterial heavy metal resistance: new surprises. *Annu. Rev. Microbiol.* 50:753–89

102. Smart CJ, Garvin DF, Prince JP, Lucas WJ, Kochian LV. 1996. The molecular basis of potassium nutrition in plants. *Plant Soil* 187:81–89

103. Stephan UW, Schmidke I, Stephan VW, Scholz G. 1996. The nicotianamine molecule is made-to-measure for complexation of metal micronutrients in plants. *BioMetals* 9:84–90

104. Supek F, Supekova L, Nelson H, Nelson N. 1996. A yeast manganese transporter related to the macrophage protein involved in conferring resistance to mycobacteria. *Proc. Natl. Acad. Sci. USA* 93:5105–10

105. Supek F, Supekova L, Nelson H, Nelson N. 1997. Function of metal-ion homeostasis in the cell division cycle, mitochondrial protein processing, sensitivity to mycobacterial infection and brain function. *J. Exp. Biol.* 200:321–30

106. Sussman MR. 1994. Molecular analysis of proteins in the plant plasma membrane. *Annu. Rev. Plant Physiol. Plant Mol. Biol.* 45:211–34

107. Tabata K, Kashiwagi S, Mori H, Ueguchi C, Mizuno T. 1997. Cloning of a cDNA encoding a putative metal-transporting P-type ATPase from *Arabidopsis thaliana. Biochim. Biophys. Acta* 1326:1–6

108. Takahashi M, Yamaguchi H, Nakanishi H, Kanazawa K, Shioiri T, et al. 1997. *Cloning and sequencing of nicotianamine aminotransferase gene (Naat)—a key enzyme for the synthesis of mugineic acid-family phytosiderophores.* Presented at Int. Symp. Iron Nutr. Interact. Plants, 9th, Stuttgart, Ger.

109. Tang HX, Vasconcelos AC, Berkowitz

GA. 1995. Evidence that plant K$^+$ channel proteins have two different types of subunits. *Plant Physiol.* 109:327–30

110. Tang HX, Vasconcelos AC, Berkowitz GA. 1996. Physical association of KAB1 with plant K$^+$ channel α subunits. *Plant Cell* 8:1545–53

111. Thiel G, Wolf AH. 1997. Operation of K$^+$-channels in stomatal movement. *Trends Plant Sci.* 2:339–45

112. Uozumi N, Gassmann W, Cao Y, Schroeder JI. 1995. Identification of strong modifications in cation selectivity in an *Arabidopsis* inward rectifying potassium channel by mutant selection in yeast. *J. Biol. Chem.* 270:24276–81

113. Wimmers LE, Ewing NN, Bennett AB. 1992. Higher plant Ca^{2+}-ATPase: primary structure and regulation of mRNA abundance by salt. *Proc. Natl. Acad. Sci. USA* 89:9205–9

114. Yamaguchi H, Shioiri T, Mori S. 1997. *Isolation* and *characterization of a barley cDNA clone which restores growth defect of yeast ferrous uptake mutant*, ctr1. Presented at Int. Symp. Iron Nutr. Interact. Plants, 9th, Stuttgart, Ger.

115. Yao W, Hadjeb N, Berkowitz GA. 1997. Molecular cloning and characterization of the first plant K(Na)/proton antiporter (Abstract No. 999). *Plant Physiol.* 114: S200

116. Yi Y. 1995. Iron uptake in *Arabidopsis thaliana*. PhD thesis. Dartmouth Coll., Hanover, NH

117. Yi Y, Guerinot ML. 1996. Genetic evidence that induction of root Fe(III) chelate reductase activity is necessary for iron uptake under iron deficiency. *Plant J.* 10: 835–44

118. Yuan DS, Stearman R, Dancis A, Dunn T, Beeler T, Klausner RD. 1995. The Menkes/Wilson disease gene homologue in yeast provides copper to a ceruloplasmin-like oxidase required for iron uptake. *Proc. Natl. Acad. Sci. USA* 92:2632–36

119. Zhao H, Eide D. 1996. The yeast *ZRT1* gene encodes the zinc transporter of a high affinity uptake system induced by zinc limitation. *Proc. Natl. Acad. Sci. USA* 93:2454–58

120. Zhao H, Eide D. 1996. The ZRT2 gene encodes the low affinity zinc transporter in *Saccharomyces cerevisiae. J. Biol. Chem.* 271:23203–10

121. Zhen R-G, Hirschi KD, Chapman D, Fink GR, Rea PA. 1997. Cax1p: an H$^+$/Ca^{2+}, H$^+$/Cd^{2+} and/or Na$^+$/H$^+$ antiporter from *Arabidopsis*? (Abstract No. 51). *Plant Physiol.* 114:S28

122. Zhou B, Gitschier J. 1997. *hCTR1*: a human gene for copper uptake identified by complementation in yeast. *Proc. Natl. Acad. Sci. USA* 94:7481–86

Annu. Rev. Plant Physiol. Plant Mol. Biol. 1998. 49:697–725

CALMODULIN AND CALMODULIN-BINDING PROTEINS IN PLANTS

Raymond E. Zielinski

Department of Plant Biology and the Physiological and Molecular Plant Biology Program, 1201 W. Gregory Drive, University of Illinois, Urbana, Illinois 61801; e-mail: rez@uiuc.edu

KEY WORDS: calcium binding, gene expression, protein-protein interaction, regulation, signal transduction

ABSTRACT

Calmodulin is a small Ca^{2+}-binding protein that acts to transduce second messenger signals into a wide array of cellular responses. Plant calmodulins share many structural and functional features with their homologs from animals and yeast, but the expression of multiple protein isoforms appears to be a distinctive feature of higher plants. Calmodulin acts by binding to short peptide sequences within target proteins, thereby inducing structural changes, which alters their activities in response to changes in intracellular Ca^{2+} concentration. The spectrum of plant calmodulin-binding proteins shares some overlap with that found in animals, but a growing number of calmodulin-regulated proteins in plants appear to be unique. Ca^{2+}-binding and enzymatic activation properties of calmodulin are discussed emphasizing the functional linkages between these processes and the diverse pathways that are dependent on Ca^{2+} signaling.

CONTENTS

1040-2519/98/0601-0697$08.00

INTRODUCTION

A vast array of cellular responses to external stimuli in eukaryotes involve second messenger Ca^{2+} signals. Transducing these signals, integrating their effects with those of other signaling pathways, and maintaining a homeostatic balance of Ca^{2+} to minimize its cytotoxic effects are initiated by a specialized group of cellular proteins, the EF-hand calcium-modulated proteins. Calmodulin (CaM) is the most widely distributed member of this family of proteins and is thought to be a primary intracellular Ca^{2+} receptor in all eukaryotes. The hallmark of CaM's mechanism of action is that it transduces second messenger Ca^{2+} signals by binding to and altering the activity of a variety of other proteins. Two decades have passed since the description of NAD kinase activation in peas (91), which led to the discovery of CaM in plants (2). In the ensuing 20 years, Ca^{2+} has been firmly established as a second messenger in plants, and CaM has been accepted as a primary intracellular receptor for Ca^{2+} and a multifunctional regulator of protein activity. But the question "Just what functions does CaM regulate in plants?" has been more difficult to answer. In recent years, calmodulin genes (*Cam*) have been isolated and used as tools to perturb cellular Ca^{2+} homeostasis and to identify components of the signal transduction pathways that act downstream from the initial perception of second messenger Ca^{2+} by CaM. However, considerable work remains to be carried out to understand how these pathways are integrated into the whole cellular framework. As was the case when the area was last reviewed in this series (120), the preponderance of information about CaM and the EF-hand family of calcium-modulated proteins, their structures, and physiological roles has come from studies on animal species. However, since that last review, considerable information has emerged on the structures of plant CaMs and the genes encoding them, as well as the identities of the protein targets upon which they exert their regulatory effects. At the same time, significant progress has been made in understanding the mechanisms by which Ca^{2+}/CaM-mediated signaling pathways are integrated into the physiological responses of both plants and animals to external stimuli. Accordingly, the primary focus of this review is an examination of the more recent cellular and molecular approaches to understanding the structure-function relationships of CaM, its mechanisms of interaction with other proteins, the identities and functions of plant proteins whose activities are regulated by CaM, and the physiological implications for Ca^{2+}-mediated signaling this information holds.

CALMODULIN: STRUCTURES, EXPRESSION, AND MODIFICATION

Calmodulin Structures

The primary structures of CaM proteins are highly conserved across all lines of eukaryotic phylogeny (reviewed in 67). However, a surprising development is the discovery that numerous isoforms of CaM may occur within a single plant species. Plant CaMs possess four functional EF-hand Ca^{2+}-binding domains. These domains are numbered I through IV, beginning from the amino-terminus. Figure 1 shows the amino acid sequence of a typical angiosperm CaM and aligns it with the sequences of vertebrate (45) and yeast (36) CaMs. Because of the large number of plant CaM sequences that have been deduced by cDNA cloning, one plant sequence, the isoform encoded by *Arabidopsis thaliana Cam-2, -3,* and *-5* (49, 64, 80, 106) is used as a representative plant sequence for comparison. Positions of frequent substitutions in the plant sequences, arbitrarily defined as occurring in >5% of the sequences, are indicated in Figure 1. Five general conclusions can be drawn from this sequence alignment. 1. Apart from a few exceptional sequences that contain small amino- or carboxy-terminal extensions, plant CaMs are ~16.7–16.8 kDa, highly acidic proteins. 2. CaM sequences are strongly conserved across all species: Plant sequences share 91% and 61% identity with those from vertebrates and yeast, respectively, and sequence conservation among plant and algal species ranges from 84 to 100%. 3. With the exception of Cys 27, which appears to be a hallmark of higher plant CaM sequences, all Ca^{2+}-coordinating residues are conserved in plant and vertebrate CaMs. 4. Sequence substitutions in non-Ca^{2+}-coordinating residues usually are conservative and preserve regions of charged or hydrophobic character. 5. Sequence substitutions occur throughout the molecule but are most common in domain IV. Domain IV of CaM contains almost as many positions of frequent sequence substitution as domains I, II, and III combined, and it displays the largest degree of variation among CaM isoforms expressed in any one species, as well as among the plant, vertebrate, and yeast sequences. Domain IV functions in concert with the most conserved region of the molecule among plant species, domain III, to form the high affinity sites for Ca^{2+} binding (82) and to initiate interaction between CaM and its target proteins (137). The sequence variability found in this region of the molecule may function to facilitate interaction between CaM and different target protein sequences, but this idea has not been tested experimentally.

The sequence alignment shown in Figure 1 also suggests the importance of hydrophobic interaction in the mechanism of CaM function. The positions of the Phe and Met residues are conserved in all angiosperm and algal CaMs. Conservation of these residues is maintained between plant and vertebrate CaMs,

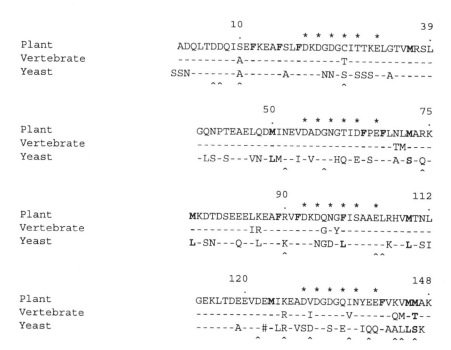

Figure 1 Comparison of the primary structures of angiosperm, vertebrate, and yeast calmodulin proteins. Protein sequences shown in this comparison were deduced from the nucleotide sequences of *Arabidopsis thaliana Cam2* (80), human (52), and *Saccharomyces cerevisiae CMD1* (36) and are displayed to show the sequence relationships of the four EF-hand domains. Amino acids are indicated in *one-letter* IUPAC nomenclature, and *dashes* indicate identical residues. Residues marked by an * indicate Ca^{2+}-binding ligands. Highly conserved phenylalanine (F) and methionine (M) residues are highlighted in *bold typeface*. The residue marked # in the yeast sequence is a deletion introduced for alignment purposes. A caret (ˆ) indicates a residue in plant calmodulins in which substitutions occur in >5% of the reported sequences. Lysine 115 (K115) is trimethylated in spinach and wheat and is presumed to be similarly modified in other plant species.

Calmodulin sequences from the following plant and algal species (and their GenBank Accession Nos.) used in this comparison are as follows: *A. thaliana* (M38379, M38380, M73711, Z12022, D45848, Z12024); *B. pilosa* (X89890); *B. juncea* (M88307); *B. napus* (U10150); *B. dioica* (L14071); *C. annuum* (X97558, X98404, U83402); *C. reinhardtii* (M20729); *D. carota* (X59751); *D. salina* (U62865); *F. sylvatica* (X97612); *G. max* (L01430, L01431, L01432, L01433, L15359); *H. annuus* (U79736); *H. vulgare* (M27303); *L. longiflorum* (L18912, Z12839); *L. esculentum* (M67472); *M. pyrifera* (X85091); *M. domestica* (X60737); *M. sativa* (X52398); *O. sativa* (L18913, L18914, X65016, Z12827, Z12828); *P. hybrida* (M80831, M80832, M80836); *P. patens* (X90560); *P. sativum* (U13882); *S. tuberosum* (J04559, U20291, U20292, U20293, U20294, U20295, U20296, U20297); *S. oleracea* (A03024); *T. aestivum* (U48242, U48688, U48689, U48690, U48691, U48692, U48693, U49103, U49104, U49105); *V. radiata* (L20507, L20691, S81594); and *Z. mays* (X74490, X77396, X77397).

with the exception of the MetMet dipeptide at positions 145/146 in plant CaMs, which is displaced one residue compared with the vertebrate proteins. With the exception of Phe 99, the location of all Phe residues is also conserved between plant and yeast CaMs. Furthermore, regions in plant CaM comprising the remaining hydrophobic residues—Leu, Ile, and Val—are highly conserved in their relative locations in all CaMs. This pattern of conservation is consistent with the finding that about 80% of the contacts between CaM and its target proteins are hydrophobic interactions rather than charge-charge interactions (reviewed in 34).

It should be noted that the criteria for comparing the proteins listed in Figure 1 as CaMs rather than CaM-like proteins are arbitrary at the present time. Proteins were eliminated from the list if they diverged by more than 20% with respect to the "nonvariant" residues in the plant CaM sequence shown in Figure 1 and included significant (>10–15 residues), unrelated amino acid sequence domains that extended from the amino- or carboxy-terminus of their CaM-like domain. By these criteria, the soybean sequences SCaM-4 and SCaM-5 (75) are near the boundary separating CaMs from CaM-like proteins. These proteins differ from the example plant CaM sequence by 20.9% and 22.3%, respectively, but by less than 20% if only the residues showing less than 5% sequence substitution are considered. The Petunia protein encoded by cDNA clone CAM53 (47) shares a high level of sequence identity with the plant sequence shown in Figure 1 but encodes a carboxy-terminal extension of 35 amino acids. As more EF-hand proteins are characterized in plants, the distinction between CaM and CaM-like proteins is likely to become less clear.

The three-dimensional structure of Ca^{2+}-replete CaM deduced by X-ray crystallography (5, 72) revealed an unusual shape for the protein as shown in Figure 2A. The EF-hand Ca^{2+}-binding domains occur in pairs embedded within two separate globular regions of the molecule, which is consistent with the cooperative kinetics of Ca^{2+} binding by CaM (82). An unexpected feature revealed in these structures, however, is the extended, solvent-exposed α-helical region joining the globular domains, which imparts a dumbbell-like shape to the molecule. This feature differs from the results of NMR (7) and biochemical studies (108), which indicated CaM has a more globular shape. Recent structures of apo-CaM solved by NMR (73, 148) (Figure 2B) show that in the absence of Ca^{2+}, CaM is considerably more globular than was suggested by the X-ray structures of the Ca^{2+}-loaded protein. In particular, the central portion of the α-helical linker domain region (designated LD in Figure 2A,B) is more flexible in the apo-CaM structure. The current view is that Ca^{2+}-loaded and apo-CaM can adopt a variety of conformations that are determined by the shape of the central α-helical linker, which acts as a region of variable expansion and contraction to allow CaM to bind different protein targets (73, 108, 148). Thus,

A   B

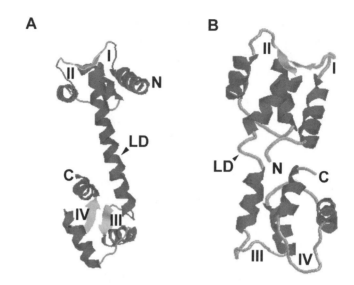

Figure 2   Structures of (*A*) Ca²⁺ calmodulin determined by X-ray crystallography (5) and (*B*) apo-calmodulin determined by NMR spectroscopy (148). The Ca²⁺-binding loops of the EF-hands are numbered (I–IV), and the amino- (N) and carboxy- (*C*) termini of the protein as well as the central linker domain (LD) are indicated in the figures. The structures were reproduced from atomic coordinates deposited in the Brookhaven Protein Data Bank. Structure codes 3CLN and 1CFD for the Ca²⁺-calmodulin and apo-calmodulin structures, respectively, are available via the World Wide Web at http://www.pdb.bnl.gov/ and were visualized using the program Rasmol (available at http://www.umass.edu/microbio/rasmol/).

CaM's three-dimensional structure should be thought of as lying somewhere between the two extreme conformations shown in Figure 2*A,B*.

Posttranslational Modifications

Plant CaMs that have been examined directly are posttranslationally trimethylated at Lys 115 (120). In all other plant species, Lys 115 is conserved and is assumed to be trimethylated. Interest in the methylation state of this residue arises from reports that CaMs lacking Lys 115 trimethylation are readily conjugated with ubiquitin in vitro (55, 104), they hyperactivate NAD kinase in vitro (125), and they accumulate differentially during development in pea roots (96) and during the culture cycle of carrot suspension cells (97). In roots and cultured cells, higher levels of CaM lacking Lys 115 trimethylation were found in cell populations containing high proportions of dividing cells. However, very low levels of the unmodified protein were detected in more highly differentiated regions of roots and in older cell cultures containing fewer dividing cells. Measurements of *Cam* mRNA and protein levels and net CaM synthesis as a function of cell culture age indicated that CaM turnover rates are higher in younger, more rapidly dividing cell populations (107). Taken together, these studies establish a correlation between low levels of CaM Lys 115 trimethylation and high rates of CaM turnover, but whether the mechanism of turnover involves ubiquitination of Lys 115 remains to be determined. The more physiologically important question, however, is what functional significance the lower levels of CaM methylation and higher rates of synthesis and turnover have in younger, more rapidly dividing populations of cells. Roberts and coworkers approached this problem by examining the effects of ectopic expression of synthetic genes encoding CaM, VU-1, and VU-3 in transgenic plants (118). VU-1 encodes Lys at position 115 and can be methylated in vitro by a CaM-specific N-methyltransferase, while VU-3 encodes a form of the protein that contains an Arg for Lys substitution at position 115 (Lys 115 Arg) and cannot be trimethylated (121). In vitro the mutant Lys 115 Arg protein is a more potent activator of NAD kinase isolated from pea, but it activates cyclic nucleotide phosphodiesterase to normal levels (121). These CaMs were expressed in transgenic tobacco under the control of the cauliflower mosaic virus 35S promoter. Steady state levels of total CaM increased approximately 1.5- to twofold over the levels in untransformed plants, with tobacco CaM expressed at near normal levels and the synthetic *transgene* expression accounting for the remainder. Transgenic lines expressing the VU-1 gene product grew normally, but in contrast, lines expressing the VU-3 (Lys 115 Arg) gene displayed shortened internodes, reduced seed production, and reduced seed and pollen viability. These results may indicate that reduced CaM methylation alters the activation of proteins that play key roles in the signaling steps required for organ differentiation. Alternatively, disruption in the pattern of CaM methylation may perturb the sequence of events of DNA replication or chromosome partitioning in meiosis, resulting in phenotypes similar to those observed as artifacts resulting from plant regeneration from tissue culture (30). This possibility is a concern because specifically timed changes in CaM gene expression occur in meiosis (124). If altered levels of CaM methylation and accumulation disrupted meiotic progression in the tobacco plants expressing VU-3 CaM, however, the effect was subtle because no differences in chromosome number were observed in the transformed plants used in this study (118).

Calmodulin Expression and Subcellular Localization

While CaM expression is ubiquitous among eukaryotic cells, concentrations of the protein can vary widely in specific cell types. CaM protein or mRNA levels generally are highest in proliferating populations of cultured cells and in plant meristematic regions (105). Higher concentrations of CaM on a total cellular protein basis have been reported in the apices of pea shoots and roots, compared with more mature regions of these tissues. Immunolocalization studies indicate that root cap cells as well as meristematic zones are specific sites of increased CaM accumulation (reviewed in 110, 120). Similarly, increased steady state levels of *Cam* mRNA have been measured in the intercalary meristematic zone of barley leaves (149), apical meristems (27, 28), differentiating siliques compared with mature leaves and roots (80), and in stolon tips (129). Studies with transgenic plants harboring chimaeric genes consisting of a plant *Cam* promoter fused to a bacterial β-glucuronidase (GUS) reporter gene are consistent with the steady state measurements of *Cam* mRNAs described above. Takezawa et al (129) found that the promoter of potato *PCM1* drives GUS expression in transgenic potato at high levels in meristematic tissues, including the shoot apex, stolon tip, and vascular tissues. Similar results were observed in transgenic tobacco using *Cam* promoters from Arabidopsis (131) and rice (27). Timme et al (131) detected *Cam3*-driven GUS activity in the root apical meristem even though *Cam3* mRNA levels were below the limits of detection in RNA gel blots and RT-PCR assays of Arabidopsis whole root RNA fractions (106). It should be noted, however, that comprehensive examinations of the cell-type patterns of expression for an entire *Cam* family in one plant species have not been reported. Thus, it cannot be excluded that there may be instances of *Cam* sequences whose expression is limited to either specific differentiated cell types or to nonproliferating tissues.

Overexpression of *Cam* sequences also has been used as a means to perturb the steady state accumulation of CaM and evaluate the protein's role in regulating plant growth and development (111). In transgenic potato plants a number of phenotypic effects were observed that correlated with the expression of a CaM-encoding *trans*-gene. The phenotypes included loss of apical dominance, shortened internodes, and the formation of tubers on aerial portions of the plants. Unfortunately, because of the variety of phenotypes observed in these plants, it is difficult to evaluate which process(es) regulated by CaM were critically affected that altered normal development.

There are few quantitative measurements of CaM concentration in vivo in either animal or plant systems. Ling & Assmann (79) examined the distribution of CaM in different organs, tissues, and protoplast types from *Vicia faba* by gel densitometry and enzyme activation assays. From their data, it can be calculated that CaM levels are in the range of 1 to 2 μM, based on the total volumes of protoplasts isolated from mesophyll, epidermal, and guard cells. Thus, CaM concentrations in the cytosol are likely to range from 5 to 20 μM. This estimate is comparable to the levels of CaM, 4 μM and 39 μM, measured in carrot cell suspensions (46) and bovine tracheal smooth muscle cells (87), respectively. Although concentrations of CaM in the cytosol are in the μM range and the K_ds for CaM interaction with its target proteins are in the nM range (34, 43, 103),

CaM accumulates at subsaturating concentration compared with the concentration of its binding sites on CaM-regulated proteins (87, 122). Therefore, competition for CaM among different CaM-binding proteins is likely to play an important role in determining the cellular responses to increases in cytosolic Ca^{2+}.

Calmodulin Genes, Ca^{2+} Signaling, and the Nucleus

Genes encoding CaM have been isolated from several plant species, including apple (142), *Arabidopsis* (23, 64, 106), potato (129), rice (27), wheat (145), and the alga *Chlamydomonas reinhardtii* (150). The higher plant genes all appear to be members of surprisingly large families of at least 6 to 12 members, while *Chlamydomonas* has a single *Cam* sequence. In all known higher plant *Cam* sequences, the coding region consists of two exons separated by a single intron that splits the codon encoding Gly 26 within the Ca^{2+}-binding loop of domain I. This intron position is conserved in *Chlamydomonas*, but here the gene contains five additional introns within the coding region. Promoter regions of the plant genes generally contain AT-rich regions and recognizable TATA box motifs, in contrast to vertebrate *Cam* 5′ flanking regions, which are GC-rich and do not contain TATA boxes (94). In vertebrates, there appear to be a limited number of *Cam* sequences (three in mouse, rat, and human) that encode multiple mRNA species, but identical polypeptides (45, 94). In these systems, the "multiple genes–one protein" hypothesis postulates that the function of the gene family is to fine tune expression to meet the specific needs of different tissues and organs for CaM. The expression pattern at the cellular level for an entire *Cam* family has not been examined for any plant species to date, and the presence of CaM isoform proteins in plants may make this analysis more complex than in animals. However, measurements of *Cam* mRNA expression in plants at the organ level (49, 80, 106, 129) are consistent with the idea that multiple genes are needed fine tune *Cam* expression in plants.

CaM is most commonly viewed as a cytosolic protein, but attention has recently been focused on CaM found in the nucleus (6), where it specifically associates with certain transcription factors (31, 128), an NTPase of unknown function (26), and histone H1 (115). Disruption of normal growth and development in transgenic mice expressing a peptide inhibitor of CaM function in the nucleus supports the physiological relevance of CaM-nuclear protein interaction and provides in vivo evidence for CaM activity in the nucleus (139). Luby-Phelps et al (87) reported that signaling levels of Ca^{2+} caused translocation of CaM into the nucleus in approximately half the smooth muscle cells loaded with fluorescently labeled CaM they imaged. Significant levels of CaM in the nuclei of barley aleurone protoplasts have been observed by immunofluorescence microscopy (123). Changes in the levels of *Cam* mRNA and protein were measured in the aleurone cells in response to treatments with GA and ABA. However, these hormone treatments had no obvious effect on the relative distribution of CaM in the cytosol and nucleus. An exciting challenge for future work lies in understanding the function of nuclear CaM and determining whether CaM pools in the nucleus and cytosol are exchangeable and respond to changes in Ca^{2+} concentration.

Signal transduction pathways involving Ca^{2+} or CaM have been implicated in changes in nuclear gene expression in response to auxin (100), gibberellic acid (24, 123), light (15, 74, 93), mechanical perturbation (14, 17, 48, 64, 95, 106, 129), and wounding (95). In many instances, changes in the expression of *Cam* sequences themselves are elicited by stimuli that trigger changes in cytosolic Ca^{2+}. It should be noted, however, that although changes in *Cam* mRNA levels have been measured in response to these stimuli, the only instance in which accompanying increases in CaM protein has been measured is in response to gibberellic acid treatment (123). The finding that *Cam* expression is enhanced by stimuli that raise cytosolic Ca^{2+} levels seems paradoxical: Why should expression of a gene encoding a receptor for a signaling molecule be induced by the molecule its gene product is produced to detect? No definitive explanation is available to answer this question, but at least two possibilities can be entertained based on the mechanisms of CaM function. CaM is needed in stoichiometric amounts to activate target proteins but appears to be produced in limiting quantities under resting conditions (87, 122). Short-term application of a stimulus may potentiate CaM accumulation in preparation for repeated or long-term stimuli that require higher levels of the protein. Alternatively, newly synthesized CaM is not methylated at Lys 115, and as such, it may selectively activate certain target proteins [e.g. NAD kinase (121)] and be turned over rapidly (107).

How could changes in intracellular Ca^{2+} concentration that are generated by extracellular stimuli lead to changes in gene expression? One mechanism of Ca^{2+} signaling in animal systems that can alter patterns of gene expression is through the transcription factors CREB (41, 90), ERK, JNK, NFAT, and NFκB (39). Activities of these transcriptional regulators are modified by phosphorylation/dephosphorylation by a CaM kinase cascade (41, 42) and the CaM-dependent protein phosphatase, calcineurin (59). At this time, however, regulation of plant transcription factor activity by CaM-dependent protein kinases or phosphatases has not been described. A second mechanism for coupling changes in cytosolic Ca^{2+} to alterations in transcription involving CaM also has been described. Several transcription factors of the basic-helix-loop-helix family bind CaM in the presence of Ca^{2+}, which inhibits their binding to DNA as homodimers and their ability to *trans*-activate gene expression (31). In plants, interaction between CaM and DNA-binding proteins, including the bZIP transcription factor TGA3 and related proteins, has been reported (128). In this case, Ca^{2+}-dependent interaction between CaM and the DNA-binding proteins enhanced their ability to bind a region of the Arabidopsis *Cam3* promoter in vitro. The challenge for future work on CaM in the plant nucleus is to correlate CaM interaction with nuclear proteins with changes in the expression of specific genes.

MECHANISMS OF CALMODULIN-MEDIATED Ca^{2+} SIGNALING

A key feature of the four EF-hands of CaM is that they respond to changes in cytosolic Ca^{2+} concentration, and in doing so allow CaM to act as a molecular switch. A crucial and often unappreciated aspect of this response mechanism is that it is dependent upon CaM's interaction with target proteins. These interactions have important consequences for the fine tuning of signal transduction and make it more appropriate to think of CaM as a tunable switch rather than a simple on and off switch.

Ca^{2+} Binding and Target Protein Interaction: The Chicken or the Egg?

Ca^{2+}-mediated signaling through CaM is commonly described as an ordered process. That is, Ca^{2+} enters the cytosol in response to a signal, it is bound by CaM, and the Ca^{2+}-CaM complex binds and activates a collection of target proteins leading to a physiological response. In this view, it is difficult to envision how Ca^{2+}/CaM can regulate such a wide array of cellular processes: The initial rate of all responses would be defined by the K_d of Ca^{2+} binding by CaM and the rate of diffusion of the complex through the cytosol. However, differential amplitudes, patchy cellular distributions, and temporal differences are now recognized to endow Ca^{2+} signals with considerable complexity (22, 134, 135). In a similar fashion, Ca^{2+}-dependent differences in the interactions between CaM and its target proteins also contribute variety and complexity to the physiological responses to increases in cytosolic Ca^{2+}.

The idea that second-messenger Ca^{2+} is perceived by CaM acting in concert with its target proteins is supported by several lines of evidence. First, saturation of the four Ca^{2+}-binding sites on CaM in vitro under conditions of physiological pH, ionic strength, and Mg^{2+} concentration requires nonphysiologically high levels of Ca^{2+} (82). Storm and colleagues observed, however, that in the presence of a target protein, Ca^{2+} binding by CaM saturates at physiologically relevant concentrations of the ion (101, 102). Under these conditions all four EF-hands bind Ca^{2+} in a cooperative fashion (63, 137) and CaM's Ca^{2+}-binding curve becomes sharply focused over the physiological range of 10^{-7} to 10^{-6} M

(Figure 3). A similar effect is reflected in the Ca^{2+} dependence of NAD kinase (77) and glutamate decarboxylase (GAD) (126) activity. Enhanced binding of Ca^{2+} by CaM also is observed in the presence of peptides comprising the CaM-binding domain of target proteins (56). The effect of CaM-binding peptides on CaM's ability to bind Ca^{2+} is so potent, in fact, that they restore the ability of proteins with mutant EF-hand Ca^{2+}-binding domains to bind the ion with normal affinity (44, 56). Since $K_d = K_{off}/K_{on}$, the effect of target proteins on Ca^{2+} binding by CaM might be to enhance K_{on} or to decrease K_{off} of the ion. Recent studies show that peptides derived from CaM-binding proteins reduce CaM's K_d for Ca^{2+} by both mechanisms (65). A crucial finding is that the enhancing effects of different target proteins on Ca^{2+} binding by CaM are not identical.

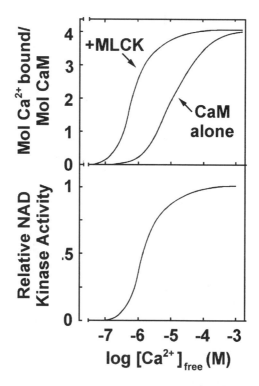

Figure 3   Calmodulin-binding proteins enhance the ability of calmodulin to bind Ca^{2+} and activate enzymes. The *top panel* shows that in the presence of myosin light-chain kinase (MLCK), Ca^{2+} binding by calmodulin occurs within the concentration range characteristic of second messenger signals (adapted from 101). The *bottom panel* shows that activation of NAD kinase occurs in a similar, but not identical, range of Ca^{2+} concentrations as Ca^{2+} binding by calmodulin-MLCK (adapted from 77).

Interaction of CaM with myosin light chain kinase (MLCK), troponin I, and cyclic nucleotide phosphodiesterase yields protein-CaM complexes having different K_ds for Ca^{2+} (102). Furthermore, kinetic studies of CaM association and dissociation with peptides derived from CaM-regulated proteins (66, 109) and model CaM-binding peptides (21) demonstrate that different CaM complexes with target protein have different association and dissociation rates. In summary, interactions between different target proteins and CaM have a variable effect on the range of Ca^{2+} concentration to which CaM responds, how rapidly different Ca^{2+}/CaM-regulated proteins are activated when a stimulus initiates a Ca^{2+} signal, and how rapidly the activities are down-regulated when the Ca^{2+} signal is attenuated. Thus, the complex interplay between Ca^{2+}, CaM, and target proteins can account for a significant portion of the diversity attributed to CaM-mediated signal transduction.

There also are well-documented Ca^{2+}-independent interactions between CaM and certain proteins. These interactions have been demonstrated for target proteins that require Ca^{2+} for activation, such as higher plant glutamate decarboxylase (3), as well as ones whose activities are down-regulated by CaM (50, 92), and ones that do not require Ca^{2+} for activation (54). Ca^{2+}-independent association of a major fraction, if not all, of the CaM in the cell with other proteins also is supported by fluorescence recovery after photobleaching experiments in smooth muscle cells (87), where it was shown that CaM within the cytosol is not freely diffusable at either resting or signaling levels of Ca^{2+}. Finally, based on thermodynamic considerations of the rates of diffusion in the cytosol, it has been suggested that CaM-target protein complexes must exist at resting Ca^{2+} concentrations to account for rapid activation in response to a signal (69).

These observations suggest that Ca^{2+} signaling through CaM is not triggered by a single step, whose initial kinetics are determined by Ca^{2+} binding by apo-CaM and Ca^{2+}/CaM diffusion through the cytosol. Rather, the available evidence indicates that Ca^{2+}-mediated signaling through CaM is initiated via several independent pathways. In this scenario, each pathway is activated with kinetics that depend on specific interactions between Ca^{2+}, CaM, and a CaM-regulated target protein. Different CaM-Ca^{2+}-target protein associations would, then, greatly increase the potential for cell-, tissue-, or organ-specific responses to stimuli, if CaM-regulated binding proteins are expressed differentially. Furthermore, the CaM-target protein interactions described above also provide a molecular mechanism for selectively decoding frequency-modulated and amplitude-modulated Ca^{2+} signals (reviewed in 12). Identifying the proteins regulated by CaM, describing their patterns of expression, and defining the Ca^{2+}-dependence of their interaction with CaM and its effect on their activities are the linchpins to understanding how so many different stimuli utilize Ca^{2+} signals and yet elicit distinctive physiological responses.

Calmodulin-Target Protein Interaction

Molecular recognition and biochemical regulation by CaM are dictated by its interaction with short peptide sequences in target proteins. An unusual feature of this interaction is that it occurs exclusively by interactions of amino acid side chains in CaM and the proteins to which it binds (34). CaM binds with high affinity to peptides ranging from 17 to 25 amino acids in length having a basic, amphiphilic α-helical structure (103). Binding is accomplished by a change in conformation of the central α-helix of CaM to a random coil. This conformational change permits the globular lobes of the protein to engulf the target peptide (reviewed in 34). When this interaction is initiated, the CaM-binding peptide can be displaced from an autoinhibitory site on the protein, as is the case with glutamate decarboxylase (126). CaM-activation also can involve other conformational changes such as multimerization (9, 58). Peptides having structures other than basic amphiphilic α-helices have recently been identified as binding targets for CaM by screening peptide display libraries (37). However, it is not clear whether any of the synthetic peptides bound by CaM in this study are representative of structures found in CaM-binding proteins.

The predominant interaction in the association between CaM and basic, amphiphilic α-helical peptides is hydrophobic (reviewed in 34, 133, 138). This mechanism was elegantly exploited using yeast genetics to show that different combinations of residues in CaM are critically required for different functions (98, 99). This conclusion is supported by in vitro studies that have examined target protein activation by site-directed mutants of CaM (138). Hydrophobic interaction is significant because it allows for tremendous primary structure flexibility, and thus, it accounts for the wide sequence variation in CaM-binding domains. No discernible amino acid sequence motif comprising a consensus CaM-binding domain has emerged from comparisons of the sequences of CaM-binding peptides and CaM-binding domains within CaM-regulated proteins (34, 103). While frustrating for the molecular biologists trying to identify and characterize CaM-binding proteins, this mechanism of interaction is crucial from the physiological perspective of creating signaling pathways having distinct and characteristic responses to second messenger Ca^{2+}.

PLANT CALMODULIN-BINDING PROTEINS

Because CaM is ubiquitously expressed and has no enzymatic activity of its own, the signaling pathways initiated by CaM and the physiological responses they elicit are derived from the expression patterns and activities of the proteins regulated by CaM. Identifying these proteins represents an area of intensive current interest in all eukaryotes. Table 1 lists the proteins in plants that have been shown to bind or whose activities have been demonstrated by some means

Table 1 Identities of known and putative plant calmodulin-regulated proteins and primary structures of their calmodulin-binding domains

Protein	Method of identification[a]	Sequence of CaM-binding domain[b]	Method of identifying CaMBD[c]	Reference[d]
Glutamate decarboxylase[d]	A, B, C	HKKTDSEVQLEMITAWKKFVEEKKKK VKKSDIDKQRDIITGWKKFVADRKKT	E S	3, 8, 147
NAD kinase	A, B	n.d.	—	2, 75, 77, 119, 121
Apyrase (Nuclear NTPase)	A, B	FNKCKNTIRKALKLNY	E	26, 61
Superoxide dismutase	B	n.d.	—	53
Kinesin heavy chain–like protein	A, B, C	ISSKEMVRLKKLVAYWKEQAGKK	E	116, 127
Elongation factor 1α	A, B	n.d.	—	40
Myosin V homolog (MYA1)	D	IQRQFRTCMAR[e]	S	68, 70
Heat-shock repressed protein	B, C	GWKIKAAMRWGFFVRKKA	E	84
Vacuolar Ca^{2+}-ATPase	A, B	ARQRWRSSVSVIVKNRARRFRMISNL	E	89
Plasmalemma Ca^{2+}-ATPase	A, B	n.d.	—	113, 114
ER Ca^{2+}-ATPase	A	n.d.	—	4, 52, 62
Slow vacuolar cation-channel	A	n.d.	—	13
CaM-dependent protein kinase II	B, C	ATPLKRLALKALSKALSEDELL	S, E	85, 143
Ca^{2+}/CaM-activated protein kinase	A, B	VVSRLRSFNARRKLRAAAIASVSLSS	E	105, 130
Phosphoprotein phosphatase 2B	A	n.d.	—	1, 86
Transcription factor TGA3	A, B	LKMLVDSCLNHYANLFRMK	C	128
FK506-binding protein	D	KIKEINKKDAKFYSNMFSKM	S	10
Root tip protein	C	GKAVVGWKIKAAMRWGIFVRKKAA	B	117
Multidrug resistant protein homolog	C	n.d.	—	141

[a] Indicates the methods that have been used to demonstrate regulation of protein function by calmodulin or calmodulin binding. The general methods are: A, activity assay; B, binding assay; C, cloning via ligand binding; D, deduced from similarity to known calmodulin-binding homologs.

[b] Amino acid sequences, given in one-letter IUPAC code, of regions demonstrated or predicted to be regions for calmodulin binding.

[c] Indicates the methods used to determine or predict the calmodulin-binding domain. The general methods are: B, predicted ability to form a basic, amphiphilic α-helix; E, experimentally determined using recombinant protein or peptide; C, computer prediction using the criteria of hydrophobic moment and mean hydrophobicity as described in Reference 43; S, sequence similarity with a calmodulin-binding domain from a homologous protein.

[d] The sequences for the calmodulin-binding domain of GAD are from Petunia and Arabidopsis.

[e] One representative example of the six potential IQ motifs in MYA1 is shown.

to be regulated by CaM. At first glance the number of plant proteins that have been identified as CaM-binding or -regulated proteins since the field was last reviewed in this series (120) is impressive. However, it should be noted that the evidence for including some of the proteins on this list is tenuous. Accordingly, particular attention should be paid to the method(s) of inferring CaM regulation. The more complementary or independent methods employed to identify a protein's interaction with or regulation by CaM, the more reliable is its listing in Table 1. As is the case in animal systems, a wide variety of enzymes, transport proteins, and cytoskeletal elements are represented among plant CaM-binding proteins. However, this list is not simply a collection of homologs to CaM-regulated proteins in animal systems. Glutamate decarboxylase (GAD) is a particularly striking illustration of this point. Alignment of the amino acid sequences of GAD from plant, animal, and bacterial sources reveals commonly shared elements in the catalytic portions of the enzymes, but no identity or similarity in the region of the CaM-binding domain, which is present as a carboxy-terminal extension on the plant enzyme (3, 8).

Conspicuously underrepresented, relative to animal systems, are CaM-regulated protein kinases in plants. These enzymes are prominent among the protein kinases in animal systems and include a kinase cascade in which all members are regulated by Ca^{2+} and CaM (42). Interactive cloning has produced reasonable evidence for plant homologs of the multifunctional Ca^{2+}/CaM-dependent protein kinase II (85, 143), but no enzymatic data have been published to confirm that CaM, in fact, regulates the activities of the proteins encoded by these cDNA clones. A novel finding in plants, however, is the Ca^{2+}/CaM-regulated protein kinase described by Poovaiah and colleagues (105, 130). This protein resembles the CaM-like domain protein kinase (CDPK) (120) in its overall structure, but unlike CDPK, requires CaM for its activity (130). Expression of the Ca^{2+}/CaM-regulated protein kinase at the level of mRNA is spatially limited, with transcripts detectable only in developing anther tissues (105). What role this kinase plays in anther development or microsporogenesis, however, remains to be determined. Pharmacological evidence implicates the activity of a CaM kinase II homolog in the mechanism of stomatal closure (32) as well as in the mechanism of growth reorientation in response to gravity (83). Ca^{2+} signaling has been demonstrated to play an important role in both of these plant responses (22). Unfortunately, substrates for either the CaM kinase II homologs or the Ca^{2+}-CaM-dependent kinase have yet to be identified. The guard cell system has also yielded tantalizing results in the search for a potential CaM-regulated protein phosphatase activity homologous to the type IIB phosphatase, calcineurin. Pharmacological evidence implicates the activity of a CaM-regulated phosphatase activity in regulating ion fluxes across both the plasma membrane (86) and the tonoplast (1) of guard cells. However, biochemical identification of a such an activity has been elusive (125).

Biochemical studies have provided evidence for CaM-regulated Ca^{2+}-ATPase activities in plant cells. Ca^{2+}-transport activities have been characterized on various membrane fractions, but whether the ATPases driving the transport are regulated by CaM and the identities of the membranes from which the CaM-regulated activities are derived have been controversial. It seems most likely that there are tissue and species differences in the numbers, locations, and mechanisms of regulation of this family of transport proteins. Ca^{2+}-ATPases play an important role in Ca^{2+} homeostasis by restoring concentrations of the ion to resting levels following a signal-induced increase in cytosolic Ca^{2+} (reviewed in 19). This transport drives Ca^{2+} extrusion from the cytosol against a large difference in electrochemical potential. Current evidence supports a widespread distribution of CaM-regulated Ca^{2+}-ATPase activities in plant cells, including transporters on the plasmalemma (113, 114), tonoplast (89), and ER (4, 52, 62) membranes. Evidence that CaM-regulated activities play a role in maintaining cytosolic Ca^{2+} levels comes from the effects of a CaM inhibitor on cytosolic Ca^{2+} concentrations measured by imaging (51). Treatment with CaM inhibitors resulted in a slow rise in cytosolic Ca^{2+}, consistent with the idea that CaM-stimulated Ca^{2+}-pumping is required to maintain low resting levels of the ion. Therefore, it appears that, although the locations of the Ca^{2+}-transport proteins may differ somewhat in different plant tissues, plants possess a CaM-regulated system of Ca^{2+} homeostasis similar to that of animals.

There is a wide array of protein sequences that are known, or presumed, to bind CaM in plant proteins (Table 1). Identification of these domains have produced some surprising findings. (*a*) Two cDNA clones have been isolated encoding proteins with nearly identical CaM-binding domains. Unfortunately, the functions of the tobacco heat shock-repressed protein (84) and the maize root tip protein (117) are unknown. Based on the previous discussion of the influence of target proteins on Ca^{2+} binding by CaM, these two proteins should trigger signaling pathways initiated by similar types of Ca^{2+} signals. (*b*) GAD and the transcription factor TGA3 possess acidic residues in their CaM-binding domains, which have been demonstrated directly and predicted by computer, respectively. The region from GAD is not predicted to be a particularly strong α-helix former. Nevertheless, Vogel and co-workers have presented preliminary evidence that a peptide comprising the CaM-binding domain of GAD binds CaM with a 1:1 stoichiometry and is induced to form an α-helix upon binding Ca^{2+}/CaM (147). (*c*) Trp residues, long thought to be one of the hallmarks, although not absolute requirements, for CaM binding (103), are found in only about half of the CaM-binding domains of the plant proteins.

While CaM regulation of the 19 proteins listed in Table 1 ranges in certainty from tenuous to well documented evidence, protein gel blot assays using labeled CaM as a ligand probe against proteins extracted from a variety of plant (78, 79) and algal (18) tissues indicate the number of CaM-binding proteins is considerably larger. The challenge in the field over the next few years will be to determine how many proteins are regulated by CaM. Determining whether CaM functions in a specific signaling pathway or regulates key activities leading to a particular physiological response presents a special challenge because of the multifunctional nature of CaM, its ubiquitous distribution, and the large family of CaM-like Ca^{2+}-binding proteins in eukaryotes. Identifying specific proteins that either bind CaM or whose activities are modulated by CaM in vitro is one of three strategies that have been used to elucidate CaM function. Of the two remaining approaches, pharmacological strategies that utilize inhibitors of CaM function have been reviewed extensively (110, 120) and will not be addressed here. The last section of this review focuses on the use of molecular tools to manipulate or visualize CaM activity or the activity of CaM-regulated proteins.

Molecular Dissection of Calmodulin Function

Cloning of sequences encoding CaM and CaM-binding proteins has facilitated the development of molecular tools for dissecting CaM-mediated signaling pathways in plants and the molecular mechanisms of CaM action. The ability to produce tens of milligrams of purified, recombinant CaM, CaM isoforms, and site-directed mutants (reviewed in 78) has expanded the analyses of CaM structure-function relationships and greatly facilitated the identification of CaM-regulated proteins by interactive cloning. Expression of CaM-binding proteins in vitro and in vivo in transgenic plants is allowing further dissection of CaM-mediated signaling pathways. This section describes several approaches to elucidating CaM function focusing on three different targets of CaM regulation.

NAD KINASE   NAD kinase has neither been purified to homogeneity nor cloned, which would permit the structure of its CaM-binding domain to be ascertained. Nevertheless, this enzyme has served an important role in examining CaM's structure-function relationships because it is sensitive to a number of sequence changes in CaMs isolated from different organisms (119) and in recombinant plant CaM isoforms (75, 77) and to site-directed mutations in different domains of recombinant CaM (33, 76, 77), including the nonmethylatable Lys 115 Arg mutation (121). Perhaps most surprising is the finding that amino acid sequence changes and deletions in the central α-helical linker domain of CaM were shown to have a large effect on NAD kinase activation but little impact on its ability to activate cyclic nucleotide phosphodiesterase and MLCK (33). At the same time, NAD kinase activation is also sensitive to mutations in both domain I (76) and the extreme carboxy-terminal region of domain IV (77) of CaM. These studies demonstrate a sensitivity of the protein to sequence changes throughout the CaM molecule that are unprecedented in other CaM-regulated proteins. An explanation for the broad sensitivity of NAD kinase activity to changes in CaM may be that the flexible linker domain between domains II and III of CaM (108) is maximally extended to accommodate interaction with NAD kinase. From a CaM structure-function perspective, future studies of CaM-NAD kinase interaction once the sequence of this target protein's CaM-binding domain is determined should be exciting because they may begin to define precisely some of the structural limits governing CaM's interaction with other proteins.

Although NAD kinase has been an informative enzyme from the perspective of CaM-target protein interaction, getting a clear handle on its physiological role and why its activity should be tightly regulated has been more difficult. Developmental changes and environmental signals, such as fertilization and pathogen attack, that increase metabolic demand for NADP have been proposed as physiological rationale for the regulation of NAD kinase activity (reviewed in 120). However, no clear-cut relationship between NAD kinase activity and Ca^{2+} signaling had been observed in a plant system. Recently, however, Roberts and co-workers provided evidence that NAD kinase activation plays an important role in the generation of active oxygen species (57). NADPH is needed for the production of superoxide and hydrogen peroxide, which are believed to be used to combat invading pathogens (16, 38). To test the idea that CaM-regulated NAD kinase provides the NADP needed for NADPH production, cultured tobacco cells expressing the VU-3 (Lys 115 Arg) form of CaM were tested for their ability to generate active oxygen species (57). The cells were challenged with stimuli known to increase cytosolic Ca^{2+} and active oxygen species production, and their response was compared with controls expressing only normal CaM. As predicted from the in vitro activation of NAD kinase (121), cells expressing the VU-3 CaM displayed an active oxygen burst that was more rapid and more intense than normal control cells challenged with the same stimuli. Larger increases in NADPH level in the VU-3-expressing cells that coincided with the onset of the active oxygen species burst also were measured. These data strongly support the operation of a Ca^{2+}-signaling pathway involving CaM and NAD kinase that activates a plant defense response pathway.

The cellular responses that balance the accumulation of active oxygen species may also involve at least one other CaM-regulated enzyme activity. At least one form of superoxide dismutase (SOD) was tentatively identified as a CaM-regulated protein based on the Ca^{2+}-dependent, quantitative retention of SOD activity extracted from germinating maize seeds on CaM-Sepharose affinity columns (53). Unfortunately, however, the effect of Ca^{2+} and CaM on SOD activity was not described in this report. A previous study of Ca^{2+} signaling during oxidative stress described a down-regulation of SOD activity, but the mechanism accounting for the SOD activity decrease was not explored (112). Thus, the involvement of CaM in regulating SOD activity in response to stress (16) is an interesting field for future study.

GLUTAMATE DECARBOXYLASE (GAD)   GAD catalyzes the formation of γ-amino butyric acid (GABA) and CO_2 from glutamic acid. GABA is well known as an inhibitory neurotransmitter in animals, but its role in plant metabolism is unclear. GABA is produced in response to a variety of stresses and has been hypothesized to participate in a variety of cellular functions (9, 20, 81). In contrast, the observations that GAD mRNA and protein levels are developmentally regulated in different organs of Petunia (25) and in the developing roots and shoots of fava bean seedlings (81) have been suggested as evidence that GAD plays a role in plant development (9). Cloned cDNAs encoding GAD have been used to demonstrate that Ca^{2+} and CaM play an important role in regulating GAD activity and GABA accumulation in vivo. Activity of recombinant GAD purified to apparent homogeneity was stimulated nearly 100-fold by Ca^{2+} and nM levels of CaM (126). Perhaps a more significant finding, however, was that the stimulatory effect of CaM is most pronounced in the physiological pH range. Previously, it had been difficult to explain why GAD activity is optimal at nonphysiologically acidic pH values, where the activity is essentially CaM-independent. This observation led to the suggestion that GAD activity is triggered in response to cytosolic acidification as part of a mechanism to restore pH balance (20). However, CaM's effect on the activity of recombinant GAD argues against a role for GABA accumulation strictly in response to changes in pH, but rather, supports the idea that GAD is activated by signals that trigger changes in cytosolic Ca^{2+} that may not involve changes in pH. To explore what the developmental roles of GAD are in plants, From and coworkers constructed transgenic tobacco that overexpress both wt GAD and a truncated form of the enzyme (GADΔC) that lacks its CaM-binding domain under the direction of the CaMV 35S promoter (9). Their experiments showed a relationship between the metabolic pools of glutamate and GABA and established GAD as an important enzyme in controlling the flux of carbon and nitrogen between these pools. For both the wt GAD and GADΔC, overexpression of the protein resulted in increased GABA levels with concomitant decreases in glutamate accumulation. This effect was especially pronounced in plants overexpressing GADΔC, where the pool of glutamate was reduced to 2 to 10% of the wt level. Together with co-immunoprecipitation and enzyme activation experiments using extracts from normal and transgenic plant lines, comparison of the wt GAD and GADΔC activities offers strong evidence that CaM is the endogenous regulator of GAD activity. At the level of plant growth and development, the GADΔC-expressing plants displayed a number of abnormalities. However, whether the developmental effects observed in these experiments were a consequence of signaling or enhanced accumulation of GABA, or a consequence of the depletion of the glutamate pool cannot be determined precisely. Nevertheless, the approach of uncoupling CaM regulation from the activity of CaM-regulated proteins offers a powerful avenue for establishing the in vivo relevance of CaM-target protein interaction and dissecting the physiological roles of CaM-binding proteins.

CYTOSKELETAL PROTEINS   A variety of critical cellular functions, including vesicle trafficking, chromosome segregation, tip growth, and cell shape changes, involve the cytoskeleton. Ca^{2+} is thought to be an important regulator in orchestrating the behavior of both microtubules and actin microfilaments, and a growing body of evidence implicates CaM as an important receptor linking changes in Ca^{2+} with cytoskeletal function. Three sites of interaction between CaM and the plant cytoskeleton have been directly established or inferred by homology with animal systems. CaM has been shown to associate with plant mitotic spindles (136, 144), although the proteins mediating CaM interaction with these structures remain unidentified as are the effects of CaM on chromosome segregation. CaM also is associated with cortical microtubule components of the plant cytoskeleton (35). The Ca^{2+}-dependence of CaM's microtubule association suggests an amplitude-modulated interplay between CaM and at least two different sets of CaM-binding microtubule associated proteins (MAPs) (40, 46). One MAP target of CaM regulation is elongation factor (EF) 1-α, a protein that displays multiple functions beyond its role as a regulator of translation. CaM binds EF1α directly and inhibits its ability to bundle microtubules in vitro (40). A second class of MAPs bind CaM only in the presence of relatively low (<320 nM) levels of Ca^{2+} including several proteins ranging from 29 to 80 kDa (46). These proteins appear to be MAPs that, together with CaM, stabilize microtubules. Clearly, identifying CaM-binding MAPs represents an important step that will facilitate further analyses of the effects of CaM on microtubule function. Interestingly, one of the multiple functions demonstrated for EF1α is as a regulator of phosphatidylinositol 4-kinase in carrot (146). This enzyme is responsible for the formation of phosphatidylinositol 4-monophosphate (PI-4-P). PI-4-P is the precursor to phosphatidylinositol 4,5-bisphosphate, the source of the second messengers inositol 1,4,5-trisphosphate and diacylglycerol (11). A key question is whether all EF1α proteins can enhance translation, interact with the cytoskeleton and catalyze phosphorylation, and perhaps more importantly, whether CaM plays a role in determining EF1α's commitment to these alternative activities.

Recently, a novel link between CaM and motor proteins associated with the plant cytoskeleton was identified. This involves a kinesin heavy chain–like protein possessing a CaM-binding domain, which appears to be unique to the plant kingdom (116, 140). In vitro microtubule gliding assays indicated that the CaM-binding kinesin heavy chain functions as a minus-end directed microtubule motor whose activity, like the microtubule bundling activity of EF-1α, is inhibited by CaM (127). Unfortunately, in this study the effect of Ca^{2+} on gliding activity was observed at concentrations higher than those determined to be critical for microtubule stabilization (46). The question of Ca^{2+}-dependence in this system is interesting because the effect of CaM on microtubule gliding mediated by the CaM-binding kinesin heavy chain could be similar to the effect of CaM on myosin I (and V) motor activity on actin microfilaments. Myosins I and V bind CaM as their regulatory light chains by a mechanism similar to the interaction of CaM with basic, amphiphilic α-helices involving a structural motif in myosin known as the IQ domain (60). For myosin I, increases in Ca^{2+} cause CaM dissociation from the IQ domain. This dissociation inhibits heavy chain ATPase activity and motility (29). Plant homologs of nonmuscle myosins have been identified in actin-based motility assays (71) and by cloning (68, 70). The deduced amino acid sequences of plant myosins contain several potential IQ domains. However, no reports have appeared describing whether these proteins interact with CaM or CaM-like proteins from plants.

Although they are in their early stages, these studies indicate that the interactions between Ca^{2+}/CaM and the plant cytoskeleton are complex, and exciting progress is being made in identifying the proteins through which CaM regulation of cytoskeletal function is mediated. The interactions between MAPs and CaM and the action of CaM on motor activity suggest that the cytoskeleton may represent an important system for examining amplitude-modulated signaling in plant cells. Growing pollen tubes represent a particularly attractive system for examining CaM regulation of cytoskeletal function, since Ca^{2+} is known to play a key regulatory role (88), and CaM has been shown to be localized near the growing tube tip (132).

SUMMARY AND FUTURE PERSPECTIVES

In spite of the inherent difficulties in studying the function of a pleiotropic regulatory protein, insights into the functions of CaM in plants are emerging. Distinctive differences have been described between the plant and animal complements of CaM isoforms, the locations of CaM-regulated Ca^{2+}-ATPases, and the identities of many of the downstream targets of CaM regulation. Accordingly, there are several priorities for progress in plant research on CaM-mediated

Ca^{2+} signaling. First, because CaM has no activity of its own, and even Ca^{2+} binding by CaM in the physiological range depends upon its interaction with target proteins, identification of more of the targets of CaM regulation in plants is of paramount importance. Second, the Ca^{2+} sensitivities of CaM-target protein interaction hold valuable information on how Ca^{2+} signals can be transduced to give diverse physiological responses. Unfortunately, information on the sensitivities of specific target protein–CaM complexes for Ca^{2+} is sparse in plant systems. Coupled with continued progress in imaging Ca^{2+} signals in plants, this information should provide new insights into the mechanisms of amplitude and frequency-modulated signaling. Third, it is still unclear how large the families of genes encoding CaM and CaM-like proteins are in the plant kingdom. Identifying and expressing CaM isoforms and CaM-like proteins, determining their patterns of expression, and measuring their abilities to bind different protein substrates and Ca^{2+} will not only identify all the players in the game but provide a better understanding of how these protein-protein interactions are integrated into the framework of plant growth and development. Fourth, genetic strategies based on expression of target proteins whose interaction with CaM is perturbed or eliminated should be pursued further to determine the in vivo relevance of specific CaM-mediated pathways in different physiological responses. A major challenge in this field in the next few years will be to elucidate the mechanisms by which CaM-mediated Ca^{2+} signaling cross talks and is integrated with the multitude of other signaling pathways that are active in any given plant cell.

Visit the *Annual Reviews home page* at
http://www.AnnualReviews.org.

Literature Cited

1. Allen GJ, Sanders D. 1995. Calcineurin, a type 2B protein phosphatase, modulates the Ca^{2+}-permeable slow vacuolar ion channel of stomatal guard cells. *Plant Cell* 7:1473–83
2. Anderson JM, Charbonneau H, Jones HP, McCann RO, Cormier MJ. 1980. Characterization of the plant nicotinamide adenine dinucleotide kinase activator protein and its identification as calmodulin. *Biochemistry* 19:3113–20
3. Arazi T, Baum G, Snedden WA, Shelp BJ, Fromm H. 1995. Molecular and biochemical analysis of calmodulin interactions with the calmodulin-binding domain of plant glutamate decarboxylase. *Plant Physiol.* 108:551–61
4. Askerlund P, Evans DE. 1992. Reconstitution and characterization of a calmodulin-stimulated Ca^{2+}-pumping ATPase purified from *Brassica oleracea* L. *Plant Physiol.* 100:1670–81
5. Babu YS, Bugg CE, Cook WJ. 1988. Structure of calmodulin refined at 2.2A resolution. *J. Mol. Biol.* 204:191–204
6. Bachs O, Agell N, Carafoli E. 1994. Calmodulin and calmodulin-binding proteins in the nucleus. *Cell Calcium* 16:289–96
7. Barbato G, Ikura M, Kay LE, Pastor RW, Bax A. 1992. Backbone dynamics of calmodulin studied by ^{15}N relaxation using inverse detected two-dimensional NMR spectroscopy: the central helix is flexible. *Biochemistry* 31:5269–67
8. Baum G, Chen YL, Arazi T, Takatsuji

H, Fromm H. 1993. A plant glutamate decarboxylase containing a calmodulin-binding domain: cloning, sequence, and functional analysis. *J. Biol. Chem.* 268: 19610–17

9. Baum G, Levyadun S, Fridmann Y, Arazi T, Katsnelson H, et al. 1996. Calmodulin binding to glutamate decarboxylase is required for regulation of glutamate and GABA metabolism and normal development in plants. *EMBO J.* 15:2988–96

10. Belcher O, Erel N, Callebaut I, Aviezer K, Breiman A. 1996. A novel plant peptidyl-prolyl-cis-trans-isomerase (PPIase): cDNA cloning, structural analysis, enzymatic activity and expression. *Plant Mol. Biol.* 32:493–504

11. Berridge MJ. 1987. Inositol trisphosphate and diacylglycerol: two interacting second messengers. *Annu. Rev. Biochem.* 56:159–93

12. Berridge MJ. 1997. The AM and FM of calcium signaling. *Nature* 386:759–60

13. Bethke PC, Jones RL. 1994. Ca^{2+}-calmodulin modulates ion channel activity in storage protein vacuoles of barley aleurone. *Plant Cell* 6:277–85

14. Botella JR, Arteca RN. 1994. Differential expression of two calmodulin genes in response to physical and chemical stimuli. *Plant Mol. Biol.* 24:757–66

15. Bowler C, Neuhaus G, Yamagata H, Chua N-H. 1994. Cyclic GMP and calcium mediate phytochrome phototransduction. *Cell* 77:73–81

16. Bowler C, Van Montagu M, Inzé D. 1992. Superoxide dismutase and stress tolerance. *Annu. Rev. Plant Physiol. Plant Mol. Biol.* 43:83–116

17. Braam J, Davis RW. 1990. Rain-, wind-, and touch-induced expression of calmodulin and calmodulin-related genes in Arabidopsis. *Cell* 60:357–64

18. Brawley SH, Roberts DM. 1989. Calmodulin binding proteins are developmentally regulated in gametes and embryos of fucoid algae. *Dev. Biol.* 131:313–20

19. Briskin DP. 1990. Ca^{2+}-translocating ATPase of the plant plasma membrane. *Plant Physiol.* 94:397–400

20. Brown A, Shelp BJ. 1989. The metabolism and physiological roles of 4-aminobutyric acid. *Biochem. Life Sci. Adv.* 8:21–25

21. Brown SE, Martin SR, Bayley PM. 1997. Kinetic control of the dissociation pathway of calmodulin-peptide complexes. *J. Biol. Chem.* 272:3389–97

22. Bush DS. 1995. Calcium regulation in plant cells and its role in signaling.

Annu. Rev. Plant Physiol. Plant Mol. Biol. 46:95–122

23. Chandra A, Thungapathra M, Upadhyaya KC. 1994. Molecular cloning and characterization of a calmodulin gene from *Arabidopsis thaliana. J. Plant Biochem. Biotechnol.* 3:31–35

24. Chen XF, Chang MC, Wang BY, Wu R. 1997. Cloning of a Ca^{2+}-ATPase gene and the role of cytosolic Ca^{2+} in the gibberellin-dependent signaling pathway in aleurone cells. *Plant J.* 11:363–71

25. Chen YL, Baum G, Fromm H. 1994. The 58-kilodalton calmodulin-binding glutamate decarboxylase is a ubiquitous protein in petunia organs and its expression is developmentally regulated. *Plant Physiol.* 106:1381–87

26. Chen Y-R, Datta N, Roux SJ. 1987. Purification and partial characterization of a calmodulin-stimulated nucleoside triphosphatase from pea nuclei. *J. Biol. Chem.* 262:10689–94

27. Choi YJ, Cho EK, Lee SI, Lim CO, Gal SW, et al. 1996. Developmentally regulated expression of the rice calmodulin promoter in transgenic tobacco plants. *Mol. Cells* 6:541–46

28. Chye ML, Liu CM, Tan CT. 1995. A cDNA clone encoding Brassica calmodulin. *Plant Mol. Biol.* 27:419–23

29. Collins K, Sellers JR, Matsudaira P. 1990. Calmodulin dissociation regulates brush border myosin I (110-kD-calmodulin) mechanochemical activity in vitro. *J. Cell Biol.* 110:1137–47

30. Contolini CS, Hughes KW. 1989. Reciprocal differences in intraspecific crosses of tobacco result from embryo death. *Am. J. Bot.* 76:6–13

31. Corneliussen B, Holm M, Waltersson Y, Onions J, Thornell A, Grundstrom T. 1994. Calcium/calmodulin inhibition of basic helix-loop-helix transcription factor domains. *Nature* 368:760–64

32. Cotelle V, Forestier C, Vavasseur A. 1996. A reassessment of the intervention of calmodulin in the regulation of stomatal movement. *Physiol. Plant.* 98:619–28

33. Craig TA, Watterson DM, Prendergast FG, Haiech J, Roberts DM. 1987. Site-specific mutagenesis of the α-helices of calmodulin; effects of altering a charge cluster in the helix that links the two halves of calmodulin. *J. Biol. Chem.* 262:3278–84

34. Crivici A, Ikura M. 1995. Molecular and structural basis of target recognition by calmodulin. *Annu. Rev. Biophys. Biomol. Struct.* 24:85–116

35. Cyr R. 1991. Calcium/calmodulin affects

microtubule stability in lysed protoplasts. *J. Cell Sci.* 100:311–17

36. Davis TN, Urdea MS, Masiarz FR, Thorner J. 1986. Isolation of the yeast calmodulin gene: calmodulin is an essential protein. *Cell* 47:423–31

37. Dedman JR, Kaetzel MA, Chan HC, Nelson DJ, Jamieson GA. 1993. Selection of targeted biological modifiers from a bacteriophage library of random peptides: the identification of novel calmodulin regulatory peptides. *J. Biol. Chem.* 268:23025–30

38. Doke N, Miura Y, Sanchez LM, Park HJ, Noritake T, et al. 1996. The oxidative burst protects plants against pathogen attack—mechanism and role as an emergency signal for plant bio-defense—a review. *Gene* 179:45–51

39. Dolmetsch RE, Lewis RS, Goodnow CC, Healy JI. 1997. Differential activation of transcription factors induced by Ca^{2+} response amplitude and duration. *Nature* 386:355–58

40. Durso NA, Cyr RJ. 1994. A calmodulin-sensitive interaction between microtubules and a higher plant homologue of elongation factor-1α. *Plant Cell* 6:893–905

41. Enslen H, Sun PQ, Brickley D, Soderling SH, Klamo E, Soderling TR. 1994. Characterization of Ca^{2+}/calmodulin-dependent protein kinase IV: role in transcriptional regulation. *J. Biol. Chem.* 269:15520–27

42. Enslen H, Tokumitsu H, Soderling TR. 1995. Phosphorylation of CREB by CaM-kinase IV activated by CaM-kinase IV kinase. *Biochem. Biophys. Res. Commun.* 207:1038–43

43. Erickson-Viitanen S, DeGrado WF. 1987. Recognition and characterization of calmodulin-binding sequences in peptides and proteins. *Methods Enzymol.* 139:455–78

44. Findlay WA, Martin SR, Beckingham K, Bayley PM. 1995. Recovery of native structure by calcium binding site mutants of calmodulin upon binding of sk-MLCK target peptides. *Biochemistry* 34:2087–94

45. Fischer R, Koller M, Flura M, Mathews S, Strehler-Page MA, et al. 1988. Multiple divergent mRNAs code for a single human calmodulin. *J. Biol. Chem.* 263:17055–62

46. Fisher DD, Gilroy S, Cyr RJ. 1996. Evidence for opposing effects of calmodulin on cortical microtubules. *Plant Physiol.* 112:1079–87

47. Fromm H, Carlenor E, Chua N-H.

1991. Molecular characterzation of petunia cDNAs encoding calmodulins and a calmodulin-related protein. GenBank Access. Nos. M80831, M80832, M80836

48. Galaud JP, Lareyre JJ, Boyer N. 1993. Isolation, sequencing and analysis of the expression of Bryonia calmodulin after mechanical perturbation. *Plant Mol. Biol.* 23:839–46

49. Gawienowski MC, Szymanski D, Perera IY, Zielinski RE. 1993. Calmodulin isoforms in Arabidopsis encoded by multiple divergent mRNAs. *Plant Mol. Biol.* 22:215–25

50. Gerendasy DD, Herron SR, Jennings PA, Sutcliffe JG. 1995. Calmodulin stabilizes an amphiphilic α-helix within RC3/neurogranan and GAP-43/neuromodulin only when Ca^{2+} is absent. *J. Biol. Chem.* 270:6741–50

51. Gilroy S, Hughes WA, Trewavas AJ. 1987. Calmodulin antagonists increase free cytosolic calcium levels in plant protoplasts in vivo. *FEBS Lett.* 212:133–37

52. Gilroy S, Jones RL. 1993. Calmodulin stimulation of unidirectional calcium uptake by the endoplasmic reticulum of barley aleurone. *Planta* 190:289–96

53. Gong M, Li Z-G. 1995. Calmodulin-binding proteins from *Zea mays* germs. *Phytochemistry* 40:1335–39

54. Greenlee DV, Andreasen TJ, Storm DR. 1982. Calcium-independent stimulation of *Bordetella pertussis* adenyl cyclase by calmodulin. *Biochemistry* 21:2759–64

55. Gregori L, Marriott D, Putkey JA, Means AR, Chau V. 1987. Bacterially synthesized vertebrate calmodulin is a specific substrate for ubiquitination. *J. Biol. Chem.* 262:2562–67

56. Haiech J, Kilhoffer M-C, Lukas TJ, Craig TA, Roberts DM, Watterson DM. 1991. Restoration of the calcium binding activity of mutant calmodulins toward normal by the presence of a calmodulin binding structure. *J. Biol. Chem.* 266:3427–31

57. Harding SA, Oh S-H, Roberts DM. 1997. Transgenic tobacco expressing a foreign calmodulin gene shows an enhanced production of active oxygen species. *EMBO J.* 16:1137–44

58. Hellermann GR, Solomonson LP. 1997. Calmodulin promotes dimerization of the oxygenase domain of human endothelial nitric oxide synthase. *J. Biol. Chem.* 272:12030–34

59. Hiraga K, Suzuki K, Tsuchiya E, Miyakawa T. 1993. Identification and characterization of nuclear calmodulin-binding proteins of *Saccharomyces cerevisiae*. *Biochim. Biophys. Acta* 1177:25–30

60. Houdusse A, Cohen C. 1995. Target sequence recognition by the calmodulin superfamily: implications from light chain binding to the regulatory domain of scallop myosin. *Proc. Natl. Acad. Sci. USA* 92:10644–47

61. Hsieh H-L, Tong C-G, Thomas C, Roux SJ. 1996. Light-modulated abundance of an mRNA encoding a calmodulin-regulated, chromatin-associated NTPase in pea. *Plant Mol. Biol.* 30:135–47

62. Hsieh W-L, Pierce WS, Sze H. 1991. Calcium-pumping ATPases in vesicles from carrot cells. Stimulation by calmodulin or phosphatidylserine and formation of a 120 kilodalton phosphoenzyme. *Plant Physiol.* 97:1535–44

63. Ikura M, Hasegawa N, Aimoto S, Yazawa M, Yagi K, Hikichi K. 1989. Cadmium-112 NMR evidence for cooperative interactions between amino and carboxy terminal domains of calmodulin. *Biochem. Biophys. Res. Commun.* 161:1233–38

64. Ito T, Hirano M, Akama K, Shimura Y, Okada K. 1995. Touch-inducible genes for calmodulin and a calmodulin-related protein are located in tandem on a chromosome of *Arabidopsis thaliana. Plant Cell Physiol.* 36:1369–73

65. Johnson JD, Snyder C, Walsh M, Flynn M. 1996. Effects of myosin light chain kinase and peptides on Ca^{2+} exchange with the N- and C-terminal Ca^{2+}-binding sites of calmodulin. *J. Biol. Chem.* 271:761–67

66. Kasturi R, Vasulka C, Johnson JD. 1993. Ca^{2+}, caldesmon, and myosin light chain kinase exchange with calmodulin. *J. Biol. Chem.* 268:7958–64

67. Kawasaki H, Kretsinger RH. 1994. Calcium-binding proteins 1: EF-hands. *Protein Profile* 1:343–517

68. Kinkema M, Wang HY, Schiefelbein J. 1994. Molecular analysis of the myosin gene family in *Arabidopsis thaliana. Plant Mol. Biol.* 26:1139–53

69. Klee CB. 1988. Interaction of calmodulin with Ca^{2+} and target proteins. In *Calmodulin, Molecular Aspects of Cellular Regulation*, ed. P Cohen, CB Klee, 5:35–56. Elsevier: Amsterdam

70. Knight A, Kendrick-Jones J. 1993. A myosin-like protein from a higher plant. *J. Mol. Biol.* 231:148–54

71. Kohno T, Okagaki T, Kohama K, Shimmen T. 1991. Pollen tube extract supports the movement of actin filaments in vitro. *Protoplasma* 161:75–77

72. Kretsinger RH, Rudnick SE, Weisman LJ. 1986. Crystal structure of calmodulin. *J. Inorg. Biochem.* 28:289–302

73. Kuboniwa H, Tjandra N, Grzesiek S, Ren H, Klee CB, Bax A. 1995. Solution structure of calcium-free calmodulin. *Nat. Struct. Biol.* 2:768–76

74. Lam E, Benedyk M, Chua N-H. 1989. Characterization of phytochrome-regulated gene expression in a photoautotrophic cell suspension: possible role for calmodulin. *Mol. Cell. Biol.* 9:4819–23

75. Lee SH, Kim JC, Lee MS, Heo WD, Seo HY, et al. 1995. Identification of a novel divergent calmodulin isoform soybean which has differential ability to activate calmodulin-dependent enzymes. *J. Biol. Chem.* 270:21806–12

76. Lee SH, Seo HY, Kim JC, Heo WD, Chung WS, et al. 1997. Differential activation of NAD kinase by plant calmodulin isoforms: the critical role of domain I. *J. Biol. Chem.* 272:9252–59

77. Liao B, Gawienowski MC, Zielinski RE. 1996. Differential stimulation of NAD kinase and binding of peptide substrates by wild-type and mutant plant calmodulin isoforms. *Arch. Biochem. Biophys.* 327:53–60

78. Liao B, Zielinski RE. 1995. Production of recombinant plant calmodulin and its use to detect calmodulin-binding proteins. *Methods Cell Biol.* 49:481–94

79. Ling V, Assmann SM. 1992. Cellular distribution of calmodulin and calmodulin-binding proteins in *Vicia faba* L. *Plant Physiol.* 100:970–78

80. Ling V, Perera IY, Zielinski RE. 1991. Primary structures of Arabidopsis calmodulin isoforms deduced from the sequences of cDNA clones. *Plant Physiol.* 96:1196–202

81. Ling V, Snedden WA, Shelp BJ, Assmann SM. 1994. Analysis of soluble calmodulin binding protein from fava bean roots: identification of glutamate decarboxylase as a calmodulin-activated enzyme. *Plant Cell* 6:1135–43

82. Linse S, Forsen S. 1995. Determinants that govern high-affinity calcium binding. *Adv. Second Messenger Phosphoprotein Res.* 30:89–151

83. Lu Y-T, Feldmann LJ, Hidaka H. 1993. Inhibitory effects of KN-93, an inhibitor of Ca^{2+}/calmodulin-dependent protein kinase II, on light-regulated root gravitropism in maize. *Plant Physiol. Biochem.* 31:857–62

84. Lu Y-T, Harrington HM. 1994. Isolation of tobacco cDNA clones encoding calmodulin-binding proteins and characterization of a known calmodulin-binding domain. *Plant Physiol. Biochem.* 32:413–22

85. Lu Y-T, Hidaka H, Feldmann LJ. 1996.

Characterization of a calcium/calmodulin-dependent protein kinase homolog from maize roots showing light-regulated gravitropism. *Planta* 199:18–24

86. Luan S, Li WW, Rusnak F, Assmann SM, Schreiber SL. 1993. Immunosuppressants implicate protein phosphatase regulation of K+ channels in guard cells. *Proc. Natl. Acad. Sci. USA* 90:2202–6

87. Luby-Phelps K, Hori M, Phelps JM, Won D. 1995. Ca^{2+}-regulated dynamic compartmentalization of calmodulin in living smooth muscle cells. *J. Biol. Chem.* 270:21532–38

88. Malho R, Trewavas AJ. 1996. Localized apical increases of cytosolic free calcium control pollen tube orientation. *Plant Cell* 8:1935–49

89. Malmstrom S, Askerlund P, Palmgren MG. 1997. A calmodulin-stimulated Ca^{2+}-ATPase from plant vacuolar membranes with a putative regulatory domain at its N-terminus. *FEBS Lett.* 400:324–28

90. Matthews RP, Guthrie CR, Wailes LM, Zhao XY, Means AR, McKnight GS. 1994. Calcium/calmodulin-dependent protein kinase types II and IV differentially regulate CREB-dependent gene expression. *Mol. Cell. Biol.* 14:6107–16

91. Muto S, Miyachi S. 1977. Properties of a protein activator of NAD kinase from plants. *Plant Physiol.* 59:55–60

92. Neel VA, Young MW. 1994. igloo, a GAP-43–related gene expressed in the developing nervous system of *Drosophila*. *Development* 120:2235–43

93. Neuhaus G, Bowler C, Kern R, Chua N-H. 1993. Calcium/calmodulin-dependent and calcium/calmodulin-independent phytochrome signal transduction pathways. *Cell* 73:937–52

94. Nojima H. 1989. Structural organization of multiple rat calmodulin genes. *J. Mol. Biol.* 208:269–82

95. Oh S-A, Kwak JM, Kwun IC, Nam HG. 1996. Rapid and transient induction of calmodulin-encoding gene(s) of *Brassica napus* by a touch stimulus. *Plant Cell Rep.* 15:586–90

96. Oh S-H, Roberts DM. 1990. Analysis of the state of posttranslational calmodulin methylation in developing pea plants. *Plant Physiol.* 93:880–87

97. Oh S-H, Steiner H-Y, Dougall DK, Roberts DM. 1992. Modulation of calmodulin levels, calmodulin methylation, and calmodulin binding proteins during carrot cell growth and embryogenesis. *Arch. Biochem. Biophys.* 297:28–34

98. Ohya Y, Botstein D. 1994. Diverse essential functions revealed by complementing yeast calmodulin mutants. *Science* 263:963–66

99. Ohya Y, Botstein D. 1994. Structure-based systematic isolation of conditional lethal mutations in the single yeast calmodulin gene. *Genetics* 138:1041–54

100. Okamoto H, Tanaka Y, Sakai S. 1995. Molecular cloning and analysis of the cDNA for an auxin-regulated calmodulin gene. *Plant Cell Physiol.* 36:1531–39

101. Olwin BB, Edelman AM, Krebs EG, Storm DR. 1984. Quantitation of energy coupling between Ca^{2+}, calmodulin, skeletal muscle myosin light chain kinase, and kinase substrates. *J. Biol. Chem.* 259:10949–55

102. Olwin BB, Storm DR. 1985. Calcium binding to complexes of calmodulin and calmodulin binding proteins. *Biochemistry* 24:8081–86

103. O'Neil KT, DeGrado WF. 1990. How calmodulin binds its targets: sequence independent recognition of amphiphilic α-helices. *Trends Biochem. Sci.* 15:59–64

104. Parag HA, Dimitrovsky D, Raboy B, Kulka RG. 1993. Selective ubiquitination of calmodulin by UBC4 and a putative ubiquitin protein ligase (E3) from Saccharomyces cerevisiae. *FEBS Lett.* 325:242–46

105. Patil S, Takezawa D, Poovaiah BW. 1995. Chimeric plant calcium/calmodulin-dependent protein kinase gene with a neural visinin-like calcium-binding domain. *Proc. Natl. Acad. Sci. USA* 92:4897–901

106. Perera IY, Zielinski RE. 1991. Structure and expression of the Arabidopsis CaM-3 calmodulin gene. *Plant Mol. Biol.* 19:49–64

107. Perera IY, Zielinski RE. 1992. Synthesis and accumulation of calmodulin in suspension cultures of carrot (*Daucus carota* L.). *Plant Physiol.* 100:812–19

108. Persechini A, Kretsinger RH. 1988. The central helix of calmodulin functions as a flexible tether. *J. Biol. Chem.* 263:12175–78

109. Persechini A, White HD, Gansz KJ. 1996. Different mechanisms for Ca^{2+} dissociation from complexes of calmodulin with nitric oxide synthase or myosin light chain kinase. *J. Biol. Chem.* 271:62–67

110. Poovaiah BW, Reddy ASN. 1993. Calcium and signal transduction in plants. *CRC Crit. Rev. Plant Sci.* 12:185–211

111. Poovaiah BW, Takezawa D, An G, Han TJ. 1996. Regulated expression of a

calmodulin isoform alters growth and development in potato. *J. Plant Physiol.* 149:553–58

112. Price AH, Taylor A, Ripley SJ, Griffiths A, Trewavas AJ, Knight MR. 1994. Oxidative signals in tobacco increase cytosolic calcium. *Plant Cell* 6:1301–10

113. Rasi-Caldogno F, Carnelli A, De Michelis MI. 1993. Controlled proteolysis activates the plasma membrane Ca^{2+} pump of higher plants. A comparison of the effects of calmodulin in plasma membrane from radish seedlings. *Plant Physiol.* 103:385–90

114. Rasi-Caldogno F, Carnelli A, De Michelis MI. 1995. Identification of the plasma membrane Ca^{2+}-ATPase and of its autoinhibitory domain. *Plant Physiol.* 108:105–13

115. Rasmussen C, Garen C. 1993. Activation of calmodulin-dependent enzymes can be selectively inhibited by histone H1. *J. Biol. Chem.* 268:23788–91

116. Reddy ASN, Safadi F, Narasimhulu SB, Golovkin M, Hu X. 1996. A novel plant calmodulin-binding protein with a kinesin heavy chain motor domain. *J. Biol. Chem.* 271:7052–60

117. Reddy ASN, Takezawa D, Fromm H, Poovaiah BW. 1993. Isolation and characterization of two cDNAs that encode calmodulin-binding proteins from corn root tips. *Plant Sci.* 94:109–17

118. Roberts DM, Besl L, Oh S-H, Masterson RV, Schell J, Stacey G. 1993. Expression of a calmodulin methylation mutant affects the growth and development of transgenic tobacco plants. *Proc. Natl. Acad. Sci. USA* 89:8394–98

119. Roberts DM, Burgess WH, Watterson DM. 1984. Comparison of the NAD kinase and myosin light chain kinase activator properties of vertebrate, higher plant, and algal calmodulins. *Plant Physiol.* 75:796–98

120. Roberts DM, Harmon AC. 1992. Calcium-modulated proteins: targets of intracellular calcium signals in higher plants. *Annu. Rev. Plant Physiol. Plant Mol. Biol.* 43:375–414

121. Roberts DM, Rowe PM, Siegel FL, Lukas TJ, Watterson DM. 1986. Trimethyllysine and protein function: effect of methylation and mutagenesis of lysine-115 of calmodulin on NAD kinase activation. *J. Biol. Chem.* 261:1491–94

122. Romoser VA, Hinkle PM, Persechini A. 1997. Detection in living cells of Ca^{2+}-dependent changes in the fluorescence emission of an indicator composed of two green fluorescent protein variants linked by a calmodulin-binding sequence. *J. Biol. Chem.* 272:13270–74

123. Schuurink RC, Chan PV, Jones RL. 1996. Modulation of calmodulin mRNA and protein levels in barley aleurone. *Plant Physiol.* 111:371–80

124. Slaughter GR, Means AR. 1989. Analysis of expression of multiple genes encoding calmodulin during spermatogenesis. *Mol. Endocrinol.* 3:1569–78

125. Smith RD, Walker JC. 1996. Plant protein phosphatases. *Annu. Rev. Plant Physiol. Plant Mol. Biol.* 47:101–25

126. Snedden WA, Koutsia N, Baum G, Fromm H. 1996. Activation of a recombinant petunia glutamate decarboxylase by calcium/calmodulin or by a monoclonal antibody which recognizes the calmodulin binding domain. *J. Biol. Chem.* 271:4148–53

127. Song H, Golovkin M, Reddy ASN, Endow SA. 1997. In vitro motility of AtKCBP, a calmodulin-binding kinesin protein of Arabidopsis. *Proc. Natl. Acad. Sci. USA* 94:322–27

128. Szymanski DB, Liao B, Zielinski RE. 1996. Calmodulin isoforms differentially enhance the binding of cauliflower nuclear proteins and recombinant TGA3 to a region derived from the Arabidopsis CaM-3 gene promoter. *Plant Cell* 8:1266–73

129. Takezawa D, Liu ZH, An G, Poovaiah BW. 1995. Calmodulin gene family in potato: developmental and touch-induced expression of the mRNA encoding a novel isoform. *Plant Mol. Biol.* 27:693–703

130. Takezawa D, Ramachandiran S, Paranjape V, Poovaiah BW. 1996. Dual regulation of a chimaeric plant serine/threonine kinase by calcium and calcium/calmodulin. *J. Biol. Chem.* 271:8126–32

131. Timme MJ, Szymanski DB, Zielinski RE. 1996. Expression of calmodulin-GUS gene fusions in transgenic tobacco. *Plant Physiol. Suppl.* 111:156 (Abstr. 717)

132. Tirlapur UK, Cresti M. 1992. Computer-assisted video image analysis of spatial variations in membrane-associated Ca^{2+} and calmodulin during pollen hydration, germination and tip growth in *Nicotiana tabacum* L. *Ann. Bot.* 69:503–8

133. Torok K, Whitaker M. 1994. Taking a long, hard look at calmodulin's warm embrace. *BioEssays* 16:221–24

134. Trewavas A, Knight M. 1994. Mechanical signaling, calcium and plant form. *Plant Mol. Biol.* 26:1329–41

135. Trewavas AJ, Malho R. 1997. Signal perception and transduction: the origin of the phenotype. *Plant Cell* 9:1181–95

136. Vantard M, Lambert A-M, De Mey J, Piquot P, Van Eldik LJ. 1985. Characterization and immunocytochemical distribution of calmodulin in higher plant endosperm cells: localization in the mitotic apparatus. *J. Cell Biol.* 101:488–99

137. Vogel HJ. 1994. Calmodulin: a versatile calcium mediator protein. *Biochem. Cell Biol.* 72:357–76

138. Vogel HJ, Zhang M. 1995. Protein engineering and NMR studies of calmodulin. *Mol. Cell. Biochem.* 149/150:3–15

139. Wang J, Campos B, Jamieson GA Jr, Kaetzel MA, Dedman JR. 1995. Functional elimination of calmodulin within the nucleus by targeted expression of an inhibitor peptide. *J. Biol. Chem.* 270:30245–48

140. Wang W, Takezawa D, Narasimhulu SB, Reddy ASN, Poovaiah BW. 1996. A novel kinesin-like protein with a calmodulin-binding domain. *Plant Mol. Biol.* 31:87–100

141. Wang W, Takezawa D, Poovaiah BW. 1996. A potato cDNA encoding a homologue of mammalian multidrug resistant P-glycoprotein. *Plant Mol. Biol.* 31:683–87

142. Watillon B, Kettmann R, Boxus P, Burny A. 1992. Cloning and characterization of an apple (*Malus domestica* (L) Borkh) calmodulin gene. *Plant Sci.* 82:201–12

143. Watillon B, Kettmann R, Boxus P, Burny A. 1993. A calcium/calmodulin-binding serine/threonine protein kinase homologous to the mammalian type II

144. Wick SM, Muto S, Duniec J. 1985. Double immunofluorescence labeling of calmodulin and tubulin in dividing plant cells. *Protoplasma* 126:198–206

145. Yang TB, Segal G, Abbo S, Feldman M, Fromm H. 1996. Characterization of the calmodulin gene family in wheat: structure, chromosomal location, and evolutionary aspects. *Mol. Gen. Genet.* 252:684–94

146. Yang WN, Burkhart W, Cavallius J, Merrick WC, Boss WF. 1993. Purification and characterization of a phosphatidylinositol 4-kinase activator in carrot cells. *J. Biol. Chem.* 268:392–98

147. Yuan T, Vogel HJ. 1997. The calmodulin-binding domain of plant glutamate decarboxylase. *Plant Physiol. Suppl.* 114:273 (Abstr. 1416)

148. Zhang M, Tanaka T, Ikura M. 1995. Calcium-induced conformational transition revealed by the solution structure of apo calmodulin. *Nat. Struct. Biol.* 2:758–67

149. Zielinski RE. 1987. Calmodulin mRNA in barley (*Hordeum vulgare* L.): apparent regulation by cell proliferation and light. *Plant Physiol.* 84:937–43

150. Zimmer WE, Schloss JA, Silflow CD, Youngblom J, Watterson DM. 1988. Structural organization, DNA sequence, and expression of the calmodulin gene. *J. Biol. Chem.* 263:19370–83

calcium/calmodulin-dependent protein kinase is expressed in plant cells. *Plant Physiol.* 101:1381–84

Annu. Rev. Plant Physiol. Plant Mol. Biol. 1998. 49:727–60

FROM VACUOLAR GS-X PUMPS TO MULTISPECIFIC ABC TRANSPORTERS

Philip A. Rea, Ze-Sheng Li, Yu-Ping Lu, and Yolanda M. Drozdowicz

Plant Science Institute, Department of Biology, University of Pennsylvania, Philadelphia, Pennsylvania 19104; e-mail: parea@sas.upenn.edu

Enrico Martinoia

Institut de Botanique, Université de Neuchâtel, 2007 Neuchâtel, Switzerland

KEY WORDS: detoxification, glutathione, oxidative stress, xenobiotics

ABSTRACT

While the concept of H^+-coupling has dominated studies of energy-dependent organic solute transport in plants for over two decades, recent studies have demonstrated the existence of a group of organic solute transporters, belonging to the ATP-binding cassette (ABC) superfamily, that are directly energized by MgATP rather than by a transmembrane H^+-electrochemical potential difference. Originally identified in microbial and animal cells, the ABC superfamily is one of the largest and most widespread protein families known. Competent in the transport of a broad range of substances including sugars, peptides, alkaloids, inorganic anions, and lipids, all ABC transporters are constituted of one or two copies each of an integral membrane sector and cytosolically oriented ATP-binding domain. To date, two major subclasses, the multidrug resistance-associated proteins (MRPs) and multidrug resistance proteins (MDRs) (so named because of the phenotypes conferred by their animal prototypes), have been identified molecularly in plants. However, only the MRPs have been defined functionally. This review therefore focuses on the functional capabilities, energetics, organization, and regulation of the plant MRPs. Otherwise known as GS-X pumps, or glutathione-conjugate or multispecific organic anion Mg^{2+}-ATPases, the MRPs are considered to participate in the transport of exogenous and endogenous amphipathic anions and glutathionated compounds from the cytosol into the vacuole. Encoded by a multigene family and possessing a unique domain organization, the types of processes

1040-2519/98/0601-0727$08.00

that likely converge and depend on plant MRPs include herbicide detoxification, cell pigmentation, the alleviation of oxidative damage, and the storage of antimicrobial compounds. Additional functional capabilities might include channel regulation or activity, and/or the transport of heavy metal chelates. The identification of the MRPs, in particular, and the demonstration of a central role for ABC transporters, in general, in plant function not only provide fresh insights into the molecular basis of energy-dependent solute transport but also offer the prospect for manipulating and investigating many fundamental processes that have hitherto evaded analysis at the transport level.

CONTENTS

INTRODUCTION

According to the prevailing chemiosmotic model for energy-dependent solute transport by plant cells, the primary energizers for solute transport are pumps that mediate electrogenic H^+-translocation across the membranes concerned to generate a gradient of electrical potential ($\Delta\psi$) and H^+ chemical potential (ΔpH) (86, 99). Depending on the secondary transporter concerned, the motive force for solute transport is derived from the ΔpH and/or $\Delta\psi$ components of the H$^+$-electrochemical potential difference ($\Delta\bar{\mu}_{H^+}$). However, while the literature on solute transport in plants is large and a wide range of transport processes have been found to be H$^+$-coupled by electrophysiological and biochemical criteria (11, 105), surprisingly few studies have addressed the transport of substances other than amino acids, sugars, and inorganic ions (see 105). As a result, the spectrum of candidate transport substrates investigated has not been sufficiently exhaustive for alternate, more direct modes of energization to be excluded a priori.

In retrospect, two classes of observation were decisive in reorienting attention away from chemiosmotic coupling toward more direct mechanisms for the energization of organic solute transport in plants. The first was the molecular cloning of a multidrug resistance- (MDR-) like gene (*AtPGP1*) from *Arabidopsis thaliana* (22) and the independent isolation of two similar but nonidentical cDNAs, *AtMDR1* and *AtMDR2*, from the same organism (KT Howitz, A Menkens, J Darling, EJ Kim, AR Cashmore, & PA Rea, unpublished results). All three genes encode polypeptides bearing strong sequence similarities to one another and to the human and mouse *MDR* gene products. Given that the mammalian MDRs belong to a superfamily of ATP-binding cassette (ABC) transporters (33), most of which utilize ATP as a direct energy source for organic solute transport, the existence of genes encoding homologs in Arabidopsis, and their subsequent demonstration in other plant species (potato, 115; barley, 17), implied that analogous ATP-dependent, primary active transport functions are operative in plants. The second critical observation was that intact vacuoles and vacuolar membrane vesicles isolated from plants are capable of MgATP-dependent, H$^+$-ATPase-independent accumulation of glutathione *S*-conjugates (GS-conjugates) (65, 75). This finding was seminal in two respects. Not only did it demonstrate that MgATP could act as a direct energy source for organic solute transport across plant membranes, but since several of the GS-conjugates studied were those of herbicides, a transport pathway capable of contributing to the vacuolar sequestration, and possibly to the detoxification, of xenobiotics seemed to have been identified.

Here we examine these and more recent findings to explore the characteristics of ABC transporters (otherwise known as "traffic ATPases") and their established and speculated functional capabilities in plants. At the time of writing, however, only two subclasses of ABC transporter have been investigated in any detail in plants: the multidrug resistance-associated proteins (MRPs) and their distant cousins, the MDRs. Of these, only the MRPs have been defined biochemically. Therefore, by necessity, our focus is restricted to the MRPs. Only in passing will some of the principles learned from studies of the MRPs be extended to other plant ABC transporters.

Notwithstanding the fact that only a handful of ABC transporters have been characterized molecularly and fewer still have been characterized functionally in plants, the existence of 29 ABC protein genes in the yeast (*Saccharomyces cerevisiae*) genome alone (19) indicates that many more remain to be discovered in plants.

ABC TRANSPORTERS: AN OVERVIEW

Distribution and Transport Functions

The ABC transporter superfamily is the largest protein family known, and its members are capable of a multitude of transport functions (32). In excess of 130 representatives have been identified in species ranging from archaebacteria to humans (33, 113). Included among the transport capabilities assigned to ABC transporters from nonplant sources are those for alkaloids, lipids, peptides, steroids, sugars, inorganic anions, and heavy metal chelates. With the exception of inorganic anions, the transport of all of these substances is directly energized by MgATP.

The most thoroughly investigated ABC family members in eukaryotes include the MDRs, some of which are implicated in resistance to antitumor drugs (24, 29) and others in lipid translocation (91); the pleiotropic drug resistance protein (PDR5) (5, 6) and STE6 peptide-mating pheromone transporter of *S. cerevisiae* (78); the *Schizosaccharomyces pombe* heavy metal tolerance protein (HMT1) (80, 81); the cystic fibrosis transmembrane conductance regulator (CFTR) chloride channel (2, 90), mutation of which is associated with cystic fibrosis in humans; the malarial, *Plasmodium falciparum*, chloroquine transporter (PfMDR1) (119); and the MHC transporters (TAPs), responsible for peptide translocation and antigen presentation in T lymphocytes (47).

Motifs and Domains

All ABC transporters are constituted of one or two copies each of two basic structural elements—a hydrophobic, integral transmembrane domain (TMD) containing multiple (usually four or six) transmembrane helices, and a cytoplasmically oriented ATP-binding domain (nucleotide binding fold, NBF) (33, 39). The TMDs span the membrane to form the pathway for solute movement across the bilayer (or between bilayer leaflets) and, in those proteins investigated in sufficient detail, determine the substrate specificity (or selectivity) of the transporter. The cytosolically oriented NBFs couple ATP hydrolysis to solute movement by an unknown mechanism. The NBFs, the most diagnostic structural feature of ABC transporters, are 30–40% identical between family members over a span of about 200 amino acid residues, and each encompasses one copy each of three idiotypic sequence motifs. These motifs are a Walker A box (G-X(4)-G-K-[ST]) and a Walker B box ([RK]-X(3)-G-X(3)-L-[hydrophobic](3)), separated by approximately 120 amino acids (114), and an ABC signature (or C motif), situated between the two Walker boxes, consisting of the amino acids [LIVMFY]-S-[SG]-G-X(3)-[RKA]-[LIVMYA]-X-[LIVMF]-[AG] (3). Walker motifs are found in other nucleotide-binding proteins, for example cation ATPases, myosin, adenylate kinase, phosphofructokinase, and ATP/ADP exchangers (114), but the C motif is unique to the NBFs of ABC transporters.

Modular Construction

A feature of ABC transporters, evident from a survey of the superfamily, is their modular construction. The four core domains—two NBFs and two TMDs—may be expressed as separate polypeptides or as multidomain proteins (39). In some ABC transporters (usually those from bacterial sources) the four domains reside on different polypeptides. An example is the oligopeptide permease of *Salmonella typhimurium*, which is constituted of four polypeptides, Opp B, C, D, and F (39). In others, the domains are fused in various combinations. In the ribose system of *Escherichia coli*, the two NBFs are fused (39); in the mammalian TAP1-TAP2 peptide transporter (47), and the *HMT1* gene product of *S. pombe* (80), one TMD is fused with an NBF; and in the mammalian MDRs and CFTR, and *S. cerevisiae* STE6 and PDR5 proteins, all four domains are fused into a single polypeptide (24, 54, 90). In those proteins in which the four core domains are fused into a single polypeptide, a TMD1-NBF1-TMD2-NBF2 configuration is found in some (e.g. MDR, STE6, CFTR) while an NBF1-TMD1-NBF2-TMD2 configuration (e.g. PDR5) is found in others.

VACUOLAR GS-X PUMPS

Considerable effort has been made in investigating the roles played by the tripeptide, glutathione (GSH, γ-Glu-Cys-Gly), and glutathione *S*-transferases (GSTs) in the detoxification of exogenous toxins (xenobiotics) (reviewed in 71), but it is only in the past few years that the significance of ABC transporters for GSH-dependent detoxification has been appreciated.

Phases I, II, III, and IV of Detoxification

The detoxification of exogenous and endogenous cytotoxins is the culmination of a multiphase process (42, 43, 53). The first phase (activation) is the exposure or introduction of functional groups of the appropriate reactivity for phase II enzymes. Cytochrome P450-dependent monooxygenases and mixed function oxidases, responsible for the hydroxylation of aromatic rings, alkyl groups, or heteroatom release, are examples of phase I enzymes. In phase II (conjugation) the activated derivative is conjugated with GSH (animals and plants), glucose (plants), glucuronic acid (animals and possibly some plants), or sulfate (animals) by GSTs, UDP:glucosyltransferases, UDP:glucuronyltransferases, or sulfotransferases, respectively. In phase III (elimination) the conjugates are transported out of the cytosol into intracellular compartments and/or the extracellular space. In phase IV (transformation) the transported conjugates are further substituted and/or degraded to yield transport-inactive derivatives.

Whereas the cytochromes P450, a group of five-liganded heme proteins encoded by a large multigene family (over 60 isoforms have been identified in Arabidopsis) (9) prepare cytotoxins for detoxification, the reactions these enzymes catalyze are activating (generate electrophilic centers) and seldom decrease toxicity. Hence, the requirement for the reactions of phase II, which serve to block the reactive groups generated in phase I and confer negative charges and/or increased water solubility on the compounds concerned.

In the case of phase II of the GSH-dependent pathway for herbicide detoxification by plants, for example, numerous studies have established the importance of GSTs comprising a superfamily of isoforms possessing varying degrees of substrate specificity for various compounds. There are many instances where herbicide selectivity is attributable to the formation of herbicide-GS conjugates in the resistant but not in the susceptible cultivars or species, and of agents (chemical "safeners") that selectively induce GST activity, thereby diminishing herbicide injury (reviewed in 52). Similarly, during anticancer drug detoxification by mammals, an association between cellular GSH levels and GST activity has long been recognized (reviewed in 31). Stable overexpression of individual GSTs and γ-glutamylcysteine synthetase (γ-GCS), a rate-limiting enzyme for GSH synthesis, has been observed in many tumors, and in a few cases, gene transfection has established a causal connection between increased GST levels and resistance to a range of anticancer agents (reviewed in 43).

Phase III transport is considered to rid the cytosol of GS-conjugates that would otherwise end product–inhibit GSTs (and inhibit GSH reductases) and impede sustained detoxification. In the special case of GS-conjugates that are themselves toxic, phase III may be critical for detoxification per se. In plants, the herbicide tridiphane is known to be converted to its corresponding GS-conjugate in situ to generate a potent inhibitor of the metabolism of the chloroacetanilide herbicide, atrazine (56). In mammals, dihaloethanes and the anticancer alkylating agent, alkyl-N-nitro-N'-nitroguanidine, undergo enormous increases in cytotoxicity and genotoxicity as a direct result of GST-catalyzed glutathionation (15).

A Phase III Transporter from Plants

It was against this background and through investigations of the uptake of the two model GS-conjugates, N-ethylmaleimide-glutathione (NEM-GS) and S-(2,4-dinitrophenyl)-glutathione (DNP-GS) (Figure 1), and the glutathionated

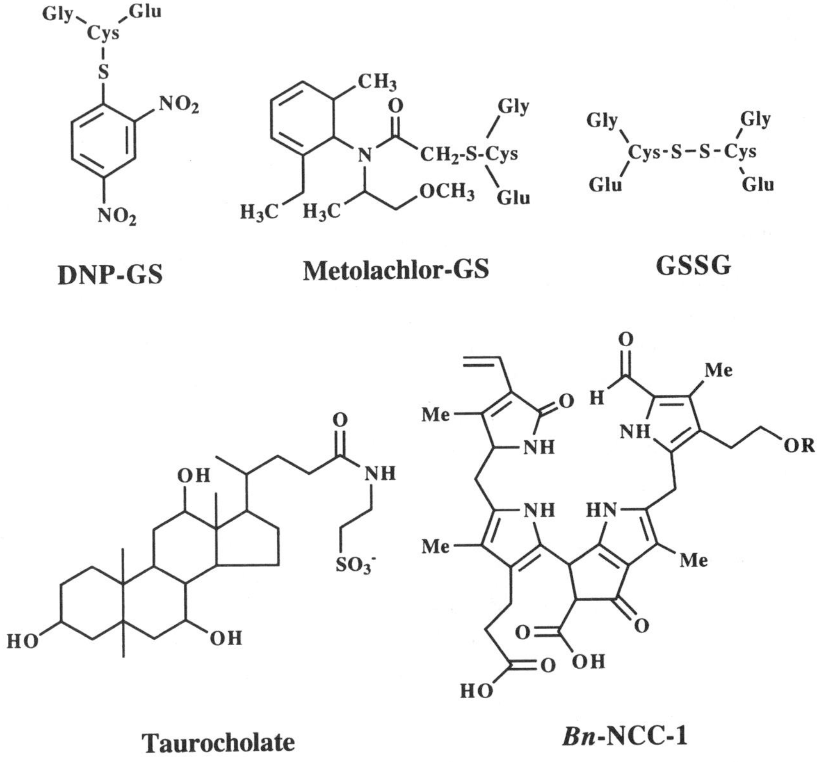

Figure 1 Examples of some of the compounds known to be transported by vacuolar ABC transporters in plants. Shown are the model GS-X pump substrate, *S*-(2,4-dinitrophenyl)-glutathione (DNP-GS), glutathionated metolachlor (metolachlor-GS), oxidized glutathione (GSSG), taurinylated cholate (taurocholate), and the *Brassica napus* chlorophyll catabolite, *Bn*-NCC-1. R = malonyl group.

chloroacetanilide herbicide, metolachlor (metolachlor-GS) (Figure 1) by isolated vacuoles (75) and vacuolar membrane vesicles (65), that ABC transporter activity was first demonstrated in plants.

The salient features of vacuolar GS-conjugate uptake are fourfold. 1. Direct energization by MgATP: Uptake is driven by MgATP but not by free ATP or nonhydrolyzable ATP analogs such as $5'$-(β, γ-imino)triphosphate (AMP-PNP). Other nucleoside triphosphates, for example GTP or UTP, can partially substitute for ATP, but inorganic pyrophosphate, the primary energy source for the vacuolar H^+-pyrophosphatase (V-PPase), cannot. 2. Insensitivity to the transmembrane H^+-electrochemical potential difference: Elimination of the inside-acid ΔpH and inside-positive $\Delta \psi$ that would otherwise be established by the vacuolar H^+-ATPase (V-ATPase) and/or V-PPase coresident on the vacuolar membrane through the addition of protonophores, ionophores, or H^+-phosphohydrolase-specific inhibitors does not inhibit GS-conjugate uptake. 3. Exquisite sensitivity to vanadate: Micromolar concentrations of the phosphoryl transition-state analog, vanadate, a compound employed to inhibit transporters that form an acylphosphate during their catalytic cycles, strongly inhibits GS-conjugate uptake. 4. Lack of activity toward GSH: The vacuolar GS-conjugate pump transports GS-conjugates, GSSG, and several nonglutathionated compounds (below) but not GSH.

A Vacuolar Transporter of High Accumulative Capacity

While the estimated accumulation ratios established by vacuolar GS-conjugate pumps are several orders of magnitude lower than the theoretical upper limit of 4×10^8 predicted for a 1ATP:1GS-conjugate ATPase (53), they are well in excess of or close to in excess of what is thermodynamically feasible for $\Delta \psi$-driven GS-conjugate anion uniport (≤ 3.2) and $1H^+$:1GS-conjugate antiport (≤ 100) (53, 87). Measurements of DNP-GS uptake by vacuolar membrane vesicles and of the distribution of the glutathionated herbicide, alachlor-GS, between the vacuolar and nonvacuolar compartments of barley leaves yield accumulation ratios in excess of 50 (65, 120). The mechanism of energization has not been investigated, but the pronounced sensitivity of GS-conjugate transport to inhibition by vanadate suggests that a critical step in the reaction cycle is transfer of the γ-phosphoryl group of ATP to the enzyme concomitant with the formation of an aspartyl- or glutamyl-phosphate intermediate (65).

MRP-SUBCLASS ABC TRANSPORTERS

Convergence of Three Research Areas

The convergence of three ostensibly disparate research areas—mammalian tumor biology, microbial heavy metal tolerance, and plant herbicide detoxification—has revealed a new subclass of ABC transporters to which the vacuolar GS-conjugate pump belongs: the MRP subclass.

The vacuolar GS-conjugate pump bears a close functional resemblance to the GS-conjugate transporting Mg^{2+}-ATPase (GS-X pump) (42) of mammalian cells. Membrane vesicles purified from erythrocytes (55), liver canaliculus (1, 48, 51), lung epithelium (58), and myocytes (41) mediate GS-conjugate transport. In all these preparations, transport is directly energized by MgATP, selective for (but not limited to) GS-conjugates and GSSG, but not GSH, and strongly inhibited by micromolar concentrations of vanadate (reviewed in 43).

The multidrug resistance associated protein (MRP1), encoded by the human *MRP1* (*HmMRP1*) gene, catalyzes the transport of leukotriene C_4 (a glutathionated arachidonic acid derivative), DNP-GS, and related GS-conjugates (59, 79).

Originally isolated by differential screens of antitumor drug-sensitive and drug-resistant small cell lung carcinoma (SCLC) cell lines (13), this ABC transporter gene when transfected into transport-deficient HeLa or SCLC cell lines restores MgATP-energized, vanadate-inhibitable GS-conjugate transport.

The yeast (*S. cerevisiae*) cadmium factor 1 gene (*ScYCF1*), which was isolated by screening a yeast genomic library for the ability of multicopy DNA fragments to confer resistance to cadmium salts in the growth medium (103), encodes a vacuolar GS-X pump resembling HmMRP1. Direct comparisons between *S. cerevisiae* (*ycf1*Δ) mutants harboring a deletion of the *YCF1* gene reveal that YCF1 is required for alleviation of the cytotoxic effects of the DNP-GS precursor, 1-chloro-2,4-dinitrobenzene (CDNB) and the MgATP-energized, uncoupler-insensitive, vanadate-inhibitable transport of DNP-GS (64). YCF1-dependent GS-conjugate transport in vitro strictly copurifies with the vacuolar fraction, intact wild type, but not *ycf1*Δ cells, mediate vacuolar accumulation of the fluorescent GS-conjugate, bimane-GS, and introduction of plasmid-borne *YCF1* or *HmMRP1* into *ycf1*Δ cells alleviates the CDNB-hypersensitive phenotype concomitant with restoration of the capacity of vacuolar membrane vesicles isolated from such cells for MgATP-energized GS-conjugate transport (64, 106).

Canalicular Multispecific Organic Anion Transporter

Another MRP-like GS-X pump from mammalian sources that has since been identified molecularly, and will be referred to below, is the canalicular multi-specific organic anion transporter (cMOAT) (10, 82). cMOAT and MRP1 have equivalent domain organizations (below), but their sequence identities with respect to each other (48%) are not much greater than those between MRP1 (or cMOAT) and ScYCF1, indicating that divergence of these two mammalian transporters antedated the appearance of mammals. Studies of mutant (TR$^-$) hyperbilirubinemic rats in which cMOAT is not functional suggest that it has a substrate specificity similar to that of MRP1 (10, 48). MRP1 and cMOAT, however, differ markedly in their tissue distributions. cMOAT has the characteristics of a liver (canalicular membrane)-specific isoform (reviewed in 43) whereas MRP1 has the characteristics of a lung-specific isoform (13, 102).

Molecular Cloning of GS-X Pumps from Arabidopsis

Through the identification of ScYCF1 as a vacuolar ortholog of HmMRP1, and by implication the suitability of the *ScYCF1* gene and yeast *ycf1*Δ mutants, respectively, as probes for and null backgrounds against which the corresponding genes from plants may be identified and tested functionally, several genes encoding GS-X pumps have been cloned from Arabidopsis. To date, four unique classes of clone, designated *AtMRP1* (68), *AtMRP2* (67), *AtMRP3* (108), and

AtMRP4 (93), have been characterized molecularly. *AtMRP1* and *AtMRP2* have been isolated as both full-length genomic and full-length cDNA clones, functionally expressed in yeast and physically mapped (to chromosomes 1 and 2, respectively) (67, 68); *AtMRP3* has been isolated as a full-length cDNA alone and functionally expressed in yeast (108); and *AtMRP4* has been isolated as a full-length genomic clone (93).

The sequences of the four 170–180-kDa deduced AtMRP polypeptides exhibit greater than 33% identity (55% similarity) to ScYCF1, HmMRP1, and cMOAT (67, 68, 93, 108), and when subjected to phylogenetic analysis using parsimony, group with LePGP1 and the rabbit epithelial conductance regulator (RbEBCR—see below) in addition to ScYCF1, HmMRP1, and cMOAT. This subclass has a common branch point with the CFTR cluster but is remote from those incorporating the MDRs, PDR5, TAPs, SpHMT1, and STE6. The rank order of sequence homology between the four AtMRPs [AtMRP1/AtMRP2 (85% identity, 94% similarity) > AtMRP3/AtMRP4 (41% identity, 63% similarity) > AtMRP1 (or AtMRP2)/AtMRP3 (or AtMRP4) (34% identity, 56% similarity)] suggests a two-pair grouping: AtMRP1 with AtMRP2 and AtMRP3 with AtMRP4.

AtMRP1, AtMRP2, and AtMRP3 Are Functional GS-X Pumps

AtMRP1, *AtMRP2*, and *AtMRP3* encode functional GS-X pumps. Transformation of *S. cerevisiae ycf1*Δ strains from which more than 95% of the coding sequence of the *YCF1* gene has been deleted (103) and high-affinity, MgATP-energized vacuolar GS-conjugate transport is abolished (64, 106), with expression vector containing the entire open reading frames of *AtMRP1*, *AtMRP2*, or *AtMRP3* is sufficient to restore GS-conjugate transport (67, 68, 108). However, while in vitro kinetic analyses of heterologously expressed AtMRP1 and AtMRP2—the two AtMRPs for which detailed and quantitative transport data are currently available—show both to be capable of transporting DNP-GS, GSSG, and glutathionated derivatives of the anthocyanin, cyanidin-3-glucoside (C3G), and the herbicide, metolachlor (67, 68), there are striking differences between the two pumps with respect to transport capacity and substrate preference. The overall transport capacity of AtMRP2 exceeds that of AtMRP1 by three- to sixfold, and the substrate specificity of AtMRP1 better approximates that of the endogenous vacuolar GS-X pump than does AtMRP2. Whereas the rates of uptake of DNP-GS, GSSG, metolachlor-GS and C3G-GS by AtMRP1 fall in the same rank order as for mung bean vacuolar membrane vesicles (65) (C3G-GS > metolachlor-GS > DNP-GS $\geq$ GSSG; uptake ratio for AtMRP1 = 2.7:1.5:1.0:0.7; uptake ratio for mung bean = 5.5:3.9:1.0:0.8), that of AtMRP2 is quite distinct (metolachlor-GS > C3G-GS $\gg$ GSSG > DNP-GS; uptake ratio = 7.8:6.4:2.2:1.0) (67, 68). Moreover, AtMRP2, unlike AtMRP1, has the facility for the high-affinity, high-capacity transport of nonglutathionated chlorophyll catabolites (below) (67). With the exception of GSSG, whose K_m for AtMRP1-mediated uptake (ca 220 μM) is three times greater than that for AtMRP2-mediated uptake (ca 66 μM), the K_m values of AtMRP2 and AtMRP1 are very similar for all of the glutathionyl derivatives examined (60–75 μM).

The substrate range of AtMRP3 is apparently similar to that of AtMRP2, inasmuch as this pump is competent in the transport of DNP-GS and chlorophyll catabolites and transport of the former is competitively inhibited by metolachlor-GS and GSSG (108), but the transport characteristics of AtMRP4 have not been determined (93).

MRP Structure

Computer-assisted hydropathy and sequence alignments between the AtMRPs and other members of the MRP subclass reveal a fundamental equivalence of domain organization (67, 68, 93, 108). The domains identified include two NBFs (NBF1 and NBF2), each of which contains the Walker A and B motifs and C motif characteristic of ABC transporters, two TMDs (TMD1 and TMD2), and two subclass-specific structures: a putative "regulatory" (R) or "connector" domain contiguous with NBF1, rich in charged amino acid residues, common to the MRP and CFTR subclasses but truncated in the former class (117–161 versus more than 250 amino acids) and an approximately 200 amino acid residue hydrophobic N-terminal extension, containing five hydrophilicity minima, absent from the CFTR subclass but present in all MRP-subclass members (Figure 2).

A tentative topological model of the MRPs showing the relative dispositions of these domains is depicted in Figure 2. Although derived largely from analyses of HmMRP1 and its murine ortholog, this model is probably as applicable to the AtMRPs as it is to the mammalian MRPs given the almost exact equivalence of the hydropathy profiles of all four AtMRPs and their close correspondence with those of other members of the MRP-subclass (67).

The structure shown consists of a cytosolically oriented R domain, two cytosolically oriented NBFs, a five-span N-terminal extension containing an extracytoplasmic N-terminus, one TMD (TMD1) containing six transmembrane spans and a second (TMD2) containing four or six spans.

A cytosolic disposition for NBFs 1 and 2 and the R domain is inferred from in situ immunolocalization and epitope-mapping experiments, while investigations of the 14 N-glycosylation sequons of HmMRP1 imply an odd number of transmembrane spans in the N-terminal extension. Immunostaining of permeabilized but not intact cells with antisera and mAbs directed against NBF1 and NBF2 (26), mAb QCRL-1 directed against the R domain epitope

AtMRP2

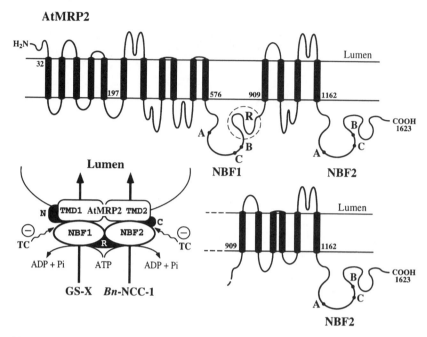

Figure 2 Tentative topological model of AtMRP2 and schematic diagram depicting parallel transport of GS-conjugates (GS-X) and *Bn*-NCC-1 by AtMRP2. The topological model shown, which is equally applicable to AtMRP1, AtMRP3, and AtMRP4, is based on that deduced for HmMRP1 by multiple sequence alignment of the human and murine MRPs, rabbit cMOAT, and ScYCF1 and profile-based neural network topology analysis (36). A model of TMD2 with a more conventional ABC transporter configuration of six transmembrane spans is also shown. The *numbers* delimit the starts of the membrane-spanning section of the N-terminal extension, TMD1, NBF1, TMD2, and NBF2. *A, B, C,* and *R* denote the positions of the Walker A boxes, Walker B boxes, ABC signature motifs, and "regulatory" or "connector" domain, respectively. In the schematic diagram of AtMRP2-mediated transport, it is proposed that GS-conjugates are transported by one NBF-TMD pair (e.g. NBF1-TMD1) while *Bn*-NCC-1 is transported by the other NBF-TMD pair (e.g. NBF2-TMD2). Taurocholate inhibits transport through both NBF-TMD pairs without itself undergoing transport.

[918]SSYSGDI[924] (37) and mAb-MRP1, whose epitope is encompassed by amino acid residues 194–360 of HmMRP1 (4), localize NBF1, together with its contiguous R domain, and NBF2 to the cytosolic face of the membrane. Parallel protein *N*-glycosidase F digestion and site-directed mutagenesis experiments demonstrate that Asn[19] and Asn[23] in the N-terminal extension and Asn[1006] in TMD2 are the sole sites in HmMRP1 modified with *N*-linked oligosaccharides, thus localizing the N-terminus to the extracytoplasmic face of the membrane and necessitating an odd number of spans, probably five on the basis of the

number of hydrophilicity minima of the appropriate length (19–21 residues) in the N-terminal extension (36, 67).

The precise topology of TMD2 is uncertain. The results of comparative hydropathy analyses and sequence alignments are equally consistent with six (102, 110) or four transmembrane spans in this region (13, 36). This is not a problem unique to the MRPs, however. A reevaluation of the results of previous studies of MDR, the archetypal 6 + 6 protein, show them also to be consistent with four or six transmembrane spans in the region corresponding to TMD2 of the MRPs (97, 121). For this reason and because the data and reagents available prohibit resolution of a 6 + 6 from a 6 + 4 model for MRPs—for instance, the established extracytoplasmic orientation of Asn[1006] in TMD2 of HmMRP1 (36) does not distinguish between different even-span models, only between even- and odd-span models—both topologies are incorporated into Figure 2.

The two most distinctive structural features of the MRPs, the CFTR-like truncated R domain and N-terminal extension, are implicated in regulation by kinases and the recognition of glutathionyl moieties, or some other common structural determinant, characteristic of their transport substrates, respectively. ScYCF1 contains a protein kinase A (PKA) dibasic consensus motif (RRAS[908]) in the R domain, equivalent to those of CFTR, which have an established regulatory role, which when S908A-substituted confers a cadmium-hypersensitive phenotype and abolishes MgATP-energized DNP-GS transport (Z-S Li, M Szczypka, DJ Thiele, & PA Rea, unpublished results). ScYCF1 NΔ208 mutants lacking the MRP-subclass-specific N-terminal extension, though unimpaired in vacuolar targeting, are no longer able to confer cadmium resistance (116).

Excretion versus Storage Excretion

It is not known whether GS-X pump activity exclusively localizes to the vacuolar membrane of plant cells. On the one hand, Coleman et al (14) have elegantly demonstrated in situ glutathionation of monochlorobimane to bimane-GS and accumulation of the latter in the vacuoles of protoplasts and cell suspension cultures, and Wolf et al (120) find that exposure of barley leaves to alachlor results in massive vacuolar accumulation of alachlor-GS. On the other hand, it should be appreciated that some herbicide metabolites derived from the GSH- and glucose-dependent detoxification pathways are deposited as "bound residues" in the extracellular matrix (57, 94) and most plant membrane density gradient fractionation procedures have an inherent bias in favor of the isolation of sealed (transport-competent) vacuolar membrane vesicles by comparison with transport-competent plasma membrane vesicles (88). Consequently, while measurements on density gradient fractionated plant membranes demonstrate an exclusive association of GS-X pump activity with the vacuolar membrane fraction (65, 66), the question of whether the same pump and/or one or more of its isoforms is also active in the discharge of GS-conjugates across the plasma membrane into the apoplast awaits the application of alternative fractionation procedures, for example the polyethylene glycol/dextran two-phase method, to the isolation of transport-competent plasma membrane vesicles and/or the results of Western analyses or in situ immunolocalizations using subclass- or isoform-specific MRP antibodies.

The need for caution is reinforced by the functional contiguity between the GS-X pump activities on the endomembranes and exomembranes of some mammalian cell types (reviewed in 43), and possibly yeast (101) (discussed in 64), and the fact that the membrane locations of none of the four isoforms of the GS-X pump from Arabidopsis has yet been determined using preparations from plant sources.

Conjugate Pumps as Targets for Manipulating Plant Xenobiotic Resistance

The high capacity of AtMRP2 and moderate capacity of AtMRP1 for the transport of metolachlor-GS (67, 68) and by implication GS-conjugates of other herbicides such as alachlor, atrazine and symetryn, for which glutathionation in vivo (52, 120) and transport by the endogenous vacuolar GS-X pump in vitro and in vivo (75, 120) are also demonstrable, is consistent with the molecular identification of transporters capable of removing these and related compounds from the cytosol. The AtMRPs and their orthologs, particularly AtMRP2 whose V_{max} for metolachlor-GS transport (136 ± 28 nmol/mg/10 min) is severalfold higher than that reported for any other MRP-subclass transporter/substrate couple (67), are therefore attractive targets for manipulating plant responses to herbicides and other xenobiotics.

By comparison with the closest known similar technologies—for instance, the breeding or genetic engineering of plants in which the target for xenobiotic action is no longer sensitive, the generation of mutants with elevated cellular GSH levels or increased levels of the appropriate GSTs, or the application of chemical safeners that elevate GSH levels or GST activities—manipulation of a GS-X pump has several advantages. First, the utility of mutated target gene products is limited in its application to those xenobiotics that directly interact with the target, whereas GS-X pumps are of broad substrate selectivity. Second, technologies based on elevated GSH levels and/or an increase in phase II enzyme catalytic competence are likely limited ultimately by the capacity of cells to transport conjugates out of the cytosol and consequently by the activity of GS-X pumps. Third, manipulations of the vacuolar GS-X pump, unlike the other technologies, directly exploit the large volume of the plant vacuole (typically 40–70% of the total intracellular volume of a mature cell; 87) and the potential it offers for high levels of xenobiotic accumulation on a tissue fresh weight basis.

Technologies of this type are still some way from implementation, but the increased resistance of transgenic Arabidopsis plants expressing ScYCF1 to both organic and inorganic xenobiotics, as exemplified by CDNB and cadmium, respectively (Y-P Lu, Z-S Li, & PA Rea, unpublished results), clearly illustrates the potential of this approach and others based on GS-X pump-catalyzed vacuolar compartmentation.

Endogenous Glutathionated Transport Substrates

The capacity of the endogenous GS-X pump(s) and heterologously expressed AtMRPs (67, 68, 108) for the transport of glutathionated herbicides is of agronomic interest, but it does not explain the existence of these transporters in the first place. It is untenable that the primary function of pumps of this type is the extrusion and/or sequestration of artificial compounds whose manufacture and application has been widespread for only about 50 years unless compounds like them are to be found endogenously or in the natural habitat.

An impediment to the identification of natural substrates for GS-X pumps is the extraordinary range of GS-conjugates GSTs are capable of synthesizing. The common chemical determinant, a carbon-carbon double bond adjacent to an electron-withdrawing group ($CH_2=CH-Z$ or $Z'-CH=CH-Z$) (a Michael acceptor) shared by all GST inducers and substrates (104), is widely distributed among biological compounds. Notable examples of compounds containing this determinant are anthocyanins, indole-3-acetic acid, base propenols (toxic products of oxidative DNA degradation), cinnamic acid, hydroxyalkenals (products of lipid peroxidation), and salicylic acid (see 72).

Having made these cautionary remarks and knowing that at least some GS-X pumps are able to transport amphipathic organic anions other than GS-conjugates (below), there are nevertheless several cases, based on physiological necessity or chance genetic observations, of compounds that are strong candidates as endogenous substrates.

A well-documented class of likely endogenous transport substrates for the GS-X pumps of mammals are the leukotrienes, primarily leukotriene C_4 (LTC_4). Not only is LTC_4 transported with a K_m of 0.2 μM, which is an order of magnitude lower than that for any other compound screened against HmMRP1 and the endogenous mammalian GS-X pump (44, 59, 79), but its extrusion from the cell is a physiological necessity. The cysteinyl leukotrienes (C_4, D_4, E_4) are mediators of inflammatory responses, LTD_4 and LTE_4 are derived from LTC_4 after its export from the cytosol into the extracellular space by the GS-X pump, and LTD_4 and LTE_4 exert the majority of their effects extracellularly (82).

In plants, the vacuolar sequestration of pigments, antimicrobial compounds, and natural herbicides (allelochemicals) seems likely. Of these, the vacuolar accumulation of anthocyanins is the perhaps best defined.

It has been known for some time that the bronze coloration of maize *Bronze-2* (*bz2*) mutants is due to accumulation of anthocyanin, C3G, derivatives in the cytosol. In wild-type plants, anthocyanins are accumulated in the vacuole as purple or red derivatives, but in *bz2* mutants they are restricted to the cytosol, where they undergo oxidation and cross-linking to generate brown products. However, the biochemical basis of this lesion was not known. Incisive are experiments showing that *Bz2* encodes a GST capable of conjugating C3G with GSH (73). CDNB-conjugating GST activity is diminished in *bz2* versus *Bz2* plants, and transient expression of *Bz2* results in increased extractable GST activity concomitant with an enhancement of vacuolar anthocyanin accumulation (73). A rational basis for the *bz2* phenotype therefore follows. Being defective in the glutathionation of anthocyanins, *bz2* mutants are unable to pump these pigments from the cytosol into the vacuole via the vacuolar GS-X pump. In agreement with this conclusion, exposure of wild-type protoplasts to the potent GS-X pump inhibitor, vanadate, phenocopies the *bz2* mutation (73), and of the various GS-conjugates tested as in vitro substrates for the vacuolar GS-X pump, C3G-GS is one of the most efficacious (Z-S Li, M Alfenito, PA Rea, & V Walbot, unpublished results). Whereas unconjugated C3G undergoes negligible uptake into vacuolar membrane vesicles isolated from roots of maize or hypocotyls of mung bean, C3G-GS is subject to high rates of MgATP-dependent, uncoupler-insensitive uptake. The K_m values for uptake are in the micromolar range (40–45 μM), the V_{max} values (45–80 nmol/mg/10 min) are 7- to 40-fold greater than those measured with the model transport substrate, DNP-GS, and the concentrations of DNP-GS and vanadate required for 50% inhibition of C3G-GS uptake are commensurate with those defined previously for the transport of other GS-conjugates by the vacuolar GS-X pump (65, 66).

The facility of both AtMRP1 and AtMRP2 for the high-efficiency (low K_m/high V_{max}) transport of C3G-GS (67, 68) and the restoration of petal pigmentation consequent on introduction of the *Bz2* gene by particle bombardment into anthocyanin sequestration (GST-deficient) *an9* mutants of petunia and *fl3* mutants of carnation (M Alfenito, E Sover, CD Goodman, R Buell, J Mol, R Koes, & V Walbot, unpublished results) indicate that GS-X pump-mediated vacuolar anthocyanin-GS uptake is a mechanism of wide distribution.

Another group of compounds for which there is an association between vacuolar accumulation and glutathionation are the isoflavonoid phytoalexins of legumes. Constitutive isoflavonoids are known to occur as vacuolarly localized malonyl glycosides in tissues and cell cultures of several species including chickpea, soybean, and alfalfa (70). Elicitation of infection is associated with both de novo synthesis of isoflavonoid phytoalexins from distant precursors and/or release of isoflavone precursors from vacuolar stores (21, 83). An increase in GST transcripts and activities coincides with the production of isoflavonoid phytoalexins in elicitor-treated or infected cells (23, 60). Medicarpin, the major isoflavonoid phytoalexin of alfalfa and several other leguminous species, is amenable to GST-catalyzed glutathionation in vitro, and the product of this reaction, medicarpin-GS, is the most efficacious (glutathionated) substrate for the plant vacuolar GS-X pump known. Its K_m (<20 μM) is more than 2- to 25-fold lower than that estimated for DNP-GS and C3G-GS in the same system, and its V_{max} (78 nmol/mg/10 min) is similar to that for C3G-GS (62). It has therefore been suggested that one function for the GSTs induced following the hypersensitive response to avirulent fungal pathogens might be to facilitate the vacuolar storage of antimicrobial compounds in the healthy cells surrounding the hypersensitive lesion (62). It is notable that in some plant species, anthocyanins, for example 3-deoxyanthocyanidin in sorghum, are themselves phytoalexins and accumulate in subcellular inclusions upon pathogen attack (98). These pathogen-related processes and those associated with cell pigmentation apparently deploy similar biochemical machineries.

A connection between the involvement of the vacuolar GS-X pump in (artificial) herbicide detoxification and the sequestration of endogenous toxins is indicated by the allelopathic (natural herbicidal) properties of plant phenolics such as cinnamic acid and its derivatives. Although there are no published reports of the transport of glutathionated cinnamic acid by the vacuolar GS-X pump, the GST responsible for its glutathionation (glutathione S-cinnamoyl transferase, GSCT)—the only GST other than BZ2 for which the endogenous substrate is known in plants—is induced after the exposure of cell-suspension cultures to fungal elicitors and activated by p-coumaric acid and 7-hydroxycoumarin (18), both of which are themselves allelochemicals (61). Notwithstanding our ignorance of the physiological meaning of these reactions, the possibility that GSCT catalyzes the conversion of either exogenous (allelopathic) or endogenous (pathogen-elicited) $trans$-cinnamic acid, or related compounds, into a form that can be transported into the vacuole by the GS-X pump for the purposes of detoxification and/or long-term storage is attractive in that it provides a mechanistic link between plant responses to artificial herbicides, natural herbicides, and pathogens.

The results of these studies, especially those of vacuolar anthocyanin accumulation, illustrate an important principle: namely, that more compounds than would be predicted from measurements of the steady state levels of glutathionyl derivatives in plant extracts may be subject to glutathionation and transport by the GS-X pump. On the one hand, the investigations of maize $bz2$ mutants and the results of in vitro transport assays demonstrate the necessity of glutathionation for vacuolar anthocyanin compartmentation. On the other hand, measurements of the steady state levels of anthocyanins in intact plants demonstrate that vacuolar anthocyanins are not stored in appreciable quantities as their glutathionated derivatives but rather as their malonylated derivatives (72). It is therefore probable that while C3G-GS is a necessary intermediate for vacuolar anthocyanin compartmentation, it is short-lived and not a terminal product of this process. If this is true of anthocyanins, it may also be true of many other compounds—for example, phytohormones—that induce GSTs and satisfy the chemical requirements of GST substrates (72) but have not been detected as GS-conjugates in plant extracts.

Oxidative Stress as a Unifying Principle?

In considering transport substrates for GS-X pumps, the status of GSSG as an endogenous GS-conjugate (of GSH with itself) and the involvement of GSH, GSSG, and GSTs in cellular responses to active oxygen species (AOS) should not be overlooked.

GSH is a protectant employed by all aerobic organisms to guard against "electrophilic attack" whether this be by endogenous or exogenous toxins or the products of AOS action (89). Xenobiotics are toxic because they contain centers of low electron density. Cytotoxicity or genotoxicity ensues when xenobiotics covalently bond nucleophilic sites containing nonbonded electron pairs or π bonds. AOS, such as superoxide radicals (O_2^-), hydroxyl radicals (OH·) and peroxide are toxic because they are strongly nucleophilic and react with biomolecules to generate reactive electrophiles. The importance of GSH derives from its sulfhydryl group, which confers on this compound the facility to act as both a reducing agent, protecting the cell from AOS, and as a nucleophile protecting the cell from the electrophilic products of AOS action.

That GSH and cellular catalysts such as GS-X pumps and GSTs arose initially from a need to combat the toxic action of an oxidizing atmosphere is suggested by the tight phylogenetic association between GSH and deployment of oxygen as a terminal acceptor in respiration. GSH is found in the majority of eukaryotes, where it is poised at a concentration of between 1 and 10 mM and accounts for more than 90% of nonprotein sulfur (see 43), but in prokaryotes (eubacteria) it appears to be restricted to the cyanobacteria and purple bacteria (25). If, as is generally supposed, cyanobacteria were the first group of organisms capable of oxygenic photosynthesis and these and the purple bacteria gave rise to chloroplasts and mitochondria, respectively, GSH biosynthesis may have emerged at about the same time as oxygenic metabolism, approximately 2.0 to 3.5×10^9 years ago (25, 43).

The general applicability of the idea of a direct connection between oxidative stress and GS-X pump function has yet to be addressed systematically, but a number of disparate observations can, at least to a first approximation, be accommodated by such a scheme.

Most of the factors known to induce GSTs in plants—organic xenobiotics, heavy metals, pathogen attack, wounding and ethylene—promote AOS production (reviewed in 40), and several of these, as well as agents that influence the efficacy with which AOS are scavenged, modulate GS-X pump expression and/or activity. The model cytotoxin, CDNB, which is metabolized exclusively via the GSH-dependent detoxification pathway, elicits a two- to threefold increase in the activity of the vacuolar GS-X pump of mung bean (66). The steady state levels of transcripts derived from *AtMRP2*, *AtMRP3*, and *AtMRP4* (probed using ESTs 39B12T, ATTS1601, 107J19T7, and 147I122T7, respectively) are elevated by the addition of CDNB, metolachlor, primisulfuron, or IRL 1803 to Arabidopsis root cultures (109). *AtMRP3* and *AtMRP4* transcript levels are elevated by more than fivefold by addition of the catalase inhibitor, 3-amino-1,2,4-triazole, and the superoxide inhibitor, menadione, respectively, to Arabidopsis cell cultures (93).

All functionally characterized MRP-subclass transporters, including endogenous vacuolar GS-X pumps (65, 107), AtMRP1 (68), AtMRP2 (67), AtMRP3 (108), and ScYCF1 (64), recognize GSSG, a product of antioxidant action, as a substrate. In the case of AtMRP2, GSSG is transported at higher capacity and at comparable affinity to DNP-GS (67).

In *S. cerevisiae* overexpression of YAP1, a bZIP transcription factor, activates not only the *YCF1* and *GSH1* genes (117), the latter of which encodes γ-GCS, but also a multitude of oxidoreductases (20). Of the 17 genes whose mRNA levels are increased by more than threefold by YAP1 when screened against yeast genomic DNA microarrays, more than two thirds contain canonical upstream YAP1-binding sites, five bear homology to aryl-alcohol oxidoreductases, and four belong to the general class of dehydrogenases/oxidoreductases (20). When the capacity of YAP1 overexpression to confer increased resistance to hydrogen peroxide, *o*-phenanthroline, and heavy metals, and the susceptibility of *YCF1* to YAP1-dependent induction by both CDNB and cadmium (63) are taken into account, the fact that an appreciable fraction of the YAP1-regulated target genes identified against the yeast genome project sequence database are putative oxidoreductases and regulated in parallel with both *YCF1* and *GSH1*, suggests that all play a role in protection from oxidative stress. The extent to which these findings apply to the regulatory networks of plant MRPs is not known, but it is noteworthy that the *AtMRP2* gene promoter contains a putative bZIP recognition sequence while the *AtMRP1* promoter contains putative xenobiotic regulatory elements and antioxidant response elements (Y-P Lu, & PA Rea, unpublished results).

It is intriguing to consider, since two particularly harmful and early effects of AOS production are membrane lipid peroxidation and oxidative DNA damage, in which fatty acids are converted to 4-hydroxyalkenals (46) and DNA bases are converted to base propenols (7)—compounds that are established substrates for mammalian GSTs (7, 16) and subject to high-efficiency transport by mammalian GS-X pumps (41)—that the various physiological roles ascribed to GS-X pumps may have evolved from an ancestral GS-X pump function: that of protection from the damaging effects of lipophilic electrophiles, such as the α, β-unsaturated aldehydes, generated by AOS action. Hydroxyalkenals, base propenols, and their GS-conjugates have not been tested as plant GS-X pump substrates, though decyl-GS, an alkyl S-derivative, is a potent inhibitor of the vacuolar GS-X pump (8) and LTC$_4$, a lipid GS-conjugate, is transported by HmMRP1 with an affinity several orders of magnitude greater than that reported for any plant GS-X pump transport substrate (59, 79).

OTHER VACUOLAR ABC TRANSPORTER ACTIVITIES

GS-conjugate transport was the first and is the most rigorously characterized ABC transporter activity identified on the plant vacuolar membrane, but several others have since been detected. Three of these are implicated in the transport of large amphipathic anions (chlorophyll catabolites, bile acids, and glucuronate conjugates), while the other two are implicated in the transport of glucose conjugates and phytochelatins. All five transport functions, like GS-conjugate transport, are directly energized by nucleoside triphosphates, insensitive to agents that dissipate transmembrane H^+ gradients and strongly inhibited by micromolar concentrations of vanadate.

Glucose Conjugates

Comparisons of the energy requirements for uptake of isovitexin, a C-glucosy-lated endogenous flavonoid, and hydroxyprimisulfuron-glucoside (HPS-gluco-side), a sulfonylurea herbicide O-glucoside, by barley vacuoles implicate two transport mechanisms for glucosides: direct energization by MgATP and H^+-antiport, respectively (50). Isovitexin uptake is abolished by bafilomycin A$_1$ and protonophores, as would be expected for a V-ATPase-dependent (118), ΔpH-driven process, whereas HPS-glucoside uptake is more than 60% insensitive to these agents but directly inhibited by vanadate. In view of the susceptibility of the AtMRP3 gene (EST 107J19T7) of Arabidopsis to strong induction by primisulfuron (109) and the capacity of herbicide safeners to increase HPS-glucoside transport activity in barley vacuoles (27), there is a possibility that the transporter responsible for MgATP-energized, vanadate-inhibitable uptake of HPS-glucoside into barley vacuoles is an ortholog of AtMRP3.

Chlorophyll Catabolites

Conversion of mesophyll chloroplasts into gerontoplasts during leaf senescence and yellowing is accompanied by the disassembly of thylakoid pigment-protein complexes and breakdown of chlorophyll into phytol and water-soluble por-phyrin derivatives (76). The water-soluble porphyrins generated are exported from the gerontoplasts and accumulated ultimately in the vacuoles of senescing cells (77), where they remain until the end of the senescence period. Three of the principal vacuolar tetrapyrroles of rape—(*Brassica napus*)—*Bn*-NCC-1, *Bn*-NCC-2, *Bn*-NCC-1—have been purified and structurally characterized, and one, *Bn*-NCC-1 (Figure 1), has been shown to undergo MgATP-energized, vanadate-inhibitable uptake into barley vacuoles (35).

There is some doubt about whether the in vivo substrates for this transport activity are the nonfluorescent substituted end products (the NCCs) or their fluorescent precursors (the FCCs). If vacuolar tetrapyrrole sequestration serves to protect the cytosol from photooxidative damage consequent on the production of free radicals by illuminated unsubstituted (fluorescent) tetrapyrroles, it might be expected that these, rather than their substituted derivatives, undergo transport. The rather high K_m for *Bn*-NCC-1 uptake by barley vacuoles (110 μM) and the potency of the fluorescent barley chlorophyll catabolite, *Hv*-FCC-2, as a competitive inhibitor ($K_i = 60$ μM) (35) is consistent with this possibility.

Bile Acids

An ABC transporter activity associated with the plant vacuolar membrane, whose in vivo function is not known, is the transport of bile acids such as taurocholate, a taurinylated derivative of cholate (Figure 1) (8, 38). Although the initial identification of this activity, which on the basis of its oligomycin sensitivity was inferred to be distinct from that responsible for the transport of GS-conjugates, contributed to the notion of a cMOAT-like activity in plants, in that cMOAT-catalyzed transport of bile acids across the canalicular membrane of hepatocytes is critical for lipid metabolism in animals (reviewed in 82), there is no precedent for an equivalent transport function in plants. There are no reports of bile acids in plants, and the most closely related known plant products, steroid saponins, such as digitonin, do not markedly inhibit taurocholate transport except at concentrations sufficient to destabilize the vacuolar membrane (38). Presumably the activity measured as bile acid uptake by vacuoles is responsible for the transport of sterol or lipid derivatives that have yet to be identified in plants, or this transporter recognizes a structural determinant on taurocholate that is shared by an otherwise unrelated group of compounds.

Glucuronate Conjugates

The biosynthesis and transport of glucuronate conjugates in animals, where they participate in the degradation of heme and its cMOAT-mediated secretion into bile as bilirubin diglucuronide, has been studied extensively (see 82), but in plants it has generally been assumed that conjugates of this type play only a minor role (if any) in detoxification by comparison with their glutathionyl and glucosyl counterparts. While this is probably true of most plants, one notable exception is rye, which produces large quantities of glucuronides and for which MgATP-energized, vanadate-inhibitable vacuolar uptake of the model glucuronate conjugate, β-estradiol 17-β-glucuronide (β-EG) is demonstrable in vitro (49). The activity responsible is competent in the transport of endogenous glucuronides and capable of distinguishing these from heterologous glucuronides in that the former but not the latter compete with β-EG for uptake.

Phytochelatins

It is generally accepted that plants respond to heavy metal stress by the elaboration of a class of peptides (phytochelatins, PCs) consisting of repeating units of γ-glutamylcysteine followed by a C-terminal glycine [poly-(γ-Glu-Cys)$_n$-Gly peptides] (85, 100). Derived from GSH by PC synthases (γ-glutamylcysteine dipeptidyl transpeptidases), which transfer a γ-glutamylcysteine moiety from GSH to a second molecule of GSH or to a previously synthesized PC molecule, these peptides complex heavy metals, such as Cd^{2+}, with high affinity, localize together with Cd^{2+} to the vacuole of intact cells (112), and contribute to Cd^{2+} detoxification (85).

The molecular basis of PC-associated vacuolar Cd^{2+} sequestration is best defined in the fission yeast S. pombe. Cadmium-sensitive mutants capable of synthesizing PCs but deficient in the accumulation of Cd^{2+}-PC (Cd.PC) complexes are mutated in a 90.5-kDa vacuolar ABC transporter, designated HMT1, which unlike most other eukaryotic representatives of the superfamily, consists of only one TMD and one NBF in a TMD-NBF configuration (80). HMT1 has the characteristics of a phytochelatin-selective Mg^{2+}-ATPase (81): it catalyzes the MgATP-energized, vanadate-inhibitable uptake of both Cd.PCs and apo-PCs into vacuoles of wild-type but not $hmt1\Delta$ cells (81).

HMT1 homologs have not been isolated from plants or identified in any of the plant EST sequence databases, but vacuolar membrane vesicles isolated from oat roots do mediate an MgATP-energized transport reaction resembling that identified in S. pombe (92).

Despite the superficial phenotypic similarity of S. pombe $hmt1\Delta$ mutants to S. cerevisiae $ycf1\Delta$ mutants, SpHMT1 and ScYCF1 are distinct. SpHMT1 are no more sequence-related than any other pair of eukaryotic ABC transporters and their transport activities appear not to overlap. Vacuolar membranes from S. pombe catalyze GS-conjugate transport, but as indicated by the inability of DNP-GS to compete with PCs for uptake and the retention of GS-conjugate transport in membranes from $hmt1\Delta$ strains, this activity is not attributable to HMT1 but instead to an S. pombe ortholog of ScYCF1 (12, 81). Reciprocally, PC fractions from S. pombe show no evidence of YCF1-dependent transport by vacuolar membrane vesicles purified from *S. cerevisiae* regardless of whether the PC fractions are complexed with Cd^{2+} or not (Z-S Li, & PA Rea, unpublished results).

Transport of Nonglutathionated Compounds by the GS-X Pump

Accumulating evidence from studies of MRP-subclass members from nonplant sources reveals that the group of transporters formerly referred to as GS-X pumps because of their affinity toward GS-conjugates, GSSG, and cysteinyl leukotrienes do not transport GS-conjugates exclusively (43). HmMRP1 is able to transport glucuronidated and sulfated compounds such as glucuronosyl estradiol and sulfatolithocholyltaurine (45). cMOAT is able to transport bile acids (10), bilirubin glucuronide (82), and unconjugated organic acids such as bromosulfphthalein and indocyanine green (96).

Against this background and in view of the existence of vacuolar ABC transporter activities for compounds other than GS-conjugates, several of which are also bulky amphipathic anions, the question of whether some plant GS-X pumps are able to satisfy more than one of these transport functions must be addressed.

In the case of AtMRP2 and AtMRP3, the answer to both of these questions is yes. Not only does AtMRP2 transport GS-conjugates with an order of preference distinguishable from that of AtMRP1, but it has the ability, which AtMRP1 lacks, for high-affinity transport of *Bn*-NCC-1. In fact, of all the compounds found to be transported by AtMRP2, *Bn*-NCC-1 is transported with the highest affinity ($K_m = 15 \ \mu M$) (67). Likewise, while the precise kinetics of *Bn*-NCC-1 transport by AtMRP3 have not been defined, it too is able to transport this compound as well as GS-conjugates (108).

It was initially surprising to find that AtMRP2 can transport both *Bn*-NCC-1 and GS-conjugates, because the functionality, and by implication the pump, responsible for MgATP-energized *Bn*-NCC-1 uptake by barley vacuoles had earlier been concluded to be different from that responsible for GS-conjugate uptake because GS-conjugates do not compete with *Bn*-NCC-1 for uptake (35). A reconciliation of the seeming discrepancy between the results reported for barley vacuoles (35) and those for heterologously expressed AtMRP2 is, however, provided by the unusual kinetics of AtMRP2. AtMRP2 is competent in the semi-autonomous transport of *Bn*-NCC-1 and GS-conjugates; it has the facility for the transport of both classes of compound without either interfering with the transport of the other (67).

The simplest interpretation of this phenomenon is that two functionally distinguishable domains reside on AtMRP2 (Figure 2). One domain transports GS-conjugates, the other transports compounds such as *Bn*-NCC-1, and transport through one domain is largely independent of transport through the other.

When account is taken of the modular construction of ABC transporters (39) and the fact that AtMRP2, and other members of the MRP subclass, contain two homologous halves, each half containing an NBF located on the cytosolic face of the membrane and a TMD containing multiple transmembrane spans, it is conceivable that one NBF-TMD pair can transport one class of compounds (e.g. GS-conjugates) across the membrane while the other NBF-TMD pair transports another class of compounds (e.g. chlorophyll catabolites) (Figure 2).

Whether the same principle applies to AtMRP3 is not known, but there are indications of a similar phenomenon for the endogenous glucuronide transport activity of rye vacuoles. Rather than being inhibited or unaffected by the addition of GS-conjugates to the assay medium, β-EG uptake by rye vacuoles is stimulated by GS-conjugates in direct proportion to their efficacies as transport substrates in this preparation (49). Apparently GS-conjugates allosterically activate β-EG transport in this system. Assuming that a distinct GS-conjugate transporter does not positively modulate the glucuronide transporter of rye vacuoles in proportion to its substrate occupancy, a model similar to that proposed for AtMRP2 may apply, except that in the case of the rye transporter, transport through one NBF-TMD pair promotes transport through the other.

The contrary situation, antagonism of transport by a compound that is not itself transported, applies to taurocholate. Neither AtMRP2 nor AtMRP1 transports taurocholate, but transport of both *Bn*-NCC-1 and GS-conjugates by AtMRP2 is strongly inhibited by this compound (67).

The high sensitivity of *Bn*-NCC-1 and GS-conjugate transport by AtMRP2 and the comparative insensitivity of AtMRP1-mediated transport to inhibition by taurocholate shows that, while this bile acid is relatively specific for AtMRP2 versus AtMRP1, it is capable of inhibiting both of AtMRP2's transport activities. If the idea that one NBF-TMD pair of AtMRP2 transports GS-conjugates while the other transports chlorophyll catabolites is valid, taurocholate is apparently capable of interacting unproductively with both of these or of binding some other component of AtMRP2 whose interaction with this compound blocks transport through both NBF-TMD pairs (Figure 2). It is likely that the potency of free estradiol, which is not itself transported, as an inhibitor of β-EG uptake into rye vacuoles (49) has a similar mechanistic basis.

In view of the multispecificity of such transporters, the susceptibility of *AtMRP3* to induction by primisulfuron (109), a herbicide metabolized via the glucose-dependent pathway, raises the possibility that the transporter encoded may transport not only GS-conjugates and chlorophyll catabolites but also glucosides such as HPS-glucoside.

Two methodological corollaries applicable to analyses performed on preparations likely containing several transporters with overlapping substrate specificities (native membrane vesicles or intact organelles) follow from these considerations. Lack of competition between a candidate transport substrate and an established substrate does not automatically imply that the former is not, itself, transported. Tests of the capacity of a candidate substrate to inhibit uptake of an established substrate are no substitute for direct measurements of transport of the candidate, because some compounds that are potent competitors (and established transport substrates for other members of the MRP subclass) are not transported by all MRPs.

The MRP Subclass and Heavy Metals

A consequence of the demonstrated identity of ScYCF1 with the vacuolar GS-X pump of *S. cerevisiae* is the question of the relationship between organic GS-conjugate transport and the ability of *ScYCF1* to confer resistance to Cd^{2+}. While *ScYCF1* was originally isolated according to its ability to confer resistance to cadmium salts (103), its mode of interaction with Cd^{2+} was unknown. Through investigations of the substrate requirements, kinetics, and Cd:GS stoichiometry of Cd^{2+} uptake and the measured molecular weight of the transport-active complex, it is now known that ScYCF1 catalyzes the MgATP-energized transport of *bis*(glutathionato) cadmium ($Cd.GS_2$) (63). *ycf1*Δ mutants are hypersensitive to Cd^{2+} or CDNB in the growth medium, and both hypersensitivities are alleviated by transformation with plasmid-borne *ScYCF1*. Vacuolar membrane-enriched vesicles purified from *ycf1*Δ cells are grossly impaired not only in MgATP-energized organic GS-conjugate but also in GSH-promoted Cd^{2+} uptake. ScYCF1-dependent Cd^{2+} uptake by membrane vesicles has a specific requirement for GSH—related compounds such as GSSG, *S*-methylglutathione, and cysteinylglycine do not promote uptake. The transport-active $Cd.GS_2$ complex and the model organic GS-conjugate, DNP-GS, compete with each other for uptake.

From these characteristics and ScYCF1's membership in the MRP-subclass, it might be anticipated that HmMRP1 and the AtMRPs are also competent in the transport of $Cd.GS_2$ and possibly other heavy metal–glutathione complexes. Despite initial enthusiasm for this possibility, more detailed investigations suggest that not all MRP-subclass members confer resistance to Cd^{2+} and that even those that do may do so by a mechanism distinct from that of ScYCF1.

Transformation of yeast *ycf1*Δ strains with *HmMRP1* restores organic GS-conjugate transport and alleviates the cadmium-hypersensitive phenotype, but all attempts to measure GSH-promoted Cd^{2+} transport by membranes isolated from such transformants have either failed (106) or yielded rates markedly lower than those seen when the same strains are transformed with *ScYCF1* (Z-S Li, YM Drozdowicz, & PA Rea, unpublished results). The situation with the

AtMRPs is less clear, but a pattern more reminiscent of HmMRP1 than ScYCF1 seems to be emerging. AtMRP1 and AtMRP2 neither confer resistance to Cd^{2+} nor mediate GSH-dependent transport of Cd^{2+} (67, 68), and although AtMRP3 relieves the Cd^{2+}-hypersensitivity of yeast $ycf1\Delta$ strains (108), it is not known if this pump directly mediates $Cd.GS_2$ transport.

There are at least three alternate explanations for these findings. The first is that suppression of the Cd^{2+}-hypersensitivity of $ycf1\Delta$ mutants by heterologous MRPs is not attributable to vacuolar heavy metal sequestration per se, but instead to alleviation of the consequences of heavy metal toxicity. For instance, some heterologous MPRs might ameliorate Cd^{2+} toxicity by vacuolarly sequestering the products of its action rather than Cd^{2+} itself. The involvement of MRP-subclass members in cellular responses to oxidative stress has already been discussed. The second possibility is that ScYCF1, and possibly AtMRP3, are special cases within the MRP subclass. That is, there may be a basic functional bifurcation within the MRP subclass such that only certain representatives are competent in the transport of Cd.GS complexes. If so, ScYCF1 may have specific structural characteristics, absent from other MRP-subclass members, that enable its interaction with $Cd.GS_2$. It is known, for example, that ScYCF1, unlike AtMRP2 (67) and AtMRP3 (108), does not transport Bn-NCC-1. The third explanation is that in the case of those transporters that suppress the $ycf1\Delta$ mutation but do not mediate GSH-promoted Cd^{2+} transport in vitro (HmMRP1 and possibly AtMRP3), a coreactant (or coreactants) in addition to or instead of GSH present in the intact yeast cell but absent from the in vitro assays is required for efficient transport by these but not by ScYCF1. Attractive is the possibility that chaperones, with capabilities similar to those proposed for the cadmium-elicited 14 kDa alternative splicing product of BZ2 (74), are required for the transport of $Cd.GS_2$ or even Cd.PC complexes by AtMRP3. Pertinent perhaps is the formal structural resemblance of bis(glutathionato) cadmium [$Cd.GS_2 = Cd$-$(\gamma$-Glu-Cys-Gly$)_2$] to the low molecular weight $Cd.PC_2$ complex [= Cd-$(\gamma$-Glu-Cys$)_2$-Gly] and the fact that although the molecular identity of $SpHMT1$ has been known for several years (80) there are no reports in the literature of the cloning of plant homologs. Could the lack of plant HMT1 homologs and the structural resemblance between $Cd.GS_2$ and $Cd.PC_2$ mean that vacuolar PC transport in plants is catalyzed by an AtMRP/ScYCF1-like transporter with the facility for recognizing either PCs or $Cd.GS_2$ after their association with some other component?

Intrinsic and Associated Channel Activities?

That CFTR is a PKA-regulated Cl^- channel and members of the MRP -subclass most closely resemble the CFTRs raises the question "Do any MRP-subclass members have channel activity or the potential to modulate channel activity?"

This question is worthy of investigation for two reasons. First, precedent for such a phenomenon—namely, an organic solute pump capable of modulating anion channel activity—is provided by another member of the ABC transporter superfamily, the MDR P-glycoprotein (P-gp) (see 34 for discussion of this and other examples). P-gp exists in two interconvertible states: one in which it is competent in MgATP-energized drug transport and another in which it mediates ligand (ATP, AMP-PNP, or AMP-PC) activated protein kinase C-dependent activation of swelling-activated Cl^- channels in a variety of mammalian cell types (28, 30, 69). The second reason is the molecular cloning of the epithelial basolateral cAMP-regulated Cl^- channel conductance regulator (EBCR) from rabbit (RbEBCR), which exhibits approximately 50% sequence identity to HmMRP1, ScYCF1, cMOAT and the AtMRPs and 30% identity to CFTR (111). It is not known if the translation product of *RbEBCR* has GS-X pump activity, but microinjection of *Xenopus* oocytes with capped *RbEBCR* RNA confers a cAMP-activated Cl^- conductance that is blocked by the anion channel blockers niflumic acid, 5-nitro-2-(3-phenylpropylamine)benzoic acid and 4,4'-diisothiocyanato-stilbene-2,2'-disulfonic acid. Thus, RbEBCR has the characteristics of a Cl^- channel (or channel activator) bearing a greater overall similarity to the MRP subclass than the CFTR subclass of ABC transporters. If similar principles were to apply to plants, there is a possibility that the AtMRPs and/or other MRP-subclass members define a new category of channel proteins or channel regulators.

CONCLUDING REMARKS

Research on plant ABC transporters is still in its formative phases, but from what we have learned of the one category investigated in any detail, the MRP subclass, it is clear that this area is going to be of intense interest in the next few years.

At an academic level, many of the key questions remain to be answered. Do the AtMRPs, and their equivalents from plants other than Arabidopsis, actually alleviate the toxic effects of xenobiotics and/or oxidative damage? When, where, and under what circumstances are these transporters deployed? What intracellular signals modulate their expression and/or activity? What structural characteristics determine the substrate preferences of different isoforms? Do all plant MRP-subclass members localize to the vacuolar membrane, or are other membranes involved? What features of ScYCF1 do the AtMRPs and their orthologs from other plants lack that would otherwise enable them to catalyze high-efficiency $Cd.GS_2$ transport? Are some or all of the AtMRPs able to transport glucose conjugates and glucuronate conjugates in addition to GS-conjugates and chlorophyll catabolites? Are glutathionated base propanols and/or hydroxyalkenals high-efficiency substrates for the AtMRPs? Are any of the AtMRPs functional anion channels or active in the regulation of channel activity?

Elucidation of the in vivo functions of the AtMRPs and related transporters will probably depend on the isolation and characterization of null mutants for the genes encoding these proteins, an objective partially fulfilled by the recent isolation of T-DNA insertional mutants for *AtMRP1* and *AtMRP2* (Y-P Lu, & PA Rea, unpublished results). However, in the light of the material discussed in this review—the functional divergence between the AtMRPs; the facility for some to transport compounds other than GS-conjugates, for example chlorophyll catabolites; the capacity of the endogenous vacuolar GS-X pump for high-efficiency transport of glutathionated isoflavonoid phytoalexins; the amenability of C3G-GS, a plant pigment, and GSSG, a product of peroxide detoxification and protein thiol reduction, to transport by the AtMRPs—the processes that might be affected by any one or combination of these null mutations could encompass (but may not be limited to) xenobiotic detoxification, chlorophyll catabolism, antimicrobial compound storage, cell pigmentation, and protection from oxidative stress.

At an agrotechnical level, MRP-subclass plant ABC transporters hold much promise. In a "brown fields" context, through an increase in our understanding of the involvement of GS-X pumps and related transporters in xenobiotic detoxification in plants, novel strategies for manipulating herbicide resistance (even perhaps pathogen resistance) and engineering or selecting plants for phytoremediation applications may be developed. The demonstrated capacity of ScYCF1 for the GSH-dependent transport of Cd^{2+}, as well as GS-conjugable cytotoxins, and the generation of xenobiotic-resistant Arabidopsis *ScYCF1* transgenics are examples of the applications of such technology. In a "green liver" context (95), the toxicological relevance of studies of plant MRPs may be considerable. If excretion storage is a general phenomenon in plants, the susceptibility of many of the compounds generated in phase II to mammalian digestive enzymes (95) and the potential for liberation of the parent compound or a toxic, possibly genotoxic, derivative (84) in the digestive tract will have profound repercussions for the consumers of plants. Critical, therefore, for the rational design of consumer-safe plant or soil applications will be information on their amenability to, and fate after, GS-X pump-mediated delivery into the vacuole.

What is written here is no more than a preface to studies of non-chemiosmotic, energy-dependent organic solute transporters in plants. Undoubtedly investigations of the functional capabilities of other plant ABC transporters, for example the MDRs, whose roles are likely to be equally as significant and varied as those of the MRPs, and include alkaloid transport, the maintenance of membrane bilayer asymmetry and signal peptide perception, will form the basis of future reviews.

ACKNOWLEDGMENTS

This work was funded by grants from the Department of Agriculture (National Research Initiative Competitive Grants Program grant 93-37304-8932) and Department of Energy (DE-FG02-91ER20055) awarded to PAR. YMD is a Triagency (Department of Energy/National Science Foundation//United States Department of Agriculture) Plant Training Grant Fellow. PAR extends his love and thanks to Jenny, Amy, and Emily for their immense patience and understanding during the completion of this and several other equally demanding projects during the latter half of 1997.

> **Visit the *Annual Reviews* home page at**
> **http://www.AnnualReviews.org.**

Literature Cited

1. Akerboom TP, Narayanaswami V, Kunst M, Sies H. 1991. ATP-dependent *S*-(2,4-dinitrophenyl)glutathione transport in canalicular plasma membrane vesicles from rat liver. *J. Biol. Chem.* 266:13147–52

2. Anderson MP, Rich DP, Gregory RJ, Smith AE, Welsh MJ, 1991. Generation of cAMP activated chloride currents by expression of CFTR. *Science* 251:679–82

3. Bairoch A. 1992. PROSITE: a dictionary of sites and patterns in proteins. *Nucleic Acids Res.* 20:2013–18

4. Bakos E, Hegedus T, Hollo Z, Welker E, Tusnady GE, et al. 1996. Membrane topology and glycosylation of the human multidrug resistance-associated protein. *J. Biol. Chem.* 271:12322–26

5. Balzi E, Goffeau A. 1994. Genetics and biochemistry of yeast multidrug resistance. *Biochim. Biophys. Acta* 1187:152–62

6. Balzi E, Wang M, Leterme S, VanDyck L, Goffeau A. 1994. PDR5, a novel yeast multidrug resistance conferring transporter controlled by the transcription regulator PDR1. *J. Biol. Chem.* 269:2206–14

7. Berhane K, Widersten M, Engstrom A, Kozarich JW, Mannervik B. 1994. Detoxification of base propenals and other α, β-unsaturated aldehyde products of radical reactions and lipid peroxidation by human glutathione *S*-transferases. *Proc. Natl. Acad. Sci. USA* 91:1480–84

8. Blake-Kalff MMA, Coleman JOD. 1996. Detoxification of xenobiotics by plant cells: characterisation of vacuolar amphiphilic organic anion transporters. *Planta* 200:426–31

9. Bolwell GP, Bozak K, Zimmerlin A. 1994. Plant cytochrome P450. *Phytochemistry* 37:1491–1506

10. Buchler M, Konig J, Brom M, Kartenbeck J, Spring H, et al. 1996. cDNA cloning of the hepatocyte canalicular isoform of the multidrug resistance protein, MRP, reveals a novel conjugate export pump deficient in hyperbilirubinemic mutant rats. *J. Biol. Chem.* 271:15091–98

11. Bush DR, 1993. Proton-coupled sugar and amino acid transporters in plants. *Annu. Rev. Plant Physiol. Plant Mol. Biol.* 44:513–42

12. Chauduri B, Ingavale S, Bachhawat AK. 1996. *apd1+*, a gene required for red pigment formation in *ade6* mutants of *Schizosaccharomyces pombe*, encodes an enzyme required for glutathione biosynthesis: a role for glutathione and a glutathione-conjugate pump. *Genetics* 145:75–83

13. Cole SP, C. Bhardwaj G, Gerlach JH, Mackie JE, Grant CE, et al. 1992. Overexpression of a transporter gene in a multidrug-resistant human lung cancer cell line. *Science* 258:1650–54

14. Coleman JOD, Randall R, Blake-Kalff MMA. 1997. Detoxification of xenobi-

otics in plant cells by glutathione conjugation and vacuolar compartmentalization: a fluorescent assay using monochlorobimane. *Plant Cell Environ.* 20:449–60

15. Commandeur JNM, Stijntjes GJ, Vermeulen NPE. 1995. Enzymes and transport systems involved in the formation and disposition of glutathione *S*-conjugates: role in bioactivation and detoxification mechanisms of xenobiotics. *Pharmacol. Rev.* 47:271–330

16. Danielson UH, Esterbauer H, Mannervik B. 1987. Structure-activity relationships of 4-hydroxyalkenals in the conjugation catalyzed by mammalian glutathione *S*-transferases. *Biochem. J.* 247:707–12

17. Davies TGE, Theodoulou FL, Hallahan DL, Forde BG. 1997. Cloning and characterization of a novel P-glycoprotein homologue from barley. *Gene* 199:195–202

18. Dean JV, Machota JH. 1993. Activation of corn glutathione *S*-transferase enzyme by coumaric acid and 7-hydroxycoumarin. *Phytochemistry* 34:361–65

19. Decottignies A, Goffeau A. 1997. Complete inventory of the yeast ABC proteins. *Nat. Genet.* 15:137–45

20. DeRisi JL, Iyer VR, Brown PO. 1997. Exploring the metabolic and genetic control of gene expression on a genomic scale. *Science* 278:680–86

21. Dixon RA, Paiva NL, 1995. Stress-induced phenylpropanoid metabolism. *Plant Cell* 7:1085–97

22. Dudler R, Hertig C. 1992. Structure of an *mdr*-like gene from *Arabidopsis thaliana*: evolutionary implications. *J. Biol. Chem.* 267:5882–88

23. Edwards R, Dixon RA. 1991. Glutathione *S*-cinnamoyl transferases in plants. *Phytochemistry* 30:79–84

24. Endicott JA, Ling V. 1989. The biochemistry of P-glycoprotein-mediated multidrug resistance. *Annu. Rev. Biochem.* 58: 137–71

25. Fahey RC, Sundquist AR. 1991. Evolution of glutathione metabolism. *Adv. Enzymol.* 64:1–53

26. Flens MJ, Izquierdo MA, Scheffer GL, Fritz JM, Meijer CJLM, et al. 1994. Immunochemical detection of the multidrug resistance protein MRP in human multidrug-resistant tumor cells by monoclonal antibodies. *Cancer Res.* 54:4557–63

27. Gaillard C, Dufaud A, Tommasini R, Kreuz K, Amrhein N, Martinoia E. 1994. A herbicide antidote (safener) induces the activity of both the herbicide detoxifying enzyme and of a vacuolar transporter for the detoxified herbicide. *FEBS Lett.* 352:219–21

28. Gill DR, Hyde SC, Higgins CF, Valverde MA, Mintenig GM, Sepulveda FV. 1992. Separation of drug transport and chloride channel functions of the human multidrug-resistance P-glycoprotein. *Cell* 71:1–10

29. Gottesman M, Pastan I. 1993. Biochemistry of multidrug resistance mediated by the multidrug transporter. *Annu. Rev. Biochem.* 62:385–427

30. Hardy SP, Goodfellow HR, Valverde MA, Gill DR, Sepulveda FV, Higgins CF. 1995. Protein kinase C-mediated phosphorylation of the human multidrug resistance P-glycoprotein regulates cell volume-activated chloride channels. *EMBO J.* 14: 68–75

31. Hayes JD, Pulford DJ. 1995. The glutathione *S*-transferase superfamily: regulation of GST and the contribution of isoenzymes to cancer chemoprotection and drug resistance. *CRC Crit. Rev. Biochem. Mol. Biol.* 30:445–600

32. Henikoff S, Greene EA, Pietrokovski S, Bork P, Attwood TK, Hood L. 1997. Gene families: the taxonomy of protein paralogs and chimeras. *Science* 278:609–14

33. Higgins CF. 1992. ABC transporters: from microorganisms to man. *Annu. Rev. Cell Biol.* 8:67–113

34. Higgins CF. 1995. The ABC of channel regulation. *Cell* 82:693–96

35. Hinder B, Schellenberg M, Rodon S, Ginsburg S, Vogt E, et al. 1996. How plants dispose of chlorophyll catabolites: directly energized uptake of tetrapyrrolic breakdown products into isolated vacuoles. *J. Biol. Chem.* 271:27233–36

36. Hipfner DR, Almquist KC, Leslie EM, Gerlach JH, Grant CE, Deeley RG, Cole SPC. 1997. Membrane topology of the multidrug resistance protein (MRP). A study of glycosylation-site mutants reveals an extracytoplasmic NH$_2$-terminus. *J. Biol. Chem.* 272:23623–30

37. Hipfner DR, Almquist KC, Stride BD, Deeley RG, Cole SPC. 1996. Location of a protease-hypersensitive region in the multidrug resistance protein (MRP) by mapping of the epitope of MRP-specific monoclonal antibody QCRL-1. *Cancer Res.* 56:3307–14

38. Hörtensteiner S, Vogt E, Hagenbuch B, Meier PJ, Amrhein N, Martinoia E. 1993. Direct energization of bile acid transport into plant vacuoles. *J. Biol. Chem.* 268: 1844–449

39. Hyde SC, Emsley P, Hartshorn MJ, Mimmack ML, Gileadi U, et al. 1990. Structural model of ATP-binding proteins

associated with cystic fibrosis, multidrug resistance and bacterial transport. *Nature* 346:362–65

40. Inzé D, Van Montagu M. 1995. Oxidative stress in plants. *Curr. Opin. Biotechnol.* 6:153–58

41. Ishikawa T. 1989. ATP/Mg^{2+}-dependent cardiac transport system for glutathione S-conjugates. *J. Biol. Chem.* 264:17343–48

42. Ishikawa T. 1992. The ATP-dependent glutathione S-conjugate export pump. *Trends Biochem. Sci.* 17:433–38

43. Ishikawa T, Li Z-S, Lu Y-P, Rea PA. 1997. The GS-X pump in plant, yeast and animal cells: structure, function and gene expression. *Biosci. Rep.* 17:189–207

44. Ishikawa T, Mueller M, Kluenemann C, Schaub T, Keppler D. 1990. ATP-dependent primary active transport of cysteinyl leukotrienes across liver canalicular membrane: role of the ATP-dependent transport system for glutathione S-conjugates. *J. Biol. Chem.* 265: 19279–86

45. Jedlitschky G, Leier I, Buchholz U, Barnouin K, Kurz G, Keppler D. 1996. Transport of glutathione, glucuronate and sulfate conjugates by the *MRP* gene-encoded conjugate export pump. *Cancer Res.* 56:988–94

46. Kappus H. 1985. Lipid peroxidation: mechanisms, analysis, enzymology and biological relevance. In *Oxidative Stress*, ed. H Sies, pp. 273–310. London:Academic

47. Kelly A, Powis SH, Kerr LA, Mockridge I, Elliott T, et al. 1992. Assembly and function of the two ABC transporter proteins encoded in the human major histocompatibility complex. *Nature* 355:641–44

48. Kitamura T, Jansen P, Hardenbrook C, Kamimoto Y, Gatmaitan Z, Arias IM. 1990. Defective ATP-dependent bile canalicular transport of organic anions in mutant (TR$^-$) rats with conjugated hyperbilirubinemnia. *Proc. Natl. Acad. Sci. USA* 87:3557–61

49. Klein M, Martinoia E, Weissenböck G. 1998. Directly energized uptake of' β-estradiol 17-(β-glucuronide) in plant vacuoles is strongly stimulated by glutathione conjugates. *J. Biol. Chem.* 273: 262–70

50. Klein M, Weissenböck G, Dufaud A, Gaillard C, Kreuz K, Martinoia E. 1996. Different energization mechanisms drive the vacuolar uptake of a flavonoid glucoside and a herbicide glucoside. *J. Biol. Chem.* 271:29666–71

51. Kobayashi K, Sogman Y, Hayashi K, Nicotera P, Orrenius S. 1990. ATP sti-

mulates the uptake of S-dinitrophenylglutathione by rat liver plasma membrane vesicles. *FEBS Lett.* 240:55–58

52. Kreuz K. 1993. Herbicides safeners: recent advances and biochemical aspects of their mode of action. In *Proc. Brighton Crop Prot. Conf., Weeds.* pp. 1249–58, Brighton, UK

53. Kreuz K, Tommasini R, Martinoia E. 1996. Old enzymes for a new job: herbicide detoxification in plants. *Plant Physiol.* 111:349–53

54. Kuchler K, Sterne RE, Thorner J. 1989. *Saccharomyces cerevisiae STE6* gene product: a novel pathway for protein export in eukaryotic cells. *EMBO J.* 8:3973–84

55. LaBelle EF, Singh SV, Srivastava SK, Awasthi YC. 1986. Evidence for different transport systems for oxidized glutathione and S-dinitrophenyl glutathione in human erythrocytes. *Biochem. Biophys. Res. Commun.* 139:538–44

56. Lamoureux GL, Rusness DG. 1986. Tridiphane [2-(3,5-dichlorophenyl)-2-(2,2,2-trichloroethyl)oxirane] an atrazine synergist: Enzymatic conversion to a potent glutathione S-transferase inhibitor. *Pestic. Biochem. Physiol.* 26:323–42

57. Lamoureux GL, Shimabukuro RM, Frear DS. 1991. Glutathione and glucoside conjugation in herbicide selectivity. In *Herbicide Resistance in Weeds and Crops*, ed. JC Caseley GW Cussan RK Atkin, pp. 227–61. Oxford:Butterworth-Heinemann

58. Lautier D, Canitrot Y, Deeley RG, Cole SPC. 1996. Multidrug resistance mediated by the multidrug resistance protein (MRP) gene. *Biochem. Pharmacol.* 52: 967–77

59. Leier I, Jedlitschky G, Buchholz U, Cole SPC, Deeley RG, Keppler D. 1994. The *MRP* gene encodes an ATP-dependent export pump for leukotriene C$_4$ and structurally related conjugates. *J. Biol. Chem.* 269:27807–10

60. Levine A, Tenhaken R, Dixon R, Lamb C. 1994. H$_2$O$_2$ from the oxidative burst orchestrates the plant hypersensitive disease resistance response. *Cell* 79:583–93

61. Li HH, Inoue M, Nishimura H, Mizutani J, Tsuzuki E. 1993. Interactions of transcinnamic acid, its related phenolic allelochemicals, and abscisic acid in seedling growth and seed germination of lettuce. *J. Chem. Ecol.* 19:1775–87

62. Li Z-S, Alfenito M, Rea P, Walbot V, Dixon RA. 1997. Vacuolar uptake of the phytoalexin medicarpin by the glutathione conjugate pump. *Phytochemistry* 45: 689–93

63. Li Z-S, Lu Y-P, Zhen RG, Szczypka M, Thiele DJ, Rea PA. 1997. A new pathway for vacuolar cadmium sequestration in *Saccharomyces cerevisiae*: YCF1-catalyzed transport of *bis*(glutathionato)cadmium. *Proc. Natl. Acad. Sci. USA* 94:42–47

64. Li Z-S, Szczypka M, Lu Y-P, Thiele DJ, Rea PA. 1996. The yeast cadmium factor protein (YCF1) is a vacuolar glutathione *S*-conjugate pump. *J. Biol. Chem.* 271:6509–17

65. Li Z-S, Zhao Y, Rea PA. 1995. Magnesium adenosine 5′-trisphosphate-energized transport of glutathione *S*-conjugates by plant vacuolar membrane vesicles. *Plant Physiol.* 107:1257–68

66. Li Z-S, Zhen RG, Rea PA. 1995. 1-Chlorodinitrobenzene-elicited increase in vacuolar glutathione *S*-conjugate transport activity. *Plant Physiol.* 109:177–85

67. Lu Y-P, Li Z-S, Drozdowicz YM, Hortensteiner S, Martinoia E, Rea PA. 1998. AtMRP2, an *Arabidopsis* ATP-binding cassette transporter able to transport glutathione *S*-conjugates and chlorophyll catabolites: functional comparisons with AtMRP1. *Plant Cell* 10:1–18

68. Lu Y-P, Li Z-S, Rea PA. 1997. *AtMRP1* gene of *Arabidopsis* encodes a glutathione *S*-conjugate pump: isolation and functional definition of a plant ATP-binding cassette transporter gene. *Proc. Natl. Acad. Sci. USA* 94:8243–48

69. Luckie DB, Krouse ME, Law TC, Sikic BI, Wine JJ. 1996. Doxorubicin selection for MDR1/P-glycoprotein reduces swelling-activated K^+ and Cl^- currents in MES-SA cells. *Am. J. Physiol.* 270:C1029–36

70. Mackenbrock U, Gunia W, Barz W. 1993. Accumulation and metabolism of medicarpin and maackiain malonylglucosides in elicited chickpea (*Cicer arietinum* L.) cell suspension cultures. *J. Plant Physiol.* 142:385–91

71. Mannervik B, Danielson UH. 1988. Glutathione transferases: structure and catalytic activity. *CRC Crit. Rev. Biochem.* 23:283–37

72. Marrs KA. 1996. The functions and regulation of glutathione *S*-transferases in plants. *Annu. Rev. Plant Physiol. Plant Mol. Biol.* 47:127–57

73. Marrs KA, Alfenito MR, Lloyd AM, Walbot V. 1995. A glutathione *S*-conjugate transferase involved in vacuolar transfer encoded by the maize *Bronze-2*. *Nature* 375:397–400

74. Marrs KA, Walbot V. 1997. Expression and RNA splicing of the maize glutathione *S*-transferase *Bronze2* gene is regulated by cadmium and other stresses. *Plant Physiol.* 113:93–102

75. Martinoia E, Grill E, Tommasini R, Kreuz K, Amrhein N. 1993. An ATP-dependent glutathione *S*-conjugate "export" pump in the vacuolar membrane of plants. *Nature* 364:247–49

76. Matile P. 1992. In *Crop Photosynthesis: Spatial and Temporal Determinants*, ed. NR Baker, H Thomas, pp. 413–40. Amsterdam: Elsevier Sci.

77. Matile P, Ginsburg S, Schellenberg M, Thomas H. 1988. Catabolites of chlorophyll in senescing barley leaves are localized in the vacuole of mesophyll cells. *Proc. Natl. Acad. Sci. USA* 85:9529–32

78. McGrath JP, Varshavsky A. 1989. The yeast *STE6* gene encodes a homologue of the mammalian multidrug resistance P-glycoprotein. *Nature* 340:400-4

79. Müller M, Meijer C, Zaman GJR, Borst P, Scheper RJ, et al. 1994. Overexpression of the gene encoding the multidrug resistance-associated protein results in increased ATP-dependent glutathione *S*-conjugate transport. *Proc. Natl. Acad. Sci. USA* 91:13033–37

80. Ortiz DF, Kreppel L, Speiser DM, Scheel G, McDonald G, Ow DW, 1992. Heavy metal tolerance in the fission yeast requires an ATP-binding casette-type membrane transporter. *EMBO J.* 11:3491–99

81. Ortiz DF, Ruscitti T, McCue KF, Ow DW. 1995. Transport of metal-binding peptides by HMT1, a fission yeast ABC-type vacuolar membrane protein. *J. Biol. Chem.* 270:4721–28

82. Oude Elferink RPJ, Frijters CMG, Paulusma C, Groen AK. 1996. Regulation of canalicular transport activities. *J. Hepatol.* 24:94–99

83. Park H-H, Hakamatsuka T, Sankawa U, Ebizuka Y. 1995. Rapid metabolism of isoflavonoids in elicitor-treated cell suspension cultures of *Pueraria lobata*. *Phytochemistry* 38:373–80

84. Plewa MJ, Wagner ED. 1993. Activation of promutagens by green plants. *Annu. Rev. Genet.* 27:93–113

85. Rauser WE. 1990. Phytochelatins. *Annu. Rev. Biochem.* 59:61–86

86. Rea PA, Poole RJ. 1993. Vacuolar H^+-translocating pyrophosphatase. *Annu. Rev. Plant Physiol. Plant Mol. Biol.* 44:157–80

87. Rea PA, Sanders D. 1987. Tonoplast energization: two H^+ pumps, one membrane. *Physiol. Plant.* 71:131–41

88. Rea PA, Turner JC. 1990. Tonoplast adenosine triphosphatase and inorganic pyrophosphatase. *Methods Plant Biochem.* 3:385–405

89. Rennenberg H. 1982. Glutathione metabolism and possible biological roles in higher plants. *Phytochemistry* 21:2778–81

90. Riordan JR, Rommens JM, Kerem B, Alon N, Rozmahel R, et al. 1989. Identification of the cystic fibrosis gene: cloning and characterization of complementary DNA. *Science* 245:1066–80

91. Ruetz S, Gros P. 1994. Phosphatidylcholine translocase: a physiological role for the *mdr2* gene. *Cell* 77:1071–81

92. Salt DE, Rauser WE. 1995. MgATP-dependent transport of phytochelatins across the tonoplast of oat roots. *Plant Physiol.* 107:1293–1301

93. Sanchez-Fernandez R, Ardiles-Diaz W, Van Montagu M, Inze D, May MJ. 1998. Cloning and expression analyses of the *Arabidopsis thaliana* glutathione-conjugate transporter AtMRP4. *FEBS Lett.* In press

94. Sandermann H. 1992. Plant metabolism of xenobiotics. *Trends Biochem. Sci.* 17:82–84

95. Sandermann H. 1994. Higher plant metabolism of xenobiotics: the 'green' liver concept. *Pharmacogenetics* 4:225–41

96. Sathirakul K, Susuki H, Yamada T, Hanano M, Sugiyama Y. 1993. Kinetic analysis of hepatobiliary transport of organic anions in Eisai hyperbilirubinemic mutant rats. *J. Pharmacol. Exp. Ther.* 268:65–73

97. Skach WR, Calayag MC, Lingappa VR. 1993. Evidence for an alternate model of human P-glycoprotein structure and biogenesis. *J. Biol. Chem.* 268:6903–8

98. Snyder BA, Nicholson RL. 1990. Synthesis of phytoalexins in sorghum as a site-specific response to fungal ingress. *Science* 248:1637–39

99. Spanswick RM. 1981. Electrogenic ion pumps. *Annu. Rev. Plant Physiol.* 32:267–89

100. Steffens JC. 1990. The heavy metal-binding peptides of plants. *Annu. Rev. Plant Physiol. Plant Mol. Biol.* 41:553–75

101. St-Pierre MV, Ruetz S, Epstein LF, Gros P, Arias IM. 1994. ATP-dependent transport of organic anions in secretory vesicles of *Saccharomyces cerevisae*. *Proc. Natl. Acad. Sci. USA* 91:9476–79

102. Stride BD, Valdimarsson G, Gerlach JH, Cole SPC, Deeley RG. 1996. Structure and expression of the mRNA encoding the murine multidrug resistance protein

(MRP), an ATP-binding cassette transporter. *Mol. Pharmacol.* 49:962–71

103. Szczypka MS, Wemmie JA, Moye-Rowley WS, Thiele DS. 1994. A yeast metal resistance protein similar to human cystis fibrosis transmembrane conductance regulator (CFTR) and multidrug resistance-associated protein. *J. Biol. Chem.* 269:22853–57

104. Talalay P, De Long M, Prochaska HJ. 1998. Identification of a common chemical signal regulating the induction of enzymes that protect against chemical carcinogenesis. *Proc. Natl. Acad. Sci. USA* 85:8261–65

105. Tanner W, Caspari T. 1996. Membrane transport carriers. *Annu. Rev. Plant Physiol. Plant Mol. Biol.* 47:595–626

106. Tommasini R, Evers R, Vogt E, Mornet C, Zaman GJR, et al. 1996. The human multidrug resistance-associated protein (MRP) functionally complements the yeast cadmium resistance factor (YCF1). *Proc. Natl. Acad. Sci. USA* 93:6743–48

107. Tommasini R, Martinoia E, Grill E, Dietz KJ, Amrhein N. 1993. Transport of oxidized glutathione into barley vacuoles: evidence for the involvement of the glutathione *S*-conjugate ATPase. *Z. Naturforsch. Teil C* 48:867–71

108. Tommasini R, Vogt E, Fromenteau M, Hörtensteiner S, Matile P, et al. 1998. An ABC-transporter of *Arabidopsis thaliana* has both glutathione-conjugate and chlorophyll catabolite transport activity. *Plant J.* 13:In press

109. Tommasini R, Vogt E, Schmid J, Fromenteau M, Amrhein N, Martinoia E. 1997. Differential expression of genes coding for ABC-transporters after treatment of *Arabidopsis thaliana* with xenobiotics. *FEBS Lett.* 411:206–10

110. Tusnady GE, Bakos E, Varadi A, Sarkadi B. 1997. Membrane topology distinguishes a subfamily of the ATP-binding cassette (ABC) transporters. *FEBS Lett.* 402:1–3

111. van Kuijck MA, van Aubel RAMH, Busch AE, Lang F, Russel FGM, et al. 1996. Molecular cloning and expression of a cyclic AMP-activated chloride conductance regulator: a novel ATP-binding cassette transporter. *Proc. Natl. Acad. Sci. USA* 93:5401–6

112. Vögeli-Lange R, Wagner GJ. 1990. Subcellular localization of cadmium and cadmium binding peptides in tobacco leaves. *Plant Physiol.* 92:1086–93

113. Volkl P, Markiewicz P, Baikalov C, Fitz-Gibbon A, Stetter KO, Miller JH. 1996. Genomic and cDNA sequence tags of the

hyperthermophilic Archaeon *Pyrobacu-lum aerophilum*. *Nucleic Acids Res.* 24: 4373–78

114. Walker JE, Saraste M, Runswick MJ, Gay NJ. 1982. Distantly related sequences in α- and β-subunits of ATP synthase, myosin, kinases and other ATP-requiring enzymes and a common nucleotide binding fold. *EMBO J.* 1:945–51

115. Wang W, Takezawa D, Poovaiah BW. 1996. A potato cDNA encodes a homologue of mammalian multidrug resistance P-glycoprotein. *Plant Mol. Biol.* 31:683–87

116. Wemmie JA, Moye-Rowley WS. 1997. Mutational analysis of the *Saccharomyces cerevisiae* ATP-binding cassette transporter protein Ycf1p. *Mol. Microbiol.* 25: 683–94

117. Wemmie JA, Szczypka MS, Thiele DJ, Moye-Rowley WS. 1994. Cadmium resistance mediated by the yeast AP-1 protein

requires the presence of an ATP-binding cassette transporter-encoding gene, *YCF*. *J. Biol. Chem.* 269:32592–97

118. Werner C, Matile P. 1985. Accumulation of coumarylglucosides in vacuoles of barley mesophyll protoplasts. *J. Plant Physiol.* 118:237–49

119. Wilson CM, Serrano AE, Wasley A, Bogen Schultz MP, Shankar AH, Wirth DF. 1989. Amplification of a gene related to mammalian *mdr* genes in drug-resistant *Plasmodium falciparum*. *Science* 244:1184–86

120. Wolf AE, Dietz K-J, Schröder P. 1996. Degradation of glutathione *S*-conjugates by a carboxypeptidase in the plant vacuole. *FEBS Lett.* 384:31–34

121. Zhang JT, Ling V. 1991. Study of membrane orientation and glycosylated extracellular loops of mouse P-glycoprotein by *in vitro* translation. *J. Biol. Chem.* 266: 18224–32

AUTHOR INDEXES

761

SUBJECT INDEX

A

A23187 calcium ionophore
  hormone-induced signaling
    during moss development
    and, 511
ABC model
  ovule development and, 16
ABC transporters
  multispecific, 727–55
ABF-1 nuclear factor
  light-regulated promoter
    evolution and, 532
Abiotic environmental stresses
  abscisic acid signal transduction
    and, 199
  ascorbate and glutathione,
    254–56
Abscisic acid
  signal transduction and
    ABA perception, 209–10
    cis-acting promoter
      elements, 203–4
    dormancy, 200–2
    desiccation tolerance, 202–6
    downstream signaling
      elements, 211–13
    germination, 200–2
    guard cell nucleus, 213
    intermediates, 207–8
    introduction, 200
    ion channels, 210
    perspectives, 213–15
    promoters, 207
    regulation of gene
      expression, 207–9
    reserve accumulation, 202–6
    second messengers, 210–11
    seeds, 200–6
    single-cell systems, 208–9
    stomatal aperture, 209–13
    stress response, 206–13
    trans-acting promoter
      elements, 204–5
    upstream signaling elements,
      205–6
Acacia senegal
  cell wall proteins and, 298
Acacia sp.
  meiotic chromosome
    organization and
    segregation, 375
ACA genes
  cation transport and,
    679

ACBP protein
  phytoremediation and, 654
Acer pseudoplatanus
  lignin biosynthesis and, 602
Acetabularia acetabulum
  body plan and phase change in
    apices, 177–82
  classic developmental
    system, 192–93
  functional significance of
    phases, 182
  future research, 191–92
  life cycle, 175
  morphogenesis, 186–91
  reasons for studying, 174–77
  removing and adding body
    regions, 182–86
  spatial control, 188–91
  temporal control, 188–91
  translational control,
    186–88
Acetabularia sp.
  carotenoid biosynthesis and,
    575
Acetylenase
  fatty acid modifications and,
    634–35
Acid patch
  photosynthetic cytochromes *c*
    and, 409
Active oxygen species
  ascorbate and glutathione,
    249–69
Acyl-ACP desaturases
  fatty acid modifications and,
    621–25, 631–32
Acyl chains
  fatty acid modifications and,
    611, 615–18, 623–25
Adenine
  hormone-induced signaling
    during moss development
    and, 510–11
Adenosine triphosphate (ATP)
  thylakoid protein transport and,
    109–14
  vacuolar GS-X pumps and
    multispecific ABC
    transporters, 727, 729–30,
    733, 735
Adonis palaestina
  carotenoid biosynthesis and,
    572
Aequoria victoria
  pre-mRNA splicing and, 88

AG genes
  ovule development and, 6
AG-rich elements
  pre-mRNA splicing and,
    77
Agrobacterium aurantium
  carotenoid biosynthesis and,
    572
Agrobacterium tumefaciens
  ovule development and,
    11
Agropyron desertorum
  phytoremediation and,
    660
AKT genes
  cation transport and, 672,
    676–77
Alanine
  fatty acid modifications and,
    625
Alaria esculenta
  photosynthetic cytochromes *c*
    and, 402–3
Alcaligenes PC-1
  carotenoid biosynthesis and,
    572
Algae
  Acetabularia body plan and
    phase change, 173–93
  boron and, 483
  calmodulin and
    calmodulin-binding
    proteins, 705
  photosynthetic cytochromes *c*
    and, 397–420
  phytoremediation and, 657
  sulfoquinovosyl diacylglycerol
    and, 53, 55, 60–61, 68, 69,
    70
Alicyclobacillus sp.
  sulfoquinovosyl diacylglycerol
    and, 56
Alkane ω-hydroxylase
  fatty acid modifications and,
    629–30
Alkanes
  fatty acid modifications and,
    626, 633
Alkenes
  fatty acid modifications and,
    633
Allene oxide synthase
  cytochrome P450-dependent
    monooxygenases and, 321,
    322, 323

CUMULATIVE INDEXES

CONTRIBUTING AUTHORS, VOLUMES 39–49

CHAPTER TITLES, VOLUMES 39–49

833

Protein Structure/Function/Regulation/Synthesis

GENETICS AND MOLECULAR BIOLOGY

Structure and Function of Nucleic Acids